Introduction to Bioorganic Chemistry and Chemical Biology

Introduction to Bioorganic Chemistry and Chemical Biology

David Van Vranken and Gregory Weiss

CRC Press
Taylor & Francis Group
Boca Raton London New York

CRC Press is an imprint of the
Taylor & Francis Group, an **informa** business

David Van Vranken earned his PhD in chemistry from Stanford University. He is a professor of chemistry at UC Irvine, where he carries out research on chemical structure and reactivity in peptides, bioactive natural products, and transition metal catalyzed reactions.

Gregory Weiss earned his PhD in chemical biology from Harvard University. He is a professor of chemistry, molecular biology, and biochemistry at UC Irvine, where his laboratory focuses on the interface between chemistry and biology, including combinatorial chemistry applied to antiviral drug discovery, membrane proteins, and bioelectronics.

CRC Press
Taylor & Francis Group
6000 Broken Sound Parkway NW, Suite 300
Boca Raton, FL 33487-2742

First issued in paperback 2019

ISBN-13: 978-0-8153-4214-4 (pbk)

Illustrator, Cover, and Text Design: Matthew McClements, Blink Studio, Ltd.

Visit the Taylor & Francis Web site at
http://www.taylorandfrancis.com

and the CRC Press Web site at
http://www.crcpress.com

Preface

This textbook provides a chemical blueprint with which to understand the synthesis and function of molecules that comprise living cells. In order to tackle chemical biology—a field of vast breadth and depth—as it continues to expand and evolve, we focus on the organic chemistry of biooligomers. Biooligomers are responsible for every main function of the cell, including control, communications, and manufacturing. The central dogma of molecular biology serves as an organizing principle for students embarking on a tour of these molecules. Starting at the level of genes and advancing through each class of biooligomer—DNA, RNA, proteins, glycans, polyketides, and terpenes—we study their chemistry, structure, and interactions with other molecules. Ultimately, we return to the topic of chemical control over gene expression. Although the majority of studies during the past millennium were based on model organisms, we concentrate on human cells as much as possible.

Introduction to Bioorganic Chemistry and Chemical Biology is written for upper-division undergraduate students and graduate-level students who have completed a year of organic chemistry. No initial knowledge of biology is assumed, but some familiarity is beneficial. The book is also appropriate for advanced students in the health sciences. A series of chemical biology courses we teach at the University of California, Irvine, inspired the development of this book. Every student in our courses speaks a common language—organic chemistry—but there was no single textbook to provide an adequate description of the cell at the level of atoms, bonds, and arrow-pushing reaction mechanisms. Starting with foundational and simple concepts, the book gradually builds in complexity, across sections and chapters. For the undergraduate course, we simply skip the more advanced topics in both the reading assignments and lectures. This format offers flexibility in coverage, depending on the length of the course and the level of student.

This book is organized according to the central dogma of molecular biology. Chapters 1 and 2 cover the fundamentals of chemical biology and the chemical origins of biology, respectively. Each subsequent chapter (3–8) focuses largely on one type of biooligomer in a human cell: DNA, RNA, proteins, glycans, polyketides, and terpenes. We steer clear of the metabolic and biosynthetic pathways, which have become synonymous with introductory biochemistry courses. Chapter 9 catapults the student beyond the central dogma to explain mechanisms of cellular control over the production of biooligomers. This final chapter will bring students closest to an understanding of modern biology, physiology, and medicine. We recognize the relentless pace of new technology to emerge and quickly become antiquated; therefore, we limit our discussion of instrumentation and laboratory techniques. Since *Introduction to Bioorganic Chemistry and Chemical Biology* is not intended to be comprehensive, many well-known examples and techniques from the field of chemical biology are omitted. Furthermore, we fully expect that instructors will select from and supplement the material in the book with additional information that best meets the needs of their students.

From experience, we know that composing moving, informative lectures constitutes the fun part of teaching. In this vein, each chapter begins with a snapshot of a significant discovery by luminaries such as Dorothy Crowfoot Hodgkins, Phoebus Levene, and Hermann Emil Fischer. We tell the stories of early examples of chemical biology illustrated in the *Iliad* and ancient Egyptian pharmacopeias. Our book combines rich, full color imagery with relevant biological examples to engage the reader as the combinatorial molecular architecture of life is revealed. Our choice of colors is deliberate: in general, we use red for DNA, green for RNA, blue for proteins, purple for glycans, and brown for polyketides and terpenes. Therefore, when biooligomers are represented by abstract shapes, the color of the shape offers an additional level of biochemical information. We employ modern depictions of organic structures and mechanistic arrow-pushing in a way that will be familiar to all students who have taken an introductory course in organic chemistry. Biological macromolecules are rendered to reveal their secondary structures and PDB codes are included to allow students to interact directly using the Protein Data Bank website. In-text problems are situated throughout the chapters, contextualized for greater understanding; additional problems are found at the end of each chapter. We have starred (*) those problems best suited to self-study, and solutions are available online at the Student Resource Site. Chapters also begin with a list of general Learning Objectives and end with a list of specific Learning Outcomes to focus students on important topics of comprehension.

David L. Van Vranken and Gregory A. Weiss

ONLINE RESOURCES

Accessible from www.garlandscience.com/bioorganic-chembio, the Student and Instructor Resource Websites provide learning and teaching tools created for *Introduction to Bioorganic Chemistry and Chemical Biology*. The Student Resource Site is open to everyone, and users have the option to register in order to use book-marking and note-taking tools. The Instructor Resource Site requires registration and access is available to instructors who have assigned the book to their course. To access the Instructor Resource Site, please contact your local sales representative or email science@garland.com. Below is an overview of the resources available for this book. On the Website, the resources may be browsed by individual chapters and there is a search engine. You can also access the resources available for other Garland Science titles.

For students

Flashcards
Each chapter contains a set of flashcards that allow students to review key terms from the text.
Glossary
The complete glossary from the book and can be searched and browsed as a whole or sorted by chapter.
References
For each chapter, a list of references is organized by concept headings.
Problems
Solutions to all in-text problems as well as all starred (*) end-of-chapter problems are available to students.

For instructors

Figures
The images from the book are available in two convenient formats: PowerPoint® and JPEG. They have been optimized for display on a computer. Figures are searchable by figure number, figure name, or by keywords used in the figure legend from the book.
Solutions
Solutions to all problems (in-text and end-of-chapter) are available to qualified adopters.

Acknowledgments

First and foremost, without the unwavering love and support of our families, this book wouldn't have become a reality. Our love and deep appreciation are given to Maureen, Julia, and Kim.

This book also owes a debt of gratitude to a cast of thousands—the researchers in the field who did the work we have had the privilege of describing. It's impossible to single out every one of these deserving scientists for acknowledgement. Limitations of space in this introductory text imposed agonizing brevity. We omitted important contributions made by key players, and we mentioned seminal discoveries without attribution. We greatly appreciate the efforts of those who drove the field forward.

We thank the following authors for writing textbooks we found particularly inspirational: Bruce Alberts, Ian Fleming, Clayton Heathcock, Alexander Johnson, Jack Kyte, Julian Lewis, Martin Raff, Keith Roberts, Richard Silverman, Wolfram Saenger, Andrew Streitwieser, Christopher Walsh, and Peter Walter. We also appreciate the countless students at UC Irvine who patiently endured our test drives of the material during our teaching. We are also grateful for the friendship and support of our colleagues at UC Irvine.

The Garland Science team made this challenging project manageable. We would like to thank our editor, Summers Scholl, who managed this project with panache; talented developmental editor John Murdzek; the unflappable senior editorial assistant Kelly O'Connor who kept the project on track and watched over a million details; and production editor Emma Jeffcock for expert layout and other crucial matters. Additionally, the illustrations benefited from the wonderful artistry of Matt McClements; and Becky Hainz-Baxter sifted through thousands of possibilities to find perfect photos and permissions.

The following people provided valuable commentary as readers, reviewers, and advisors during the development of the project:

Peter A. Beal (University of California, Davis); Annette Beck-Sickinger (University of Leipzig); Danielle Dube (Bowdoin College); Marina Gobbo (University of Padova); David Grayson (Trinity College, Dublin); Paul Harrison (McMaster University); Jessica Hollenbeck (Trinity University, Texas); Gerwald Jogl (Brown University); Andrej Lupták (University of California, Irvine); José Luis Mascareñas (University of Santiago); Christian Melander (North Carolina State University); Nicola Pohl (Indiana State University); James Redman (Cardiff University); Carmelo J. Rizzo (Vanderbilt University); Erland P. Stevens (Davidson College); Terry Smith (University of St. Andrews); Ali Tavassoli (University of Southampton); Doug Tobias (University of California, Irvine).

Dedication

This book is dedicated to our graduate mentors, Pete and Stuart.

Contents

Detailed Contents

The Fundamentals of Chemical Biology

1

LEARNING OBJECTIVES

- Provide a working definition of chemical biology.
- Apply the central dogma of molecular biology as an organizing principle in chemical biology.
- Describe key features of genes, transcripts, proteins, and the associated genomes, transcriptomes, and proteomes.
- Explain the importance of evolution in chemical biology experiments.
- Introduce the most important model organisms, and provide examples of their applications in chemical biology experiments.
- Provide a brief introduction to viruses and plasmids.

Why organize a book on chemical biology around biooligomers?

Living organisms must obey the rules of chemistry. **Chemical biology** applies the rules of chemistry—at the level of atoms and bonds—to biological systems. Chemistry is an expansive set of subdisciplines, each with limited dynamic range, that offer insight into the behavior of molecules. Among those subdisciplines, organic chemistry offers a unique compromise between qualitative and quantitative approaches, and is ideal for explaining chemical reactions and nonbonding interactions that generate the molecular diversity needed for evolution.

According to Charles Darwin, selection for beneficial traits is a key step in the evolution of the various species (**Figure 1.1**). The success of selection as an evolutionary tool depends on a diverse population. If a field of plants, say Cape cowslip lilies, possesses a drought-resistant plant, a flood-resistant plant, a cold-resistant plant, and a heat-resistant plant, the chances are good that at least one member of the population will survive a single climate catastrophe. But what happens to the population if there are a series of different climate catastrophes, such as a drought this summer and a flood the next summer? To survive a *second* type of climate catastrophe, the surviving plants must propagate in a way that re-establishes a diverse population. Thus, evolution is not just driven by survival of the fittest; it is also driven by the propagation and diversification of the fittest.

Figure 1.1 Charles Darwin. Charles Darwin at age 6 holding a pot of Cape cowslips. By the time he published *On the Origin of Species* at age 50, Darwin understood the importance of diverse populations at the organismal level. The molecular basis of that diversity would not be revealed for almost one hundred years. (Courtesy of J. van Wyhe, ed., The Complete Work of Charles Darwin Online, http://darwin-online.org.uk, 2002.)

Generating diverse populations is a key ingredient in the recipe of evolution. Living organisms are engines for the assembly of diverse populations from a limited set of subunits: diverse populations of molecules, diverse populations of cells, diverse populations of organisms, and diverse ecosystems. This modular approach to assembly can be described as **combinatorial**. We can appreciate the flexibility of this approach by comparing it to writing. Writing is combinatorial at every level: letters can be combined to form words; words can be combined to form sentences; sentences can be combined to form paragraphs; and the meaning of letters, words, and paragraphs is dependent on sequence. In the right combination, letters can express ideas that organize societies, mobilize armies, solidify the bonds of love, and even explain our universe. Words, sentences, and paragraphs are oligomers. They are formed by connecting a limited set of subunits in a nonrepeating fashion. We have rules for each level of writing: rules of spelling for words, rules of grammar for sentences, and rules of composition for paragraphs. Those rules allow us to encode powerful ideas in a compact form, to store them, to transmit them, to duplicate them, and later to decipher them without ambiguity.

Chemical biologists seek to explain the combinatorial origin of diversity at the molecular level of detail, atom by atom and bond by bond. Most of the molecules in your cells are biooligomers made up of molecular subunits, and each of those molecular subunits is a combinatorial oligomer of atoms. **Biooligomers**, including proteins, DNA, RNA, glycans, lipids, and terpenes, account for more than 90% of the dry weight of the human cell. To a first-order approximation, such biooligomers *are* the cell—just add water. Oligomeric architectures are ideally suited to the synthesis of diverse molecules. The goal of this book is to explain the rules of chemical grammar that govern the assembly of diverse biooligomers and, in so doing, to illuminate the workings of the human cell. Because the central dogma of molecular biology governs the assembly of biooligomers we use it as the thematic organization for this book.

Problem 1.1

If a thousand chimpanzees typed for a thousand years, is it likely that one of them would turn out a work of Shakespeare? (Courtesy of New York Zoological Society.)

1.1 THE CENTRAL DOGMA OF MOLECULAR BIOLOGY

The central dogma of molecular biology is an organizing principle for chemical biology

Cells can be simpler to understand than writing because the sequence of every biooligomer is ultimately determined by the sequence of other biooligomers. In the case of DNA, RNA, and proteins, the correspondence between sequences is straightforward. The potential relationships between those biooligomers were first outlined by Francis Crick, a pioneer in determining the structure of DNA. Crick noted a hierarchical flow of information from DNA to other biooligomers and titled it **the central dogma of molecular biology** (**Figure 1.2**). In that hierarchy, DNA can be used as a template to direct the synthesis of RNA and vice versa. RNA can be used as a template to direct the synthesis of proteins, but not the other way around. The structure of proteins makes it difficult to use as a template for directing the biosynthesis of RNA, DNA, or any type of oligomer. The flow of information is not completely unidirectional. Proteins do affect the synthesis of DNA and RNA, but not through a direct encoding mechanism. We will examine those indirect mechanisms in later chapters.

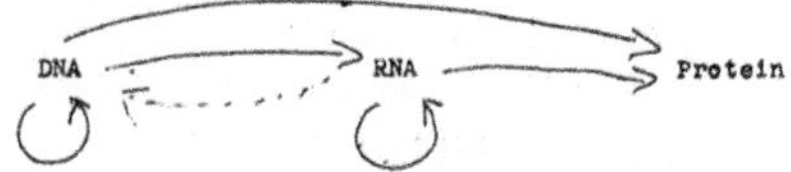

Figure 1.2 The original central dogma. This first rendition of the central dogma (in the figure above) included both hypothetical and known patterns of information transfer. In its humblest form, Crick's central dogma stated that "Once information has got into a protein it can't get out again." (Courtesy of Wellcome Library, London.)

The biopolymer DNA provides a master blueprint for the cell and organism. To follow this blueprint, we will need to master three highly specific biochemical terms: **replicate**, **transcribe**, and **translate** (**Figure 1.3**). Most of the reactions taking place in biology are catalyzed by proteins, called **enzymes**. DNA is replicated by the enzyme DNA polymerase, using each strand of DNA as a template for the synthesis of new strands. RNA is transcribed by RNA polymerase, using one of the two strands of DNA as a template. RNA carries out diverse functions in the cell, but the function heralded by the central dogma is that it serves as a template for protein synthesis. Proteins are translated from a messenger RNA (mRNA) template by the ribosome. In organisms, proteins contribute diverse roles. For example, proteins called enzymes can catalyze the formation of other biopolymers, such as oligosaccharides, lipids, and terpenes.

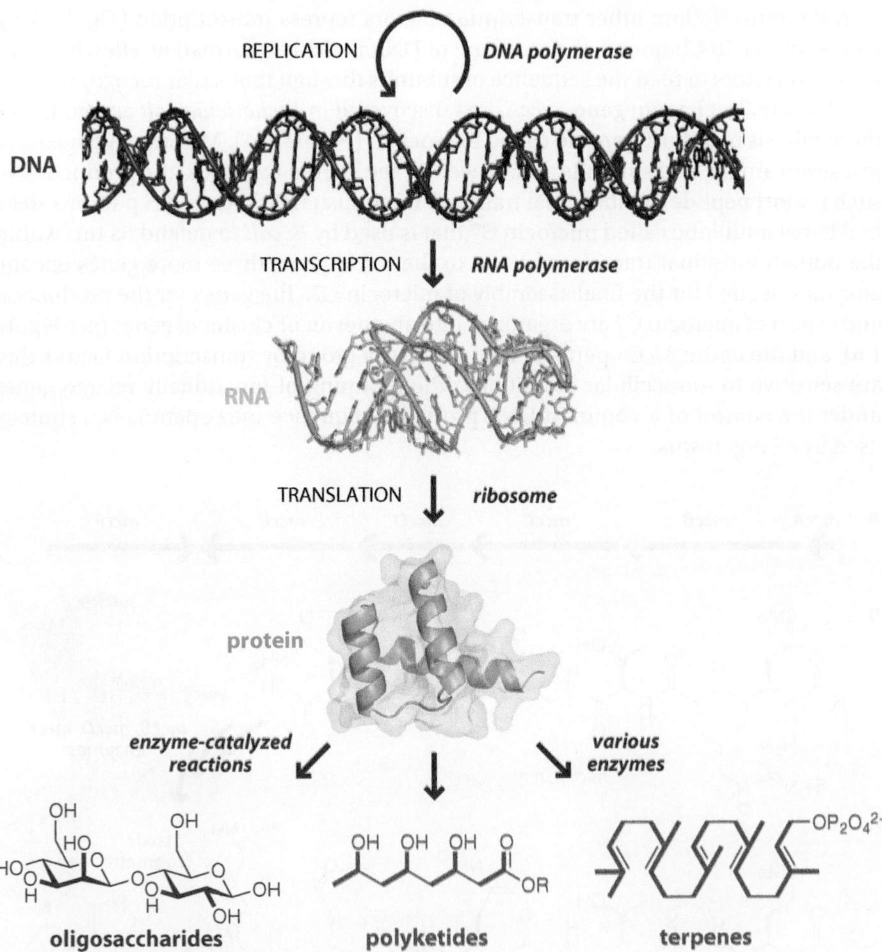

Figure 1.3 The central dogma of molecular biology, expanded. The arrows indicate the flow of information as biooligomers of one type serve as templates or catalysts for the synthesis of other types of biooligomers.

1.2 GENES

A gene is made up of a promoter and a transcribed sequence

In the cell, the transcription of RNA is carefully orchestrated. RNA polymerase does not begin transcription at random places in the genome or even at random genes. At any point in time, RNA polymerase transcribes only specific genes in the genome. Proteins called **transcription factors** bind to a specific DNA sequence called a **promoter**. The combination of a promoter and a DNA sequence that encodes an RNA sequence composes a simple **gene** (**Figure 1.4**). The promoter and its associated transcription factors control gene expression. Some transcription factors recruit RNA polymerase and

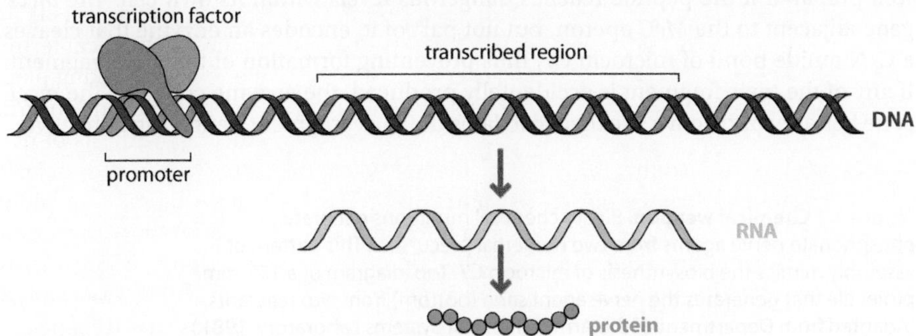

Figure 1.4 Gene expression. DNA appears uniform when rendered as a cartoon, but when viewed at the level of atoms and bonds each nanometer of sequence contains rich information. All genes consist of a promoter sequence and a transcribed sequence. Transcription factors bind to the promoter and recruit RNA polymerase to transcribe RNA.

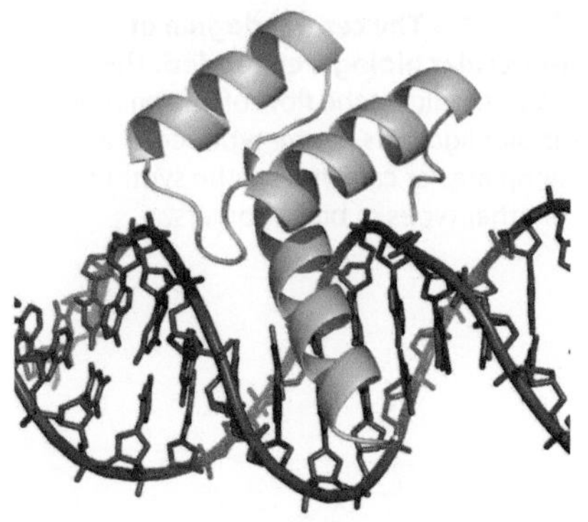

Figure 1.5 A transcription factor binding to DNA. Proteins called transcription factors (blue) turn genes on and off, by binding to promoters, which are specific DNA sequences (red). In this depiction, the protein is shown as a ribbon tracing the connections between amino acids. DNA is depicted as sticks between the non-hydrogen atoms, and a ribbon traces the phosphodiester backbone of DNA. More information about DNA and protein structure is provided in later chapters of this book.

activate transcription; other transcription factors repress transcription (**Figure 1.5**). As we will see in Chapter 3, the structure of DNA is richly informative, allowing transcription factors to read the sequence of subunits through molecular recognition.

The smallest known gene, *mccA*, was discovered in *Escherichia coli* and codes for the synthesis of a short **peptide** called microcin A (**Figure 1.6**). Microcin A consists of just seven amino acid subunits and, as we will see later in this book, the production of such a short peptide by ribosomal translation is unusual. Microcin A is part of a sleek and lethal antibiotic called microcin C7 that is used by *E. coli* to defend its turf within the human intestinal tract. In addition to the *mccA* gene, three more genes encode enzymes needed for the final assembly of microcin C7. The genes for the production and export of microcin C7 are organized into an **operon** or cluster of genes (see Figure 1.6), and the entire *McC* operon is controlled as a group by transcription factors that are sensitive to intracellular conditions. The grouping of functionally related genes under the control of a common DNA promoter sequence into operons is a strategy used by all organisms.

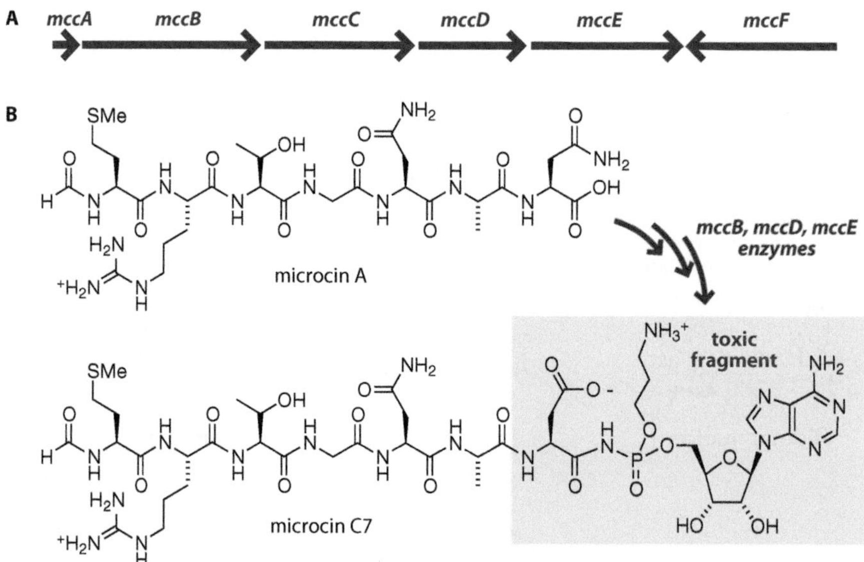

Figure 1.6 The smallest gene makes a chemical weapon. (A) Microcin A is encoded by the tiny *mccA* gene grouped together with related genes in the *McC* operon. These genes are made up of DNA. (B) The peptide microcin A is assembled into a sleek chemical weapon. Me is a commonly used abbreviation for a methyl group (CH_3).

The *McC* operon has one more gene, which codes for a protein transporter that aggressively expels microcin C7 from the producing cell. Microcin C7 resembles the binary chemical weapons invented by humans but is much more sophisticated (**Figure 1.7**). Microcin C7 is a booby trap. Hapless strains of *E. coli* see the tasty peptide and actively import it. When those bacteria try to strip it for usable parts, one of the fragments turns out to be toxic, targeting an enzyme that is essential for protein synthesis, called aspartyl tRNA synthetase. The bacterium that produces microcin C7 is well prepared if the peptide reaches dangerous levels within its own cell. The *mccF* gene adjacent to the *McC* operon, but not part of it, encodes an enzyme that cleaves a C–N amide bond of microcin C7, thus preventing formation of the toxic fragment. If any of the toxic fragment is accidentally produced, the enzyme coded by the *mccE* gene has a second active site that acetylates the toxic fragment, rendering it inactive.

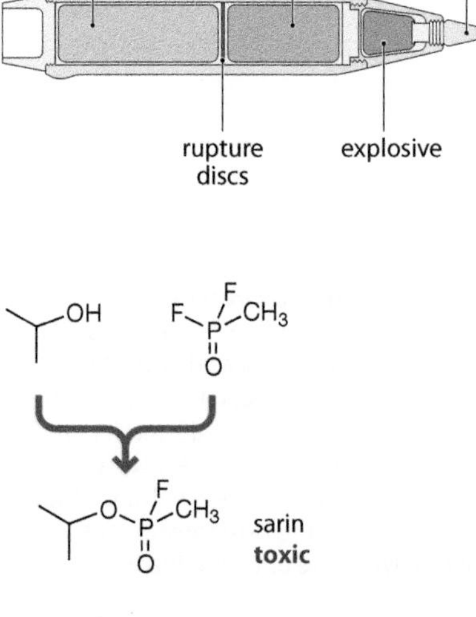

Figure 1.7 Chemical weapon. Binary chemical munitions generate phosphonate nerve agents from two nonlethal precursors. This pattern of assembly mimics the biosynthesis of microcin C7. Top: diagram of a 155-mm projectile that generates the nerve agent sarin (bottom) from two reactants. (Adapted from Department of the Army, Chemical Systems Laboratory, 1981.)

1.3 GENOMES

We have sequenced the human genome and many others. Now what?

The turn of the millennium saw one of the greatest scientific achievements in the history of humankind—the structural elucidation of the human genome. A **genome** is a collection of all of the DNA in an organism. The human genome carries instructions for all human biosynthesis: DNA, RNA, proteins—everything. Because DNA is a linear oligomer of four chemical subunits, the process of structure elucidation involves determining the sequence of the subunits. We will refer to DNA subunits as **base pairs**, or **bases** for short, for reasons that will become apparent in Chapter 3. Technology for genomic sequencing was first applied to a diminutive bacterium, *Haemophilus influenzae,* and within a few years a rough draft of the human genome was available. The ability to sequence entire genomes has given us unprecedented access to the genetic differences that distinguish bacteria from yeast, yeast from worms, worms from flies, flies from mice, mice from humans, and healthy humans from diseased humans.

Scientists routinely study simpler organisms to understand important principles about human biology. We still endeavor to understand the minimum requirements for life. The smallest known genome, with about 200 genes, belongs to the microorganism *Carsonella rudii,* but *C. rudii* is a symbiont that cannot live outside its insect host. The smallest known genome for a *free-living* organism is that of the bacterium *Mycoplasma genitalium,* which sometimes infects the human respiratory or genital tract (**Figure 1.8**). The *M. genitalium* genome has only 521 genes efficiently coded within 582,970 DNA bases; 482 of those genes code for proteins. From the gene sequences, we can predict the chemical structures of the resulting RNA transcripts and, ultimately, the chemical structures of the proteins. Only 382 of those proteins are essential for life. However, the bacterium probably would not be viable if all of the 100 nonessential genes were removed at once. J. Craig Venter, one of the leaders of the human genome project, recently generated a viable cell from a chemically synthesized genome of *M. genitalium*—the first synthetic organism! It would seem that we are well on our way to understanding a minimal organism—how it survives, evolves, and responds to widely varying conditions—but we are not. Scientists have been staring at the chemical structures of the *M. genitalium* proteins for almost two decades, and we still cannot figure out what one-quarter of the essential proteins do.

Why is there a yawning gap between protein sequence and protein function? There are two main reasons. First, the function of many of those proteins relates to other types of biooligomers (such as glycans, lipids, terpenes, and metabolites) that are not easy to study. Second, the protein may only serve a relevant function under a certain set of conditions. This is why an alien making a brief visit to your home might not understand the function of the ceiling fire alarm. Furthermore, the function of many proteins can be understood only when observed within a dynamic system. For example, it is virtually impossible to predict the function of a synchronizer gear once it is removed from a 1967 four-speed Toploader transmission. As long as the top is bolted on the transmission, you cannot see what the synchronizer is doing as you shift gears (**Figure 1.9**). Similarly, no one can see what is going on in *M. genitalium.* Chemical biologists are uniquely equipped for such problems. They are trained to think at the level of atoms and bonds, and can design molecular tools to probe and peer into complex systems such as cells.

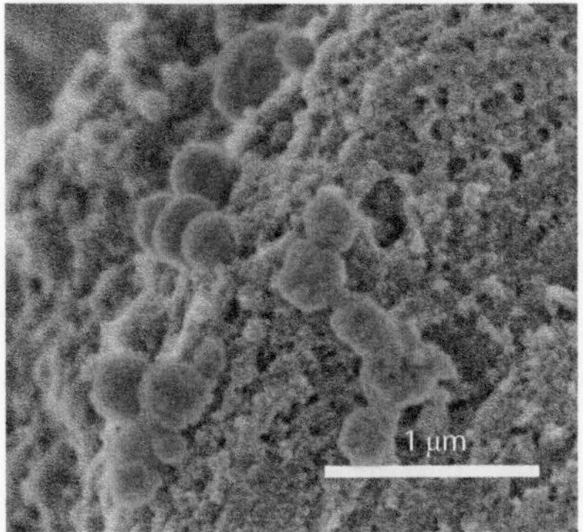

Figure 1.8 The smallest genome. *Mycoplasma genitalium* has the smallest genome of any free-living organism. This image was taken by scanning electron microscopy and then colored. Optical microscopy does not have sufficient resolution to resolve the features of *M. genitalium.* (From C. McGowin et al., *BMC Microbiol.* 9:139, 2009. With permission from BioMed Central Ltd.)

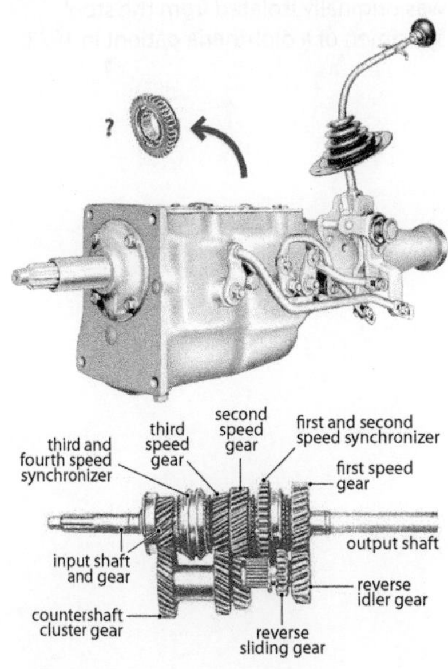

Figure 1.9 What does it do? Sometimes the function of a component—such as a synchronizer from an automobile transmission—is difficult to discern if you cannot see inside. Even with an accurate static diagram, the roles of the components are cryptic if you cannot see the system respond dynamically. (Courtesy of Roland Dudley and Mark Olson, eds., Classic Tiger. http://www.classictiger.com)

We are far from understanding cells that we understand the best—*Escherichia coli*

Escherichia coli is the best-understood species of organism, and it is larger and more complex than a mycoplasmic bacterium. Some strains of *E. coli* live harmoniously within the human gastrointestinal tract, and even thrive there; one-third of the dry weight of human feces is *E. coli* cells. The task of understanding *E. coli* is sometimes complicated by the fact that there are numerous strains. The common laboratory strain, *E. coli* K-12, is harmless, but other strains of *E. coli* are associated with disease.

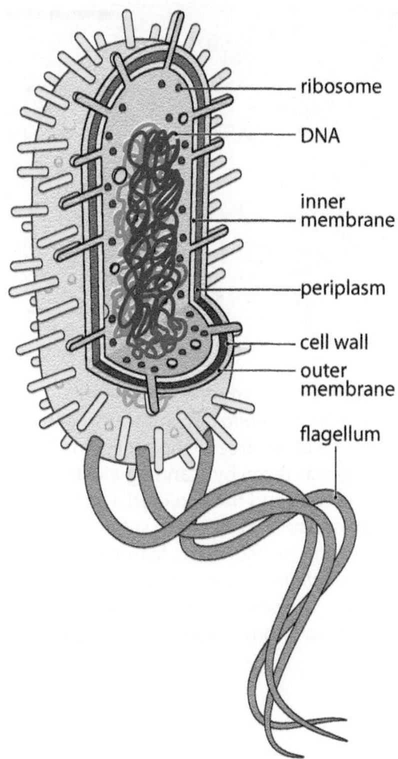

Figure 1.10 Inside a bacterium. Diagram of *E. coli* K-12 showing the arrangement of membranes and the location of genomic DNA and ribosomes. *E. coli* K-12 was originally isolated from the stool specimen of a diphtheria patient in 1922.

Uropathogenic strains such as CFT073 sometimes lead to human bladder infections. The enterohemorrhagic strain O157:H7 sometimes contaminates the beef sold in grocery stores.

The bacterium *E. coli* is a small rod-shaped organism (**Figure 1.10**), typically a little more than 1 μm in diameter. It has a complex coating composed of two fluid lipid membranes. A tough, web-like network, called the cell wall, is sandwiched in the periplasmic space between the two membranes. Most of the bacterial contents, such as the DNA, are housed within the inner membrane as opposed to the cramped periplasmic space. *E. coli* K-12 is capable of a solitary itinerant existence. It is equipped with rotating whip-like flagella that propel it toward nutrients and away from toxins. In spite of its itinerant tendencies, *E. coli* is also quite sociable. Under nutrient-rich conditions, *E. coli* undergoes binary fission to generative cooperative colonies akin to superorganisms. Chemical signaling within the colony maximizes the use of environmental resources (related to invasiveness) and inhibits overpopulation. Each member of the colony is a genetic clone, equally fit yet equally unfit to survive. *E. coli* cells rely on promiscuity to improve genetic diversity and increase the chances of survival. The bacterial capsule is covered with thin hair-like projections—much smaller than the flagella—that allow the bacteria to exchange genetic material with other strains.

Many genotypic variants of K-12 have been engineered for laboratory experiments such as viral transfection, selecting clones, and producing proteins. The genome of the MG1655 variant of *E. coli* K-12 is composed of 4377 genes, encoded within 4,639,221 bases of DNA. Both the cell and the genome of *E. coli* are about 10 times larger than that of *M. genitalium*. Like *M. genitalium*, more than 95% of the *E. coli* K-12 genes encode proteins. Advances in DNA sequencing technology have reduced the sequencing of bacterial genomes to an expensive exercise, and the genomes of more than 50 strains of *E. coli* have now been sequenced. You would be hard pressed to distinguish the various strains of *E. coli* by looking at micrographs (**Figure 1.11**).

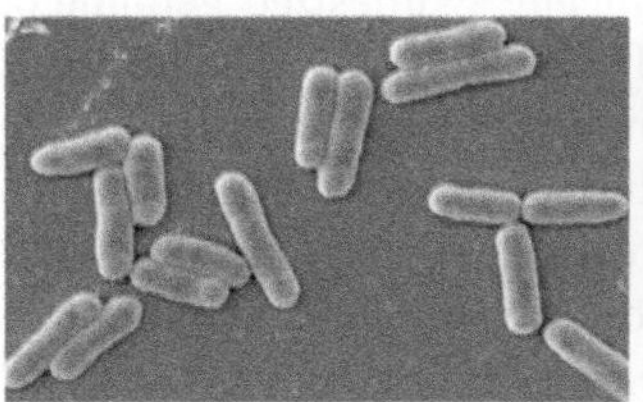
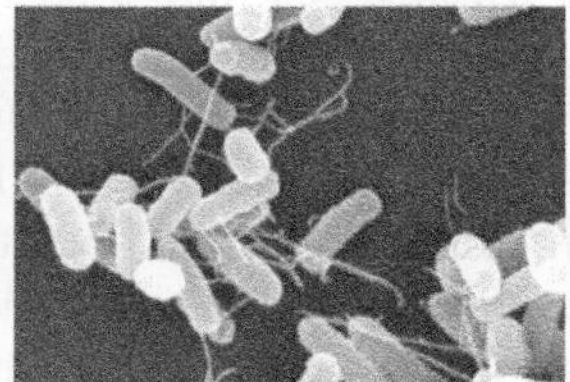
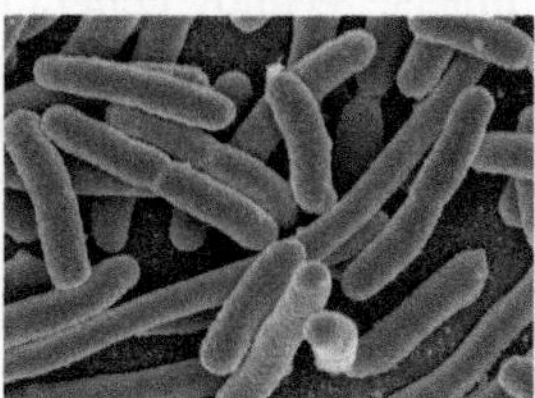

Figure 1.11 What is inside is what counts. Strains of *Escherichia coli* appear similar under a scanning electron microscope, but share only 20% genetic identity. Left to right: *E. coli* strains K-12 MG1655, O157:H7, and CFT073. (Left, from S. Kar et al., *Proc. Natl. Acad. Sci. USA* 102:16397–16402, 2005. With permission from the National Academy of Sciences; middle, from S. Suwalak and S.P. Voravuthikunchai, *J. Elec. Microsc.* 58:315–320, 2009. With permission from Oxford University Press; right, courtesy of Rocky Mountain Laboratories, NIAID, NIH.)

We are far from understanding any one strain of *E. coli* and much further from understanding all the different strains. As with *M. genitalium*, we do not know the function of about 20% of the genes in *E. coli* K-12. Surprisingly, only about 20% of the *E. coli* genes are conserved among all of the strains. The genetic diversity of *E. coli* is further increased by the presence of smaller circles of DNA called **plasmids**. The plasmids contained within bacteria vary, even within the same strain. Plasmids typically contain one or several genes, usually beneficial. Bacteria trade plasmids during their conjugal visits. There is only one copy of the bacterial genome per cell, but many copies of each of the plasmids.

We are even farther from understanding human cells

Human cells are about a thousand times larger than *E. coli* cells and have about a thousand times more DNA. Like *E. coli*, human cells are topologically separated by two lipid membranes (**Figure 1.12**). The outer membrane, called the **plasma membrane**,

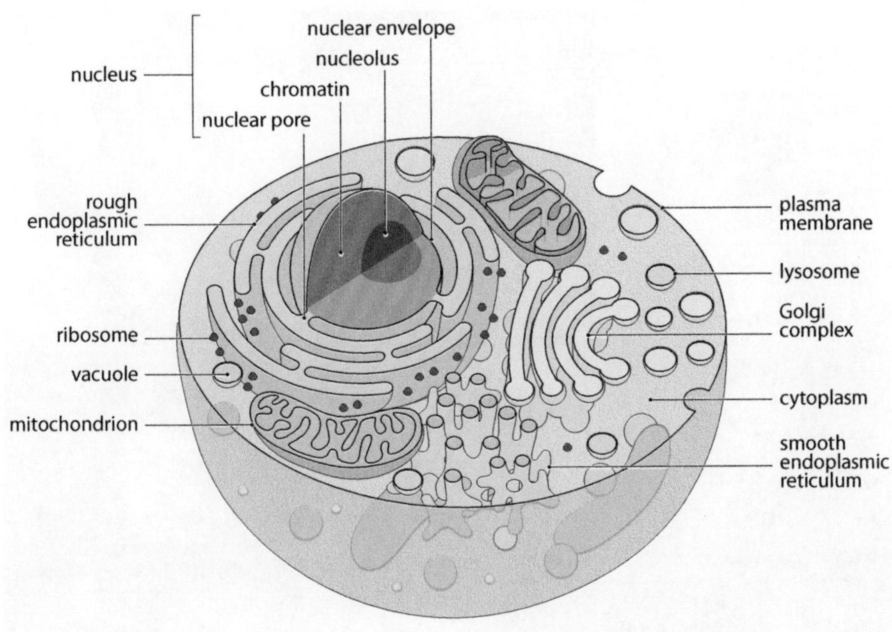

Figure 1.12 Inside a human cell. A human cell is compartmentalized by various membranes, allowing the segregation of biochemical processes.

is exposed to the extracellular milieu. The nuclear envelope is actually a double membrane, with a thin space between the layers; it houses the DNA. Human cells have nothing like a bacterial cell wall. Instead, the shape of human cells is maintained by internal protein scaffolding. There is plenty of room in the cytoplasm, between the nuclear membrane and the plasma membrane, for the things that distinguish one human cell type from another. In addition, the cytoplasm also contains other membrane-enclosed organelles. Among the most distinctive organelles are the mitochondria, which produce chemical energy for the cell. Mitochondria are like symbiotic organisms, replicating independently of the cell. They have two membranes, and within the inner membrane mitochondria have their own DNA, their own enzymes, and their own ribosomes. In addition to mitochondria, microscopy also reveals a network of bubble-like vesicles (the endoplasmic reticulum and Golgi complex) that carry protein and glycan products to the exterior of the cell.

Human DNA exists in two cellular locations: the nucleus and the mitochondria. The genomic DNA within the nucleus is divided between 23 pairs of homologous chromosomes. The mitochondrial DNA is a smaller circle of DNA, much like the *E. coli* genome. Whereas most of the DNA in *E. coli* encodes proteins, only a small fraction (1.5%) of the DNA in human cells encodes proteins. The human genome encodes only about 20,000–25,000 transcribed proteins—only five times as many proteins as one finds in *E. coli*. Thus, humans, accustomed to thinking of themselves as tremendously more complex than unicellular life, have a surprisingly modest parts list. How can a mere 25,000 genes give rise to a multicellular organism as complex as a human being (**Figure 1.13**)? To understand this complexity, we will need to further explore the origins of molecular diversity at the level of RNA, proteins, and beyond.

Figure 1.13 Millions of parts. A Boeing 747–400 aircraft is built from over 6 million parts. In contrast, humans require only about 20,000 different genes, which encode perhaps 100,000 different proteins or parts. (Courtesy of Wikimedia Commons.)

Figure 1.14 Humans are 99.9% identical at the genetic level, even though different people of different origins and different ages have markedly different appearances. (A, courtesy of Michael Schwart, CDC; B, courtesy of Nancy Michael, Florida Photographic Collection http://www.floridamemory.com; C and D, courtesy of Amanda Mills, CDC; F, courtesy of CDC.)

You cannot judge a cell by its genome

Strains of *E. coli* have similar structures but very different genomes, whereas the opposite is true for mammals such as mice and humans, which share 99% of the same genes. If *E. coli* were geneticists, they would easily mistake you for a mouse. All humans of the same sex, regardless of age, race, or ethnic origin, are 99.9% identical (**Figure 1.14**). The wide variation in human appearance and disease susceptibilities results from differences in a mere 0.1% of the human genome. Ultimately, analysis of genome sequences from many different humans may reveal genetic differences that underlie conditions such as Alzheimer's disease, osteoporosis, and diabetes. Many of these differences will lead to targets for the next generation of pharmaceuticals.

Each human individual is made up of roughly 220 different cell types, which have a dazzling array of sizes, shapes, and functions. With a few special exceptions, all of the somatic cells in your body share 100% genetic identity: nerve cells, muscle cells, white blood cells, photoreceptor cells, excretory cells, fat cells, bone-making cells, hair-producing cells, and skin cells (**Figure 1.15**). A major focus of **developmental biology** is to understand how the various human cell types arise during embryonic cell division. Ultimately, the differences between these cells are attributable not to differences in the genomic DNA sequences but to differences in gene expression—that is, which of these genes are "turned on" and which are "turned off."

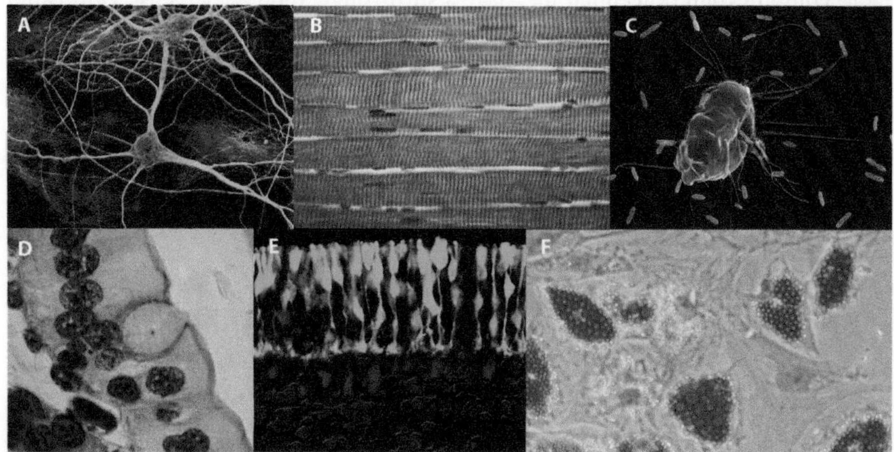

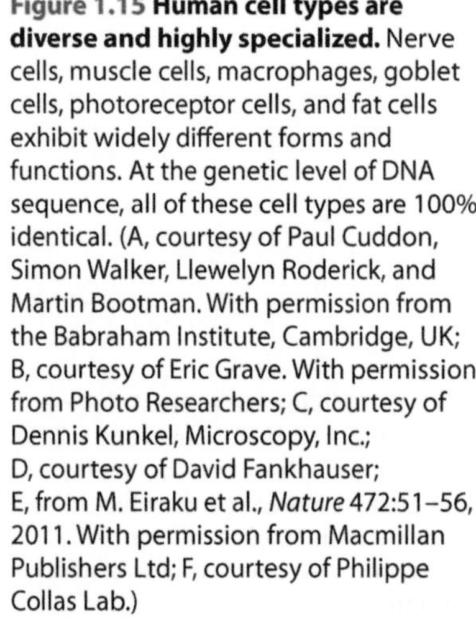

Figure 1.15 Human cell types are diverse and highly specialized. Nerve cells, muscle cells, macrophages, goblet cells, photoreceptor cells, and fat cells exhibit widely different forms and functions. At the genetic level of DNA sequence, all of these cell types are 100% identical. (A, courtesy of Paul Cuddon, Simon Walker, Llewelyn Roderick, and Martin Bootman. With permission from the Babraham Institute, Cambridge, UK; B, courtesy of Eric Grave. With permission from Photo Researchers; C, courtesy of Dennis Kunkel, Microscopy, Inc.; D, courtesy of David Fankhauser; E, from M. Eiraku et al., *Nature* 472:51–56, 2011. With permission from Macmillan Publishers Ltd; F, courtesy of Philippe Collas Lab.)

The observable phenotype belies the hidden genotype

Each gene or combination of genes, referred to as a **genotype**, results in an observable set of characteristics, called the **phenotype** (such as red hair or antibiotic resistance). Resistance to the antibiotic erythromycin is a phenotype, but there are many genetic variations that can confer resistance to erythromycin. Bacteria with the *mef* gene (abbreviated *mef*$^{+}$ and spoken mef positive) have a protein pump that exports erythromycin. Bacteria with the *ermA*$^{+}$ and *ermC*$^{+}$ genotypes produce an enzyme that methylates a single atom in the ribosome, preventing the binding of erythromycin (**Figure 1.16**). A **mutation** to the DNA sequence within an existing gene can result in erythromycin resistance. Mutation of a single DNA subunit in the gene that codes for the ribosomal RNA (for example a mutation termed A2058G) also makes bacteria resistant to erythromycin. Mutations to the gene that encodes the ribosomal protein L4 (for example a mutation termed K63E) can make bacteria resistant to erythromycin.

Cancer is a phenotype of human cells that arises from a combination of genetic changes. The phenotype of cancer is invasive, uncontrolled cell proliferation. Each cancer cell line usually exhibits a unique genotype. In the previous millennium, most of the drugs prescribed for leukemia targeted the phenotype of uncontrolled cell growth in the fervent hope that they would kill the cancer faster than the patient. That strategy often failed. Newer anticancer drugs are intended to distinguish cancer cells from normal cells at the level of genotypic variation, resulting in fewer side effects.

The field of chemical biology has entered this millennium armed with an abundance of genomic information, but we cannot yet make useful predictions about the workings of cells based solely on DNA structure. Genomic structure is essential knowledge for understanding cellular properties, but it does not enable us to predict the molecular workings of even the simplest cell. To use this information to the fullest extent, we will need to master the chemical principles that underlie all of the processes embodied in the central dogma and obeyed by each and every molecule in the universe.

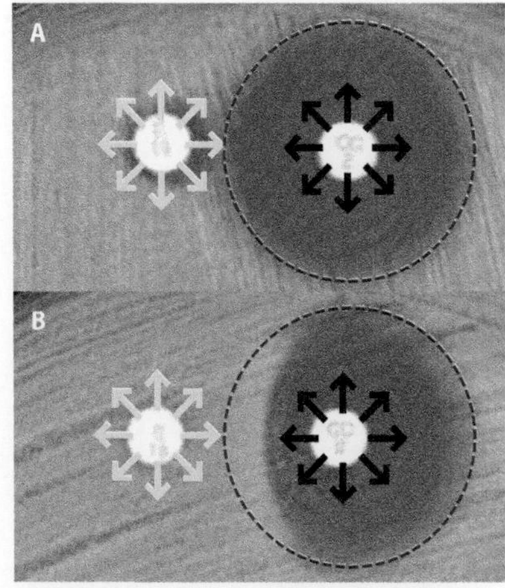

Figure 1.16 The phenotype can reveal an underlying genotype. Two different strains of *Staphylococcus aureus* exhibit different phenotypic responses to antibiotics. Antibiotics are diffusing out of these paper disks (arrows) creating potential zones of death within the bacterial lawn on the Petri dish. (A) An *ermC*$^{+}$ strain of *S. aureus* almost resistant to erythromycin (left disk) but highly susceptible to clindamycin (right disk). (B) A different strain of *S. aureus* is completely resistant to erythromycin (left disk), but highly susceptible to clindamycin (right disk). Notably, in the zone between both disks, the presence of erythromycin is inducing resistance to clindamycin, leading to a phenotypic D-shaped zone of death. This phenotype results from a special form of the *ermC* gene that can be translated by stalled ribosomes. (From K.R. Fiebelkorn et al., *J. Clin. Microbiol.* 41:4740–4744, 2003. With permission from the American Society for Microbiology.)

1.4 SOURCES OF DIVERSITY BEYOND GENOMES

The transcriptome is the collection of all of the RNA transcripts in a cell

In Chapter 4 we will immerse ourselves in the world of RNA, where we will meet an ancient and secretive class of molecules whose chemical similarity to DNA belies a pervasive set of functions. The central dogma (see Figure 1.3) emphasizes the role of RNA as a messenger, but the structural diversity available to RNA molecules allows it to exert control over gene expression at every level.

The complete collection of all RNA found in a cell is called a **transcriptome** (**Figure 1.17**). The transcriptional profile of a human cell is a direct readout of the active genes and thus depends on the tissue type and many other variables, such as the nutrients, extracellular signaling molecules, and temperature. The transcriptional profiles vary between cells, even cells from the same tissue. Thus, the exact details of the sample source and preparation become vitally important. By comparing the transcriptional profiles of different cells, we gain insight into which genes are being expressed and how the resulting RNA transcripts are being processed. For example, growing human cells in the laboratory at a slightly elevated temperature markedly alters the transcriptome, as the cells scramble to adjust for the shock of the higher temperature. The condition, termed **heat shock**, equips the cell with new specialized

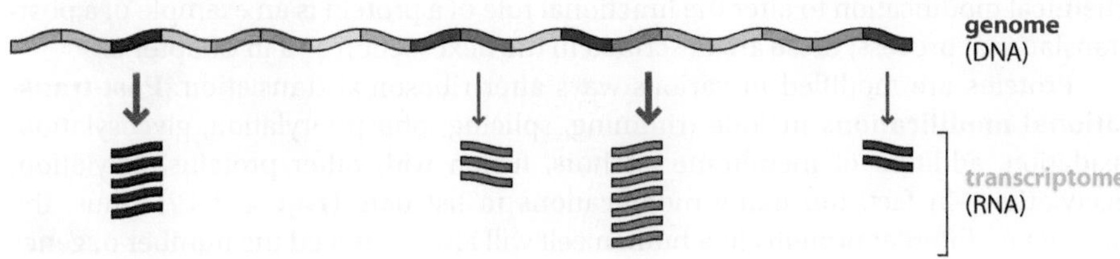

Figure 1.17 A simplified picture of the transcriptome. Transcription of selected genes, each at a different rate, leads to a collection of RNA transcripts called the transcriptome. Actual transcriptomes consist of thousands of types of RNA transcripts, each present in varying amounts.

mRNA transcripts; in addition, the levels of the standard mRNA transcripts are altered. Some genes will be transcribed at lower levels, but others critical for cell stability at the higher temperature will be produced more often. For example, some proteins fall apart at slightly elevated temperatures (increases of just 5 °C in some cases), and other proteins are deployed to stabilize and help remove such breakdowns. Thus, the identity and concentrations of individual mRNA transcripts within the transcriptome control the cell's future fate and current activities.

RNA splicing amplifies the diversity of the transcriptome

In bacterial cells, mRNA is immediately seized by ribosomes for translation into protein, but in human cells, RNA transcripts undergo substantial editorial processing, called **splicing** (**Figure 1.18**), that removes various stretches of RNA before it is ready to serve as mRNA. We cannot yet predict which pieces of human RNA will be spliced out by looking at the original DNA sequence. Because we cannot predict which parts of an RNA transcript will be spliced, it is impossible to predict protein sequences based solely on DNA gene sequences. Our predictive power is further diminished by the fact that splicing depends on environmental conditions. For example, the transcription factor ATF3 is spliced in two different ways depending on whether growth factors are present in the culture medium. Differential splicing allows a single human gene to code for more than one protein and amplifies the diversity of the human genome.

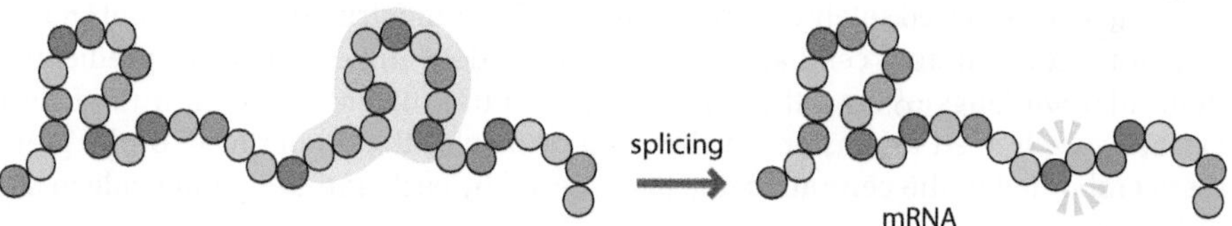

Figure 1.18 Splicing. Splicing refers to the removal of an internal segment, highlighted in yellow, from a biopolymer. In humans, RNA splicing leads to mRNA molecules that cannot be predicted from the original DNA gene sequences.

Post-translational modification of proteins amplifies the diversity of the proteome

Proteins are assembled from 20 universal building blocks called amino acids. Amino acids have a much broader range of chemical functionality than the nucleotide subunits that make up DNA and RNA. Folded proteins accomplish the bulk of activity in the cell, including the catalysis of specific reactions and formation of structural supports. Analogous to the transcriptome, the **proteome** is the complete collection of all proteins in a cell, organism, or tissue sample. Simple methods are available for isolating and sequencing mRNA and the proteins they encode, but it is still difficult to predict the full range of proteins within a human cell.

Analogously to the regulation of gene transcription by transcription factors, translation of mRNA into proteins is regulated by multiple translational control proteins. First, protein translation begins at a specific sequence near the beginning of the mRNA transcript. In humans, the eIF-2 protein assists in recognition of this start sequence. Illustrating the complexity inherent in the regulation of protein production, levels of protein production in eukaryotic cells can be decreased by covalent modification of eIF-2. This attenuation can take place in response to viral attack or heat shock. This chemical modification to alter the functional role of a protein is an example of a post-translational process; these are described in the next section and in Chapter 6.

Proteins are modified in various ways after ribosomal translation. **Post-translational modifications** include trimming, splicing, phosphorylation, glycosylation, oxidation, addition of membrane anchors, fusion with other proteins, alkylation, acetylation—in fact, too many modifications to list here (**Figure 1.19**). Thus, the number of different proteins in a human cell will always exceed the number of genes

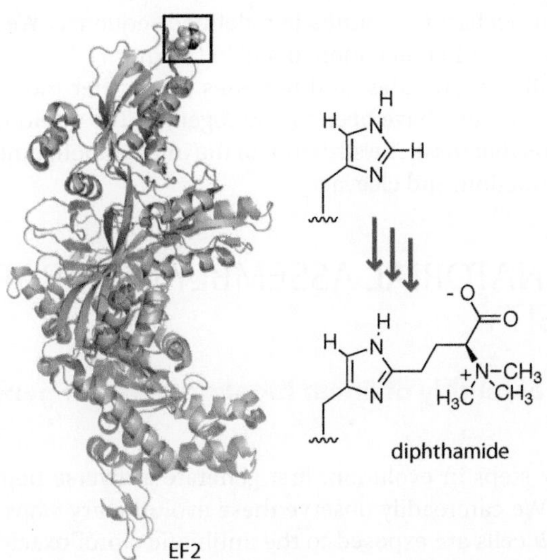

Figure 1.19 Post-translational modification. A single amino acid subunit (boxed) in elongation factor 2 (EF2) is modified post-translationally to produce a new subunit called diphthamide. EF2, with the modified diphthamide residue, is essential for protein synthesis. *Corynebacterium diphtheriae* shuts down human protein synthesis by producing toxin that seeks out EF2 and chemically modifies the unique diphthamide residue. (PDB 3B8H)

being expressed. The potential diversity of the human proteome greatly exceeds the approximately 23,000 genes in the genome, as a result of splicing and the potential for post-translational modifications. Like transcriptomes, proteomes vary considerably between different tissue samples from the same organism. Chemical biology seeks to understand such differences; for example, characterizing differences between seemingly identical cells from young and old humans could solve many fundamental questions in biology.

Beyond template-directed synthesis of biooligomers

Enzymes catalyze the synthesis of the three other types of biooligomers—namely, polyketides, oligosaccharides, and terpenes. For these three types of biooligomers, each subunit is added by a unique enzyme that does not use another biooligomer as a template, in contrast with the way in which RNA polymerase can transcribe DNA or the ribosome can translate mRNA. In general, the order of subunits in polyketides, oligosaccharides, and terpenes does not correlate with the order of the genes that encode the biosynthetic enzymes. Some polyketide synthase genes are important exceptions, and we discuss them in Chapter 8.

Many oligosaccharides are nonlinear and/or branched; thus, the sequence of subunits cannot correlate in a simple way with a linear DNA sequence or protein sequence. For example, the synthesis of a branched oligosaccharide with 11 types of bonds could require 11 different enzymes, encoded by 11 different genes (**Figure 1.20**). The genes can be in any order within the genome, even if the enzymes

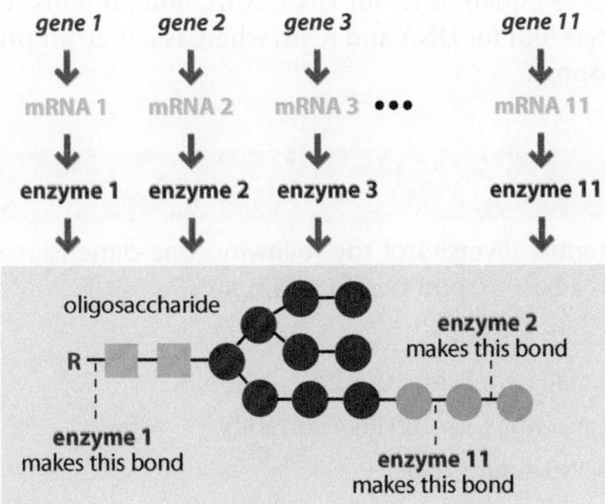

Figure 1.20 Oligomerization without a template. Oligosaccharides are constructed by many enzymes—one for each type of linking bond—without a biooligomer template. The order of genes that encode the enzymes is unrelated to the structure of the final oligosaccharide.

assemble the oligosaccharide subunits in a defined sequence. We discuss the structures of oligosaccharides in much more detail in Chapter 7.

Polyketides, oligosaccharides, and terpenes are further modified by additional enzymes after the subunits have been linked together. The additional enzymes catalyze many reactions that mask the structure of the initial biooligomer, such as cyclization, oxidation, reduction, and cleavage.

1.5 COMBINATORIAL ASSEMBLY GENERATES DIVERSITY

Combinatorial assembly of linear biooligomers can generate massive diversity

There are two key steps in evolution: first generate a diverse population, and then select for fitness. We can readily observe these evolutionary steps in *E. coli* populations. When *E. coli* cells are exposed to the antibiotic ciprofloxacin, the bacteria initiate an SOS response characterized by a 10,000-fold increase in the rate of genetic mutations. Almost all of the resulting mutants are less fit for survival, even under ideal growth conditions. A few mutants, however—and it only takes one—are resistant to the antibiotic. The mutation rates subside as these healthy, resistant bacteria begin to thrive and regenerate healthy colonies.

All organic molecules have combinatorial architectures based on atomic subunits. For example, 31 stable compounds can be assembled from four carbon atoms, eight hydrogen atoms, and one oxygen atom. However, these varying molecules cannot be accessed through a single efficient synthetic route. The origin of diverse populations can be found at the molecular level of detail in the one-dimensional oligomeric structures of DNA, RNA, proteins, oligosaccharides, polyketides, and terpenes. One-dimensional oligomeric architectures make it possible to generate massive, diverse collections of molecules from a few simple joining reactions and a small number of subunits (**Figure 1.21**). Like words in the English language, the meaning (that is, the function) of a biooligomer depends on both the constitution and the order of the letters. The total number of possible "words" of a particular length depends on the number of letters in the alphabet. In a similar way, one can calculate the number of possible oligomers based on the the number of types of subunits and the number of subunits in the oligomer (Equation 1.1). With the 26 letters of the English alphabet, one could construct 26^4, or 456,976, four-letter words. The order of the letters is essential, too; the letters "N-O-I-T-U-L-O-V-E" have no particular meaning in the English language, but spelled backward they produce a word with a well-defined meaning: evolution. As with spoken words, the order of subunits in biooligomers determines their functions.

$$\text{number of possible oligomers} = (\text{number of types of subunits})^{\text{length of oligomer}} \tag{1.1}$$

We can customize Equation 1.1 for DNA, RNA, and proteins. That is, the number of types of subunits is four for DNA and RNA, whereas it is 20 for proteins synthesized by the human ribosome.

Figure 1.21 Child's play. More than 300 million unique chains, 11 beads long, can be assembled from just six types of children's toy beads. (Courtesy of Edushape.)

Problem 1.2

Calculate the potential diversity of the following one-dimensional constructs and give the answer as a base 10 number (for example 3.7×10^4).

A English words with 10 letters.

B An unusually small gene with 1000 DNA subunits.

C A microRNA transcript with 100 RNA subunits.

D A protein with 100 amino acids.

Combinatorial synthesis can be used to synthesize DNA libraries

The principles of combinatorial assembly can be used to synthesize libraries of biooligomers. In Chapter 3 we describe the chemical synthesis of DNA. The chemistry is so highly optimized that DNA oligomers with more than 50 subunits can be reliably synthesized by machines on a 50 nmol scale. That may not sound like much, but 50 nmol is 3×10^{16} molecules. If you carry out each coupling step with an equal mixture of all four DNA subunits, you will generate a combinatorial library of DNA molecules in which each strand has the same length but a different sequence. The potential diversity of a DNA strand with 50 subunits is 4^{50}, or about 10^{30} molecules. If you somehow made all of the possible DNA 50-mers, collectively they would weigh more than 25,000 kilotons. Thus, automated DNA synthesis makes it easy to assemble combinatorial libraries of DNA, but practical considerations usually keep us from preparing DNA samples that contain all of the possible variants. The tools of molecular biology dramatically amplify the power of DNA libraries. If you have a DNA library, you can use RNA polymerase to generate a complementary RNA library. Similarly, that RNA library can be used as a template for ribosomal translation to generate a protein library.

Problem 1.3

What length of DNA (based on the number of subunits) has a potential diversity close to 3×10^{16}?

Modular architecture lends itself to the synthesis of non-natural chemical libraries

For example, a non-natural class of biooligomers known as peptoids can be assembled by organic synthesis. Up to 8000 different trimeric peptoids can be created from 20 different subunits through the iterative application of just one efficient bond-forming sequence (**Figure 1.22**). Collections of molecules are often referred to as libraries. When a library of 5120 peptoid trimers was screened for activity against apoptosis, a form of cell suicide, potent inhibitors were identified. We discuss apoptosis in more detail in Chapters 6 and 9. The apoptosis inhibitor was found to bind selectively to the protein Apaf-1 *in vitro*, but the inhibitor was unable to rescue living cells because it could not cross cell membranes. Fortunately, a cyclic, membrane-permeable version of the inhibitor, with fewer hydrogen-bonding groups, was shown to act on living cells.

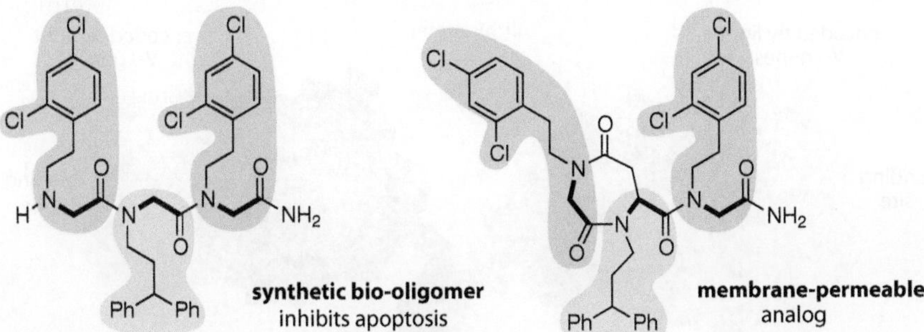

Figure 1.22 Synthetic peptoid oligomers. An inhibitor of cell suicide was identified from a combinatorial library of 5120 synthetic peptoids; the individual peptoid subunits are highlighted. The compound binds selectively to the protein Apaf-1. A membrane-permeable version of the inhibitor was active against U937 lymphoma cells. Ph is a commonly used abbreviation for a phenyl group (C_6H_5).

The concepts of combinatorial architecture can be applied to any type of molecule that is assembled from modular subunits, even when the subunits are connected using completely different types of chemical reactions (**Figure 1.23**). For example, a

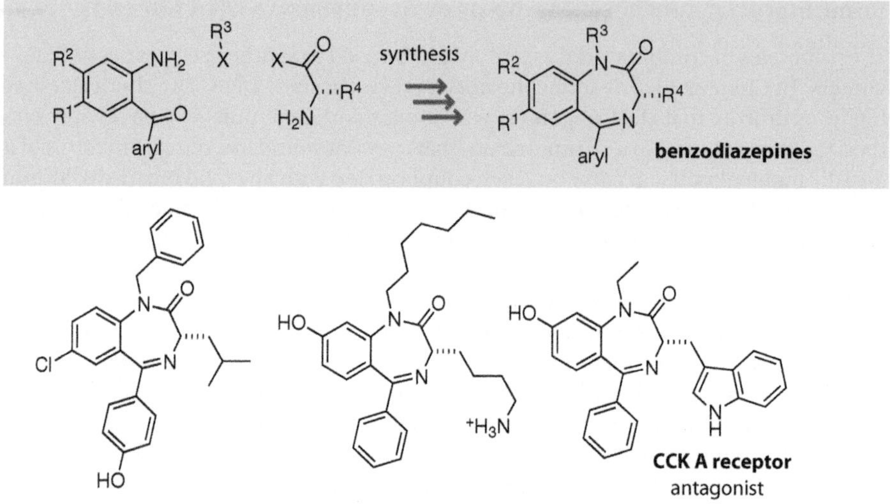

Figure 1.23 Combinatorial synthesis of nonoligomeric small molecules. Parallel synthesis can be used to generate diverse libraries of 1,4-benzodiazepines, all of which have drug-like structural characteristics—namely, low molecular mass, membrane permeability, and so on.

synthetic route involving an acylation reaction followed by an imine condensation followed by an S_N2 reaction could be used to make large numbers of synthetic molecules from a small set of starting materials. As long as the reactions are high yielding, with no side reactions, solid-phase synthesis (discussed further in Chapters 3 and 5) and robotics can be used to carry out the synthesis. The important advantage of chemical synthesis is that it offers direct access to small molecules with drug-like properties, such as oral bioavailability, low immunogenicity, and slow metabolism. Since the introduction of parallel synthesis in the 1990s, the synthesis and screening of synthetic libraries has become a standard tool in the development of pharmaceuticals.

The human immune system uses combinatorial biosynthesis

Your immune system synthesizes combinatorial libraries of **antibody** proteins (sometimes called immunoglobulins) through the combinatorial assembly of genetic modules. This allows your body to generate libraries of different B lymphocytes, ready to fend off a wide range of infections (**Figure 1.24**). Each antibody can potentially bind to a unique nonhuman molecule, such as those found on viral and bacterial pathogens. When an influenza virus binds to a surface-bound antibody on a B lymphocyte, the binding event turns on the genes for cell proliferation (**Figure 1.25**). The transduction of extracellular signals into changes in transcription are fascinating, and we discuss

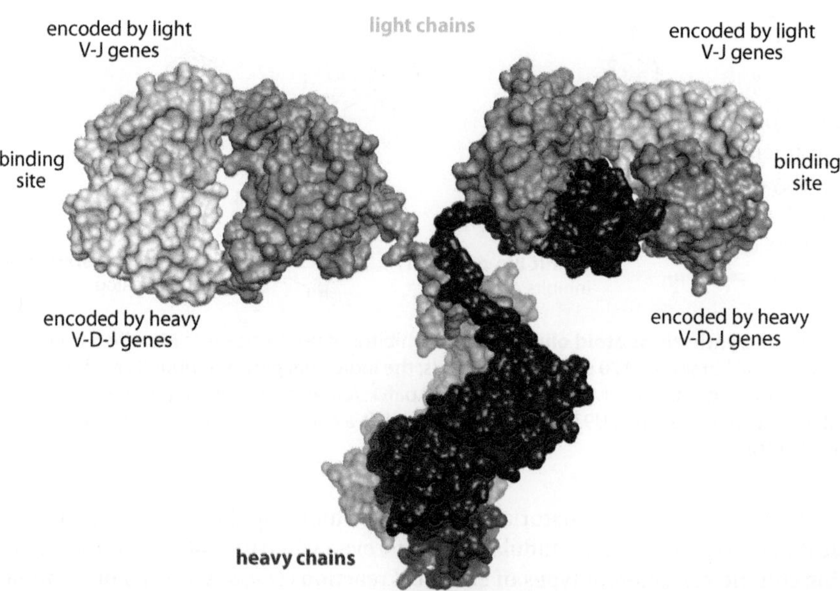

Figure 1.24 A typical immunoglobulin protein. This antibody is made up of two light chains (green) and two heavy chains (blue). The variable regions of the antibody, encoded by the V-J and V-D-J genetic modules, contain the binding sites that recognize foreign proteins. (PDB 1IGT)

many more examples in Chapter 9. As the B lymphocytes proliferate, they generate additional variations of the antibody gene, some that bind viruses more tightly and some that bind viruses less tightly. The tighter binders continue to proliferate and ultimately release antibodies into the extracellular medium. The human B-lymphocyte response is very much a model for evolution, involving diversity-generating steps and selection steps. A similar system of combinatorial gene assembly is used to generate libraries of T lymphocytes, each displaying a different T-cell receptor, except that T-cell receptors are not released into the surrounding medium.

An antibody is composed of four peptide chains: two identical heavy chains and two identical light chains. The gene encoding the heavy chain is constructed from three genetic modules, each selected from a pool. There are 40 variable (V) modules, 25 diversity (D) modules, and six joining (J) modules, offering a theoretical diversity of $40 \times 25 \times 6 = 6000$ heavy chains (**Figure 1.26**). There are two types of light chains, each constructed from two genetic modules. Kappa (κ) light chains arise from 40 variable (V) modules and five joining (J) modules; lambda (λ) light chains arise from 30 variable (V) modules and four joining (J) modules. When summed, there are $(40 \times 5) + (30 \times 4) = 320$ possible types of light chains. Overall, there are almost 2 million ways to combine the heavy chains with the light chains. The human immune system has harnessed an additional source of diversity that scares away most chemists: it is sloppy. Instead of joining genetic modules in precise ways, it does so in an imprecise way that further amplifies the potential diversity by a factor of more than 10 million. B lymphocytes employ one additional diversity-generating trick. When they proliferate in response to antibody binding, the V, D, and J modules are much more subject to point mutations than the rest of the genome, about one mutation per module per cell division. This hypermutation accesses further genetic variation that was not present in the original genome.

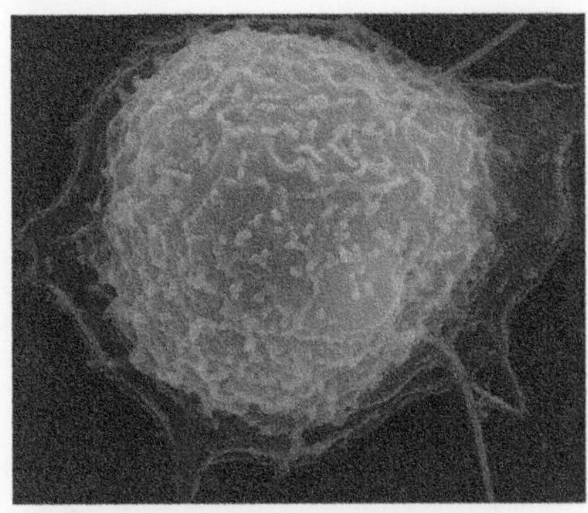

Figure 1.25 Antibody factory. The surface of B lymphocytes is dotted with antibodies (too small to see) that serve as receptors against foreign molecules. (Courtesy of Louisa Howard, Dartmouth College.)

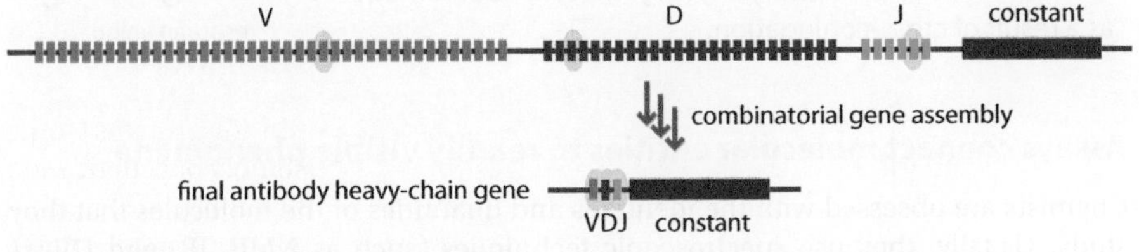

Figure 1.26 Combinatorial antibody gene assembly. Combinatorial assembly of V, D, and J gene modules with a constant gene module generates a final antibody heavy-chain gene.

1.6 SOME COMMON TOOLS OF CHEMICAL BIOLOGY

Chromophores reveal invisible molecules

Most biological molecules are colorless, making them difficult to quantify or locate by light microscopy. Molecules that absorb or emit light at visible wavelengths are particularly valuable for assays and microscopy. Cross-conjugation is a common feature of molecules with high extinction coefficients, particularly when charge separation competes with aromaticity. For example, the *p*-nitrophenolate anion is stabilized by resonance delocalization of the negative charge, but delocalization of the negative charge into the nitro group generates a cross-conjugated nonaromatic π system (**Figure 1.27**). Solutions of *p*-nitrophenolate strongly absorb violet light (405 nm) and, when you subtract violet light from white light, the material appears yellow.

When visible chromophores absorb a photon of light, they shake off the excess energy through bond vibrations. When a fluorescent chromophore absorbs a photon, only a small amount of the energy is lost to vibrations; the remaining energy is emitted as a photon of lower energy (and higher wavelength) than the photon that was absorbed. Detecting emission at a different wavelength from excitation minimizes background caused by reflections and scattering. Many fluorescent chromophores exhibit frustration between resonance and aromaticity (**Figure 1.28**). Fluorescent

A

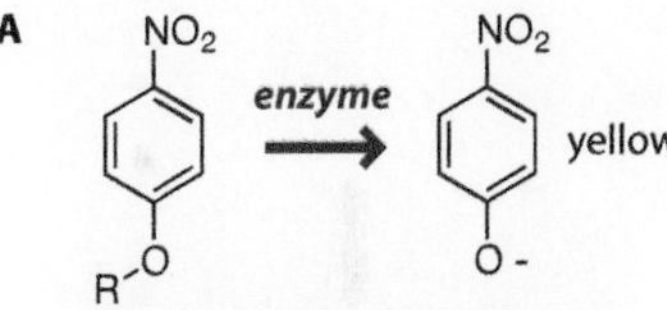

B

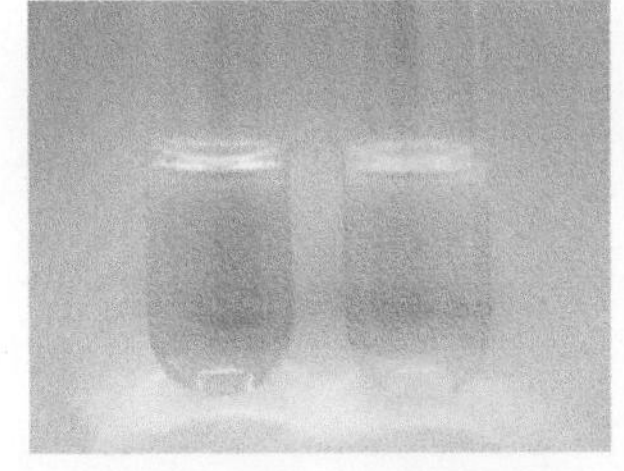

Figure 1.27 A simple color test. (A) A colorimetric substrate releases yellow *p*-nitrophenolate anion, revealing the presence of the enzyme α-galactosidase. (B) The intensity of the *p*-nitrophenolate anion is attributable to frustration between charge delocalization and aromaticity. (A, from S.-F. Chien et al., *J. Nanomaterials* Article ID 391497:1–9, 2008.)

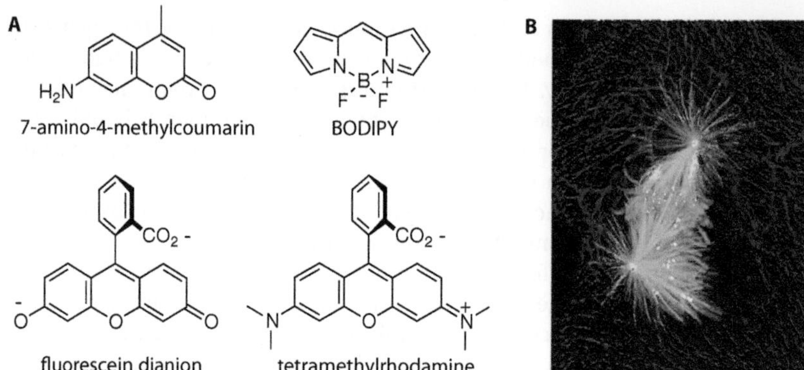

Figure 1.28 Brilliant molecules. (A) Common fluorophores used in biological assays and fluorescence microscopy. (B) A newt lung cell undergoing mitosis after staining with fluorescently labeled antibodies. (B, courtesy of Alexey Khodjakov.)

chromophores such as aminocoumarins absorb ultraviolet photons (which you cannot see) and emit violet photons (which you can see). Fluorescein absorbs cyan light and emits green light. Derivatives of tetramethylrhodamine emit red light. BODIPY derivatives are available with a wide range of emission colors. Most of the fluorescent micrographs of cells that show dazzling colors were generated with chemical derivatives of these types of **fluorophores.** The process involves highly specific antibodies that have been chemically linked to synthetic fluorophores.

Problem 1.4

Draw all of the resonance structures for the resorufin anion that satisfy the octet rule (that is, no carbocations). In each resonance structure, identify any rings that are nonaromatic as a result of cross-conjugation.

resorufin anion

Assays connect molecular entities to readily visible phenomena

Chemists are obsessed with the identities and quantities of the molecules that they study. Usually, they use spectroscopic techniques (such as NMR, IR, and UV-vis spectroscopy) or mass spectrometry to establish the structural identity of pure molecules and then employ assays to determine the concentration within mixed samples—some assays are even based on spectroscopy. Selective detection of molecules in biological samples is difficult because the biological molecules rarely have unique chromophores; they are also present in small amounts and exist as mixtures of chemical homologs. By coupling chemical entities such as genes, mRNA, enzymes, and carbohydrates to observable phenomena, we can infer their presence or absence, even in a system as complex as a cell. Typical examples of observable phenomena are precipitation, colony growth, and absorption of light. Recall from Figure 1.16 that bacterial growth on agar plates is a reliable indicator of the presence or absence of antibiotic resistance genes.

Antibodies are a useful class of reagents for the detection of specific biological molecules. Hemagglutination assays, for example, are based on the ability of specific antibodies to crosslink erythrocytes or to prevent viruses from crosslinking erythrocytes. Cellular crosslinking is easily detected by the precipitation of a cellular mass. In the early 1950s the Hungarian physician Gyula Takátsy initiated the modern form of the high-throughput assay by developing a method for accurately dispensing small amounts of serum. To make the best use of his dispensing technique, he machined plates with 8 rows and 12 columns of miniature wells so that up to 96 hemagglutination assays could be carried out in parallel. The 96-well microplate is now a standard tool in chemical biology.

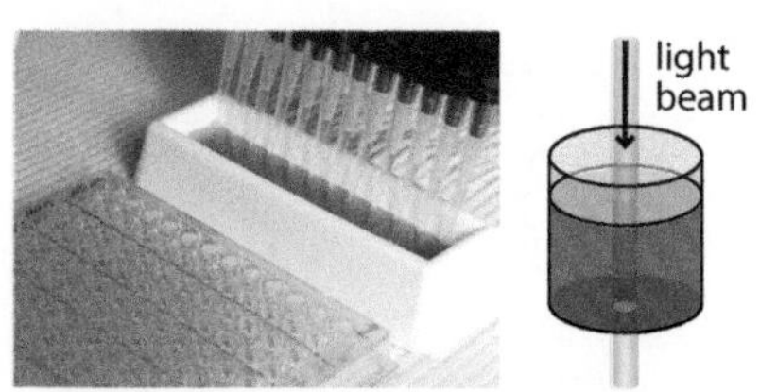

Figure 1.29 Parallel assays. Multichannel micropipettors can simultaneously dispense liquid into multiple wells of a 96-well microplate with high accuracy. Each well acts as both a reaction vessel and a cuvette. The absorbance of the solution in each and every well can be read in seconds with a microplate reader. (Left, courtesy of Linda Bartlett. With permission from the National Cancer Institute.)

The essential advance for high-throughput assays came from the ability to manufacture uniform plates with flat, optically transparent bottoms, allowing the absorbance of the solution in each well to be measured with a vertical beam of light (**Figure 1.29**). Each well of a 96-well plate can serve as a reaction vessel, a cuvette, or even a Petri dish. Microplate readers can measure the absorbance in every well of a 96-well microplate in a few seconds. Plates with 6, 24, 96, 384, or even 1536 (16 × 96) wells are widely available, all with the same 8.5 cm × 12.8 cm footprint. Multichannel pipettors allow multiple wells of a 96-well plate to be filled simultaneously. Commercial robotic systems can fill all the wells of a microplate simultaneously. In the quest for drug development leads, pharmaceutical companies use automated systems to screen tens of thousands of compounds in 96-well or 384-well format to test their effects on enzyme-catalyzed reactions or living cells.

Powerful microbiological screens reveal interesting chemical phenomena

Robotics and miniaturization have greatly increased the capacity of high-throughput assays, but they do not represent a truly parallel approach. Each compound to be tested must be aliquoted into an individual well, and most plate readers measure absorbance one well at a time, albeit very fast. If you want to screen truly massive collections of molecules—say, a billion—you cannot do anything one at a time or the logistics would be unmanageable. For example, testing a billion compounds in 384-well microplates would require enough plates to cover four soccer fields. Screening high-diversity collections requires parallel assays on a miniature scale. Cells are nearly ideal for small-scale work because a single tiny cell can proliferate to produce large numbers of identical clones, enough to determine the structures of the molecules inside. Auxotrophic bacterial selections are based on strains of bacteria that are lacking a gene necessary for growth. For example, when a clonal population of Δ*fes E. coli*, lacking an enzyme needed for the release of iron, was spread onto nutrient agar, the bacteria failed to generate colonies (**Figure 1.30**). However, when a diverse library of 1.5 million different variants of Δ*fes E. coli*, each expressing a different test protein, were plated on the same medium, some of the bacteria benefited from the test protein and thrived. These types of auxotrophic selections, where one screens for cell growth/division, are extremely space-efficient because the nonviable bacteria take up virtually no room on the plate. A single milliliter of bacterial culture readily holds more than a billion bacteria and it is straightforward to fit up to a billion different bacteria on a single Petri dish.

The classical Ames test for chemical mutagens is a screen that tests the ability of chemicals to induce genetic mutations into *Salmonella typhimurium* bacteria. The auxotrophic T100 *Salmonella* strain used in the Ames test has a single mutation in the gene that encodes an enzyme essential for histidine biosynthesis. Transcription and translation of the mutant gene leads to a defective enzyme. When plated on histidine-deficient nutrient agar, T100 cells cannot proliferate unless they undergo a DNA reversion mutation back to the correct coding sequence. Chemical mutagens increase the rate of random mutations, including the mutation that rescues histidine biosynthesis. The Ames test has often been used as evidence for the potential carcinogenicity of synthetic chemicals. However, in large-scale random tests of natural and synthetic compounds that repel insects from crops, Ames found that natural compounds were just as likely as synthetic compounds to have potential carcinogenic effects.

The major weakness of bacterial selections is that cell populations are resilient, and natural mutations offer many unexpected mechanisms for surviving stringent selections. Imagine creating a library of enzymes (in a large population of *E. coli*) in the hope that one will catalyze the hydrolysis of phosphorus–nitrogen bonds. If you plate the library of bacteria on nutrient agar containing microcin C7, a small number of the bacteria would probably generate colonies. However, the bacteria that grow into colonies would probably do so for reasons unrelated to P–N bond cleavage, and it would require years of effort to figure out the strategy that a colony of bacteria used to survive. Some of the survivors will have mutant enzymes that acetylate the microcin C7. Other survivors will produce more pumping proteins that export microcin C7 faster than it enters the cell. Other survivors will produce vast amounts of natural enzymes that have weak phosphoramide-cleaving activity.

One can readily screen bacterial libraries for phenotypic traits other than survival. For example, moving the biochemical machinery for the production of lycopene (a

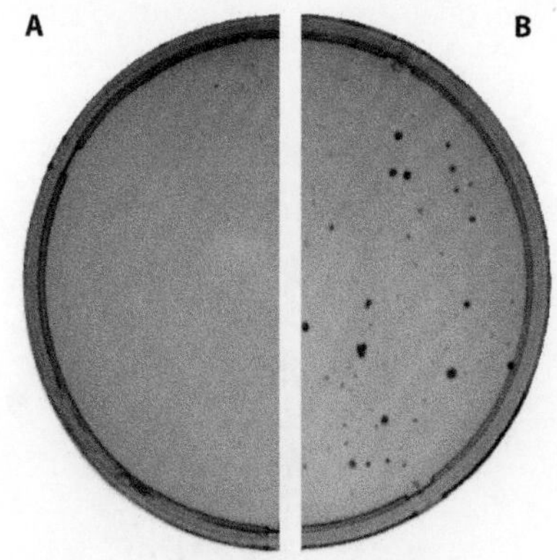

Figure 1.30 Powerful auxotrophic selection. (A) *fes*⁻ *E. coli* do not generate colonies (shadow spots), because they lack the ability to metabolize iron. (B) When each member of the *fes*⁻ bacterial population expresses a different test protein (out of 1.5 million variants), some of the bacteria thrive and generate colonies. (From M.A. Fisher et al., *PLoS ONE* 6:e15364, 2011.)

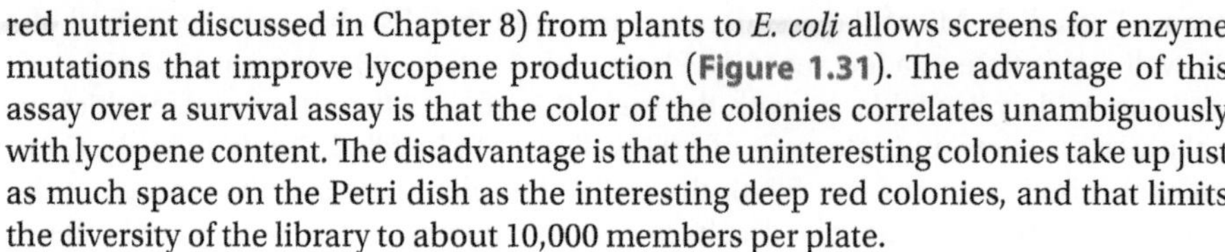

red nutrient discussed in Chapter 8) from plants to *E. coli* allows screens for enzyme mutations that improve lycopene production (**Figure 1.31**). The advantage of this assay over a survival assay is that the color of the colonies correlates unambiguously with lycopene content. The disadvantage is that the uninteresting colonies take up just as much space on the Petri dish as the interesting deep red colonies, and that limits the diversity of the library to about 10,000 members per plate.

Figure 1.31 Screening colonies. Bacteria are easily screened for the production of lycopene, an organic pigment found in tomatoes, simply by looking at the color of the bacterial colonies. (Courtesy of Harris Wang, Harvard University. From M. Baker, *Nature* 473:403–408, 2011.)

Viruses deliver genes efficiently

Viruses are small packages of genetic material that hijack the biosynthetic machinery of a cell to make additional copies of the virus. Some contain DNA; others contain RNA. They are usually extremely specific to one cell type. For example, the human immunodeficiency virus targets only one type of T lymphocyte in the human body. It does not target other cells and it does not target other organisms. Viruses are well known as the pathogenic agents that cause all manner of human diseases, such as Ebola, smallpox, AIDS, severe acute respiratory syndrome, papillomas, herpes, influenza, and the common cold. They also attack crops, livestock, and pets—even the gastrointestinal bacteria that assist human digestion. As far as we know, viruses infect every type of free-living organism. They also infect computers. Computer viruses are aptly named because they mimic biological viruses in almost every respect except for a general lack of evolutionary capability.

In spite of their notoriety, viruses have several important properties that make them useful for molecular biology. First, they deliver genetic material *efficiently* to cells (**Figure 1.32**). Viral gene delivery has been ruthlessly honed by evolution. So far, all of the nonviral methods (such as carrier molecules, heating, electrocution, and ballistics) that have been developed for delivering genes into cells are both disruptive and inefficient. A second important property of viruses is that they encode enzymes that are better translated, fold more efficiently, and catalyze better than the enzymes of the host cell. Viral enzymes have to be superior for the virus to take over the cell. Some of the most efficient enzymatic tools used by molecular biologists were obtained from viruses. Third, viruses are lean and simple. Many human viruses gain added complexity by coating themselves in membrane material from the host cell (a complex mixture of lipids, proteins, and glycans), but ultimately all viruses encapsulate relatively small genetic payloads that encode a small repertoire of proteins. Some viruses are small enough for us to model the positions of all of the atoms in the virus. No living cell has that level of simplicity.

Viruses that infect bacteria, called **bacteriophage**, or phage for short, are particularly useful, because the bacterial hosts are cheap and easy to grow. Phages are valuable for packaging DNA and making combinatorial libraries.. Packaging a synthetic DNA library into phage is inefficient, but once accomplished, each member of the phage library is capable of efficiently transfecting a bacterium to produce a limitless supply of copies (**Figure 1.33**). Importantly, you can package libraries of phage genes

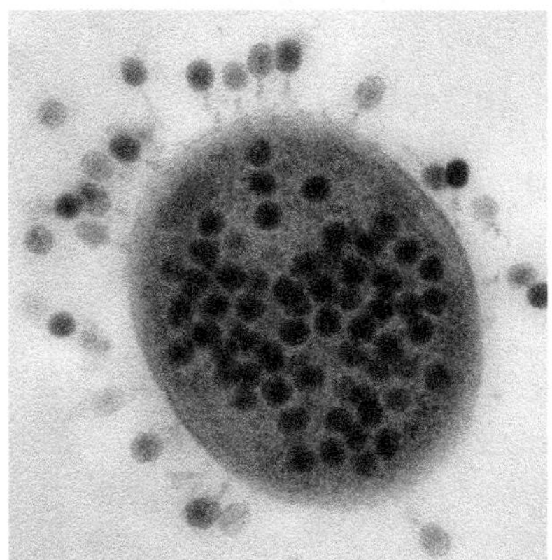

Figure 1.32 Transfection. This scanning electron micrograph of T bacteriophages attacking *E. coli* reveals the relative sizes of bacterium and virus. It takes only one phage to transform the bacterium into a virus factory. (Courtesy of John Wertz, Yale University.)

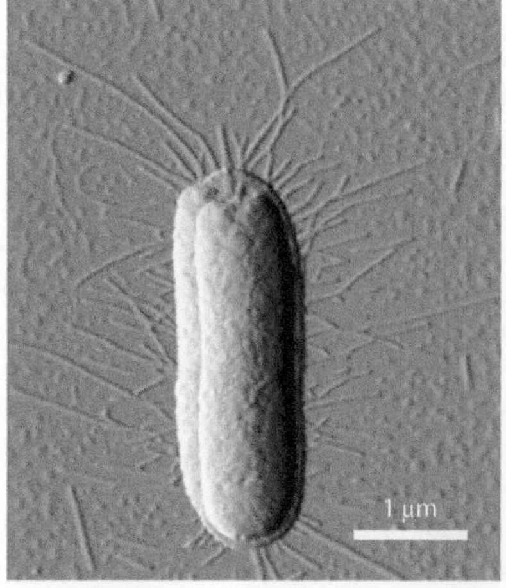

Figure 1.33 Amplification. Following transfection, atomic force microscopy reveals the extrusion of many M13 filamentous phage from an *E. coli* bacterium. (From M. Ploss and A. Kuhn, *Phys. Biol.* 7:045002, 2010. With permission from IOP Publishing.)

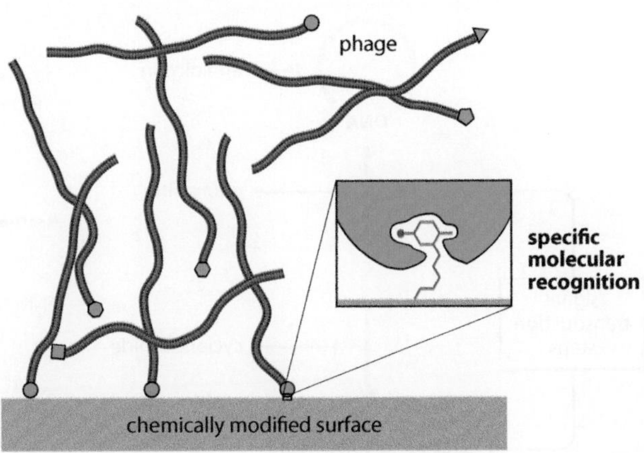

Figure 1.34 Virus libraries. Libraries of millions of filamentous phage, each with a different protein at the tip, can be screened for the ability to recognize molecules immobilized on a surface.

in phage that display the corresponding proteins on the viral surface. Libraries of proteins displayed on bacteriophages are a powerful tool for selection in chemical biology.

Vast libraries of proteins can be screened *in vitro* using bacteriophages

In vitro selections involving viruses accommodate even greater diversity than selections involving living cells. For example, one can conveniently generate and work with solutions containing more than 10 billion infective bacteriophages per milliliter. Like a bacterium, a single virus has the potential to proliferate to give a vastly larger population of identical clones. Wild-type filamentous phages have proteins designed to bind tightly and specifically to bacteria, but one can alter the viral genome so that novel proteins are displayed on the virus. Using this phage display technique, one can conveniently screen vast libraries of viruses, each displaying a novel protein, for the ability to cling to a surface. Thus, if you wanted to screen a library of proteins displayed on phages for the ability to bind tightly to estrogen, you would chemically attach estrogen molecules to the surface of, for example, a microplate well (**Figure 1.34**). Bacteriophages with proteins that bind to estrogen would stick to the derivatized well, whereas other bacteriophage would not bind at all. The unbound phages are readily rinsed away. Once the selection is complete, the phages that bound to the well could be detached and used to infect bacteria, re-establishing a vast population of clones. Many companies sell microplates with chemically reactive surfaces for use in these kinds of experiments.

In vitro screens of DNA and RNA push the limits of library diversity

The largest libraries that have been created and screened are based on DNA. Combinatorial chemical synthesis can conveniently be used to create libraries containing more than 10^{13} DNA molecules. Some DNA molecules fold into unique three-dimensional shapes that are capable of binding tightly and specifically to other biomolecules or even catalyzing chemical reactions, much like protein-based enzymes. DNA libraries are usually screened for the ability to bind to chemically immobilized molecules, much like phage libraries. Those that bind tightly and specifically can be amplified into larger quantities with DNA polymerase. DNA and RNA libraries are usually much larger than phage-based protein libraries; however, DNA possesses a limited repertoire of chemical functionality. DNA libraries can be transcribed with RNA polymerase to generate RNA libraries, which can be screened and amplified in a manner similar to DNA (**Figure 1.35**).

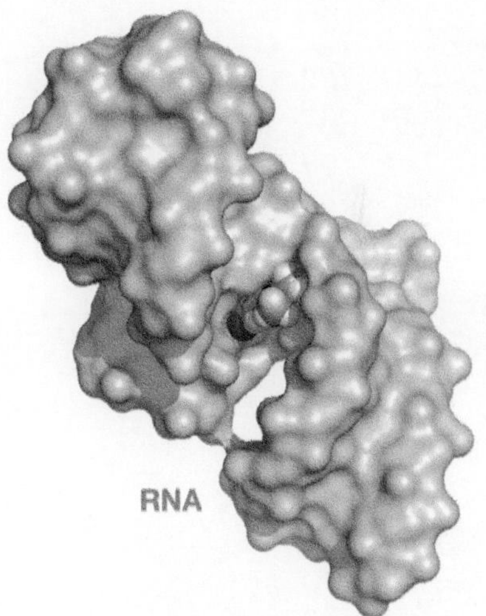

Figure 1.35 The needle in the haystack. This RNA molecule was selected from a pool of 5×10^{15} RNA molecules. It folds into a unique shape that binds to the fluorescent dye *N,N,N',N'*-tetramethylrosamine. (PDB 1FIT)

Small molecules take control

Chemical biologists prefer to use small molecules to control cell function and to report on cellular processes. The main advantage of small-molecule effectors is that they can

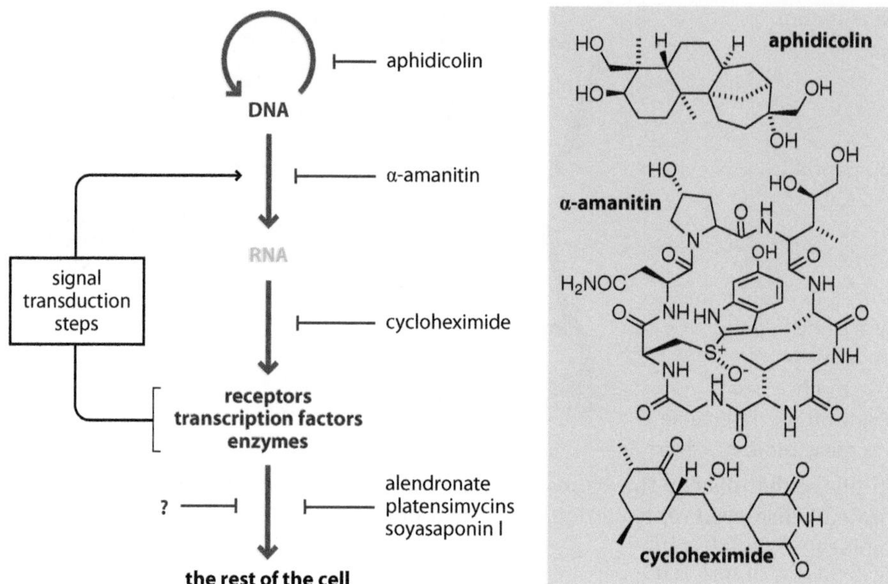

Figure 1.36 Limited control. We have small-molecule inhibitors of many steps in the central dogma, but the signal transduction pathways that control transcription have been difficult to sort out and target selectively. A line with a bar at the end symbolizes inhibition.

be membrane permeable. It is difficult to overstate the advantages of a molecule that can slip undetected through the defensive system of membranes that protect all living cells, gaining access to high-value targets that are deep inside the cell. In contrast, few biological macromolecules can cross cell membranes without some kind of clumsy assistance.

The main disadvantage of small-molecule effectors is that it is hard to find ones with high selectivity. High potency is also desirable, but it is not as important as selectivity. Potency is less important than selectivity because, in theory, you can always increase the concentration of a weak inhibitor, provided it is selective. In practice, solubility limits the concentrations that can be achieved with small molecules. Selectivity is paramount if you want to use a small molecule as a tool to understand cell biology. For example, if you want to know whether the phosphorylation of the transcription factor STAT1 is essential for macrophage activation, you could add staurosporine, a potent inhibitor of protein phosphorylation, and see whether macrophages respond to bacteria. Unfortunately, macrophages contain hundreds of phosphorylating enzymes that target hundreds of proteins, and staurosporine inhibits most of those phosphorylating enzymes. If you use a nonselective inhibitor such as staurosporine, you cannot draw any specific conclusions, even if the compound inhibits macrophage activation.

Small molecules have already been identified that inhibit the main steps in biooligomer assembly (**Figure 1.36**). Most were isolated from organisms that use them to fend off predators, but synthetic chemistry is beginning to overtake nature as a source of selective inhibitors. The fungal natural product aphidicolin, for example, is a selective inhibitor of DNA polymerase α and ε. The mushroom toxin α-amanitin is a selective inhibitor of RNA polymerase II. The bacterial natural product cycloheximide inhibits ribosomal protein synthesis (in humans, but not in bacteria). These inhibitors may seem large and complex compared with the molecules one typically sees in an introductory organic chemistry class; however, as we will see shortly, they are minute compared with typical biooligomers such as DNA, RNA, and proteins. There are many enzymes involved in the construction of terpenes, polyketides, glycans, and other molecules that make up the rest of the cell. A few can be inhibited selectively with small molecules. For example, the osteoporosis drug alendronate is a specific inhibitor of terpene assembly. The bacterial natural product platensimycin inhibits polyketide assembly. A soy natural product, soyasaponin I, selectively inhibits the glycosylation of proteins.

Arguably, the most interesting steps in the control of biooligomer synthesis are those that control the transcription of genes—specifically, the process by which extracellular signals are transduced into changes in gene expression. These signal transduction pathways determine the differences between various cell types and between healthy cells and diseased cells. Decades of convoluted experiments have outlined the rough details of the seven main signal transduction pathways, and we discuss them in more detail in Chapter 9. Many of the steps are still poorly understood. Selective inhibitors of signal transduction pathways would allow biologists to test hypotheses better and offer medicinal chemists ideal leads for drug design.

Short RNA molecules silence gene expression

Short RNA molecules, about 20 subunits in length, elicit a destructive response in human cells that leads to the cleavage of related RNA sequences. This phenomenon prevents the translation of mRNA transcripts and is called **RNA interference**. We will talk much more about RNA interference in Chapter 4, but RNA interference merits attention here because it has the potential to displace small molecules as a tool for chemical biology. Short synthetic RNA molecules can enter human cells grown in laboratory culture, and that ability was once thought to be limited to small molecules. Moreover, it is very easy to design and synthesize short RNA molecules without developing new synthetic strategies, as is required for small drug-like molecules.

Will RNA interference replace small molecules as drugs and biological tools? Universal application of RNA interference has been hindered by three major obstacles. First, the process of RNA cell entry is poorly understood and not completely general. When injected directly into the eye, small RNA molecules are effective against viruses or over-vascularization, but the same injection approach is unlikely to work on other human organs. Second, because RNA interference directs the cleavage of both identical and similar sequences, one has to check the RNA transcriptome carefully to anticipate and avoid collateral damage. Finally, RNA interference does not inhibit proteins that have already been translated; it only prevents new proteins from being translated. Until these problems are addressed, small molecules are likely to remain the central tool in modern chemical biology.

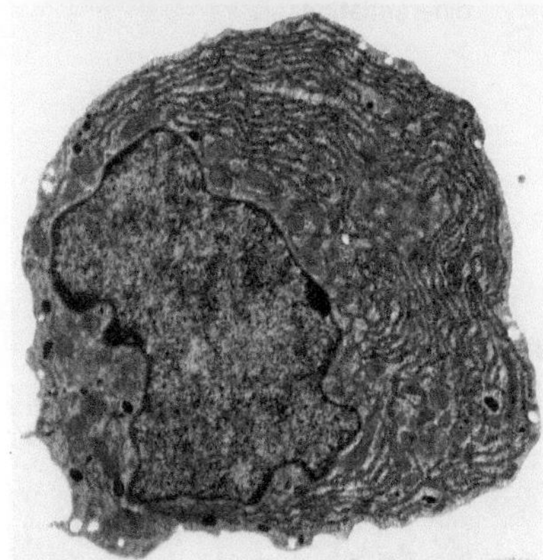

Figure 1.37 Mammalian cell culture. Mammalian cell lines can be grown in the laboratory under controlled conditions (top), typically in an atmosphere of carbon dioxide that creates a carbon dioxide/bicarbonate buffer system in the growth medium. The media is usually supplemented with phenol red as a harmless pH indicator. Hybridoma cell lines combine the immortality of cancer cells and the antibody-producing ability of an evolved B cell (bottom). This false color image of a hybridoma reveals the nucleus in blue and the extensive vesicular network that is important for trafficking antibodies out of the cell. (From A. Karpas et al., *Proc. Natl. Acad. Sci. USA* 98:1799–1804, 2001. With permission from the National Academy of Sciences.)

Monoclonal antibodies bind specifically

Recall that the human immune system can rapidly evolve highly specific antibodies that are capable of binding foreign molecules. The same is true for other mammals, including laboratory strains of mice. When mice are challenged with foreign proteins, or proteins that have been chemically derivatized with foreign molecules, the mice generate B cells that produce highly specific antibodies against the foreign molecules. If you remove B cells from a mouse spleen and try to culture them in the laboratory, they will eventually die. This is because most differentiated mammalian cells have a limited capacity for cell division; however, by fusing antibody-producing B cells with immortal mouse myeloma cells, one can obtain hybrid cell lines called hybridomas that live forever in culture yet produce antibodies (**Figure 1.37**). Monoclonal antibodies against many important proteins are widely available for laboratory use and can be chemically derivatized to create highly selective reagents for chemical biology experiments. The human immune system would react violently to antibodies from a mouse, but clever techniques have enabled human-like monoclonal antibodies against disease proteins to be obtained and used as human drugs. Most clinically used protein drugs are based on antibodies, including Herceptin™ (breast cancer), Humira™ (autoimmune diseases), Erbitux™ (colorectal cancer), Rituxan™ (non-Hodgkin's lymphoma), and Avastin™ (macular degeneration). Growing bacteria is easy. In contrast, culturing mammalian cells in the laboratory is a costly endeavor that requires special nutrients and highly controlled, sterile conditions. Thus, protein drugs obtained from mammalian cells command a premium.

Immortal cancer cell lines serve as mimics of human organs

Mammalian cells require controlled conditions for laboratory culture—namely sterile conditions, special carbon dioxide incubators, and complex mixtures of growth factors. A wide range of immortal mammalian cell lines are available for laboratory

Table 1.1 Examples of immortal mammalian cell lines.

Cell line	Tissue
Jurkat cells	Human T-cell lymphoma
HeLa	Human cervical cancer
MCF-7	Human breast adenocarcinoma
HepG2	Human hepatocarcinoma
NIH 3T3	Embryonic mouse fibroblast
MDCK	Madin–Darby canine kidney
CHO	Chinese hamster ovary

studies (**Table 1.1**). Thus, much of our understanding of mammalian cell function has come from laboratory studies of cancer cell lines with cryptic designations such as HeLa cells, MCF-7 cells, and Jurkat cells. The inner workings of these cells are not necessarily the same as the workings of cells in healthy tissue, but they are often the best systems for study and testing, short of living human or animal subjects. A few other common cell lines have been derived from noncancerous tissue. The immortality of NIH 3T3 cells arose spontaneously while culturing embryonic fibroblasts from a Swiss mouse. The Chinese hamster ovary (CHO) cell line is widely used for the expression of protein pharmaceuticals that require the unique machinery of mammalian cells for biosynthesis.

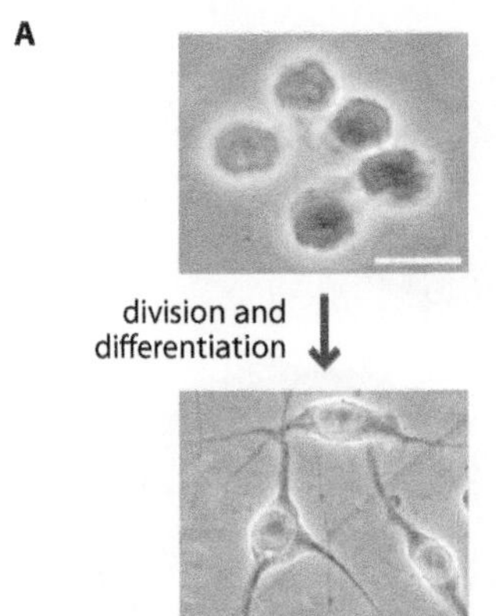

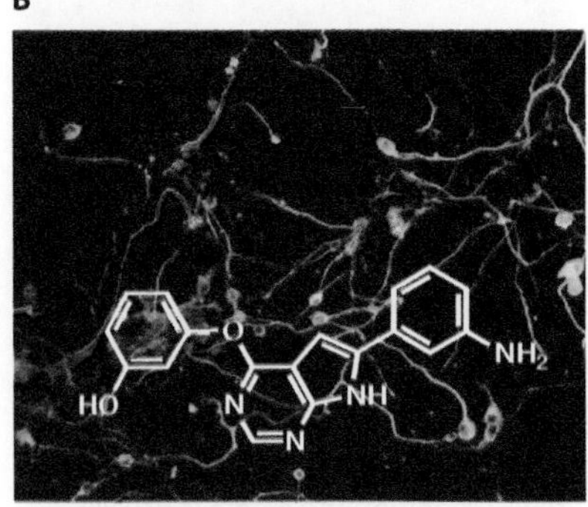

Figure 1.38 Controlling differentiation. (A) Small molecules can control differentiation. (B) A small-molecule inhibitor (TWS119) of the enzyme GSK-3β induces a mouse embryonic cell line to differentiate into neurons. (A, from K. Miyake and K. Nagai, *Neurochem. Int.* 50:264–270, 2007. With permission from Elsevier; B, from S. Ding et al., *Proc. Natl. Acad. Sci. USA* 100:7632, 2003. With permission from the National Academy of Sciences.)

Human stem cells are highly valuable tools for research and medicine

Many of the cells in the developing human embryo retain their capacity to regenerate most or all of the various tissue types. These **stem cells** may have the capacity to replace injured or diseased organs such as spinal cords, hearts, kidneys, eyes, and skin. It is the dream of chemical biologists to use small molecules to control the differentiation of stem cells (**Figure 1.38**). The most powerful stem cells, capable of regenerating virtually any type of tissue, must be harvested early in embryonic development and therefore require one to destroy the embryo. The use of such embryonic stem cells is mired in moral controversy, so tremendous effort has been invested in the development of stem-cell lines that are obtained without sacrificing an embryo or fetus.

The various cells of the bone marrow differentiate throughout human life through the process of hematopoiesis. We discuss the signal transduction pathways that control this differentiation later, in Chapter 9. Pluripotent hematopoietic stem cells divide and differentiate to form red blood cells, T lymphocytes, B lymphocytes, macrophages, platelets, neutrophils, and eosinophils as needed. Differentiation and proliferation along these different pathways occurs in response to various protein hormones. In most cases of human cell differentiation, the expression of genes changes, but the genotype does not. In response to viral infection, antibody-producing B lymphocytes undergo a dramatic process of differentiation, at both the genetic level and the cellular level. Thus, the genotype of mature B lymphocytes is permanently altered through this special type of differentiation.

Model organisms teach us about humans

The study of human physiology, human cell biology, and human development is driven not merely by curiosity but also by the desire to improve the human condition. However, just as human cells are much more difficult to grow in the laboratory than bacterial cells, human beings create significant logistical and ethical challenges as organisms of study. Thus, most of the scientific literature is focused on simpler organisms. These range from simple, single-celled organisms, such as bacteria and yeast, to

Table 1.2 Commonly studied model organisms.

Organism	Common name	Generation time	Primary importance
Escherichia coli	Bacterium	20 minutes	Production of DNA and proteins
Saccharomyces cerevisiae	Yeast	2 hours	Genetics, production of complex human-like proteins
Caenorhabditis elegans	Roundworm	36 hours	Genetic model for cell development/differentiation
Drosophila melanogaster	Fruit fly	10 days	Genetic model for organismal development/differentiation
Mus musculus	Mouse	3 months	Model for human physiology
Homo sapiens	Human	14 years	Target organism for the development of new medicines

complex animals, such as mice (**Table 1.2** and **Figure 1.39**). The DNA, RNA, and proteins generated by bacteria are very similar to their counterparts found in human cells. Bacterial cells are useful for the production of DNA and proteins in the laboratory, but are a bit too simple to be good models for human cells. At a cellular and genetic level, yeast cells work in a way that is much more relevant than *E. coli*. Moreover, humans have been working with yeast for thousands of years—for as long as we have been baking bread and brewing beer. Furthermore, some genera of yeast cause harmful infections, particularly in immunosuppressed patients.

Microscopic soil worms and fruit flies seem to be extravagant research targets. They are not pathogenic, so why study them? Worms and flies serve as economical models for the study of human biology, physiology, and behavior. They are faster, easier, and more ethical to breed than humans. In general, the reproduction times for model organisms tend to scale inversely with the simplicity of the organism. "Simple"

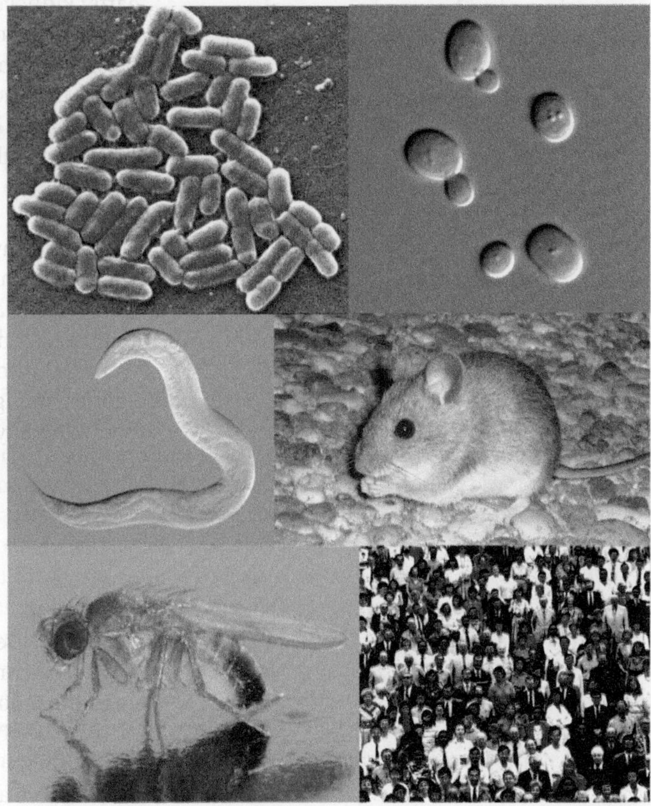

Figure 1.39 Model organisms. Model organisms help us understand humans. Clockwise from the upper left: *E. coli* bacteria, *Saccharomyces* yeast, a mouse, humans, *Drosophila*, and *C. elegans*. (*E. coli*, courtesy of Janice Haney Carr; yeast, courtesy of Masur, Wikimedia Commons; mouse, courtesy of Florean Fortescue, Wikimedia Commons; humans, from National Heart, Lung, and Blood Institute, 1987; *Drosophila*, courtesy of André Karwath, Wikipedia; *C. elegans*, courtesy of Bob Goldstein, UNC Chapel Hill.)

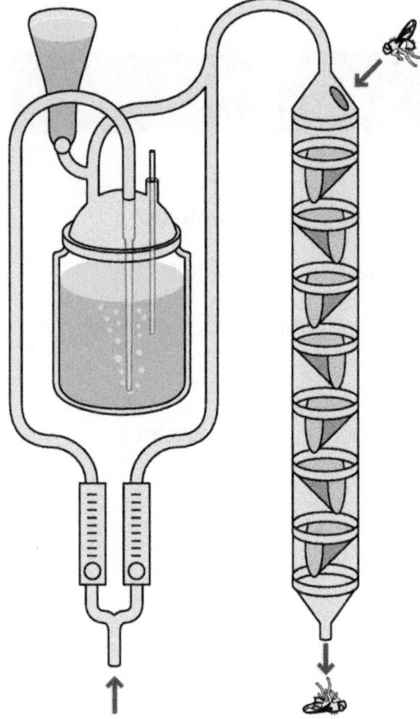

Figure 1.40 An inebriometer. Alcohol-soaked air is pumped through the stack of mesh funnels to which fruit flies cling. As the flies succumb to the alcohol, they fall from one mesh to the one below, eventually collecting at the bottom of the apparatus in a drunken stupor. By collecting flies that fell through the inebriometer at a faster rate, genes associated with alcohol susceptibility were identified. (Adapted from U. Heberlein et al., *Integr. Comp. Biol.* 44:269–274, 2004. With permission from Oxford University Press.)

organisms are not that simple: *Drosophila* and *Caenorhabditis elegans* pack an impressive number of genes, about 14,000 and about 20,000, respectively. As a consequence of evolution, worms, flies, and humans share similar genes, similar proteins, and similar biochemical mechanisms.

The soil worm *C. elegans* is transparent, allowing one to see the cells and the organs. Developmental biologists have traced the development of every one of the 959 cell types in *C. elegans* starting from a single cell. Fruit flies are not transparent, but the stages of embryological development more closely mimic that of humans than do microscopic soil worms. In particular, the use of irradiation to induce distinct phenotypic mutations motivated our early understanding of the relationship between genes and physiological development. Had it not been for *Drosophila,* we would not have a molecular-level understanding of how humans develop the basic body plan of two arms and two legs. Model organisms can also allow us to study genes that relate to human behavior. On ingestion of alcohol, fruit flies experience hyperactivity, decreased coordination, and ultimately sleepiness, much like their human counterparts. An inebriometer was devised to screen a diverse population of fruit flies, to identify mutations that would make them more susceptible to alcohol (**Figure 1.40**). Using the inebriometer, researchers demonstrated that increased sensitivity to alcohol can result from mutation to a gene they named *cheapdate.* Mutations to the gene *cheapdate* affect levels of a molecule called cyclic AMP (cAMP) in neuronal cells.

Finally, the resilient house mouse has become indispensable for studies of mammalian physiology, metabolism, immunity, cognition, and behavior. In the drug development process, as one progresses from cellular models to human subjects, mice are usually the first animal model in which drugs are tested. No other mammal has been subjected to as much study. Thus we have cured cancer, obesity, heart disease, bacterial infections—all manner of diseases—in mice. New, useful strains of mice have been bred in the laboratory. Many knockout strains, lacking a particular gene, have been created to provide a better understanding of the importance of defective proteins in humans. Mice with defective immune systems have been particularly important because they can be used as hosts for the growth of either diseased or normal tissue from other mammals (**Figure 1.41**).

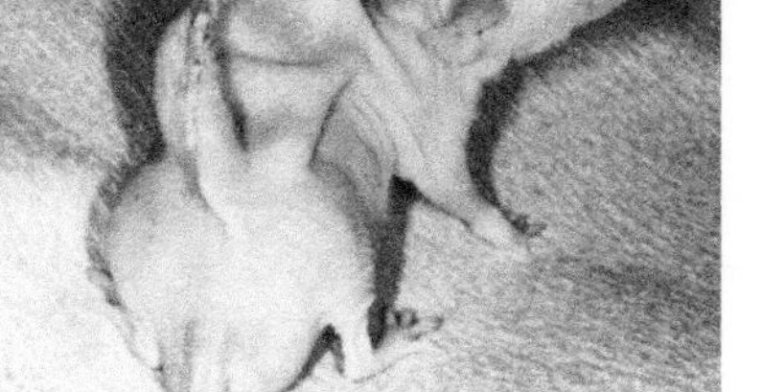

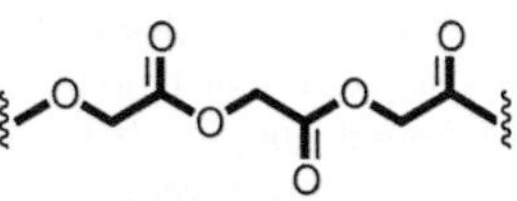

biodegradable polymer

Figure 1.41 Tissue engineering. Cow cartilage cells were injected into an immunodeficient mouse strain and grown around an ear-shaped mold made from a biodegradable polymer. Mice with a healthy immune system would have rejected the foreign tissue. (Courtesy of Charles A. Vacanti.)

1.7 SUMMARY

Combinatorial assembly readily accesses unlimited diversity from limited sets of building blocks, and the creation of diversity is central to evolution. It is possible to see the power of combinatorial assembly at all levels of life. The biomolecules that make us human are constructed from a limited set of atom types—mostly hydrogen, oxygen, carbon, nitrogen, phosphorus, and sulfur. The biooligomers that make up cells—DNA, RNA, proteins, oligosaccharides, polyketides, and terpenes—are each assembled from a limited set of subunits, about 40 in total. Further assembly of cells to organisms, to populations, and to ecosystems is accompanied by a marked increase in complexity.

As chemical biologists, we are chiefly concerned with atoms and molecules, and in this book we focus primarily on organic molecules. The central dogma serves as a blueprint for understanding the combinatorial assembly of the biooligomers that make up human cells. The structure of every protein is encoded in a simple way by an RNA molecule and, ultimately, by a fragment of the DNA that makes up the genome. The combinatorial assembly of other biooligomers is more powerfully encrypted. Glycans, lipids, fats, and steroids each require unique sets of catalytic protein modules for their assembly.

LEARNING OUTCOMES

- Contrast the architectural efficiency of one-dimensional oligomers such as DNA, RNA, and proteins with multidimensional oligomers such as organic molecules.
- Identify each type of biooligomer in the central dogma and the flow of chemical information.
- Diagram the elements of a gene and how the expression of genes is controlled.
- Contrast the size and structure of the human genome with that of the bacterium *E. coli.*
- Contrast the size and morphology of human cells with those of bacterial cells.
- Understand the nongenomic origins of the diversity of transcriptomes and proteomes.
- Contrast the template-directed biosynthesis of DNA, RNA, and proteins with the nontemplated biosynthesis of oligosaccharides, polyketides, and terpenes.
- Calculate the potential diversity of biooligomers of a given length.
- Understand the combinatorial genetic architecture of antibodies.
- Recognize the chemical structures of common fluorophores.
- Diagram the process of selection and amplification using a phage library.
- Contrast the advantages of small organic molecules over large biooligomers, such as antibodies, as medicines and tools in chemical biology.
- Contrast the advantages and disadvantages of immortal mammalian cell lines and bacteria for understanding normal human cells.
- Identify major applications of common model organisms, such as *E. coli*, *Saccharomyces*, *C. elegans*, *Drosophila*, and mouse.

PROBLEMS

Solutions to all in-text problems as well as all starred (*) end-of-chapter problems are available on the Student Resource Site at: www.garlandscience.com/bioorganic-chembio.

1.5 Throughout this book we will refer to macromolecular structures whose atomic coordinates can be accessed through the Protein Data Bank (PDB) Web site. The Web site has numerous built-in applications that allow you to view and render molecules of DNA, RNA, and proteins. Use the PDB Web site to answer the following questions:

A How many phosphorus atoms are present in the DNA dodecamer (PDB 1BNA)?

B How many monomeric RNA subunits are present in the crystal structure of the hepatitis B virus encapsidation signal (PDB 2IXY)?

C How many iron atoms are present in the enzyme aconitase (PDB 1ACO)?

***1.6**

A How many membranes separate the DNA of *E. coli* from the extracellular environment?

B How many membranes separate the nuclear DNA of human cells from the extracellular environment?

C How many membranes separate the mitochondrial DNA of human cells from the extracellular environment?

1.7 Use the Internet to determine the major organelle that is missing from mature human erythrocytes (red blood cells).

***1.8** Show the starting materials and steps (in the correct order) that would be needed to synthesize the ester shown below. Using that synthetic route, how many different esters could be synthesized using the starting materials shown below? Me, Et, *n*-Bu, and Ph are abbreviations for methyl, ethyl, *n*-butyl and phenyl substituents, respectively.

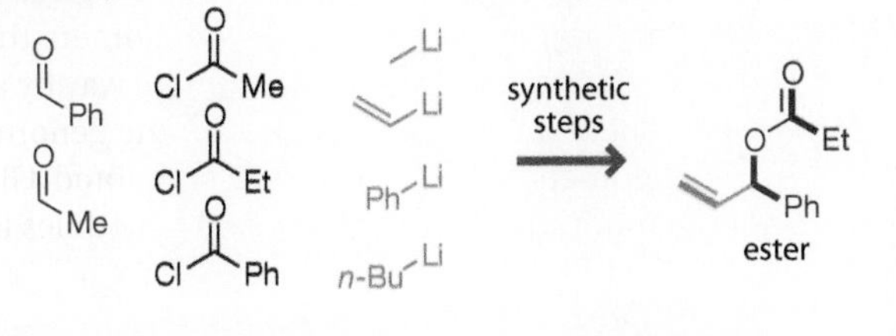

synthetic steps → ester

1.9 If one synthesized every possible DNA molecule, each with 20 subunits, how much would your collection of molecules weigh? The average molecular mass of a DNA subunit is 165 g mol^{-1}.

1.10 Use the Internet to identify the organism and tissue of origin of the following laboratory tumor-cell lines:

A COS7

B LNCaP

C LA-4

D Sf21

1.11 Biologists often use Latin animal adjectives (for example human or feline) to describe the origin of tissues, cells, and molecules. For each of the following viruses, identify the common name of the animal it infects.

Avian influenza
Bovine respiratory syncytial virus
Canine parvovirus
Caprine arthritis–encephalitis virus
Equine infectious anemia
Feline panleukopenia
Abelson murine leukemia virus
Ovine pulmonary carcinoma
Piscine mycobacteriosis
Porcine circovirus disease
Simian immunodeficiency virus

The Chemical Origins of Biology 2

LEARNING OBJECTIVES

- Draw arrow-pushing mechanisms to depict chemically reasonable reaction mechanisms.
- Identify frontier orbitals involved in bond forming and breaking mechanistic steps.
- Describe the chemical basis, strengths, and geometries of hydrogen bonds and proton transfers.
- Provide and discuss equations approximating noncovalent interactions.
- Discuss and quantify the wide range of noncovalent interactions critical to chemical biology.
- Understand how entropy affects systems that can exist in a multitude of atomic arrangements.

In his book *What is Life?*, the physicist Erwin Schrödinger (**Figure 2.1**) argued that all living things are governed by the same physical laws as those encountered in everyday life, such as the laws of thermodynamics and Newton's laws of motion. His book appeared in the 1940s and influenced a generation of physicists to explore the physical and molecular mechanisms responsible for life. At the time, the molecules at the heart of life were essentially invisible, because spectroscopy and crystallography were not developed for another decade. No exceptions to Schrödinger's argument have ever been found. Life with all of its surprises, diversity, and complexity seems entirely bounded by the rules of chemistry and physics.

Figure 2.1 Erwin Schrödinger. The quantum physicist Erwin Schrödinger was a proponent of the idea that complex living organisms are beholden to elementary scientific principles. (Photograph by Francis Simon. With permission from Photo Researchers, Inc.)

Diversity is a fundamental requirement for Darwinian evolution. At the molecular level, oligomerization reactions provide a simple mechanism for generating diverse, structurally related molecules. These reactions can be understood using mechanistic arrow-pushing, which is based on quantum mechanical principles. Using arrow-pushing, we will show that simple reactions, applied combinatorially, can explain the formation of all the major classes of biological molecules and the molecular diversity necessary for the emergence of life. The complex assemblies of large molecules found inside the cell are held together by noncovalent interactions, in addition to **covalent bonds**. To understand the driving forces for such interactions requires a nuanced portrait of the subtle interplay of atomic distances, solvent, and other environmental effects.

2.1 MECHANISTIC ARROW-PUSHING IS AN EXPRESSION OF MOLECULAR ORBITAL THEORY

Three properties control chemical reactivity

Arrow-pushing is our most powerful tool for connecting chemical reactivity with biological phenomena. It was first introduced by Robert Robinson and Arthur Lapworth in 1922, who were probably unaware of the quantum mechanical underpinnings (**Figure 2.2**). To understand arrow-pushing and apply it to its fullest potential, we need to understand its strong connection to molecular orbital theory. Every time we choose one set of curved arrows over another, we are concluding that one

$^{+}CH_2$:CH:CH:CH:CH:CH_2^{-}

Figure 2.2 The origin of arrow-pushing. This first application of curved arrows was used to represent resonance in a conjugated molecule. Curved arrows were introduced to mechanistic organic chemistry before line depictions of bonds were widely adopted. (From W.O. Kermack and R. Robinson, *J. Chem. Soc. Trans.* 121:427–440, 1922. With permission from the Royal Society of Chemistry.)

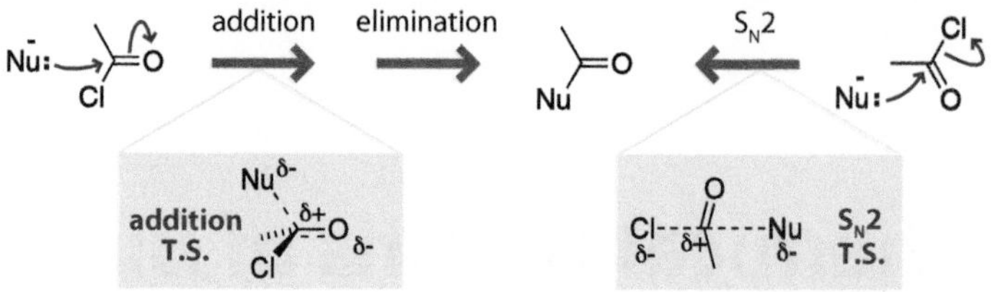

Figure 2.3 Why push arrows? Arrow-pushing helps us evaluate possible transition states (T.S.) for nucleophilic substitution of acid chlorides.

transition state is lower in energy. For example, it is well understood that acid chlorides do not undergo substitution through a concerted S_N2 mechanism (**Figure 2.3**). Instead, substitution of acid chlorides proceeds through a two-step addition–elimination mechanism. The addition step is slower than the elimination step, yet both of the steps in the addition–elimination mechanism are faster than the alternative S_N2 displacement of chloride.

How can we quantify the transition-state energies? The energetics of all the important reactions in biology can be evaluated by an equation that has roughly three terms (**Equation 2.1**): interactions between charges, repulsive interactions, and attractive interactions. Chemical biologists favor different forms of this equation depending on whether they are interested in bonding or nonbonding interactions.

$$\text{Energy} \propto \text{charge interactions} + \text{repulsive interactions} - \text{attractive interactions} \qquad (2.1)$$

Perturbational molecular orbital theory connects arrow-pushing with quantum mechanics

Chemists who are interested in chemical reactions prefer a form of Equation 2.1 that is based on perturbational molecular orbital theory (the quantum mechanical interaction of two orbitals). If we had mathematical expressions that described the molecular orbitals of the nucleophile and the acid chloride, we could calculate all of the possible transition-state energies with high precision. Applying perturbational molecular orbital theory realizes a direct connection between quantum mechanics and chemical reactivity. Sadly, the mathematics would be overwhelming for all except the simplest molecules.

Organic chemists prefer to use arrow-pushing as a proxy for mathematical equations. That connection between arrow-pushing and orbital interactions was popularized by Ian Fleming in a classic handbook, *Frontier Orbitals and Organic Chemical Reactions*. Along the lines of Equation 2.1, Fleming reminded us that we could compare any two transition states through a qualitative expression of perturbational molecular orbital theory involving three terms (**Equation 2.2**): 1. Coulomb's law for interacting charges, 2. sterics, and 3. the favorable overlap of filled orbitals with unfilled orbitals. By recognizing those three terms we will resist the temptation to explain chemical phenomena on the basis of either "sterics" or "electronics"; there are three terms, not two, in Equation 2.2. The first two terms of Equation 2.2 are simple to comprehend because they correlate with macroscopic phenomena and are widely taught in introductory organic chemistry courses. The third term of Equation 2.1 is likely to be unfamiliar.

$$\text{Reaction energy} \propto \text{Coulomb's law} + \text{sterics} - \text{filled–unfilled orbital overlap} \qquad (2.2)$$

Coulombic effects in Equation 2.2 can be repulsive or attractive. We frequently reinforce these notions of charge with colloquial phrases such as "opposites attract" and "familiarity breeds contempt." Happily, social forces are merely a mirror of the atomic world, where opposite charges attract and like charges repel. Anyone who has used a hammer can appreciate the steric term in Equation 2.2. The steric term results from the interaction of filled orbitals with filled orbitals. This interaction is always repulsive, and such repulsion explains the sluggish rates of S_N2 attack at tertiary centers and the equatorial preference of a methyl group on a cyclohexane ring.

A wrong: HO^- $H_3C-\overset{+}{N}(CH_3)_3$ — right: HO^- $H_3C-\overset{+}{N}(CH_3)_3$

B *anti* conformer ⇌ *syn* conformer (methyl formate, H–C(=O)–O–CH_3)

Figure 2.4 Why can we not simply attribute all effects to sterics and electronics? (A) Charge is usually important in organic reactions, but using formal charges for arrow-pushing is often misleading. (B) Sterics cannot explain why the *syn* conformer of methyl formate is favored.

Although clearly important, charge and sterics are overworked concepts in undergraduate organic chemistry textbooks. Worse yet, the formal charges in Lewis structures are often misleading (**Figure 2.4A**). For example, if we consider the reaction of a hydroxide anion with a tetramethylammonium cation, the attack of hydroxide on the positively charged nitrogen atom has an irresistible appeal. However, the interaction of charges depicted in Figure 2.4 is wrong at nearly every level of detail. The biggest problem is that such an attack would lead to a fifth bond to nitrogen—a violation of the octet rule. Hydroxide is least likely to attack nitrogen and more likely to attack the carbon atoms or protons. If we really want to understand the reaction of hydroxide anion with a tetramethylammonium ion, we need to embrace the third term of Equation 2.2, for it alone serves as the underpinning for mechanistic arrow-pushing. Similarly, sterics can not explain why the *syn* conformer of methyl formate is much more stable than the *anti* conformer (Figure 2.4B). That preference is best illuminated by considering orbital interactions. To understand the interaction of filled orbitals with unfilled orbitals, we need to review the various types of orbitals used by organic chemists, for it is these orbitals that ultimately determine molecular structure and reactivity.

Six canonical frontier orbitals can be used to predict reactivity

There are three important types of orbitals: atomic orbitals (such as *s*, *p*, *d*, *f*), hybrid atomic orbitals (such as sp^3, sp^2, *sp*), which are built as combinations of atomic orbitals on the same atom, and finally molecular orbitals (such as σ, π, ψ_n) which are built by combining the atomic orbitals of different atoms within the molecule. Electrons in 2*s* atomic orbitals are closer to the nucleus than electrons in 2*p* atomic orbitals. This proximity makes electrons in 2*s* orbitals more stable and less reactive (**Figure 2.5**). Thus, *s* character confers stability to all orbitals. Conversely, *p character confers nucleophilicity and basicity.*

Subtle differences in the *p* character of hybrid atomic orbitals or molecular orbitals translate into extraordinary differences in reactivity (**Table 2.1**). For example, in considering the basicity of carbanions, $R_3C\text{-}CH_2\!:^-$ anions are 10^{26} times more basic than $RC{\equiv}C\!:^-$ anions, primarily owing to differences in the *p* character of the two lone pairs. The effect of *p* character is also useful for predicting aqueous acidity, which is the opposite of aqueous basicity: $RC{\equiv}NH^+$ (pK_a -10) > $R_2C{=}NH_2^+$ (pK_a 5-7) > $R_3C\text{-}NH_3^+$ (pK_a 10).

Table 2.1 *p* Character and basicity.

pK_a'	lone pair	%*p*
50	sp^3	75
41	sp^2	67
24	*sp*	50

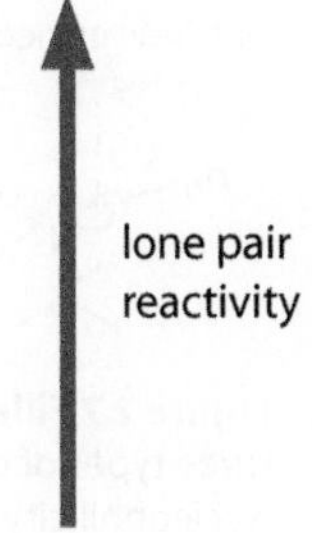

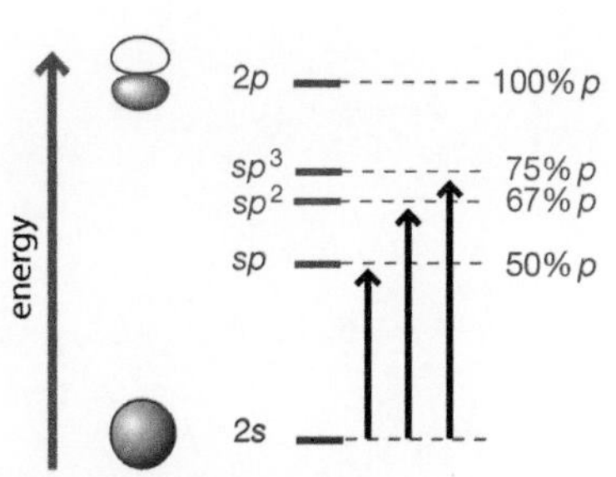

Figure 2.5 Always note the hybridization. Electrons in orbitals with higher *p* character are more reactive than electrons in orbitals with lower *p* character.

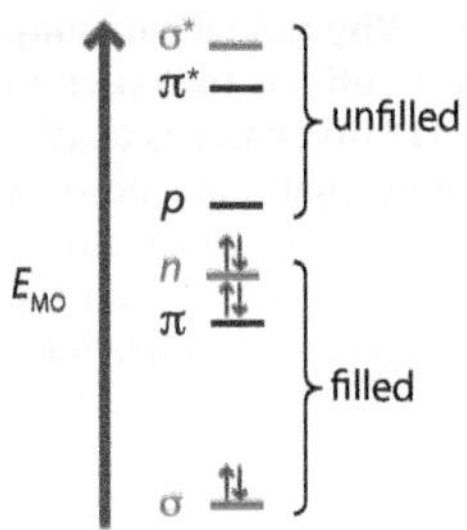

Figure 2.6 Frontier orbitals. There are six types of canonical frontier molecular orbitals: three filled and three unfilled.

Problem 2.1

For each pair of compounds below, predict the hybridization of the lone pair, and specify the percentage of *s* and *p* character. Then rank the compounds in order of decreasing basicity.

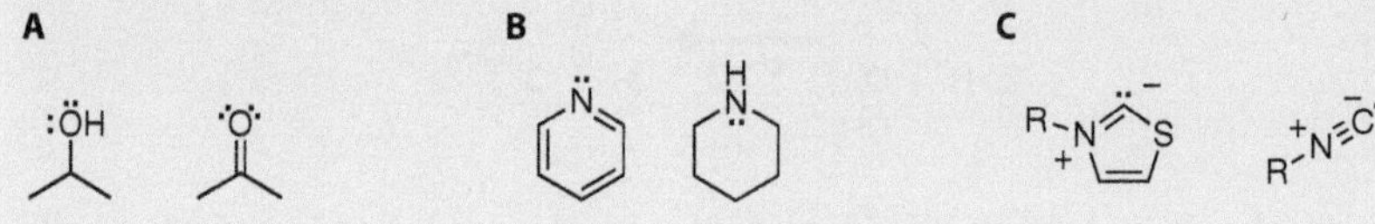

Lewis structures depict only the valence electrons. For accounting purposes, we ignore the 1*s* electrons of boron, carbon, nitrogen, and oxygen, and concern ourselves solely with the 2*s* and 2*p* electrons. Similarly, rather than consider all of the molecular orbitals, and they are legion, we will consider only the most reactive orbitals, namely the **frontier molecular orbitals.** Usually, the frontier molecular orbitals of a molecule correspond to the highest occupied molecular orbital (HOMO) and lowest unoccupied molecular orbital (LUMO). We can explain most of the important reactions in biology by using six canonical types of frontier molecular orbitals: σ, π, nonbonded lone pairs (*n*), empty *p* orbitals, π*, and σ*, as shown in **Figure 2.6**. One can rationalize the relative energies of these orbitals by understanding that they arise from *s* and *p* orbitals, but for now you should permanently commit the relative energies of these frontier orbitals to memory.

The filled frontier molecular orbitals—*n*, π, and σ—should be familiar because they correspond to the lines and dots in Lewis structures. The energies of these filled frontier molecular orbitals are important because *electrons in higher-energy filled orbitals are more reactive.* Thus, the general ordering of reactivity is: lone pairs > pi bonds > sigma bonds (**Figure 2.7**).

The unfilled frontier molecular orbitals—*p* orbitals, π*, and σ*—may seem less familiar because they are not depicted in traditional Lewis structures. Thankfully, we can rely on a simple relationship to visualize these invisible antibonding orbitals: for every pi bond there is a corresponding π* antibonding orbital, and for every sigma bond there is a corresponding σ* antibonding orbital. The third type of unfilled frontier orbital is the empty *p* orbital. Empty *p* orbitals are relatively easy to pinpoint on second-row atoms; just look for trivalent boron or carbocations. The energies of these unfilled orbitals are as important as the energies of the filled orbitals, *because lower-energy unfilled orbitals are more reactive toward nucleophiles.* If you were a pair of electrons, would you not want to go into the most stable orbital? The lowest-energy orbital is the most stable.

Thus, if sterics and charge are equal, then nucleophiles should add to empty orbitals in the following order: empty *p* orbitals > π* orbitals > σ* orbitals (**Figure 2.8**). Add to carbocations first, then to C=X double bonds. Only when we exhaust these alternatives would we consider an S_N2 reaction on a C–X bond, which requires attack on a σ* orbital. Unfortunately, in most undergraduate textbooks, S_N2 reactions at carbon are usually discussed first, promoting them to an undeserved level of importance in the chemical pantheon.

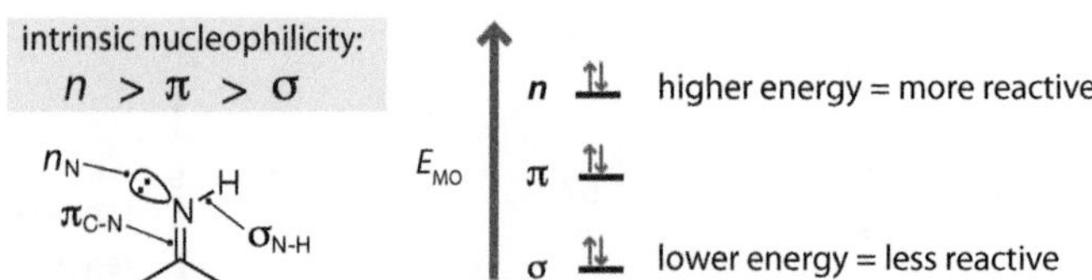

Figure 2.7 Filled frontier orbitals. The relative energies of the three types of canonical filled orbitals are a good predictor of the nucleophilicity of the electrons that occupy them.

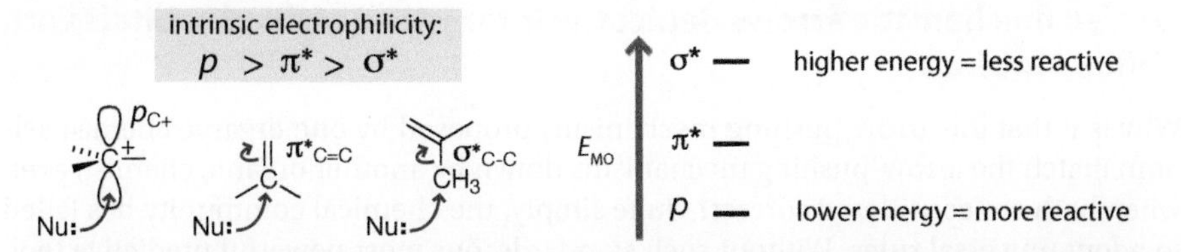

intrinsic electrophilicity:
$p > \pi^* > \sigma^*$

σ^* — higher energy = less reactive
π^* —
p — lower energy = more reactive

Figure 2.8 Unfilled frontier orbitals. The relative energies of the three types of canonical unfilled frontier orbitals are a good predictor of electrophilicity.

Problem 2.2

For each compound below, identify the atom most likely to be attacked first by a nucleophile and identify the unoccupied frontier orbital (for example p, π^*, or σ^*) associated with that atom.

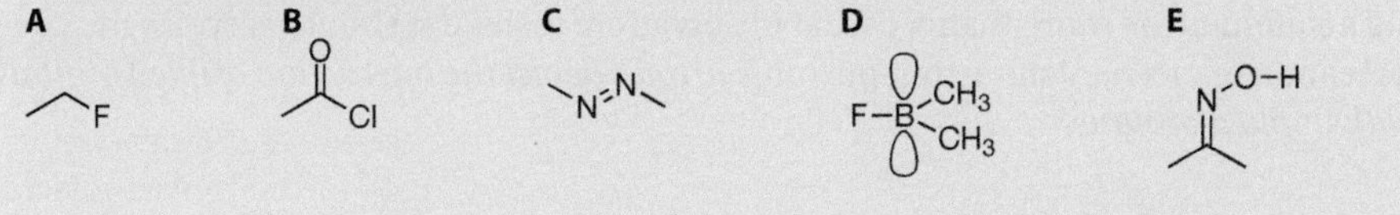

A B C D E

Electronegativity affects both frontier orbitals and Coulombic interactions

Electronegativity affects frontier orbital interactions, the third term in Equation 2.2. When you replace a carbon atom with a more electronegative atom such as nitrogen or oxygen, all of the orbitals are lowered in energy, including the energies of the frontier orbitals. For example, lone pairs on H_3N are higher in energy and more nucleophilic than lone pairs on H_2O, because O is more electronegative than N (**Figure 2.9**). Similarly, S_N2 reactions are faster on CH_3–F than CH_3–CH_3 because the σ^*_{C-F} orbital is lower in energy than the σ^*_{C-C} orbital.

Electronegativity also affects Coulombic interactions, the first term in Equation 2.2. Electronegative atoms (such as F, O, N, and Cl) tend to have partial negative charge that gives them a Coulombic advantage when they act as nucleophiles. Electronegativity predicts partial charge, even when the formal charges on Lewis structures are misleading. For example, in a hydronium ion (H_3O^+) the protons, not the oxygen atom, bear most of the positive charge. Formal charges are still useful, particularly for bonds to negatively charged atoms; B–H bonds in the BH_4^- anion (σ_{H-B^-}) are much more nucleophilic than B–H bonds in BH_3 (σ_{H-B}).

How do we weigh the two opposing effects of electronegativity on reactivity, namely decreased frontier orbital energies versus enhanced Coulombic interactions? Two broad generalizations will cover most of the reactions in this book. For reactions that involve an attack of nucleophiles on carbon atoms, we should focus on the fact that frontier orbitals are lower in energy when they involve electronegative atoms. For reactions that involve an attack of nucleophiles on protons and cations, we should focus more on the fact that electronegative atoms have more negative charge.

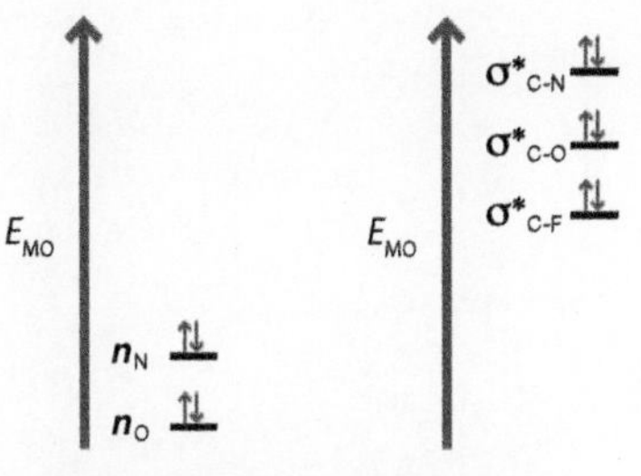

Problem 2.3

For each compound below, identify the most reactive pair of electrons (either a bond or a lone pair) and specify the corresponding type of frontier orbital (namely n, π, or σ).

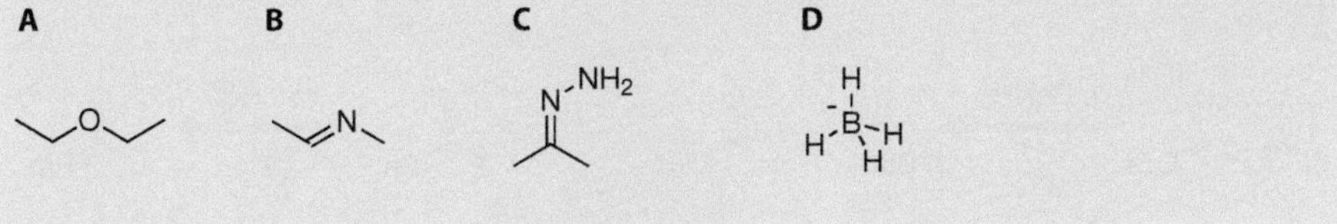

A B C D

E_{MO} n_N n_O

E_{MO} σ^*_{C-N} σ^*_{C-O} σ^*_{C-F}

Figure 2.9 Frontier orbitals are lower in energy when they involve electronegative atoms. The more electronegative the atom, the lower the energy of the orbital.

Curved mechanistic arrows depict the interaction of filled orbitals with unfilled orbitals

Why is it that the arrow-pushing mechanisms proposed by one organic chemist seldom match the arrow-pushing mechanisms drawn by another organic chemist, even when both are considered correct? Quite simply, the chemical community has failed to adopt universal rules. Without such standards, our most powerful predictive tool, arrow-pushing, is more art than science.

Equation 2.2 tells us that the relative *rates* of chemical reactions are certainly determined by both orbital interactions and Coulombic interactions, yet arrow-pushing cannot, and should not, be used to depict both. As stated by Fleming, "*Curly arrows*, when used with a molecular description of bonding, work as well as they do simply because they *illustrate the electron distribution in the frontier orbital*, and for reaction kinetics it is the frontier orbital that is most important." We should treat Fleming's statement as more than a casual observation; instead it should serve as the central canon of mechanistic arrow-pushing: *arrows depict the interaction of filled orbitals with unfilled orbitals.*

There are three basic rules for mechanistic arrow-pushing

Mechanistic Arrow-Pushing
Arrows depict the interaction of filled orbitals with unfilled orbitals

Rules for Arrow-Pushing
1. Arrows never start with atoms
2. Arrows never start or end on charges
3. Arrows begin with lone pairs, pi bonds, or sigma bonds, and end on unfilled orbitals

To follow the convention suggested by Fleming, we must adhere to three basic arrow-pushing rules:

1. Arrows never indicate the motion of atoms.
2. Arrows never originate from or terminate on a charge; charges do not form bonds—electrons do.
3. Arrows begin with lone pairs, pi bonds, or sigma bonds, and end on unfilled orbitals.

To briefly demonstrate the application of these rules and the predictive power of frontier molecular orbitals, let us consider the trivial substitution reaction of **Figure 2.10A**. If our goal were simply to follow the three arrow-pushing rules then we might be inclined to choose the tidy S_N2 mechanism depicted in Figure 2.10B. However, our choice of mechanisms should be guided by our knowledge of frontier orbital reactivity. Recall that sigma bonds are the worst nucleophiles, whereas lone pairs are the best; thus, it is the lone pair of nitrogen that will attack the β carbon and not the N-H bond. Frontier orbital theory also helps us to sort out the behavior of the electrophile. Recall that addition to π* orbitals is more favorable than addition to σ* orbitals. In the

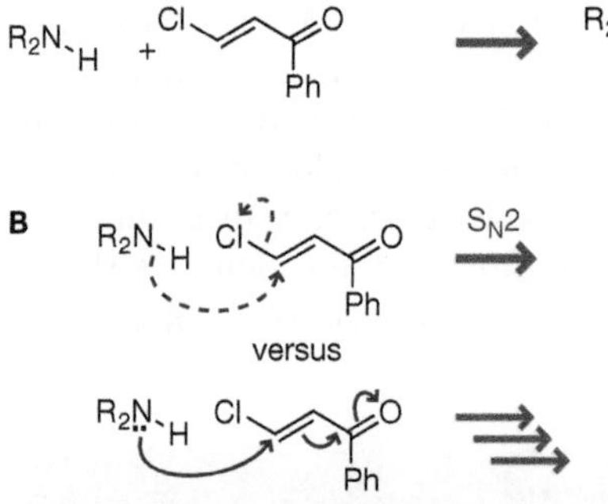

Figure 2.10 Using curved arrows to distinguish mechanistic pathways. (A) A substitution reaction. (B) The tidy one-step S_N2 mechanism is implausible, whereas the three-step addition–elimination reaction is plausible.

Problem 2.4

For each transformation below, suggest a plausible arrow-pushing mechanism.

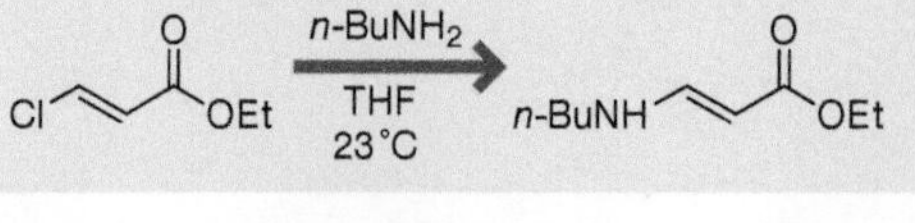

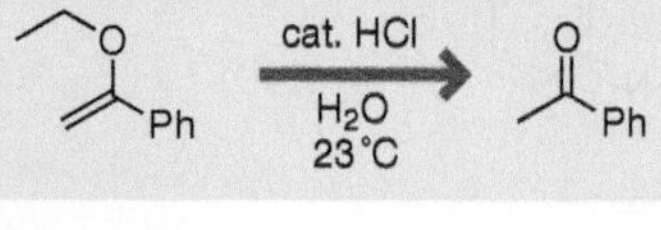

absence of mechanistic data to the contrary, we would shun the one-step S_N2 mechanism involving the addition of a $\sigma_{N\text{-}H}$ to $\sigma^*_{C\text{-}Cl}$ and favor a three-step addition–elimination mechanism in which the first step involves addition of the nonbonding lone pair n_N to π^* of the enone.

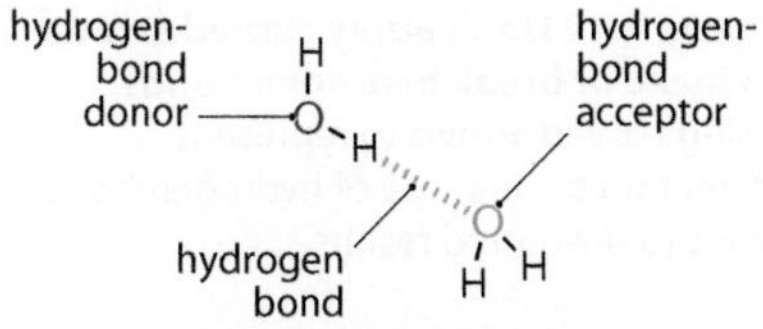

Figure 2.11 The geometry of the ideal hydrogen bond. The shared proton knits together a hydrogen bond donor and acceptor. The angle between the acceptor and donor results from the lone pair occupying a sp^3-hybridized orbital.

2.2 HYDROGEN BONDS AND PROTON TRANSFERS

Hydrogen bonds involve three atoms

Life as we know it would be inconceivable without hydrogen bonds. Hydrogen bonds hold water onto the surface of the Earth. Hydrogen bonds provide the glue to unite DNA strands, and thus provide the basis for genetic heredity. Hydrogen bonds hold proteins in defined conformations.

The hydrogen bond can be considered a special type of interaction between three atoms: a donor atom, an acceptor atom, and a proton (**Figure 2.11**). The interaction is mainly electrostatic; thus the Coulombic term in Equation 2.1 is particularly large for a hydrogen-bonding interaction. Hydrogen bonds are strongest when they involve highly electronegative donor and acceptor atoms such as oxygen and nitrogen. In this arrangement, the electronegative atoms have partial negative charge and the proton has a partial positive charge, resulting in the strongest possible hydrogen bond; the distinction between partial charges and formal charges from Lewis structures is discussed below. Thus, carbon atoms do not form strong hydrogen bonds, because they are not highly electronegative.

Because hydrogen bonds are mainly a Coulombic interaction, their strengths are very sensitive to the dielectric constant or polarity of the environment. Hydrogen bonds are weakest in environments with a high dielectric constant (such as water) and strongest in environments with a low dielectric constant (such as a lipid membrane or the interior of a protein). Good hydrogen bonds are slightly less than 2 Å in length—about twice the length of an O–H or N–H bond. Hydrogen bonds are shortest and strongest when they are linear, and most hydrogen bonds observed in crystal structures have donor–proton–acceptor bond angles of about 180°. Furthermore, the charge state of the hydrogen bond acceptor can markedly affect the strength of the hydrogen bond, as greater electron density results in a stronger interaction with the donated proton; carboxylates, for example, which are charged at the physiological pH of 7, can form very strong hydrogen bonds.

Problem 2.5

Identify the hydrogen bond donors and acceptors in the following molecules.

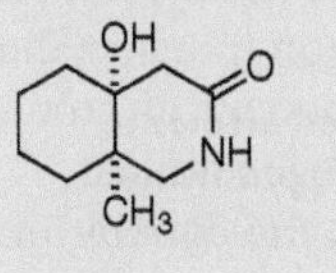

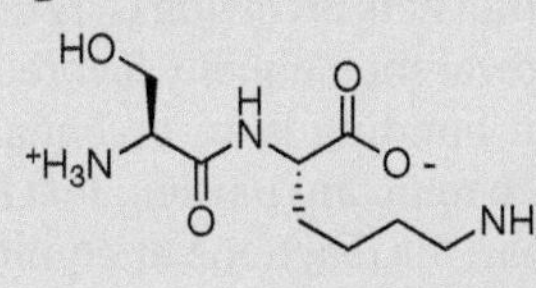

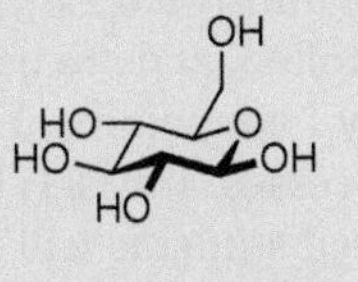

Problem 2.6

Draw the structure of the polymer Kevlar™ (used in bulletproof vests), and depict the hydrogen bonds between polymer strands.

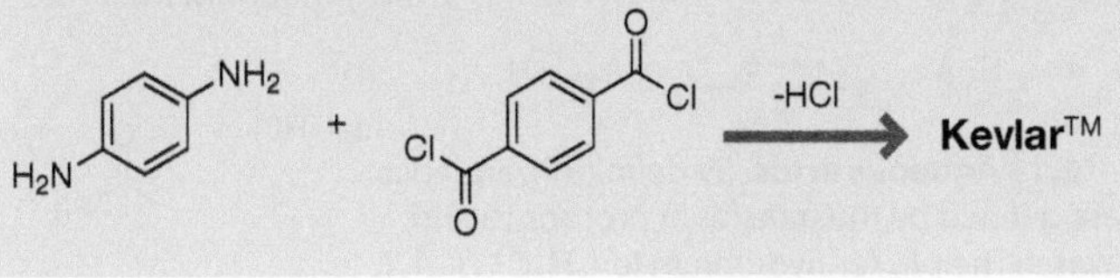

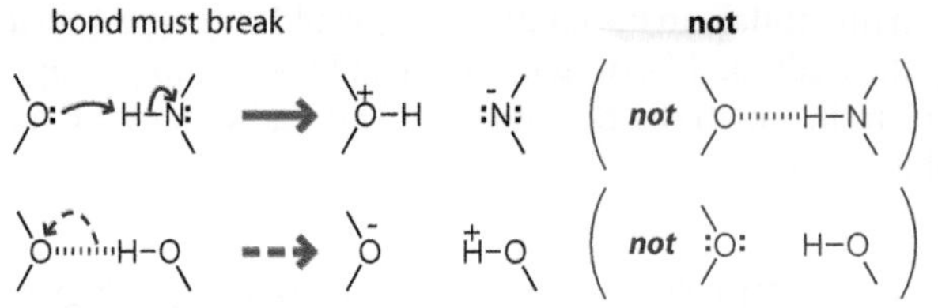

Figure 2.12 Don't apply curved arrows to make or break hydrogen bonds. Using curved arrows to represent formation or cleavage of hydrogen bonds leads to unwanted results.

The hydrogen bond is represented with a hashed or dashed line. This is not part of the Lewis formalism for drawing chemical structures, which means we should never use curved arrows to show the formation or cleavage of hydrogen bonds (**Figure 2.12**). For example, if we use a lone pair to attack a proton that is bonded to nitrogen, we are putting electrons into a $\sigma^*_{H\text{-}N}$ antibonding orbital and the bond must break. This is why it is called an antibonding orbital. Conversely, if we tried to use a curved arrow to represent cleavage of a hydrogen bond, we would be obliged to add a negative charge to the acceptor atom and a positive charge to the donor atom.

Hydrogen bonding plays a key mechanistic role in many enzyme mechanisms, facilitating the attack of nucleophiles on carbonyls and the expulsion of leaving groups in substitution reactions. These are two-step processes, and one step involves cleavage of a hydrogen bond. We must either omit the arrow-pushing for the proton transfer step or omit the formation of the obvious hydrogen bond (**Figure 2.13**).

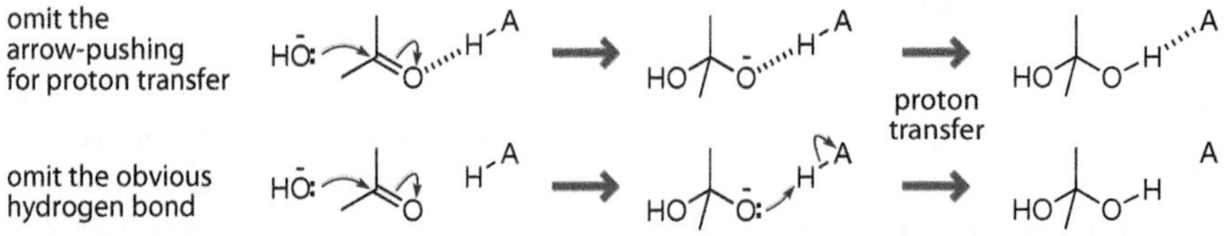

Figure 2.13 Depicting mechanisms that involve hydrogen bonds. There are two ways to represent a proton transfer that involves a hydrogen bond: omit the curved arrows or omit the hydrogen bond.

Proton transfers to and from heteroatoms are usually very fast

Proton transfers to and from oxygen, nitrogen, and sulfur are usually very fast. For strongly favorable proton transfers, such as when the pK_a of the acid is at least 3 units higher than the pK_a' of the base, the proton transfer will be diffusion-controlled. In diffusion-controlled processes, essentially every collision leads to a successful reaction. The rate of diffusion sets the ultimate speed limit for any bimolecular reaction. In practice, diffusion-controlled rates for molecules other than water are about 10^9 M^{-1} s^{-1} in aqueous solution at 25 °C. As is evident from the aqueous-phase proton transfers (**Figure 2.14** and **Table 2.2**), proton transfers involving C–H bonds are usually not diffusion controlled, and are usually slower than equally favorable proton transfers that involve only heteroatoms. Hydrogen bonding is an essential component of proton transfer processes, and C–H bonds are reluctant participants in hydrogen bonds.

Advanced students will recall that, for reactions at equilibrium, the equilibrium constant is equal to the ratio of the forward and reverse rate constants (**Figure 2.15**). One can calculate the equilibrium constant for acid-base reactions by using pK_a values. Because proton transfer in the favorable direction is often diffusion controlled we can also calculate the rate constant for proton transfer in the unfavorable direction! Note that the diffusion-controlled rate constants in Table 2.2 are about 1000-fold faster

$$H_2O + H\text{-}A \underset{k_{-1}}{\overset{k_1}{\rightleftharpoons}} H_2\overset{+}{O}\text{-}H + A^-$$

Figure 2.14 Aqueous acids. By definition, aqueous pK_as are defined by the transfer of protons to and from water. The pK_a for hydronium ion, H_3O^+ is –1.7.

Table 2.2 Rate constants for proton transfers to water (see Figure 2.14).

HA	pK_a	k_1 ($M^{-1}s^{-1}$)	k_{-1} ($M^{-1}s^{-1}$)
HF	3.2	10^8	10^{11}
AcOH	4.7	10^6	10^{11}
H_2S	7.2	10^4	10^{11}
$MeCOCH_2CO_2Et$	9.0	10^{-3}	6×10^7
NH_4^+	9.3	25	~10^{11}
CH_3NO_2	10.2	10^{-8}	6×10^2

Problem 2.7

For each of the proton transfers below, estimate the forward equilibrium constant, and the forward and reverse rate constants. Hint: Note the relative consistency of the rate constant given as k_{-1} in Table 2.2. Could this consistency provide a simplifying assumption?

A

$$\underset{pK_a\ 10}{PhOH} + H_2O \underset{k_{-1}}{\overset{k_1}{\rightleftharpoons}} PhO^- + \underset{pK_a\ -1.7}{H_3O^+}$$

B

$$\underset{pK_a\ 38}{H_3N} + H_2O \underset{k_{-1}}{\overset{k_1}{\rightleftharpoons}} H_2N^- + \underset{pK_a\ -1.7}{H_3O^+}$$

$$K_{eq} = \frac{k_{forward}}{k_{reverse}}$$

Figure 2.15 The equilibrium constant for a reaction is the ratio of the rate constants for the forward and reverse reactions.

than typical diffusion-controlled rates (10^9 M^{-1} s^{-1}) because one of the reactants (the base) is water. This ability to estimate absolute rate constants from thermodynamic constants, in this case pK_a values, is an exceptional instance.

Linear geometries are preferred for proton transfers

Throughout this book we will depict proton transfers as a one-step concerted displacement even though ultra-fast kinetic studies have revealed that hydrogen bonding precedes and follows the key proton transfer step. Given the importance of hydrogen bonding in proton transfers, it makes sense that proton transfers are fastest when they occur through linear trajectories and are slower when they require bent trajectories. This view is supported by kinetic isotope effects; deuteron transfers are much slower than corresponding proton transfers when the trajectories are linear.

The need to avoid proton transfers through square transition states with 90° angles will add a lot of ink to our arrow-pushing mechanisms. For example, consider the mechanism for imine formation, one of the most important fundamental reactions in chemical biology (**Figure 2.16**). Because one of the reactants is an amine, it may come

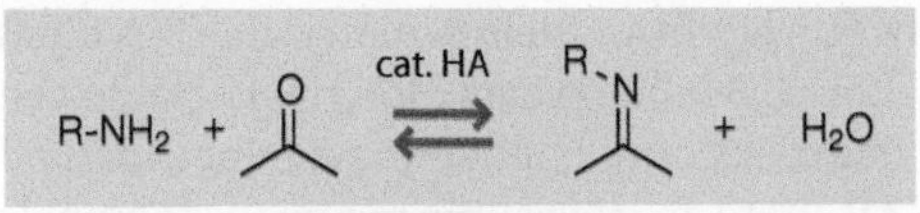

R-NH₂ + O=C(Me)₂ ⇌ (cat. HA) R-N=C(Me)₂ + H_2O

mechanism:

not concerted

iminium ion

imine

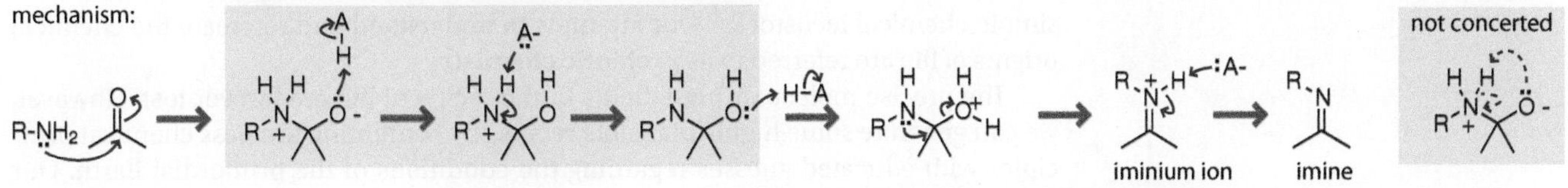

Figure 2.16 The mechanism for iminium ion formation involves a two-step proton transfer in a tetrahedral intermediate. The two-step mechanism is preferred over a one-step proton transfer through a four-atom transition state.

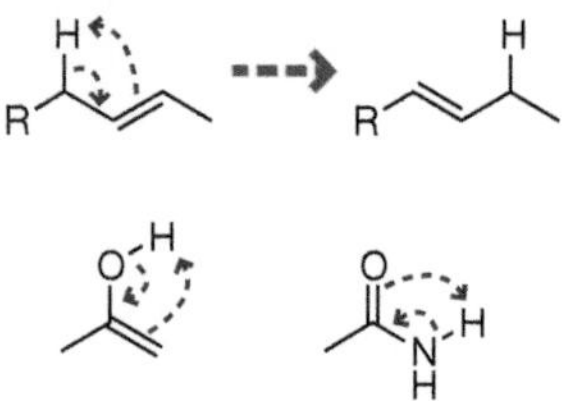

Figure 2.17 1,3-Sigmatropic rearrangements are implausible. When tautomerizations occur, they involve acid or base catalysis.

as a surprise that the reaction is fastest under slightly acidic conditions. Between pH 5 and 6, the amine adds directly to the carbonyl group. Then, an acidic species (symbolized with H–A) donates a proton to the alkoxide group and subsequently removes a proton from the ammonium ion. This two-step proton transfer is faster than a one-step proton transfer from ammonium to alkoxide. Even if the reaction were carried out under basic conditions, the medium would contain some species capable of rapidly protonating the alkoxide. The rate-determining step in imine formation is protonation of the hydroxyl group, which then leaves as water, generating an iminium ion. Deprotonation of the iminium ion generates an imine. The iminium ion is highly reactive and appears in many mechanisms throughout this book.

New students might also be inclined to draw tautomerizations (such as keto–enol tautomerization) as one-step processes that proceed through four-membered transition states (**Figure 2.17**). However, these one-step depictions are completely implausible, for reasons much deeper than poor trajectory. One-step tautomerizations that involve a pi bond and a four-member transition state, as shown below, are classified as 1,3-sigmatropic shifts. As far as chemical biologists are concerned 1,3-sigmatropic shifts of protons can not be concerted, because of a mismatch in the symmetry of pi and sigma orbitals. It is fortunate that 1,3-sigmatropic shifts of protons are not facile; if they were, double bonds would migrate relentlessly within simple alkenes.

Reluctantly, we must forego a deeper examination of orbital symmetry—a Nobel prize-winning concept. For now, we should take stock in the higher plausibility of two-step proton transfers relative to one-step proton transfers that involve four-membered transition states.

Problem 2.8

Suggest a plausible arrow-pushing mechanism for the following tautomerization reactions.

A O, NH_2 —cat. HA→ OH, NH

B H–O, NH_2 —cat. B→ ^-O, NH_3^+

Students may find that some mechanisms published in the chemical literature do not follow the simple rules that connect their arrows to frontier orbitals. Students should not think that a mechanism is acceptable simply because it is printed in a journal. Our rules for mechanistic arrow-pushing are simple and powerful. The best organic chemists follow these rules intuitively, and beginning students are admonished to follow suit. Now that we have a common language for chemistry that is based on a quantum mechanical description of atoms and molecules, we are now empowered to grapple with a fundamental question: How did life arise on the planet Earth?

2.3 PREBIOTIC CHEMISTRY

HCN and CH_2O are key ingredients in the primordial soup

We humans are a relatively new addition to life on planet Earth, having evolved from *Homo sapiens neanderthalensis* as recently as 100,000 years ago. Life itself is much older and arose relatively quickly from lifeless materials. The oldest fossils, derived from cyanobacteria, are about 3.6 billion years old, a respectable age considering that the oldest rocks on Earth date back only 4 billion years. What enabled living organisms to evolve with such haste? Is there an irresistible recipe that compels life to form from simple chemical feedstocks? Our attempts to understand and re-create the chemical origins of life are referred to as **prebiotic chemistry**.

The precise molecular ingredients in the recipe of life are forever lost. However, we can generate some highly plausible recipes by combining timeless chemical principles with educated guesses regarding the conditions of the primordial Earth. Our

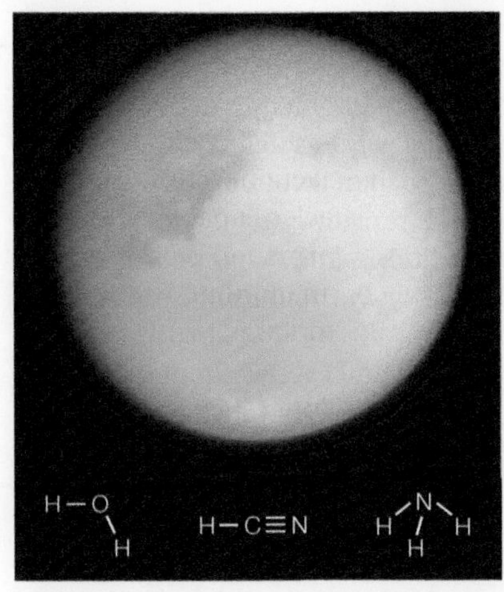

Figure 2.18 Celestial building blocks. Many of the essential prebiotic building blocks can be found on Titan, the largest moon of Saturn. This false-color image of Titan is based on ultraviolet and infrared wavelengths. (Image courtesy CICLOPS and the Cassini Imaging Team. With permission from NASA/JPLCaltech/SSI.)

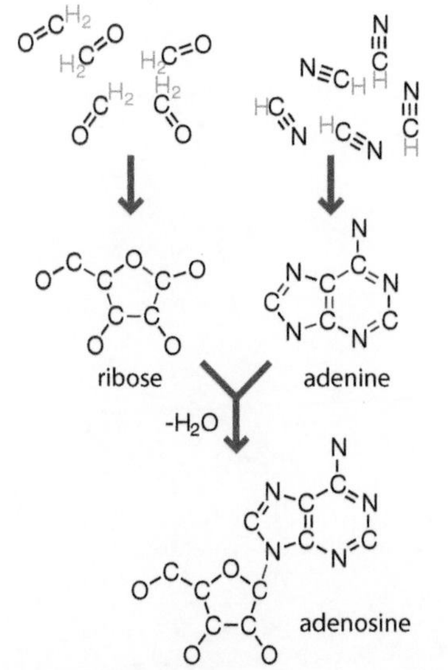

ribose

adenine

$-H_2O$

adenosine

Figure 2.19 DNA from oligomerization of simple molecules. The subunits of DNA, such as adenosine, are simple oligomers of HCN and CH_2O.

guesses about the conditions of primordial Earth are guided by spectroscopic studies of substances present in interstellar space. There is general agreement that the surface of primordial Earth was endowed with simple molecular building blocks: water, ammonia, hydrogen cyanide, acetonitrile, acrylonitrile, cyanogen (NC–CN), and cyanoacetylene. These celestial molecules can be seen right here in our solar system on the surface of Titan, a moon of Saturn (**Figure 2.18**). Molecular oxygen is not an obligate component of planetary atmospheres. The Earth probably remained relatively free of molecular oxygen until the evolution of photosynthetic bacteria around 2.7 billion years ago.

What can we make from aqueous solutions of these primordial building blocks in an oxygen-free environment? Nucleic acids—the building blocks of DNA and RNA! Ribose and adenine are merely pentamers of formaldehyde and hydrocyanic acid, respectively. These two pentamers condense to form adenosine (**Figure 2.19**).

Aqueous solutions of hydrocyanic acid engage in carbon–carbon bond-forming reactions at room temperature, ultimately forming a tetramer, diaminomaleonitrile (DAMN) and larger oligomers. After hydrolysis of the oligomers, adenine can be isolated in 0.03–0.04% yield, without divine intervention. Many different recipes have now been identified for the formation of DNA subunits from primordial molecules, inspiring confidence that the formation of DNA subunits may be an inevitable consequence of the intrinsic reactivity of the primordial molecules.

Solutions of HCN contain both nucleophile and electrophile at pH 9.2

A plausible mechanism for formation of DAMN, based on the available mechanistic data, is shown in **Figure 2.20**. The reaction is fastest at pH 9.2. At this pH, which is the pK_a for HCN, hydrocyanic acid and cyanide anion are present in a 1:1 ratio. Cyanide anion can attack HCN, forming a new carbon–carbon bond. Protonation of the imine would then generate a highly electrophilic iminium ion, making it susceptible to attack by another cyanide anion. A third addition of cyanide anion would complete the construction of a four-carbon fragment that could tautomerize to give the DAMN, which is stabilized by resonance. Under ideal conditions, DAMN can be formed in up to 30% yield by the following mechanism (see Figure 2.20).

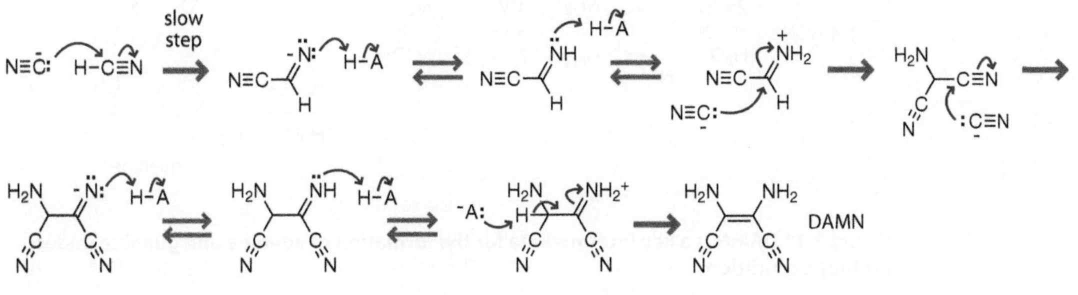

slow step

DAMN

Figure 2.20 Prebiotic carbon–carbon bond formation. Under basic conditions, cyanide can serve as both a nucleophile and an electrophile, ultimately leading to DNA building blocks like DAMN.

An important characteristic of a rigorous arrow-pushing reaction mechanism is that *each set of curved arrows corresponds to a single elementary reaction with one transition state.* Curved arrows are meant to indicate the interaction of filled orbitals with unfilled orbitals in the transition state. One simple way to safeguard against mistakes is to follow the *three-arrow rule.* If you draw more than three curved arrows in a single step, it probably does not correspond to a single transition state. When the horizontal reaction arrow is written with reagents or conditions, and no curved arrows are drawn, the horizontal arrow represents a transformation with no implications about the number of steps in the reaction mechanism.

The Henderson–Hasselbalch equation is a powerful tool even without a slide rule or calculator: $\log([HA]/[A^-]) = pK_a - pH$. Once you know the pH at which an acid and conjugate base are present in a 1:1 ratio, it is easy to estimate the ratio of acid and conjugate base around physiological pH. As we just saw, HCN and CN^- are present in a 1:1 ratio at pH 9.2. Therefore the two species must be present at a 10:1 ratio at pH 8.2, where the medium is 10 times more acidic. Similarly, HCN and CN^- must be present in a 100:1 ratio at pH 7.2, and increasing amounts of HCN are present with decreasing pH. Thus, if we know the pK_a for any compound, we can estimate the ratio (or percentage) of protonated and deprotonated species at physiological pH. Ratios are much more intuitive than equilibrium constants.

Problem 2.9

For each of the following molecules, predict the ratio of protonated acid to deprotonated conjugate base at pH 7.2.

A	B	C
HCN (pK_a 9.2)	NH_4^+ (pK_a 9.2)	$PhCO_2H$ (pK_a 4.2)

HCN forms purines and pyrimidines under prebiotic conditions

Returning to the mechanism for prebiotic synthesis of DNA, we should note that the Sun was, and still is, an abundant energy source. Without an ozone layer to shield it, the primordial Earth was undoubtedly exposed to high-energy ultraviolet radiation. The key to understanding the prebiotic formation of DNA subunits was the finding that ultraviolet radiation converts DAMN to the imidazole derivative aminoimidazole carbonitrile (AICN) (**Figure 2.21**). When the photochemical conversion is performed in the absence of oxygen, the yield of AICN is nearly quantitative. AICN can react further with HCN to form the DNA base adenine, or with urea to form the DNA base guanine. The exact mechanism for this transformation is left to the reader's study.

Uracil can be obtained in very low yield from the hydrolysis of HCN oligomers. A recipe for the formation of cytosine involves the reaction of cyanoacetylene with

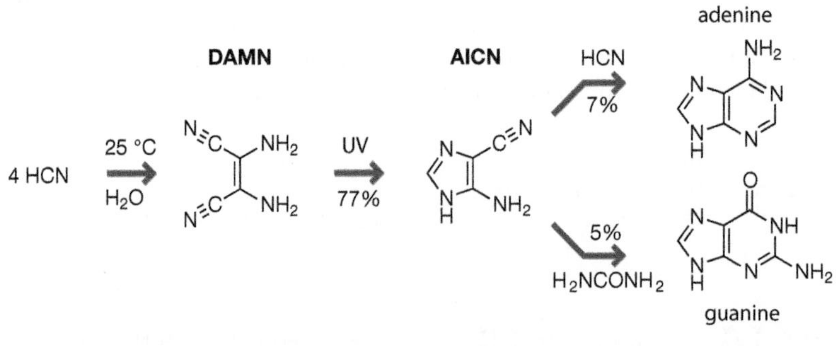

4 HCN → (25 °C, H_2O) DAMN → (UV, 77%) AICN → (HCN, 7%) adenine; AICN → (H_2NCONH_2, 5%) guanine

Figure 2.21 DAMN is a key intermediate for the formation of adenine and guanine under prebiotic conditions.

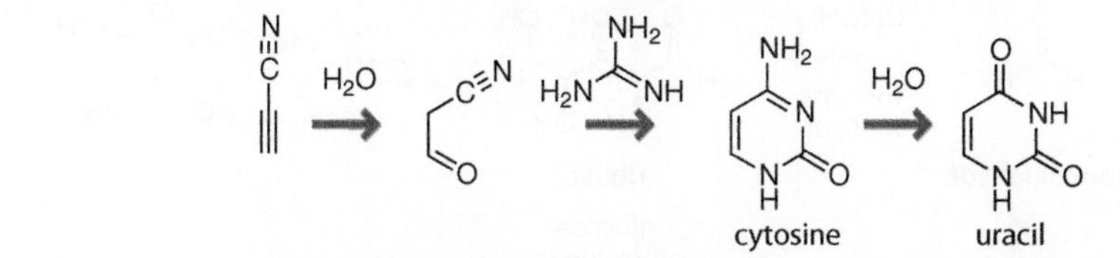

Figure 2.22 **Cyanoacetylene is a precursor in the formation of pyrimidine bases.**

water and guanidine as depicted in **Figure 2.22**. Hydrolysis of cytosine would then form uracil, a base found in RNA. We discuss these DNA subunits in significant detail in the next chapter.

The mechanism for the formation of cyanoacetaldehyde from cyanoacetylene involves the conjugate addition of water to cyanoacetylene to form an enol. The enol could undergo acid-catalyzed tautomerization to the keto form (**Figure 2.23**).

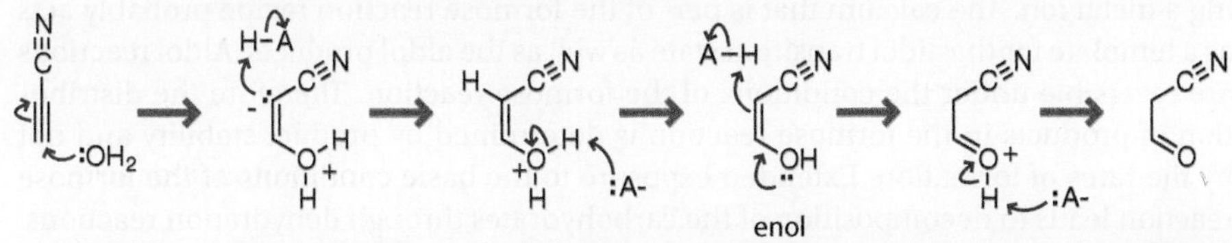

Figure 2.23 **Conjugate addition of water to cyanoacetylene leads to cyanoacetaldehyde.**

The mechanism for the condensation of cyanoacetaldehyde with guanidine is quite lengthy (17 steps), consisting mostly of proton transfers. The key transformations are depicted in abbreviated form in **Figure 2.24**. Note that it is perfectly acceptable to omit some of the obvious steps by stacking elementary reaction arrows (**Figure 2.25**), but if we leave out the intermediate structures, we must leave out the curved arrows. *In this book, we will abbreviate three or more skipped elementary mechanistic steps with three stacked reaction arrows.* Condensation of cyanoacetaldehyde and guanidine could produce an iminium ion. Cyclization of the guanidino group onto the nitrile would generate a heterocyclic ring. Hydrolysis of the imino tautomer would lead to the DNA base cytosine.

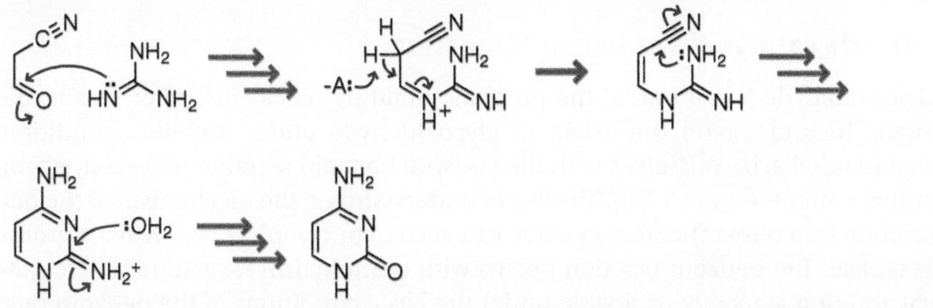

Figure 2.24 **An abbreviated mechanism for formation of cytosine.** To save space, we depict only the curved arrows for the first of several mechanistic steps.

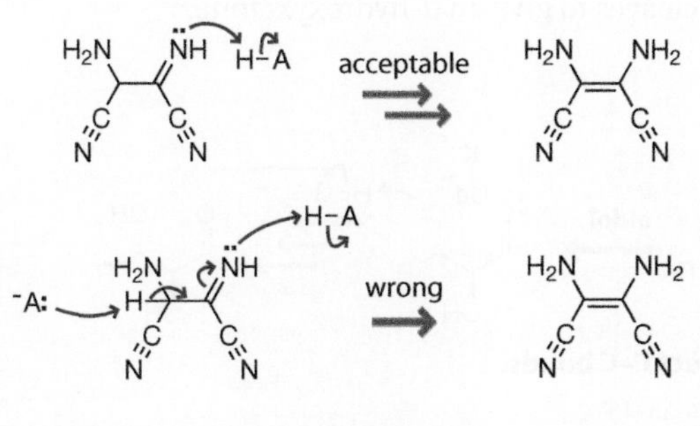

Figure 2.25 **A shortcut for experts.** Skipping obvious elementary reaction steps in a mechanism is acceptable as long as the horizontal reaction arrows are retained. However, it is misleading to combine arrows for multiple transition states into a single step.

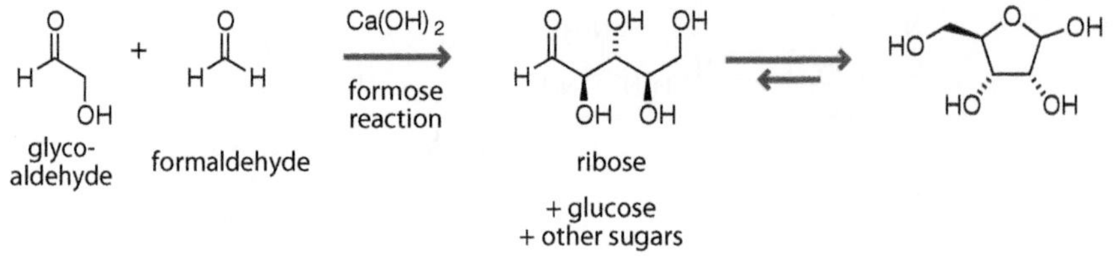

Figure 2.26 Prebiotic formation of carbohydrates. The formose reaction generates carbohydrates like ribose and glucose under prebiotic conditions.

glyco-aldehyde + formaldehyde → $Ca(OH)_2$ formose reaction → ribose + glucose + other sugars

Aldol reactions with formaldehyde generate carbohydrates

The "formose reaction" of formaldehyde and glycoaldehyde, first described in 1861, generates a mixture of virtually all possible carbohydrates, including D-ribose, which is found in DNA and RNA (**Figure 2.26**). The formose reaction works best when glycoaldehyde and formaldehyde react in a 3:1 ratio in aqueous calcium hydroxide.

The formose reaction generates carbohydrates through a series of aldol reactions. The best conditions involve calcium, which is abundant in many minerals. A plausible mechanism for the key aldol reaction leading to formation of ribose is shown below (**Figure 2.27**). Most aldol reactions proceed through chair-like transition states involving a metal ion. The calcium that is part of the formose reaction recipe probably acts as a template for the aldol transition state as well as the aldol products. Aldol reactions are reversible under the conditions of the formose reaction. Therefore the distribution of products in the formose reaction is determined by product stability and not by the rates of formation. Extended exposure to the basic conditions of the formose reaction leads to decomposition of the carbohydrates through dehydration reactions. However, carbohydrate products can be stabilized by the presence of borates, which form cyclic borate esters with vicinal hydroxyls.

Problem 2.10

Suggest a plausible arrow-pushing mechanism involving a retro-aldol reaction for the following interconversion under the conditions of the formose reaction (namely aqueous calcium hydroxide).

Cyanide catalyzes the benzoin reaction

Glycoaldehyde is not one of the prebiotic building blocks identified in space or on Titan. To understand the origin of glycoaldehyde under prebiotic conditions we should familiarize ourselves with the classical benzoin reaction of benzaldehyde and cyanide anion (**Figure 2.28**). The key to understanding the mechanism of the benzoin reaction is to remember that cyanide can act as a nucleophile, as a leaving group, and as a base. The benzoin reaction begins with cyanohydrin formation; in fact, the initial addition is readily reversible under the basic conditions of the benzoin reaction. Because nitriles stabilize carbanions, the protons alpha to the nitrile can be deprotonated, and the resulting carbanion can attack another aldehyde group. The resulting cyanohydrin readily cleaves to give an α-hydroxyketone.

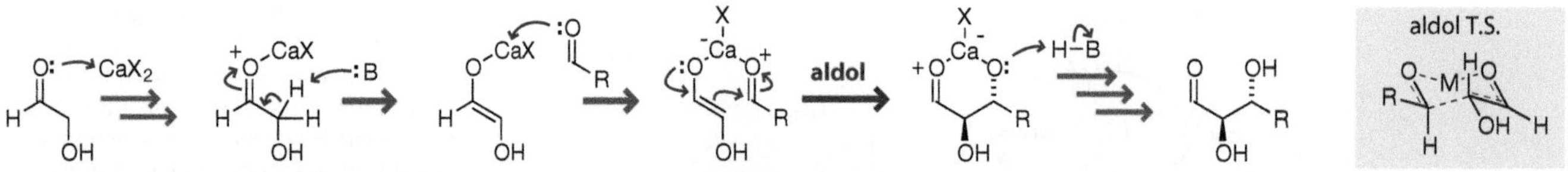

CaX_2 → aldol → aldol T.S.

Figure 2.27 The formose reaction involves aldol reactions to construct C–C bonds.

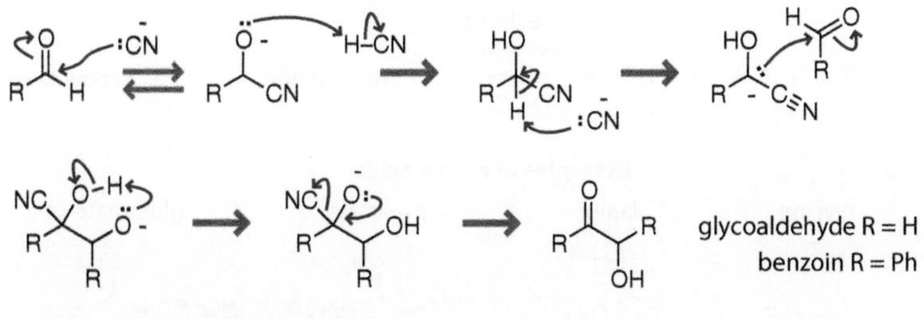

glycoaldehyde R = H
benzoin R = Ph

Figure 2.28 The benzoin reaction. The mechanism of the benzoin reaction involves a carbanion stabilized by a nitrile group.

The recipe for formation of the nucleoside adenosine is ultimately quite simple: five molecules of formaldehyde, five molecules of HCN, and some light. However, DNA and RNA are oligomers of phosphate esters. By invoking inorganic phosphates (plentiful in minerals) as chemical participants we gain a fuller picture of the potential chemical origin of DNA, the basic molecule of life on Earth.

Did we arise from a primordial RNA world?

Unfortunately, the previous discussion of prebiotic building blocks best explains the formation of RNA, not DNA. Although structural differences between RNA and DNA are subtle, those differences have profound effects on reactivity that we will address later. For now, we are faced with a conundrum: DNA is the quintessential genetic material from which all living organisms make RNA, yet prebiotic chemistry readily explains the formation of RNA rather than DNA. Two key findings support the hypothesis that RNA was the seminal biomolecule in the primordial world. The first key finding was the discovery of the viral enzyme reverse transcriptase in 1971. Reverse transcriptase catalyzes the formation of DNA from an RNA template. A second key finding is that RNA itself can catalyze specific chemical reactions, including the cleavage and formation of the bonds that hold RNA together. Without much difficulty, one can imagine a prebiotic world dominated by RNA. As we will see shortly, however, RNA is a poor choice for long-term information storage.

Amino acids arise spontaneously under prebiotic conditions

The field of prebiotic chemistry can be traced back to the famous Miller–Urey experiment performed in the early 1950s (**Figure 2.29**). The experiment was designed to

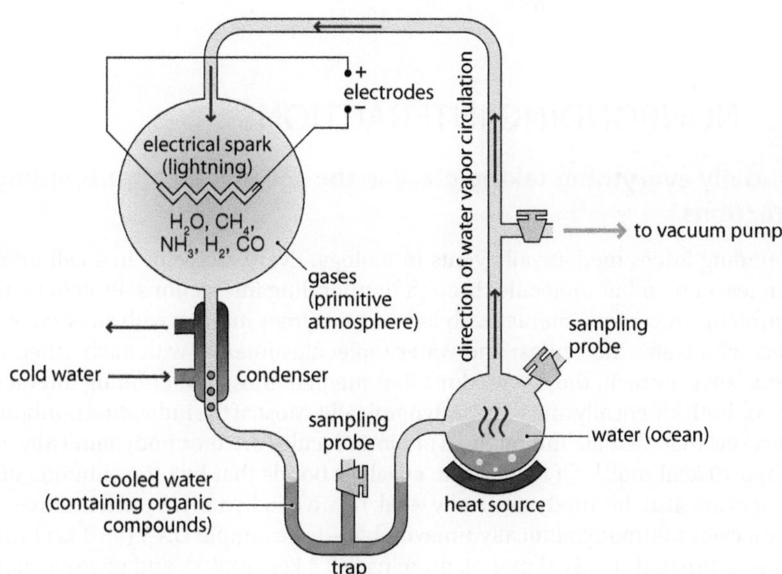

Figure 2.29 The Miller–Urey experiment. In the classic Miller–Urey experiment, amino acid building blocks can be formed from prebiotic starting materials and environmental conditions. (Adapted from W. Schwemmler, Reconstruction of Cell Evolution: a Periodic System. Boca Raton: CRC Press, 1984.)

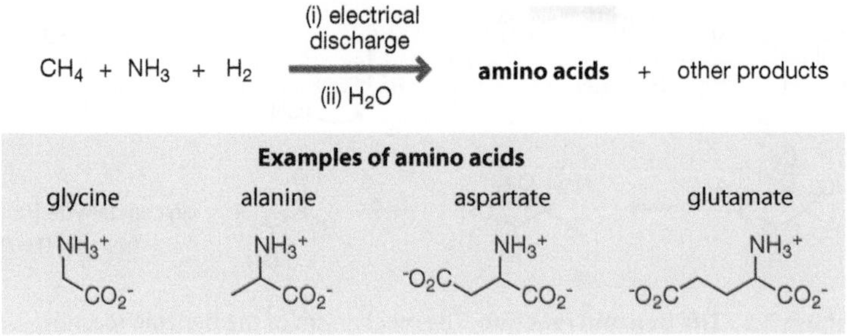

Figure 2.30 The Miller–Urey experiment showed that amino acids are generated under prebiotic conditions.

$$CH_4 + NH_3 + H_2 \xrightarrow[\text{(ii) } H_2O]{\text{(i) electrical discharge}} \textbf{amino acids} + \text{other products}$$

mimic the effects of lightning on a primordial atmosphere consisting of methane, ammonia, hydrogen, and water vapor, but in place of lightning a simple electrical discharge was used. The reactive intermediates produced by the electric discharge were allowed to react in a mock "ocean." Surprisingly, several naturally occurring amino acids arose under these conditions: glycine, alanine, aspartic acid, and glutamic acid (**Figure 2.30**). We will revisit these amino acids in more detail in Chapter 5.

The formation of amino acids in the Miller–Urey experiment was probably related to the classical two-step Strecker synthesis of amino acids. The first step in the Strecker method involves the condensation of ammonia, HCN, and aldehydes to form α-aminonitriles (**Figure 2.31**). The aminonitriles can be hydrolyzed under strongly acidic conditions to form α-amino acids.

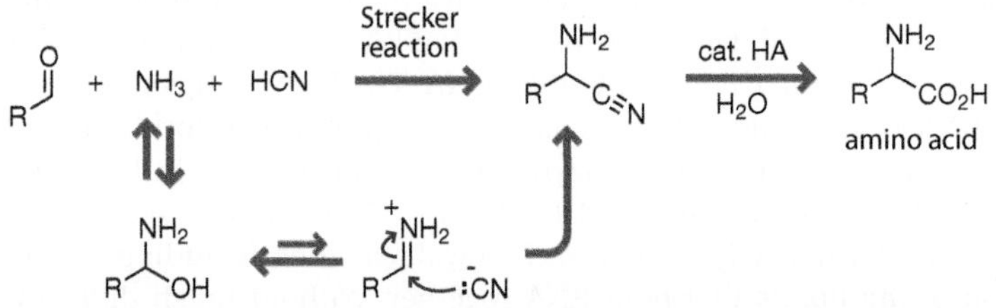

Figure 2.31 The Strecker reaction. Formation of amino acids under prebiotic conditions involves addition of a cyanide anion to highly reactive iminium ion.

Problem 2.11

Suggest a complete arrow-pushing mechanism for the transformation of acetaldehyde to α-aminopropionitrile.

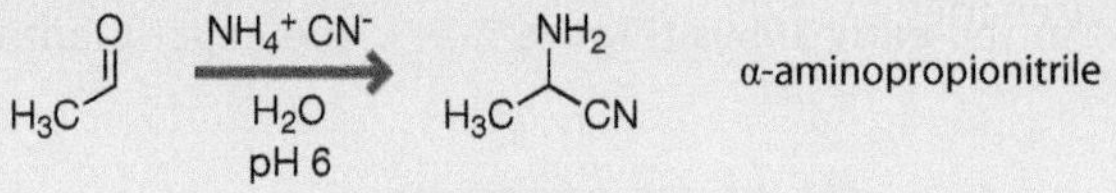

2.4 NONBONDING INTERACTIONS

Essentially everything taking place in the cell involves nonbonding interactions

Nonbonding forces mediate all events in biology. Every molecule in a cell interacts with at least one other molecule through nonbonding interactions. Proteins interact with proteins; receptors interact with ligands; enzymes interact with substrates; ions interact with water molecules; and water molecules interact with each other. In an aqueous environment, the interactions that mediate these nonbonding interactions are weak, both kinetically and thermodynamically. Most of the individual nonbonding interactions that mediate interactions of biomolecules are thermodynamically worth less than 10 kcal mol^{-1}. Of course, the covalent bonds that link the subunits of bio-oligomers are also thermodynamically weak relative to hydrolysis, and in some cases they are even thermodynamically unfavorable—for example DNA (–5.3 kcal mol^{-1}), RNA (an estimated –5.3 kcal mol^{-1}), proteins (+0.4 kcal mol^{-1}), and oligosaccharides

(+5.5 kcal mol^{-1}). Yet, if you pull on the ends of a folded protein in water, the nonbonding interactions will always be disrupted before the covalent bonds break. Ultimately, it is kinetic stability, not thermodynamic stability, that sets covalent bonds apart from the nonbonding interactions that dominate biology. We will address the **lability** of covalent bonds and nonbonding interactions later in this book. For now, it is essential for us to rank the kinetically labile nonbonding interactions and identify those that are highly dependent on the local dielectric constant. Once we have gained a picture of the underlying forces that mediate those interactions we can use them to discuss complex phenomena such as ligand binding, enzymatic catalysis, and protein folding.

Figure 2.32 van der Waals interaction. The weak attraction between two methane molecules is not easily quantified by perturbational molecular orbital theory that we used to think about bonding interactions (Equation 2.2).

The weak energies of nonbonding interactions are not easily calculated using perturbational molecular orbital theory

When we think about chemical reactions that form bonds, the perturbation treatment of molecular orbital theory fits well with the mechanistic depictions used by organic chemists. Please go back and review the three components of Equation 2.2. Often, either Coulombic effects or orbital effects dominate bond-forming events. When electropositive atoms interact with electronegative atoms, as in K–Cl, the Coulombic effects dominate Equation 2.2 and we describe the resulting bond as **ionic**. When the bonded atoms have similar electronegativity, as in H_3C–H or H_3C–CH_3, the orbital effects dominate Equation 2.2 and we describe the resulting bond as covalent. Bonds that are somewhere between these extremes are often referred to as **polar covalent** bonds, as in RO–Li.

But what happens if we try to analyze the interaction of two methane molecules with perturbation theory (**Figure 2.32**)? Neither carbon nor hydrogen has a significant charge so the Coulombic effects will be small. The covalent component of this interaction would also be almost negligible. In theory, each molecule of methane has filled orbitals that can interact with the unfilled orbitals of the other methane, but none of the filled orbitals in CH_4 are very high in energy and none of the unfilled orbitals in CH_4 are very low in energy. We might see some bonding interactions if we got close—extremely close—but, at such close distances, steric repulsion would cancel out the bonding interactions. Ultimately, the interaction between two methane molecules might be slightly favorable but would not make or break covalent bonds. Chemists refer to these nonbonding interactions as **van der Waals interactions.** Such interactions are vitally important to all events in biology, including ligand binding, protein-protein interactions, transport through membranes, chemical reactions, and protein folding. The ubiquity of these nonbonding interactions justifies a version of Equation 2.1 that is more user-friendly than the perturbational molecular orbital treatment that we invoked for chemical reactions.

For nonbonding interactions, the energies can be fitted to a simplified equation

If you measured the distance dependence of the potential energies that result from nonbonding interactions, you could roughly fit the energetics to an equation with three mathematical terms (**Equation 2.3; Figure 2.33**). The fitted Equation 2.3 looks similar to the perturbation treatment of molecular orbital theory (Equation 2.2) with terms that correspond to charge–charge, repulsive sterics, and attractive energy terms.

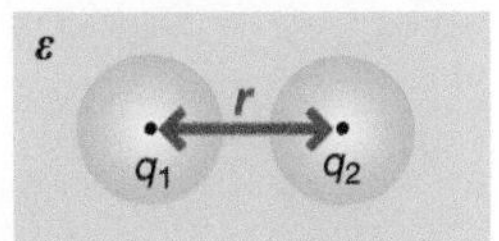

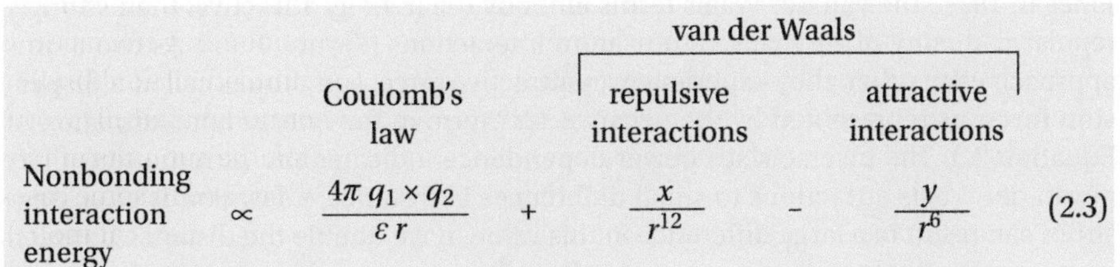

van der Waals
Coulomb's law
repulsive interactions
attractive interactions
$$\text{Nonbonding interaction energy} \propto \frac{4\pi\, q_1 \times q_2}{\varepsilon\, r} + \frac{x}{r^{12}} - \frac{y}{r^6} \qquad (2.3)$$

Figure 2.33 Three important contributions to nonbonding interactions. Nonbonding interactions can be quantified by an equation involving electrostatic interactions and both attractive and repulsive van der Waals interactions.

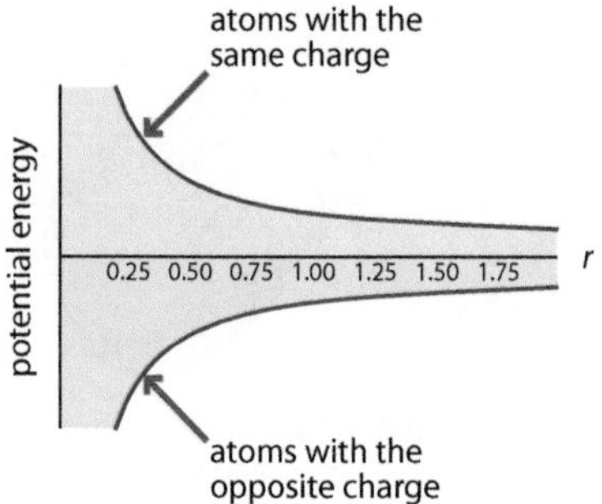

Figure 2.34 The Coulombic potential. As distance between the atoms (r in van der Waals radii) increases, the potential energy for the interaction approaches zero.

The equation provides an approximation for the interaction energy of a noncovalent interaction involving either electrostatic interactions or the nonbonding interactions that resemble the weak methane–methane interaction. A negative potential energy for the interaction indicates a favorable interaction, and a positive potential energy indicates an unfavorable interaction. The charges (q_1 and q_2) and the numerators in the van der Waals terms (x and y) are constants based on the van der Waals radii of the two atoms. The dielectric constant ε describes how effectively the environment screens the two charges. The distance r between the nuclei is determined by the atomic positions. Humans are good at looking up constants and plugging them into simple equations such as Equation 2.3; computers are even better, and computational modeling can provide insight into protein folding, receptor–ligand interactions, and other processes in chemical biology.

The Coulombic effects are the same as those we previously considered in Equation 2.2, except that in Equation 2.3 we reveal the full form of the equation (see Figure 2.33). The charges q_1 and q_2 on the two interacting atoms will always be partial charges determined by electronegativity. If the interacting atoms or functional group have the same charges, the Coulombic potential returns a value greater than 0, which describes the expected repulsion of the two atoms. However, the interaction of two oppositely charged atoms results in a negative value for the potential, which provides the expected attractive potential. Thus, the plot of this potential has two lines depending on the charges of the two atoms (**Figure 2.34**).

The partial charges on the individual atoms within a functional group or molecule are often the opposite of the formal charges drawn in the Lewis structure. For example, the nitrogen in NH_4^+ has a partial negative charge, –0.98; each of the four protons has a partial positive charge, +0.5 (**Figure 2.35**). A negatively charged chloride ion would be attracted by the partial positive charges on the protons but repelled by the negatively charged nitrogen atom. Lewis structures correctly predict the partial negative charge on oxygen, but they do not reveal that the α carbons have about the same partial negative charge as the oxygen. Fortunately, the net effect of these Coulombic interactions based on partial atomic charges is usually the one we would have predicted on the basis of the formal Lewis structure charge.

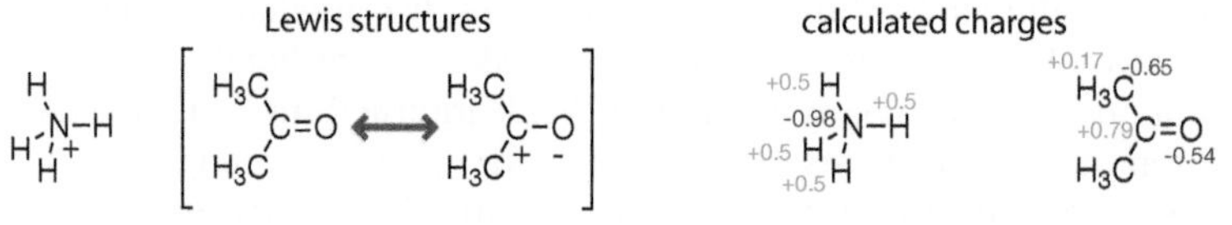

Lewis structures | calculated charges

Figure 2.35 Where is the charge? The formal charges in Lewis structures offer a misleading picture of the partial charges on the atoms.

The value of ε, the dielectric constant, is highly dependent on context, and it can change enormously depending on the surrounding environment. Such changes can in turn drastically alter the strength of the electrostatic interaction. A more polar environment results in a higher value for this dielectric constant, which weakens the attraction between two oppositely charged atoms. The very polar solvent water, for example, has a high dielectric constant value of about 78 at room temperature. In water, identically charged atoms repel each other much less strongly than expected, and water is sometimes described as having a charge-shielding effect. Furthermore, oppositely charged atoms in water attract each other far less strongly than two charged atoms in a vacuum.

van der Waals interactions can be described by the Lennard-Jones potential

The second two terms in Equation 2.3 were empirically formulated by John Lennard-Jones in 1924. The van der Waals terms embody the initially attractive, then strongly repulsive, duality of very close atom–atom interactions (**Figure 2.36**). As two atoms approach each other, they experience an attractive force, sometimes called a **dispersion force**, as represented by the negative $1/r^6$ term in the energy approximation of Equation 2.3. The inverse sixth power dependence indicates the extreme sensitivity of van der Waals attractions to small differences in distance; a few tenths of an ångström can result in a large difference in this value. If we double the distance between interacting methyl groups in a receptor–ligand interaction, the attractive forces will decrease by a factor of 64. Proteins tend to fold and interact with ligands in ways that avoid leaving empty space, because empty space represents a lost opportunity for dispersive attractions.

The positive $1/r^{12}$ term of the Lennard-Jones potential approximates the steric repulsion experienced by two atoms pushed too close to each other. The inverse twelfth power dependence on distance strongly penalizes a close approach of nonbonding atoms. Each time we cut the distance in half, the repulsions increase by a factor of more than 4000. Forces go from attractive to repulsive at a distance σ, which is proportional to the van der Waals radii for the two atoms. When you see space-filling models of molecules, atoms are usually drawn as spheres based on these van der Waals radii; such depictions thus provide the distance to which another molecule can approach before experiencing the tremendously repulsive force embodied by the Lennard-Jones equation.

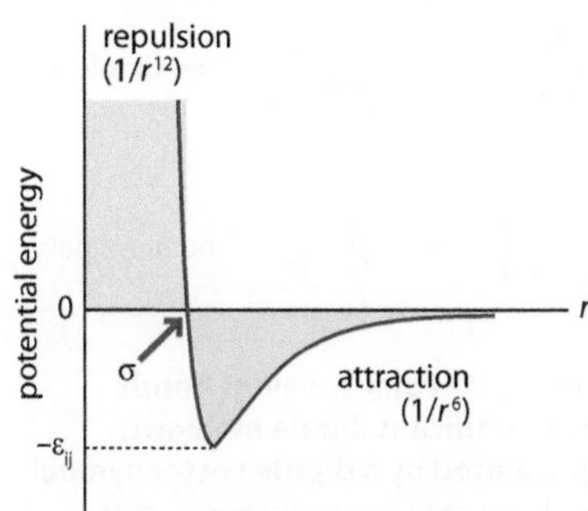

Figure 2.36 The Lennard-Jones potential. As the distance r between the atoms increases, the attraction between the two atoms (blue) approaches zero. Strong repulsion (yellow) results when the two atoms are too close.

Problem 2.12

A Two negatively charged functional groups are spaced 2 Å apart. If the two atoms move to 3 Å away, by what percentage does their repulsive interaction energy change?

B Two neutral atoms move from 2 Å to 3 Å apart. By what percentage does their attractive interaction energy change?

It is helpful to distinguish reversible from irreversible interactions

Chemical biologists usually distinguish interactions that form highly covalent bonds (such as C-C, C-O, O-P, and C-S) from interactions that do not. More importantly, we tend to group interactions on the basis of kinetics rather than thermodynamics. Under biological conditions, bonds to carbon and phosphorus tend to be kinetically stable and irreversible. Conversely, dispersive interactions, Coulombic interactions, dipole–dipole interactions, and hydrogen bonds tend to be kinetically unstable and reversible. Such readily reversible interactions are important in receptor–ligand binding, protein folding, and other events in chemical biology.

The readily reversible interactions in **Figure 2.37** are each worth less than 10 kcal mol^{-1} relative to dissociation in the gas phase. Individual dispersion interactions between atoms or functionalities are weakest—usually less than 1 kcal mol^{-1}. However, such individual interactions can add up, as is evident in the π-stacking interactions of arene rings, which require multiple contacts between atoms.

Two geometries dominate the interactions of aromatic rings. Rings can interact in parallel or T-shaped structures with roughly equivalent interaction strengths. For the parallel interactions of π-stacked aromatic rings, the arenes prefer an offset or displaced orientation to position regions of electron density over the comparatively

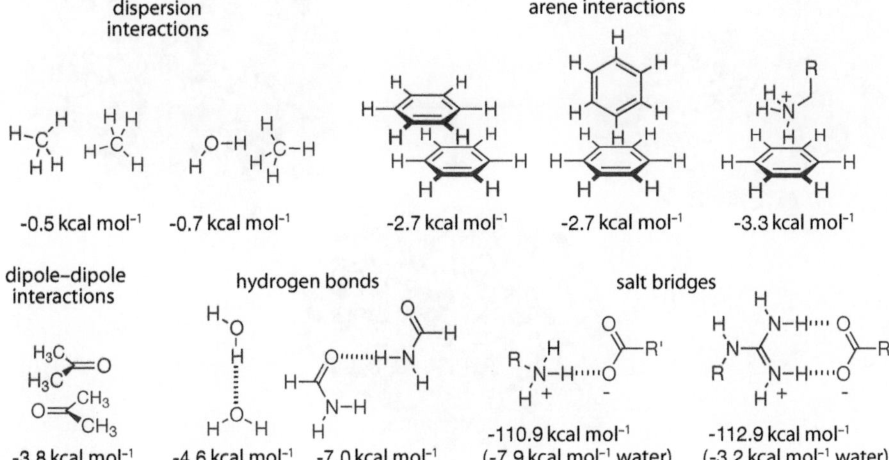

Figure 2.37 Readily reversible gas-phase interactions between molecules and functional groups are only modestly favorable, except for salt bridges. Dipole–dipole interactions, hydrogen bonds, and salt bridges are weaker in an environment with a high dielectric constant (such as water) because it screens the charges; thus, salt bridges are much weaker in an aqueous environment.

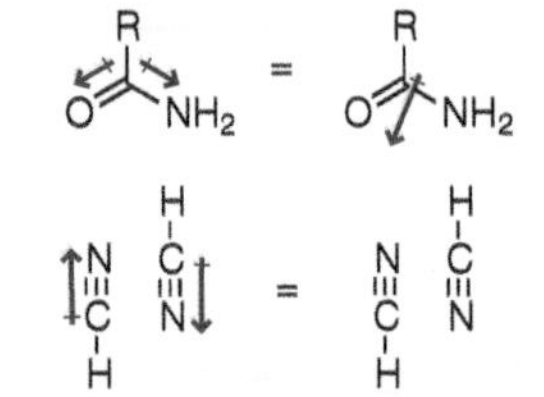

net dipole

no net dipole

Figure 2.38 Polar covalent bonds have significant dipole moments, represented by a dipole vector symbol. Dipole moments are additive vector quantities that can augment or cancel each other. Because a dipole moment corresponds to charge separation, opposing dipoles that cancel each other are strongly favored in nonpolar environments.

electron-deficient centers of the rings. Aromatic rings also interact with cations, C–H, and the lone pairs of oxygens. Found widely in protein structures, cation–π interactions are important in receptor–ligand binding, protein folding, and the interactions of membrane proteins with the phospholipid headgroups of the membrane.

Dipole–dipole interactions, which are Coulombic in nature, are more stabilizing than dispersion interactions. The acetone–acetone interaction, for example, benefits from the interaction of carbonyl dipoles (see Figure 2.37). Dipoles prefer to line up in opposite directions, with the negative aligned with positive ends of the dipole. Thus, the lowest-energy configuration cancels out the individual dipoles (**Figure 2.38**).

Salt bridges, which are interactions between oppositely charged protein side chains, are markedly favorable in the gas phase, where it takes a great deal of energy—more than 100 kcal mol^{-1}—to pull oppositely charged functional groups apart. However, water destabilizes salt bridges in two ways. Most importantly, the high bulk dielectric constant of water (78) screens the charges, reducing their Coulombic interaction by up to two orders of magnitude. In addition, water competes for such interactions and can form stabilizing hydrogen bonds with most cations and anions. Thus, although ionic interactions have immense capacity for repulsion and attraction, that capacity is rarely maximized in an aqueous environment.

Problem 2.13

Identify the functional groups in the following molecules that can engage in noncovalent (hydrogen-bonding and nonbonding) interactions, and describe the type of interaction.

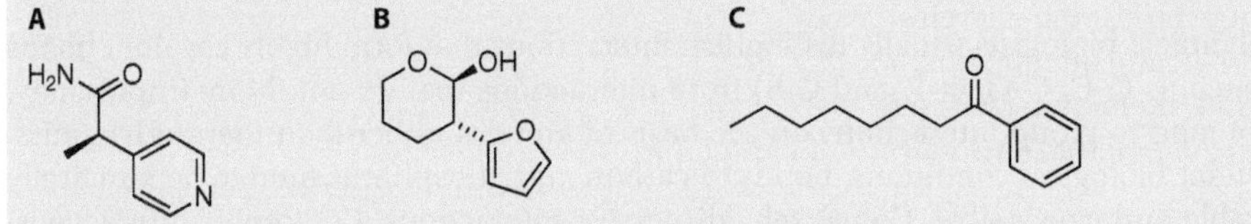

A **B** **C**

As described above, interactions between biomolecules are usually driven by a large number of small noncovalent forces. Collectively, these smaller forces add up to the net potential energy for the interaction. Thus, the interaction energy for any noncovalent interaction is the sum total of the interaction energies contributed by every charge–charge, arene, dipole–dipole, and dispersion interaction and also hydrogen bonds at the interface between the two molecules.

The complexity of this summation is apparent if we closely examine the structures of the enzyme lysozyme and a tightly binding ligand (**Figure 2.39**). In this example,

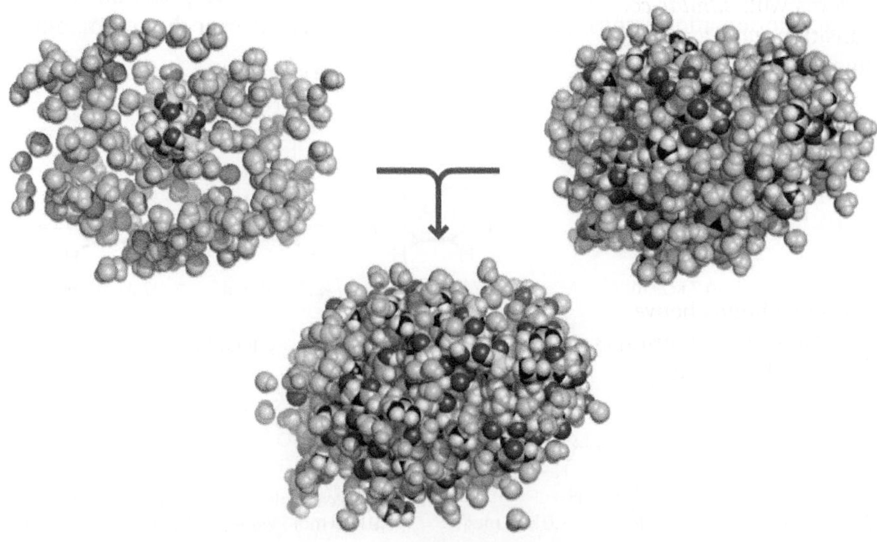

Figure 2.39 Both a ligand (yellow) and a protein (gray) must give up many favorable interactions with water (green) to form a protein–ligand complex. (PDB 1HEW and 1IO5.)

the ligand is actually the proteoglycan substrate, which is hydrolyzed by lysozyme. Studded with potential hydrogen bond donors and acceptors and various other functionalities, the ligand can readily access a large number of potential binding modes to lysozyme and other molecules such as water.

To understand the high affinity of this interaction we must ask a deeper question: Because both the protein and the ligand can make many favorable interactions with water molecules, why should they prefer to associate with each other? An answer to this question requires us to consider carefully the advantages and disadvantages of interactions with water.

Entropy makes it difficult to identify favorable states among seemingly endless possibilities

We must be careful how we apply the energies tabulated in Figure 2.39. We never showed the alternative with which those states were compared—with the two molecules separated by an infinite distance in a vacuum. To illustrate the danger in misusing those numbers, let us consider the hydrogen-bonded water dimer (**Figure 2.40**). The ideal hydrogen-bonded, dimerized configuration is more stable than the two water molecules held an infinite distance apart; the energetic difference is 4.6 kcal mol^{-1}. However, many configurations are preferred over infinite separation. If we compare the one ideal hydrogen-bonded configuration with the multitude of alternative configurations, we find that at room temperature only 1 out of 16 pairs of water molecules will exist in the ideal hydrogen-bonded configuration; the other 15 pairs will adopt some other configuration. This plethora of possibilities is the unsettling effect of entropy, and it is this effect that will prevent us from summing up a bunch of small energies to make accurate predictions about the energetics of receptor–ligand interactions. In other words, focusing simply on the enthalpy of some idealized interaction (ΔH) neglects the entropic ($T\Delta S$) term composing the free energy of the interaction ($\Delta G = \Delta H - T\Delta S$, the Gibbs–Helmholtz equation).

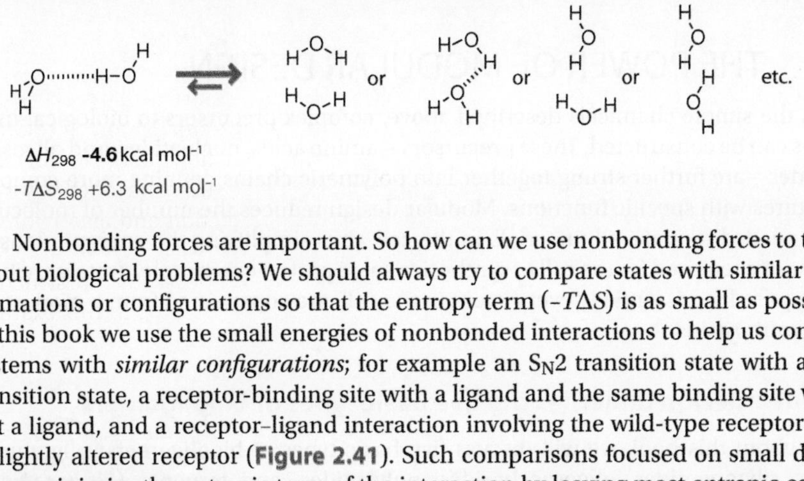

or or or etc.

ΔH_{298} **-4.6** kcal mol^{-1}

$-T\Delta S_{298}$ +6.3 kcal mol^{-1}

Figure 2.40 Entropy complicates hydrogen-bonding. An ideal hydrogen bond between water molecules is thermodynamically unfavorable relative to the multitude of other possibilities.

Nonbonding forces are important. So how can we use nonbonding forces to think about biological problems? We should always try to compare states with similar conformations or configurations so that the entropy term ($-T\Delta S$) is as small as possible. In this book we use the small energies of nonbonded interactions to help us contrast systems with *similar configurations*; for example an S_N2 transition state with an E2 transition state, a receptor-binding site with a ligand and the same binding site without a ligand, and a receptor–ligand interaction involving the wild-type receptor with a slightly altered receptor (**Figure 2.41**). Such comparisons focused on small differences minimize the entropic term of the interaction by leaving most entropic contributions unchanged.

The hydrophobic effect results from a balance between attractive forces and entropy

Hydrogen bonds between water molecules are directional and incur a high entropic cost. Most of this entropic cost is recovered when water molecules form energetically favorable hydrogen bonds. The dispersive interactions between water and alkanes are so weak that they do not compensate for the entropic cost of the interaction. As a result, water tends to maximize interactions with itself and minimize interactions with hydrophobic solutes; hydrophobic solutes are left to associate with themselves through meek dispersive interactions. We call this tendency the **hydrophobic effect**. The hydrophobic effect motivates proteins to adopt globular shapes in water.

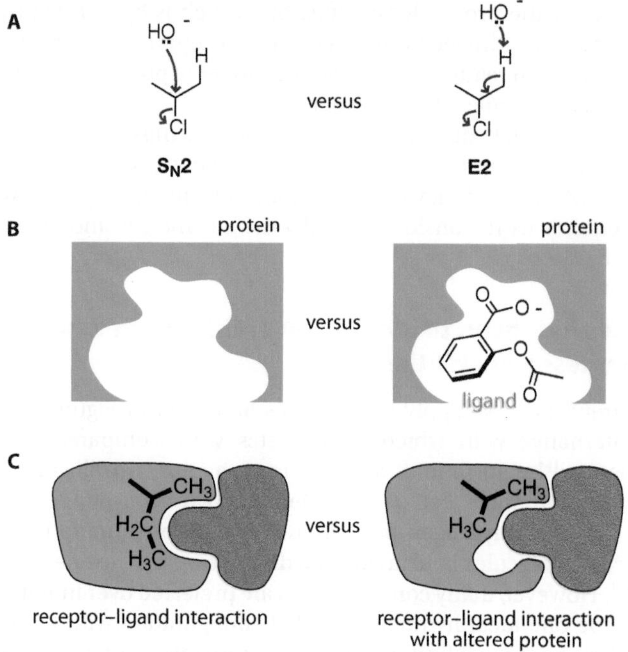

Figure 2.41 It is best to use small nonbonding energies to analyze systems with similar configurations. Comparing systems with similar configurations minimizes the entropy differences: (A) comparing an S_N2 substitution with the competing E2 elimination, (B) comparing an enzyme active without a ligand and with a ligand, and (C) the interaction ligand with a wild-type protein and a mutant protein.

Unlike hydrogen bonds, dispersive interactions do not show strong orientational preferences, so several small dispersive interactions can sum to a large attractive interaction without a large entropic penalty. The hydrophobic effect applies to hydrophobic surfaces or ligand-binding pockets of proteins in aqueous solution. Thus, a small hydrophobic patch on the surface of an otherwise hydrophilic large protein can become a beacon for protein–protein interactions (**Figure 2.42**).

A

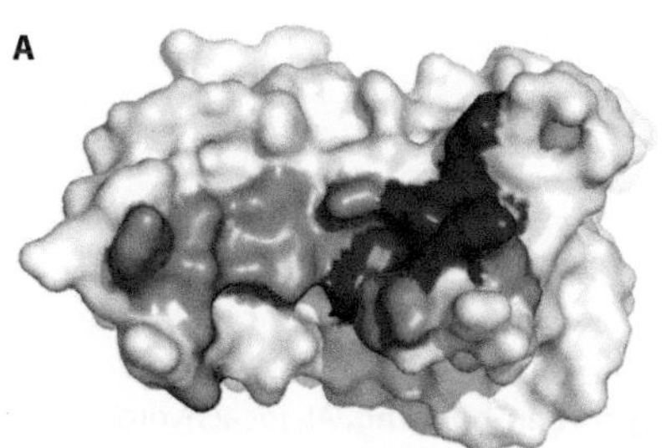

B

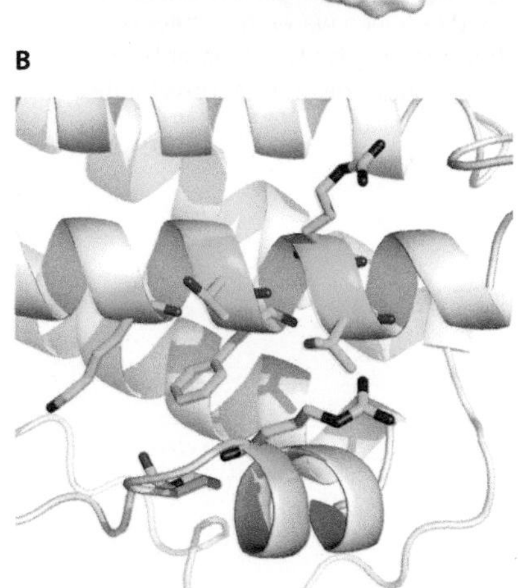

Figure 2.42 Protein–protein interactions often involve interfaces composed of hydrophobic surfaces. (A) The surface colored in red and blue is buried when the protein human growth hormone (hGH) binds to its receptor, hGH-binding protein. (B) A close-up of the red patch, which contributes most of the interaction potential, reveals a small patch of hydrophobic functionalities surrounded by hydrophilic functionalities. (PDB 3HHR.)

2.5 THE POWER OF MODULAR DESIGN

From the simple chemicals described above, complex precursors to biological molecules can be constructed. These precursors—amino acids, nucleotides, and oligosaccharides—are further strung together into polymeric chains defining more complex structures with specific functions. Modular design reduces the number of molecular parts needed to make all materials in the cell, thus simplifying the energy-intensive operations required for the cell to synthesize spare parts. For example, proteins, which are required for essentially every event in the cell, are constructed from a modest set of 20 amino acids.

Modular design underlies the five basic types of biooligomers

Throughout this book we will discuss five basic types of biooligomers: oligonucleotides, oligopeptides, oligosaccharides, polyketides, and terpenes (**Figure 2.43**). The functional groups that characterize the five types of biooligomers confer special shapes and special patterns of chemical reactivity. For now, we should learn to recognize these functional groups; in later chapters we will learn more about their structures, reactivities, and functions.

The bonds that join oligonucleotides, oligopeptides, and oligosaccharides are ultimately formed through the loss of water or broken through the addition of water, respectively. This is a particular advantage for bond cleavage, because water is a plentiful reagent. Carboxylic ester-based biooligomers are conspicuously absent from our list. Although some microorganisms produce polyhydroxyalkanoates, these ester oligomers have no known role in signaling, catalysis, or information storage. However, they have found use as biodegradable plastics. In Chapter 8 we explore the importance of nonoligomeric esters in lipids. For now, we will consider why evolution selected against highly labile carboxylic esters for information storage.

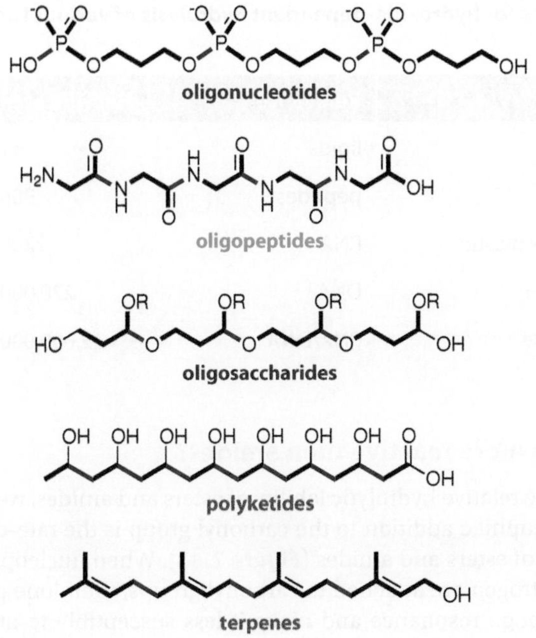

Figure 2.43 Modular design. The biooligomers shown can be constructed through repetitive bond-forming steps from simple building blocks.

Lability correlates inversely with information longevity

A cell is a dynamic system composed of short-lived and long-lived molecules, present in different amounts. Each biosynthetic step in the central dogma is accompanied by an amplification: each DNA sequence can be transcribed into multiple mRNA molecules; each mRNA molecule can be translated into several identical enzymes; each enzyme can catalyze the formation of vast numbers of product molecules (**Figure 2.44**). DNA molecules are rare and precious, whereas (small-molecule) enzyme products tend to be numerous and disposable.

The labilities of the various bio-oligomer functional groups are tabulated in **Table 2.3**. It is satisfying to note that carboxylic ester bonds are fragile, whereas the all-important genetic material, DNA, is assembled using robust phosphodiester groups. However, the reactivity of a phosphate ester is dependent on the neighboring functional groups. The transient nature of mRNA is better suited to the more labile ribose phosphodiester backbone. Amide bonds have kinetic stabilities in between phosphate esters and carboxylic esters. Because these hydrolyzable functional groups have a key role in chemical biology, we will discuss the interplay between biooligomer structure and reactivity.

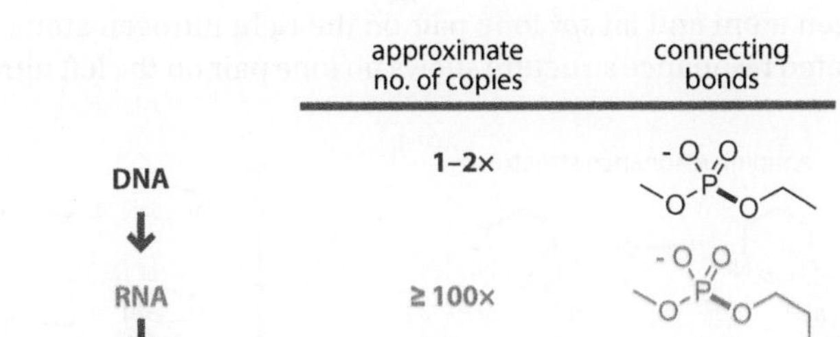

approximate no. of copies
connecting bonds
DNA
1–2×
RNA
≥ 100×
protein (enzymes)
>>1000×
substrate
product
1,000,000×

Figure 2.44 Numerical amplification in biosynthesis.

Table 2.3 Half-lives for hydroxide-dependent hydrolysis of various functional groups in neutral water.

Functionality	Relevance	Half-life at pH 7 (years)
carboxylic ester	lipids	<1
carboxylic amide	peptides	300
ribose phosphate diester	RNA	2200
phosphate diester	DNA	220,000
β-glucofuranoside	RNA/DNA	22,000,000

Why are esters more reactive than amides?

To understand the relative hydrolytic lability of esters and amides, we should first recognize that nucleophilic addition to the carbonyl group is the rate-determining step in the hydrolysis of esters and amides (**Figure 2.45**). When nucleophilic atoms such as oxygen and nitrogen are attached to carbonyl groups, their lone pairs donate into the carbonyl through resonance and make it less susceptible to attack by external nucleophiles such as hydroxide (**Figure 2.46**). Because nitrogen lone pairs are more nucleophilic than oxygen lone pairs, the amide is made less reactive than the ester toward external nucleophiles. Aldehydes are highly reactive because a hydrogen atom attached to a carbonyl is incapable of donating into the π^*_{CO}.

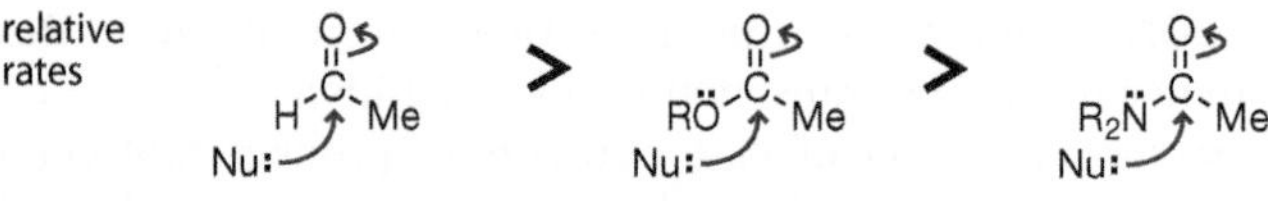

Figure 2.45 The relative reactivity of carbonyl compounds is determined by donation of the neighboring atom into π^*_{CO}.

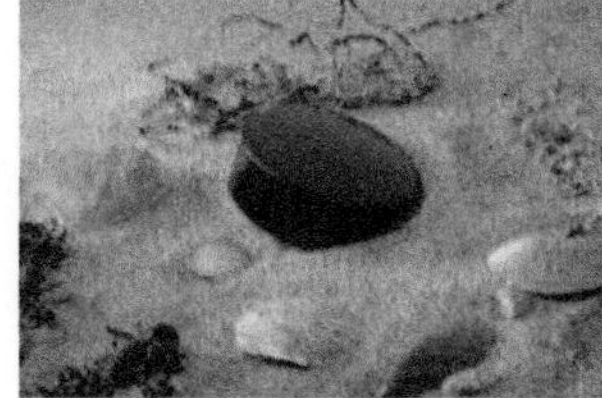
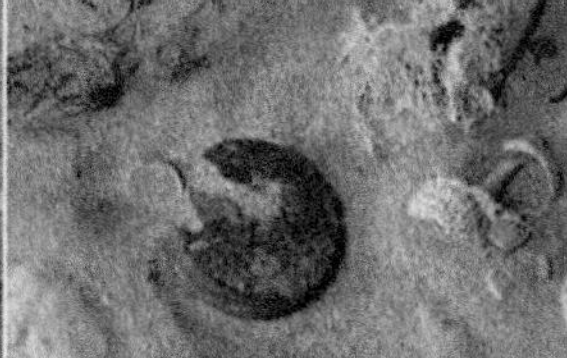
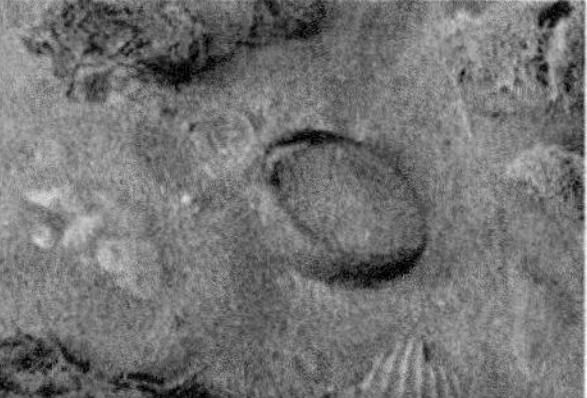
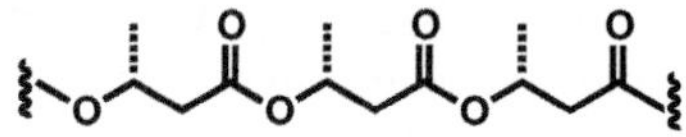

Figure 2.46 Lability of esters. Polyesters produced by bacteria are used to make the bioplastic Mirel, which is stable enough for processing into useful components, such as the black cosmetic case shown above. The material is fully biodegradable under a wide range of environmental conditions including in the oceans, because the ester bonds of the polymer at the right are susceptible to enzymatic hydrolysis by organisms present in the environment. (Courtesy of Metabolix.)

Recall that lone pair reactivity correlates with the degree of p character. However, traditional Lewis structures will often lead to erroneous predictions if we fail to consider important resonance contributors. For example, the typical Lewis structure depiction of an amidine in **Figure 2.47** suggests the presence of an sp^3 lone pair on the left nitrogen atom and an sp^2 lone pair on the right nitrogen atom. However, the charge-separated resonance structure shows no lone pair on the left nitrogen and two

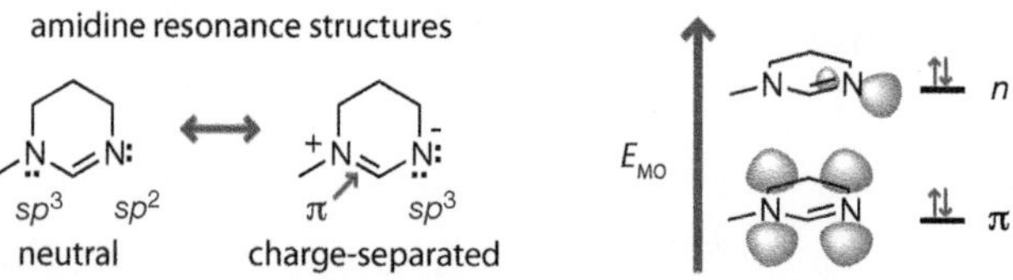

Figure 2.47 Consider resonance structures in assessing hybridization. In order to determine which nitrogen atom in the amidine above is most nucleophilic, one has to consider both the neutral resonance structure and the charge-separated resonance structure in order to arrive at an accurate picture of hybridization.

sp^3 lone pairs on the negatively charged right nitrogen. In fact, the truth lies somewhere between these two extremes. The most reactive electrons—that is, those in the highest occupied molecular orbital—spend their time centered on the right nitrogen atom in a nonbonding molecular orbital that looks much like an sp^2 lone pair. The second most reactive pair of electrons spend their time almost equally distributed on the left and right nitrogen in an orbital with π symmetry.

Charge-separated resonance structures correctly predict the reactivity of amidines, imidazole, and guanidine (**Figure 2.48**). Resonance structures also predict why protonation of the amide carbonyl is a billion times more favorable than protonation of an amide nitrogen (see Figure 2.48). The preference is even greater for esters: protonation of the carbonyl oxygen is about 10^{13} times more favorable than protonation of the carboxyl oxygen. There is only exceptional circumstance in which *N*-protonation of an amide has been observed (**Figure 2.49**). As we will see many times in the coming chapters, ultra-precise placement of functionality can facilitate unfavorable chemical events.

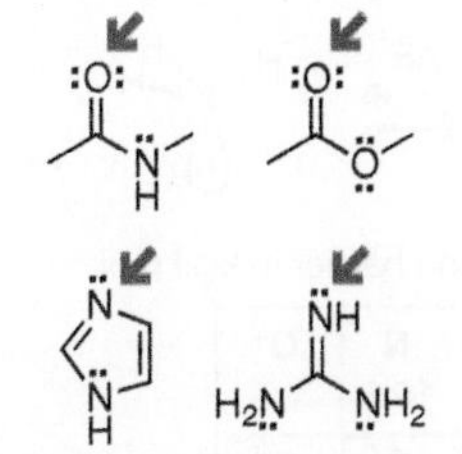

Figure 2.48 The effects of resonance. The atoms indicated by the arrows are the preferred sites of protonation. The connection between reactivity and hybridization is not apparent for these neutral resonance structures.

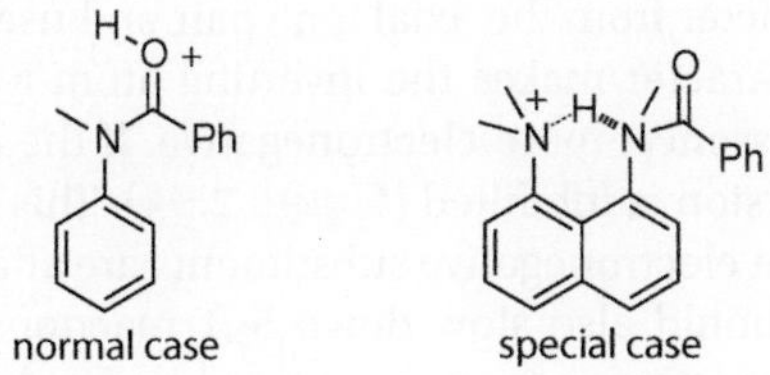

normal case

special case

Figure 2.49 A rare case of *N*-protonation of an amide.

Why are phosphate esters less reactive than carboxylic esters?

Resonance structures do not explain why nucleophilic attack at phosphate esters is quite slow relative to carboxylic esters (**Figure 2.50**). Before we delve into the details of phosphodiester reactivity, we should evaluate the importance of the P=O pi bond. Second-row atoms such as C, N, O, and F form stable pi bonds because the atom–atom distances are short and the 2*p* orbitals are just the right size to overlap. This explains why *the best resonance structures generally have the most bonds.* However, higher-level *p* orbitals (3*p*, 4*p*, 5*p*, and so on) do not overlap effectively to create strong π bonds—not with 2*p* orbitals and not with each other. Thus, the charge-separated depiction of a phosphate ester, $P^+–O^-$, is just as good, and perhaps better, than the double-bond depiction, P=O. Organic chemists are probably too comfortable with drawing double bonds to phosphorus to give up the practice. Similarly, we will continue to draw double bonds to phosphorus and sulfur, but we should never forget the low degree of π character in such bonds.

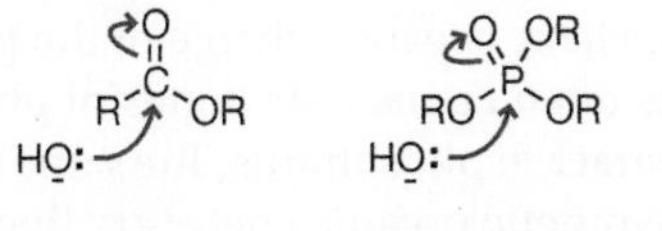

Figure 2.50 Substitution of esters. Nucleophilic attack on a carboxylic ester is much faster than attack on a phosphate ester.

The hydrolysis and transesterification of phosphate esters is a two-step reaction. The initial step involves an attack at phosphorus to form a pentavalent phosphorane intermediate. Pentavalent phosphoranes are analogous to the obligatory tetrahedral intermediates formed during the substitution of carboxylic acid derivatives (**Figure 2.51**).

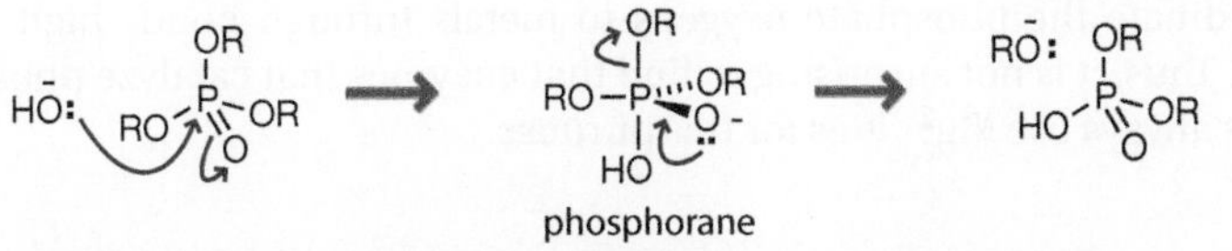

phosphorane

Figure 2.51 Pentavalent phosphorane. Reactions of phosphate esters involve a trigonal bipyramidal phosphorane intermediate with five bonds to phosphorus.

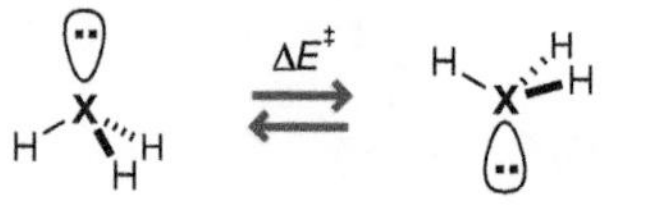

$:XH_3$ inversion barrier in kcal mol^{-1}

C^-	N	O^+
13.0	16.5	1.7
Si^-	P	S^+
34.7	37.8	32.4

Figure 2.52 Third-row atoms versus second-row atoms. Pyramidal inversion through trigonal bipyramidal transition states is slower for third-row atoms than for second-row atoms.

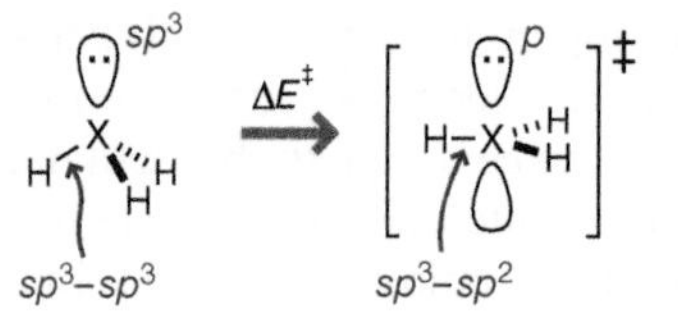

Figure 2.53 Trigonal bipyramidal structures. The transition from tetrahedral to trigonal bipyramidal geometry is accompanied by changes in hybridization.

Phosphate esters resist nucleophilic attack because of the high cost of rehybridization. One way in which to appreciate the cost of this rehybridization is to examine the analogous process of pyramidal inversion. As suggested by the calculated inversion barriers in **Figure 2.52**, second-row atoms such as carbon, nitrogen, and oxygen undergo rapid pyramidal inversion below room temperature. In contrast, third-row atoms such as silicon, phosphorus, and sulfur are reluctant to invert, even at elevated temperatures.

The planar transition states for pyramidal inversion mirror the transition states for addition to tetrahedral atoms (**Figure 2.53**). To accommodate this planar arrangement, the inverting atom must take *s* character from the axial lone pair and use it for the equatorial bonds. This increased *s* character makes the inverting atom a more attractive environment for electrons—in essence, more electronegative. If the equatorial atoms are also electronegative, inversion is inhibited (**Figure 2.54**). This effect explains why S_N2 reactions are slower when electronegative substituents are attached to the carbon under attack. This effect should also slow down S_N1 reactions, but S_N1 reactions involve carbocations. Such reactions are so dominated by Coulombic (charge-charge) interactions that the effect of rehybridization is not apparent. By analogy, nucleophilic addition to phosphorus should be additionally slowed by the oxygen substituents.

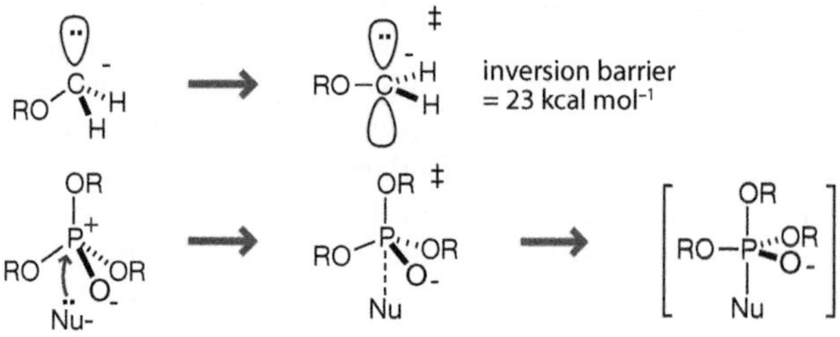

Figure 2.54 Slow attack on phosphate esters. Equatorial electronegative substituents slow the transition from tetrahedral to trigonal bipyramidal geometries.

Every factor seems to conspire against the addition of a nucleophile to a phosphate group. In particular, the high cost of rehybridization, the presence of electronegative alkoxy substituents, and the nucleophile-repellent negative charge of the phosphate ester slow the reaction. In fact, nucleophiles often attack C–O bonds of phosphates, leading to S_N2 cleavage in competition with attack at phosphorus. The same trends are played out in sulfonate esters. Recall from synthetic organic chemistry that tosylates and mesylates are great leaving groups, precisely because it is easier to attack a $\sigma^*_{C\text{-}O}$ orbital than to force the sulfur atom to rehybridize (**Figure 2.55**).

Charge has an important role in the reactions of phosphate esters. The negative charge on phosphate oxygens tends to repel negatively charged nucleophiles such as hydroxide ion. For example, trimethylphosphate reacts with hydroxide 200,000 times faster than dimethyl phosphate anion (**Figure 2.56**). One way in which to mitigate the repulsive effects in the attack of hydroxide on a negatively charged phosphate diester is to coordinate the phosphate oxygens to metals through bonds high in covalent character. Thus, it is not surprising to find that enzymes that catalyze phosphodiester hydrolysis always use Mg^{2+} ions for this purpose.

Figure 2.55 Slow attack on sulfonate esters. Nucleophiles prefer to attack the $\sigma^*_{C\text{-}O}$ orbital, which makes mesylates and tosylates great leaving groups and not electrophiles.

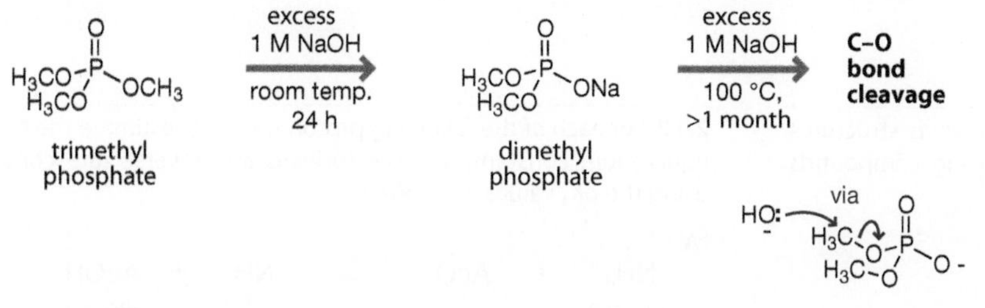

Figure 2.56 The resiliency of phosphate esters. Phosphate esters resist hydrolysis quite readily when incubated with sodium hydroxide at room temperature. However, elevated temperatures and long time periods can hydrolyze the phosphate ester, through nucleophilic attack on the electrophilic methyl group.

2.6 SUMMARY

Chemical biologists think about molecules at the level of atoms and bonds. We have established two different ways of thinking about the interactions of molecules. Both involve Coulombic interactions, repulsive steric interactions, and attractive interactions. For chemically reactive molecules the attractive interactions lead to bonding, and we use arrow-pushing to describe them. When used correctly, arrow-pushing has a direct connection to molecular orbital theory. Reactive nucleophiles have high-energy frontier orbitals that look like lone pairs. Reactive electrophiles have low-energy frontier orbitals that look like empty p orbitals or π^* orbitals.

Hydrogen bonding has a central role in chemical biology and it arises whenever there are polar bonds to hydrogen atoms. Because water is the preferred solvent for life on the planet Earth, hydrogen bonding is pervasive. We can describe hydrogen bonds by using molecular orbitals but they are mainly a Coulombic interaction; thus, the energetics are governed by the dielectric constant of the medium.

Nonbonding forces are omnipresent. They are involved in every interaction—both bonding and nonbonding—between molecules. Nonbonding interactions are readily reversible and tend to be energetically weak; but when you sum the vast numbers of weak interactions they help explain the solubility of molecules in water, reaction selectivity, the binding of small-molecule ligands to proteins, the binding of substrates to enzymes, the self-assembly of membranes, protein folding, and the effects of mutations on protein stability.

Humans do not think naturally in terms of logarithms or exponents. We think in terms of numerical ratios—1:1, 10:1, 1000:1—and Gibbs free energy is our access to ratiometric thinking because each -1.4 kcal mol^{-1} in free energy corresponds to a factor of 10, either in rates or in equilibrium populations. Unfortunately, entropy is the *bête noire* of free energy. Whenever a multitude of atomic arrangements are possible, as in the case of macromolecules or an interactive solvent such as water, it becomes virtually impossible to compare all the possibilities. However, as long as we are careful about our comparisons, we can invoke a range of powerful tools such as arrow-pushing, hydrogen bonding, Coulombic interactions, and van der Waals interactions to explain life on the planet Earth.

LEARNING OUTCOMES

- Understand the effect of hybridization on basicity and nucleophilicity.
- Use frontier orbitals to make comparisons of reactivity.
- Hypothesize plausible arrow-pushing mechanisms in which curved arrows represent the interaction of filled orbitals with unfilled orbitals.
- Draw the mechanism for
  - HCN oligomerization to produce DNA bases
  - benzoin reaction
  - aldol reaction
  - Strecker synthesis.
- Predict the relative rates of hydrolysis ($^-$OH-dependent) of different functional groups.
- Describe, compare, and contrast mechanisms for generating chemical diversity in nature and in the laboratory.
- List the various noncovalent interactions important to chemical biology.
- Rank and explain the relative strengths of noncovalent interactions.
- Use equations to estimate the affinity of atom–atom interactions.
- Estimate the strengths of noncovalent interactions from the functional groups involved.
- Explain the basis for affinity and specificity in noncovalent interactions.

PROBLEMS

***2.14** All arrow-pushing must begin with correct Lewis structures. Draw valid Lewis structures for each of the following compounds. Include all nonbonding lone pairs.

A Acetic acid (CH_3CO_2H)
B Acetaldehyde (CH_3CHO)
C Acrolein ($H_2C{=}CHCHO$)
D Hydrogen peroxide (H_2O_2)
E Carbon monoxide (CO)
F Methyl formate (CH_3O_2CH)
G Benzaldehyde (PhCHO)
H Methyl isonitrile (CH_3NC)

2.15 Phosphorus has an essential role in chemical biology, so we need to familiarize ourselves with the structures of phosphorus-based compounds.

A Draw the Lewis structure for phosphoric acid (H_3PO_4). Include all nonbonding lone pairs.
B Draw the Lewis structure for phosphorus tribromide (PBr_3). Include all nonbonding lone pairs.

***2.16** For each of the species below, circle the most reactive lone pair or bond and indicate the type of frontier orbital that it represents (n, π, or σ).

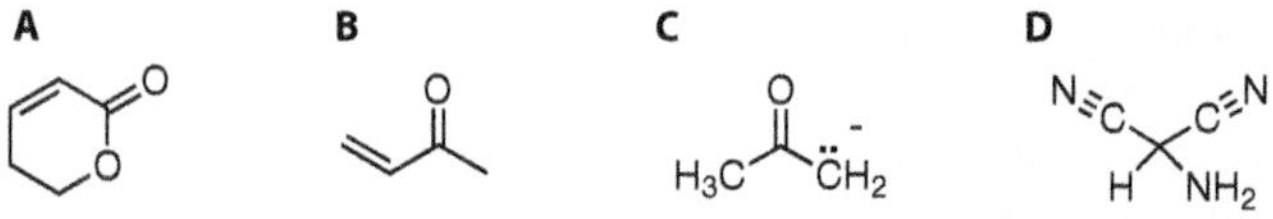

A B C D

H_3C $\ddot{C}H_2$ -

N≡C C≡N H NH_2

***2.17** For each species, draw a second resonance depiction that best satisfies the following, in order: 1) the octet rule, 2) aromaticity, 3) charge separation.

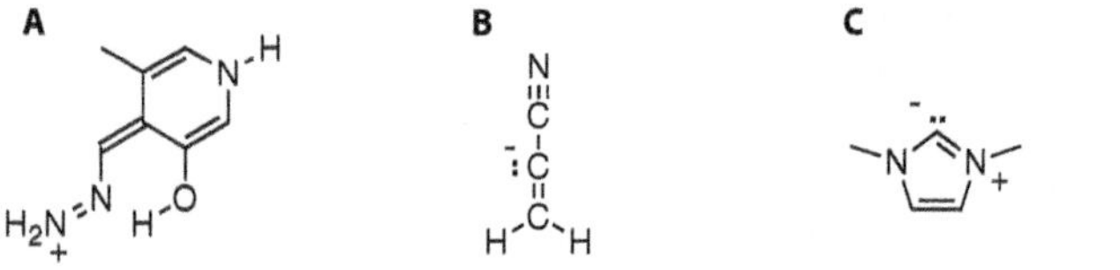

A B C

N-H H_2N=N + H-O

N≡C–:C=C H H

N N +

***2.18** When we write arrow-pushing mechanisms involving proton transfer, it is often difficult to keep track of acids and bases. In general, when we write arrow-pushing mechanisms that occur under acidic conditions, it is best to represent the acid and conjugate base as H–A and A:⁻, respectively. Conversely, for mechanisms that occur under basic conditions, it is best to represent the base and conjugate acid as B: and B–H⁺. For reactions that occur under neutral conditions, you can choose either H–A/A:⁻ or B:/B–H⁺.

A Suggest a plausible arrow-pushing mechanism for the following reaction, using H–A as the acid and A:⁻ as the conjugate base.

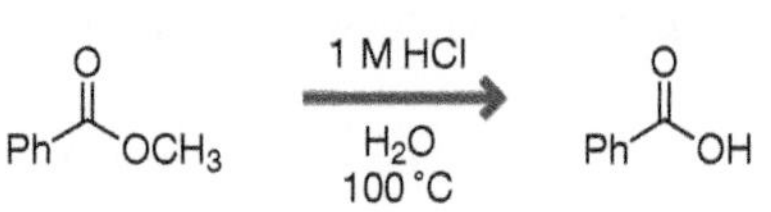

Ph–C(=O)–OCH_3 → (1 M HCl, H_2O, 100 °C) → Ph–C(=O)–OH

B Suggest a plausible arrow-pushing mechanism for the following reaction, using B: as the base and +B–H as the conjugate acid.

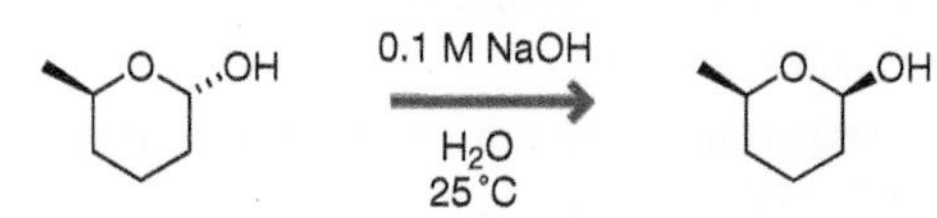

(0.1 M NaOH, H_2O, 25 °C)

2.19 For each of the following proton transfers, estimate the forward equilibrium constant, and the forward and reverse rate constants, using the pK_a values in **Table 1**.

***A**

NH_4^+ + AcO^- ⇌ NH_3 + AcOH

pK_a 9.3 pK_a 4.7

B

NH_2^+ (H_2N, NH_2) + NH_3 ⇌ NH (H_2N, NH_2) + NH_4^+

pK_a 13.6 pK_a 9.3

C

O + HO - ⇌ O - + H_2O

pK_a 20 pK_a 15.7

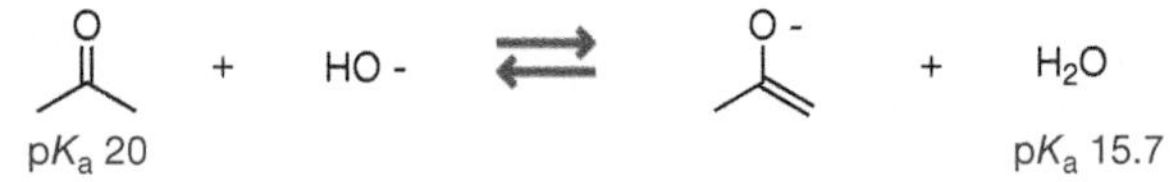

***2.20** Identify the functional groups in the morphine analog below that would be ionized (either protonated or deprotonated) between pH 2 and 12. Draw each of those functional groups in the correct ionic form at pH 2, 7, and 12.

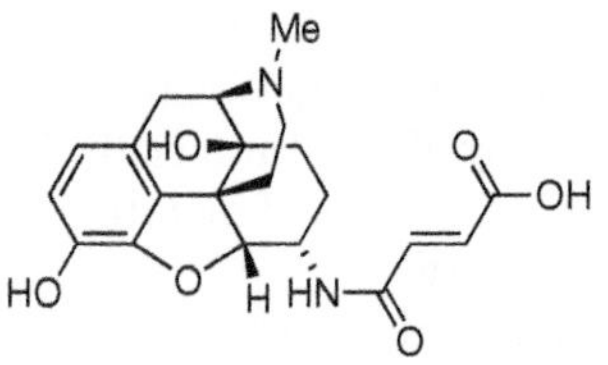

Me N HO O OH HO O H HN O

***2.21** Use Table 1 in Problem 2.19 to answer the following.

A Assign a pK_a to the most acidic functional group in each of the molecules in the figure below.
B Assign a pK_a' to the most basic functional group. The pK_a' is the pK_a of the protonated form, also called the conjugate acid.
C Redraw each structure in the preferred ionic form (with protonated or deprotonated functional groups) at pH 7.2.

pinnaic acid
inhibitor of phospholipase A2

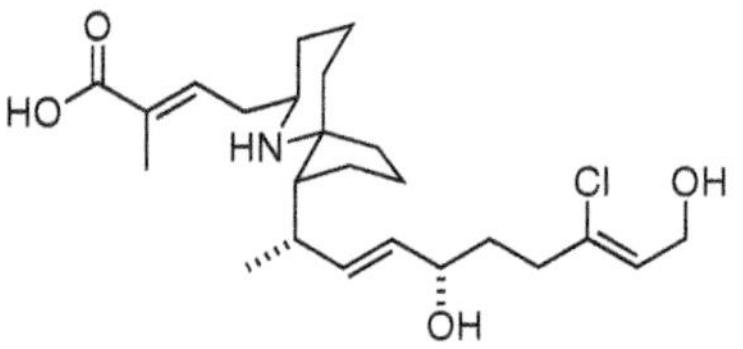

O HO HN Cl OH OH

nipecotic acid derivative
anticonvulsant

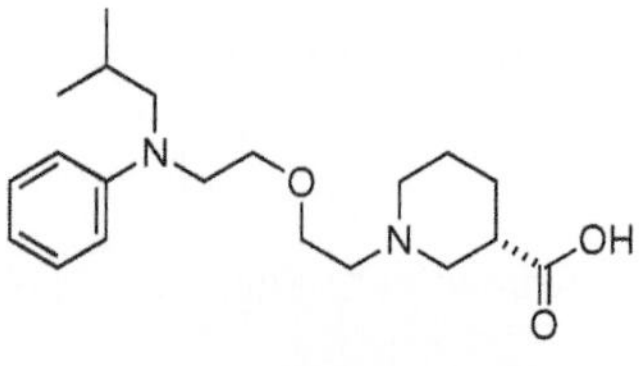

N O N OH O

quinine
antimalarial

OMe N OH N

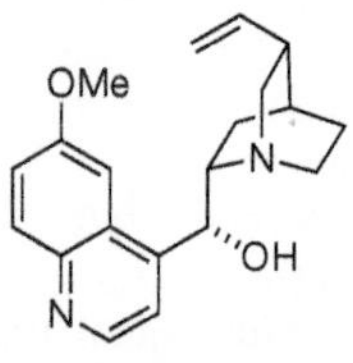

Table 1 (For Problem 2.19)

1. Electronegativity		2. Hybridization		4. Charge		5. Row	
$H_3C{-}CH_3$	**50**	$H_3C{-}NH_3^+$	**10**	HO–H	**16**	$H_3CO{-}H$	**16**
$H_3C{-}NH_2$	**~36**	$H_2C{=}NH_2^+$	**~5**	$H_2\overset{+}{O}{-}H$	**-2**	$H_3CS{-}H$	**9**
$H_3C{-}OH$	**16**	$HC{\equiv}NH^+$	**-10**	$H_3C\overset{H}{\underset{+}{O}}{-}H$	**-4**	F–H	**3**
				O=C(CH₃)OH	**5**	Cl–H	**-8**
				OH^+=C(CH₃)OH	**-6**	Br–H	**-9**
				C₆H₅–NH_3^+	**5**	$(^-O)_2P(=O)O{-}H$	**12**
						$HO(^-O)P(=O)O{-}H$	**7**

3. Resonance			
CH₂=C(CH₃)–CH₃	**41**	O=C(CH₃)–CH₃	**~20**
CH₂=C(CH₃)–NH₂	**~28**	O=C(CH₃)–NH₂	**~17**
CH₂=C(CH₃)–OH	**10**	O=C(CH₃)–OH	**5**

2.22 A good rule of thumb for free energies (at 25 °C) is that each -1.4 kcal mol^{-1} of free energy is worth a factor of 10 in the equilibrium constant.

***A** Estimate the equilibrium binding constant for the association of a receptor with a ligand with a $\Delta G_{\text{association}}$ of -9.8 kcal mol^{-1}.

B Is this rule of thumb good at both room temperature and the physiological temperature of 37 °C?

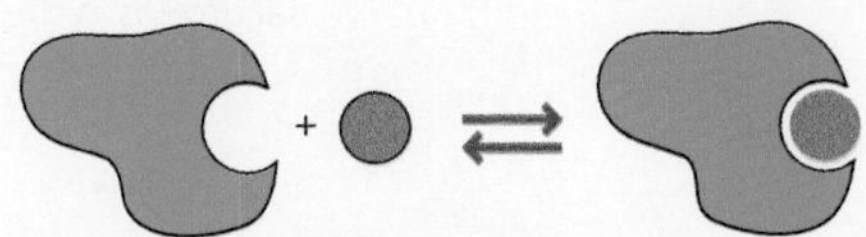

$\Delta G = -9.8$ kcal mol^{-1} at 25 °C

2.23 By what percentage do the repulsion and attraction terms of the Lennard-Jones potential change when two atoms move 0.1 Å closer? What percentage change in repulsion and attraction results when the two atoms move 0.1 Å further apart? Assume that the two atoms are initially 1.5 Å apart.

***2.24** The hydrophobic effect works best with large, hydrophobic functionalities. For each of the molecules shown below, identify the functionality that contributes the most to the hydrophobic effect experienced by the molecule.

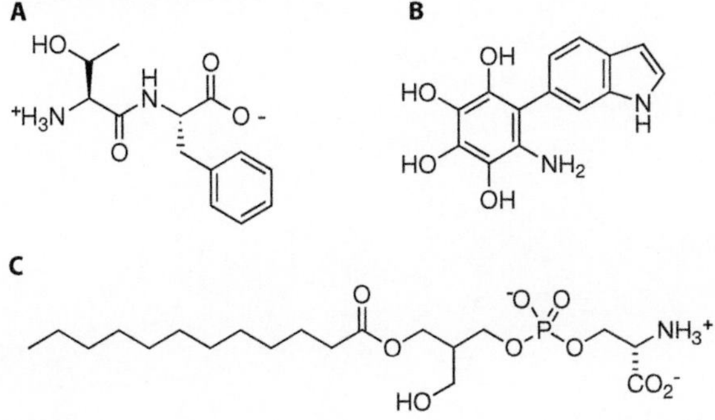

2.25 Which of the following molecules can dimerize through hydrogen-bonding interactions?

A *o*-Nitrotoluene
B Methanol
C Acetone
D Chloroform
E Acetic acid

2.26 The protein fragment below can, in theory, generate two different types of intramolecular hydrogen bonds. Draw depictions of the two hydrogen bonded conformations and determine which is more stable.

2.27 Suggest a plausible arrow-pushing mechanism for the formation of adenine from 4-aminoimidazole-5-carbonitrile (AICN) and HCN.

***2.28** In aqueous solution, the open-chain form of ribose exists in equilibrium with a cyclic hemiacetal form called ribofuranose. Note that ribofuranose exists as a mixture of diastereomers called anomers. Suggest a plausible arrow-pushing mechanism for the formation of both anomers of ribofuranose.

ribose ⇌ (pH 6) ribofuranose (anomeric center)

2.29 Ribose can also form a six-membered ring called ribopyranose. Draw a structure for both anomers of ribopyranose and suggest a mechanism for the cyclization of the open-chain form of ribose to α-D-ribopyranose.

α-D-ribopyranose   β-D-ribopyranose

***2.30** The formose reaction is not stereoselective, and all possible stereoisomers of ribose are formed.

A How many (total) stereoisomers of acyclic ribose are possible?

B Draw the enantiomer of the acyclic open-chain form of D-ribose.

2.31 Suggest a plausible arrow-pushing mechanism for the formation of cyanamide (aminonitrile) from cyanogen.

$$N{\equiv}C{-}C{\equiv}N \xrightarrow{NH_3} H_2N{-}C{\equiv}N + HCN$$

cyanogen → aminonitrile

2.32 Carbodiimide is a tautomer of aminonitrile. Carbodiimides are powerful reagents for the formation of peptide bonds. Draw a structure for carbodiimide and a two-step mechanism for acid-catalyzed tautomerization.

***2.33** Aminonitrile can react with ammonia to generate guanidine. Suggest a plausible arrow-pushing mechanism.

$$H_2N{-}C{\equiv}N \xrightarrow{NH_3} H_2N{-}C(=NH){-}NH_2 \quad \text{guanidine}$$

2.34 Aminonitrile can react with water to generate urea. Suggest a plausible arrow-pushing mechanism.

$$H_2N{-}C{\equiv}N \xrightarrow[\text{pH 6}]{H_2O} H_2N{-}C(=O){-}NH_2 \quad \text{urea}$$

2.35 Urea condenses with AICN to form guanine in 5–10% yield. Guanine is one of the four heterocyclic bases that are found in DNA. Suggest a plausible arrow-pushing mechanism.

2.36 Predict the product of the following Strecker reaction.

$$^{+}H_3N{-}CH_2CH_2CH_2CH_2{-}CH{=}O \xrightarrow{CN^-}$$

DNA

3

LEARNING OBJECTIVES

- Become conversant with the nucleotide nomenclature and atom numbering.
- Recognize the chemical landscape of the canonical DNA double helix.
- Predict folding and hybridization based on DNA sequence.
- Draw arrow-pushing mechanisms involved in the chemical synthesis of DNA.
- Contrast biological synthesis of DNA with chemical synthesis of DNA.
- Design experiments using the rudimentary tools of molecular biology.
- Recognize molecules that are likely to damage DNA.
- Propose mechanisms for small-molecule reactivity with DNA.

The biggest questions we face are often the simplest. What are we made of? Why are we here? What is the future of our species? From a reductionist's perspective, it seems that all forms of life on our planet are merely machines for the propagation of selfish genes. Thus, the big questions we face, questions about life and the human species, can be resolved at the molecular level of detail.

By the 1940s it was clear that deoxyribonucleic acid, or DNA, was the transforming principle that conveyed heritable traits from parent organisms to offspring. However, although it was recognized that DNA must have an intrinsic capacity for replication, the structural basis for DNA replication remained a tantalizing puzzle. The frenetic race to solve this puzzle, arguably the most important structural problem in the field of natural products, is now the stuff of history.

In 1953, using Rosalind Franklin's highly accurate X-ray diffraction data, James D. Watson and Francis Crick published a correct interpretation of the structure of DNA—a complementary double helix held together by hydrogen bonds (**Figure 3.1**). Their dryly written paper ends with an incandescent understatement: "It has not escaped our notice that the specific pairing we have postulated immediately suggests a possible copying mechanism for the genetic material." Indeed, the complementary double-helical model of Watson and Crick provides a chemical basis for the storage of genetic information, for replication, and for mutation—key requirements for evolution (**Figure 3.2**). All known examples of evolution, either found in Nature or developed in the laboratory, are based on the Watson–Crick paradigm. Of course, the importance of DNA transcends the ability to store information faithfully. DNA holds a unique position at the headwaters of biomolecular synthesis, and thus it controls all chemistry within living cells. This chapter is devoted to the chemistry of DNA, not at the level of molecular biology but at the deeper chemical level of atoms and bonds.

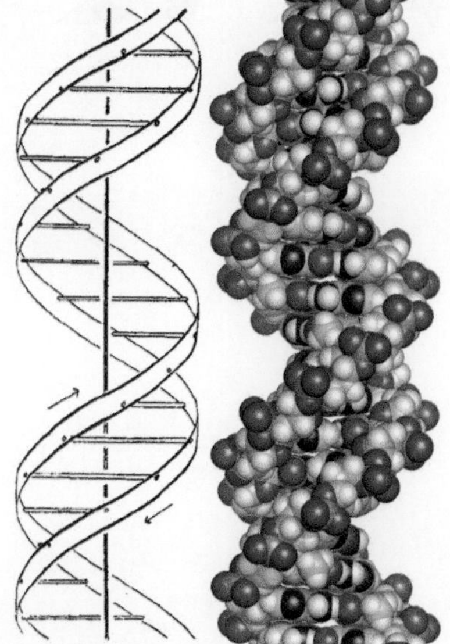

Figure 3.1 The Watson–Crick model for DNA. The double-helix model was revolutionary, consisting of complementary strands rather than identical strands. The double helix is now an icon for molecular biology. The drawing of DNA on the left is from the seminal Watson and Crick paper first reporting the double-helix structure of DNA. (From J.D. Watson and F.H.C. Crick, *Nature* 171:737–738, 1953. With permission from Macmillan Publishers Ltd.)

3.1 FORMS OF DNA

The canonical double helix is one of several forms of DNA

Double-stranded nucleic acids generally adopt one of three types of helical conformations, namely A, B, or Z. The A form is shorter and fatter and is usually found in RNA/RNA or DNA/RNA duplexes. The B form is the canonical conformation proposed by

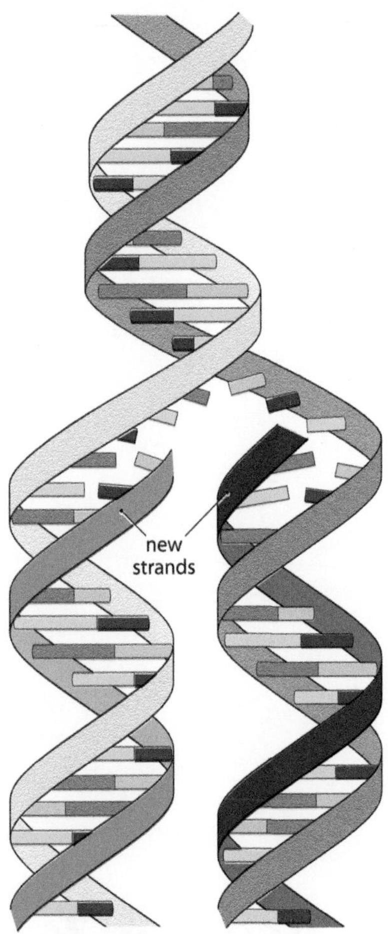

Figure 3.2 Complementary strands. Strand complementarity provides an obvious mechanism for copying at the molecular level. (Adapted from the National Institute of General Medical Sciences.)

Watson and Crick and is the structure we normally associate with double-stranded DNA. Both A-form helices and B-form helices have a right-handed helical twist, but in alcohol or unusually high concentrations of salt, DNA can adopt the Z form, with a left-handed helical twist.

The two strands of the type B DNA double helix are joined by hydrogen bonds rather than covalent bonds. The double helix has a pitch of about 3.4 Å per rung on the ladder—the thickness of an aromatic ring. Each strand of the DNA double helix has an opposing orientation, referred to as antiparallel. As a result of the pseudo-C_2 symmetry of the double helix, when viewed perpendicular to the major axis, it looks the same when it is twisted through 180°. The double helix has two nonidentical grooves (**Figure 3.3**): a wider major groove and a narrower minor groove. The shapes of those two grooves dictate the types of interactions that allow small molecules and large proteins to specifically recognize DNA.

Under physiological conditions, the floor of the deep minor groove is lined with fidgety sodium cations topped by a layer of water molecules that form hydrogen bonds with the phosphate backbone. Water is readily visible in X-ray crystal structures of DNA, but sodium is not. There are several reasons. First, sodium does not prefer a specific coordination geometry. Second, water often adopts fixed positions in DNA crystals because it can act as both a hydrogen-bond donor and a hydrogen-bond acceptor; sodium can act only as a Lewis acid. Additionally, it is difficult to distinguish sodium ions from water molecules in low-resolution crystal structures, because they contain the same number of electrons.

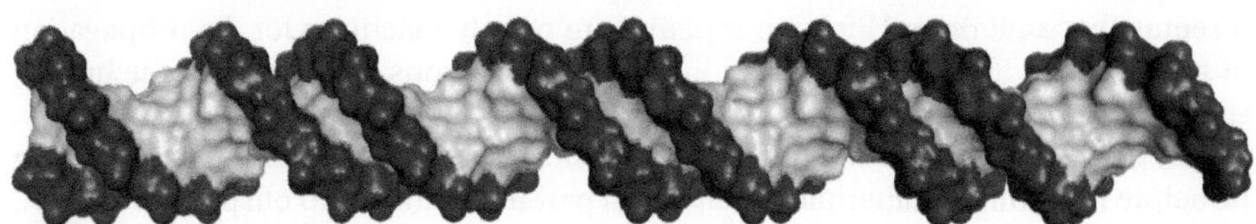

Figure 3.3 Groovy. A surface representation of B DNA showing the major (blue) and minor (green) grooves.

The organization of genomic DNA molecules depends on the type of organism

In all cellular organisms, the appropriate form of DNA for information storage is the double helix. Recall from Chapter 1 that in bacteria such as *Escherichia coli* all of the molecular components mingle together in the cytoplasm: for example, the genetic material (DNA), the protein synthesis machinery, protein modification factories, and the power plants (**Figure 3.4**). The *E. coli* genome is a cyclic DNA molecule in which the ends of the double helix are joined to make a ring. Similarly, bacteria sometimes contain smaller cyclic **plasmid** DNA molecules, about 1000 times smaller than the genome. Plasmids contain short sequences that trick the host into making copies. As an exception to the double-stranded genomes, some compact viruses convey a single-stranded DNA genome from one host to another.

Human cells are more than 1000 times more complex than bacteria, and many of the components are segregated by membranes. Human DNA is present in two different compartments—the mitochondria and the nucleus. The 16,569 subunits of mitochondrial DNA exist as a cyclic molecule, like the genome of *E. coli*, suggestive of its origin as a stand-alone microorganism. In the nucleus, the 3 billion subunits of the human genome are divided among 23 massive chromosomes. From the nuclear command center the DNA dictates the production and import of all the molecules that make up the cell, and ultimately, the very existence of the cell.

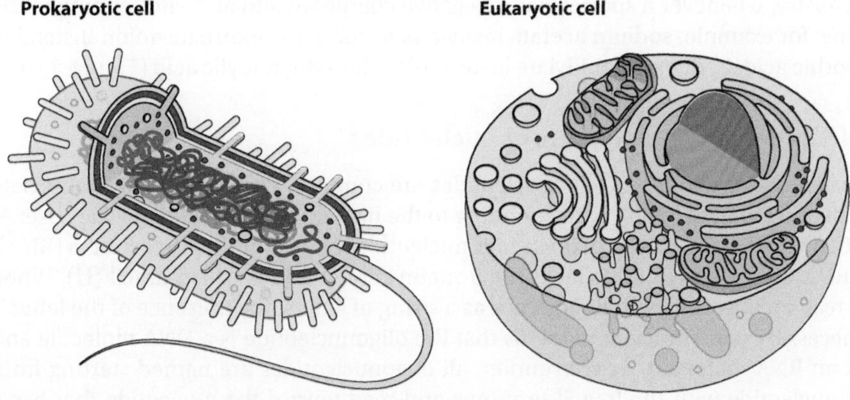

Figure 3.4 Compartmentalization of DNA. DNA and other components of human cells are spatially segregated by membranes. Bacteria and other prokaryotes lack these kinds of compartments. These cells are not drawn to scale. A red blood cell has 50–100 times the volume of an *E. coli* bacterium.

3.2 THE RIBONUCLEOTIDE SUBUNITS OF DNA

Nucleotides are phosphate esters

We ask the reader to make a sizable investment by memorizing the names of the monomers that make up DNA, as well as the numbering of the atoms in those monomers. That precise vocabulary will prove to be essential as we resort to one-letter codes and atom numbers to describe reactions of DNA molecules that are unwieldy to draw. The ability to think chemically, at the level of atoms and bonds, is central to the fields of chemical biology and bioorganic chemistry.

Each subunit of DNA, called a **nucleotide**, is a five-membered ring carbohydrate with two important functional groups: a phosphoryl group and a nitrogen heterocycle. The four nucleotides in DNA are distinguished by the nitrogen heterocycle substituent: adenine, cytosine, guanine, and thymine (**Figure 3.5**). Cytosine and thymine belong to a class of six-membered ring heterocycles called **pyrimidines**, whereas adenine and guanine belong to a class of bicyclic heterocycles called **purines**. It is accepted practice to abbreviate the nucleotide subunits as dA, dC, dG, and dT ("d" stands for 2′-deoxy), but when it is clear that one is referring to DNA, the shorter abbreviations A, C, G, and T are more common. By convention, the atoms in the heterocyclic base are numbered from 1 to 9, whereas the carbons in the ribose sugar are numbered from 1′ to 5′. Even though DNA monomers are monoesters of phosphoric acid, they are rarely handled as acids. Instead they are sold, bought, and handled as the anions. In organic

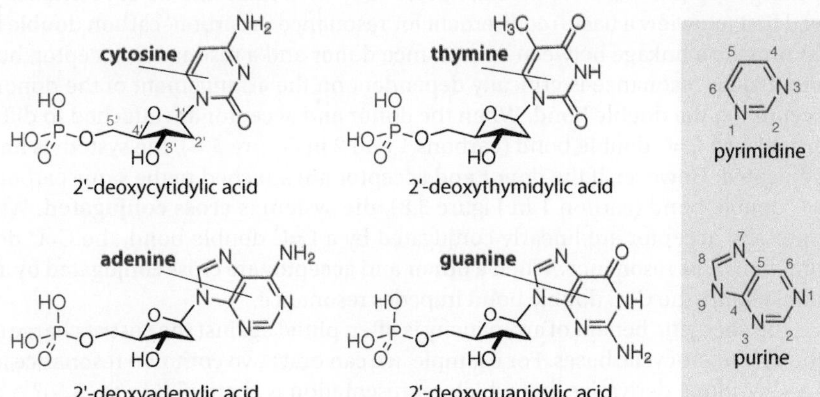

Figure 3.5 Nucleotide anatomy. Nucleotides are phosphate monoesters of ribose, each with a unique heterocyclic base substituent.

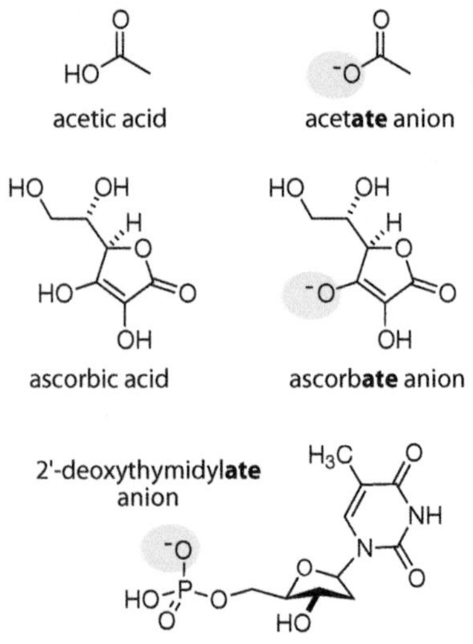

Figure 3.6 Carboxylic acids and phosphoric acids exist in a deprotonated anionic form at physiological pH. The anionic form is distinguished by the suffix *-ate.*

chemistry, whenever a species has a negative charge we add an "-ate" ending to the name: for example, sodium acetate instead of acetic acid, ascorbate anion instead of ascorbic acid, 2′-deoxythymidylate instead of 2′-deoxythymidylic acid (**Figure 3.6**).

DNA and RNA are polymers of nucleotides

Because the structures of oligonucleotides are complex, we will typically abbreviate them as a string of letters corresponding to the first letter of the heterocyclic base: A, C, G, or T. Similar versions of these four nucleotides appear in ribonucleic acid (RNA). In RNA the base thymidine lacks the 5-methyl group and is called uracil (U). When we read an oligonucleotide sequence as a string of letters, the presence of the letter T is necessary and sufficient to tell us that the oligonucleotide is a DNA molecule and not an RNA molecule. By convention, all oligonucleotides are named starting from the nucleotide with the free 5′ terminus and read toward the nucleotide that has a free 3′ terminus (**Figure 3.7**). The order is therefore important, and ACGT is neither an enantiomer nor a diastereomer of TGCA; at best, the tetranucleotides ACGT and TGCA are constitutional isomers with the same molecular formulae. The presence of a phosphate ester at either the 5′ end or the 3′ end of the oligonucleotide is indicated with a lower-case letter p (see Figure 3.7).

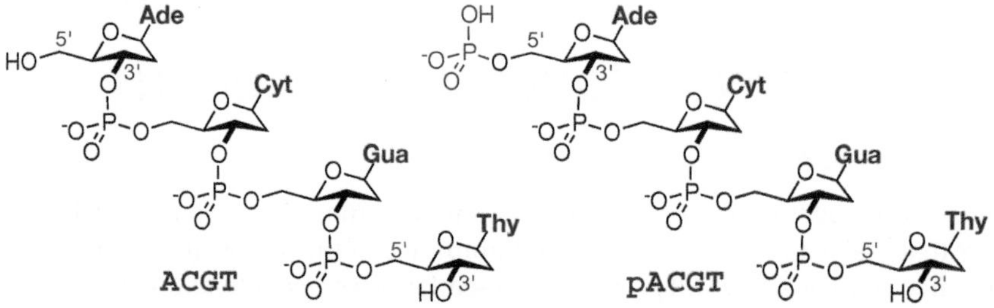

Figure 3.7 Reading a DNA strand. By convention, oligonucleotides are read from the 5′ end to the 3′ end, using the letter p to indicate a phosphate group.

Problem 3.1

Draw the complete chemical structure of the oligonucleotide pTGCAp, using dash/wedge notation to indicate stereochemistry.

Are the heterocyclic DNA bases aromatic?

The DNA bases and related heterocycles enjoy aromaticity, but not to the extent of a simple benzene ring. In many heterocycles, aromaticity is balanced against the cost of charge separation. To appreciate more fully the requirements of aromaticity, we need first to review a basic requirement for resonance. A carbon–carbon double bond can provide a linkage between a resonance donor and a resonance acceptor, but the potential for resonance is critically dependent on the arrangement of the donor and acceptor on the double bond. When the donor and acceptor are attached to different carbons of a C=C double bond (carbons 1 and 2 in **Figure 3.8**), the system is linearly conjugated. However, if the donor and acceptor are attached to the same carbon of a C=C double bond (carbon 1 in Figure 3.8), the system is **cross conjugated.** When a donor and acceptor are linearly conjugated by a C=C double bond, the C=C double bond transmits resonance. When a donor and acceptor are cross conjugated by a C=C double bond, the C=C double bond impedes resonance.

The energetic benefit of aromaticity is often pitted against the cost of charge separation in heterocyclic bases. For example, we can draw two common resonance forms of a 4-pyridone derivative, but which representation is "better" (**Figure 3.9**)? A naive

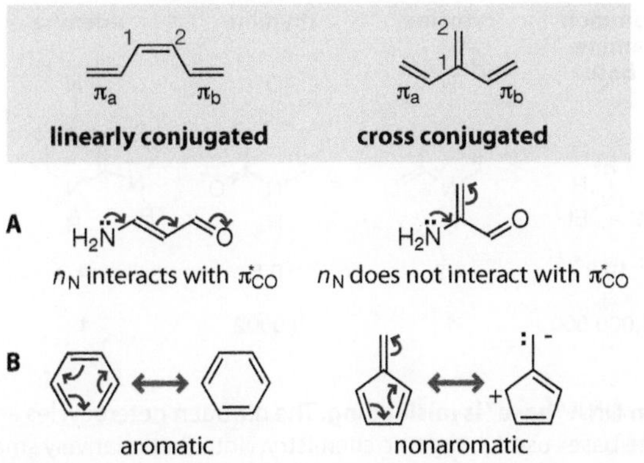

Figure 3.8 Arrow-pushing helps to reveal the importance of linear conjugation versus cross conjugation. (A) When linearly conjugated, a C=C double bond transmits resonance between the nonbonding lone pair on the amino group (n_N) and a carbonyl. When cross conjugated, the C=C double bond impedes resonance. (B) Benzene is aromatic, whereas the cross-conjugated isomer, fulvene, is not aromatic.

bond-strength analysis should lead you to favor the keto resonance depiction with a C=O π bond, which is over 15 kcal mol^{-1} more stable than a C=N π bond. However, the keto resonance depiction is nonaromatic (because it is cross conjugated), whereas the pyridinium resonance structure is aromatic. Aromaticity can be worth as much as 36 kcal mol^{-1}. Unfortunately, the pyridinium depiction suggests substantial charge separation, with a formal positive charge on the ring nitrogen atom and a negative charge on the exocyclic oxygen 4.0 Å away. Recall from Equation 2.2 in Chapter 2 that it costs energy to separate charges. Because separation of charges is inversely dependent on the dielectric constant of the medium, we should expect the charge-separated aromatic representation to have greater importance in a polar environment such as water.

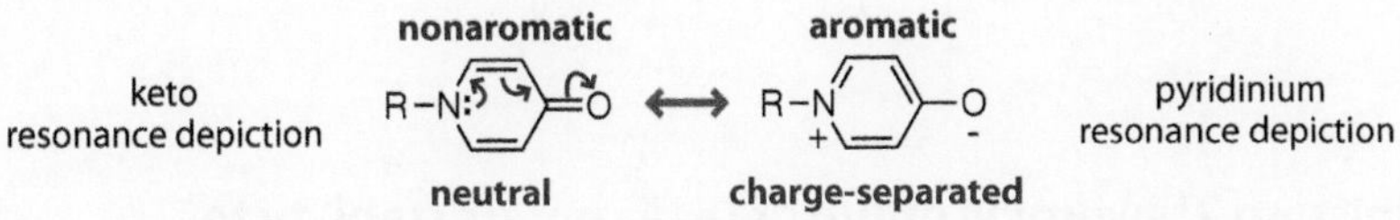

Figure 3.9 How much aromaticity? In heterocycles related to 4-pyridone, the beneficial effects of resonance conjugation (aromaticity) are balanced against the destabilizing effects of charge separation.

Problem 3.2

The cost of charge separation depends dramatically on the dielectric constant. Using the integrated form of Coulomb's law [energy = $(k_e \cdot q_1 \cdot q_2)/(\varepsilon \cdot \text{distance})$], calculate the cost of discrete charge separation in 4-pyridone (4.0 Å) in the gaseous phase ($\varepsilon = 1.0$) and in water ($\varepsilon = 78$). Use $k_e = 9.0 \times 10^9$ J m C^{-2}; the charge on an electron is 1.6×10^{-19} C. Is the cost of charge separation greater than or less than the aromaticity of a simple benzene ring?

Nucleic acids are not acidic, and DNA bases are not basic

In the context of DNA, the terms base and nucleic acid belie the true reactivity of the functional groups under buffered physiological conditions. The pK_a of a phosphate diester is typically between 1 and 2. Thus, at pH 7 fewer than 1 in 100,000 phosphate groups in a DNA strand will have a proton to donate. Therefore a nucleic acid is not much of an acid under physiological conditions. Most DNA phosphate linkages will exist as phosphate anions. Similarly, referring to the purine and pyrimidine heterocycles of DNA as "bases" exaggerates their reactivity. The sp^2 hybridized nitrogen atoms of adenine, guanine, and cytosine are, at best, weakly basic (**Figure 3.10**). In fact, thymine accepts protons with great reluctance and is only protonated to a significant extent under strongly acidic conditions.

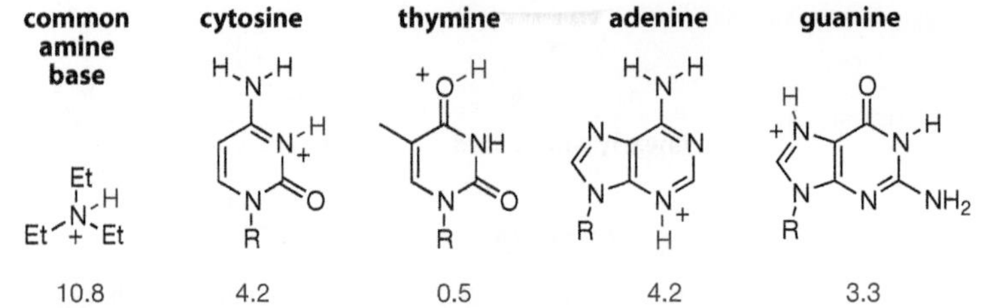

	common amine base	cytosine	thymine	adenine	guanine
pK_a	10.8	4.2	0.5	4.2	3.3
relative basicity of conj. base	**4,000,000**	**1**	**0.0002**	**1**	**0.1**

Figure 3.10 The term DNA "base" is misleading. The nitrogen heterocycles are much less basic than traditional amine bases used in organic chemistry. Note that relatively small differences in logarithmic pK_a values translate into large differences in equilibrium (base 10).

Problem 3.3

Because charge has a role in most organic reactions, it is important to recognize the distribution of charge in common functional groups. Amides and amidines are the most important reactive functional groups in DNA and RNA. Draw the most important alternative resonance structure of each functional group and, on the basis of the location of the charge, predict which atom in the structure would be most basic (and therefore most nucleophilic).

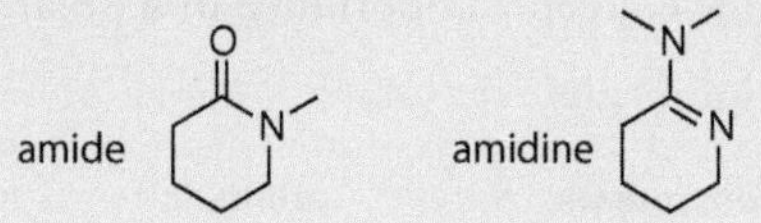

amide amidine

The missing 2′-hydroxyl group of DNA confers stability to phosphodiester hydrolysis

An essential difference between DNA and RNA is that the 2′-hydroxyl of ribose is missing from DNA nucleotides—hence the name deoxyribonucleic acid. As we noted in the previous chapter, the absence of this hydroxyl group makes the phosphate esters of DNA more resistant to base-promoted hydrolysis than the phosphate esters of RNA. However, 2′-deoxyribose derivatives lose the heterocyclic base through acid-catalyzed S_N1 solvolysis 1000 times faster than the corresponding ribonucleotides of RNA (**Figure 3.11**). More stable products do not always form faster, but they do in the case of carbocation formation. S_N1 ionization of a base from RNA leads to a carbocation that is destabilized by the electronegative oxygen atom at the 2′ position. This destabilization is commonly referred to as either an inductive effect or, less commonly, a

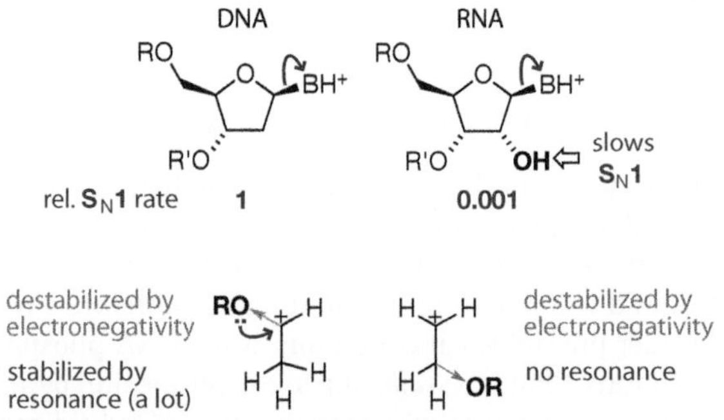

	DNA	RNA
rel. S_N1 rate	1	0.001

slows S_N1

destabilized by electronegativity / stabilized by resonance (a lot)

destabilized by electronegativity / no resonance

Figure 3.11 Neighboring groups affect SN1 ionization of nucleic acid bases. The 2′-hydroxyl cannot stabilize the carbocation through resonance but still exerts a slight destabilizing inductive effect.

Problem 3.4

Suggest a plausible arrow-pushing mechanism for hydrolysis of 2′-deoxyadenosine under acidic conditions.

polar effect. Inductive (polar) effects should be distinguished from effects based on resonance. Thus, the presence of the 2′ hydroxyl in RNA makes the phosphodiester bonds about 100 times more labile, whereas the absence of the 2′ hydroxyl in DNA makes the heterocyclic base about 1000 times more labile.

Modifications to DNA bases are as important as the nucleotide DNA sequence

In the earliest days of molecular biology it was recognized that viral DNA that had been propagated through one strain of *E. coli* was not always effective at infecting other strains of *E. coli*. The origin of this restriction is twofold. First, bacteria possess DNA endonucleases that target non-native DNA sequences; these "restriction endonucleases" have become widely used tools in molecular biology. Second, bacteria contain enzymes that chemically methylate adenine and cytosine bases in newly synthesized DNA, making it resistant to the vigilant restriction endonucleases. In *E. coli*, the enzyme DNA adenine methyltransferase specifically methylates the 6-amino group of adenine in GATC sequences. The enzyme DNA cytosine methyltransferase methylates the C5 position of the second cytosine in CCGG sequences. Modern strains of *E. coli* used for cloning lack the genes for these DNA methylation enzymes. Methyltransferases that modify the 4-amino group of cytosine are found in only a few bacteria.

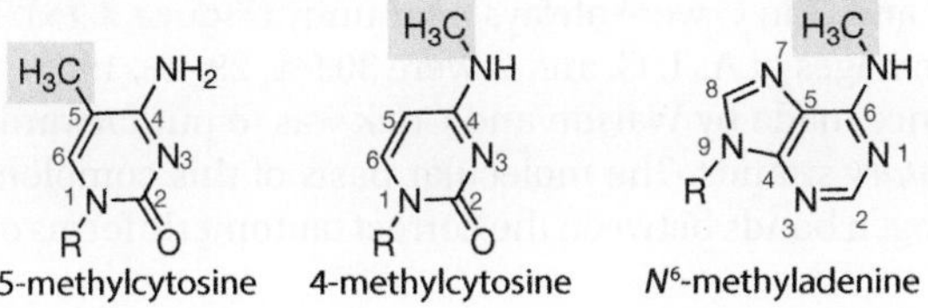

Figure 3.12 Adding to the four bases in DNA. Methyl groups are added to DNA bases after replication resulting in more than four types of heterocyclic bases.

Methylation of DNA bases affects sequence recognition, both by endonucleases and by transcription factors (**Figure 3.12**). Thus, the methylation state of DNA is just as important as the nucleotide sequence in determining gene expression. Humans have three genes for cytosine methyltransferases, which can affect the transcription of genes. As we will see in Section 3.4, human DNA is wound tightly around special protein spools which are also subject to various modifications; such modifications can control the transcription of the associated DNA sequences. The modifications can also confer heritable traits, because modified DNA and DNA-associated proteins end up in daughter cells after mitotic cell division. The field of **epigenetics** is chiefly concerned with heritable factors that influence gene transcription but are not determined by nucleotide sequence. Methylation of cytosine, for example, is a key epigenetic modification that has long eluded study. Methods are now available to detect 5-methylcytosine in human DNA on the basis of its resistance to bisulfite (HSO_3^-) addition. Cytosine undergoes conjugate addition by bisulfite in single-stranded DNA, but 5-methylcytosine does not. The resulting bisulfite adduct is highly susceptible to spontaneous hydrolysis of the 4-amino group. This hydrolysis ultimately generates the base uracil, which base pairs with adenine but not guanine. The bisulfite can be removed through an E1cB mechanism under basic conditions (**Figure 3.13**). There

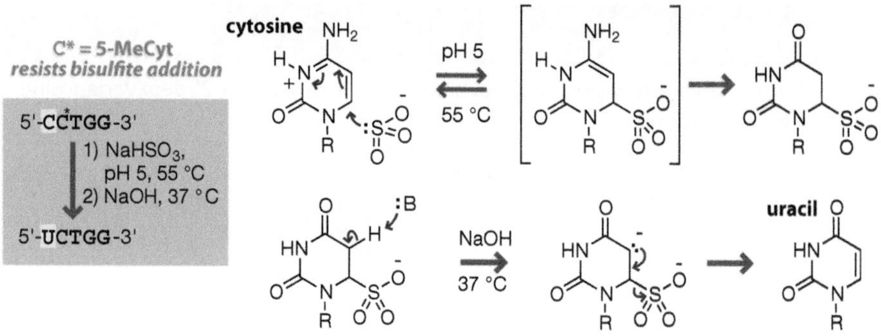

Figure 3.13 Cytosine is susceptible to a two-step bisulfite-promoted hydrolysis. 5-Methylcytosine resists this hydrolytic transformation.

C* = 5-MeCyt
resists bisulfite addition

5'-CC*TGG-3'
1) $NaHSO_3$, pH 5, 55 °C
2) NaOH, 37 °C
5'-UCTGG-3'

cytosine
pH 5, 55 °C
NaOH, 37 °C
uracil

are now a number of economical, high-throughput methods for sequencing DNA into nucleotide sequences made up of A, C, G, and T. None of the high-throughput methods of DNA sequencing distinguish between methylated and non-methylated bases.

3.3 ELEMENTARY FORCES IN DNA

Base pairing knits together the two strands of DNA

It is increasingly difficult in this millennium to recall a time when the double-helical structure of DNA was not a well-known icon of molecular biology. However, to appreciate the insight afforded by the DNA double helix we should look at the state of knowledge in the decades leading up to Watson and Crick's landmark 1953 paper. During that time it was widely held that all four DNA bases were present in equal amounts. Furthermore, the tautomeric forms of the DNA bases were not known. Finally, even when X-ray diffraction data were made available, it was unclear whether the normal structure of DNA should involve two, three, or four strands (**Figure 3.14**). Understandably, the most obvious models for DNA as a self-replicating oligomer necessarily involved *identical* strands.

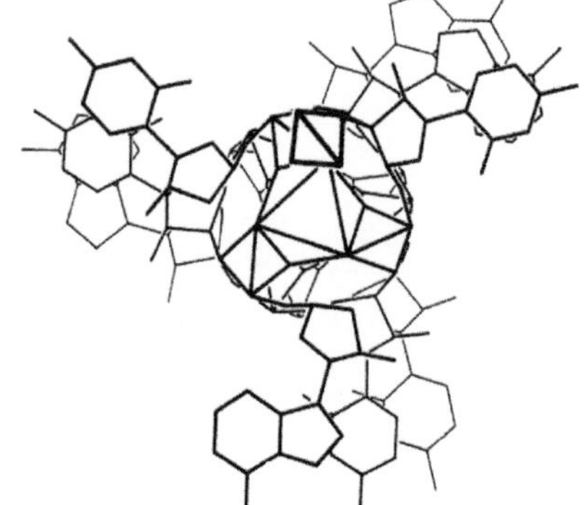

Figure 3.14 Base pairing was not always an obvious concept. An incorrect structure of DNA proposed by Linus Pauling and Robert Corey in early 1953 involves a triple helix composed of three identical strands and no base pairing. In this incorrectly hypothesized structure, the phosphodiester backbones of each strand appear in the center of the structure and the bases around the edges. (From L. Pauling and R.B. Corey, *Proc. Natl. Acad. Sci. USA* 39: 84–97, 1953. With permission from the Oregon State University Libraries Special Collections.)

An essential clue to the true structure of DNA came from the meticulous measurements of a biochemist named Erwin Chargaff, who showed using DNA isolated from thymus glands and other organs that the ratio of A to G and C to T varied, whereas the ratios of A to T and G to C were always near unity (**Figure 3.15**). For example, in human DNA, percentages of A, T, G, and C were 30.9%, 29.4%, 19.9%, and 19.8%. The monumental advance made by Watson and Crick was to put forward a model based on two *complementary* strands. The molecular basis of this complementarity lies in the pattern of hydrogen bonds between the correct tautomeric forms of the four bases.

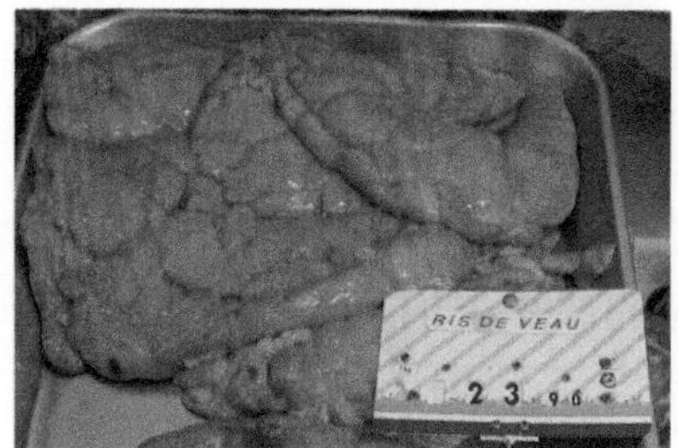

Figure 3.15 The thymus has the highest DNA content of any mammalian organ—more than 10% of the dry weight of the tissue. In culinary circles, calf thymus is known as "sweetbread." (Courtesy of Jennifer McLagan.)

For each base, these tautomeric preferences lead to a distinctive pattern of hydrogen-bond donors and hydrogen-bond acceptors (**Figure 3.16**). As a result, adenine can only pair with thymine, and guanine can only pair with cytosine—A•T and G•C. RNA molecules have adenine-uracil (A•U) base pairs instead of A•T base pairs. The structures of the Watson–Crick base pairs are essential for an understanding of DNA recognition and many forms of DNA reactivity. They should be committed to memory. The two antiparallel phosphoribose backbones that form the rails of the DNA ladder are oriented about 130° apart from each other relative to the helical axis

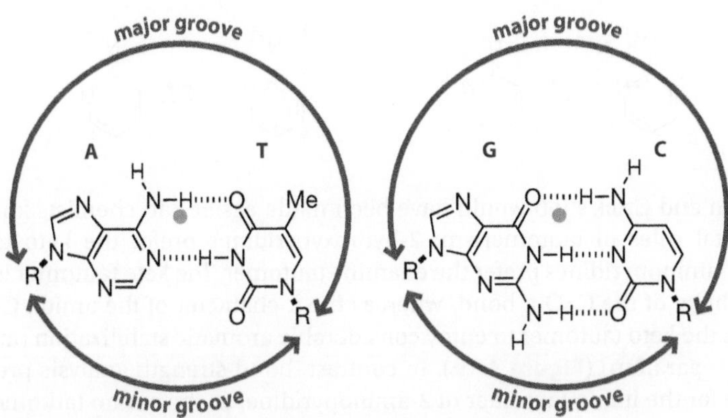

Figure 3.16 The hydrogen bonds in Watson–Crick base pairs are the basis of life on the planet. The gray dot represents the helical axis. The ribosyl groups are not oriented symmetrically (180°) from the helical axis; that asymmetry leads to major and minor grooves.

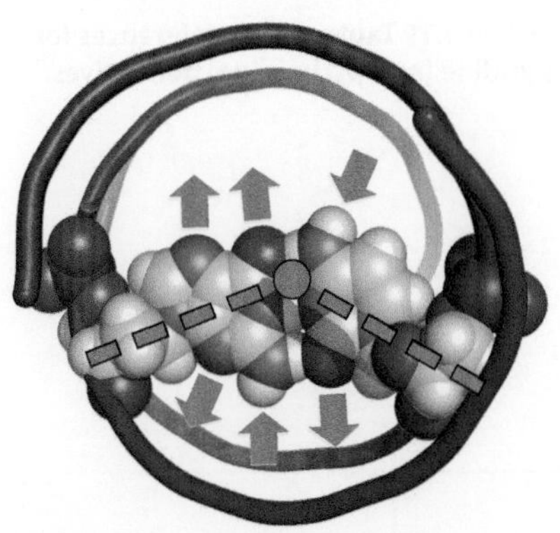

Figure 3.17 Reading DNA. Each edge of a G•C base pair exhibits a unique pattern of hydrogen-bond donors (arrows pointing away from the bases) and acceptors (arrows pointing towards the bases) that is distinguishable from the edges of the other three possible base pairs, C•G, A•T, and T•A.

(see Figure 3.16). This is the origin of the differential sizes of the major and minor grooves of B DNA.

The edges of each of the four possible base-pair arrangements (A•T, T•A, G•C, and C•G) display a unique arrangement of hydrogen-bond donors and hydrogen-bond acceptors (**Figure 3.17**). Thus, the edge of an A•T base pair is not superimposable on the edge of a T•A base pair, generated by flipping it over like a pancake. Those information-rich edges are most accessible in the major groove. A protein or small molecule that binds to the major groove of DNA can easily distinguish the pattern of functional groups in an A•T base pair from the pattern of functional groups in a G•C base pair. Transcription factors (such as engrailed in Figure 1.5)) generally read specific DNA sequences by using helices to read the Braille-like code of hydrogen-bonding groups within the major groove (**Figure 3.18**). Smaller molecules can also bind with some specificity; the antibiotic chromomycin binds selectively to the minor groove of DNA.

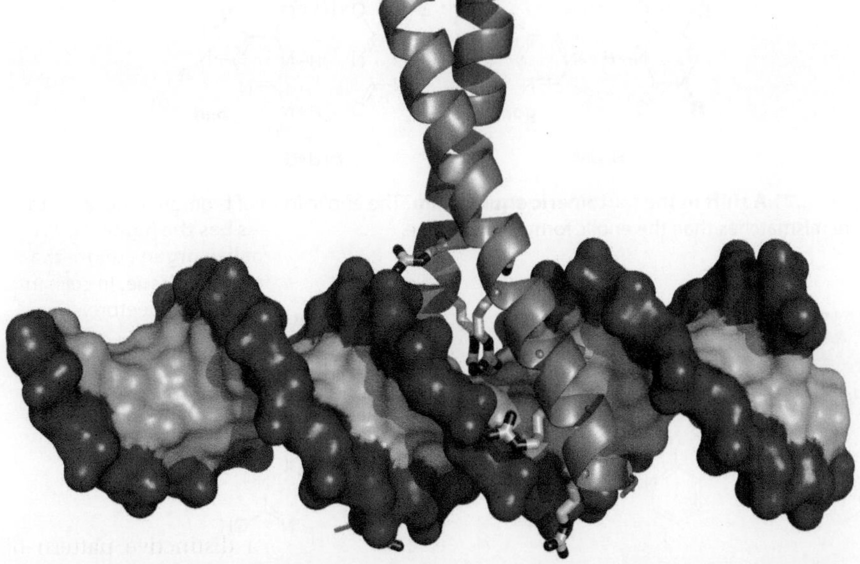

Figure 3.18 Sequence-specific DNA binding. The transcription factors Fos (sky blue at the back) and Jun (purple-blue in the front) recognize the sequence TGAGTCA by probing the major groove of DNA. (PDB 1FOS)

Problem 3.5

Draw a Watson–Crick base pair between adenine and a thymidine with an R group at the 5-position of the base. In which groove of DNA, major or minor, does the R group reside?

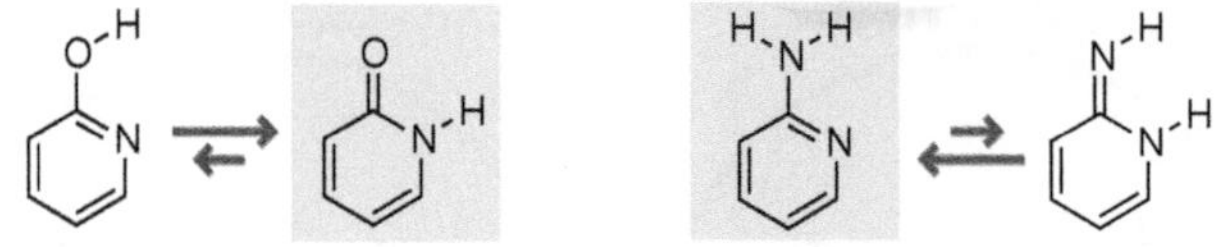

Figure 3.19 Tautomeric preferences for pyridine (and pyrimidine) derivatives.

Watson and Crick's job would have been made easier had chemists known two fundamental rules of tautomerism: 2-hydroxypyridines prefer the keto tautomer, whereas 2-aminopyridines prefer the enamine tautomer. The keto tautomer is favored by the stability of the C=O π bond, whereas the π character of the amide C-N bond still allows the keto tautomer to enjoy considerable aromatic stabilization (at the cost of charge separation) (**Figure 3.19**). In contrast, bond-strength analysis predicts no advantage for the imino tautomer of 2-aminopyridine, so the amino tautomer is preferred.

Whereas thymine and uracil heavily favor the keto tautomer, 5-bromouracil (BrU) shows a significantly higher proportion (up to 1%) of the aberrant enol tautomer. This subtle change makes 5-bromouracil highly mutagenic. Because bromine is about the size of a methyl group, 5-bromouracil is incorporated into DNA in place of thymine by the cell. When DNA polymerase adds a complementary nucleotide to a growing strand it misincorporates G opposite bromouridine instead of A. These mutations can overwhelm the normal repair systems, leading to effects that are apparent at the molecular level and beyond. For example, chromosomes from cells exposed to bromodeoxyuridine show obvious defects in packaging (**Figure 3.20** and **Figure 3.21**). When pregnant mice are injected with bromodeoxyuridine, most of the offspring exhibit polydactyly (extra digits on the hands and feet).

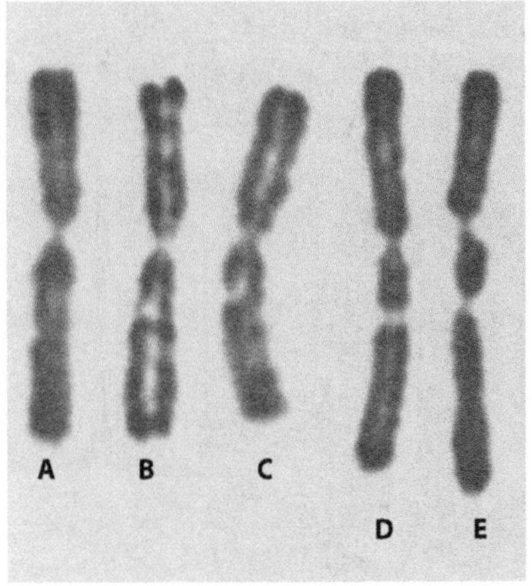

Figure 3.20 Chromosome 1 from hamster cells exposed to bromodeoxyuridine. (A) Normal chromosome. (B–E) Aberrant chromosomes. (From T.C. Hsu and C.E. Somers, *Proc. Natl. Acad. Sci. USA* 47: 396–403, 1961. With permission from the MD Anderson Cancer Center.)

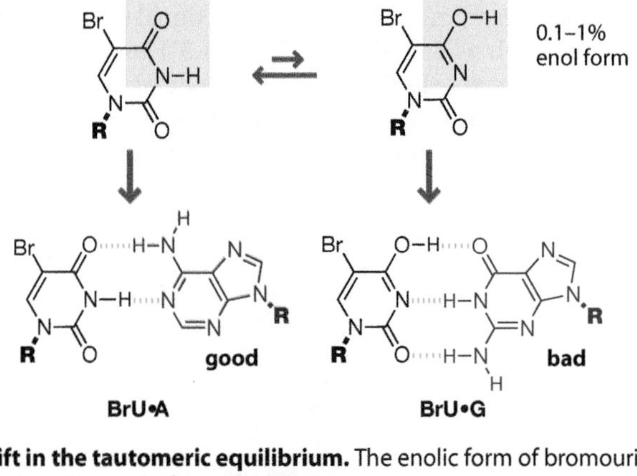

0.1–1% enol form

good

bad

BrU•A

BrU•G

Figure 3.21 A shift in the tautomeric equilibrium. The enolic form of bromouridine leads to more mismatches than the enolic form of thymidine.

Problem 3.6

Predict the preferred tautomer of the following heterocycles.

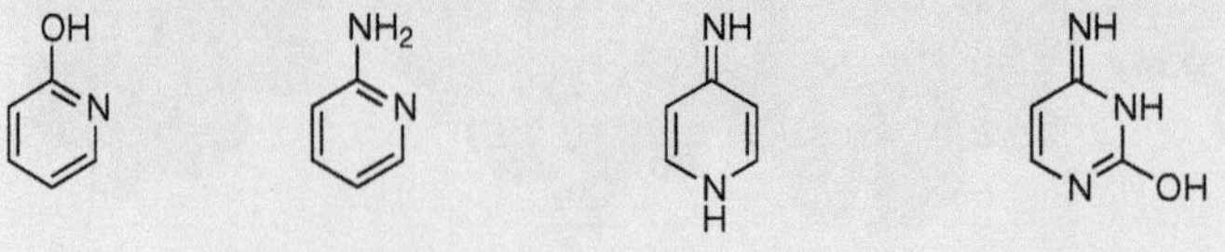

Some non-natural, isomeric bases form effective base pairs

Of the four types of pyrimidines that are readily accessible from HCN, ammonia, and water under prebiotic conditions, Nature has chosen only two pyrimidines for specific base pairing: thymine and cytosine (**Figure 3.22**). Why is there not a third or fourth

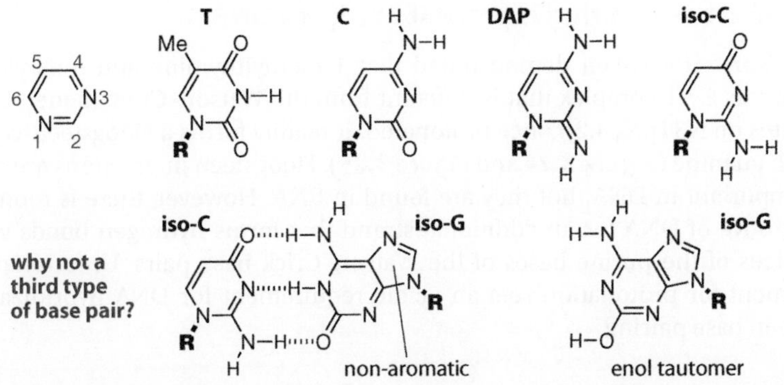

Figure 3.22 Examples of pyrimidine isomers and isomeric base pairs. iso-C can form a unique base pair with iso-G. Unfortunately, the enol tautomer of iso-G is present in unacceptably high amounts.

type of unique base pair based on 2,4-diaminopyrimidine (DAP) or iso-cytosine (iso-C)? Unfortunately, DAP can readily form a tautomer that would base pair with guanine.

Nucleosides based on the complementary bases iso-cytosine and iso-guanosine have been synthesized and incorporated into DNA and RNA strands. By all accounts, iso-C and iso-G form specific base pairs much like their natural counterparts C and G or A and T. In fact, RNA molecules containing iso-C and iso-G are accepted by the normal ribosomal protein synthesis machinery. Unfortunately, iso-C and iso-G show problems. First, cross conjugation in the keto tautomer of iso-G prevents the five-membered imidazole ring from enjoying aromaticity, and that causes the enolic tautomer of iso-G to be present to a much larger extent than the enolic tautomer of guanine (see Figure 3.22). The enol tautomer of iso-G is complementary to thymine, not cytosine. Second, and less obviously, the 2-amino group of iso-C is much more susceptible to hydrolysis than the 4-amino group of cytosine. Thus, any primitive organism attempting to generate a third base pair using iso-C and iso-G would not have been able to generate complementary strands of DNA with the exquisite levels of fidelity needed for the preservation of genetic information.

Other nucleotides have been synthesized by joining the pyrimidine to the ribose sugar through carbon–carbon bonds, but such nucleotides would not readily arise through prebiotic condensation of ribose with a heterocyclic base. Nature has, it seems, converged on an ideal set of nucleotide bases optimized for complementarity, stability, ease of synthesis, and information content.

Hydrogen bonds are not absolutely essential for complementary base pairing

It is certainly true that hydrogen bonds ensure the stability of the DNA double helix, but we should ask whether they are essential for the complementarity of each and every base pair. This question has been addressed through the synthesis of oligonucleotides that incorporate bases that lack the capacity to form hydrogen bonds (**Figure 3.23**). Surprisingly, the enzyme DNA polymerase, which is responsible for the replication of DNA (described in further detail below), can selectively incorporate these nucleotides, which are based on sterics rather than hydrogen bonding. For example, a nucleotide possessing a pyrenyl group in place of a heterocyclic base (P) is selectively incorporated across from a nucleotide lacking a base (ϕ). Other base analogs, such as the thiopyridone SNICS, are preferentially incorporated opposite each other. This kind of homo-pairing is completely different from the hydrogen-bond-based complementary pairing system used by Nature.

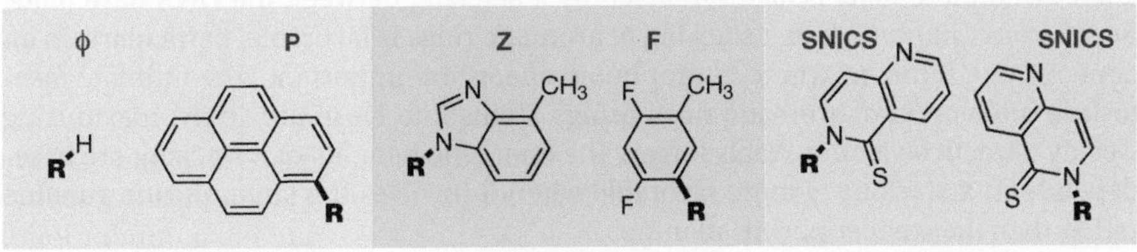

Figure 3.23 Base pairs without hydrogen bonds. Nucleoside substituents without hydrogen-bonding groups can act as complementary "bases."

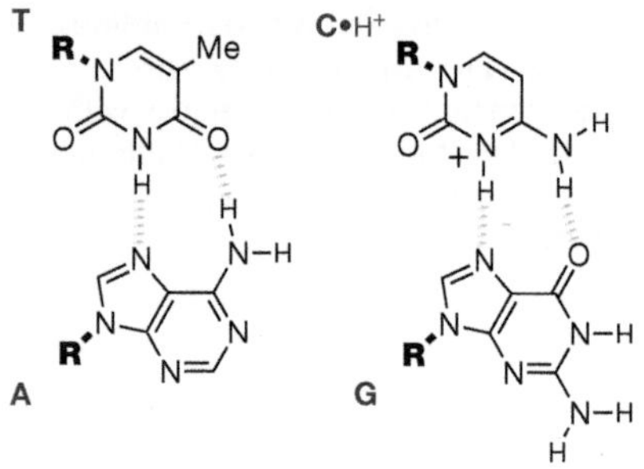

Figure 3.24 Hoogsteen base pairs. Hoogsteen base pairs use a different edge of the purine from a Watson–Crick base pair.

Hoogsteen base pairing is present in triplex DNA

In 1963 Karst Hoogsteen demonstrated that 1-methylthymine and 9-ethyladenine crystallize as a 1:1 complex that is different from the Watson–Crick model. Cytosine protonates on N3 (pK_a 4.2); once protonated, it readily forms a Hoogsteen-type base pair with guanine (**Figure 3.24** and **Figure 3.25**). Hoogsteen interactions are not normally important in DNA, but they are found in RNA. However, there is room in the major groove of DNA for an additional strand that forms hydrogen bonds with the open edges of the purine bases of the Watson–Crick base pairs. Unfortunately the requirement for protonation sets an acidic requirement for DNA hybridization by Hoogsteen base pairing.

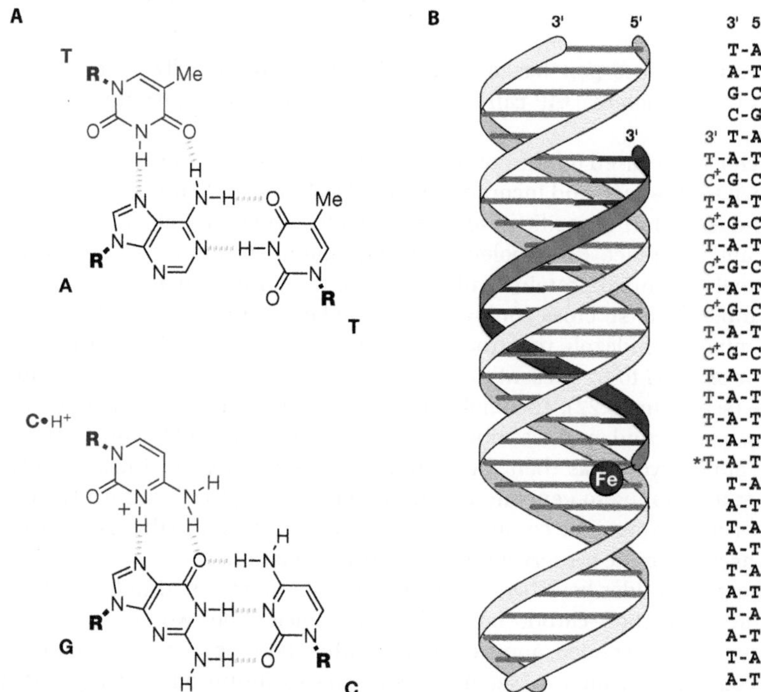

Figure 3.25 Triple-stranded DNA. (A) Complementary DNA base triplets TAT and C^+GC involve both Hoogsteen and Watson–Crick base pairing. (B) Single-stranded oligonucleotides can be designed that bind selectively to the major groove of DNA sequences by Hoogsteen base pairing. (B, adapted from H.E. Moser and P.B. Dervan, *Science* 238: 645–650, 1987. With permission from AAAS.)

Hoogsteen's work emphasized the fact that purine bases have two "edges" for base pairing: one edge for Watson–Crick base pairing and another edge for Hoogsteen base pairing. Because there is room in the major groove for an additional strand, it is possible to design oligonucleotides that specifically recognize polypurine sequences (A and G) through DNA triplex formation. The third strand fits best when it runs parallel to the polypurine strand with which it forms Hoogsteen base pairs. Triplex recognition has been used to carry reactive molecules to specific sites in DNA. For example, Dervan and coworkers were the first to show that triplex formation could be used to direct the iron-mediated cleavage of DNA to specific sites (see Figure 3.25B).

Aromatic π stacking stabilizes the DNA double helix

The DNA double helix is also stabilized by π stacking between the DNA base pairs. Recall from Chapter 2 that π stacking of aromatic rings is favorable, particularly in an aqueous environment where hydrophobic effects are important. The intimate face-to-face interaction of aromatic rings brings atoms into close proximity, maximizing weakly attractive van der Waals forces. The energetic benefits of π stacking are base-dependent: π stacking is more favorable when it involves the larger purine guanine rather than the smaller purine adenine.

The DNA double helix is a dynamic structure. The bases are not rigidly fixed by the hydrogen bonds that form Watson–Crick base pairs. The stacking interactions do not demand perfect coplanarity between the base pairs. Crystal structures of B-form DNA reveal significant distortions from the ideal of coplanar base pairs (**Figure 3.26**).

Figure 3.26 Stacking up. The DNA double helix is stabilized by π stacking between DNA base pairs, rendered here as spheres. (PDB 1BNA)

Intercalation between DNA base pairs involves π stacking

The sugar-phosphate backbone of DNA can untwist slightly, allowing flat aromatic molecules to "**intercalate**" between the base pairs through π-stacking interactions (**Figure 3.27**). A wide range of antitumor natural products target DNA by intercalating between DNA base pairs; examples are daunomycin and adriamycin, which are used in the treatment of cancer (**Figure 3.28**). The best intercalators tend to be polycyclic aromatic compounds with a positive charge. The drug ditercalinium binds tightly to double-stranded DNA and was designed with a tether long enough to span two base pairs. As the concentration of the intercalator is increased, the DNA will stiffen and lengthen to accommodate an increasing number of intercalating molecules. However, DNA has a limited capacity for such intercalation, up to one intercalator for every two base pairs.

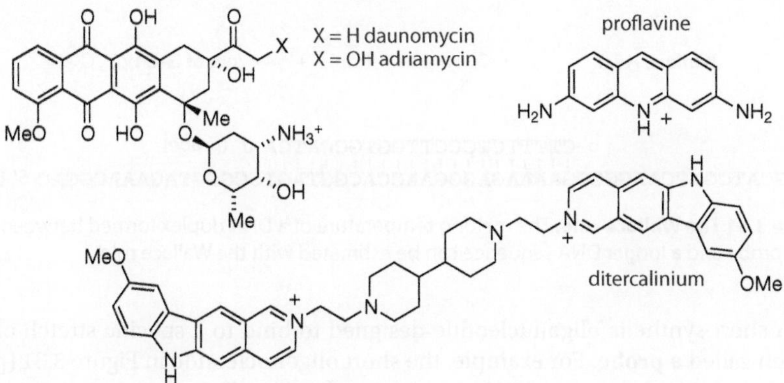

Figure 3.27 Intercalators. Small molecules that intercalate between DNA base pairs usually have flat polycyclic aromatic structures and cationic functional groups.

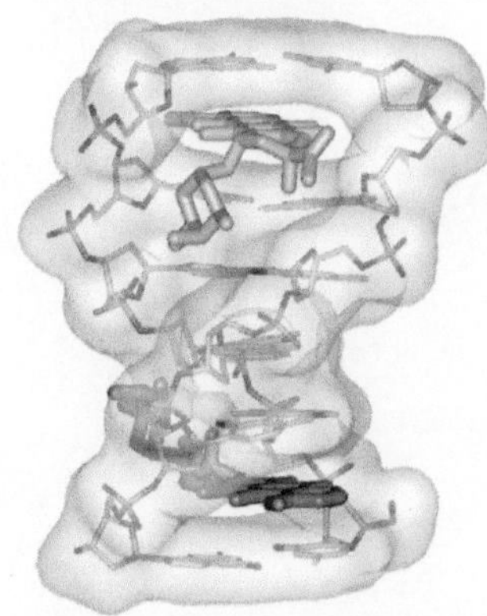

Figure 3.28 Flat aromatic molecules like daunomycin intercalate between DNA base pairs. The result is a sandwich in which the base pairs are the bread and the intercalator is the meat. This crystal structure of a DNA hexamer has two molecules of daunomycin bound through intercalation. (PDB 1D11)

Double-stranded DNA undergoes reversible unfolding and refolding

Like ex-girlfriends and ex-boyfriends at a high school reunion, the bases G and C (or A and T) will inevitably seek each other out. The association of DNA strands with complementary sequences is spontaneous and rapid under physiological conditions. The process is often referred to as **hybridization**. The stability of the resulting duplex DNA is usually characterized by the melting temperature (T_m). At the T_m for duplex DNA, half of the DNA is double-stranded, held together by π stacking and Watson-Crick base pairs, and the other half is in single-stranded form. The ultraviolet absorbance at 260 nm increases slightly as DNA unwinds (**Figure 3.29A**). By measuring the

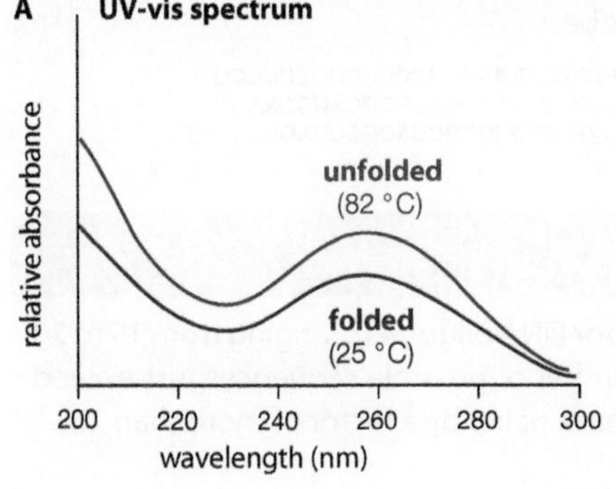

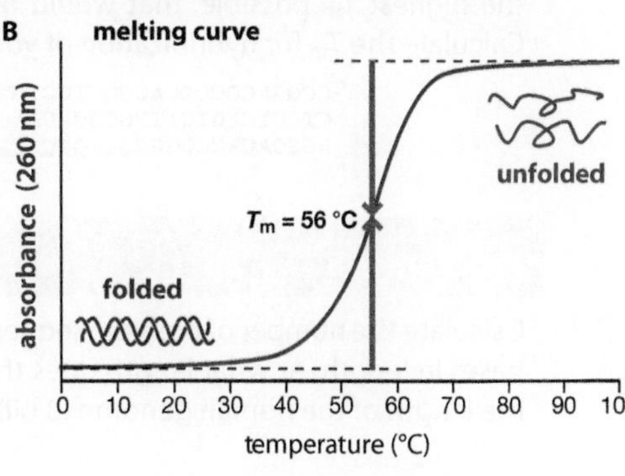

Figure 3.29 Melting DNA. (A) The UV-vis spectra for folded and unfolded forms of DNA reveal a significant difference in absorption at 260 nm. (B) The melting curve for a DNA duplex at increasing temperature shows a point of inflection at the melting temperature (T_m).

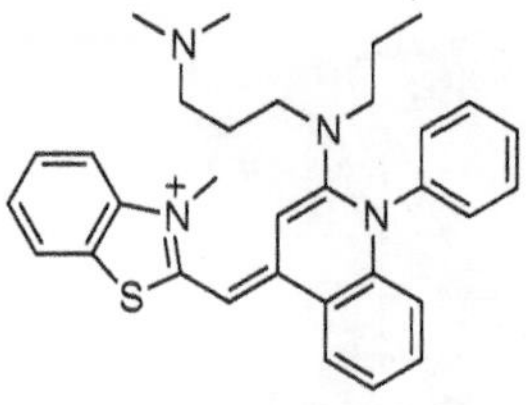

SYBR Green I

Figure 3.30 SYBR Green I becomes more fluorescent when it binds to duplex DNA.

absorbance at 260 nm at different temperatures, it is easy to determine the T_m of a double-stranded DNA molecule by using a UV-vis (ultraviolet and visible light) spectrophotometer. In this experiment, a solution of duplex DNA is simply heated while the absorbance at 260 nm is monitored. The temperature at which the absorbance is halfway between the minimum and maximum is the T_m (Figure 3.29B). Some dyes exhibit enhanced fluorescence when they bind to double-stranded DNA relative to single-stranded DNA (**Figure 3.30**). One such dye, SYBR Green I, is used for highly sensitive quantification of double-stranded DNA.

Duplex DNA with high CG content tends to be more stable than duplex DNA with high AT content because G•C base pairs have more hydrogen bonds and enjoy more stable π stacking. Surprisingly, the complex phenomenon of DNA folding can be predicted with a simple equation. Thus, knowing the number of A•T base pairs (A•T bp, below) and G•C base pairs (G•C bp, below) allows a simple estimation of the T_m for the sequence by using a rudimentary equation called the Wallace rule (**Figure 3.31**). The Wallace rule is useful for all experiments requiring manipulation of DNA. Students therefore should commit the formula to memory. The Wallace rule is most accurate for the hybridization of short (14–20 bp) oligonucleotides binding to much longer pieces of DNA. More accurate versions of the Wallace rule take into account variables such as ion concentrations and co-solvents.

Wallace rule: $T_m = 2 \times (\text{no. of A-T bp}) + 4 \times (\text{no. of G-C bp})$ °C

```
      5'-CTTTTCTCCCTTGGTGCCATCA-3' (probe)
         ||||||||||||||||||||||
3'-TGCATGGACCAAGGGGGAAAAGAGGGAACCACGGTAGTGCGGGTAGAAACGGAC-5' (MBL2)
```

Figure 3.31 The Wallace rule. The melting temperature of a DNA duplex formed between a short probe and a longer DNA sequence can be estimated with the Wallace rule.

A short synthetic oligonucleotide designed to bind to a specific stretch of DNA is often called a **probe**. For example, the short oligonucleotide in Figure 3.31 (probe) would bind to a sequence of DNA containing the complementary sequence (*MBL2*, below) with a T_m of $(2 \times 11 + 4 \times 11)$ °C = 66 °C. The DNA sequence above is part of the gene that encodes the protein mannose-binding lectin 2 (MBL2), an important component of innate immunity. Note that gene names are italicized, whereas protein names are not. A single nucleotide mutation at the G base highlighted in green can make a child susceptible to infection. Because the incorrect base pair at this position will result in the loss of three hydrogen bonds (between the G•C base pairs), a mutation in the highlighted position will result in a slightly lower T_m for the interaction between the probe and the mutant genome. Methods to detect this difference can be used to diagnose such mutations and the resultant disease.

Problem 3.7

A deletion of the 14 underlined bases from the 5′ end of the *RFXANK* gene has been linked to bare lymphocyte syndrome type II. Design an 18-base probe sequence, with the highest T_m possible, that would be complementary to the 14-base sequence. Calculate the T_m for hybridization of your probe.

5'-CCGGACGCCGCACGGCTCCTGTTCCGGTGTCAGAGGGCCCGCCCTCCCCG
CTCCTCAGTCTTTGCGGACAAGAAAGGGGCTGTGTGAGACGCAGGGAAGG
AGGCACACCCGGG<u>GGTGGCGCAGTGAG</u>GAGGGGGCGCGACGGCCAGG...-3'

Problem 3.8

Calculate the number of possible sequences for DNA oligomers ranging from 15 to 20 bases in length. At what length does the number of possible sequences just exceed the length of the human genome (3 billion base pairs) by a factor of more than 20?

Oligonucleotides can still form duplexes even when there are a few mismatches between bases that prevent Watson–Crick base pairing. However, each mismatched base decreases the stability of the duplex. The instability induced by a base mismatch can be used in genomics. Humans vary widely in appearance and susceptibility to diseases even though they share 99.9% homology in their DNA. The differences are even smaller between fraternal siblings. Most of the genetic differences between individuals are due to **single nucleotide polymorphisms** (SNPs)—differences of just a single base pair—spread throughout the genome of each individual. Many of the diagnostic tools that have been developed to identify SNPs take advantage of the differences in stability between a fully matched DNA probe and a probe with a single mismatch. Mismatches usually decrease the melting temperature by several degrees. For example, a G1947A SNP (mutation of guanidylate 1947 to adenylate) in a 110 bp fragment of the *COMT* (catechol-*O*-methyltransferase) gene decreases the melting temperature by 3.8 °C. The *COMT* gene affects the biosynthesis of the neurotransmitter dopamine. DNA can still hybridize when one of the strands has an additional nucleotide but, as expected, the stability of the duplex suffers. The presence of an additional nucleotide subunit results in a motif known as a bulge (**Figure 3.32**). Bulges significantly reduce the stability of DNA duplexes.

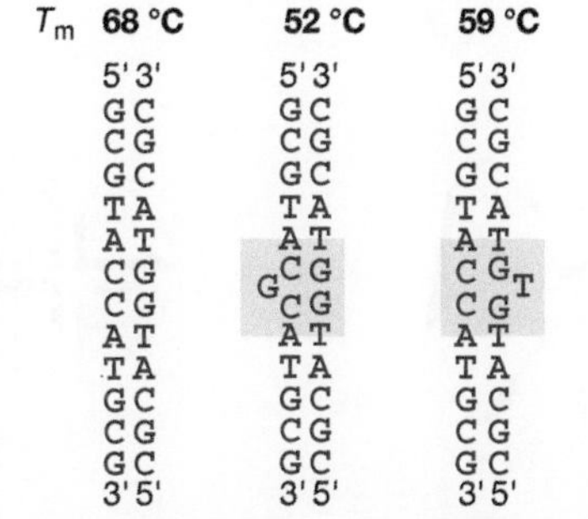

Figure 3.32 Bulges are destabilizing to duplex DNA, reducing the T_m.

Complementarity drives self-assembly of DNA

When one of the strands of a DNA duplex is longer than the other it leaves a **sticky end**. Two pieces of double-stranded DNA that have complementary sticky ends will associate to generate a single long duplex (**Figure 3.33**). However, the resulting duplex will have **nicks**—sites where the adjacent nucleotides are not covalently bonded. Unless the sticky ends are very long, the resulting duplex will be relatively labile. Enzymes called **DNA ligases** can repair nicks by forming a phosphate ester from the 3′ hydroxyl of one strand and the 5′ phosphate of another (**Figure 3.34**); in these ligation reactions a complementary strand is necessary, serving as a splint to hold the two reactive functional groups in proximity.

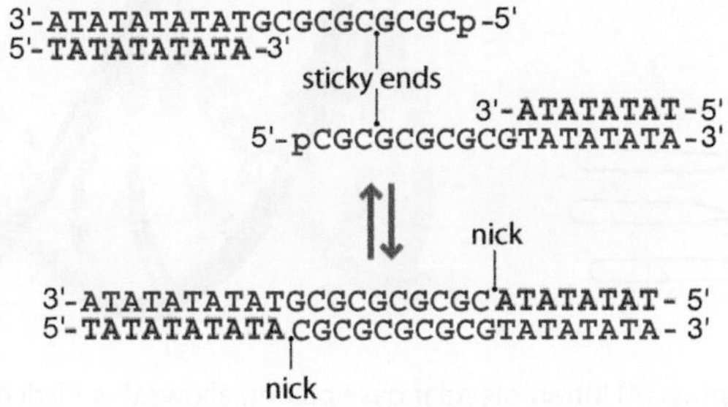

Figure 3.33 Sticky situation. Two complexes of double-stranded DNA will self-associate if they possess complementary sticky ends.

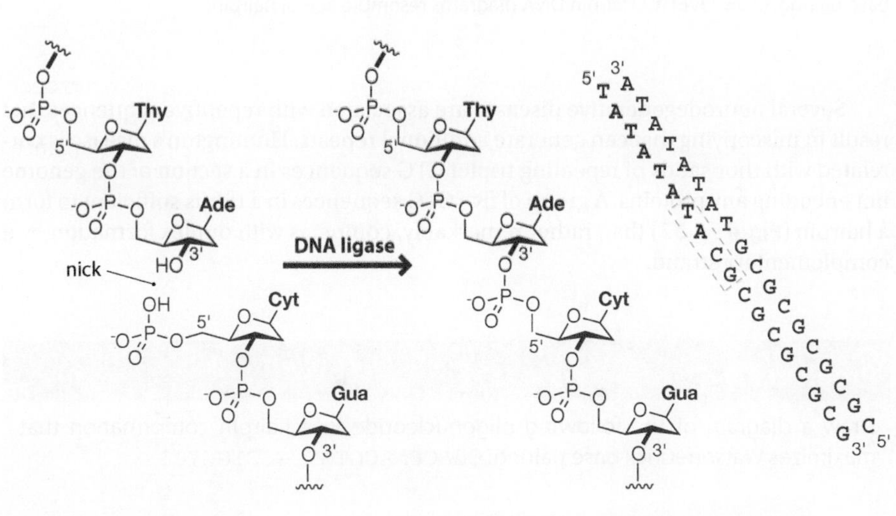

Figure 3.34 DNA ligation. Nicked DNA, generated through the association of sticky ends, can be rejoined with DNA ligase.

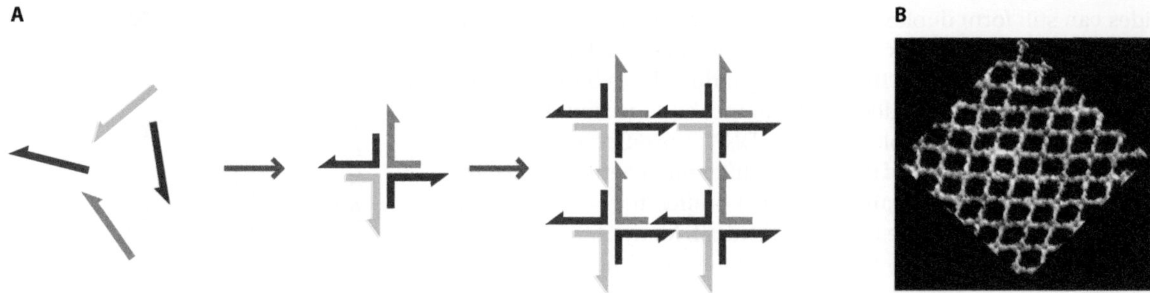

Figure 3.35 DNA materials.
(A) Oligonucleotides can be designed with sequences that self-associate with other strands in a controlled orientation, leading to multidimensional materials.
(B) A DNA nanogrid generated through self-assembly of oligonucleotides.
(B, from H. Yan et al., *Science* 301: 1882–1884, 2003. With permission from AAAS.)

The ability of DNA strands with complementary sticky ends to self-assemble is an immensely powerful engineering tool that allows molecular biologists to join pieces of DNA together. The power of these self-assembly processes for nano-engineering was recognized even before it was possible to obtain custom oligonucleotides readily. Oligonucleotides have been designed that hybridize to three, four, and more strands, resulting in two-dimensional lattices, cubes, and three-dimensional networks (**Figure 3.35**).

Short stretches of DNA can fold into hairpins

Short segments of DNA can fold back on themselves forming hairpins as long as they have complementary sequences. Folded DNA is inaccessible to hybridization with probes. For example, some DNA sequences can form well-defined, conformational folds termed **secondary structure** (named in deference to the primary structure, which is the DNA sequence) by folding onto themselves. The resultant short stretches of double-stranded DNA can block hybridization by added oligonucleotides (**Figure 3.36**).

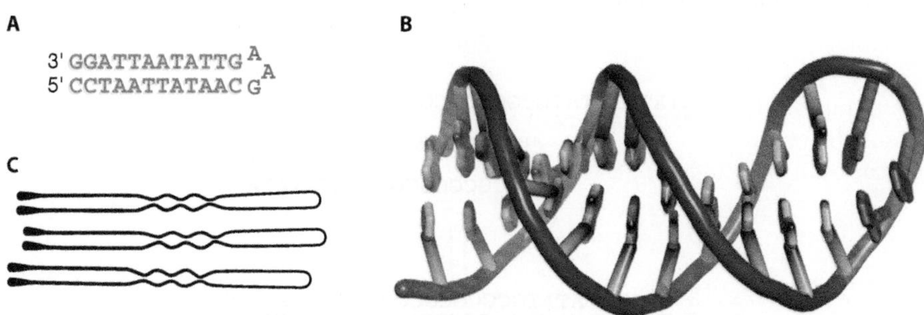

Figure 3.36 DNA hairpins. (A) Intramolecular base pairing allows this AT-rich oligonucleotide to form a stable hairpin structure. (B) The three residues in purple are not involved in Watson–Crick base pairing. (PDB 1JVE) (C) Hairpin DNA diagrams resemble actual hairpins.

Several neurodegenerative diseases are associated with repetitive sequences that result in miscopying that can generate additional repeats. Huntington's disease is correlated with thousands of repeating triplet CTG sequences in a section of the genome not encoding any proteins. A group of five CTG sequences in a row is sufficient to form a hairpin (**Figure 3.37**) that, rather remarkably, competes with duplex formation by a complementary strand.

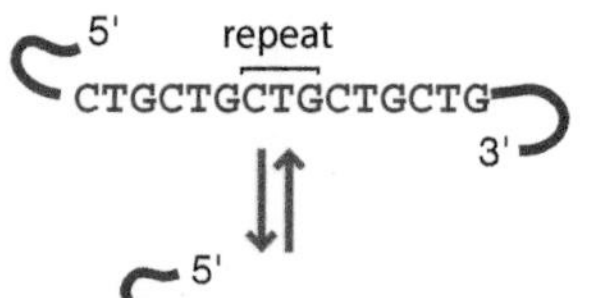

5' repeat
CTGCTGCTGCTGCTG
3'
5'
CTGCTGCT
GTCGTCG hairpin
3'

Figure 3.37 The danger of triplet repeats. In Huntington's disease, CTG repeats form hairpins that compete with duplex formation.

Problem 3.9

Draw a diagram of the following oligonucleotide in a hairpin conformation that maximizes Watson–Crick base pairing: CGACCAACGTGTCGCCTGGTCG.

Most mischievous DNA sequences are culled from genomes by natural selection. Short DNA probes designed by molecular biologists lack this advantage and can sometimes suffer from kinetic problems due to misfolding and hybridization to sequences with imperfect complementarity. Products that form the fastest (kinetic products) are not always the most stable (thermodynamic products). To overcome kinetic hybridization problems, DNA can be heated to a high temperature (above the T_m for misfolded structures) and then slowly cooled. This method, called **annealing**, provides enough energy for the system to reach thermodynamic equilibrium.

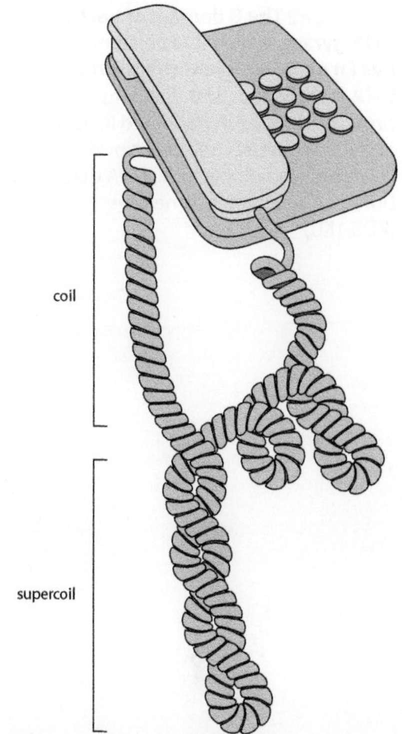

Figure 3.38 Supercoiling. The topological phenomenon of supercoiling in DNA is similar to that in a molded phone cord. (Adapted from D.L. Nelson and M.M. Cox, Lehninger Principles of Biochemistry, 5th ed. W.H. Freeman, 2008.)

3.4 DNA SUPERSTRUCTURE

Double-stranded DNA forms supercoils

Without careful management, the DNA of a cell would be a tangled mess. Genomic DNA is always long. If extended, the DNA from a single human cell would be about 2 m in length. Stringy objects with an intrinsic twist are highly susceptible to **supercoiling**, a topological concept familiar to everyone who has experienced the annoying propensity of coiled telephone cords to twist themselves. Both DNA and phone cords have an intrinsic twist. The twist of DNA is imposed by the chiral ribophosphate backbone; the twist of a phone cord is imposed by the molded plastic. If a phone handset is turned over several times, it introduces enough torsional strain in the coiled cord to induce supercoiling (**Figure 3.38**). After repeated use, the tangles and twists of supercoiling can become so pronounced that it is difficult to pull the cord away from the phone. Supercoiling can be either positive or negative. Supercoiling does not change the bonding relationship between the atoms in DNA; it merely changes the topology.

Supercoiling of DNA prevents the transcriptional machinery from teasing apart the individual strands to make RNA (**Figure 3.39**). Some method is needed to remove supercoiling, allowing DNA to relax. It is easy to remove supercoiling from a strand of DNA simply by twisting either of the ends, but a cyclic DNA molecule does not have ends. A supercoiled phone cord presents a similar problem. It is impossible to remove the supercoiling from a phone cord while the handset is in the cradle, but once the handset is picked up you can remove the supercoiling by turning the handset (and the end of the cord to which it is attached). If the phone cord were 66,732 km in length, it would be very difficult to remove any supercoiling in the middle of the phone cord. Similarly, it is impractical to remove supercoiling from the middle of a human chromosome by twisting the ends.

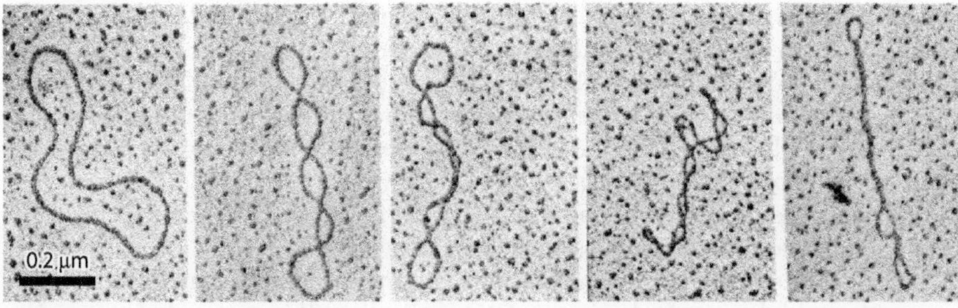

Figure 3.39 Supercoiled plasmids. These DNA plasmids have the same chemical structure but different degrees of supercoiling, from lowest (left) to highest (right). (Courtesy of Laurien Polder.)

Topoisomerases resolve topological problems with DNA

Instead of attempting to work out the kinks in supercoiled DNA by rotating the ends, cells employ two types of **topoisomerase** enzymes to introduce transient breaks

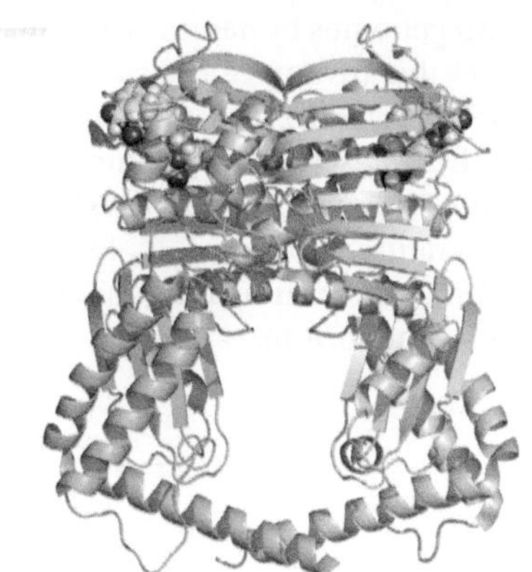

Figure 3.40 The B domain of bacterial DNA gyrase, a type II topoisomerase, has an obvious hole through which DNA is threaded. Both the natural product novobiocin (rendered in yellow) and the synthetic antibiotic ciprofloxacin selectively inhibit bacterial DNA gyrase but not human topoisomerases. (PDB 1KIJ)

into the DNA (**Figure 3.40**). Type I topoisomerases cleave a single strand, allowing the other strand to relax through rotation about individual bonds. Then the enzyme rejoins the strands. Some topological problems of DNA cannot be resolved by mere twisting. In eukaryotic cells, the mitochondrial DNA is present as a mass of many copies of the circular genome called the kinetoplast. The kinetoplast DNA found in trypanosomal parasites exists as a necklace of catenated circles. Type II DNA topoisomerases are required for the decatenation of DNA necklaces. Type II topoisomerase cleaves both DNA strands while holding precariously onto the ends. The enzyme then allows another section of the double-stranded DNA to pass through the break. Then the topoisomerase stitches the two strands of DNA back together, using an ATP-dependent mechanism analogous to DNA ligase, discussed later in this chapter. Many thermodynamically unfavorable enzymatic transformations are driven by coupling the reaction to the thermodynamically favorable hydrolysis of a phosphate triester bond in ATP. Type II topoisomerases burn two molecules of ATP per strand-passage event.

When topoisomerases are prevented from rejoining DNA strands, this results in cell death. Rapidly dividing cells such as cancer cells and bacteria are especially sensitive to topoisomerase inhibitors (see Figure 3.40). Both natural and synthetic topoisomerase inhibitors are notable for their extreme toxicity.

Bacterial plasmids are rings of DNA

Bacterial cells often contain small circular pieces of DNA plasmids. Plasmids are not part of the bacterial genome but they usually carry at least one gene that confers a selective advantage on the host bacterium. Plasmids are generally present in multiple copies, increasing the likelihood that they will be passed on to each daughter cell during cell division. Plasmids are readily isolated in pure form from bacteria by lysing the cell membrane and trapping the plasmid DNA on silica gel. After new genes have been spliced into plasmids they can be reintroduced back into bacteria and expressed as proteins. The field of molecular biology emerged from the ability to carry out reliable chemistry with plasmid molecules as opposed to living organisms.

All plasmids, and even genomic DNA, have an essential sequence called the **origin of replication** (**ORI** or **ori**) (**Figure 3.41**). The origin of replication instructs cellular enzymes to make a copy of the plasmid. Some chromosomes have multiple ORIs to provide multiple spots for beginning DNA replication. This strategy helps speed up replication of the chromosome. Some plasmid ORIs can lead to rapid DNA synthesis and consequent high copy numbers; others result in slow synthesis and low copy numbers.

As carriers of instructions for producing proteins and DNA, plasmids are often introduced into *E. coli* to turn the bacteria into tiny, high-efficiency factories. However, why should the *E. coli* bother making the plasmid or following its instructions, especially if the foreign DNA drains resources away from the bacteria? Useful plasmids carry a gene that confers resistance to antibiotics such as tetracycline (*Tet*) or ampicillin (*Amp*). If tetracycline is included in the growth medium, only bacteria carrying Tet^+ plasmids will grow.

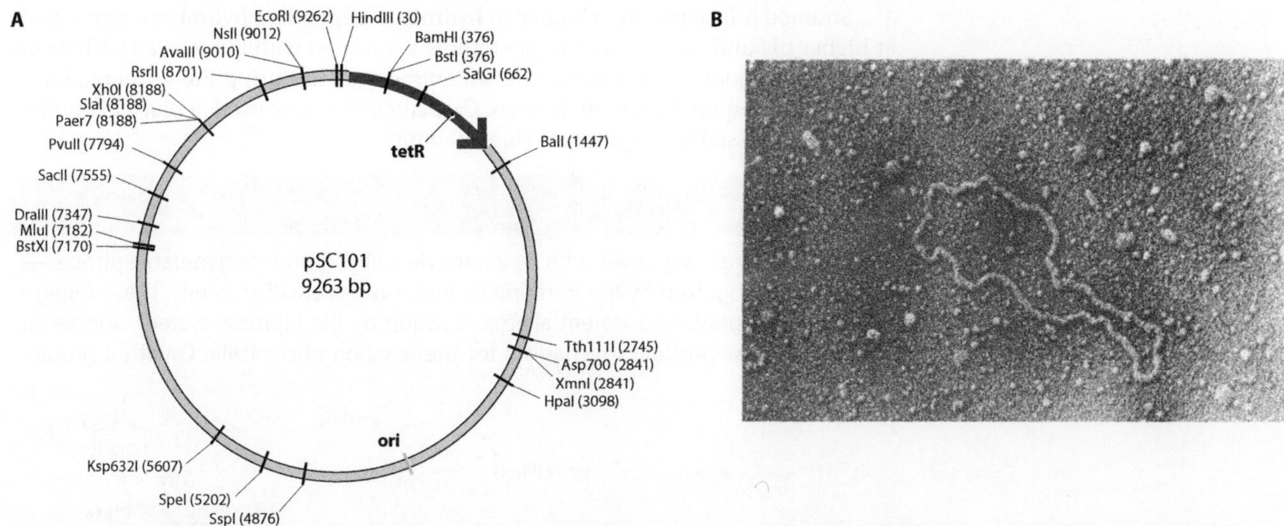

Figure 3.41 A plasmid. (A) Plasmid map. The pSC101 plasmid consists of 9263 base pairs. Restriction sites (lines) are labeled with the corresponding restriction enzyme. The red arrow indicates the location of the gene conferring resistance to the antibiotic tetracycline. (B) Electron micrograph of the pSC101 plasmid. (B, courtesy of S.N. Cohen and A.C. Chang, from *Proc. Natl. Acad. Sci. USA* 70:1293–1297, 1973.)

Problem 3.10

The circular single-stranded genome of the bacterial virus Fd has an origin of replication. Once the viral DNA enters a bacterium, a bacterial RNA polymerase starts to transcribe a complementary RNA strand but gets stuck as a result of folded structure in the Fd DNA. Draw a plausible structure, using base pairing, for the folded region of the Fd ORI.

```
                              folded
5'-GGGTGATGGTTCACGTAGTGGGCCATCGCCCTGATAGACGGTTTTTCGCCCTT -3' Fd DNA
                                3'-ACUAUCUGCCAAAAAGCGGGAA -5' new RNA
                                 |<--- E. coli RNA polymerase
```

Plasmids contain genes that confer advantageous traits

Antibiotics target processes required for life. The most useful antibiotics selectively kill bacteria or human cancer cells, but in medicine the term is usually restricted to small molecules that target bacteria. Some antibiotics inhibit topoisomerase enzymes; some block protein synthesis; others lead to leaky cell walls. Penicillin and other β-lactam antibiotics disrupt synthesis of the bacterial cell wall—a semi-rigid outer coat of some bacteria. Compounds that disrupt the integrity of cell walls are particularly attractive for treating bacterial infections because human cells do not have cell walls. Early studies of antibiotic resistance revealed plasmids that carry genes (such as *Amp*) encoding enzymes that cleave the strained ring of β-lactam antibiotics (**Figure 3.42**). Plasmids with the *Amp* gene are widely used in molecular biology (**Figure 3.43**).

β-lactam γ-lactam δ-lactam

penicillin G R = H
ampicillin R = NH_3^+
carbenicillin R = CO_2^-

Figure 3.42 β-Lactam antibiotics have a distinctive four-membered cyclic amide, called a lactam. Ampicillin and carbenicillin are analogs of penicillin G, one of the earliest β-lactam antibiotic drugs. Carbenicillin finds widespread application in the laboratory in conjunction with plasmids that confer resistance to β-lactams.

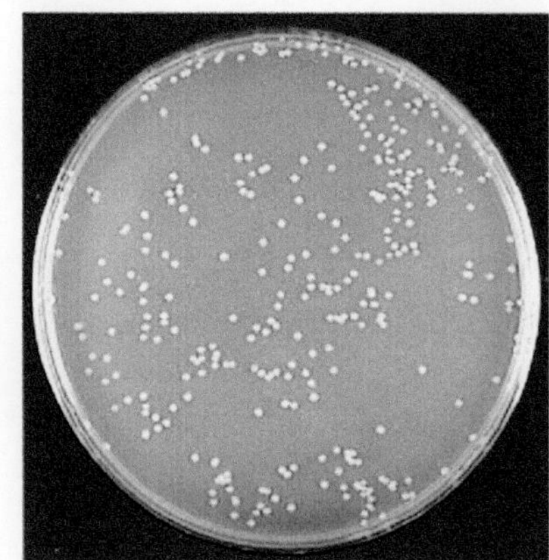

Figure 3.43 Use of a genetic selection marker. Only bacteria containing Amp^+ plasmids can grow into colonies on agar containing carbenicillin. (From G. Karimova et al., *Proc. Natl. Acad. Sci. USA* 95: 5752–5756, 1998. With permission from the National Academy of Sciences.)

Strained β-lactams are sensitive to hydroxide-dependent hydrolysis, particularly at higher pH and the elevated temperatures associated with autoclave sterilization. Additional aliquots of antibiotics are sometimes added during the fermentation of bacteria that require long growth times. Carbenicillin can be used in place of ampicillin and is more stable to hydrolytic ring opening.

Problem 3.11

Human proteins react slowly with β-lactams via amino groups to generate epitopes—structures recognized by the immune system. Even in small amounts, these foreign epitopes can provoke a violent allergic reaction by the immune system. Suggest a plausible arrow-pushing mechanism for the reaction of penicillin G with a protein side chain.

protein NH_2 + penicillin G $\xrightarrow[H_2O]{pH\ 7.2}$ protein (N-H amide product: HN, S, H, O, N-H, Ph, Me, Me, ^-O_2C)

Eukaryotic DNA is coiled around histone proteins

The human genome—all 6 billion base pairs of the diploid genome—fits concisely in human cells with room to spare. Imagine packing 9 km of thread into a soccer ball while keeping it sorted, free of knots, and with every micrometer of thread completely accessible. This insane scenario is business as usual for a human cell. The human genome is not one long string of DNA; the human genome is diploid—two copies of each gene—divided between 23 chromosomes, each consisting of a pair of homologous DNA molecules (**Figure 3.44**). Normally chromosomal DNA is stored as tightly

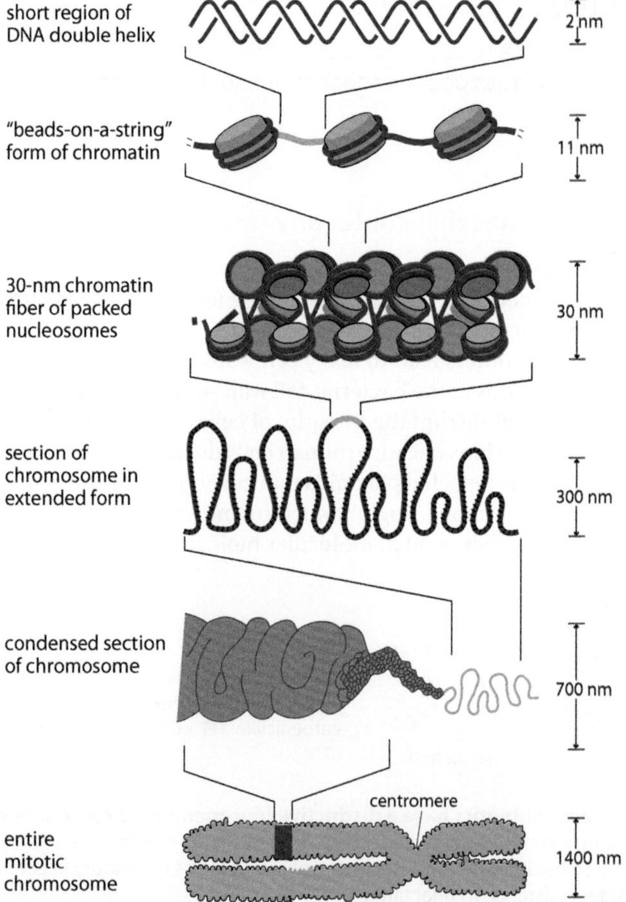

Figure 3.44 Packing of human DNA in chromosomes. (Adapted from B. Alberts et al., Essential Cell Biology, 3rd ed. New York: Garland Science, 2009.)

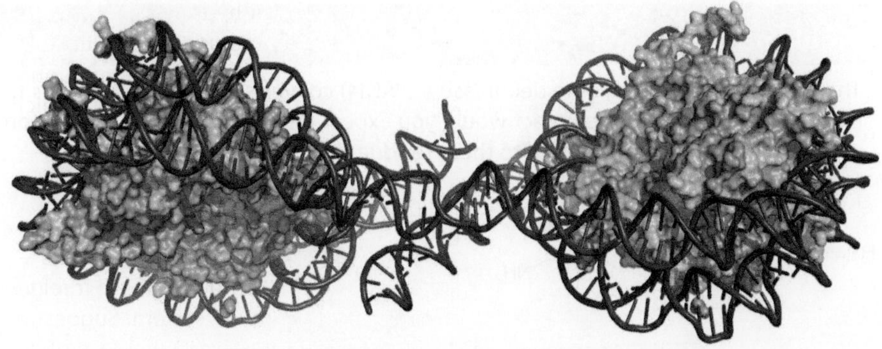

Figure 3.45 Electrostatic hooks. Positively charged lysine and arginine side chains (rendered in green) keep the negatively charged DNA bundled around the histones. They also help nucleosomes to associate closely together.

wrapped chromatin (a mixture of proteins and DNA). Transcription of genes requires the unraveling of specific regions of chromatin; cell division requires the unraveling of all of the chromatin. The building blocks of chromatin are nucleosomes, in which DNA, about 150 bp, is neatly spooled around a complex of eight histone proteins.

Histones interact with the negatively charged DNA backbone through specific lysine and arginine side chains that are protonated and positively charged at physiological pH (**Figure 3.45**). During cell division, enzymes acetylate the ε-amino groups of lysine on histone proteins. Without the positively charged ammonium groups, the tight grip of histones on DNA is lost. Once freed from its histone core, the DNA adopts an extended flexible form that can be transcribed. Histone acetylation and deacetylation, key steps in cell division, are carried out by enzymes called **histone acetyltransferases** (HATs) and **histone deacetylases** (HDACs), respectively. Regulating the structure of nucleosomes through histone modification provides a powerful mechanism for the cell to control transcription, simply by making the DNA accessible or inaccessible for transcriptional machinery. For example, methylation of Lys4 and Lys79 of histone H3 is associated with gene expression. In contrast, methylation of Lys9 and Lys27 is associated with silent genes. A wide range of histone modifications are now known to occur on the lysine, arginine, and serine side chains: methylation, acetylation, phosphorylation, citrullination, ADP-ribosylation, and conjugation to the proteins ubiquitin and SUMO (**Figure 3.46**). Because histones are passed on to daughter cells during cell division, histone modifications are epigenetic, meaning that their effect on transcription is heritable and as important as the sequence of nucleotides.

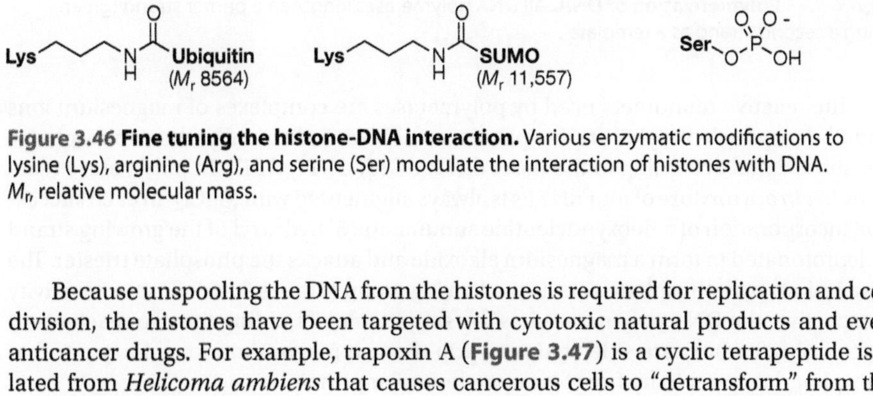

Figure 3.46 Fine tuning the histone-DNA interaction. Various enzymatic modifications to lysine (Lys), arginine (Arg), and serine (Ser) modulate the interaction of histones with DNA. M_r, relative molecular mass.

Because unspooling the DNA from the histones is required for replication and cell division, the histones have been targeted with cytotoxic natural products and even anticancer drugs. For example, trapoxin A (**Figure 3.47**) is a cyclic tetrapeptide isolated from *Helicoma ambiens* that causes cancerous cells to "detransform" from the

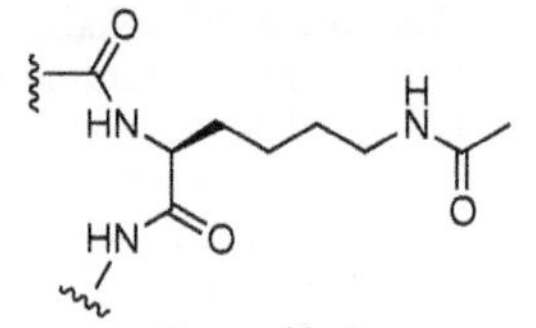

N-acetyl lysine

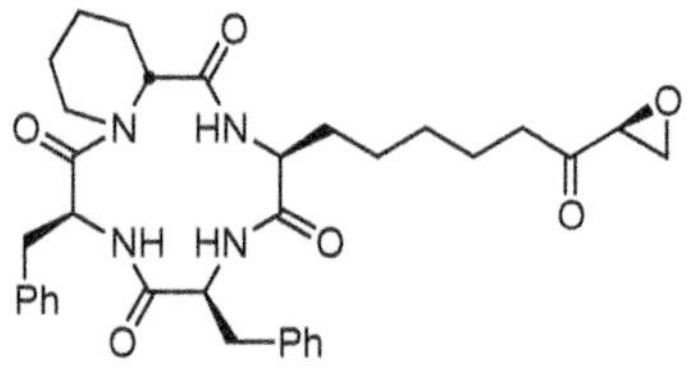

trapoxin A

Figure 3.47 A chemical fishing lure. Trapoxin A is a covalent inhibitor of HDAC-1. The epoxy ketone functional group of trapoxin A mimics the *N*-acetyl lysine side chains of histones.

Problem 3.12

The enzyme peptidyl arginine deiminase 4 (PADI4) converts arginine side chains to citrulline side chains. What impact would you expect this transformation to have on the transcription of genes bound to these residues?

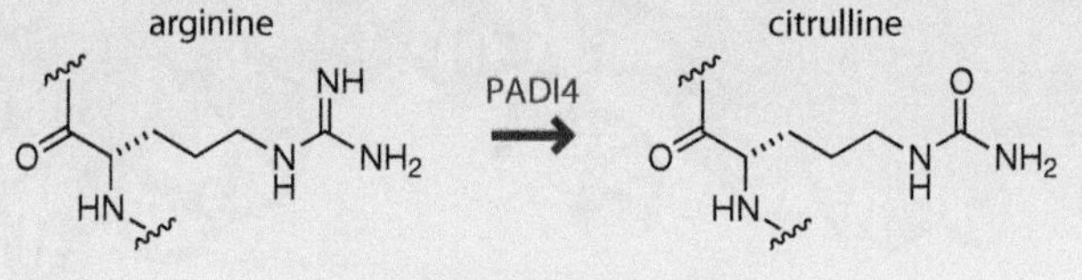

cancerous phenotype to a shape associated with normal cells. By investigating the biological mechanism of trapoxin, Yoshida and coworkers determined that trapoxin inhibits histone deacetylase activity in the cell. Used as a chemical fishing lure by Schreiber and coworkers, trapoxin was found to bind covalently and directly inhibit HDAC-1.

3.5 THE BIOLOGICAL SYNTHESIS OF DNA BY POLYMERASE ENZYMES

DNA polymerases lengthen existing strands

The key enzymes involved in the replication of DNA are polymerases. DNA polymerases lengthen existing strands but do not start new strands. DNA polymerases add deoxynucleotidyl subunits to the 3′ end of a primer strand, using a template strand to determine complementarity (**Figure 3.48**). Human cells have several different types of DNA polymerases: some fill in gaps and make repairs, whereas others copy the entire genome. The DNA polymerase complexes Pol δ and Pol ε are the main polymerases involved in copying the genome before cell division; the complexes are composed of multiple protein subunits.

template strand 3'-TTTTTTAAAAGTTACTTATTTTACGACGTAAGATATCCAATAGTTAAA...
primer strand 5'-AAAAAATTTTCAATG-3'

DNA polymerase enzyme ↓ **nucleotide monomers**

template strand 3'-TTTTTTAAAAGTTACTTATTTTACGACGTAAGATATCCAATAGTTAAA...
elongated strand 5'-AAAAAATTTTCAATGAATAAAATGCTGCATTCTATAGGTTATCAATTT...

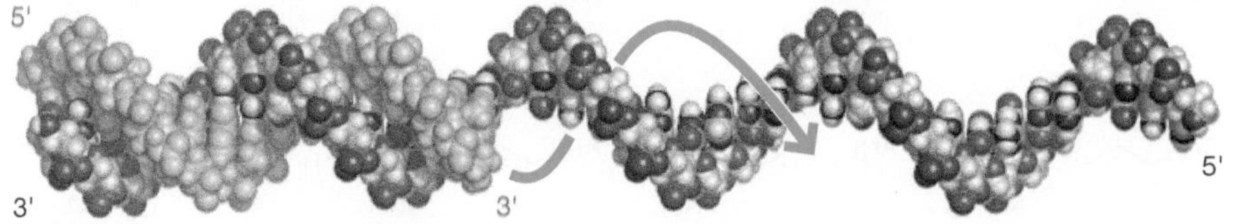

Figure 3.48 Polymerization of DNA. All DNA polymerases lengthen a primer strand (green) using a second strand as a template.

The reactive monomers used by polymerases are complexes of magnesium ions and 2′-deoxynucleotidyl 5′-triphosphates, or dNTPs. Individually, the triphosphates are abbreviated according to the base: dATP, dCTP, dGTP, dTTP. In polymerase reactions *in vitro*, a mixture of four dNTPs is always augmented with $MgCl_2$. In each successive incorporation of a deoxynucleotide subunit, the 3′ hydroxyl of the growing strand is deprotonated to form a magnesium alkoxide and attacks the phosphate triester. The magnesium pyrophosphate acts as the leaving group (**Figure 3.49**). From a reactivity perspective, the DNA polymerase has the simple role of holding the two reactants in place, using carboxylate side chains to grip the two magnesium ions. Enzymes that phosphorylate protein hydroxyl groups use similar principles.

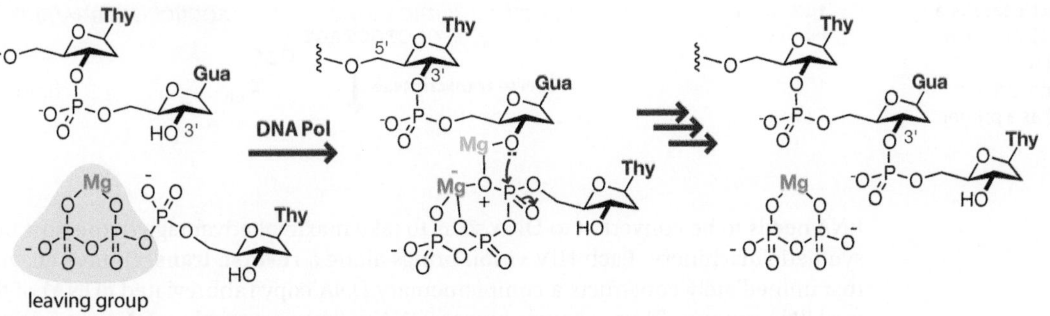

Figure 3.49 Magnesium requirement. The chemical mechanism of human DNA polymerase η (DNA Pol) involves two Mg^{2+} ions.

DNA polymerases copy with high fidelity

The true magic of DNA polymerases is the ability to incorporate complementary nucleotides with high fidelity based on a template strand. DNA polymerase grips the growing duplex at just the right place, using a structure that resembles a half-open right hand (**Figure 3.50**). The short stretch of double-stranded DNA formed by the primer, followed by the single-stranded DNA, enters at the crook of the hand, between the "thumb" and "forefinger." As bases are added to the strand, the upper index finger pushes on the resultant double-stranded structure, testing it for correct Watson–Crick base pairing. Incorrectly base-pairing dNTPs are pushed out of the active site. When correctly base-pairing dNTPs are found, the fingers touch the thumb, closing the active site around the nascent double strand and the incoming dNTP. A second enzymatic domain (the **exonuclease**) checks the quality of the work before the newly added nucleotide moves from the active site. If the newly added nucleotide does not form a perfect Watson–Crick base pair, the exonuclease cleaves the aberrant nucleotide, giving the polymerase the opportunity to introduce the correct subunit. This proof-reading reduces the overall error rate to 1 out of 10^8 bases synthesized. On average, DNA polymerases lengthen DNA at a rate of 1000 nucleotides per minute—an impressive rate of speed considering the fastidious nature of the enzyme!

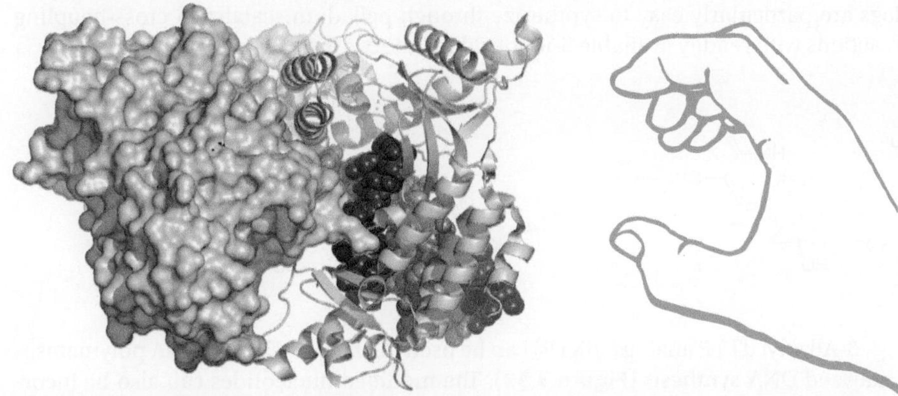

Figure 3.50 The right hand model (cyan) is conserved for all known DNA polymerase structures. This structure of DNA polymerase δ from yeast (PDB 3IAY) also includes an exonuclease domain, rendered in green. The template strand is rendered in orange and the growing DNA strand is rendered in red.

Reverse transcriptase lengthens existing DNA strands on an RNA template

The human genome is stored as a double-stranded DNA template, but some viruses, termed **retroviruses**—including HIV (human immunodeficiency virus)—have genomes composed of single-stranded RNA. Once it enters a human T cell the HIV

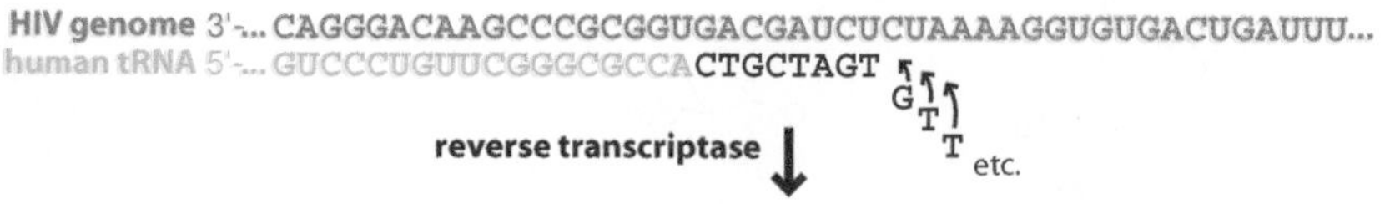

HIV genome 3'-... CAGGGACAAGCCCGCGGUGACGAUCUCUAAAAGGUGUGACUGAUUU...
human tRNA 5'-... GUCCCUGUUCGGGCGCCACTGCTAGT G T T etc.

reverse transcriptase

Figure 3.51 Reverse transcriptase is a type of DNA polymerase. HIV reverse transcriptase generates a stable cDNA copy of the HIV RNA genome, using a human transfer RNA (tRNA) as a primer.

RNA needs to be converted to DNA so as to take maximal advantage of the host biosynthetic machinery. Each HIV virion brings along a **reverse transcriptase** enzyme that immediately constructs a complementary DNA copy (abbreviated cDNA) of the viral RNA genome. Thus, whereas human DNA polymerases make a DNA strand from a complementary DNA template, reverse transcriptases make a DNA strand from a complementary RNA template. The overall process is the reverse of human RNA transcription. In human cells, HIV reverse transcriptase borrows a human transfer RNA molecule, which we discuss in the next chapter, and uses it as a primer (**Figure 3.51**). Reverse transcriptases will lengthen either DNA or RNA primers.

Reverse transcriptases have become indispensable tools for molecular biology. Reverse transcriptase has a few features that are unusual for a DNA polymerase. In addition to the synthesis of one strand of cDNA complementary to the genomic RNA, reverse transcriptase can destroy the RNA template, and even synthesize the second strand of cDNA, using the first strand as a template. A subunit of reverse transcriptase, RNase H, catalyzes the breakdown of the RNA template. Such multifunctional abilities are characteristic of viruses, which typically have small genomes; minimization of viral genome size confers the benefit of easier replication and consequently improved evolutionary fitness. Thus, each member of the HIV genome, including reverse transcriptase, must perform multiple functions.

DNA polymerase incorporates modified thymidylate residues

Given the high fidelity of DNA polymerases, it is surprising that many will incorporate nucleotides with modified bases. Some DNA polymerases are more tolerant than others, but a key requirement is that the modified bases must have the ability to form Watson–Crick base pairs. 5-Alkynylthymidine analogs and 5-alkenylthymidine analogs are particularly easy to synthesize through palladium-catalyzed cross-coupling reactions with readily available 5-iodouridine (**Figure 3.52**).

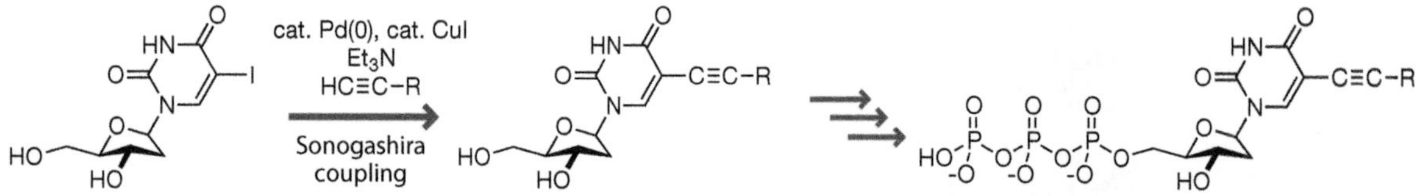

cat. Pd(0), cat. CuI
Et_3N
HC≡C–R
Sonogashira coupling

Figure 3.52 Adding chemical functionality to DNA bases. dTTP analogs can be readily synthesized by using palladium-catalyzed cross-coupling.

5-Alkynyl dTTP analogs (dXTP) can be used in place of dTTP in DNA polymerase-catalyzed DNA synthesis (**Figure 3.53**). The modified nucleotides can also be incorporated even when the modified bases are already present in the template strand, leading to DNA in which both strands have modified bases. The flexibility of DNA polymerase to incorporate modified bases also enables DNA sequencing through the incorporation of unique dye-labeled analogs for each DNA base.

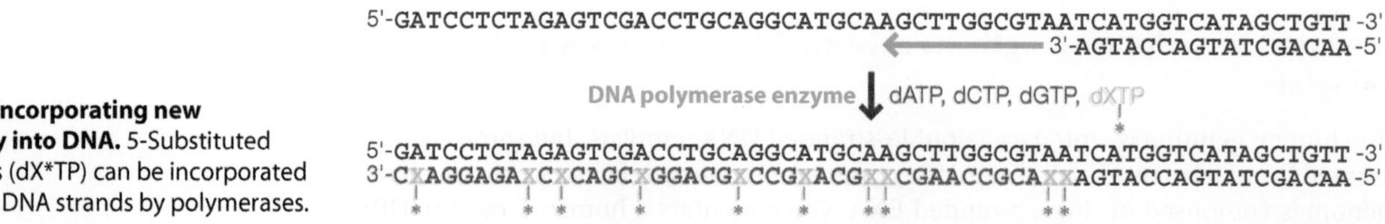

Figure 3.53 Incorporating new functionality into DNA. 5-Substituted dTTP analogs (dX*TP) can be incorporated into growing DNA strands by polymerases.

The polymerase chain reaction amplifies DNA through iterative doubling

DNA polymerase finds numerous applications in chemical biology laboratories. To illustrate, the polymerase chain reaction (PCR) is described here. Other applications for DNA polymerase include DNA sequencing (Section 3.7) and mutagenesis, and these are described later in this chapter. PCR can be used to amplify essentially any sequence of DNA; it is based on Le Chatelier's principle of using excess reagents to drive a thermodynamic equilibrium.

PCR consists of essentially two stages: (1) equilibration of double-stranded DNA with two sets of 5′ primers, each in excess, and (2) extension with DNA polymerase (**Figure 3.54**). In practice, the equilibration stage involves two 1-minute steps. Because the process involves temperatures close to 100 °C, a heat-stable DNA polymerase from a thermophilic bacterium is used; high temperatures cause most bacterial or human enzymes to misfold. The thermostable DNA polymerase *Taq* polymerase has an ideal operating temperature around 72 °C; it does not work if the reaction is too hot or too cold. Each cycle of PCR doubles the number of strands, and typically 25 or more cycles are used to amplify DNA. Each cycle of PCR consists of three 1-minute steps. The initial reaction mixture includes DNA polymerase, dNTPs, $MgCl_2$, target DNA, and two sets of 5′ primers that flank the target sequence. No reagents are added during PCR. The only variable is temperature. The reaction is run in a special thin-walled Eppendorf tube. The reaction is conducted in a computer-controlled heating block called a thermal cycler.

In the first step of PCR, the reaction mixture is heated to 95 °C, facilitating equilibration between the longer duplexes and the abundant shorter primers. In the second step, the temperature of the solution is lowered to 55 °C to allow the primers to hybridize specifically to the DNA at flanking 3′ ends of the double-stranded target sequence. The sequences of these two PCR primers exactly match the sequences at the 5′ ends of the double-stranded target. In the third step, the mixture is heated to 72 °C, the optimal temperature for *Taq* polymerase. This three-step process—heating to dissociate, cooling to anneal, and warming to polymerize—is repeated over and over. After each cycle, more target DNA accumulates. PCR is efficient enough to amplify even a single

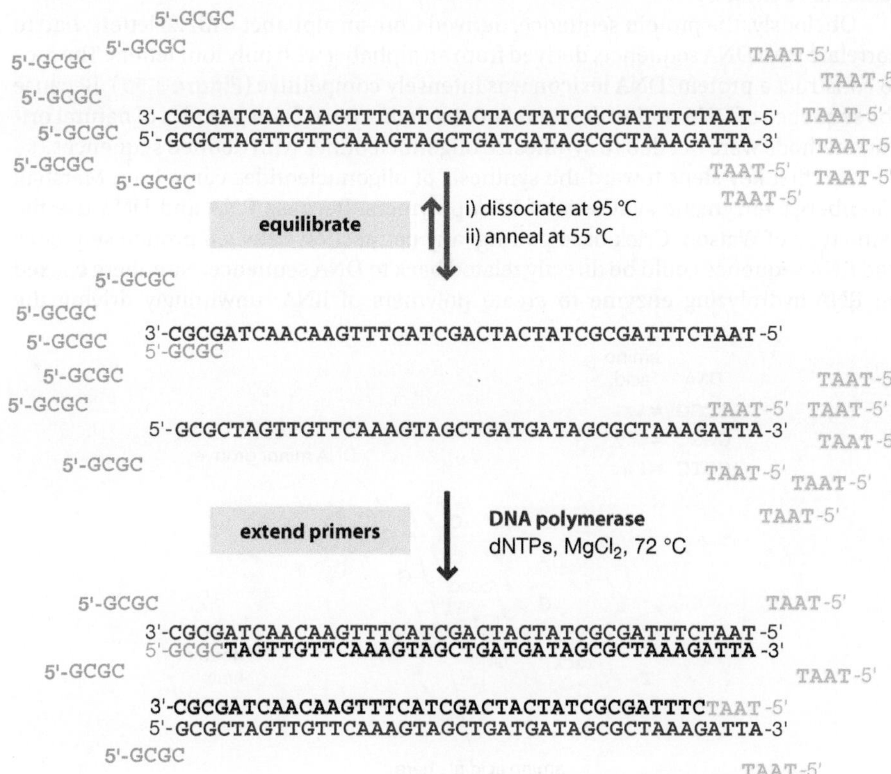

Figure 3.54 Each cycle of PCR involves a two-step equilibration process followed by a polymerization step. For simplicity, primers of only four base pairs are shown, although much longer primers are typically used to ensure high specificity of the amplification.

molecule of target DNA. Because the annealing process is dependent on the T_m of the primer, the technique can be used to detect single base differences within complex sequences. This ability to amplify a specific gene within an entire genome from even a single molecule has revolutionized biology, forensics, anthropology, and many other fields. Such power also finds daily use in most chemical biology laboratories.

Problem 3.13

Design a pair of oligonucleotide primers, each 21 base pairs in length, that could be used to amplify the following gene.

```
5'-CCATGCCTATGTTCATCGTGAACACCAATGTTCCCCGCGCCTCCGTGCCAGAG
GGGTTTCTGTCGGAGCTCACCCAGCAGCTGGCGCAGGCCACCGGCAAGCCCGCAC
AGTACATCGCAGTGCACGTGGTCCCGGACCAGCTCATGACTTTTAGCGGCACGAA
CGATCCCTGCGCCCTCTGCAGCCTGCACAGCATCGGCAAGATCGGTGGTGCCCAG
AACCGCAACTACAGTAAGCTGCTGTGTGGCCTGCTGTCCGATCGCCTGCACATCA
GCCCGGACCGGGTCTACATCAACTATTACGACATGAACGCTGCCAACGTGGGCTG
GAACGGTTCCACCTTCGCTTGAGTCCTGGCCCCACTTACCTGCACCGCTGTTC-3'
```

3.6 THE CHEMICAL SYNTHESIS OF DNA

The race to crack the genetic code drove the development of DNA synthesis

In 1950 Per Edman developed a method for sequencing proteins, which are much more complex than DNA; however, reliable methods for sequencing DNA did not become available until 1977. Before 1977, the only information that could be gained for any DNA sample was the nucleotide composition, that is A, G, C, and T percentages. Watson and Crick's 1953 discovery offered tremendous insight into heredity and also bolstered the one-gene/one-enzyme hypothesis. The one-gene/one-enzyme hypothesis demanded a correlation between DNA sequence and protein sequence, referred to as the **genetic code**. Without a way to sequence DNA, the genetic code remained a mystery.

Obviously, the protein sequence, derived from an alphabet with 20 letters, had to correlate with DNA sequence, derived from an alphabet with only four letters. The race to construct a protein/DNA lexicon was intensely competitive (**Figure 3.55**). Because the sequence of DNA molecules could not be determined from samples of natural origin, methods were needed to synthesize oligonucleotides with defined sequences.

The first key steps toward the synthesis of oligonucleotides came from Marshall Nirenberg's enzymatic synthesis of RNA polymers. Because RNA and DNA use the same type of Watson–Crick base pairing, any connection between protein sequence and RNA sequence could be directly related back to DNA sequence. Nirenberg coaxed an RNA-hydrolyzing enzyme to create polymers of RNA, unwittingly driving the

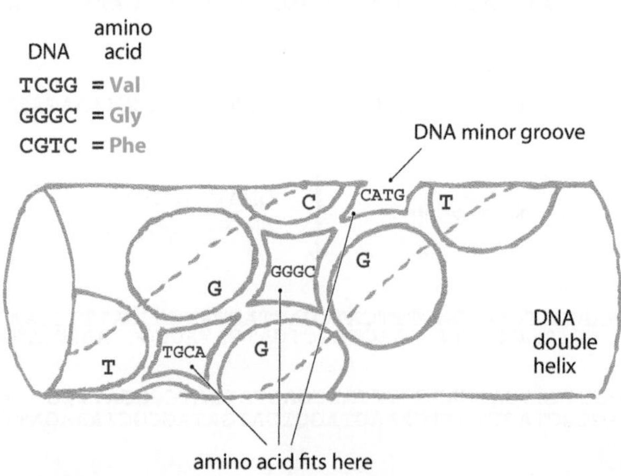

Figure 3.55 The diamond code hypothesis. In 1954 the physicist George Gamow proposed that groups of four bases created pockets in the minor groove that were complementary to amino acid side chains. The flawed diamond code hypothesis—based on a four-base codon—was difficult to prove or disprove.

biological reaction in reverse. Using a large excess of individual nucleotidyl triphosphate building blocks, Nirenberg was able to create random RNA polymers that contained, for example, only A, only C, or mixtures of A and C. These random RNA polymers were translated into proteins by using cellular extracts containing ribosomes and many other molecules necessary for protein synthesis (**Figure 3.56**). Nirenberg used his random polymers of RNA to show that triplet "codons" specify amino acids during ribosomal translation, and he further identified 50 (out of a possible 64) three-nucleotide sets that coded for amino acids. However, to assign the order of nucleotides unambiguously in many of these three-base codons and complete missing pieces of the code, oligonucleotides of precisely defined sequence were required.

Ultimately, the goal of cracking the genetic code drove the early development of DNA synthesis. Modern oligonucleotide synthesis is a triumph for synthetic organic methodology. In response to the need of biologists for short to medium-sized oligonucleotides (up to 100 bp), chemists developed chemistry so efficient and so robust that yields per step typically exceed 99%.

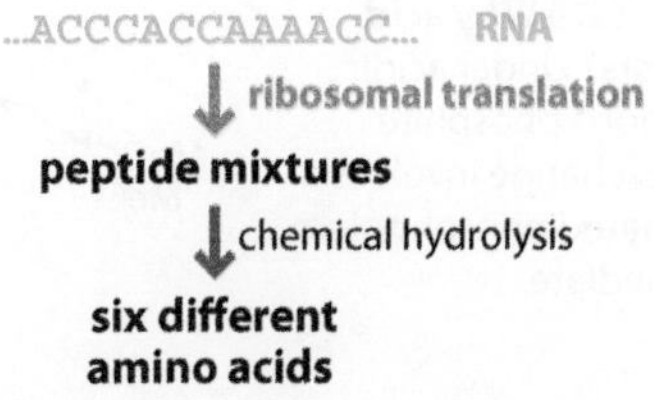

...ACCCACCAAAACC... RNA
↓ ribosomal translation
peptide mixtures
↓ chemical hydrolysis
six different amino acids

Figure 3.56 Cracking the code. Random RNA polymers were synthesized and translated into proteins in order to investigate the genetic code.

The Khorana method of DNA synthesis relies on phosphate coupling chemistry

The pioneering synthetic work of the chemist H. Ghobind Khorana led to the completion of the genetic code. Khorana developed a practical method for the synthesis of simple oligonucleotides that were used as templates in protein synthesis. The original Khorana method involved the direct formation of phosphate esters by attack of the 3′ hydroxyl on an activated phosphate group (**Figure 3.57**). Similar chemical methods of activation are used in peptide synthesis (Chapter 5). This phosphate ester chemistry mirrors the enzymatic reaction catalyzed by DNA polymerase except that the 3′ position of each monomer is protected as an acetate ester. After each coupling, sodium hydroxide is used to saponify the acetate ester and hydrolyze the phosphate triester to the phosphate diester. A major disadvantage of this chemistry is that it uses a hindered secondary hydroxyl as a nucleophile, and yields are never above 80% for each coupling step. Purification is an even greater problem. For example, after 10 nucleotide subunits have been added, it can be difficult to separate the decamer from the nonamer.

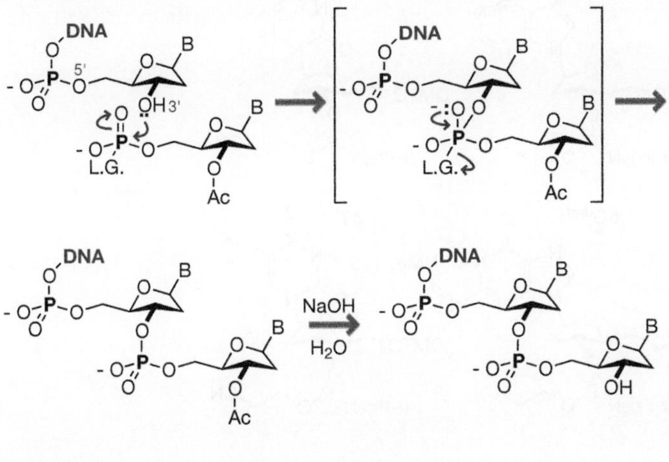

Figure 3.57 An early approach to DNA synthesis. The original Khorana method for DNA synthesis involved activated phosphate esters.

	phosphate	phosphite
k_{rel} (base)	1	1000
k_{rel} (acid)	1	1,000,000,000,000

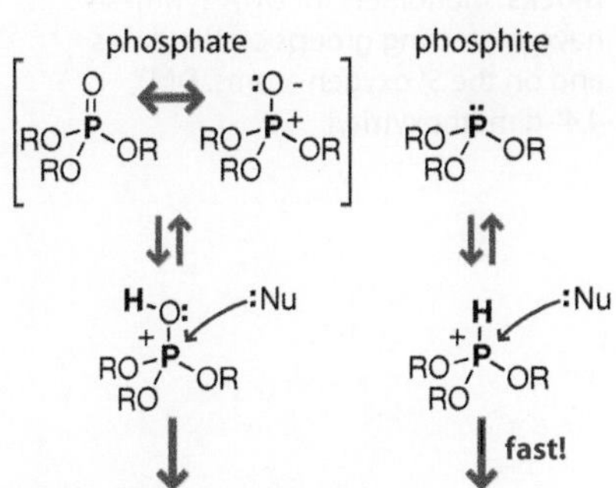

Figure 3.58 Phosphites versus phosphates. Phosphites are much more reactive than phosphates, particularly under acidic conditions.

Letsinger recognized the speed and efficiency of phosphite couplings

Recognition of the ease of phosphite ester formation relative to phosphate ester formation resulted in a major advance in oligonucleotide synthesis. To understand the difference in reactivity between phosphates and phosphites one should consider the effect of the oxygen lone pairs (**Figure 3.58**). The oxygen lone pairs donate into

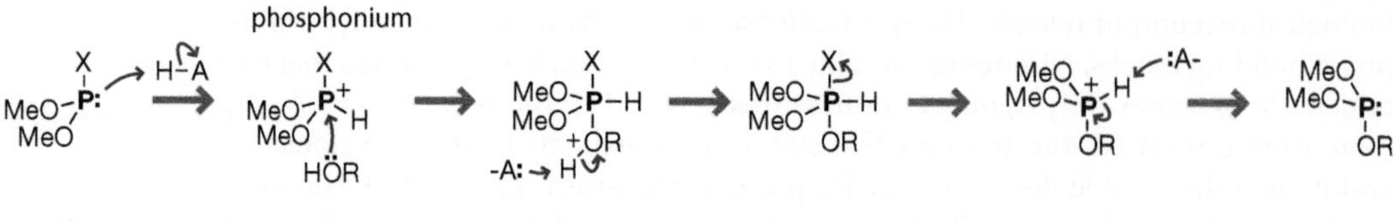

Figure 3.59 Why acid catalysis? Under acidic conditions phosphite ester exchange involves a protonated phosphonium intermediate.

empty orbitals on phosphorus, competing with incoming nucleophiles. Because a protonated phosphite has one less oxygen substituent than a protonated phosphate the protonated phosphite is much more electrophilic. Thus, the key step in the mechanism of phosphite ester formation is the formation of a protonated phosphonium intermediate (**Figure 3.59**). Under acidic conditions, phosphites hydrolyze 10^{12} faster than phosphates. Recall that C–O bond cleavage competes with O–P bond cleavage under acidic conditions.

Caruthers synthesized DNA by using phosphoramidites on solid phase

Marvin Caruthers made two key advances in the synthesis of oligonucleotides. First, he extended the Letsinger phosphite chemistry to solid phase, making it possible to purify intermediates by rinsing away the unbound reagents. The second key advance was the development of stable **phosphoramidite** building blocks. This combination of solid-phase synthesis and phosphoramidite building blocks is the basis of modern oligonucleotide synthesis. A modern phosphoramidite building block has two essential features: an acid-labile dimethoxytrityl (DMT) protecting group on the 5′ oxygen and a relatively stable cyanoethyl phosphoramidite (**Figure 3.60**). The mechanism for phosphoramidite coupling involves protonation of the phosphoramidite with a mild acid to generate the activated phosphonium salt (**Figure 3.61**). The resulting phosphite is extremely sensitive to water, but oxidation to the corresponding phosphate provides a stable linkage.

Three of the four nucleotide bases have nucleophilic amino groups requiring protection: cytosine, adenine, and guanine. The protecting groups are removed with

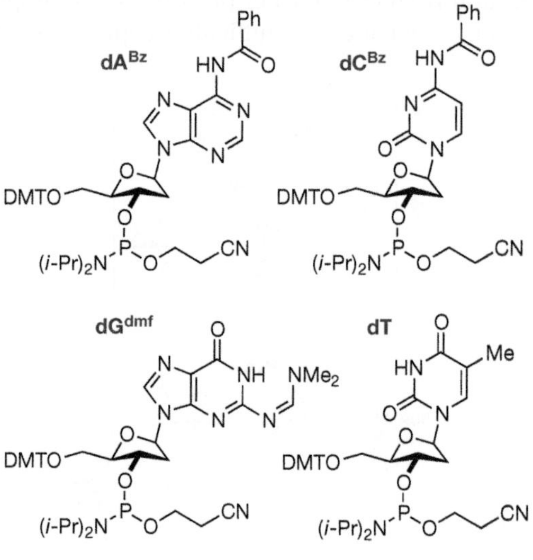

Figure 3.60 Protection of DNA building blocks. Monomers for DNA synthesis have protecting groups on the bases and on the 5′ oxygen atoms. DMT, 4,4′-dimethoxytrityl.

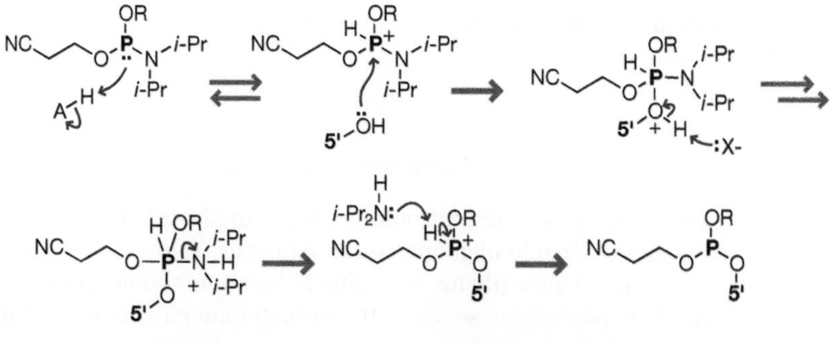

Figure 3.61 Why phosphoramidites? Protonation of the diisopropylamino group turns it into a good leaving group. The attacking 5′-hydroxyl group is abbreviated as 5′-OH.

concentrated (15 M) ammonium hydroxide (Figure 3.71), or mixtures of ammonium hydroxide and aqueous methylamine, at elevated temperature. Aqueous ammonium hydroxide solutions have three advantages over sodium hydroxide: ammonia and methylamine are volatile (whereas sodium hydroxide is not), the free amines are nucleophilic, and the presence of ammonium counterions facilitates acid-catalyzed reaction steps. When guanosine is protected with an N^2-benzoyl group, it is too easily deprotonated under the deprotection conditions, so a dimethylformamidine (dmf) protecting group is used to protect the N^2-amino group of guanine.

Problem 3.14

Deprotection of N^4-benzoyldeoxycytidine with a mixture of aqueous methylamine and ammonia leads to the formation of an *N*-methylated side product in 10% yield. Suggest a plausible arrow-pushing mechanism to account for this side product.

NHBz → ($8.5\ M\ NH_3$, $6.5\ M\ MeNH_2$, H_2O, 65 °C, 5 min) → NH_2 90% + $NHCH_3$ 10%

Automated oligonucleotide synthesis is performed on glass particles

The iterative synthesis of long bio-oligomers is best performed on a solid support. To achieve fast coupling rates, reagents are generally used in excess (rate = k[ROH][reagent]). Unfortunately, a large excess of chemical reagents is difficult to remove from the desired product. By attaching the substrate to a solid support, the by-products and reagents can be rinsed away from the product, just like washing a car. This important conceptual advance, initially pioneered by Robert Merrifield for peptide synthesis, has made possible the automated synthesis of oligonucleotides, oligopeptides, and oligosaccharides.

Tiny polystyrene beads are the most common solid supports for most types of solid-phase chemical synthesis, but controlled-pore glass (CPG) is also used as a solid support for oligonucleotides with fewer than 50 bases. We will examine the topic of synthesis on polystyrene supports further when we discuss peptide synthesis in Chapter 5. CPG is a porous borosilicate glass with a high surface area. The pore diameter of CPG is usually 500–1000 Å, and the particle size can range from 3 μm to 3 mm. The initial nucleoside is connected to the CPG support by first converting the CPG surface to an aminopropylsilane surface by condensing with 3-aminopropyl triethoxysilane and heating to drive off ethanol. Similar siloxane chemistry can be used to derivatize microscope slides or SiO_2 on semiconductor chips. There are no known S_N2 reactions at silicon; substitution reactions involving silicon always proceed through addition–elimination reactions via a pentacoordinate siliconate intermediate. Once the glass surface has been carpeted with amino groups, they can be acylated by a nitrophenyl ester derived from the first nucleoside (**Figure 3.62**). The final deprotection with

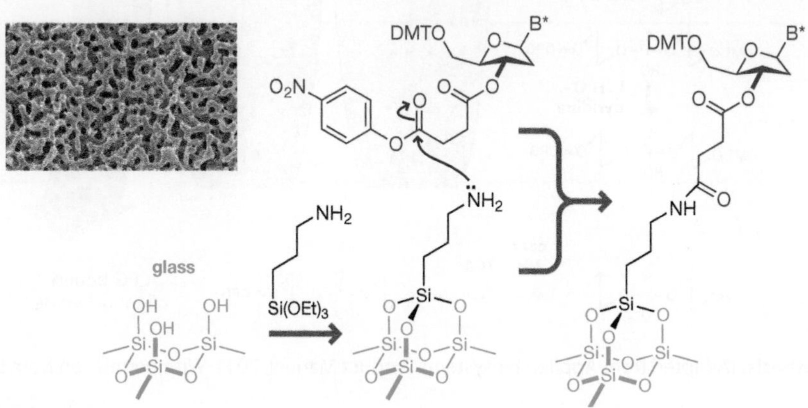

Figure 3.62 Siloxane condensation can be used to link the first nucleoside to the surface of controlled-pore glass. Left: an electron micrograph of controlled-pore glass reveals the high internal surface area. (Image from V. Beschieru and B. Rathke, *Microporous Mesoporous Mat.* 125:63–69, 2009. With permission from Elsevier.)

Figure 3.63 Pre-derivatized CPG columns for an automated DNA synthesizer. These plastic cartridges contain controlled-pore glass particles, pre-derivatized with one of the four DNA nucleotides. (From Applied Biosystems Product Bulletin, 2011. With permission from Life Technologies Corporation.)

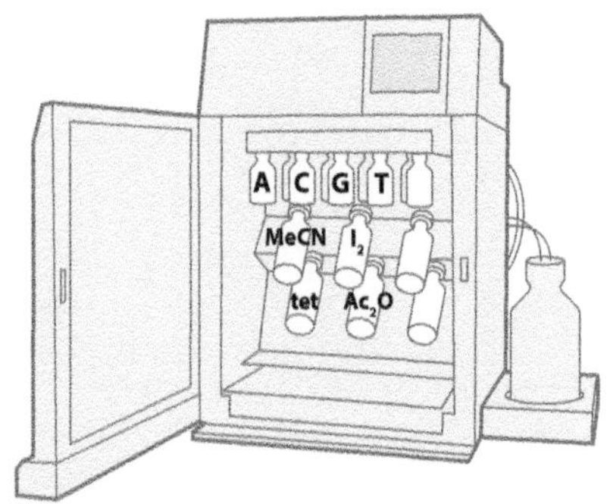

Figure 3.64 PE Applied Biosystems Expedite® DNA Synthesizer. (From Applied Biosystems Product Bulletin, 2011. With permission from Life Technologies Corporation.)

ammonium hydroxide cleaves the 3′ ester bond. The CPG-nucleoside conjugates are encapsulated in flow-through columns that are connected to an automated synthesizer (**Figure 3.63**). Because the 3′ oligonucleotide can possess any one of the four bases, four different columns are sold, each with a different base attached to the solid support.

Modern automated DNA synthesis involves repetitive four-step cycles

Most laboratories order custom-synthesized oligonucleotides over the Internet. These oligonucleotides are made on automated DNA synthesizers (**Figure 3.64**). Each cycle of automated DNA synthesis involves four steps, repeated over and over (**Figure 3.65**): 5′ deprotection, phosphoramidite coupling, capping of unreacted 5′-hydroxyl groups, and oxidation of phosphites to phosphates.

The 4,4′-dimethoxytrityl group is deprotected through an S_N1 reaction

Triphenylmethyl (trityl) groups are commonly used to protect alcohols in DNA and peptide synthesis. Under acidic conditions, DMT groups are more easily removed than trityl groups because the methoxy groups stabilize the buildup of positive charge during formation of the trityl cation. DMT groups are removed through an S_N1 reaction, using undiluted trichloroacetic acid (pK_a -0.5), which is about 100,000 times more acidic than acetic acid (**Figure 3.66**).

The dimethoxytrityl cation imparts a deep orange color to the waste effluent from the deprotection reaction. The absorbance at 498 nm can be used to estimate the overall yield of the synthesis. For example, if the A_{498} of the trichloroacetic acid effluent is 1.2 after the first round but only 0.8 after 20 rounds, then the overall yield is about 67%, or an average of 98% yield for each round.

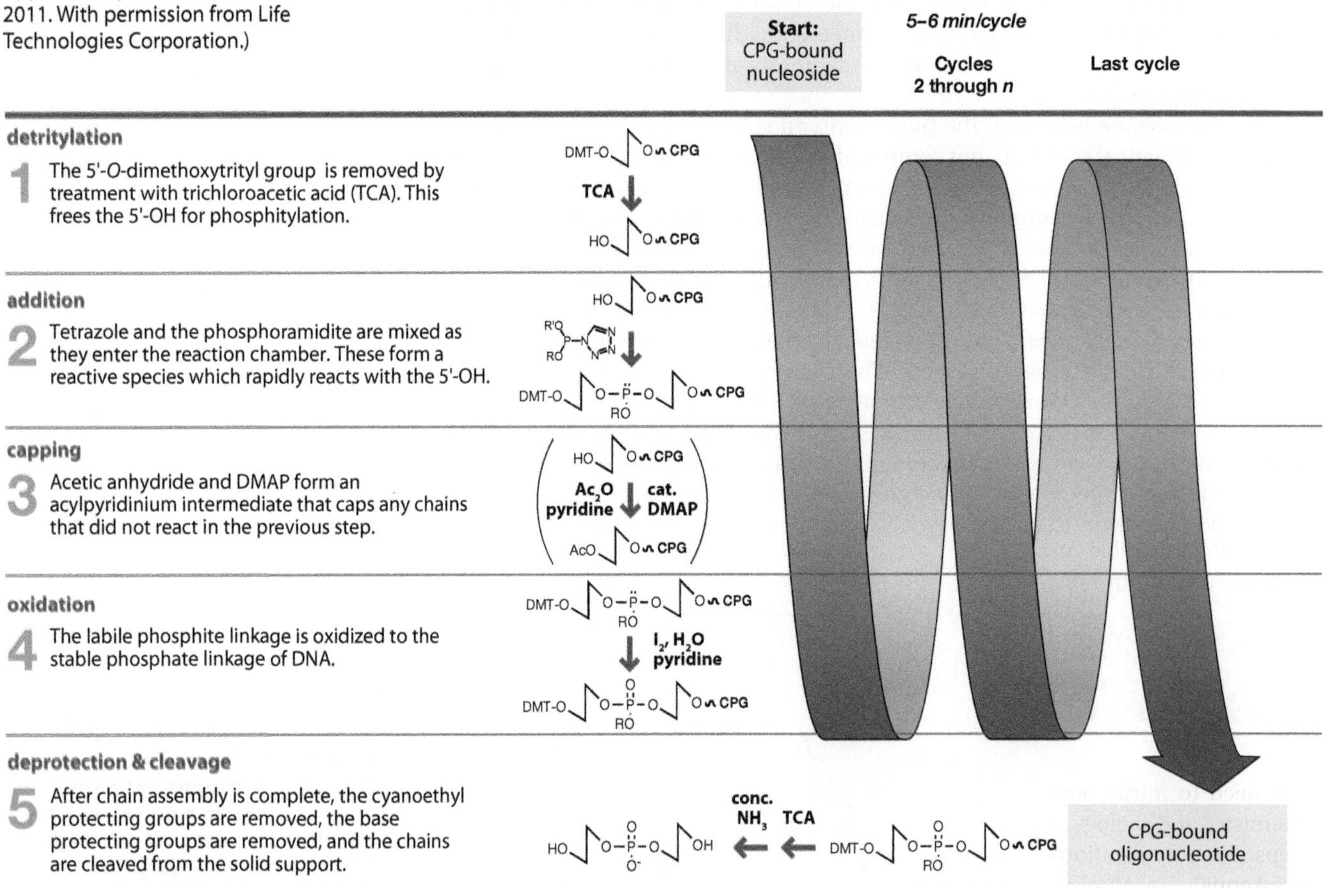

Start: CPG-bound nucleoside

5–6 *min/cycle*

Cycles 2 through *n*

Last cycle

detritylation

1 The 5′-*O*-dimethoxytrityl group is removed by treatment with trichloroacetic acid (TCA). This frees the 5′-OH for phosphitylation.

DMT-O … O CPG → (TCA) → HO … O CPG

addition

2 Tetrazole and the phosphoramidite are mixed as they enter the reaction chamber. These form a reactive species which rapidly reacts with the 5′-OH.

HO … O CPG → DMT-O … O–P–O … O CPG (RO)

capping

3 Acetic anhydride and DMAP form an acylpyridinium intermediate that caps any chains that did not react in the previous step.

HO … O CPG → (Ac_2O, pyridine, cat. DMAP) → AcO … O CPG

oxidation

4 The labile phosphite linkage is oxidized to the stable phosphate linkage of DNA.

DMT-O … O–P–O … O CPG (RO) → (I_2, H_2O, pyridine) → DMT-O … O–P(=O)–O … O CPG (RO)

deprotection & cleavage

5 After chain assembly is complete, the cyanoethyl protecting groups are removed, the base protecting groups are removed, and the chains are cleaved from the solid support.

CPG-bound oligonucleotide: DMT-O … O–P(=O)–O … O CPG (RO) → (TCA) → (conc. NH_3) → HO … O–P(=O)(O⁻)–O … OH

Figure 3.65 The chemistry of automated DNA synthesis. (Adapted from Applied Biosystems Product Manual, 2011. With permission from Life Technologies Corporation.)

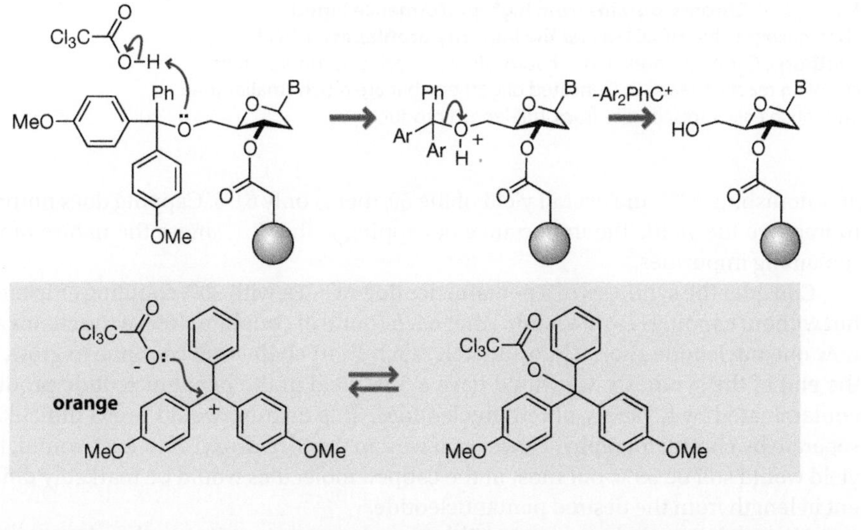

Figure 3.66 Healthy orange color. During the solvolysis of the 5′ DMT group the solution is orange as a result of the resonance-stabilized triarylmethyl cation.

Tetrazole serves as an acid catalyst in phosphoramidite couplings

The coupling reaction is performed in acetonitrile by using a very special acid, 1H-tetrazole, that has about the same acidity as acetic acid. If tetrazole is premixed with the phosphoramidite before it is added to the alcohol, it will initially form a *P*-tetrazolyl intermediate (**Figure 3.67**). Activated intermediates also participate in capping.

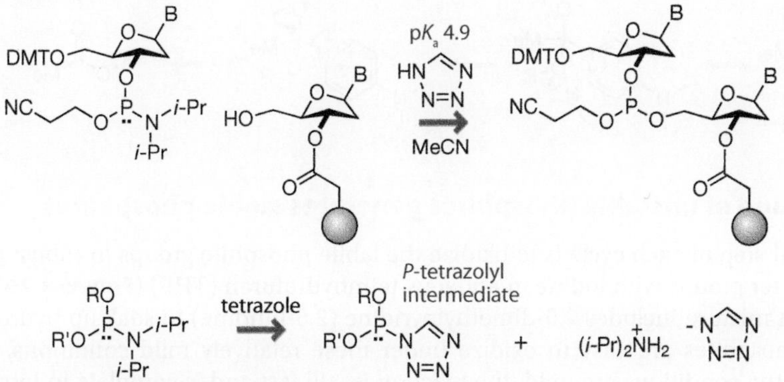

Figure 3.67 Secret ingredient. Phosphoramidite coupling uses the highly specialized acid tetrazole to generate a *P*-tetrazolyl intermediate.

Problem 3.15

Suggest a plausible arrow-pushing mechanism for the formation of a *P*-tetrazole intermediate from a phosphoramidite.

Table 3.1 Effect of coupling yield on overall yield of a 50-mer.

Yield per step (%)	Overall yield (%)
95	8
98	36
99	61
99.5	78

Capping unreacted 5′-hydroxyl groups prevents the propagation of mistakes

The need to purify products from impurities is of paramount importance to both chemistry and biology, particularly for a synthesis involving hundreds of sequential steps. Most applications of oligonucleotides require very little material, but purity is essential. For an oligonucleotide 50 bases in length (referred to colloquially as a 50-mer), a yield of 95% per cycle would afford only an 8% yield of the desired oligonucleotide (**Table 3.1**). Under typical conditions, where the average coupling yield is

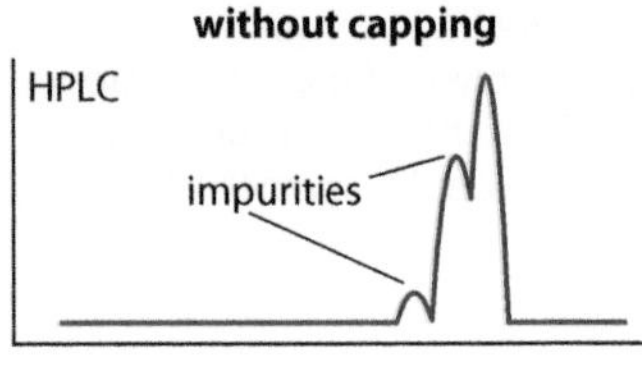

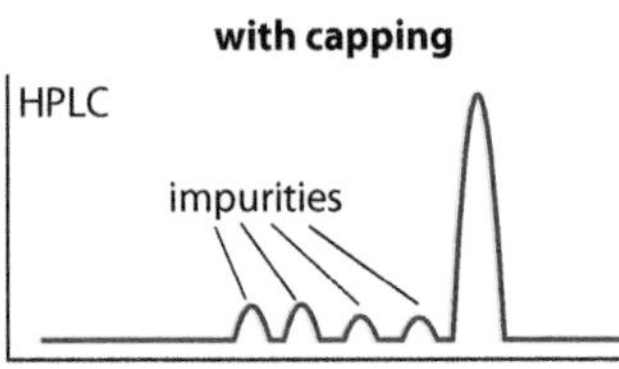

Figure 3.68 Chromatograms from high-performance liquid chromatography (HPLC) reveal the impurity profiles after DNA synthesis. Capping unreacted 5′ hydroxyls after each phosphoramidite coupling reaction leads to truncated oligomers that are much smaller than, and much easier to separate from, full-length product.

an astonishing 99%, the overall yield of the 50-mer is only 61%. Capping does nothing to improve the yield. The importance of capping is that it changes the nature of the remaining impurities.

Consider the synthesis of a pentanucleotide AGCTA with 90% coupling efficiency, but without capping (**Figure 3.68**). After each round of coupling, 10% of the chains are now one nucleotide short. Unfortunately, such short chains will continue to grow. By the end of the synthesis, we would have a 59% yield of the pentanucleotide product contaminated by a variety of tetranucleotides. This mixture would prove difficult to separate by chromatography. However, if we cap the 5′ hydroxyl after each round, the yield would still be 59%, but most of the capped molecules would be markedly different in length from the desired pentanucleotide.

After each round, the unreacted 5′-hydroxyl groups, usually less than 1%, must be capped off by acetylation with acetic anhydride in pyridine (**Figure 3.69**). The reaction is catalyzed by 4-dimethylaminopyridine—a powerful acyl transfer catalyst—and proceeds via an activated *N*-acylpyridinium intermediate. These acetylation conditions are even more efficient than phosphite formation. When used in conjunction with a stoichiometric base such as triethylamine, 10 mol% 4-dimethylaminopyridine can accelerate acylation reactions by up to four orders of magnitude.

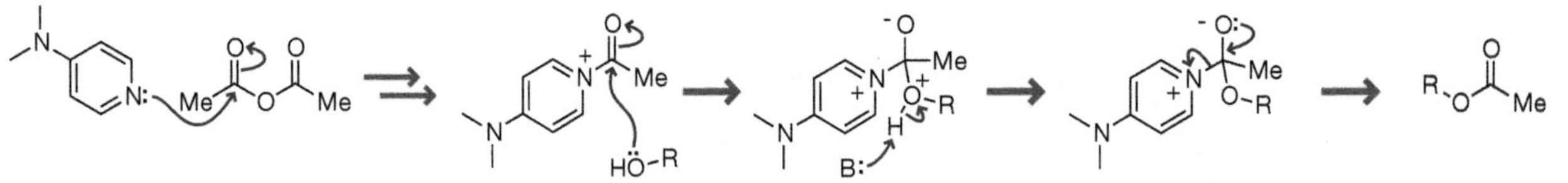

Figure 3.69 Nucleophilic catalysis. 4-Dimethylaminopyridine catalyzes the acylation of naked 5′-hydroxyl groups through nucleophilic catalysis, a mechanistic strategy common to many enzymes.

Oxidation of unstable phosphites generates stable phosphates

The final step of each cycle is to oxidize the labile phosphite groups to robust phosphodiester groups with iodine in aqueous tetrahydrofuran (THF) (**Figure 3.70**). The reaction mixture includes 2,6-dimethylpyridine (2,6-lutidine) to soak up hydroiodic acid. Phosphites are easy to oxidize under these relatively mild conditions. Even though the conditions are mild, the reaction is still fast and is complete in less than 30 seconds.

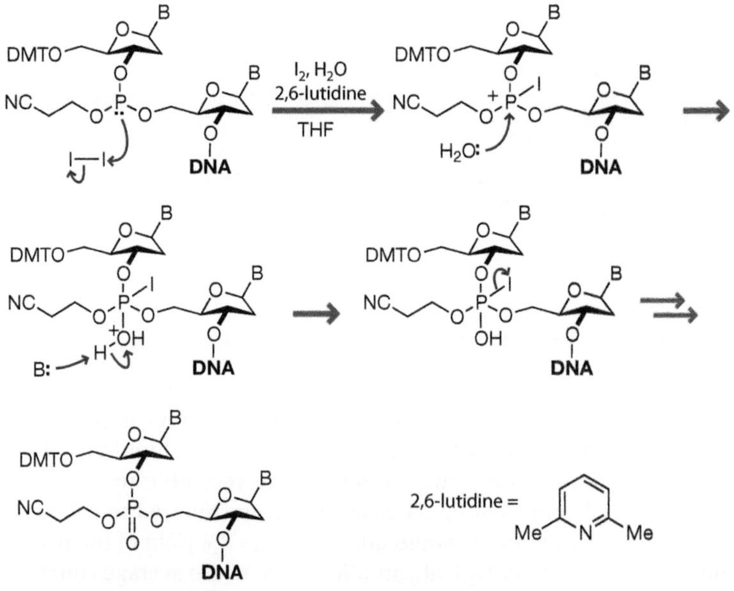

Figure 3.70 Phosphite esters are readily oxidized by a wide variety of reagents. Iodine oxidizes labile phosphites via pentavalent phosphorane intermediates.

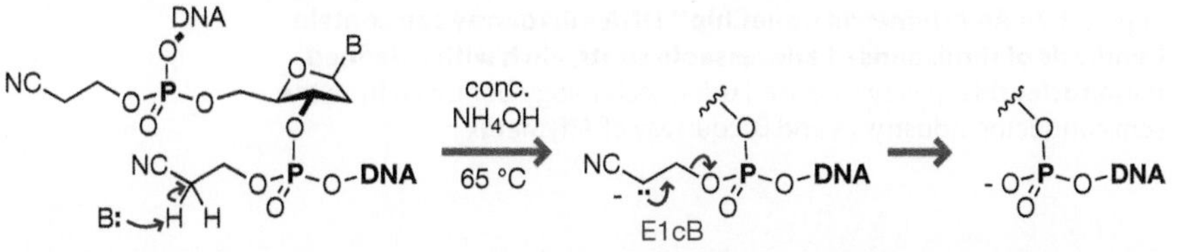

Figure 3.71 E1cB elimination mechanism for the deprotection of cyanoethyl esters.

Aqueous ammonium hydroxide cleaves and deprotects synthetic DNA

After the synthesis is complete, the cartridge containing the solid support is sealed in a sample vial containing concentrated aqueous ammonium hydroxide (17 M). The reaction is heated for 2 hours to effect nucleophilic cleavage of the base protecting groups, elimination of the cyanoethyl groups, and aminolysis of the ester linker. The ammonium hydroxide deprotection removes the cyanoethyl groups by an E1cB mechanism (**Figure 3.71**). This elimination mechanism involves the formation of a nitrile-stabilized carbanion followed by elimination of the phosphate anion leaving group.

Microarrays of DNA facilitate screening

DNA microarrays have been at the vanguard of genomic analysis. A DNA microarray can have thousands of different oligonucleotide probes immobilized in known positions on a small chip that can be digitally imaged to detect the binding of fluorescent DNA molecules. DNA microarrays exemplify the ingenuity and breadth of thinking in chemical biology. The simplest and cheapest microarrays are produced from collections of purified oligonucleotides by robotic microspotters. However, inkjet technology is now being used to create arrays of DNA molecules with extraordinary resolution. The most advanced method for the production of DNA microarrays involves the synthesis of the DNA molecules on a glass chip with the use of techniques borrowed from the semiconductor industry.

Integrated circuits are produced with a technique known as photolithography. The basic idea behind silicon-based integrated circuits is that doped silicon is a conductor, whereas silicon dioxide is an insulator. Thus, patterns of doped silicon on silicon dioxide can function as coupled electronic devices. Photolithographic etching involves four basic steps. The first step is to spread a thin, uniform layer of photoreactive compound, called photoresist, to the surface of an SiO_2 water. Next, a glass mask is placed over the surface and the masked wafer is irradiated with ultraviolet light (**Figure 3.72**). One of the earliest versions of photoresist contains diazonaphthoquinone. The photoresist exposed to the ultraviolet radiation undergoes a photochemical reaction called the Wolff rearrangement to generate a highly reactive ketene that generates a water-soluble carboxylate on exposure to hydroxide anion (**Figure 3.73**).

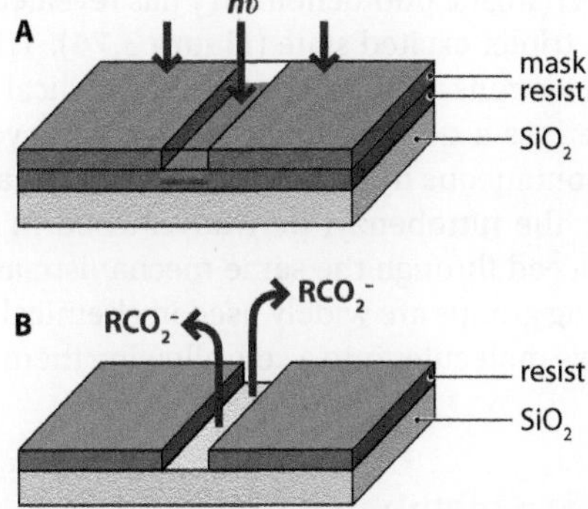

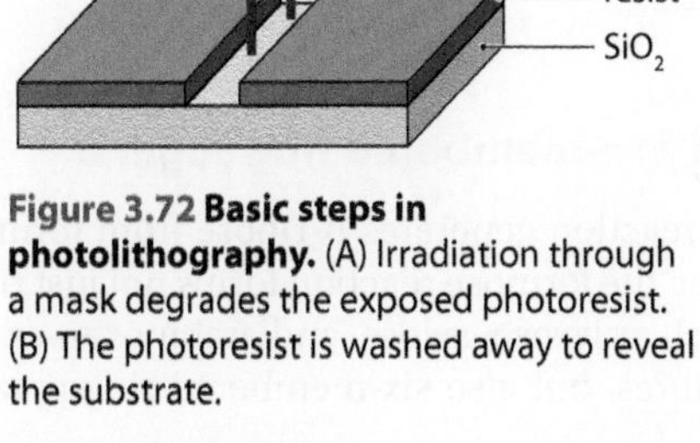

Figure 3.72 Basic steps in photolithography. (A) Irradiation through a mask degrades the exposed photoresist. (B) The photoresist is washed away to reveal the substrate.

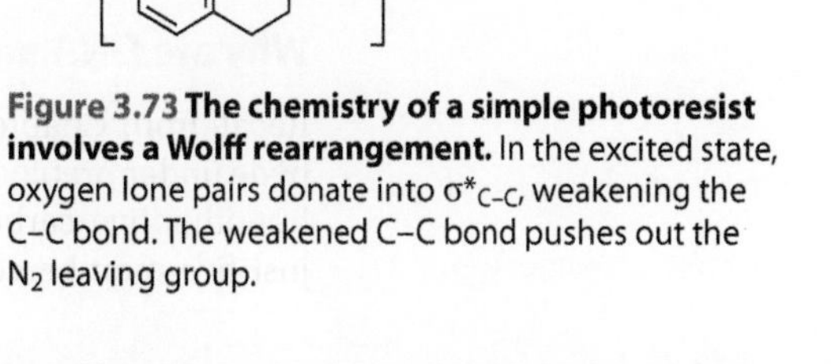

Figure 3.73 The chemistry of a simple photoresist involves a Wolff rearrangement. In the excited state, oxygen lone pairs donate into σ^*_{C-C}, weakening the C–C bond. The weakened C–C bond pushes out the N_2 leaving group.

A

B

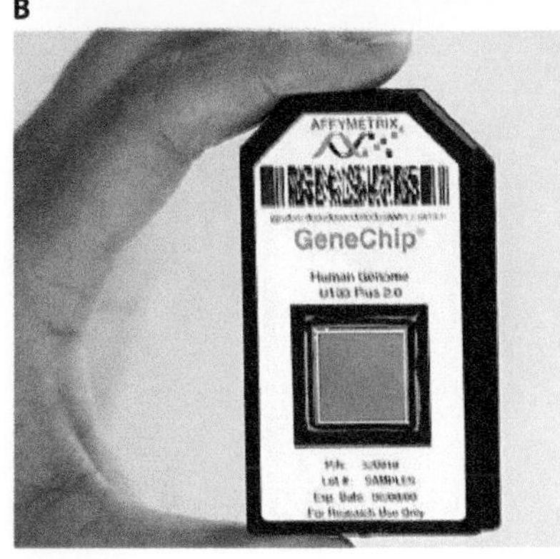

Figure 3.74 An Affymetrix GeneChip™ DNA microarray can contain hundreds of thousands of addressable spots, each with a defined oligonucleotide. It is synthesized using technology borrowed from the semiconductor industry. (A and B, courtesy of Affymetrix.)

When the surface is rinsed with aqueous base, the photoexposed areas wash away, whereas the unexposed regions remain impervious to the aqueous solution. Finally, the exposed SiO_2 surface is etched with hydrofluoric acid to produce the final pattern.

Similar masking and irradiation technology allows the spatially addressable synthesis of oligonucleotides on a glass surface (**Figure 3.74**). A photolabile 5′-*O*-nitrobenzylcarbonate protecting group is used instead of acid-labile 5′-DMT groups. The molecules are attached to the glass surface by using the same types of linkers developed for CPG. Synthesis typically involves iterative deprotection, phosphoramidite coupling, and phosphite oxidation (**Figure 3.75**). Initially, using a specially designed mask, certain areas of the surface are selectively deprotected and then coupled with dA phosphoramidites. With the use of a second specially designed mask, a different set of areas are selectively deprotected and then coupled to dC phosphoramidites. By repeating this process over and over with increasingly complex masks, arrays of hundreds of thousands of DNA molecules can be created on spots only 5 μm wide (see Figure 3.75). When a sample containing fluorescently labeled DNA molecules is washed over the chip, a highly sensitive charge-coupled device (CCD) detector, like those found in sensitive microscopes or telescopes, can be used to detect hybridization. The data are digitized, stored, and analyzed with a computer.

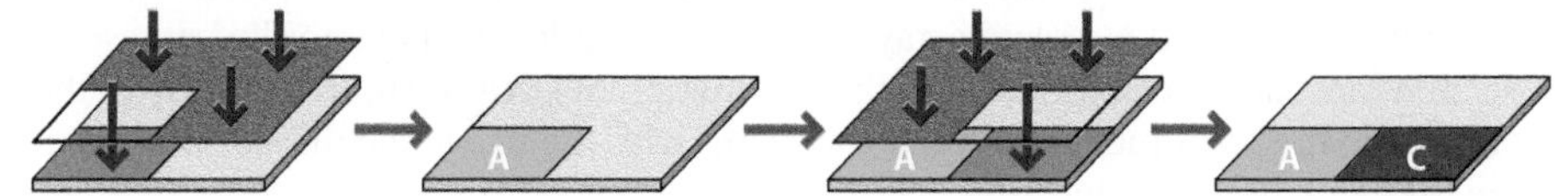

Figure 3.75 Masking. Iterative photodeprotection and coupling chemistry adds new nucleotides to precisely defined positions as determined by a patterned mask.

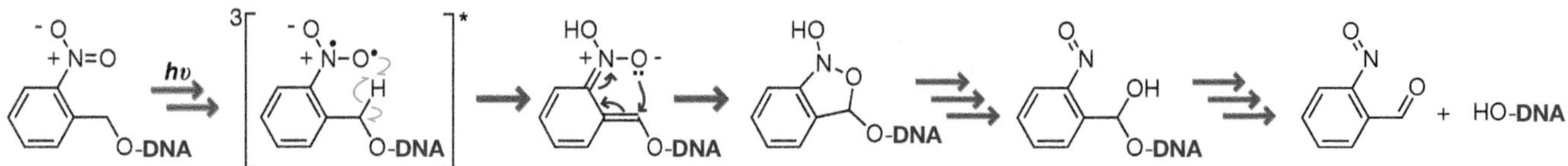

Figure 3.76 Photodeprotection. Excitation of nitrobenzyl derivatives generates a triplet excited state intermediate. The intermediate rearranges to generate a labile hemiacetal.

The key to using photoresist technology for the synthesis of oligonucleotide arrays lay in the availability of photolabile protecting groups, first developed by the synthetic organic chemist Robert B. Woodward. Mechanistic photochemistry has revealed that the irradiation of nitroarenes generates a triplet excited state (**Figure 3.76**). Triplets have unpaired spins on the most reactive electrons and must proceed via radical reactions. Hydrogen-atom transfer then generates a quinoid intermediate that cyclizes to form an acetal with full aromaticity. Spontaneous hydrolytic cleavage of the acetal proceeds via a hemiacetal intermediate. The nitrobenzyl derivatives used in DNA synthesis have more substituents, but proceed through the same mechanism as that shown in Figure 3.76. Photolabile protecting groups are widely used in chemical biology to produce "caged" versions of bioactive molecules into a cell, allowing them to be instantly released by light.

Why are DNA and RNA made up of five-membered ring sugars?

Recall from Chapter 2 that the formose reaction generates D-ribose from formaldehyde under prebiotic conditions. However, the formose reaction forms not just ribose, but other five-carbon sugars as well—such as lyxose, xylose, and arabinose—and not just five-membered ring furanose structures, but also six-membered ring pyranose structures (**Figure 3.77**). Why did Nature choose a five-membered ring ribofuranose backbone for DNA and RNA? It is tempting to speculate that the ribofuranose backbone is uniquely suited for Watson–Crick base pairing. That would be wrong.

A wide range of other six-membered ring nucleoside analogs have been synthesized and assembled into octanucleotides using phosphoramidite chemistry. Surprisingly, most octanucleotides based on six-membered ring sugars were capable of duplex formation. The same is also true for deoxynucleotides based on six-membered ring sugars. However, oligonucleotides based on the six-carbon sugars formed weak duplexes, and those based on glucopyranose did not form duplexes at all. The bigger surprise is that alternative pentopyranose backbones form duplexes that are more stable than the naturally occurring duplex (**Figure 3.78**).

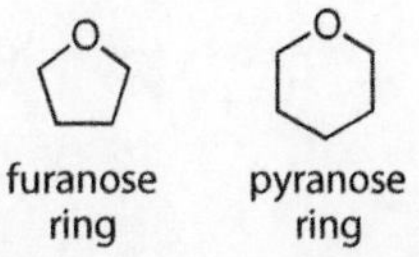

furanose ring

pyranose ring

Figure 3.77 Furanose and pyranose rings.

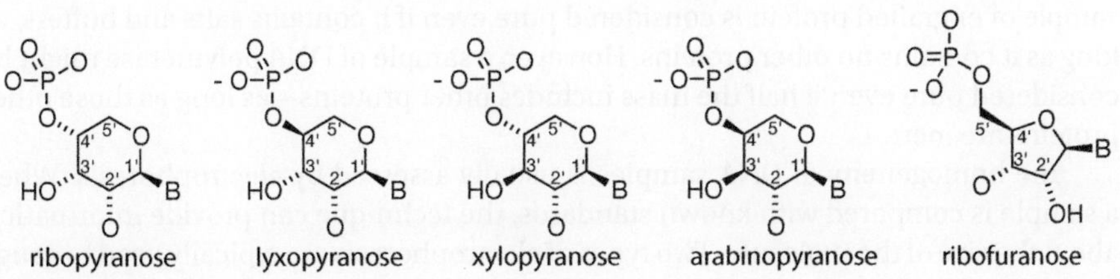

ribopyranose lyxopyranose xylopyranose arabinopyranose ribofuranose

4'-AAAAAAAA-2' 2'-TTTTTTTT-4'	T_m (°C)
ribopyranose	**46.0**
lyxopyranose	**51.0**
xylopyranose	**47.3**
arabinopyranose	**79.7**
RNA (ribofuranose)	**16.3**

Figure 3.78 Comparing oligonucleotides from alternate universes. Alternative double-stranded oligonucleotides (based on monosaccharides other than ribofuranose), assembled using phosphoramidite chemistry, are more stable than normal 5′–3′ RNA.

Clearly, Nature's choice of ribofuranose was not based solely on the stability of duplex RNA or DNA, because oligonucleotides based on other polyols can form much more stable duplexes. Unlike furanosyl RNA duplexes, pyranosyl RNA duplexes are predicted to have only a slight helical twist—about one-third of the pitch that one finds in DNA duplexes (**Figure 3.79**). Ultimately, the superior fitness of our D-ribofuranose DNA may be attributable to lability—allowing it to unfold for replication or transcription—and the deep grooves it presents to other molecules, such as transcription factors.

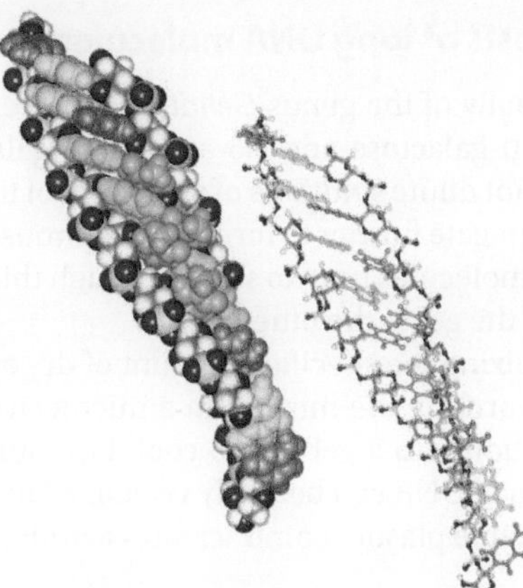

Figure 3.79 Is ribofuranose special? Two different views of a model of a pyranosyl-RNA 4′-CGAATTCG-2′ duplex, showing the subtle twist and shallow grooves. (Courtesy of Romain M. Wolf.)

3.7 SEPARATION OF DNA MOLECULES BY ELECTROPHORESIS

Scientists use different criteria for the purity of biological macromolecules versus small, organic molecules

Anyone who studies or creates molecules must have a way to establish the identity (that is, the chemical structure) of the sample. For small molecules these properties can be rigorously established by using a combination of spectroscopic methods, usually ^{1}H NMR, ^{13}C NMR, infrared spectroscopy, and sometimes UV-vis spectroscopy, together with chromatography techniques. Unfortunately, these spectroscopic techniques are challenging to apply to massive bio-oligomers such as oligonucleotides, proteins, and oligosaccharides; for such molecules, spectroscopic techniques can be

Figure 3.80 **A jelly is extracted from dried *Gelidium*, a seaweed, with boiling water.** It can be extruded to make edible noodles for the Japanese dish tokoroten or purified to make laboratory grade agar or agarose. (Top, courtesy of M.D. Guiry, AlgaeBase; bottom, courtesy of Studio Eye, Corbis Images.)

less informative or very time-consuming to yield structural information. Therefore a variety of indirect or destructive methods, to be discussed later, are used to establish the identity of biological oligomers. Mild methods for suspending large molecules in the gas phase have revolutionized mass spectrometry and its application to biological macromolecules. Even though mass spectrometry is intrinsically destructive, modern instruments require only minute quantities of sample.

Most chemists and biochemists spend more time purifying their molecules than they spend making them. Two of the key commandments of enzymology noted by the Nobel prize winner Arthur Kornberg relate to purity. First, do not waste clean thinking on dirty enzymes. Second, do not waste clean enzymes on dirty substrates. (Yes, the converse to the first rule is also considered true.) Chemists usually use the term purity to mean mass fraction. A sample is 99% pure when 99% of the sample mass is one compound. In contrast, biologists usually use the term purity to mean homogeneity. A sample of engrailed protein is considered pure even if it contains salts and buffers, as long as it contains no other proteins. However, a sample of DNA polymerase might be considered pure even if half the mass includes other proteins—as long as those other proteins are inert.

The homogeneity of DNA samples is usually assessed by electrophoresis. When a sample is compared with known standards, the technique can provide information about the size of the molecule. Two types of electrophoresis are typically used to gauge homogeneity: gel electrophoresis and capillary electrophoresis. The idea behind electrophoresis is simple: *negatively charged anions move toward a positively charged anode.* A gel matrix serves as a physical barrier to the motion of molecules. The rate at which biomolecules move through a gel depends mostly on molecular size, but also on shape and charge. There are three basic types of gel electrophoresis: agarose gel electrophoresis is used to separate large pieces of DNA, polyacrylamide sequencing gels are used to separate short pieces of DNA, and sodium dodecyl sulfate-polyacrylamide gel electrophoresis (SDS-PAGE) is used to separate proteins. Polyacrylamide DNA-sequencing gels have been largely supplanted by capillary electrophoresis.

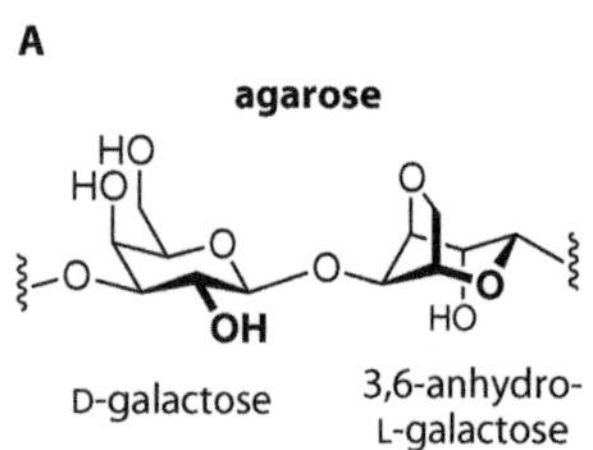

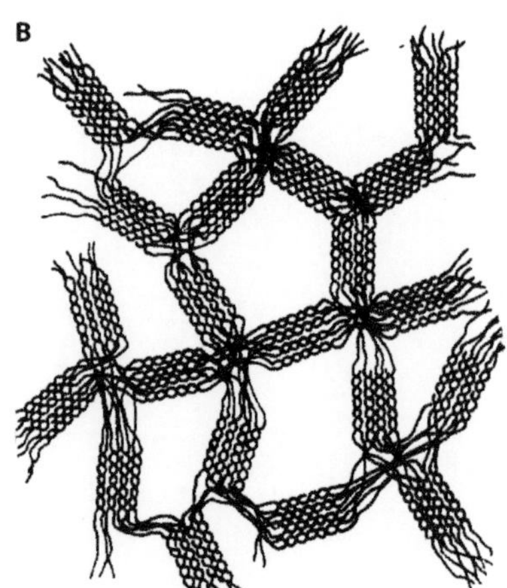

Figure 3.81 **The structure of agar.** (A) Agarose is a polymer made up of a repeating disaccharide unit. (B) Depiction of an agarose gel network. (B, from S. Arnott et al., *J. Mol. Biol.* 90: 269–284, 1974. With permission from Elsevier.)

Agarose gel is used for electrophoresis of long DNA molecules

Agarose is isolated from red seaweeds, usually of the genus *Gelidium* (**Figure 3.80**). It is a repeating disaccharide made up of D-galactose and 3,6-anhydro-L-galactose. Agarose chains form a double helix. When hot dilute solutions of agarose cool to room temperature the agarose double helices aggregate further to form macroporous three-dimensional networks (**Figure 3.81**). DNA molecules have to snake through this complex network as they move from one end of the gel to the other.

Agarose gels are usually prepared by mixing the specified amount of dry agarose powder with buffer solution (**Table 3.2**), warming the mixture in a microwave oven until dissolved, and then pouring the solution into a gel box to cool. Ethidium bromide is included in the gel so that the bands of DNA can be easily visualized on top of a fluorescent light. As the gel cools, a removable plastic "comb" creates a uniform line

Table 3.2 **Agarose content for separation of various lengths of DNA.**

Agarose (%)	DNA size (bp)
0.3	60,000–5000
0.6	20,000–1000
0.7	10,000–800
0.9	7000–500
1.2	6000–400
1.5	4000–200
2.0	3000–100

Problem 3.16

Seaweed from the genus *Gracilaria* is easier to grow than seaweed from the genus *Gelidium*, but the agar is of lower quality as a result of the presence of ionic 1,4-L-galactose-6-sulfate residues within the agarose polymer. Treatment with alkali has been shown to improve the quality of agarose from *Gracilaria*. Suggest a plausible arrow-pushing mechanism for the reaction.

1.5 M NaOH, H_2O, 80 °C, 2 h → agarose

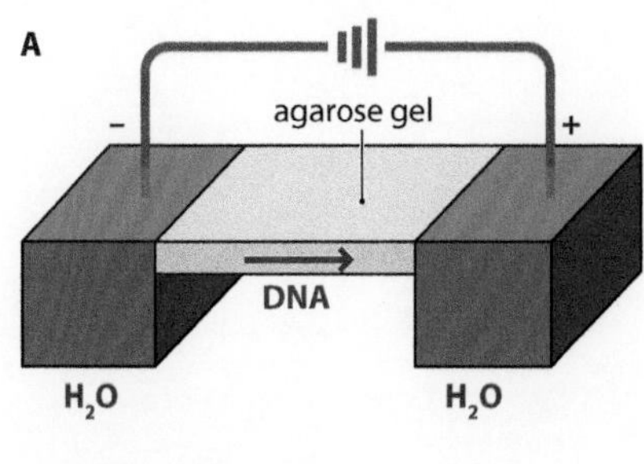

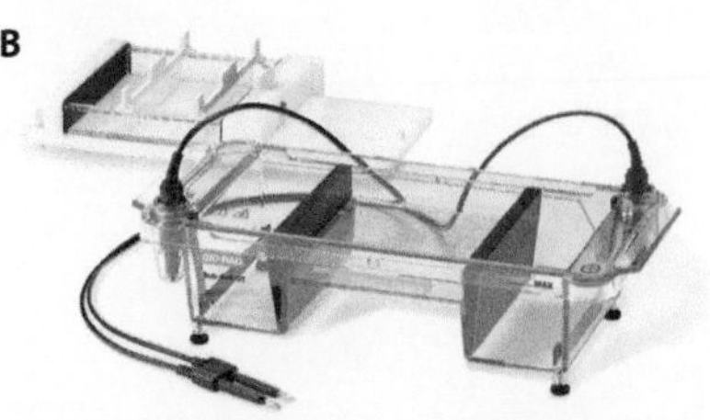

Figure 3.82 A horizontal gel rig. The agarose gel sits between two buffer tanks, one at negative potential and the other at positive potential. Negatively charged molecules such as DNA move slowly through the gel toward the positive buffer. (B, courtesy of Bio-Rad Laboratories, Inc., 2012.)

of rectangular holes in the gel for loading the samples. Buffer solutions on each end of the gel (**Figure 3.82**) provide uniform contact with the positive and negative terminals of a power supply. In the presence of the electric field, the negatively charged DNA molecules move toward the buffer pool held at positive potential.

The distance traveled by DNA in a gel depends on both applied voltage and duration, so every gel must include a "marker" lane, usually containing a mixture of DNA molecules with known size (**Figure 3.83**). Gels are always oriented with the faster migrating bands toward the bottom and the more slowly migrating bands toward the top.

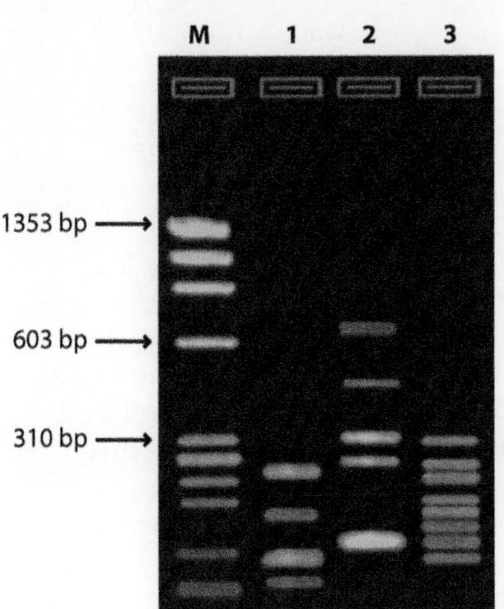

Figure 3.83 Shorter DNA moves faster. Shorter pieces of DNA migrate faster through an agarose gel. Lane M contains a mixture of DNA size "markers": 72, 118, 194, 234, 271, 281, 310, 603, 872, 1078, and 1353 bp. Lanes 1–3 contain PCR amplification products from different clinical specimens of *Mycobacterium tuberculosis*. The DNA sample is loaded into "wells," which are the rectangular boxes at the top of each lane. (From A.M. Kearns et al., *J. Clin. Pathol.* 53: 122–124, 2000. With permission from BMJ Publishing Group Ltd.)

The shape and rigidity of DNA molecules also affect migration through agarose. Tautly supercoiled plasmids migrate faster than plasmids with a single-strand cut (called a nick), which allows the plasmid to relax in a manner similar to that of a topoisomerase I-DNA complex. A double-strand cut linearizes the plasmid and also reduces the rate of migration, but the linearized plasmid still migrates faster than the relaxed plasmid (**Figure 3.84**).

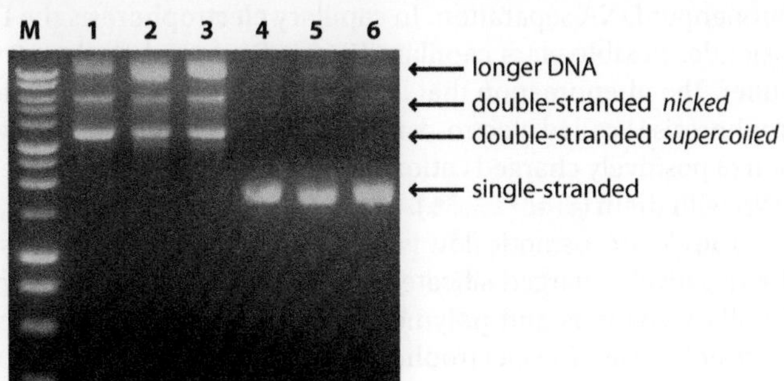

Figure 3.84 Both size and shape affect DNA mobility. A single-stranded plasmid moves faster than a double-stranded plasmid of equal length. The migration of the double-stranded plasmid depends on its topology and shape. A nicked, relaxed plasmid migrates slowly relative to the supercoiled plasmid, which is more compact (lanes 1–3). Lane M contains the DNA size markers. In the other lanes, the DNA is approximately 6200 bases (single-stranded DNA) or base pairs (double-stranded DNA). (Courtesy of Gregory Weiss and Phillip Y. Tam.)

Problem 3.17

Six samples of identical plasmid DNA were treated with DNA topoisomerase I, which allows the supercoiled plasmid to relax. Each sample was treated with a different amount of topoisomerase inhibitor. Which band corresponds to supercoiled plasmid and which band to fully relaxed plasmid? Which sample had the least inhibitor and which had the most inhibitor?

sample 1 2 3 4 5 6

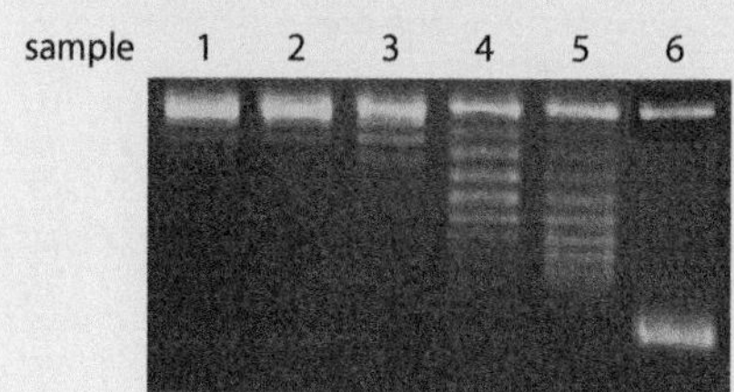

(Adapted from I. Larosche et al., *J. Pharmacol. Exp. Ther.* 321: 526–535, 2007. With permission from the American Society for Pharmacology and Experimental Therapeutics.)

The DNA bands are only visible when the ethidium-laced gel is visualized over an ultraviolet light. Ethidium is excited by fluorescent light, but water molecules normally quench the excited state of ethidium before it can emit visible photons. However, when ethidium is intercalated between the base pairs of DNA, the aqueous quenching is suppressed and the DNA/ethidium complex shows strong orange fluorescence (**Figure 3.85**).

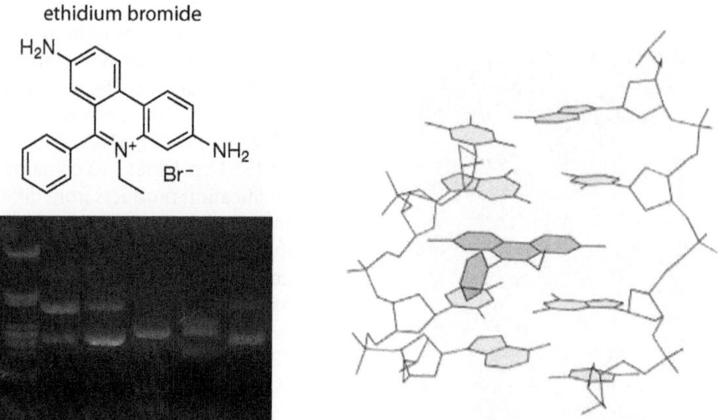

Figure 3.85 Ethidium bromide. Ethidium bromide is a fluorescent DNA intercalator that is used to stain DNA in agarose gels. Ethidium is a mutagen, but is sold as homidium to treat trypanosomal infections in cattle. (Gel image courtesy of Markus Nolf; model on right adapted from K.V. Miroshnychenko and A.V. Shestopalova, *Int. J. Quantum Chem.* 110: 161–176, 2010. With permission from John Wiley and Sons.)

Capillary electrophoresis is used for analytical separation of short DNA molecules

The first methods of DNA sequencing made use of thin vertically oriented, polyacrylamide gels sandwiched between glass plates. Polyacrylamide gels readily resolve short DNA molecules (up to a few hundred bases) that differ by just a single base. Unfortunately, these DNA-sequencing gels are laborious to prepare. The human genome project spurred the development of capillary electrophoresis for automated high-throughput DNA separation. In capillary electrophoresis the DNA is separated within a slender flexible glass capillary (inner diameter less than 0.1 mm) filled with ionic buffer. The phenomenon that leads to separation of DNA molecules in capillary electrophoresis is called electro-osmotic flow. Electro-osmotic flow arises from the migration of positively charged cations toward the negatively charged cathode, dragging the DNA with them (**Figure 3.86**).

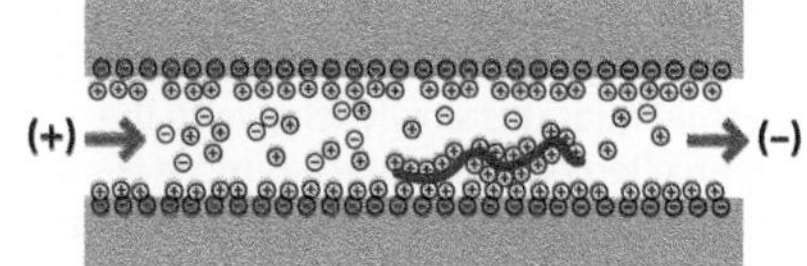

Figure 3.86 Electro-osmotic flow. Buffer ions and DNA molecules (red) in a thin glass capillary column migrate toward the negatively charged cathode.

The electro-osmotic flow is also influenced by interactions between the buffer and the negatively charged silicate groups (Si–O$^-$) on the capillary surface (Figure 3.86). Capillary coatings and polymer additives improve the resolving ability of capillary electrophoresis. The electrophoretic movement of short (less than 500 nucleotides in length) single-stranded DNA molecules through glass microcapillaries is extremely

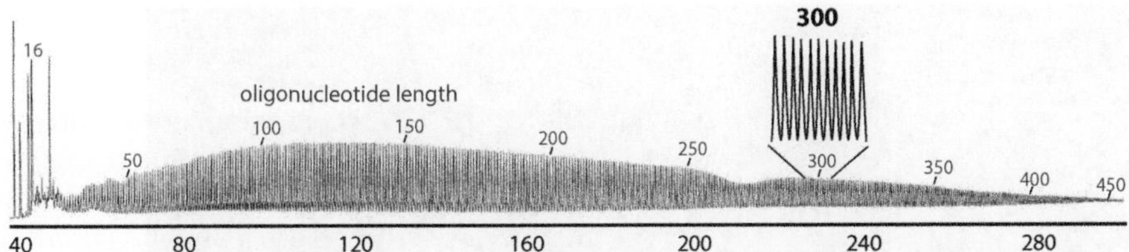

Figure 3.87 High resolution. Capillary electrophoresis leads to well-resolved peaks for a series of fluorescently labeled oligonucleotides $(dT)_n$ ranging in length from 16 to 500 nucleotides. (Adapted from T. Manabe et al., *Anal. Chem.* 66: 4243–4252, 1994. With permission from the American Chemical Society.)

sensitive to size and charge. Oligonucleotides that differ by a single base are easily resolved, even when the oligonucleotides are hundreds of nucleotides in length (**Figure 3.87**).

Because the capillaries used in capillary electrophoresis are fine, the amount of sample needed is necessarily small. To detect those small amounts of material, the analytes must be highly fluorescent. A precisely aimed laser detects the fluorescent molecules before they escape from the end of the capillary (**Figure 3.88**). Most of the techniques for the highly sensitive detection of DNA involve fluorescently labeled DNA incorporated during DNA polymerization, as described next.

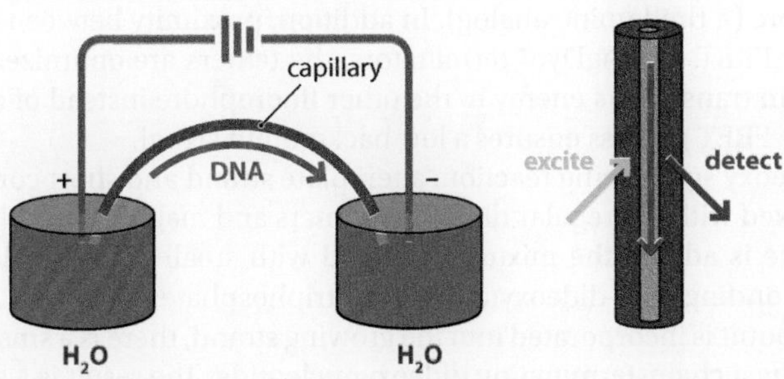

Figure 3.88 Capillary electrophoresis. Fluorescently labeled DNA molecules are dragged through a flexible capillary and detected with a laser.

DNA dideoxy sequencing capitalizes on the tolerance of DNA polymerase

The human DNA polymerases that replicate the human genome are fastidious, checking before and after they add each nucleotide subunit. In contrast, viral polymerases such as HIV reverse transcriptase are less meticulous in their rush to outpace the host machinery. HIV reverse transcriptase will accept nucleotides with alternative bases such as inosine with a 2-keto group, missing 3′-hydroxyl groups, or even bizarre substitutions such as a 3′ azido group (**Figure 3.89**). If a polymerase incorporates a nucleotide that lacks a 3′-hydroxyl, it leads to chain termination because the 3′-hydroxyl group is the nucleophile that makes each stable phosphodiester bond. The ability of 2′,3′-dideoxynucleotides to induce chain termination by polymerases was central to the sequencing of the human genome.

In 1977 the first methods for DNA sequencing were described: Maxam–Gilbert sequencing and Sanger dideoxy sequencing. Dideoxy sequencing involves DNA synthesis with a promiscuous DNA polymerase and fluorescently labeled (as indicated with an asterisk) chain-terminating 2′,3′-dideoxynucleotidyl triphosphates

ddI ddC AZT

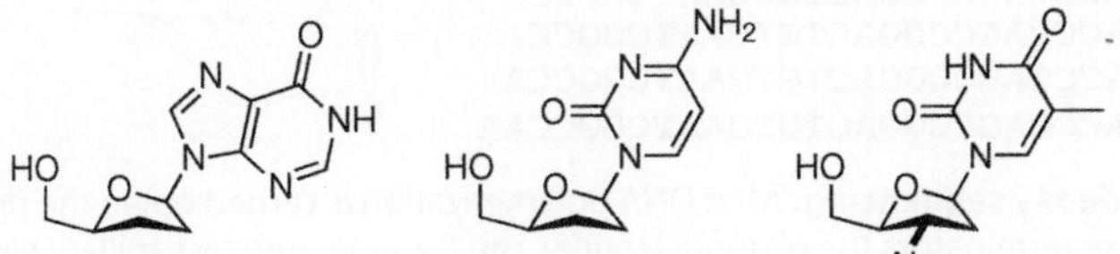

Figure 3.89 Dideoxynucleosides as drugs. These nucleosides without the 3′ hydroxyl are converted into nucleotidyl triphosphates by human enzymes. The triphosphates are eschewed by human DNA polymerases but incorporated by HIV reverse transcriptase, leading to chain termination. They are the basis for antiretroviral drug therapy.

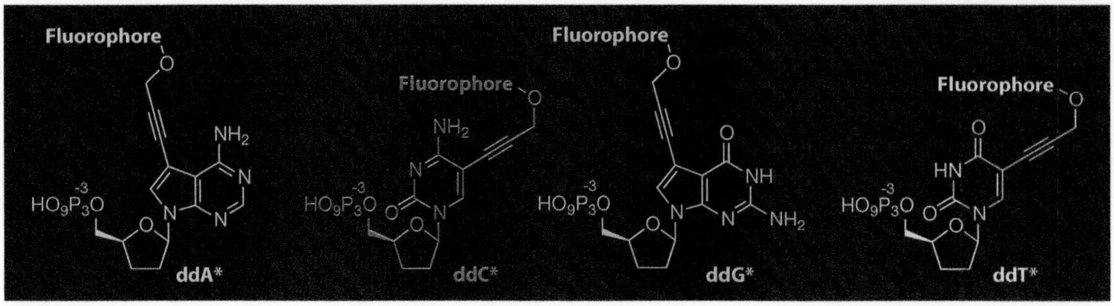

Fluorophore
ddA*
Fluorophore
ddC*
Fluorophore
ddG*
Fluorophore
ddT*

Figure 3.90 Fluorescent terminators for DNA sequencing. Fluorescence resonance energy transfer (FRET) fluorophores are attached to the bases of the BigDye™ dideoxy terminators.

(abbreviated ddN*): ddA*, ddC*, ddG*, and ddT* (**Figures 3.90 and 3.91**). Importantly, each of these dideoxynucleotides is also substituted with a fluorophore that emits light at a unique wavelength. The most advanced fluorescent dideoxy terminators take advantage of fluorescence resonance energy transfer (FRET). All four dideoxynucleotidyl triphosphates incorporate a fluorescence acceptor based on the dye carboxyfluorescein, which is excited by an argon laser. In each of the four dideoxy terminators the fluorescein is attached to a different acceptor fluorophore with a unique emission wavelength: approximately, 542 nm (ddG), 568 nm (ddA), 596 nm (ddT), and 622 nm (ddC). FRET is most efficient when the emission maximum of the first fluorophore (carboxyfluorescein) is matched to the excitation maximum of the second fluorophore (a rhodamine analog). In addition, proximity between fluorophores is required for FRET. For BigDye™ terminators, the tethers are optimized so that carboxyfluorescein transfers its energy to the other fluorophore instead of emitting light at 517 nm. The FRET process ensures a low background signal.

In the dideoxy sequencing reaction, a template strand and short complementary primer are mixed with the regular dNTP monomers and magnesium chloride. Before the polymerase is added, the mixture is doped with small amounts (less than 1%) of the corresponding 2′,3′-dideoxynucleotidyl triphosphates (**Figure 3.92**). As each nucleotide subunit is incorporated into the growing strand, there is a small probability of incorporating a chain-terminating dideoxynucleotide. The result is a series of fragments, each ending with a distinctive fluorescent nucleotide that reports on the last base in the sequence. When the mixture is analyzed by capillary electrophoresis, the DNA sequence can be read directly from the capillary electrophoresis trace.

emits at 596 nm
energy transfer
excite at 377 nm
517 nm
ddT*

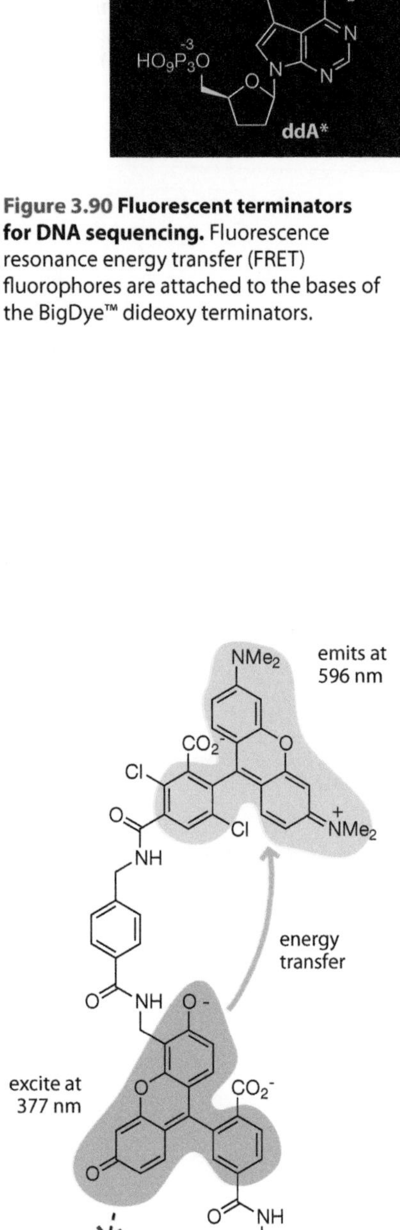

Figure 3.91 FRET requirements. FRET is very efficient when the emission of one fluorophore is matched with the excitation of another fluorophore, and when the two fluorophores are close in space.

Large-scale sequencing methods avoid the need for electrophoresis

The desire to customize medicine to an individual's genome has driven the development of other technologies for cost-effective full-genome sequencing. Each innovation

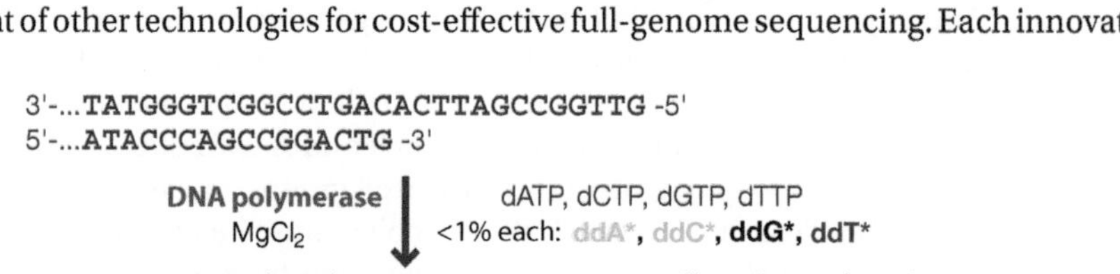

3'-...TATGGGTCGGCCTGACACTTAGCCGGTTG -5'
5'-...ATACCCAGCCGGACTG -3'
DNA polymerase
$MgCl_2$
dATP, dCTP, dGTP, dTTP
<1% each: ddA*, ddC*, ddG*, ddT*
truncated products
capillary electrophoresis trace
5'-...ATACCCAGCCGGACTGT*
5'-...ATACCCAGCCGGACTGTG*
5'-...ATACCCAGCCGGACTGTGA*
5'-...ATACCCAGCCGGACTGTGAA*
5'-...ATACCCAGCCGGACTGTGAAT*
5'-...ATACCCAGCCGGACTGTGAATC*
5'-...ATACCCAGCCGGACTGTGAATCG*
5'-...ATACCCAGCCGGACTGTGAATCGG*
5'-...ATACCCAGCCGGACTGTGAATCGGC*
5'-...ATACCCAGCCGGACTGTGAATCGGCC*
5'-...ATACCCAGCCGGACTGTGAATCGGCCA*
5'-...ATACCCAGCCGGACTGTGAATCGGCCAA*
5'-...ATACCCAGCCGGACTGTGAATCGGCCAAC*
5' T G A A T C G G C C A A C 3'

Figure 3.92 Dideoxy sequencing. After DNA polymerization is carried out in the presence of BigDye™ dideoxy terminators the oligonucleotides can be separated by capillary electrophoresis. The sequence can be directly read from the trace.

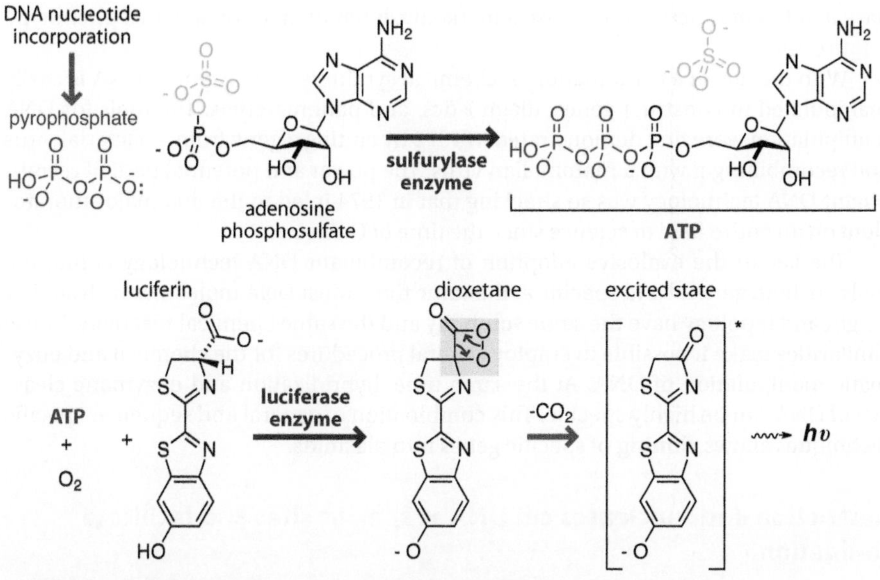

DNA nucleotide incorporation

pyrophosphate

adenosine phosphosulfate

sulfurylase enzyme

ATP

luciferin

luciferase enzyme

ATP + O_2

dioxetane

$-CO_2$

excited state

$h\nu$

Figure 3.93 Pyrosequencing involves the very sensitive detection of pyrophosphate through a coupled enzyme assay. Pyrophosphate is converted into ATP by sulfurylase; the ATP serves as a substrate for luciferase, which generates light.

Figure 3.94 The enzyme luciferase was originally isolated from the common eastern firefly *Photinus pyralis*. Assays coupled to luciferase are commonly employed because they involve low levels of background light that lead to high sensitivity. (Courtesy of Gail Shumway. With permission from Getty Images.)

in DNA sequencing has expanded the throughput of the technique; in turn, this has extended the field into new technological realms. There are now a variety of methods for sequencing capable of ultra-high throughput, none of which involve capillary electrophoresis. One such method, called **pyrosequencing**, is based on the very sensitive detection of pyrophosphate, the by-product that is formed when polymerase adds a nucleotide subunit to DNA. In pyrosequencing, the DNA is immobilized in a tiny (picoliter-volume) well along with all of the necessary reagents for the assay, with one exception—the dNTPs used by DNA polymerase during polymerization. Each of the four dNTPs is introduced one at a time until pyrophosphate is detected.

Pyrosequencing takes advantage of two sequential enzyme-catalyzed reactions of pyrophosphate, ultimately leading to the production of light (**Figure 3.93**). The first reaction involves pyrophosphate and adenosine phosphosulfate, forming ATP. In the presence of ATP and oxygen, the enzyme luciferase catalyzes the transformation of luciferin to a strained dioxetane that undergoes a [2+2] retro-cycloaddition that is forbidden by rules of orbital symmetry unless it produces an excited state. The excited state of the product then relaxes to the ground state by emitting a photon of light (**Figure 3.94**). A very sensitive camera is used to observe the flashes of light that indicate the successful incorporation of a specific DNA base. A single CCD camera can image thousands to hundreds of thousands of beads at once, offering massive scalability of DNA sequencing.

3.8 RECOMBINANT DNA TECHNOLOGY

Molecular biology connects DNA molecules to biological phenotypes

New tools can redefine or even revolutionize a scientific field. Biology, for example, has undergone a transition from a purely observational art to a science, leveraging the tools of synthesis for direct hypothesis testing. Biologists no longer have to wait for many generations of an organism to grow before finding a mutant with a particular attribute. Instead, biological synthesis equips the experimenter with tools to tinker under the hood of the organism—to make changes to molecular structure and see what happens. Perhaps not surprisingly, most analogs of biological macromolecules begin with custom fabrication of DNA. When you alter DNA, you alter the corresponding mRNA, leading ultimately to an altered protein. In turn, this protein analog can

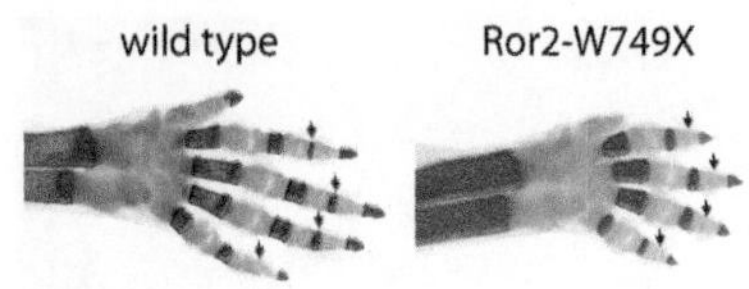

Figure 3.95 A revolution in biology. The synthesis approach to developmental biology targets a receptor protein encoded by the *ROR2* gene. Truncation of the mouse *ROR2* gene leads to the synthesis of a truncated protein, and ultimately to shortened fingers (arrows indicate missing bone sections). (Courtesy of Sigmar Stricker.)

have a different function and cause a markedly different, new, organismal phenotype (**Figure 3.95**).

With the use of a combination of chemical synthesis and enzymes, DNA is easily manipulated to construct genes, diagnostics, and nanomaterials. The tools for DNA manipulation were first demonstrated in 1972 by cutting a gene from a bacterial virus and recombining it with a mammalian virus. The power and potential peril of recombinant DNA technology was so shocking that in 1974 it led to the first major moratorium on an entire field of science since the time of Galileo.

The key to the explosive adoption of recombinant DNA technology is that the tools are both general and specific at the same time. Most DNA molecules with similar length and topology have the same solubility and the same chemical reactivity. Those similarities make it possible to employ general procedures for the chemical and enzymatic manipulation of DNA. At the same time, hybridization and enzymatic cleavage of DNA can be highly specific. This combination of general and sequence-specific techniques allows **cloning** of specific genes into plasmids.

Restriction endonucleases cut DNA at specific sites and facilitate re-ligation

Most restriction endonucleases consist of two identical subunits paired together as a homodimer (**Figure 3.96**). Because the subunits have identical active sites, restriction enzymes act on palindromic sequences of DNA. A palindrome has the same meaning when read forward and backward ("Able was I ere I saw Elba" or "I prefer Pi"); DNA palindromes consist of two complementary strands with the same sequence when read in the 5′ to 3′ direction (**Figure 3.97**). Restriction endonucleases are selective for specific sequences up to eight bases in length. One of the best-known of these, restriction endonuclease I from *E. coli* (EcoRI, for short), cuts DNA at the palindromic sequence GAATTC, at the shortest distance across the major groove making a staggered cut. Other common restriction enzymes make similar staggered cuts. A staggered cut leaves self-complementary "sticky ends." The sticky ends hybridize weakly to each other, allowing them to be rejoined by the enzyme DNA ligase.

Importantly, if you cut a segment of frog DNA with the enzyme NotI, the overhanging bases, 3′-CCGG-5′, will be complementary to human DNA that has also been cut with NotI. These sticky ends make it easy to ligate DNA cut from one genome, into DNA cut from any other genome. If you want to insert a newly synthesized gene into a double-stranded plasmid, restriction sites can provide a specific site for the insertion. Fortunately, a wide range of plasmids are commercially available that have been sequenced and mapped to show the sites at which common restriction endonucleases will cut.

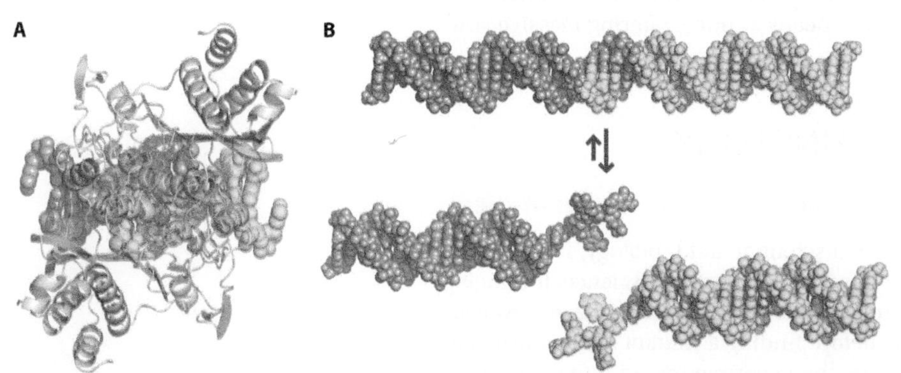

Figure 3.96 Staggered cuts leave sticky ends. (A) EcoRI makes staggered cuts at the two sites occupied by magnesium ions (green). (B) Making staggered cuts across the major groove leaves self-complementary sticky ends with phosphate groups at the dangling 5′ ends.

restriction enzyme	**targeted sequence**
EcoRI	3'-NNNCTTAAGNNN-5' 5'-NNNGAATTCNNN-3'
HindIII	3'-NNNTTCGAANNN-5' 5'-NNNAAGCTTNNN-3'
BamHI	3'-NNNCCTAGGNNN-5' 5'-NNNGGATCCNNN-3'
NotI	3'-NNCGCCGGCGNN-5' 5'-NNGCGGCCGCNN-3'
BglI	3'-NCGGNNNNNCCGN-5' 5'-NGCCNNNNNGGCN-3'
SmaI	3'-NNCCCGGGNN-5' 5'-NNGGGCCCNN-3'

Figure 3.97 Common restriction endonucleases make staggered cuts, indicated by the lines. Most leave sticky ends but a few, such as SmaI, do not.

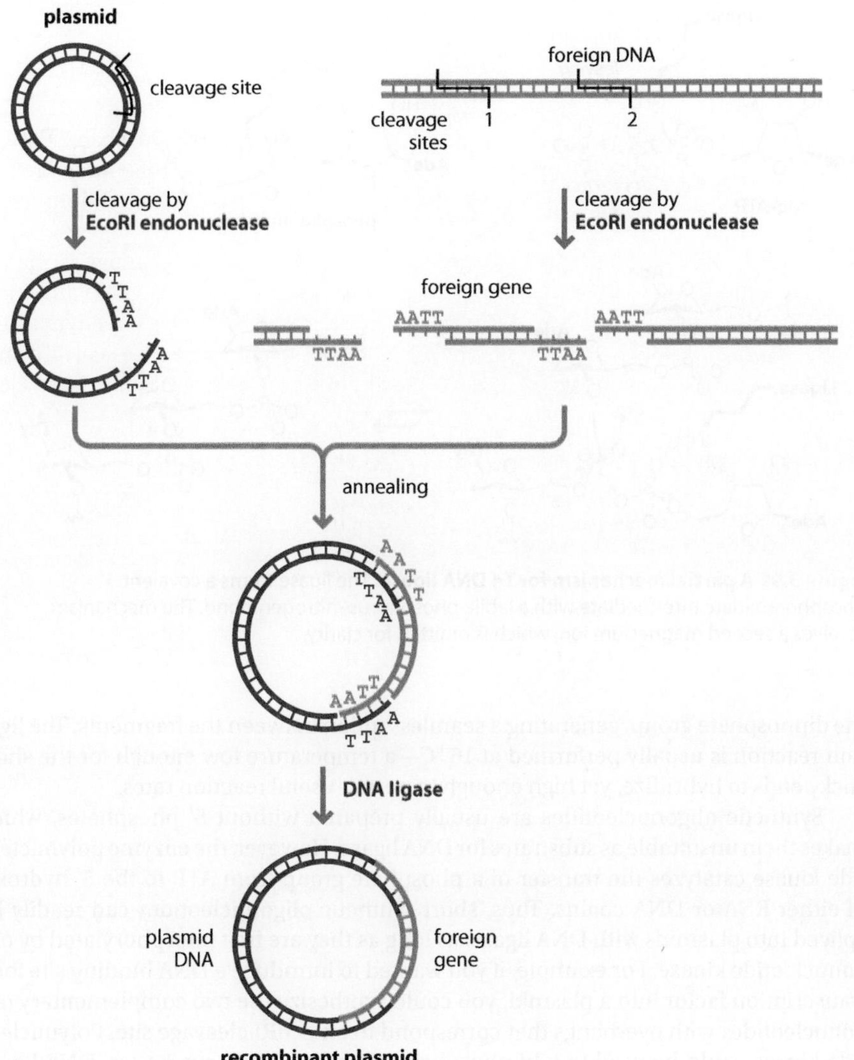

Figure 3.98 Recombinant DNA technology. A combination of restriction enzymes and DNA ligase can be used to clone DNA into plasmid vectors. Many protein pharmaceuticals based on human proteins include the name "recombinant" to indicate that they were produced in non-human cells using recombinant DNA technology. For example rhGH stands for recombinant human Growth Hormone.

If a single EcoRI site is available in the correct location in a plasmid, a new piece of DNA can be readily inserted (**Figure 3.98**). First, a gene bounded on both sides by EcoRI cleavage sites is excised from a piece of foreign DNA. Then the plasmid is digested with EcoRI resulting in the creation of complementary sticky ends. When the foreign gene and plasmid are mixed together, the sticky ends anneal through base-pairing. Finally, DNA ligase can be added to fill in the gaps in the phosphodiester backbone of DNA, providing a perfectly intact plasmid with the new piece of DNA added at the EcoRI restriction site. Note that this method adds an additional EcoRI site to the plasmid. The new piece of DNA is therefore flanked by EcoRI restriction sites. Digestion of the new plasmid with EcoRI, rehybridization of the plasmid sticky ends (after separation of the digested plasmid from the inserted DNA by purification), and re-ligation can remove the inserted DNA.

Recall that the enzyme DNA ligase (see Figure 3.34) can stitch together nicks in the backbone of the DNA. The most commonly used DNA ligase is an extremely aggressive variant from the bacterial virus T4. Recall, too, that DNA polymerase uses a pyrophosphate leaving group when it unites the 3′ end of DNA with a dNTP. To rejoin nicked DNA, T4 DNA ligase needs to add a leaving group to the 5′ phosphate group. It does so by attaching an adenosyl monophosphate group to the 5′ end, making a labile diphosphate bond (**Figure 3.99**). An amino side chain in the ligase enzyme first attacks Mg•ATP to generate a labile phosphoramidate, which is readily attacked by the 5′ phosphate at the nick site. The 3′ hydroxyl on the other side of the nick then attacks

Figure 3.99 A partial mechanism for T4 DNA ligase. The ligase forms a covalent phosphoramidate intermediate with a labile phosphorus–nitrogen bond. The mechanism involves a second magnesium ion, which is omitted for clarity.

the diphosphate group, generating a seamless union between the fragments. The ligation reaction is usually performed at 16 °C—a temperature low enough for the short sticky ends to hybridize, yet high enough to permit useful reaction rates.

Synthetic oligonucleotides are usually prepared without 5′ phosphates, which makes them unsuitable as substrates for DNA ligase. However, the enzyme polynucleotide kinase catalyzes the transfer of a phosphate group from ATP to the 5′ hydroxyl of either RNA or DNA chains. Thus, short synthetic oligonucleotides can readily be spliced into plasmids with DNA ligase, as long as they are first phosphorylated by oligonucleotide kinase. For example, if you wanted to introduce a DNA binding site for a transcription factor into a plasmid, you could synthesize the two complementary oligonucleotides with overhangs that correspond to an EcoRI cleavage site. Polynucleotide kinase could be used to add phosphate groups to the 5′ overhangs. DNA ligase could then be used to splice the resulting short duplex into a plasmid, or any other DNA, that had been cut with EcoRI (**Figure 3.100**).

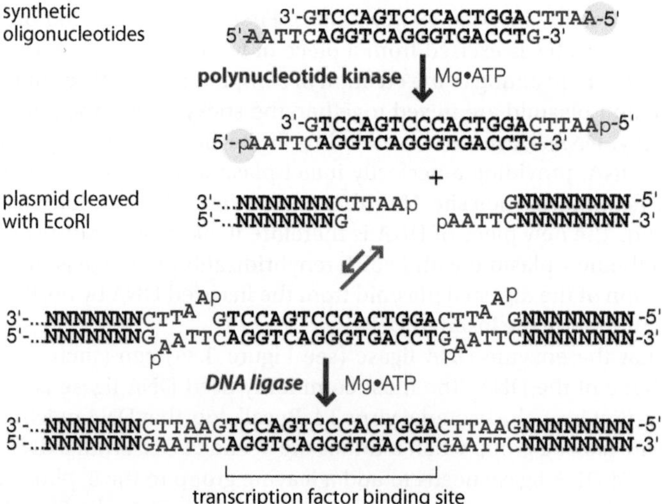

Figure 3.100 Ligation of synthetic DNA. Synthetic oligonucleotides are less expensive when they are synthesized with 5′ hydroxyl groups instead of 5′ phosphates. Polynucleotide kinase prepares synthetic oligonucleotides for ligation by adding the 5′ phosphates.

Problem 3.18

A panoply of different restriction enzymes is available, and each has a particular sequence preference. For example, the sequence G↓GATCC is cut by BamHI at the backbone position indicated by the arrow.

A Rewrite the sequence AACTGAATTTCAGGGGGATCCGCATGGCGT as a double-stranded DNA sequence, circle the nucleotides that make up the BamHI restriction site, and indicate the sites of strand cleavage.

B Describe the steps that would be required to splice a double-stranded oligonucleotide GCATGGGTTTCAAATC into the BamHI site of the sequence in question A, if it is part of a plasmid vector.

C What happens to the sequence GCATCCTTAAGTGGTGGGATCCCCATTCGGATCC after digestion with BamHI? After digestion with BamHI and ligation with DNA ligase, what sequence(s) result?

Mutations in DNA can lead to changes in expressed proteins

With robust chemical tools for synthesizing oligonucleotides and agile biological tools for cutting and ligating DNA, it is relatively straightforward to translate changes in DNA to changes in protein structure and ultimately to changes in biological function. Mutations in DNA can be made randomly (replacing A with A, G, C, or T) or specifically (for example replacing A with G). Similarly, mutations can be made at random sites or specific locations.

Chemical mutagens show no sequence selectivity. Recall that 5-bromouridine causes low-frequency T•A to C•G transversions in living cells, but with no sequence selectivity. Error-prone PCR is one of the most effective approaches for mutating a specific gene. If a denaturing agent such as dimethylsulfoxide (DMSO; $(CH_3)_2S{=}O$) is added to the PCR mixture, the amplification will take place in an error-prone manner with a large number of incorrectly incorporated bases. DMSO disrupts hydrogen bonding and interferes with high-fidelity base pairing during DNA synthesis. Substituting the transition metal Mn^{2+} for Mg^{2+} also decreases the fidelity of DNA polymerase. Error-prone PCR can only introduce a low rate of mutations in the targeted gene—a maximum of 2%. For mutation rates as high as 5%, a plasmid is usually amplified in a special strain of *E. coli* that has a careless DNA polymerase. Afterward, the target gene can be amplified by PCR.

Random mutagenesis usually leads to misfolded, nonfunctional proteins and is a risky strategy for improving the fitness of an organism. Sexual organisms achieve higher success rates through the recombination of homologous genes. The *engrailed* gene from any human male can hybridize with the *engrailed* gene of any human female because the sequences are highly homologous. Importantly, any region of the *engrailed* gene from a human sperm can hybridize with the corresponding region of *engrailed* gene from a human egg. The resulting *engrailed* genes are likely to generate correctly folded engrailed protein because the progenitor proteins folded correctly and were very similar. The technique of **DNA shuffling** mimics homologous recombination. It involves mixing any number of homologous genes from different persons—even different species, such as human, monkey, lemur—then randomly fragmenting them. The mixture of fragments, which act as primers, is subjected to PCR, resulting in vast numbers of new genetic combinations (**Figure 3.101**). DNA shuffling leads to a high proportion of functional proteins because they incorporate only successful mutations that were present in each of the original genes.

Site-specific mutagenesis gives chemical biologists the power to formulate and test specific hypotheses about protein structure. Why should a particular Leu to Phe substitution turbo-charge an enzyme's activity? Why should a mutation well away from the active site cause a loss of protein function? Such questions cannot be readily answered with the broad ax of random mutagenesis, and they motivate techniques for making mutations at specific locations in a protein, for example residues in the flexible loop of an antibody, the residues of a transcription factor that make contacts with DNA, or a nucleophilic residue in an enzyme's active site.

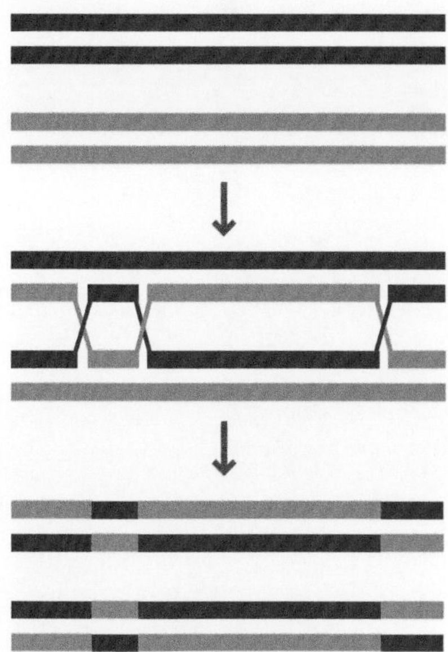

Figure 3.101 The principle of homologous recombination is that homologous regions of genes readily hybridize, which can facilitate genetic recombination. The biochemical mechanisms are more involved, and are not discussed here.

Site-directed mutagenesis involves labile plasmid templates

The most common methods of site-directed mutagenesis are based on the ability of DNA polymerase to extend oligonucleotide primers on a plasmid template, which works even when the oligonucleotide has mismatched bases. Oligonucleotide-directed mutagenesis requires both a synthetic oligonucleotide primer that has the mutant DNA sequence and a template plasmid containing the gene to be mutated. Oligonucleotide-directed mutagenesis leads to two similar plasmids—the original and the mutant—that are not easily separated by electrophoresis or any other technique. Modern methods of oligonucleotide-directed mutagenesis take advantage of bacteria to degrade DNA selectively (**Figure 3.102**).

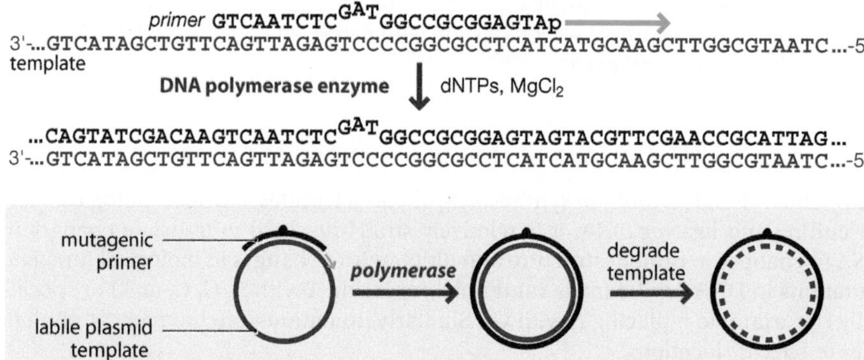

Figure 3.102 Oligonucleotide-directed mutagenesis. A synthetic oligonucleotide primer with a mutagenic codon is extended around a labile plasmid template. After the labile template is degraded only the mutant form of the plasmid remains.

Recall that bacteria aggressively target non-native DNA for destruction by using, for example, restriction endonucleases to cleave DNA, and DNA methyltransferases to protect their own DNA from the restriction endonucleases. In 1985, Thomas Kunkel reported a method for creating labile plasmids that involves growing special single-stranded plasmids in a special strain of *E. coli* that misincorporates uridylate residues (at low frequency) instead of thymidylate. After performing extension of a mutagenic primer *in vitro*, the mixture of labile template strand and full-length mutant strand is amplified as double-stranded plasmids in a strain of *E. coli* that degrades DNA containing small proportions of uridylate.

A more recent variant of the Kunkel method called Quikchange™ mutagenesis uses a special DNA polymerase to amplify double-stranded plasmid DNA from almost any laboratory strain of *E. coli* by PCR. It avoids the Kunkel requirement for special single-stranded plasmids and special strains of *E. coli*, and you get more plasmid DNA than you start with. Most laboratory strains of *E. coli* generate methylated plasmid DNA (for example N^6-methylguanosine), whereas PCR with dNTPs does not generate DNA with modified bases. Therefore, after performing mutagenic PCR on a plasmid, you can treat the mixture with a special restriction endonuclease that targets only methylated template DNA and not the newly amplified DNA from PCR; the endonuclease will remove the original plasmid template.

3.9 NUCLEIC ACID PHOTOCHEMISTRY

Ultraviolet radiation promotes [2+2] photodimerization of thymine and uracil bases

The π-stacking arrangement of pyrimidine bases makes them susceptible to photochemical crosslinking. Ultraviolet radiation can promote [2+2] photocycloaddition reactions between thymine bases in 5′-TT-3′ sequences, leading to intrastrand crosslinks made up of cyclobutane rings. In particular, these lesions are produced by UV-B and UV-C radiation with wavelengths below 320 nm.

The excitation step in a photochemical reaction promotes an electron from a bonding orbital to an antibonding orbital. Unfortunately, we have no way to depict structures with partly occupied bonding and antibonding orbitals by using traditional Lewis structures. The situation is further complicated by the fact that the electrons in singly occupied orbitals can now have one of two spin states: a singlet state with opposing spins and a triplet state with like spins. Pyrimidine photodimerizations are believed to proceed through a singlet excited state, consistent with a concerted pericyclic reaction. However, most solution-phase photochemical reactions of enones involve triplet states and involve radical intermediates. One way to represent an excited state is to place the structure in brackets and denote the excited state with an asterisk (**Figure 3.103**). In so doing, we imply nothing about the type of antibonding orbital occupied by the excited electron. Indicating the singlet state with a superscript provides an additional level of information.

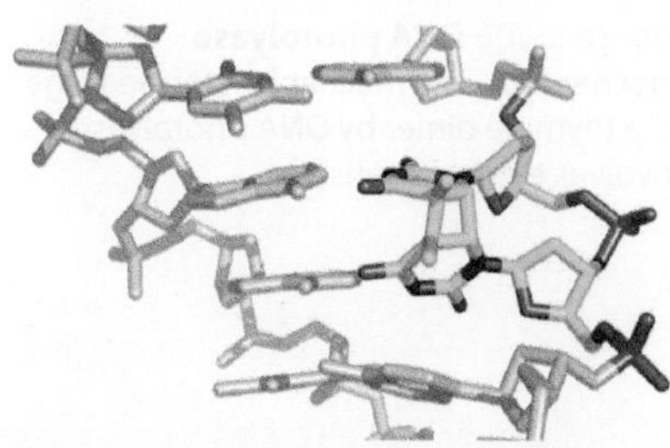

Figure 3.104 Intrastrand crosslinking. The thymine dimer is formed between adjacent thymine bases in a DNA strand. (PDB 1N4E)

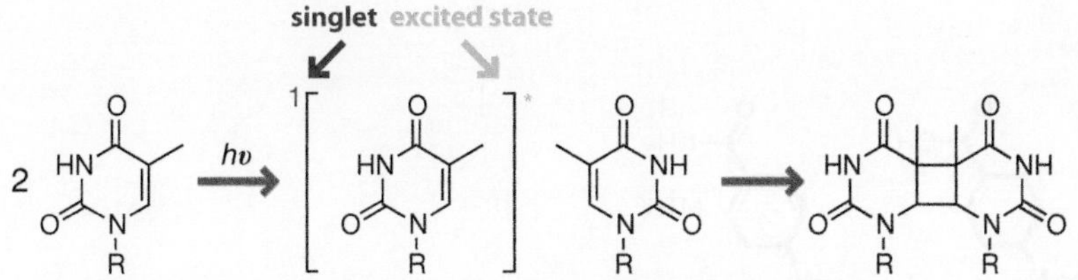

singlet excited state

Figure 3.103 Thymine photodimers. Ultraviolet light promotes [2+2] cycloaddition of thymine bases via a singlet excited state.

Fortunately, the thymine bases are not well aligned for cycloaddition in B-form DNA. However, in A-form DNA the thymines are well aligned to form an adduct (**Figure 3.104**). Thus, at any point in time, some of the base pairs of a typical genome will exist in conformations amenable to pyrimidine photodimerization.

Thymine dimers in DNA can be repaired

There are two types of repair mechanisms that remove thymine dimers from DNA: a simple light-dependent process and a complex multienzyme process. Lower organisms have a light-dependent enzyme called DNA photolyase that seeks out thymine dimers, flips the dimer out of the double helix, and holds it directly in contact with a protein cofactor—reduced flavin adenine dinucleotide (FADH) (**Figure 3.105**). Once excited, FADH transfers an electron to the thymine dimer, generating a radical anion (**Figure 3.106**). The radical anion of the thymine dimer undergoes a stepwise cleavage and then returns the electron to the FADH.

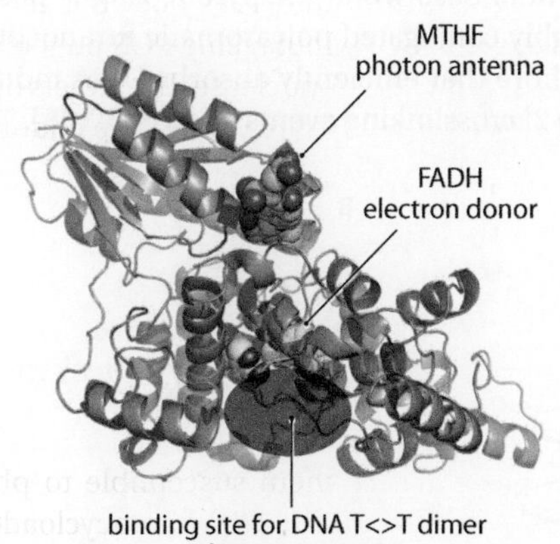

Figure 3.105 Solar-powered DNA repair. *Escherichia coli* DNA photolyase recruits two additional molecules to cleave thymine dimers: a powerful chromophore to absorb light (methylene tetrahydrofolate, MTHF), and a redox active molecule to donate and accept an electron. Bactericidal ultraviolet lamps emit light at 254 nm, which is ideal for thymine dimerization, but not for repair by DNA photolyase.

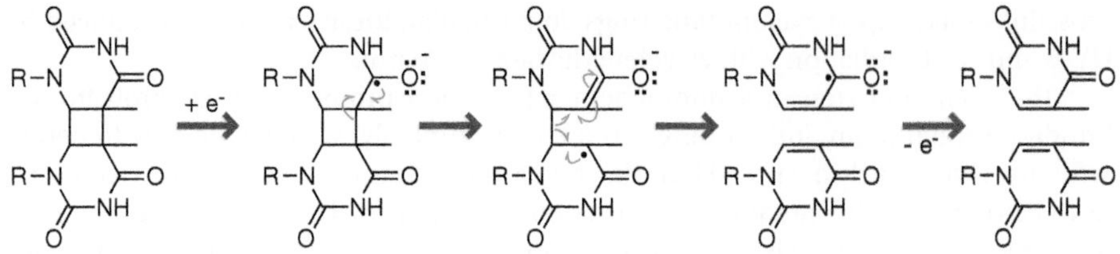

Figure 3.106 DNA photolyase mechanism. The mechanism for cleavage of a thymine dimer by DNA photolyase involves electron transfer.

At this stage we need not worry about the structure of FADH except to recognize that FADH is not particularly good at using long-wavelength ultraviolet light with wavelengths between 315 and 400 nm. *E. coli* DNA photolyase employs an extra photon-capture device, methylene tetrahydrofolate (MTHF), to absorb photons in this region of the ultraviolet spectrum; the excited MTHF then transfers excitation to FADH. Ironically, one subunit of the MTHF chromophore is *para*-aminobenzoic acid (PABA), which is widely used in ultraviolet-absorbing sunscreens (**Figure 3.107**).

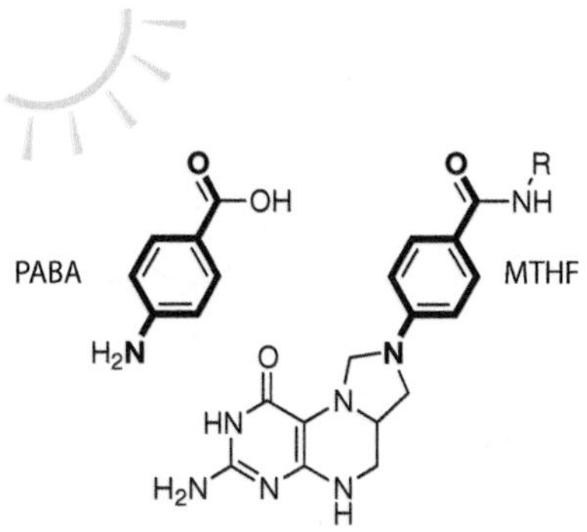

Figure 3.107 Protective sunscreens use the same chromophore as *E. coli* DNA photolyase. PABA, *p*-aminobenzoic acid; MTHF, methylene tetrahydrofolate.

Human cells do not express a DNA photolyase enzyme. Instead, they rely on a complex set of enzymes that perform "nucleotide excision repair." As the name implies, the enzymes involved in excision repair recognize damage, cut out a large section of the damaged strand, and fill in the spaces with new nucleotides. The advantage of excision repair systems is that they do not require light and can repair many types of DNA damage. Mutations in the enzymes in the excision repair system are associated with the genetic disorder xeroderma pigmentosum. Individuals with xeroderma pigmentosum must avoid direct sunlight and have a roughly 1000–2000-fold higher risk of skin cancer than normal individuals.

Psoralens intercalate between DNA base pairs and photocrosslink opposing strands

Psoralens are a class of natural products that generate covalent crosslinks between opposing strands of duplex DNA on exposure to ultraviolet radiation. Psoralens intercalate between DNA base pairs. When they intercalate between 5′-TA-3′ sequences, the furan ring is positioned under one thymine base while the pyrone ring is positioned above another thymine base. The highly conjugated polyaromatic furanocoumarin ring system is an excellent chromophore that efficiently absorbs UV-A radiation, ultimately leading to two sequential [2+2] crosslinking events (**Figure 3.108**).

A

5,8-dimethoxypsoralen

B

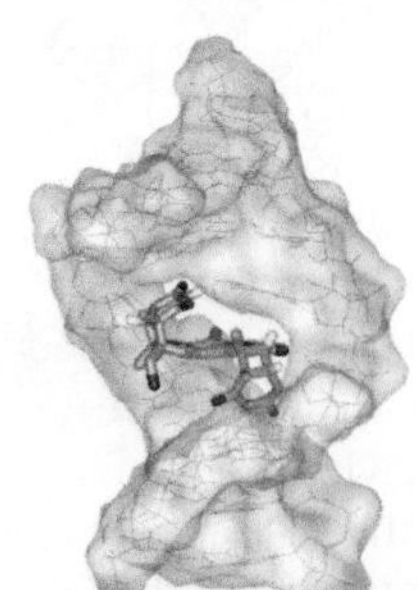

Figure 3.108 Crosslinking DNA strands with psoralens. (A) The oils in the rind of Persian limes are rich in psoralens like 5,8-dimethoxypsoralen. Psoralens can intercalate between base pairs. (B) Each of the photoreactive double bonds of the psoralen can engage in photochemical [2+2] cycloadditions with pyrimidines in double-stranded DNA.

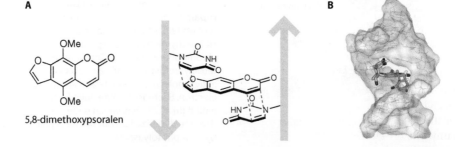

Figure 3.109 Bizarre streaks and linear erythematous vesicular plaques with hyperpigmentation on the abdomen. From lime juice squeezed on the body while sunbathing. (Reproduced with permission from J.C. Cather et al., *Proc. Bayl. Univ. Med. Cent.* 13:405–406, 2000.)

Lime oil is a rich source of psoralens, and cases of phototoxic dermatitis are commonly observed in persons working with limes who are subsequently exposed directly to sunlight. One of the largest outbreaks of phototoxic dermatitis involved children at a day camp in Owings Mills, Maryland. About one-sixth of the 622 children reported skin eruptions, dark vesicular patches, and rashes (**Figure 3.109**). The outbreak was eventually traced to the preparation of pomander balls by sticking fragrant cloves into Persian limes and then engaging in outdoor activities. The skins of Persian limes are rich in 5,8-dimethoxypsoralen. Some proteins prevent the intercalation of psoralens and the subsequent crosslinking reaction. Psoralen photocrosslinking of these protein–DNA complexes leaves a chemical "footprint" where the protein was bound.

Psoralen photocrosslinking ultimately leads to hyperpigmentation of skin. The use of psoralen-containing plant extracts to treat vitiligo (loss of skin pigmentation) was reported as early as 1400 BC. In the 1950s the writer John Howard Griffin used psoralen photochemistry to darken his skin so that he could experience at first hand what it meant to be a black man in the prejudiced environment of the deep South of the United States. Psoralen phototreatment is now listed as a known human carcinogen.

3.10 DNA AS A TARGET FOR CYTOTOXIC DRUGS

Cell division is highly controlled in normal human cells

The process of cell division is highly regulated in simple and complex organisms. For example, *E. coli* bacteria divide only in uncrowded environments when nutrients are plentiful. As long as these conditions are met, bacterial cell division is boundless. Most human cells are not engaged in frequent mitotic cell division, even when nutrients are abundant. For example, in adult humans, mitosis is rare among most cell types, such as skeletal muscle cells, kidney cells, and neurons. A few human cell types—epithelial cells (gastrointestinal tract, skin, hair, and fingernails) and hematopoietic cells (related to blood)—are engaged in slow constant cell division but are poised to undergo highly conditional, explosive mitotic proliferation: white blood cells proliferate in response to infection; skin cells proliferate in response to a wound; breast and uterine cells proliferate during pregnancy; gonadal cells proliferate during puberty. Not surprisingly, these cell types are most prone to form various human cancers: leukemia, lung cancer, skin cancer, breast cancer, ovarian cancer, testicular cancer, and prostate cancer.

Dividing human cells must pass through checkpoints, or die

The decision to engage in mitotic cell division has serious implications for a human cell. It requires immense investment of resources to generate enough components—ribosomes, lipids, histones, and so on—for two daughter cells. It also requires the construction of complex machinery needed to sort out the cell components and orchestrate mitotic division. All the while, the cell is in a precarious state, poised to commit suicide at the slightest sign of trouble.

When a cell receives the appropriate mitotic signals, it proceeds through a series of four stages that culminate in cell division (cytokinesis). These four phases of cell division are usually described by a clock diagram referred to as the cell cycle (**Figure 3.110**). There are two major phases in the cell cycle: DNA synthesis, and mitosis. Each of these events is preceded by a preparatory phase known as a gap phase, which involves intensive biosynthetic activity. A cell that is big enough to divide and is receiving the right chemical signals can irreversibly leave the resting state and enter the first gap phase, G_1. During this phase the cell must produce enzymes that can unwind, un-knot, repair, and copy the genome. During S (synthesis) phase, these

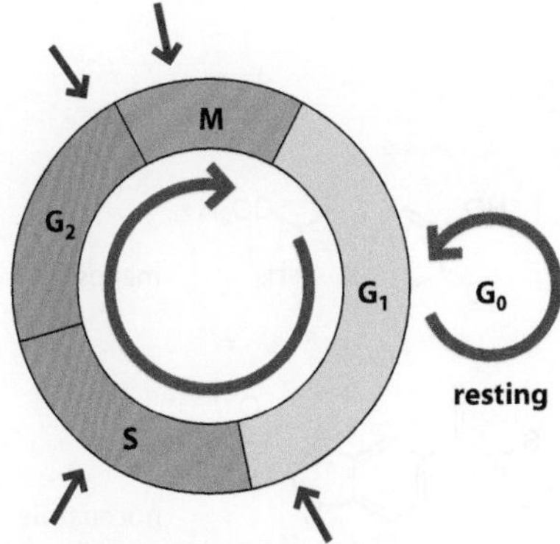

Figure 3.110 The cell cycle. The cell cycle is a clock with preparatory stages for cell division. Red arrows indicate key checkpoints. (Courtesy of Eishi Noguchi.)

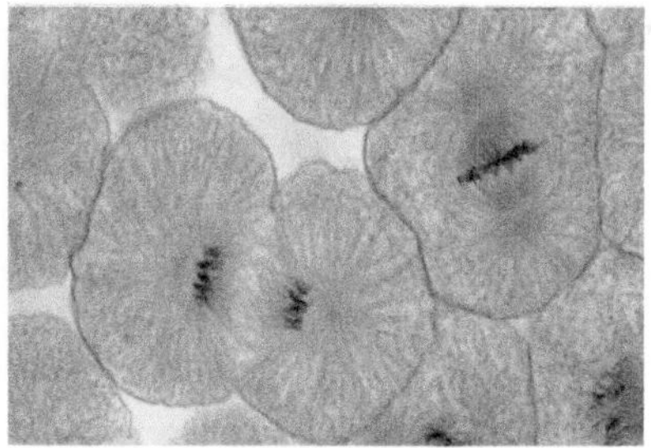

Figure 3.111 Embryonic whitefish blastula cells in various stages of mitosis. The cell at the lower left is in the metaphase stage of mitosis. (Courtesy of Alexander Cheroske.)

enzymes then must replicate 3 billion base pairs with no mistakes! In the next gap phase, G_2, the cell must produce a protein scaffold for sorting the 23 pairs of chromosomes, dividing them equally, and dragging them to opposite sides of the cell. Cells in G_1, S, and G_2 phases are not readily distinguished under a microscope, but during the final phase, mitosis (M), drastic changes occur that are readily visible under a light microscope (**Figure 3.111**). During mitosis, the nuclear membrane dissolves, and chromosomes line up on scaffolds and are then pulled to opposite sides of the cell.

The molecular changes in each phase of the cell cycle are extensive, so that once a cell has committed itself to cell division and has entered the cell cycle, it cannot directly return to a non-mitotic state; it must proceed correctly through all four phases until two daughter cells are produced. If irreparable problems are detected, the cell commits suicide in a programmed process of cell death known as **apoptosis**, which in Greek means "falling off" (**Figure 3.112**). The decision to proceed, repair, or die is controlled by cell-cycle **checkpoints**: the G_1 checkpoint, the G_2/M checkpoint, and the metaphase checkpoint. At these various checkpoints the cell decides whether the environment is favorable, the cell is big enough, the DNA has been replicated accurately, or the chromosomes are lined up. If DNA damage is detected at any one of these checkpoints it leads to apoptosis.

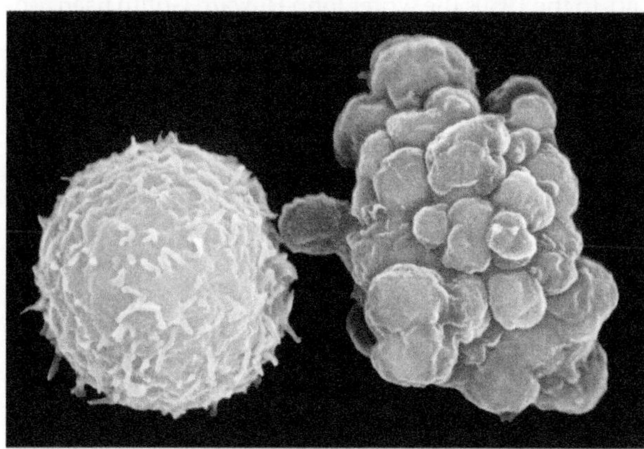

Figure 3.112 A normal white blood cell (left) and an apoptotic white blood cell (right). The apoptotic cell shows characteristic "blebbing" of the cell membrane in the final throes of death. (Courtesy of Photo Researchers.)

Cultured cells can be synchronized in S phase by treatment with the natural product mimosine (**Figure 3.113**). Mimosine inhibits the initiation of DNA replication, arresting cells in late G_1 phase. If cultured cells are exposed to mimosine, all of the cells end up synchronized at the same stage of the cell cycle, somewhat like having runners line up at the starting line before a race. Remarkably, brief exposure does not induce apoptosis. Once the mimosine-containing medium is replaced with mimosine-free medium, all of the cells restart the cycle synchronized at the S phase. Similarly, nocodazole inhibits cells during mitosis, just before they reach the metaphase checkpoint.

Traditional chemotherapy targets DNA in rapidly dividing cells, cancerous or not

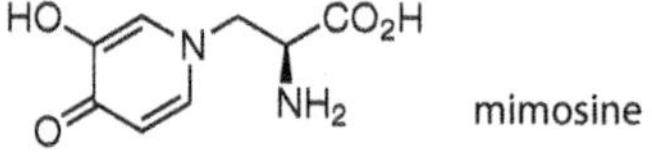

Figure 3.113 Cell-cycle synchronizers. Mimosine and nocodazole arrest cells at specific points in the cell cycle without inducing apoptosis.

Most of the clinically used chemotherapeutic agents target DNA, so it is easy to fall into the trap of thinking that damaging DNA kills cells. DNA damage is not lethal to most cells; it is only lethal to cells that are participating in the cell cycle. Recall that most cells in the human body—muscle cells, neurons, and so on—are not engaged in the cell cycle, do not proceed through checkpoints, and do not undergo apoptosis in response to molecular damage. Ironically, the fastest growing tumors, if caught early, can be easier to treat with drugs that trigger apoptosis at cell-cycle checkpoints (**Table 3.3**).

DNA was the main target for cancer chemotherapy for most of the previous century. The method is brutal because it also targets healthy tissue in a constant state of cell division, such as the immune system, fingernails, and gastrointestinal tract. After chemotherapy, patients are usually isolated to prevent infections; they lose their hair;

Table 3.3 Growth rate versus curability with traditional chemotherapy.

Cancer	Doubling time (days)	Curability
Burkitt's lymphoma	1	+
Choriocarcinoma	2	+
Acute lymphocytic leukemia	3	+
Hodgkin's lymphoma	4	+
Testicular embryonic carcinoma	5	+
Colon cancer	80	–
Lung cancer	90	–
Breast cancer	100	–/+

Figure 3.114 Effects of chemotherapy. (A) Leukonychia (fingernail irregularities) after multiple rounds of chemotherapy. (B) Hair loss following chemotherapy 1 month after treatment. (C) Regrowth of hair 6 months after treatment. (A, courtesy of Kelly Joppa; B and C, courtesy of Jenny Mealing.)

the fingernails stop growing; and they develop intense nausea as the cells lining the gastrointestinal tract die out (**Figure 3.114**). Newer methods of pharmaceutical intervention that specifically target the signaling pathways of the cancer cells have none of the common side effects that we associate with chemotherapy.

In the following sections we describe three main ways of targeting DNA that lead to apoptosis in human cells: (1) inhibition of thymine biosynthesis, (2) inhibition of DNA replication, and (3) inhibition of mitosis.

Inhibition of thymine biosynthesis triggers apoptosis during the S phase of the cell cycle

Synthesis of the genome creates a huge demand for deoxynucleotidyl triphosphates during the S phase of the cell cycle. Thus, inhibiting the biosynthesis of these subunits will cause cells to stall during S phase and ultimately lead to apoptosis at the S-phase checkpoint. Unfortunately, inhibiting the biosynthesis of nucleosides also affects non-mitotic cells, because nucleosides are necessary for mRNA production. However, there are two key differences between the nucleosides used to make RNA and the nucleosides used to make DNA. First, the nucleosides used to make RNA are derivatives of ribose, whereas the nucleosides used to make DNA are derivatives of 2′-deoxyribose. Second, in DNA the heterocyclic base complementary to adenine is thymine, whereas in RNA the heterocyclic base complementary to adenine is uracil, which lacks a methyl group. Thus, inhibiting the synthesis of 2′-deoxynucleosides or thymidine should selectively affect human cells making DNA and passing through cell-cycle checkpoints.

2′-Deoxythymidylate (TMP) is synthesized from uridine diphosphate (UDP) in several steps (**Figure 3.115**): removal of the 2′-hydroxyl group by ribonucleotide

UDP

1) Remove 2′ OH
2) Remove 5′ diphosphate
3) Add 5-CH_3

TMP

Figure 3.115 The first and third steps in thymidylate biosynthesis are major targets of chemotherapy.

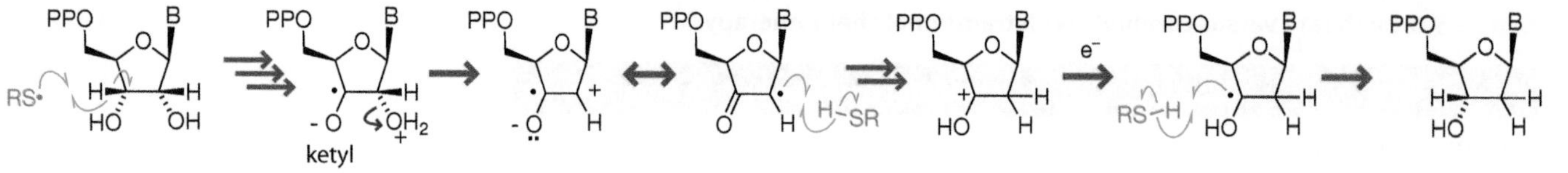

Figure 3.116 **Enzymatic deoxygenation.** Ribonucleotide reductase generates DNA subunits by removing the 2′-hydroxyl group of nucleoside 5′ diphosphates.

reductase, two-step conversion of the 5′ diphosphate to a 5′ monophosphate, and addition of the methyl group by thymidylate synthase. The first and last steps are targets for chemotherapy.

Ribonucleotide reductase is an iron-dependent enzyme that catalyzes the deoxygenation of nucleoside diphosphates such as ADP to generate the corresponding 2′-deoxynucleoside diphosphates such as dADP. These 2′-deoxynucleoside diphosphates are then converted to triphosphate building blocks for DNA synthesis. The key step in the mechanism of ribonucleotide reductase is the formation of an electron-rich ketyl anion radical that facilitates loss of the neighboring water molecule (**Figure 3.116**). The effect of the ketyl radical on the neighboring leaving group can be understood through the less favorable resonance form that places the unpaired electron on oxygen and the anionic lone pair on carbon. Note that arrow-pushing, which is usually effective for generating resonance structures, cannot be used to interconvert between resonance structures of the ketyl radical or any radical stabilized by an adjacent lone pair (**Figure 3.117**).

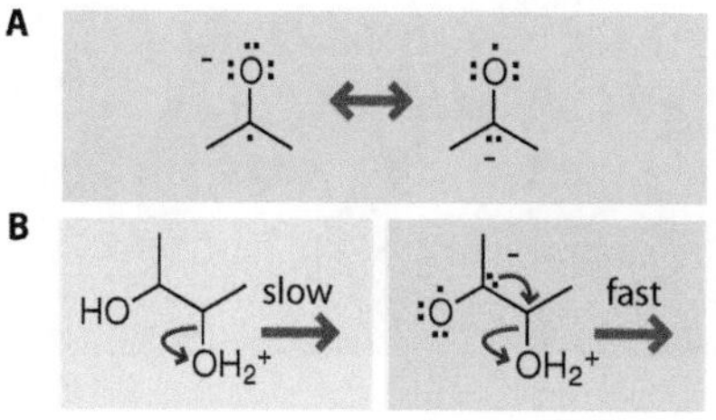

Figure 3.117 **Ketyl radicals.** (A) Arrow-pushing cannot be used to interconvert between resonance forms of a ketyl radical. (B) The ketyl radical has nucleophilic character on the carbon atom, which facilitates expulsion of the neighboring leaving group.

The clinically used drug gemcitabine is an inhibitor of ribonucleotide reductase with two fluorine atoms that resist radical-atom abstraction. Gemcitabine is a prodrug that is phosphorylated *in vivo* to generate the active diphosphate form that has high affinity for the enzyme (**Figure 3.118**).

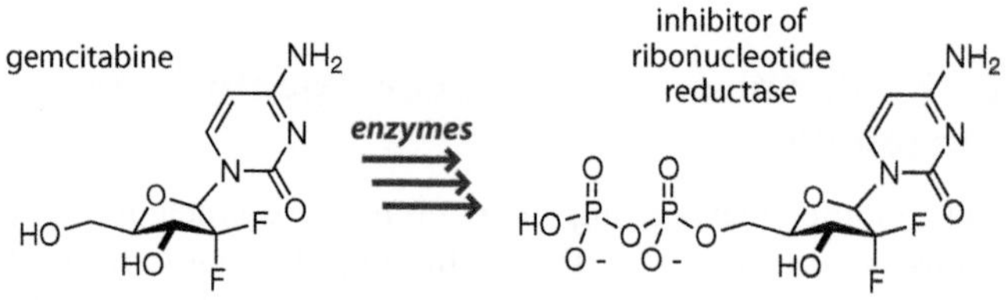

Figure 3.118 **Prodrug activation.** Like most prodrugs, gemcitabine is metabolically converted to an active form. The diphosphate of gemcitabine can inhibit ribonucleotide reductase.

Adding the methyl group to thymine is essential for DNA synthesis

The nucleoside thymidine monophosphate is biosynthesized from 2′-deoxyuridine monophosphate and methylene tetrahydrofolate by the action of the enzyme thymidylate synthase. Methylene tetrahydrofolate is the enzyme cofactor derived from folic acid that donates a single carbon atom from an iminium ion functional group to each of the thymine bases of DNA (**Figure 3.119**). Recall that methylene tetrahydrofolate was also the handy chromophore employed by DNA photolyase. Folic acid has a key role in DNA biosynthesis and cell proliferation. Pregnant women are urged to eat foods rich in folic acid before and during pregnancy to ensure an ample supply for proliferating fetal cells.

Figure 3.119 Folic acid is converted into a key metabolic reagent (methylene tetrahydrofolate) that is essential for rapidly dividing cells. Folic acid supplements are recommended during pregnancy as insurance, because fetal development involves DNA synthesis on a massive scale. The March of Dimes Folic Acid for a Healthy Pregnancy campaign logo is no longer active. (Courtesy of March of Dimes Foundation.)

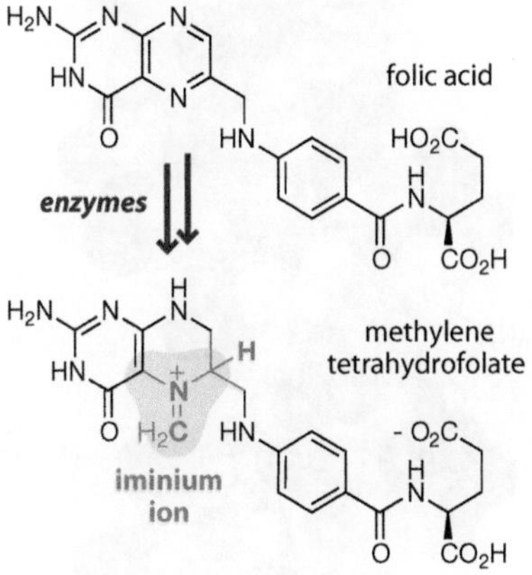

The mechanism of thymidylate synthase involves an initial conjugate addition of an active-site thiolate anion to the uracil ring followed by addition of the enolate intermediate to the iminium form of methylene tetrahydrofolate. E1cB elimination of the folate group completes the transfer of the one-carbon unit. The final stage of the reaction involves a remarkable hydride transfer from the tetrahydrofolate to the enone, regenerating an enolate that will eliminate the enzyme thiolate (**Figure 3.120**). The prototropic states of the intermediates remain to be fully elucidated because X-ray crystallography does not easily resolve the positions of protons, particularly for unstable intermediates along a reaction pathway. The by-product of this reaction is methylene dihydrofolate, which is reduced back to methylene tetrahydrofolate by the enzyme dihydrofolate reductase. Dihydrofolate reductase is as important for DNA synthesis as thymidylate synthase and is also a target for chemotherapy.

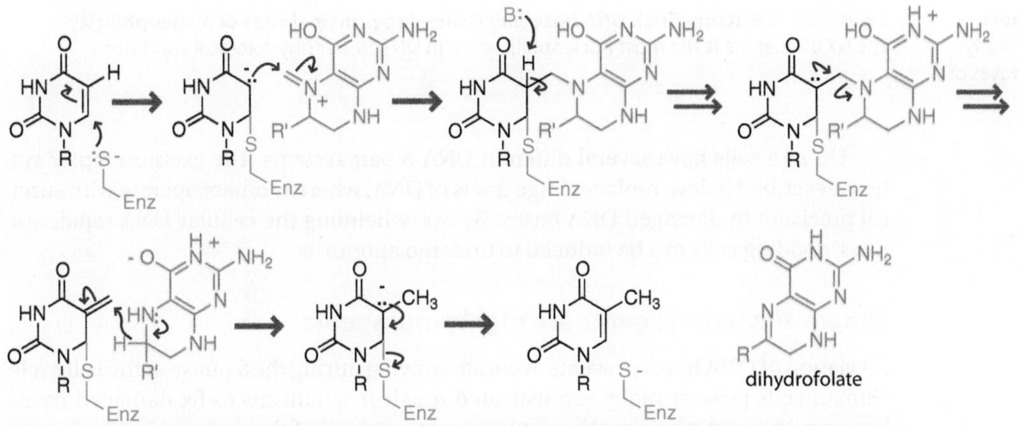

Figure 3.120 Thymidylate synthase mechanism. Thymidylate synthase transfers a carbon atom from methylene tetrahydrofolate to the 5 position of uridine.

5-Fluorouracil is a highly effective inhibitor of thymidylate synthase because it is transformed into the 5-fluoro analog of deoxyuridine monophosphate. Once 5-fluorouridylate enters the active site of thymidylate synthase, it follows the same reaction path as deoxyuridine monophosphate. However, the E1cB elimination step cannot proceed with the fluorine atom in place of a hydrogen atom, so the substrate remains covalently bound to the enzyme (**Figure 3.121**). Mechanism-based inhibitors such as 5-fluorouracil that react with enzymes are sometimes called suicide inhibitors.

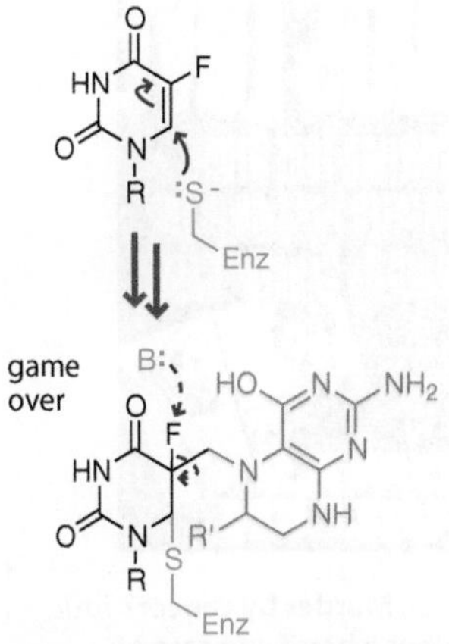

Figure 3.121 Suicide inhibitor. 5-Fluorouridylate forms a covalent intermediate with thymidylate synthase that cannot undergo E1cB elimination.

Problem 3.19

Suggest a plausible arrow-pushing mechanism for the formation of a covalent adduct between the suicide inhibitor 5-trifluoromethyluridylic acid and thymidylate synthase that involves the active-site thiolate.

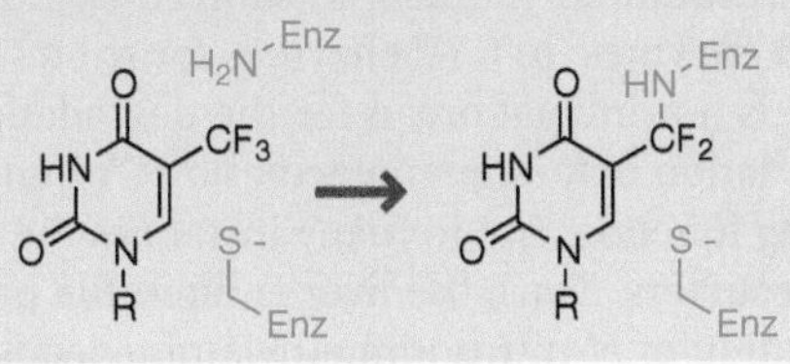

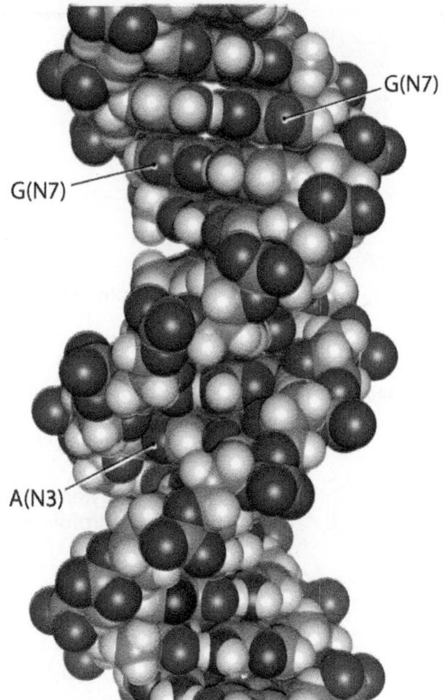

Figure 3.122 Nucleophilic hotspots. Nucleophilic nitrogen atoms are clearly visible in the major and minor grooves of double-stranded DNA.

DNA is a nucleophile

DNA-damaging agents offer a conceptually simple approach to the treatment of cancer. DNA is a nucleophile, which can react with electrophilic drugs, and some types of alkylating events are particularly cytotoxic. The principal nucleophilic sites on DNA are the nitrogen atoms of the heterocyclic bases. At this point you may want to go back and review the atom numbering of pyrimidine and purine bases that we introduced in Figure 3.5. When reacted with methyl methanesulfonate (methyl mesylate), the relative nucleophilicity of atoms in double-stranded DNA is **G**(N7) > **A**(N3) >> **A**(N1) > **C**(N3) (**Figures 3.122** and **Figure 3.123**). This order reflects the ability of lone pairs to act as nucleophiles versus participating in hydrogen bonds or aromaticity. For single-stranded DNA, the N1 position of adenine is more nucleophilic than the N3 position of adenine because it is not protected by base pairing.

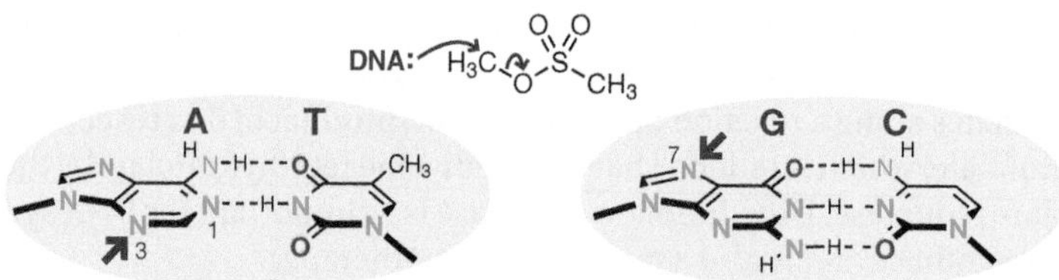

Figure 3.123 Watson–Crick DNA base pairs, showing principal sites of nucleophilicity. The N7 of guanine is the most nucleophilic atom in DNA under physiological conditions.

Human cells have several different DNA repair systems. The excision repair system, described below, replaces huge tracts of DNA, whereas others operate with surgical precision in damaged DNA bases. By overwhelming the cellular DNA repair systems, dividing cells can be induced to undergo apoptosis.

Simple alkylating agents are highly mutagenic

Alkylation of DNA bases prevents accurate copying during the S phase of the cell cycle. Human cells possess many sophisticated repair mechanisms to fix damaged bases, but even the best repair mechanisms cannot catch all of the mutations. Most mutations are likely to be silent, because only about 1% of the human genome is expressed as protein. In other cases, the daughter cells will be nonviable. In rare cases, mutations will cripple the control systems that prevent uncontrolled cell division. The accumulation of such mutations can lead to cancer. DNA is not the most nucleophilic nor the most accessible molecule within a human cell, so special tricks are needed to bring an electrophile through a cytoplasmic sea of thiolate and amine nucleophiles and into close proximity to DNA.

One method of generating highly reactive alkylating agents inside of cells takes advantage of *N*-nitrosoamines. The book *Toxic Love* by Tomas Guillen recounts a 1978 incident of chemical poisoning by Steven Roy Harper, who attempted to give cancer to his former girlfriend and her family by spiking their refrigerator drinks with a known carcinogen, *N*-nitrosodimethylamine (**Figure 3.124**). Although several of the family members died of acute poisoning, none of the affected individuals immediately contracted cancer. This disturbing incident reminds us that cells have many control systems to resist transformation into cancer cells and that cancer is a result of many low-probability mutations.

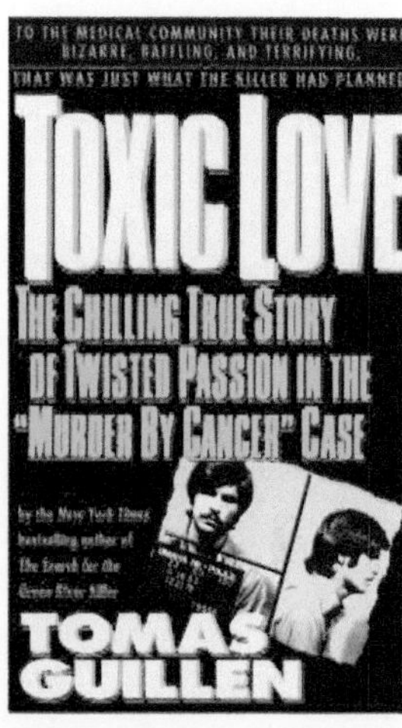

Figure 3.124 Murder by cancer? *Toxic Love* describes a horrifying case of poisoning with a carcinogenic agent. (From T. Guillen, Toxic Love. Bantam Books, 1995. With permission from Random House, Inc.)

In mammals, *N*-nitrosodimethylamine is oxidized by iron-dependent enzymes called cytochrome P450 enzymes, or CYP enzymes for short. The human liver is rich with CYP enzymes and is a principal organ for the degradation and removal of foreign molecules. The oxidation of foreign molecules by CYP enzymes normally serves an important detoxifying function, but in many instances the oxidation products are more toxic than the precursors. Thus, the liver is often the principal site of damage by various toxins. Oxidation of *N*-nitrosodimethylamine leads to demethylation and

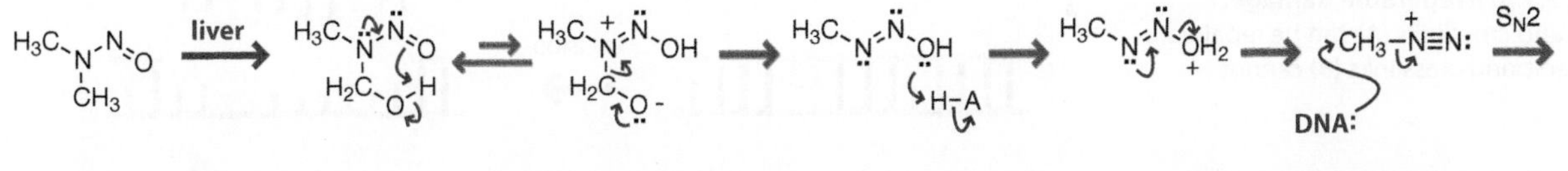

Figure 3.125 **The danger of *N*-nitroso compounds.** Metabolism of *N*-nitrosodimethylamine generates methyl diazonium ion, a potent alkylating agent.

ultimately generates a highly reactive alkylating agent, methyl diazonium ion, that reacts with DNA (**Figure 3.125**). Many alkylating agents are known to cause cancer in laboratory animals and are therefore *suspected* human carcinogens. However, relatively few of these agents have undergone sufficient study to be classified as *known* human carcinogens.

N-Nitrosoamines are prepared with sodium nitrite and acid, which generate nitrous acid (HO–N=O). Sodium nitrite is a preservative in cured meats that prevents the growth of deadly *Clostridium botulinum* bacteria. However, you might wonder what happens when the sodium nitrite that is used to cure bacon enters the acidic environment of your stomach. Fortunately, nitrite reacts quickly in meats and very little makes it into your stomach. In theory, nitrite has the opportunity to generate dangerous *N*-nitrosoamines by reacting with amines in bacon meat. The most popular brands of cured meats include sodium ascorbate (vitamin C) or the diastereomer sodium erythorbate along with the sodium nitrite. Ascorbate and erythorbate react with nitrous acid faster than amines, inhibiting the buildup of dangerous *N*-nitrosoamines in cured meats (**Figure 3.126**).

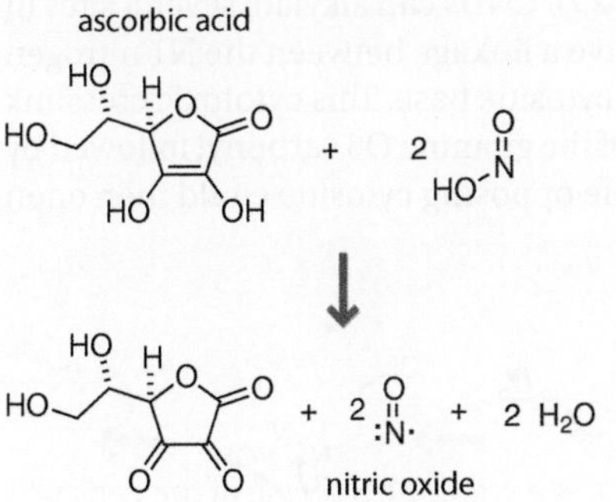

Figure 3.126 **Saving your bacon.** Ascorbic acid reacts faster than amines with nitrous acid. Instead of forming *N*-nitroso compounds, the nitrous acid is converted to nitric oxide.

Problem 3.20

In addition to generating dangerous *N*-nitroso compounds, nitrous acid can also react directly with DNA. Suggest a plausible mechanism for the formation of a 2-diazonium guanine intermediate and the subsequent formation of a guanine–guanine crosslink.

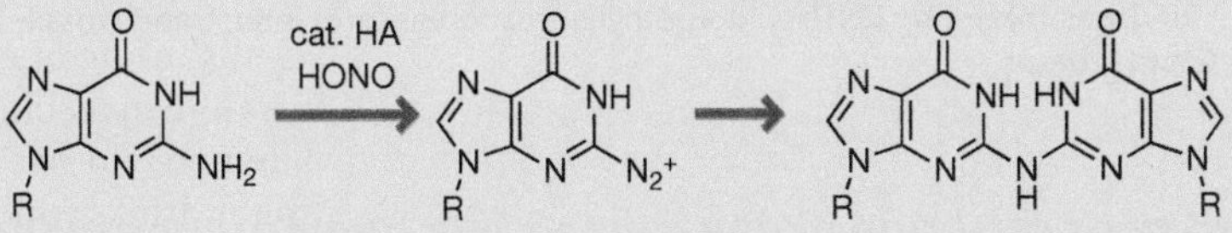

Bifunctional alkylating agents that crosslink DNA are highly cytotoxic

Bifunctional alkylating agents can be highly cytotoxic to dividing cells and are widely used in the treatment of cancer. One clinically used bifunctional alkylator is busulfan, which is shockingly similar to the mutagenic alkylating agent methyl mesylate (**Figure 3.127**). Bifunctional alkylators are highly cytotoxic to rapidly growing cancer cells, but they unfortunately increase the risk of future cancers.

Human cells have several mechanisms for the repair of DNA damage. Nucleotide excision repair is the most powerful because it can readily remove damaged sections of DNA. Like thymine photodimers, DNA bases that are crosslinked by a bifunctional

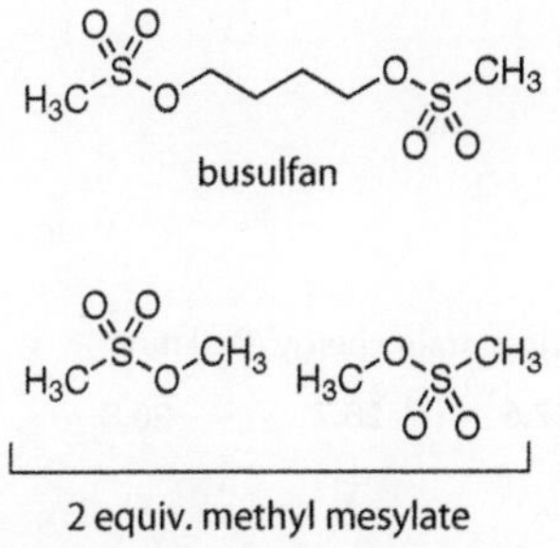

Figure 3.127 **Cytotoxicity versus mutagenicity.** The simple bis-mesylate busulfan is far more cytotoxic than two molecules of methyl mesylate because it can crosslink DNA strands.

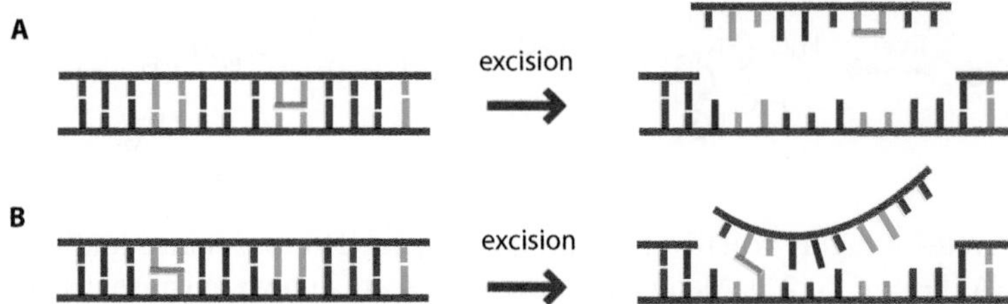

Figure 3.128 Irreparable damage. Intrastrand crosslinks (A) can be repaired, but interstrand crosslinks (B) cannot.

alkylating agent such as busulfan can readily be removed as long as the crosslinked bases are present on the same strand. However, if the crosslinked bases are on opposite strands, the DNA lesion is irreparable because the excised strand remains covalently bound to the template strand (**Figure 3.128**). Crosslinks between strands can lead to apoptosis during either the G_1/S checkpoint or the G_2/M checkpoint. This cell-cycle dependence of bifunctional alkylating agents confers selective cytotoxicity toward dividing cells over non-dividing cells. One problem with bifunctional alkylating agents is that the second alkylation event is not very efficient. So the majority of alkylation events will affect only one strand and will either be repaired or ignored or will lead to heritable mutations in the cell lineage.

Chloroethyl nitrosoureas (CNUs) are a general class of bifunctional alkylators used in the treatment of brain tumors. Like *N*-nitrosodimethylamine, diazonium chemistry has an important role in the mechanism of CNUs, but generation of the key diazonium ion is base-catalyzed (**Figure 3.129**). CNUs can alkylate several sites in DNA. The cytotoxic crosslink is believed to involve a linkage between the N1 nitrogen of guanine and the N3 nitrogen of the opposing cytosine base. This cytotoxic crosslink is believed to arise through an initial alkylation of the guanine O6 carbonyl followed by a slower intramolecular cyclization. The N3 of the opposing cytosine could then open this ring in an S_N2 reaction.

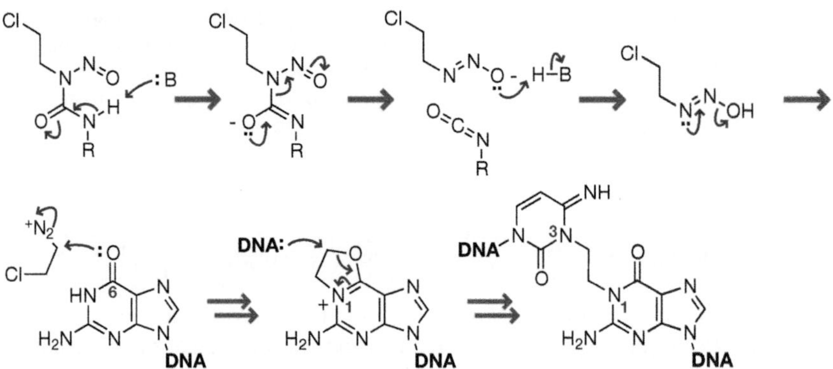

Figure 3.129 Crosslinking by CNUs. Chloroethyl nitrosoureas form interstrand crosslinks between G and C bases.

Strained rings can bring highly reactive functional groups to DNA

Strained three-membered rings offer, in a compact package, a means of delivering highly reactive functionality to the DNA that is protected within the nucleus of human cancer cells. Three-membered rings such as cyclopropanes, oxiranes (epoxides), and aziridines have two important properties. First, they are significantly strained, with about 27 kcal mol^{-1} of ring strain relative to similar unstrained compounds (**Figure 3.130**). Second, the three bonds inside the strained ring are vastly more nucleophilic than unstrained bonds. It is easy to rationalize the considerable nucleophilicity of the bonds in three-membered rings by looking at the direct correlation between bond angles and hybridization. Recall that smaller angles are made by mixing *p* character into orbitals. Thus, tighter bond angles require more *p* character. For instance, a carbon atom that forms a 180° CCC bond angle is essentially *sp* hybridized; a tetrahedral

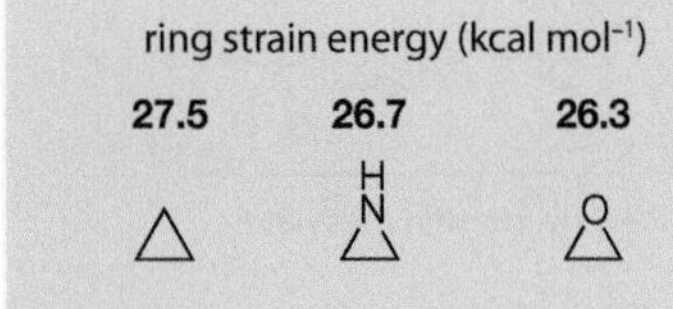

Figure 3.130 Strained relationship. Three-membered rings composed of second-row atoms exhibit comparable levels of ring strain.

Figure 3.131 Bent out of shape. Angle strain corresponds to *p* character, which ultimately corresponds to bond nucleophilicity. The three bonds within a cyclopropane ring are more nucleophilic than unstrained carbon–carbon bonds.

bond angle	180°	109.5°	60°
hybridization	*sp*	sp^3	"sp^5"
%*p*	50	75	84

carbon atom that forms a 109.5° CCC bond angle is sp^3 hybridized. As the bond angles decrease, the molecular orbitals corresponding to the bonds incorporate more *p* character (**Figure 3.131**). For a cyclopropane ring, each carbon–carbon bond in the ring is essentially made from sp^5 hybridized orbitals. Recall from the last chapter that small increases in *p* character correlate with vast increases in nucleophilicity. A particularly important consequence of the nucleophilic character of cyclopropane ring bonds is that they can donate into and stabilize carbocations. For example, one cyclopropane ring is more effective at stabilizing an adjacent carbocation than the resonance donation from three phenyl rings!

Epoxide alkylators of DNA are highly mutagenic

In 1775 a London surgeon named Percival Pott noted the high incidence of "soot wart" among chimney sweeps. These scrotal tumors are now known to arise from exposure to molecules with fused benzene rings called polycyclic aromatic hydrocarbons (PAHs). PAHs are formed through the incomplete combustion of coal, petroleum, foods, and other organic substances. PAHs with a characteristic cleft called a "bay region" are oxidized by mammalian metabolic enzymes to form carcinogens (**Figure 3.132**). For example, two particular cytochrome P450 enzymes, CYP1A1 and CYP1A2, are known to oxidize benzo[*a*]pyrene (see Figure 3.132) to benzo[*a*]pyrene diol epoxide (BPDE). Because BPDE alkylates the 2-amino group of guanine bases it must be binding selectively to the minor groove (otherwise it would alkylate the more reactive N7 nitrogen of guanosine). As a result of this alkylation the strand is miscopied, leading to G to T

bay region
benzo[*a*]pyrene
CYP1A
HO
OH
CYP1A
O
HO
OH
BPDE
DNA:
DNA
HO
HO
OH

Figure 3.132 Metabolic generation of a carcinogen. Oxidation of benzo[*a*]pyrene by CYP1A enzymes generates the carcinogen BPDE.

transversions. In 1996 it was demonstrated that BPDE leads to heritable mutations in the *p53* gene, a key apoptotic gene that responds to DNA damage. Mutations in *p53* are oncogenic, because they allow further DNA damage to occur unchecked (**Figure 3.133**). For many decades, tobacco companies denied that there was any scientific proof that cigarettes cause cancer. Because smoke from a typical cigarette contains 20–40 ng of benzo[*a*]pyrene, the evidence for the carcinogenic potential of cigarettes could no longer be plausibly denied.

Nature produces carcinogens just as dangerous as benzo[*a*]pyrene. Aflatoxin B1 is a natural product produced by the fungus *Aspergillus flavus*, which infects peanuts

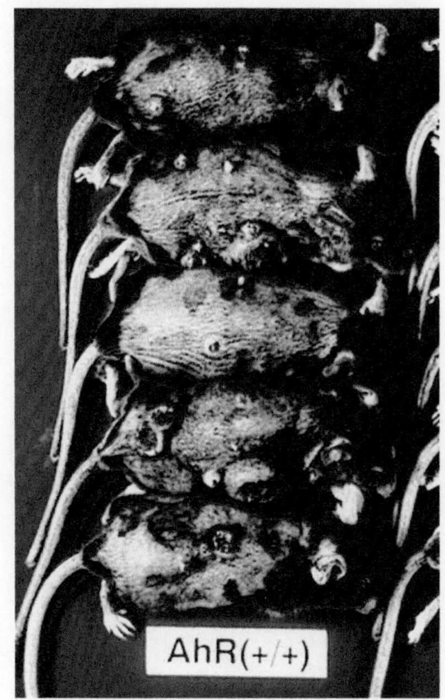

Figure 3.133 Tumorigenicity. Repeated applications of benzo[*a*]pyrene can lead to eruptive skin tumors in mice. (From Y. Shimizu et al., *Proc. Natl. Acad. Sci. USA* 97: 779–782, 2000. With permission from the National Academy of Sciences.)

Figure 3.134 All-natural carcinogens. The mold *Aspergillus flavus*, which produces aflatoxins, infects many types of foods, such as peanuts and corn. (Courtesy of The State of Queensland, Department of Agriculture, Fisheries and Forestry, Australia, 2010.)

(**Figure 3.134**). Aflatoxin B1 is highly carcinogenic and can cause tumors at concentrations below one part per billion. Like benzo[*a*]pyrene, aflatoxin is a substrate for CYP1A enzymes, which convert it into a highly mutagenic epoxide. The resulting tumors tend to be focused in the liver, which is the most prominent site for CYP enzymes. Once formed, the highly reactive aflatoxin epoxide does not get very far without reacting with something. In the nucleus it intercalates between DNA base pairs and then reacts selectively with the N7 atom of guanidylate residues (**Figure 3.135**). The reaction with DNA is stereospecific; the diastereomeric epoxide (on the top face instead of the bottom) does not alkylate DNA.

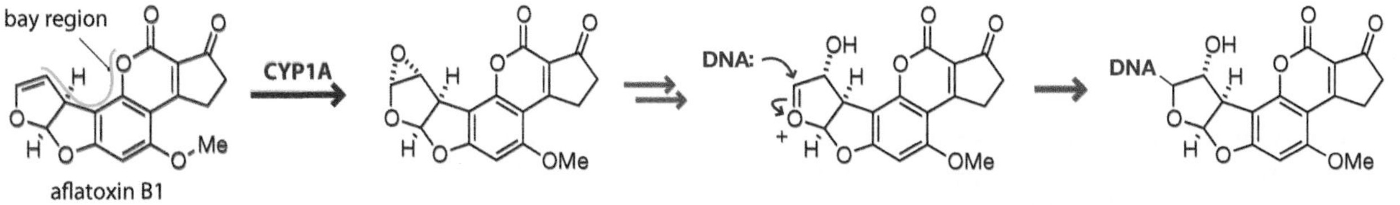

bay region

CYP1A

DNA:

aflatoxin B1

Figure 3.135 Metabolic activation of aflatoxin B1. Metabolic epoxidation of aflatoxin B1 leads to a highly reactive epoxide.

Aziridinium rings are relatively selective alkylators of DNA

Much of what we know about the reaction of epoxides with DNA comes from carcinogenic epoxides generated by CYP enzymes. In contrast, much of what we know about the reactions of anticancer aziridines with DNA comes from synthetic compounds. The development of chemotherapeutic aziridine drugs had an inauspicious beginning. During World War I, bis-chloroethyl sulfide, also known as mustard gas, was used as a chemical warfare agent (**Figure 3.136**). Mustards undergo nucleophilic substitution reactions through **neighboring group participation**. The eyes, respiratory tract, and mucous membranes are acutely susceptible to mustard gas because, in these moist environments, the initial reaction is hydrolysis to produce hydrochloric acid. However, mustard gas also induces longer-term damage to cells by reacting with proteins and DNA.

Substitution reactions of mustards involve neighboring group participation. The intramolecular S_N2 displacement of the chloride leaving group by sulfur is extremely fast—much faster than any intermolecular S_N2 reaction (**Figure 3.137**). The resulting episulfonium ion is highly reactive and rapidly reacts with nucleophiles, even modest nucleophiles such as water. After an accidental release of mustard gas in World War II, mustard agents were observed to inhibit the proliferation of white blood cells

Figure 3.136 Chemical mustards as weapons. An observer working with the Royal Australian Artillery examines an unexploded mustard gas shell. (From the Collection Database of the Australian War Memorial, 1943.)

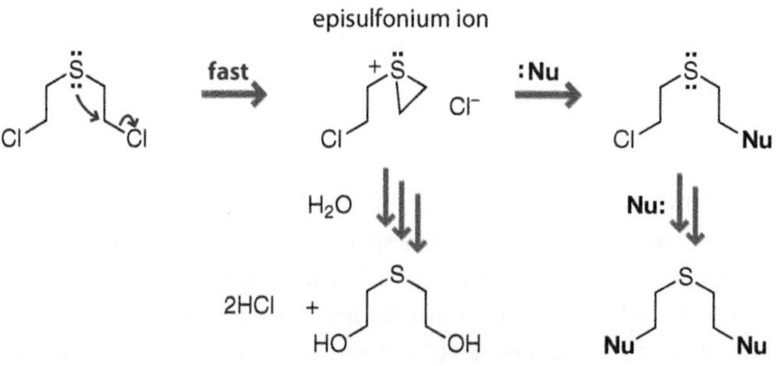

Figure 3.137 Sulfur mustards. Sulfur mustards react via strained episulfonium ions. The episulfonium ions react rapidly with water to produce hydrochloric acid, but they can also react with proteins or DNA.

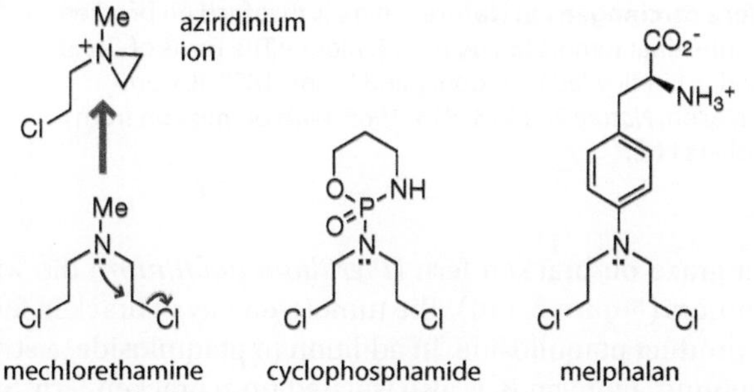

Figure 3.138 **Nitrogen mustards.** Nitrogen mustards used in the treatment of cancer generate strained aziridinium ions.

in the victims, suggesting a treatment for leukemia. These observations were held in secrecy until the close of the war. Nitrogen mustards cyclize to form aziridinium ions almost 10 times more slowly than sulfur mustards. In addition, they are more readily absorbed through the skin than sulfur mustards. Thus, nitrogen mustards were pushed into development as anticancer drugs (**Figure 3.138**). Some examples of aziridinium alkylators include mechlorethamine, cyclophosphamide, chlorambucil, and melphalan.

N-Ethylaziridine is a selective electrophile, reacting with double-stranded DNA seven times faster than it reacts with single-stranded DNA, and 50 times faster than the nucleoside 2-deoxyguanosine (**Figure 3.139**). Because *N*-ethylaziridine exists primarily in the protonated form at pH 7 it is reasonable to suspect that part of the selectivity is due to the electrostatic attraction with the polyanionic backbone of DNA and the positively charged aziridinium ion. The enhanced reactivity of double-stranded DNA over single-stranded DNA is harder to explain.

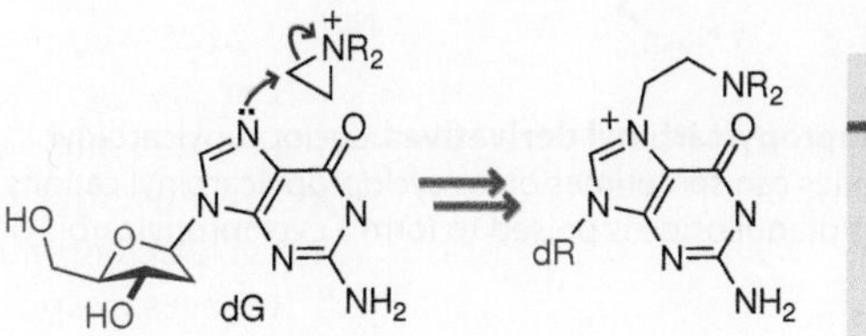

substrate	k_{rel}
dG	1
ssDNA	7
dsDNA	50

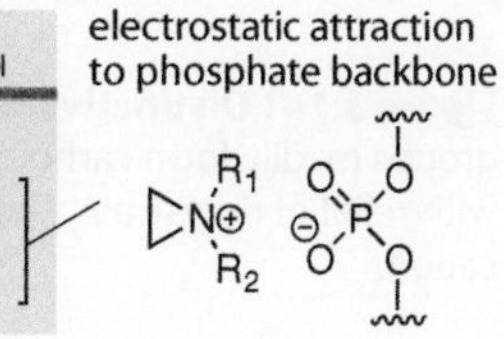

Figure 3.139 **Highly attractive DNA.** Aziridinium ions react faster with nucleic acids than with free nucleosides as a result of electrostatic localization of the reagent.

Problem 3.21

Suggest a plausible arrow-pushing mechanism for the formation of the DNA adduct shown below by the marine natural product fascicularin.

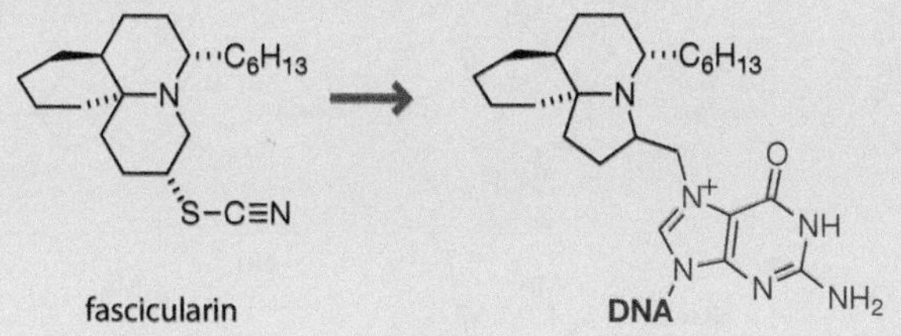

Cyclopropane rings can serve as spring-loaded electrophiles

Epoxides and aziridinium ions are readily opened by nucleophiles, but cyclopropanes are not directly susceptible to nucleophilic attack unless combined with other structural elements. Three important cyclopropane natural products that attack DNA are ptaquiloside, CC-1065, and duocarmycin.

A

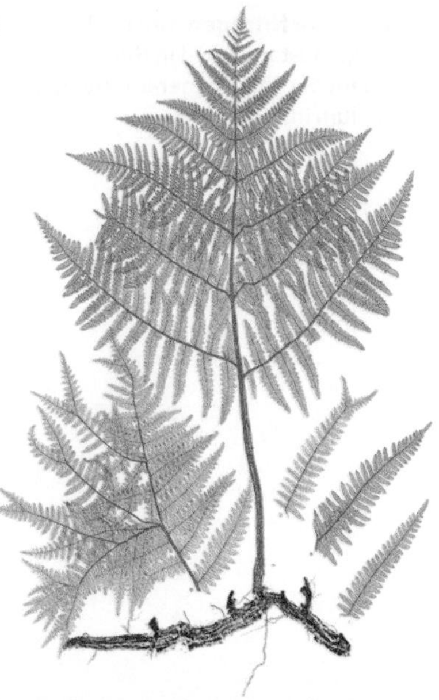

B

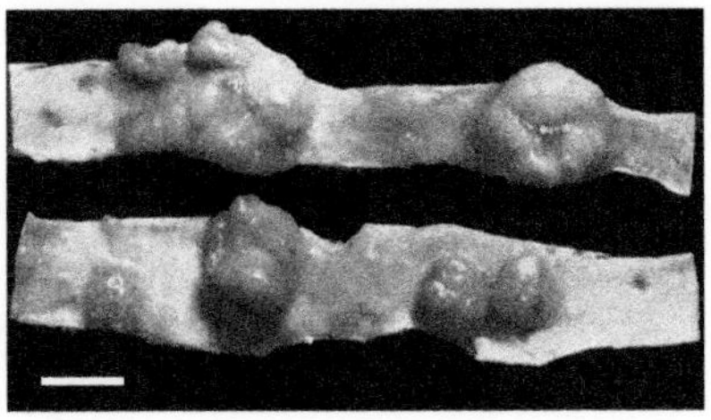

Figure 3.140 More carcinogens in Nature. Animals that feed on bracken fern (A) develop intestinal tumors (B). (A, from T. Moore: The Ferns of Great Britain and Ireland, J. Lindley (ed.). Bradbury and Evans, 1857; B, from I.A. Evans and J. Mason, *Nature* 208:913–914, 1965. With permission from Macmillan Publishers Ltd.)

Cattle that graze on bracken fern (*Pteridium aquilinum*) die with massive gastrointestinal tumors (**Figure 3.140**). The tumorigenicity of bracken fern is attributable to the natural product ptaquiloside. In addition to ptaquiloside, a structurally related aromatic compound, pterosin B, is also isolated from bracken fern. Subsequent studies demonstrated that pterosin B is formed from ptaquiloside in mildly basic aqueous solution. The cyclopropane ring is a key structural element for the mechanism of action of ptaquiloside (**Figure 3.141**). Recall that bonds in cyclopropane rings are highly nucleophilic relative to unstrained carbon–carbon bonds and that this nucleophilicity facilitates the formation of adjacent carbocations, referred to as cyclopropylcarbinyl cations.

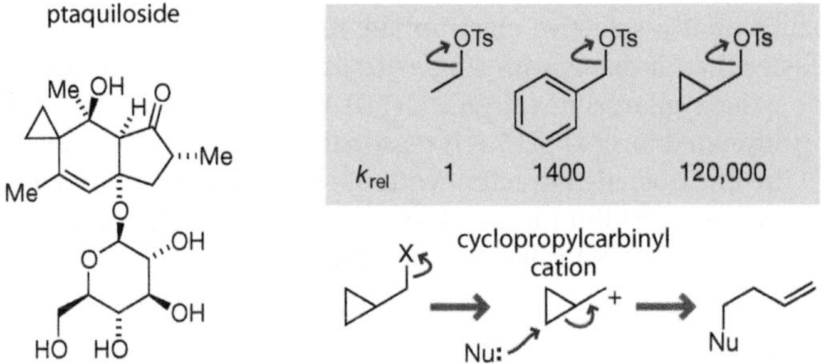

Figure 3.141 Distinctive reactivity of cyclopropylcarbinyl derivatives. Cyclopropylcarbinyl groups readily form carbocations. Nucleophiles can sometimes open cyclopropylcarbinyl cations with relief of ring strain. The natural product ptaquiloside is poised to form a cyclopropylcarbinyl cation.

Even though the cyclopropane ring of ptaquiloside is a focal point for its reactivity, bioactivation requires an initial base-catalyzed elimination of the β-glucoside through an E1cB mechanism (**Figure 3.142**). At this point the tertiary alcohol is perfectly set up to ionize because the resulting carbocation is stabilized by donation from the highly nucleophilic bonds of the cyclopropane ring, the neighboring π bond of the diene, and the neighboring C–H bonds of the methyl group.

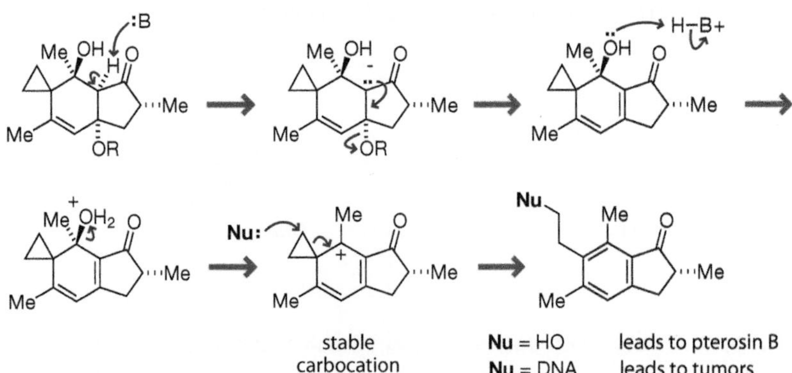

Figure 3.142 Ptaquiloside activation mechanism. E1cB elimination and S_N1 ionization lead to a cyclopropylcarbinyl cation that alkylates DNA. The overall reaction is thermodynamically driven by relief of ring strain and gain of aromaticity.

CC-1065 was initially identified by the Upjohn Company in a screen for anticancer compounds. CC-1065 binds to the deep narrow minor groove of DNA. A variety of other synthetic and natural molecules bind to the minor groove: the antimicrobial drug pentamidine, the antibiotics netropsin and distamycin, and the fluorescent DNA dye Hoechst 33258 (**Figure 3.143**). The best minor-groove binders are generally oligomers of flat molecules with rotatable hinges connected at five-membered rings (**Figure 3.144**); such molecules adopt a curvature that is naturally suited to the DNA double helix.

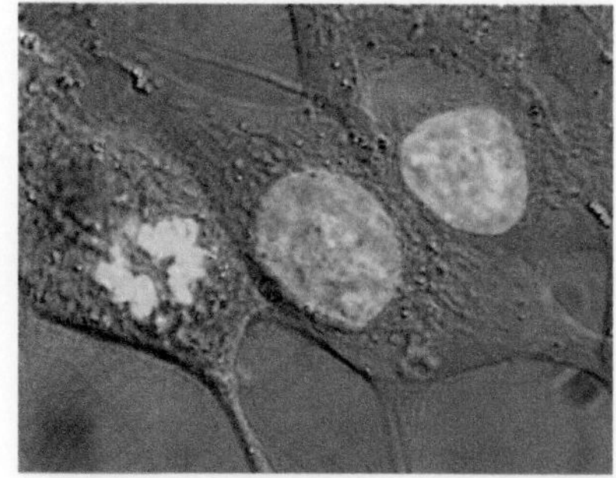

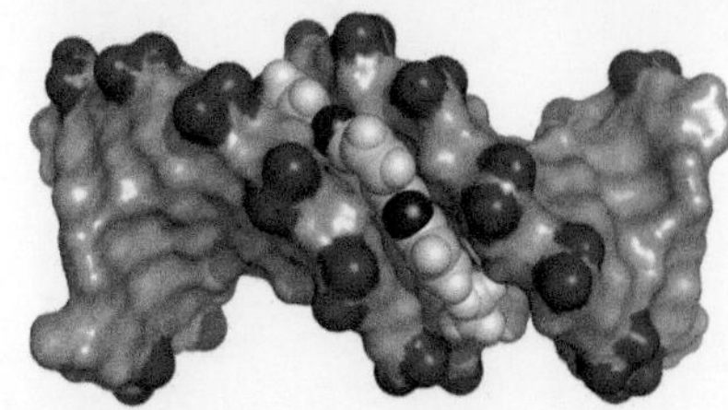

Figure 3.143 Nuclear stain. (A) Hoechst 33258 is commonly used to stain nuclei of mammalian cells. (B) Hoechst 33258 is fluorescent when bound to the minor groove of DNA. (From Wikimedia Commons.)

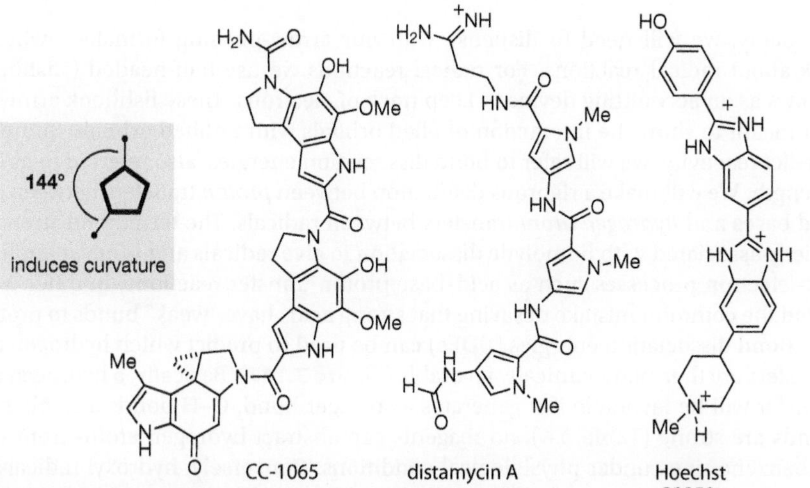

144° induces curvature

CC-1065 distamycin A Hoechst 33258

Figure 3.144 Common features of minor-groove binders. Many minor-groove binders contain five-membered rings and rotatable bonds that help them adopt flat, curved shapes.

Once snugly fitted within the minor groove of DNA, CC-1065 is positioned to react with the N3 position of adenine (**Figure 3.145**). When the adenine attacks the cyclopropane it relieves ring strain and generates an aromatic ring. However, these advantages must be weighed against the loss of a stable C=O π bond with substantial resonance stabilization. Even though CC-1065 binds tightly to DNA, the adduct is still kinetically reversible.

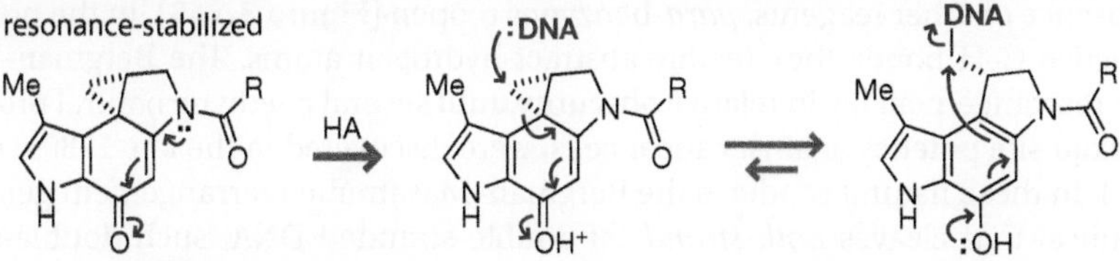

Figure 3.145 Reversible DNA alkylation. CC-1065 forms a tight but reversible covalent adduct within the minor groove of DNA.

Free radicals and oxygen conspire to cleave DNA sugars

Strained-ring alkylators make formidable weaponry, but the principal problem with all electrophilic reagents is that they are also susceptible to attack by thiolates, amines, hydroxyl groups, and other competitive nucleophiles within the nuclear compartment. Not surprisingly, Nature has found other reactions that can damage DNA. It is not immediately obvious that the 2′-deoxyribose backbone should be susceptible to attack. Any reagent vicious enough to attack the ribose skeleton should almost certainly react with water—unless that reagent is a reactive radical. Heteroatoms stabilize carbon-centered radicals. This makes C_1', C_3', C_4', and C_5' of the ribosyl moiety a target for H-atom abstraction by unstable radicals (**Figure 3.146**). Through careful orchestration of radical reactions, the ribose backbone is as easily targeted as the DNA bases.

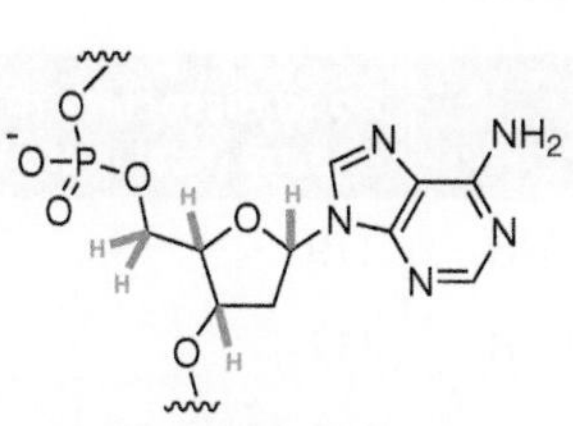

relatively weak C–H bonds **≤93 kcal mol^{-1}**

Figure 3.146 Weak bonds. The terms weak bond and strong bond refer only to radical reactions. Carbon–hydrogen bonds in the ribose backbone of DNA are susceptible to hydrogen-atom abstraction.

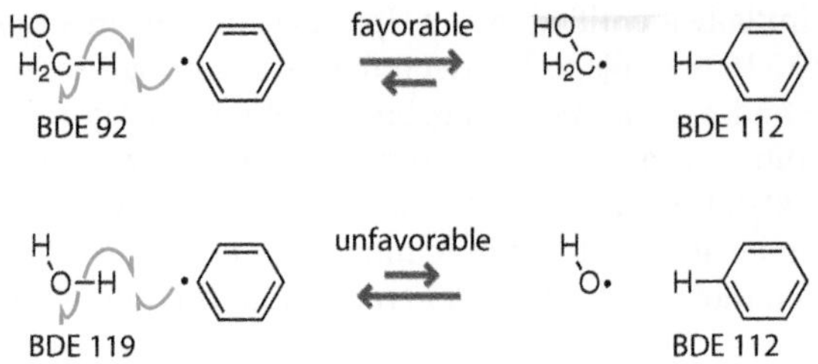

Figure 3.147 Hydrogen atom transfers are favorable when they generate stronger bonds (with higher bond dissociation energies, BDEs). In theory, homolysis of a C–H bond would give H˙ even though free hydrogen atoms are never present under physiological conditions.

Sadly, we will need to dispense with our arrow-pushing formalism when we talk about radical reactions. For radical reactions we use half-headed ("fishhook") arrows as an accounting device to keep track of electrons. These fishhook arrows are not meant to show the interaction of filled orbitals with unfilled orbitals. To help us predict reactivity, we will refer to bond dissociation energies, also referred to as bond strength. We will make a rigorous distinction between *proton* transfers between acids and bases and *hydrogen atom* transfers between radicals. The term bond strength is strictly associated with homolytic dissociation to give radicals and is never applied to two-electron processes such as acid-base proton transfer reactions; in other words, avoid the common mistake of saying that strong acids have "weak" bonds to protons.

Bond dissociation energies (BDEs) can be used to predict which hydrogen atom transfers are thermodynamically favorable (**Figure 3.147**). Basically, a hydrogen atom transfer will be favorable if it generates a stronger bond. O–H bonds and phenyl–H bonds are strong (**Table 3.4**); no reagents can abstract hydrogen atoms from water or benzene rings under physiological conditions. Conversely, hydroxyl radicals and phenyl radicals aggressively abstract hydrogen atoms from susceptible C–H bonds, like those on the DNA backbone. Such reactions are fast when they involve bimolecular collisions and even faster when the reactive radical is held in the proximity of a weak C–H bond. Whenever two carbon radicals recombine with other radicals they allow the carbon atom to achieve a filled octet. As a result, you can expect radicals to recombine at rapid rates.

Enediyne antitumor antibiotics cleave both strands of DNA via *para*-benzyne diradicals

In 1971 Bergman showed that electrocyclic rearrangements of enediynes, first reported by Masamune, proceed through a *para*-benzyne diradical intermediate. In the absence of other reagents, *para*-benzynes reopen (**Figure 3.148**). In the presence of reactive C–H bonds they readily abstract hydrogen atoms. The Bergman–Masamune rearrangement lay in relative obscurity until several enediyne natural products, with exquisite potency against cancer cells, were discovered in the late 1980s (**Figure 3.149**). In these natural products the Bergman–Masamune rearrangement generates a diradical that cleaves *both strands* of double-stranded DNA; such double-strand breaks cannot be repaired.

The enediyne natural products are sophisticated molecular devices with homing devices, triggers, and warheads. The enediyne moiety of dynemicin is prevented from

Table 3.4 Bond dissociation energies of X–H bonds.

Bond	Bond dissociation energy (kcal mol^{-1})
HO–H	119
Ph–H	112
H_3CO–H	104
R_2CH–H	98
$HOCH_2$–H	92
$PhCH_2$–H	90
RS–H	88

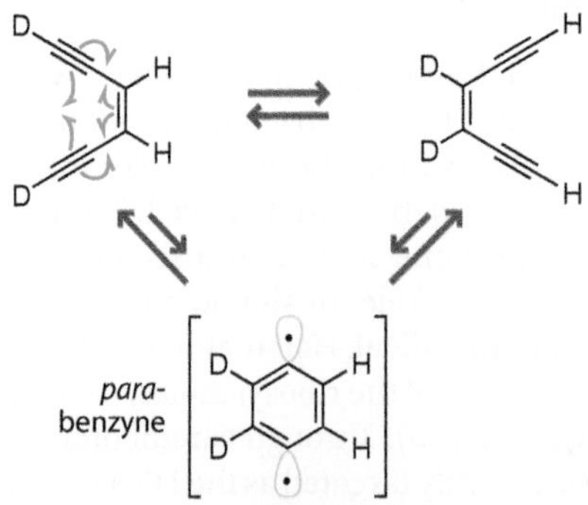

Figure 3.148 The Bergman–Masamune rearrangement. This electrocyclic reaction, often referred to simply as a Bergman rearrangement, involves a *para*-benzyne intermediate with diradical character.

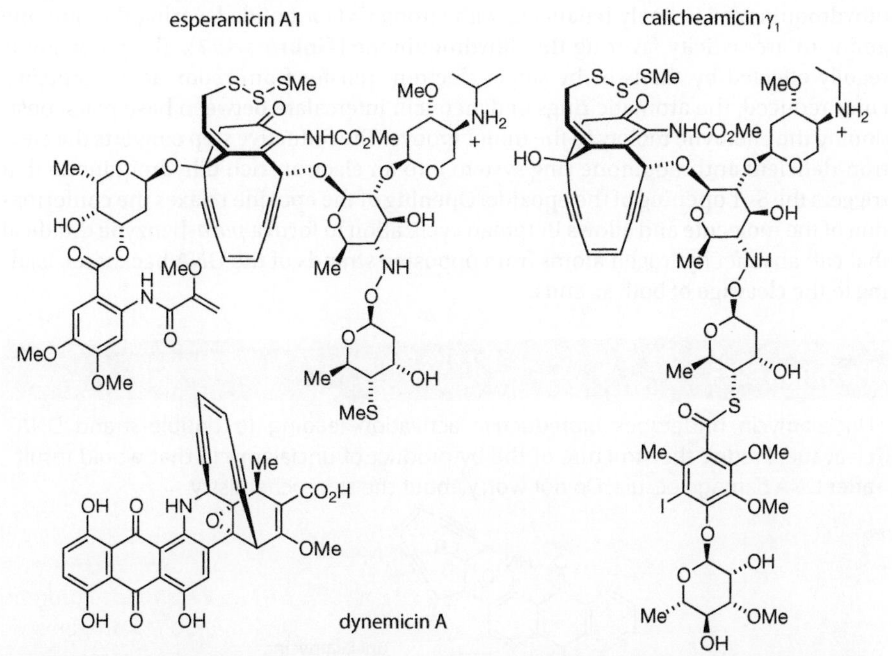

Figure 3.149 Enediyne natural products. The distinctive enediyne functional group of these natural products suggests the possibility of a Bergman rearrangement.

Problem 3.22

1,4-Cyclohexadiene is sometimes used as a surrogate for DNA when studying reactions of reactive diradicals. Show a half-arrow (fishhook) arrow-pushing mechanism for the formation of benzene in the following reaction.

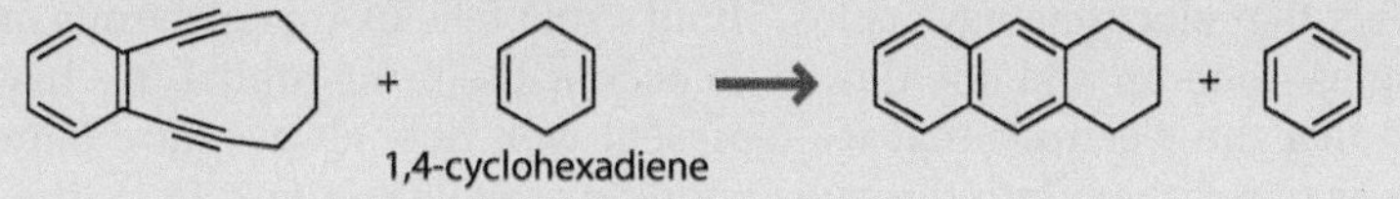

engaging in a Bergman–Masamune rearrangement by the epoxide moiety. Once the epoxide is opened, a subtle bond rotation brings the ends of the alkyne into proximity, leading to a Bergman–Masamune rearrangement (**Figure 3.150** and **Figure 3.151**). Hal Moore first proposed that many drugs require **bioreductive activation** before engaging in chemical reactions with biological targets. Quinones such as dynemicin are readily reduced in human cells. The redox equilibrium between quinones and

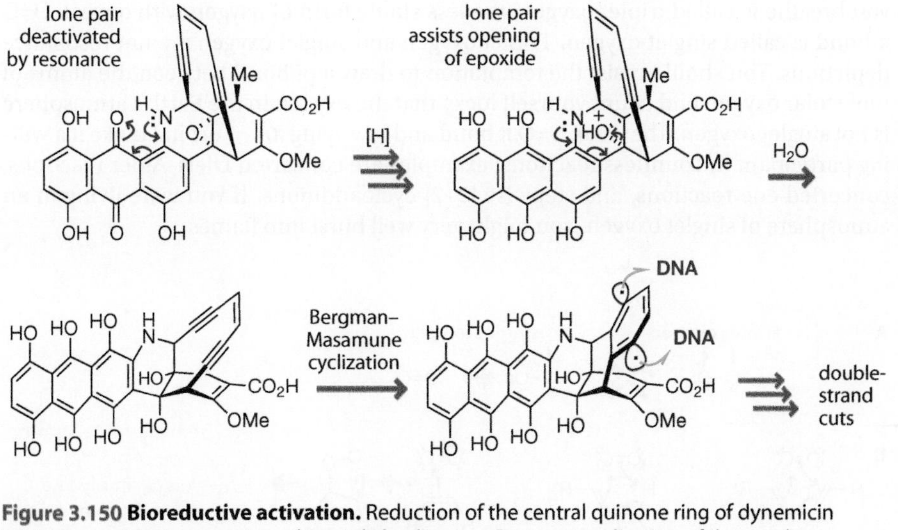

Figure 3.150 Bioreductive activation. Reduction of the central quinone ring of dynemicin initiates epoxide ring opening, ultimately leading to a Bergman cyclization of the enediyne moiety.

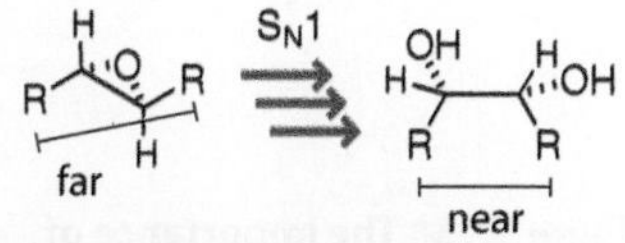

Figure 3.151 Enediyne trigger. S_N1 opening of the epoxide in dynemicin allows the reactive ends of the enediyne moiety to react.

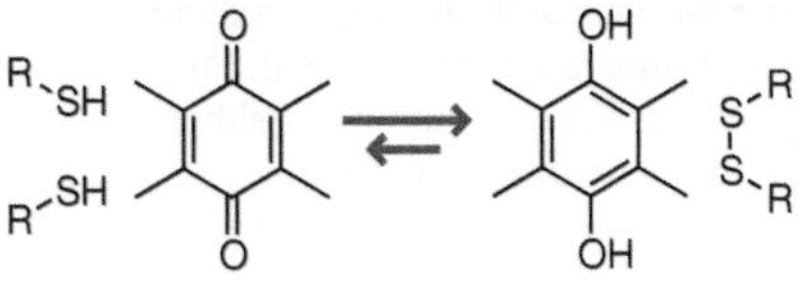

quinone dihydroquinone

Figure 3.152 Quinone reduction is a common trigger in bioreductive activation. Thiols and other reducing agents readily reduce quinones.

dihydroquinones is finely balanced, with strong C=O π bonds favoring the quinone and with aromaticity favoring the dihydroquinone (**Figure 3.152**). The reduction is readily effected by thiols or by single-electron transfer from redox active proteins. Once reduced, the aromatic rings of dynemicin intercalate between base pairs, positioning the enediyne moiety in the minor groove. The reductive step converts the electron-deficient anthraquinone ring system into an electron-rich dihydroquinone that triggers the S_N1 opening of the epoxide. Opening of the epoxide relaxes the conformation of the molecule and allows Bergman cyclization to form a *para*-benzyne diradical that can abstract hydrogen atoms from opposing strands of the DNA backbone, leading to the cleavage of both strands.

Problem 3.23

Uncialamycin undergoes bioreductive activation leading to double-strand DNA cleavage. Predict the structure of the by-product of uncialamycin that would result after DNA damage occurs. Do not worry about the stereochemistry.

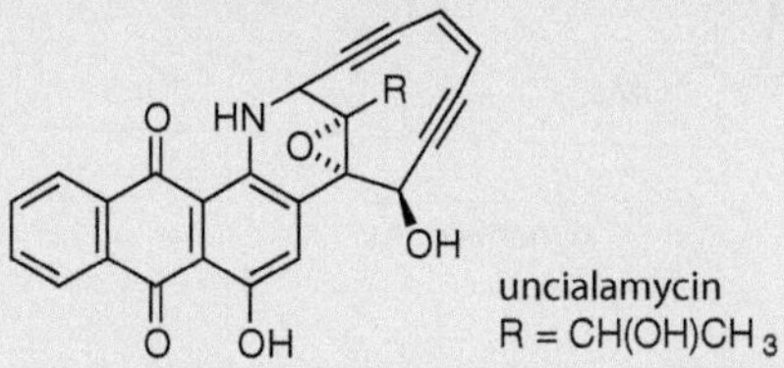

uncialamycin
R = CH(OH)CH$_3$

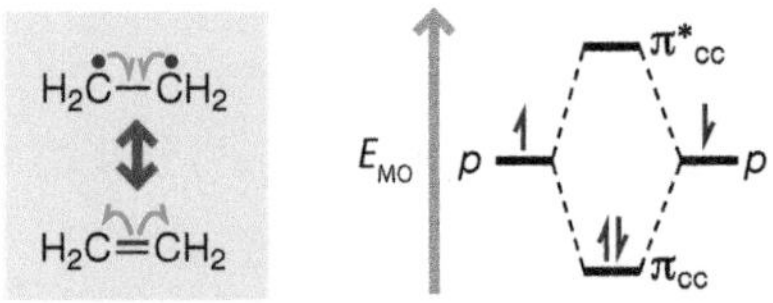

Figure 3.153 Spin city. Adjacent radicals can form bonds if and only if they have paired spins—one up and one down.

To understand the mechanism by which hydrogen-atom abstraction leads to DNA cleavage we will need to consider more carefully the character of the oxygen that you are breathing as you read this book. In general, we are well served by ignoring the fact that electrons have spins. Hund's rule tells us that electrons must have paired spins—one up and one down—to occupy the same orbital. We have simply assumed that the electrons that we represent with dots always have paired spins, allowing us to put them into the same bonding orbital (**Figure 3.153**). The assumption of paired spins underlies the notion that the most stable resonance structures (involving second-row atoms) will always have the most bonds, regardless of whether we discuss one-electron or two-electron processes. The assumption of paired spins breaks down for molecular oxygen. The two most energetic electrons in an oxygen molecule have the same spin and cannot be used to form a bond. Thus, oxygen is a diradical. There is a form of oxygen with an oxygen–oxygen double bond, but it is more than 20 kcal mol^{-1} higher in energy. Chemists use spin quantum numbers to distinguish between the two states (**Figure 3.154**): the more stable form of the oxygen that you breathe is called triplet oxygen; the less stable form of oxygen with a weak O=O π bond is called singlet oxygen. Triplet oxygen and singlet oxygen are not resonance depictions. You should resist the temptation to draw a pi bond between the atoms of molecular oxygen and count yourself lucky that the oxygen in the Earth's atmosphere is not singlet oxygen. The weak O=O π bond and low-lying π_{O-O}^* orbital make it a willing participant in countless reactions; examples are concerted Diels–Alder reactions, concerted ene reactions, and stepwise [2+2] cycloadditions. If you were living in an atmosphere of singlet oxygen, you might very well burst into flames.

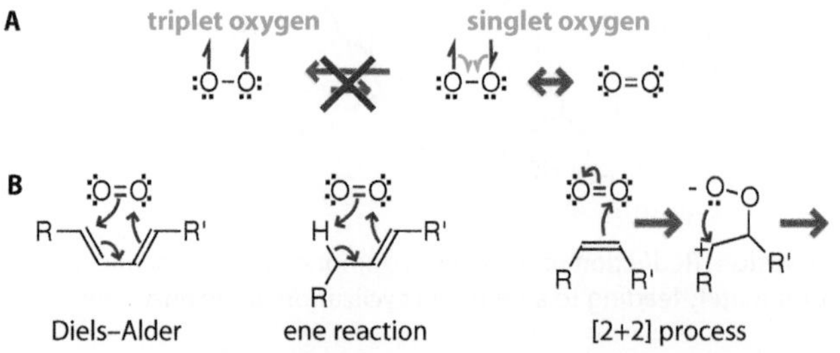

A triplet oxygen singlet oxygen

B Diels–Alder ene reaction [2+2] process

Figure 3.154 The importance of electron spin in O$_2$. (A) The diradical triplet form of oxygen is much more stable than singlet oxygen. (B) Singlet oxygen engages in many reactions.

Given the instability and extraordinary reactivity of singlet oxygen, we will restrict all further discussion of oxygen to the triplet state. Each breath you inspire is a cloud of reactive diradicals. You need them. Aerobic organisms use the reactivity of oxygen diradicals to power their biochemical machinery. However, the unpaired electrons of triplet oxygen are desperate for a partner (just not each other). Carbon-based radicals always recombine at phenomenal rates because it allows them to form a stable bond and satisfy the octet rule. Oxygen-based radicals do not recombine as readily because this leads to unstable O–O bonds. Imagine the enthusiasm with which carbon-based radicals react with molecular oxygen! If you generate a carbon radical on the DNA backbone, no matter how stable, every collision with molecular oxygen will generate a stable carbon–oxygen bond. The resulting oxygen-centered peroxy radical is stabilized by lone pairs on the adjacent oxygen, but it will rapidly abstract a hydrogen atom from several functional groups, including thiols and phenols (**Figure 3.155**).

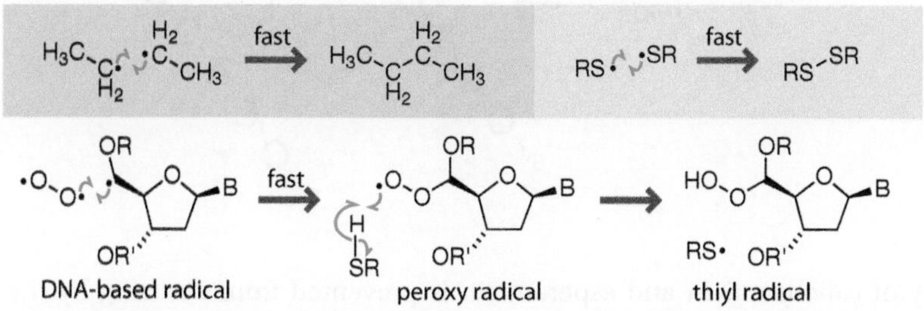

Figure 3.155 Radical recombination is fast if it forms stable C–C, S–S, or C–O bonds. Thus, DNA-based radicals recombine rapidly with oxygen. The peroxy radicals can abstract hydrogen atoms from thiols to generate thiyl radicals that recombine to form disulfides.

para-Benzyne diradicals always abstract hydrogen atoms from carbons that are adjacent to oxygen. When those carbon radicals react with O_2, they generate peroxyacetal functional groups. Acetals are labile, particularly if one of the oxygens is part of a good leaving group such as phosphate. Thus, when reactive radicals abstract hydrogen atoms from the 5′ carbon of DNA, it leads to direct strand cleavage (**Figure 3.156**). Hydrogen-atom abstraction from the 4′ carbon, 3′ carbon, and 1′ carbon also leads to strand cleavage via peroxyacetals, but the mechanisms are more complex.

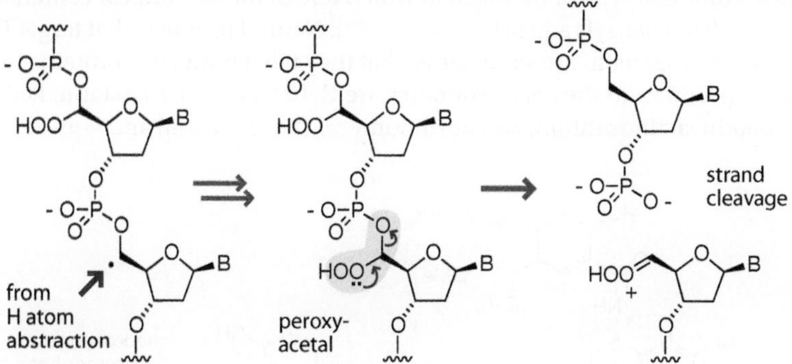

Figure 3.156 Carbon-based radicals recombine with O_2. When molecular oxygen recombines with DNA-based radicals it can generate labile functional groups like peroxyacetals.

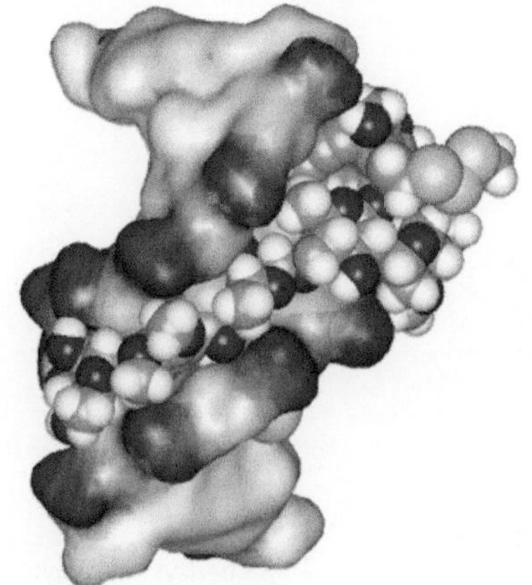

Figure 3.157 Snug fit. The oligosaccharide moiety of calicheamicin binds the minor groove of DNA. The trisulfide trigger (yellow) of calicheamicin is readily accessible to other thiols like glutathione.

The oligosaccharide groups dangling from calicheamicin and esperamicin bind selectively to the minor groove of double-stranded DNA (**Figure 3.157**). For calicheamicin, this binding event positions the enediyne moiety so that the resulting diradical efficiently abstracts hydrogen atoms from opposing strands. Calicheamicin and esperamicin are triggered in a different way from dynemicin. The enediyne

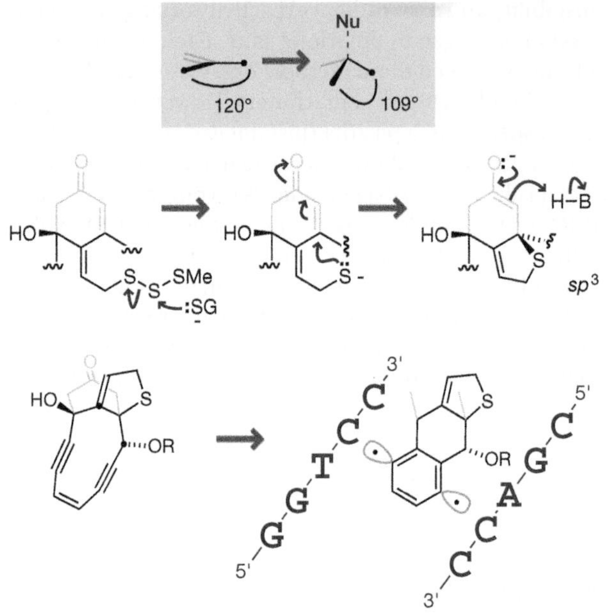

Figure 3.158 The trigger mechanism for calicheamicin involves pyramidalization. Calicheamicin binds selectively to the dsDNA sequence 5′-CCTCC-3′.

moiety of calicheamicin and esperamicin is prevented from cyclizing by the rigid bicyclic structure. Pyramidalization of a single carbon atom in the 10-membered ring serves as the conformational release that allows the enediyne cyclization to occur (**Figure 3.158**). The trigger mechanism for calicheamicin involves reductive cleavage of a trisulfide, to reveal a thiolate that undergoes conjugate addition to the cyclohexenone. The resulting conformational change and Bergman–Masamune rearrangement unleashes the reactive *para*-benzyne diradical. Calicheamicin is too toxic to be useful on its own, but a derivative of calicheamicin, chemically linked to an antibody that binds to leukemia cells, is sold under the trade name Mylotarg®.

Some highly reactive enediyne natural products are protected by protein delivery vehicles

Enyne allenes can also undergo electrocyclic ring closures related to the Bergman–Masamune rearrangement. This analogous process, called the Myers rearrangement, generates a different type of diradical in which one of the two radical centers is a resonance-stabilized benzyl radical (**Figure 3.159**). Natural products that target DNA via the Myers rearrangement are so sensitive that they need protective proteins to deliver the natural products to the cells; examples are C-1027, neocarzinostatin, kedarcidin, maduropeptin, actinoxanthin, and auromomycin (**Figure 3.160** and **Figure 3.161**).

Figure 3.159 The Myers rearrangement. The electrocyclic ring closure of enyne allenes, named after Andy Myers, generates a reactive species with two types of radicals, a benzyl radical that is stabilized by resonance and a phenyl radical that is not.

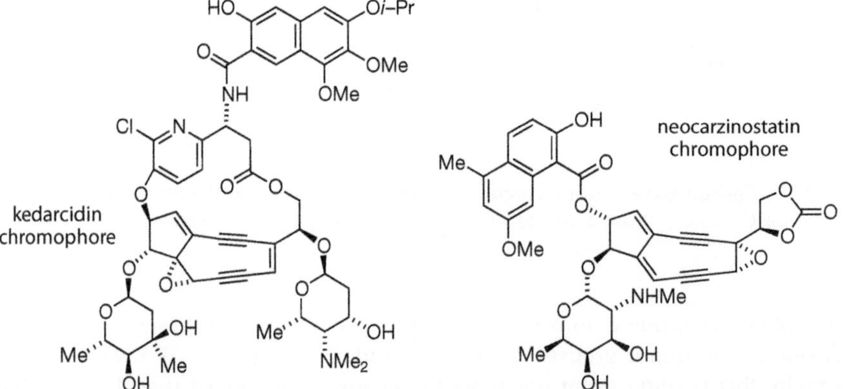

Figure 3.160 Naked enediynes. These highly sensitive enediynes are normally complexed with protective proteins.

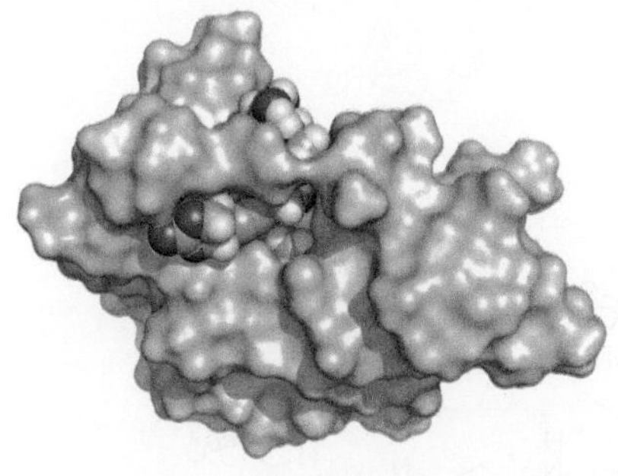

Figure 3.161 A Trojan horse. The small molecule chromophore of neocarzinostatin is delivered by a protein. As described in Virgil's poem *The Aeneid*, Greek warriors hid inside a wooden horse that was carried into the besieged city of Troy by unsuspecting Trojans. (Image from R. Robbins, The World Displayed in its History and Geography, W.W. Reed, 1830. Left, PDB 1NCO.)

The enyne allene ligand of neocarzinostatin is bound very tightly and specifically to the protein, with a dissociation constant of 10^{-10} M. The protein carrier acts like the fabled Trojan horse, protecting the lethal cargo until it is delivered to the target cell (see Figure 3.161). The small organic ligands in these complexes are often referred to as the chromophores because the ligands exhibit prominent ultraviolet absorption spectra very different from those of proteins. The **apoproteins** (that is, proteins without ligand) have no activity.

The mechanism of action of these enediyne relatives involves an initial activation by glutathione, an abundant cellular thiol that will be discussed more extensively in Chapter 5, followed by Myers cyclization that generates a reactive diradical. Both of the radicals are sp^2 hybridized and orthogonal to the benzene pi system, preventing resonance stabilization (**Figure 3.162**). Intercalation of the naphthalene ring of neocarzinostatin between DNA base pairs is believed to precisely place the highly reactive diradical for H-atom abstraction on opposing strands.

Figure 3.162 Neocarzinostatin mechanism. When bound to DNA, epoxide opening and thiolate addition lead to a Myers rearrangement of the neocarzinostatin chromophore. The resulting diradical is poised to abstract hydrogen atoms from opposite strands of DNA. (Model, from A. Galat and I.H. Goldberg, *Nucleic Acids Res.* 18: 2093–2099, 1990. With permission from Oxford University Press.)

Problem 3.24

The following model of neocarzinostatin reacts at temperatures above –60 °C. Suggest a plausible fishhook arrow-pushing mechanism.

Bleomycin catalyzes the formation of reactive oxygen species

Nature has developed even more devious mechanisms for abstracting hydrogen atoms from the DNA backbone. The bleomycins are a closely related set of natural

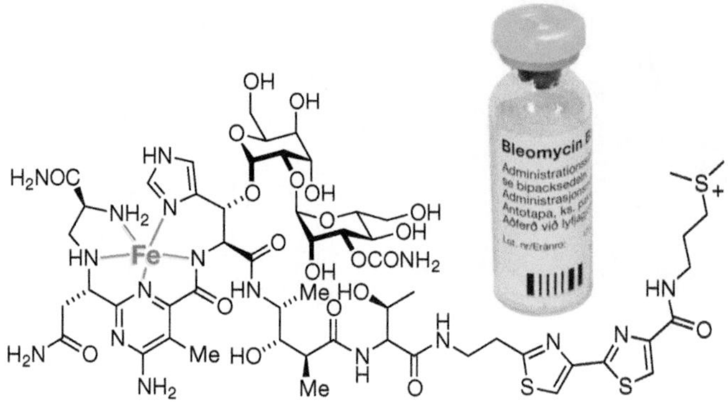

Figure 3.163 Tight grip. In this proposed structure for the active complex of bleomycin A_2, nitrogen atoms tightly grip a catalytically active iron atom, bringing it into close proximity with the DNA backbone. Bleomycin is used clinically in the treatment of cancer. (Courtesy of Haukeland University Hospital.)

products isolated from the mycobacterium *Streptomyces verticillus* (**Figure 3.163**). They enter cells, stealing iron atoms on their way to the minor groove of DNA, where the iron complex nestles comfortably. The bleomycin•Fe(II) complex then catalyzes the reduction of oxygen to generate reactive oxygen species, such as the insanely reactive HO•, that abstract hydrogen atoms from the DNA backbone. However, there is also evidence that points to an oxoferryl (Fe=O) species as the key intermediate leading to the insertion of oxygen atoms into C–H bonds. In this respect, the mechanism of action of Fe(II) bleomycin resembles that of the CYP enzymes, which harness an Fe(II)•porphyrin complex to epoxidize polycyclic aromatic hydrocarbons.

3.11 SUMMARY

DNA is the most precious molecule in the cell. It provides the blueprint for all biosynthesis in the cell and is the molecular basis for heredity. The DNA double helix is held together by a combination of π stacking and Watson–Crick base pairing. The asymmetry inherent in Watson–Crick base pairs leads to the presence of two grooves in the double helix: a wide, information-rich major groove and a deep, narrow minor groove. Most transcription factors and restriction endonucleases recognize specific sequences by forming hydrogen bonds with functional groups in the major groove. Small organic molecules tend to interact with DNA by intercalating between base pairs or snuggling within the minor groove. Watson–Crick base pairing drives the highly predictable self-assembly of duplex structures. The pairing of A with T and G with C is so predictable that chemical structures are unnecessary for most experiments involving DNA. However, a modern view of DNA goes beyond the nucleotide sequence. Human DNA is reprogrammed by enzymes that methylate the DNA bases and modify the histone proteins. The heritability of these modifications is driving the growing field of epigenetics.

A variety of chemical and biological tools have converged to empower chemical biologists. DNA polymerases are among the most important because they allow one to amplify oligonucleotides. The mechanism of DNA polymerases exhibits little variation among species. All DNA polymerases use Mg•NTPs as substrates. All DNA polymerases add new nucleotide subunits to the 3′ end of the growing strand. All DNA polymerases require a primer strand, which they lengthen, and a template strand. Restriction endonucleases and DNA ligases make it possible to easily cut genes from one genome and insert them into another genome, or into bacterial plasmids for fast inexpensive protein expression. Phosphoramidite chemistry has been optimized so successfully that oligonucleotides are readily synthesized by machines. Those oligonucleotides can be used as primers for PCR amplification, as probes for hybridization experiments, as tools for mutagenesis, and as building blocks for nanoengineering. Scientists have recently used synthetic oligonucleotides as the starting material for the creation of the first synthetic organism.

Electrophoresis can be used to separate DNA on the basis of size. Agarose gel electrophoresis is best for long DNA strands, whereas polyacrylamide gel electrophoresis is best for short strands, like those generated in Sanger dideoxy sequencing. For high-throughput sequencing, two-dimensional gels have been replaced by capillary columns. However, the field of genomics has driven the development of sequencing methods that obviate the need for electrophoresis.

Cell-cycle checkpoints make DNA the Achilles' heel of cancers. Cells have a wide range of repair systems that can mend damage to a single strand, but reactions that crosslink or cut both strands usually lead to apoptosis. Special tricks are required to bring reactive species into proximity with DNA. The most successful DNA-damaging drugs use strained rings to alkylate DNA bases or generate radicals that abstract hydrogen from the ribose backbone. Unfortunately, DNA damage is indiscriminate and does not focus on the molecular mechanisms that make cancer cell lines abnormal. To understand those differences we will need to delve deeper into the workings of human cells, following the path from DNA to transcription and ultimately to the pathways that control transcription.

LEARNING OUTCOMES

- Become conversant with nucleotide nomenclature and numbering.
- Translate between letter sequences and chemical structures.
- Identify major and minor grooves and the functional groups they contain.
- Apply the Wallace rule for DNA melting temperatures to oligonucleotides.
- Predict folding and hybridization on the basis of DNA sequence.
- Design reactions involving DNA polymerase, including PCR and dideoxy sequencing.
- Draw arrow-pushing mechanisms involved in the chemical synthesis of DNA.
- Design experiments using rudimentary tools of molecular biology: DNA polymerase, restriction endonucleases, and DNA ligases.
- Understand which parts of DNA are susceptible to chemical damage.
- Recognize molecules that are likely to damage DNA: strained rings and radical precursors.

PROBLEMS

3.25 For each of the following heterocycles, decide whether the resonance structure drawn represents an aromatic ring system.

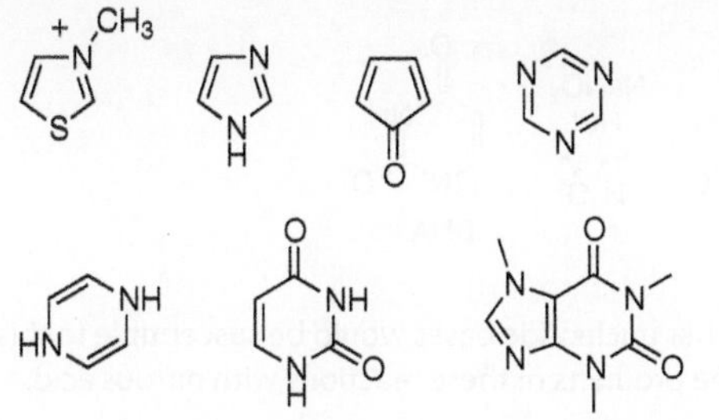

***3.26** Draw chemical structures for each of the following oligonucleotides. Pay attention to clues that distinguish DNA from RNA. You may abbreviate the heterocyclic bases as Ade, Cyt, Gua, Thy, Ura.

A d(GACA)
***B** pTATA
C GUCU

***3.27** The designed nucleoside shown engages in a specific Watson–Crick base pair with one of the four DNA bases. Draw the Watson–Crick base pair, clearly indicating the hydrogen bonds.

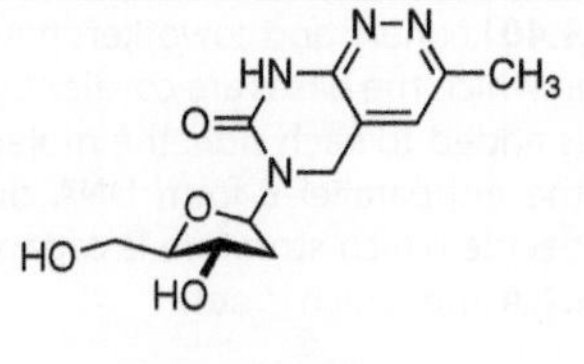

3.28 What is the most precise relationship (for example none, constitutional isomers, diastereomers, or enantiomers) between the following pairs of dsDNA?

5'-CGCGCG-3'
3'-GCGCGC-5'
vs.
5'-GCGCGC-3'
3'-CGCGCG-5'

3.29 Write out the DNA sequences that are complementary to the following single-stranded DNA sequences.

A 5′-TATAATCGTTACTGAATGTCTT-3′ a Pribnow box sequence
***B** 5′-TATAAAAGTCTTTGTAACCTTG-3′ a Hogness/TATA box sequence
C 5′-GGCCAATCTTGTGCTTCTAGAT-3′ a CAAT box sequence

***3.30** The anticancer drug cisplatin, $(H_3N)_2PtCl_2$, coordinates tightly to double-stranded DNA. Access the crystal structure for the duplex d(CCTCTGGTCTCC)•cisplatin complex (PDB 1AIO) at the Protein Data Bank Web site and examine the interaction between platinum and DNA.

A Which atoms in which bases are crosslinked by the platinum atom?
B Which groove of B DNA is occupied by the platinum atom?
C Does the platinum generate an interstrand crosslink (between opposite strands) or an intrastrand crosslink (within a single strand)?

3.31 Write out the mRNA sequence (in the 5′ to 3′ direction) that is complementary to each of the following single-stranded DNA sequences.

***A** 3′-CATAGCTGTCCTCCT-5′ a Shine–Dalgarno sequence
B 3′-CATGGTGGT-5′ a Kozak sequence

3.32 Write out the complementary DNA sequence for each of the following RNA molecules.
***A** 5′-UCGAAUGCAUUAUUCGU-3′
B 5′-GCUUUACGUUGUCAAUG-3′

***3.33** When the following DNA oligomers were subjected to a special nickel oxidant system, only the underlined guanosine residues were susceptible to oxidation. (When two oligonucleotides are shown, the second was not tested for oxidation.)

i. 5′-CATGCGTTCCCGTG-3′
ii. 5′-CATGCGTTCCCGTG-3′ + 5′-CACGGGAACGCATG-3′
iii. 5′-AGTCTAGTAGACT-3′
iv. 5′-ACGTCAGGTGGCAT-3′ + 3′-TGCAGTCACCGTGA-5′
v. 5′-AGTCTATGGGTTAGACT-3′

A For each of the substrates i–v, predict whether strands can fold to form a stable hairpin or associate to form a stable double-stranded duplex.
B Suggest a plausible rationale for why some base pairs are susceptible to oxidation whereas other base pairs are not.

3.34 Chiron Corporation patented the following DNA sequence because it selectively hybridizes with DNA from hepatitis C virus. What is the melting temperature (T_m) of the complex that would form by hybridization with a much longer strand?

5′-CCTGGTTGCTCTTTCTCTATCT-3′

3.35
A Predict the sequence of the product(s) that would be formed if a DNA polymerase reaction were performed on the following single-stranded oligonucleotide (at a 1 μmol scale) using DNA polymerase, excess primer, a mixture of dNTPs, and $MgCl_2$.
ssDNA: 3′-ACTGGCCGTCGTTTTACAACGTACGTACGTACGTACGT ACGTACGTACGT-5′
Primer: 5′-TGACCGGCAGCAAAATGT-3′
B What product would you expect if the dATP was replaced with 2′,3′-dideoxyadenosine triphosphate (ddATP)?

Mg•ddATP

C List the sequences and yields of the four major products you would expect if one-quarter of the dATP was replaced with 2′,3′-dideoxyadenosine triphosphate (ddATP).

3.36 Peptide nucleic acids (PNAs) are chemical analogs of oligonucleotides based on amide bonds. They can hybridize with DNA or RNA and exhibit better membrane permeability than either DNA or RNA. PNAs are more flexible than DNA. Identify the freely rotatable bonds in each subunit of DNA and each subunit of PNA.

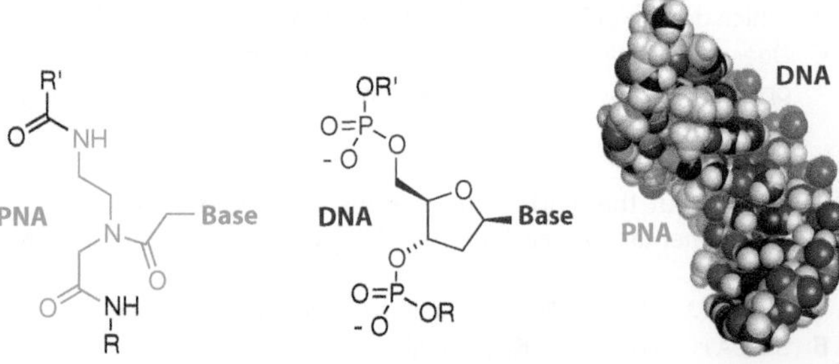

***3.37** Photolabile protecting groups can be used to trigger the instantaneous release of bioactive molecules inside cells. Predict the two organic fragments that would result during photodeprotection of this caged form of calcium ions.

$h\nu$, H_2O → Ca^{2+}

***3.38** As we saw from the chemistry of *N*-nitrosodimethylamine, *N*-nitroso compounds can lead to methylation of DNA through the formation of diazonium ion intermediates. Diazonium ion intermediates can have other roles in the chemistry of DNA.
A Diazonium salts are generally made using nitrous acid (HO–N=O), which must be prepared *in situ* from sodium nitrite in aqueous acid. Suggest a mechanism for the formation of benzenediazonium chloride in aqueous solution. You may have to go back to an undergraduate textbook to refresh your understanding of this reaction.

$NaNO_2$, HCl, H_2O

B Deoxycytidylate residues hydrolyze at very slow rates into deoxyuridine. This chemical change could result in a mutation from a CG base pair to an AT base pair, but most cells have an efficient repair system that searches dsDNA for dU residues and replaces them with dC. Nitrous acid converts deoxycytidylate residues to deoxyuridylate residues too fast for the normal DNA repair systems to find and correct all the changes. Suggest a mechanism for this chemical reaction.

$NaNO_2$, HCl, H_2O

C What other nucleotide bases would be susceptible to this reaction? Show the products of these reactions with nitrous acid.

3.39 When DNA is treated with busulfan (sold as Myleran®), 1,4-bis(7-guanyl)butane crosslinks are formed. Suggest a plausible arrow-pushing mechanism for the formation of these crosslinks between guanine bases.

busulfan

3.40 Leonard and coworkers have prepared a mimic of an A•T base pair in which the bases are covalently fused together. When one nucleotide is added to each side, the molecule can form a π-stacked segment of the antiparallel B-form DNA duplex. For the two structures below, decide which structure is capable of forming a π-stacked segment of DNA and which is not.

A•T mimic

A•T mimic

***3.41**

A Diazomethane reacts instantaneously with carboxylic acids to give the corresponding methyl esters in quantitative yields. Suggest a mechanism for this esterification reaction.

diazomethane

$H_2C{=}N_2$ + $HO{-}C({=}O){-}R$ ⟶ $H_3C{-}O{-}C({=}O){-}R$

B Diazomethane is often generated from a rather strange-looking compound, *N*-methyl-*N*-nitrosoguanidine, using aqueous sodium hydroxide below room temperature. Suggest a plausible arrow-pushing mechanism for the formation of diazomethane.

pK_a 7.7

cat. base

N-methyl-*N*-nitrosoguanidine

***3.42** A library of 2×10^{11} DNA molecules 54 nucleotides long was synthesized by incorporating a random mixture of the four phosphoramidites (A, C, G, and T) into the 30 middle positions. The fixed sequences on the ends were included to facilitate PCR amplification.

5'-GGGAGAATTCCCAGACCNNNNNNNNNNNNNNNNNNNNNNNNNNNNNNCTGAGGGAAATTCTCCC-3'

30 nucleotides

A What is the theoretical diversity of this DNA library (in base 10)? In other words, how many members of this library are possible?

B What are the sequences of the two PCR primers that would be used to amplify any member(s) of the DNA library?

C A selection was then performed (do not worry about how) to identify DNA molecules that would selectively bind the amino acid arginine. This sequence was found to fold into a hairpin through intramolecular Watson–Crick base pairing. Sketch the structure of this hairpin, clearly indicating which nucleotides are involved in base pairs.

DNA sequence that binds arginine

5'-**GGGATCGAAACGTAGCGCCTTCGATCCC**-3'

D The three-dimensional structure of the hairpin was determined by NMR, revealing a deep pocket that bound the arginine side chain. The arginine side chain formed a strong interaction with one of the cytidine residues that was not involved in Watson–Crick base pairing. Sketch a plausible structure for this interaction between a cytosine base and an arginine side chain.

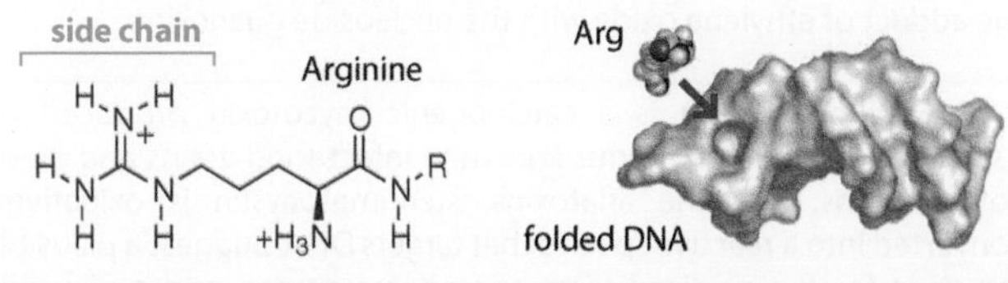

3.43 Cyclophosphamide is the most commonly prescribed alkylating agent used in the treatment of cancer. Like benzo[*a*]pyrene, cyclophosphamide is activated by a cytochrome P450 enzyme. The phosphoramide mustard has been shown to crosslink dG to dG in double-stranded GAC sequences. Suggest an arrow-pushing mechanism for the formation of crosslinks starting from the hydroxylated cyclophosphamide intermediate. Do not worry about the acrolein by-product.

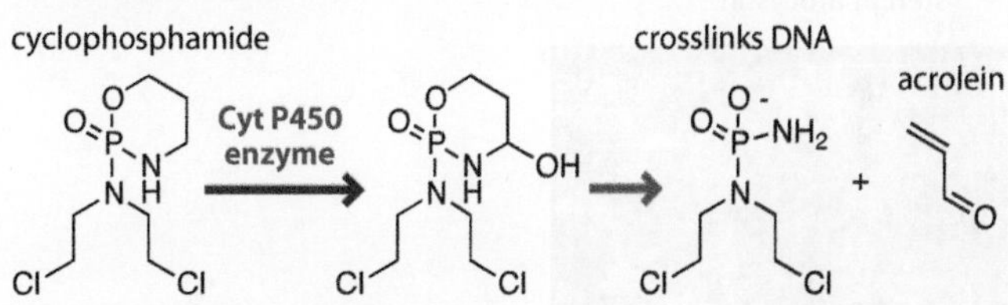

***3.44** Illudin S is a DNA-alkylating agent from the jack-o-lantern mushroom. Suggest a plausible arrow-pushing mechanism for the activation of illudin S with thiols and the subsequent alkylation of DNA. (Image, courtesy of Jason Hollinger.)

RSH
pH 7
illudin S

3.45 Both temozolomide and mitozolomide form the same active intermediate, MITC, in the human body. Suggest a plausible arrow-pushing mechanism for the alkylation of DNA by MITC.

temozolomide

> pH 7

MITC

+ DNA methylation

metabolism

mitozolomide

3.46 The epoxide ethylene oxide (C_2H_2O) is an Occupational Safety and Health Administration class III carcinogen, yet it is used to sterilize medical equipment and food spices. Ethylene oxide reacts with a variety of protein functional groups and with DNA. The principal site of attack on DNA is the N7 position of guanine bases. Draw a structure of the adduct of ethylene oxide with the nucleoside guanosine.

3.47 Sterigmatocystin is a carcinogenic mycotoxin produced by *Aspergillus versicolor*, a fungus known to infect food grains and green coffee beans. Like the aflatoxins, sterigmatocystin is oxidatively converted into a reactive species that targets DNA. Suggest a plausible structure for the oxidized form of sterigmatocystin and a plausible arrow-pushing mechanism involving acid catalysis for the reaction of the metabolite with DNA. (Image, courtesy of Dennis Kunkel Microscopy, Inc.)

sterigmatocystin —[O]→ sterigmatocystin metabolite —DNA→ adduct

***3.48** Adozelesin preferentially alkylates at the 3′ end of the trinucleotide sequences AAA, TAA, TTA, and ATA in double-stranded DNA. Draw a chemical structure of the adenine/adozelesin adducts that result from the alkylation reaction.

adozelesin

3.49 The antitumor agent bizelesin was found to form interstrand crosslinks at the underlined positions within the double-stranded sequence G<u>T</u>ACTA<u>A</u>GT in the *p53* gene. Suggest a plausible arrow-pushing mechanism for the formation of DNA crosslinks and include the structure of the final crosslinked adenine bases.

bizelesin

3.50 Suggest a plausible arrow-pushing mechanism for the following Bergman reactions.

***A**

+ CH_3OH —heat→

B

+ CH_3OH —180 °C, sealed tube→

***3.51** Suggest a plausible arrow-pushing mechanism for the Myers reaction using 1,4-cyclohexadiene as a hydrogen atom donor.

benzene, 80 °C

3.52 Myers rearrangement in a polar solvent gives two types of products, A and B. Product A can be explained on the basis of a diradical intermediate; however, product B cannot be explained on the basis of a diradical intermediate. Offer an alternative way to push arrows that leads to an alternative "resonance structure" of the reactive intermediate to explain the formation of product B.

+ CH_3OH —CH_3OH, 80 °C→ **A** + **B**

3.53 The following model of neocarzinostatin reacts at temperatures above −60 °C. Suggest a plausible mechanism.

> -60 °C

***3.54** The marine natural product shishijimicin C recently isolated from the ascidian *Didemnum proliferum* is structurally related to the known enediyne natural products namenamicin and calicheamicin.

A Shishijimicin C is active at about a 10-fold lower concentration than namenamicin. Suggest a specific explanation for the enhanced potency of shishijimicin C relative to namenamicin.

B Draw the aromatic product that would be formed from shishijimicin C if it cleaves double-stranded DNA by a mechanism similar to that of calicheamicin.

namenamicin   shishijimicin C

3.55 The following rearrangement of ketenes is referred to as a Moore rearrangement. Suggest a plausible arrow-pushing mechanism for the second part of the reaction. Do not worry about the mechanism of the Wolff rearrangement.

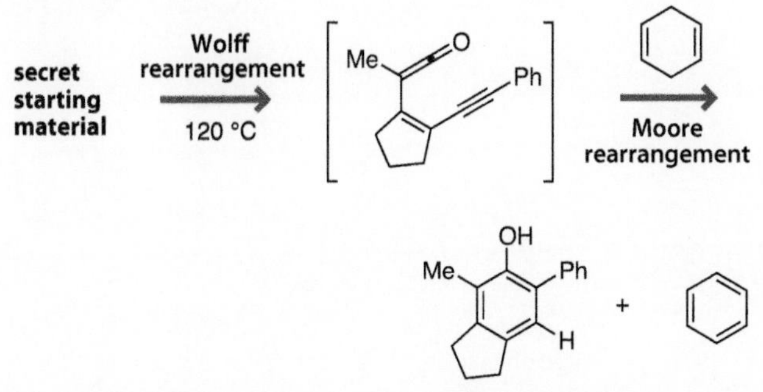

secret starting material —**Wolff rearrangement**, 120 °C→ [Me, O, Ph] —**Moore rearrangement**→ OH, Me, Ph, H +

3.56 When heated at 50 °C, the enediyne below reacts with 1,4-cyclohexadiene through a Bergman cyclization. Predict the products of the reaction with cyclohexadiene.

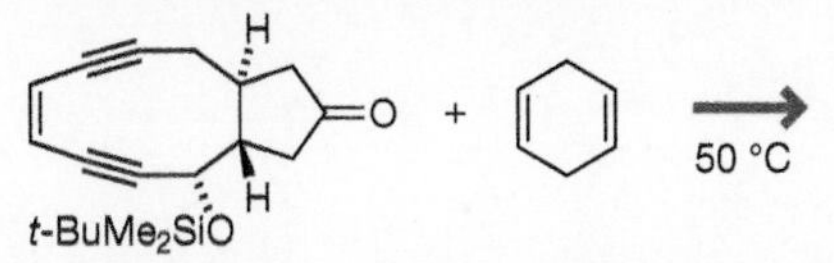

H, O, H, *t*-$BuMe_2SiO$ + → 50 °C

***3.57** The oligosaccharide moiety of calicheamicin $\gamma_1{}^{I}$ binds selectively to dsDNA at the sequence 5′-TCCT-3′. By dimerizing the calicheamicin $\gamma_1{}^{I}$ with a special linker ($-OCH_2CH_2OCH_2CH_2O-$) that spans any two nucleotides (NN), what 10-base dsDNA sequence would you expect to be targeted?

synthetic dimer of oligosaccharide from calicheamicin $\gamma_1{}^{I}$

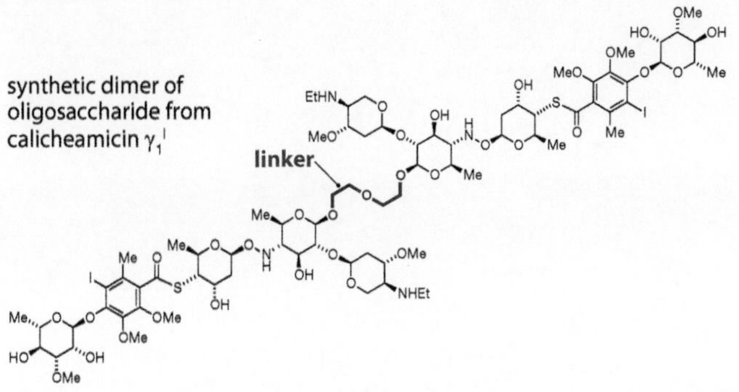

linker

***3.58** Cyanosporasides A and B have been isolated from the culture broth of a marine organism. It was hypothesized that the cyanosporasides derive from a Bergman cyclization of an enediyne precursor that reacts further through ionic reactions instead of radical reactions. Suggest a structure of the enediyne precursor that might lead to the cyanosporasides.

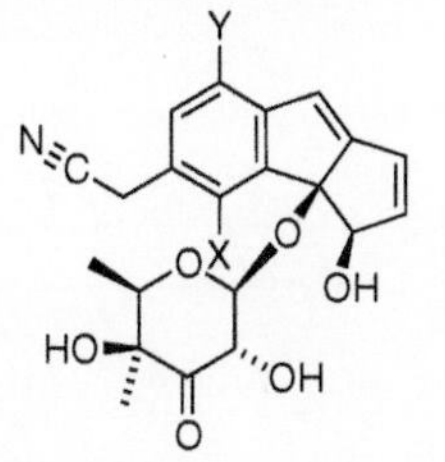

cyanosporaside A X=Cl,Y=H
cyanosporaside B X=H,Y=Cl

3.59 Tallimustine is a synthetic antitumor compound that preferentially generates double-stranded DNA crosslinks in the sequence 5′-ATTTTGAT-3′.

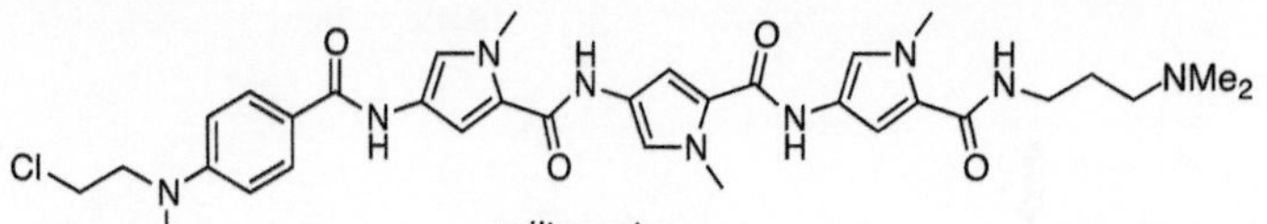

tallimustine

A Predict the conformation of tallimustine that would form the most specific interaction with double-stranded DNA.

B Propose a plausible arrow-pushing mechanism for the crosslinking reaction.

C Predict which sites on DNA should be most reactive.

RNA

4

LEARNING OBJECTIVES

- Understand the chemical and structural differences between RNA and DNA.
- Describe the features of a DNA gene sequence that determine when genes will be expressed and what part of the gene will be transcribed.
- Draw a scheme showing the synthesis, processing, and translation of RNA in eukaryotic cells.
- Explain how RNA interference can be used to control gene expression in cells.
- Describe the features of an mRNA sequence that control ribosomal binding and what part of the mRNA will be translated.
- Explain the role of tRNA and elongation factors in translation.
- Describe the steps needed to produce a combinatorial library of proteins.

Phoebus Aaron Theodore Levene (**Figure 4.1**) studied both RNA and DNA, and contributed key insights into the structures of both biopolymers. When Levene began his research at the Rockefeller Institute for Medical Research, precipitation techniques had been used to isolate two different types of nucleic acids—one from yeast and the other from calf thymus. In 1909, Levene showed that the carbohydrate in the nucleic acids isolated from yeast was the pentose sugar ribose, correctly identifying the material from yeast as RNA. Twenty years later, Levene also correctly identified the equivalent carbohydrate in material extracted from calf thymus as deoxyribose. Additionally, Levene deduced key structural elements of nucleic acids; for example, he coined the term "nucleoside" to describe a carbohydrate linked through a glycosidic bond to the base of a nucleic acid. With *tour de force* experimentation, he also discovered that the phosphodiesters link to the 3′ and 5′ carbons of the nucleic acid carbohydrate. Thus, Levene earned a spot in the pantheon of chemical biology through key insights into the chemical structures of RNA and DNA.

When Francis Crick first proposed the central dogma in 1956, RNA was considered a minor league player charged merely with transducing information. The attentions of biochemists remained firmly fixed on DNA and proteins. However, information about and appreciation for the myriad roles of RNA have exploded over the past 25 years. RNA molecules assume diverse roles in the cell, including courier, craftsman, interpreter, and sentry. Given the chemical similarity between DNA and RNA, it may seem strange that DNA is assigned one principal role while RNA is assigned many diverse roles. The great agility of cellular RNA relative to cellular DNA is principally due to the fact that RNA is single-stranded, and any single-stranded oligonucleotide will tend to hybridize with itself to form a three-dimensional shape. In contrast, cellular DNA is generally double-stranded, leading to a monotonous unidimensional form.

It has been hypothesized that life began with RNA molecules rather than DNA or proteins. The idea of RNA as the primordial molecule is compelling because we can explain the formation of RNA from prebiotic building blocks such as HCN and formaldehyde. In addition, RNA exhibits key capabilities, including information storage and catalysis. Proteins have no role in cell-free information storage. Similarly, DNA has no catalytic role in the chemistry underlying life processes. RNA is a jack-of-all-trades but master of none. A primordial RNA world would have needed to evolve biooligomers better suited for information storage, structure, and catalysis—namely, biooligomers

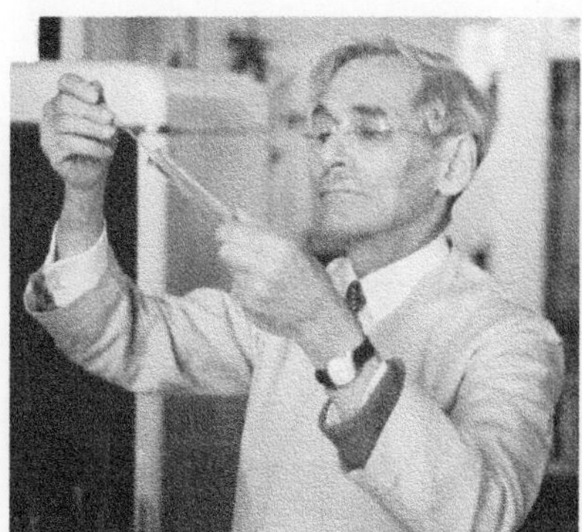

Figure 4.1 RNA pioneer. Phoebus Levene (1869–1940) defined the basic structure of oligonucleotides, deoxyribose, and both DNA and RNA, as phosphate esters of ribose. (Courtesy of the Rockefeller Archive Center.)

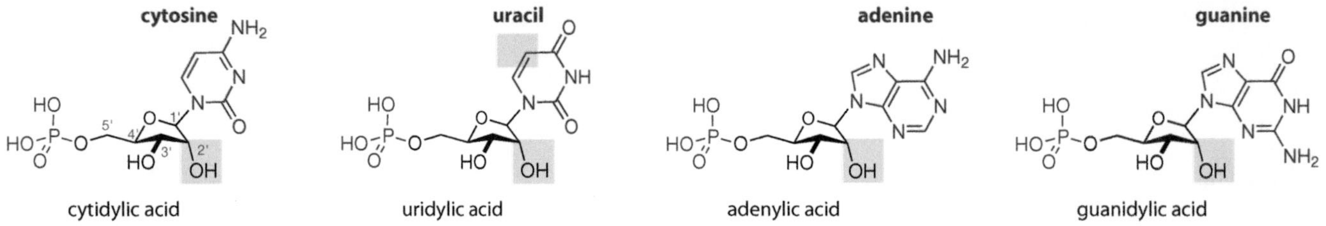

cytidylic acid uridylic acid adenylic acid guanidylic acid

Figure 4.2 How does RNA differ from DNA? RNA is composed of four types of nucleotide subunits. Structural features that distinguish RNA subunits from DNA subunits are highlighted in green.

such as DNA and proteins. Are the diverse roles of mammalian RNA vestigial roles from an RNA world or roles earned through the meritocracy of evolution? The answer to this question may reveal itself in another four billion years. To help you consider such questions this chapter surveys RNA structure, synthesis, and functions and its many roles in the central dogma from transcription to translation.

4.1 RNA STRUCTURE

The nucleotide subunits of RNA are subtly different from those of DNA

RNA is made up of four nucleotide subunits that are analogous to the nucleotide subunits in DNA. Those similarities give RNA strands the ability to hybridize with complementary DNA strands through Watson–Crick base pairing. Two structural differences between the nucleotide subunits of RNA and the nucleotide subunits of DNA result in major functional consequences for RNA. First, the nucleotide subunits of RNA are complete ribosyl sugars with a hydroxyl substituent at the 2′ position (**Figure 4.2**). In contrast, 2′-deoxyribonucleic acid lacks this 2′-hydroxyl group. Second, the heterocyclic substituent that base pairs with adenine is uracil, not thymine. The only difference between uracil and thymine is that uracil lacks the 5-methyl substituent of thymine.

The 2′-OH of RNA confers high chemical reactivity

Recall from Chapter 2 that the presence of the 2′-hydroxyl group in RNA makes the phosphodiester bonds about 100 times more susceptible to base-promoted hydrolysis. To understand why this group confers lability, it is instructive to recall the difficulty of attacking at an anionic phosphate diester, highlighted in Chapter 2. Attack on the phosphorus atom of dimethyl phosphate is so slow that half of the hydrolyzed product actually arises through an S_N2 attack on one of the unhindered methyl groups, leading to C–O bond cleavage (**Figure 4.3**). In contrast, a 2-hydroxyethyl substituent makes the phosphate diester highly unstable toward base, even at room temperature.

The 2′-hydroxyl group of an RNA nucleotide is about 1000 times more acidic than the hydroxyl group of a typical alcohol. Once deprotonated, the 2′-hydroxyl group is poised to attack the phosphate group, generating a five-membered ring cyclic phosphate diester (**Figure 4.4**). Even though the cyclic phosphate forms rapidly, it

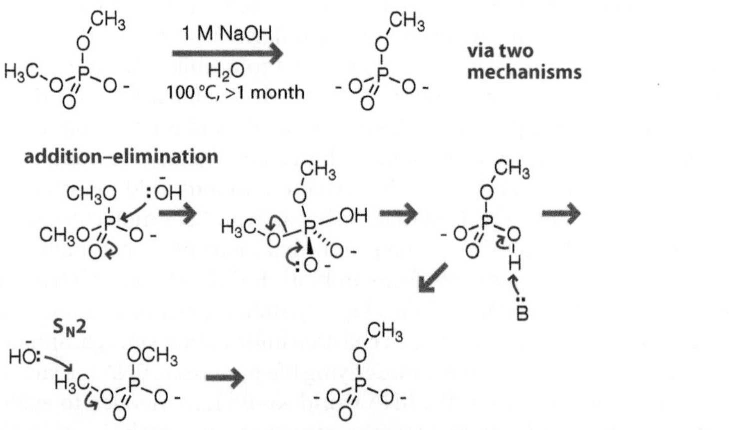

Figure 4.3 Phosphodiester cleavage. Most simple phosphate diesters, like those in DNA, resist intermolecular attack at phosphorus.

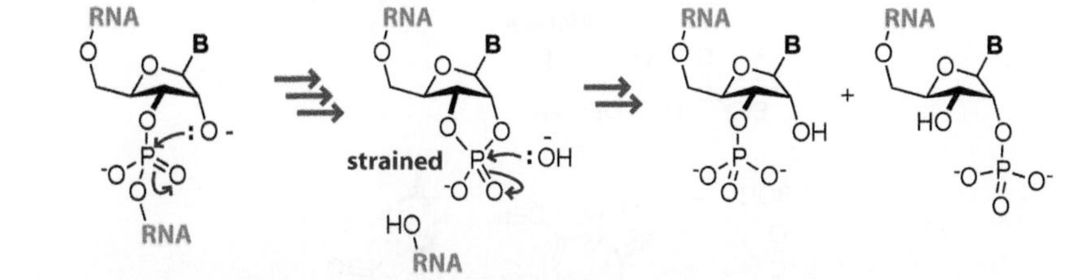

Figure 4.4 Assistance from the 2′-hydroxyl group. The 2′-hydroxyl group of RNA facilitates chemical cleavage in aqueous sodium hydroxide by forming a cyclic phosphate ester.

is actually strained, and attack on this cyclic ester occurs a million times faster than attack on a simple acyclic phosphate diester, ultimately leading to a mixture of 2′- and 3′-phosphates.

Problem 4.1

Which of the following phosphodiesters will undergo hydrolysis more rapidly in aqueous sodium hydroxide? *Hint:* Draw chair conformations of the cyclohexane ring.

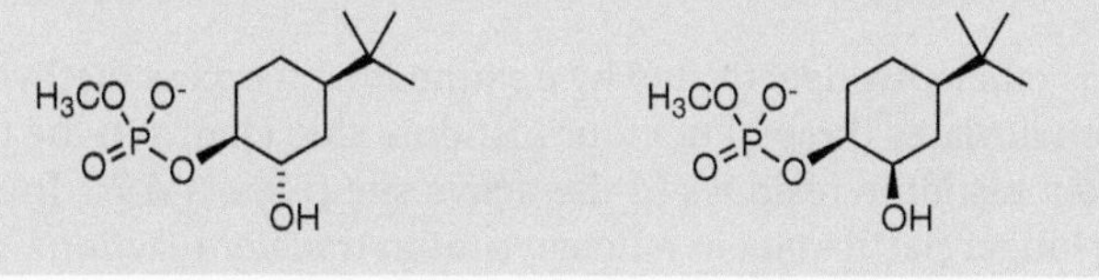

Ubiquitous ribonucleases rapidly degrade RNA

Humans have at least seven different extracellular ribonucleases, probably as a part of the innate immunity against potential retroviral invaders. Because human skin is rife with ribonucleases they end up everywhere and contaminate everything. Extreme care must be taken when handling RNA samples or any laboratory equipment that will come into contact with RNA samples. RNA-hydrolyzing enzymes such as ribonuclease A (RNase A) take advantage of the 2′-hydroxyl group to effect the rapid cleavage of RNA molecules. RNase A was one of the earliest enzymes to receive a thorough mechanistic study. In the 1940s the hot dog company Armour Inc. isolated and purified more than 1 kg of pancreatic RNase A and made it readily available to researchers in 10 mg batches (**Figure 4.5**).

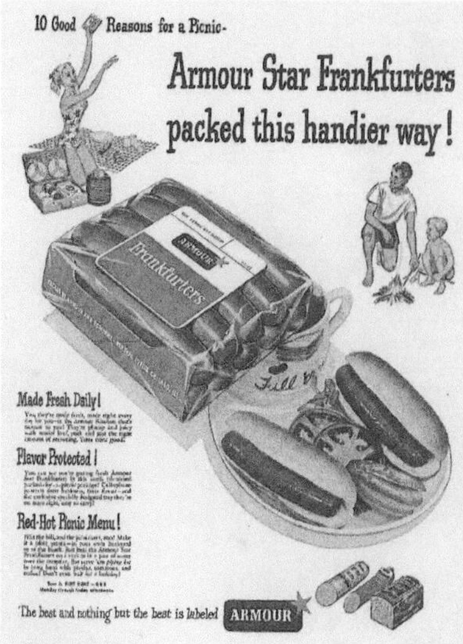

Figure 4.5 A 1948 advertisement for Armour frankfurters. This appeared in *American Home* in September 1948. Before the development of recombinant DNA technology, slaughterhouses were the most abundant source of bovine (cow) and porcine (pig) enzymes, including RNase. (Courtesy of Armour and Company.)

Several residues hold the substrate in the active site. A cationic lysine side chain positions the anionic phosphate group with surgical precision. In the active site of RNase A, two catalytic imidazoles—the side chains of histidine—are exquisitely positioned to act as acid and base (**Figure 4.6**). The substrate and imidazoles are held

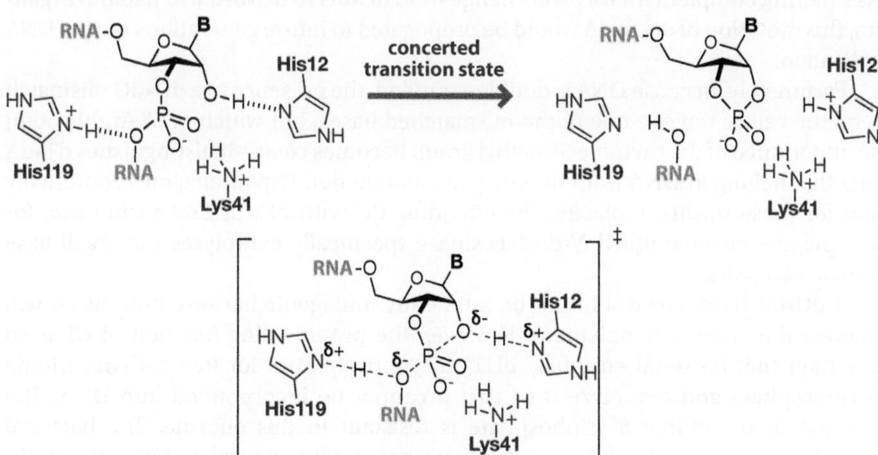

Figure 4.6 The role of catalytic histidine residues in the active site of ribonuclease A. The substrate and two histidine side chains are held in such perfect alignment that addition of the nucleophile and expulsion of the leaving group are concerted.

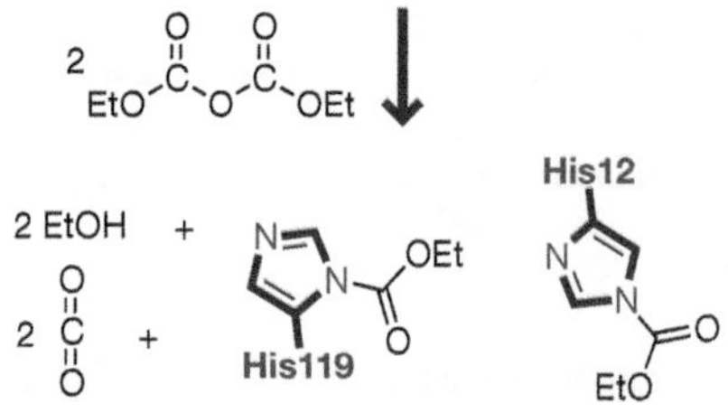

Figure 4.7 Deactivating pesky RNases. Diethyl pyrocarbonate (DEPC) acylates the active-site histidines of RNase, deactivating the enzyme.

with such perfect alignment that detailed studies have failed to provide evidence for a five-coordinate phosphorane intermediate, implying a concerted S_N2-like phosphoryl transfer mechanism. Imidazoles are ideal for acid–base catalysis at physiological pH, because they have a pK_a close to 7. That means that imidazoles are equally happy in the conjugate base and conjugate acid forms at pH 7. One imidazole, from His12, abstracts a proton from the 2′-hydroxyl group, facilitating attack on the phosphorus atom, *anti* to the leaving group. At the same time, the leaving group oxygen in the pentavalent phosphorane intermediate is then protonated by the second imidazole from His119, facilitating expulsion of the neighboring 3′-hydroxyl group. The same catalytic imidazoles are then poised to catalyze cleavage of the cyclic phosphate by a water molecule.

RNase A is efficiently deactivated by treatment with excess diethyl pyrocarbonate (DEPC), which acylates three of the four histidine side chains in the protein, including the two key histidine residues in the active site (**Figure 4.7**). The acylated histidines can no longer participate in efficient proton transfer reactions, and the protein becomes considerably less stable. The DEPC reagent is typically added to all aqueous solutions (0.1% v/v, 1 hour, 37 °C) used for RNA experiments. The remaining DEPC is hydrolyzed by autoclaving the solution. Without deactivation of the ubiquitous RNase enzymes, experiments involving RNA would be prohibitively difficult.

The 5-methyl group of thymine is a form of chemical ID

The cytosine base of cytidine hydrolyzes to uracil at a slow but appreciable rate (**Figure 4.8**). When cytosine is converted to uracil, the base-pairing complementarity changes from C•G to U•A. At the level of messenger RNA (mRNA), such a point mutation could lead to incorporation of the wrong amino acid by the ribosome and ultimately to a misfunctional protein. Fortunately, both mRNA and proteins have relatively short half-lives in most cells, so the effect of a mutation from C to U would be short-lived. In addition, new correct copies of mRNA could be made on demand from the correct genomic DNA template.

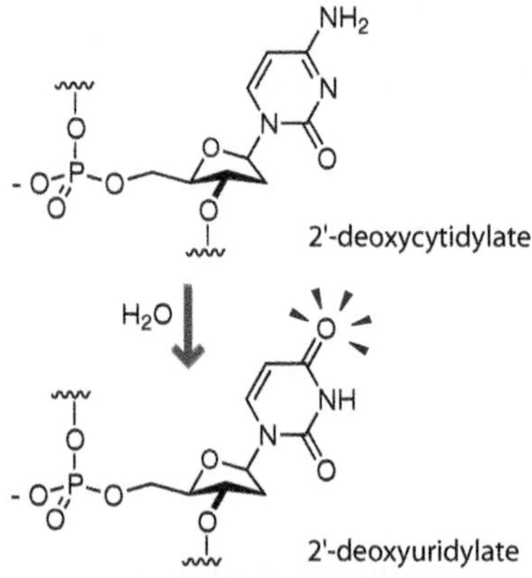

Figure 4.8 Hydrolysis as a source of oligonucleotide mutations. Hydrolysis of 2′-deoxycytidine in DNA generates an aberrant nucleotide. An enzymatic surveillance system replaces the uridine bases in DNA with cytosine.

In contrast, the consequences of cytosine hydrolysis could be disastrous if they occur in DNA. If the cytosine base of 2′-deoxycytidine is hydrolyzed to uracil, then the base-pairing complementarity will change from dC•dG to dU•dA. In a haploid organism, this mutation of dG to dA would be propogated to future generations during DNA replication.

Fortunately, because DNA is double-stranded, the presence of a dU•dG mismatch alerts the cell to remove one of the mismatched bases, but which one? At this point the importance of the thymine 5-methyl group becomes clear—it distinguishes dT•dA pairs that belong in DNA from dU•dG pairs that do not. Repair enzymes continually hunt for these misfits, replacing the offending dU with dC. In *Escherichia coli*, for example, the enzyme uracil *N*-deglycosidase specifically hydrolyzes the uracil base from dU•dG pairs.

A dU•dA base pair would not be inherently mutagenic because both bases will generate the correct complement. However, the proofreading function of dT is so important that bacterial enzymes (dUTPase) actively hunt for free 2′-deoxyuridine 5′-triphosphate and hydrolyze it so that it cannot be incorporated into DNA. The RNA precursor uridine 5′-triphosphate is resistant to this enzyme. The bacterial strain *E. coli* CJ236 lacks the genes for these two quality-control systems; the strain frequently misincorporates dU into new DNA strands, and then leaves it there. This type of DNA, riddled with dU•dA base pairs, is used to introduce mutations at specific sites by Kunkel mutagenesis (Chapter 3).

Problem 4.2

The hydrolytic deamination of cytosine is slow, even under harsh chemical conditions. Suggest a plausible arrow-pushing mechanism for the following deamination.

NH_2 — 4 M LiOH; CH_3OH/H_2O, reflux, 12 h → NH, O

RNA adopts globular shapes because it is single-stranded

In comparison with the basically extended rod structure of DNA, RNA often folds into complex globular structures (**Figure 4.9**). The distinction between rod-like DNA and globular RNA results from the fact that DNA is essentially always double-stranded, whereas RNA is typically single-stranded. The bases of the single-stranded RNA desperately seek partners for base pairing, which causes the RNA to fold in upon itself and form compact structures. In addition to the Watson–Crick base pairing featured in DNA, RNA structure makes frequent use of Hoogsteen base pairs, which were described in Chapter 3.

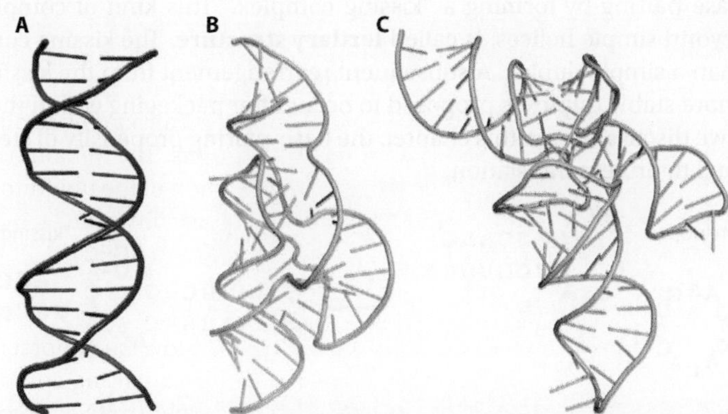

Figure 4.9 DNA structure versus RNA structure. A comparison of the structure of (A) double-stranded DNA (PDB 3BSE) with single-stranded RNA of (B) the hammerhead ribozyme (PDB 2GOZ) and (C) transfer RNA (PDB 3RG5).

RNA secondary structure that arises through base pairing is often easy to predict with computer programs that rank the probabilities of various folding patterns. RNA structure consists of sections of double helix held in place by loops. Hairpins, sometimes called stem loops, are the simplest structures. The four-nucleotide sequence UNCG is a particularly good loop for stabilizing hairpins (**Figure 4.10**). As a result of the favorability of intramolecular base pairing, the Wallace formula for estimating the melting temperature (T_m) of a DNA probe does not work for hairpin formation. Tetraloops similar to GAAA are also common, allowing the unpaired adenosines to interact with other parts of the RNA molecule. Indeed, tertiary contacts involving adenosines are the most common in structured RNAs.

5'-**GGAC**UUCG**GUCC**-3'

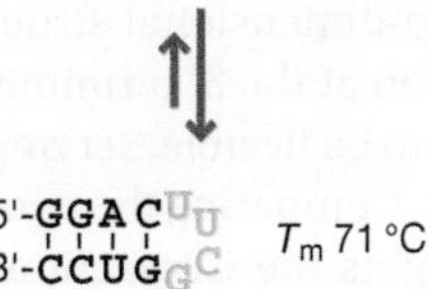

Figure 4.10 A small RNA hairpin. The RNA sequence UNCG is particularly good at stabilizing hairpin conformations, where N indicates any of the four RNA bases. Note the requirement for palindromic sequences flanking the loop for the stem loop to form.

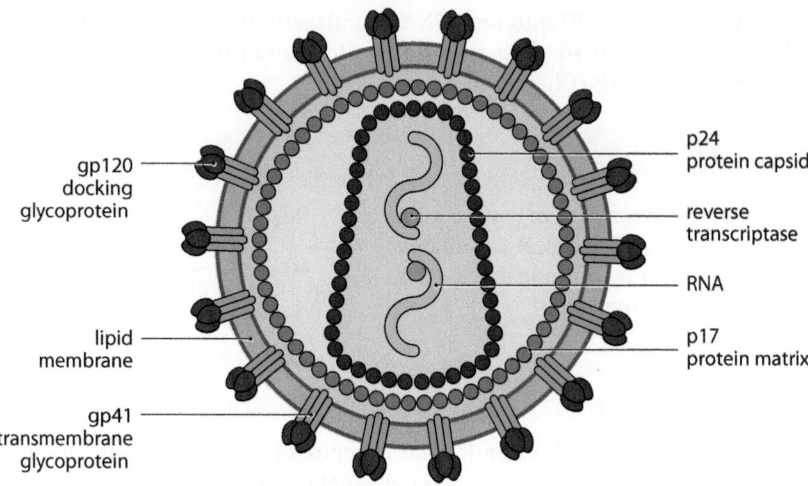

Figure 4.11 An RNA genome. The RNA genome of HIV is packaged as a dimer.

All retroviruses package exactly two copies of their RNA genome into the viral capsid, allowing them to modify both copies of the host's diploid genome. The dimerization of the HIV-1 genome (9150 nucleotides) involves stem loops (**Figure 4.11**). When the HIV genome is initially transcribed, intermolecular dimerization is disfavored by the low concentration of the transcript. A naked stem loop in the dimer initiation site seeks other bases with which to form base pairs (**Figure 4.12**). These stem loops engage in base pairing by forming a "kissing complex." This kind of complex RNA structure, beyond simple helices, is called **tertiary structure**. The kissing complex is less stable than a simple duplex. A subsequent rearrangement from the kissing complex to the more stable duplex is proposed to occur after packaging within the capsid proteins. As we discuss later in this chapter, the base-pairing propensity of stem loops is fundamental to protein translation.

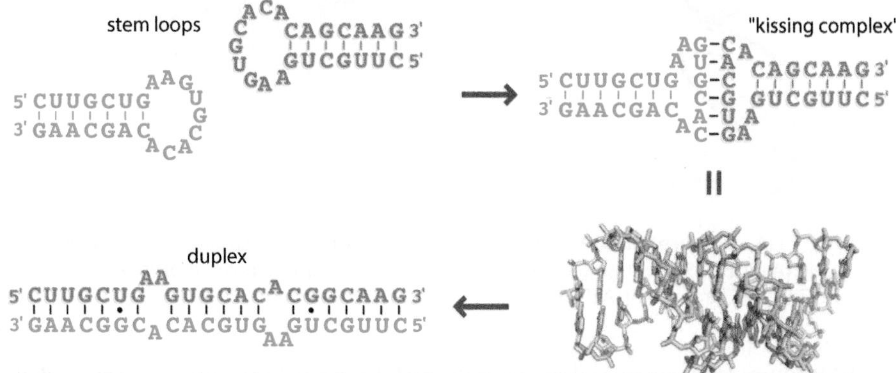

Figure 4.12 Kiss me. Stem loops in the HIV-1 dimerization initiation site initially form a "kissing complex" that rearranges to a more stable RNA duplex. (PDB 2D19)

Problem 4.3

Propose a base-pairing pattern in a stem-loop kissing complex that would form from the following two RNA sequences:

5′-GAGCCCUGGGAGGCUC-3′ and 5′-GCUGUUCCCAGACAGC-3′

Two-dimensional depictions of base-pairing patterns are helpful in seeing loop structures in folded RNA molecules. For example, the two-dimensional structure of transfer RNA (tRNA) reveals loops and an unpaired region at the 3′ terminus (**Figure 4.13**). Segments that do not engage in base pairing tend to be flexible, serving as hinges for folding or sites of interaction with other nucleotides. Comparing the two-dimensional and three-dimensional structures of tRNA highlights the interactions

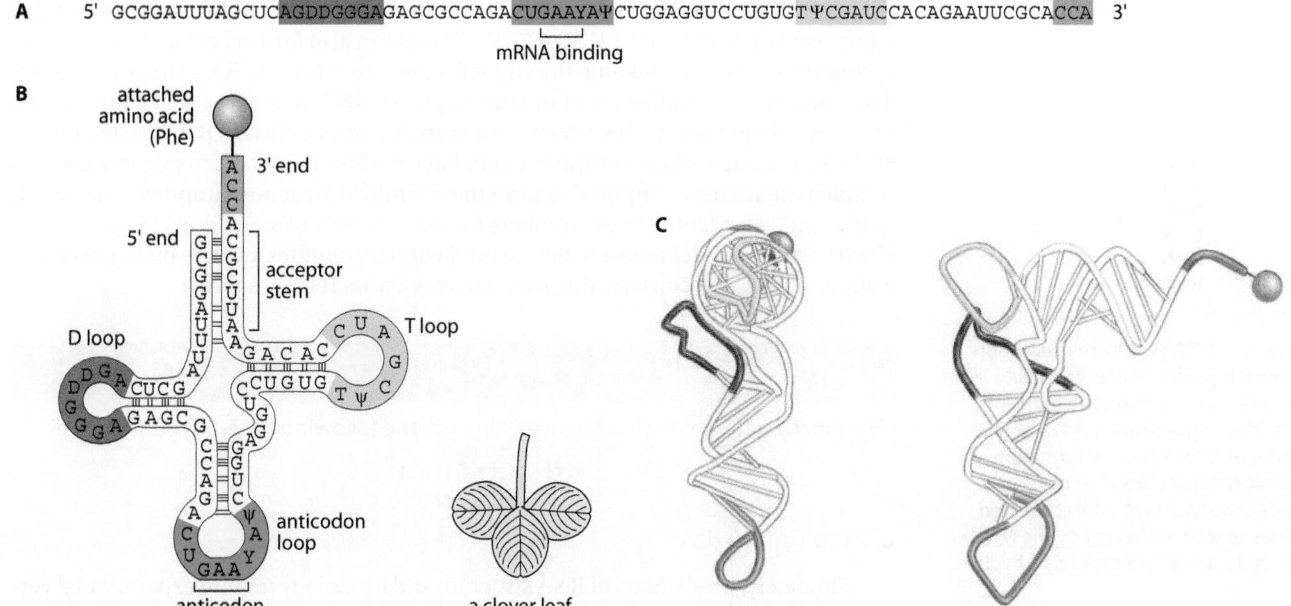

between stem loops. In the two-dimensional depiction, tRNA looks like a cloverleaf, but two-dimensional depictions of RNA usually belie the compact three-dimensional shapes of RNA molecules. Hydrogen bonds between stem loops in the two-dimensional cloverleaf structure of tRNA fold the structure into a more compact L-shape, seen in the three-dimensional structure.

Figure 4.13 Primary, secondary, and tertiary structure of tRNA. (A) In the primary structure, the sequence includes unusual bases produced by modification reactions after tRNA synthesis. Symbols other than A, C, G or U refer to enzymatically modified nucleosides. (B) The secondary structure resembles a clover leaf; highlights indicate loops and an unpaired region associated with mRNA and protein binding. (C) The tertiary structure reveals interactions between elements of secondary structure such as contacts between the D loop and T loop. (Adapted from B. Alberts et al., Molecular Biology of the Cell, 5th ed. New York: Garland Science, 2008.)

Within larger RNA structures, Watson–Crick base pairs can be disrupted or distorted. Other secondary structure elements can push on the helices, breaking up their hydrogen bonding. In addition, internal loops consist of runs of Watson–Crick base pairing interrupted by loops lacking the base pairs. In such loops, non-Watson–Crick base pairing can also contribute essential hydrogen-bonding interactions. A large number of different RNA tertiary structures have been described, including the hook-turn, kink-turn, tetraloop receptor, and lone-pair triloop interactions. Almost all RNA structures bind specifically to Mg^{2+} at the physiological concentrations of 0.5–1 mM.

Despite the complications imposed by disrupted base pairing and the large potential diversity of tertiary structures, RNA structures are relatively well understood at the level of base pairing. Furthermore, RNA folding patterns are even predictable from short RNA sequences by using free Web-based prediction programs. For more accurate structure prediction, computational folding predicts structures by modeling potential base pairs, loops, and other structures. The free energy expected for each model structure is calculated from the sum of the favorable (for example hydrogen bonds from base pairing) and the unfavorable (for example steric clash) interactions. The model with the lowest free energy is considered the best prediction of the RNA fold.

Problem 4.4

Use a free Web-based RNA structure prediction program (such as RNAfold) to determine the structure of the $tRNA^{Phe}$ from *Thermus thermophilus*:
5′-GCCGAGGUAGCUCAGUUGGUAGAGCAUGCGACUGAAAAUCGCAGUGUCGGCGGUUC GAUUCCGCUCCUCGGCACCA-3′.

Sequence alignment works well for finding highly conserved DNA and RNA sequences in databases, but it is challenging to implement when searching for conserved RNA folding patterns. For example, the sequence 5′-GGACUUCGGUCC-3′ readily folds into a hairpin because the underlined sequences are complementary,

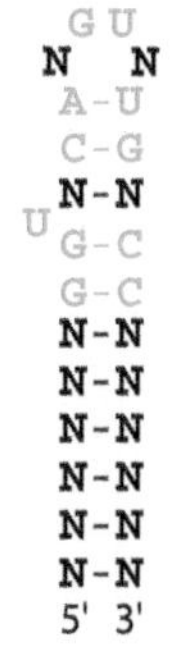

Figure 4.14 RNA structure prediction by searching with conserved RNA sequences. In this RNA stem loop, a few residues (green) are used to define requirements for structure formation. Genomic database searches can then identify sequences with the conserved residues spaced as defined by the N bases, which base pair as indicated by a hyphen.

but searching an RNA database for this exact sequence would miss the other 255 sequences (such as 5′-GAGGUUCGCCUC-3′) that can also form a hairpin held together by four base pairs. To illustrate the strengths and limitations of RNA structure prediction, consider the challenges of uncovering short RNA sequences that form hairpin structures. Sequences with such structures are found in a class of RNAs called microRNAs and are described in further detail later in this chapter. Focusing searches on complementary bases required to form the microRNA structures improves the search results and can identify large numbers (at least 30,000) of stem loops (**Figure 4.14**). Much better approaches use structure prediction techniques and use biological information, such as the conservation of sequences across related species.

Problem 4.5

How many different RNA sequences can form the following base-paired structure?

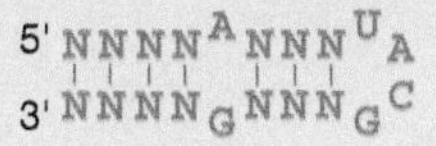

Ultimately, predictions of RNA structure and homology require experimental verification. X-ray crystallography provides the gold standard for determining the structure of RNA sequences, although RNA structures compose only a minor percentage of all biopolymer structures deposited in the Protein Data Bank. Crystallization of RNA, as in the early years of protein crystallography, is a difficult, idiosyncratic art. Chemical derivatization can also contribute key insights into RNA structure. For example, the SHAPE technique uses a reactive electrophile that preferentially modifies more flexible and more readily accessible 2′-OH groups of RNA (**Figure 4.15**). The resultant esters at the 2′ position stall transcription by reverse transcriptase; the sites of blocked DNA synthesis are then identified by DNA sequencing. The technique is particularly useful for very large RNA structures (such as the RNA core of HIV). Such analysis highlights unstructured regions of HIV RNA required for RNA splicing and hypervariation; structured regions include sequences encoding loops between domains of HIV proteins. When these considerations are taken together, analysis of the structure of RNA by SHAPE suggests yet another level of genomic regulation—namely, that the structure of RNA has evolved to regulate translation to proteins and levels of sequence variation.

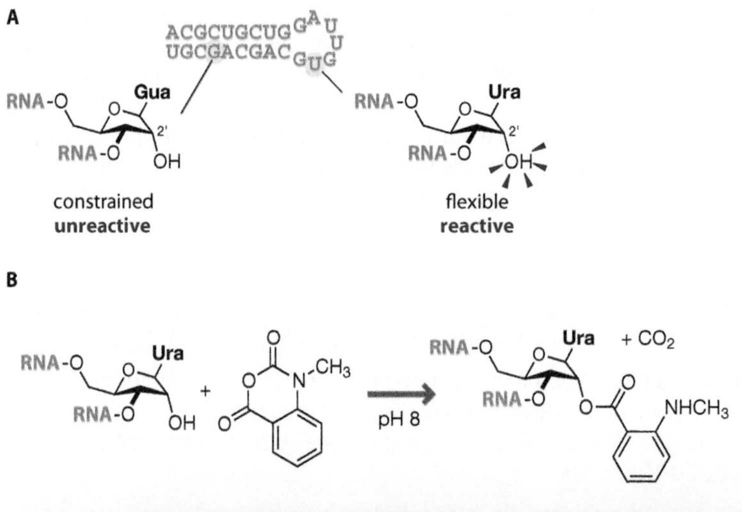

Figure 4.15 SHAPE analysis of RNA structure. (A) Less flexible or constrained regions of RNA structure block accessibility of the 2′-OH for reactions with electrophiles. (B) The more flexible regions of RNA structure are more accessible and allow reaction between the 2′-OH and the added anhydride.

Problem 4.6

Suggest a plausible arrow-pushing mechanism for the base-catalyzed acylation used in RNA SHAPE analysis (Figure 4.15B).

4.2 RNA SYNTHESIS

RNA polymerases create new strands of RNA

The human genome encodes several RNA polymerases that transcribe DNA into functional RNA molecules. RNA polymerase I transcribes ribosomal RNA (rRNA); RNA polymerase III transcribes transfer RNA (tRNA). These two polymerases account for most of the RNA in the cell; 80% of the RNA in growing mammalian cells is rRNA and 15% is tRNA. RNA polymerase II transcribes the remaining 5% of RNA found in human cells, types of RNA that are related to conditional gene expression: mRNA, microRNA, and small nuclear RNA. Conditional gene expression by RNA polymerase II is the major feature that distinguishes one gene from another, and ultimately one human cell from another. The transcriptional activity of RNA polymerase II is dependent on a tightly choreographed interaction between the polymerase, phosphorylating enzymes, and a host of transcription factor proteins. Without these attendant transcription factors, RNA polymerase II is quiescent. RNA polymerization from a double-stranded DNA template relies on many independent subunits. Working together, the machine must recognize the start of the gene, separate the strands of DNA, catalyze the formation of a complementary strand of RNA, pull apart the RNA–DNA double helix, and rejoin the DNA strands. The resulting RNA strand resembles one of the DNA strands, called the **sense strand**, and is complementary to the DNA template strand, called the **antisense strand**.

RNA polymerases I and II are massive protein complexes formed from a dozen subunits with a combined molecular mass greater than 500 kDa. The two RNA polymerases and their bacterial ortholog adopt a jaw-like structure, with the DNA template clamped between the two halves of the jaw (**Figure 4.16**). As the RNA polymerase moves along the double-stranded DNA template, strand separation occurs, and a bridge helix forces the DNA template through the active site with a ratcheting motion. A positively charged funnel guides incoming negatively charged nucleotide triphosphates (NTPs) into the active site. When the correct NTP forms a base pair with the DNA template, a short loop below the bridge helix closes the breech, ensuring the formation of a new phosphodiester bond. Notably, although the shape of RNA polymerase differs from that of DNA polymerase (see Figure 3.50), all nucleotide polymerases on the planet operate through the same chemical mechanism, involving two Mg^{2+} ions, that we discussed in Section 3.5.

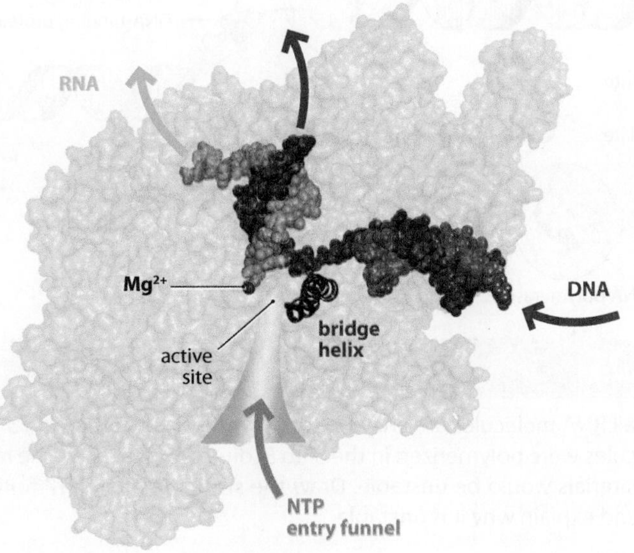

Figure 4.16 Seeing inside RNA polymerase II. The double-stranded DNA is drawn into the polymerase (transparent blue). A bridge helix helps to drive apart the two DNA strands so that new RNA subunits can form Watson–Crick base pairs with the antisense strand (red). New nucleotides are added to the 5′ end of the mRNA strand, which is shown snaking out of the active site. A key active-site Mg^{2+} ion is rendered in purple. (PDB 2O5I)

α-amanitin

Figure 4.17 Mushroom toxin. α-Amanitin, from the deathcap mushroom (*Amanita phalloides*) is a potent inhibitor of mammalian RNA polymerase II, exerting the most lethal effects within the liver. Structurally, α-amanitin is a cyclic peptide with an unusual transannular tryptathionine crosslink. (Top, courtesy of George Chernilevsky.)

Inhibition of RNA polymerase can be deadly. For example, the deathcap mushroom produces two cyclic peptide natural products called phalloidin and α-amanitin. Phalloidin binds specifically to actin filaments, whereas α-amanitin binds to and inhibits RNA polymerase II (**Figure 4.17**). Cell death quickly ensues, because transcription is required for the maintenance, growth, and division of the cell.

Problem 4.7

Draw a depiction of an RNA polymerase active site, based on its similarity to the active site of DNA polymerase. Your active site should include the NTP, two Mg^{2+} ions, and a GATC template strand. Propose an arrow-pushing mechanism for the key O–P bond-forming step in the reaction catalyzed by RNA polymerase.

DNA primase is just another RNA polymerase

RNA polymerases can create new RNA strands from scratch, but DNA polymerases cannot create new strands of DNA—they can only lengthen an existing strand, referred to as the primer strand. To satisfy the requirement for an oligonucleotide primer, some viruses recruit random snippets of RNA found inside the cell as primers, or even use a protein as a primer. In human cells, DNA strand synthesis involves the initial action of an RNA polymerase that generates a short RNA primer, and that primer is then elongated by a DNA polymerase. The tiny RNA primers are later removed and the gap is filled in with DNA.

The classical model of DNA replication at a DNA replication fork works well using the leading strand as a template because the complementary strand can be lengthened in the 5′ to 3′ direction. However, DNA synthesis using the lagging strand as a template presents a problem because no known polymerase can lengthen in the 3′ to 5′ direction. The biological solution to this challenge is the existence of an RNA polymerase called **DNA primase** (**Figure 4.18**), which creates short RNA primers on the lagging DNA strand. These primers are lengthened by DNA polymerase to create DNA fragments with RNA primers at the 5′ ends. These fragments are eventually ligated and the RNA is removed and replaced by DNA. As an RNA polymerase, DNA primase is noticeably more error-prone than other nucleotide polymerases. Like all such polymerases, DNA primase uses the same conserved mechanism, and has an active site shaped like a right hand.

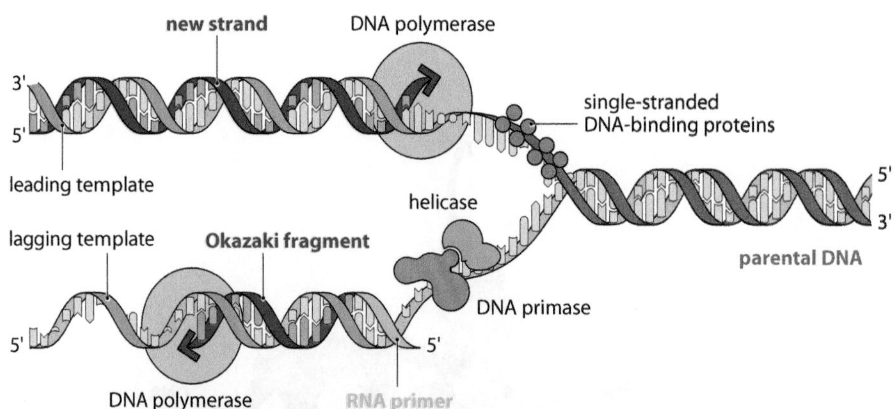

Figure 4.18 RNA primers at the DNA replication fork. DNA primase is an RNA polymerase that creates short RNA primers that are elongated by DNA polymerase. The primers allow the lagging strands to be elongated in the 5′ to 3′ direction.

Problem 4.8

In nature, all RNA molecules are enzymatically synthesized in the 5′ to 3′ direction. If RNA molecules were polymerized in the 3′ to 5′ direction, however, the required NTP starting materials would be unstable. Draw the structure of an NTP for 3′ to 5′ RNA synthesis and explain why it is unstable.

4.3 TRANSCRIPTIONAL CONTROL

DNA sequences determine start sites and stop sites for RNA polymerase

Transcription of DNA into RNA requires the equivalent of punctuation marks. Indentation, capitalization, and periods identify the beginnings and ends of sentences, and thus make the message more readable. Similarly, RNA polymerases require cues within the DNA sequence to signal the start and stop sites for RNA transcription. Sentence punctuation can also moderate the tone of the sentence; an exclamation point, for example, provides an insistent emphasis. Similarly, some start sites for RNA transcription are stronger than others.

Start and stop sites for the transcription of RNA are called **promoters** and **terminators**, respectively. As expected, the promoter appears near the start or 5′ end of the gene, and the terminator is found at the 3′ terminus. Promoters consist of readily recognizable sequences of DNA rich in T and A. In bacterial genes, variants of the promoter sequence TATAAT (called a Pribnow box) are found about 10 base pairs upstream (toward the 5′ end of the DNA) of the transcriptional start site. In human cells, variants of the promoter sequence 5′-TATAAA-3′ (called the TATA box) are typically found about 25 base pairs upstream of where RNA polymerase begins transcription.

Prokaryotic RNA polymerases recognize the transcriptional start sites through direct interactions between protein transcription factors and RNA polymerase. Similarly, in human cells a protein called the TATA-binding protein (TBP) binds to the TATA box and drastically bends the DNA into a roughly 100° angle (**Figure 4.19**). The weak π stacking of the A•T base pairs found in the TATA box makes this bending possible. The TBP uses DNA flexibility as one criterion for rapidly identifying the TATA box sequence. Additionally, the β strands in the DNA major groove recognize the specific sequence of the TATA box. The embrace between TBP and DNA provides a focal site for recruiting the complex of transcription factors and RNA polymerase to initiate transcription.

The DNA sequence also designates where transcription ends. Some bacterial and bacteriophage terminators have a self-complementary G•C-rich sequence followed by a run of As. Driven by the strength of the G•C base pair, the resultant transcript forms an RNA hairpin, which rips the relatively weak A•U base pairs out of the DNA-RNA hybrid; the RNA hairpin also binds to the RNA polymerase, slowing its forward momentum. Transcription thus screeches to a halt. Other forms of transcription termination use an ATP-hydrolyzing protein called Rho. In Rho-dependent termination, Rho binds to the new RNA transcript and destabilizes the DNA–RNA hybrid, pulling RNA out of the RNA polymerase active site.

The three different eukaryotic RNA polymerases have different mechanisms for termination. RNA polymerase I focuses on the synthesis of most of the rRNA, and stops when it reaches an 18-base-pair DNA sequence that is repeated eight times. Unlike the Rho/RNA-dependent termination of transcription in bacteria, the human protein TTF-1 binds directly to the DNA and blocks RNA polymerase I. Transcription of mRNAs, which are by definition destined for translation into proteins, is the domain of RNA polymerase II; some small regulatory RNAs are also transcribed by RNA polymerase II. The 3′ termini of mRNA-encoding genes typically include a signal for the incorporation of a poly(A) tail. The exact mechanism by which RNA polymerase II stops RNA synthesis and allows polyadenylation remains unclear. RNA polymerase III focuses on the pedestrian RNAs, including small RNAs (microRNAs, or miRNAs), tRNAs, and some rRNAs. Such RNAs are required for everyday housekeeping in the cell, and they also must be stockpiled for cell division. A short stretch of poly(T) sequence terminates transcription by RNA polymerase III.

Viruses, such as bacteriophages, have evolved very effective promoters and RNA polymerases to guarantee that the viral genome gets the highest priority during transcription. For example, the RNA polymerase from bacteriophage T7 will transcribe any double-stranded DNA with the following 23-base-pair T7 promoter sequence in the template strand: 3′-**ATTATGCTGAGTGATAT**CCCTCT...-5′. The last six nucleotides

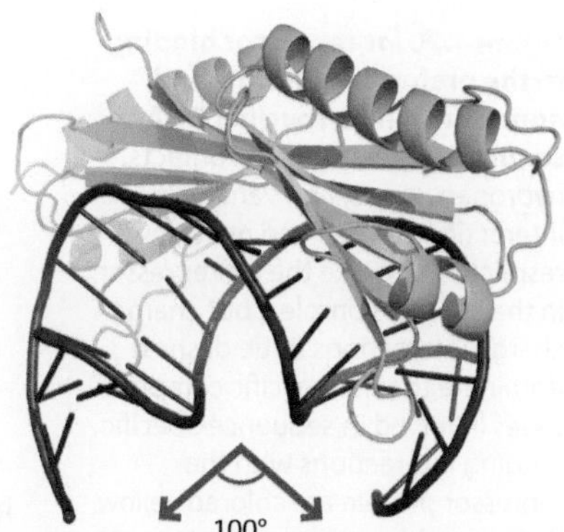

Figure 4.19 TATA-binding protein (TBP) binding and bending a TATA box. TBP (blue) has been described as a saddle sitting on DNA (red) with two β turns providing the stirrups to force the DNA into a roughly 100° bend. (PDB 1CDW)

in this sequence are transcribed; RNA will then start with the sequence 5′-GGGAGA... -3′. The indifference to other sequence characteristics makes T7 RNA polymerase a widely used experimental tool for RNA synthesis. Furthermore, the T7 promoters are often used for protein overexpression in *E. coli.*

Transcription factors bind to DNA with exquisite sequence specificity

The tightly regulated synthesis of RNA is principally controlled by protein–DNA interactions featuring transcription factors bound to specific DNA sequences. Nonspecific binding to a highly charged molecule such as DNA is easy—one simply needs to accumulate enough positively charged functional groups to stick to the anionic phosphodiester backbone of the DNA. This strategy is used by histones to provide a spool for winding up supercoiled DNA in chromosomes. However, an overreliance on Coulombic interactions will prevent sequence-specific binding, which is essential for transcription factors. A careful charge balancing is observed in proteins that are required to interact specifically with DNA—too much positive charge poisons specificity, and too little abrogates binding.

Transcription factors read out sequences in a manner analogous to reading Braille. The first problem is locating the Braille writing, just as the protein needs to first find the DNA. Complementary charge–charge interactions provide a long-range attractive force for drawing the transcription factor toward DNA. Then the protein feels the bumps, hydrogen bonds, and van der Waals interactions of the major groove in an attempt to read out the sequence (**Figure 4.20**). If the DNA sequence provides a complementary set of interactions to the protein surface, the transcription factor binds. If not, the protein either glides further along the DNA, as seen *in vitro*, or floats away to find another sequence of DNA to interrogate. The latter is a more likely scenario in the densely packed interior of the cell, where DNA is more likely to be found in a folded and supercoiled state. Oriented water molecules, sandwiched between the transcription factor and DNA, have a curious and vital role in molecular recognition. Displacing water molecules from DNA would impose a steep cost to the binding of a transcription factor. Instead, transcription factors often use these bound water molecules to communicate with DNA.

The music played by the transcriptional orchestra has a remarkably chaotic tone. Proteins jump onto a large complex regulating RNA polymerase, only to leap off a

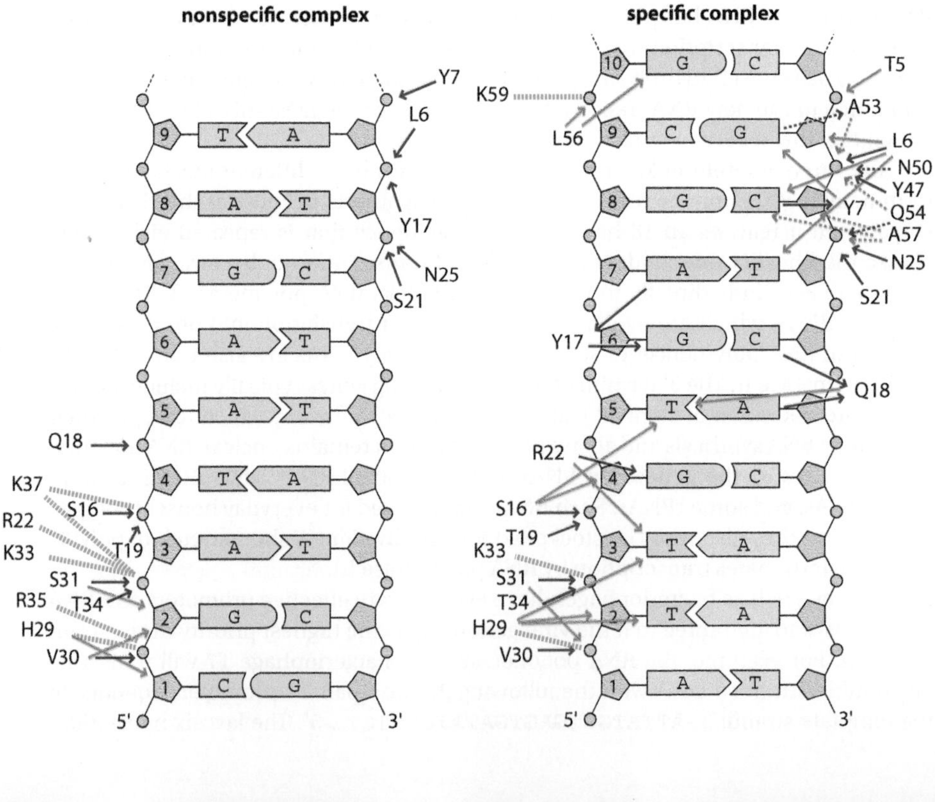

Figure 4.20 *lac* repressor binding to the preferred (specific) and nonoptimal (nonspecific) DNA sequence. A plethora of contacts, hydrogen bonds, and van der Waals interactions (green and red arrows, respectively) brace the *lac* repressor in the specific complex, but charge–charge interactions (blue dashes) dominate the nonspecific complex. Bases involved in sequence-specific binding interactions with the repressor protein are colored yellow. Protein residues are designated with one-letter amino acid codes, followed by protein sequence numbers. One-letter amino acid codes will be introduced in Chapter 5. (Adapted from C.G. Kalodimos et al., *Science* 305: 388–389, 2004. With permission from AAAS.)

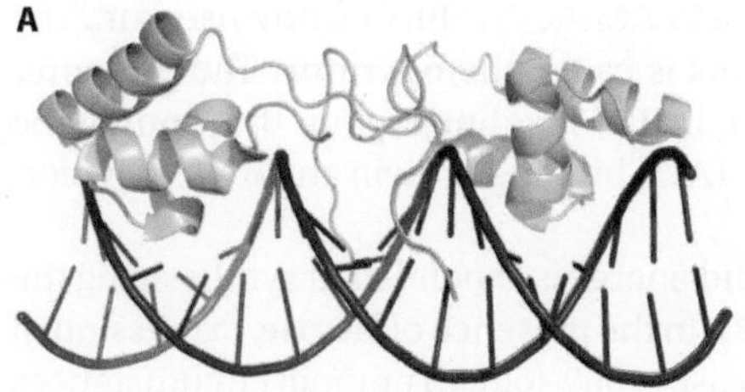

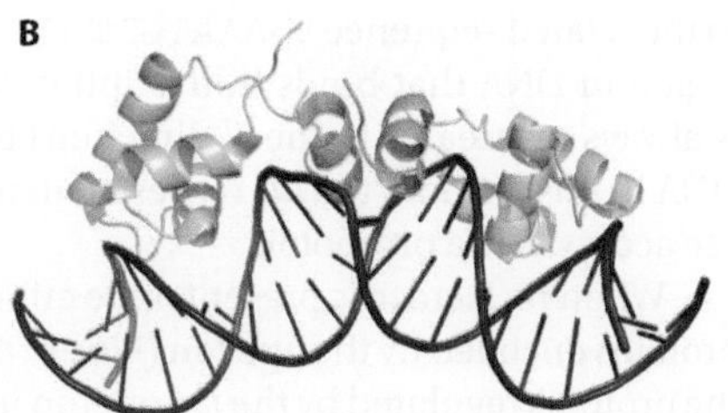

Figure 4.21 Sequence-specific binding of the *lac* repressor. (A) The hinge region of the DNA binding domain of the *lac* repressor remains unstructured in the complex with a nonoptimal DNA sequence (PDB 1OSL). (B) On binding to the preferred DNA sequence of DNA, the hinge region forms an α-helix (green), bending the DNA. (PDB 1JWL)

fraction of a second later. A closer look at the interactions between DNA and a bacterial transcription factor reveals an intricate dance (**Figure 4.21**). In the next section we explain how the *lac* repressor protein controls the transcription of several genes in *E. coli*. This control begins at the level of DNA binding. Fluttering through multiple conformations, the *lac* repressor protein alights on the DNA and wriggles into the major groove. Positively charged residues grab the rails of the phosphodiester backbone to guide the protein into the proper position with the residues critical to determining specificity available to interrogate the DNA. When binding to a DNA sequence not recognized by the transcription factor, the protein clings to the phosphodiester backbone and cannot find useful contacts with bases in the major groove. However, binding to the correct DNA sequence through side-chain contacts with the DNA bases, or to water aligned in the major groove, can cause a major change in the structure and orientation of the *lac* repressor. This gross change in protein structure and interactivity accompanies a huge gain in binding affinity. The protein–DNA complex is about 100 million times more likely to form with the correct sequence than with the incorrect sequence. Often, on binding to the transcription factor, the DNA also undergoes a drastic change in structure, such as bending. The tolerance of DNA to such manipulations provides an additional, subtle, indirect readout for the DNA sequence. Firmly π-stacked DNA sequences, for example, will be less willing to undergo bending. This pinching of the DNA sequence is also used for the recognition of DNA damage before the excision repair of alkylated DNA, discussed in Chapter 3.

Transcription can be controlled by small molecules

With such a diverse portfolio of functions, random synthesis of different RNAs would quickly prove disastrous for the cell. Instead, transcription by RNA polymerase is a tightly regulated event, dependent on the orchestration of numerous proteins and DNA sequences. Transcriptional regulation responds to two major drives: (1) the drive to take in nutrients, metabolize them, and excrete waste, and (2) the drive to grow, develop, and reproduce. These cellular imperatives are tightly controlled in all cells and in all organisms. Avoiding false alarms and providing specificity for such signals can involve a complex series of protein–protein interactions; however, the binding of transcription factors to DNA can also be directly regulated by small-molecule ligands. Precise control over transcription remains a significant goal of many research laboratories. This holy grail of chemical biology could allow physicians to shut off cancerous cells or turn on the transcription of genes necessary for growing a new arm or leg.

To understand how small molecules can control transcription in useful ways we will closely examine how *E. coli* controls gene expression on the basis of the presence of the milk sugar galactose. *E. coli* cells readily multiply when fed with a diet of essentially any carbohydrates, except when the carbohydrate is exclusively lactose. To metabolize the lactose, the bacteria need two key proteins: a permease that allows lactose to enter the cell, and a hydrolytic enzyme called β-galactosidase that cleaves the disaccharide lactose into digestible monosaccharide subunits.

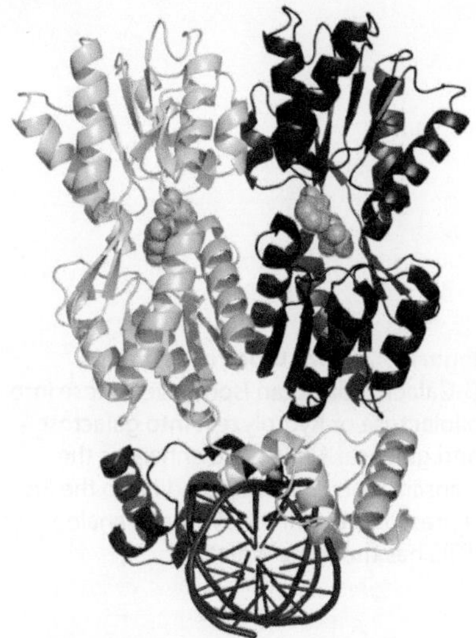

Figure 4.22 Squatting on a promoter. In this crystal structure of the *lac* repressor dimer (PDB 1JWL), a fucosyl sugar derivative, which does not prevent DNA binding, rests in the allolactose site (green).

The expression of β-galactosidase is inhibited by the *lac* repressor protein, which is expressed constitutively (that is, it is synthesized at a constant rate in the cell). The *lac* repressor binds to DNA as a tetramer, in which each subunit binds to DNA through a domain similar to that found in engrailed (**Figure 4.22**). The *lac* repressor tetramer binds tightly to the promoter at the sequence 5′-`AATTGTGAGCGGATAACATT`-3′ and to the related sequence 5′-AAATGTTGAGCGAGTAACAACC-3′ further downstream. The region of DNA that binds transcription factors is called the **operator**. The promoter is always upstream (in the 5′ direction) from the RNA-coding region. By gripping the DNA in this fashion, the *lac* repressor denies TATA-binding protein and RNA polymerase access to the promoter.

When no lactose is present, little effort and energy is expended in synthesizing the proteins encoded by the **operon** (**Figure 4.23**). In the presence of lactose, expression of the proteins regulated by the *lac* operon increases 1000-fold. In optimal circumstances β-galactosidase levels can approach up to 3% of all proteins in the cell. β-Galactosidase is a massive tetrameric enzyme that generates two different types of products from lactose (**Figure 4.24**). When the concentrations of lactose and β-galactosidase are low, the dimeric form of the enzyme isomerizes lactose into allolactose; such conditions occur initially when very little β-galactosidase has been produced through leaky transcription of its encoding gene. Allolactose binds tightly to the *lac* repressor, causing it to release its grip from DNA. Once the promoter sequence has been revealed, RNA polymerase transcribes the *lacZ* gene, which encodes β-galactosidase. At high concentrations of β-galactosidase, the enzyme forms a tetramer that efficiently cleaves the disaccharide lactose into galactose and glucose.

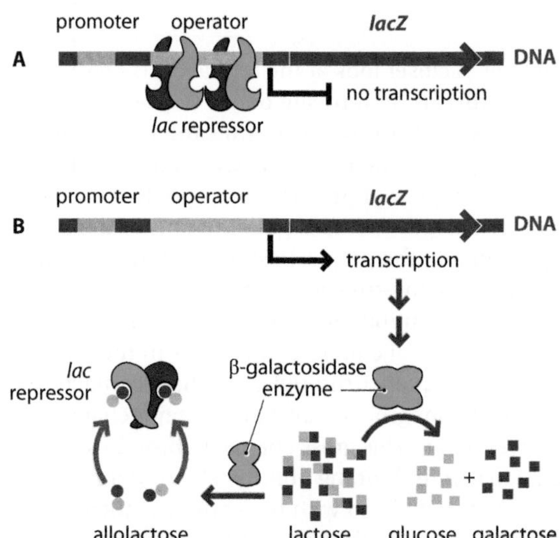

Figure 4.23 Repressing the repressor. (A) The *lac* repressor binds to the *lac* operator, inhibiting transcription. (B) At low concentrations, β-galactosidase generates allolactose, which knocks the repressor off the promoter, allowing expression of *lacZ*. Expression of β-galactosidase at higher concentrations leads to the formation of a tetramer that efficiently hydrolyzes lactose into glucose and galactose.

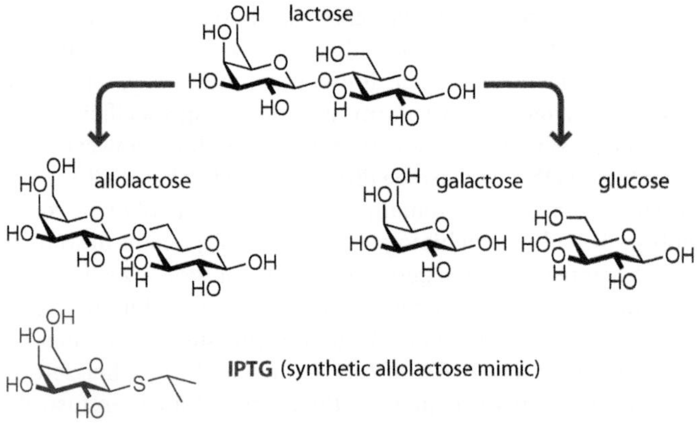

lactose
allolactose
galactose
glucose
IPTG (synthetic allolactose mimic)

Figure 4.24 Two types of products. β-Galactosidase can isomerize lactose into allolactose or hydrolyze it into galactose and glucose. Allolactose enhances the transcription of *lacZ* by binding to the *lac* repressor protein; the synthetic analog IPTG has the same effect.

Non-natural derivatives of lactose can also induce the release of the *lac* repressor, providing small-molecule control over transcription. For example, the synthetic galactose derivative isopropyl β-D-thiogalactoside (IPTG) effectively dissociates the *lac* repressor from its operator sequence. A poor substrate for β-galactosidase, IPTG is not rapidly degraded (see Figure 4.24). In place of β-galactosidase, other genes can be inserted into the *lac* operon. After allowing bacteria to multiply, one can add IPTG to turn on the overexpression (or high-level production) of the inserted open reading frame. Because *E. coli* cells only activate transcription of the *lac* operon in the presence of galactose derivatives such as allolactose, the growth of bacteria on a diet of sucrose will effectively shut off transcription of the inserted open reading frame. This system is used routinely in most chemical biology laboratories for the production of proteins by bacteria.

Transcription of mRNA in human cells involves many proteins and many regions of DNA

The bacterial *lac* operon offers just a glimpse of the diverse control mechanisms that control gene expression. Eukaryotic transcription adds several layers of complexity. First, as has been noted previously, eukaryotes store their DNA in chromosomes within a nucleus. Second, jump-starting eukaryotic transcription generally requires the binding of many proteins to DNA (**Figure 4.25**). The complex that initiates the transcription of mRNA by RNA polymerase II involves about 60 different proteins. The names of the proteins on that roster are not important here. The important point is that the transcriptional initiation complex must grip the DNA strand in many places to initiate mRNA synthesis. A final level of complexity arises from covalent modification of transcriptional proteins; for example, phosphorylation of RNA polymerase causes the protein to change conformation markedly, opening up the active site for RNA polymerization.

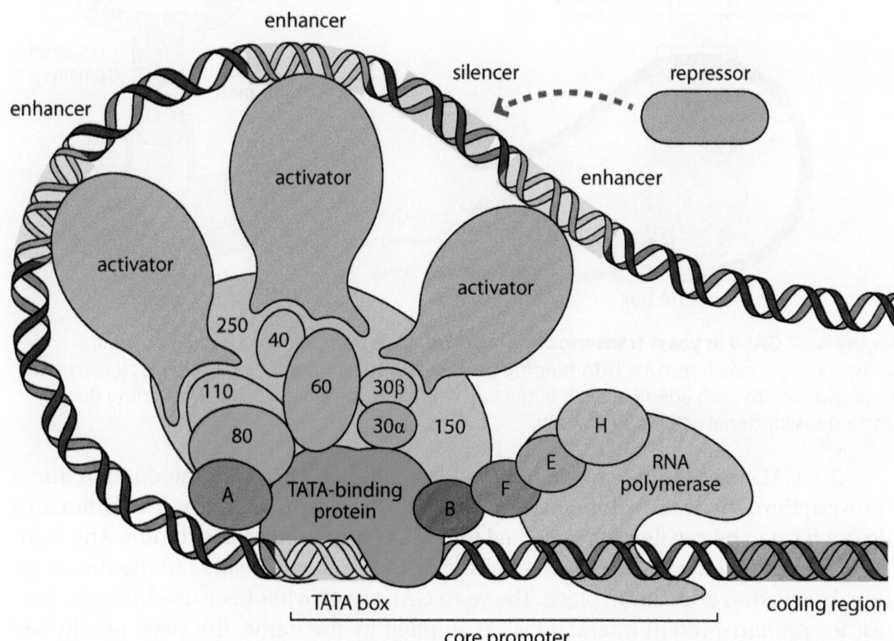

Figure 4.25 The transcriptional initiation complex is... complex. The multiprotein complex that initiates transcription of human genes involves dozens of proteins and grips the DNA in many different locations. (Adapted from P.H. Raven et al., Biology, 6th ed. New York: McGraw-Hill, 2002. With permission from the McGraw-Hill Companies, Inc.)

Transcription factors can either activate or repress transcription and are called **activators** and **repressors**, respectively. These transcription factors often bind to DNA control sequences, called enhancers or silencers, far away from the RNA-coding region of DNA. Some transcription factors can act as a repressor for some operons and as an activator for others. Mutations in such pivotal proteins can often cause cancer. For example, the Myc family of transcription factors binds weakly to DNA as both repressors and activators for a number of different genes. Myc proteins are called **oncoproteins**, because mutations in this gene family are associated with cancer.

Transcription factors are potential targets for small drug-like molecules. In a typical chemical genomics experiment, a synthetic library of 6504 compounds, spotted on a glass slide, was screened for binding to the yeast transcription factor Hap3p, a homolog of the mammalian transcription factor that binds the DNA sequence CCAAT. The Hap3p protein was fused to the common stably folded protein glutathione S-transferase (GST) and applied to the library of compounds (**Figure 4.26**). After the slide was stained with a fluorescent dye that binds to GST, two spots were found

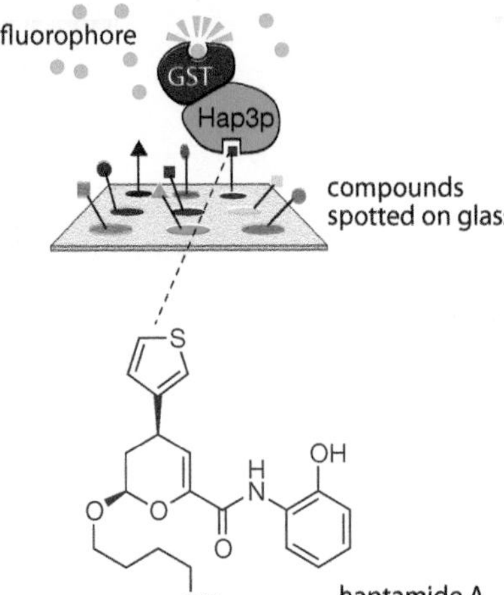

Figure 4.26 Screening a library of compounds. A protein hybrid composed of Hap3p and glutathione S-transferase (GST) was washed over a glass slide spotted with synthetic compounds. Compounds that bound tightly to Hap3p were identified by staining with a fluorophore that binds to GST.

to fluoresce. One compound was found to bind to the GST alone and was excluded from further study. The other compound, called haptamide A, was found to bind to Hap3p. A soluble version of haptamide A was found to inhibit transcription by the Hap transcription complex. An isomeric form of haptamide A, named haptamide B, was shown to bind even more tightly to Hap3p and to inhibit transcription at even lower concentrations.

The yeast two-hybrid system provides a transcription-based tool to identify protein–protein interactions

Protein-protein interactions are essential for eukaryotic transcription. Molecular biologists have harnessed eukaryotic transcription as a sensitive readout of protein–protein interactions. The **yeast two-hybrid system** was designed to show whether two proteins can bind specifically to each other inside the cell. The key to this technology is a yeast-derived transcription factor called GAL4. When GAL4 binds to the upstream activator sequence 5′-CGGNNNNNNNNNNNCCG-3′ it recruits TATA-binding protein (TBP), and other elements of the transcriptional machinery, to the downstream DNA-coding sequence, a term used to describe a DNA sequence that encodes a protein (**Figure 4.27**A). When GAL4 drives expression of the enzyme β-galactosidase, the resulting enzymatic activity is easily detected.

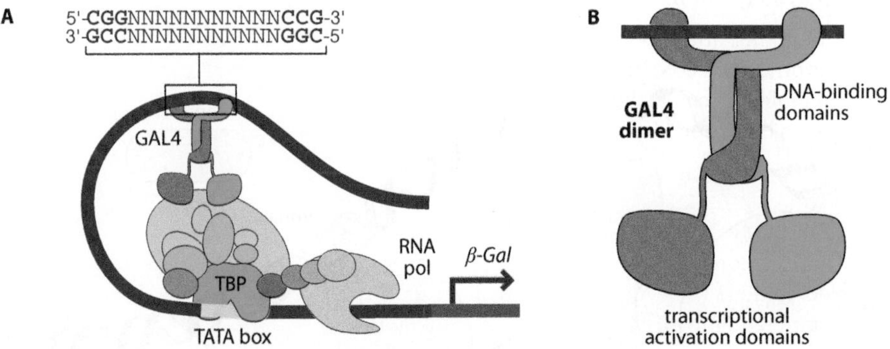

Figure 4.27 GAL4 in yeast transcriptional control. (A) When GAL4 binds to an upstream activator sequence, it recruits TATA-binding protein (TBP) and other transcription factors to start transcription. (B) Each (identical) half of the GAL4 dimer is composed of a DNA-binding domain and a transcriptional activation domain.

The GAL4 protein has two functional domains—a DNA-binding domain and a transcriptional activation domain (see Figure 4.27B). Surprisingly, these two domains do not have to be covalently connected for GAL4 to promote transcription. Any high-affinity interaction can be used to hold the DNA-binding domain and the transcriptional activation domains in place. The yeast GAL4 system has been used as a sensitive test for protein–protein interactions. As implied by the name, the yeast two-hybrid test requires two hybrid proteins, one connected to the bait and one connected to the prey. For example, if the GAL4 DNA-binding domain were connected to the human protein axin and the GAL4 activation domain were attached to the human protein β-catenin, the interaction between axin and β-catenin would lead to the assembly of a functional GAL4. The interaction between axin and β-catenin would then lead to the expression of β-galactosidase (**Figure 4.28**).

When set up properly, transcription promoted by GAL4 produces an enzyme called β-galactosidase, described previously as part of the *lac* operon in *E. coli.* β-Galactosidase can convert a colorless compound (X-Gal) into an indigo (intensely blue-colored) dye. Thus, testing whether two proteins interact inside the yeast cell is simply a matter of constructing yeast plasmids with the two separate GAL4 domains fused to the two potential binding partners. This yeast two-hybrid test can easily be converted to a screen of large numbers of proteins for the ability to bind to a single protein of interest. One protein with unknown binding partners, called the **bait**, is attached to the GAL4 DNA-binding domain. Next, a library of mutant proteins, called the **prey**, is fused to the activation domain. After the transformation of plasmids

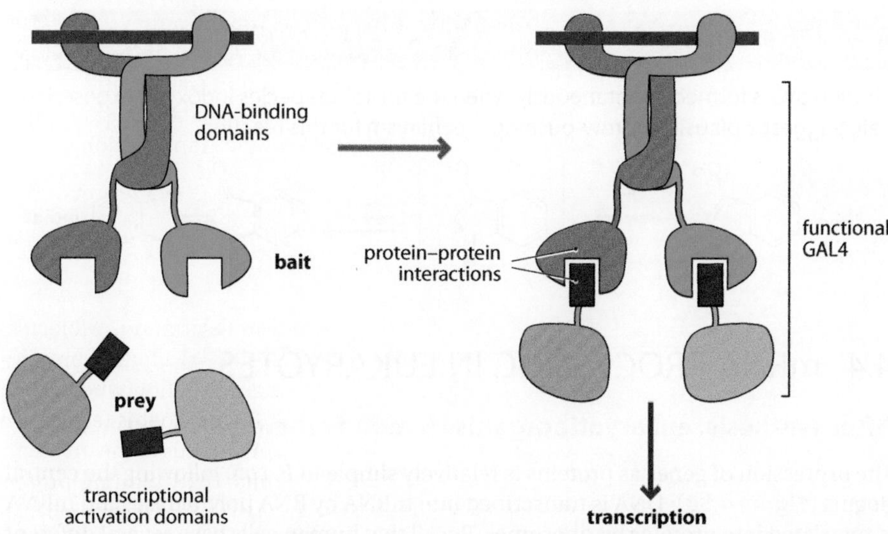

Figure 4.28 Yeast two-hybrid test. The yeast two-hybrid test involves two different hybrid proteins. One hybrid is the GAL4 DNA-binding domain fused to a bait protein. The other hybrid is the GAL4 transcriptional activation domain fused to a prey protein. When the prey takes the bait, it leads to transcription of the downstream reporter gene.

encoding the bait-fused binding domain and prey-fused activation domain, each yeast cell (and resulting colony) will have a different gene from the library of prey. Binding between bait and prey will bring together the GAL4 DNA-binding and activation domains, and this will give the yeast colony a blue color (**Figure 4.29**). Many variations on this basic scheme can be used to test binding. For example, to test whether a small molecule can bring together two protein partners, the yeast three-hybrid system has been developed (two proteins + one small molecule = three hybrids).

Problem 4.9

Using cartoon representations, design a pair of GAL4 hybrids that could be used to select for RNA molecules that bind simultaneously to the MS2 viral coat protein and the iron regulatory protein.

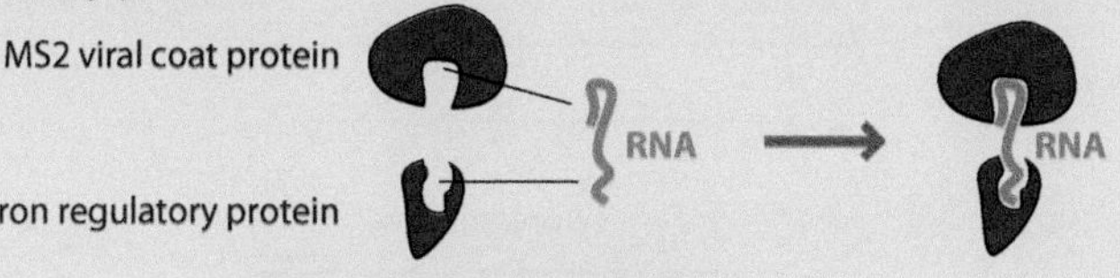

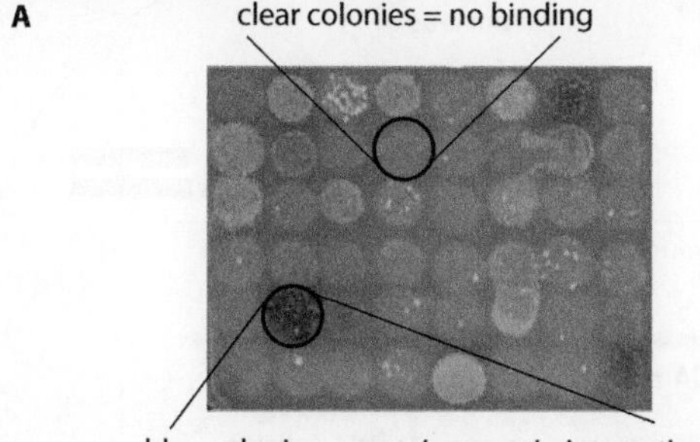

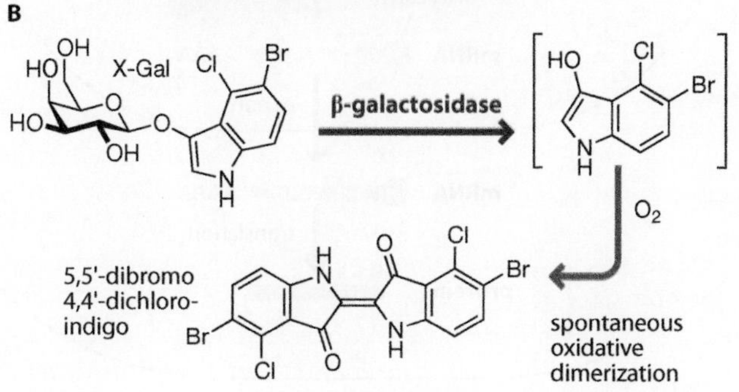

Figure 4.29 A yeast two-hybrid screen. (A) Actual data from a yeast two-hybrid screen. Blue colonies indicate successful binding between bait and prey. (B) When the prey binds the bait it leads to the expression of β-galactosidase, which converts X-Gal in the agar into an intensely blue-colored indigo dye. (A, from J. Zhong et al., *Genome Res.* 13:2691–2699, 2003. With permission from Cold Spring Harbor Laboratory Press.)

Problem 4.10

Indigo dye is formed spontaneously when the unstable species indoxyl is exposed to air. Suggest a plausible arrow-pushing mechanism for this reaction.

indoxyl $\xrightarrow{\bullet OH}$ [indoxyl radical] ⇉ indigo

4.4 mRNA PROCESSING IN EUKARYOTES

After synthesis, eukaryotic organisms modify their mRNA extensively

The expression of genes as proteins is relatively simple in *E. coli*, following the central dogma (**Figure 4.30**): DNA is transcribed into mRNA by RNA polymerase, and mRNA is translated into proteins by ribosomes. Recall that human cells have several different RNA polymerases. Genes that code for proteins are transcribed by RNA polymerase II. RNA transcripts generated by other RNA polymerases are destined for other roles in the cell.

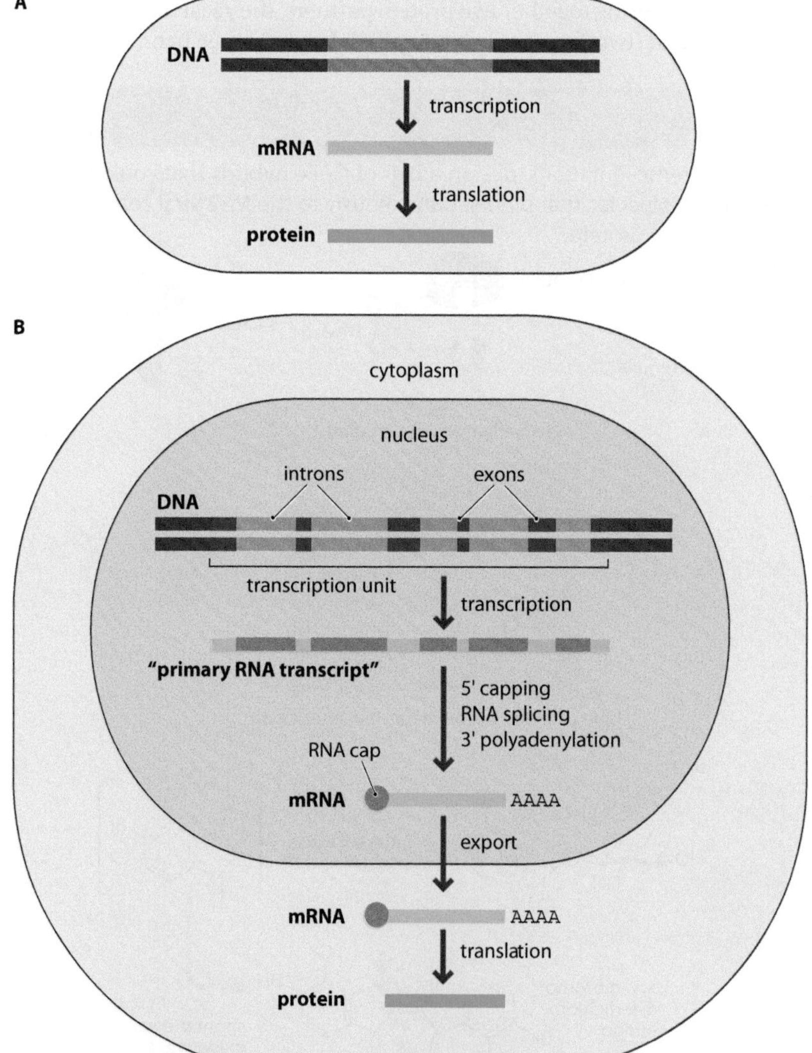

Figure 4.30 mRNA processing in eukaryotes adds complexity to gene expression. Comparison of (A) bacterial gene expression with (B) human gene expression. (Adapted from B. Alberts et al., Molecular Biology of the Cell, 5th ed. New York: Garland Science, 2008.)

The nascent RNA transcripts generated by RNA polymerase II go through a complex series of processing steps that begins during the process of transcription. Both the 5′ end and 3′ end of the RNA transcript are modified, and sections of the middle are removed. These fully processed transcripts, then called mRNA, are ready for export from the nucleus.

The ends of the mRNA are capped and polyadenylated

The first modification to the RNA transcript in eukaryotes is the addition of a conspicuous GTP cap to the 5′ end of the RNA transcript through the action of several enzymes. A phosphatase removes the terminal γ-phosphate from the 5′ end, and the resultant diphosphate is joined to GMP to form an unusual 5′-to-5′ triphosphate linkage (**Figure 4.31**). A methyltransferase enzyme catalyzes S_N2 methylation of the N7 position of the guanine by a sulfonium salt called *S*-adenosylmethionine, or SAM for short (**Figure 4.32**). SAM is a much more common reagent for biosynthetic methylations than the methylene tetrahydrofolate that was used to produce thymidylate from uridylate. A second methyltransferase catalyzes the methylation of the 2′-hydroxyl group of the penultimate nucleoside. The GTP cap is an essential feature of mRNA because it interacts with a protein called eukaryotic initiation factor 4E that helps load the mRNA onto the ribosome.

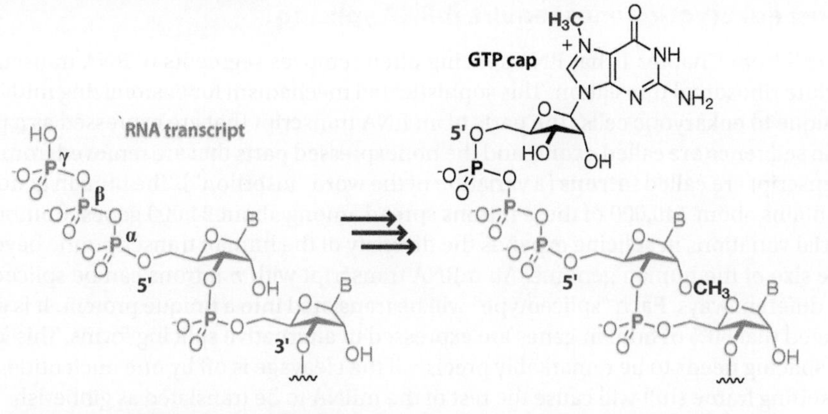

Figure 4.31 mRNA capping in eukaryotic cells. In humans, strands of mRNA undergo substantial modification at the 5′ end.

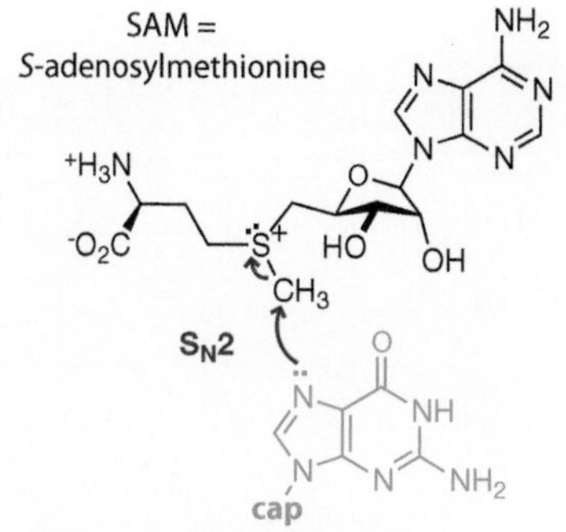

Figure 4.32 S_N2 reaction. The mechanism for methylation of the GTP cap by SAM involves an S_N2 reaction on a sulfonium ion.

RNA transcripts destined for ribosomal translation end in sequences similar to AAUAAA, making them substrates for the addition of a poly(A) tail, 50–200 nucleotides long, to the 3′ end. Poly(A)-binding protein binds to the poly(A) tail of mRNA and directs it to the ribosome for translation. As we will see in the next chapter, the RNA-binding domain is one of the most common structures in the human proteome.

mRNA can be easily separated from other oligonucleotides in mammalian cells by passing them through a matrix of cellulose derivatized with oligo(dT). The mRNAs with poly(A) tails hybridize strongly with the oligo(dT), whereas other oligonucleotides pass through the column. Oligo(dT)-cellulose is made through the chemical polymerization of deoxythymidylate using DCC (dicyclohexylcarbodiimide) followed by coupling to the hydroxyl groups of cellulose (**Figure 4.33**). The DCC polymerization procedure was developed before the advent of phosphoramidite chemistry. We discuss the chemistry of DCC and other carbodiimides in greater detail in Chapter 5.

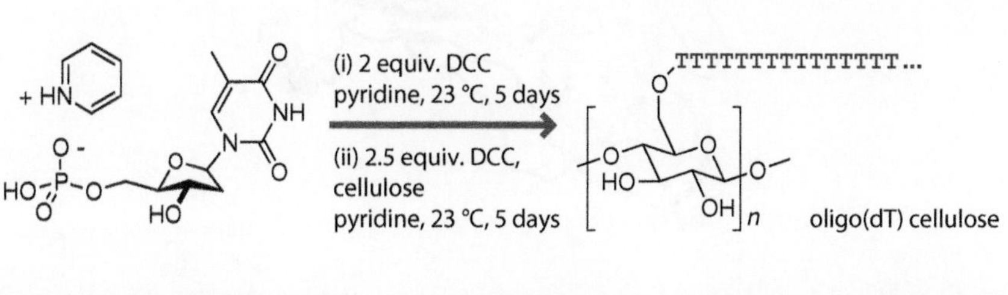

Figure 4.33 A trap for polyadenylated mRNA. Oligo(dT) cellulose is made through uncontrolled polymerization of deoxythymidylate followed by conjugation to polymeric cellulose.

Problem 4.11

S-Adenosylmethionine (SAM), prepared enzymatically from L-methionine and ATP, is an effective substrate for methyltransferases such as mRNA:guanine-N7 cap methyltransferase. SAM prepared through S-methylation with methyl iodide has only half of the activity of SAM isolated from cells, even if the anion is exchanged. Suggest a plausible explanation for the difference.

ATP

L-methionine

S-adenosyl-homocysteine

L-methionine adenosyltransferase

enzymatic S_N2 reaction

SAM = *S*-adenosylmethionine

chemical S_N2 reaction

Most eukaryotic genes require mRNA splicing

Recall from Chapter 1 that RNA splicing often removes segments of RNA transcripts before ribosomal translation. This sophisticated mechanism for customizing mRNA is unique to eukaryotic cells. The parts of an RNA transcript that are expressed as a protein sequence are called **exons**, and the nonexpressed parts that are removed from the transcript are called **introns** (a variation of the word "insertion"). The human genome contains about 140,000 of these introns spread among about 24,000 genes. Combinatorial variations in splicing expands the diversity of the human transcriptome beyond the size of the human genome. An mRNA transcript with n introns can be spliced in 2^n different ways. Each "spliceotype" will be translated into a unique protein. It is estimated that 60% of human genes are expressed in alternative splicing forms. This kind of splicing needs to be remarkably precise; if the cleavage is off by one nucleotide, the resulting frame shift will cause the rest of the mRNA to be translated as gibberish.

In human cells, the removal of introns involves a multiprotein–RNA complex known as the **spliceosome**. The spliceosome is made up of a strand of RNA about 150 nucleotides long, intimately folded together with as many as 10 other small nuclear ribonucleoproteins (snRNPs). The small nuclear RNA (snRNA) strand serves as a targeting device, hybridizing with complementary sequences in the pre-mRNA strand. The other snRNPs pile on to generate a functional spliceosome (**Figure 4.34**). One snRNP recognizes a GU sequence at the 5′ end of the intron, while another snRNP recognizes an AG sequence at the 3′ end of the intron.

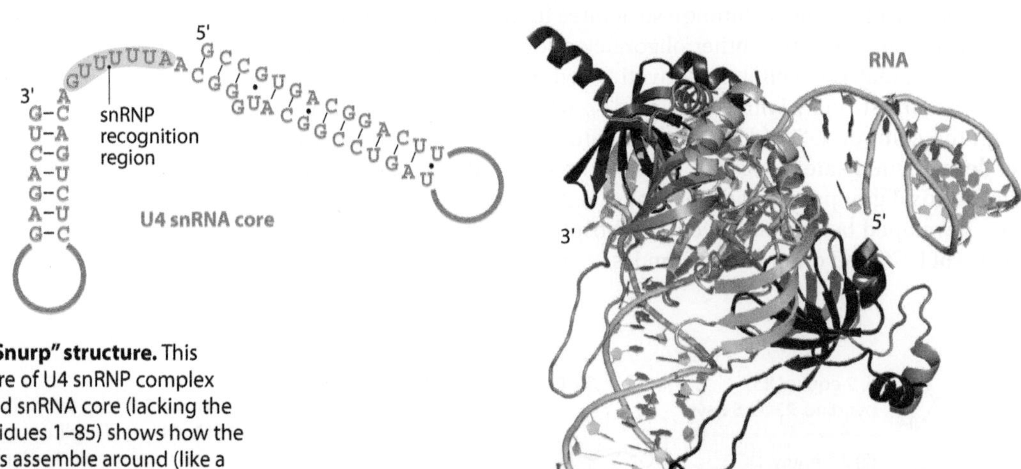

Figure 4.34 "Snurp" structure. This crystal structure of U4 snRNP complex with a modified snRNA core (lacking the 5′ terminal residues 1–85) shows how the snRNP proteins assemble around (like a donut) and recognize the central single-stranded region of the snRNA.

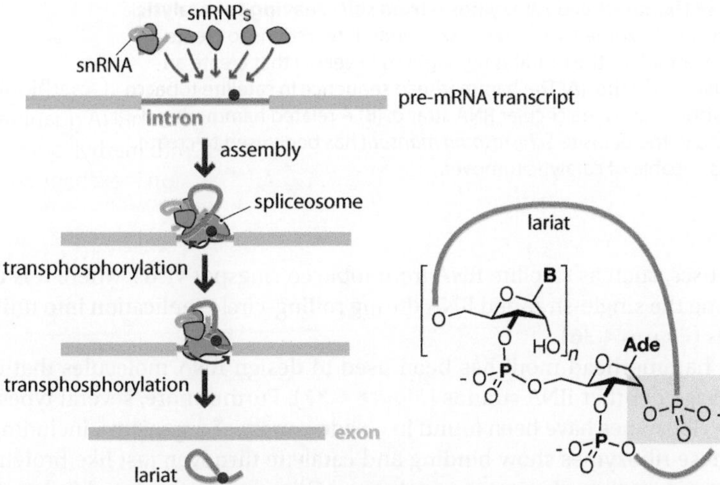

Figure 4.35 Assembly of the spliceosome. An snRNA strand helps the direct assembly of the spliceosome around an intron sequence. The spliceosome catalyzes two transphosphorylation reactions. The first transphosphorylation involves the 2′ hydroxyl group of an adenylate residue and generates the loop of the intron lariat. The second transphosphorylation seamlessly rejoins the exon.

The fully assembled spliceosome catalyzes two sequential transphosphorylation reactions (**Figure 4.35**). In the first reaction, a 2′-hydroxyl group on a specific adenine cleaves the 5′ end of the intron. In the second reaction, the newly freed 3′-hydroxyl group of the exon attacks the 3′ end of the intron, releasing the intron RNA as a lariat. The lariat-shaped RNA intron, knotted by a single adenylate residue, is released and then degraded.

Problem 4.12

The central adenylate of intron lariats has three phosphate bonds. Which P–O bond of this adenylate would be most susceptible to chemical hydrolysis with aqueous NaOH?

Some RNA introns undergo self-splicing without a spliceosome

Some RNA transcripts in lower organisms undergo self-splicing without the action of a spliceosomal complex. In these transcripts, a defined sequence has been shown to be sufficient for cleavage in the presence of Mg^{2+} ions. Before the discovery of self-splicing RNA, it was widely believed that only proteins could exhibit enzymatic behavior. The fact that RNA sequences, called **ribozymes**, can catalyze biological reactions has been hailed as evidence in support of RNA as the molecular progenitor of life on Earth. In general, ribozymes use similar mechanisms to protein-based enzymes, but as a result of their floppier structures and limited functionality they catalyze reactions with less efficiency than enzymes do. Despite this caveat, ribozymes have important, unique roles in biology.

Self-splicing RNA sequences consist of two large families, group I and group II. The group I ribozymes vary in size from about 200 to thousands of nucleotides. They form compact structures that can fold very fast (within tens of milliseconds) and catalyze transesterification reactions using Mg^{2+} ions as cofactors. The group I ribozymes also use guanosine as a cofactor. Group II self-splicing introns resemble the eukaryotic spliceosome, having all catalytic components of the spliceosome in one chain. The group II ribozymes also require Mg^{2+} for folding and catalysis. A self-splicing RNA motif known as the hammerhead ribozyme has been identified in several transcripts

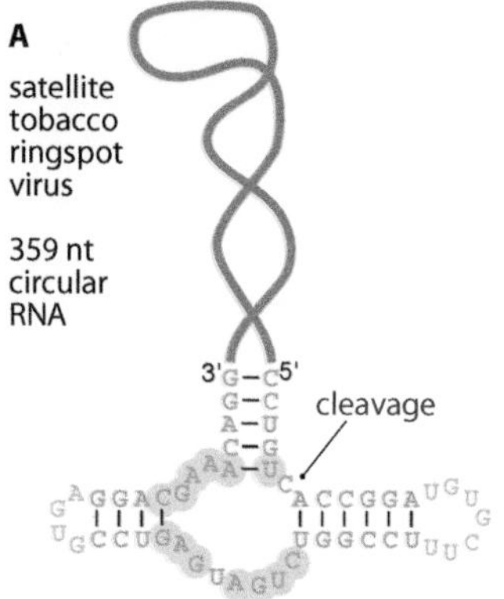

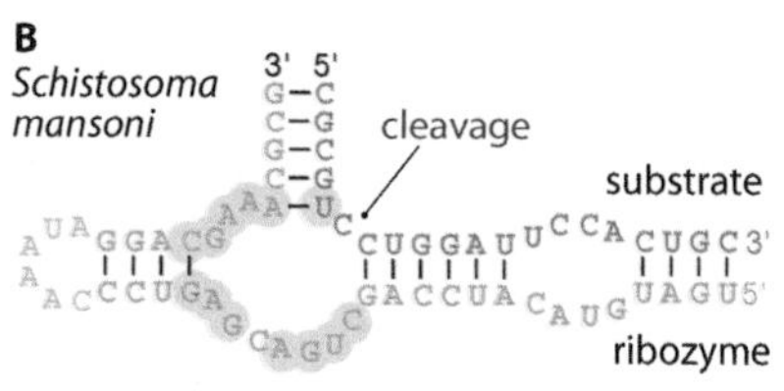

Figure 4.36 Hammerhead ribozymes—from self-cleaving to catalytic. Hammerhead ribozome motifs use base-pairing interactions to create a structural scaffold for the residues highlighted in yellow that create an enzyme-like active site. (A) The hammerhead sequence in satellite tobacco ringspot virus cleaves the circular RNA strand. (B) A related hammerhead motif found in the parasite *Schistosoma mansoni* has been used to create ribozymes capable of catalytic turnover.

from viruses, such as satellite RNA from tobacco ringspot virus, where it is used for processing the single-stranded RNA during rolling-circle replication into unit-length genomes (**Figure 4.36**).

The hammerhead motif has been used to design RNA molecules that catalyze the cleavage of other RNA strands (**Figure 4.37**). Furthermore, several types of self-cleaving ribozymes have been found in a wide variety of organisms, including mammals. These ribozymes show binding and catalytic turnover, just like protein-based enzymes except that ribozymes use Watson–Crick base pairing to bind to the substrate. The mechanism of ribozyme catalysis is believed to involve the attack of a 2′-hydroxyl group on a neighboring 3′-phosphate group. The requirement for divalent cations is believed to involve the stabilization of the negative charge in the phosphate leaving group (**Figure 4.38**).

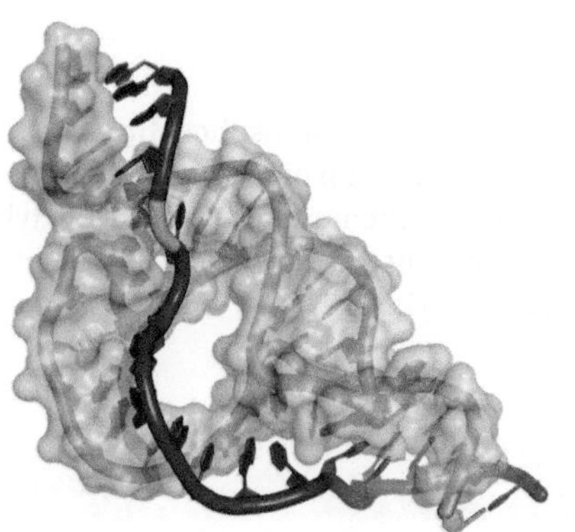

Figure 4.37 A hammerhead ribozyme-substrate complex. A crystal structure based on the *S. mansoni* hammerhead reveals the complex embrace of ribozyme (green) and substrate (brown). Conserved hammerhead residues are highlighted in yellow. The two residues with the scissile bond are colored red. (PDB: 2GOZ)

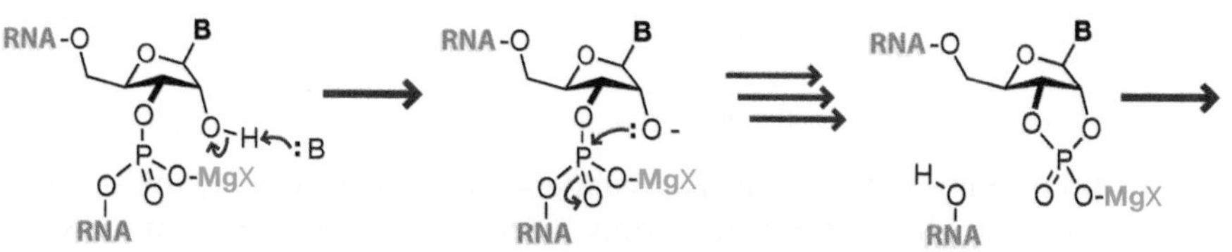

Figure 4.38 Mechanism for RNA cleavage by a ribozyme. MgX^+ indicates Mg^{2+} with a halide counterion. Typically, RNA bases and Mg^{2+} ions contribute additional stabilization to the transition state and leaving groups.

4.5 CONTROLLED DEGRADATION OF RNA

Ribonuclease H degrades RNA•DNA duplexes

Recall that during DNA replication, DNA primase generates short RNA primers that are extended by DNA polymerase (see Figure 4.18). What happens to all of those short RNA primers that end up embedded in the DNA? The RNA primers are removed by the nuclear enzyme ribonuclease H (RNase H), which recognizes RNA•DNA duplexes and cleaves the RNA strand. The gaps are then filled in by a DNA polymerase.

The cleavage of RNA•DNA duplexes by RNase H is remarkably general. Synthetic single-stranded DNA oligonucleotides will induce cleavage of the mRNA by RNase H.

Because the sequence of those oligonucleotides matches the antisense strand of the encoding gene, the method is referred to as antisense DNA, or more generally as antisense. Antisense deoxyoligonucleotides offer a potentially general way to silence genes. For them to work, the antisense deoxyoligonucleotides must hybridize effectively with mRNA in order to target the mRNA for degradation by RNase H (**Figure 4.39**). Because the DNA strand is not affected, it acts as a catalyst. The immense power of antisense technology is that it trivializes the design of the therapeutic because anyone can design a complementary DNA sequence if the mRNA sequence is known. In contrast, designing and synthesizing small-molecule inhibitors of proteins is a challenging endeavor. In practice, the sequence of antisense oligonucleotides must be carefully chosen to minimize hybridization with similar, off-target mRNA molecules and to ensure that neither the antisense DNA nor the mRNA target is sequestered as a hairpin.

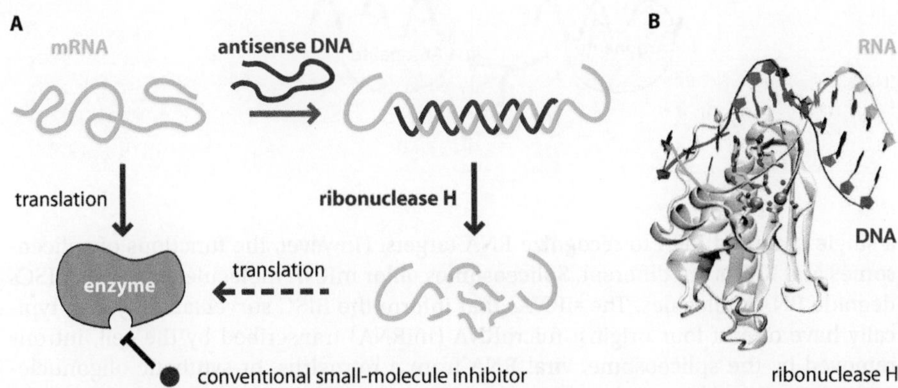

Figure 4.39 Comparison of antisense strategies with conventional small-molecule approaches to inhibition of enzymes. (A) In conventional therapeutics, small-molecule inhibitors disrupt protein function. (B) Antisense ssDNA (red) hybridizes to an mRNA strand (green) and targets the mRNA for destruction by RNase H. (Adapted from M. DeVivo et al., *J. Am. Chem. Soc.* 130:10955–10962, 2008. With permission from the American Chemical Society.)

Oligonucleotides have poor membrane permeability and are susceptible to metabolic degradation. A wide range of chemical modifications have been explored to overcome those challenges. In 1998 the first antisense oligonucleotide drug, fomivirsen (Vitravene™), was approved in the United States for the treatment of cytomegalovirus (CMV) retinitis, an opportunistic infection that causes blindness in patients with AIDS. Fomivirsen is a phosphorothioate deoxyoligonucleotide that is complementary to the mRNA that regulates viral gene expression (**Figure 4.40**). Phosphorothioates resist nuclease cleavage and are readily taken up by cells.

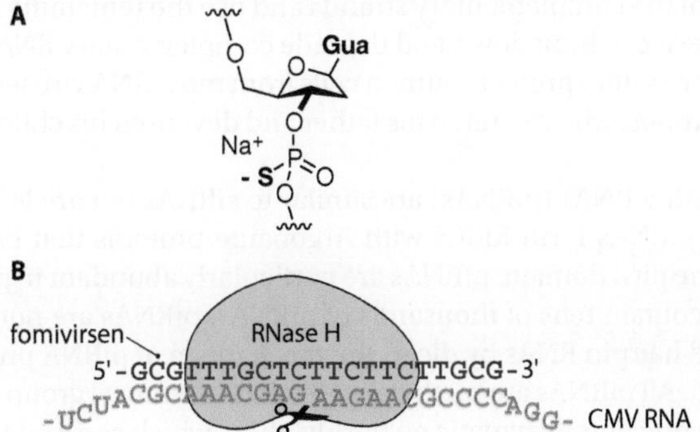

Figure 4.40 Therapeutic antisense. (A) Fomivirsen is a synthetic oligonucleotide linked by phosphorothioates. Each phosphorothioate is a stereogenic center, so the drug is actually a combinatorial library of diastereomers. (B) Fomivirsen leads to the cleavage of viral mRNA by ribonuclease H. (CMV, cytomegalovirus.)

RNA-induced silencing complexes target specific mRNA sequences

In human cells a constellation of small RNAs micromanage protein synthesis by culling the pool of mRNA. These small RNAs generate transient RNA•RNA duplexes that are cleaved through a system that is independent of RNase H. Short double-stranded RNA molecules with fewer than 35 nucleotides, also called small interfering RNA (siRNA), nucleate the assembly of a multiprotein RNA-induced silencing complex (RISC), so named because RISCs degrade mRNA transcripts, effectively "silencing" gene expression. Both RISCs and spliceosomes are multiprotein complexes that use

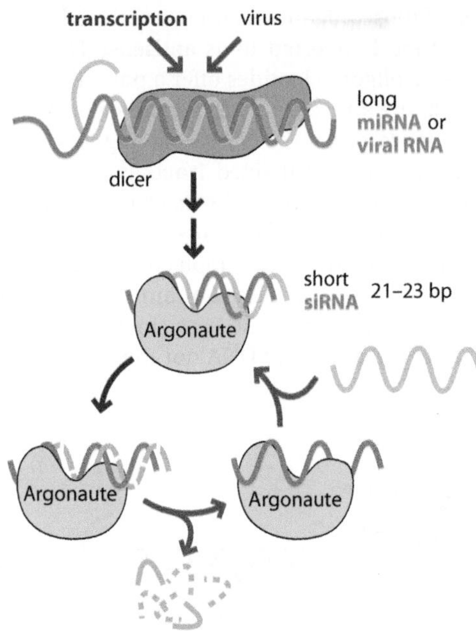

Figure 4.41 Loading the RISC. Dicer generates siRNAs of defined length. The siRNAs are seized by Argonaute and one of the strands is cleaved, leaving the other strand to serve as a template for the destruction of complementary RNA strands.

a single-stranded RNA to recognize RNA targets. However, the functions of spliceosomes and RISCs are different. Spliceosomes tailor mRNA molecules, whereas RISCs degrade RNA molecules. The siRNAs that inform the RISC surveillance system typically have one of four origins: microRNA (miRNA) transcribed by the cell, introns removed by the spliceosome, viral RNA from a retrovirus, or synthetic oligonucleotides introduced by an experiment.

First discovered in *Caenorhabditis elegans* in the early 1990s, miRNAs have since been found in many different animals, plants, and viruses. It has been estimated that up to 60% of human genes are potential targets for RNA silencing by miRNAs. MicroRNAs are transcribed and modified much like other mRNA molecules that are destined for translation, to generate a pre-miRNA with a 5′ cap and a 3′ poly(A) tail. The miRNA transcript folds into a hairpin, which can then leave the nucleus for further processing (**Figure 4.41**). In the cytoplasm, the ribonuclease **dicer** whittles the pre-miRNA down to an siRNA with 21–23 base pairs. The siRNA is seized by the enzyme **Argonaute**, nucleating formation of a multiprotein RISC. Argonautes are usually ribonucleases that cleave one of the complementary strands and use the remaining strand as a reusable targeting device to hunt down and degrade complementary RNA sequences. The RISC surveillance system protects human cells from most RNA viruses. siRNAs act like the Greek god Kronos, who castrated his father and devoured his children as they were born.

Piwi-interacting RNAs (piRNAs) are similar to siRNAs but are longer: 24–32 base pairs in length. piRNAs form RISCs with Argonaute proteins that have a distinctive domain called the piwi domain. piRNAs are particularly abundant in germline testicular cells, which contain tens of thousands of piRNAs. piRNAs are not generated from double-stranded hairpin RNAs by dicer; the mechanism of piRNA production has yet to be established. All piRNAs are methylated on the 2′-hydroxyl group of the 3′ residue; this methylation prevents enzymatic polyuridylation, which can lead to degradation of the RNA sequence (**Figure 4.42**). Some siRNAs are also *O*-methylated at the 3′ nucleotide. piRNA is believed to target retrotransposons for degradation. Retrotransposons are mobile DNA elements that can copy themselves through RNA intermediates, often

Figure 4.42 piRNA. Modifications to piwi-interacting RNA include a 5′ phosphate and a 2′ *O*-methyl group at the 3′ end.

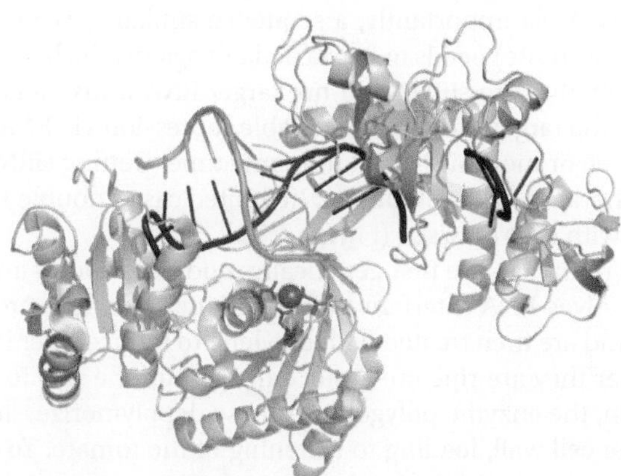

Figure 4.43 Structure of an Argonaute protein from *Thermus thermophilus*. This protein (blue) has an active site analogous to that of RNase H, with Mg^{2+} ion chelated to carboxylate-bearing side chains (purple). They guide siRNA (red) base pairs to the targeted mRNA (green). (PDB 3F73)

encoding a reverse transcriptase. More than 40% of the human genome seems to be composed of vestigial retrotransposons. Most, but not all, are inactive as a result of mutation and/or epigenetic modifications.

Reflecting its functional charge of cleaving RNA, Argonaute proteins structurally resemble RNase H (**Figure 4.43**). Members of the family charged with cleaving RNA have the full complement of active-site residues (two aspartic acids and one glutamic acid) required to bind divalent metal ions for the hydrolysis of the RNA. The siRNA binds some distance from the active site and forms Watson–Crick base pairs with the targeted mRNA. The phosphodiester bond slated for hydrolysis dangles in the active site about 10 or 11 nucleotides from the start of the complementary base pairs. The most important feature of this protein is its ability to bind many different siRNA sequences, which can direct the protein to chop apart different mRNA targets. Some Argonaute proteins derail protein translation by simply removing the 7-methylguanine cap on the 5′ end of the mRNA.

RNA interference is a useful laboratory tool

RNA interference (RNAi) provides an easy method for creating the equivalent of a genetic protein knockout. Implementation requires only a knowledge of the mRNA sequence. Creation of knockouts by more conventional genetic deletion is laborious and often lethal to the developing organism. Small interfering RNAs can be double-stranded or short hairpins. They can be prepared in the lab and injected into cells, or engineered into plasmids and used to transform cells (**Figure 4.44**). Plasmids expressing short hairpin siRNAs against many genes are commercially available. In practice, some sequences lead to more effective RNA interference than others. The best siRNA sequences tend to start with AA and end with TT, such as AA(N_{19})TT, and have less than 50% GC content. Avoiding GC-rich sequences helps to minimize nonspecific hybridization and the consequent off-target effects. Additionally, the sequence should lack

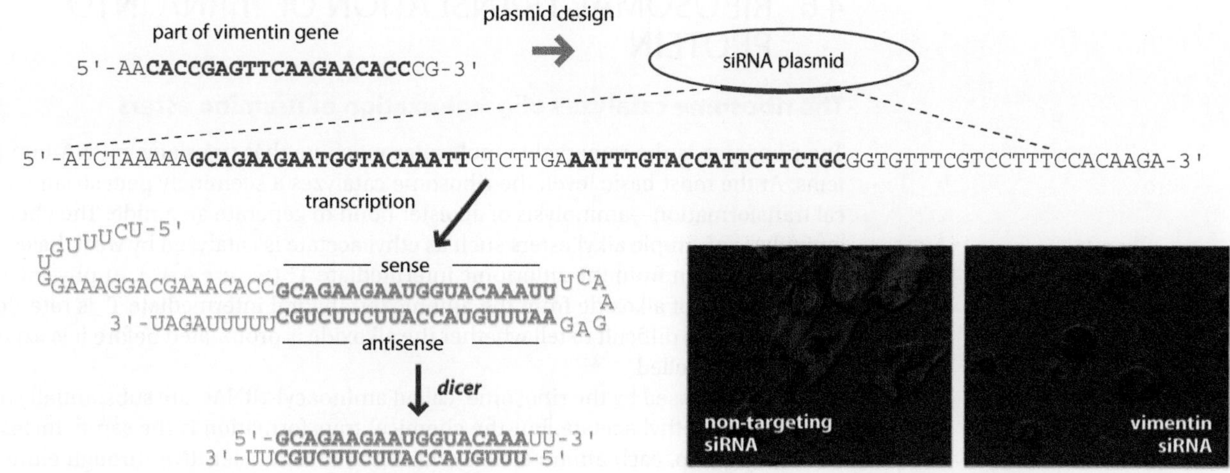

Figure 4.44 Plasmid-encoded siRNA. Cells derived from a papillary thyroid carcinoma were treated with siRNA plasmids and then stained with a fluorescent antibody against vimentin and with 4′,6-diamidino-2-phenylindole (DAPI), a fluorescent DNA stain. In cells treated with a nontargeting siRNA, both nuclear DNA and cytoskeletal vimentin were visible; in cells treated with siRNA against vimentin, only the nuclei were visible. (From V. Vasko et al., *Proc. Natl. Acad. Sci. USA* 104:2803–2808, 2007. With permission from the National Academy of Sciences.)

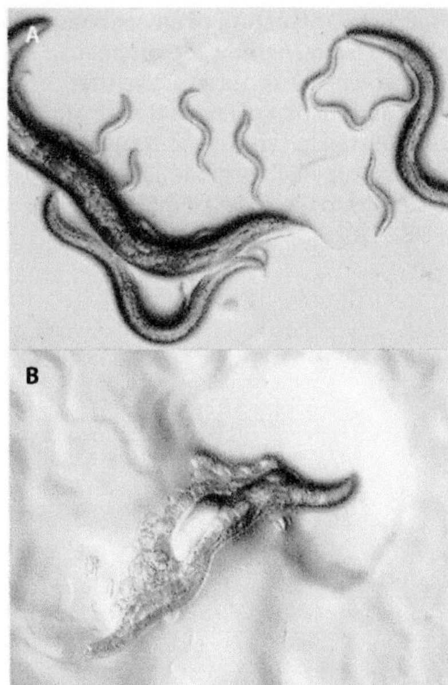

Figure 4.45 Genetic knockouts using RNAi. RNAi was used to understand the importance of an enzyme that crosslinks tyrosine residues. (A) Wild-type *C. elegans.* (B) Double-stranded RNAs based on the *Duox* gene were made *in vitro* and then injected into the gonads of *C. elegans,* leading to bizarre phenotypes in the progeny. (From W.A. Edens et al., *J. Cell Biol.* 154:879–891, 2001. With permission from Rockefeller University Press.)

nucleotide repeats. Most importantly, a sequence similarity search (such as BLAST from the PubMed Web site) needs to be carried out against the human transcriptome. This step can verify that the siRNA will not target RNA transcripts other than those corresponding to the target protein. For reliable repression via RNAi, most investigators introduce three or more siRNAs at the same time. Getting siRNA into cells is the major hurdle in therapeutic development. In limited cases, double-stranded RNA can simply be mixed with cells or tissue (**Figure 4.45**).

The Flavr Savr® tomato, the first genetically modified food, is an example of gene silencing that involves RNA interference. Traditional tomatoes are harvested while green and firm, and are then treated with ethylene to induce ripening. Tomatoes that are harvested after they are ripe are easily damaged on the way to the supermarket. As tomatoes ripen, the enzyme polygalacturonase depolymerizes an oligosaccharide component of the cell wall, leading to softening of the tomato. To decrease the rate of this softening, molecular biologists introduced a new gene into tomatoes that produced an antisense RNA strand that would hybridize to polygalacturonase mRNA, and lead to RNA interference. This strategy worked: these genetically modified tomatoes produced only 10% of the normal amount of the polygalacturonase enzyme. Flavr Savr® tomatoes could be picked after ripening on the vine without the traditional sensitivity to damage during harvest and transport. Unfortunately, the Flavr Savr® tomato could not live up to the commercial hype and, in the face of public skepticism over genetically modified foods, production was stopped in 1997 after only 2 years on the market (**Figure 4.46**).

Figure 4.46 RNAi in the Flavr Savr® tomato. The first genetically modified produce, tomato CGN-89564-2, was created by introducing a gene encoding an RNA molecule that hybridized with mRNA for the enzyme polygalacturonase. (Courtesy of Anthony Freeman. With permission from Photo Researchers.)

Problem 4.13

On the basis of Figure 4.44, design a DNA sequence that could be introduced into a plasmid for the production of a hairpin siRNA against the following sequence in rhodopsin mRNA: 5′-AACUACAUCCUGCUCAACCUA-3′.

4.6 RIBOSOMAL TRANSLATION OF mRNA INTO PROTEIN

The ribosome catalyzes oligomerization of α-amino esters

The ribosome is the remarkable molecular machine that translates mRNA into proteins. At the most basic level, the ribosome catalyzes a seemingly pedestrian chemical transformation—aminolysis of an ester bond to generate an amide. The chemical aminolysis of simple alkyl esters such as ethyl acetate is catalyzed by weak bases that remove a proton from the zitterionic intermediate $T^{\pm}$ (**Figure 4.47**). At physiological pH, expulsion of alkoxide from the anionic tetrahedral intermediate T^{-} is rate-determining, but it is difficult to tell whether the alkoxide is protonated before it is expelled or after it is expelled.

The esters used by the ribosome, called aminoacyl-tRNAs, are substantially more complex than ethyl acetate, but the chemical transformation is the same. Instead of an ethoxy group, each amino acid is connected to a transfer RNA through either the

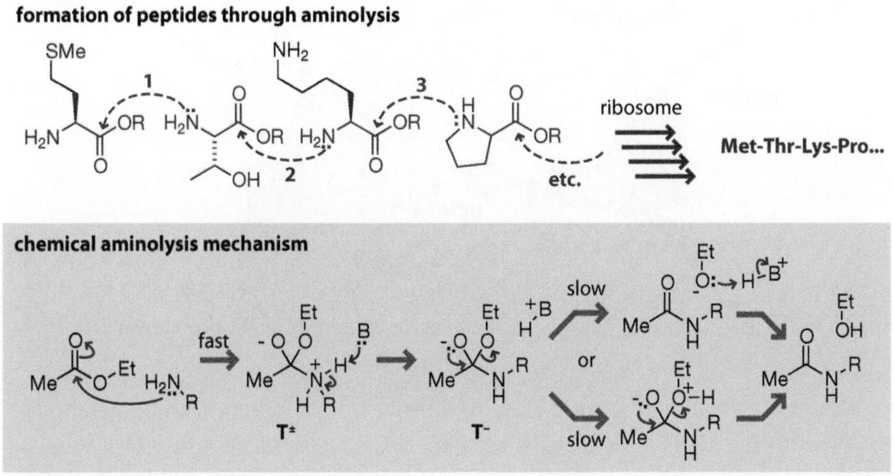

Figure 4.47 The ribosome catalyzes aminolysis. The ribosome assembles peptides through sequential aminolysis reactions. Bases can catalyze the aminolysis of simple alkyl esters by removing a proton and putting it back. Under base-catalyzed conditions, the rate-determining step becomes the breakdown of the tetrahedral intermediate T^{-}.

2′-hydroxyl or 3′-hydroxyl group (**Figure 4.48**). These two isomers equilibrate rapidly, about 5000 times per second at 25 °C, but only a 3′ ester is involved in the aminolytic formation of a peptide bond. The ribosome accelerates the formation of peptide amide bonds from aminoacyl-tRNAs by a factor of more than 10^7 over the background rate. This is a modest rate acceleration by the standards of most enzymes. The extraordinary feat achieved by the ribosome is that it controls the sequence of the newly formed peptide chain.

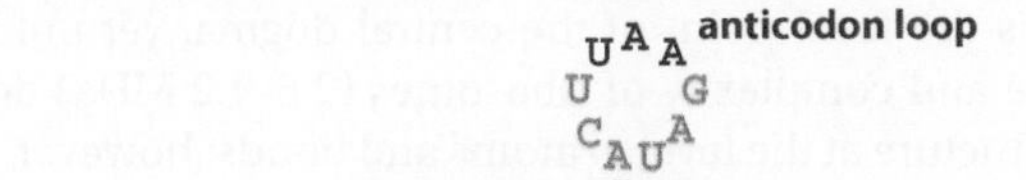

Figure 4.48 The chemical structure of a human leucyl-tRNA. Transfer RNA molecules are about 80 nucleotides in length. The amino acid is attached through either the 2′- or 3′-hydroxyl group of the nucleotide at the 3′ end through a carboxylic ester bond. The two esters equilibrate rapidly through an acyl transfer reaction.

The ribosome is a massive molecular machine, half protein and half RNA

A typical human cell contains around 4 million ribosomes, accounting for about 80% of the total RNA and 5–10% of the total protein in the cell. Mitochondrial organelles express their own ribosomes; they are smaller than cytoplasmic ribosomes but larger than the ribosomes found in *E. coli.* Structurally, mitochondrial ribosomes are better understood than cytoplasmic ribosomes. Ribosomes are conglomerates—about half protein and half RNA. A typical human ribosome is composed of four RNA chains and about 80 proteins. Human cytoplasmic ribosomal components assemble in the

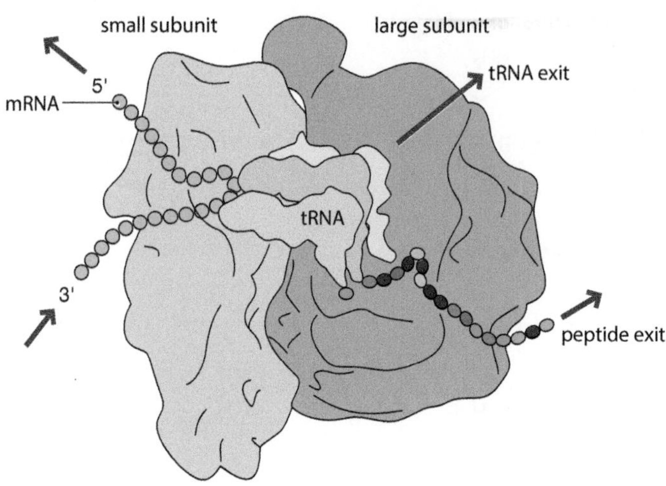

Figure 4.49 The ribosome at work. This structure highlights the key regions of the ribosome (white surface) and other molecules during peptide synthesis. (Adapted from D.L.J. Lafontaine and D. Tollervey, *Nat. Rev. Mol. Cell Biol.* 2:514–520, 2001. With permission from Macmillan Publishers Ltd.)

nucleus into two massive ribosomal subunits, one large and one small. The large ribosomal subunit consists of three RNA chains and 49 proteins. The small ribosomal subunit consists of one RNA chain and 33 proteins. Once exported to the cytoplasm, the two subunits bind to the mRNA for protein translation, through sequential aminoacylation reactions fed by aminoacyl-tRNA esters (**Figure 4.49**). Such translation requires initiation and elongation factors, which are described further below.

The ribosome is the focal point of the central dogma, yet until the turn of the millennium the size and complexity of ribosomes (2.6–4.2 MDa) defied attempts to resolve ribosome structure at the level of atoms and bonds; however, in 2000, in a *tour de force* achievement, structural biologists solved the X-ray crystal structure of a bacterial ribosome at high enough resolution (2.4 Å) to observe the atomic details of the catalytic machinery (**Figure 4.50**).

All ribosomes—mitochondrial, cytoplasmic, or prokaryotic—catalyze amide bond formation through the same chemical mechanism, by supplying a catalytic base. Deep within the active site of all ribosomes, a single lone pair on a single nitrogen atom on a single adenine carries out the critical deprotonation that lowers the barrier to amide bond formation. The lone pair on N3 of the catalytic adenine in the bacterial ribosome (A2541) has been estimated to have a pK_a' of about 7.6—just right for base catalysis at physiological pH, but about 10,000 times more basic than the N3 in free adenosine. The structural characterization of the ribosome proves that it is a ribozyme; the proteins are mere scaffolding. Enzymatic catalysis was once thought to be the exclusive provenance of proteins, but the enzymatic capabilities of RNA are now firmly rooted in the central dogma. With firm evidence that protein synthesis is catalyzed by RNA, the idea that all life began with RNA molecules has gained considerable traction.

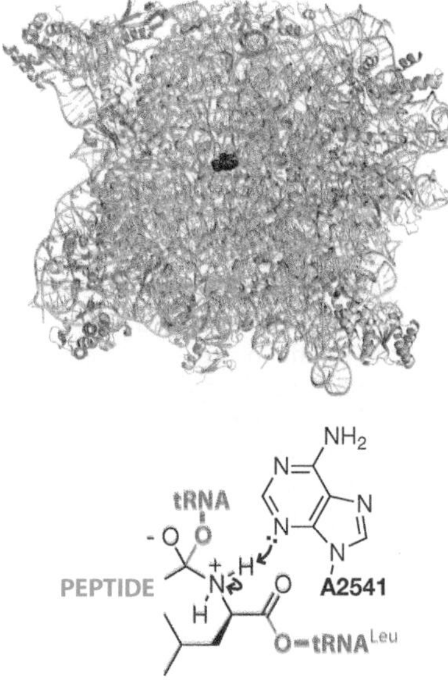

Figure 4.50 The structure of an archaebacterial ribosome. Proteins (blue) have a peripheral role in peptide-bond-forming reactions at the ribosome's active site (purple), which is composed mainly of RNA (green). The purple van der Waals spheres indicate atoms in the catalytic adenylate base. (PDB 1S72)

Problem 4.14

The phosphate group of adenine 2450 has been proposed to increase the basicity of the N3 lone pair of A2451 in the bacterial ribosome. Draw a tautomer of the A2451/G2447/A2450 triad, with hydrogen bonds, that justifies the hypothesis.

A2451 G2447 A2450

Figure 4.51 tRNA shape. All tRNA molecules, like tRNA^{Phe} from yeast, fold into a bracket shape. The three nucleotides in the anticodon loop are rendered as spheres. The amino acid (not shown) is normally attached to the 3′ adenylate residue. Mg^{2+} ions are shown as gray spheres. (PDB 1EHZ)

tRNA molecules are heavily processed and adopt fixed shapes

All living organisms on the planet Earth can translate proteins with a common set of 20 amino acids, which will be introduced in Chapter 5 along with their abbreviations. Each of these amino acids is carried by one or more types of tRNA molecules. There are a total of 64 possible tRNA molecules, each with a three-nucleotide anticodon tag that allows the ribosome to translate from a three-nucleotide codon into an amino acid. Thus, the mRNA sequence 5′-UUC-GUA-CAC-3′ uniquely encodes the amino acid sequence phenylalanine-valine-histidine.

Each of the tRNA molecules is about 80 nucleotides long but has a different nucleotide sequence, yet all tRNA molecules are tightly folded into a common bracket shape that allows them to fit within the ribosomal active sites (**Figure 4.51**). tRNA molecules are held together by a combination of Watson–Crick base pairing—including mismatches, Hoogsteen base pairing, and even some triple-strand interactions. The tRNA molecules in most organisms have a conserved trinucleotide sequence CCA at the 3′ terminus. When not encoded by the gene, the three nucleotides are added one at a time by a single nucleotidyltransferase enzyme. The amino acid is attached to the 3′ nucleotide at one end of the bracket; the anticodon nucleotides are located at the loop on the other end of the bracket.

tRNA molecules are heavily modified from the original transcripts. Some transcripts undergo splicing; others are modified by enzymes. About 100 different covalent modifications to the RNA bases are known (**Figure 4.52**). In some cases, the modified

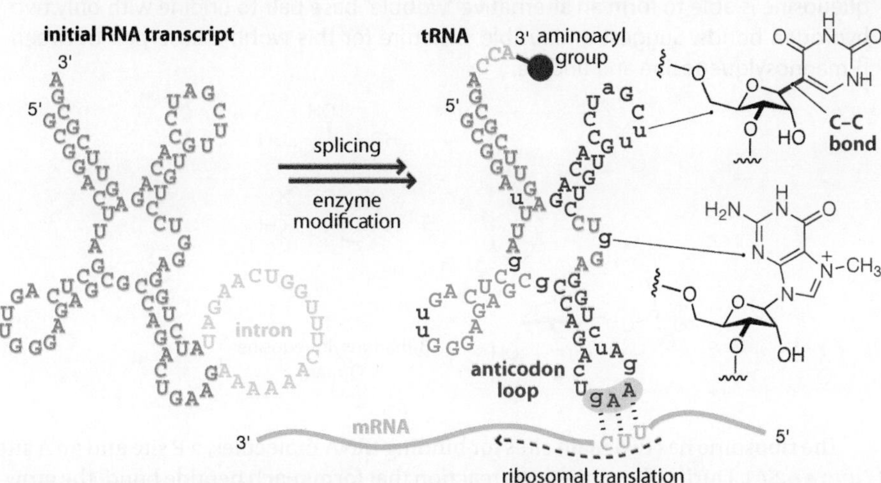

Figure 4.52 Post-transcriptional processing of tRNA. The initial transcript of tRNA^{Phe} from yeast chromosome 4 is heavily modified through splicing and enzymatic replacement of RNA bases with unusual bases (represented by lower-case red letters).

bases are relatively similar to traditional nucleotide bases, for example 7-methylguanine or 4,5-dihydrouracil. In other cases, the modifications are quite drastic (**Figure 4.53**). For example, the base pseudouridine is attached to ribose through a carbon-carbon bond. The modified bases can have an important role in insuring the fidelity of the translation. When replaced with other bases, the tRNA can either lose its specificity for the correct amino acid or even mis-hybridize with the RNA sequence to cause a frame shift, which produces the wrong protein. Such "shifty tRNAs" can result from an innate immune response to viral infection, as the human host responds to

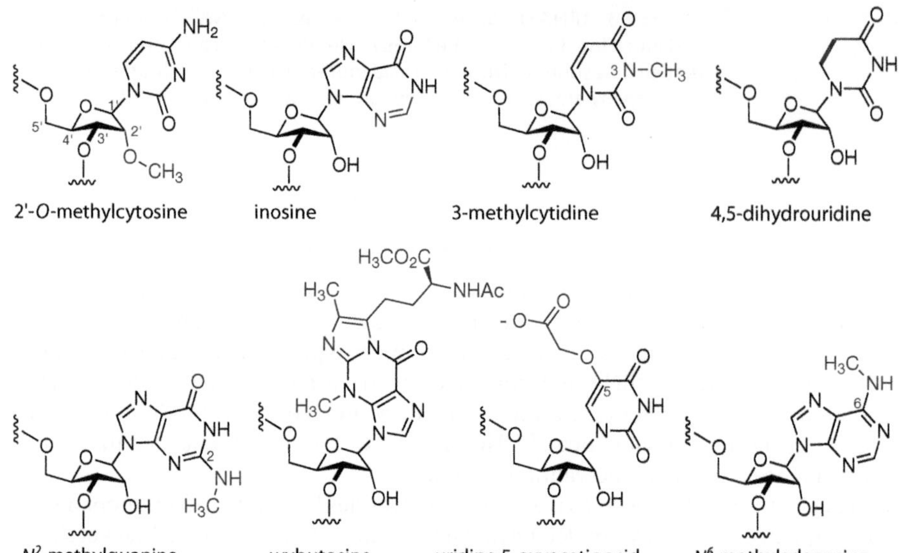

Figure 4.53 Modified nucleosides found in yeast tRNA^{Phe}. Some variations arise from modifications of the nucleoside hydroxyl group; others involve removal and replacement of the base.

2′-*O*-methylcytosine · inosine · 3-methylcytidine · 4,5-dihydrouridine

N^2-methylguanine · wybutosine · uridine 5-oxyacetic acid · *N*^{6}-methyladenosine

retroviral invasion with decreased levels of the tRNA modification enzymes required for ensuring the production of the correct protein.

Problem 4.15

The anticodon loop of tRNA^{Asp} contains the hypermodified base β-mannosyl-queuosine in the sequence 5′-Q(Man)UC-3′. tRNA^{Asp} recognizes the codons 5′-GAC-3′ and 5′-GAU-3′ equally well because the β-mannosylqueuosine residue destabilizes Watson–Crick base pairing between mannosylqueuosine and cytosine. The queuosine is able to form an alternative "wobble" base pair to uridine with only two hydrogen bonds. Suggest a plausible structure for this wobble base pair between β-mannosylqueuosine and uridine.

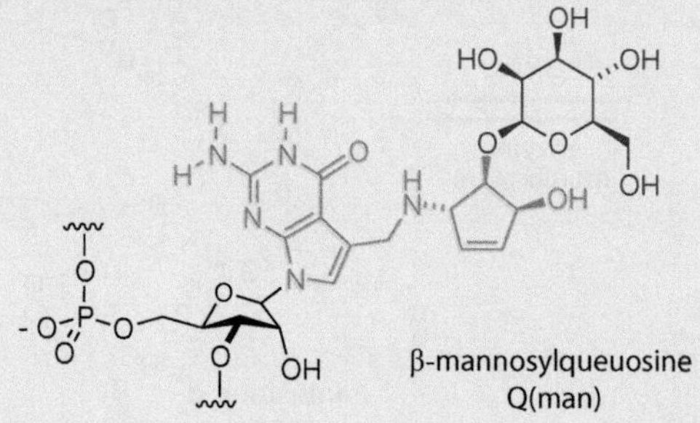

β-mannosylqueuosine
Q(man)

The ribosome has two main sites for binding tRNA molecules, a P site and an A site (**Figure 4.54**). During the aminolysis reaction that forms each peptide bond, the growing peptide chain is transferred from the tRNA in the P site to the aminoacyl-tRNA in

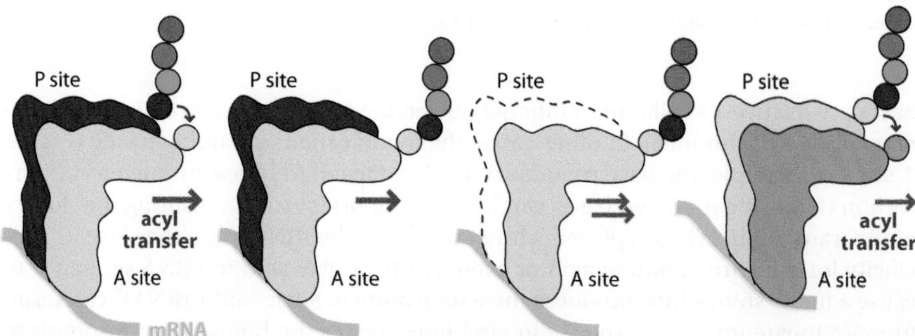

Figure 4.54 Transfer of the peptidyl chain during translation. The aminolysis reaction leads to acyl transfer of the growing peptide chain from one tRNA to another.

the A site. The tRNA molecule in the P site then leaves the ribosome. The tRNA in the A site, which now carries the peptide chain as an ester, translocates to the P site, leaving the A site vacant so that a new aminoacyl-tRNA can bind in the A site, ready for another round of elongation.

The genetic code allows one to translate from mRNA sequence into protein sequence

With four different RNA bases, the 64 possible three-base codons encode the 20 universal amino acids and three stop codons that tell the ribosome to stop translation. Some of the more commonly used amino acids are encoded by more than one codon. The **genetic code** (**Figure 4.55**) translates triplet codons into specific amino acids. Although the genetic code remains essentially invariant across all life on the planet, different organisms have slightly different codon preferences for each amino acid. For example, the most common codon for alanine in *E. coli* K12 is GCG, whereas in humans it is the least common codon for alanine; similarly, the most common codon for arginine in humans is AGA, whereas in *E. coli* K12 it is the least common codon for arginine. *E. coli* cells transformed with a human gene sometimes do not express the human protein very effectively as a result of the differences between the codon preferences of humans and *E. coli*. Resynthesis of the human gene to replace uncommon codons with the codons preferred by bacteria can sometimes solve this problem.

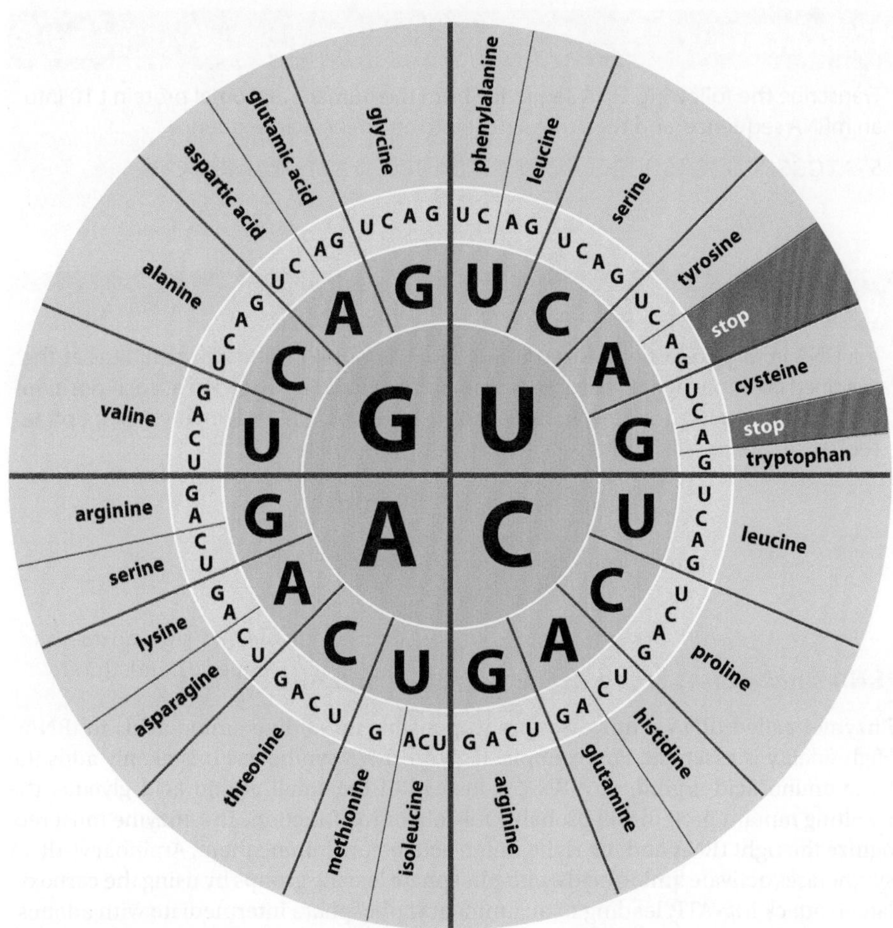

Figure 4.55 The genetic code. This wheel diagram of the genetic code allows one to translate mRNA sequence into protein sequence by reading from the inside of the wheel to the outside, big to small. For example, at 4 o'clock on the genetic code diagram, CUG encodes the amino acid leucine.

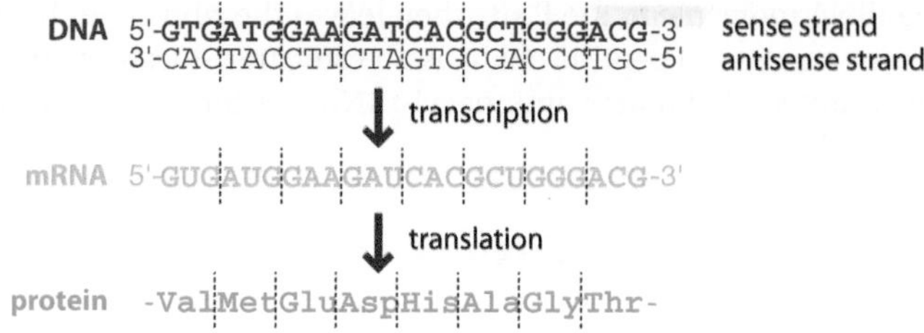

Figure 4.56 Decoding. The DNA sequence in the coding strand (top) is easily transcribed by replacing T with U. By reading the mRNA sequence in the 5′ to 3′ direction, one can use the genetic code to predict the sequence of the translated protein.

mRNA is translated by the ribosome in the 5′ to 3′ direction (**Figure 4.56**). Each aminoacyl-tRNA adds a new amino acid to the C terminus of the growing peptide chain. In other words, proteins are synthesized in the N to C direction. For example, the mRNA sequence 5′-AGUUAC-3′ would code for the dipeptide sequence Ser-Tyr as read from the N terminus to the C terminus. The final codon in any open reading frame is one of three stop codons—namely UAG, UAA, or UGA. The stop codon UAG is rarely used in *E. coli*. It is more difficult to go backward and predict the sequence of the gene that gave rise to a given protein sequence, because amino acids are usually encoded by more than one codon, and—more importantly—exons may be missing from eukaryotic mRNA. Many Web-based programs will translate DNA-coding sequences or mRNA sequences into protein sequences, but you should practice translating DNA and RNA sequences into protein sequences manually.

Problem 4.16

Transcribe the following DNA sequence from the human ribosomal protein L10 into an mRNA sequence, and then translate it into an amino acid sequence.

5′-ATGGGCCGCCGCCCCGCCCGTTGTTACCGGTATTGTAAGAAC-3′

Problem 4.17

If a DNA library were synthesized using equal mixtures of phosphoramidites at the specified positions, what mixtures of amino acids would be encoded at each position if the mixture of sequences were ribosomally expressed? How many hexapeptide sequences are possible from such a mixture of oligonucleotides?

A A A G G C
5′-G A CTG TG T CAG T-3′
C G C C C T

stacked letters indicate an equal mixture of nucleotides

tRNA synthetases recognize amino acids and nucleotides

Enzymes called tRNA synthetases are responsible for adding amino acids to tRNAs. High fidelity is essential. For example, if glycyl-tRNA synthetase mistakenly adds the large amino acid arginine to $tRNA^{Gly}$ instead of the small amino acid glycine, the resulting mutant protein will probably misfold or malfunction. The enzyme must recognize the right tRNA and the right amino acid before joining them. Aminoacyl-tRNA synthetases activate amino acids with phosphate leaving groups by using the carboxylate to attack Mg•ATP, leading to an aminoacyl phosphate intermediate with adenosine monophosphate poised to act as a leaving group (**Figures 4.57** and **4.58**). The synthetase checks its work. If the wrong amino acid is attached to AMP, the enzyme rapidly hydrolyzes the acyl phosphate through the action of an additional editing domain. If the correct amino acid is attached, the enzyme allows the bound tRNA to attack the acyl phosphate, generating the aminoacyl-tRNA ester. The combination of careful

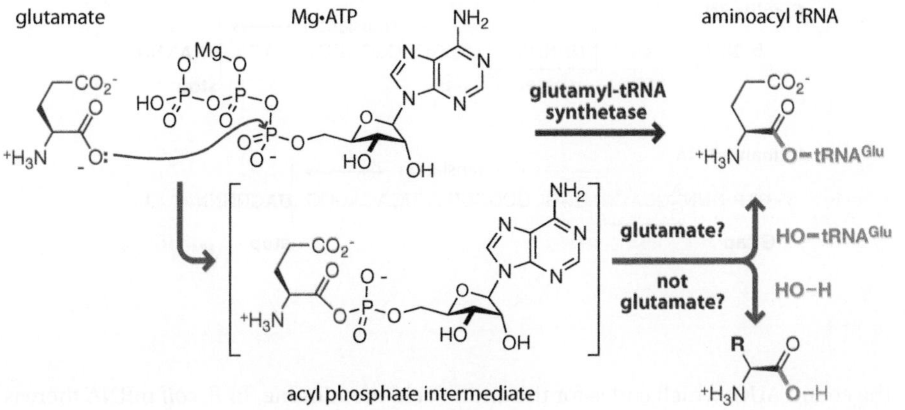

Figure 4.57 Proofreading. Glutamyl-tRNA synthetase generates a highly reactive acyl phosphate intermediate and then checks its work. If the aminoacyl group corresponds to glutamate, it allows $tRNA^{Glu}$ to react, forming an ester. If the aminoacyl group belongs to any other amino acid, water reacts, erasing the mistake.

activation and proofreading leads to high fidelity. For example, given both valine and isoleucine—which differ by a single methylene unit—isoleucine tRNA synthetase will make a mistake only once for every 10,000–100,000 correctly loaded tRNAs.

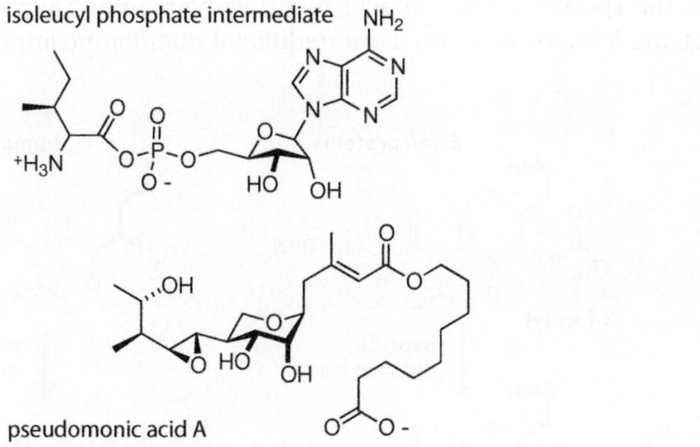

Figure 4.58 Chemical mimicry. The natural product pseudomonic acid A is a component of antibacterial ointments such as Bactroban™. Pseudomonic acid A blocks the active site of isoleucyl-tRNA synthetase by mimicking the aminoacyl phosphate intermediate.

How does glutaminyl-tRNA synthetase correctly distinguish $tRNA^{Gln}$ from the other tRNA molecules such as $tRNA^{Glu}$? Aminoacyl-tRNA synthetases cannot discriminate on the basis of shape, because all tRNAs that carry amino acids have the same tightly folded bracket shape. Moreover, the tightly packed structure of the tRNA precludes the typical helix–major groove interactions that mediate the recognition of B-form DNA. Some aminoacyl-tRNA synthetases make contacts with the anticodon nucleotides, but others do not. All aminoacyl-tRNA synthetases bind to a large portion of the tRNA surface, making it possible to select just the right set of tRNA molecules. To meet the molecular recognition challenge, the tRNA synthetase relies on induced fit, adjusting the conformation of both tRNA and protein during enzymatic loading of the tRNA (**Figure 4.59**). Each amino acid has a single tRNA synthetase, which can recognize all tRNAs having those anticodons associated with the amino acid. Thus, all six leucyl-tRNAs are loaded by a single synthetase, even though their anticodon loops bind different codons, namely UAG, GAG, CAG, AAG, UAA, and CAA.

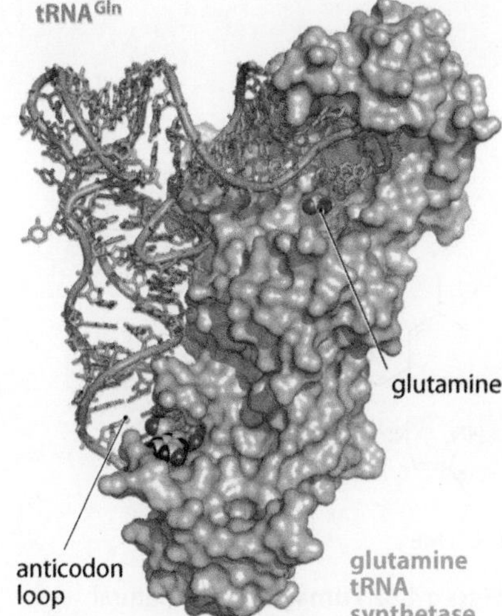

Figure 4.59 tRNA synthetase. *E. coli* tRNA synthetase binds to both ends of $tRNA^{Gln}$ (green), recognizing a large portion of tRNA, including the nucleotides CUG in the anticodon loop (spheres). The amino acid glutamine is snuggled deep within the active site, close to the 3′ end of the tRNA. (PDB 1ZJW)

What controls the beginning and end of translation?

All mRNA molecules have structural features that tell the ribosome where to begin and end translation. A short nucleotide sequence called the ribosomal binding site (RBS) hybridizes with a complementary strand on the ribosome (**Figure 4.60**). The RBS is always upstream of the coding sequence. In *E. coli* mRNA, the RBS sequence is usually similar to AGGAGGU; in human mRNA, the RBS is similar to ACCACC or GCCACC. Recall that human mRNA strands are more complex than bacterial mRNA, with a 5′ G-cap and a poly(A) tail at the 3′ end. All mRNA coding sequences start with

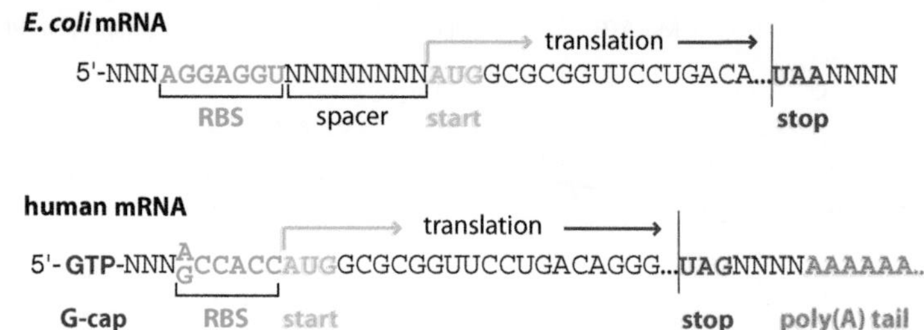

Figure 4.60 Features of prokaryotic and eukaryotic mRNA sequences. Bacterial mRNAs have a spacer between the RBS and the protein-coding sequence, but otherwise have fewer features than eukaryotic mRNAs.

the codon AUG, which codes for the amino acid methionine. In *E. coli* mRNA there is a short nucleotide spacer, about eight bases in length, between the RBS and the AUG start codon. In human mRNA the AUG start codon is directly adjacent to the RBS. Most mRNA coding sequences also have one of three different stop codons named after mineral gems: UAA (ochre), UAG (amber), and UGA (opal). The tRNAs that bind to these codons, called suppressor tRNAs, carry no amino acid so that water eventually hydrolyzes the peptidyl ester bond, releasing the protein.

In prokaryotes, such as *E. coli*, a special initiator tRNA starts protein synthesis with the specialized amino acid *N*-formylmethionine (**Figure 4.61**); in eukaryotic cells, the initiator tRNA carries a traditional methionine aminoacyl group. Enzymes customize the termini by cleaving off the formyl group (in bacterial proteins) or the methionine residue (in human proteins). Other enzymes called aminopeptidases nibble away accessible amino acids from the starting end of the protein, often referred to as the N terminus (**Figure 4.62**). The removal of amino acids from the N terminus extends the control of protein sequences beyond that available from ribosomal translation. Trimming *N*-formyl groups, methionine, and other amino acids from the N terminus of enzymes is often a critical step in revealing the unique reactivity of the terminal amino group or an amino acid that is essential for catalytic activity. Unfortunately, when non-bacterial proteins are overexpressed in *E. coli*, the aminopeptidases often generate mixtures of proteins that differ by only one or two amino acids, making it impossible to separate the full-length protein from proteins lacking one or two amino acids.

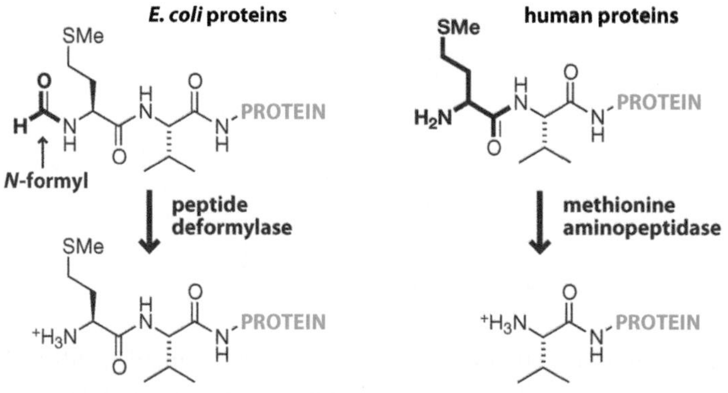

Figure 4.61 Editing the first few amino acids. In both prokaryotic and eukaryotic cells, the N terminus of the protein is subject to enzymatic cleavage to reveal amino acids other than methionine.

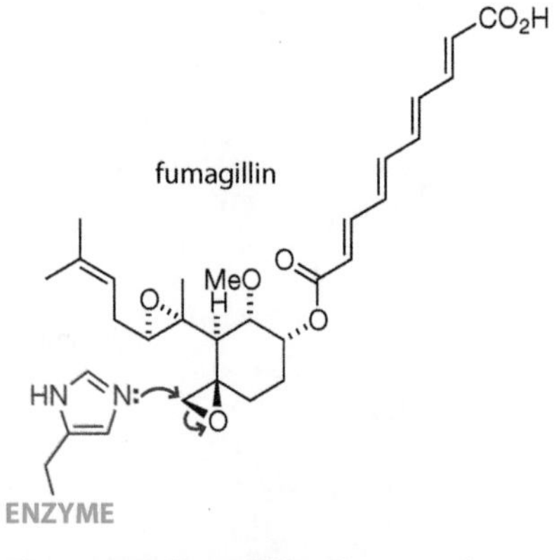

Figure 4.62 Fumagillin. The natural product fumagillin isolated from *Aspergillus fumigatus* is a selective inhibitor of the eukaryotic enzyme methionine aminopeptidase 2. It forms a reversible covalent adduct with an imidazole in the active site of human methionine aminopeptidase 2, and inhibits angiogenesis—the growth of blood vessels.

The G-cap helps direct the human ribosome to the RBS so that translation can start with AUG. Less commonly, an internal ribosomal entry sequence (IRES) can direct the start of translation, bypassing the RBS and AUG start codon. Many eukaryotic viruses and some human proteins can be translated, even when the cell is trying to shut down translation by inhibiting the traditional initiation of protein synthesis.

Problem 4.18

The natural antibiotic actinoin binds tightly to a zinc cation in the active site of peptide deformylase. On the basis of the structure of the actinoin–enzyme complex shown below, draw the structure of the natural substrates bound to the active site in a way that explains the role of the zinc ion in catalysis.

Translational initiation is a focal point for control of protein synthesis

The initiation of protein translation is the slow, rate-limiting step in translation of mRNA. By controlling translational initiation, cells exert tight control over protein synthesis. For example, many proteins misfold at elevated temperature, which is great for frying eggs but disastrous for cells. *E. coli* constitutively transcribes mRNA for heat-response proteins, but they are only translated at elevated temperatures that allow them to unfold and reveal the RBS (**Figure 4.63**A). This elegant temperature sensor provides an immediate, transcription-independent response to the elevated temperature, producing proteins that degrade or refold misfolded proteins. When temperatures return to a normal range, heat response proteins are no longer translated. Proteins can facilitate or inhibit translation by binding to the mRNA. Proteins can promote translation by binding to mRNA segments that mask the RBS (Figure 4.63B) or, more commonly, can inhibit translation by binding to the RBS. Small regulatory RNAs and proteins can bind to the untranslated regions of the mRNA to mask or unmask the RBS (Figure 4.63B, 4.63C, and 4.63D). In *E. coli*, for example, the threonyl-tRNA synthetase can bind to the mRNA encoding its synthesis, thus blocking the synthesis of additional copies of itself. This mechanism provides feedback, preventing the synthesis of excess quantities of the enzyme. Such binding typically takes place at a specific sequence toward the 5′ end of the mRNA. The complex can either unmask a RBS to trigger translation, or block the RBS to prevent translation.

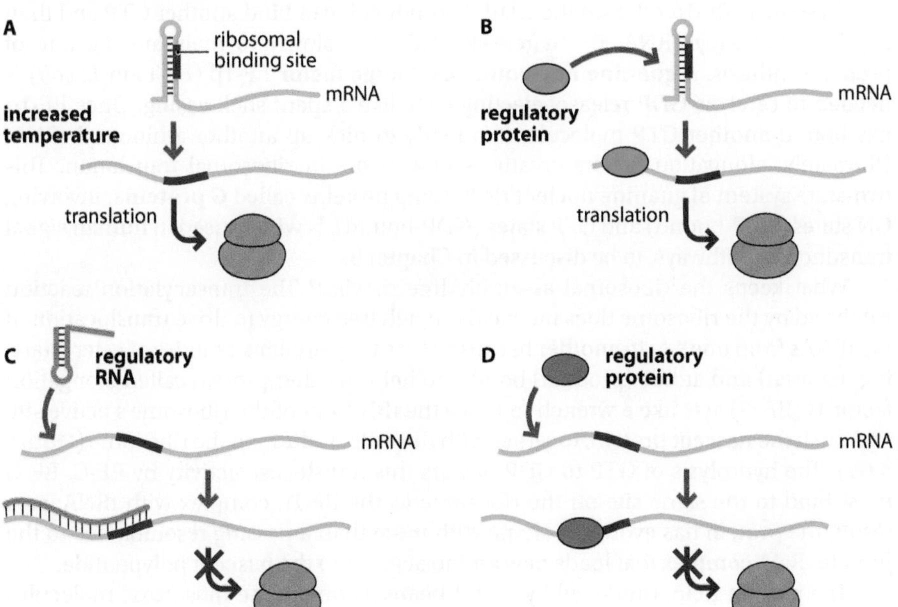

Figure 4.63 Mechanisms for controlling translational initiation. The mRNA ribosomal binding site (RBS) is a focal point for control of translational initiation. (A) At elevated temperatures, mRNA can unfold to reveal the RBS. (B) Proteins can complex with the mRNA to reveal the RBS. (C) Regulatory RNAs can hybridize to and mask the RBS. (D) Proteins can bind to and mask the RBS. (Adapted from P. Romby and M. Springer, *Trends Genet.* 19:155–161, 2003. With permission from Elsevier.)

A protein escorts each aminoacyl-tRNA to the ribosome for fidelity testing

Aminoacyl-tRNAs do not engage the ribosome on their own. They are *escorted* to the ribosome by an error-checking protein called elongation factor-1α, called EF-Tu in *E. coli* (**Figure 4.64**). The escort function is so critical that EF-Tu makes up 5% of total protein content in *E. coli*. In human cells EF-1α uses a bound GTP molecule as part of a two-state control mechanism. The ternary EF-1α–GTP–aminoacyl-tRNA complex binds reversibly to the ribosome. If the anticodon loop of the aminoacyl-tRNA matches the mRNA codon, EF-1α hydrolyzes the GTP to GDP and releases its grip on the aminoacyl-tRNA, allowing it to deliver the amino acid. This mechanism of coupling chemical energy to physical movement, through hydrolysis of a high-energy phosphate bond and subsequent conformational change, is a common theme in biology; for example, ATP-driven motor proteins undergo similar conformational rearrangements, which allow extension and worm-like walking.

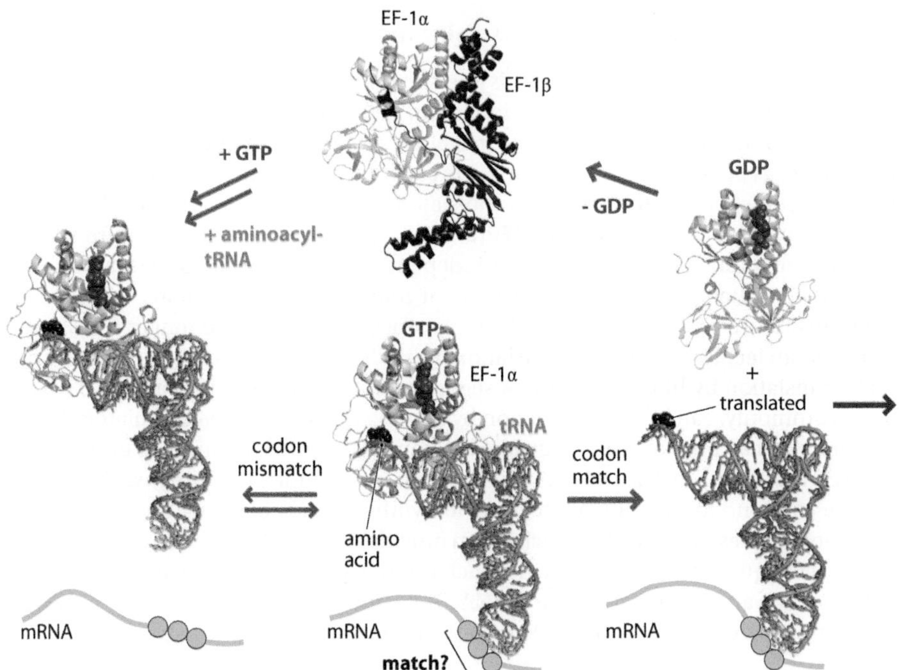

Figure 4.64 The bacterial version of the EF-1α cycle. A ternary complex between EF-1α, GTP, and an aminoacyl-tRNA binds to mRNA within the ribosome. If the tRNA does not match the codon, the ternary complex dissociates from the ribosome. If the codon matches, GTP is hydrolyzed to GDP, and EF-1α releases the tRNA so that the new amino acid can be incorporated. Dissociation of GDP is catalyzed by EF-1β, allowing a new aminoacyl-tRNA to bind. (PDB 1OB2, 1D8T, and 1EFU)

As soon as EF-1α releases the GDP by-product it can bind another GTP and then another aminoacyl-tRNA. EF-1α releases GDP too slowly to maintain the rate of protein synthesis. A **guanine nucleotide exchange factor** EF-1β (EF-Ts in *E. coli*) is needed to catalyze GDP release, ejecting GDP like a spent shell casing. Once EF-1α has bound another GTP molecule, it is ready to pick up another aminoacyl-tRNA. Ultimately, elongation factors greatly reduce errors in ribosomal translation. This two-state system of guanine nucleotide binding proteins called **G proteins**, involving ON states (GTP-bound) and OFF states (GDP-bound), is widely used in human signal transduction pathways, to be discussed in Chapter 9.

What keeps the ribosomal assembly line moving? The transacylation reaction catalyzed by the ribosome does not yield enough free energy to drive translocation of the tRNAs from one site to another because of the very similar stabilities of ester (starting material) and amide (product) bonds. To help, another protein called elongation factor-G (EF-G) acts like a wrench to move the tRNA out of the ribosome's active site and push the nascent protein, together with the mRNA, through the ribosome (**Figure 4.65**). The hydrolysis of GTP to GDP powers this translocase activity by EF-G. EF-G must bind to the same site on the ribosome as the EF-Tu complex with tRNA; as a result, the protein has evolved a shape with more than a passing resemblance to the protein–RNA complex that loads new amino acids onto the nascent polypeptide.

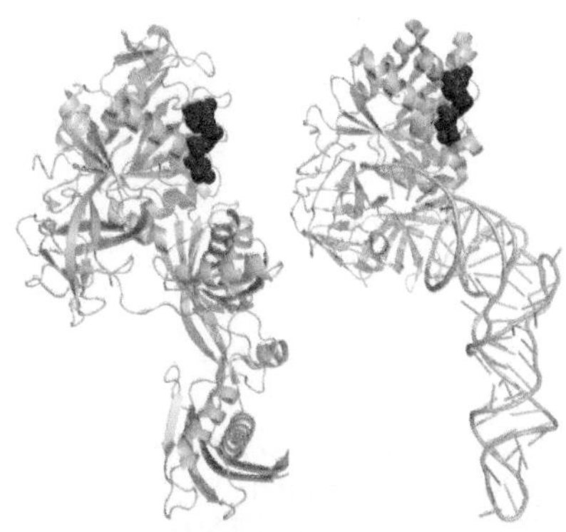

Figure 4.65 Structural mimicry by EF-G. The protein EF-G (left) pushes the nascent protein through the ribosome during biosynthesis. Remarkably, the structure of EF-G mimics the structure of EF-Tu bound to an aminoacyl-tRNA (right). Both structures must bind to the same site on the ribosome, and bind to GTP (red, van der Waals spheres). (PDB 2J7K and 1OB2)

The protein ricin, produced by castor beans, is one of the most toxic molecules on the planet. Ricin is a complex between two protein molecules, called the A chain

Problem 4.19

A EF-2, the eukaryotic homolog of EF-G, has a modified amino acid known as diphthamide (shown below). Diphtheria toxin, the causative agent of whooping cough, catalytically conjugates NAD^+ to the diphthamide residue of EF-2, rendering it ineffective. Suggest a plausible arrow-pushing mechanism for this ADP ribosylation reaction.

NAD+ — diphthamide — EF-2

diphtheria toxin or exotoxin A

B Exotoxin A from *Pseudomonas aeruginosa* also catalyzes the ADP ribosylation of EF-2 as shown in the crystal structure (PDB 2ZIT) of exotoxin A (purple), EF-2 (blue), and NADH (red). Mechanistically, what entropic and enthalpic barriers are likely to be lowered by the enzyme?

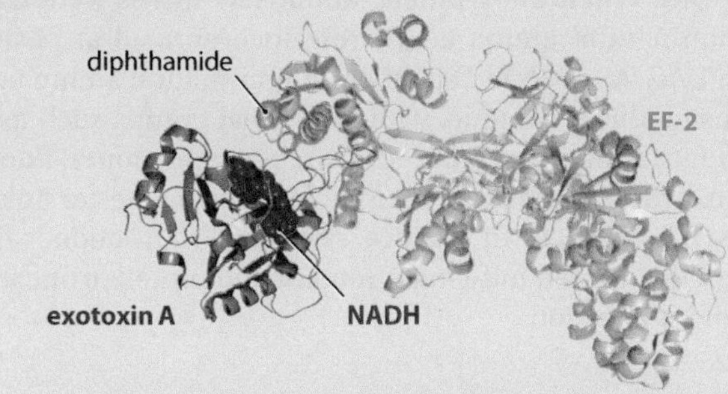

and the B chain. The A chain is a highly selective enzyme and the B chain is a delivery vehicle that ensures efficient entry into mammalian cells by binding to carbohydrates found on the surface of the cell. The A chain selectively depurinates a single adenine base from a single nucleotide (adenylate 4324) in the ribosome. A4324 is part of the ribosomal loop that binds to the elongation factors EF-1 and EF-2. Without that adenine, the entire ribosome is rendered into a pile of molecular junk. A single molecule of the ricin A chain can deactivate about 1500 ribosomes per minute.

The genetic code can be expanded beyond 20 amino acids

Every organism on the planet can incorporate a common set of 20 amino acids through ribosomal translation. Many of the codons in the genetic code are redundant, raising questions on the limits of ribosomal translation. Why are there 64 triplet codons to code for just 20 amino acids? Do we really need six codons for leucine? Can a cell get by with just two different stop codons?

Many species, including humans and *E. coli*, can incorporate a 21st amino acid, selenocysteine (abbreviated Sec). All mammals express enzymes containing selenocysteine, but only about 20% of bacteria do it. The human proteome codes for only 25 proteins containing selenocysteine. Selenium is a third-row atom, in the same group as sulfur and oxygen. At physiological pH, more than 90% of selenols are deprotonated and the large anionic selenide atom exhibits awesome nucleophilicity. The tRNA that carries selenocysteine binds to the stop codon UGA. In the presence of cysteine, the

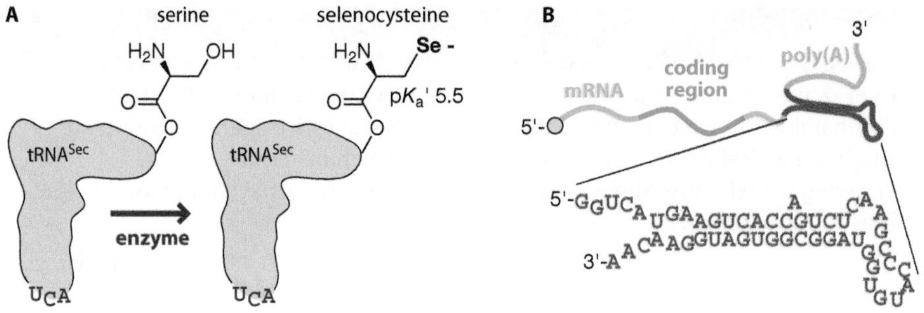

Figure 4.66 Selenocysteine incorporation. Two key components are needed for the ribosomal incorporation of selenocysteine. (A) An enzyme converts serine-$tRNA^{Sec}$ to selenocysteinyl-$tRNA^{Sec}$. (B) In human mRNAs, such as the mRNA encoding thioredoxin, a selenocysteine insertion sequence (brown) in the mRNA induces the ribosome to accept $tRNA^{Sec}$.

$tRNA^{Sec}$ is aminoacylated with serine, which is enzymatically converted to selenocysteine while connected to the tRNA (**Figure 4.66**). The presence of a UGA stop codon in the mRNA normally leads to translational termination, but a special mRNA structural element, called a selenocysteine insertion sequence, causes the ribosome to accept the selenocysteinyl-$tRNA^{Sec}$.

Long before the discovery of ribosomal selenocysteine incorporation, chemical biologists imagined reorganizing the genetic code of a living cell so as to translate proteins composed of more than 20 amino acids. The earliest efforts to incorporate non-natural amino acids site-specifically involved growing cells under normal conditions, and then isolating the protein synthesis machinery (ribosomes, tRNAs, elongation factors, and so on) for *in vitro* protein production. Synthetic amino acids were then chemically attached to an amber suppressor tRNA, which recognizes the least common stop codon UAG. When the synthetic aminoacyl-tRNAs were used for *in vitro* translation, the non-natural amino acids were incorporated at positions encoded by the stop codon UAG. More than 100 different non-natural amino acids have been incorporated by using this technique, with functional groups such as fluorophores, azides, alkynes, monosaccharides, boronic acids, benzophenones, fluorine, bromine, and iodine. Even α-hydroxy acids can be used, leading to an ester linkage within the protein backbone (**Figure 4.67**). In practice, ribosomal translation is finicky and can lead to a mixture of the desired full-length mutant and protein truncated at the position specified by the stop codon.

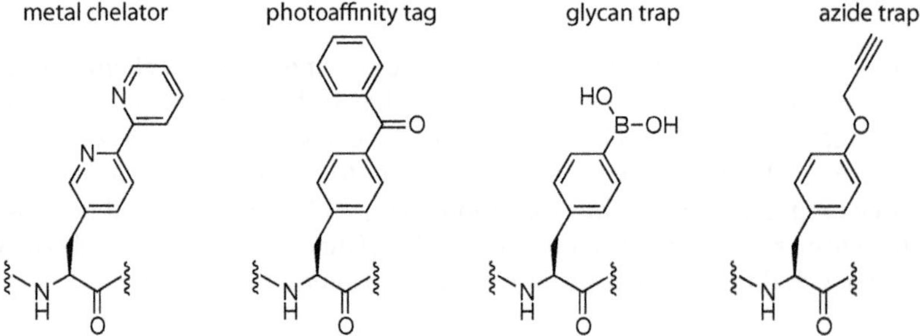

Figure 4.67 New protein functionality. A variety of functional amino acids have been ribosomally incorporated into proteins.

Incorporating non-natural amino acids into proteins in living cells is much more complex than *in vitro* overexpression experiments, because the cell has to avoid misincorporation of non-natural amino acids into its own proteins. In spite of the obstacles, Peter Schultz accomplished this feat by modifying a tRNA and aminoacyl synthetase from the archaebacterium *Methanococcus jannaschii* and introducing it into *E. coli.* The anticodon loop of the archaebacterial $tRNA^{Tyr}$ was changed from AUA to the stop anticodon CUA (**Figure 4.68**). Then the archaebacterial tyrosyl-tRNA synthetase was re-engineered to accept the novel $tRNA_{CUA}{}^{Tyr}$ and only the amino analog of tyrosine, but not tyrosine itself. When these genes were introduced into *E. coli,* the bacterium introduced the tyrosyl analog at positions specified by the amber stop codon UAG.

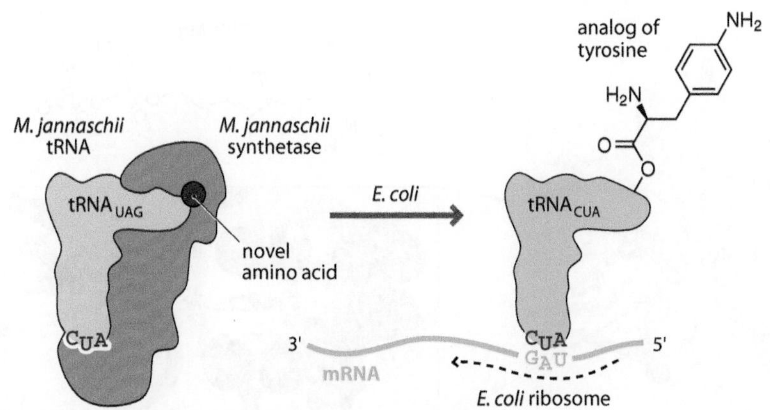

Figure 4.68 An orthogonal tRNA–synthetase pair for the delivery of non-natural amino acids. A tRNA and synthetase from *M. jannaschii* were modified to accept only *p*-aminophenylalanine. The tRNA was redesigned with the stop anticodon CUA and then introduced into *E. coli*. Instead of stopping at UAG, the bacterium incorporated *p*-aminophenylalanine.

Ligand-dependent riboswitches control protein expression

The idea that ligand–receptor interactions ultimately control gene expression is fundamental to our modern understanding of cell biology and will be discussed in Chapter 9. Until recently, the term receptor has been taken to mean a protein-based receptor. Recently it was discovered that RNA molecules can serve as receptors for small molecules, modulating various stages of gene expression. Some small-molecule metabolites can bind to structured regions of RNA, called **riboswitches**, that affect transcription, splicing, or translation. The *thiM* riboswitch controls the transcription of RNA (**Figure 4.69**). Some fungal and plant riboswitches direct alternative splicing of the mRNA, ultimately affecting the sequence of the expressed protein. The *GlmS* riboswitch binds to glucosamine 6-phosphate, which then acts as a ribozyme cofactor for self-cleavage. The amine group of the glucosamine 6-phosphate provides a Brønsted acid to protonate the leaving group during hydrolysis. The enzyme cofactor thiamine pyrophosphate, to be discussed in Chapter 6, binds to two different RNA riboswitches. When the RNA sequence *thiM* is present on an RNA transcript, it forms a sensor that can induce premature termination of transcription on binding of TPP. When the *thiC* gene is present at the appropriate site of an mRNA, it forms a TPP sensor that masks or reveals the Shine–Dalgarno sequence (AGGAGGU) that is essential for ribosomal binding and translation.

In Chapter 6 we will consider the function of vitamins as enzyme cofactors. All of the vitamin cofactors seem to have functionality with no obvious catalytic role. The presence of nucleotides within the structures of vitamins is particularly hard to ignore; on closer inspection of each cofactor, one can readily identify features that should facilitate interactions with nucleic acids through base pairing, π stacking, metal phosphates, and other interactions (**Figure 4.70**). Examples of metabolites that bind to riboswitches include thiamine pyrophosphate, flavin mononucleotide, cobalamin, *S*-adenosylmethionine (SAM), purines, glucosamine 6-phosphate, lysine, and glycine.

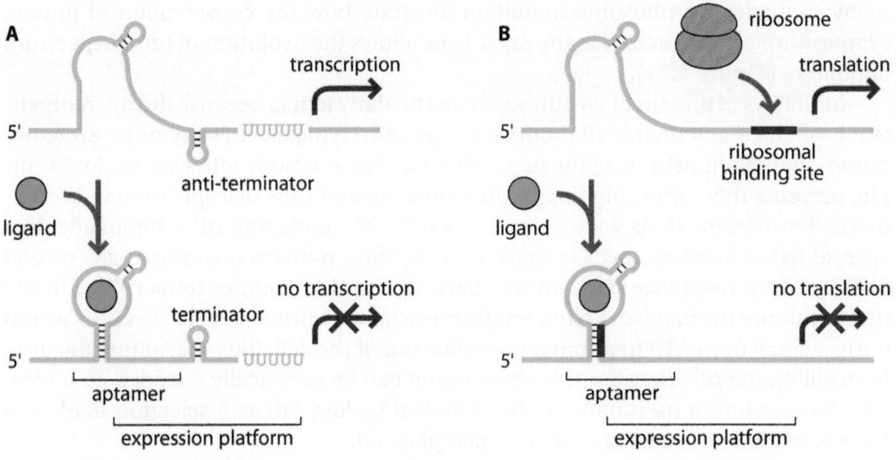

Figure 4.69 Regulation of gene expression by riboswitches. The small molecule can stabilize folded RNA structures that mask or unmask functional RNA sequences. (A) Binding of a ligand to a riboswitch can mask the transcriptional anti-terminator region and reveal a transcriptional terminator. (B) Binding of a ligand to a riboswitch can mask the ribosomal binding site, preventing the mRNA from associating with the ribosome. (Adapted from J.N. Kim and R.R. Breaker, *Biol. Cell* 100:1–11, 2008. With permission from Portland Press Ltd.)

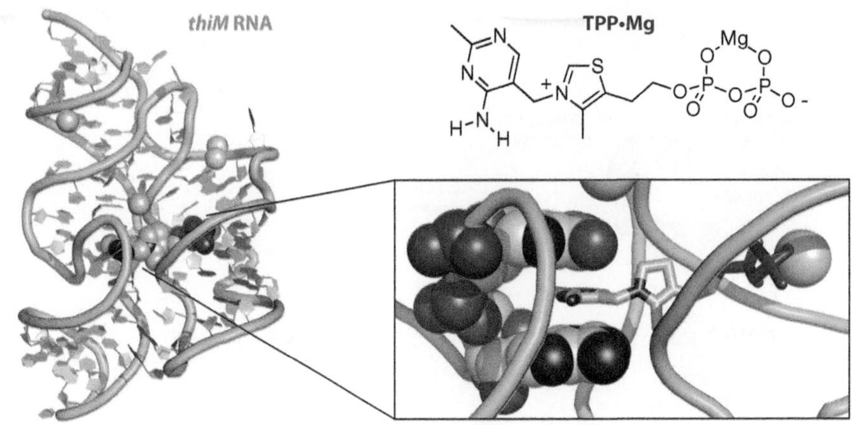

Figure 4.70 A riboswitch bound to a ligand. Structure of the *thiM* riboswitch bound to thiamine pyrophosphate•Mg^{2+} (PDB 2CKY); a close-up view reveals π stacking of the pyrimidine substituent of thiamine and the coordination of the Mg^{2+} ion.

As far as we know, every vitamin cofactor binds to an RNA riboswitch. Riboswitches can provide immediate feedback control over the biosynthesis of key metabolites such as DNA bases. Purines can bind to and trigger riboswitches within the mRNAs that encode enzymes required for purine synthesis and transport. When multiple riboswitches are present in tandem in an RNA transcript, they can result in a sophisticated control over the logic of gene expression.

Riboswitches have also emerged as potential targets for drug development. Their specificity for binding to a particular small-molecule effector suggests that antibiotics could also bind with high specificity to riboswitches. For such antibiotic applications, riboswitches can also present intriguing targets if found exclusively in bacteria and fungi. Thus, the specificities of riboswitches, combined with their powerful mechanisms for short-circuiting protein translation, will continue to interest and challenge medicinal chemists.

Many antibiotics target bacterial protein synthesis

The long-running chemical war between different microorganisms, competing for limited resources, has resulted in the evolution of clever strategies for attack, defense, and counterattack. Shutting down protein biosynthesis is one method for eliminating the competition, just as the bombing of factories is a strategy used in large-scale human wars. The ribosome presents an attractive target for such chemical intervention. First, the ribosome is universally required by all organisms for protein biosynthesis. Second, the key ribosomal features (such as ribosomal adenine 2451) are conserved across essentially all organisms, which means that an inhibitor against a particular fungus might also work against bacteria encroaching on the same turf. The different strategies for inhibiting the ribosome illustrate the diversity of solutions possible for just one problem in molecular recognition. Although all of these molecules bind to the ribosome, some to the same spot, each has a very different structure. The many strategies for ribosome inhibition illustrate how the conservation of protein synthesis machinery across many organisms allows the evolution of broad-spectrum antibiotics (**Figure 4.71**).

Inhibitors of protein biosynthesis are particularly lethal, because the microorganism is often caught unable to produce the protein (typically an enzyme or excretory pump) that might neutralize the toxic offender. The antibiotic tetracycline, for example, prevents tRNA from binding to the ribosome and thus disrupts protein biosynthesis. Tetracycline finds wide usage in modern life, including sub-therapeutic dosages added to livestock feeds to boost growth. Thus, perhaps unsurprisingly, several different drug resistance mechanisms have emerged to counter tetracycline. In the most common mechanism for tetracycline resistance, a protein called an **efflux pump** burns energy from ATP to pump tetracycline out of the cell, thus preventing the drug from killing the cell. Because this efflux pump can be genetically encoded on a plasmid, tetracycline is used often in the chemical biology lab as a selection marker to force *E. coli* to carry and act on a particular plasmid.

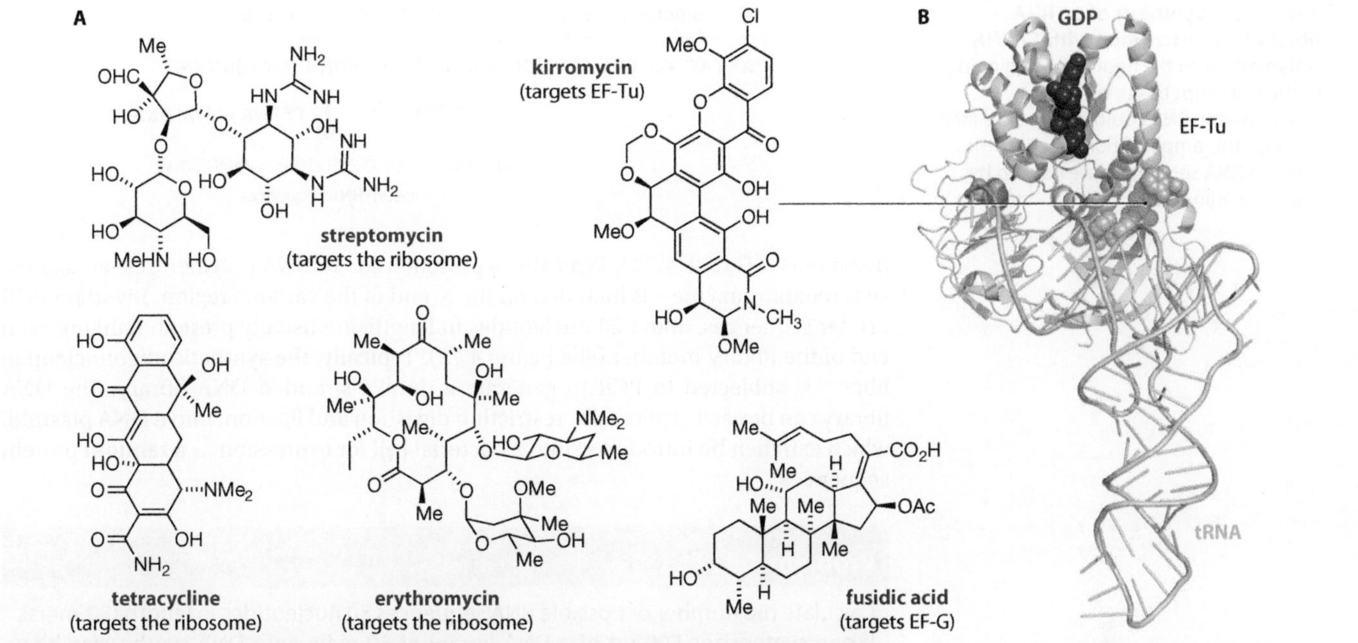

Figure 4.71 Antibiotic inhibitors of protein synthesis. (A) These polyketide and oligosaccharide (streptomycin) compounds kill cells by binding to conserved components required for protein synthesis by the cell. (B) For example, kirromycin (gray) binds to EF-Tu, locking the protein in a state similar to the GTP- and tRNA-bound conformation. The resultant complex forms fibers in the cell; this glorified precipitation makes the key protein EF-Tu unavailable for protein synthesis.

Problem 4.20

When competent bacterial cells are briefly heated at 42 °C with a plasmid containing a gene for chloramphenicol resistance, a small fraction of the cells take up the plasmid. Initially, none of the cells are resistant to chloramphenicol, but if they are allowed to grow for about an hour and then treated with chloramphenicol, the cells with the plasmid remain viable. Why are the cells that take up the plasmid not immediately resistant?

4.7 FROM OLIGONUCLEOTIDE LIBRARIES TO PROTEIN LIBRARIES

Automated oligonucleotide synthesis facilitates generation of both DNA and RNA oligonucleotide libraries

Automated DNA synthesizers make it easy to create mixtures of DNA molecules simply by using mixtures of all four phosphoramidites (A, C, G, and T) in each of the coupling steps to generate a combinatorial library of DNA oligonucleotides. One micromole of DNA—a typical synthesis scale—is enough to contain 100 copies of every possible 25-mer; but potential diversity increases rapidly with sequence length. There are more than 10^{18} possible DNA 30-mers; a micromole of DNA could consist of only about half of the possible sequences.

When designing oligonucleotide libraries for selection experiments, nonvariant sequences are usually included on the 5′ and 3′ ends of the oligonucleotides to facilitate PCR amplification (to generate double-stranded DNA), transcription into RNA, or cloning into plasmids with restriction endonucleases. The automated synthesis of combinatorial DNA libraries ultimately provides efficient access to libraries of DNA, RNA, or proteins. One can screen single-stranded DNA libraries for the ability to fold into defined shapes, driven by Watson–Crick base pairing, that bind specifically to other molecules (**Figure 4.72**). These types of functional oligonucleotides are sometimes called **aptamers**.

Single-stranded DNA aptamers that fold into functional shapes are unlikely to be biologically relevant, because DNA is generally double-stranded within cells; however, RNA is single-stranded, and aptamers selected from artificial RNA libraries are sometimes found to mimic their natural counterparts, the riboswitches, in affinity and specificity. A DNA library can be transcribed into an RNA library by using RNA

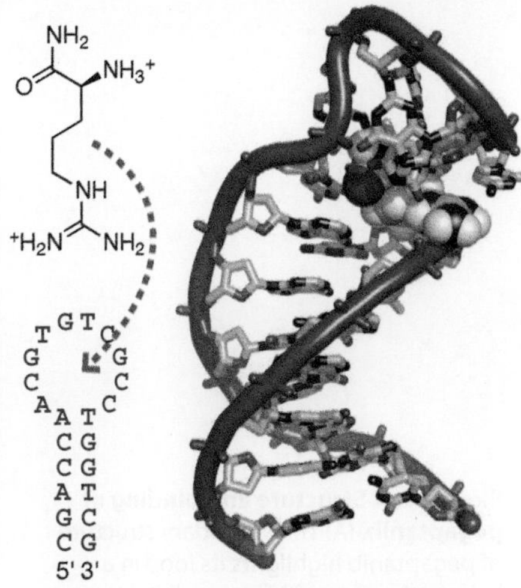

Figure 4.72 A DNA aptamer that binds arginine. An arginine-binding DNA motif was identified by screening a pool of more than 100 billion single-stranded DNA molecules. An amide derivative of the amino acid arginine binds to the 10-base loop. (PDB 1DB6)

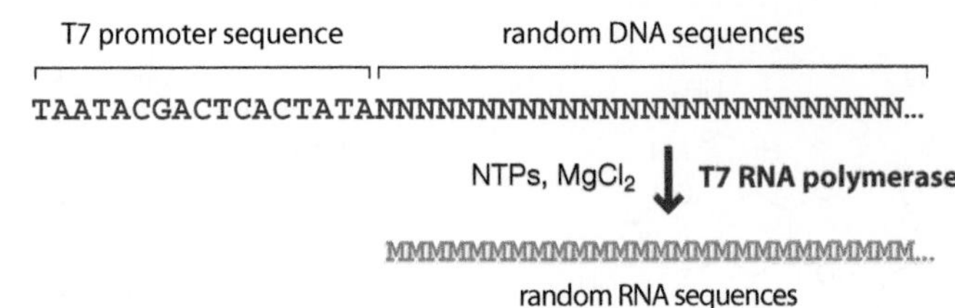

Figure 4.73 Synthesis of an RNA library by transcription with T7 RNA polymerase. In the figure, the beginning of the transcript has to be known, otherwise the RNA could not be amplified; through this amplification reaction, the random RNA sequences are flanked by primer-binding sites.

polymerase (**Figure 4.73**). Typically, a promoter for T7 RNA polymerase—an aggressive, reliable enzyme—is included on the 5′ end of the variable region. Invariant PCR primer sequences, about 20 nucleotides in length, are usually present, flanking each end of the library member (see Figure 4.73). Typically, the synthetic oligonucleotide library is subjected to PCR to generate a double-stranded DNA library. The DNA library can be easily spliced, by restriction digestion and ligation, into a DNA plasmid, which can then be introduced into a bacterial cell for expression as a random protein sequence.

Problem 4.21

Calculate the number of possible RNA sequences 50 nucleotides in length (50-mers). If one synthesizes 500 μg of a DNA library of 50-mers on a DNA synthesizer, how many individual DNA molecules does it contain? If each DNA molecule encodes a random 50-mer, what fraction of the potential 50-mers are contained within the library? (The average molecular mass for a nucleotide subunit of DNA is 305.6 g mol^{-1}.)

For example, the synthesis of and selection from oligonucleotide libraries was used in the development of the drug pegaptanib (**Figure 4.74**), which is used to treat macular degeneration. The discovery of pegaptanib involved synthesizing the following combinatorial DNA oligonucleotide composed of a variable region flanked by a T7 RNA polymerase promoter sequence to allow RNA polymerization: 5′-*TAATACGACTCACTATA*GGGAGGACGATGCGG**NNNNNNNNNNNNNNNNNNNNNNNNNNNNNNN**CAGACGACTCGCCCGA-3′, where N indicates any of the four DNA bases. Next, the library of DNA sequences was transcribed into an RNA library with T7 RNA polymerase. To improve the stability of the resultant RNA against hydrolytic cleavage, structural 2′-fluoro analogs of the NTPs were used. The resultant RNA library was mixed with the protein vascular endothelial growth factor (VEGF), which triggers the growth of blood vessels. The protein was isolated on a nitrocellulose membrane along with the tight-binding RNA molecules from the library. The RNA molecules that bound to VEGF were reverse-transcribed into DNA by using reverse transcriptase and amplified by PCR. The DNA molecules were spliced into plasmids and transformed into *E. coli* for ease of growth, cloning, and propagation. DNA sequencing revealed a consensus binding sequence, CGGAAUCAGUGAAUGCUUAUACAUCCA, that served as the basis for the development of the drug Macugen® (pegaptanib sodium). When a photosensitive iodouridine was incorporated at one of the pegaptanib loops, it was

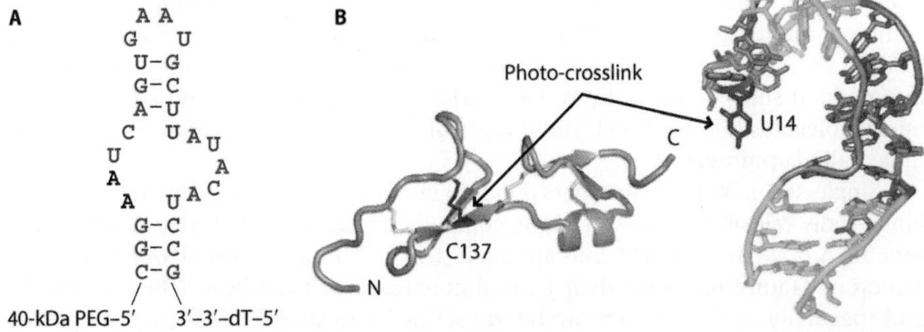

Figure 4.74 Structure and binding of pegaptanib. (A) The secondary structure of pegaptanib highlights its loop in a hairpin structure. (B) This model of the structure of pegaptanib binding to VEGF shows how the two molecules interact, as determined by photocrosslinking experiments. (B, adapted from E.M.W. Ng et al., *Nat. Rev. Drug Discovery* 5:123–132, 2006. With permission from Macmillan Publishers Ltd.)

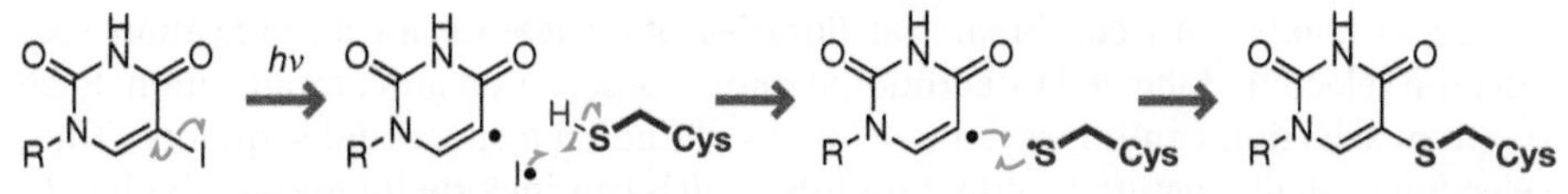

Figure 4.75 Carbon–iodine bonds are weak and can be cleaved with ultraviolet radiation to generate reactive radicals. Protein–DNA interactions can sometimes be confirmed by replacing thymidine with iodouridine and irradiating the protein–DNA complex.

shown to generate a crosslink with a thiol group on Cys137 of VEGF when irradiated at 308 nm (see Figure 4.74 and **Figure 4.75**).

RNA libraries have been used to identify RNA catalysts for a wide range of reactions, such as hydrolysis, ligation, Diels-Alder reactions, and atropisomerization, which is the ability to isomerize sterically hindered rotation around a single bond. Selection from molecular libraries always requires some kind of clever trick for selecting members of the library that have the desired properties. In addition, the selection hinges on the conversion of the RNA into DNA by the enzyme reverse transcriptase. The DNA can then be amplified by PCR and reconverted to RNA by RNA polymerase. This sequence makes it possible to amplify a small number of RNA molecules selected by the selection conditions. The amplified DNA can be used either for sequencing or for another round of selection.

RNA libraries can be screened for ribozymes

One example of selection from RNA libraries involves the identification of RNA molecules that can catalyze the stereoselective S_N1 glycosylation of uracil by 1-ribosyl pyrophosphate to produce the nucleoside uridine (**Figure 4.76**). Such catalysts may have had a role in the prebiotic origin of life. A pool of 1.5×10^{15} different RNA molecules was constructed by the transcription of a synthetic DNA library. Because each member of the library had 76 random nucleotides, the potential diversity was 6×10^{45}. A ribosyl-1′-diphosphate was added enzymatically to the 3′ end of every RNA molecule. The pool of RNA molecules was then allowed to react with 4-thiouracil, an analog of uracil with a high affinity for mercury. A small number of the RNA molecules catalyzed the desired glycosylation reaction, which requires dephosphorylation of the pyrophosphate and bringing the thiouracil into close proximity with the resultant oxocarbenium ion. The RNA molecules capable of undergoing the reaction were isolated by electrophoresis on a polyacrylamide gel derivatized with phenylmercuric chloride, which bonds tightly to sulfur-containing functional groups. Such covalent bonds slow the migration of the reactive RNAs through the gel, making it possible to isolate the sequences capable of catalyzing a reaction from the very many unreactive RNA sequences in the library.

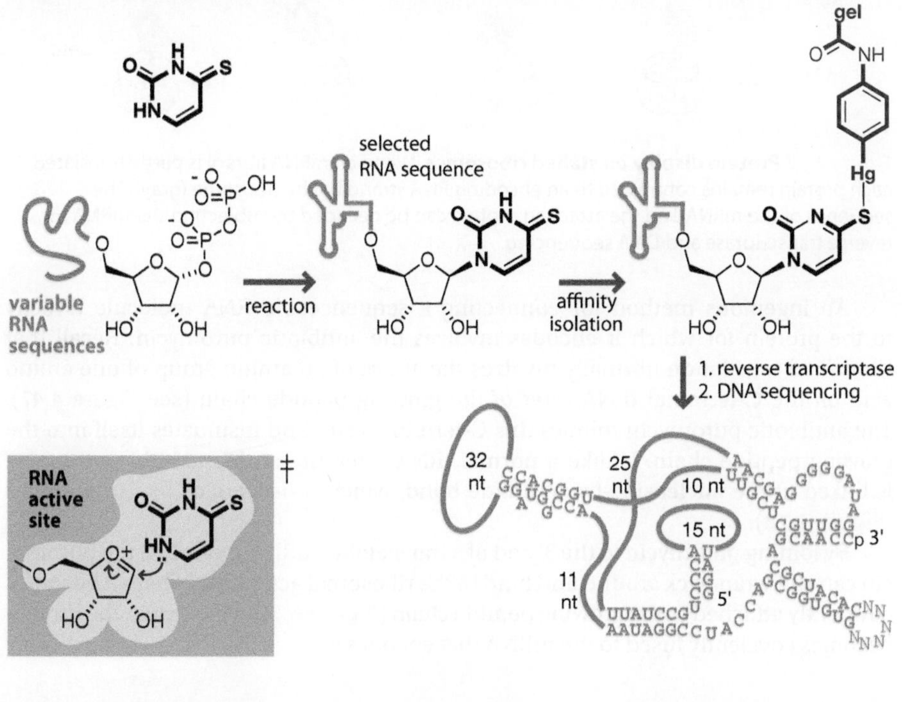

Figure 4.76 Selection for catalysis. To identify RNA motifs that catalyze ribonucleoside assembly, a library of RNA molecules with a ribosyl-1′-diphosphate at the 3′ end was allowed to react with 4-thiouracil. RNA molecules that reacted quickly with the 4-thiouracil were isolated on the basis of their slow migration through a polyacrylamide gel derivatized with thiophilic mercury groups.

Experiments with combinatorial libraries often feature an amplification step after the selection, followed by additional rounds of selection and amplification. Each round of selection can improve the chances of finding a successful sequence. Thus, selections usually feature multiple rounds, which can include increasing the harshness or stringency of the conditions to discriminate between great and merely good binders. For example, shorter binding times can be used to identify sequences with faster on-rates. The RNA sequences isolated through slow migration during gel electrophoresis were reverse transcribed to produce DNA, which could then be amplified by PCR. The product of the PCR was then transcribed into RNA sequences for additional rounds of selection and amplification. The best RNA catalysts identified from this library accelerated the reaction by a factor of 10 million over the background rate, which is quite a significant rate of acceleration.

mRNA libraries can be expressed as protein libraries

Oligonucleotides bear a sparse repertoire of functionality relative to proteins. Proteins have a broader range of functional properties than oligonucleotides, but proteins cannot be sequenced with the exquisite sensitivity of oligonucleotides. The ideal way to screen vast molecular libraries of proteins is to connect the protein to the mRNA that encoded it. RNA can be transcribed back into DNA with reverse transcriptase, amplified through PCR, and then sequenced, revealing the sequence of the protein.

One way to convert mRNA libraries into protein libraries is to display them on stalled ribosomes, a technique analogous to phage display (**Figure 4.77**). Ribosomal translation can be halted through the addition of the antibiotic chloramphenicol, leaving the partly translated mRNA and the partly synthesized protein dangling from the ribosome. The same effect is achieved by cooling the translation reaction and adding high concentrations of Mg^{2+} ions. By stalling translation, any protein that has interesting chemical or biological properties carries with it an RNA transcript that can be readily transcribed back to DNA by reverse transcriptase, amplified by PCR, and sequenced. Libraries of 10^{13} RNA molecules have been used to generate very large libraries of peptides and proteins.

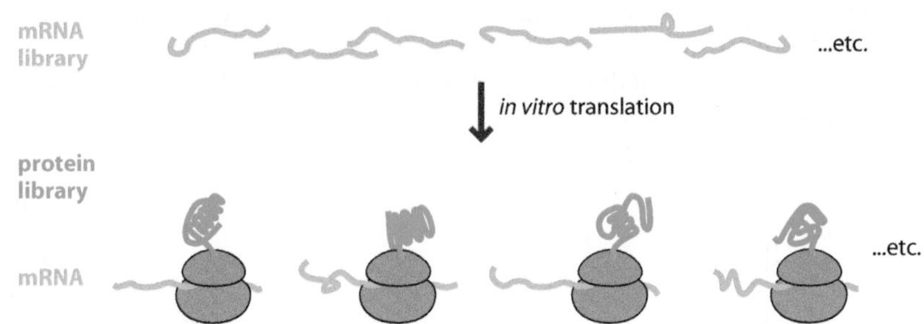

Figure 4.77 Protein display on stalled ribosomes. When an mRNA library is partly translated, each protein remains connected to an encoding RNA strand by the ribosome (gray). The sequence of the mRNA and the attached protein can be decoded by subjecting the mRNA to reverse transcriptase and DNA sequencing.

An ingenious method for connecting a sequenceable RNA molecule directly to the protein for which it encodes involves the antibiotic puromycin. Recall that ribosomal translation normally involves the attack of an amino group of one amino acid on the C-terminal tRNA ester of the growing peptide chain (see Figure 4.47). The antibiotic puromycin mimics this C-terminal ester and insinuates itself into the growing peptide chain. Unlike a normal tRNA ester, the amino acid in puromycin is linked to the nucleoside by an amide bond, which resists attack by nucleophiles (**Figure 4.78**).

By joining puromycin to the 3′ end of a translatable mRNA molecule, the puromycin can still wrap back around and bind to the ribosomal active site, where it becomes covalently attached to the growing peptide chain (**Figure 4.79**). As a result, the protein becomes covalently fused to the mRNA that encodes it.

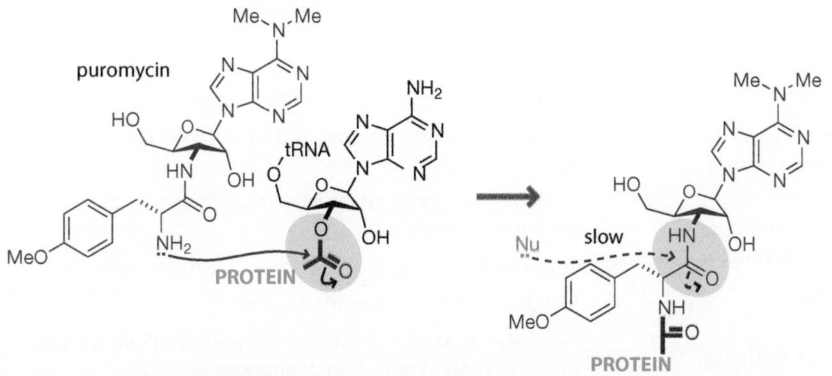

Figure 4.78 Puromycin stops protein synthesis. The antibiotic puromycin mimics the 3′-nucleotide of an aminoacyl-tRNA except that the amino acid is connected through an unreactive amide bond. Once the ribosome has transferred the growing peptide to puromycin, elongation cannot continue.

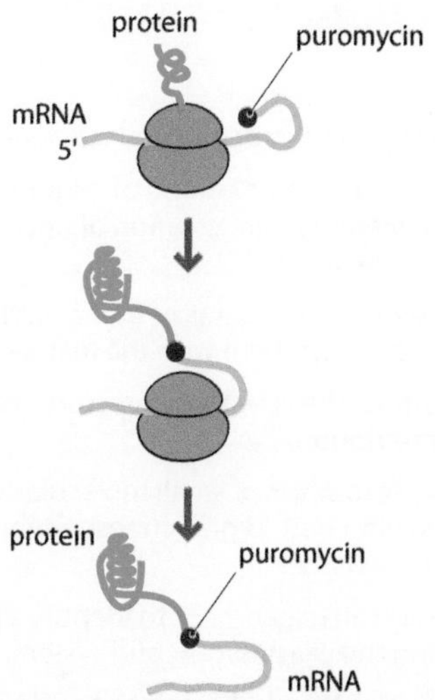

Figure 4.79 Puromycin mRNA–protein display. When puromycin is attached to the 3′ end of an mRNA, the protein chain ends up covalently attached to the mRNA, which can be sequenced.

4.8 SUMMARY

Tinker, tailor, soldier, spy... catalyst, regulator, and messenger—RNA fills a large number of different roles for the cell. The messenger role of RNA is most important to the central dogma, allowing DNA to be translated as proteins; but RNA is more than just the message. In retroviruses such as HIV, RNA is *the* genetic material. In living cells, RNA makes up the very machinery of protein production, both the ribosome, which is essentially a ribozyme, and the aminoacyl-tRNA building blocks. Moreover, many of the regulatory roles once thought to be mediated only by proteins are now known to be under the partial control of RNA. RNA molecules serve as small-molecule receptors that control transcription and translation. Finally, RNA serves as a homing device for RISC, which culls the intracellular pool of RNA.

Structurally, RNA is distinct from DNA because it is generally single-stranded, allowing it to adopt diverse structures driven by a wide range of hydrogen-bonding and π-stacking interactions, not just the canonical Watson–Crick base pairs. Chemically, RNA is distinct from DNA because the phosphodiester bonds of RNA are labile as a result of the 2′-hydroxyl group of each ribose subunit. RNA is not an ideal system for long-term information storage.

Scientists have made great strides in harnessing and extending the power of RNA. The central steps of protein synthesis—aminoacylation of tRNAs and ribosomal translation—have been modified for highly subversive activities, allowing expansion of the genetic code beyond the 20 universal amino acids. RNA molecules are becoming useful tools for both biology and medicine. Small interfering hairpin RNAs direct the RNA interference system to cleave complementary mRNAs, silencing gene expression. The ability to synthesize DNA libraries makes it easy to prepare complementary RNA libraries. RNA libraries can be screened for functional RNA molecules composed of both non-natural and natural nucleotides. Such efforts have identified ribozymes, aptamers, and RNA drugs. Vast libraries of RNA molecules can be expressed and screened as proteins.

It is time to return to the question we raised at the beginning of this chapter: Did life begin with RNA molecules? Just as computers came before hard drives and robotics it should now be easy for you to imagine that life began with RNA followed later by DNA and proteins. For many decades, DNA has garnered most of the attention; but prebiotic chemistry is most consistent with the initial production of RNA, not DNA. Moreover, RNA can store information, transmit information, catalyze reactions, sense metabolites, control transcription, trim itself, cull its population, catalyze protein synthesis, and control translation. To understand why many of these functions have been largely supplanted by proteins we will need to venture further into Chapters 5 and 6, where the architectural control and functional finesse of proteins come into full view.

LEARNING OUTCOMES

- Contrast the structure and reactivity of RNA with DNA.
- Distinguish lengthening of oligonucleotide primers by DNA polymerases from creation of new oligonucleotide strands by RNA polymerases.
- Describe the features of human DNA that define a promoter and contrast them with the features of bacterial promoters.
- Contrast the effect of repressors and activators on transcription.
- Offer examples of small molecule control over transcription involving both protein transcription factors and RNA microswitches.
- Design an experiment to identify a protein binding partner using the yeast two-hybrid system.
- Describe the modifications to RNA transcripts that prepare human mRNA for translation.
- Design a hairpin siRNA from a DNA coding sequence.
- Explain how heat, proteins, and small molecules can affect the accessibility of RBSs.
- Describe the composition of human ribosomes in terms of RNA and proteins.
- Translate RNA coding sequences into protein sequences.
- Sketch the role of elongation factors.
- List the non-natural components needed for site-specific incorporation of non-natural amino acids.
- Describe some of the mechanisms by which antibiotics can inhibit protein translation.
- Design a selection that could be used to screen an RNA library for new functions.

PROBLEMS

***4.22** Write out the mRNA sequence that is complementary to the following single-stranded DNA sequences.

A 3′-CATAGCTGTCCTCCT-5′ (a Shine–Dalgarno sequence)

B 3′-CATGGTGGT-5′ (a Kozak sequence)

4.23 Sketch the folding pattern for the following RNA sequences. You may use a Web-based RNA-folding prediction engine such as MFOLD or RNAFOLD.

***A** A polyadenylation signal from HIV:
5′-CACUGCUUAAGCCUCAAUAAAGCUUGCCUUGAGUGCUUC AAGUAGUG-3′

B A folded structure necessary for the efficient encapsidation of spleen necrosis virus retroviral RNA:
5′-GGAGGAUCGGAGUGGCGUGACGCUGCCGCUCCACCUCCG CUCAGCAGGGGACGCCCUGGUCUGAGC-3′

4.24 Predict the product or products of the following nonenzymatic hydrolysis reactions.

***A** (phosphodiester, ONa) → 1 M NaOH, H_2O, 100 °C

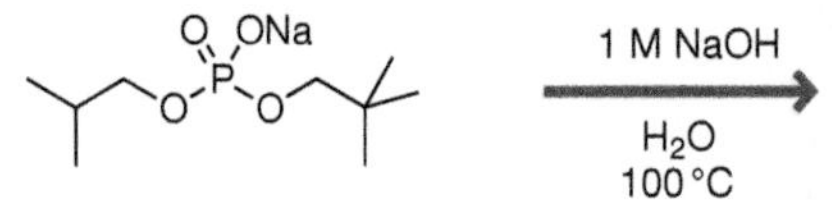

***B** (phosphodiester, ONa, OH) → 1 M NaOH, H_2O, 100 °C

C (dinucleotide: OH, NaO–P–O, Ura, Ade, OH, HO, OH) → (excess) 1 M NaOH, H_2O, 100 °C

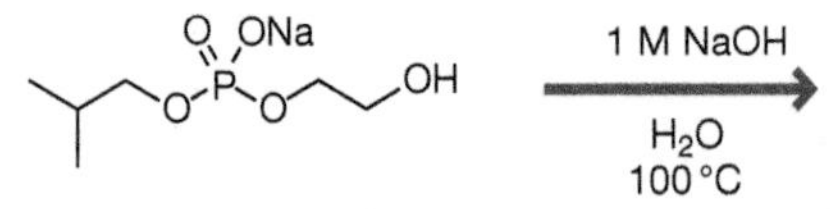

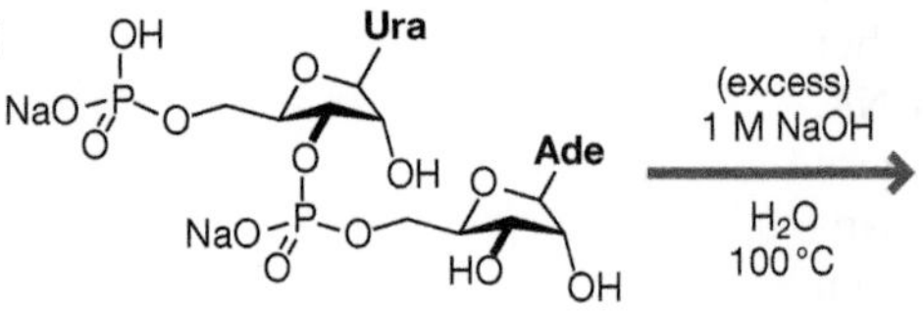

***4.25** The following RNA molecule folds into a three-dimensional shape that binds to and hydrolyzes the shorter RNA molecule.

ribozyme 5′-GGCGACCCUGAUGAGGCCGAAAGGCCGAAACCGU-3′

hydrolyzes here ↓

substrate 5′-ACGGUCGGUCGCC-3′

A Show schematically how the smaller RNA molecule might bind to the catalytic RNA in a way that maximizes the number of base pairs.

B What ribozyme sequence would efficiently catalyze the hydrolysis of the following substrate: 5′-GCCGUCGCUGCGC-3′?

4.26 Recall that bacteria that have been transformed with plasmids conferring chloramphenicol resistance require a growth induction period before the antibiotic is added. Which of the following antibiotics would also require a growth induction period after plasmid transformation before the addition of antibiotic?

Antibiotic	Target	Common resistance protein
Ampicillin	Cell wall synthesis	β-Lactamase enzyme
Tetracycline	Ribosome	Tetracycline efflux pump protein
Kanamycin	Ribosome	Kanamycin:phosphoryltransferase
Fosfomycin	Cell wall biosynthesis	Fosfomycin epoxide hydrolase
Mupirocin	Leucyl-tRNA synthetase	Resistant leucyl-tRNA synthetase

4.27 The mRNA sequence for the tachykinin precursor protein is given below with the translated region capitalized:

```
gcgccgcaaggcacugagcaggcgaaagagcgcgcucggaccuccuucccggcgg
cagcuaccgagagugcggagcgaccagcgugcgcucggaggaaccagagaaacuc
agcaccccgcgggacuguccgucgcaaaauccaacAUGAAAAUCCUCGUGGCCUU
GGCAGUCUUUUUUCUUGUCUCCACUCAGCUGUUUGCAGAAGAAAUAGGAGCCAAU
GAUGAUCUGAAUUACUGGUCCGACUGGUACGACAGCGACCAGAUCAAGGAGGAAC
UGCCGGAGCCCUUUGAGCAUCUUCUGCAGAGAAUCGCCCGGAGACCCAAGCCUCA
GCAGUUCUUUGGAUUAAUGGGCAAACGGGAUGCUGAUUCCUCAAUUGAAAAACAA
GUGGCCCUGUUAAAGGCUCUUUAUGGACAUGGCCAGAUCUCUCACAAAAUGGCUU
AUGAAAGGAGUGCAAUGCAGAAUUAUGAAAGAAGACGUUAAUAAacuaccuaaca
uuauuuauucagcuucauuugugucaaugggcaaugacagguaaauuaagacaug
cacuaugaggaauaauuauuuauuuaauaacaauuguuuggguugaaaauucaa
aaaguguuuauuuuucauauugugccaauauguauuguaaacauguguuuuaauu
ccaauaugaugacucccuuaaaauagaaauaagugguuauuucucaacaaagcac
aguguuaaaugaaauuguaaaaccugucaaugauacagucccuaaagaaaaaaaa
ucauugcuuugaagcaguugugucagcuacugcggaaaaggaaggaaacuccuga
cagucuugugcuuuuccuauuuguuuucauggugaaaauguacugagauuuuggu
auuacacuguauuuguaucucugaagcauguuucauguuuugugacuauauagag
auguuuuuaaaaguuucaaugugauucuaaugucuucauuucauuguaugaugug
uugugauagcuaacauuuuaaauaaaagaaaaaauaucuug
```

A Label the start codon and stop codon.
B Underline the ribosomal binding site.
C Note the point of attachment of the G-cap.
D Label the site of attachment of the poly(A) tail.
E List the sequence of the first five amino acids translated by the ribosome from this mRNA.

***4.28** The *ENOD40* (early nodulin 40) gene is expressed during the early development of peas, soybeans, alfalfa, and tobacco plants and can generate two different peptides, A (12 amino acids) and B (24 amino acids), that are believed to bind to the enzyme sucrose synthase. The open reading frames for peptide A (in green capital letters) and peptide B are overlapping.

```
  1 gaauuccgcu aaaccaaucu aucaaguccu gauuaaucug gugagcAUGG AGCUUUGUUG
 61 GCUCACAACC AUCCAUGGUU CUUGAagaag cuuggagaga aaggggugug agaggagagg
121 gugcucacuc cucacacucc cucacuuaaa acaguuuguu uuggcuuagc uuuggcuucu
181 cugaucaaca agggaugugu ucuaacauuc uuucuugagu ggcggaagca gauacacauu
241 cuccgacgga ggagaggcuu ggcuacagcc uggcaaaccg gcaagucaca aaaaaggcaa
301 uggacuccau uggggucucu auggcuaugu agugcucaug uaguucuucu ugcuguagaa
361 uguaauaaua aacaaaguug gucuuccuuu ugagaaguua ccagcuuuug cuguccaaaa
421 uuacucaauu ugcagcugac uagaauuccu uucucucuuc aguuucugca gaugaguagg
481 uaggcaauuu gugaucacuc ccuucccuuu ucaugucuuc uguguucccu uuuccaugcu
541 uguuuguguu guuaguuaug accuuaugag gaaauaaaag aauaguacaa uucuagucc
601 ucaguuuagg auuguauucu auugaacuuu auuagaaaag uuuccagagu ccuuucuaaa
661 aaaaaaaaaa aaaaaaaaa
```

A Translate the mRNA sequence for the open reading frame for peptide A into the peptide sequence.
B Identify the overlapping open reading frame for peptide B and translate it into the 24 amino acid sequence. *Hint:* Look for a start codon.

4.29 A mutant form of T7 RNA polymerase has been developed that incorporates 2′-*O*-methyl nucleotidyl triphosphates. Compare the reactivity of 2′-*O*-methyl RNA with wild-type RNA toward chemical hydrolysis with aqueous NaOH and toward enzymatic hydrolysis by RNase A.

2′-*O*-methyl NTPs —(T7 RNA polymerase, DNA template, $MgCl_2$)→ 2′-*O*-methyl RNA

4.30 Double-stranded RNA-activated protein kinase is activated by double-stranded RNA at least 33 bp in length, or by shorter stem loops with 5′-triphosphates on flanking single-stranded regions. Show the base-pairing structure of the following activator of (double-stranded RNA)-activated protein kinase, taking into account the potential for wobble base pairs (G•U).

5'-ppp GAGGUCACUGACUAAGUUGGUGAAAUCUUGAUUUAUCAGUGACAAG-3'

kinase-activating RNA

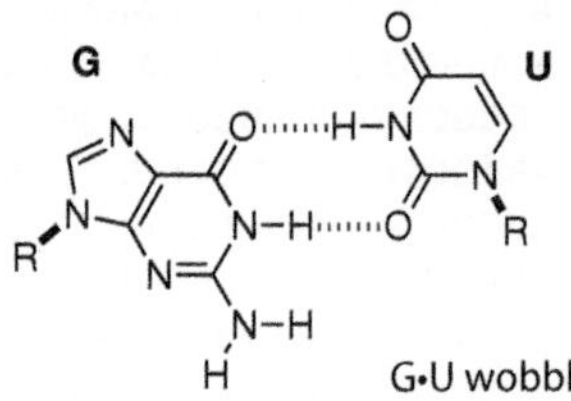

G•U wobble base pair

***4.31** DEPC is not very effective at deactivating ribonucleases in solutions buffered with Tris. Use the Web to find the structure of Tris and then suggest a plausible rationale for why Tris interferes with the inactivation of ribonucleases.

Peptide and Protein Structure

5

LEARNING OBJECTIVES

- Become fluent in the basic vocabulary of protein structure: the common ribosomal amino acids and their one-letter codes.
- Recognize hydrophobic interactions as a major contributor to protein folding and binding.
- Learn the chemical mechanisms upon which chemical peptide synthesis is based.
- Understand the elementary chemical forces that control protein conformation.
- Distinguish patterns of hydrogen bonding that distinguish helices, sheets, and turns.
- Associate amino acids with canonical secondary structure preferences.
- Recognize hierarchical structural patterns in folded proteins.

Reflecting the centrality of proteins in biology, their name is derived from the Greek word πρωτειοσ, which means "of primary importance." In 1838 the Dutch chemist Gerrit Jan Mulder conferred this name to recognize the requirement for proteins in the diet of animals. Berzelius later found that all proteins have similar empirical formulas. Proteins are oligomers of amino acids joined together by amide bonds. Composed of 20 different building blocks, proteins display a much more varied array of functional groups and structures than either DNA or RNA, which have only four building blocks each. For DNA, structure came first; sequencing came later. The opposite is true for proteins because a process known as the **Edman degradation** method made it possible to elucidate the sequence of amino acid residues in proteins long before the first protein structures were determined.

The amide backbone of proteins offers greater hydrogen-bonding capacity than any other polymer backbone. To satisfy this demand for hydrogen bonding while facilitating hydrophobic interactions, proteins fold into a wide array of well-defined shapes. The ability to fold into specific shapes, displaying a wide range of chemical functionality, makes proteins far more versatile than oligonucleotides. Indeed, proteins serve as catalysts, force-transducing machines, cellular gatekeepers, structural materials, and environmental sensors; they also have hundreds of other key roles in the cell.

Unlike the monotonous double-helical conformation of DNA, the folded conformation of a protein is highly dependent on the sequence. Most proteins adopt unique conformations out of a nearly limitless set of possibilities. Thus, a true understanding of proteins and enzymes had to await the advent of methods that could reveal the folded structures of proteins. The structural characterization of proteins progressed rapidly after the crystallization of proteins emerged as an effective method for the purification of proteins in the early twentieth century. Few believed that proteins would exhibit the kind of ordered packing that would be required for X-ray structure determination, but in 1934 Dorothy Crowfoot Hodgkin and J. D. Bernal showed that crystals of the enzyme pepsin could diffract X-rays. Throughout the 1930s and 1940s Hodgkin used X-ray crystallography to determine the structures of biomolecules with dozens of atoms, but proteins containing thousands of atoms seemed unassailable. Max Perutz and John Kendrew had the audacity to apply X-ray crystallography to two related metalloproteins, hemoglobin and myoglobin, and by 1959 the two had successfully elucidated the first three-dimensional atomic models of folded globular proteins (**Figure 5.1**). X-ray crystallography has now been used to solve the structure of thousands of proteins.

Figure 5.1 Seeing proteins. John Kendrew (left) and Max Perutz shared the 1962 Nobel Prize in Chemistry for determining the first three-dimensional structures of proteins. In the pioneering days of structural biology, scientists crafted their molecular models by hand. (Courtesy of MRC Laboratory of Molecular Biology.)

5.1 AMINO ACIDS AND PEPTIDES

The standard ribosomal amino acids include a broad range of functionalities

Twenty amino acids are commonly incorporated into proteins by ribosomes in all living organisms. As shown in **Figure 5.2**, these amino acids can be categorized on the basis of acidity, basicity, aromaticity, hydrophobicity, polarity, sulfur content, and flexibility. Although glycine is achiral, the other 19 are chiral, and all of the chiral amino acids except cysteine have the *S* configuration at the α center, according to the Cahn–Ingold–Prelog rules. Cysteine has the *R* configuration because the sulfur atom gives the side chain priority over the carboxylic acid. Threonine and isoleucine are unique because they have a second stereogenic center. The configuration of the threonine β-carbon is *R*; the configuration of the isoleucine β-carbon is *S*.

Human cells can biosynthesize 11 of the 20 amino acids, but the remaining 9 amino acids are a nutritional requirement. These "essential amino acids" are histidine, lysine, methionine (or cysteine), phenylalanine (or tyrosine), threonine, tryptophan, and valine. Under conditions of nutrient deprivation, cysteine can be biochemically converted into methionine, and tyrosine can be biochemically converted into phenylalanine. Lysine and tryptophan, two of the essential amino acids, are poorly represented in most plant proteins. Thus, strict vegetarians must take special care to ensure that their diet contains sufficient amounts of these two amino acids.

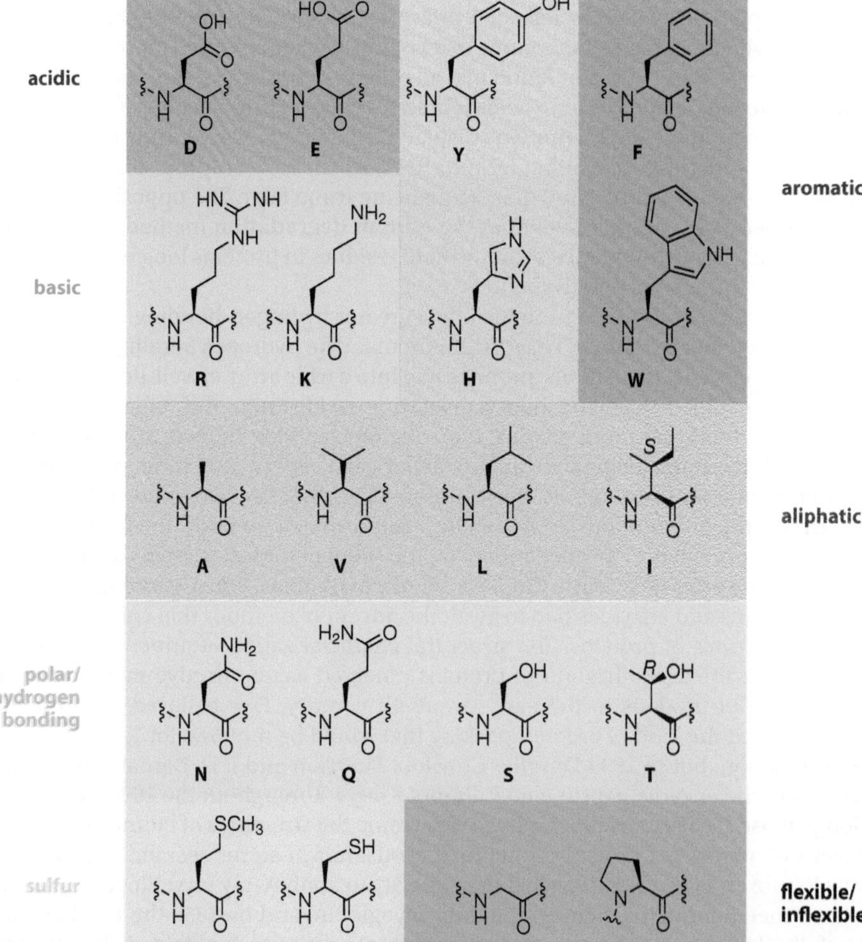

Figure 5.2 The 20 ribosomally incorporated amino acids. These amino acids, used by every ribosome on the planet, can be grouped according to their intrinsic properties.

Amino acids are polymerized into peptides and proteins

Peptides are generally less than 5 kDa in molecular mass, or about 40 amino acid residues in size. Peptides are typically unstructured and sweep through a large number of different conformations, although many examples of rigid peptides exist. The backbone of a linear peptide has two ends. The end with the amino group is referred to as the **N terminus**, whereas the end with the carboxy group is referred to as the **C terminus**. By convention, peptides are described with one-letter or three-letter codes (listed in **Table 5.1**) starting from the N terminus and reading to the C terminus. The three-letter abbreviation usually corresponds to the first three letters of the amino acid (with the exception of Asn, Gln, and Trp). The concise one-letter code may be harder to remember but is more commonly used. Amino acid residues in proteins are usually labeled using a one-letter code followed by the sequence number. For example histidine 171 would be abbreviated H171 and aspartate 167 would be labeled D167 (**Figure 5.3**). You should memorize both the three-letter and the one-letter abbreviations listed in Table 5.1. Like the base abbreviations A, G, C, T, and U, these codes are part of the fundamental alphabet of bioorganic chemistry and chemical biology.

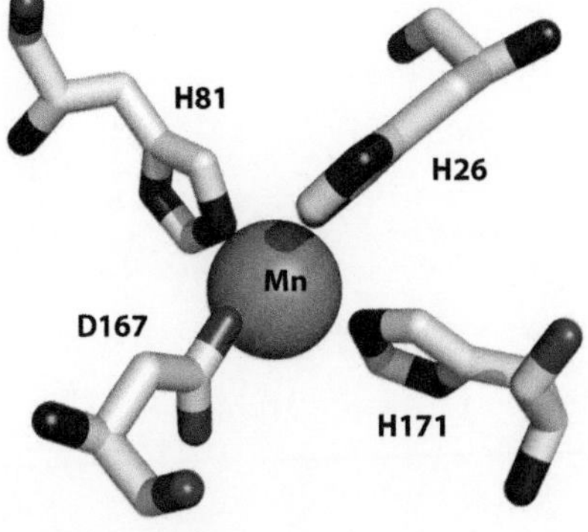

Figure 5.3 Shorthand. One-letter abbreviations followed by residue numbers are used to label protein residues. (PDB 1IXB)

Table 5.1 The 20 most common amino acids in human proteins.

Name	Three-letter code	One-letter code	Side chain pK_a	Frequency[a]	Desolvation cost (kcal mol^{-1})[b]
Leucine	Leu	L	–	10.02	–2.3
Serine	Ser	S	13.0	8.11	+5.1
Alanine	Ala	A	–	6.94	–1.9
Glutamic acid	Glu	E	5.3	6.86	+9.4
Glycine	Gly	G	–	6.59	–
Proline	Pro	P	–	6.12	–
Valine	Val	V	–	6.07	–2.0
Arginine	Arg	R	12.5	5.67	+19.9
Lysine	Lys	K	10.8	5.63	+9.5
Threonine	Thr	T	13.0	5.32	+4.9
Aspartic acid	Asp	D	3.9	4.69	+11.0
Glutamine	Gln	Q	–	4.66	+10.2
Isoleucine	Ile	I	–	4.43	–2.2
Phenylalanine	Phe	F	–	3.79	–0.8
Asparagine	Asn	N	–	3.61	+9.7
Tyrosine	Tyr	Y	10.1	2.75	+6.1
Histidine	His	H	6.0	2.60	+10.3
Cysteine	Cys	C	8.3	2.32	+1.2
Methionine	Met	M	–	2.20	+1.5
Tryptophan	Trp	W	–	1.32	+5.9

[a]Percentage occurrence based on codon usage in human cDNA (about 2008).
[b]Free energy to move a neutral side chain from water to the gas phase; the cost to desolvate a charged side chain is too large to measure directly.
(The format of this table is taken from Creighton, Proteins: Structures and Molecular Properties, 2nd ed. New York: W.H. Freeman, 1993.)

One-letter codes make it relatively easy to describe the most common protein mutations, namely substitutions and deletions. Such altered proteins are typically referred to as variants, a nomenclature that reserves the word mutant for changes at the DNA level. For example, a variant of superoxide dismutase in which aspartate 167 (D167) is replaced with a glutamate (E) would be described as the D167E variant of superoxide dismutase. A variant protein with a missing residue, moreover, is described with the Greek capital delta (Δ). For example, the ΔF508 variant of the transporter protein that causes cystic fibrosis is lacking phenylalanine 508.

Problem 5.1

Use a computer rendering program to look at the crystal structure of PDB 3ZNF. What are the identities (one-letter code plus residue number) for the four amino acids that make contact with the zinc atom?

As with oligonucleotides, a nonpalindromic peptide sequence is completely different if the order of substituents is reversed. For example, the peptide component of the antibiotic microcin C7, MRTGNAD, is neither an enantiomer nor a diastereomer of the inverted sequence DANGTRM (**Figure 5.4**). The closest definable relationship is that of constitutional isomers, which have the same molecular formula, $C_{28}H_{49}N_{11}O_{12}S$. Given the differences in chemical structure, it is unlikely that these two peptides will share the same biological activity.

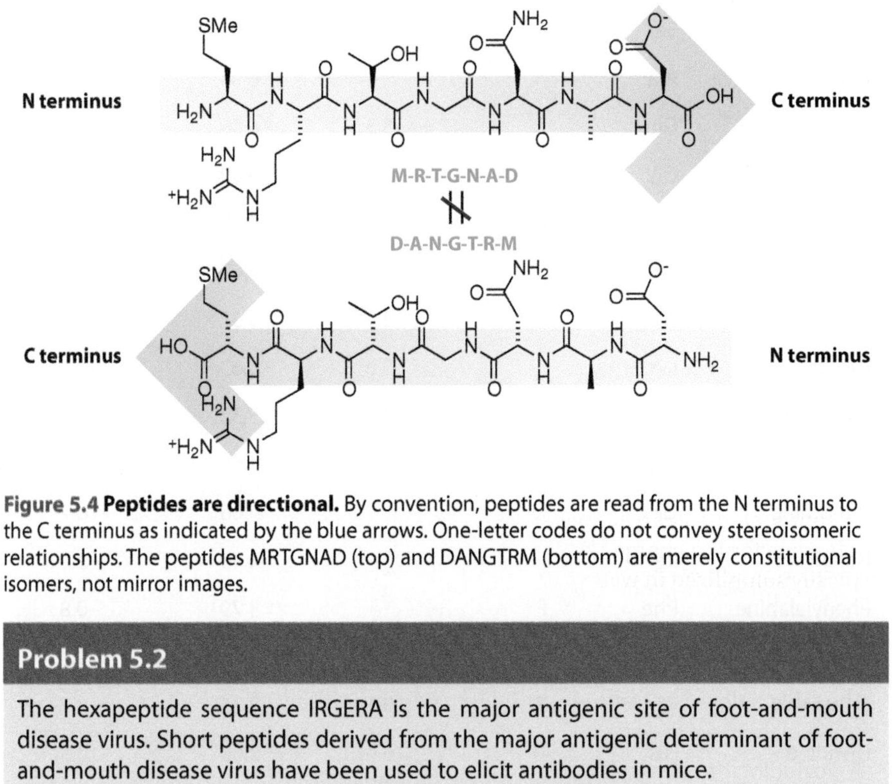

Figure 5.4 Peptides are directional. By convention, peptides are read from the N terminus to the C terminus as indicated by the blue arrows. One-letter codes do not convey stereoisomeric relationships. The peptides MRTGNAD (top) and DANGTRM (bottom) are merely constitutional isomers, not mirror images.

Problem 5.2

The hexapeptide sequence IRGERA is the major antigenic site of foot-and-mouth disease virus. Short peptides derived from the major antigenic determinant of foot-and-mouth disease virus have been used to elicit antibodies in mice.

A Draw the hexapeptide IRGERA in an extended conformation.

B Draw the enantiomer of IRGERA (ent-IRGERA) in an extended conformation.

C Draw the inverted peptide AREGRI in an extended conformation.

D Draw the enantiomer of the inverted sequence (ent-AREGRI) in an extended conformation.

E Which of the three non-natural analogs of IRGERA would you expect to bind best to antibodies elicited against the natural peptide?

Amino acid side chains have predictable protonation states

Before exploring the protonation state of amino acid side chains, first consider the pH found inside the cell. Usually, pH 7.2 is taken to be a good estimate for pH of the cytoplasm, which is buffered by various carboxylates and inorganic phosphate. pH values between 7.2 and 7.4 are commonly referred to as physiological pH. You might recall from the Henderson–Hasselbalch equation that, for any functional group, the ratio of the protonated forms to deprotonated forms is equal to $10^{(pK_a-pH)}$. At pH 7.2, any functional group with a pK_a below 7.2 will tend to be protonated at physiological pH, whereas any group with a pK_a above 7.2 will tend to be deprotonated (**Figure 5.5**). With the exception of histidine, the acidic and basic amino acids are generally charged at physiological pH, which contributes to the solubility of peptides and proteins in water. More than 7% of cysteine thiols are deprotonated at pH 7.2. The exceptional nucleophilicity of cysteine is due in part to this anionic form. The C and N termini of proteins are unusually acidic as a result of inductive effects of the backbone amides. The backbone carbonyl exerts a powerful polar effect on the N terminus, making the N-terminal ammonium group 1000–10,000 times more acidic than the ε-ammonium of the lysine side chain. The backbone nitrogens exert a less potent effect on the carboxylic acid group of the C terminus, acidifying it by less than a factor of 10 relative to a glutamic acid side chain.

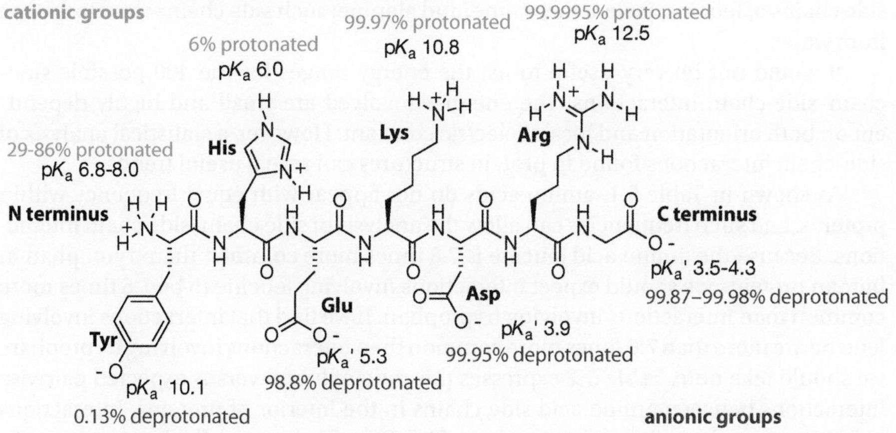

cationic groups

His: 6% protonated, pK_a 6.0; 29-86% protonated, pK_a 6.8-8.0

Lys: 99.97% protonated, pK_a 10.8

Arg: 99.9995% protonated, pK_a 12.5

N terminus

C terminus: pK_a' 3.5-4.3, 99.87–99.98% deprotonated

Asp: pK_a' 3.9, 99.95% deprotonated

Glu: pK_a' 5.3, 98.8% deprotonated

Tyr: pK_a' 10.1, 0.13% deprotonated

anionic groups

Figure 5.5 The pK_a of the protein side chains determines the ratio of protonated forms to deprotonated forms. The extent of protonation and deprotonation for each functional group is calculated for a pH of 7.2.

The use of the terms *acidic* and *basic* to describe side chains can be confusing. At physiological pH, the solvent-exposed side chains of the *acidic* residues Asp and Glu are deprotonated and cannot deliver a proton. Similarly, at physiological pH, the side chains of the *basic* residues Lys and Arg are protonated and cannot act as a base to accept another proton. The situation is often quite different in the interior of a protein. When modeling the protein interior, one typically uses effective dielectric constants that range from 4 to 20 (versus 78 for water and 1 for the gas phase). In this interior environment, the charged form of protein side chains may be disfavored, unlike the preference for the charged state at the surface of the protein, where the side chain is typically solubilized in water.

Problem 5.3

The table below lists three peptides derived from larger proteins. Each fragment can cause platelet aggregation, a key step in blood clotting.

A Including the termini, what is the net charge of the following three peptides, at pH 7.3?

Protein	Function	Sequence
ARL6IP5	Glutamate transport	NKDVLRRMKK
PVRL3	Cell adhesion	RFRGDYFAK
CLCN7	Chloride channel	KGNIDKFTEK

B Will the first peptide bind better to a positively or negatively charged surface at pH 7.3? Why?

Table 5.2 Representative side chain–side chain interactions within proteins.

Side chain	Side chain	Ratio of found to expected
Phe	Phe	4.30
Ile	Ile	3.76
Met	Met	3.73
Leu	Leu	3.57
Ile	Leu	3.28
Leu	Trp	3.04
Tyr	Tyr	2.64
Val	Leu	2.62
Val	Val	2.45
Leu	Tyr	2.35
Met	Cys	1.65
Arg	Asp	1.60
Tyr	Pro	1.60
Lys	Glu	1.27
Ala	Leu	1.26
Asp	His	1.13
Leu	Thr	1.03
Thr	Thr	0.89
Gln	Asn	0.86
Thr	Ser	0.61
Leu	Gly	0.59

Amino acid side chains mediate protein–protein interactions

Interactions between protein side chains determine the ability of a protein to adopt a uniquely favorable folded conformation. The same kinds of side chain–side chain interactions that stabilize folded proteins also mediate protein–protein interactions, receptor–ligand interactions, and enzyme–substrate interactions. To provide insights into amino acid chemistry, we will examine the side chain–side chain interactions that are most important within and between proteins.

We ask the reader to compare the side-chain desolvation energies in the final column of Table 5.1 with the chemical structures of the amino acid side chains in Figure 5.2. That essential exercise will reveal that it is hard to strip the water away from polar side chains such as those of arginine, glutamic acid, aspartic acid, glutamine, asparagine, and histidine, even in their neutral forms. Indeed, desolvation energies are too large to measure for positively charged or negatively charged side chains. Even though an interaction between an aspartate carboxylate ion and a lysine ammonium ion is highly favorable, it comes at the cost of highly favorable interactions with the aqueous environment. Similarly, even though a hydrogen-bonding interaction between serine and threonine side chains is highly favorable, it comes at the expense of highly favorable hydrogen-bonding interactions with water. Desolvation of hydrophobic side chains is less costly. In fact it is thermodynamically favorable to desolvate the aliphatic side chains of leucine, isoleucine, valine, and alanine; such side chains want to escape from water.

It would not be very useful to list the energy ranges for the 400 possible side-chain–side-chain interactions. The energies involved are small and highly dependent on both orientation and local dielectric constant. However, a statistical analysis of side-chain interactions found in protein structures can reveal useful trends.

As shown in Table 5.1, amino acids do not appear with equal frequency within proteins, and such frequencies can allow the analysis of side chain–side chain interactions. Because the amino acid leucine is 7.6 times more common than tryptophan in human proteins we should expect interactions involving leucine to be 7.6 times more common than interactions involving tryptophan. If we find that interactions involving leucine are more than 7.6 times more common than interactions involving tryptophan, we should take note. **Table 5.2** expresses this ratio of found versus expected pairwise interactions between amino acid side chains in the interior of proteins. Interactions between nonpolar side chains (such as Phe–Phe, Ile–Ile, and Ile–Leu) are significantly more common than expected (**Figure 5.6**). Cation-π interactions are also more common than expected, although numbers are not available for Table 5.2. Ionic salt bridges (such as Asp–Arg and Glu–Lys) are only slightly more common than expected in the interiors of proteins. Hydrogen-bonding interactions between side chains (such as Ser–Thr and Asn–Gln) are less common than expected within proteins. Clearly, the costs of desolvation penalize interactions between polar side chains.

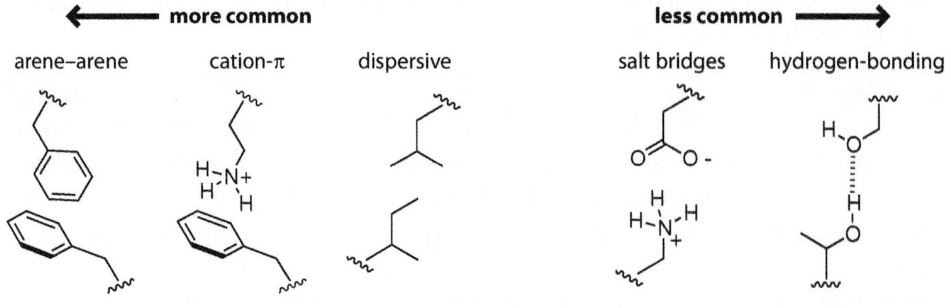

Figure 5.6 Deep inside. Within the interior of a protein, or at protein–protein interfaces, interactions between hydrophobic groups tend to be more common than interactions involving polar groups.

These trends also hold true for high-affinity interactions between proteins and even between proteins and small molecule ligands. At the protein-protein interface, the most deeply buried side chain–side chain interactions usually involve amino acids such as leucine, isoleucine, valine, alanine, phenylalanine, and methionine (**Figure 5.7**). Polar amino acids at the protein-protein interface will tend to be closer to bulk solvent.

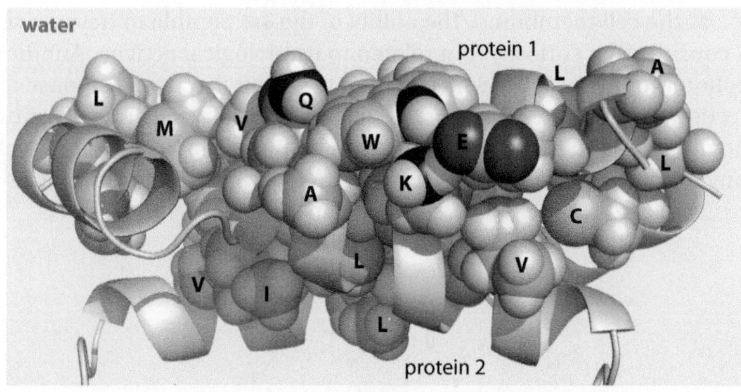

Figure 5.7 Hydrophobic interactions at a protein–protein interface. Amino acid side chains (rendered as spheres) make intimate contacts between the α helix of protein 1 (tan) and the groove of protein 2 (cyan; only the top portion of the protein is shown). Polar atoms (blue and red) tend to be exposed to solvent. (PDB 3IM4)

In membrane proteins, the aromatic amino acids such as tyrosines have major roles. These bulky and hydrophobic side chains direct the membrane protein out of the aqueous environment and into the vastly different, oily confines of the membrane. Furthermore, such residues also drive both the folding of membrane proteins and the interactions between membrane proteins. Contrary to expectations, interactions involving hydrophilic side chains, such as hydrogen bonds, contribute only modestly to the stabilization and interactions of membrane proteins.

Problem 5.4

Analyze the crystal structure of the complex between the protein MDM2 and an inhibitor peptide (PDB 3JZR). Identify the hydrophobic amino acids in the protein and the peptide (using the one-letter code and residue number, for example K42) that engage in the most significant side chain–side chain interactions.

5.2 SOLID-PHASE PEPTIDE SYNTHESIS

Peptides can be used as pharmaceuticals

Most of the interactions that control cell function occur between proteins. Logically, short peptides derived from protein–protein interfaces should be capable of inhibiting the interactions of the corresponding full-length proteins. In theory, it should be possible to create peptide drugs that inhibit cellular pathways. Unfortunately, most prescription pharmaceuticals are taken orally, and the amide bonds of peptides are rapidly hydrolyzed by enzymes called proteases in the stomach, and peptides are poorly absorbed through the gastrointestinal lumen. Even if injected directly into the bloodstream, peptides must face a host of proteases that cleave peptide bonds. Because of these proteases, most peptides have a serum half-life shorter than 1 hour. For some peptides, such as the endorphin Leu-enkephalin (YGGFL), the half-life is less than 1 minute. However, subtle modifications to the peptide backbone can lead to marked increases in half-life. These changes include the reversal of stereochemical configuration, alkylation of the backbone amide nitrogens, replacement of the charged C terminal carboxylate with a neutral carboxamide, and addition of an uncharged pyroglutamyl residue to the N terminus (**Figure 5.8**).

Membrane permeability is a final challenge for the development of therapeutic peptides that act on intracellular targets. Unfortunately, membrane permeability decreases with the number of hydrogen-bonding sites that bind strongly to water molecules. In general, peptides are not membrane permeable. The exceptions include peptides with large numbers of arginines, such as RKKRRQRRR, a short peptide sequence first identified in the HIV Tat protein. Not only does this short peptide cross cell membranes, it can also be attached to larger molecules to make it possible for

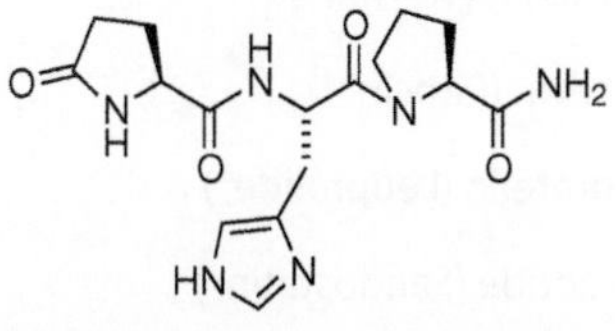

Figure 5.8 A natural peptide drug. Thyrotropin releasing hormone has sufficient serum half-life to be used as a prescription pharmaceutical.

them to cross the cell membrane. The ability of the Tat peptide to traverse cell membranes is considered a curiosity, not related to its biological activity. Another important exception is cyclosporin A, which is used to inhibit the T-cell response that leads to the rejection of transplanted organs. Cyclosporin A can be taken orally, crosses biological membranes, and has significant serum half-life. The backbone of cyclosporin A has many *N*-methyl amides, which contribute to its long serum half-life and membrane permeability (**Figure 5.9**).

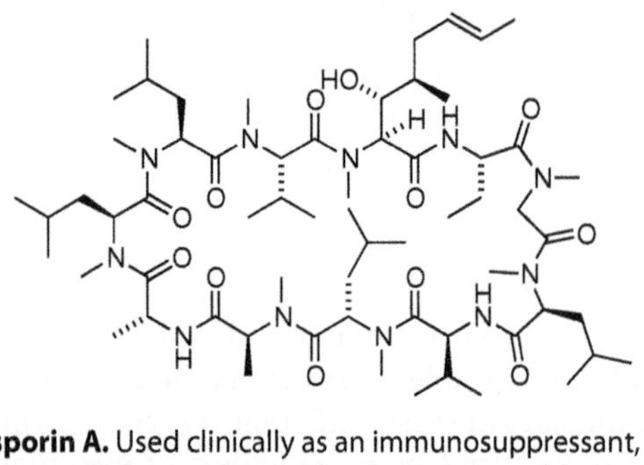

Figure 5.9 Cyclosporin A. Used clinically as an immunosuppressant, this cyclic peptide natural product differs structurally from polypeptides that are assembled by a ribosome. Cyclosporin A is both membrane permeable and resistant to proteases.

In spite of the potential instability and membrane impermeability, a wide range of peptide drugs have been approved by the U.S. Food and Drug Administration (FDA) (**Table 5.3**). Most of these peptide drugs are modified versions of naturally occurring peptide hormones that bind to cell-surface receptors without entering cells.

Often, peptide leads are used for the development of bioavailable peptides or nonpeptide drugs. Captopril is an example. Initially, the peptide teprotide was isolated from the anti-clotting venom of the lance-headed pit viper *Bothrops*

Table 5.3 Peptide drugs.

Peptide/drug name	Sequence[a]	Peptide length (residues)
Corticotropin releasing factor (Xerecept®)	SEEPPISLDLTFHLLREVLEMARAEQLAQQAHSNRKLMEII-NH_2	41
Calcitonin (Miacalcin®)	CSNLSTCVLGKLSQELHKLQTYPRTNTGSGTP-NH_2	32
Growth hormone releasing factor (1–29) (Sermorelin®)	YADAIFTNSYRKVLGQLSARKLLQDIMSR-NH_2	29
Thymosin-α-1 (Zadaxin®)	Ac-SDAAVDTSSEITTKDLKEKKEVVEEAEN	28
Adrenocorticotropic hormone (1–24) (Acthar®)	SYSMGHFRWGKPVGKKRRPVKVYP	24
Bivalirudin (Angiomax®)	FPRPGGGGNGDFEEIPEEYL	20
Somatostatin	AGCKNFFWKTFTSC	14
Daptomycin (Cubicin®)	WNDTGKDADGSIJ	13
Terlipressin (Glypressin®)	GGGCYFQNCPKG-NH_2	12
Triptorelin (Trelstar®)	pEHWSYWLRPG-NH_2	10
Oxytocin (Pitocin®)	CYIQNCPLG-NH_2	9
Leuprorelin (Leuprolide®)	pEHWSYLLRP-NHEt	9
Octreotide (Sandostatin®)	FCFWKTCT*ol*	8
Eptifibatide (Integrilin®)	CRGDWPC-NH_2	7
Thyrotropin releasing hormone (Relefact®)	pEHP-NH_2	3

[a]Lines connecting residues indicate crosslinking, such as disulfide bonding in cystine residues.

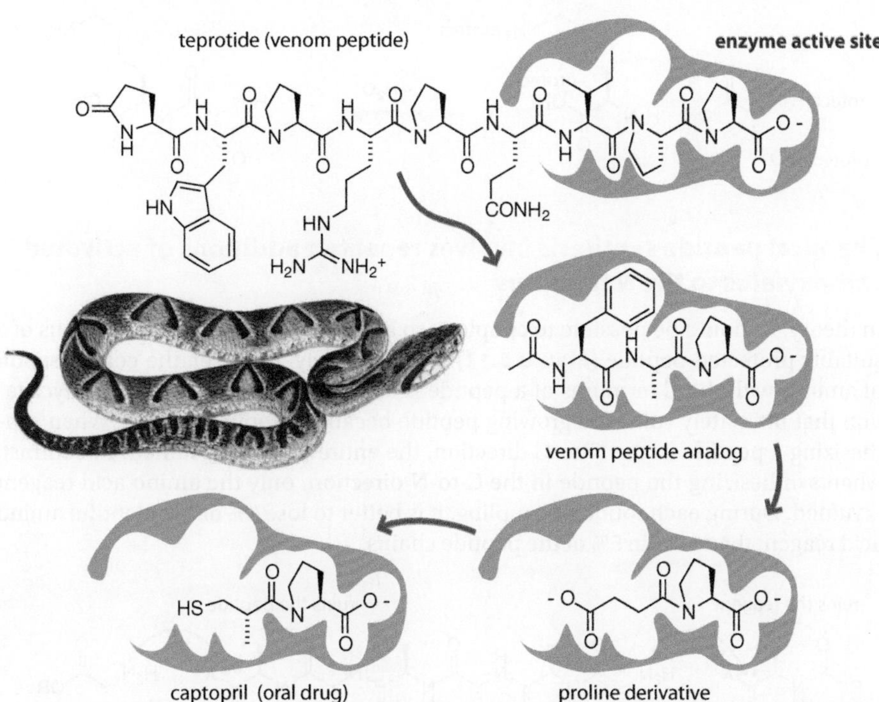

Figure 5.10 From peptide to oral drug. The peptide teprotide, from the lance-headed pit viper (*Bothrops jararaca*), inhibits blood clotting. A series of design changes led from a peptide venom to a small, orally available drug. (Image from http://www.Infobibos.com.)

jararaca, and it was chosen as a lead because it remained active for a long time *in vivo* (**Figure 5.10**). Structure–activity relationship studies revealed that a Phe-Pro-Ala sequence was the minimum pharmacophore to fit a hypothetical substrate-binding site on the targeted angiotensin converting enzyme. A thiol group was eventually added to improve interactions with the zinc ion in the enzyme's active site. The resulting drug, captopril, reached more than 1 billion US dollars in annual sales by the mid-1990s, and a related drug, enalapril, also based on teprotide, reached around 0.5 billion US dollars in worldwide sales in 2010.

Problem 5.5

On the basis of the crystal structure of angiotensin converting enzyme (PDB 1UZF), which amino acids make contacts with the key zinc atom in the active site?

Excess reagents and optimized chemistry allow high-throughput peptide synthesis

In its modern form, **solid-phase peptide synthesis** (SPPS) is the most complex and heavily optimized area of synthetic organic chemistry. In practice, SPPS can be used to construct peptides about 50 amino acid residues long. Much larger peptides (that is, proteins) are best prepared using bacterial overexpression. The common term *automated peptide synthesis* is misleading because it suggests that machines can design and execute peptide synthesis. In fact, in the hands of an amateur, a peptide synthesizer is more likely to deliver a library of compounds than a single product.

The diversity of peptide functional groups, relative to that of nucleotides, greatly complicates chemical synthesis. Recall that the ribosome deftly forms peptide amides with perfect chemoselectivity, but chemical peptide synthesis may require the protection of as many as nine different types of functional groups. Protection of the N and C termini prevents polymerization, whereas protection of the side chains prevents branching and other undesired reactions (**Figure 5.11**). Successful peptide synthesis requires a deep understanding of structure, reactivity, and mechanism so as to anticipate problems and design conditions accordingly. Protecting groups for the N or C termini must be labile enough to remove without affecting the side-chain protecting groups. The side-chain protecting groups must be robust enough to withstand round after round of acidic or basic reactions.

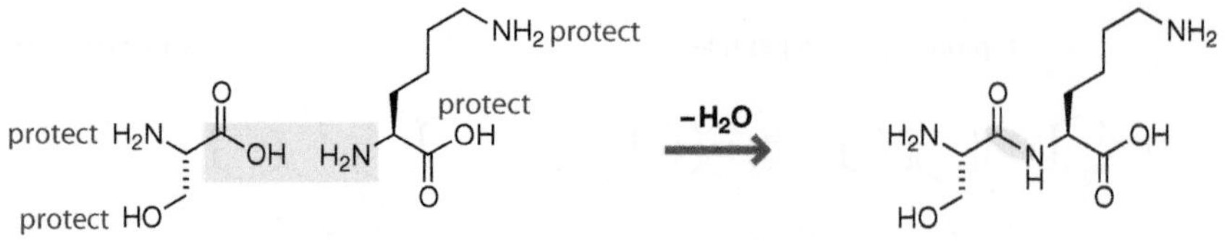

Figure 5.11 Selective protection in peptide synthesis. Protecting the indicated functional groups is required to avoid side reactions during amide bond formation.

Chemical peptide synthesis involves repeated additions of activated carboxylates to the N terminus

In theory, it should be possible to couple each amino acid at the N or C terminus of a suitably protected peptide (**Figure 5.12**). Unfortunately, however, the condensation of amines with the C terminus of a peptide is accompanied by a competing cyclization that ultimately ruins the growing peptide because of epimerization. When synthesizing a peptide in the N-to-C direction, the entire peptide is ruined. In contrast, when synthesizing the peptide in the C-to-N direction, only the amino acid reagent is ruined. During each round of coupling, it is better to lose 5% of the plentiful amino acid reagent than to ruin 5% of the peptide chains.

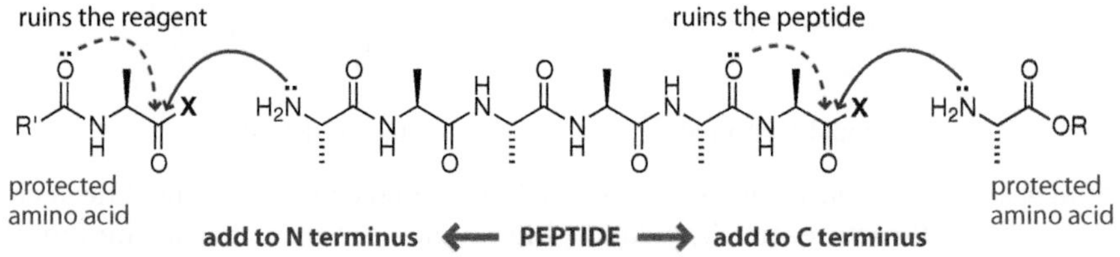

Figure 5.12 Peptides can be elongated by adding a protected amino acid to either the N terminus or the C terminus. Once a carboxylic acid is activated, nearby carbonyl groups can participate in an unwanted cyclization (indicated with a dashed arrow). Depending on the strategy, the cyclization either ruins the amino acid reagent or ruins the growing peptide chain. Ruining the reagents is better than ruining the entire peptide.

In modern peptide synthesis, the cyclization of the amino acid reagent is decreased by protecting the N_α group (see Figure 5.12) as a carbamate. Carbamate protecting groups were found empirically to give lower levels of the cyclization side reaction than amide protecting groups. After amide bond formation, the temporary carbamate protecting group can be removed selectively without disturbing the side chain or the C-terminal protecting groups.

The need to remove excess reagents and chemical by-products drove the development of solid-phase peptide synthesis

To synthesize long oligopeptides, rapid reaction rates are essential. During coupling reactions, amide bond formation competes with side reactions that deplete one or more of the reagents. Insufficient reagents lead to less peptide, and more side reactions lead to more impurities. The impurities must be laboriously separated during the final purification. If the impurities cannot be removed, the synthesis must be repeated. Thus, each coupling step is a race between amide bond formation and undesired reactions. One way to win each race is to use higher concentrations of reagents. In fact, the use of excess reagents at each step is essential for achieving the rapid coupling rates that are necessary to outpace potential side reactions.

Unfortunately, the excess reagents used to ensure efficient coupling and deprotection are not easy to separate from the product. The revolutionary solution to this problem came from R. Bruce Merrifield, who single-handedly conceived and proved the idea of SPPS. Merrifield linked the peptide to an insoluble polymer so that excess reagents could be rinsed away without losing the product. The strategy is similar to washing a car, in which excess reagents and by-products (that is, soap and dirt) are rinsed away with water.

Either acid- or base-labile carbamates are used for the temporary protection of the N$_\alpha$ group

The *t*-butoxycarbonyl (***t*-Boc**) group is widely used for the protection of amines and is often abbreviated as Boc. In the earliest versions of SPPS, Boc groups were used to protect the α-amino groups. It is still used for the protection of the amino groups of lysine side chains when the α amines are protected by base-labile protecting groups. Boc groups are quickly removed under mildly acidic conditions, such as 1:1 trifluoroacetic acid (TFA)/dichloromethane. Under these deprotection conditions the amine is protonated to give the ammonium salt. A separate neutralization step is required to convert the protonated ammonium group to the free base. The deprotection reaction is an S_N1 reaction in which the *t*-butyl cation is trapped by trifluoroacetate anion (**Figure 5.13**). E1 reactions compete with S_N1 reactions, but the olefin product isobutylene that is generated reversibly through E1 elimination is readily reprotonated (**Figure 5.14**)—the equilibrium constant for isobutylene + TFA = *t*-butyl trifluoroacetate is 600 M^{-1}.

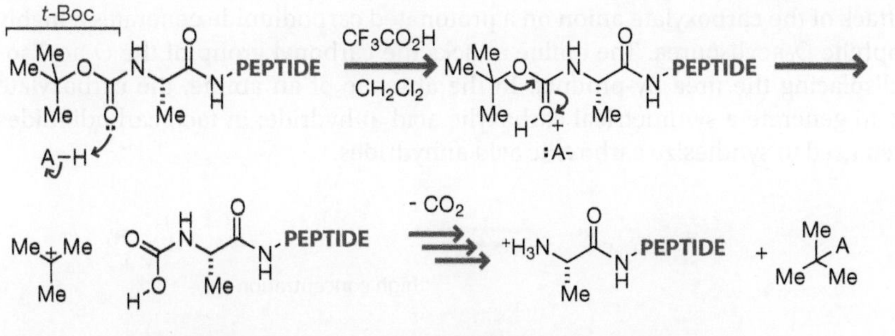

Figure 5.13 *t*-Boc deprotection proceeds through an S_N1 mechanism. The carbamic acid (bottom left) rapidly decarboxylates.

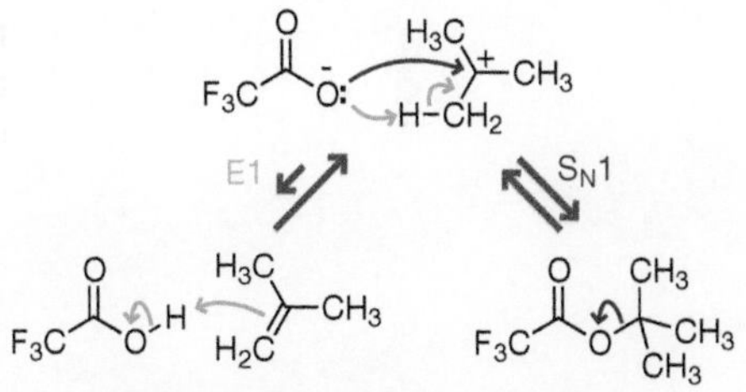

Figure 5.14 S_N1 versus E1. The *t*-butyl carbocation undergoes reversible E1 and S_N1 reactions during Boc deprotection with trifluoroacetic acid. The equilibrium favors the S_N1 substitution product *t*-butyl trifluoroacetate.

The fluorenylmethoxycarbonyl (Fmoc) group is quickly removed under mildly basic conditions, such as 1:4 (v/v) piperidine/*N,N*-dimethylformamide (DMF). The reaction involves an E1cB mechanism in which the deprotonation of the fluorenyl group leads to an aromatic anion that undergoes β-elimination (**Figure 5.15**). Excess piperidine adds to the fulvene intermediate. Generally, carbamic acids do not decarboxylate under highly basic conditions, but the piperidinium ion is sufficiently acidic to induce loss of carbon dioxide. In contrast with Boc chemistry, the removal of Fmoc leads directly to the free base, which is ready for the next amino acid coupling reaction.

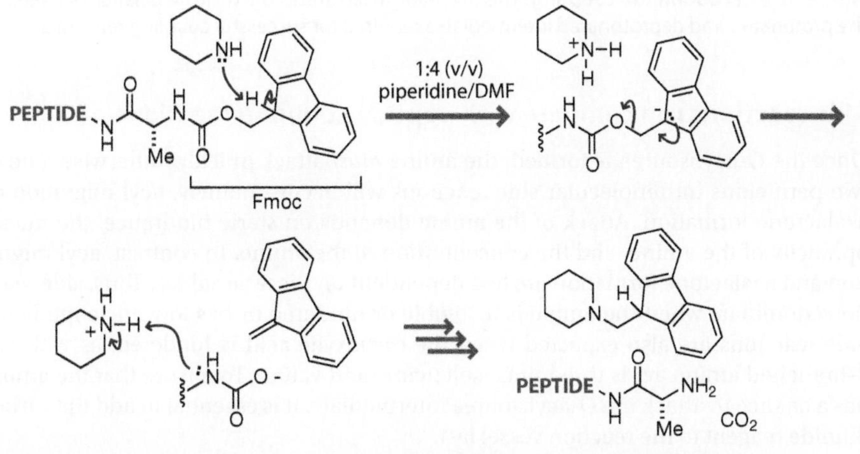

Figure 5.15 Deprotection of Fmoc proceeds through an E1cB mechanism. The decarboxylation of carbamic acids is generally disfavored under highly basic conditions, but the piperidinium ion can protonate the carbamic acid to promote decarboxylation (bottom).

Carbodiimides drive condensation to form peptide bonds

A dehydrating reagent is required to drive the condensation of amines and carboxylic acids. Generally, the thermodynamic driving force for this process is the formation of a stable C=O or P=O bond. Carbodiimides are the prototypical coupling reagents, yielding urea by-products. Carbodiimides are rarely used for amide bond formation in modern peptide synthesis but still find wide use in chemical biology for reactions forming amide bonds. Several carbodiimides are commonly used, depending on whether one wants the urea by-product to be soluble or insoluble. In peptide synthesis the most common carbodiimides are DIC (*N,N′*-diisopropylcarbodiimide) and DCC (*N,N′*-dicyclohexylcarbodiimide). Laboratory exposure to carbodiimides such as DCC can lead to extreme hypersensitivity to the reagent.

The mechanism for carbodiimide coupling (**Figure 5.16**) involves an initial acid-base equilibrium between the carboxylic acid and the amine. Protonation of the carbodiimide facilitates attack by the carboxylate anion, which is present in high concentrations. Because most of the amine is protonated, it does not compete with the carboxylate anion in nucleophilic reactions. Carbodiimide coupling is slow under strongly basic conditions because the concentration of protonated carbodiimide is low. Attack of the carboxylate anion on a protonated carbodiimide generates a highly electrophilic *O*-acylisourea. The amine attacks the carbonyl group of the *O*-acylisourea, displacing the urea by-product. In the absence of an amine, the carboxylate attacks to generate a symmetrical carboxylic acid anhydride; in fact, carbodiimides are often used to synthesize carboxylic acid anhydrides.

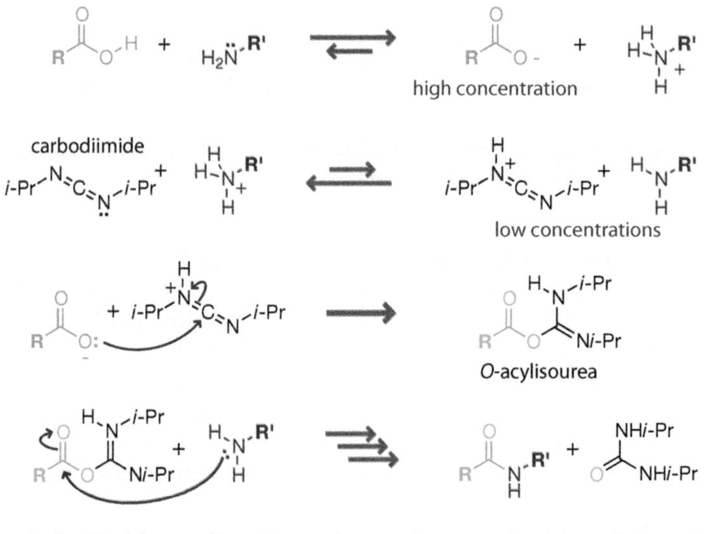

Figure 5.16 Carbodiimide coupling. This mechanism illustrates the delicate balance between the protonated and deprotonated intermediates required for successful coupling reactions.

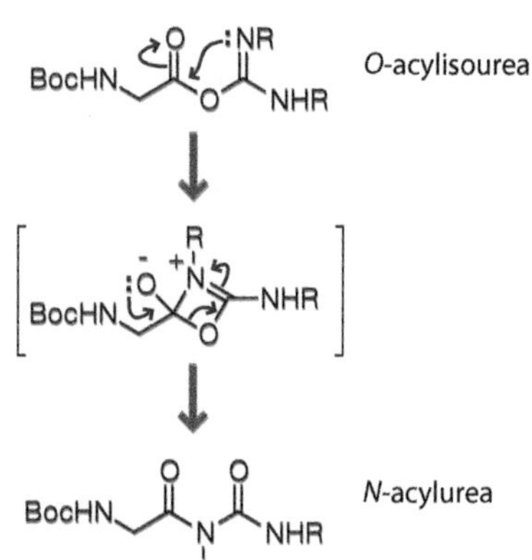

Figure 5.17 *O*-to-*N* acyl migration. This side reaction ruins the carboxylic acid component during carbodiimide coupling reactions.

Side reactions can compete with peptide coupling reactions

Once the *O*-acylisourea is formed, the amine *must* attack quickly; otherwise, one of two pernicious intramolecular side reactions will occur—namely, acyl migration or azalactone formation. Attack of the amine depends on steric hindrance, the nucleophilicity of the amine, and the concentration of the amine. In contrast, acyl migration and azalactone formation are less dependent on these variables. Thus, side reactions dominate when the amine is insoluble or hindered or has low nucleophilicity. Side reactions are also expected when the carboxylic acid is hindered, as with the β-branched amino acids threonine, isoleucine, and valine. To ensure that the amine has a chance to attack the *O*-acylisourea intermediate, it is essential to add the carbodiimide reagent to the reaction vessel last.

Acyl migration forms an unreactive *N*-acylurea. Surprisingly, kinetic studies support a mechanism involving a strained four-membered ring intermediate (**Figure 5.17**). The use of excess amino acids and coupling reagents ensures that the acylation of the amine outpaces the side reactions. In solution-phase synthesis, acyl migration is more problematic, because the *N*-acylurea by-products are not easily removed.

The second major side reaction of the *O*-acylisourea is azalactone formation (often spelled *azlactone*; **Figure 5.18**). Azalactones can still react with the amine, but epimerization usually competes with amide bond formation. The epimerization of full-length peptides is generally not a concern because of the low kinetic acidity of the protons α to amide functional groups. In contrast, azalactones are very acidic, because the resulting enolate is aromatic. Azalactone formation is particularly insidious because the epimeric products are generally impossible to separate from the desired product. Azalactone formation is faster when the acyl group is an amide, such as a peptide chain, and slower when the acyl group is a carbamate, such as *t*-Boc or Fmoc. This is why peptide synthesis is not generally conducted by addition to the C terminus of a peptide (that is, in the N to C terminus direction).

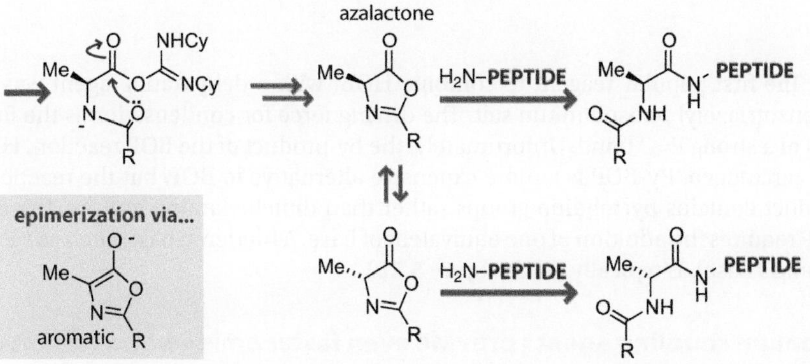

azalactone
NHCy
Me
NCy
H₂N–PEPTIDE
PEPTIDE
epimerization via...
aromatic
Me
H₂N–PEPTIDE
PEPTIDE

Figure 5.18 Side reaction. Azalactonization can cause epimerization of C_α backbone carbons.

HOBt minimizes side reactions in carbodiimide couplings

One way to slow acyl migration and azalactone formation is to convert the *O*-acylisourea to a less promiscuous acylating agent *in situ*. Through empirical studies, 1-hydroxybenzotriazole (HOBt) was found to have fast exchange rates, and the resulting HOBt esters resist azalactone formation. Generally, one equivalent of HOBt is added relative to the carboxylic acid and the coupling agent. Even if the amine couples sluggishly, the *O*-acylisourea can be quantitatively converted to the benzotriazolyl ester, which will patiently wait for attack by the amine (**Figure 5.19**).

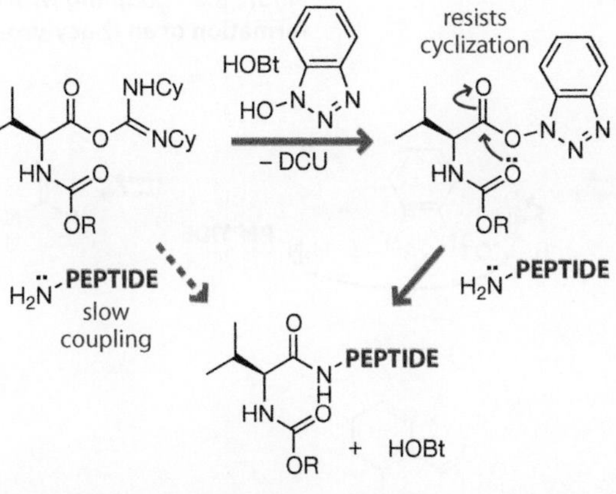

HOBt
– DCU
resists cyclization
PEPTIDE
slow coupling
PEPTIDE
PEPTIDE
+ HOBt

Figure 5.19 Activated ester formation. Slow formation of the azalactone decreases epimerization at C_α. However, the benzotriazolyl ester reacts readily with the N terminus of the peptide to form the desired amide bond.

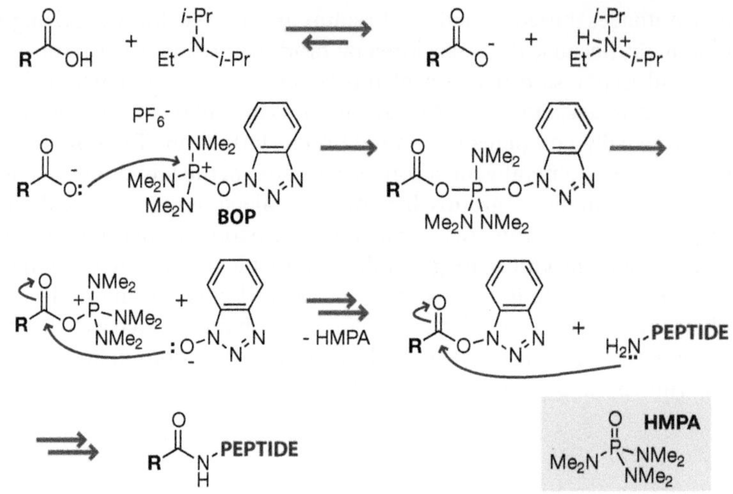

Figure 5.20 BOP coupling mechanism. This dual reagent carries both the dehydrating agent and the acyl transfer catalyst.

The first popular reagent to combine HOBt with a dehydrating agent was BOP, a benzotriazolyl phosphonium salt. The driving force for condensation is the formation of a strong P=O bond. Unfortunately, the by-product of the BOP reaction, HMPA, is a carcinogen. Py-BOP is a more expensive alternative to BOP, but the reaction by-product contains pyrrolidine groups rather than dimethylamino groups. The use of BOP requires the addition of one equivalent of base. A hindered base, such as i-Pr_2NEt (Hünig's base), is typically used (**Figure 5.20**).

Uronium coupling agents provide even faster amide bond formation

Benzotriazolyluronium salts are newer coupling agents related to BOP. Like BOP, HBTU (**Figure 5.21**) combines a dehydrating agent with HOBt, but unlike BOP, the uronium moiety is attached to the nitrogen atom of HOBt. *N,N,N′,N′*-tetramethylurea is the by-product. Like BOP, HBTU requires an additional equivalent of base to prevent the amine from being protonated by the carboxylic acid. One of the most advanced coupling reagents currently in use is HATU (**Figure 5.22**). HATU is an aza analog of HBTU, and gives faster coupling rates. HATU is presumed to couple faster because the nitrogen atom can lower the energy of the tetrahedral transition state by hydrogen bonding to the amine as it attacks. In cases where racemization is a problem, the reagent DEPBT is exceptionally resistant to racemization via base-catalyzed azalactonization of the activated ester intermediate.

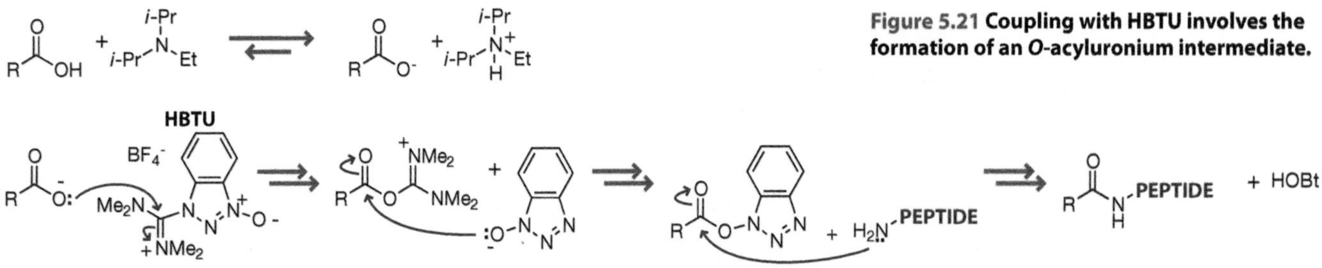

Figure 5.21 Coupling with HBTU involves the formation of an *O*-acyluronium intermediate.

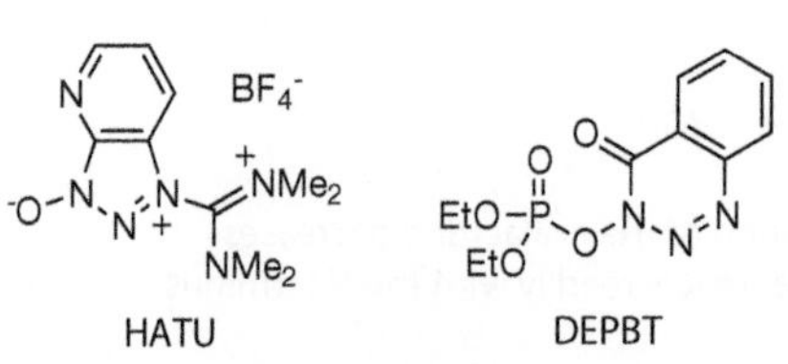

Figure 5.22 Advanced coupling reagents. HATU and DEPBT are more advanced versions of peptide coupling reagents than HBTU.

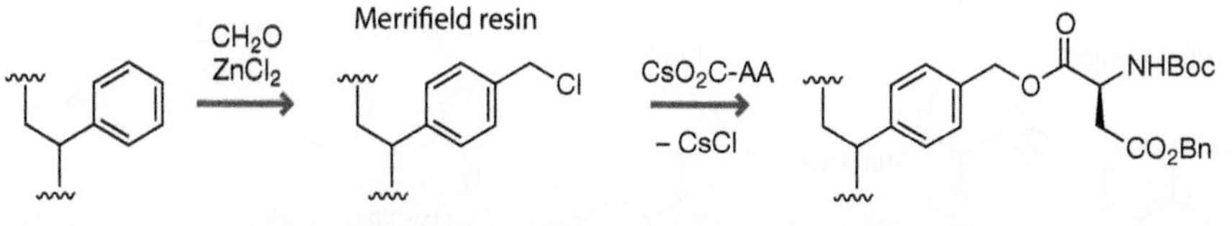

Figure 5.23 Functionalization of polystyrene. The attachment to the resin both provides a solid support for efficient synthesis and protects the C-terminal carboxylate.

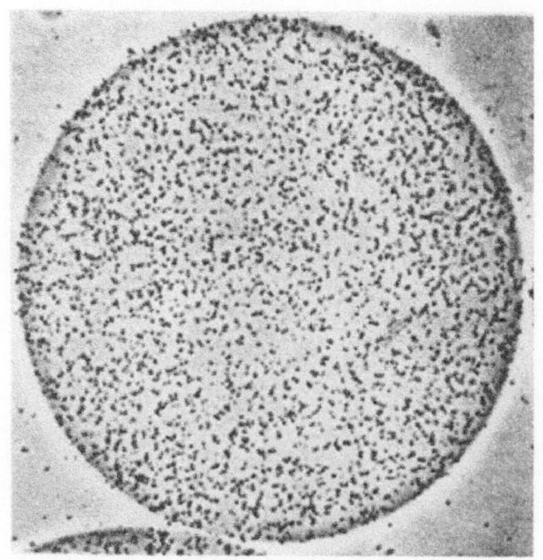

Figure 5.24 Inside and out. An ultrathin slice of 1% divinylbenzene/polystyrene functionalized with a radiolabeled peptide reveals the relatively uniform distribution of peptide chains throughout the bead. (From R.B. Merrifield, *Intra-Sci Chem. Rept.* 5:184–198, 1971.)

Resins for solid-phase peptide synthesis are made of plastic

The traditional resin for SPPS is polystyrene. Polystyrene is a hard, colorless plastic, like the kind used to make the cases for compact discs. Expanded polystyrene, also known as styrofoam, is used for insulation and packing materials. The polystyrene beads used for SPPS are prepared by the radical polymerization of polystyrene with a 1-divinylbenzene added as a crosslinking agent. Traditionally, beads are about 0.1–0.2 mm in size—smaller than grains of beach sand. Each bead can hold about 50–100 pmol of peptide, so 1 g of resin can hold about 0.5 mmol of peptide.

Polystyrene is functionalized with formaldehyde and zinc chloride to give chloromethylated polystyrene, also known as Merrifield resin (**Figure 5.23** and **Figure 5.24**). The first amino acid is added by an S_N2 reaction, leading to an ester linkage. Thus, the solid support serves as a C-terminal protecting group for the first amino acid. It is generally cheaper to purchase resin with the first amino acid attached than to functionalize it yourself.

The term *solid-phase* is somewhat misleading in that it implies immobility. Instead of an immobile solid, though, try to think of SPPS beads as liquid spheres. Once resin beads swell with solvent, the peptides and polystyrene chains are highly mobile, allowing reagents to diffuse in and out. Because of this mobility, there is no significant barrier to interactions between peptides bound to a solid-phase resin. Thus, peptides can aggregate during peptide synthesis, which reduces the efficiency of the amino acid coupling reaction.

For some nonpeptide reactions, or on-bead assays, hydroxylic solvents may be desirable. Unfortunately, polystyrene is not solvated by water or alcohols, so peptide chains inside the bead are inaccessible to reagents. If the polystyrene is functionalized with poly(ethylene glycol) (PEG), however, the beads are then solvated by a wide range of solvents, including alcohols and water. Protein drugs modified with long hydrophilic PEG chains (**Figure 5.25**) often have longer half-lives in the human body, meaning that injections are needed less frequently.

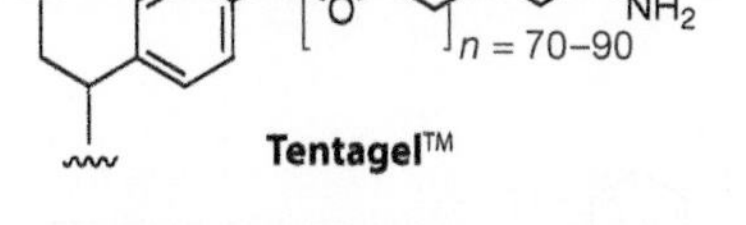

Figure 5.25 Tentagel™ is a grafted poly(ethylene glycol)/polystyrene resin. The PEG linkers provide additional solubility and capability with hydroxylic solvents (such as water).

Cleavable linkers between the synthesized peptide and solid support provide stable, yet reversible, attachments

When performing SPPS, the peptide–resin linkage must be stable throughout all the peptide elongation steps, yet easily cleaved at the end of the synthesis. This concept is known as orthogonal protection. Linkers are functional groups that provide a more stable or more labile connection between the peptide and the solid support, as needed.

For SPPS using Boc-protected amino acids, it is essential that the linkage to the solid support be stable to round after round of treatment with TFA. When using Merrifield resin, about 0.65% of the peptide falls off the resin during each round of Boc deprotection. In an effort to decrease this side reaction, the peptide can be attached through a phenylacetamidomethyl (PAM) linker. Because amides ($pK_a' = 0.0$) are a million times more basic than esters ($pK_a' = -6.5$), the PAM linker preferentially protonates before the ester, making the benzyl ester linkage about 100 times more stable and resistant to ionization (**Figure 5.26**).

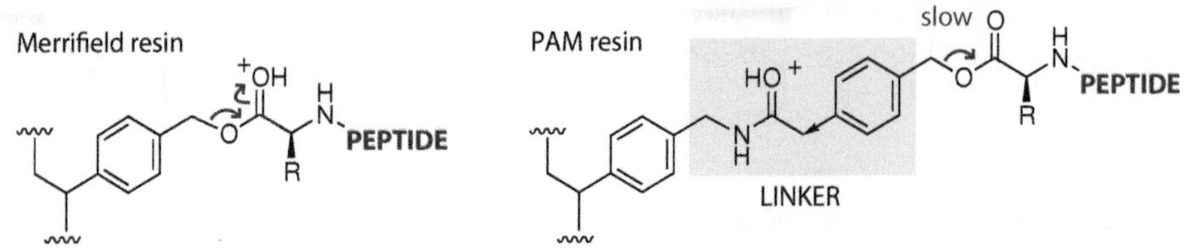

Figure 5.26 Premature cleavage. With Merrifield resin, S_N1 solvolysis of the peptide chains decreases chemical yield after many rounds of Boc deprotection with CF_3CO_2H. With PAM resin, protonation of the PAM linker inhibits S_N1 solvolysis of the peptide chain.

Peptides are cleaved from Merrifield and PAM resin under strongly acidic conditions. The best conditions for S_N2 cleavage involve anhydrous HF, condensed from the gas phase with special apparatus that is unavailable in most laboratories. HF dissolves both glass and human bones. The weak acidity of HF allows it to penetrate deeply into tissues, making it an extremely dangerous chemical. Trifluoromethanesulfonic acid (TFMSA) is sometimes used as an alternative: CF_3SO_3H/PhSMe/1,2-ethanedithiol (EDT) (2:2:1), 25 °C, 1 hour. The choice of cleavage and deprotection conditions depends greatly on the constitution of the peptide. Typical conditions always employ nucleophilic scavengers to react with benzyl and *t*-butyl cations. The low–high conditions ((i) CF_3SO_3H/TFA/Me_2S/*m*-cresol (1:5:3:1), 0 °C, 3 hours; (ii) CF_3SO_3H/TFA/PhSMe/EDT (1:10:3:1), 25 °C, 30 min) prevent alkylation of tyrosine and acylation by glutamine side chains.

A carboxamide group is often desired at the C terminus rather than a carboxylate group. For example, if synthesizing a peptide fragment of a larger protein, the carboxamide avoids introducing the negative charge inherent in a carboxylate at the C terminus. It is impossible to selectively perform chemistry on the C terminus once the side chains have been deprotected, and the peptide is cleaved from the resin. In fact, cleavage and side-chain deprotection usually occur in the same step. Benzhydrylamine resins such as MBHA (methyl benzhydrylamine resin) were developed to allow facile S_N1 cleavage of carbon–nitrogen bonds, providing a carboxamide at the C terminus (**Figure 5.27**). A second aryl group provides sufficient activation for efficient ionization using anhydrous HF or TFMSA.

When SPPS is performed with base-labile, Fmoc-protected amino acids, peptides go through round after round of coupling and base deprotection; the growing peptide chain does not see acid at any point during peptide elongation. As a result, highly acid-labile linkers can be used that allow the final deprotection and cleavage to be performed under much milder conditions than peptides constructed from N_α-Boc-protected amino acids. For example, typical conditions use the much safer TFA (CF_3CO_2H) in place of the highly toxic anhydrous HF. One set of conditions is TFA/*i*-Pr_3SiH/H_2O (95:2.5:2.5), 25 °C, 1–10 hours. Two types of resin that are commonly used for the synthesis of peptides during Fmoc chemistry are chlorotrityl resin and Rink amide resin (**Figure 5.28**). Chlorotrityl resin generates a stable triphenylmethyl

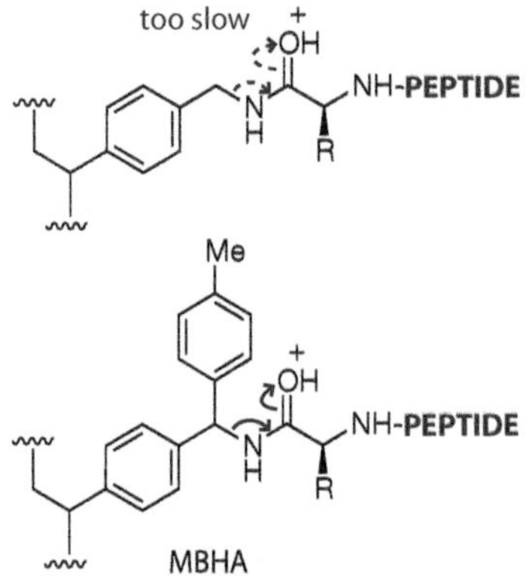

Figure 5.27 Making C-terminal amides. MBHA resin is stable to the repeated exposure to TFA used in Boc chemistry, but can be cleaved by HF or TFMSA to give peptides with carboxamide termini.

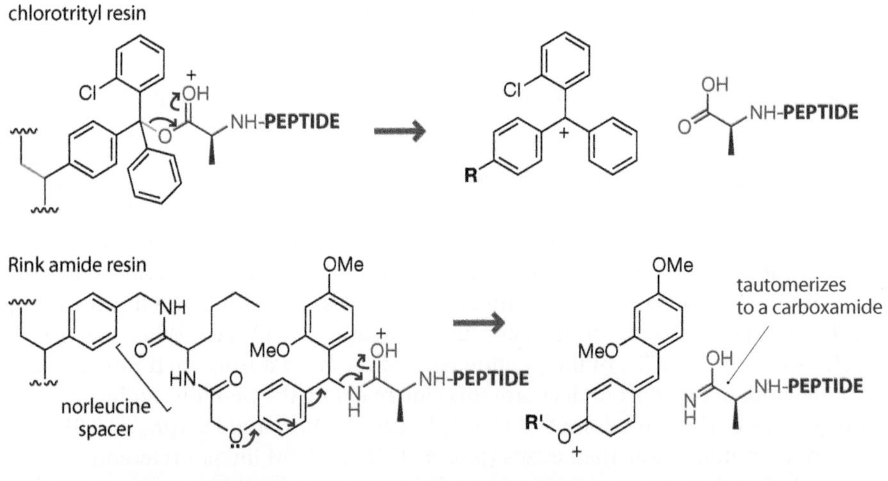

Figure 5.28 Common resins for Fmoc synthesis. Chlorotrityl resin and Rink amide resin are designed to generate highly stable carbocations, resulting in rapid S_N1 cleavage.

cation. The chlorine atom subtly attenuates the rate of S_N1 ionization. To facilitate the cleavage of a peptide with a carboxamide terminus, a Rink amide linker is used. The Rink linker generates a highly stabilized carbocation, making up for the slow S_N1 ionization of carbon–nitrogen bonds.

Side-chain protecting groups come off under acidic conditions

Side-chain protecting groups are designed to come off when immersed in the acidic cocktail that cleaves the peptide from the resin. The synthesis strategy dictates the choice of amino acid starting materials (**Figure 5.29**). When peptides are synthesized with N_α-Boc-protected amino acids, the side-chain protecting groups must be stable

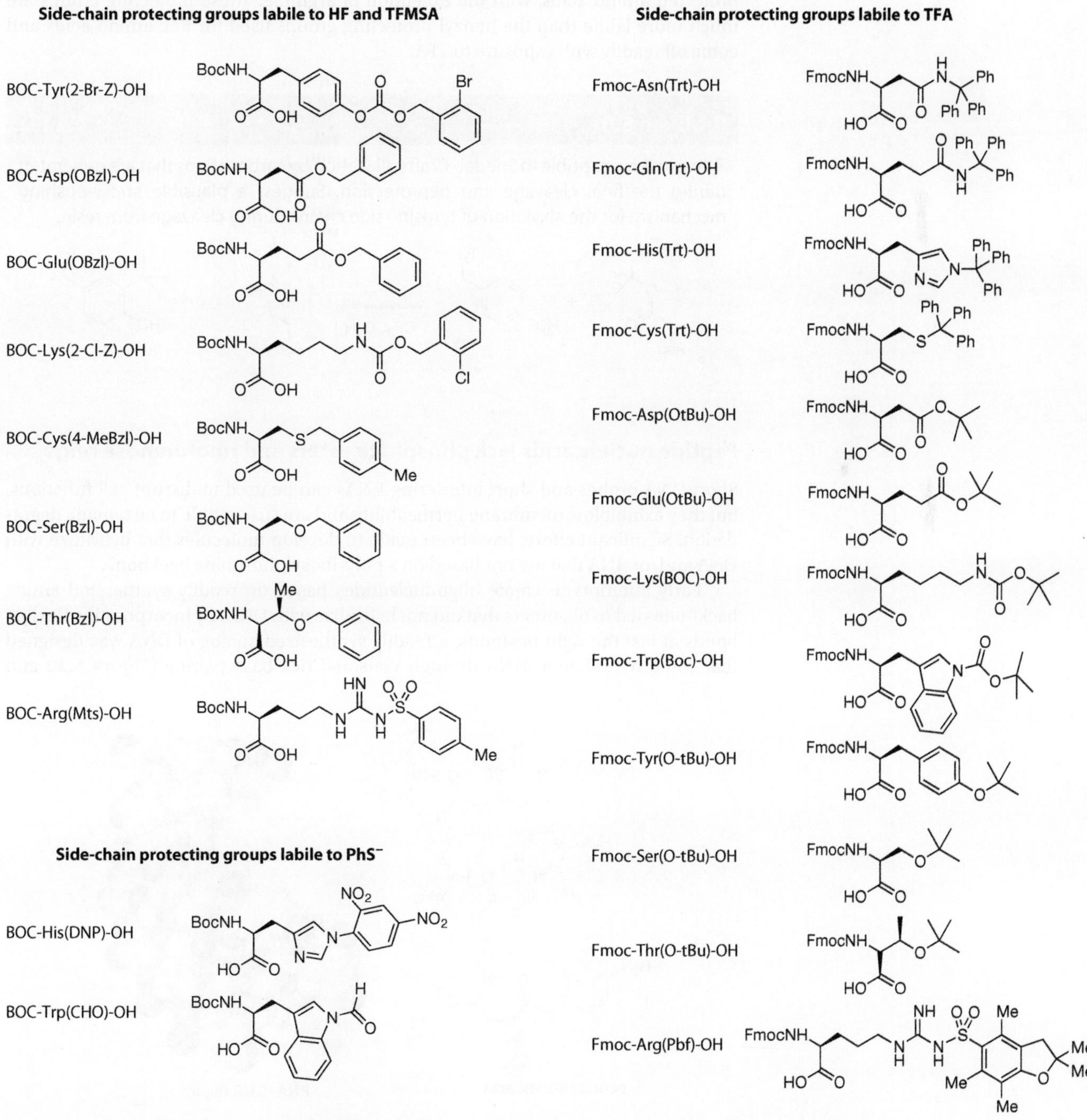

Side-chain protecting groups labile to HF and TFMSA

BOC-Tyr(2-Br-Z)-OH

BOC-Asp(OBzl)-OH

BOC-Glu(OBzl)-OH

BOC-Lys(2-Cl-Z)-OH

BOC-Cys(4-MeBzl)-OH

BOC-Ser(Bzl)-OH

BOC-Thr(Bzl)-OH

BOC-Arg(Mts)-OH

Side-chain protecting groups labile to PhS⁻

BOC-His(DNP)-OH

BOC-Trp(CHO)-OH

Side-chain protecting groups labile to TFA

Fmoc-Asn(Trt)-OH

Fmoc-Gln(Trt)-OH

Fmoc-His(Trt)-OH

Fmoc-Cys(Trt)-OH

Fmoc-Asp(OtBu)-OH

Fmoc-Glu(OtBu)-OH

Fmoc-Lys(BOC)-OH

Fmoc-Trp(Boc)-OH

Fmoc-Tyr(O-tBu)-OH

Fmoc-Ser(O-tBu)-OH

Fmoc-Thr(O-tBu)-OH

Fmoc-Arg(Pbf)-OH

Figure 5.29 Amino acids used for peptide synthesis. Amino acid reagents are designed for either Boc chemistry (repeated exposure to mild acid) or Fmoc chemistry (repeated exposure to mild base).

to round after round of immersion in TFA. After all the peptide bonds are formed, harsh acidic conditions such as anhydrous HF or CF_3SO_3H are then used for both the resin cleavage and side-chain deprotection. Benzyl protecting groups are generally used, but the reactivity of some benzyl protecting groups has been tuned with halogen substituents. Arg(Mts) has a sulfonyl protecting group that comes off under acidic conditions. His(DNP) and Trp(CHO) are special exceptions that require an additional nucleophilic deprotection step. These amino acids possess protecting groups for nitrogen atoms that are removed by thiols.

When peptides are synthesized with N_α-Fmoc-protected amino acids, they are exposed to round after round of basic deprotection conditions. The nascent peptides do not experience acidic conditions until the end of the synthesis. Triphenylmethyl (trityl) and *t*-butyl protecting groups are used for all the side chains of N_α-Fmoc-protected amino acids, with the exception of arginine. These protecting groups are much more labile than the benzyl protecting groups used for Boc amino acids and come off readily with exposure to TFA.

Problem 5.6

Tyrosine is susceptible to Friedel–Crafts alkylation by carbocations that are generated during the final cleavage and deprotection. Suggest a plausible arrow-pushing mechanism for the alkylation of tyrosine side chains during cleavage from resin.

Asp + Tyr (HO) $\xrightarrow{CF_3CO_2H}$ Asp (OH) + Tyr (HO)

Peptide nucleic acids lack phosphate esters and ribofuranose rings

Short DNA probes and short interfering RNAs can be used to disrupt cell functions, but they exhibit low membrane permeability and are susceptible to enzymatic degradation. Significant efforts have been made to develop molecules that hybridize with DNA and/or RNA that are not based on a polyribosylphosphate backbone.

Early attempts to create oligonucleotides based on readily synthesized amide backbones led to oligomers that did not hybridize with DNA. By incorporating flexible bonds at just the right positions, a readily synthesized analog of DNA was designed that could hybridize to DNA through Watson–Crick base pairing (**Figure 5.30** and

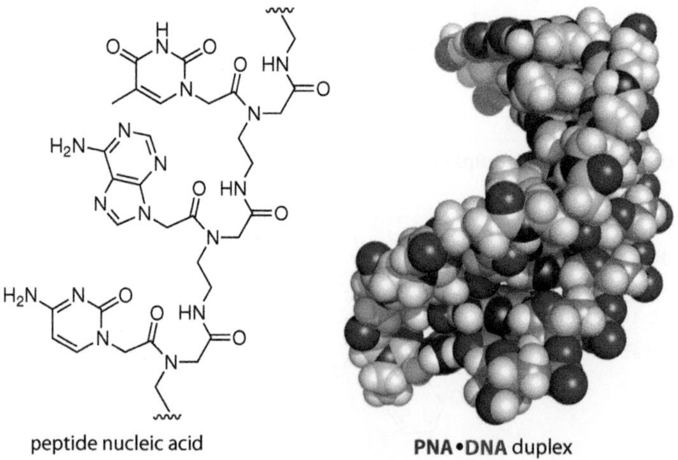

Figure 5.30 Peptide nucleic acids can hybridize with DNA. One helical turn of a PNA•DNA duplex consists of 18 base pairs. (PDB 1PDT)

Figure 5.31). These new DNA analogs, called **peptide nucleic acids (PNAs)**, are resistant to intracellular enzymes that cleave phosphodiesters and also to intracellular enzymes that cleave amide bonds. Lacking the negative charge of the phosphodiester backbone of DNA, PNAs are not subjected to the charge–charge repulsion inherent in the hybridization of two strands of DNA. Thus, PNAs bind so strongly to complementary DNA sequences that they can insinuate themselves into a long double-stranded DNA molecule and displace one of the native DNA strands. Furthermore, PNAs bind strongly to genomic DNA in cells, and have been used to inhibit transcription. Unfortunately, like oligonucleotides, PNAs have low solubility and low intrinsic membrane permeability. Although various tricks can be used to induce the uptake of PNAs into cells grown in plastic dishes, the prospects for the use of PNAs to control gene expression in humans seem dim. Clearly, new methods are needed to carry bioactive molecules efficiently across biological membranes.

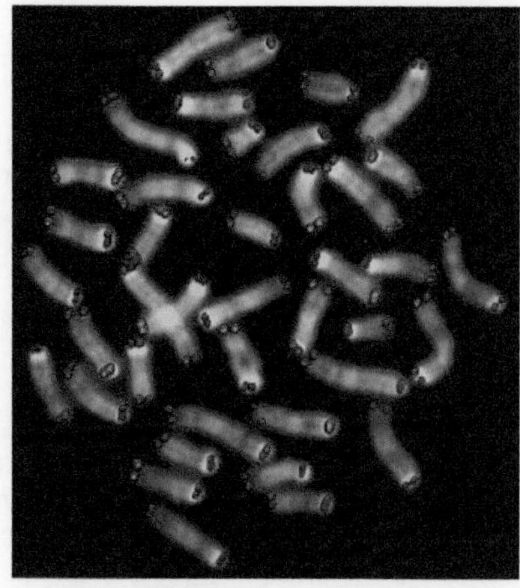

Figure 5.31 Using peptide nucleic acids. A fluorescently labeled peptide nucleic acid $(CCCTAA)_3$ binds specifically to the telomeric DNA sequences at the ends of chromosomes. (From J.M. Zijlman et al., *Proc. Natl. Acad. Sci. USA* 94:7423–7428, 1997. With permission from the National Academy of Sciences.)

Problem 5.7

Draw the structure of a monomeric analog of thymidylic acid that would be used for Fmoc synthesis of a peptide nucleic acid.

Native chemical ligation generates cysteinyl amides through aminolysis of thiol esters

Full-length proteins are usually too long for an iterative synthesis on a peptide synthesizer. Strategically, it is better to synthesize long peptides by uniting shorter peptide segments; unfortunately, protected peptides have low solubility in both organic solvents and water. **Native chemical ligation** (**Figure 5.32**) was developed as an efficient method for joining long unprotected peptide fragments in aqueous solution. The method requires that one peptide have a cysteine at the N terminus; the other peptide must have a non-natural C-terminal thioester.

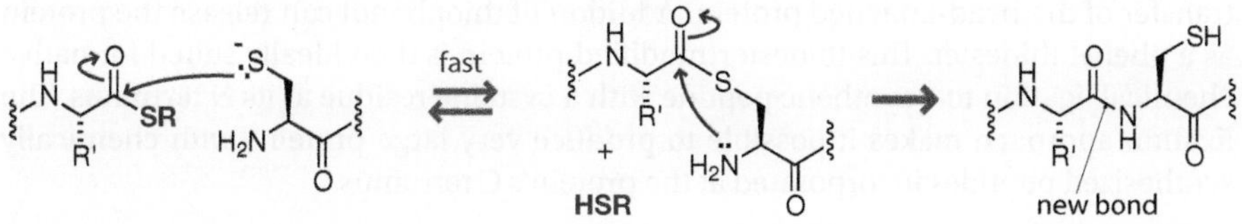

Figure 5.32 Native chemical ligation. Thiol esters undergo facile exchange under physiological conditions. Thiol esters derived from cysteine are poised to undergo an S to N acyl migration, generating a new peptide bond.

Native chemical ligation takes advantage of the unique reactivity of thiol esters and the low pK_a of cysteine thiols (about 9.0). Like traditional carboxylic acid esters thiol esters are relatively stable toward hydrolysis at pH 7. Unlike carboxylic acid esters, thiol esters exchange rapidly at physiological pH in the presence of excess thiol. When N-terminal cysteine residues form thiol esters they undergo a facile N to S acyl migration via a five-membered ring tetrahedral intermediate. The resulting amide bond is indistinguishable from cysteinyl amide bonds constructed by a ribosome.

Not long after the development of native chemical ligation in the laboratory, related chemistry was discovered in a select set of proteins that undergo a splicing process analogous to mRNA splicing. By analogy with RNA introns, the excised portions of self-splicing proteins are referred to as **inteins**. Protein splicing proceeds through thioester intermediates much like the native ligation chemistry (**Figure 5.33**). Self-splicing leads to proteins with sequences that do not exactly match mRNA or gene

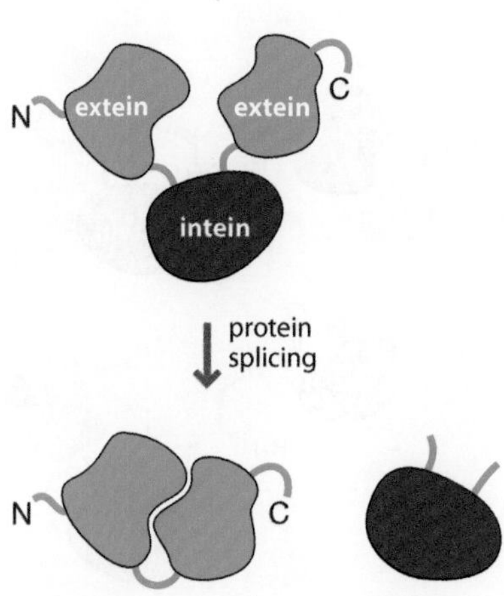

Figure 5.33 Self-splicing proteins. Inteins are removed from the middle of fully translated protein chains.

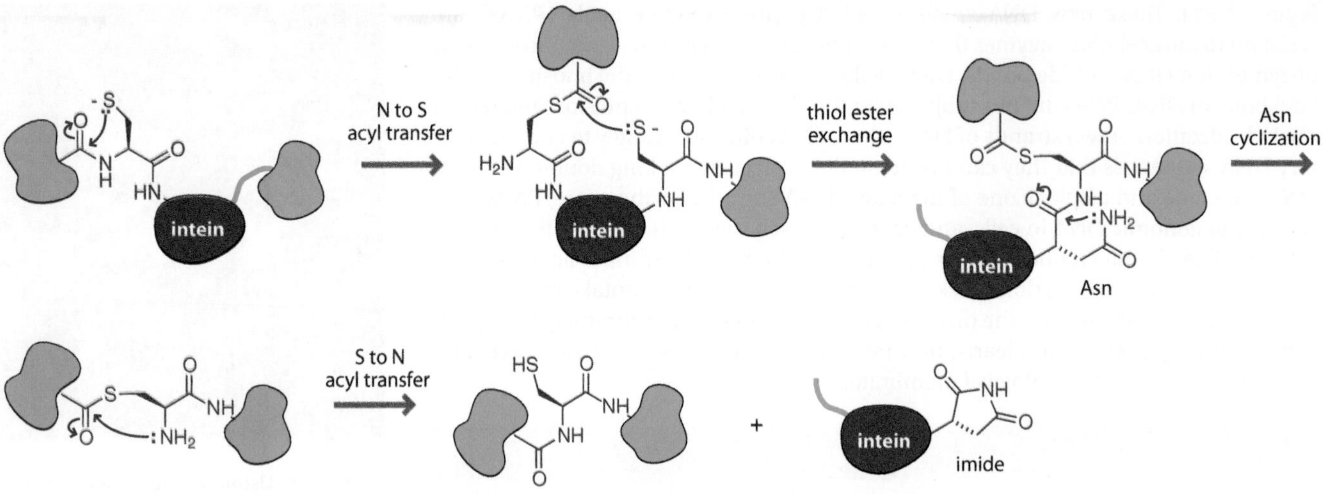

Figure 5.34 Thiol esters are key intermediates in the mechanism for intein splicing. An asparagine side chain attacks the amide backbone, cleaving the intein and revealing a free α-amino group that undergoes the final S to N acyl transfer.

sequences. As shown by the mechanism for the reaction, the reaction takes advantage of the facile exchange of thioesters. Residues flanking the intein also participate in this reaction (**Figure 5.34**).

A method called expressed protein ligation further breaks down the barrier to obtaining large proteins with unnatural elements (**Figure 5.35**). This method uses expression in *E. coli* to synthesize most of the target protein. However, the protein is expressed as a fusion or covalent attachment to an intein, which is also fused to a chitin-binding domain to simplify the purification of the recombinant protein. Beads with chitin attached will preferentially trap only the recombinant protein, and all other proteins produced by the *E. coli* are washed away. The intein drives an N to S acyl transfer of the bead-attached protein. Addition of thiophenol can release the protein as a phenyl thioester. This thioester-modified protein is then ideally suited for native chemical ligation to a synthetic peptide with a cysteine residue at its N terminus. The ligation approach makes it possible to produce very large proteins with chemically synthesized peptides incorporated at the protein's C terminus.

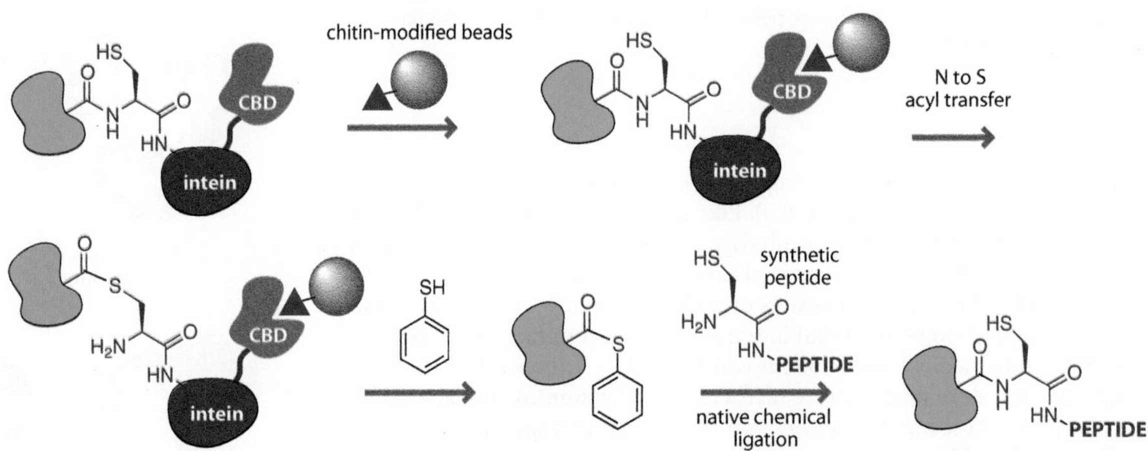

Figure 5.35 Expressed protein ligation for the synthesis of large semisynthetic proteins. Following this procedure, a synthetic peptide is attached to the C terminus of the resultant protein and can include unnatural amino acids or other modifications. CBD, chitin-binding domain.

5.3 FUNDAMENTAL FORCES THAT CONTROL PROTEIN SECONDARY STRUCTURE

Secondary structure involves different patterns of hydrogen bonding between backbone amides

The primary structure—that is, the sequence of amino acid residues—largely determines the three-dimensional structure and function of the protein. The secondary structure describes the canonical, most common folds of a protein—that is, the α helices, β sheets, and turns. Even in aqueous solution, the protein backbone is preordained to form intramolecular hydrogen bonds by two distinctive effects, namely intramolecularity and hydrophobicity, both of which are entropic effects. Each backbone amide is tethered to its neighbor by a single carbon atom. Thus, the peptide backbone can provide an entropic discount to intramolecular hydrogen bonds. In contrast, positioning a water molecule to form an intermolecular hydrogen bond with a backbone amide imposes a non-negotiable entropic cost. Hydrophobicity, in contrast, tends to minimize the volume of the protein in the same way as an oil droplet in water. Hydrophobic collapse brings together the backbone amides and further facilitates backbone hydrogen bonding. Hydrophobic collapse is widely regarded as the major driving force in protein folding, but hydrophobic interactions have little predictive value because they also drive protein misfolding.

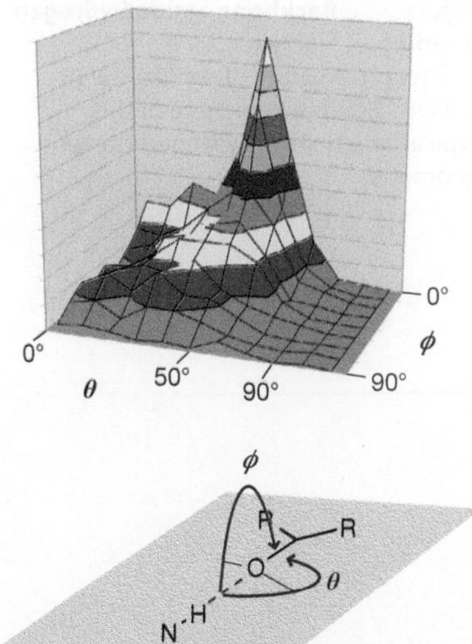

Figure 5.36 Angles of hydrogen bonds found in crystal structures. The analysis of crystal structures reveals strong preferences for hydrogen bonding between N-H and C=O groups.

The strength of a hydrogen bond depends strongly on geometry. Before we discuss intramolecular hydrogen bonds in proteins, we should establish an idealized image of a hydrogen bond between two amides. Ideal hydrogen bonds prefer linear N–H···O angles and distances of around 1.93 Å. A statistical analysis of intermolecular hydrogen bonds in the crystal structures of small organic molecules revealed that C=O···HN hydrogen bonds tend to lie in the plane of the carbonyl. The analysis also revealed that C=O···HN angles of less than 110° are strongly disfavored (**Figure 5.36**). A similar conclusion would be reached by analyzing the negative electrostatic potential for an amide, which is centered on the carbonyl oxygen (**Figure 5.37**). Both the empirical analysis and the calculated electrostatic potential correlate reasonably well with the traditional depiction of sp^2 lone pairs.

Figure 5.37 Electrostatic potential maps help explain hydrogen-bond geometry in amides. The traditional depiction of carbonyl lone pairs from valence shell electron pair repulsion theory as rabbit ears (left) correlates with an electrostatic potential map (right) at bonding (1 Å) distances but not at hydrogen-bonding (2 Å) distances.

Gas-phase calculations performed on the (unsymmetrical) formamide dimer agree with the trends observed in crystal structures—namely, ideal hydrogen bonds prefer to be linear, in the plane of the carbonyl, with an optimal C=O···H angle of 112° (see Figure 5.36). Nonlinear N–H···O angles down to 160° are nearly as favorable as angles of 180°, but the stability quickly decreases below 150°. The stability of amide hydrogen bonds is sensitive to deviations from the plane of the carbonyl. Hydrogen bonds with a 90° N–C=O···H torsion angle are over 2 kcal mol^{-1} less stable than in-plane hydrogen bonds, even when other parameters are optimized. A hydrogen bond with a C=O···HN angle of 112° is about 1.8 kcal mol^{-1} more favorable than a hydrogen bond with a 180° C=O···HN angle, so distortions toward acute angles are energetically costly. Although hydrogen bonds tend to align with oxygen lone pairs, geometries are limited by sterics; these steric costs are underestimated by studying formamide instead of true peptides (**Figure 5.38**).

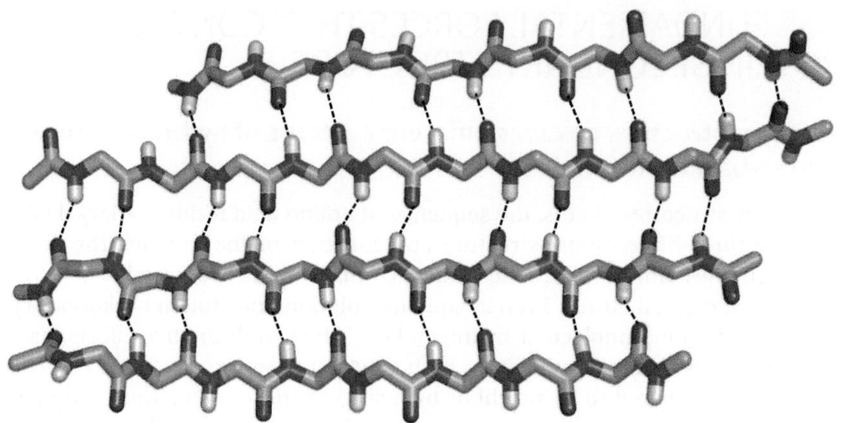

Figure 5.38 Backbone amide hydrogen bonding. A highly accurate neutron diffraction structure of concanavalin A (PDB 1XQN) reveals some of the typical variations in hydrogen-bonding geometry.

α Helices allow effective hydrogen bonding between neighboring amide N–H and C=O

In theory, the C=O of every residue could form a hydrogen bond to the N–H of the same residue, called the *i*th residue (**Figure 5.39**). Unfortunately, the resulting hydrogen bonds are bent and necessarily weak, because the resultant N–H···O angle deviates significantly from linearity. According to computational modeling, each of these distorted hydrogen bonds is worth 4.7 kcal mol^{-1} less than a corresponding hydrogen bond in a β sheet. Not surprisingly, these extended conformations are not observed in aqueous solution. Furthermore, it is not possible to form a hydrogen bond between the C=O of the *i*th residue and the N–H of the $i + 1$ residue, because they are each joined by a peptide bond. It is possible, however, to form hydrogen bonds between the C=O of the *i*th residue and the N–H of the $i + 2$ residue, resulting in a serpentine conformation of the backbone chain. This hydrogen-bonding pattern is particular to isolated γ turns but is not found as a repetitive secondary structure motif.

By twisting a peptide into a right-handed helix, it is possible to form hydrogen bonds between every *i*th and $i + 3$ residues, thus creating a strained structure called a 3_{10} helix (the subscript 10 clarifies the number atoms in the ring formed by hydrogen bonding). Not surprisingly, 3_{10} helices rarely exceed more than a single turn composed of four or five residues. Hydrogen bonds between every *i*th and $i + 4$ residues result in a much more stable and common helix, called the α helix. Every two turns of the α helix lines up the amino acid functionalities; thus, the helical repeat of the α helix is seven residues every two turns. Notably, it takes about seven turns (21 amino

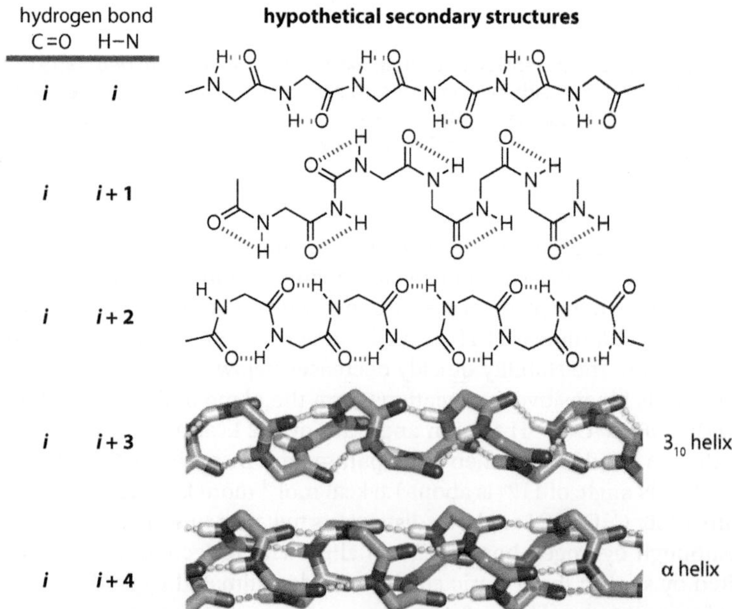

Figure 5.39 Repeating patterns of hydrogen bonds. Hydrogen bonding between the C=O of every *i* residue and the N–H of the $i + 4$ residue generates an α helix.

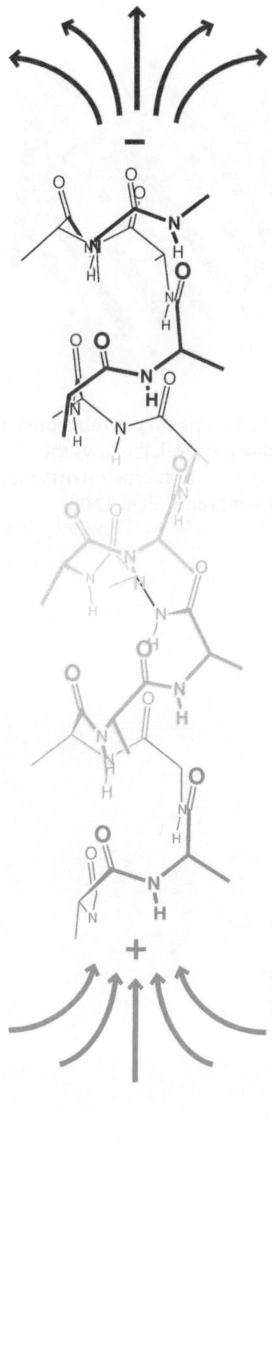

Figure 5.40 The helical macrodipole. The individual dipoles of the carbonyl groups sum up to generate a larger electrostatic field.

acid residues) to span a lipid bilayer, which is a very common arrangement for transmembrane proteins. The side chains on an α helix are barbed toward the N terminus, and all carbonyls point toward the C terminus.

Some amino acids do not cooperate with the steric requirements for the formation of an α helix. For example, proline forces turns and is a notorious α-helix breaker. In addition, the backbone nitrogen of proline is unavailable for the requisite hydrogen bonds; despite this, proline is often found within the first turn (from the N terminus) of an α helix. Moreover, the β-alkyl groups of the β-branched amino acids valine, isoleucine, and threonine can clash with the peptide backbone of the α helix.

The α helix is highly polarized, because all of the carbonyl groups from each amino acid point in the same direction (**Figure 5.40**). The additive effect of the carbonyl dipoles causes the ends of an α helix to "feel" the effects of roughly half of an unbalanced charge. The N terminus of the α helix is positive, and the C terminus is negative. Enzymes and other proteins sometimes point the N termini of α helices into active sites to help stabilize negative transition states (**Figure 5.41**).

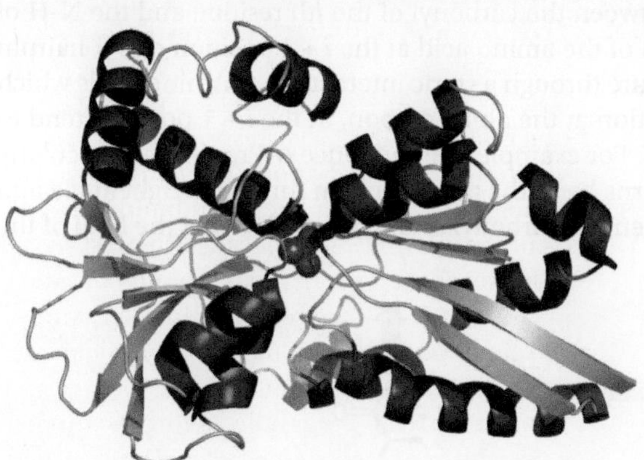

Figure 5.41 Proteins use helical macrodipoles for binding. The binding of SO_4^{2-} (red) to sulfate-binding protein (PDB 1SBP) is stabilized by the macrodipoles of three α helices rather than by positively charged groups.

β Sheets satisfy hydrogen bonding by backbone amides with linkages between different strands

Extended peptide chains present edges that can hydrogen bond in a parallel or antiparallel fashion (**Figure 5.42**). Antiparallel β sheets are more common than parallel β sheets, probably because of their more stable and more linear hydrogen bonds. In contrast with α helices, β sheets tend to favor β-branched amino acids such as valine, isoleucine, and threonine. Proline disrupts β-sheet formation, just as proline is bad for α helices.

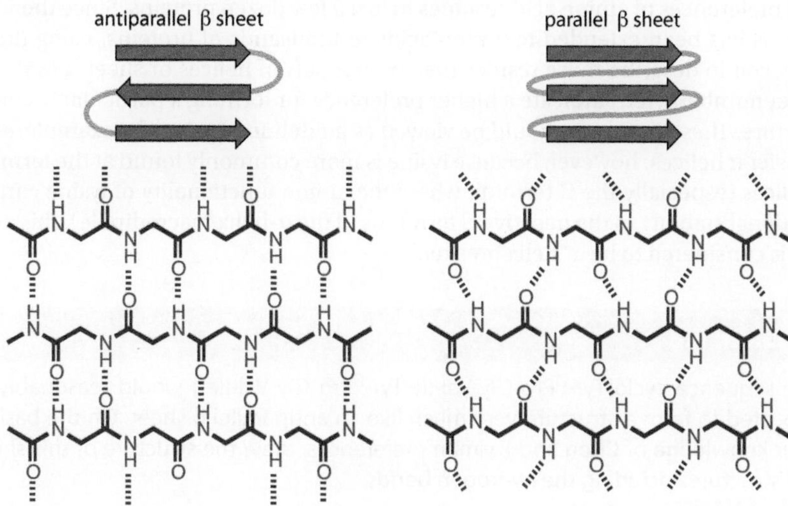

Figure 5.42 Hydrogen bonding in β sheets. The hydrogen-bonding patterns in antiparallel and parallel β sheets are distinctive.

Figure 5.43 This porin protein consists of a massive β barrel. It allows the selective passage of sucrose across the bacterial membrane. (PDB 1A0S)

β sheets are not flat. They often have a curved surface that allows them to fit snugly against α helices in the same way that a hot dog bun envelops a hot dog. The **β barrel** is a common motif for folded proteins, so named because the strands of the β sheet align like the staves of a wooden barrel (**Figure 5.43**). β sheets have large exposed surfaces that promote aggregation. Proteins that misfold into β sheets often aggregate through edge-to-edge and face-to-face interactions. The protein Aβ, for example, which is associated with Alzheimer's disease, forms insoluble aggregates that are visible by microscopy.

Turn structures have minimal hydrogen bonding between backbone amides

For proteins to adopt globular conformations, the peptide chain must change directions. The points of directional change are referred to as turn structures (**Figure 5.44**). These turn structures are often found on the exterior of proteins, and have unsatisfied hydrogen bonds. These features make them common sites for protein binding interactions, in which hydrogen bonds to the backbone can contribute to the interaction. The most abrupt turn motif is the β turn, sometimes referred to as a **β hairpin**. It involves a hydrogen bond between the carbonyl of the *i*th residue and the N–H of the $i + 3$ residue. The side chain of the amino acid at the $i + 1$ position of a β hairpin generally destabilizes the structure through a steric interaction. D-Amino acids which have a non-natural R configuration at the alpha carbon, at the $i + 1$ position tend to generate more stable β hairpins. For example, the sequence D-Pro-Gly is particularly good at favoring β hairpins. γ turns bend the peptide chain through a right angle and involve a hydrogen bond between the carbonyl of the *i*th residue and the N–H of the $i + 2$ residue.

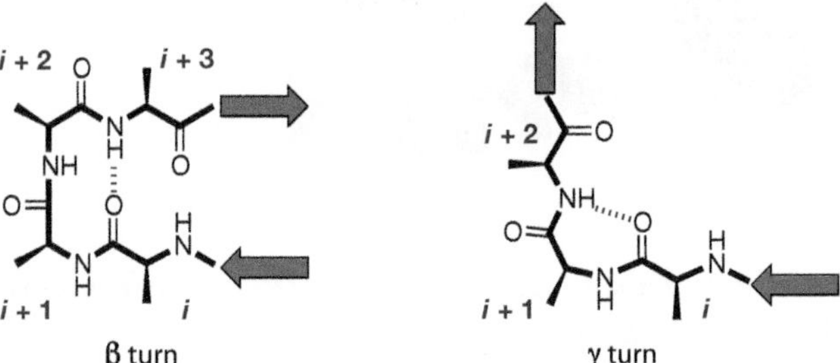

Figure 5.44 Hydrogen bonding in turn structures. Turn structures that allow the peptide backbone to change directions.

By examining the large numbers of X-ray crystal structures of proteins, P.Y. Chou and G.D. Fasman calculated the statistical likelihood of finding a given amino acid in helices, sheets, and turns. Chou and Fasman originally analyzed the secondary structural preferences of amino acid residues in just a few dozen proteins. Since then, their analysis has been extended to amino acids in thousands of proteins, using the category coil to describe those resides that are not part of helices or sheets (**Table 5.4**). Larger numbers (red) indicate a higher preference for forming a particular secondary structure. These numbers should be viewed as guidelines. Lysine, for example, seems to prefer α helices; however, because lysine is more commonly found at the termini of α helices (especially the C termini, where the amine functionality of lysine can lend additional stability to the negative C terminus of the α-helix macrodipole), this amino acid is considered to be a "helix breaker."

Problem 5.8

The sequence cyclo[Tyr-^{D}Pro-Gly-Val-Ile-Tyr-^{D}Pro-Gly-Val-Ile-] would reasonably be expected to form a structure very much like an antiparallel β sheet. On the basis of your knowledge of Chou and Fasman preferences, draw the structure of this sheet-like structure, including the hydrogen bonds.

Significant efforts have been invested in identifying short peptides that can form β hairpins in which two peptide strands adopt stable antiparallel β structure using naturally occurring amino acids. The most notable successes have come from peptides that take advantage of clusters of hydrophobic amino acids. For example, clusters of tryptophan side chains, called tryptophan zippers, provide enhanced stability to short peptides that form hairpins even in aqueous solution.

Rotation about substituted ethanes, butanes, and pentanes reveals the fundamental forces dictating protein folding

To better understand the forces that control the individual torsion angles in proteins, it is instructive to examine much simpler models first. The barrier to rotation about the C-C bond in ethane is about 3 kcal mol^{-1}, which is usually attributed to the high energy of the eclipsed conformation. The classical notion of sterics fails to account for this energy difference, however. The hydrogen atoms are too small and too far apart to clash sterically in the eclipsed ethane. Orbital interactions provide an explanation. The donation of a filled σ_{C-H} orbital into a σ^*_{C-H} should contribute a stabilizing effect, and we can find six of these potential interactions in ethane. Collectively, this sixfold donation of filled orbitals into unfilled orbitals provides 2.6 kcal mol^{-1} of the stabilization energy associated with the staggered isomer. Because the stereochemical alignment is important, we will refer to this donation as a **stereoelectronic effect**. In the eclipsed conformation, stereoelectronic effects strongly disfavor the eclipsed conformation of ethane, because the alignment forces filled σ_{C-H} orbitals to interact with filled σ_{C-H} orbitals.

Admittedly, a C-H bond is a rather poor nucleophile, and a C-H σ* orbital is a rather poor acceptor. However, if we replace two of the C-H bonds with C-O bonds, as in ethylene glycol, the effect is surprising. Although dipole repulsion favors the anti conformation in glycol, the stereoelectronic effect (two σ_{C-H} orbitals donating into two σ^*_{C-O} orbitals) wins out in polar environments, favoring the gauche conformation. Thus, **PEG**, a polymer commonly used to improve the serum half-life of protein pharmaceuticals and to induce the fusion of mammalian cells, tends to adopt a helical conformation (**Figure 5.45**).

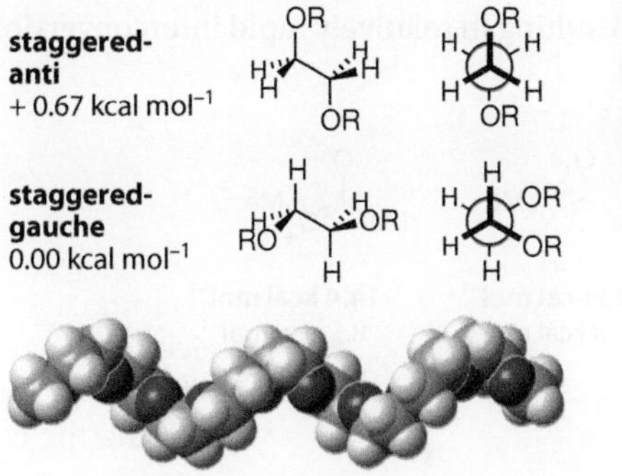

Figure 5.45 Glycols prefer gauche conformations over anti conformations. Calculations (left) reveal that 180° O-C-C-O torsion angles are disfavored relative to 60° O-C-C-O torsion angles. As a result PEG prefers a helical conformation.

Table 5.4 Conformational preferences of amino acid residues.

Residue	α helix	β sheet	coil
Ala	1.39	0.75	0.80
Glu	1.35	0.72	0.86
Leu	1.32	1.10	0.68
Gln	1.29	0.76	0.89
Met	1.21	0.99	0.83
Arg	1.17	0.91	0.91
Lys	1.11	0.83	1.00
His	0.92	0.99	1.07
Asp	0.89	0.55	1.33
Ser	0.82	0.85	1.24
Asn	0.77	0.62	1.39
Pro	0.50	0.44	1.72
Gly	0.47	0.65	1.62
Thr	0.76	1.23	1.07
Trp	1.06	1.30	0.79
Cys	0.74	1.31	1.05
Phe	1.01	1.43	0.76
Tyr	0.95	1.50	0.78
Ile	1.04	1.71	0.59
Val	0.89	1.86	0.64

(Data from S. Costantini et al., *Biochem. Biophys. Res. Commun.* 342:441–451, 2006.)

The effect that twists PEG into a helix also forces acetylcholine, an essential neurotransmitter, to adopt a gauche conformation (**Figure 5.46**). What is the effect of the positive charge on orbital energies? Atoms that have formal positive charges lead to much lower orbital energies. Thus, the σ^*_{C-N+} orbital is a better acceptor than a σ^*_{C-N} orbital. Donation from two σ_{C-H} orbitals into σ^*_{C-N+} and σ^*_{C-O} orbitals drives the preference for the gauche conformation of acetylcholine. This effect also explains why

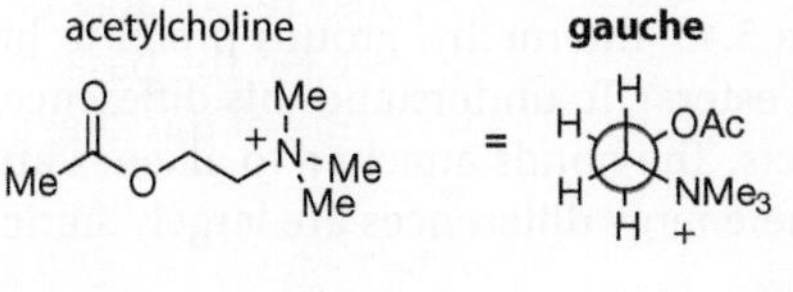

Figure 5.46 Acetylcholine prefers a gauche conformation.

protonated carbonyls react so much faster than free carbonyls; $\pi^*_{C=OH+}$ is much lower in energy than $\pi^*_{C=O}$. Stereoelectronic effects have a profound effect on carbohydrate structure and reactivity, and we return to these concepts in Chapter 7.

Staggered conformations are preferred in simple alkanes, but the energetic difference between the gauche and anti conformations is largely determined by sterics. The gauche conformation of butane is less stable than the anti conformation by 0.9 kcal mol^{-1}. This is an important number and can be used to judge the energetic differences in a variety of polycyclic hydrocarbons. The eclipsed conformation is over 4 kcal mol^{-1} less stable than either of the staggered conformations. These preferences dominate reactions that form carbon–carbon bonds. Transition states that must adopt eclipsed conformations will feel a significant amount of the 5.5 kcal mol^{-1} destabilization of eclipsed butane (**Figure 5.47**).

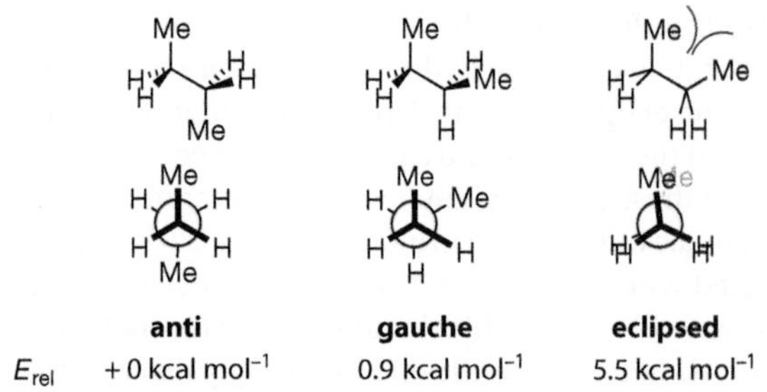

	anti	gauche	eclipsed
E_{rel}	+0 kcal mol^{-1}	0.9 kcal mol^{-1}	5.5 kcal mol^{-1}

Figure 5.47 Conformations of butane (substituted ethanes).

Stereoelectronic effects distinguish amides and esters from substituted ethenes

There is a sizable barrier (65 kcal mol^{-1}) to rotation around carbon–carbon double bonds, which prevents the interconversion of *E* and *Z* isomers. Resonance causes amides and esters to exhibit partial double-bond character about the acyl–nitrogen and acyl–oxygen bonds, respectively. The amide bond has almost 40% π character, leading to slow interconversion between the *cis* and *trans* isomers with a half-life of about an hour at physiological temperature. The double-bond character in an ester is substantially lower than that in an amide, resulting in relatively rapid interconversion between the *E* and *Z* isomers (**Figure 5.48**).

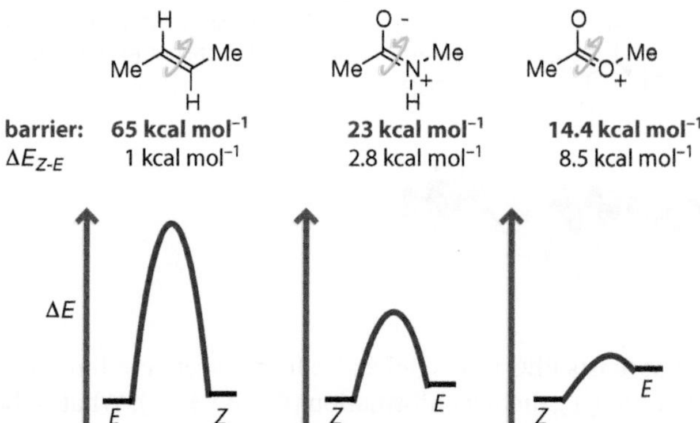

Figure 5.48 Rotational barriers of π bonds and partial π bonds in alkenes, amides, and esters. In amides and esters, the O$^-$ substituent has a higher Cahn–Ingold–Prelog priority than the methyl group, so that the "*trans*" isomer actually has a *Z* configuration.

The big surprise is that the energy difference between the *cis* and *trans* isomers should be so great. In all three cases in Figure 5.49, the methyl groups prefer to be *trans*. Why is the difference so much larger for esters? To understand this difference, we must pay attention to stereoelectronic effects. The bonds attached to alkenes are neither good acceptors nor good donors, so the energy differences are largely steric.

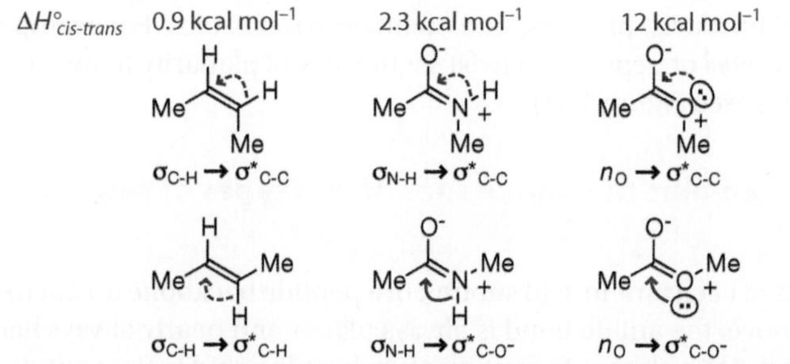

Figure 5.49 Why are *trans* amides and *trans* esters so much more favorable than *trans* alkenes? Donation of nucleophilic filled orbitals into unfilled orbitals stabilizes *trans* conformers more than *cis* conformers.

In contrast, the $\sigma^*_{C\text{-}O}$ orbital of esters and amides is a relatively good acceptor. Because an oxygen lone pair is a much better donor than an N–H sigma bond, the carboxyl lone pair prefers to be anti to the carbonyl sigma bond (**Figure 5.49**).

Interactions between allylic substituents and alkene substituents limit the conformation of substituted propenes

Allylic substituents direct conformations by exerting two possible types of strains, called $A_{1,2}$ strain and $A_{1,3}$ strain (**Figure 5.50**). $A_{1,2}$ strain discourages coplanarity between large functionalities on the allylic carbon (1) and the first carbon of the olefin (2). The effects of $A_{1,3}$ strain are more dramatic than those of $A_{1,2}$ strain. The cost of eclipsing a C–Me group and the hydrogen of a C=CHMe group is only about 0.9 kcal mol^{-1} (**Figure 5.51**). This energy is significantly lower than the 5.5 kcal mol^{-1} cost of eclipsing a C–Me group with a C–Me group in eclipsed butane, primarily because alkenes have spacious bond angles (120°) relative to alkanes (109.5°). However, the presence of a *cis* substituent (Me in Figure 5.51) on the olefin greatly magnifies the steric strain that distinguishes various allylic conformations. Any conformation that puts allylic substituents in the same plane as a *cis* olefinic substituent will be highly unfavorable.

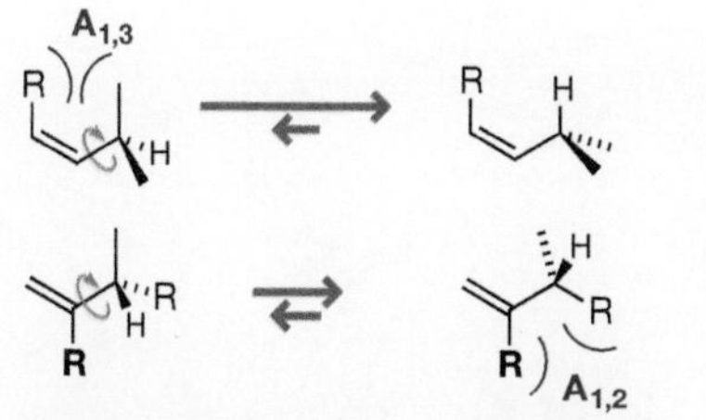

Figure 5.50 Two types of allylic strain, designated $A_{1,2}$ and $A_{1,3}$, have been defined for alkene substituents and disubstituted allylic groups. $A_{1,3}$ is the more significant.

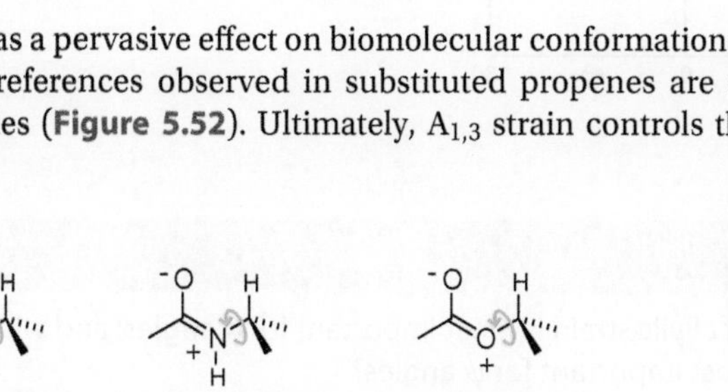

Figure 5.51 When allylic carbons are substituted, certain conformations are energetically disfavored. The effect of allylic strain is most significant when the alkene has a substituent that is *cis* to the substituted allylic carbon. Relative energies are in kcal mol^{-1}.

$A_{1,3}$ strain has a pervasive effect on biomolecular conformations. The same allylic conformation preferences observed in substituted propenes are also expressed in esters and amides (**Figure 5.52**). Ultimately, $A_{1,3}$ strain controls the conformations

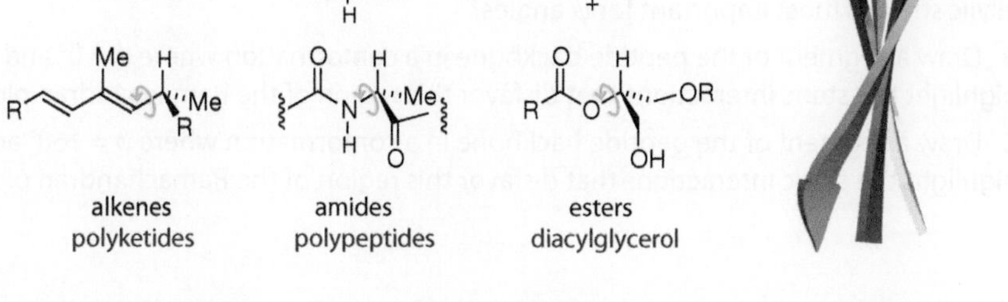

Figure 5.52 Allylic strain controls conformations in a wide range of biomolecules, including amides and esters, which have only partial double-bond character. The characteristic curvature of protein β sheets is attributable to $A_{1,3}$ strain. (PDB 2F73)

of many natural products, proteins, and membrane molecules. For example, the tendency of the ϕ angles of peptides to twist slightly out of planarity tends to impart curvature to β sheets (see Figure 5.52).

Allylic strain explains the dominance of two types of secondary structures

The three bonds of each amino acid subunit in a peptide backbone do not rotate freely. As described above, the amide bond is always planar and nearly always has the *trans* (*Z*) configuration. Thus, the conformation of each amino acid in the peptide backbone can be described in terms of only two torsion angles, designated phi (ϕ) and psi (ψ). These angles are defined as 180° in a fully extended peptide with antiperiplanar conformations about each bond (**Figure 5.53**). Chi (χ_n) angles are used to describe the torsion angles in the side chains. The bonds in the side chains tend to adopt extended conformations.

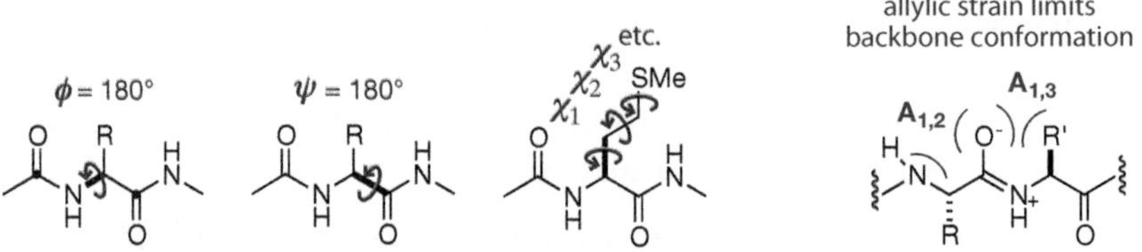

Figure 5.53 Variables used to describe torsion angles in proteins.

If all of the ϕ and ψ angles found in large numbers of different proteins are plotted on a two-dimensional map (called a **Ramachandran plot**), some regions are highly populated, whereas other regions are sparse (**Figure 5.54**). This is because the two types of allylic strain conspire to limit the available conformations. The two most heavily populated regions of the Ramachandran plot correspond to α helices and β sheets.

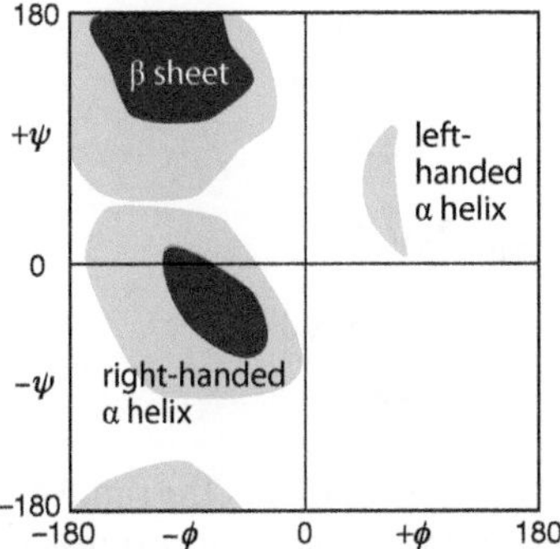

Figure 5.54 A Ramachandran plot labeled with characteristic regions of secondary structure. Blue areas indicate higher frequency of occurrence of particular ϕ and ψ angles in protein structures; the darker color indicates a larger number of examples.

Problem 5.9

A Which type of allylic strain is most important for ϕ angles; and which type of allylic strain is most important for ψ angles?

B Draw a segment of the peptide backbone in a conformation where ϕ = 0° and highlight the steric interactions that disfavor this region of the Ramachandran plot.

C Draw a segment of the peptide backbone in a conformation where ϕ = 160° and highlight the steric interactions that disfavor this region of the Ramachandran plot.

5.4 THE CHEMISTRY OF DISULFIDE CROSSLINKS

Cystine disulfides form readily under oxidative conditions

Cysteine thiols readily form disulfide crosslinks in proteins. The disulfide form of cysteine is referred to as **cystine**. All heteroatom–heteroatom bonds prefer 90° torsion angles: O–O bonds, N–O bonds, and the S–S bonds of cystines. The eclipsed and anti conformations of disulfides are strongly disfavored, by 14 and 7 kcal mol^{-1}, respectively. Disulfides are chiral and can adopt either +90° or –90° torsion angles; both are common in proteins (**Figure 5.55**).

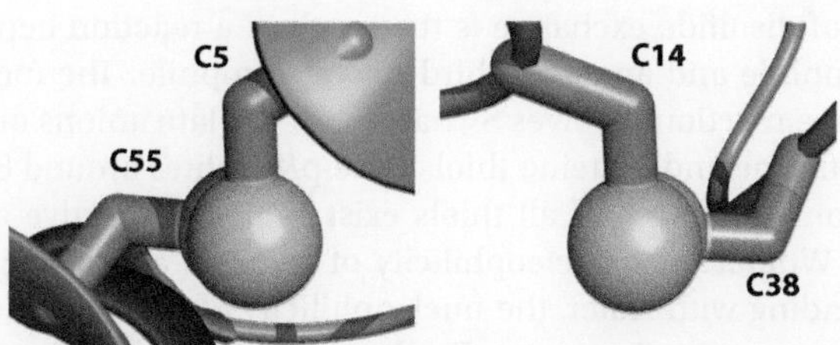

Figure 5.55 Right angles. Disulfides strongly prefer either –90° or +90° torsion angles. The C5–C55 disulfide in bovine pancreatic trypsin inhibitor (PDB 1BPI) adopts a –90° left-handed torsion angle, whereas the C14–C38 disulfide adopts +90° right-handed torsion angle.

All thiols readily form disulfides under other oxidative conditions (**Figure 5.56**). The cytoplasm of human and bacterial cells is a reducing environment and disfavors protein disulfide formation. Anywhere outside the cell, oxygen exerts a thermodynamic preference for disulfide formation. Free cysteine thiols are unusual in secreted proteins such as antibodies or in the extracellular domains of membrane-bound proteins. Peptides containing a cysteine will form disulfide bonds in the presence of air if left at room temperature overnight; the oxidation is even faster under basic conditions.

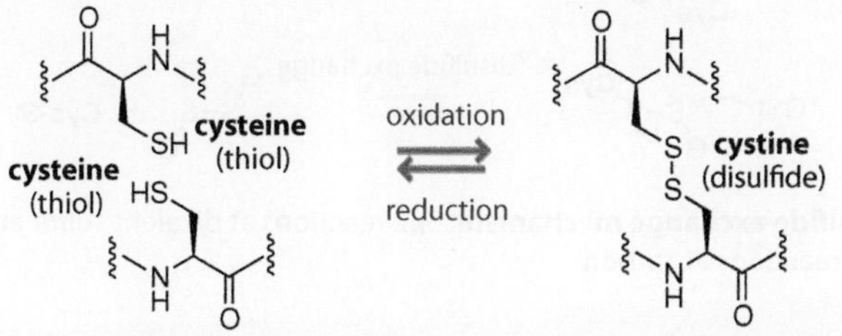

cysteine (thiol) + cysteine (thiol) ⇌ (oxidation / reduction) cystine (disulfide)

Figure 5.56 Cysteine undergoes facile oxidative dimerization to form cystine disulfide crosslinks. Reducing conditions can readily convert cystines into the free thiol-based cysteines.

Problem 5.10

One facile method for introducing disulfide bonds into a cysteine-containing peptide is to stir the peptide overnight in 20% (v/v) dimethylsulfoxide. Suggest a plausible arrow-pushing mechanism for this acid-catalyzed transformation.

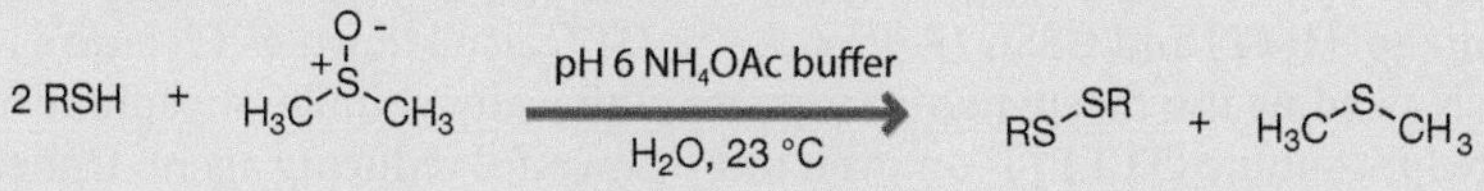

$$2\,RSH + H_3C{-}\overset{+}{S}(O^-){-}CH_3 \xrightarrow[H_2O,\ 23\ ^\circ C]{pH\ 6\ NH_4OAc\ buffer} RS{-}SR + H_3C{-}S{-}CH_3$$

Glutathione is an intracellular thiol buffer

Disulfides undergo rapid exchange with free thiols at physiological pH. Cells use thiols as convenient nucleophiles and as a buffer for soaking up oxidizing agents. Unfortunately, H_2S is too small and slippery for enzymes to hold. Instead, cells use glutathione as a steerable thiol with a unique handle (**Figure 5.57**). The handle is important for binding the cofactor to the enzyme, but not for its catalytic abilities. Reduced glutathione is commonly abbreviated **GSH**; in the oxidized, disulfide state, glutathione is abbreviated **GSSG**.

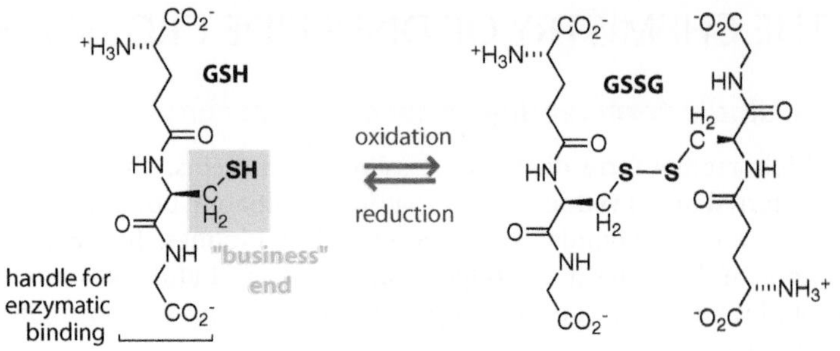

Figure 5.57 Glutathione redox equilibrium. Glutathione is an all-purpose reducing agent for the cell.

The facility of disulfide exchange is the result of a reaction between a powerful third-row nucleophile and an eager third-row electrophile. The mechanism for the disulfide exchange reaction involves S_N2 attack of thiolate anions on disulfides (**Figure 5.58**). Glutathione and cysteine thiols have pK_a values around 8.8. Under physiological conditions, up to 5% of all thiols exist as highly reactive and nucleophilic thiolate anions. Whereas the nucleophilicity of alkoxide anions is greatly tempered by hydrogen bonding with water, the nucleophilicity of thiolate anions is much less affected by an aqueous environment. Furthermore, S_N2 reactions at divalent sulfur atoms are extremely fast. In fact, sulfenic acids (RS–OH) are so susceptible to nucleophilic displacement that they cannot be isolated.

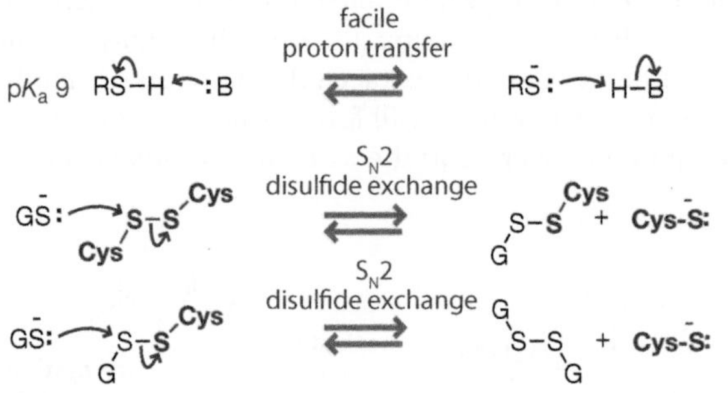

Figure 5.58 Disulfide exchange mechanism. S_N2 reactions at divalent sulfur are extremely facile, unlike S_N2 reactions at carbon.

Cystine disulfides in proteins are in equilibrium with glutathione disulfides

Native cystine crosslinks are not always thermodynamically favorable. For example, once the three disulfides of human insulin have been cleaved, the biologically active set of three disulfides cannot be selectively re-formed. Geometry places stringent demands on disulfide stability, but entropy has an equally important role. The role of entropy is exemplified by the favorable equilibrium for the reduction of GSSG by dithiothreitol (**DTT**) and GSSG (**Figure 5.59**). DTT, also known as Cleland's reagent, is usually used as the inexpensive racemate. The water-soluble phosphine tris-carboxyethylphosphine (**TCEP**) is an even more powerful reducing agent. Under aqueous conditions, the irreversible reduction of disulfides to thiols by TCEP is driven by formation of the phosphine oxide. TCEP is used widely to reduce disulfides in

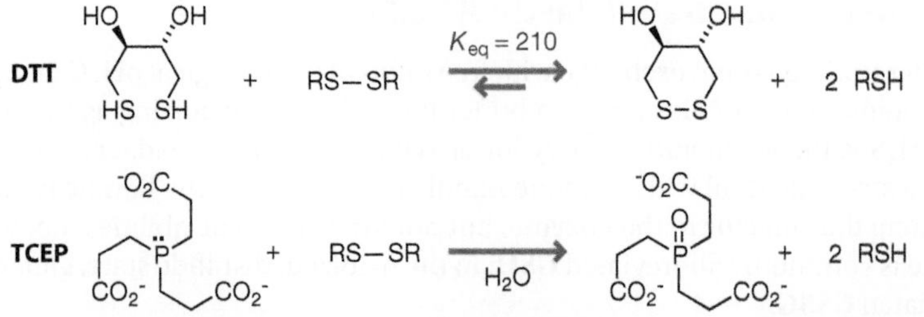

Figure 5.59 Chemical reduction of disulfides. In the laboratory, disulfide reduction can be entropically driven by the formation of a cyclic disulfide (with DTT), or enthalpically driven by the formation of a stable P=O bond (with TCEP).

Problem 5.11

Suggest a plausible arrow-pushing mechanism for the reduction of disulfides by phosphines in water. Assume that proton transfers occur through base catalysis.

$$R_3P + \text{Cys–S–S–Cys} \xrightarrow{H_2O} R_3P{=}O + 2\ \text{Cys-SH}$$

preparation for reaction with the resultant free thiol of cysteine. Unlike thiol-based reducing agents (such as DTT), TCEP does not add a competing nucleophilic thiol to the reaction conditions.

The equilibrium between intramolecular disulfide formation (for example in a protein) and GSSG is described by the equilibrium constant K_{ox}, defined below. Cysteine pairs with low K_{ox} values favor the free thiol form, whereas cysteine pairs with high K_{ox} values favor the cystine disulfide (**Figure 5.60**). The stability of protein disulfides varies widely. As shown in **Table 5.5**, some disulfides are unstable compared with GSSG (for example DsbA), whereas others are likely to form under all normal redox conditions (for example BPTI).

The facility of disulfide exchange, the geometric demands of the disulfide torsion angle, and the entropic requirement for proximity make it difficult to improve protein stability through the mutagenic incorporation of non-native cystine disulfides. Finding effective sites for stabilization by disulfide crosslinking is difficult and unpredictable. For each promising example of protein stabilization through disulfide introduction, there seems to be a disappointing counterexample of destabilization.

$$K_{ox} = \frac{[\text{PSSP}][\text{GSH}]^2}{[\text{PSH}][\text{GSSG}]}$$

Figure 5.60 K_{ox}, an equilibrium constant for glutathione-driven (GS) equilibration of free thiol and disulfide-bonded forms of a protein (P). When K_{ox} is large, the protein disulfide is stable in the presence of glutathione.

Table 5.5 Protein disulfide stabilities.

Protein	K_{ox} (M)
DsbA (reducing agent)	0.000008
DsbC (disulfide isomerase)	0.00002
Thioredoxin	10
Subtilisin C22-C87	120
RNase A C2-C10	1000
BPTI C14-C38	1500
BPTI C5-C55	11,000,000

Combinatorial crosslinking and protein misfolding can complicate attempts to produce disulfide-containing proteins

As proteins are translated by the ribosome, disulfides that form fastest are not necessarily those found in the native, active protein (**Figure 5.61**). As the number of disulfides in a folded protein increases, the number of potential incorrect disulfide isomers grows astronomically. Proteins that have the wrong disulfide crosslinks have little chance of adopting a correctly folded conformation. The refolding of tissue plasminogen activator from an aggregate formed in *E. coli*, called an inclusion body, requires the correct formation of 34 out of 35 cysteines to form 17 disulfide bonds. Only 1 in 6.3×10^{18} potential crosslinking combinations is correct. As long as the glutathione concentration is high enough to support rapid disulfide exchange, proteins can sample vast numbers of conformations on a short timescale; however, if the concentration of glutathione is too high, protein disulfides will not be stable. A class of enzymes called protein disulfide isomerases also helps to catalyze disulfide exchange, helping proteins adopt correctly folded conformations before aggregation occurs.

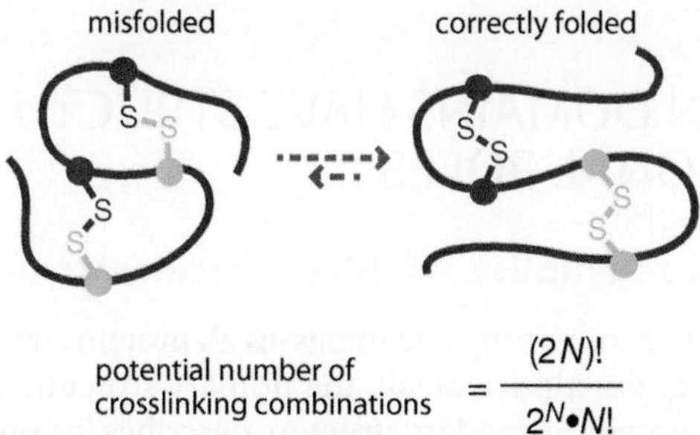

$$\text{potential number of crosslinking combinations} = \frac{(2N)!}{2^N \bullet N!}$$

Figure 5.61 If disulfide exchange is slow, proteins can become trapped in misfolded conformations as a result of the presence of incorrect disulfide crosslinks. For a protein with only $N = 2$ disulfides there are three potential crosslinking combinations.

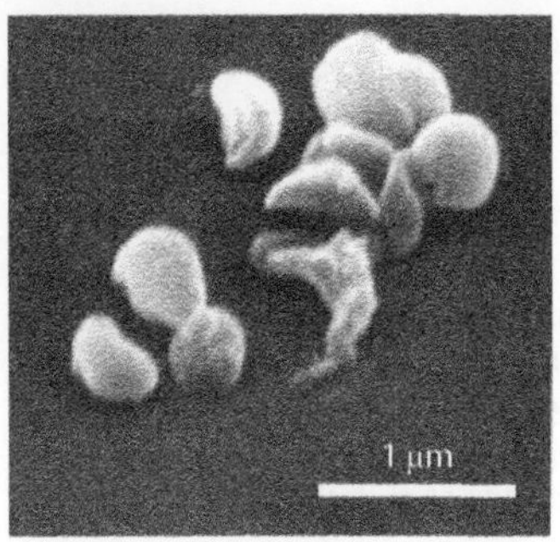

Figure 5.62 Production problems. Attempts to overexpress eukaryotic proteins in bacteria can lead to large insoluble protein aggregates called inclusion bodies. (From E. De Bernardez-Clark and G. Georgiou, *Protein refolding, ACS Symposium Series* 470, 1991. With permission from the American Chemical Society.)

Table 5.6 Empirically determined thiol ratios for protein folding.

Protein	[GSH] (mM)	[GSSG] (mM)
Bovine pancreatic trypsin inhibitor	10	0.1
Bovine pancreatic RNase	1	0.2
RNase A	2	1
E. coli alkaline phosphatase	<0.4	4–8
Bacterial TEM β-lactamase	<0.4	4–8

Heterologous expression (i.e., the production of a human protein in a bacterium) of disulfide-containing proteins in non-native organisms often leads to misfolded proteins. The problem is exacerbated in the case of overexpression, where intermolecular disulfide crosslinks can contribute to aggregation (**Figure 5.62**). In some cases, misfolded proteins can be refolded *in vitro*, but the thiol concentrations have to be determined empirically and can vary widely (**Table 5.6**). In general, extracellular eukaryotic proteins seem to fold best with glutathione concentrations around 1–5 mM and a [GSH]/[GSSG] ratio of about 1–3.

Concentrations of glutathione depend on location

The total glutathione concentration (that is, reduced + oxidized) in most cells is about 5 mM, but the *ratio* of GSH to GSSG varies markedly from 1:1 to 300:1 (**Table 5.7**). Glutathione buffering allows cells to control the degree of disulfide formation in nascent proteins. However, the concentration of glutathione depends on the location inside the cell. For example, proteins in the endoplasmic reticulum, which are destined for export outside the cell, face a more oxidizing environment. The cytoplasm of *E. coli*, moreover, is a more reducing environment than the cytoplasm of mammalian cells. As a result, cystine disulfides that are perfectly stable in mammalian cells may be disfavored in *E. coli*. The inability to form disulfides can make it difficult to express mammalian proteins in *E. coli*.

Table 5.7 Examples of cellular glutathione levels.

Cell	[GSH] (mM)	[GSSG] (mM)	GSH/GSSG ratio
Rat liver	11	0.04	300
Endoplasmic reticulum	2	1	1–3
Mammalian cytosol	2	0.05	30–100
E. coli cytoplasm	5	0.05	50–200

5.5 PROTEIN DOMAINS HAVE STRUCTURAL AND FUNCTIONAL ROLES

Biological protein assemblies exhibit hierarchical structures

A hierarchy of structural complexity encompasses all protein structure (**Figure 5.63**). At the simplest level of descriptive detail, the primary structure refers to the amino acid sequence of the protein. Secondary structure describes the protein's composition in terms of the number and location of α helices, β sheets, and turns. Earlier in this chapter we discussed the structures of amino acids and the secondary structure elements formed by short oligomers of amino acids. Secondary structures can wrap and

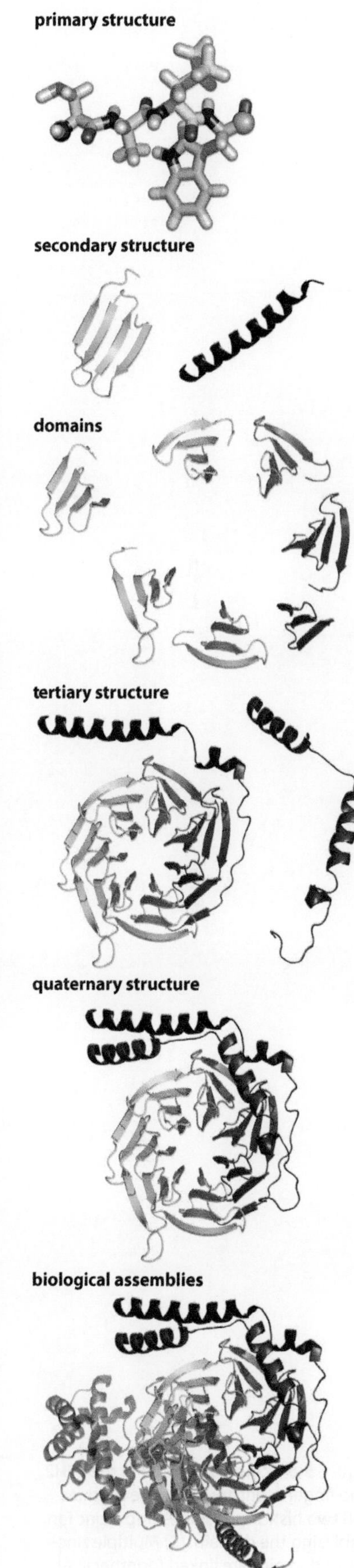

Figure 5.63 Hierarchical elements of protein structure. At the lowest level protein structure is defined by the sequence of amino acids. These generate elements of secondary structure such as helices and sheets, which interact to form domains. Multiple domains can interact to yield the independently folded tertiary structure of the protein chain. The quaternary structure of proteins is defined by the interaction of multiple folded protein chains. These fully functional quaternary structures interact with other structured macromolecules to form fully functional biological assemblies, such as the complex between a receptor and a heterotrimeric G protein. (PDB 1GOT)

fold on themselves to form the next level of complexity, called tertiary structure. In this book we use the word **peptide** to describe unstructured polypeptides; we use the word *protein* to indicate folded polypeptides with at least two elements of interacting secondary structure. Those interactions between α helices and/or β sheets ultimately lead to protein **domains**—small regions of proteins (typically with fewer than 100 amino acid residues) that are capable of folding independently when removed from the larger protein structure. The exceptions to the independent folding rule are short repeats (less than 50 amino acid residues), which stack together to provide stability to the multidomain unit. The **tertiary structure** of a folded protein chain can consist of a single domain or multiple domains.

Large proteins often consist of multiple independently folded peptide chains. This higher level of structural complexity, in which the interactions of the tertiary structures and domains form complex structural assemblies and nanometer-scale machines, is called **quaternary structure**. These quaternary structures then interact with each other either directly or indirectly to create the biological assemblies that are necessary for cellular, tissue, and organismal function.

The tertiary and quaternary structures of proteins access a wide range of different archetypal protein folds

Our discussion of protein structure will be focused mostly on protein domains. Protein domains are modular units of protein structure that consist of unique arrangements of α helices, β strands, turns, and loops. They sometimes fold into stable structures, even when they are clipped out of larger proteins. For example, the anti-inflammatory drug etanercept was engineered by joining two independent protein domains—namely, an immunoglobulin Fc domain (the constant fragment of an antibody) and an extracellular domain of the tumor necrosis factor receptor. Each domain of etanercept folds independently, and the entire folded conjugate dimerizes through the immunoglobulin domain to produce the active drug. Homologous recombination of genetic elements corresponding to protein domains has had a major role in protein evolution. Hence, protein domains tend to be reused for different functions. For example, a wide range of modern proteins harness the energy of ATP hydrolysis using a highly conserved nucleotide-binding domain, referred to at the genetic level as a Walker box. These nucleotide-binding domains drive membrane transport, microfilament assembly, and error checking during ribosomal protein synthesis.

Some of the most common protein domains serve primarily structural roles, such as those in collagen or those based on β sandwiches. Other highly common protein domains are functional, such as the seven-transmembrane domain and the protein kinase catalytic domain. Thousands of unique protein domains have been characterized and we cannot hope to discuss all of them. Instead, we will focus only on the most common protein domains in the human proteome. The 2011 Pfam database listed 4400 unique protein domains; the top 10 are listed in **Table 5.8**.

The list of most common protein domains varies from organism to organism, and even between strains of bacteria. Toward the end of the last century, our understanding of protein domains was naturally biased toward bacterial proteins because they were easier to express and crystallize than human proteins. Even though current efforts have shifted more toward mammalian proteins, the current set of protein structures is still skewed away from membrane-bound proteins, which are notoriously

Table 5.8 Common protein domains in the human proteome.

Domain	Occurrences[a]	Description
Zinc finger, C2H2 type	9398	α helix + loop + Zn^{2+}
Immunoglobulin domains	3563	Antiparallel β sandwiches
WD40 domain	2084	Twisted wedge-shaped β sheet
Fibronectin type III domain	1605	Antiparallel β sandwich
Cadherin domains	1012	Antiparallel β sandwich + Ca^{2+}
Seven-transmembrane domain	980	Seven α helices
Protein kinase	812	α helices with a β sheet lid
Sushi domain	773	β sandwich with disulfides
Collagen triple helix	751	Triple helix
RNA recognition motif	687	Two α helices + four-stranded β sheet

[a]As listed in the Pfam database (http://pfam.sanger.ac.uk), accessed on July 14, 2011.

difficult to crystallize. The statistical ranking of common human protein domains is likely to change as structural biologists extend their successes to the remainder of the human proteome. However, the emergent picture of domains as functional and structural modules will remain valid, even within an ever-changing zoo of protein structures.

Zinc-finger domains recognize DNA sequences

The most common domain in the human proteome, the "classical" zinc finger, is about 25 amino acid residues in length and is composed of an α helix and a peptide loop. The exceptional stability of the C2H2 variant of the zinc-finger domain arises from two cysteine residues on the loop and two histidine residues on the helix that maintain a tight grip on a zinc ion (**Figure 5.64**A). Recall from Chapter 4 that zinc-finger domains define an important class of transcription factors. The C2H2 zinc finger is just one of many domains that are important in DNA recognition. The predominance of DNA recognition domains indicates the importance of transcriptional control in human cells.

Each full-length zinc-finger transcription factor possesses multiple C2H2 domains. For example, each of the three zinc-finger domains of Zif268 recognizes a three-base sequence (Figure 5.64B). Six zinc-finger domains are needed to recognize

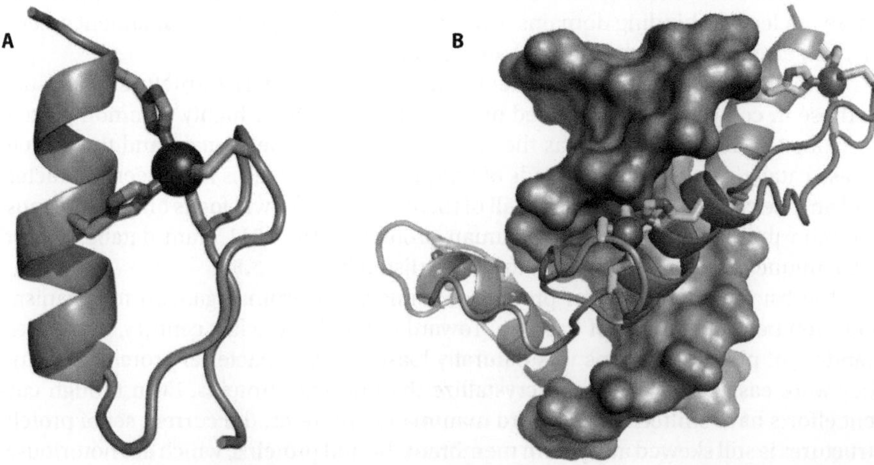

Figure 5.64 Zinc fingers. (A) In the C2H2 zinc-finger motif, two cysteine residues and two histidine residues grip a zinc ion, stabilizing the domain. (B) Multiple zinc-finger domains are linked together in a transcription factor. (PDB 1AAY)

a DNA sequence of about 18 base pairs, which defines a sequence long enough to be unique within the human genome. α Helices are just the right diameter to fit into the major groove of DNA. Zinc fingers recognize DNA through contacts between the helix and DNA bases. Zinc-finger domains have been extensively engineered and then strung together to create custom transcription factors and nucleases capable of targeting specific DNA sequences. In addition to binding to DNA, zinc-finger domains can sometimes interact with other proteins both in the presence and in the absence of DNA.

A number of common domains are based on β-sandwich architectures

Many of the most common human protein domains are composed of β-sheet sandwiches; examples are immunoglobulin domains, fibronectin type III domains, cadherin domains, and sushi domains. They vary in the number of β strands and in the order and topology of the connections between the β strands. Sandwiches of β sheets, called β sandwiches, also often occur as repeats, strung together like beads on a string.

Immunoglobulin domains are the most common β-sandwich domains. As suggested by their name, they are central to antibody structure (**Figure 5.65**), but are also present in other types of proteins. The binding specificity, customizability, and drug-like properties (low toxicity and long serum half-lives) of antibodies have led to their dominance of the present biopharmaceutical market. Recall from Chapter 1 that high-affinity type G immunoglobulins are composed of two heavy chains held together by a disulfide bond and two light chains (Figure 1.24). Each antibody chain is composed of nonidentical immunoglobulin domains connected by peptide loops. An immunoglobulin domain is composed of two face-to-face antiparallel β sheets. Typically, each sheet consists of three to four β strands. The two β sheets in an immunoglobulin domain are not independent. For example, if one labeled the β strands 1–7 in an immunoglobulin CH2 domain, starting from the N terminus, the bottom sheet would be composed of strands 1, 2, 5, and 4, and the top sheet would be composed of strands 3, 6, and 7 (see Figure 5.65). Thus, one can further categorize the nonidentical immunoglobulin domains on the basis of the topological connectivity of the individual β strands that make up each domain.

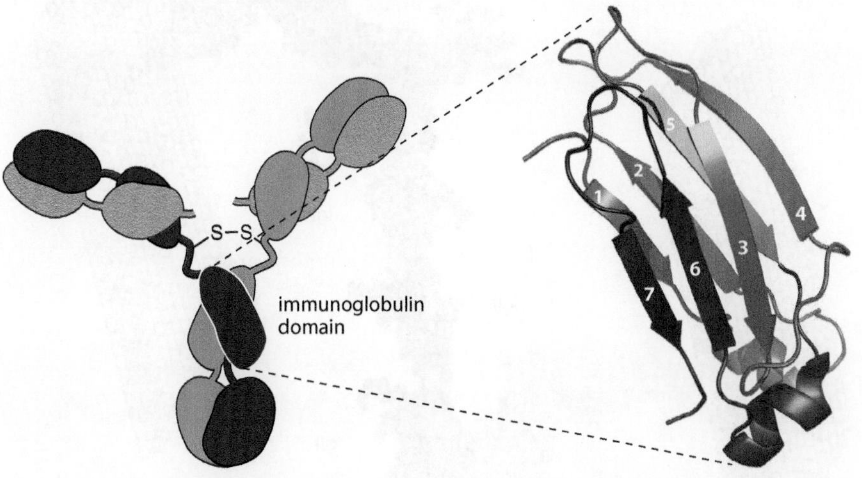

Figure 5.65 Immunoglobulin domains. Each of the four strands of a type G immunoglobulin is composed of eponymous immunoglobulin domains. Each domain is a sandwich of β sheets. The β strands 1–7 in each type of domain are connected through a distinctive topology. (PDB 1HZH)

Both electron microscopy and X-ray crystallography demonstrate the considerable flexibility of the hinge regions that connect the "arms" of immunoglobulins, which can accommodate a wide range of geometries and distances. In the hinges, long stretches of protein lack any secondary structure; they thus avoid conformation-limiting backbone hydrogen bonds. Such flexibility is especially useful for bidentate (two-pronged) binding to antigens in close proximity (for example, on the surface of a bacterial cell). The human immune system reacts violently to antibodies from mice, which are conventionally used to raise monoclonal antibodies (that is, antibodies with a single protein sequence). Thus, numerous techniques have emerged to "humanize"

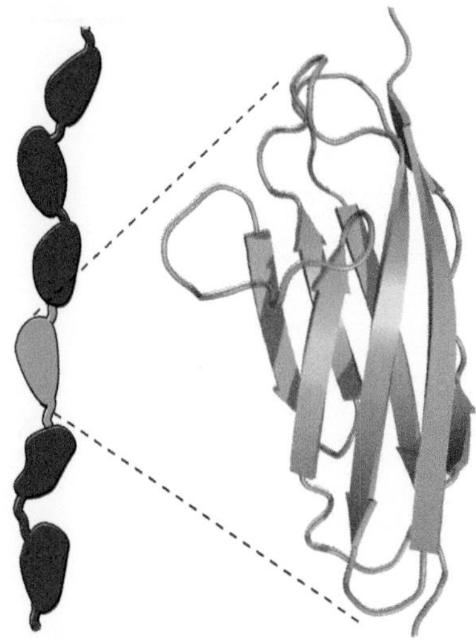

Figure 5.66 Muscle protein. The resiliency of the muscle protein titin at low loads is attributable to a string of up to 95 immunoglobulin domains. (PDB 3B43)

mouse antibodies by basically grafting complementarity-determining regions from mouse antibodies onto a human antibody framework.

Immunoglobulin domains are useful architectural subunits that are present in a wide range of other proteins. When strung together, multiple immunoglobulin domains can provide the optimal length to span precise distances or as expendable subunits in a resilient chain. For example, each domain of the muscle protein titin can unfold without allowing the protein backbone to snap (**Figure 5.66**). With a large number of built-in length extenders, titin is exceptionally stretchable yet strong.

The fibronectin type III (FNIII) domain is another antiparallel β-sandwich domain with a topological arrangement of β strands different from the immunoglobulin domains of antibodies. The fibronectin type III domain was first identified in the large extracellular protein fibronectin, which has a string of 16 FNIII domains. The FNIII domain is common among soluble extracellular proteins and the extracellular domains of membrane-bound proteins such as receptors (**Figure 5.67**).

Sushi domains are another type of β sandwich, alternatively known as short consensus repeats (on the basis of sequence homology) or complement control protein domains (on the basis of their functional context). They have been likened to a piece of Japanese nigiri sushi with a piece of fish on a box of rice. Each short consensus repeat (SCR) has about 60 residues folded into a β sandwich with two β strands atop three β strands. Two disulfide bonds hold together the SCR, a structural requirement that explains why the involved cysteine residues are conserved by essentially all SCRs. The SCR appears in many proteins responsible for regulating the complement immune response, and some complement proteins have strings of up to 30 SCRs. Typically, two to four SCRs are involved in protein–protein interactions, so the long strings of SCRs in complement proteins suggest the ability to recruit large numbers of proteins into oligomeric complexes.

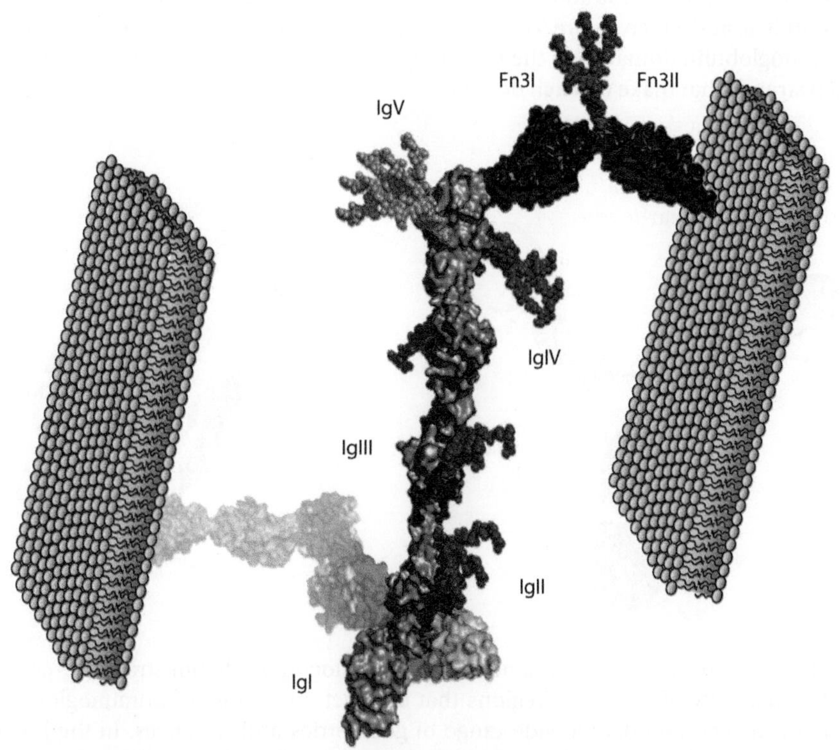

Figure 5.67 Immunoglobulin and fibronectin domains. The snake-like extracellular portions of olfactory cell adhesion molecules (OCAMs, one green and one blue) bind to each other via immunoglobulin domains, leading to cell adhesion. The domains of the blue OCAM molecule are labeled: immuglobulin domains (IgI-IgV), fibronectin type III domains (Fn3I and Fn3II). The red and orange tree-like structures are oligosaccharides. (Adapted from N. Kulahin et al., *Structure* 19:203–211, 2011. With permission from Elsevier.)

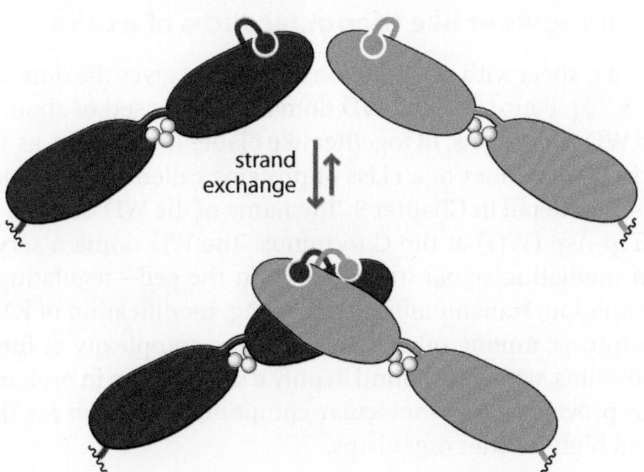

Figure 5.68 Strand exchange facilitates interactions between cadherin domains.

Calcium promotes interactions between cadherin domains

Another common β-sandwich domain in the human proteome, the cadherin domain, coordinates cell–cell adhesion through protein complexes anchored to the internal cytoskeleton of each cell. The name cadherin evokes its key structural feature and function—calcium and adhesion. The extracellular domains of C-cadherin feature five cadherin folding domains strung together, and these are anchored to the plasma membrane by a single α-helical segment. A fully folded domain on the cytoplasmic side of the membrane makes contacts with the actin cytoskeleton, responsible for structuring the cell. In C-cadherin, the domain farthest from the cell surface can dimerize with similar cadherin domains on other cells (**Figure 5.68**). When it does so, a conserved tryptophan residue from one cadherin domain fits into a hydrophobic pocket on the complementary domain, leading to an exchange of β strands—a process called strand exchange. Loss of cadherin on the surface of the cancer cells is associated with more aggressive, more invasive tumors. The cadherin domain requires bound Ca^{2+} ions for structural support; cadherin holds the calcium only weakly, because protein adhesion and function require millimolar concentrations of extracellular Ca^{2+} ions. The folded protein binds to the Ca^{2+} ions through chelation, largely by the carboxylate-bearing side chains of aspartic acid and glutamic acid (**Figure 5.69**).

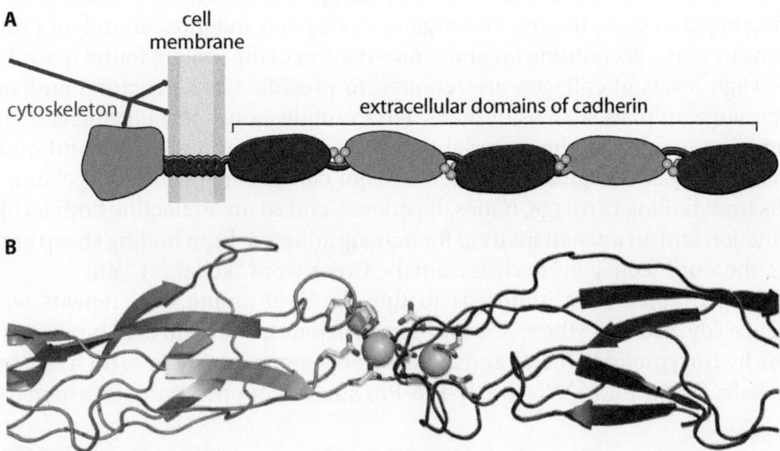

Figure 5.69 Cadherin domains. (A) The regions between cadherin domains rigidify in the presence of calcium ions (green). The rigidified protein transduces that information through the cell membrane to the actin scaffolding within the cell. (B) The three Ca^{2+} ions (green) at the domain interfaces are chelated by eight carboxylates and two carboxamides. (PDB 1L3W)

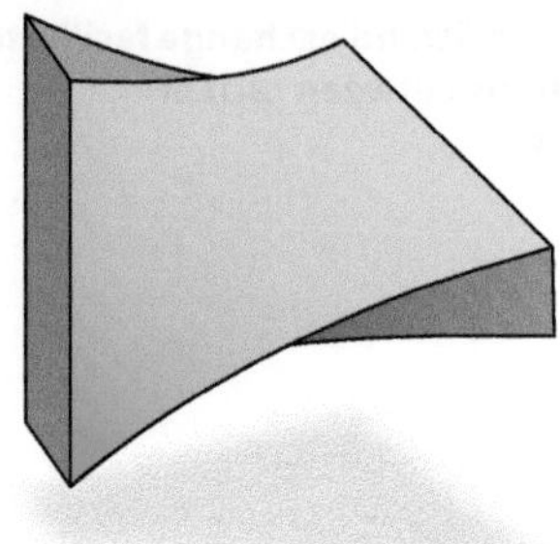

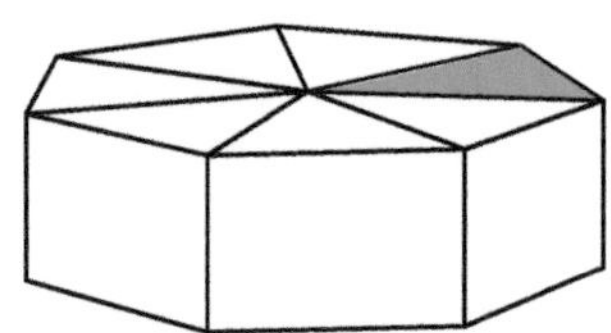

Figure 5.70 WD40 domains. Twisting of the β sheet gives the WD40 domain its wedge-like profile. The wedges fit together like slices of a cake.

WD domains fit together like triangular slices of a cake

A WD domain is a β sheet with an orthogonal twist that gives the domain a wedge-like profile (**Figure 5.70**). Four-stranded WD domains composed of about 40 amino acid residues, called WD40 domains, fit together like blades of a turbine, as shown in Figure 5.70. Transducin is a member of a class of proteins called heterotrimeric G proteins that are discussed in detail in Chapter 9. The name of the WD domain arises from the prevalence of Trp-Asp (WD) at the C terminus. The WD domain serves remarkably diverse roles in mediating signal transduction in the cell—regulating transcription, cell-fate determination, transmembrane signaling, modification of RNA, vesicle trafficking, and apoptosis, among other functions. The complexity of functions orchestrated by WD domains, which are found in only a simple form in prokaryotes, suggests that the protein provides a key molecular component required for the evolution of multicellular and higher-order organisms.

Problem 5.12

How many WD domains are present in the quaternary structure of the transcription factor Groucho/TLE (PDB 1GXR)?

Collagen is formed from a three-stranded helix

Mainly found in the extracellular matrix, collagen composes the connective tissue holding together cells, tissues, and organs. Collagen is the most abundant protein in humans by mass, accounting for about one-quarter of the protein found in the human body. High levels of collagen are required to provide a key structural and support role for large animals such as humans. For example, bones rely on collagen for their strength, and cartilage, the material that provides shape to the nose and outer ear, consists largely of collagen. Collagen is a major component of culinary gelatin, which results from boiling cartilage, bones, ligaments, and so on. Reflecting both its biological function and an ancient method for making adhesive from boiling sheep and oxen necks, the word "collagen" derives from the Greek word "kolla" for "glue."

Collagen consists of hundreds to thousands of amino acid repeats with the sequence Gly-Xaa-Yaa, where Xaa and Yaa designate any amino acid but usually proline or hydroxyproline. The strands intertwine spontaneously to form a left-handed triple helix (**Figure 5.71**). Each Gly-Pro-Pro subunit of one strand is fastened to the

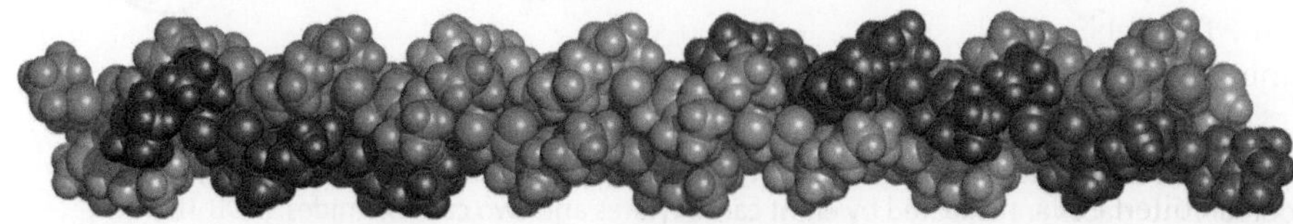

Figure 5.71 Collagen. Three protein strands intertwine in the collagen triple helix.

other two strands by a hydrogen bond (**Figure 5.72**). The glycine N–H donates a hydrogen bond to a proline carbonyl on the second strand and the second proline carbonyl accepts a hydrogen bond from the glycine N–H of the third strand. The unusual triple-helical structure of collagen relies on the very tight turns allowed by glycine and on the rigidification enforced by the abundant proline residues. Recall that the standard protein α helix consists of just a single strand. α helices can form bundles involving two or more helices; however, in those bundles, peptide backbones do not form hydrogen bonds between helices. Conversely, all of the hydrogen bonds in a collagen triple helix exist between different strands.

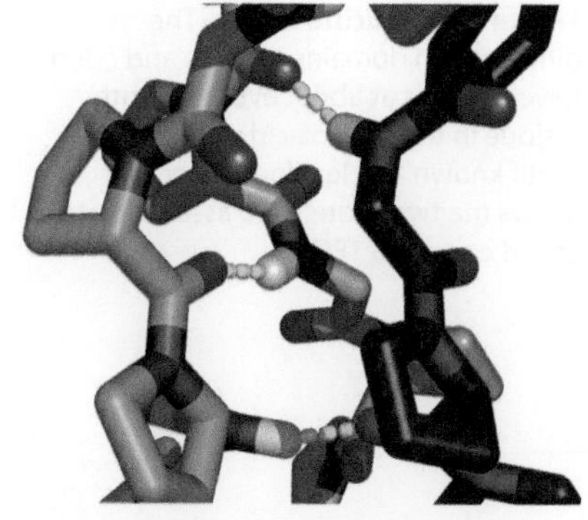

Figure 5.72 Interchain hydrogen bonds hold together the three strands of collagen. (PDB 1K6F)

Post-translational enzymatic modification of proline to form (*R*)-4-hydroxyproline stabilizes the collagen triple helix. Studies with 4-fluoroproline analogs confirm that the stabilization arises from a stereoelectronic effect and not because of hydrogen bonding from the introduced hydroxyl group. Five-membered rings lack the structural rigidity of a cyclohexane chair, but when (*R*)-4-hydroxyproline is part of a collagen triple helix, the hydroxyl group adopts a favorable gauche orientation relative to the ring nitrogen (**Figure 5.73**). In contrast, (*S*)-4-hydroxyproline, not found in collagen, leads to an unfavorable anti orientation of the hydroxyl group and the ring nitrogen.

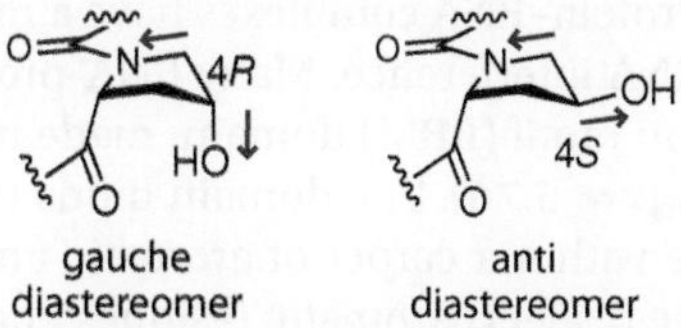

4*R* HO gauche diastereomer

4*S* OH anti diastereomer

Figure 5.73 Stereoelectronic effects in collagen. The gauche arrangement of electronegative atoms in (*R*)-4-hydroxyproline stabilizes the collagen triple helix. The anti arrangement of electronegative atoms in (*S*)-4-hydroxyproline destabilizes it.

The enzyme that catalyzes hydroxylation of collagen proline residues requires the cofactor vitamin C. Thus, too little vitamin C in the diet causes weakened collagen triple helices, which manifests as scurvy, a disorder associated with bleeding gums, loose teeth, and a violent, mutinous disposition. In the eighteenth-century British Navy, scurvy is credited with having caused more deaths than fighting with the enemy.

Protein kinase domains and seven-transmembrane domains have key roles in signal transduction

Protein kinase domains and seven-transmembrane domains are prevalent because they are common elements of the signal transduction pathways that control transcription in human cells. Protein kinases are among the most common enzymes in humans and their structures and mechanisms will be discussed in Chapter 6. **Seven-transmembrane (7TM) domain receptors** are characterized by the presence of seven α helices that sit within the cell membrane. Heptahelical is a more apt description for this important class of membrane-bound receptors because in this book, an α-helix is not defined as a protein domain. The helices of a 7TM receptor form a basket that is ideal for reversibly binding small molecules such as neurotransmitters and hormones and transducing the binding event into a conformational change within the cytoplasm. In the G protein-coupled receptor rhodopsin, the aldehyde ligand retinal is always bound to a lysine side chain of the receptor as an imine (**Figure 5.74**). Light isomerizes an olefin in retinal from a *trans* configuration to a *cis* configuration, and the resultant change in conformation is transmitted to the ends of the protein for communication with the cell. 7TM receptors interact with heterotrimeric G proteins, composed of WD domains, and the biological assembly for a 7TM receptor–G-protein complex was shown in Figure 5.63. 7TM receptors will be discussed extensively in Chapter 9, because they modulate pathways that control transcription in human cells.

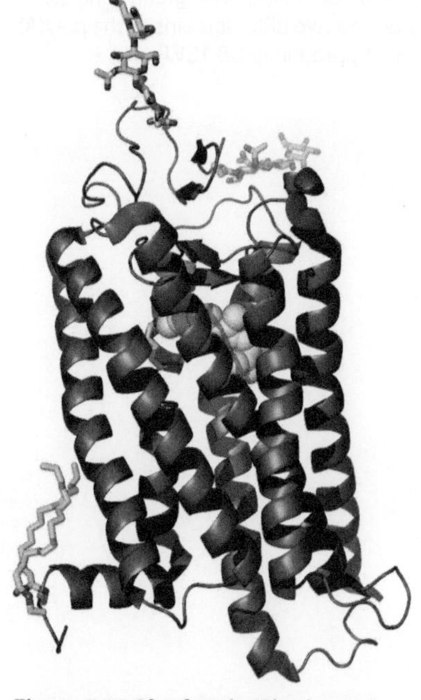

Figure 5.74 Rhodopsin. Rhodopsin is a 7TM receptor that is covalently bound to a light-sensitive carotenoid called retinal (yellow). Note the presence of two types of post-translational modifications: glycosylation of two extracellular loops and palmitoylation of two C-terminal cysteine residues as thioesters. (PDB 1GZM)

Protein domains composed of α-helical bundles are common among all organisms. The helices in these bundle domains tend to have long stripes of hydrophobic amino acids that run the length of the helix and drive protein folding in an aqueous environment. Four-helix bundles are particularly common but there is no intrinsic limit to the number of α helices that can associate side by side. Two-helix bundles, called coiled coils, need to be relatively long to fold independently. For example, the

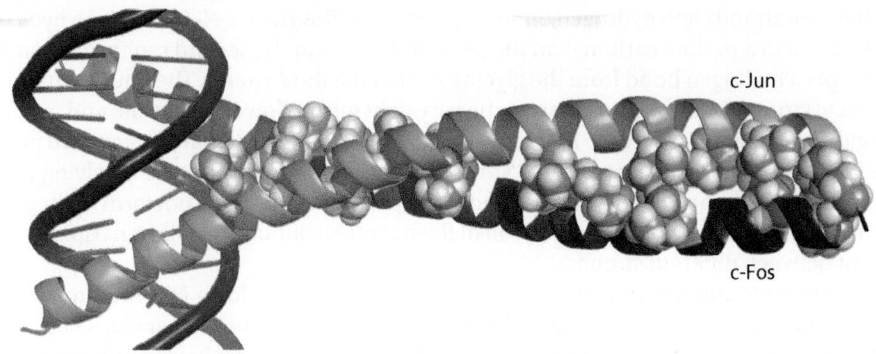

Figure 5.75 Leucine zipper. The dimerization domains of c-Fos and c-Jun have leucines at about every seventh residue in the amino acid sequence, a motif known as a leucine zipper that allows the two proteins to associate as a coiled coil. (PDB 1FOS)

DNA-binding domains of the transcription factors c-Fos and c-Jun (Figure 3.18 and **Figure 5.75**) are held together by a coiled coil of α helices.

The RNA recognition motif domain binds to single-stranded RNA

A variety of protein domains are specialized for binding particular classes of molecules, such as peptides, nucleotides, or sugars. Protein–RNA complexes have a major role in ribosomal structure, RNA splicing, and RNA interference. Many RNA–protein interactions are mediated by the RNA recognition motif (RRM) domain, made up of four antiparallel β strands and two α helices (**Figure 5.76**). The domain binds to an elongated strand of RNA. The RNA bases nestle within a carpet of aromatic amino acid side chains that coat the exposed side of the β sheet. Aromatic residues engage in π-stacking interactions; other residues make base-specific contacts that make each RRM domain specific for a different RNA sequence. Positively charged residues (arginine and lysine) make nonspecific contacts with the phosphate backbone.

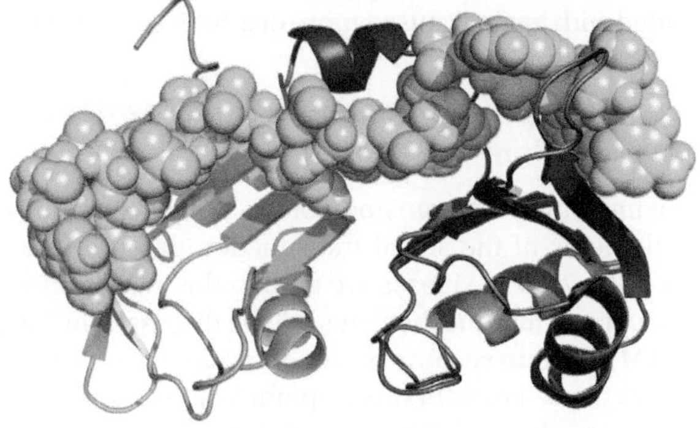

Figure 5.76 RNA recognition motif. A strand of poly(A) RNA (green spheres) bound to two RRM domains of the poly(A) binding protein. (PDB 1CVJ)

Peptide-binding domains can confer modular functions to proteins

The language of protein communication often involves the sequence-selective binding of short peptides to canonical protein motifs. SH2 (Src homology 2) domains recognize the addition of a phosphate group to tyrosine residues, a key event in human signal transduction (**Figure 5.77**). The domain consists of an antiparallel β sheet with one α helix on top and another α helix on the bottom. Like many peptide-binding domains, SH2 domains are often tethered together with other domains into much larger proteins. Phosphorylation of tyrosine residues on proteins with SH2 domains can often lead to homodimerization such that the phosphotyrosine on one protein binds to the SH2 domain on the other protein, and vice versa—like a molecular handshake.

Another modular unit for protein–protein binding, the SH3 domain, recognizes and binds specific short stretches of peptides within larger proteins (**Figure 5.78**). SH3 domains are β barrels. The preferred targets of SH3 domains are peptides with multiple proline residues, which adopt a characteristic helical conformation called a

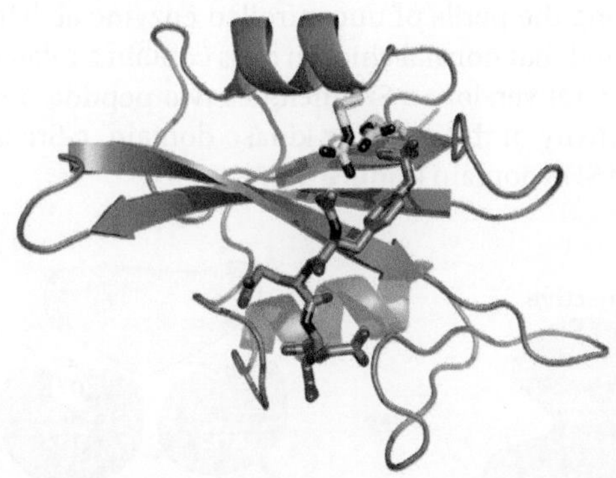

Figure 5.77 An SH2 domain. The SH2 domain (teal) in p56 Lck kinase binds to a peptide (blue) with a phosphotyrosine residue. The phosphate dianion engages in complementary electrostatic interactions with two positively charged arginine residues (yellow). (PDB 1LKL)

polyproline helix. Wound a bit tighter than the standard α helix, each turn of the polyproline helix is spaced three residues above the previous turn (as opposed to four in the standard α helix). The human signal transduction protein GRB2 is composed of one SH2 domain flanked by two SH3 domains. It binds to phosphotyrosine residues on the epidermal growth factor receptor and beckons proteins with cognate polyproline sequences.

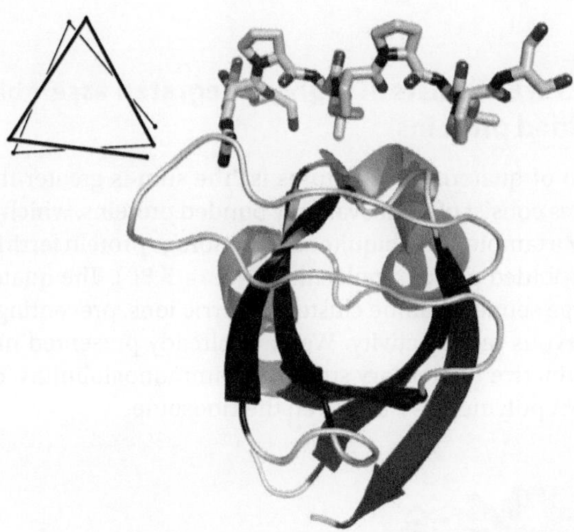

Figure 5.78 The SH3 domain. The SH3 domain of phosphoinositide 3-kinase (blue) is a β barrel that binds proline-rich peptides that form polyproline helices. The polyproline peptide adopts a helical conformation with a three-residue pitch (depicted at the top left). (PDB 3I5R)

5.6 HIGHER LEVELS OF PROTEIN STRUCTURE

The tertiary structure consists of one or more domains

The folded structure of an individual protein chain defines the tertiary structure. The structure can be a single domain; however, more often than not, it consists of multiple protein domains. In many cases the multiple domains merely reinforce a large, highly stable folded conformation. In other cases the domains add modular functions to the protein.

The tyrosine kinase Src provides a good example of a single-chain protein with multiple functional domains that must work together to regulate protein activity. The Rous sarcoma virus encodes a viral version of Src, which phosphorylates many proteins in infected chicken cells. The ensuing havoc leads to uncontrolled cell growth and tumors, illustrating the perils of uncontrolled enzyme activity. In the late 1970s investigators discovered that normal chicken cells contain a related kinase c-Src (cellular Src). Unlike the viral version, c-Src includes two peptide-binding domains that tightly control the activity of the catalytic kinase domain. c-Src is restrained by one SH2 domain and one SH3 domain (**Figure 5.79**).

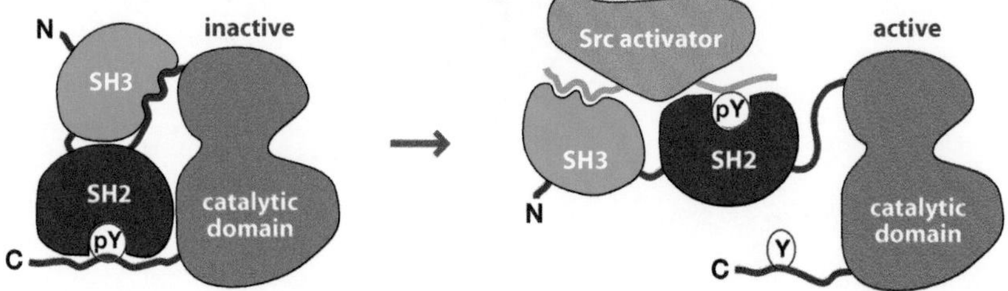

Figure 5.79 The catalytic domain of c-Src is sequestered by the internal binding of SH2 and SH3 domains. The catalytic domain can be released by dephosphorylation or by competitive displacement of the SH2 and SH3 domains by a Src activator.

The c-Src kinase domain is maintained in an inactive state through the binding of the SH2 domain to a C-terminal peptide with a phosphotyrosine residue at position 527. The inactive state is further stabilized by the weak binding of the SH3 domain to a flexible peptide linker that joins the SH2 domain to the catalytic domain. The catalytic domain can be unleashed either by dephosphorylation of Tyr527 or by a Src-activating protein with the right polyproline SH3 ligand and the right phosphotyrosine SH2 ligand.

Quaternary structure consists of highly integrated assemblies of independent, folded proteins

The unofficial motto of quaternary structures is "the sum is greater than the parts." Quaternary structures consist of noncovalently bonded proteins, which assemble into large complexes. For example, the ubiquitous iron-storage protein ferritin is a complex of 24 independently folded protein molecules (**Figure 5.80**). The quaternary assembly encapsulates large semicrystalline clusters of ferric ions, preventing the iron from engaging in mischievous redox activity. We have already presented many examples of proteins with distinctive quaternary structures: immunoglobulins, collagen, DNA gyrase, histones, DNA polymerases, and even the ribosome.

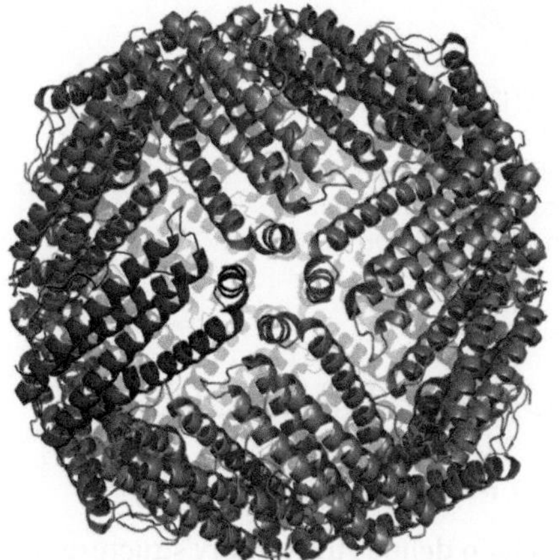

Figure 5.80 The quaternary structure of the iron-storage protein ferritin is composed of 24 identical four-helix bundles. Each ferritin complex (one highlighted in purple) can store around 4500 Fe^{3+} ions in the core. (PDB 1FHA)

Sometimes it is difficult to distinguish between quaternary structure, tertiary structure, and domain structure. For example, the collagen folding domain is composed of three independent strands and is, by definition, a form of quaternary structure. Collagen can be very long, with hundreds of Gly-Pro-Pro repeats in each strand. The bundling of these peptide chains into multistranded collagen fibrils involves organization at the level of domain interactions, tertiary folding, and quaternary assemblies (**Figure 5.81**). After translation of the collagen peptide into the endoplasmic reticulum, the C-terminal domain folds into a globular structure that trimerizes. The trimer serves to nucleate triple-helix formation, such that strands of the collagen helix are in precise registry. The *cis–trans* isomerization of the proline amide bond is the rate-determining step in triple-helix formation. Fortunately, prolyl *cis–trans* isomerase enzymes can offer roadside assistance as the collagen trimers are trafficked through the endoplasmic reticulum. As the collagen triple helices reach the Golgi network they aggregate into lateral aggregates and are then exported from the cell. Enzymes in the extracellular environment cleave the globular domains from the N and C termini, allowing the collagen helices to assemble irreversibly into resilient fibrils with each triple helix displaced by about one-quarter of its length along the fibril axis. The collagen fibrils are further strengthened by the enzymatic oxidation of lysine side chains, yielding covalent spot welds that hold together the macrostructure of collagen.

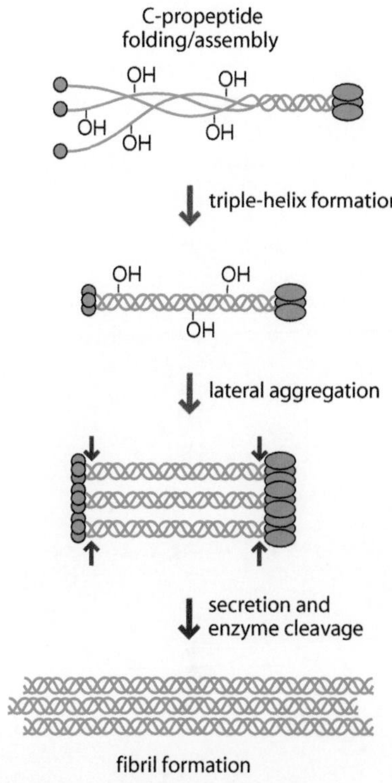

C-propeptide folding/assembly

OH OH OH OH OH

triple-helix formation

OH OH OH

lateral aggregation

secretion and enzyme cleavage

fibril formation

Figure 5.81 Controlled formation of quaternary structure. The stepwise assembly of collagen fibrils is dependent on tertiary and quaternary structure.

Problem 5.13

Lysyl oxidase catalyzes the enzymatic oxidation of lysine side chains, leading to imine intermediates.

A Suggest a plausible acid-catalyzed mechanism for the formation of an imine crosslink from an amine and imine functional group.

lysyl oxidase

R ~ NH_2 + HN= ~ R → imine crosslink

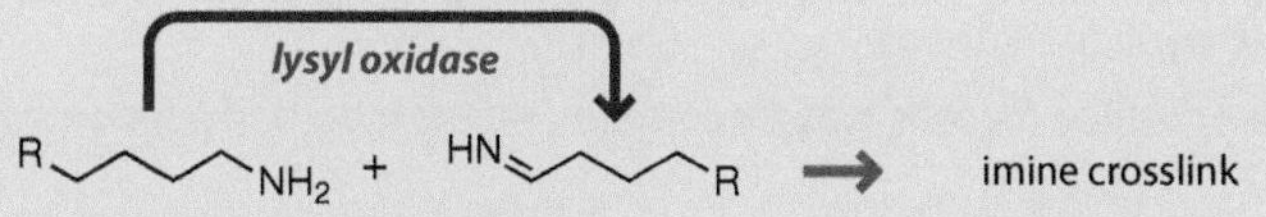

B Suggest a plausible acid-catalyzed mechanism for the formation of an aldol crosslink from two imine functional groups.

R ~ =NH + HN= ~ R → R ~ NH ~ R aldol crosslink

5.7 SUMMARY

The amino acids are like an orchestra with musical side chains conducted by the peptide backbone. Proteins have hydrophobic amino acids for a string section, aromatic amino acids for brass, polar amino acids for woodwinds, and ionic amino acids for percussion. Acting in concert, the amino acids create triumphant works of protein structure, notable for their size and complexity. The ribonucleotide subunits of RNA play more like a jazz quartet: unfettered by a complementary strand, they have the freedom to fully capitalize on the strengths of each musician. The deoxyribonucleotides of double-stranded DNA are like a classical string quartet, narrowly focused yet perfected for playing timeless masterpieces. The first time you hear an orchestral piece you will have a tendency to simply enjoy the fullness of the music, but if you focus you can pick out the dominant players at each point in the music. In so doing, you will achieve a richer appreciation of the work and the capabilities of the instruments. Similarly, the beguiling pictures of proteins in this book may at first overwhelm you with their color and complexity. Take the time to examine each structure carefully, focusing on any atoms and bonds that have been rendered. This is the level of the detail at which chemical biologists wield their greatest power.

In spite of the immense strategic challenges associated with protein synthesis, chemists have now developed robust strategies for the synthesis of peptides. Recall that the solid-phase strategy that enabled oligonucleotide synthesis was initially developed for peptides. Chemists elongate peptides in the same direction as the ribosome, adding activated amino acid esters to the N terminus of a growing peptide chain. The ribosome uses tRNA esters; chemists use *N*-protected amino acids and activate the carboxylic acid with fancy coupling reagents. Diverse protecting groups for the amino acid side chains have been designed to come off under the same S_N1 reaction conditions that lead to cleavage from the solid-phase resin.

The peptide backbone facilitates formation of intramolecular hydrogen bonding, leading to canonical structures with helices, sheets, and turns. A few simple forces, including torsional strain, allylic strain, and stereoelectronics, dictate the geometric attributes of these structures. α helices, β sheets, and turns combine to form modular protein domains that define the tertiary structure of individual protein chains. Higher levels of architectural complexity determine the function of biological assemblies. In the next chapter we will gain a deeper chemical understanding of the capabilities of proteins.

LEARNING OUTCOMES

- Learn the structures of each of the 20 common ribosomally incorporated amino acids and the distinctive chemical properties of each side chain.
- Achieve literacy with the one-letter codes for amino acids.
- Design the synthesis of a full-length peptide using Fmoc-protected amino acids using solid-phase peptide synthesis.
- Draw arrow-pushing mechanisms for common side reactions in peptide synthesis and deprotection.
- Correlate conformational preferences in peptide ϕ, ψ, and χ angles with fundamental types of torsional strain and allylic strain.
- Recognize the secondary structure preferences of each amino acid.
- Distinguish between secondary structure, tertiary structure, and quaternary structure.
- Render protein databank (.pdb) files to reveal secondary structure and individual amino acids.
- Recognize the most common protein domains.
- Identify key elements defining protein structure at the primary, secondary, tertiary, and quaternary levels.

PROBLEMS

***5.14** Draw the structure of the following peptides in the ionic form that would dominate at pH 7.

A SV40 NLS sequence: PKKKRKV

B spinorphin: LVVYPWT

C oxytocin: CYIQNCPLG

D phakellistatin 13: cyclo-[FGPTLWP]

E malformin A: cyclo-[^{D}C^{D}CV^{D}LI]

***5.15** Show the one-letter code designation for the following peptides.

A substance P

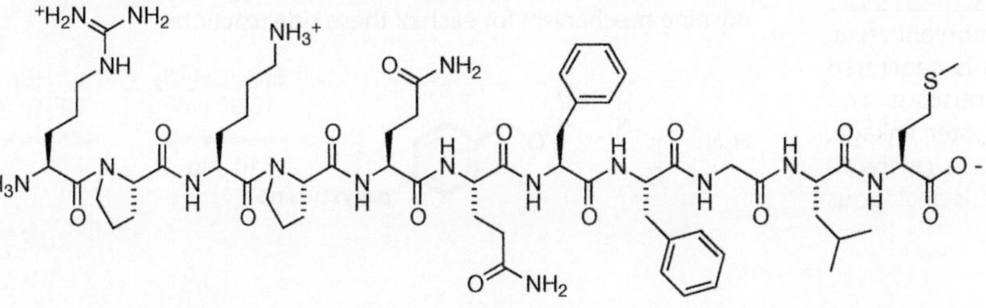

B antimicrobial peptide

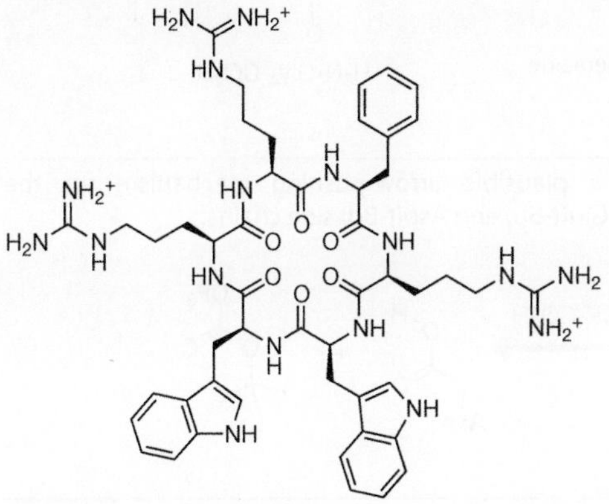

C synthetic integrin antagonist

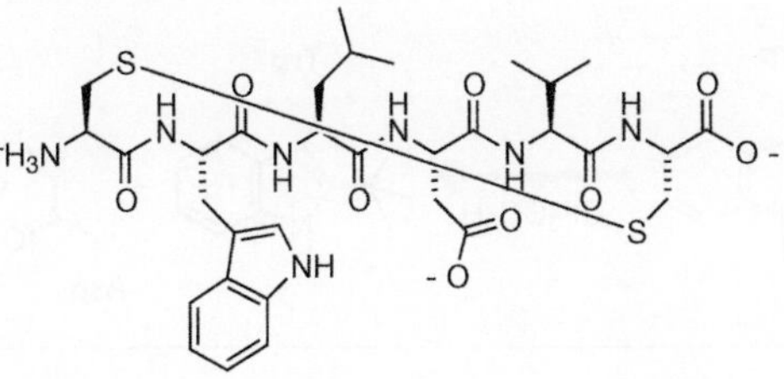

5.16 Kororamide and microcyclamide are believed to be derived from cyclic peptides. Identify each of the putative amino acid precursors in each structure.

kororamide

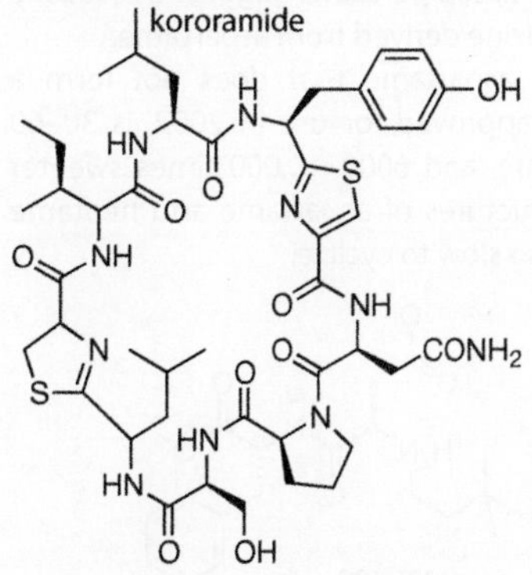

microcyclamide

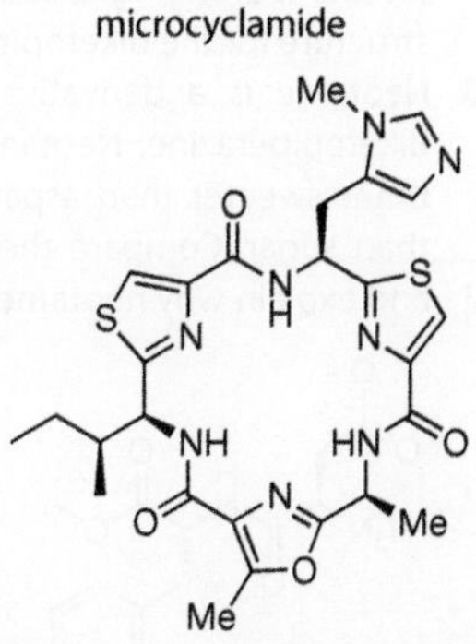

***5.17** Calculate the net charge of the following peptides at pH 7.3:

A Ovine pulmonary surfactant peptide, GADDDDD

B HIV Tat peptide, GRKKRRQRRRPPQC-NH_2

5.18

A Under neutral to basic conditions, aspartame can cyclize to form a six-membered ring called a diketopiperazine. Suggest a plausible structure for the diketopiperazine derived from aspartame.

B Neotame is a derivative of aspartame that does not form a diketopiperazine. Neotame, approved for use in 2002, is 30–60 times sweeter than aspartame, and 6000–10,000 times sweeter than sugar. Compare the structures of aspartame and neotame and explain why neotame is so slow to cyclize.

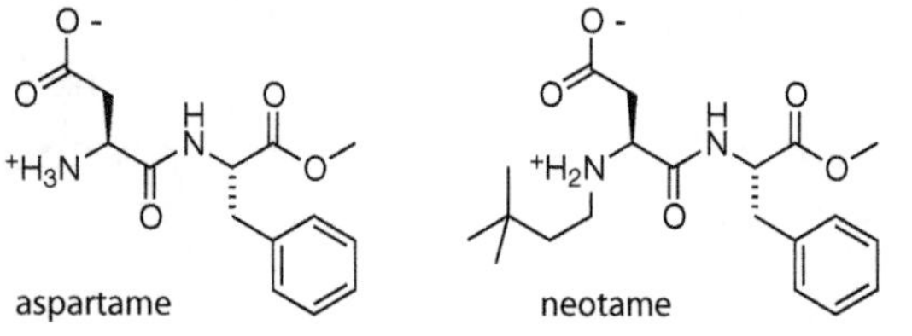

aspartame neotame

***5.19** Sonic hedgehog is a protein involved in embryonic/fetal development. The active hedgehog signaling protein is generated through a self-catalyzed cleavage that leaves the business end of hedgehog connected to cholesterol through an ester linkage. Suggest a plausible arrow-pushing mechanism for the base-catalyzed formation of the hedgehog–cholesterol conjugate that is analogous to the mechanism for intein splicing.

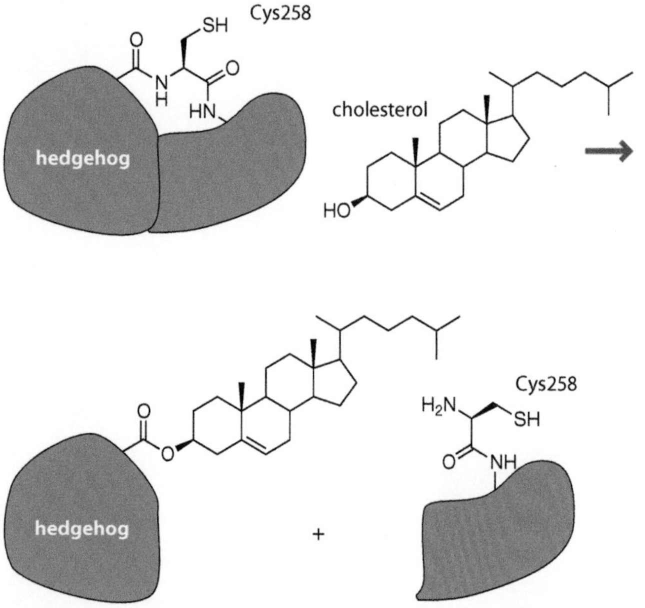

Cys258 hedgehog cholesterol

hedgehog Cys258

5.20 The drug glatiramer acetate (Copaxone™) is sold as a mixture of peptides for the treatment of multiple sclerosis. It is generated through the random polymerization of four different *N*-carboxy anhydride derivatives. Suggest a plausible arrow-pushing mechanism for the initial reaction of diethylamine with an *N*-carboxy anhydride to make an amino acid derivative with a free N terminus.

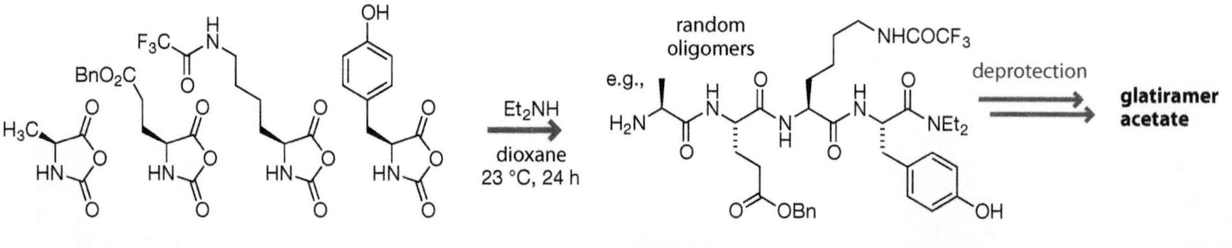

Et_2NH, dioxane, 23 °C, 24 h → random oligomers, e.g., … → deprotection → **glatiramer acetate**

5.21 Carbonyl sulfide (O=C=S) is present at up to 0.09 mol% in volcanic gases and has been proposed as a peptide coupling agent under prebiotic conditions. Suggest a plausible arrow-pushing mechanism.

8 equiv.; pH 9.6, 25 °C, 48 h; 7% yield

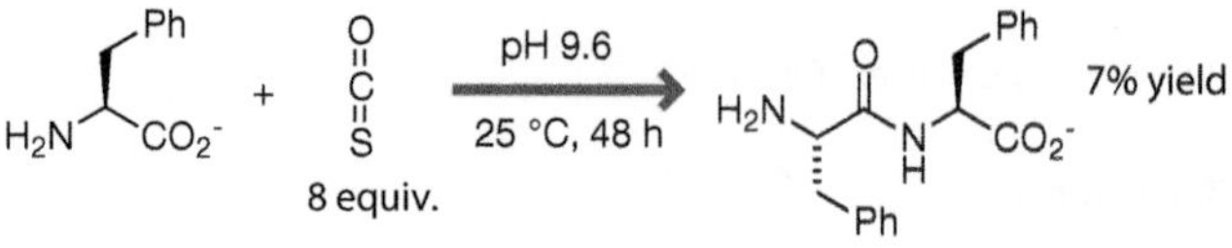

5.22 Triethylamine is often used to deprotonate the N-terminal ammonium group of peptides after acidic deprotection of an N_α-Boc group. Two types of side reactions can reduce overall peptide yield during this free-basing step. First, at the stage of dipeptide, glycine-containing and proline-containing peptides are highly susceptible to an autocleavage reaction, generating a diketopiperazine. Second, peptides of all lengths are subject to a doubling reaction. For example, if the dipeptide ^+H_3N-Gly_2-CO_2H is cleaved from resin after the final treatment with base, up to 1% of the tetramer ^+H_3N-Gly_4-CO_2H is present in the product mixture. Suggest a plausible arrow-pushing mechanism for each of these side reactions.

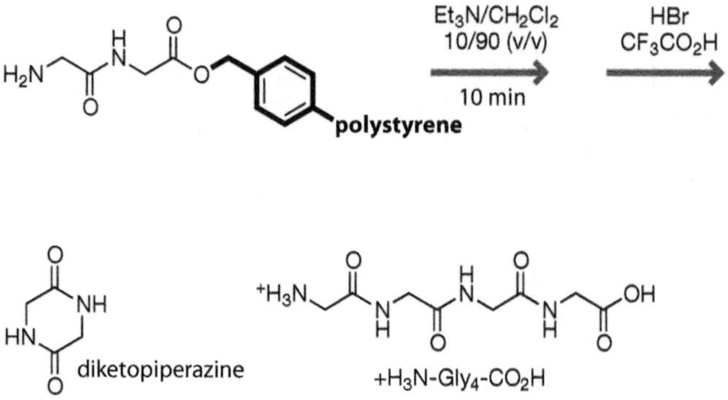

polystyrene; Et_3N/CH_2Cl_2 10/90 (v/v), 10 min; HBr, CF_3CO_2H

diketopiperazine; ^+H_3N-Gly_4-CO_2H

***5.23** Provide a plausible arrow-pushing mechanism for the deprotection of Glu(*t*-Bu) and Asp(*t*-Bu) side chains.

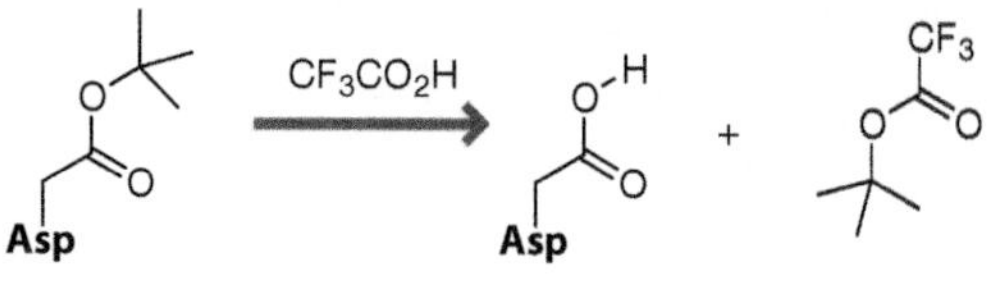

Asp → CF_3CO_2H → Asp +

***5.24** Suggest a plausible arrow-pushing mechanism for the following modification of tryptophan side chains under peptide cleavage/deprotection conditions.

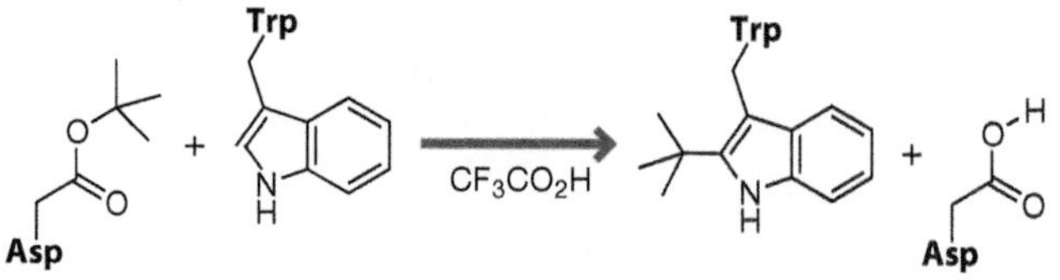

Asp + Trp → CF_3CO_2H → Trp + Asp

5.25 *t*-Butyl carbocations participate in Friedel–Crafts (and other) side reactions during peptide deprotection/cleavage. Water and triisopropylsilane are included in some peptide deprotection cocktails to scavenge carbocations. Siliconate–hydride bonds (R_3XSiH^-) are nucleophilic, whereas silicon hydride bonds (R_3SiH) bonds are not. Hydridosiliconates are highly selective and attack only carbocations.

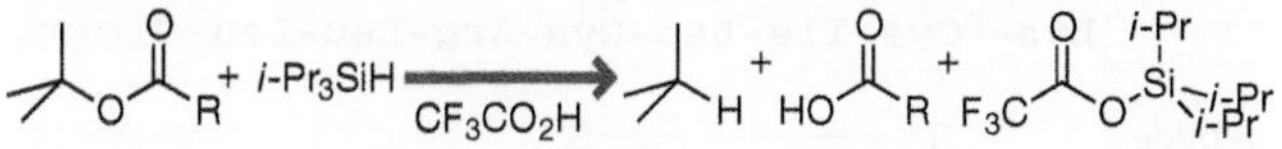

A Suggest a plausible arrow-pushing mechanism for the scavenging reaction that removes *t*-butyl carbocations.

B What is the other major carbocation present during cleavage of peptides made through Fmoc chemistry that you would expect to be reduced by triisopropylsilane?

***5.26** Suggest a plausible arrow-pushing mechanism for deprotection of the following amino acid side chains using the standard deprotection cocktail 95:2.5:2.5 (by volume) $CF_3CO_2H/(i\text{-Pr})_3SiH/H_2O$.

A

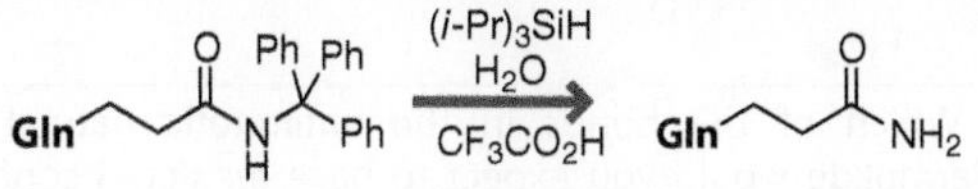

B

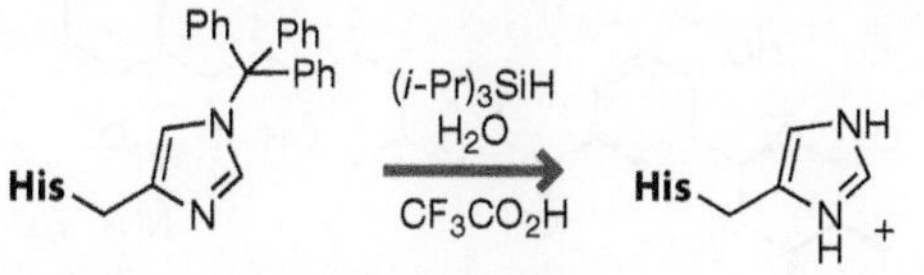

C

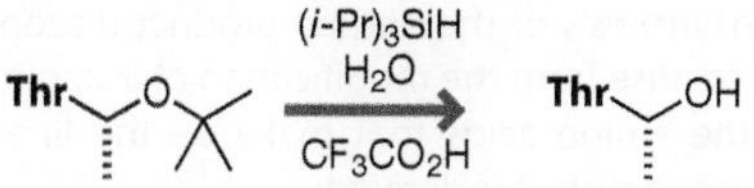

D

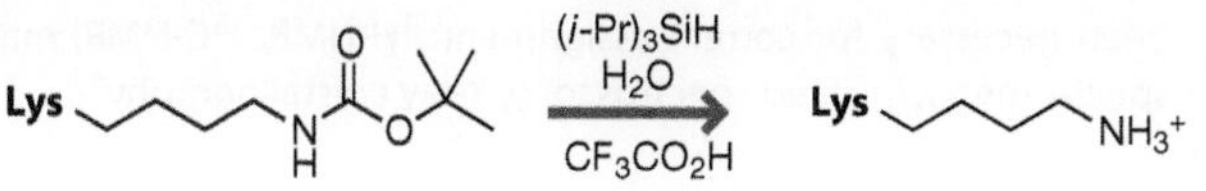

E

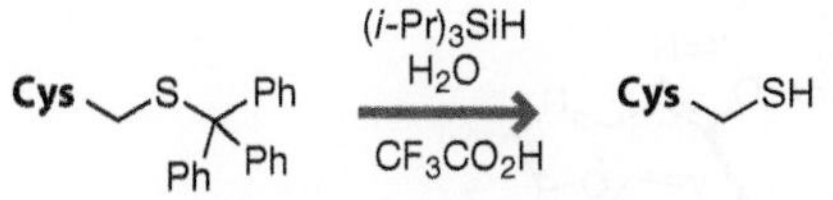

5.27 Suggest a plausible arrow-pushing mechanism for cleavage of the following tripeptide from Rink AM resin using 95:2.5:2.5 (by volume) $CF_3CO_2H/(i\text{-Pr})_3SiH/H_2O$.

ValPheLeu

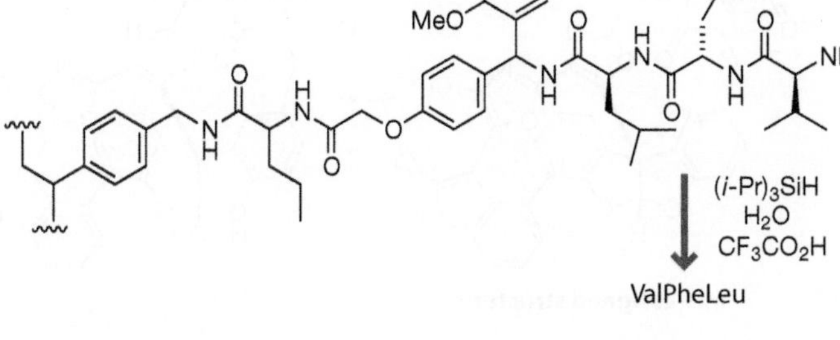

***5.28** Select from the following set of peptides to answer the following questions.

Ac-CAAAKAAAAKAAAAKA-$CONH_2$
Ac-NLEDKAEELLSKNYHLENEVARLCONH$_2$
Ac-AAAAEAAAKAAAAYR-$CONH_2$
Ac-YMSEDELKAAEAAFKRHNPT-$CONH_2$
Ac-AAQAAAAQAAAAQAAY-$CONH_2$
Ac-AEAAAKEAAAKEAAAKACONH$_2$
Ac-KIVFKNNAGFPH-$CONH_2$
Ac-KVKVKVKVKVKVK-$CONH_2$
Ac-GPPGPPGPPGPPGPPGPPGPP-$CONH_2$

A Which of the sequences above would have the highest β-sheet propensity?

B Which of the sequences above would have the least β-sheet propensity?

C Which of the sequences above would be most likely to dimerize as a coiled coil?

D Which of the sequences above would be most likely to form a triple helix?

E Which of the sequences above would have the least α-helix propensity?

5.29 The disulfide exchange equilibrium below favors the mixed disulfides, even though $\Delta H°$ is approximately zero for the equilibrium. Suggest a plausible explanation for the difference in stabilities.

$$\text{EtS–SEt} + t\text{-BuS–S}t\text{-Bu} \underset{}{\overset{K_{eq} = 25}{\rightleftharpoons}} t\text{-BuS–SEt} + t\text{-BuS–SEt}$$

5.30 Which amino acid would you expect to most favor ϕ angles around 160°?

***5.31** The EF hand is a common protein domain that folds into a helix–loop–helix motif in the presence of Ca^{2+} ions. In many calcium-sensing proteins, EF hands dimerize after binding Ca^{2+}, leading to marked changes in tertiary structure. The twelve-amino acid calcium-binding loop of the EF-hand motif forms a neutral hexacoordinate complex with the calcium ion.

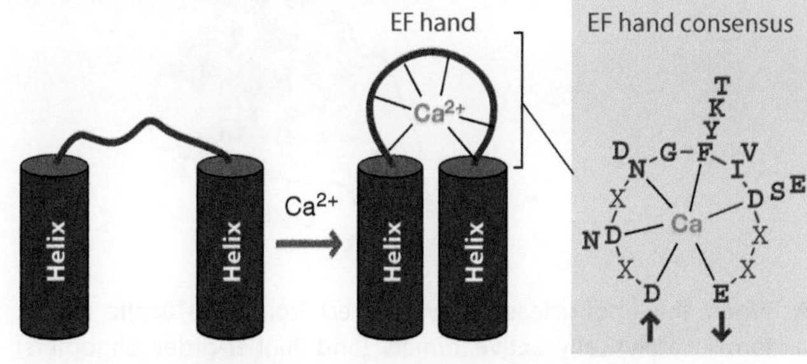

A The calcium signaling protein calmodulin has four EF hands. On the basis of the consensus amino acid sequence of the calcium-binding loop of the EF hand, identify each of the putative calcium-binding loops in the protein sequence for human calmodulin.

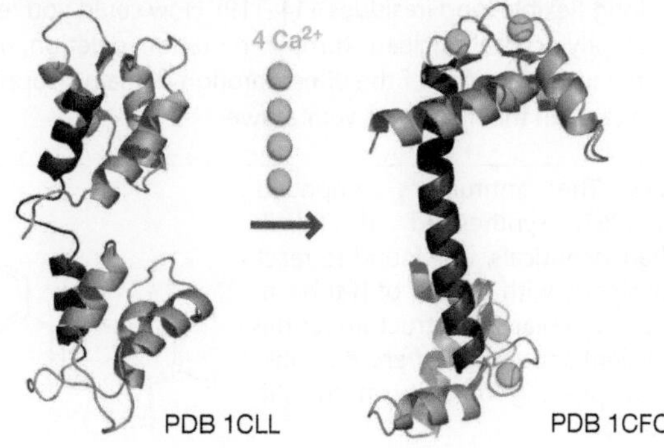

MADQLTEEQIAEFKEAFSLFDKDGDGTITTKELGTVMRSLGQNPT
EAELQDMINEVDADGNGTIDFPEFLTMMARKMKDTDSEEEIR
EAFRVFDKDGNGYISAAELRHVMTNLGEKLTDEEVDEMIREAD
IDGDGQVNYEEFVQMMTAK

B On the basis of the consensus amino acid sequence shown to coordinate to calcium (above), draw a coordination diagram that indicates which atoms in the *first* EF hand of human calmodulin are ligated to the Ca^{2+} ion. Indicate clearly which atoms bind as anionic ligands (using solid lines) and which atoms bind as neutral ligands (using dashed lines).

C Two of the conserved positions in the EF-hand motif are not ligated to the Ca^{2+} ion. What particular properties of these conserved residues are likely to be important for the structure of the loop?

***5.32** Some proteins have small domains or helices that can be removed without unfolding the protein. For example, selective proteolysis of ribonuclease A by subtilisin cleaves a flexible loop (residues 15–22), leading to the dissociation of a helix (residues 1–14, referred to as the S-peptide). The truncated protein is still folded but is catalytically inactive.

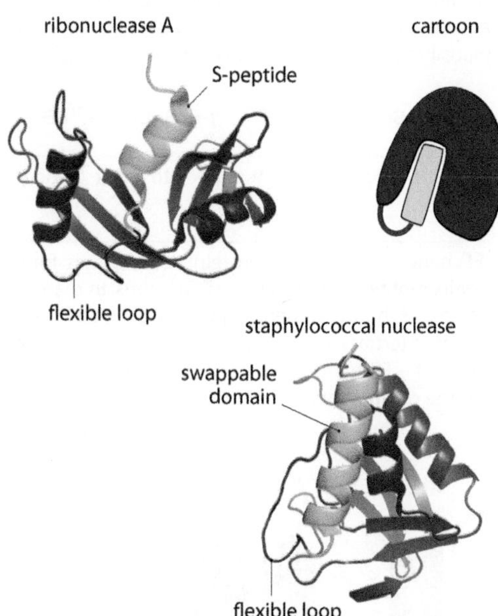

A When the ribonuclease is lyophilized from 40% acetic acid it forms catalytically active dimers (and higher-order oligomers) through domain swapping of the S-peptide between ribonuclease molecules. Draw a cartoon of a domain-swapped dimer of ribonuclease involving the S-peptide domain.

B Like ribonuclease A, staphylococcal nuclease also has a swappable helical domain (residues 120–141) connected by a long flexible loop (residues 114–119). How could you re-engineer staphylococcal nuclease through mutation, deletion, or addition to favor formation of the dimer through domain swapping? Draw a cartoon that illustrates your answer.

5.33 The antitumor compound T138067, synthesized at Tularik Pharmaceuticals, was found to react covalently with Cys239 of β-tubulin. Suggest a plausible structure for the covalent adduct and offer a plausible arrow-pushing mechanism for the reaction.

T138067

5.34 Which of the following peptides would be most likely to adopt an α-helical structure?

peptide 1 Lys-Cys-Ile-Leu-Cys-Arg-Leu-Leu-Gln-NH_2 (S–S)

peptide 2 Lys-^{D}Cys-Ile-Leu-Cys-Arg-Leu-Leu-Gln-NH_2 (S–S)

peptide 3 Lys-Cys-Ile-Leu-^{D}Cys-Arg-Leu-Leu-Gln-NH_2 (S–S)

5.35 Assume that the conjugated π system of sclerotonin is planar and adopts a conformation that minimizes $A_{1,3}$ strain. Will the ethyl group be on the same face as the acetoxy group or on the opposite face?

sclerotioramine

ethyl group

***5.36** Which of the bonds in the antimitotic natural product discodermolide would you expect to have restricted conformation due to $A_{1,3}$ strain?

discodermolide

5.37 The biosynthesis of the natural product diazonamide A can be rationalized to arise from the modification of a single linear peptide.

A Identify the amino acids that make up the linear peptide from which diazonamide A is derived.

B The structure of diazonamide A was initially mis-assigned. Which of the following spectroscopic techniques would have been necessary for correct assignment: ^1H-NMR, ^{13}C-NMR, mass spectrometry, infrared spectroscopy, X-ray crystallography?

diazonamide A

mis-assigned structure

***5.38** The natural product helenalin inhibits the transcription factor NF-κB by reacting with Cys38 in the p65 subunit.

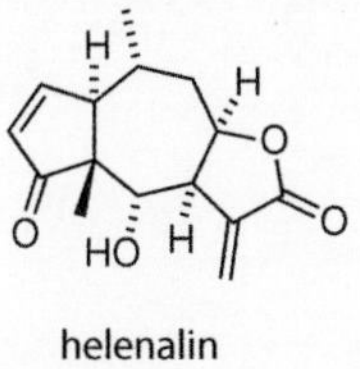

helenalin

A On the basis of the crystal structure PDB 1RAM, suggest a plausible reason why Cys38 is important for NF-κB function.

B Helenalin has been proposed to form a covalent crosslink between Cys38 and another nearby Cys residue. Identify the nearby Cys residue.

C Suggest a plausible structure for the proposed adduct of helenalin with NF-κB .

5.39 Predict which five of the common amino acids would not be isolable after protein hydrolysis at 105 °C for 10 hours in 4 M NaOH.

5.40 Ottelione A forms a covalent bond with the thiolate group of Cys239 of tubulin. Suggest a plausible structure for the covalent adduct.

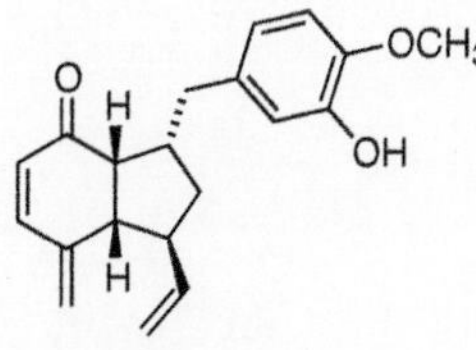

ottelione A

***5.41** Tyrosine residues are known to generate dityrosine crosslinks in the presence of radicals. Suggest a plausible fishhook arrow-pushing mechanism for the following dimerization reaction.

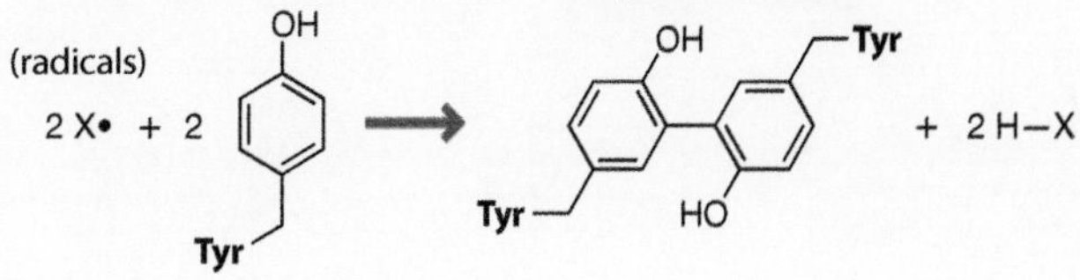

Protein Function

6

LEARNING OBJECTIVES

- Predict the toxicity or therapeutic value of a compound using a dose–response curve.
- Understand binding interactions in terms of both thermodynamics and kinetics.
- Intuit the importance of substrate binding as measured by K_m, and the importance of catalytic turnover as measured by k_{cat}.
- Contrast several different regulatory mechanisms for the control of protein kinase activity.
- Distinguish between the chemical mechanisms of serine/cysteine proteases, metalloproteases, and aspartic proteases.
- Compare several different mechanisms for the regulation of protease activity.
- Gain familiarity with techniques for protein mutagenesis and the power of selections.

Human proteins exhibit a wide range of functions in the human body, but the function of proteins that has held the most enduring attraction for chemists is catalysis—the acceleration of chemical reactions. Even the ancient Greeks marveled at the speed of enzyme action. As mentioned in *The Iliad*, stirring warm milk with a fig branch leads to rapid curdling of milk, a process that involves the catalytic action of the enzyme ficain (**Figure 6.1**). Chemists once envied the rate accelerations afforded by enzymes, but it is now commonplace for chemists to design transition-metal catalysts with equally amazing rate accelerations. In fact, the speed and efficiency of many transition-metal-catalyzed reactions—such as hydrogenation, olefin metathesis, and olefin polymerization—cannot be matched by any enzyme.

The aspect of enzyme catalysis that is still largely elusive to chemists is specificity. Moreover, chemists covet the range of specificities found within many enzyme families. For example, digestive proteases such as pepsin are necessarily promiscuous, whereas the insidious protease tetanus toxin is exquisitely selective, cleaving just a single bond of a single protein **substrate**. This level of specificity is commonplace among enzymes and receptors, particularly among membrane-bound receptors tuned to receive specific chemical messages from the extracellular environment.

Biosynthesis is an obvious role for enzymes, building the molecules that make up the cell. Since the start of the new millennium, even more interest has been focused on the role of enzymes in cellular signaling processes. Regardless of their role, most enzymes require activation through, for example, localization, cleavage, or activation by small, nonsubstrate molecules. Ultimately, mechanisms that control enzyme activity and their biological context are as interesting as the chemical transformations catalyzed by the enzyme.

Figure 6.1 The latex from a fig leaf contains the enzyme ficain, which rapidly cleaves the amide bonds of casein, initiating misfolding and aggregation. The process makes a cameo in Homer's *The Iliad*, as follows, *"As the juice of the fig-tree curdles milk, and thickens it in a moment though it is liquid, even so instantly did Paeon cure fierce Mars."* (Courtesy of Ellen Friedman.)

6.1 RECEPTOR–LIGAND INTERACTIONS

The thermodynamics and kinetics of receptor–ligand interactions govern all processes in biology

Specific noncovalent binding is essential for virtually all processes in chemical biology: the binding of signaling proteins, the binding of enzymes to substrates, and the

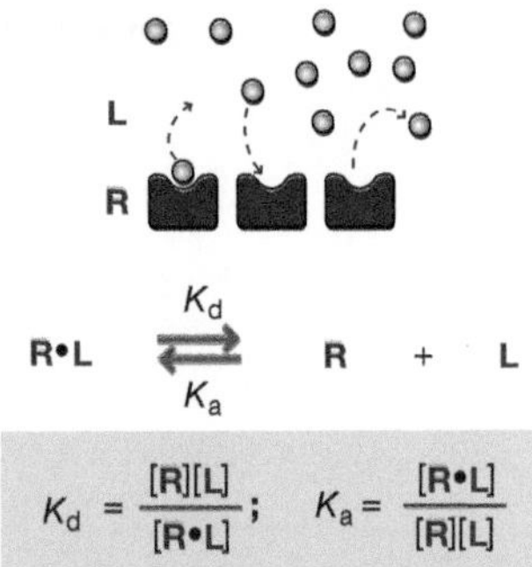

Figure 6.2 Equilibrium concentrations. The dissociation constant (K_d) can be defined in terms of concentrations. It describes the lability of a receptor–ligand complex and is measured in units of moles per liter. Smaller values of K_d correlate with tighter binding by the ligand to the receptor.

binding of receptors to ligands. Receptor–ligand interactions are the most useful context for thinking about how we quantify binding phenomena. In chemistry, the "tightness" of association is usually quantified by an equilibrium **association constant, K_a**. In biology, it is more common to use the **dissociation constant, K_d**, which quantifies lability rather than stability (**Figure 6.2**). Lower K_d values correspond to tighter binding. Most chemical biologists would describe a K_d of less than 1 nM as tight and receptor–ligand interactions with a K_d of greater than 1 mM as weak.

Knowing K_d and the ligand concentration can provide an intuitive feel for the proportions of bound and unbound receptors. If we set the ligand concentration equal to the dissociation constant ($[L] = K_d$), the concentration of bound receptor will equal the concentration of unbound receptor. Mathematically, the ratio [R•L]/[R] will be 1:1. Furthermore, when the concentration of ligand is 10 times K_d, the ratio [R•L]/[R] will be 10:1 and more than 90% of the receptors will be bound. Conversely, if the concentration of ligand is one-tenth of the K_d, the ratio [R•L]/[R] will be 1:10 and less than 10%—a small minority—of the receptors will be bound. These relationships are important because ligand concentration is one of the variables that can be easily controlled by a biologist. Obviously, K_d is an important variable to know for any and every binding interaction. K_d is also useful because it allows us to say something about the free energy of binding in the hypothetical situation in which all components are present at 1 M: $\Delta G° = -RT \ln K_{eq}$.

The experimental technique isothermal titration calorimetry (ITC) can be used for the direct measurement of the heat released when aliquots of a ligand solution are added to a solution of protein. Changes in enthalpy, entropy, and ultimately free energy can be obtained through computerized curve-fitting of data from the experiment. ITC is also a fairly sensitive technique, allowing investigators to husband their precious protein samples.

Problem 6.1

Calculate K_d for each of the following protein–protein interactions, using data derived from isothermal titration calorimetry (ITC) measurements.

Protein	Protein	ΔH (kcal mol^{-1})	ΔS (cal K^{-1} mol^{-1})
Mutant TCR β chain 8.2	*S. aureus* enterotoxin C3	−15.8	−21 at 25 °C
p67^{phox}	Rac•GTP complex	−7.3	52 at 18 °C
Iso-1-cytochrome *c*	Iso-1-*cc* peroxidase	−2.6	18.5 at 25 °C

Most freshman chemistry courses emphasize a definition of equilibria based on concentrations. However, to understand biological interactions, our thinking needs to be recast in terms of reaction rates. We use rate constants to characterize the rates of chemical reactions. Rate constants are most powerful when we apply them to elementary chemical reactions; that is, reactions that involve a single transition state. There are only two types of elementary chemical reactions that are relevant to solution-phase chemistry on the planet Earth: unimolecular reactions involving one reactant, and bimolecular reactions in which two reactants must collide to initiate a chemical event (**Figure 6.3**). Termolecular and higher-order elementary reactions are unrealistic, because it is rare for several reactants to collide simultaneously with the precise orientations and sufficient energies to form bonds or complexes.

unimolecular step — k_1 — + $^-$OR — rate = k_1[X]

bimolecular step — k_2 — rate = $k_2[HO^-][Y]$

Figure 6.3 Rate laws include rate constants and concentrations. The rate law for an elementary reaction step includes a rate constant *k*, and the concentrations of the species involved in the rate-determining step of the reaction.

The rate constants for meaningful unimolecular elementary reactions span about 20 orders of magnitude: 10^{13} s^{-1} > k_1 > 10^{-7} s^{-1}. The dissociation of receptor–ligand complexes is usually described as an approximately unimolecular process. The speed limit for the fastest unimolecular reactions is determined by the fastest rates at which atoms vibrate, and corresponds to a rate constant of about 10^{13} times per second. The lower limit is determined by practical considerations. Unimolecular reactions with rate constants lower than 10^{-7} s^{-1} would have half-lives measured in years. Few cellular reactions have the luxury of taking that long. Rate constants for meaningful bimolecular reactions span an even smaller range: 10^9 M^{-1} s^{-1} > k_2 > 10^{-7} M^{-1} s^{-1}. The association of receptors and ligands into complexes is usually described as a bimolecular process. The speed limit for the fastest bimolecular reactions is set by the rate of diffusion, which determines how often molecules can randomly collide with each other. For small molecules in water, that diffusion-controlled limit is about 10^9 M^{-1} s^{-1}. The lower limit is again determined by practical considerations. Bimolecular reactions with rate constants lower than 10^{-7} M^{-1} s^{-1} would have half-lives measured in years if the concentrations of both reactants were 1 M. Of course, concentrations of enzymes and substrates are typically in the micromolar and millimolar ranges, respectively. At those concentrations, a bimolecular reaction with a rate constant of 10^{-7} M^{-1} s^{-1} would take billions of years. Unlike in the cell, in a chemical laboratory you typically cool reactions with microsecond half-lives to prevent them from overheating. Similarly, you typically heat reactions with long half-lives so that you do not have to wait weeks or years for the reaction to take place.

Thermodynamic equilibrium is based on the equality of both forward and reverse reaction rates. For receptor–ligand interactions, the associative reaction rate is referred to as the on-rate; it is equal to k_{on}[R][L]; the dissociative reaction rate is referred to as the off-rate; it is equal to k_{off}[R•L]. The K_d can also be defined as the ratio of the off-rate constant to the on-rate constant (**Figure 6.4**). For high-affinity interactions, the off-rates are always much slower than the on-rates. The American scientist-statesman Benjamin Franklin prescribed this kind of advice for social situations: "Be slow in choosing a friend, slower in changing."

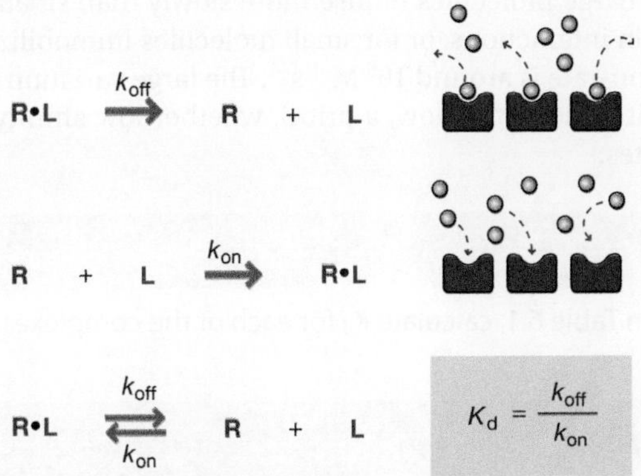

Figure 6.4 Basing equilibrium on rate constants. The equilibrium dissociation constant K_d can be defined in terms of rate constants instead of equilibrium concentrations.

In his classical text *Enzyme Structure and Mechanism*, Alan Fersht tabulated the on-rate and off-rate constants for a variety of specific interactions involving proteins (**Table 6.1**). His books on enzyme catalysis are recommended for students seeking a deeper appreciation of enzyme kinetics. A valuable trend is apparent for small-molecule ligands, as opposed to large macromolecular ligands. Small molecules tend to bind to specific proteins with fast on-rates. The theoretical maximum rate for diffusion in water is about 5×10^8 M^{-1} s^{-1}. If a ligand binds with a receptor with a rate constant $k_{on} = 5 \times 10^8$ M^{-1} s^{-1}, it means that every collision leads to a successful binding event; if $k_{on} = 5 \times 10^6$ M^{-1} s^{-1}, only 1 in 100 collisions results in the formation of a receptor-ligand complex.

Table 6.1 The kinetics of binding.

Protein	Small ligands	k_{on} ($M^{-1} s^{-1}$)	k_{off} (s^{-1})
Chymotrypsin	Proflavin	10^8	8300
Creatine kinase	ADP	0.2×10^8	18,000
Glucose 3-phosphate dehydrogenase	NAD^+	0.2×10^8	1000
Lactate dehydrogenase	NADH	$\sim 10 \times 10^8$	~10,000
Alcohol dehydrogenase	NADH	0.3×10^8	9
Lysozyme	$(N\text{-Ac-Glu})_2$	0.4×10^8	100,000
Ribonuclease	3′-UMP	0.8×10^8	11,000
Protein	**Large ligands**	**k_{on} ($M^{-1} s^{-1}$)**	**k_{off} (s^{-1})**
$tRNA^{Ser}$ synthetase	$tRNA^{Ser}$	2×10^8	11
Trypsin	Protein inhibitor	0.0007×10^8	0.0002
Insulin	Insulin	10^8	20,000
β-Lactoglobulin	β-Lactoglobulin	0.00005×10^8	2
α-Chymotrypsin	α-Chymotrypsin	0.000004×10^8	0.7

The important and powerful generalization is that small ligands bind to protein receptors at rapid rates, particularly if electrostatic interactions are involved. Thus, most of the variation in dissociation constants (K_d) arises from how fast or slowly the ligands dissociate from the receptors. On-rates tend to be slower and exhibit greater variation for interactions between two macromolecules than for interactions involving small molecules. Large molecules diffuse more slowly than small molecules; for typical protein–protein interactions, or for small molecules immobilized on surfaces, the basal diffusional on-rate is around $10^6\ M^{-1}\ s^{-1}$. The large variation in protein–protein on-rates makes it difficult to know, a priori, whether low affinity is due to slow on-rates or fast off-rates.

Problem 6.2

Using the numbers in Table 6.1, calculate K_d for each of the complexes.

Problem 6.3

What percentage of alcohol dehydrogenase is bound when the concentration of NADH is (A) 3 μM, (B) 3 nM?

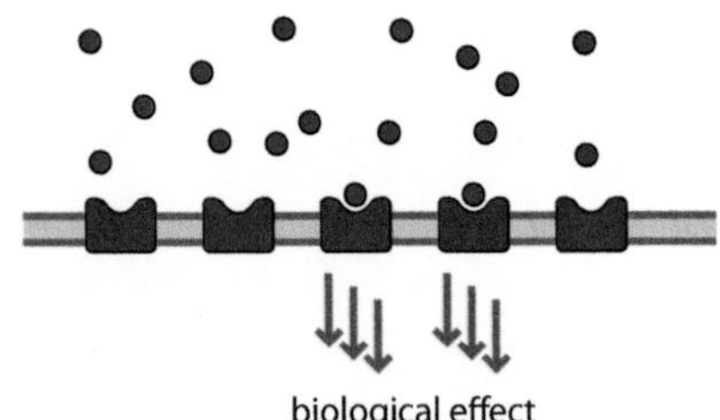

Figure 6.5 Receptor occupancy model. The biological effect of a drug is proportional to the receptor–drug concentration, not the total drug concentration.

Dose–response curves measure protein function, and correlate with affinity

According to pharmacological receptor theory, the biological effect of a drug is directly proportional to the concentration of ligand-bound receptor, and not on the concentration of unbound ligand (**Figure 6.5**). Focusing on the concentration of the receptor•ligand complex compels us to use K_d when estimating the minimum therapeutic concentration of a drug. For a drug to exert an effect, the concentration of the drug should be much higher (at least 10 times higher) than the K_d of the receptor–drug interaction. Most pills are prescribed with a regimen that ensures that the biological target is always more than 90% bound by the drug.

The profound implication of pharmacological receptor theory is that one can estimate the percentage of reversibly bound receptors, and ultimately K_d, by examining complex cellular phenomena such as cell death or bacterial swimming, or even complex physiological phenomena such as coughing frequency. If the biological effect is due to a specific interaction, the dose required to elicit 50% of the desired response (IC_{50}) should correspond to the K_d of the receptor–drug interaction. The relationship between binding and biological effect is usually plotted with a **dose–response curve** in which biological response is plotted against the logarithm of the drug concentration (**Figure 6.6**). If multiple receptors, each with a different K_d, contribute to the biological response, then the dose–response curve will represent a composite of those interactions and may lack a sharp sigmoidal shape.

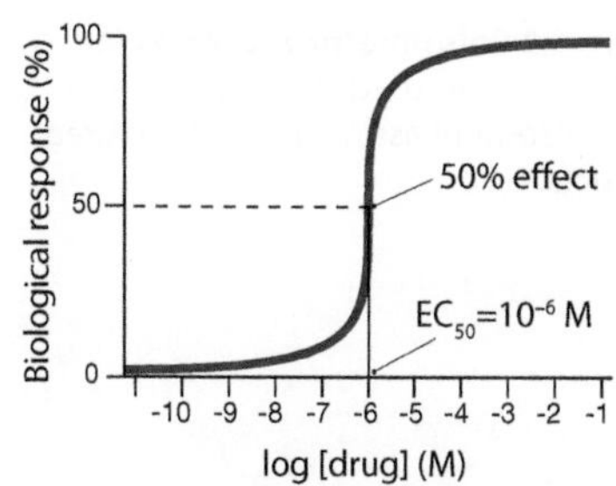

Figure 6.6 The dose–response curve. A dose–response curve explains why doubling the concentration of a drug generally does not double the effect. It explains why many toxic chemicals are nontoxic and why many "natural" cures do not cure.

The dose–response curve is probably the most important concept in chemical biology. An understanding of the dose–response relationship allows chemical biologists to set up assays at meaningful concentrations of ligand, substrate, or drug. It also allows doctors and patients to maintain dosages of medicines at levels that treat their ailments and avoid resistance of infectious diseases. A misunderstanding of the dose–response curves leads to an irrational fear of chemicals, even when they are present at harmless concentrations. The historical figure Paracelsus correctly summed up the dose–response curve in a simple way, saying that it is the dose that makes the poison (**Figure 6.7**). An ignorance of dose–response relationships can be dangerous; every year it leads sick people to spend billions of dollars on "natural" cures that have too little of the active components to offer any benefit beyond the placebo effect.

Problem 6.4

Beneficial properties of the spice turmeric have been attributed to the natural product curcumin, which is active against chronic lymphocytic leukemia cells with an EC_{50} of 5.5 μM. Assuming that the cytotoxicity of curcumin follows classical pharmacological receptor theory, fill in the table below to estimate the percentage of viable cancer cells when treated with the following concentrations of curcumin. What concentration of curcumin would be required to kill 99.9% of the cells?

curcumin

[Curcumin] (μM)	cancer cells: Dead / Live	cancer cells: Percentage viable
5.5	1 : 1	50
11	: 1	
55	: 1	

Because immunoglobulins (antibodies), which bind specifically to nearly any molecule, can be easily generated in the laboratory, it is common to use antibodies as reagents to detect the presence of other molecules and to plot the results using a dose–response curve. The method is made both quantitative and highly sensitive by chemically linking special enzymes to the antibodies. When colorless enzyme substrate molecules are added at high concentrations, the enzymes generate colored products in amounts that are proportional to the concentration of the enzyme, and hence are proportional to the concentration of the antibody. Two of the most common enzymes that are linked to antibodies are horseradish peroxidase and alkaline phosphatase. Peroxidases catalyze the decomposition of hydrogen peroxide, H_2O_2; however, the highly specific reductive decomposition of hydrogen peroxide is coupled to the relatively nonselective oxidation of other molecules such as redox proteins, dihydroquinones, or nicotinamides (discussed below).

Figure 6.7 The dose makes the poison or cure. According to Paracelsus, "All things are poison, and none without; it is the dose alone that makes a thing not poison. [*Alle Ding' sind Gift und nichts ohn' Gift; allein die Dosis macht, das ein Ding kein Gift ist.*]"—Theophrastus Philippus Aureolus Bombastus von Hohenheim, a.k.a. Paracelsus, 1493–1541. (*Portrait of Paracelsus* by Quentin Massys, 1528.)

The enzyme-linked immunosorbent assay (ELISA) provides a low-tech method for quantifying binding levels in chemical biology and diagnostics laboratories. In one variation of this commonly applied technique, the target is attached to the surface of a microtiter plate; typically, the coating of the plate uses nonspecific, but very strong, adsorption by hydrophobic regions of the target to the polystyrene surface of

Figure 6.8 Colorimetric reactions for ELISA. Enzymes used in enzyme-linked immunosorbent assays generate colored products.

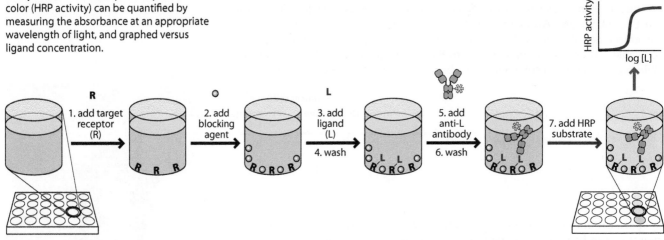

Figure 6.9 ELISA procedure. The enzyme-linked immunosorbent assay takes place in multiwell blocks called microtiter plates (upper left). In step 1, the target receptor (R) is bound to the polystyrene surface of the well, through nonspecific interactions. A blocking agent in step 2 prevents nonspecific interactions with the sticky polystyrene surface by either the ligand or the antibody. Steps 3 and 4, addition of the ligand (L) followed by washing allows the formation of the receptor–ligand interaction. A ligand-specific antibody conjugated to the enzyme horseradish peroxidase (HRP) is added before nonspecific binding is washed away in steps 5 and 6. After addition of the HRP substrate, wells with the sandwich of the R–L complex bound to antibody–HRP turn yellow. The intensity of this yellow color (HRP activity) can be quantified by measuring the absorbance at an appropriate wavelength of light, and graphed versus ligand concentration.

the plate. Any remaining hydrophobic spots on the plate are blocked through addition of a solution with an inexpensive protein in high concentration, such as casein from dairy milk; this blocking step prevents nonspecific binding to the target or plate from dominating the assay results. Next, an antibody covalently attached to one of the enzymes described above (e.g., horseradish peroxidase) is added, and the plates are washed extensively with detergent in buffer to remove any nonspecific binding.

The peroxidase activity correlates with levels of antibody binding or target levels in the example above, and is usually monitored by adding synthetic substrates to generate colored products. *Ortho*-phenylenediamine is a common component of hair dyes that forms a range of products when exposed to air. Horseradish peroxidase converts *ortho*-phenylenediamine into the orange-yellow 2,3-diaminophenazine chromophore; further oxidation leads to a brown precipitate. Alkaline phosphatase can hydrolyze *para*-nitrophenyl phosphate to generate *para*-nitrophenolate anion, which is intensely yellow (**Figure 6.8**). The quantitative nature of the ELISA has led to its widespread use in chemical biology (**Figure 6.9**).

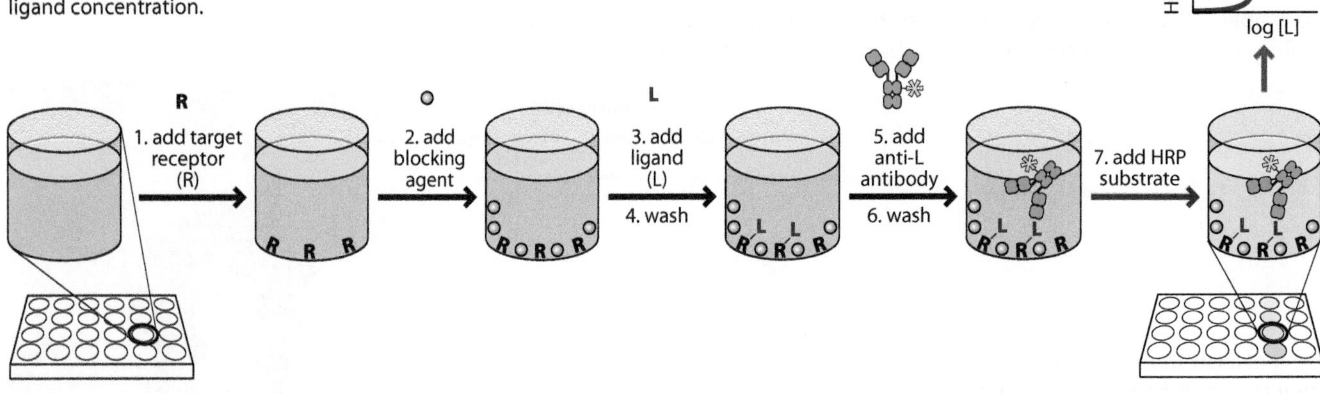

Highly specific protein–small-molecule interactions are useful

Small molecule-protein interactions are noteworthy for their specificity and tight binding. For example, the immunosuppressant rapamycin and the related natural product FK-506 both bind to the FK-506 binding protein (FKBP), as shown by Schreiber and coworkers. The FK-506–FKBP complex can then inhibit a protein (mTOR) that modulates T-cell signaling (**Figure 6.10** and **Figure 6.11**). These natural products have common elements of structure. FK-506 binds to FKBP with a dissociation constant of 0.4 nM; rapamycin binds to FKBP with a dissociation constant of 0.2 nM. As a result of their high affinity and binding specificity, the two small molecules have been extensively used in chemical biology studies to shut down cell signaling pathways associated with cell growth and protein synthesis.

The binding of the small molecule biotin to the protein avidin ($K_d = 10^{-15}$ M at pH 7) is one of the strongest known protein–ligand interactions. Animals, including humans, that eat raw egg whites over a period of months develop biotin deficiency

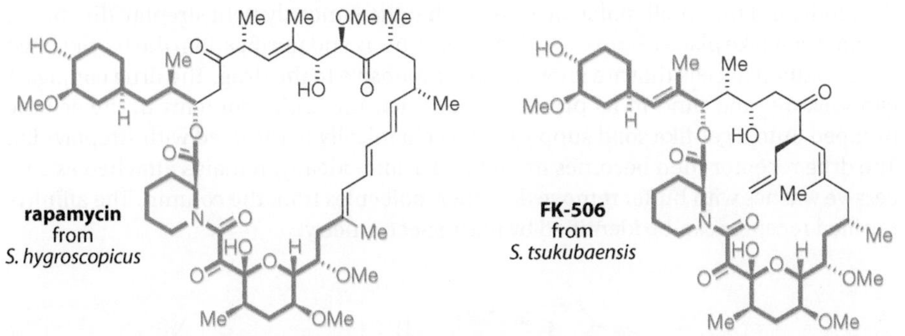

Figure 6.10 Chemical cousins. Soil microorganisms from Japan (*Streptomyces tsukubaensis*) and the remote Pacific island of Rapa Nui (*Streptomyces hygroscopicus*) produce natural products with virtually identical structural features, drawn in green.

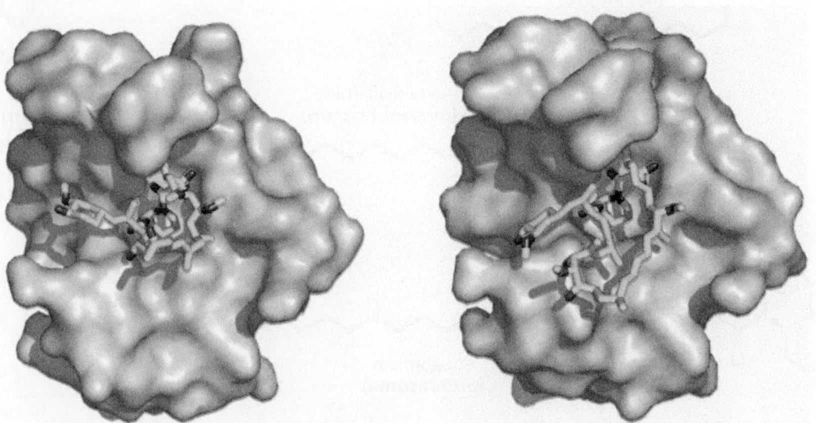

Figure 6.11 Similar modes of binding. FK-506 (PDB 1FKF, left) and rapamycin (PDB 1FKL, right) both bind to the same pocket of the protein FKBP. The conformations of the small molecules are slightly different, because the FKBP–FK-506 complex crystallized as a dimer with FK-506 at the interface.

caused by high concentrations of avidin. Avidin binds to and leaches biotin, also known as vitamin B7, preventing its adsorption in the gastrointestinal tract. Avidin is a glycoprotein and cannot be expressed in bacteria. However, *Streptomyces* produces a form of avidin known as streptavidin, which has no post-translational modifications (**Figure 6.12**).

Streptavidin can be easily expressed in bacteria and has a slightly higher specificity than avidin; thus, streptavidin is more popular for biotechnology applications. The rate of association of biotin with avidin is unimpressive: $k_{on} = 7 \times 10^7\ M^{-1}\ s^{-1}$ (at pH 5). This on-rate constant, when compared with the diffusion-controlled limit of $50 \times 10^7\ M^{-1}\ s^{-1}$, suggests that about 7 out of 50 collisions result in the formation of an avidin–biotin complex. What makes the avidin–biotin interaction so impressive is the glacial off-rate ($k_{off} = 0.00000004\ s^{-1}$ at pH 5). Dissociation of any 1:1 complex is unimolecular, and the half-life is easily calculated according to the equation $t_{1/2} = (\ln 2)/k_{off}$. At pH 7, the half-life of the avidin–biotin complex is 200 days! Tight binding is not enough to be useful in biology. Covalent bonds are much stronger than the avidin–biotin interaction (by about 10^{50}-fold). The utility of the avidin–biotin interaction in biotechnology is the extraordinary *specificity* of the interaction. If you conjugate biotin to *any* molecule, the conjugate (biotin-linked molecule) can be reliably fished out of a complex mixture with a streptavidin affinity matrix. This approach is so common that companies now sell flow-through purification columns prepacked with streptavidin–agarose and streptavidin–Sepharose (**Figure 6.13**).

This strategy for finding the targets of small molecules requires the synthesis of the small molecule conjugated to biotin. The streptavidin binding site buries the biotin in a deep cleft, which is then covered by a trapdoor. So, the drug needs some room between it and the biotin, lest binding to its receptor prevent closure of the streptavidin trapdoor. A linker of at least six atoms (such as 6-aminohexanoic acid) between

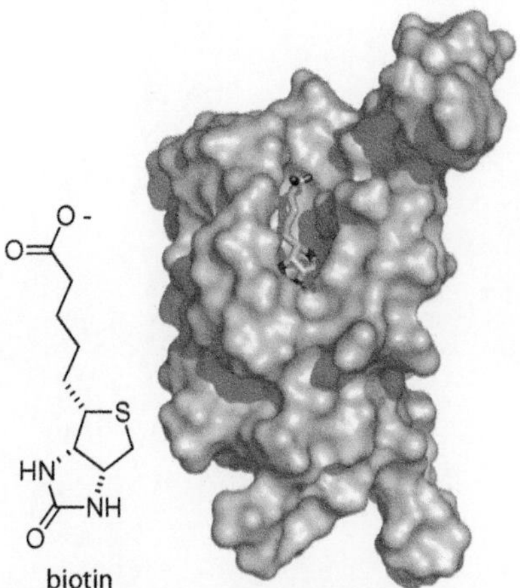

Figure 6.12 The origin of high affinity. Biotin binds snugly in the deep binding pocket of streptavidin. When streptavidin forms a tetramer of the monomer depicted here, the bicyclic core of biotin is completely encapsulated. (PDB 1STP)

the biotin and the small molecule ensures that the famously tight streptavidin–biotin binding can take place (**Figure 6.14**). After synthesis and purification, the biotinylated drug is added to cells that are known to be responsive to the drug. The drug conjugate can seek out and bind to its protein target. Then the entire contents of the cell are dumped onto a gel-like solid support that is chemically derivatized with streptavidin. The drug receptor then becomes immobilized, and, ideally, remains attached as successive washes with buffer remove the other molecules from the column. The affinity-purified receptor can be identified by mass spectrometry.

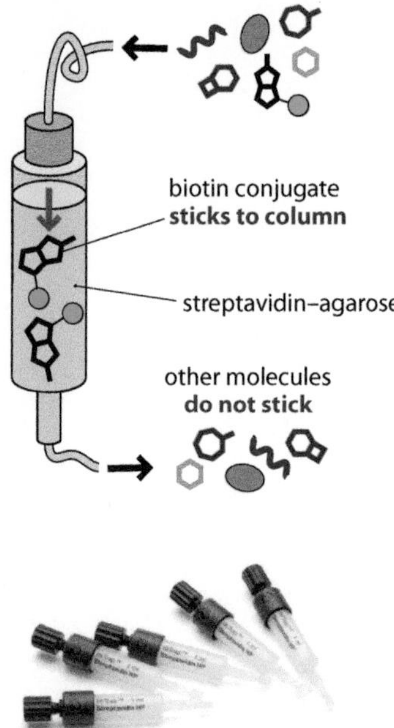

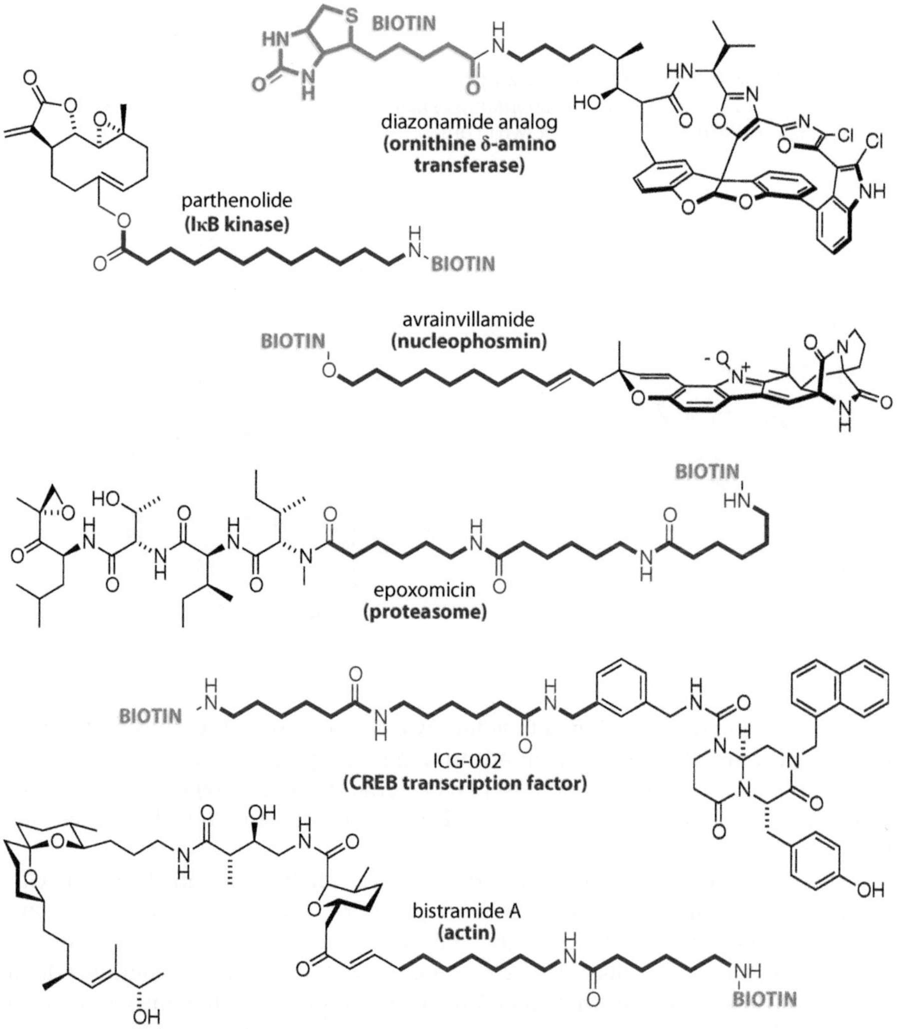

Figure 6.13 Commercial streptavidin columns. Prepackaged columns of streptavidin–agarose can isolate biotin conjugates from complex cellular mixtures. (Bottom, courtesy of GE Healthcare.)

Figure 6.14 Biotinylated small molecules. Examples of biotinylated probes used to identify their protein targets (in parentheses) in cells. Note that a flexible linker (red) is usually necessary to allow both the protein target and streptavidin to bind at the same time.

6.2 A QUANTITATIVE VIEW OF ENZYME FUNCTION

Enzymes are catalytic receptors

With a kinetic picture of receptor–ligand interactions in place, we are now prepared to understand enzymes. In his influential book *The Organic Chemistry of Drug Design and Drug Action*, Richard B. Silverman popularized the view of enzymes as catalytic receptors. By thinking of enzymes as receptors, one can capitalize on a kinetic understanding of receptor•ligand interactions. Recall that the dissociation constant for a receptor–ligand interaction is easily described by rate equations because the receptor–ligand interactions: $R + L \underset{k_{off}}{\overset{k_{on}}{\rightleftharpoons}} R{\bullet}L$

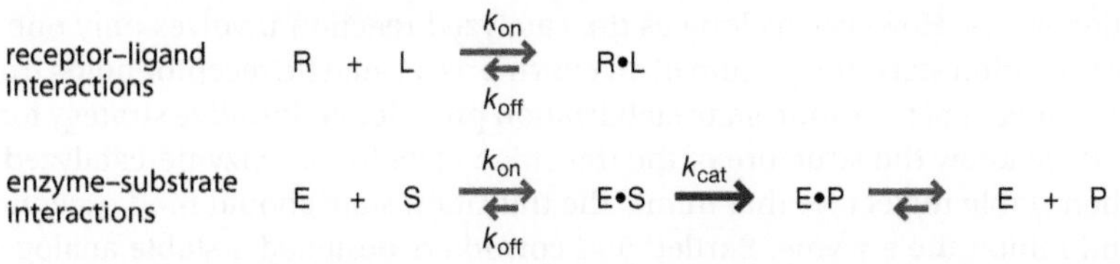

enzyme–substrate interactions: $E + S \underset{k_{off}}{\overset{k_{on}}{\rightleftharpoons}} E{\bullet}S \xrightarrow{k_{cat}} E{\bullet}P \rightleftharpoons E + P$

Figure 6.15 Enzymes as catalytic receptors. On-rates and off-rates are important for both receptor–ligand interactions and enzyme–substrate interactions. In addition, the enzyme–substrate complex (E•S) must choose between catalysis (k_{cat}) and dissociation (k_{off}).

receptor–ligand complex has only one choice—to dissociate. However, an enzyme-substrate complex (E•S) has two choices (**Figure 6.15**): to dissociate with a rate constant of k_{off} or to transform into an enzyme-product complex with a rate constant of k_{cat}.

We can recreate this picture in terms of free energy with a reaction-coordinate-energy diagram that compares the uncatalyzed reaction with the enzyme-catalyzed reaction (**Figure 6.16**). In this diagram we include the enzyme in the uncatalyzed reaction as a spectator that merely balances out the equations. Most reaction coordinate diagrams use standard-state free energies ($\Delta G°$) in which every species is present at 1 M concentration. For chemical biologists it makes more sense to think of conditions under which species are present at *physiological* concentrations. These concentrations will vary from enzyme to enzyme and from substrate to substrate, but a useful approximation is that enzymes are usually present at or below 1 μM.

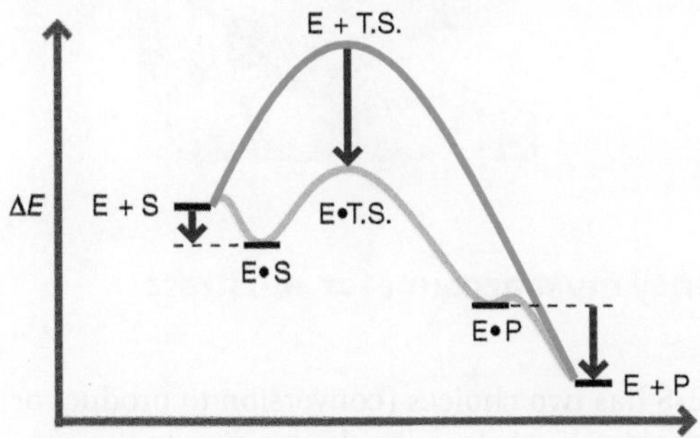

Figure 6.16 The enzyme changes the free-energy landscape. A reaction-coordinate–energy diagram allows one to compare the uncatalyzed pathway (blue) in which the enzyme is not bound to substrate, transition state (T.S.), or product with the catalyzed pathway, in which the enzyme is bound to substrate, transition state, and product.

Think about a hypothetical enzyme-catalyzed reaction with an enzyme at 1 μM and a substrate at 1 mM, and focus on the changes in free energy in the reaction-coordinate-energy diagram (see Figure 6.16). To ensure catalytic turnover, it should be energetically favorable for enzymes to bind to substrates, yet energetically unfavorable for the enzyme to remain bound to the product. More importantly, it should be *highly* energetically favorable for enzymes to bind to transition states. Because the goal of enzymes is to catalyze reactions, enzymes should bind to (and stabilize) transition states more strongly than they bind to substrates. Thus, enzymes can be thought of as receptors for transition states.

The concept of enzymes as receptors for transition states is easiest to visualize for reactions that involve a single elementary mechanistic step. For example, consider the enzyme chorismate mutase, which catalyzes the transformation of chorismate into prephenate through a pericyclic [3,3] sigmatropic rearrangement called a Claisen rearrangement (**Figure 6.17**). The enzyme binds the substrate in a conformation that

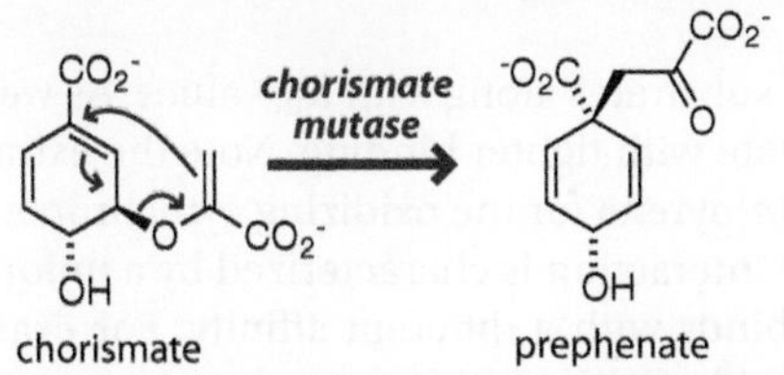

chorismate —*chorismate mutase*→ prephenate

Figure 6.17 Pericyclic reactions involve a single transition state. The transformation of chorismate to prephenate involves a Claisen rearrangement, which can be catalyzed by the enzyme chorismate mutase.

mimics the preferred chair-like transition state for the Claisen rearrangement. The vast majority of enzymes catalyze chemical reactions that have multiple steps and multiple transition states. However, as long as the catalyzed reaction involves only one high-energy transition state, the picture of an enzyme as a catalytic receptor holds true.

The concept of transition state stabilization provides an intuitive strategy for inhibition. If we know the structure of the transition state for an enzyme-catalyzed reaction, then stable molecules that mimic the transition state should bind exceptionally well and inhibit the enzyme. Bartlett and coworkers designed a stable analog of the bicyclic transition state of chorismate mutase and found it to be a potent inhibitor (**Figure 6.18**). The strategy of synthesizing a stable mimic of the transition state, termed a transition-state analog, for inhibition of a target enzyme works quite well. However, enzymes must bind to any other intermediates required for the reaction and also to the starting material. Thus, enzyme inhibitors can also mimic the structures of the starting material or even products. In addition, the crevassed, densely functionalized surfaces of enzyme active sites present numerous opportunities for high-affinity binding by compounds with little structural relationship to the enzyme's usual binding partners.

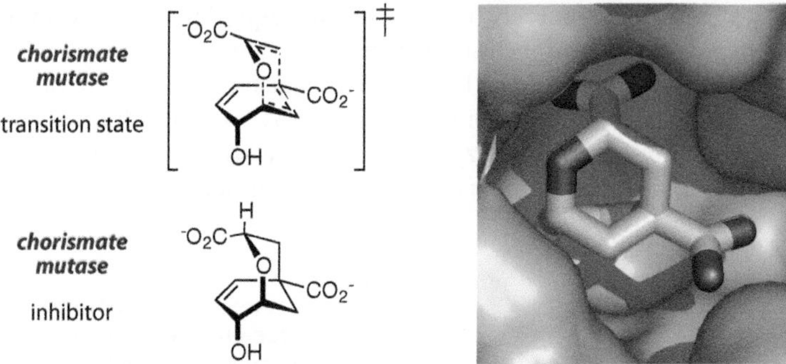

Figure 6.18 Enzyme inhibition by a transition-state analog. The bicyclic compound provides a stable mimic of the transition state for the reaction catalyzed by chorismate mutase (PDB 2CHT). Because the enzyme evolved to stabilize this transition state, the transition-state analog binds quite well, and effectively inhibits the enzyme.

Measurements of enzyme efficiency must account for substrate binding and catalysis

Since the enzyme-substrate complex E•S has two choices (conversion to product or substrate dissociation), we can no longer use K_d to describe the binding to the substrate. Furthermore, the dissociation constant K_d is defined for systems at thermodynamic equilibrium, whereas most enzyme-catalyzed reactions involve dynamic changes in concentrations. When we cannot use K_d to describe enzyme-substrate interactions, we use a very similar variable called the Michaelis constant, K_m (**Figure 6.19**). For a modest enzyme, K_m can be thought of as the dissociation constant for the enzyme and substrate; K_m tells you how tightly the enzyme binds the substrate. Enzyme efficiency is often described using the ratio k_{cat}/K_m.

$$E + S \underset{k_{off}}{\overset{k_{on}}{\rightleftharpoons}} E{\bullet}S \xrightarrow{k_{cat}} E + P \qquad K_m = \frac{k_{off} + k_{cat}}{k_{on}}$$

Figure 6.19 The Michaelis constant. K_m resembles an equilibrium constant for the formation of E•S and the resultant catalysis. K_m is similar to K_d for the substrate, particularly if k_{off} is much faster than k_{cat} (that is, for enzymes having a hard time getting a grip on their substrate and also responsible for the catalysis of a difficult reaction).

Table 6.2 gives some typical enzyme substrates along with K_m values. As we have seen for K_d values, lower numbers correlate with tighter binding. Note the extraordinary affinity of the precarcinogen benzo[*a*]pyrene for the oxidizing cytochrome P450 enzyme. Each specific substrate-enzyme interaction is characterized by a unique K_m value because each substrate (usually) binds with a different affinity. For example, from the two entries in Table 6.2, the peptide GGYAELRMGG binds to matrix metalloprotease-11 (MMP-11) over three times more tightly than GGAANLVRGG does.

We can use the dose-response curve for receptor-ligand interactions to describe classical enzyme-substrate behavior by simply renaming the axes "initial velocity"

Table 6.2 K_m values for typical enzyme–substrate interactions.

Enzyme	Substrate	K_m (μM)
aconitase	citrate HO CO_2^- HO_2C CO_2H	2900
MMP-11 (protease)	GlyGlyAlaAlaAsnLeuValArgGlyGly GlyGlyTyrAlaGluLeuArgMetGlyGly	705 210
fucosyl transferase	GDP-fucose	100
protein kinase A	Mg_2•ATP LeuArgArgAlaSerLeuGly	20 7
dUTPase	dUTP	4
Cyp450	benzo[a]pyrene	0.006

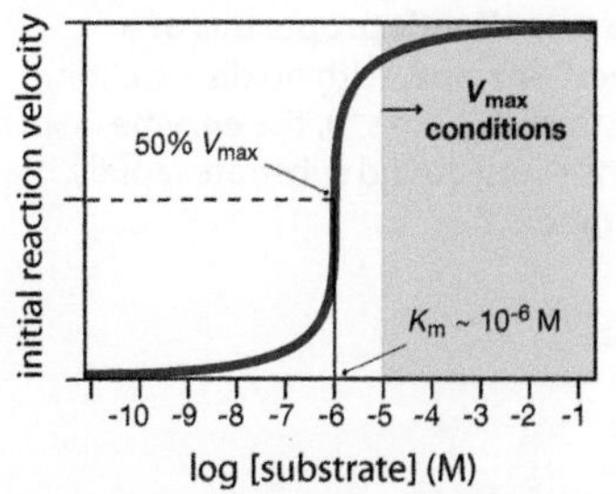

Figure 6.20 Visualizing enzyme catalysis with a dose–response curve. Dose–response curves describe enzyme–substrate interactions. The inflection point (50% of the maximum velocity) would be called K_m instead of EC_{50}. We refer to substrate concentrations of more than 10 K_m as "V_{max} conditions", because the reaction velocity is near the maximum rate.

and "log[substrate]" for the *y* and *x* axes, respectively (**Figure 6.20**). In most cases, the inflection point will be close to the K_m for the substrate. If the substrate concentration is 10 times lower than K_m, there will be negligible turnover. Thus, if K_m is known under physiological conditions, it offers some predictive value concerning the minimum concentration of substrate that one should expect to find in the cell. Conversely, when the substrate concentration is at least 10 times K_m, most of the enzyme's active sites will be bound by substrate waiting for catalytic transformation. Assay measurements made under such conditions, with a saturating concentration of substrate, are termed maximum-rate, or V_{max}, conditions. Under V_{max} conditions, the rate of the catalytic reaction is directly proportional to the concentration of enzyme. ELISA screens, or other high-throughput assays that involve enzymes, are performed under V_{max} conditions. V_{max} conditions are also used to determine the quantity of enzyme in an unknown sample.

The catalytic step(s) of enzyme catalysis are intramolecular reactions because the substrate(s) are bound to the enzyme. Most enzyme-catalyzed steps that go on inside human cells are slow relative to on-rates and off-rates because they involve the formation and cleavage of bonds. **Table 6.3** provides some typical k_{cat} values for common classes of human enzymes.

Table 6.3 Typical k_{cat} values for common enzyme classes.

Substrates	Enzyme	Products	k_{cat} (s^{-1})
H_2O + R–C(=O)–NH–R	**protease** →	R–C(=O)–O^- + H_3N^+–R	**1**
PROTEIN–OH + HO–P(=O)(O^-)–O-ADP	**kinase** →	PROTEIN–O–P(=O)(O^-)–OH + HO-ADP	**10**
H_2O + –O–P(=O)(O^-)–O–	**RNase** →	–O–P(=O)(O^-)–OH + HO–	**100**
HO–C(=O)–OH	**carbonic anhydrase** →	O=C=O + H_2O	**1,000,000**
HO–OH + HO–OH	**catalase** →	O_2 + 2 H_2O	**100,000,000**

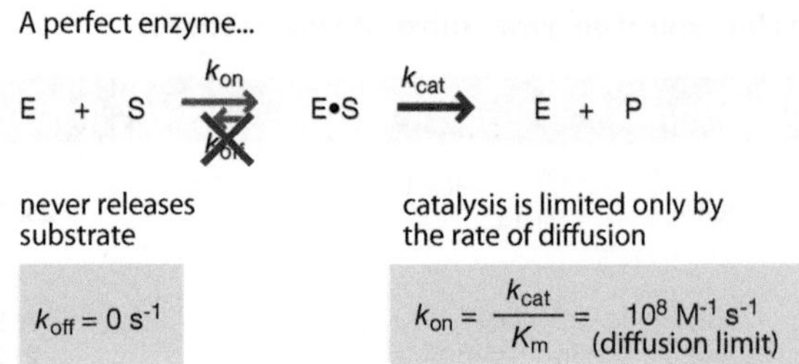

Figure 6.21 Kinetic properties of a "perfect" enzyme. With no dissociation of substrate ($k_{off} = 0\ s^{-1}$), the enzyme converts every bound substrate rapidly to the product.

For a perfect enzyme, k_{cat}/K_m is equal to the diffusion-controlled rate of between 10^8 and 10^9 M^{-1} s^{-1} (**Figure 6.21**). In this perfect enzyme, every collision between substrate and enzyme leads to the formation of a complex and the substrate is never released. Instead, the substrate is converted to product and released faster than another substrate molecule can find the enzyme. Some enzymes, such as superoxide dismutase and triosephosphate isomerase, approach these performance parameters for unrivaled catalytic abilities. Unfortunately, many of our conceptual shortcuts fail for perfect enzymes; in particular, K_m will not be a good approximation for the substrate K_d. Fortunately, enzymatic perfection is not common among human enzymes. It is tempting to think of k_{cat} and K_m as characteristics of the enzyme, simply because most enzymes have evolved to act on a single substrate. However, every possible enzyme–substrate interaction has a distinctive k_{cat} and K_m value.

Problem 6.5

In the cheese-making bacterium *Lactococcus lactis*, the enzyme galactose mutarotase catalyzes the epimerization of the hemiacetal moiety of galactose.

A Which of the three substrates below binds more tightly?

B Overall, which substrate is isomerized most efficiently by the enzyme?

C At what concentration of glucose would more than 90% of the enzyme be occupied by glucose?

galactose mutarotase

	K_m (mM)	k_{cat} (s^{-1})
galactose	20	3700
glucose	34	430
xylose	21	410

6.3 A MECHANISTIC VIEW OF ENZYMES THAT CATALYZE MULTISTEP REACTIONS

Protein kinases and proteases catalyze reactions through multistep mechanisms

The International Union of Biochemistry and Molecular Biology has designated six broad functional enzyme classes: EC 1, **oxidoreductases** that catalyze oxidation and reduction reactions; EC 2, **transferases** that catalyze the transfer of functional groups such as a phosphoryl group, a nucleotide, or a sugar; EC 3, **hydrolases** that catalyze hydrolysis of functional groups such as amides; EC 4, **lyases** that cleave bonds through mechanisms other than hydrolysis or redox reactions; EC 5, **isomerases** that catalyze isomerizations without changes in constitution (molecular formula); and EC 6, **ligases** that catalyze the ligation of two independent molecules. Solely on the basis of its hierarchical designation EC 3.2.1.142, the enzyme pullulanase can be recognized as a hydrolase that hydrolyzes *O*-glycosidic bonds.

Table 6.4 Number of human genes encoding enzymes.

Enzyme function	No. of genes	Notable examples
Hydrolases	1753	Protease, nuclease, glycosylase, lipase
Transferases	1675	Protein kinase, glycosyltransferase
Oxidoreductases	603	Cytochrome P450, alcohol dehydrogenase
Ligases	473	Ubiquitin ligase, DNA ligase
Isomerases	357	Helicase, topoisomerase, epimerase
Synth(et)ases	213	Aminoacyl-tRNA synthetase, thymidylate synthase
Lyases	157	Aldolase, decarboxylase

Some enzymes do not fall neatly into these six categories. In particular, enzymes that synthesize biomolecules through complex mechanisms are classified as **synthases** or **synthetases**; the term synthetase is usually reserved for enzymes that are driven by a high-energy molecule such as ATP. At the time of writing of this book, more than 5000 out of about 22,000 protein-encoding human genes code for enzymes of known function (**Table 6.4**), but the functions of many genes remain to be assigned.

The mechanism of chorismate mutase was readily understood by using the catalytic receptor model for enzymatic catalysis, because it catalyzed a chemical reaction with a single transition state. However, unlike chorismate mutase, most enzymes proceed through reactions that involve more than one elementary reaction step, and hence more than one transition state. This introductory book does not provide sufficient space to describe the many complex mechanisms of all human enzymes. So we will focus on the most common enzymes in humans.

Within the seven broad classes of human enzymes listed in Table 6.4, protein kinases and proteases stand out for their abundance. The human genome codes for 518 protein kinases and more than 460 catalytically active proteases. The large populations and importance of these two enzyme classes merit detailed examination. Protein kinases and proteases catalyze reactions through multistep mechanisms. Many of the steps in these enzyme-catalyzed reactions involve relatively rapid proton transfers that do not require any special feat by the enzyme other than to provide a carefully positioned acidic or basic functional group. However, the mechanisms of reactions catalyzed by kinases and proteases do involve more than one high-energy transition state—one for addition of the nucleophile and one for elimination of the leaving group. As we will see, Nature chooses catalytic pathways in which multiple high-energy transition states are nearly identical in terms of geometry and charge. Thus, even kinases and proteases can be thought of as receptors for multiple similar transition states.

Protein kinases share a common motif

Protein kinases catalyze the transfer of the γ-phosphate group of $2Mg^{2+}$•ATP to a protein functional group, usually the hydroxyl group of serine, threonine, or tyrosine. The human protein kinases can be roughly grouped into five classes based on sequence homology; however, these broad homology classes have no correlation with the roles of the kinases in human cells. All human protein kinases possess a common structure and catalytic core that is easily recognized (**Figure 6.22**).

The Mg^{2+} ions help to neutralize the charge of the triphosphate and the charge that develops in the transition state. Many of the mechanistic hypotheses have been based on the crystal structures of $2Mn^{2+}$•ATP complexes of protein kinases. It is not entirely clear how relevant such structures are in deciphering the role of Mg^{2+} ions in the mechanism of protein kinases. ANP (**Figure 6.23**) is also used in structural studies as a nonhydrolyzable analog of ATP. Our best guesses as to the mechanism

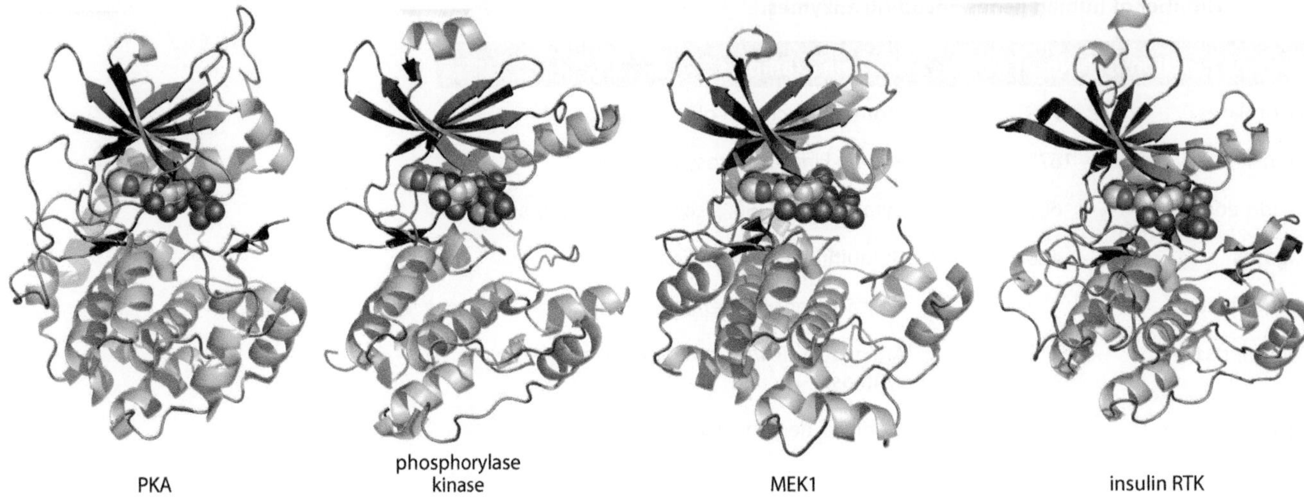

Figure 6.22 Protein kinase structures are highly conserved. A comparison of four protein kinases with ATP bound in the active site reveals similarities: PKAα (PDB 1RDQ); phosphorylase kinase (PDB 1QL6); MAP kinase/ERK kinase 1 (PDB 1S9J); and insulin receptor tyrosine kinase (PDB 1IR3).

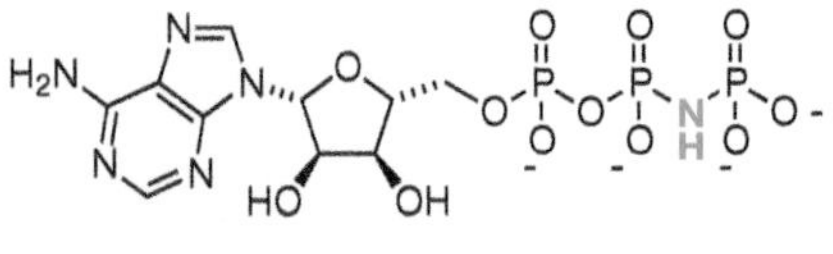

Figure 6.23 A stable analog of ATP. Crystal structures of protein kinases are often obtained with ANP, an aza-analog of ATP. ANP is not readily hydrolyzed when bound to the kinase active site.

for phosphoryl transfer from $2Mg^{2+}$•ATP to serine hydroxyl groups come from analysis of cyclic-AMP-dependent protein kinase, also known as protein kinase A (PKA), for which co-crystal structures have been obtained before and after hydrolysis of the $2Mg^{2+}$•ATP ligand (**Figure 6.24**).

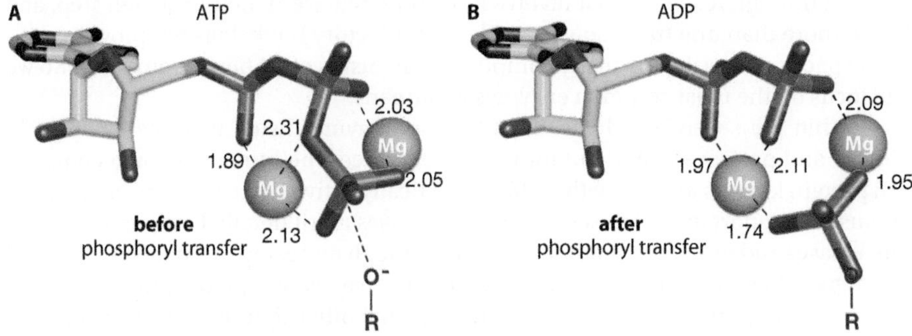

Figure 6.24 Before and after phosphoryl transfer. Crystal structures of $2Mg^{2+}$•ATP before (A) and after (B) hydrolysis offer a model for the mechanism of protein kinases (PDB 1RDQ); Mg–O distances are in ångströms.

Protein kinases possess a conserved lysine and a conserved aspartate residue in the active site. The aspartate residue helps to deprotonate the serine hydroxyl acceptor. The serine alkoxide group then attacks the phosphate, forming a stable pentavalent phosphorane intermediate, which then ejects the β-phosphate of the ADP leaving group. The Mg^{2+} ions seem to remain coordinated throughout the reaction (**Figure 6.25**). The trigonal bipyramidal intermediate looks so similar to an S_N2 transition state that many students tend to think of kinase-catalyzed phosphoryl transfer as a concerted process. However, enzymes obey the same chemical rule as nonenzymatic reactions—there are no S_N2 reactions at phosphate phosphorus atoms.

Figure 6.25 Kinase mechanism. The mechanism of phosphoryl transfer by a protein kinase (Ado = 5′-adenosyl) involves a trigonal bipyramidal intermediate. The Lewis acidic Mg^{2+} ions can coordinate to and stabilize the transferred phosphate.

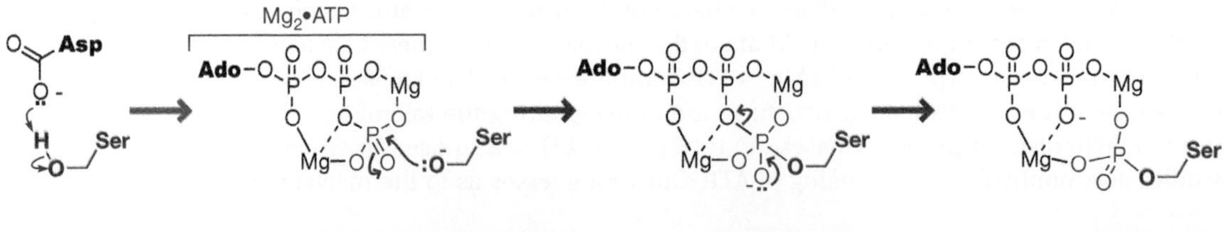

Problem 6.6

The following kinetic parameters were measured when the peptides below were phosphorylated with ATP by cAMP-dependent protein kinase.

A Which peptide substrate binds most tightly to the kinase?

B Once the substrate is bound, which one is phosphorylated fastest?

C What is the ratio of rates of phosphorylation between the best and worst substrates?

Substrate	K_m (μM)	k_{cat} (s^{-1})
LRAASLG	12,200	8.7
LHRASLG	804	19.8
LRRASLG	31	33.1

A knowledge of enzyme mechanism and structure allows chemical biologists to design reagents with unprecedented properties. For example, the Shokat group has engineered an ATP analog that crosslinks protein kinases with mutated substrates having cysteine in place of the serine acceptor residue. The reagent is a hybrid between adenosine and the fluorogenic amino acid detection reagent *ortho*-phthalaldehyde. The reagent binds in the ATP site of the kinase and the *ortho*-phthalaldehyde then condenses with the conserved lysine in the kinase active site and the cysteine residue of the mutated target protein. The resulting crosslink is a fluorescent isoindole that facilitates detection (**Figure 6.26**).

designed peptide/protein substrate

designed covalent inhibitor

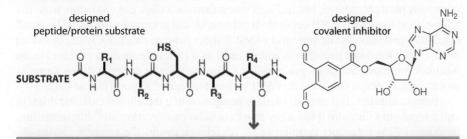

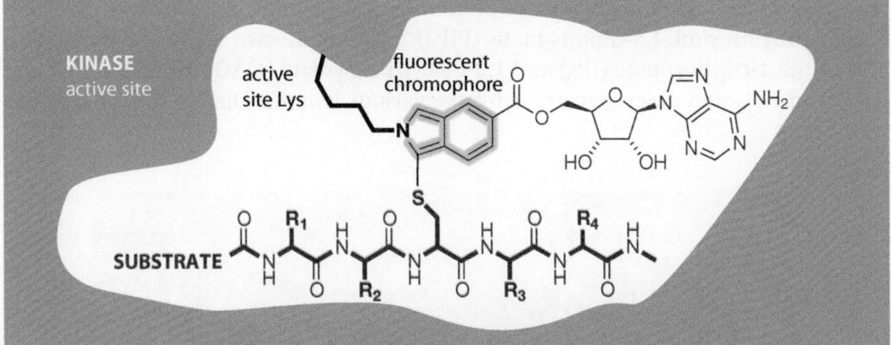

Figure 6.26 A mechanism-based reagent for trapping mutant kinase substrates. The designed peptide substrate (blue) has cysteine in place of the normal serine phosphoryl acceptor.

Problem 6.7

Suggest a plausible arrow-pushing mechanism for the following method for derivatizing amino acids.

fluorescent

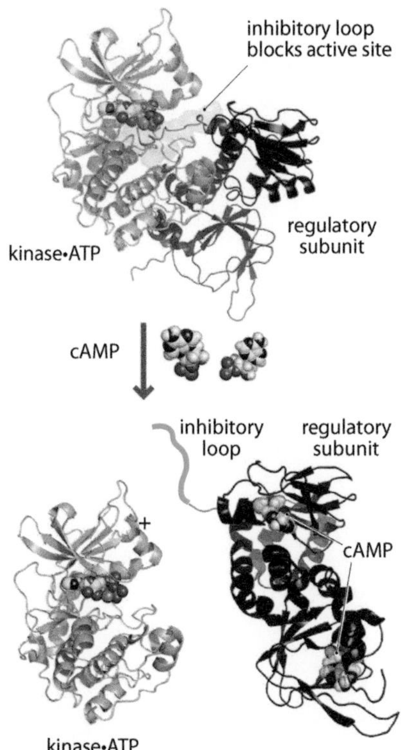

Figure 6.27 Sequestration and release of cAMP-dependent protein kinase (PKA). Normally, a regulatory subunit inhibits the activity of PKA (PDB 2QCS) by sticking protein loops (light blue) in the kinase active site. Binding of cAMP to the regulatory subunit (PDB 1NE6) of PKA releases the active kinase.

Regulation of protein kinase activity requires allosteric binding

All communication networks require the ability to switch between on and off states. For example, in early telephone networks, an operator connected callers manually. Then electromechanical switches largely replaced the human operators. In modern optical fiber networks, photonic switchers control the flow of optical signals. Faithful transmission of chemical signals is essential for the function and cooperation of human cells. In signal transduction pathways, protein kinases serve as both the switches and the conduits of information. The activity of protein kinases can be controlled through a variety of mechanisms, most commonly localization, ligand binding, and phosphorylation.

The activity of many common kinases is controlled by the binding of small messenger molecules. For example, cyclic-AMP-dependent protein kinase (PKA) normally exists as an inactive tetrameric complex composed of two kinases and two regulatory proteins. When two molecules of 3′,5′-cyclic adenosine monophosphate (cAMP) bind to each of the regulatory subunits, the kinases are released and proceed to phosphorylate a wide variety of proteins, ultimately leading to cell proliferation (**Figure 6.27**).

The human genome has genes for three different forms of cAMP-dependent protein kinases and four different forms of regulatory subunits. The regulatory subunits of cAMP-dependent protein kinase lodge an extended peptide into the active site, sealing off access (**Figure 6.28**). When two molecules of 3′,5′-cyclic adenosine monophosphate (cAMP) bind to a regulatory subunit, it undergoes a drastic conformational change that promotes the release of the kinase, allowing it to phosphorylate target proteins.

Inhibitory loops that interact with the active site of cAMP-dependent protein kinases tend to have a common sequence motif: RRXSϕ, where X is any amino acid and ϕ is a hydrophobic amino acid. In the regulatory inhibitor α_2, the extended peptide is actually phosphorylated, but unlike a true substrate it does not dissociate from the kinase. In some cells, cAMP-dependent protein kinase is strongly inhibited by small regulatory proteins that do not bind cAMP. A short peptide called PKI-tide, based on one of these inhibitor proteins, is widely used as a cAMP-dependent protein kinase inhibitor. Similarly, a heptapeptide, called Kemptide after its designer Bruce Kemp, is an efficient, selective substrate for cAMP-dependent protein kinase (**Table 6.5**).

Protein kinase C is a modular kinase composed of a traditional catalytic domain and a regulatory domain. It has a key role in cellular proliferation and differentiation. Binding of 7TM receptors coupled to G_q proteins activates the enzyme phospholipase C, which triggers activation of protein kinase C. Phospholipase C hydrolyzes phosphatidylinositol 4,5-diphosphate (PIP_2) to generate two signaling molecules: inositol 1,3,4-triphosphate (IP_3) and 1,2-diacyl-*sn*-glycerol (DAG). Note that the lipid chains on PIP_2 and diacylglycerol can have various lengths. Diacylglycerol activates

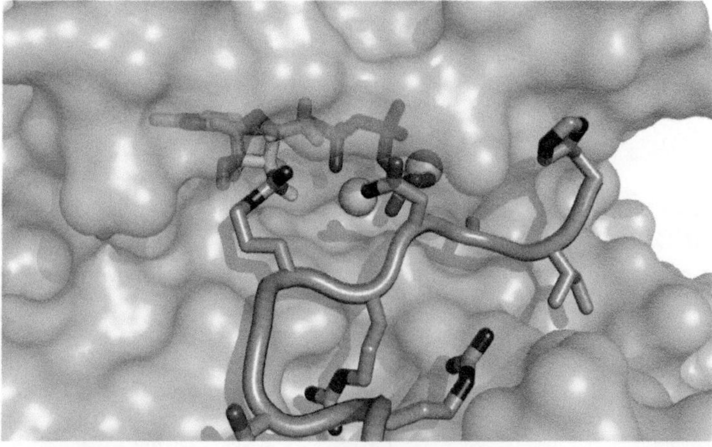

Figure 6.28 A peptide–kinase interaction. Structural basis for peptide binding to PKA. Residues RRQAIH of a PKI-tide block the active site. (PDB 3FJQ)

Table 6.5 Sequence homology among inhibitors and the Kemptide substrate of cAMP-dependent protein kinase.

Sequence	Found in
...KGRRRRGAISA...	Human regulatory inhibitor α1
...SRFNRRVSVCA...	Human regulatory inhibitor α2
...GRTGRRNAIHD...	cAMP-dependent protein kinase inhibitor
IAAGRTGRRQAIHDILVAA	PKI-tide
LRRASLG	Kemptide substrate (K_m 33 μM)

protein kinase C directly; IP_3 activates it indirectly. IP_3 opens Ca^{2+} ion channels—both in the plasma membrane and in vesicles that store Ca^{2+} ions—and the Ca^{2+} ions then act on cytosolic protein kinase C (**Figure 6.29**). The natural product staurosporine is a nonselective inhibitor of protein kinases that binds to the ATP site. Various isoforms of protein kinase C are distributed in cells throughout the body. For example, PKCβ_1 and PKCβ_2 phosphorylate proteins that ultimately control cell growth, and are active in many types of cancers. The synthetic analog ruboxistaurin is highly selective for the β isoforms of mammalian protein kinase C (**Figure 6.30**).

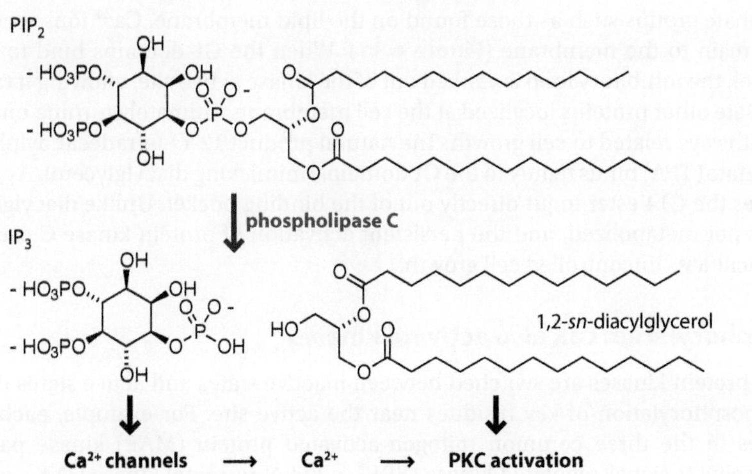

PIP₂ → phospholipase C → IP₃ + 1,2-*sn*-diacylglycerol; IP₃ → Ca^{2+} channels ⟶ Ca^{2+} ⟶ PKC activation

Figure 6.29 PKC activation pathway. Cleavage of PIP_2 releases two signaling molecules that ultimately facilitate the activation of protein kinase C (PKC).

Like cAMP-dependent protein kinase, protein kinase C is a multidomain protein with a pseudosubstrate inhibitory loop that blocks the active site. Protein kinase C is activated when the concentration of Ca^{2+} is high in the cytosol and the concentration of diacylglycerol is high in the membrane. Ca^{2+} ions have a natural affinity for anionic

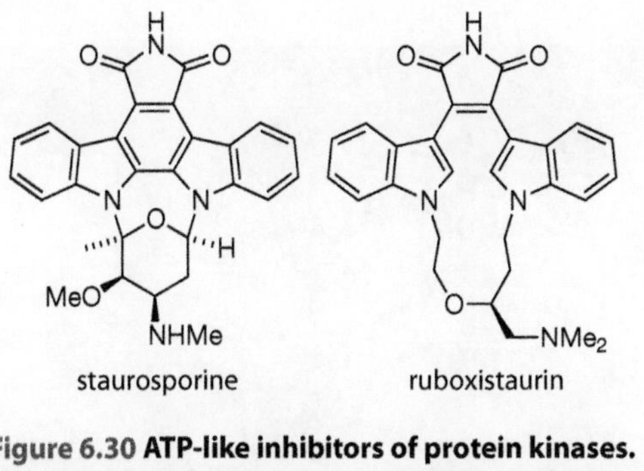

staurosporine ruboxistaurin

Figure 6.30 ATP-like inhibitors of protein kinases. The natural product staurosporine is a non-selective inhibitor of many kinases. In contrast, a synthetic analog of staurosporine, ruboxistaurin, targets the β isoforms of PKC with high selectivity.

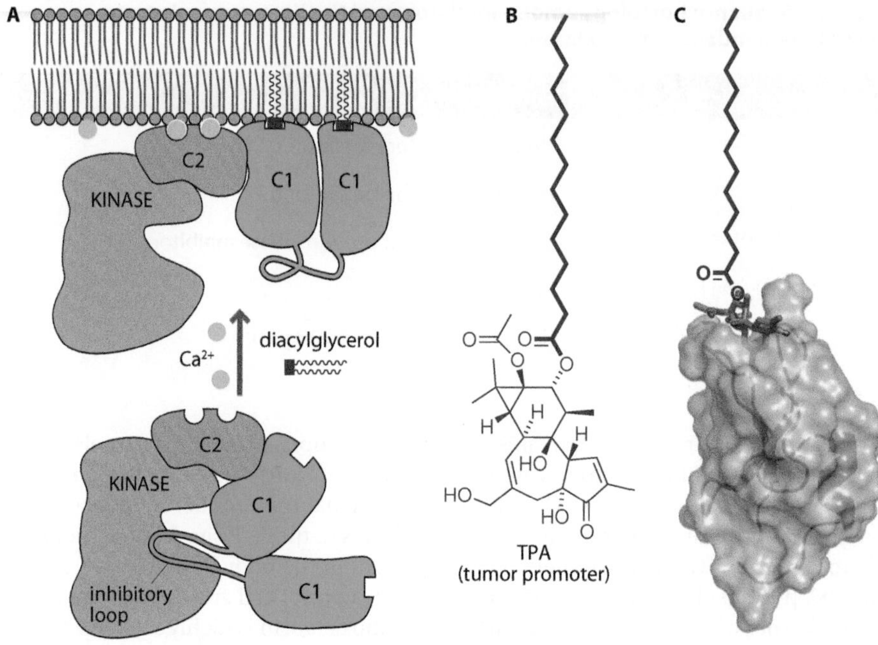

Figure 6.31 Ca^{2+} and diacylglycerol recruit and activate PKC at the cell membrane. (A) Normally, the inhibitory loop of PKC blocks the active site. Ca^{2+} and diacylglycerol mobilize the kinase at the cell membrane in a conformation that is not inhibited. (B) The tumor promoter TPA has a long fatty acid substituent that serves as a lipid membrane anchor. (C) When TPA binds to the C1 domains of PKC the lipid anchor juts outward. (Adapted from J.M. Berg et al., Biochemistry, 5th ed. New York: W.H. Freeman, 2002.)

phosphate groups such as those found on the lipid membrane. Ca^{2+} ions recruit the C2 domain to the membrane (**Figure 6.31**). When the C1 domains bind to diacylglycerol, the inhibitory loop is yanked out of the kinase active site, allowing it to phosphorylate other proteins localized at the cell membrane, ultimately turning on signaling pathways related to cell growth. The natural product 12-*O*-tetradecanoylphorbol-13-acetate (TPA) binds tightly to the C1 domains, mimicking diacylglycerol. $A_{1,3}$-strain induces the C14 ester to jut directly out of the binding pocket. Unlike diacylglycerol, TPA is not metabolized, and the persistent activation of protein kinase C caused by TPA facilitates uncontrolled cell growth.

Phosphorylation can also activate kinases

Many protein kinases are switched between inactive states and active states through the phosphorylation of key residues near the active site. For example, each of the kinases in the three common mitogen-activated protein (MAP) kinase pathways (extracellular signal-regulated kinase (ERK), c-Jun N-terminal kinase (JNK), and p38 pathways) is activated through the phosphorylation of two residues in a loop near the active site (**Figure 6.32**). The phosphorylation loops tend to have the following motifs: MAP kinases, ThrXxxTyr; MAP kinase kinases, SerXxxXxxXxxSer/Thr; MAP kinase kinase kinases, ThrXxxXxxSer/Thr (**Figure 6.33**). Without these negatively charged phosphoryl groups, the kinases will not phosphorylate their downstream targets.

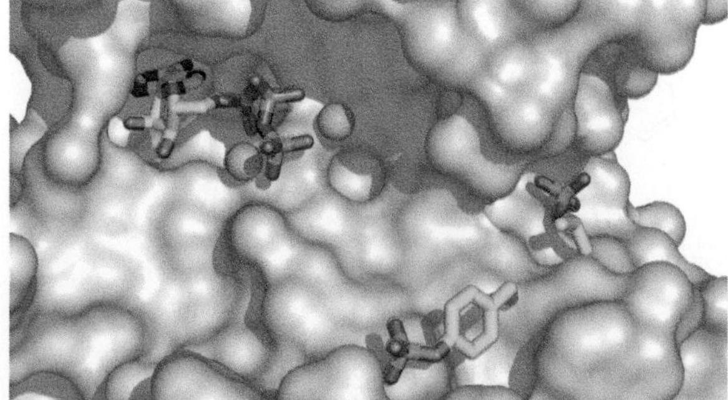

Figure 6.32 Phosphorylation to activate kinases. The phosphotyrosine and phosphothreonine residues of p38γ MAP kinase are located close to the active site (PDB 1CM8). Note the structure of the nonhydrolyzable ATP analog ANP in the active site (upper left).

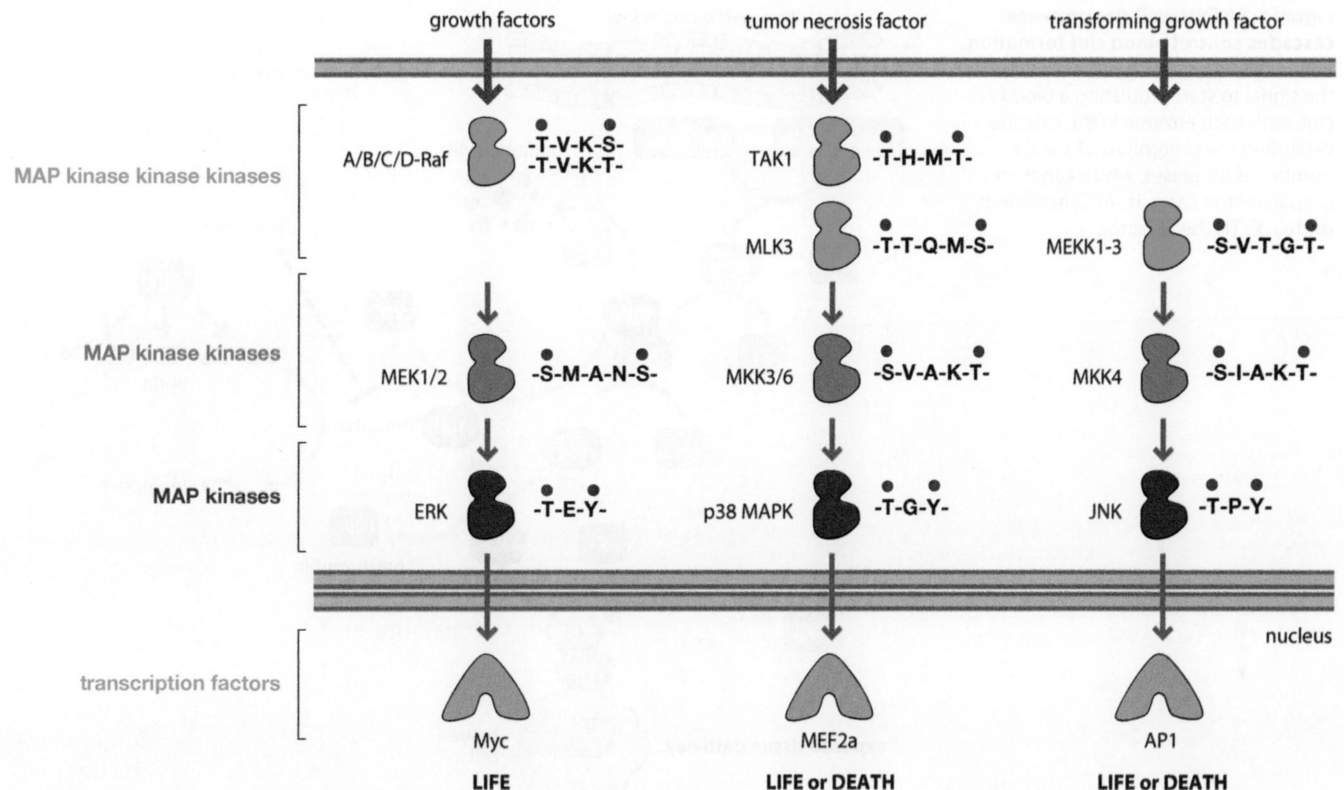

Figure 6.33 MAP kinases are activated through double phosphorylation. Double phosphorylation of short loops leads to the activation of kinases in MAP kinase pathways. Homologous MAP kinases (such as MAP kinase kinases) have similar sequence motifs in the activation loops. The red dots indicate phosphorylation sites.

Constitutively active variants of MAP kinase kinases and MAP kinase kinase kinases can often be generated by mutating phosphorylatable serine or threonine residues to glutamate. Constitutively active variants of MAP kinases cannot readily be prepared by mutagenesis, because there are no typical amino acids that mimic phosphotyrosine.

Proteases serve roles in degradation and protein signaling

Proteases are enzymes that cleave proteins and peptides hydrolytically. The sequence specificity of DNA-cleaving endonucleases is usually very high. However, many proteases target rather vague structural features such as the positively charged amino acids Lys and Arg or residues at the C terminus. Proteases such as trypsin, chymotrypsin, elastase, and carboxypeptidase act in the human digestive tract to break proteins down into peptides and amino acids. Collagenase, elastase, and cathepsin K have an essential role in tissue remodeling by hydrolyzing the extracellular polymeric proteins. Most cathepsins act inside lysosomal compartments to degrade proteins that have been taken in through endocytosis. Among the intracellular proteases, the large multisubunit proteasome has a key role in the ongoing molecular salvage operation. It degrades proteins into their constituent amino acids, ensuring that amino acid building blocks are readily available and that damaged proteins do not persist.

Sequence-selective proteases are essential in injury and signaling pathways, both between cells and within cells. For example, the remodeling of connective tissues is dependent on a series of sequence-selective proteases: tissue plasminogen activator, plasmin, stromelysin, collagenase, and gelatinase. Thrombosis (blood clotting) is another excellent example of an extracellular signaling pathway that involves a series of sequence-selective proteases. Many of these proteases retain their mysterious historical names, such as clotting factors VII–XII.

Enzymes that are activated through covalent modification are referred to as zymogens. Many kinases fall into this category, but the term zymogen is more commonly reserved for precursors to proteases, also called pro-proteases. For example, the main mechanism for initiation of the clotting protease cascade is the exposure of prekallikrein, on the surface of damaged endothelial cells, to the protease factor XII.

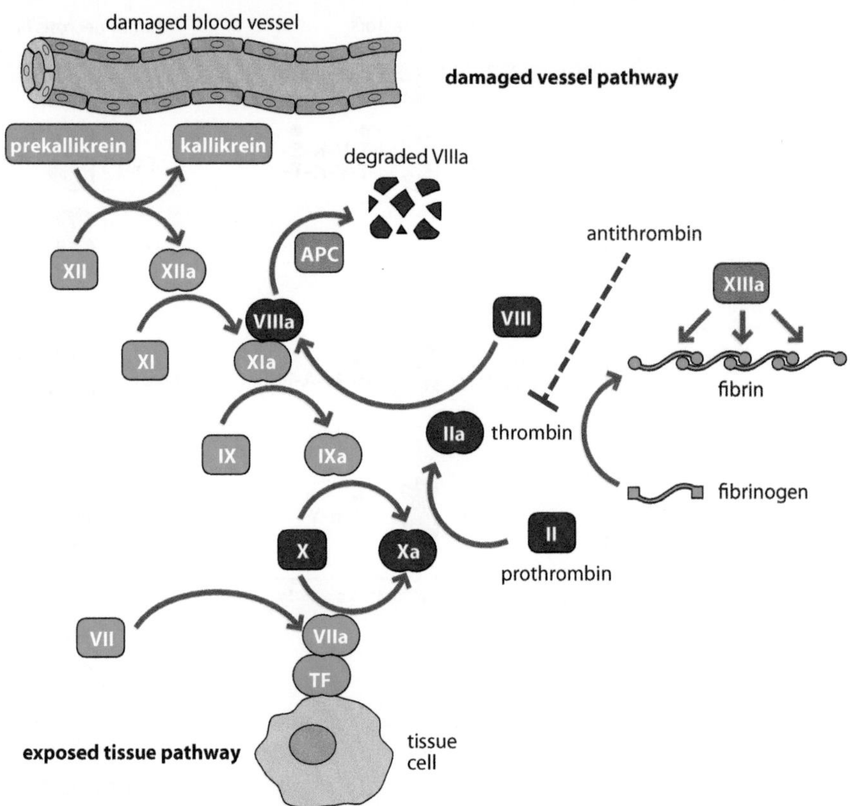

Figure 6.34 Extracellular protease cascades control blood clot formation. This complex signaling pathway amplifies the signal to start producing a blood clot, with each enzyme in the cascade catalyzing the proteolysis of a large number of proteases, which can then propagate the cascade. APC, activated protein C; TF, tissue factor.

Factor XII cleaves prekallikrein to the activated form kallikrein; and kallikrein reciprocally cleaves factor XII, converting it to the activated form, factor XIIa (the letter "a" means "activated"). Factor XIIa cleaves and activates factor XI; factor XIa cleaves and activates factor IX; factor IXa cleaves and activates factor X; factor Xa cleaves prothrombin (factor II) to generate thrombin (factor IIa). Finally, the active thrombin proteolytically activates fibrinogen into fibrin. Fibrin oligomerizes into fibrils that form noncovalent crosslinks between platelets (**Figure 6.34**). The connections between the fibrin monomers are welded together by a transglutaminase that forms crosslinks between lysine and glutamine side chains of fibrin monomers.

Each step in the pathway amplifies the signal in the cascade. For example, each molecule of factor XIIa can act on 1000 or more molecules of factor XI. Thus, such enzyme-based signaling pathways should be viewed as less like a row of dominos and more like a chain letter. Critical to control over the pathway, antithrombin acts like the person receiving the chain letter who throws it in the trash. By restricting signal propagation, antithrombin prevents the signal from raging out of control to coagulate the entire blood system's supply of fibrin.

The coagulation process is autocatalytic because factor XIa must dimerize with another protease, factor VIIIa, that is generated by thrombin. Exposed tissues also initiate coagulation. Cells that are not part of the vascular endothelium express the membrane-bound protein cryptically called tissue factor (TF). Tissue factor recruits factor VII, which can then be activated by a wide range of protease clotting factors. The activated factor VIIa then cleaves and activates factor X. The coagulation cascade is kept in check by a variety of mechanisms. One important mechanism is the **proteolytic** degradation of factor VIIIa by activated protein C (APC, not to be confused with antigen-presenting cell). As described above, antithrombin also keeps coagulation in check, by binding tightly to thrombin and blocking the active site. Most animals that suck human blood, such as leeches and ticks, produce proteins that bind to and inhibit thrombin. Anticoagulation drugs can work by inhibiting proteases in this cascade (**Figure 6.35**).

Defective enzymes in the coagulation protease pathway can lead to hemophilia—the inability to form effective clots after injury. The most common form of hemophilia, hemophilia A, is due to mutations in the factor VIII gene. The second most common

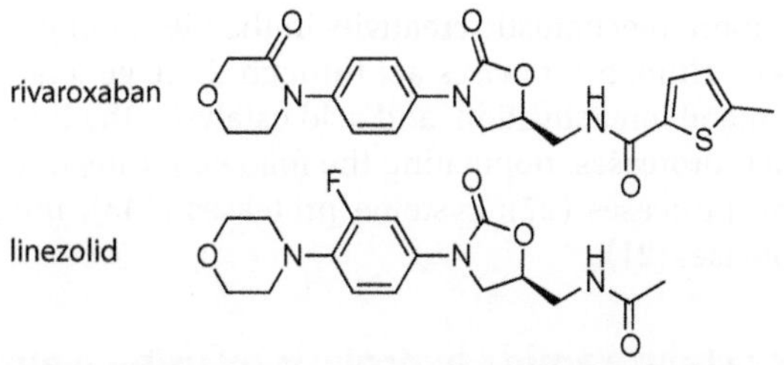

Figure 6.35 Similar structures, different targets. The potent factor Xa inhibitor rivaroxaban (IC_{50} = 0.7 nM) is similar in structure to the antibiotic linezolid, an inhibitor of ribosomal protein synthesis in Gram-positive bacteria. Surprisingly, however, rivaroxaban does not exhibit significant antibiotic activity.

Figure 6.36 Hemophilia. Czar Nicholas II's wife Alexandra passed on the gene for hemophilia to their son Alexis. (Courtesy of Hulton Archive. With permission from Getty Images.)

form of hemophilia is due to mutations in the factor IX gene (for example in the Russian royal family; **Figure 6.36**). The genes for these clotting factors are located on the X chromosome. Women rarely suffer from hemophilia A or B because they usually have a functional clotting factor gene on at least one of their two X chromosomes. Hemophilia A and B are observed almost exclusively in men because they have only one X chromosome. Males therefore have no backup system for defective genes on their X chromosomes. Although too little clotting is problematic, too much clotting is equally undesirable. Uncontrolled clotting of blood in the vessels that supply oxygen to the heart leads to heart attacks. In addition, uncontrolled clotting of blood in the vessels that supply oxygen to the brain leads to strokes. Heart disease and strokes are the two major causes of death in the United States.

The coagulation cascade is an extracellular process; however, proteases also have diverse roles in intracellular signal transduction pathways. For example, the apoptotic pathways triggered when cytotoxic T lymphocytes target cells infected with viruses or bacterial parasites. When the trimeric Fas protein, displayed on the surface of cytotoxic T lymphocytes, binds to the Fas receptor on target cells, it leads to apoptosis of the target cell. Fas ligand (FasL) binds to the Fas receptor (**Figure 6.37**). The intracellular protein Fas-associated death domain (FADD) binds to several molecules of procaspase-8, which proteolytically activate each other to generate active caspase-8. Caspase-8 can activate caspase-3 directly, but it also activates caspase-3 through the mitochondria. Caspase-8 cleaves the protein Bid, causing it to associate with the mitochondrial membrane, where it induces release of the redox protein cytochrome *c* from the mitochondria through a mechanism that has not yet been identified. When the cytochrome *c* is released from mitochondria, it binds to apoptotic protease activating factor-1 (Apaf-1) and dATP to form a complex called the **apoptosome**. The apoptosome converts procaspase-9 to caspase-9. Caspase-9 proteolytically activates procaspase-3. Caspase-3 then cleaves ICAD (inhibitor of caspase-activated DNase). Once the inhibitor has been deactivated, the DNase cleaves DNA. Other caspases are activated by the Fas–FADD complex. Caspase-10 cleaves and activates caspase-7; caspase-7 cleaves and activates the MAP kinase kinase kinase MEKK1 at the top of the JNK pathway, ultimately leading to cell death.

Proteases are impressive because the amide bond offers formidable resistance to nucleophilic attack. Two ways of facilitating amide hydrolysis are to increase the reactivity of the amide with a Lewis acid or to increase the reactivity of the nucleophile by deprotonation. In practice, both of these methods are used in the chemical hydrolysis of proteins: either acid-catalyzed hydrolysis with 6 M HCl or base-promoted hydrolysis with 4 M NaOH. However, both of these chemical methods require high temperatures. Enzymes can cleave amide bonds rapidly at low temperature because they deploy both acidic and basic functional groups simultaneously.

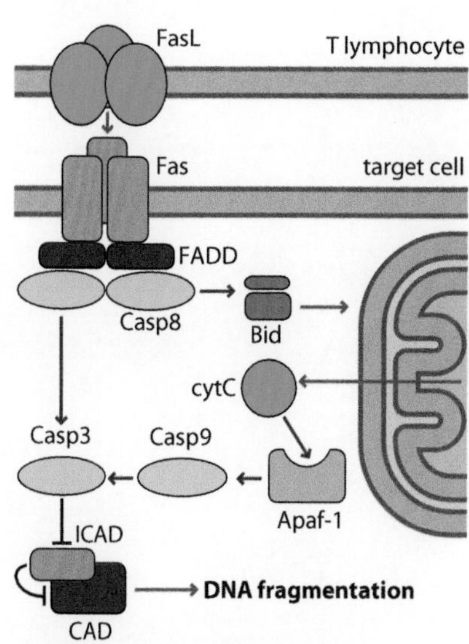

Figure 6.37 Intracellular proteases in signaling. Apoptotic pathways involve several different sequence-selective proteases called caspases.

Problem 6.8

Predict which five of the common amino acids would not be isolable after protein hydrolysis in 4 M NaOH at 105 °C for 10 h.

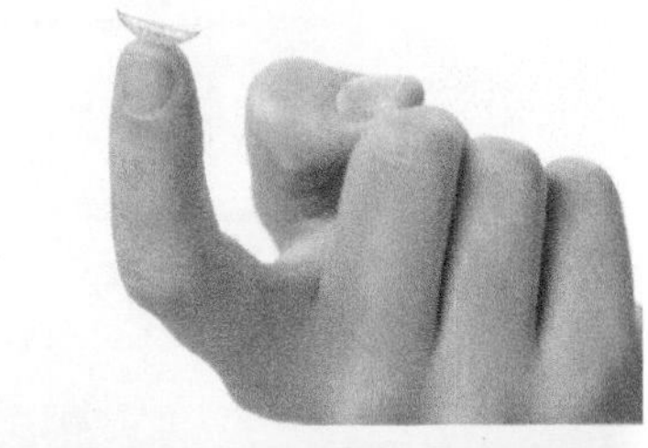

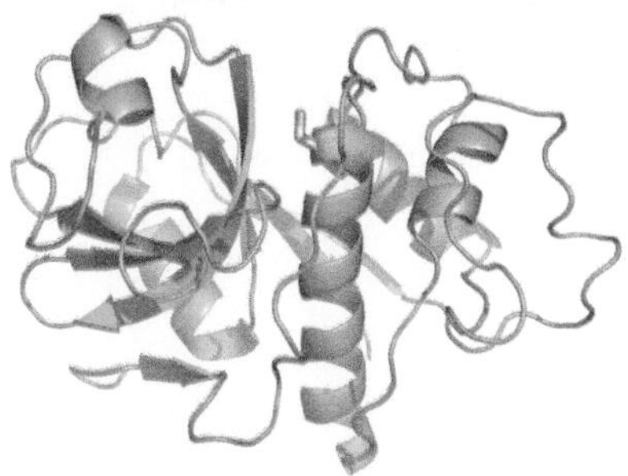

Figure 6.38 Papain. The cysteine protease papain from papaya is a common ingredient in cleaning solutions for contact lenses, breaking down protein deposits that build up on such lenses. (Top, courtesy of Amanda Mills, CDC; middle, from Wikimedia Commons.)

Nature has shown more mechanistic creativity in the design of proteases than it has with protein kinases. Human proteases act through three general mechanisms: nucleophilic catalysis, metal ion activation, and acid catalysis. The human proteome codes for more than 500 proteases, populating the following categories: serine proteases (176), threonine proteases (27), cysteine proteases (143), metalloproteases (186), and aspartic proteases (21).

Cysteine proteases catalyze amide hydrolysis by using a nucleophilic cysteine thiolate

Cysteine thiolates are the most nucleophilic functional groups commonly found on proteins. It is therefore not surprising to find that many proteases use cysteine nucleophiles to attack peptide carbonyl groups (**Figure 6.38**). For example, caspase-1 converts pro-interleukin-1β into the active cytokine interleukin-1β by using a cysteine thiolate to cleave off the first 116 residues at the sequence -YVHD↓APVR-, where the symbol ↓ indicates the site of amide bond cleavage. The aspartate residue at the cleavage site is essential for binding because it makes a critical salt bridge with an arginine side chain of the protease. Caspase-1 is also activated by proteolytic cleavage by an enzyme complex referred to as an inflammasome.

Before we begin drawing mechanisms for proteases, we must return to a common problem in bioorganic chemistry: How does one formulate a plausible arrow-pushing enzyme mechanism based on a crystal structure? X-ray crystal structures are static and usually do not resolve the positions of hydrogen atoms. Let us start by analyzing a common mechanistic depiction of cysteine proteases (**Figure 6.39**).

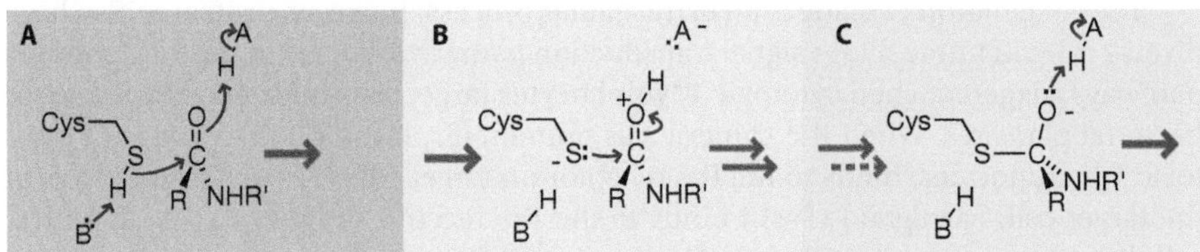

Figure 6.39 Which mechanism is correct? Often, several plausible steps can be drawn for a hypothetical enzyme mechanism. (A) Does the enzyme deprotonate the thiol and protonate the carbonyl group as the sulfur atom attacks the carbonyl group? (B) Does the enzyme protonate the carbonyl group before the thiolate attacks? (C) Does the enzyme protonate the carbonyl group after the thiolate attacks?

The mechanism for formation of a tetrahedral intermediate in Figure 6.39A looks quite plausible but raises several troubling questions. First, does the S–H sigma bond really attack the carbonyl group? Probably not. Sulfur lone pairs are much more nucleophilic than S–H sigma bonds, and it would be very difficult for an enzyme to change this order of reactivity. Similarly, it is unlikely that the acid protonates the π bond of the carbonyl group because the lone pair on the carbonyl oxygen is much more basic than the π bond.

The mechanism depicted in Figure 6.39A raises a third troubling question: Is every step *concerted*, as indicated in the mechanism? If the steps are concerted, then the base must be pulling off a proton *as* the sulfur is attacking and *as* the oxygen is being protonated. To be more precise, the B–H distance must be getting shorter *as* the A–H distance is getting longer. This is certainly possible, but proton transfers such as these are usually more than 1000 times faster than the attack of thiolates on amide carbonyl groups. Thus, on average, we should expect the base to remove, and then replace, the thiol proton 1000 times before the sulfur atom attacks the carbonyl group. If we are confident that proton transfers precede attack on the carbonyl group, then we might choose to depict only the arrow-pushing for the nucleophilic attack and assume that the reader is familiar with the roles of Brønsted bases and acids (Figure 6.39B). But what if we aren't sure about the nucleophilic attack? It would be much safer, and no more burdensome, to use a dashed reaction arrow to indicate that we are not sure how many mechanistic steps are involved: one step or two steps (Figure 6.39C).

Proton transfers will create mechanistic headaches for arrow-pushers unless we realize that we are taking a shortcut. The legendary chemist Manfred Eigen won the 1967 Nobel Prize in Chemistry for measuring the rates of fast chemical reactions—in particular, proton transfers (see Section 2.2). Surprisingly, Eigen showed that fast proton transfers are not like S_N2 reactions, because they involve three mechanistic steps instead of one: (1) association to form a hydrogen bond, (2) bond vibration, and (3) dissociation to break the hydrogen bond (**Figure 6.40**). Including these laborious hydrogen-bonding steps can take much of the joy out of mechanistic arrow-pushing, but omitting these hydrogen-bonding steps will make some steps of enzymatic catalysis seem confusing. For example, an alkoxide leaving group will look much more attractive if it is hydrogen-bonded to an imidazolium ion. Similarly, a water molecule will seem to be much more nucleophilic if it is hydrogen-bonded to an amine base. It is the duty of bioorganic chemists and chemical biologists to understand the benefits of hydrogen bonding in chemical mechanisms even if we choose not to depict it.

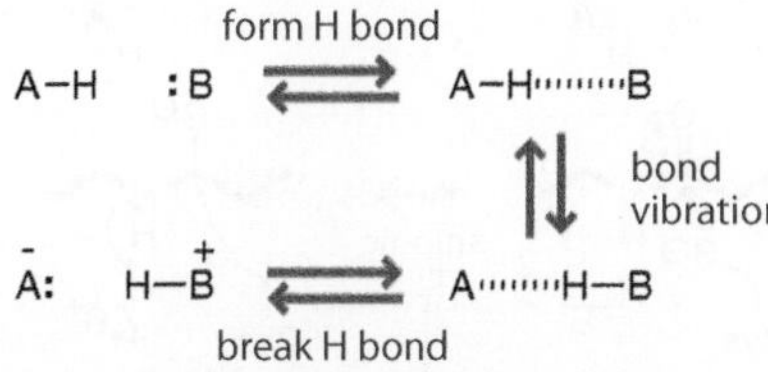

Figure 6.40 Proton transfers involve three mechanistic steps.

Enzymes proceed via mechanisms with the minimum number of different types of transition states

The mechanism of action of caspase-1 involves two key residues: a nucleophilic thiol group and a basic histidine imidazole moiety (**Figure 6.41**). It is unclear whether the carbonyl group is protonated by the enzyme. The imidazole side chain deprotonates the cysteine thiol, and the resulting thiolate then attacks the amide (step *a*). The imidazolium side chain then protonates the amine leaving group (step *b*), leading to collapse of the tetrahedral intermediate with ejection of the leaving group as a neutral amine (step *c*). Next, a water molecule binds in place of the amine (step *d*). Water then attacks the thioester (step *e*), and the imidazole group deprotonates the OH_2^+ group to generate an anionic tetrahedral intermediate (step *f*). Collapse of the anionic tetrahedral intermediate releases the carboxylic acid and regenerates the nucleophilic thiolate (step *g*).

Although the mechanism of caspase-1 involves many steps and many transition states, the energetically challenging steps are similar and involve the formation of tetrahedral intermediates through the addition of nucleophiles to carbonyl groups or the breakdown of tetrahedral intermediates through the expulsion of leaving groups. Thus, a single active site that is tuned to stabilize anionic tetrahedral intermediates can catalyze more than one step in a reaction. In general, when postulating mechanisms for enzyme-catalyzed reactions, the most plausible mechanism will involve the fewest types of high-energy transition states. This provides an important guide for students proposing arrow-pushing mechanisms when no experimental data are

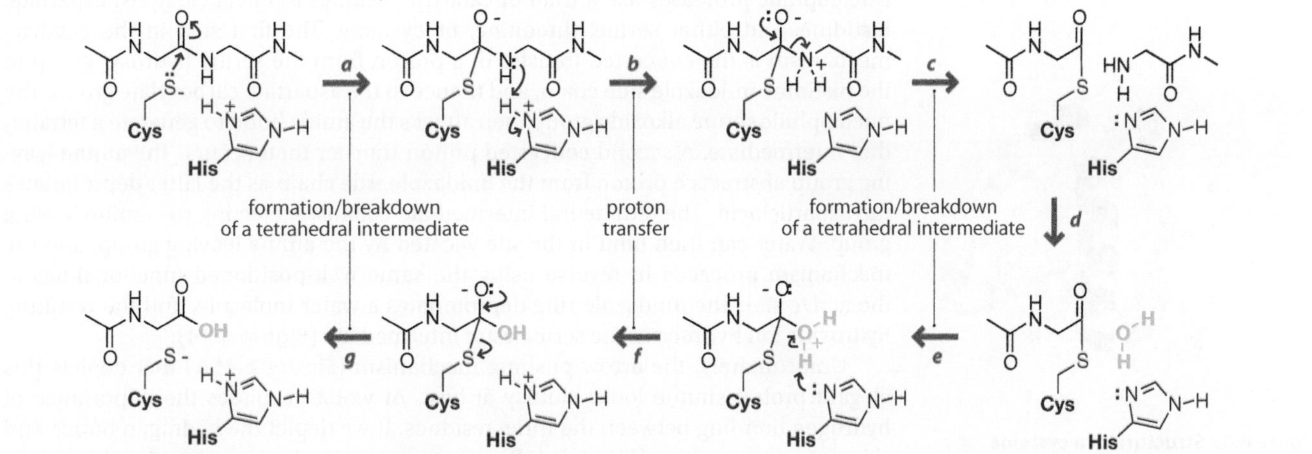

Figure 6.41 Enzyme mechanisms often involve many similar transition states. When applied to cysteine protease interleukin-1β-converting enzyme, the principle of minimum transition states suggests that the transition states for the indicated steps will be rather similar.

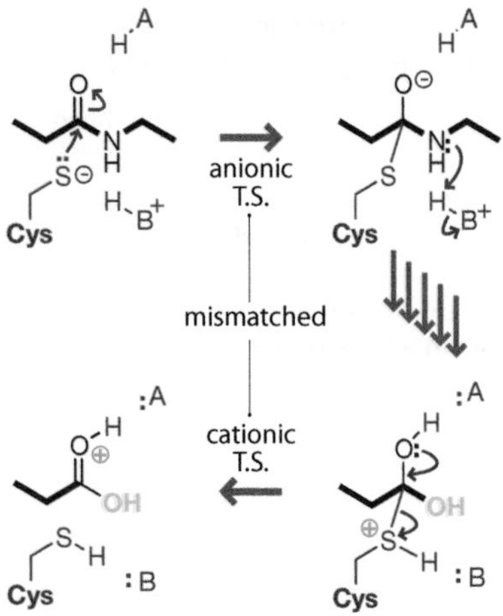

Figure 6.42 Avoid mismatched transition states. If you propose anionic transition states in the first step, try not to switch to cationic transition states later in the mechanism.

available: *minimize the number of different types of transition states that need to be stabilized by the enzyme* (**Figure 6.42**).

The oxyanion transition states for steps *a*, *c*, *e*, and g in Figure 6.41 have a similar tetrahedral geometry. If the thiolate attacks as the anion in step *a*, then it must leave as an anion through the same trajectory in step *g*. However, we should exercise similar logic if the enzyme also stabilizes the transition states for aminolysis of the thioester intermediate (steps *c*, *d*, and *e*). If we propose that the scissile amine group leaves as a neutral amine in step *c*, we are compelled to consider the attack of a neutral water molecule through the same trajectory in step *e*. Hydroxide anion is certainly the nucleophilic species in the uncatalyzed hydrolysis of thioesters at pH 7, but one would be hard pressed to explain how an enzyme could comfortably fit an anionic hydroxide nucleophile in the same space that was suited to a cationic ammonium group just moments before. Unfortunately, it is very difficult to unambiguously establish each of the elementary steps in an enzyme mechanism. To understand why, the reader should reflect on the similarity between an ammonium group that is hydrogen-bonded to an imidazole ring and an amine that is hydrogen-bonded to an imidazolium ion, or between a hydroxide anion that is hydrogen-bonded to an imidazolium ion and a water molecule that is hydrogen-bonded to an imidazole ring.

Problem 6.9

The enzyme alkaline phosphatase, commonly used in ELISAs, employs a key serine residue in the active site to generate a phosphoserine intermediate. Suggest a plausible four-step arrow-pushing mechanism for the hydrolysis of phosphate esters that proceeds via the key phosphoserine intermediate.

phosphoserine intermediate

Recall that an α helix with just a few turns can create an electrostatic field that is the equivalent of about half a positive charge. The N termini of α helices are often trained on the active sites of proteases to stabilize the buildup of negative charge associated with the formation of various intermediates. Such α helices are readily seen in the crystal structure of cathepsin B with an epoxide inhibitor covalently bound in the active site (**Figure 6.43**).

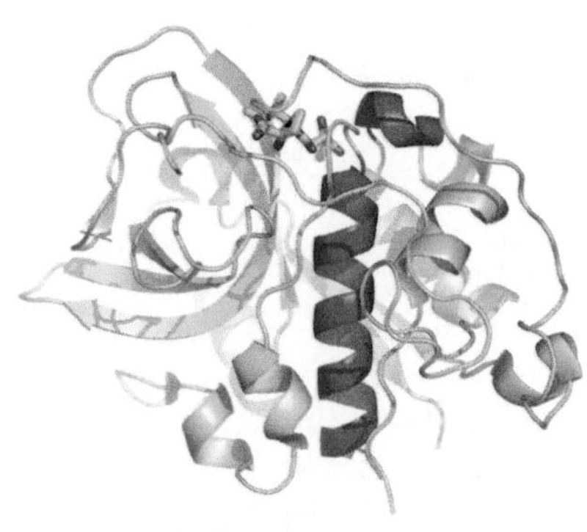

Figure 6.43 Structure of a cysteine protease. α Helices converge on the active site of a cysteine protease cathepsin B. (PDB 1QDQ)

Serine proteases cleave amides by using an alkoxide nucleophile

Nucleophilic proteases use a triad of catalytic residues to effect catalysis: aspartate, histidine, and either serine, threonine, or cysteine. The first step in the catalytic mechanism is the concerted transfer of a proton from the serine hydroxyl group to the histidine imidazole side chain, and thence to the aspartate carboxylate group. The nucleophilic serine alkoxide group then attacks the amide bond to generate a tetrahedral intermediate. A second concerted proton transfer then occurs. The amine leaving group abstracts a proton from the imidazole side chain as the latter deprotonates the aspartic acid. The tetrahedral intermediate collapses, ejecting the amine leaving group. Water can then bind in the site vacated by the amine leaving group, and the mechanism proceeds in reverse using the same well-positioned functionalities at the active site. The imidazole ring deprotonates a water molecule, and the resulting hydroxide ion hydrolyzes the serine ester intermediate (**Figure 6.44**).

Unfortunately, the arrow-pushing mechanism (**Figure 6.45A**) that depicts this elegant proton shuttle looks clumsy at best. At worst, it misses the importance of hydrogen bonding between the three residues. If we depict the hydrogen bonds and skip the arrow-pushing (Figure 6.45B), we can more clearly appreciate that this proton transfer involves a pair of concerted bond vibrations, like the motion of pistons in a fast-revving V8 engine.

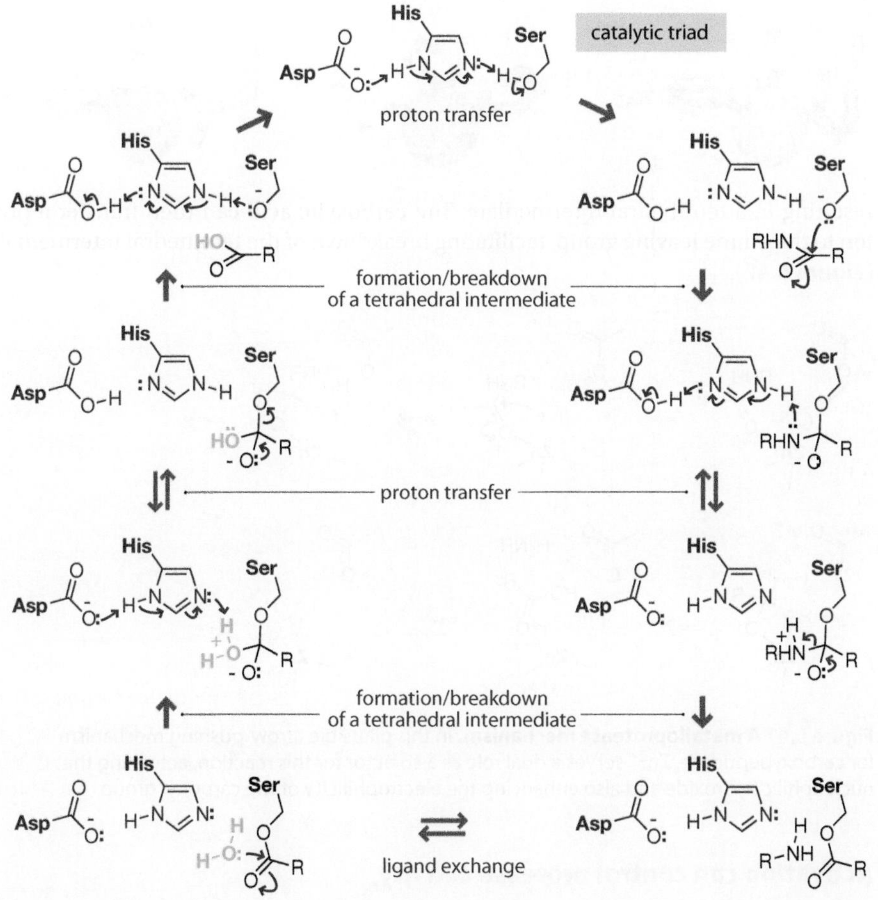

Figure 6.44 **A catalytic triad.** Three residues in serine proteases work together to generate an anionic alkoxide nucleophile. The enzyme stabilizes four similar zwitterionic tetrahedral transition states.

There is some variation in the residues activating the nucleophilic hydroxyl group. Proteases with Asp-Asp-Ser, His-His-Ser, and even Lys-Ser are known. The **proteasome** is a large multisubunit protein complex with a key role in protein turnover, degrading proteins so that they can be synthesized anew. The proteasome has a catalytic grouping slightly different from that found in most serine/threonine proteases. The nucleophilic threonine is located at the N terminus, and it is the free N-terminal amino group that deprotonates the nucleophilic hydroxyl moiety.

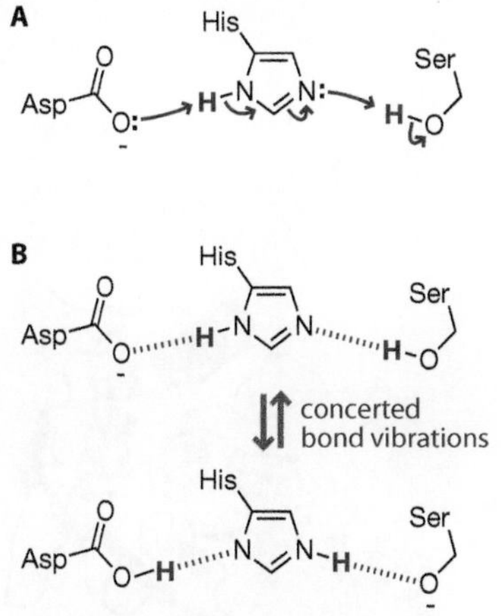

Metalloproteases use Zn^{2+} ions to activate the nucleophilic water and stabilize the tetrahedral intermediate

Hundreds of human enzymes depend on Zn^{2+} ions either to maintain structure or to participate directly in catalysis. For the latter, Zn^{2+} is an ideal Lewis acid; the Zn^{2+} ion is not redox active and has no ligand field. Lack of a ligand field means that Zn^{2+} ion can take as many ligands as it likes in any geometry that is convenient. Thus, the Zn^{2+} ion is not rigidly constrained to a particular coordination geometry, such as forming a tetrahedral or octahedral complex. The Zn^{2+} ion of metalloproteases is usually gripped by three side chains in the protease active site with nanomolar affinity (**Figure 6.46**). In human metalloproteases, the Zn^{2+} ligands are usually the imidazole side chains of histidine or the carboxylate side chains of aspartate or glutamate. When a water molecule binds, the coordination geometry is roughly tetrahedral. Zn^{2+} is by far the most common metal in human metalloproteases, although some involve other metals such as Mn^{2+} (for example methionine aminopeptidase 2).

The mechanism for zinc metalloproteases is believed to involve coordination of water to the Zn^{2+} ion. The amide carbonyl group probably coordinates to the Zn^{2+} ion as well. Deprotonation of the water molecule by a basic carboxylate group generates a zinc hydroxide that can attack the carefully positioned amide carbonyl group,

Figure 6.45 **Choose depictions.** You cannot push arrows with hydrogen bonds, so chemical biologists have to make an agonizing choice between depicting the interaction of filled orbitals with unfilled orbitals with an arrow pushing mechanism (A) or depicting hydrogen bonds (B).

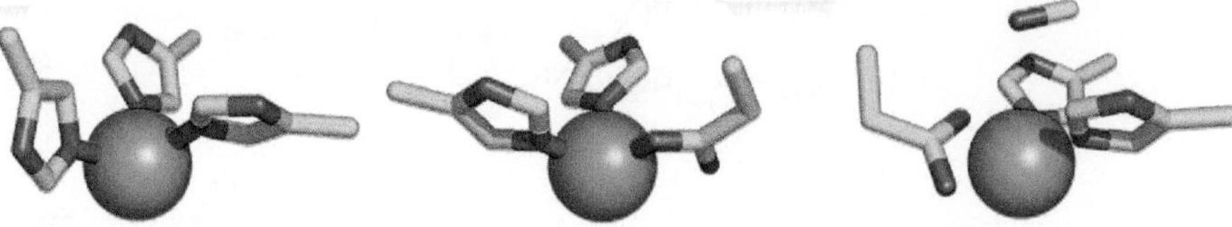

Figure 6.46 Metalloproteases grip metal ions. Examples of Zn^{2+} ions in the active sites of (left to right) matrix metalloprotease-1 (PDB 2J0T), neprilysin (PDB 1R1J), and neurolysin (PDB 1L1L). Water coordinates to the fourth position of Zn^{2+}.

resulting in a tetrahedral intermediate. The carboxylic acid can then transfer a proton to the amine leaving group, facilitating breakdown of the tetrahedral intermediate (**Figure 6.47**).

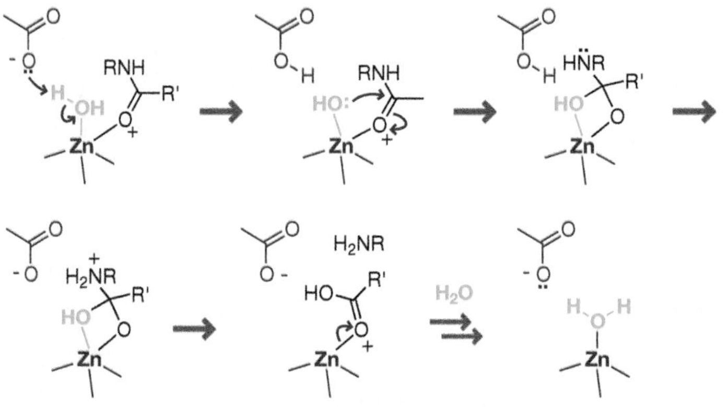

Figure 6.47 A metalloprotease mechanism. In this plausible arrow-pushing mechanism for carboxypeptidase, Zn^{2+} serves a dual role as a cofactor for this reaction, activating the nucleophilic hydroxide and also enhancing the electrophilicity of the carbonyl group.

Activation can control protease activity

Recall that sequence-selective proteases are important in clot formation and intracellular signaling pathways. Proteases are kept at bay through a variety of mechanisms. For example, proteases can be expressed as inactive zymogens that require proteolytic activation. In some cases, proteolytic activation of such pro-proteases results in a conformational change. In other cases, proteolytic activation leads to the loss of an inhibitory fragment that is blocking access to the active site. Alternatively, the proteases can be inhibited by endogenous protein inhibitors that bind with high affinity. In addition, proteases can be sequestered in locations remote from their cognate substrates. All of these inhibitory mechanisms are employed to control caspases in apoptotic pathways, a necessity for limiting cell suicide to critical situations only (see Figure 6.37).

The cleavable segments that restrain pro-proteases range from short peptides to large protein subunits. For example, the activation of the zymogen prochymase to generate the active protease chymase is effected by the removal of the N-terminal dipeptide $^{+}H_3N$–Gly–Glu– by the nonspecific protease dipeptidyl protease I. Typically, the cleavable segments are longer than just a few amino acids. The Zn^{2+}-dependent matrix metalloprotease stromelysin-1 (also known as MMP-3) is expressed as a peptide chain 477 amino acids in length. The first 17 N-terminal residues are a signal sequence, which is cleaved before the zymogen is exported from the cell. Active stromelysin-1 is generated from prostromelysin-1 through cleavage at His82 by enzymes such as elastase (**Figure 6.48**). The disposable inhibitory segment forms a folded structure that places an extended strand in the active site in the reverse orientation (N to C) of the normal

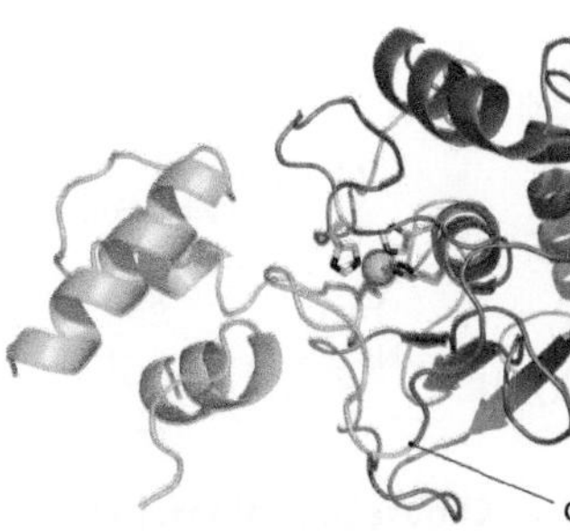

Figure 6.48 Protease activation. Prostromelysin-1 is activated by cleavage of the yellow residues at the site indicated. Missing residues indicated by the dashed line were disordered in the crystal structure. (PDB 1SLM)

Problem 6.10

Treatment of prostromelysin with 4-aminophenylmercuric acetate brings about autoproteolysis at Glu74-Val75, leading to an active form of stromelysin. Suggest a plausible arrow-pushing mechanism to explain how $ArHg^+$ induces cleavage of prostromelysin-1.

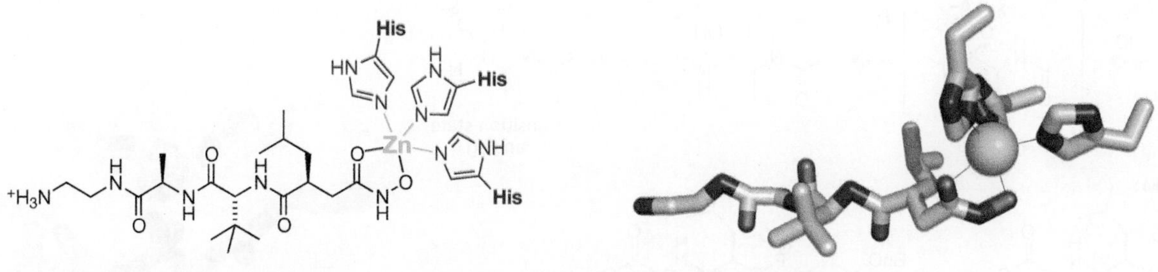

Figure 6.49 Hydroxamic acids inhibit zinc proteases. Two views of the active site of TNFα-converting enzyme with a hydroxamic acid inhibitor bound to the active Zn^{2+} ion. (PDB 1BKC)

substrate. Cys75, in the disposable segment, also coordinates with the active-site Zn^{2+} ion, ensuring that the enzyme does not cleave the propeptide to activate itself.

Three major biopharmaceuticals inhibit the interaction of tumor necrosis factor α (TNFα) binding to its cell-surface receptor. Remicade® and Humira® are antibodies that bind to TNFα, thereby preventing it from acting on its receptor; Enbrel® is a chimeric protein generated by replacing the two antigen-binding arms of an antibody with a soluble form of the TNFα receptor. Like Remicade and Humira, it binds to TNFα. Because anti-TNFα therapies have proven so successful for immunological diseases such as rheumatoid arthritis, many companies have sought to prevent the formation of TNFα by targeting TNFα-converting enzyme, the protease that generates TNFα from its precursor protein. TNFα-converting enzyme is a membrane-bound metalloprotease that cleaves the TNFα precursor at the sequence LeuAlaGlnAla↓ValArgSerSer (the down arrow indicates the cleavage site) to generate the active 17 kDa protein (**Figure 6.49**).

Caspases recognize tetrapeptide sequences ending in aspartic acid: XXXD. For example, caspase-8 and caspase-9 cleave the dimer of procaspase-3 at -GIETD↓SGVDD-. Once cleaved, caspase-3 undergoes a self-cleavage event to produce the active caspase-3 (**Figure 6.50**). The cleavage site is located on a long peptide loop at the protein–protein interface. One end of the cleaved strand folds back to complete the active site.

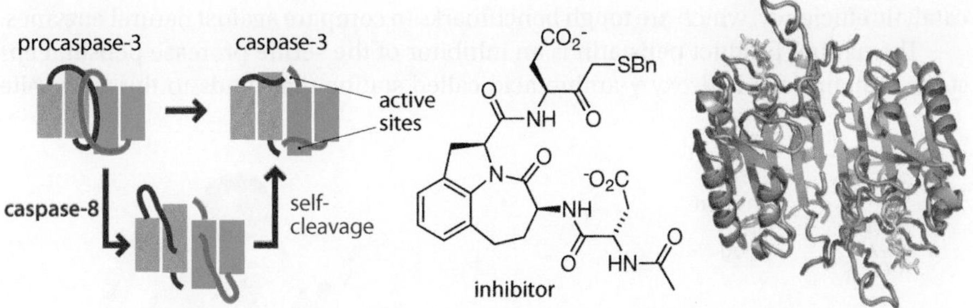

Figure 6.50 Caspase activation. The proteolytic activation of the procaspase-3 dimer generates the active enzyme caspase-3 as a homodimer. A crystal structure of caspase-3 bound by a peptidomimetic inhibitor (yellow) reveals the active sites. (PDB 1RHU)

Reversible enzyme inhibitors include transition-state analogs with very high affinity

All proteases stabilize transition states leading to and from tetrahedral intermediates. Phosphonamidates are excellent mimics of such transition states because P–O bonds are longer than C–O bonds and charge is delocalized on two oxygen atoms of the phosphonamidate (**Figure 6.51**). For example, the bacterial zinc metalloprotease thermolysin efficiently cleaves short peptides such as Boc-Phe-Leu-Ala-CO_2H

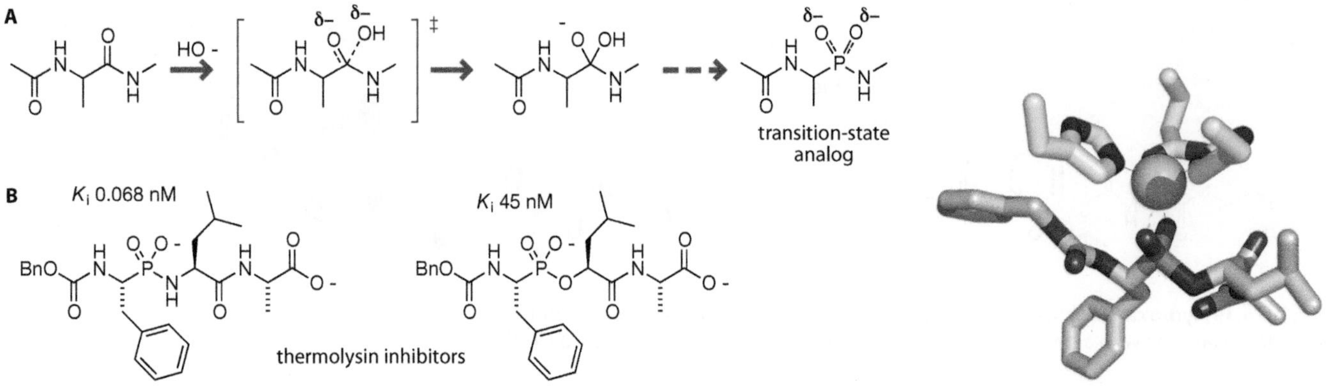

A

transition-state analog

B K_i 0.068 nM

K_i 45 nM

thermolysin inhibitors

Figure 6.51 Phosphonamidate inhibitors of proteases. (A) The protease transition state can be well mimicked by a (B) phosphonamidate (left) and the less effective phosphonate (right) inhibitors of the zinc metalloprotease thermolysin. (C) The crystal structure of the phosphoramidite bound to Zn^{2+} in the thermolysin active site (PDB 4TMN).

($k_{cat}/K_m = 6.7 \times 10^5\ M^{-1}\ s^{-1}$). The phosphoramidate analog shown in Figure 6.51 is an exquisite inhibitor, with an **inhibitory constant**, K_i of 0.068 nM. In the crystal structure of this inhibitor, one of the phosphoramidate oxygens is bound to the Zn^{2+} atom at the active site, just like the putative hydroxide nucleophile. The corresponding phosphonate is almost three orders of magnitude less efficient. In another example from earlier in this chapter, a transition-state analog was a very potent inhibitor of chorismate mutase.

In the 1960s William P. Jencks noted, "If complementarity between the active site and the transition state contributes significantly to enzymatic catalysis, it should be possible to synthesize an enzyme by constructing such an active site. One way to do this is to prepare an antibody to a haptenic group which resembles the transition state of a given reaction." Because phosphonates are good transition-state analogs for ester hydrolysis, antibodies elicited against phosphonates should be capable of catalyzing the hydrolysis of the corresponding esters. It took two decades to reduce this prognostication to practice by eliciting antibodies against phosphonate esters, which mimic the transition state for a hydrolysis reaction (**Figure 6.52** and **Figure 6.53**). Subsequently, it was shown that antibodies elicited against a wide range of transition-state analogs catalyze the corresponding reactions, for example amide formation, hydride reduction, Diels–Alder reactions, and transamination. Key issues for developing such artificial enzymes include the number of reactions catalyzed before inactivation and catalytic efficiency, which are tough benchmarks to compare against natural enzymes.

The natural product pepstatin is an inhibitor of the serine protease pepsin. Pepstatin contains a β-hydroxy γ-amino acid called statine that binds to the active site

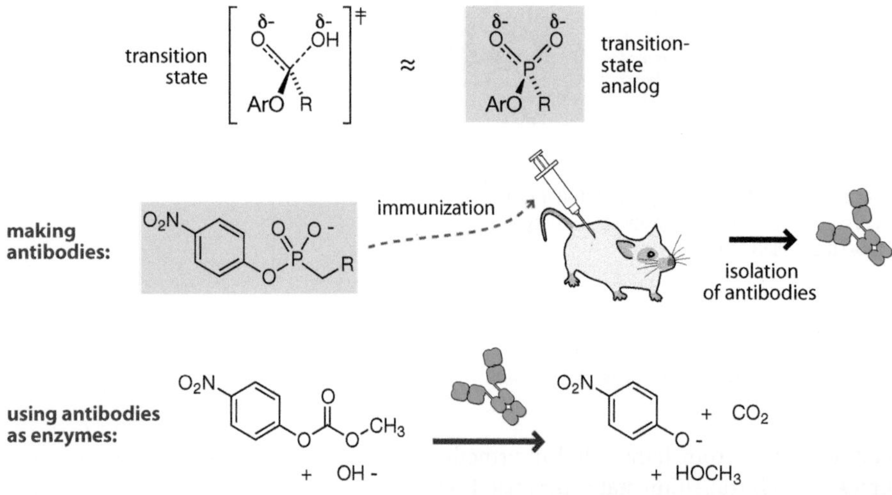

Figure 6.52 Catalytic antibodies. Phosphonates are good transition-state analogs for ester and carbonate hydrolysis. When mice are immunized with phosphonates, they produce antibodies that bind tightly to phosphonates and catalyze the corresponding chemical reactions.

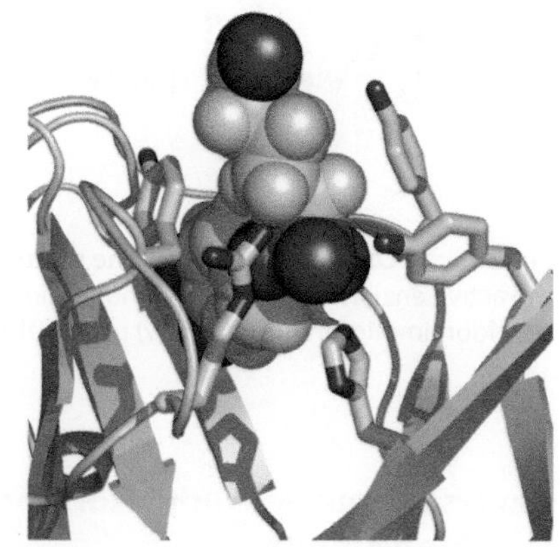

Figure 6.53 Structure of a catalytic antibody. Crystal structure of a nitrophenylphosphonate ligand bound to the active site of a catalytic antibody. (PDB 1GAF)

statine statine

pepstatin

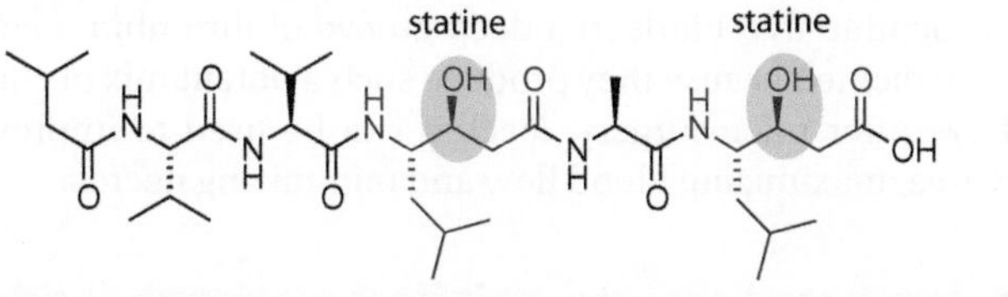

Figure 6.54 Statine-containing peptides. The two statine subunits of pepstatin allow it to inhibit a wide range of proteases. Statine subunits are dipeptide isosteres with a carbinol center (blue) that mimics the tetrahedral transition state for amide hydrolysis.

of the protease. The statine subunit can be embedded in other peptides to produce highly effective inhibitors (**Figure 6.54**).

The peptide inhibitor bortezomib forms a covalent bond with the active-site threonine of the proteasome. In spite of the robust appearance of the covalent adduct, bortezomib is a reversible inhibitor (**Figure 6.55**). Thus, a readily reversible covalent adduct will be a reversible inhibitor. Bortezomib has been shown to be effective against some myelomas. It is believed that the proteasome degrades pre-apoptotic factors in those myelomas, conferring immortality.

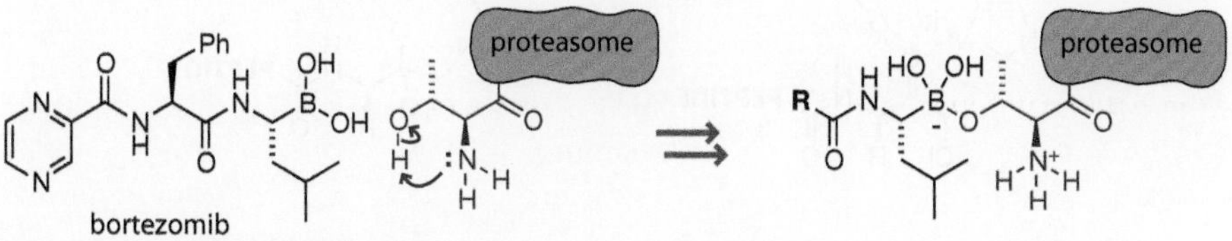

proteasome proteasome

bortezomib

Figure 6.55 Inhibition of the proteasome by bortezomib. Boronic acids are electrophilic because boron wants to form an octet. Bortezomib reversibly forms a covalent bond with the proteasome.

Full-length protein inhibitors of proteases are common, and they are produced both by the organism that makes the protease and by organisms that do not. For example, the mammalian enzyme pancreatic trypsin is inhibited by a small folded protein, pancreatic trypsin inhibitor. Bovine pancreatic trypsin inhibitor (BPTI, **Figure 6.56**) is one of the best-studied proteins, because it is a small (58 amino acids) stably folded protein. Leeches, ticks, and vampire bats produce potent inhibitors of proteases in the blood clotting cascade. Medicinal leeches (*Hirudo medicinalis*) produce a powerful anticoagulant that prevents blood from clotting when they take in a meal from their host. Hirudin is an inhibitor of the serine protease thrombin (**Figure 6.57**). It binds with a phenomenal affinity (K_d = 20–200 fM), unfurling an

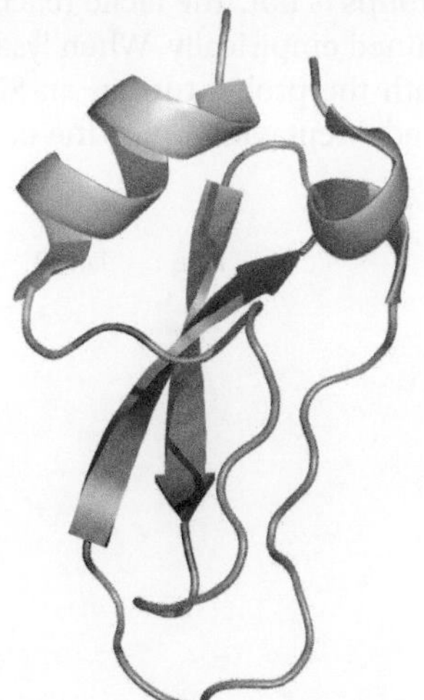

Figure 6.56 Protein folding model. Bovine pancreatic trypsin inhibitor (BPTI) is a small, robust, well-behaved protein that folds and unfolds reversibly. It has become a widely adopted model for studies of protein folding. (PDB 9PTI)

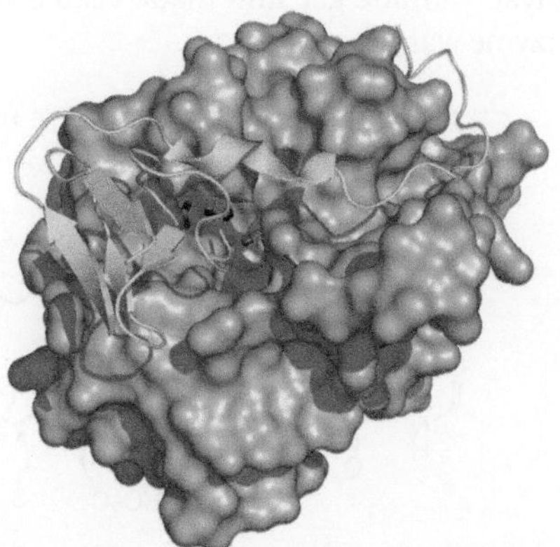

Figure 6.57 Leech protein. Structure of the complex between thrombin (blue) and hirudin (yellow). The first three residues of hirudin are thrust deep into the catalytic pocket, making contact with the serine of the catalytic triad. (PDB 4HTC)

extended peptide that binds in a deep groove of thrombin. Leeches still find use in modern medicine because they produce such a potent mix of compounds to improve blood flow. After microsurgery, leeches can be used to improve blood flow to the affected area, maximizing blood flow and minimizing necrosis.

Mechanism-based enzyme inhibitors react with residues at the active site

S_N2 reactions on α-haloketones are generally faster than the S_N2 reactions on other haloalkanes, because the adjacent carbonyl group lowers the energy of the C–X σ* orbital. α-Halomethylketones are potent inhibitors of serine proteases such as the digestive enzymes trypsin and chymotrypsin. Both the active-site histidine and the active-site serine end up covalently bonded to the inhibitor (**Figure 6.58**).

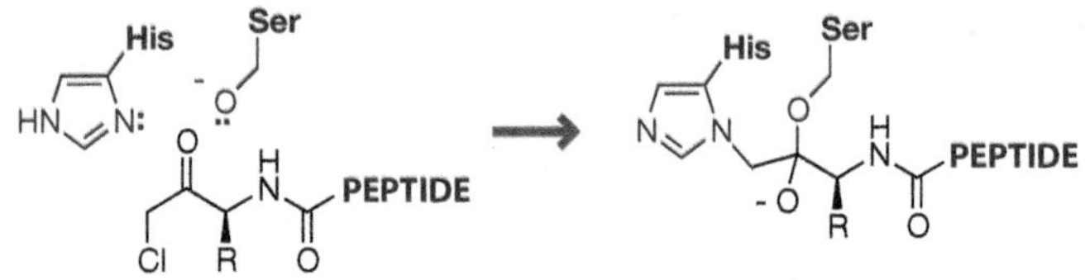

Figure 6.58 Covalent protease inhibitor. α-Chloromethylketones alkylate active-site imidazoles and form hemiketals with the carbonyl group.

Many proteases are expressed as zymogens, and it is difficult to tell which proteases are active at any point in time in a particular cell line. Proteases that are specifically upregulated in diseased cells such as cancer cells may be targets for drug development. Activity-based protein profiling allows one to identify proteases that are upregulated specifically in cancer cells. A knowledge of chemical mechanism allows one to design molecules that form covalent bonds with specific classes of enzymes. For example, when α-haloketones, α-acyloxyketones, or α,β-unsaturated esters are attached to peptides, they react selectively with cysteine proteases such as caspases. If these peptides are also labeled with a biotin tag or imaging label, it is possible to isolate the enzyme or make it visible.

For example, the electrophilic probe in **Figure 6.59** was designed to react specifically with interleukin-1β-converting enzyme, a serine protease that cleaves a target sequence YVHD↓APVR. There is no reason to expect that α-acyloxyketones should be good electrophiles in S_N2 reactions, because addition to C=O π* is undoubtedly faster than S_N2 addition to C–O σ*. However, addition of thiolates to ketones is readily reversible, whereas S_N2 displacement of acyloxy groups is not. The facile reaction of α-acyloxyketones with serine proteases was determined empirically. When lysates from a monocytic leukemia cell line were treated with the probe, run on an SDS-polyacrylamide gel, and made visible with radiolabeled streptavidin, only the target enzyme was labeled.

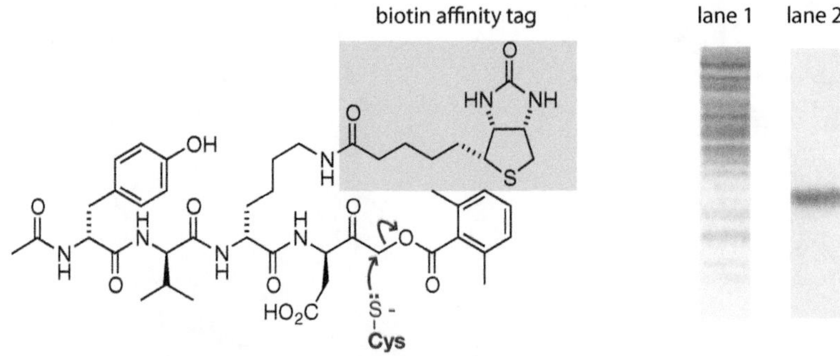

Figure 6.59 Highly selective protease trap. A proteomics probe for the identification of active IL-1β-converting enzyme (ICE) with a biotin affinity tag. Cell lysates treated with the affinity tag reveal the protease(s) that react covalently. Lane 1: nonselective labeling with a silver stain shows that ICE is not the major protein. Lane 2: when visualized with radioactive streptavidin, only ICE gives a detectable band. (Right, adapted from N.A. Thornberry et al., *Biochemistry* 3:3934–3940, 1994. With permission from the American Chemical Society.)

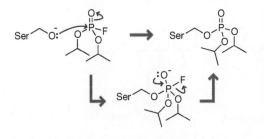

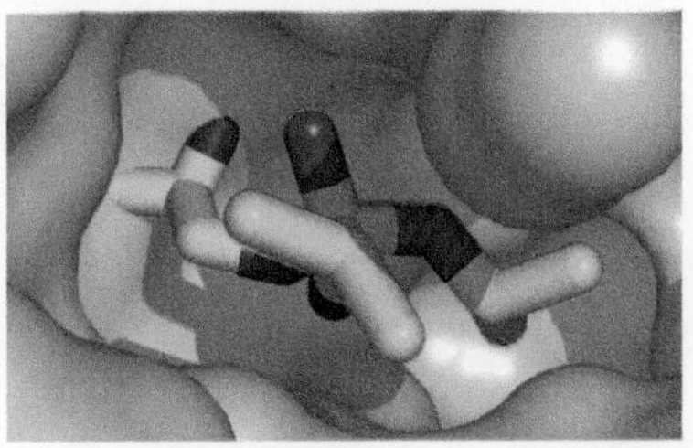

Figure 6.60 Fluorophosphate inhibition. Structure of an active-site serine residue in the protease "factor D" covalently bound to a diisopropylphosphoryl group. (PDB 1DFP)

The human immune system targets pathogenic bacteria for destruction through the complement pathway. The pathway is named for a phenomenon recognized by Paul Erhlich in the late nineteenth century. The complement pathway involves a cascade of extracellular proteases that culminates in the formation of a massive porthole in the bacterial membrane, allowing bacterial contents to leak out of the cell. One of the proteases at the top of the complement cascade, called factor D, is a serine protease with an Asp-His-Ser catalytic triad. Diisopropylfluorophosphate covalently deactivates factor D, forming a stable phosphate triester (**Figure 6.60**).

Electrophilic phosphate derivatives are potent inhibitors of various serine proteases and esterases, which act through analogous mechanisms. Among the most toxic phosphorus compounds are those that deactivate the enzyme cholinesterase. Cholinesterase hydrolyzes the ester group of the neurotransmitter acetylcholine (ACh). Acetylcholine is essential because it transmits signals between nerve cells and at the neuromuscular junction. When cholinesterase is inhibited, acetylcholine builds up at the junction between nerve cells, effectively preventing the transmission of signals. Nerve agents (sarin, tabun, and VX) are all potent inhibitors of the esterase cholinesterase (**Figure 6.61**).

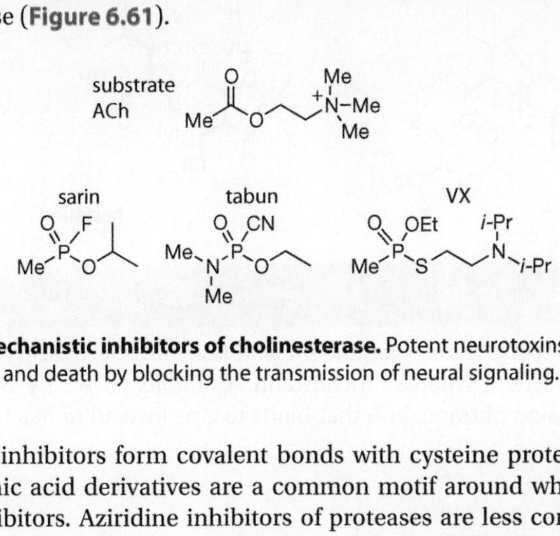

substrate ACh

sarin tabun VX

Figure 6.61 Mechanistic inhibitors of cholinesterase. Potent neurotoxins, these molecules cause paralysis and death by blocking the transmission of neural signaling.

Epoxide inhibitors form covalent bonds with cysteine proteases (**Figure 6.62**). Epoxy succinic acid derivatives are a common motif around which one can design protease inhibitors. Aziridine inhibitors of proteases are less common, but the natural product miraziridine A is an inhibitor of serine proteases, cysteine proteases, and aspartyl proteases. Peptides containing α,β-unsaturated γ-amino esters are also

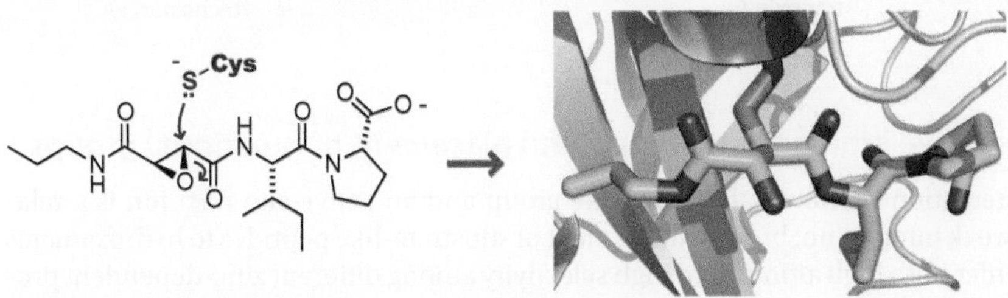

Figure 6.62 Covalent inhibition of nucleophilic proteases. Epoxide inhibitors are readily opened by nucleophilic proteases such as cathespin B. (PDB 1QDQ)

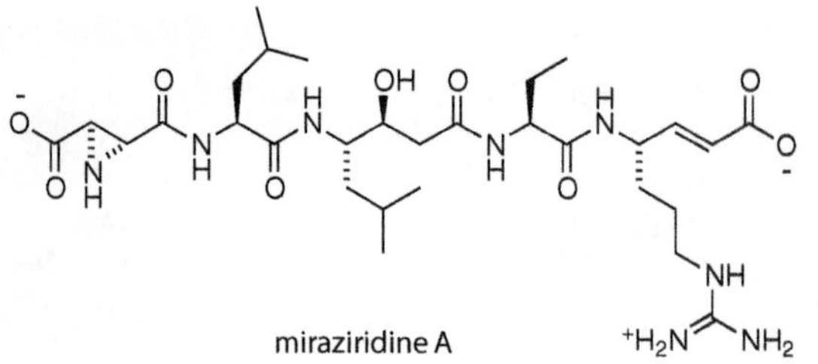

Figure 6.63 Catch-all protease inhibitor. Miraziridine A incorporates several protease inhibitor functional groups: an enone Michael acceptor for thiol proteases, a statine subunit that mimics a tetrahedral transition state, and a potentially electrophilic aziridine.

miraziridine A

susceptible to attack by cysteine proteases. Thus, miraziridine A can covalently inhibit nucleophilic proteases through a variety of mechanisms (**Figure 6.63**).

Most inhibitors of zinc metalloproteases contain hydroxamic acid functional groups, which bind tightly to the zinc atoms (**Figure 6.64**). For example, galardin is an inhibitor of the zinc-dependent metalloproteases MMP-1 and MMP-3; matlystatin A is an inhibitor of MMP-2 and MMP-9. Hydroxamic acids are weak acids ($pK_a = 9.3$) and the hydroxamate anion tightly chelates Zn^{2+} through a bidentate interaction (see Figure 6.64). However, other chelating functional groups can also bind zinc, like the β-amino-α-hydroxy amide moiety of bestatin, which binds two metal ions in some aminopeptidases. If the inhibitor makes sufficient specific contacts with the protease, then only a single thiolate or carboxylate group is needed to coordinate to zinc.

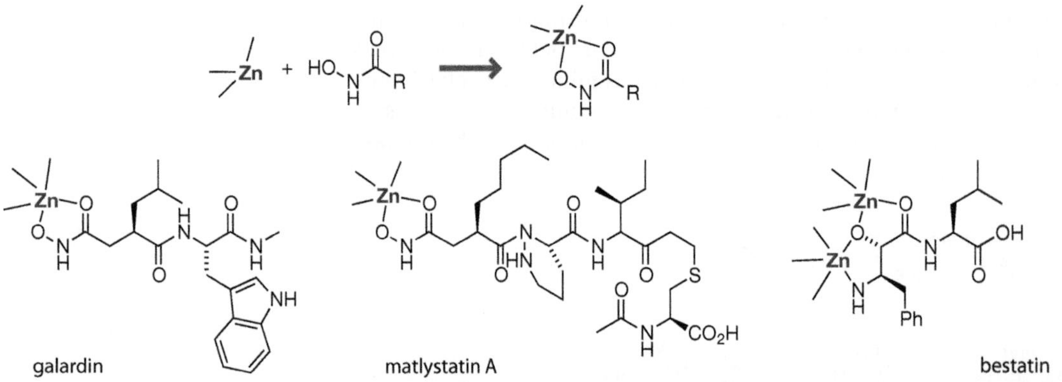

galardin

matlystatin A

bestatin

Figure 6.64 Hydroxamic acids as zinc metalloprotease inhibitors. Hydroxamic acids bind effectively to zinc and are a common feature of natural and synthetic zinc metalloprotease inhibitors. An amide bond resonance structure reveals significant negative charge on the carbonyl oxygen, explaining its affinity for the Zn^{2+} atom.

Problem 6.11

Trapoxin A deactivates the zinc-dependent amidase histone deacetylase (HDAC) through an S_N2 reaction, whereas trichostatin A inhibits HDAC by binding zinc. Design a hybrid analog of trapoxin A that binds to zinc instead of reacting through an S_N2 mechanism.

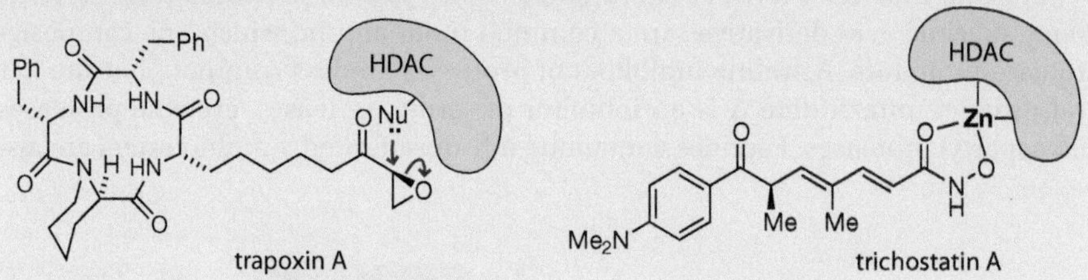

trapoxin A

trichostatin A

Cooperative binding requires careful placement of functional groups

The interaction between a hydroxamate group and an active-site Zn^{2+} ion is a relatively weak interaction, but the attachment of substrate-like peptides to hydroxamates can confer both high affinity and high selectivity among different zinc-dependent proteases. Whenever two ligands are tethered together with a covalent linker, the structure of the linker is critical in determining whether the free energy of binding is higher or lower than the additive free energies of the individual ligands. For example, in the

Table 6.6 Cooperativity in binding: free energy of ligand interactions with the zinc-dependent protease stromelysin.

Ligand	ΔG (303 K)
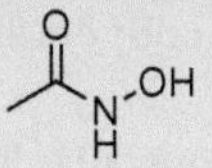	−2.3
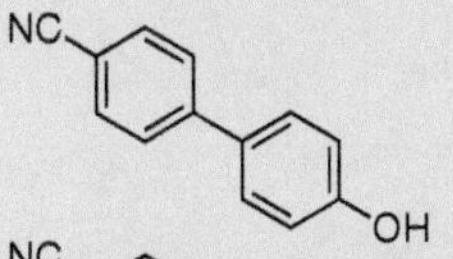	−4.0
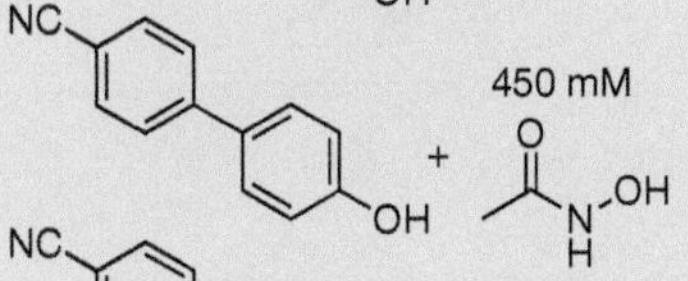 450 mM +	−5.4
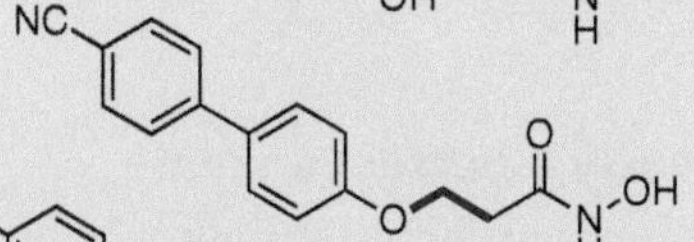	−10.0
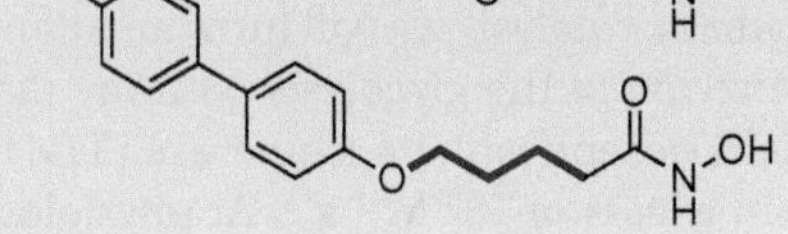	−6.6

design of inhibitors of the zinc-dependent protease stromelysin, acetylhydroxamic acid was found to bind to the active-site Zn^{2+} ion with a free energy of −2.3 kcal mol^{-1} (**Table 6.6**). A simple biaryl group was found to bind within the substrate cleft with a free energy of −4.0 kcal mol^{-1}. However, the presence of the hydroxamate group facilitated the binding of the biaryl group. When the acetylhydroxamate was present at a sufficiently high concentration to ensure that the active sites were completely occupied, the biaryl group bound with even greater affinity, −5.4 kcal mol^{-1}, perhaps as the result of an induced change in protease conformation. If acetylhydroxamate binds to stromelysin with a free energy of −2.3 kcal mol^{-1} and the biaryl binds with a free energy of −5.4 kcal mol^{-1} (when acetylhydroxamate is present), then what would be the free energy of binding when the two ligands are covalently linked? The net effect can be either greater than or less than the sum of the two interactions, −7.7 kcal mol^{-1}. For example, a three-carbon linker is clearly too long and the free energy of binding is only −6.6 kcal mol^{-1} (1.1 kcal mol^{-1} less than the free energy of the two ligands mixed in solution). The +1.1 kcal mol^{-1} penalty may arise from strain, sterics, or the cost of removing bound water from the active site so as to stuff in a hydrophobic linker. In contrast, if the linker has just the right length, then the free energy of binding is enhanced. For example, the one-carbon linker raises the free energy of binding to −10.0 kcal mol^{-1}. This enhancement of −2.3 kcal mol^{-1} may arise from entropy—that is, the preorganization of the two species into a bound conformation—or from removing bound water molecules from the ligands before they enter the active site.

Pairing up has an entropic cost. For example, the pairing of ballroom dancers has a higher entropic cost than a random mix of dancers in a nightclub. The entropic advantage of perfect preorganization can be substantial. In a classic analysis of the cost of bimolecular association in a Diels–Alder cycloaddition, it was estimated that the entropic cost of simply pairing a mole of diene with a mole of dienophile, before bond formation, could be more than 14 kcal!

Of course, the Diels–Alder cycloaddition is entropically more demanding than most reactions that involve the formation of a single bond, such as proton transfers or carbonyl additions. Before bond formation and cleavage occur in a Diels–Alder cycloaddition, the two reactants must adopt the correct distances and orientations. This kind of preorganization is similar to that required to bind two ligands in the active site of stromelysin. Obviously, the one-carbon linker does not provide a 14 kcal mol^{-1} cooperative advantage. Although it does hold the biaryl group close to the hydroxamate group, it does little to restrict torsion angles or the relative orientations of the functional groups.

Problem 6.12

Two ligands bind to nearby sites in a receptor, each with a dissociation constant of 1 μM. What dissociation constant would you expect if you were to join the two compounds together with a perfectly cooperative linker?

cooperative linker

K_d 10^{-6} M K_d 10^{-6} M K_d ?

Triosephosphate isomerase is nearly a perfect enzyme

To understand better the power of enzymatic catalysis we will turn our attention from kinases and proteases to a common enzyme in the glycolysis pathway that human cells use to produce ATP from glucose. Triosephosphate isomerase (TIM) is nearly a perfect enzyme with a k_{cat}/K_m ratio in excess of 10^8 M^{-1} s^{-1}. At physiological concentrations of dihydroxyacetone phosphate (DHAP), TIM binds substrate with nearly every collision, catalyzes the isomerization, and releases the product (*R*)-glyceraldehyde 3-phosphate before it bumps into another substrate (**Figure 6.65**).

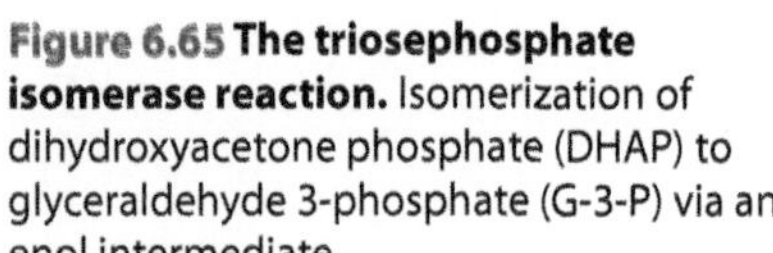

Figure 6.65 The triosephosphate isomerase reaction. Isomerization of dihydroxyacetone phosphate (DHAP) to glyceraldehyde 3-phosphate (G-3-P) via an enol intermediate.

Triosephosphate isomerase (TIM) is a canonical β barrel (sometimes referred to as a TIM barrel). The inner staves of the barrel are made up of parallel β strands and the outer staves are made up of α helices. The crystal structure of TIM with a phosphoglycolate inhibitor bound in the active site (**Figure 6.66**) has given considerable insight into the tricks used by enzymes to achieve efficient catalysis. TIM sequesters the highly reactive enolate intermediate with a flexible peptide loop (colored yellow in the figure) in a conformation that prevents side reactions. The slow step in the uncatalyzed reaction is the deprotonation of an α proton to form an enol.

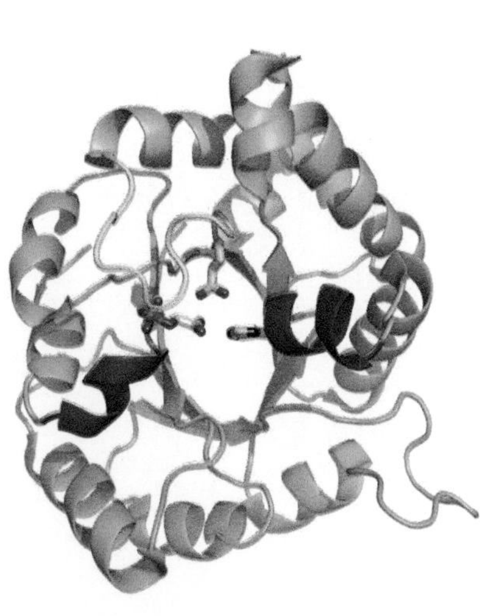

Figure 6.66 Operating room. A crystal structure of triosephosphate isomerase bound by a phosphoglycolate inhibitor reveals an array of helices and catalytic groups aligned with surgical precision. (PDB 2YPI)

To achieve this astonishing rate acceleration, TIM holds the substrate in a conformation that aligns the labile C–H bond of DHAP with the π* orbital of the carbonyl group. A glutamate side chain is poised above the substrate ready to abstract a proton. The imidazolium side chain of a histidine residue protonates the carbonyl group, facilitating the formation of the enolate. TIM has two α helices whose macrodipoles are precisely aligned. One helix stabilizes the binding of the negatively charged phosphate. The other helix destabilizes the protonated form of the histidine side chain, making it a more effective acid (see Figure 6.66). After the glutamate carboxylate group has abstracted a proton, the carboxylic acid merely needs to swivel around a single bond to place the acidic O–H bond above the basic C2 carbon atom (**Figure 6.67**). Simple buffers such as imidazole can also catalyze this transformation, but TIM is 10 billion times more efficient than imidazole.

Unlike imidazole or other base catalysts, TIM holds the phosphate group, which is a good leaving group, in the same plane as the enolate. This conformation prevents bond rotation that would facilitate the alignment of the electron-rich π system of the

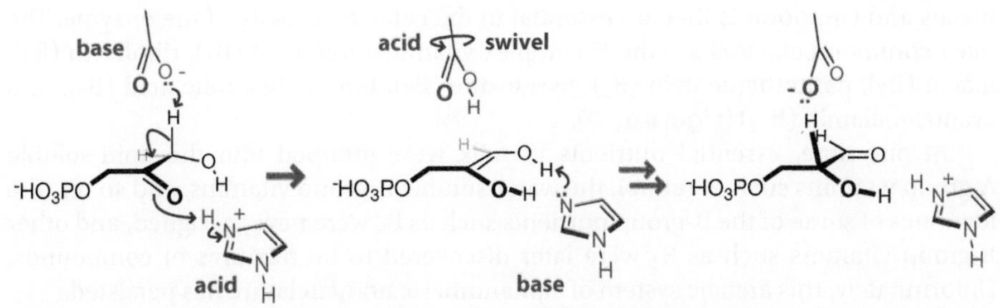

Figure 6.67 Rapid catalysis involving a bond rotation. A glutamate side chain can readily deprotonate and reprotonate the substrate through a simple bond rotation.

enolate with the $\sigma_{C\text{-}O}{}^*$ of the phosphate leaving group (**Figure 6.68**). The hold on the phosphate therefore prevents elimination of the phosphate, shutting down this undesirable side reaction.

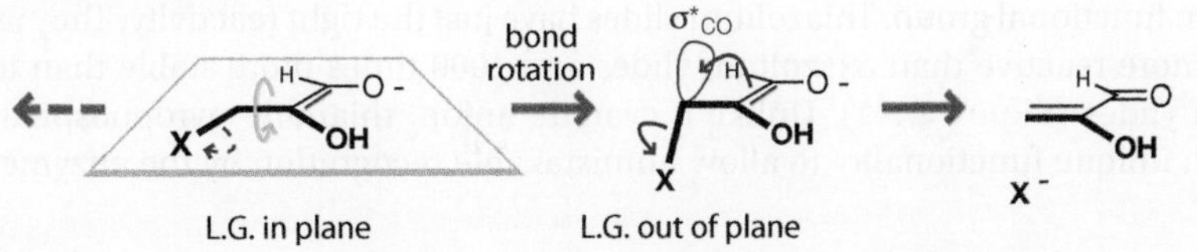

Figure 6.68 Safekeeping. TIM holds the reactive enolate in a conformation that prevents loss of the allylic phosphate group X. L.G., leaving group.

6.4 ENZYMES THAT USE ORGANIC COFACTORS

Enzyme cofactors extend the capabilities of enzymes

Because ribosomes incorporate only 20 amino acids into proteins, the number of potential functional groups in newly expressed enzymes is rather limited. Many enzymes recruit additional molecules, known as cofactors, to increase their functionality and catalytic capabilities (**Figure 6.69**). Common cofactors include various

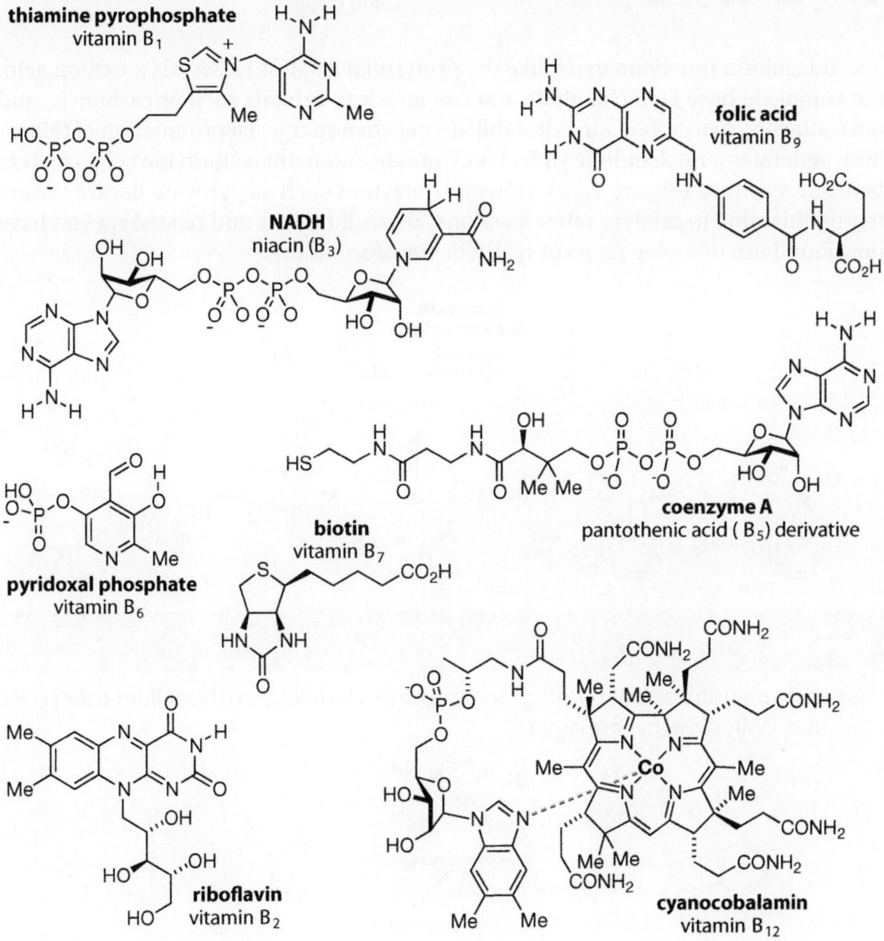

Figure 6.69 Enzyme prosthetic groups. B-group vitamins greatly extend the chemical abilities of enzymes.

Supplement Facts
Serving Size 1 tablet

Amount Per Tablet %DV		Amount Per Tablet %DV	
Vitamin A 2500 I.U,	50%	Biotin 30 mcg	10%
60% as Beta Carotene		Pantothenic Acid 10 mg	100%
Vitamin C 180 mg	300%	Calcium 162 mg	16%
Vitamin 031000 I,U,	250%	Iron 18 mg	100%
Vitamin E 50 I,U,	167%	Iodine 150 mcg	100%
Vitamin K 80 mcg	100%	Magnesium 100 mg	25%
Thiamin 1.5mg	100%	Zinc 15 mg	100%
Riboflavin 1.7 mg	100%	Selenium 70 mcg	100%
Niacin 20 mg	100%	Copper 2 mg	100%
Vitamin B6 2 mg	100%	Manganese 4 mg	200%
Folic Acid 400 mcg	100%	Chromium 120 mcg	100%
Vitamin B12 6 mcg	100%	Molybdenum 75 mcg	100%

Daily Value (DV)

Figure 6.70 Vitamins as cofactors. Most of the B-group vitamins that are taken as nutritional supplements are enzyme cofactors. (Courtesy of Pharmavite.)

metals and compounds that are essential to the catalytic activity of the enzyme. The most common cofactors are the B-complex vitamins: thiamine (B_1), riboflavin (B_2), niacin (B_3), pantothenic acid (B_5), pyridoxine (B_6), biotin (B_7), folic acid (B_9), and cyanocobalamin (B_{12}) (**Figure 6.70**).

At one time, essential nutrients in milk were grouped into the lipid-soluble A-group vitamins such as retinol, the water-soluble B-group vitamins, and so on. The identities of some of the B-group nutrients such as B_9 were never assigned, and other B-group vitamins such as B_4 were later discovered to be mixtures of compounds. Unfortunately, this archaic system of alphanumeric nomenclature has persisted.

Thiamine pyrophosphate provides a stabilized ylide

The enzyme cofactor thiamine pyrophosphate (vitamin B_1) possesses a key *N*-alkylthiazolium functional group. Thiazolium ylides have just the right reactivity. They are 100 times more reactive than oxazolium ylides and 1000 times more stable than imidazolium ylides (**Figure 6.71**). Unlike a cyanide anion, thiamine pyrophosphate has enough unique functionality to allow unmistakable recognition by the enzymes that need it.

pK_a 18

relative reactivity: 1, 100, 100,000

Figure 6.71 Comparing thiazolium reactivity with analogs. Thiazolium ions have just the right mix of nucleophilicity and stability.

Thiazolium functions much like the primordial catalyst HCN: it is a carbon acid; the conjugate base is nucleophilic and can attack π^* orbitals such as carbonyls; and once attached to a carbon atom it stabilizes negative charge. Deprotonation of thiazolium generates a nucleophilic ylide. Like cyanide anion, thiazolium ions can catalyze benzoin reactions (**Figure 6.72**). However, enzymes such as pyruvate decarboxylase recruit thiamine to catalyze other reactions. Thiazolium ions and related species have now found use in a wide range of synthetic transformations.

pyruvate decarboxylase

Figure 6.72 Mechanism of thiazolium catalysis. As a cofactor for pyruvate decarboxylase, thiazolium provides a stabilized ylide.

Problem 6.13

Suggest a plausible arrow-pushing mechanism for the following thiazolium-catalyzed reaction. DMF, dimethylformamide.

cat. thiazolium salt (Cl^-), Et_3N, DMF, 25 °C — 86%

The dihydropyridine group of niacin (vitamin B3) provides a reactive hydride

Enzymes use a dihydropyridine, nicotinamide adenine dinucleotide (NADH), as a hydride donor, generating the pyridinium analog, NAD^+. The reverse of this reaction is used by the human body to oxidize ethanol to acetaldehyde. Donation of a hydride group by NADH is facilitated by the gain in aromaticity of the pyridinium ring. Even before the structure of NADH was correctly assigned as a 1,4-dihydropyridine, Westheimer and coworkers conducted an elegant study with deuterium-labeled substrates to show that hydride donation was stereospecific (**Figure 6.73**).

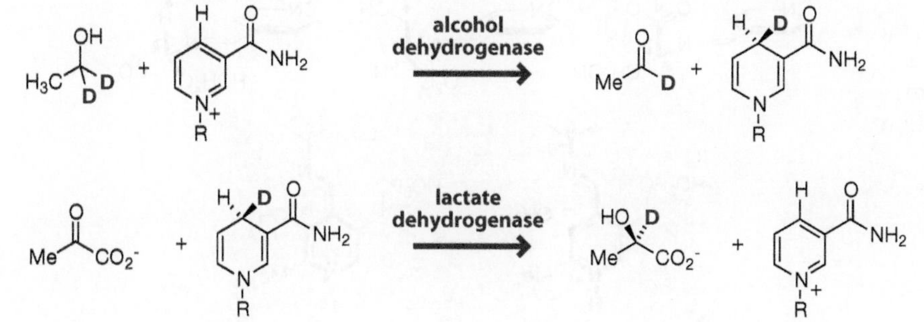

Figure 6.73 Elegant chemical proof. Westheimer's experiments to establish the stereospecificity of hydride transfers. The first step (top) produces a single enantiomer of deutero-substituted NADH. This reagent is then used (bottom) to transfer exclusively a deuteride to reduce pyruvate. Taken together, the experiments demonstrate that hydride extraction and donation take place in the enzymes' active sites with perfect stereochemical fidelity.

It has been suggested that the dihydropyridine ring of NADH is subtly bent into a boat conformation, with the nitrogen lone pair and pro-*R* hydride in pseudoaxial conformations (**Figure 6.74**). The nitrogen lone pair is aligned to donate into the π^* orbital of the adjacent double bonds, weakening the C–H bond of the axial pro-*R* hydride, making it more nucleophilic.

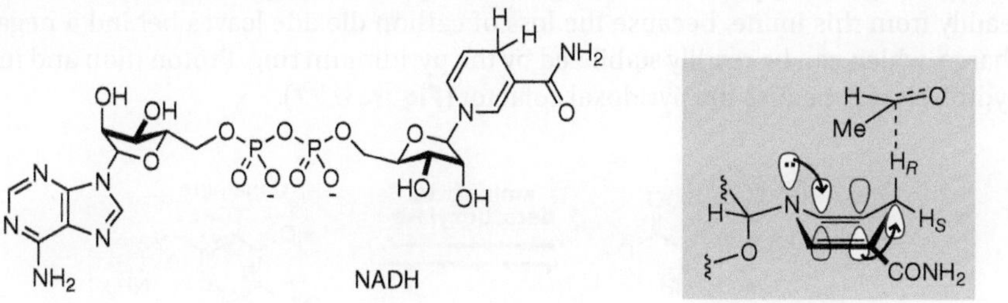

Figure 6.74 The structural basis for stereochemical fidelity in hydride transfer by NADH. The enzyme's active site positions NADH to allow only the pro-*R* hydride to reduce the carbonyl group. The nitrogen lone pair in the dihydropyridine ring activates the pro-*R* hydride.

Alcohol dehydrogenase 1B (ADH1B), expressed predominantly in the liver, is the key enzyme for the metabolism of ethanol in humans. The active site of ADH1B has two cofactors, a zinc atom and NAD^+ (**Figure 6.75**). The zinc atom is a powerful Lewis acid that favors binding of the ethoxide anion. The lone pairs of the anionic ethoxide ligand donate into the adjacent $\sigma_{C\text{-}H}^*$ orbitals, weakening the C–H bonds and making them more nucleophilic (see Figure 6.75).

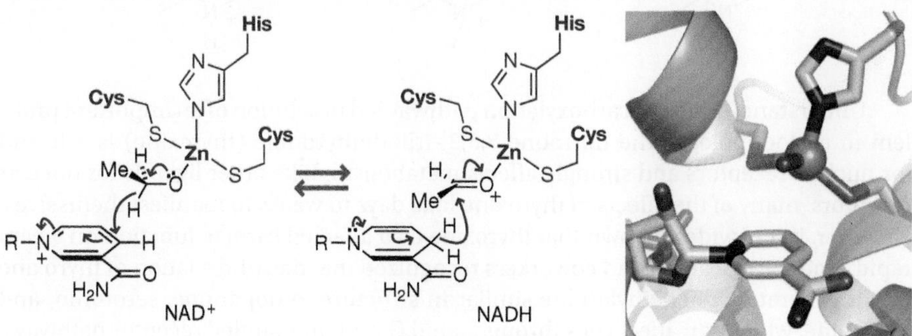

Figure 6.75 Lewis acid activation. Zinc activates the alcohol in the active site of alcohol dehydrogenase. In the crystal structure of human alcohol dehydrogenase 1B, a water molecule is bound to the active-site zinc atom in place of an alcohol. (PDB 1HSZ)

The pyridoxal cofactor serves as an electron sink

The pyridinium ring of protonated pyridoxal phosphate (PLP; vitamin B_6) can serve as an electron sink for a variety of catalytic and noncatalytic transformations (**Figure 6.76**). All reactions of pyridoxal involve the formation of an imine with a primary amine. The key requirement for bond activation of bonds adjacent to the amine nitrogen is that the bond must be aligned with the π^* system of the planar iminopyridinium ring system.

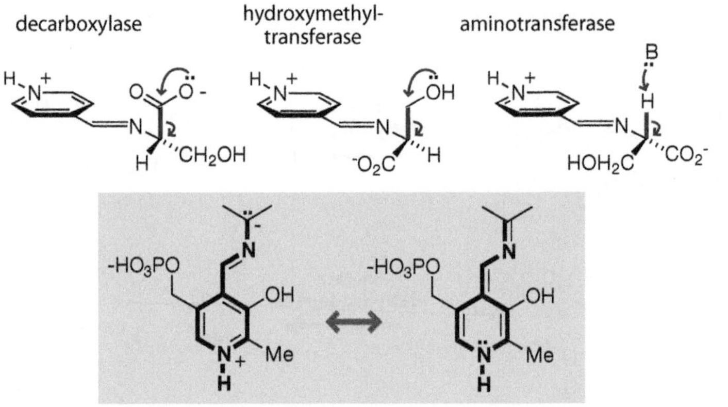

Figure 6.76 Pick a bond ... any bond. When pyridoxal forms an imine with an amino acid, it can activate any of the three bonds to the α carbon.

Aromatic amino acid decarboxylases use pyridoxal as a cofactor for the decarboxylation of amino acids. Key neurotransmitters such as dopamine, serotonin, and histamine arise through this type of decarboxylation reaction. The mechanism begins with the formation of an imine between the amino acid and pyridoxal phosphate. In its protonated form, the positively charged pyridinium ring serves as an electron sink that can stabilize negative charge; a resonance structure, for example, can move a negatively charged lone pair to the nitrogen of the pyridine ring. Decarboxylation occurs readily from this imine, because the loss of carbon dioxide leaves behind a negative charge, which can be readily stabilized by the pyridinium ring. Protonation and imine hydrolysis regenerates the pyridoxal cofactor (**Figure 6.77**).

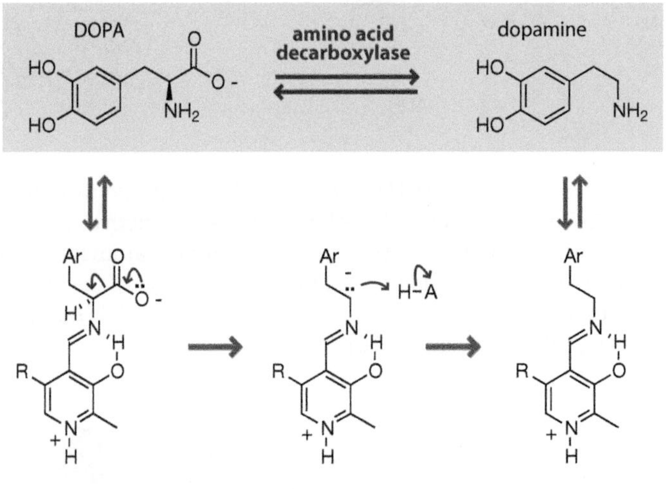

Figure 6.77 Amino acid decarboxylase mechanism. Decarboxylation generates a resonance-stabilized carbanion. DOPA, L-3,4-dihydroxyphenylalanine.

Understanding this decarboxylation pathway led to solution of an important problem in endocrinology. The hormone 3,5,3′-triiodothyronine (thyroxine) is a ligand for nuclear receptors and strongly affects metabolism. Like other ligands for nuclear receptors, many of the effects of thyroxine take days to weeks to manifest themselves. However, it was widely known that thyroxine also affected cardiac function on a very rapid timescale. Scanlan and coworkers recognized that decarboxylation of thyroxine might generate a phenethylamine similar in structure to dopamine, serotonin, and histamine, which can affect cells through rapid G protein-coupled receptor pathways.

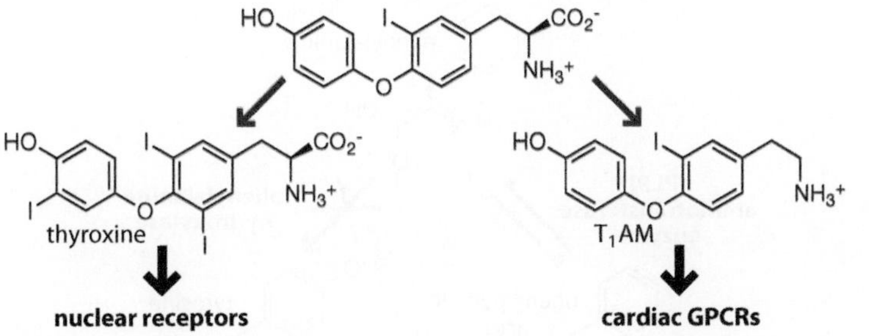

Figure 6.78 New versions of old hormones. 3-Iodothyronamine (T_1AM) is generated through decarboxylation mediated by pyridoxal phosphate. GPCRs, G protein-coupled receptors.

Indeed, the investigators hypothesized that the rapid, and sometimes undesirable, effects on cardiac function might be mediated by a yet to be identified decarboxylated form of iodothyronines (**Figure 6.78**). They then identified 3-iodothyronamine (T_1AM) as an endogenous signaling molecule. Apparently, thyroxine and T_1AM are generated from a common precursor.

Retro-aldol reactions are another type of reaction of amino acids catalyzed by pyridoxal-dependent enzymes such as serine hydroxymethyltransferase. Deprotonation of the side-chain hydroxyl group facilitates the subsequent retro-aldol reaction to generate an enolate intermediate (**Figure 6.79**). Free formaldehyde is not released by the enzyme because a molecule of tetrahydrofolate (K_m = 0.05 mM) is poised above the serine–PLP adduct to react with the formaldehyde as it is generated.

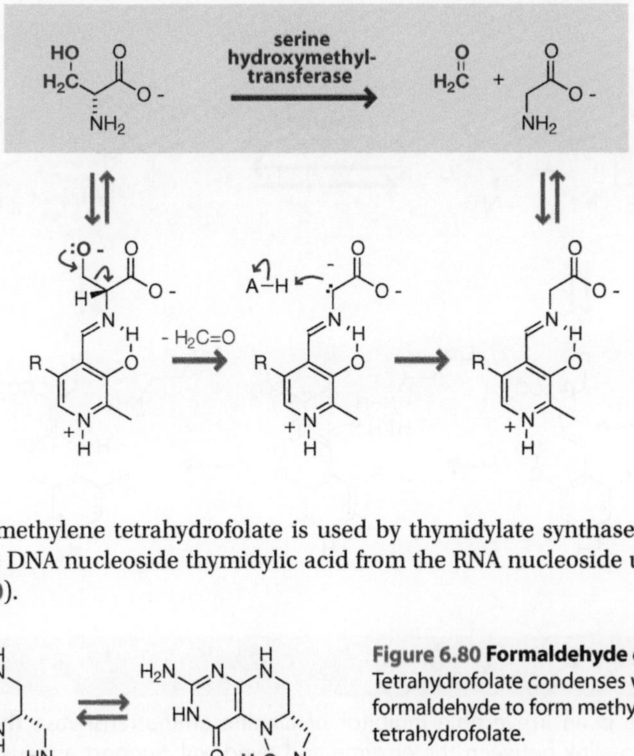

Figure 6.79 Pyridoxal catalyzed retro-aldol. Pyridoxal-dependent enzymes can catalyze enantioselective aldol reactions with glycine.

Recall that methylene tetrahydrofolate is used by thymidylate synthase in the synthesis of the DNA nucleoside thymidylic acid from the RNA nucleoside uridylic acid (**Figure 6.80**).

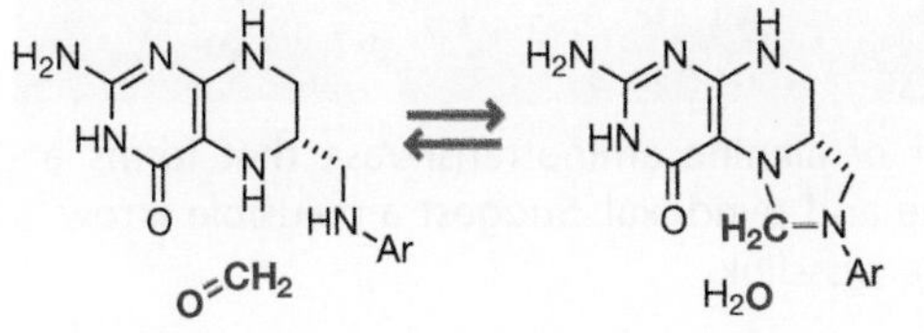

Figure 6.80 Formaldehyde carrier. Tetrahydrofolate condenses with formaldehyde to form methylene tetrahydrofolate.

The human amino acid tyrosine is generally synthesized from dietary phenylalanine through the action of phenylalanine hydroxylase (**Figure 6.81**). A mutational deficiency in phenylalanine hydroxylase prevents the conversion of dietary phenylalanine to tyrosine. The large excess of phenylalanine leads to mental retardation during childhood development. This condition is diagnosable by the presence of high concentrations of phenylpyruvic acid in the urine—a condition referred to as

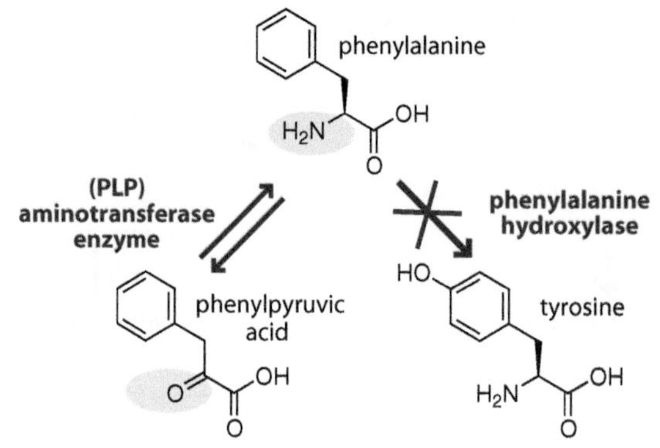

Figure 6.81 A deficiency in phenylalanine hydroxylase is diagnosable by the buildup of phenylpyruvic acid. PLP, pyridoxal phosphate.

phenylketonuria. Children diagnosed with phenylketonuria can develop normally if they adhere to a diet low in phenylalanine. Packages of the synthetic sweetener aspartame contain a warning to phenylketonurics because aspartame is a dipeptide composed of phenylalanine and aspartic acid.

The equilibrium between phenylpyruvic acid and phenylalanine is coupled to the equilibrium between pyridoxal phosphate and pyridoxylamine. This transamination reaction, catalyzed by a group of amino acid aminotransferases, is key to the biosynthesis of all amino acids. The mechanism involves the formation of an imine between the amino acid and pyridoxal phosphate, followed by a stereoselective tautomerization (**Figure 6.82**).

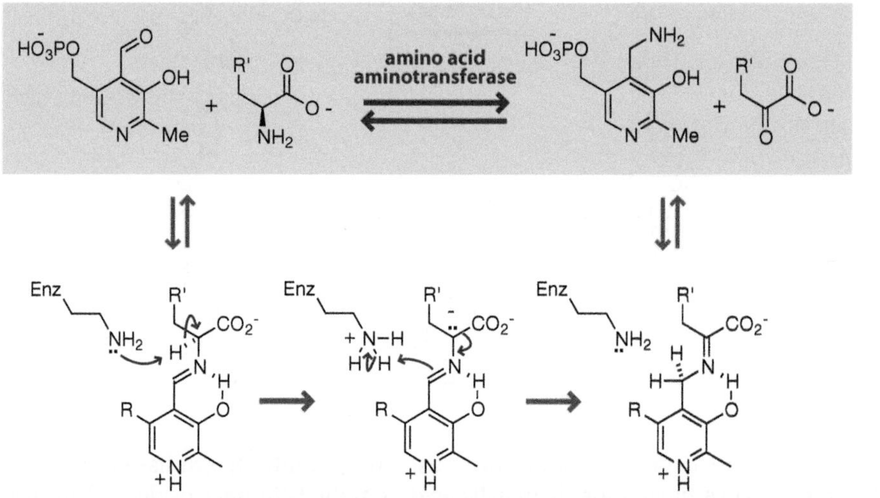

Figure 6.82 Aminotransferase mechanism. Aminotransferases catalyze tautomerizations that convert alkylamines to ketones and pyridoxal to pyridoxylamine.

Problem 6.14

Vinylglycine is an irreversible inhibitor of alanine aminotransferase that forms a covalent crosslink between the enzyme and pyridoxal. Suggest a plausible arrow-pushing mechanism for formation of the crosslink.

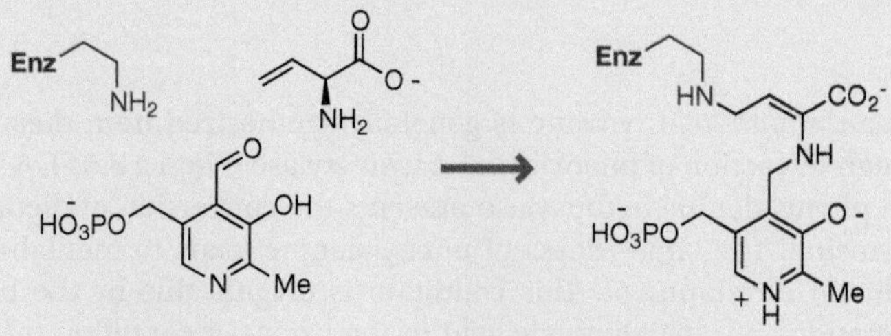

6.5 ENGINEERING IMPROVED PROTEIN FUNCTION

Protein engineering provides power tools for the dissection of protein function and the development of hyperfunctional molecules

Given the remarkable capabilities of the average protein, one might question the need to equip them with new functions. After all, each enzyme, the result of billions of years of evolution, offers a finely tuned, powerful catalyst for a particular chemical reaction. Furthermore, antibodies and hybridoma technology can provide a potentially limitless source of high-affinity receptors for binding to essentially any target. However, not all naturally occurring proteins are up to the tasks required by chemical biologists. Antibodies from mice, for example, cause immunogenic reactions in humans. Furthermore, no one wants to wait for the ideal enzyme to be identified from the proteome of, for example, the red squirrel. The quest to tinker with protein sequences to alter protein functions has also provided important insights into the mechanisms underlying protein function. A number of different computational approaches can contribute to the tailoring of protein function. Such experiments remain at the frontiers of protein engineering, and are not discussed here because of their evolving capabilities, which can quickly make even the most rudimentary discussion obsolete.

Unfortunately, the label "protein engineering" implies a much higher degree of precision than is available with current technologies. No civil engineer would plan to build a bridge or skyscraper with materials as unpredictable as proteins, which are held together and functioning primarily through the capricious properties of molecular recognition. Every change made to a protein sequence could have unpredictable consequences for the protein folding pathways, and to the ultimate conformation and function of the altered protein. In addition, proteins are large, complex machines. It is typically nonobvious where to introduce changes to obtain a desired new activity.

Many an elegant protein engineering project has foundered on misplaced assumptions and too little knowledge of the complete properties and consequences of changes to amino acid side chains. Despite such strong caveats, considerable progress has been made in the field by literally leveraging the uncertainty. In other words, if you cannot know in advance the exact properties of a particular alteration to the structure, make a large number of changes. Then, select from this collection or library for some powerful new function, and prepare (or hope?) for surprises. As we will see in this section, despite its misnomer protein engineering offers power tools for the transmogrification of protein function.

Alanine scanning assigns function to side chains and motifs

In reverse engineering, a complicated machine is taken apart piece by piece to dissect how the pieces fit together and work. This technique is often used in the computer chip and automobile industries to uncover the secrets of a competitor's product. The typical experiment begins with the thought, "Hmm... I wonder what this part does." Then, after removing it, you test how well the machine functions without it. Insight from reverse engineering can guide development of both outright copies and improved versions of a product or process. Building an improved version is an especially important part of the innovation loop, because it allows the testing of hypotheses suggested by close examination of each component of the machine.

Many experiments in protein engineering apply this reverse engineering approach. For example, both reverse engineering and protein engineering begin with a mysterious structure having a particular ability. The primary structure or protein sequence, in fact, is sometimes all that is available at the beginning of a protein engineering project. Then, changes are made to specific side chains, allowing one to deduce the contributions made to protein function by the altered side-chain functionalities. In such experiments, the naturally occurring protein found in the wild is typically referred to as the wild-type protein. In contrast, a protein that is longer, shorter, or has different amino acids as a result of genetic mutation should be referred to as a variant or a mutant protein.

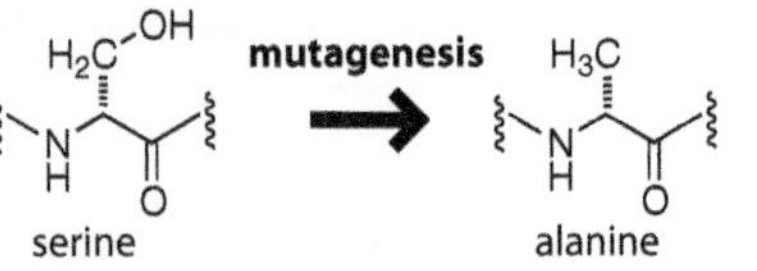

One common alteration to protein structure, replacement of a particular amino acid with alanine, effectively truncates the side chain at the β-carbon (**Figure 6.83**). Thus, the alanine-substituted protein can be tested to investigate the contributions to protein function made by the atoms past the β-carbon. For example, substitution of alanine for a serine residue allows one to test the contributions, if any, made by the hydroxyl functionality of the serine. Alanine provides a particularly useful substitution; the change clips off the side-chain functionalities without introducing excessive additional flexibility into the protein backbone, which could result from replacement with glycine.

serine **mutagenesis** alanine

Figure 6.83 Reverse engineering. The importance of a serine hydroxyl group can be revealed by mutating the serine to an alanine.

Some of the earliest peptide synthesis experiments applied this reverse engineering approach to understanding protein function. After inventing solid-phase peptide synthesis, Merrifield and coworkers chemically synthesized a wide variety of peptide hormones and proteins, including the 124-residue ribonuclease A (RNase A). In 1969, the latter *tour de force* galvanized the community around solid-phase synthesis. Overcoming objections of skeptics who held fast to the paradigm of solution-phase synthesis, Merrifield synthesized the enzyme using solid phase synthesis as described in Chapter 5. The resulting product mixture was allowed to generate the four disulfide crosslinks through air oxidation. Incorrect disulfide formation led to oligomers and misfolded protein. After isolation of monomers by gel electrophoresis and digestion of misfolded monomers with trypsin, the resultant enzyme convincingly possessed 78% of its expected enzymatic activity. In those early days of peptide synthesis, Merrifield also synthesized modified variants of RNase A with specific alterations designed to probe the functionalities required for enzymatic catalysis.

In these early mutagenesis studies, Merrifield did not have to synthesize the entire RNase A chemically to introduce each mutation. Unusually, RNase A can be split into two pieces, providing a more tractable synthetic target. The larger piece consists of RNase A digested with the highly specific protease subtilisin. The subtilisin clips off a 20-residue stretch of the N terminus (termed the S-peptide), rendering the enzyme inactive. Mixing this truncated enzyme with chemically synthesized S-peptide restores enzymatic activity to reasonable levels (more than 50%) (**Figure 6.84**). Thus, Merrifield could synthesize short peptides containing mutations and could probe the effect of each mutation on the activity of a good model for the full-length enzyme. Protein variants synthesized by this approach removed hydrogen bond-donating hydroxyl functionalities. For example, substituting an *O*-methyl ether for the hydroxyl group of a serine converts the side chain from a hydrogen-bond donor into a hydrogen-bond acceptor (**Figure 6.85**). Similar experiments form the basis for the field of protein engineering, which seeks to understand protein structure and function.

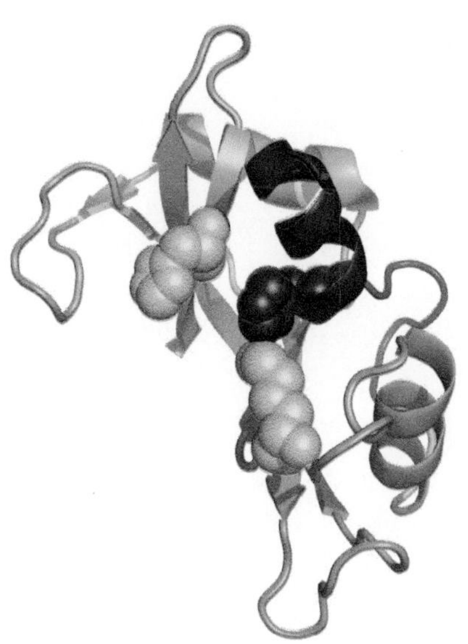

Figure 6.84 Ribonuclease A active site. The 20 N-terminal residues of the S-peptide are colored purple. In addition to His12 (purple van der Waals spheres), the active site includes Lys41 and His119 (yellow). (PDB 1FS3)

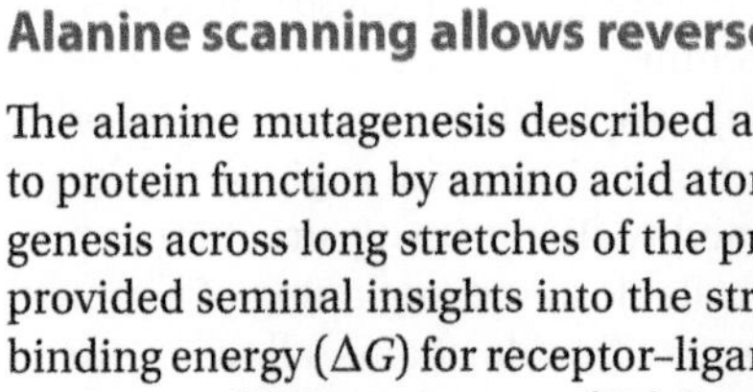

serine ? *O*-methyl serine

Figure 6.85 Replacing the serine hydroxyl group with an ether. This conversion of a hydrogen-bond donor to a hydrogen-bond acceptor requires chemical synthesis or another technique for incorporation of the non-natural amino acid *O*-methyl serine into the protein.

Alanine scanning allows reverse engineering of protein function

The alanine mutagenesis described above provides a powerful test for contributions to protein function by amino acid atoms past the β-carbon. Systematic alanine mutagenesis across long stretches of the protein sequence, termed **alanine scanning**, has provided seminal insights into the structure and function of numerous proteins. The binding energy (ΔG) for receptor–ligand interactions is typically evaluated by measuring the equilibrium constant (K_d) for the interaction and applying the following equation:

$$\Delta G = -RT \ln K_d \tag{6.1}$$

When alanine scanning a protein interface, the binding energy for the alanine-substituted variant (ΔG_{Ala}) can be measured and compared with that of the wild-type receptor (ΔG_{wt}) by the following equation, which calculates the difference in binding energy between the wild-type protein and the alanine-substituted variant:

$$\Delta\Delta G_{Ala\text{-}wt} = \Delta G_{Ala} - \Delta G_{wt} \tag{6.2}$$

This $\Delta\Delta G_{Ala\text{-}wt}$ value quantifies in kcal mol^{-1} exactly how much binding energy the side-chain atoms past the β-carbon contribute to the interaction. Large $\Delta\Delta G_{Ala\text{-}wt}$ values (more than 1.4 kcal mol^{-1}) indicate a strong contribution to the binding energy (more than 10-fold effect on the K_d). Low values (less than 0.5 kcal mol^{-1}) are often within the error of the measurement and are associated with negligible contributions to protein binding.

Human growth hormone (hGH), for example, has been extensively studied through alanine scanning to uncover how hGH binds to two hGH receptors to activate cell signaling through a JAK–STAT pathway (**Figure 6.86**). The first hGH receptor covers a large area of the hGH surface, and directly contacts 19 amino acid side chains. With so many hGH residues touching the surface of the receptor, one might expect each side chain to provide a tiny and equal fraction of binding energy to the overall interaction (like the dozens of Velcro hooks sticking into the fuzzy material, in which each weak interaction adds to the overall strength holding the Velcro together). Alanine scanning, however, definitively shows this *not* to be the case. Several contact residues provide more than 1.5 kcal mol^{-1} per residue of binding energy to the overall interaction, and many contact residues contribute essentially no energy to stabilize binding to hGH receptor.

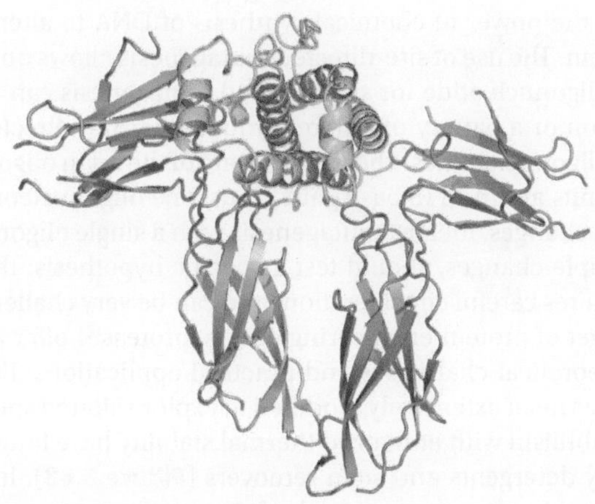

Figure 6.86 Ligand-mediated dimerization of human growth hormone receptor. Human growth hormone receptor transduces growth signals from outside the cell to the inside of the cell. Two identical receptor molecules (blue) bind to nonidentical sides of the hormone (green). Each receptor has an additional transmembrane domain on the bottom (not shown) that extends through the membrane into the cytoplasm. (PDB 3HHR)

The bulk of the binding energy gluing together hGH and hGH receptor is provided by a small, tightly clustered set of 6 out of 19 residues. This **hotspot of binding energy** looks like a core sample of a protein; the outside diameter of the hotspot has hydrophilic functionalities, which surround a hydrophobic core (**Figure 6.87**). Alanine scanning of other proteins has revealed that many, but not all, proteins bind to their targets by using a similar hotspot strategy.

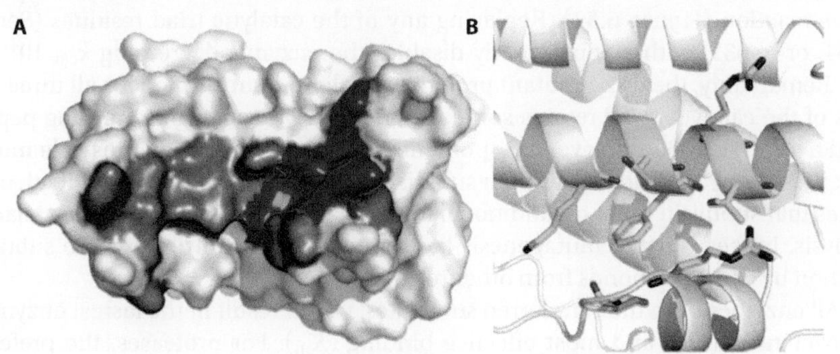

Figure 6.87 A hotspot of binding energy on the hGH surface. (A) Binding to the first human growth hormone (hGH) receptor buries 19 hGH side chains (red and blue surfaces). However, only six tightly clustered side chains contribute significantly to the binding energy (red). The other buried side chains (blue) fail to contribute much binding energy, but might provide shape requirements for further specificity in differentiating between several related protein hormones. (B) A close-up view of the hGH hotspot reveals that the side chains form a bullseye pattern with the hydrophobic aryl and alkyl carbons (green) in the center surrounded by hydrophilic functionalities with oxygen (red) and nitrogen (blue) atoms.

Figure 6.88 Thermostable proteases. Many detergents contain engineered thermostable proteases that degrade protein-based stains. (Courtesy of Proctor & Gamble.)

Problem 6.15

Shotgun alanine scanning is a combinatorial technique for alanine scanning large numbers of residues across a protein interface. After mutagenesis to introduce a 1:1 ratio of wild-type protein to Ala variants, selections (usually with phage display) are used to identify the binding members of the library, which are sequenced to identify residues that are intolerant of alanine substitutions. The site-directed mutagenesis to create the alanine shotgun scanning library requires that degenerate oligonucleotides encode the wild-type/alanine substitutions. For each of the following sequences, design degenerate oligonucleotides for alanine shotgun scanning (denote 1:1 ratios of two DNA bases as {base1/base2}). For some amino acids, up to two other residues are allowed, as a result of degeneracy in the genetic code (for example A below).

A Hexahistidine affinity tag: H_6

B Initial five residues of mannan-binding tectin serine protease 2: MRLLT

C From human leukocyte antigen: HPVSD

Protein engineering enables improvement of protein function

The different approaches to site-directed mutagenesis described in earlier chapters (Section 3.8) apply the power of chemical synthesis of DNA to alter a specific residue in a target protein. The use of site-directed mutagenesis allows one to test specific hypotheses. Each oligonucleotide for site-directed mutagenesis can encode either a single point mutation or a battery of different mutations. Site-directed mutagenesis requires synthetic oligonucleotides. The current state of the art in oligonucleotide synthesis reaches its limits around 115 base pairs. Thus, one oligonucleotide can encode up to 25 amino acid changes. Ideally, mutagenesis with a single oligonucleotide, even one encoding multiple changes, should test a specific hypothesis: the design of the mutant protein requires careful consideration, and can be very challenging.

As an early target of protein engineering efforts, proteases offer an enzyme with both interesting theoretical challenges and practical applications. For example, the enzyme subtilisin has been extensively modified to explore altered specificity and stability. Variants of subtilisin with enhanced thermal stability have long been deployed in common laundry detergents and stain removers (**Figure 6.88**). In addition, subtilisin can be expressed in large quantities by fermentation—often more than 1 g of subtilisin is obtained per liter of bacteria grown (a whopping 30% or more of the protein mass secreted by the organism). This beguiling combination of properties has prompted protein engineers to modify every aspect of subtilisin activity, including its catalytic rate, thermal stability, optimal temperature and pH for activity, and substrate specificity. The new abilities were typically introduced by rational design, and uncovered important insights into how proteases and other enzymes and proteins work.

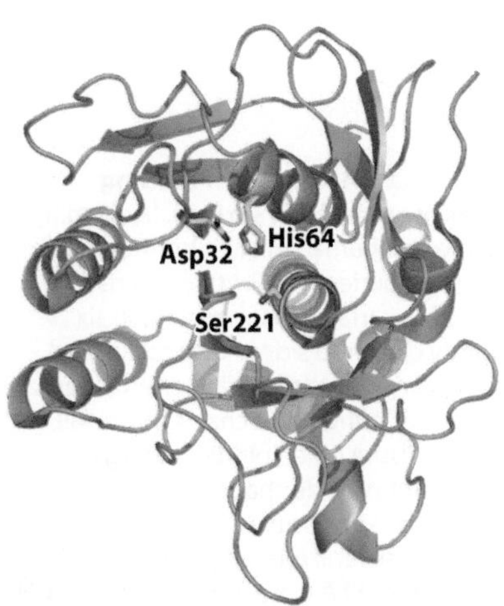

Figure 6.89 Subtilisin. The catalytic triad (sticks and labeled) accelerates peptide bond hydrolysis by using the serine protease mechanism described earlier in this chapter. (PDB 2ST1)

Given the field's uncertainties, protein engineering projects should always start with mutations aimed at testing the mechanistic contributions made by key residues. The bacterial serine protease subtilisin catalyzes the hydrolysis of protein amide bonds, and accelerates the reaction about 10^9-fold versus the uncatalyzed spontaneous reaction (**Figure 6.89**). Replacing any of the catalytic triad residues (Ser221, His64, or Asp32) with alanine nearly disables the enzyme, decreasing k_{cat} 10^4–10^6-fold. Remarkably, the triple mutant protein with alanine substituted for all three residues of the catalytic triad retains some residual catalytic activity, catalyzing peptide bond hydrolysis 1000-fold over the spontaneous reaction. Although the triple mutant protein is about 10^6-fold slower than subtilisin, the residual catalytic activity demonstrates that subtilisin relies on additional functionalities outside the catalytic triad for catalysis. Indeed, further mutagenesis has demonstrated contributions to subtilisin function by hydrogen bonds from other residues.

All enzymes have their preferred substrates, which result in the fastest enzymatic turnover rates (k_{cat}) and most effective binding (K_m). For proteases, the preferred substrates are peptide sequences that can optimally fit into the active site. Determining this specificity, the enzyme cradles substrate side chains in pockets with shapes and functionalities evolved to fit specific amino acids. The amino acids N-terminal to the cleaved amide bonds are labeled P1, P2, P3, ... *Pn*, and P1′, P2′, P3′, ..., *Pn*′ denote residues on the C-terminal side of the scissile peptide bond (**Figure 6.90**).

Subtilisin evolved to chew apart essentially any protein, exhibiting broad substrate specificity. However, by judicious substitution of the residues lining the substrate-specificity pockets, the specificity of subtilisin can be limited to specific peptide sequences. For example, engineering subtilisin to hydrolyze only the peptide bond after the sequence RAKR required the incorporation of negatively charged residues in the P4, P2, and P1 binding pockets to complement the positively charged arginine and lysine residues of the substrate. The resultant subtilisin variant then rejects sequences that the wild-type enzyme would eagerly digest. Such large changes to enzyme specificity often require additional mutations to either regulatory sequences or other residues responsible for shifting the positioning of the substrate in the active site.

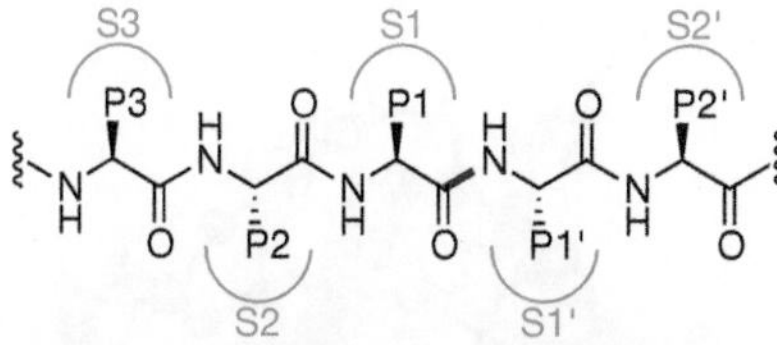

Figure 6.90 Protease specificity. The hydrolyzed bond (red) is anchored in the protease active site by amino acid side chains to either side (P1, P1′, etc.), which nestle into complementary pockets provided by the protease (S1, S1′, etc.).

Protein engineering enables a change of protein function

In addition to tinkering with the specificity and activity of proteases, similar techniques can transform a nonhydrolytic enzyme into a functional protease. Cyclophilin catalyzes the *cis–trans* isomerization of proline, a key step in protein folding, and completely lacks proteolytic activity (**Figure 6.91**). To accomplish the proline isomerization, the cyclophilin active site has evolved exquisite specificity for proline-containing peptides. Reasoning that such binding abilities would be perfect for a high-specificity protease, Quéméneur and coworkers attempted to transform cyclophilin from a *cis-trans* proline isomerase into a proline-specific protease. First, cyclophilin residues near the proline amide were mutated to serine to provide a nucleophile for serine protease activity. Mutation of Ala91 to Ser (A91S) provided some protease activity, which, although modest, represented a more than 10^5-fold gain in activity over the lowest limit of detection for the wild-type cyclophilin (**Table 6.7**). Additional mutagenesis identified positions amenable to hosting the Asp and His of the serine protease catalytic triad (F104H–N106D). The resultant triple mutant protein is a surprisingly efficient Xaa-Pro peptidase (where Xaa represents any amino acid). Despite the elegance of this effort, the catalytic efficiency and rate enhancement of a naturally occurring proline-specific peptidase remain far more impressive (prolyl oligopeptidase in Table 6.7).

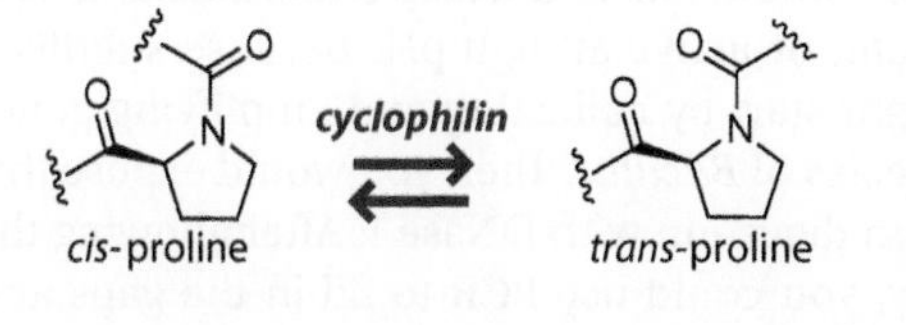

cyclophilin

cis-proline   *trans*-proline

Figure 6.91 Cyclophilin. The enzyme catalyzes rotation around the amide bond of proline. Like all enzymes, cyclophilin catalyzes both the forward and reverse reactions.

Table 6.7 Engineering protease activity into cyclophilin.

Enzyme	Efficiency (k_{cat}/K_m) (M^{-1} s^{-1})	K_m (10^{-3} M)	k_{cat} (s^{-1})	Rate enhancement (k_{cat}/k_{uncat})
Cyclophilin	$<10^{-3}$	ND		
Cyclophilin (A91S)	73.1	0.6	0.044	9×10^6
Cyclophilin (A91S-F104H-N106D)	1675	2.4	4.0	8.33×10^8
Prolyl oligopeptidase	1,000,000	0.06	60.5	1.26×10^{10} (est.)

ND, not determined; est., estimated.

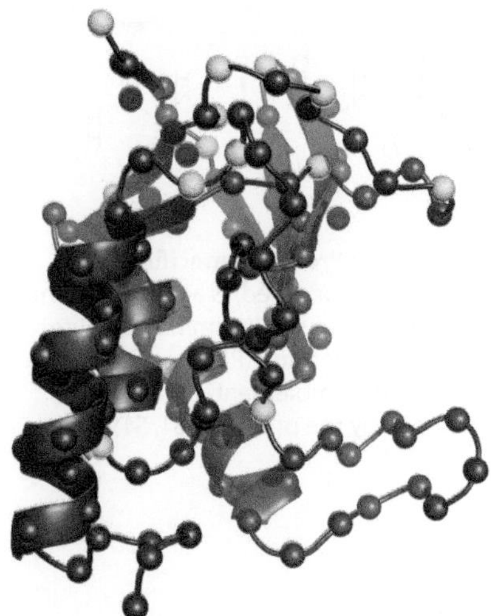

Figure 6.92 Tolerant positions. Relatively few sites in staphylococcal nuclease tolerate random mutations (yellow) that arise during error-prone PCR. Most residues (blue) are critical. Such residues stabilize secondary structure, are packed within the interior of a protein, or involve critical turn residues. Such sites do not tolerate random substitutions. (PDB 1STN)

Most random mutations debilitate rather than enhance protein function

Introducing random changes into DNA sequences is quite straightforward using **error-prone PCR**. Many error-prone PCR methods are based on the ability of additives to interfere with the proof-reading mechanisms of DNA polymerase. For example, the addition of dimethylsulfoxide to the PCR buffer disrupts base-pair hydrogen bonding, which leads to a higher rate of incorrect base incorporation. When Mn^{2+} is added to the PCR mixture, the Mn^{2+} displaces Mg^{2+} in the active site of DNA polymerase and increases the error rate through an incompletely understood mechanism. Because the G•C base pairs are stronger than A•T base pairs, altered ratios of dNTPs are often used to limit bias and favor incorporation of the less common mutations. Nucleotide mutation rates of a few percent are typical.

A few small, highly stable proteins such as egg lysozyme tolerate a wide range of random mutations. More typically, mutations lead to inactive enzymes that either do not fold or do not catalyze. For example, when random mutagenesis was applied to plasmids encoding staphylococcal nuclease, only 3% of the proteins retained enzymatic activity (**Figure 6.92**). Tightly packed side chains in the center of proteins are particularly sensitive to mutations. If you mutate an alanine to a leucine, you need to make space for the new isopropyl substituent by mutating another residue—close in space, but farther along the peptide chain—to something smaller. That is not always possible. Even if it is physically possible to make such a compensatory mutation, the chances of compensatory mutations occurring simultaneously are very low when one relies on random changes.

Recombination generates new combinations of existing mutations

Living organisms use redundancy to mitigate the effects of deleterious point mutations. A colony of staphylococci can thrive even if a single bacterium experiences a lethal point mutation. Because human cells are diploid, with two homologous copies of each gene, the cell can often tolerate a point mutation that codes for a nonfunctional protein, as long as the other copy of the gene codes for a functional protein. Most organisms, from bacteria to humans, use genetic recombination to enhance genetic diversity without the perils of random mutagenesis. Pairwise recombination of homologous genes generates new combinations of the mutations that were already present in the original strands.

Exceptionally diverse combinatorial libraries of homologous variants can be accessed by an artificial form of recombination called **DNA shuffling** to form chimeric sequences (**Figure 6.93**). Imagine that you would like to develop a thermostable version of subtilisin that is stable in hot water or active at high pH. Because subtilisin is produced by *Bacillus subtilis*, you might start by collecting and amplifying genes for homologous proteases from other species of *Bacillus*. Then you would expose this mixture of gene copies to limited, random digestion with DNase I. After allowing the gene fragments to rehybridize randomly, you could use PCR to fill in the gaps and generate a staggering number of different genes with completely new combinations of the natural mutations. With natural recombination, two strands of DNA recombine to give just two new rearranged strands. Because DNA shuffling uses many copies of the original strands, it can generate a vast number of new combinations; some of the resulting proteins are more fit and some are less fit. The diverse population of genes can be transformed into bacteria to facilitate selection and amplification.

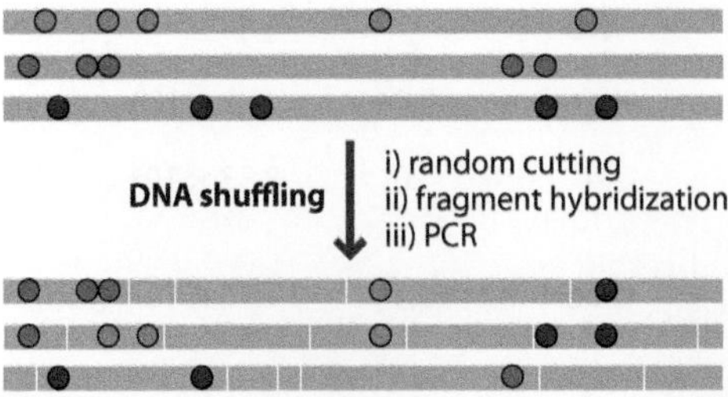

Figure 6.93 DNA shuffling. DNA shuffling breaks apart and recombines homologous DNA sequences, each with a different set of mutations. Each of the resultant chimeras contains a unique combination of the mutations that were present in the original sequences.

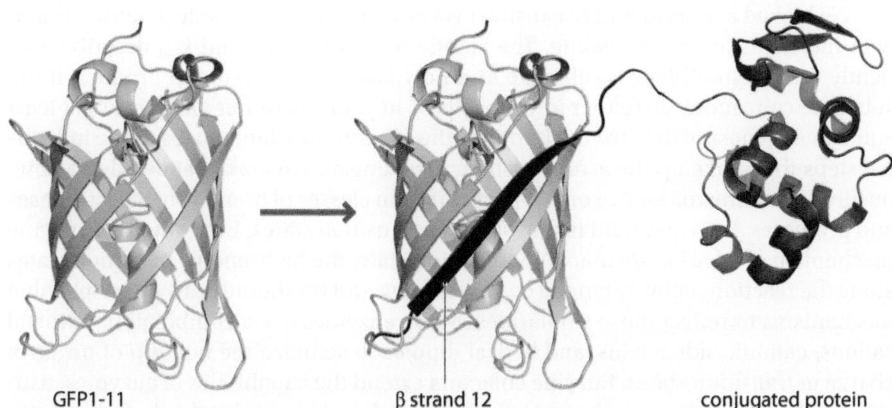

Figure 6.94 Split GFP for complementation screens. GFP1-11 (residues 1–214, in green) becomes fluorescent only when the structure is complemented by residues from the twelfth β strand (GFP12, residues 216–230, in dark blue). β Strand 12 complements GFP1-11 even when it is conjugated to another protein (light blue). (PDB 1GFL)

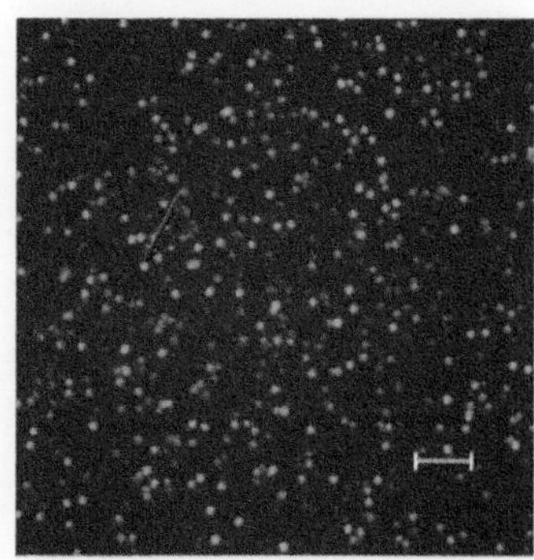

Figure 6.95 Bacterial colonies in a folded protein screen. Colonies of bacteria expressing proteins conjugated to GFP β strand 12 show varying levels of fluorescence, depending on the solubility of the protein (scale bar = 1 cm). Correctly folded, soluble proteins lead to highly fluorescent colonies. The colony with the red arrow was picked for further study. (From S. Cabantous and G.S. Waldo, *Nat. Meth.* 3:845–854, 2006. With permission from Macmillan Publishers Ltd.)

Screens work well for modest numbers of protein variants, but exceptionally diverse libraries require selections

DNA shuffling was used to improve a truncated version of green fluorescent protein (GFP) for use in **complementation assays**. The truncated form of GFP becomes fluorescent only when the C-terminal β strand 12 is present in a soluble form (**Figure 6.94**). The ability of β strand 12 to complement GFP1-11 makes the split GFP system useful for assays. For example an assay was developed using the split GFP system to distinguish soluble proteins from insoluble proteins. When a library of proteins conjugated to GFP β strand 12 was expressed in *E. coli*, the colonies expressing soluble proteins could be easily identified through fluorescence simply by holding the plate above an ultraviolet light (**Figure 6.95**). The approach allows quick screening of a large number of protein variants to identify the well-folded variants.

Screens can work well for up to a few million different protein variants. However, the theoretical diversity of combinatorial protein libraries can exceed the number of particles in the Universe; although chemical biologists do not access numbers even close to such diversity, exceptionally diverse libraries of billions to trillions are not unusual. To examine all members of such massive libraries, screens must usually be coupled to some kind of selection that amplifies only the hits. Many auxotrophic selections couple successful traits such as catalysis to bacterial survival and growth. Hits or **selectants** isolated from phage-displayed protein libraries can be transfected into *E. coli*, leading to marked amplification. Selectants from *in vitro* libraries of DNA can be easily amplified with PCR.

6.6 SUMMARY

Every chemical conversation in the human body is initiated by touch, by some kind of noncovalent interaction between molecules. Proteins bind with high affinity and specificity to a much broader range of molecules than any other type of biooligomer. Thus, proteins are capable of performing a broad range of roles in living cells. Dose-response curves offer critical insight into the effects of ligand, substrate, and drug concentration on binding, and ultimately on the effects of binding. At sufficiently low concentrations, acute toxins are not toxic and natural cures will not cure you; however, at sufficiently high concentrations, anything can be toxic, including such otherwise harmless substances as sugar and salt.

It is usually difficult to measure an association constant directly. Chemical biologists rely on IC_{50} values, EC_{50} values, and K_m values—indirect measures of lability—to describe protein binding interactions. As a consequence we are stuck with the awkward semantic situation in which lower values correspond to tighter binding.

We looked at enzymes in two distinct ways: using abstract kinetic parameters and with mechanistic arrow-pushing. The kinetic parameters K_m and k_{cat} describe how tightly the enzyme binds a substrate and how fast it is converted to product. If the substrate concentration (either in your cells or in your microtiter plate) is not at least equal to K_m, most of the enzyme is idling. The k_{cat} rate constant relates to the individual steps that make up the arrow-pushing mechanism. We looked at detailed arrow-pushing mechanisms for two of the most common classes of human enzymes: kinases and proteases. Enzymes bind to and stabilize transition states. Even when an enzyme mechanism involves more than one transition state, the high-energy transition states along the reaction pathway tend to be very similar, and we should draw arrow-pushing mechanisms to reflect those similarities. Many enzymes use a combination of metal cations, cationic side chains, and helical dipoles to stabilize the buildup of negative charge in transition states. Enzyme cofactors extend the capabilities of enzymes, usually involving covalent catalysis. Hydrogen bonding has a critical role in catalysis. Hydrogen bonds often immobilize the substrate, allowing catalytic residues to act with surgical precision. The fact that hydrogen bonding precedes proton transfers blurs the line between concerted and stepwise arrow-pushing mechanisms. If you are unsure, draw all proton transfers as distinct mechanistic steps.

Even as we endeavor to understand how proteins fold, how they assemble, how they recognize, how they cleave, graft and create, we are empowered to improve them. Site-directed mutagenesis allows us to study the roles of individual residues in enzymes or to redesign enzymes for enhanced performance. DNA-based techniques such as error-prone PCR and DNA shuffling can be used to generate vast libraries of protein mutants from which one can select proteins with improved properties, such as an enzyme with high thermal stability, or a protein drug with longer serum half-life. Now that we have a clearer view of binding and catalysis, we can explore the three remaining classes of biooligomers that are assembled by enzymes: oligosaccharides, polyketides, and terpenes.

LEARNING OUTCOMES

- Master the equations describing binding equilibrium.
- Use equations for on-rate and off-rate to relate kinetics to the equilibrium dissociation constant.
- Determine IC_{50} and EC_{50} from dose response curves.
- Based on EC_{50} or LD_{50}, determine the concentrations of a chemical at which no significant biological effect is expected.
- Use the kinetic variables K_m and k_{cat} to describe binding, catalytic turnover, and catalytic efficiency.
- Draw arrow-pushing mechanisms for serine/cysteinyl proteases, metalloproteases, and aspartyl proteases.
- Provide mechanisms for acid and base catalysis by various functionalities in enzyme active sites.
- Recognize the most common enzyme cofactors, and draw arrow-pushing mechanisms for PLP, NADH, and thiazolium.
- Diagram the dual roles of Zn^{2+} ions in enzyme structure and activity.
- Explain the mechanistic basis for the regulation of protease and kinase activities.
- Relate equilibrium constants to the thermodynamic energies governing receptor–ligand interactions.
- Sketch the steps required for the design and introduction of amino substitutions into proteins.
- Plan selections of proteins for fluorescence or other readily recognized phenotypes.

PROBLEMS

***6.16** Chorismate mutase from *E. coli* catalyzes the Claisen rearrangement of both chorismate and the *O*-methyl derivative **1** with the kinetic parameters below.

A Which substrate binds more tightly?

B Which bound substrate rearranges more rapidly?

C Overall, which substrate is rearranged more quickly by the enzyme?

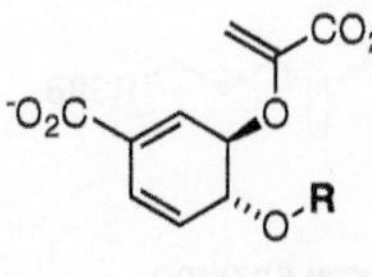

		K_m (mM)	k_{cat} (s^{-1})
Chorismate	R = H	0.14	29
Derivative **1**	R = CH_3	1.9	0.56

***6.17** In eukaryotic cells, the turnover of intracellular proteins is mediated mainly by the ubiquitin–proteasome pathway, a nonlysosomal proteolytic pathway. The 26S proteasome is a 2.5 MDa multiprotein complex found in both the nucleus and cytosol of all eukaryotic cells. The 20S particle is a threonine protease.

A The natural product lactacystin forms a strained reactive intermediate that acylates the N-terminal threonine in the active site of the 20S proteasome. Suggest a plausible structure for the reactive intermediate that modifies the enzyme and a mechanism for the acylation of the enzyme.

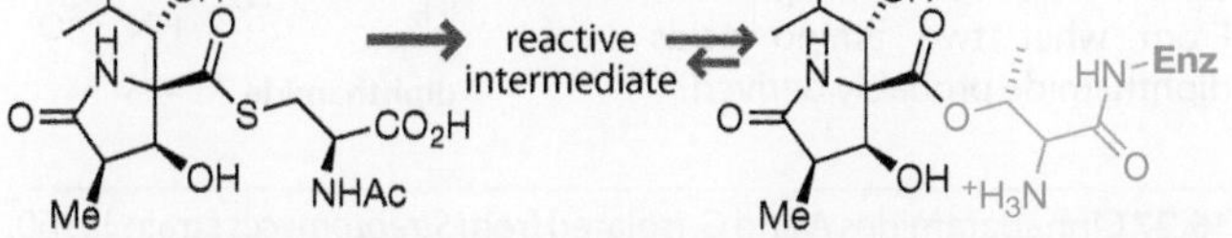

B The acylation of the proteasome by lactacystin is reversible. Suggest a plausible arrow-pushing mechanism for the reverse reaction that regenerates the reactive intermediate and the free enzyme.

C The natural product salinosporamide forms an irreversible adduct with the 20S proteasome. Suggest a plausible arrow-pushing mechanism for the reaction.

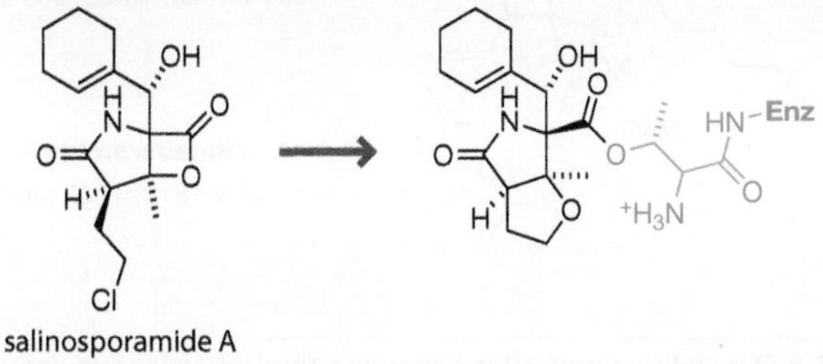

salinosporamide A

D The fluorogenic substrate below is used to monitor the activity of the proteasome. Using resonance structures, speculate on why the nonpeptide product is fluorescent but the substrate is not.

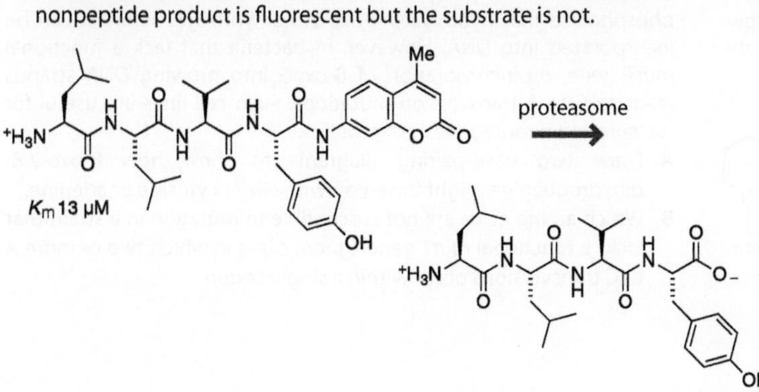

K_m 13 μM

proteasome

6.18

A Suggest a plausible arrow-pushing mechanism for the cleavage of the peptide AAPFGF by the serine protease subtilisin to give AAPF and GF.

B Mutation of the key serine in the catalytic triad of subtilisin to cysteine generates a variant enzyme that will catalyze aminolysis of esters. Draw the expected products of the following amide-bond-forming reaction catalyzed by the S221C variant of subtilisin.

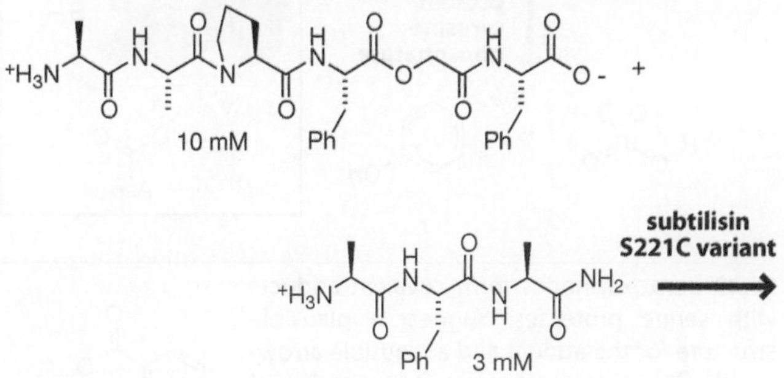

10 mM

3 mM

subtilisin S221C variant

6.19 Tryptophan synthase has a catalytic amine base in the active site to facilitate the following transformation. Suggest a plausible arrow-pushing mechanism for the reaction.

tryptophan synthase

$-H_2O$

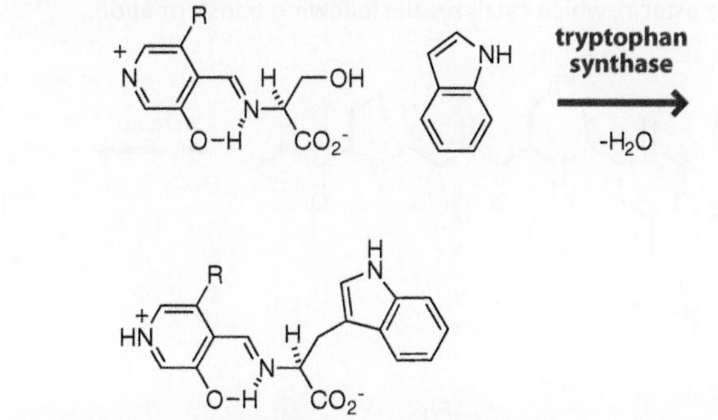

***6.20** Caspase-3 is a cysteine protease that cleaves at the sequence DEVD. Suggest a plausible mechanism for covalent inhibition of caspase-3 by the following inhibitors.

A

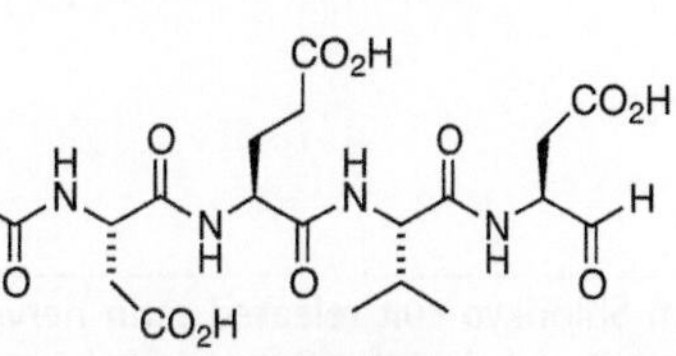

B

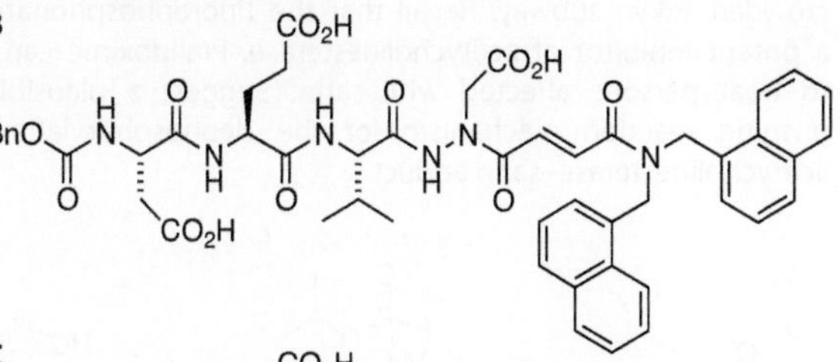

C

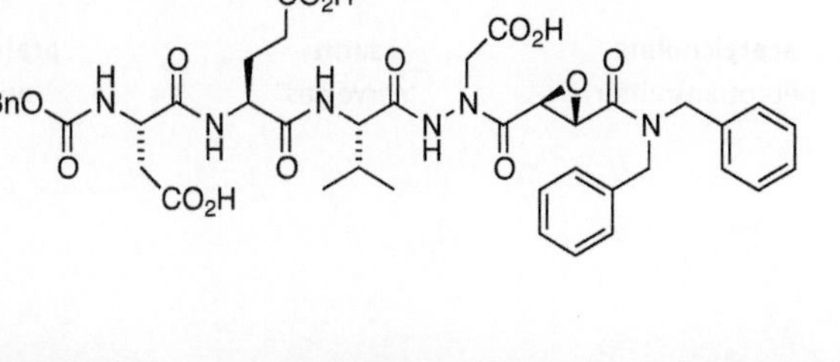

6.21 The active site of bovine protein tyrosine phosphatase has three key residues in the active site: an arginine, a cysteine, and an aspartic acid. Suggest a plausible arrow-pushing mechanism for the enzyme-catalyzed hydrolysis of tyrosine monophosphate esters that involves these three active-site residues. Expect a mechanism with at least seven steps.

protein tyrosine phosphatase

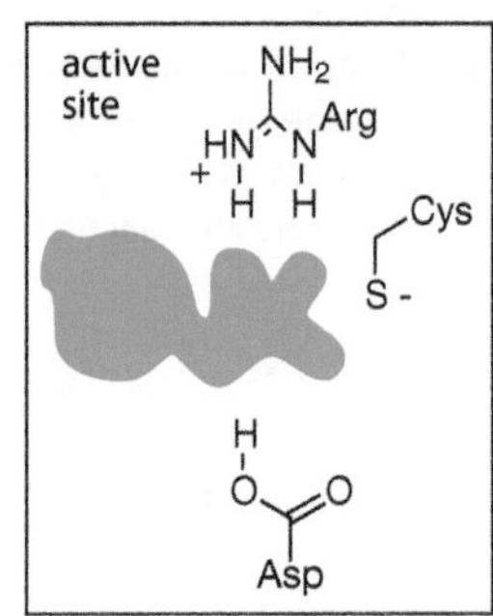

***6.22** Benzoxazinones form covalent adducts with serine proteases. Suggest a plausible structure for the adduct and a plausible arrow-pushing reaction mechanism for formation of the covalent adduct.

6.23 Design a stable transition-state analog that might inhibit the enzyme astacin, which catalyzes the following transformation.

astacin

6.24 In 1995, the Aum Shinrikyo cult released sarin nerve gas in a crowded Tokyo subway. Recall that the fluorophosphonate sarin is a potent inhibitor of acetycholinesterase. Pralidoxime can be used to treat persons affected with sarin. Suggest a plausible arrow-pushing reaction mechanism for the dephosphorylation of the acetycholinesterase–sarin adduct.

acetylcholine neurotransmitter

sarin "nerve gas"

pralidoxime antidote

***6.25** Suggest a plausible amino acid substitution in aspartate aminotransferase that would change the substrate specificity from aspartate to arginine. How many times faster does aspartate aminotransferase catalyze the transamination of aspartate relative to that of arginine?

K258 R292 D223 R386

Wild-type enzyme

Substrate	$\frac{k_{cat}}{K_m}$ ($M^{-1} s^{-1}$)
Aspartate	34,500
Arginine	0.0695

6.26 The protein elongation factor 2 is essential for ribosomal translation in all eukaryotes. It possesses a unique residue named diphthamide. From what two amino acids is diphthamide probably derived?

diphthamide

***6.27** Cinnabaramides A and G, isolated from *Streptomyces* strain JS360, are potent inhibitors of the human proteasome. Cinnabaramide G is believed to generate cinnabaramide A, with an unusual carbonyl stretch at 1812 cm^{-1}. Suggest a structure for cinnabaramide A and a plausible mechanism for covalent inhibition of the proteasome (a threonine protease).

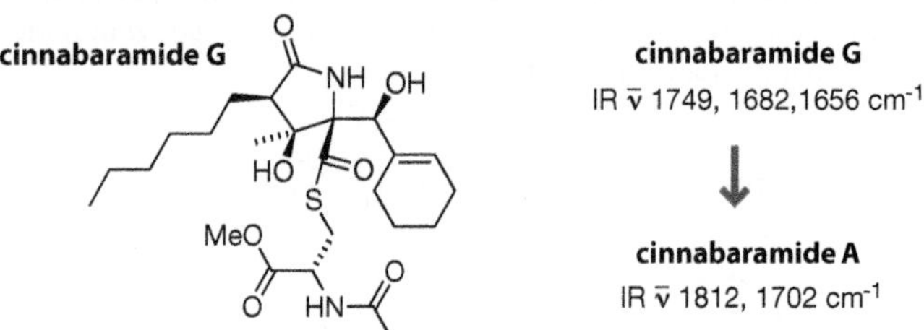

***6.28** 8-Oxo-7,8-dihydroGTP is generated through oxidative damage to guanosine. Unfortunately, 8-oxoG base-pairs with cytosine and adenine with nearly equal efficiency. The *mutT* gene encodes a phosphatase that hydrolyzes 8-oxo-7,8-dihydroGTP before it can be incorporated into DNA. However, in bacteria that lack a functional *mutT* gene, misincorporation of 8-oxoG into growing DNA strands leads to C to A transversion mutations. Such cell lines are useful for (semi)random mutagenesis of proteins.

A Draw two base-pairing diagrams to show how 8-oxo-7,8-dihydroguanine might base-pair with either cytosine or adenine.

B Which amino acids are not susceptible to mutation in a strain that lacks a functional *mutT* gene. Ignore cases in which two or more A to C transversions occur within a single codon.

Table (for Problem 6.29) Biological activity of IL-8 variants.

Variant	EC_{50} (nM)	Variant	EC_{50} (nM)
(Wild type)	5	P32A	11
E4A	>500	N36A	14
L5A	165	T37A	18
R6A	>1000	I39A	10
I10A	83	I40A	50
P16A	7	V41A	3
F17A	198	L43A	14
F21A	33	S44A	2
I22A	73	L49A	8
I28A	6	L51A	9
S30A	3	P53A	6

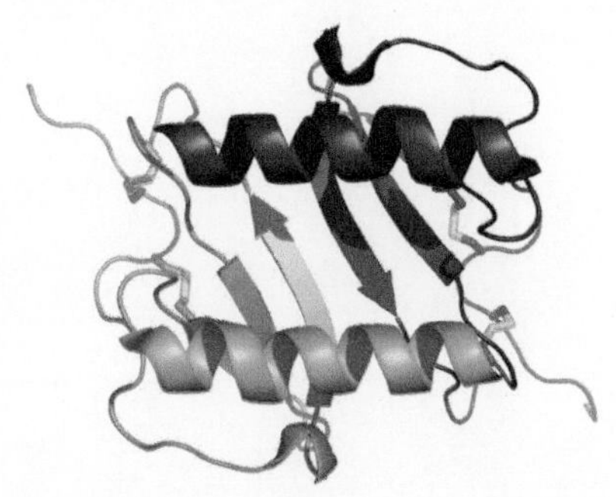

6.29 The cytokine interleukin-8 (IL-8) binds as a dimer to a pair of interleukin-8 receptor molecules. Alanine scanning mutagenesis was performed to generate a series of 21 variants that were assessed for biological activity; the results are given in the table above.

A On the basis of the solution NMR structure of IL-8 (PDB 2IL8), identify which regions of the IL-8 dimer are the most important for receptor activation.

B Is the greater activity of Phe17 relative to Phe21 due to contacts with the receptor?

***6.30** Identify the three residues that make up the catalytic triad of the serine protease neuropsin (PDB 1NPM).

Glycobiology

7

LEARNING OBJECTIVES

- Master the basic vocabulary of human glycobiology, including abbreviations and anomeric stereochemistry.
- Understand the structure and reactivity of acetals as they relate to glycans.
- Understand the basic mechanisms of enzymes that break and make glycosidic bonds.
- Differentiate the biosynthesis of O-linked glycoproteins from that of N-linked glycoproteins.
- Contrast the heterogeneity of human glycans with the homogeneity of the template-based biooligomers—DNA, RNA, and proteins.
- Distinguish the range of human glycans from those found on pathogenic organisms.
- Appreciate the role of cell-surface glycans in cell–cell communication.
- Understand the relationship between hyperglycemia and nonenzymatic glycation.

The field of glycobiology is connected to the very origins of organic chemistry through the monumental work of Hermann Emil Fischer (**Figure 7.1**). At a time when horses and trains were the common mode of transportation, Fischer worked out the relative stereochemical structures of all the common sugars, and completed the total syntheses of three stereochemically rich monosaccharides: glucose, fructose, and mannose. These achievements, made over the course of a single decade (1884–1894), are all the more remarkable when one considers that the tetrahedral nature of carbon had only been proposed 10 years earlier by van 't Hoff and Le Bel.

Despite this fast start, the field of glycobiology has been slow to mature, lagging behind the chemical biology of nucleic acids and proteins. At the time of writing of this book, there is no simple method for sequencing oligosaccharides, there is no general way to express homogeneous oligosaccharides, there is no simple way to amplify small amounts of oligosaccharides, and there is no general method for the synthesis of oligosaccharides. These inadequacies are arrayed against formidable objects of study—naturally occurring oligosaccharides are precious in quantity, dazzling in complexity, and maddening in heterogeneity. In spite of these challenges, heroic efforts of the past 50 years have elucidated the role of many of the carbohydrates in the human body, particularly in cell-cell communication and cooperation. These efforts empower chemical biologists to have an expansive view of human glycobiology, in which the oligosaccharides on the outside of cells are perhaps even more interesting than the glucose that fuels them from the inside.

Figure 7.1 Hermann Emil Fischer in his laboratory around 1900. (Courtesy of Archiv der Max-Planck-Gesellschaft, Berlin-Dahlem.)

7.1 STRUCTURE

There are 10 common monosaccharide building blocks for human glycans

The monomeric sugar building blocks, called **monosaccharides**, that make up carbohydrates are more complex than building blocks of DNA, RNA, or proteins, so we will start this chapter with an introduction to the 10 monosaccharide building blocks that are common in human **oligosaccharides**. The common term **carbohydrate** is derived from the empirical formula of monosaccharides such as ribose ($C_5H_{10}O_5$) and

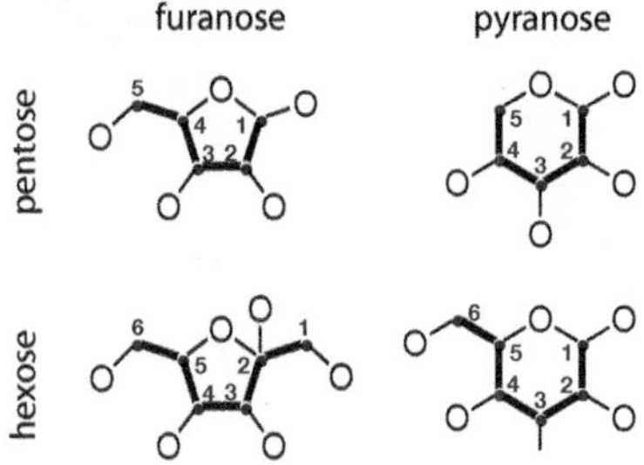

Figure 7.2 Numbering and descriptors for monosaccharides. Monosaccharide descriptors convey the number of carbons in the monosaccharide and the size of the ring.

glucose ($C_6H_{12}O_6$). A more general formula $[C(H_2O)]_n$ suggests a "hydrate of carbon." The term carbohydrate is usually used to refer to molecules or moieties composed of one or more monosaccharides. The collection of all molecules that contain carbohydrate functional groups, such as glycolipids, glycoproteins, and **polysaccharides**, are referred to as **glycans**.

The acetal carbon of monosaccharides, called the **anomeric center**, is the focal point for chemical reactivity. Carbohydrate rings are numbered such that the end of the chain closest to the anomeric center is numbered as carbon 1 (**Figure 7.2**). Only nine monosaccharides are incorporated into human glycans by **glycosyltransferases** (**Figure 7.3**), and their structures and names should be committed to memory. Without this common vocabulary, we cannot carry on a meaningful discussion of glycobiology. The actual diversity of the human **glycome** is greatly amplified by post-glycational modifications. For example, glucuronic acid is sometimes epimerized at the C5 position to generate iduronic acid. Similarly, **deacetylases** and **sulfotransferases** can drastically alter the ionic state of various sugars. Later, we will use a combination of monosaccharide names and positional numbering to abbreviate the structures of stunningly complex oligosaccharides.

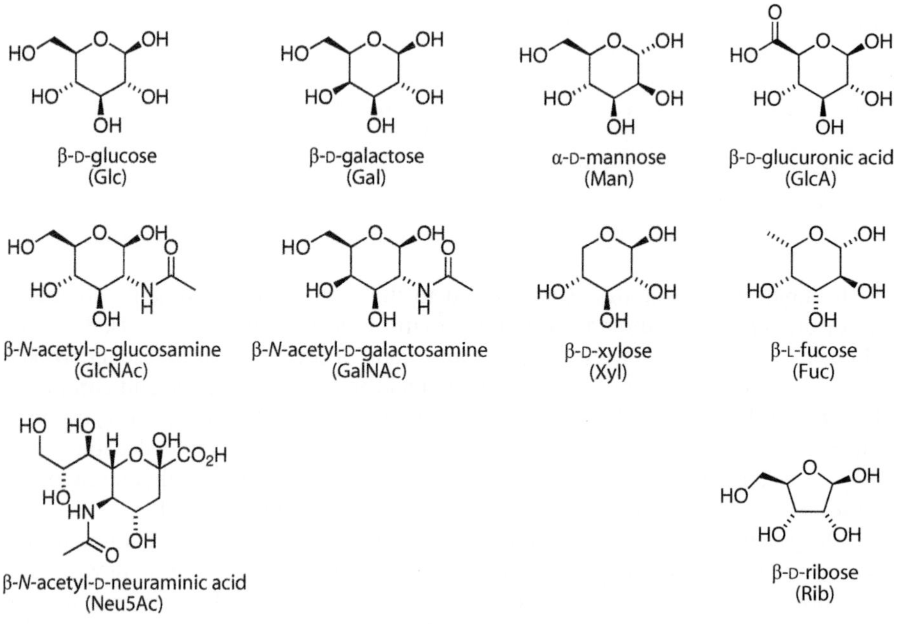

β-D-glucose (Glc)
β-D-galactose (Gal)
α-D-mannose (Man)
β-D-glucuronic acid (GlcA)
β-N-acetyl-D-glucosamine (GlcNAc)
β-N-acetyl-D-galactosamine (GalNAc)
β-D-xylose (Xyl)
β-L-fucose (Fuc)
β-N-acetyl-D-neuraminic acid (Neu5Ac)
β-D-ribose (Rib)

Figure 7.3 Key carbohydrate building blocks that are important in human oligosaccharides. Ribose, which is found in nucleic acids, is absent from human oligosaccharides.

Problem 7.1

Draw Lewis structures of the charged functional groups resulting from each of the enzymes below. Assume physiological pH (7.4).

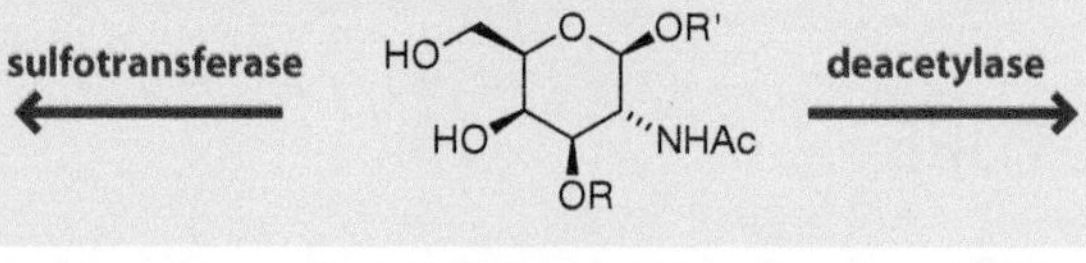

sulfotransferase ⟵ HO, OR', NHAc, HO, OR ⟶ **deacetylase**

The chemistry of ribose and its nucleoside derivatives was discussed in detail in Chapters 3 and 4. However, five-membered ring monosaccharides (furanoses) have not yet been identified in human oligosaccharides. If one considers all the organisms in Nature, the set of known oligosaccharide substituents becomes considerably more diverse. For example, in microorganisms one finds glycans containing subunits with five-membered rings, six-membered rings, and five, six, seven, eight, or nine carbon atoms. Exploiting the differences between the oligosaccharides of human cells and the oligosaccharides of pathogenic organisms is a promising avenue for the development of new vaccines and new medicines.

Problem 7.2

Draw the following sugars in the preferred chair conformation.

A β-D-Xylose

B β-D-Galactose

C α-D-Mannose

D α-L-Fucose

Glycobiology uses a compact form of nomenclature

Consider the IUPAC names that distinguish the cyclic forms of glucose and galactose: (3*R*,4*S*,5*S*,6*R*)-tetrahydro-6-(hydroxymethyl)-2*H*-pyran-2,3,4,5-tetraol and (3*R*,4*S*,5*R*,6*R*)-tetrahydro-6-(hydroxymethyl)-2*H*-pyran-2,3,4,5-tetraol. It is immediately obvious that the IUPAC system of chemical nomenclature, based on Cahn–Ingold–Prelog (*R*/*S*) assignments of stereogenic centers, is too clumsy for glycobiology. Thus, glycobiologists have retained the venerable naming system that was introduced in the 1800s. In the traditional naming system, the *relative* configurations of the stereogenic centers of monosaccharides are embodied by prefixes, such as gluco-, galacto-, manno-, and xylo-. The *absolute* configuration of the monosaccharide molecule is then designated using the (D/L) system of nomenclature (**Figure 7.4**). Usually, only a single enantiomeric form of each sugar is present in an organism (typically the D-enantiomer), and it is common practice to leave out the D or L descriptor. Configurational assignments such as D or L can be made by examining the single stereogenic carbon atom farthest from the acetal or ketal moiety. If this stereogenic center has an *R* configuration, the entire sugar is a D-sugar; if the stereogenic center has an *S* configuration, the entire sugar is an L-sugar. To maintain the analogy with the parent compounds, the uronic acids glucuronic acid and iduronic acid are not assigned stereochemistry as carboxylic acid derivatives. Instead the uronic acids are assigned D/L stereochemistry like the parent compounds, glucose and idose respectively (**Figure 7.5**).

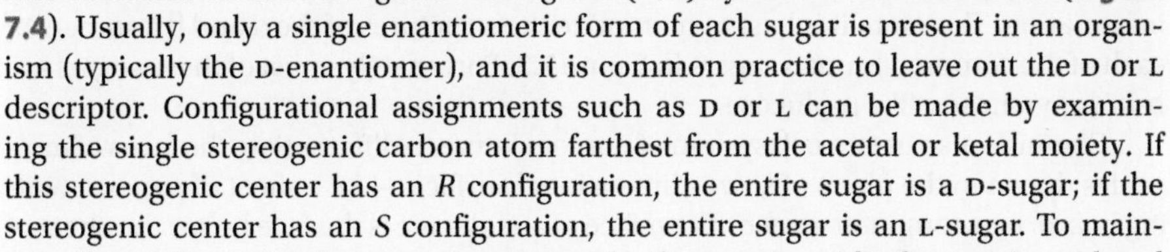

D-glucose (R) L-glucose (S) D-xylose (R) N-acetyl-D-neuraminic acid (R)

Figure 7.4 D and L configurations. For the common glycan building blocks, assignment of D and L enantiomers can be made on the basis of the stereogenic center (*R* or *S*) farthest from the acetal or ketal (indicated with a pink dot).

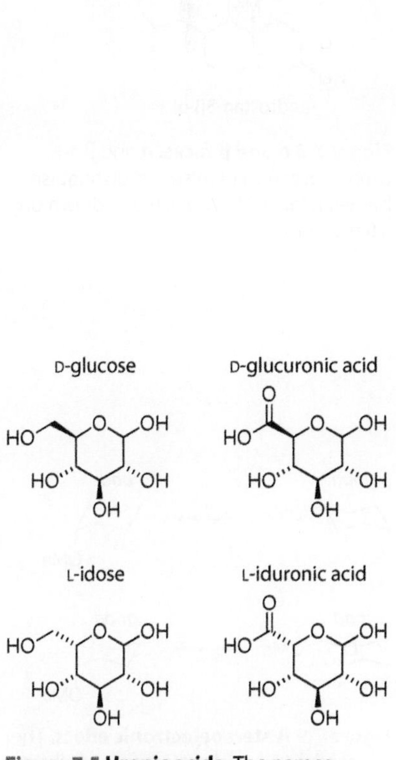

Problem 7.3

Assign the absolute configuration (D or L) of the following naturally occurring non-human monosaccharides.

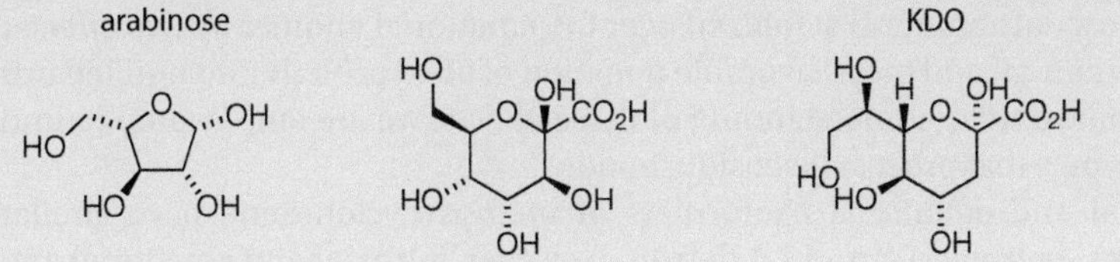

arabinose KDO

D-glucose D-glucuronic acid L-idose L-iduronic acid

Figure 7.5 Uronic acids. The names and D/L assignments of uronic acids are derived from the parent compounds.

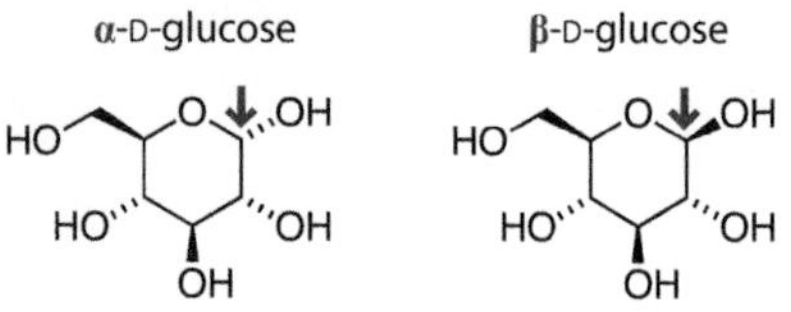

Figure 7.6 Anomers of glucose. The anomeric center is indicated with an arrow.

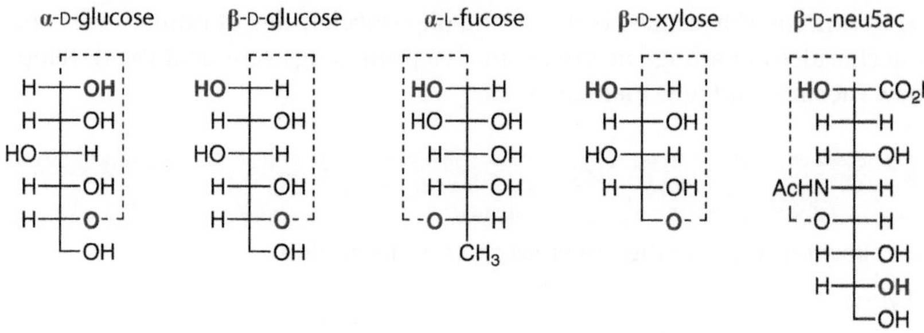

Figure 7.7 α and β configurations are based on Fischer projections. Anomeric α and β configurations are designated on the basis of the relative orientation of the anomeric hydroxyl and the substituent on the stereocenter farthest from the anomeric center. Since Fischer projections are no longer used in modern organic chemistry, it is easier simply to memorize α and β configurations for each of the monosaccharides in Figure 7.3.

The acetal or ketal carbon is a stereogenic center in all naturally occurring glycoconjugates. That means that it can exist in two different configurations, referred to as **anomers** (**Figure 7.6**). According to IUPAC rules, the two anomers are labeled as α or β (see Figure 7.6) depending on the relative orientation between the D/L reference carbon (farthest from the acetal or ketal) and the anomeric carbon in the Fischer projection representation (see Figure 7.3 and **Figure 7.7**). When the substituents at the anomeric center and the D/L reference center are on the same side of the carbon backbone, the stereoisomer is referred to as the α anomer. When the substituents at the anomeric center and the D/L reference center are on the opposite sides of the carbon backbone, the stereoisomer is referred to as the β anomer. This book does not use Fischer projections, because they do not convey meaningful spatial or conformational information. Therefore it is best to memorize the structures of the α and β anomers of each of the common human sugars in Figure 7.3. Fortunately, most of these sugars are D-hexopyranoses, like D-glucose.

Chemists also use an α/β system of nomenclature to differentiate the faces of molecules drawn in the plane of a two-dimensional surface—like a sheet of paper, a writing board, or a computer screen. The α face of a molecule is the face below the sheet of paper; the β face of a molecule is the face above the sheet of paper. A useful mnemonic device is that fish swim down (α resembles a fish) and bees fly up (β resembles the letter B). The α/β facial descriptors are highly effective whenever the molecule is drawn on a chalkboard for an audience, but are ambiguous if the molecule is not drawn. The α and β facial designations have been adopted as unambiguous descriptors for steroids because, by convention, chemists draw steroids with the A, B, C, and D rings in a left-to-right orientation (**Figure 7.8**).

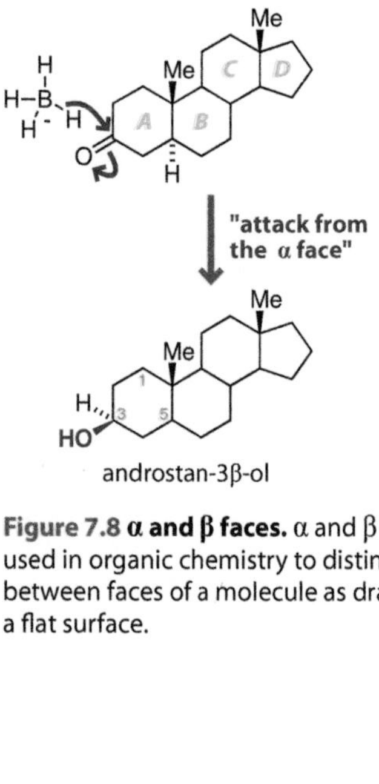

Figure 7.8 α and β faces. α and β are used in organic chemistry to distinguish between faces of a molecule as drawn on a flat surface.

Polar effects and stereoelectronic effects determine the relative stability of α and β anomers

Axial substituents are never preferred in monosubstituted cyclohexane rings. A few substituents with long bonds, such as acetoxymercurio groups (AcOHg), show no preference, but most substituents on cyclohexane rings favor equatorial conformations. In some cases the preferences are small; in other cases the preferences are large. Alkoxy groups show modest preferences for equatorial positions over axial positions; in methoxycyclohexane the equatorial isomer is favored over the axial isomer by only about 0.6 kcal mol^{-1}. However, for methoxytetrahydropyran the axial anomer is preferred over the equatorial anomer by 1.0 kcal mol^{-1} (**Figure 7.9**). Similarly, glucopyranosides such as methyl D-glucopyranose show a slight preference for the axial anomer. The axial anomer is stabilized over the equatorial anomer by two effects: a lower dipole moment and more favorable donation of filled orbitals into unfilled orbitals. By developing a fuller understanding of these effects, we are empowered to understand the enzymes that process glycosidic bonds.

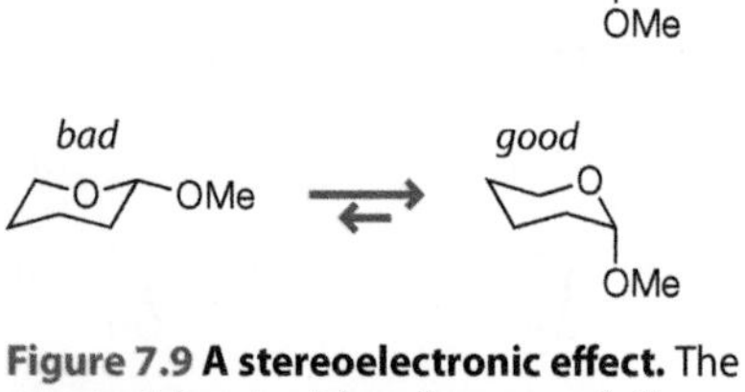

Figure 7.9 A stereoelectronic effect. The equatorial and axial preferences of alkoxy substituents are different for cyclohexane rings and pyranose rings.

Axial and equatorial conformers of methoxycyclohexane have similar dipole moments, between 1.2 and 1.3 debyes. However, the axial and equatorial anomers of

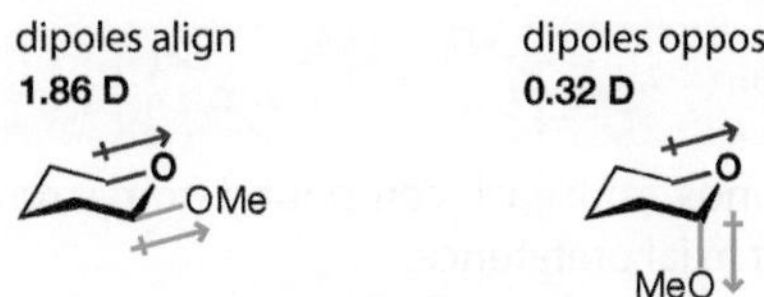

Figure 7.10 A polar effect. Axial anomers have lower dipole moments.

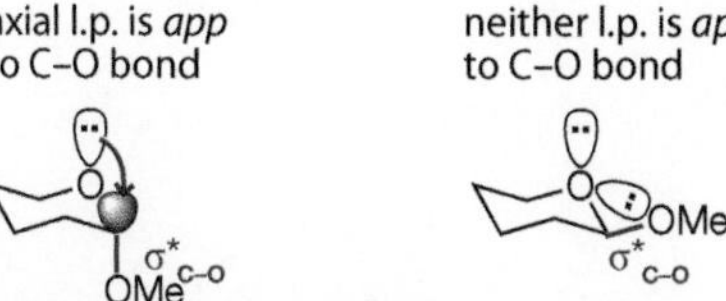

Figure 7.11 Origin of the stereoelectronic effect. Axial anomers are better aligned for donation into the σ* orbital of the anomeric substituent. *app*, antiperiplanar; l.p., lone pair.

2-methoxypyranose (IUPAC numbering) have very different dipole moments. In the equatorial anomer the dipoles align to create an overall dipole moment of 1.9 debyes (**Figure 7.10**). In the axial anomer the dipoles are almost in opposition, leading to a much lower dipole moment of 0.32 debyes. A larger dipole moment implies more charge separation, and Coulomb's equation predicts that charge separation is more costly in environments with a low dielectric constant. In nonpolar solvents, therefore, the more polar equatorial anomer is destabilized relative to the less polar axial anomer. The overall preference for axial anomers is diminished in polar solvents because the dipole penalty is decreased. For example, the axial form of 2-methoxytetrahydropyran is favored over the equatorial form by a 5:1 ratio in benzene, but in polar solvents like methanol and acetonitrile the ratio of the axial conformer to the equatorial conformer is only 2:1.

The **anomeric effect** also has a stereoelectronic component that we discussed in Chapter 5. In the axial anomer, the axial pyranose lone pair is antiperiplanar to the polar anomeric bond. Thus, the lone pair is *perfectly* aligned with the antibonding orbital of the anomeric C–O bond, and because of this alignment there is significant donation from the filled nonbonding orbital of the pyranose oxygen into the empty σ* orbital of the anomeric substituent (**Figure 7.11**). In the equatorial anomer, neither lone pair of the pyranose oxygen is aligned with the C–O bond. Note that only the hydroxyl groups (or alkoxy groups) at the 1 position of the pyranose ring can enjoy this effect; hydroxyl groups at other positions of a pyranose ring prefer conventional equatorial conformations.

The anomeric effect is evident in crystal structures in which donation into antibonding orbitals lengthens the anomeric C–O bond of axial anomers relative to equatorial anomers. The same donation that lengthens the bonds helps to push out the anomeric leaving group in S_N1 solvolysis reactions. The methyl-α-D-glucopyranoside hydrolyzes twice as fast as the β anomer. Small differences in bond length correlate with large differences in reactivity (**Figure 7.12**). A 0.01 Å increase in bond length corresponds to a 3 kcal mol^{-1} decrease in the activation energy for ionization of the anomeric leaving group.

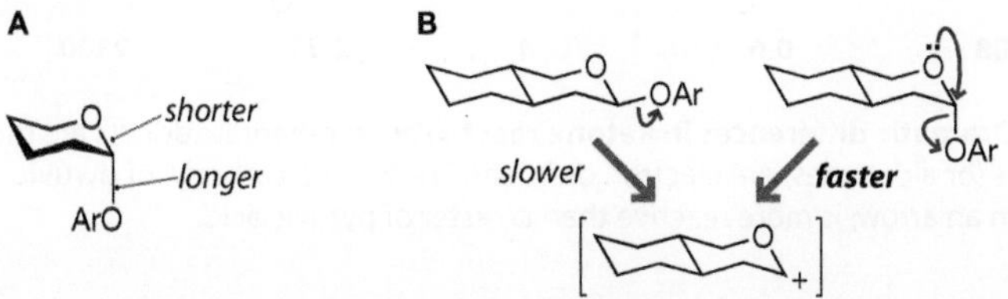

Figure 7.12 The anomeric effect in structure and reactivity. (A) Structural changes due to the anomeric effect have been observed in crystal structures. (B) The anomeric effect alters the reactivity of the anomeric carbon.

Lone-pair donation into lower-energy orbitals is more effective than donation into higher-energy orbitals. Recall that the energy of σ^*_{C-X} orbitals is affected by electronegativity. For example, S_N2 attack on $CH_3–NH_2$ is much slower than S_N2 attack on $CH_3–F$. For the same reason, the anomeric effect is most pronounced for pyranosyl fluorides, because fluorine is the most electronegative atom (**Figure 7.13**). Similarly, the anomeric effect for an alkoxy substituent should be greater when the alkoxy substituent is protonated.

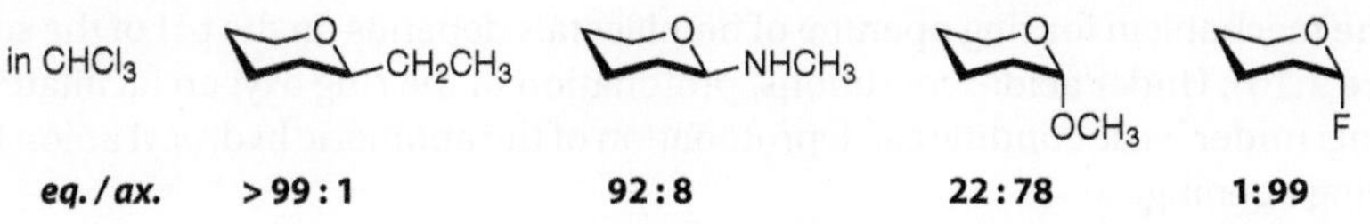

Figure 7.13 The anomeric effect depends on electronegativity. The equatorial/axial preference of anomeric substituents increases with electronegativity.

Problem 7.4

For each series of compounds, rank each compound from strongest axial preference for the OR group to lowest axial preference.

A

B

7.2 THE CHEMISTRY AND ENZYMOLOGY OF THE GLYCOSIDIC BOND

Monosaccharide carbonyl groups form hemiacetals

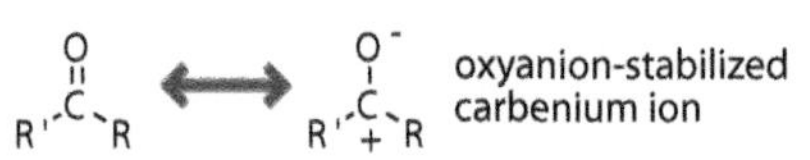

Figure 7.14 Donation. Carbenium ions are stabilized by donation of filled orbitals into the empty *p* orbital.

All carbohydrates possess latent carbonyl groups. So before we begin our discussion of carbohydrate structure, it is worth considering the reactivity of carbonyls with hydroxylic nucleophiles. Most aldehydes and ketones have a slight tendency to form tetrahedral hydrates by addition of water. To understand the factors that favor hydrate formation, it is best to treat the carbonyl group as an oxyanion-stabilized carbenium ion and recall that carbenium ions are stabilized by overlap of filled orbitals with the empty *p* orbital (**Figure 7.14**). Recall that hydrogen atom substituents on a carbenium ion have no lone pairs or bonds that can overlap with the empty *p* orbital. That is why H_3C^+ is less stable and more reactive than $(CH_3)_3C^+$. Similarly, $H_2C{=}O$ is less stable and more reactive than $(CH_3)_2C{=}O$. Therefore, the stability of hydrates is inversely correlated with the stability of the carbonyl starting material: ketone < aldehyde < formaldehyde (**Figure 7.15**). In addition, acyl groups and inductively electron-withdrawing groups such as trifluoromethyl enhance the rates and extent of hydration. In some cases, special features of the hydrate product might contribute to stability. For example, the greater tendency of pyruvic acid to undergo hydration might be attributable to the greater capacity of carboxylic groups to form hydrogen bonds than ketones.

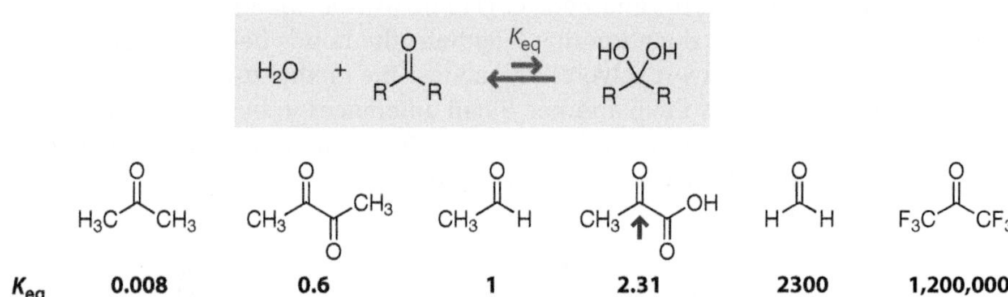

Figure 7.15 Dramatic differences in ketone reactivity. Tetrahedral adducts are more stable when ketones (or aldehydes) are electron deficient. The ketone carbonyl of pyruvic acid, indicated with an arrow, is more reactive than an ester of pyruvic acid.

Six- and five-membered ring hemiacetals are common

When ketones and aldehydes form adducts with alcohols they are referred to as hemiketals and hemiacetals, respectively. To simplify our discussion we will use the term acetals to refer collectively to acetals and ketals. Similarly, we will use the term hemiacetals to refer to either hemiacetals or hemiketals. The trends in stability for acetal and hemiacetal formation with alcohols follow the trends for hydrate formation with water shown in Figure 7.15.

The mechanism for ring opening of hemiacetals depends on the pH of the solvent (**Figure 7.16**). Under acidic conditions, protonation of the ring oxygen facilitates ring opening; under basic conditions, deprotonation of the anomeric hydroxyl anion facilitates ring opening.

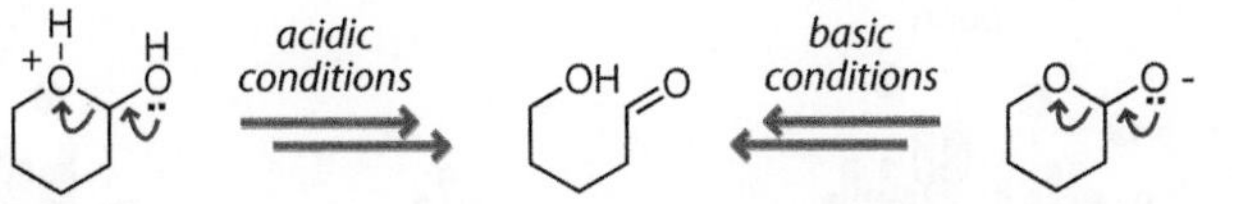

Figure 7.16 Opening cyclic hemiacetals. There are two different pH-dependent mechanisms for opening hemiacetal rings, acid-catalyzed and base-catalyzed.

Acyclic acetals and hemiacetals are thermodynamically unstable in water, but cyclic acetals and hemiacetals formed through intramolecular reactions can enjoy substantial stability (**Figure 7.17**). The equatorial hydroxyl groups of glucose give the cyclic form added stability relative to the open-chain aldehyde because of steric considerations. In aqueous solution, glucose exists almost exclusively (99.9%) in the cyclic form. However, although the concentration of the open-chain aldehyde is usually small, the reactive carbonyl of the aldehyde is often a key functional group in many reactions of carbohydrates.

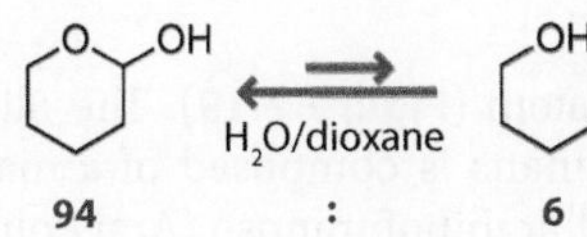

ring size	percentage closed
4	<1
5	89
6	94
> 6	≤20

Figure 7.17 Furanose and pyranose are special. Cyclic hemiacetals exhibit an intrinsic preference over acyclic hydroxyaldehydes for five- and six-membered rings.

Most naturally occurring carbohydrates have many hydroxyl groups and can, in theory, form more than one size of ring; however, among the possible cyclic forms, one is always preferred. For example, D-glucose can, in theory, exist in a five-membered ring furanose form, a six-membered ring pyranose form, or even a seven-membered ring form (**Figure 7.18**). In practice, only the glucopyranose form is observed under aqueous conditions. Only two types of cyclic acetals are found in oligosaccharides: six-membered ring pyranoses and five-membered ring furanoses. All human oligosaccharides are made up of six-membered ring monosaccharides. Oligosaccharides from other plants and microorganisms also include five-membered ring monosaccharides. No naturally occurring monosaccharides are known to prefer a seven-membered ring form.

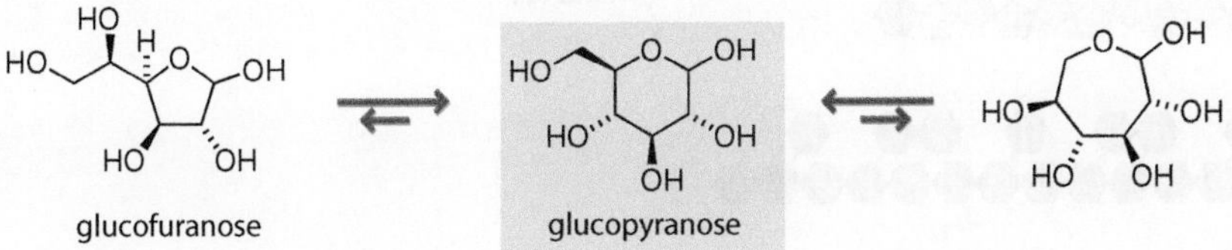

Figure 7.18 The pyranose form of D-glucose is preferred over other ring sizes. Seven-membered rings are highly disfavored because of the entropic costs required for their formation.

Problem 7.5

The drug lubiprostone has a carefully positioned difluoromethylene group (X = F). What effect would these substituents have on the equilibrium between the open-chain ketone and the cyclic hemiketal, relative to the parent compound (X = H)?

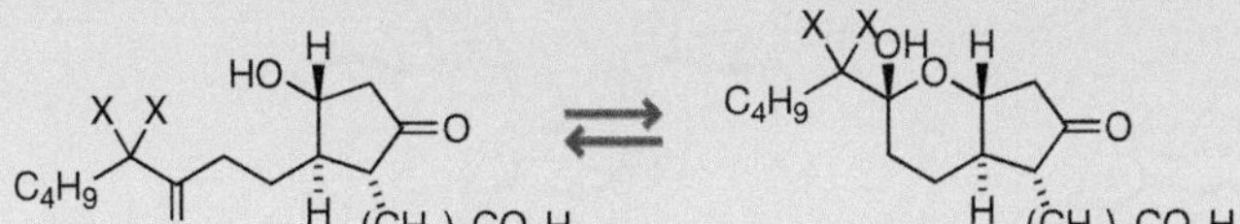

Human macrophages swallow up pathogenic organisms into phagocytic vesicles called phagosomes. When these phagosomes fuse with the protease-rich lysosomes, a lethal environment is created that attacks the pathogen. However, mycobacteria can inhibit the fusion of their phagosomal prison with lysosomes, allowing the mycobacteria to survive and replicate inside their would-be assassin. Much of the stealth of *Mycobacterium tuberculosis* is associated with its waxy, protective coat. Lipoarabinomannans are one component of the *M. tuberculosis* membrane that helps it avoid

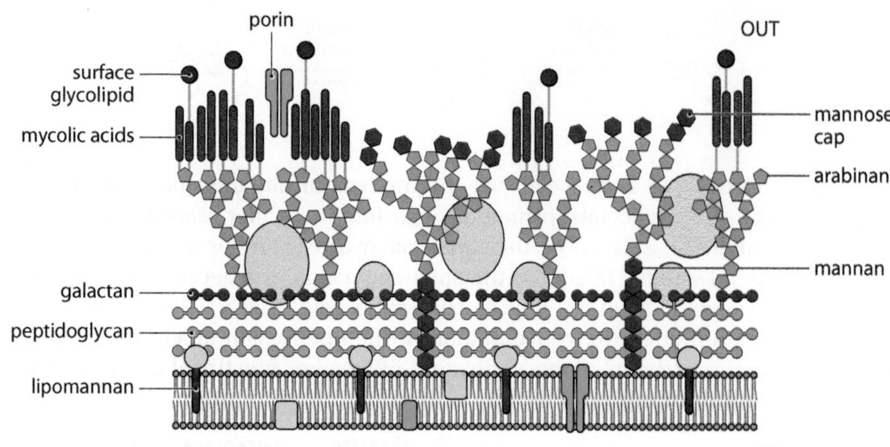

Figure 7.19 Furanose rings in mycobacterial oligosaccharides. Five-membered ring arabinofuranose sugars (pentagons) are present in two types of highly branched glycans that make up the cell wall of *Mycobacterium tuberculosis*, lipoarabinomannans and arabinogalactans. Hydrophobic long-chain mycolic acids attached to the outer branches of arabinogalactans create a waxy external barrier. (Adapted from D. Chatterjee and K.H. Khoo, *Glycobiology* 8:113–120, 1998. With permission from Oxford University Press.)

unwanted attention from the human immune system (**Figure 7.19**). The oligosaccharide portion of mycobacterial lipoarabinomannans is composed of a mannose oligomer called a mannan and a highly branched arabinofuranose (Araf) oligomer called an arabinan. Polysaccharides composed of five-membered ring furanose sugars are unique to mycobacteria (**Figure 7.20**). Furanoses are sometimes found in oligosaccharides in lower organisms, but not in human oligosaccharides. Of course, furanoses do have a role in other aspects of cell function. Recall that the backbones of DNA and RNA are based on the five-membered ring sugar ribose.

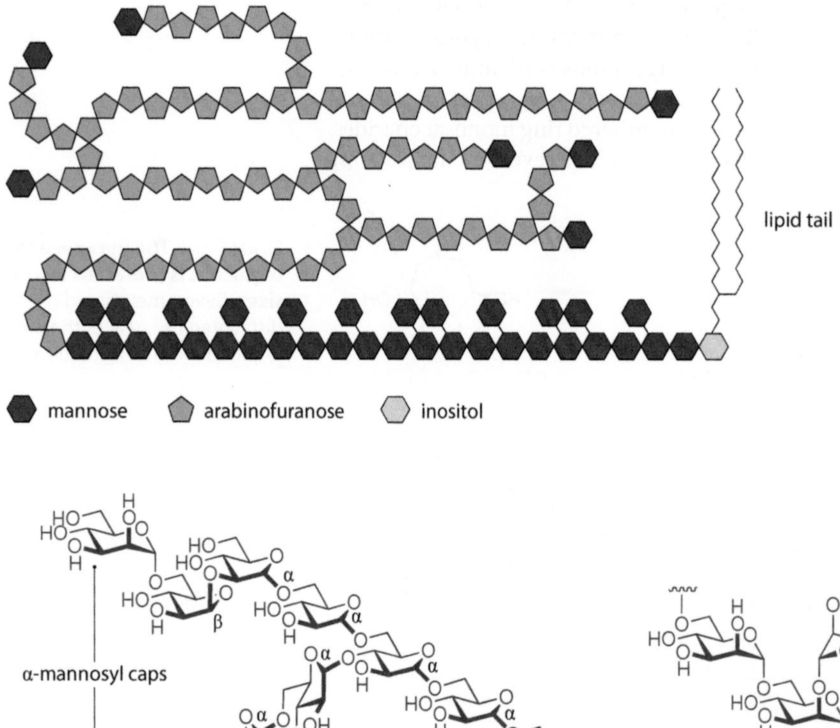

lipid tail

mannose arabinofuranose inositol

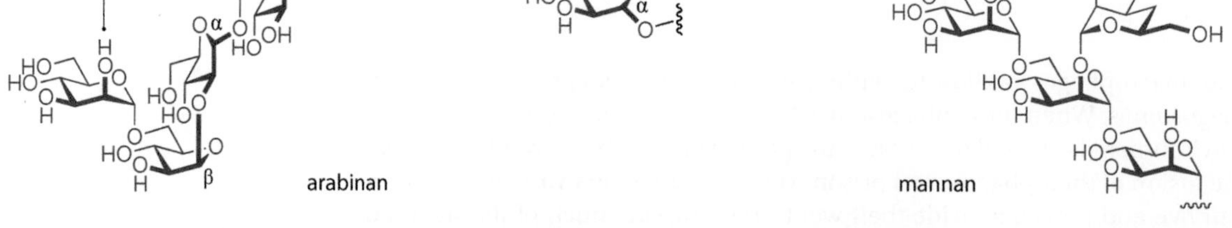

α-mannosyl caps

arabinan

mannan

Figure 7.20 Lipoarabinomannans. Mycobacterial lipoarabinomannans are tree-like structures with a lipid for roots, a mannan trunk, arabinan branches, and α-mannosyl caps for leaves. Human cells don't make oligosaccharides out of five-membered ring sugars.

Chemical hydrolysis of glycosidic bonds involves S_N1 reactions

In the laboratory, chemical cleavage of glycosidic bonds is conducted under acid-catalyzed conditions and usually involves the ionization of a leaving group to produce a carbocation intermediate. The mechanism can be drawn as an unassisted solvolysis leading to a carbocation or as an assisted solvolysis leading to an **oxonium ion** (**Figure 7.21**). Because the carbocation and the oxonium ion are resonance structures, the two mechanistic depictions are equally acceptable for use in arrow-pushing. There are two types of carbocations: trivalent carbenium ions such as R_3C^+ and pentavalent carbonium ions such as R_5C^+. Carbonium ions are protonated alkanes and are so unusual that chemical biologists usually use the terms carbocation and carbenium ion interchangeably. The term **oxocarbenium ion** is very precise; it describes a carbenium ion with an oxygen substituent. Although the oxocarbenium ion resonance structure and the oxonium ion resonance structure are equally acceptable for arrow-pushing, they are not equal. Students who are learning about the stability, conformation, and bonding of carbocations for the first time may find that the oxonium ion depiction is preferable to the oxocarbenium ion depiction. The formal charges in the oxonium ion resonance structure are wildly misleading (**Figure 7.22**). Electronic structure calculations reveal that the positive charge is distributed among the hydrogen atoms and carbenium carbon atom, whereas the oxygen atom and all of the sp^3 carbon atoms have partial negative charges. It is most often the case that students are interested in chemical reactivity and arrow-pushing, and for these applications the oxocarbenium ion is more useful than the oxonium ion resonance structure; the oxocarbenium representation emphasizes the carbenium center as the site of nucleophilic attack. Sadly, all organic chemistry textbooks cling to the oxonium ion representation even though it misleads many students to attack the positively charged oxygen atom.

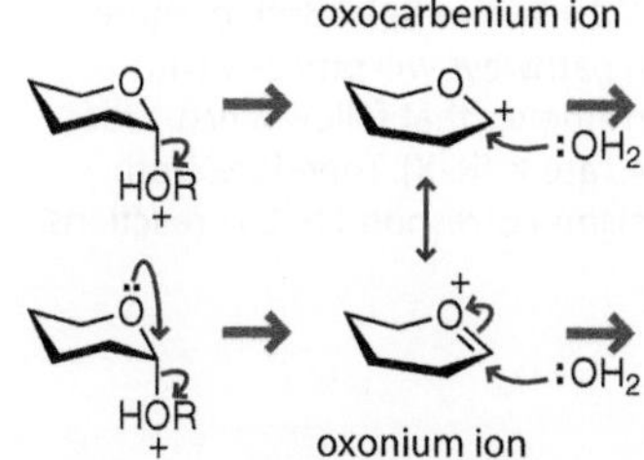

Figure 7.21 Equivalent depictions. The two arrow-pushing mechanisms above are equivalent, showing alternative resonance structures for the starting materials and intermediates.

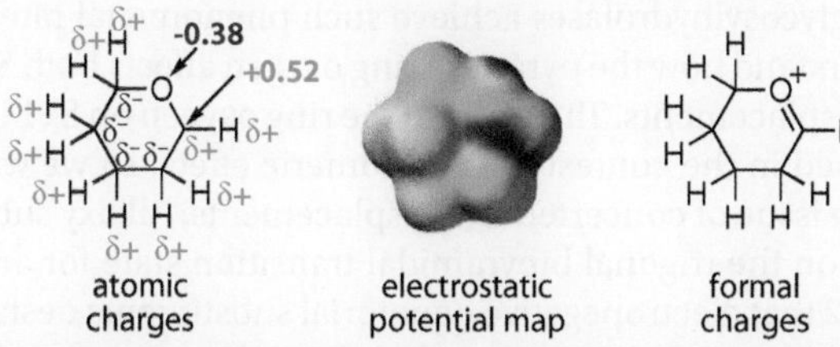

atomic charges — electrostatic potential map — formal charges

Figure 7.22 Where is the charge? The formal charges in the oxonium ion representation are at odds with the calculated atomic charges. The electrostatic potential map reveals the positive charge (colored blue) to be centered on the carbenium atom.

Problem 7.6

Identify the α and β anomers shown below. On the basis of stereoelectronic effects, predict which anomer would ionize faster in an S_N1 solvolysis reaction.

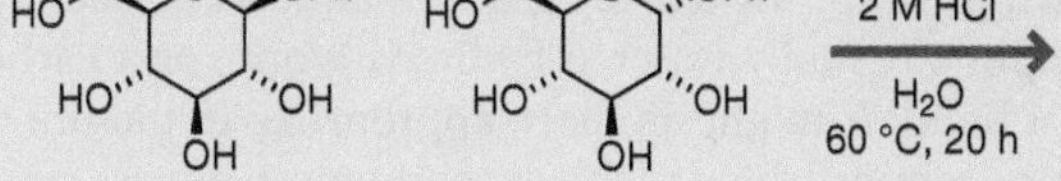

2 M HCl, H_2O, 60 °C, 20 h

The hydroxyl substituents of sugars drastically slow the rate of acid-catalyzed hydrolysis. Acid-catalyzed hydrolysis of glucosides is 10 million times slower than the hydrolysis of a tetrahydropyranoside with no hydroxyl substituents. Tetrahydropyranyl groups are commonly used as protecting groups for alcohols in organic synthesis because they are removed under such mild acidic conditions. The inductive effect of the hydroxyl substituents is most significant for hydroxyl groups at the 2 position, next to the anomeric center. Given the robustness of glycosidic bonds, it is surprising that **glycosylhydrolases** can hydrolyze them with such apparent ease. For example, enzyme-catalyzed hydrolysis rates can be up to 100,000,000,000,000,000 (10^{17}) times faster than background hydrolysis rates under neutral conditions. In all known glycosylhydrolases and glycosyltransferases, a nucleophile is poised on the side opposite the leaving group. In such an arrangement the reactants seem poised for a concerted S_N2 displacement; however, the system is also poised to form a very stable cation in an S_N1 process. The question of whether glycosylhydrolases and glycosyltransferases

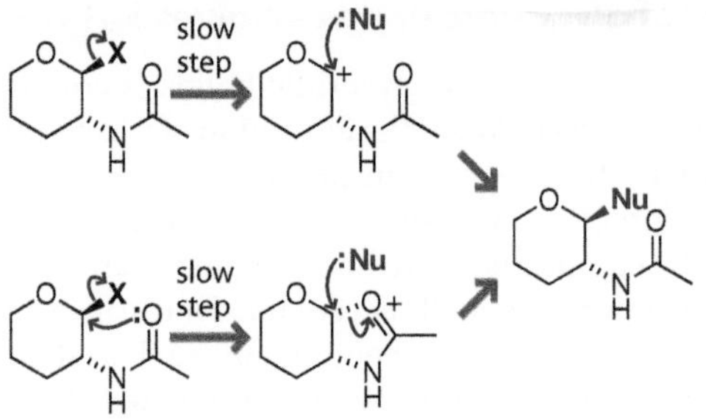

Figure 7.23 S_N1 or S_N2? Both of these reaction pathways will proceed with reaction kinetics that follow a first-order rate law: rate ∝ [R–X]. Therefore, both mechanisms correspond to S_N1 reactions.

act via concerted displacements, via discrete oxocarbenium ions, or via some hybrid mechanism is not easy to answer.

The widely used descriptions S_N1 and S_N2 can be misleading when used to label elementary reaction steps. For example, the solvolyses of 2-acetamido-2-deoxyglycosides involve S_N1 reactions (**Figure 7.23**). However, the term S_N1 does not necessarily mean that the mechanism involves a discrete carbocation. If a neighboring group pushes out the leaving group in an intramolecular S_N2 reaction to generate a highly reactive substrate for a faster intermolecular S_N2 reaction, the reaction will also obey first-order kinetics which makes it an S_N1 reaction. In fact, hydrolysis of β-GlcNAc and β-GalNAc derivatives proceeds through this neighboring-group participation mechanism. Enzymes that hydrolyze and form β-GlcNAc and β-GalNAc derivatives favor this type of neighboring-group displacement mechanism.

Enzymatic hydrolysis of glycosidic bonds involves S_N1-like S_N2 reactions

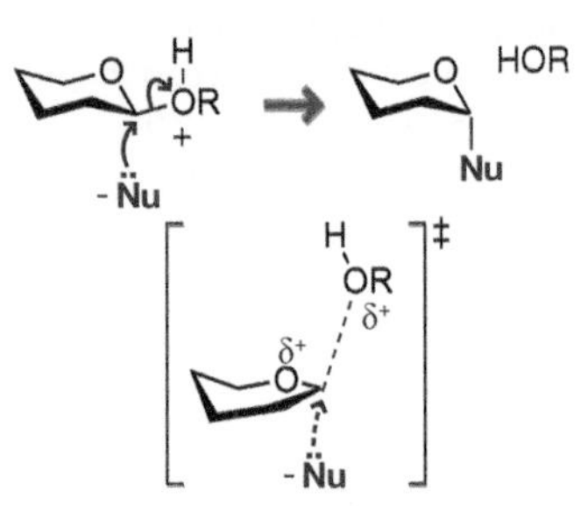

Figure 7.24 Glycosylhydrolase mechanism. The arrangement of groups in glycosylhydrolases favors an S_N2 displacement mechanism with the hallmarks of an S_N1 reaction.

To understand how glycosylhydrolases achieve such phenomenal rate accelerations, it is essential to understand how the pyranose ring oxygen affects both S_N1 ionizations and concerted S_N2 displacements. The effect of the ring oxygen on S_N1 ionizations has already been discussed in the context of the anomeric effect, so we will now turn to the more challenging issue of concerted S_N2 displacements. Alkoxy substituents have two opposite effects on the trigonal bipyramidal transition state for an S_N2 reaction. Recall from Chapter 2 that electronegative equatorial substituents destabilize trigonal bipyramidal S_N2 transition states. However, the oxygen lone pair can donate into the empty *p* orbital of the S_N2 transition state, leading to stabilization. Thus the pyranose oxygen can significantly accelerate the S_N2 displacement of anomeric substituents by S_N2 mechanisms for reasons that are analogous to the way in which it can accelerate S_N1 ionization. Electronic structure calculations suggest that glycosylhydrolases act via concerted S_N2 mechanisms (**Figure 7.24**). However, the nucleophile is not really pushing out the leaving group. The leaving group is almost completely dissociated by the time the nucleophile (usually water, a hydroxyl group, or a carboxylic acid side chain) has formed a bond. It might be more appropriate to picture the nucleophile getting sucked into the empty orbital of the oxocarbenium ion as it forms.

Members of all classes of glycosylhydrolases have two carboxylic acids in the active site

Mechanistically, glycosylhydrolases fall into two distinct classes, inverting and retaining, which reflect the stereochemistry of the product compared with the starting material (**Figure 7.25**). Both types of glycosylhydrolases involve the carefully orchestrated

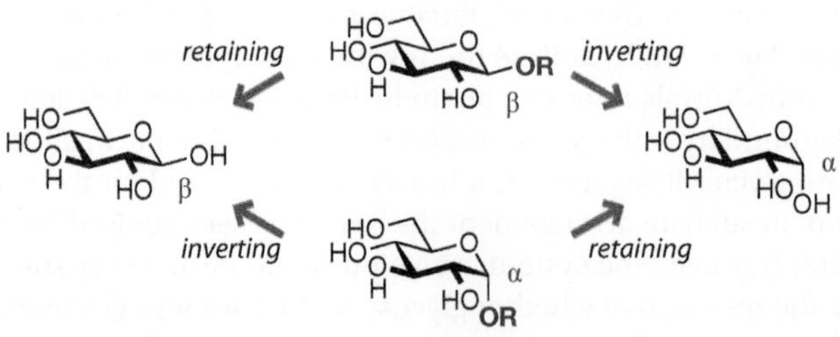

Figure 7.25 Two types of glycosylhydrolases. Glycosylhydrolase enzymes are described as "retaining" or "inverting," depending on the stereochemical outcome of the reaction.

actions of two carboxylic acid functional groups. One carboxylic acid protonates the leaving group. The role of the other carboxylate is dependent on the type of glycosylhydrolase. For inverting glycosylhydrolases, the second carboxylate deprotonates a water molecule that displaces the leaving group through an S_N2 reaction (see Figure 7.24). In retaining glycosylhydrolases, the second carboxylate acts as a nucleophile, displacing the protonated leaving group. The covalently bound carboxylate is then displaced by an attacking water molecule. Retaining glycosylhydrolases lead to net retention through double inversion.

Problem 7.7

Provide arrow-pushing mechanisms for both retaining and inverting β-D-glucosylhydrolases, each with two acidic residues in the active site. Do not worry about changes in pyranose conformation.

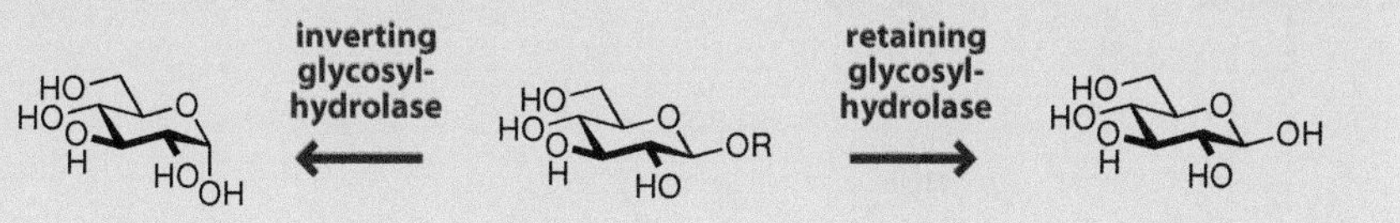

The mechanism for glycosylhydrolases involves several steps of indeterminate sequential order. As we have discussed, X-ray crystal structures usually do not reveal the positions of the protons. It is therefore difficult to know whether the nucleophilic water is deprotonated before it attacks, as it attacks, or after it attacks. Students who are looking for a single correct arrow-pushing mechanism for enzyme-catalyzed reactions may find this disheartening, but the order of proton transfers is moot if the protons are already engaged in a hydrogen bond (which is often the case in enzyme–substrate complexes). As we have discussed, proton transfers are really just ultra-fast vibrations of hydrogen bonds; the precise order of sequential ultra-fast steps is less important than the order of the slow, rate-determining steps.

Substrate distortion is important in glycosylhydrolase enzymes

What is the origin of the phenomenal rate accelerations of glycosylhydrolases? In mechanistic terms, the inverting α-glycosylhydrolases are the simplest to understand because the anomeric bond starts off in the stereoelectronically activated α configuration. For example, the yeast enzyme processing α-glucosidase I is an inverting α-glycosylhydrolase that removes glucosyl residues from the $Glc_3Man_9GlcNAc_2$ oligosaccharide that is added to many newly synthesized proteins. Removal of these exterior glucosyl residues is critical because these carbohydrates serve as signals for glycoprotein folding and degradation in the endoplasmic reticulum. The enzyme class I α1–2 mannosidase (PDB 1FO2) is also an inverting enzyme and has a role in the degradation of misfolded glycoproteins in the endoplasmic reticulum. The enzyme binds to its substrate in a skew-boat conformation as the Ca^{2+} complex (**Figure 7.26**). The Ca^{2+} ion also coordinates to the nucleophilic water molecule (**Figure 7.27**). Two glutamate residues drive the key proton transfers.

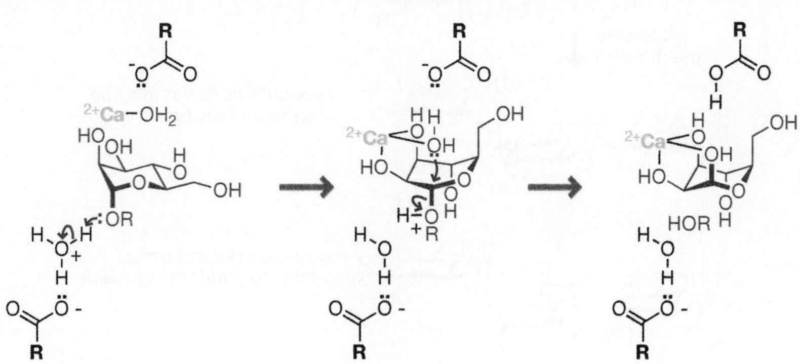

Figure 7.26 Special delivery. A Ca^{2+} ion in class I α1–2 mannosidase forms a distorted substrate complex and coordinates a nucleophilic water molecule (red) that is delivered to the anomeric carbon.

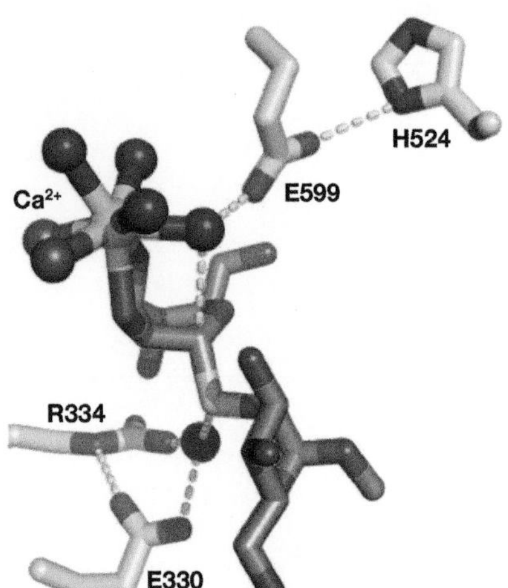

Figure 7.27 Freeze frame. Co-crystallization of class I α1–2 mannosidase (PDB 1X9D) with a nonhydrolyzable thiodisaccharide substrate reveals the precise placement of functional groups in the active site. The sulfur atom is yellow; water molecules are rendered as red spheres; a Ca^{2+} ion is rendered as a green sphere.

The mechanism of β-glycosylhydrolases is less obvious because equatorial anomeric substituents are not aligned to benefit from the stereoelectronic effect of the pyranose oxygen. However, if the ring flips into a boat conformation, an advantageous alignment of the pyranose oxygen lone pair with the antibonding orbital of the anomeric substituent can be achieved (**Figure 7.28**). Unfortunately this imposes an energetic cost on the enzyme, because boat conformations are generally less stable than chair conformations. Protonation of the anomeric substituent increases the anomeric effect markedly. In a simple pyranose ring, protonation of an anomeric hydroxyl group will help to drive the anomeric substituent into an axial conformation that is stereoelectronically favorable toward hydrolysis. Importantly, the stereoelectronic benefits of an axial leaving group are realized regardless of whether the reaction proceeds through an S_N1 or an S_N2 pathway.

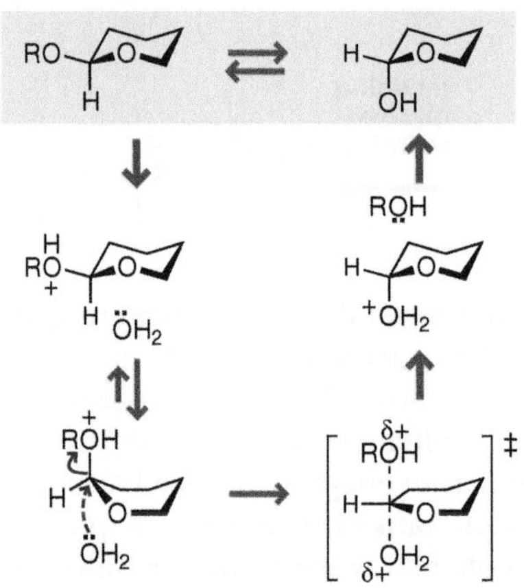

Figure 7.28 Acid catalysis. Protonation of the anomeric substituent can drive a conformational change that facilitates substitution.

Studies of glycosylhydrolases have shown that they distort β-glycosides into conformations that force the anomeric leaving group into a pseudo-axial orientation. It is rarely possible to obtain crystal structures of glycosylhydrolases bound to true reactive intermediates, but one can infer a plausible mechanism from the structures of enzymes bound to substrate analogs. A crystal structure of a fungal β-glucanase with a nonhydrolyzable thioglycoside substrate revealed that the inhibitor is distorted such that the pyranose oxygen lone pair is aligned with the leaving group, weakening the bond for cleavage in an S_N2 reaction (**Figure 7.29**). Similar results have been observed for other β-glycosylhydrolases.

Lysozyme, a retaining β-glycosylhydrolase, selectively cleaves a polysaccharide that is found in bacterial cell walls but not in humans. An abundant supply of lysozyme in chicken eggs and human tears helps to make these nutrient-rich environments inhospitable to bacteria. Lysozyme was the first enzyme for which a crystal structure was obtained, dating back to 1965. Yet it took 36 years to provide convincing evidence that the mechanism involves a double inversion and a covalent intermediate. The enzyme positions an acidic glutamate residue in a perfect position to protonate the

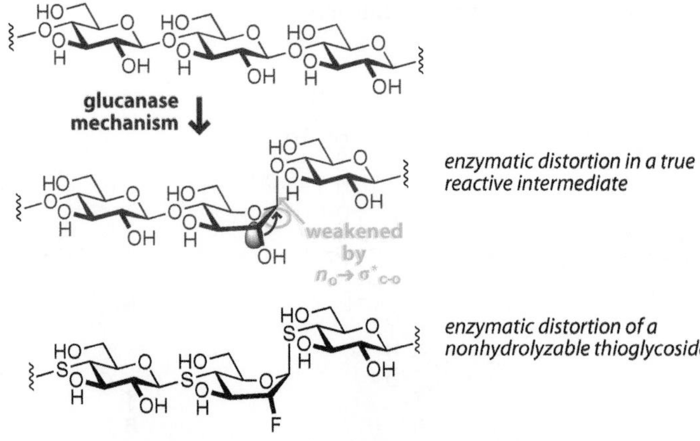

Figure 7.29 Transition state analog inhibitor. Structural studies of a synthetic analog bound to β-glucanase I from *Fusarium oxysporum* shows that substrate distortion orients and weakens the anomeric bond, facilitating S_N2 displacement of the leaving group.

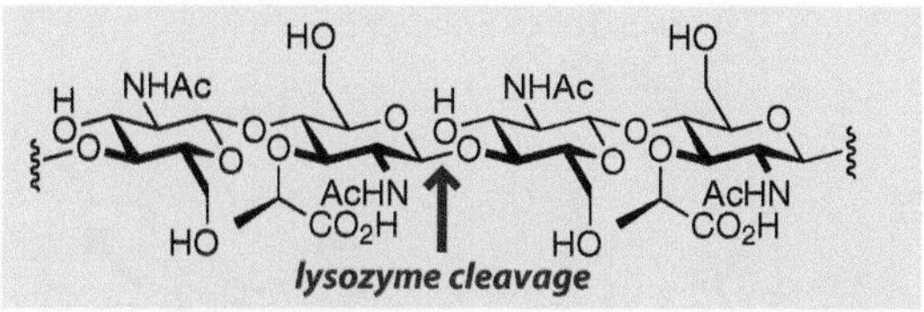

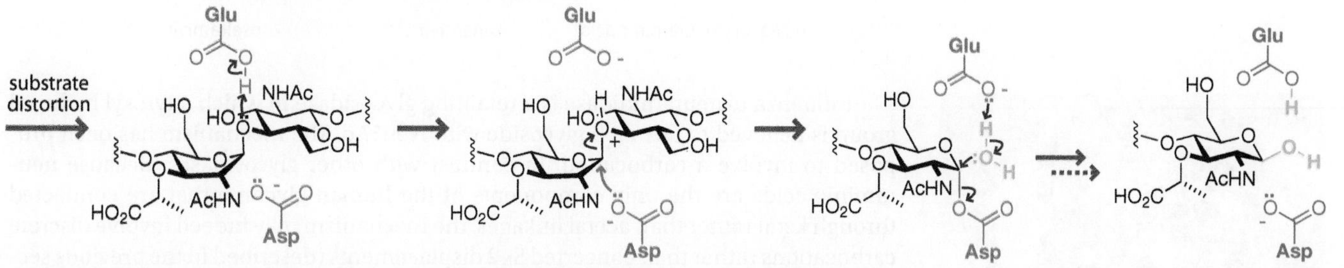

Figure 7.30 Retention mechanism. The mechanism of the retaining glycosylhydrolase lysozyme involves double inversion.

oxygen leaving group (**Figure 7.30**). Once protonated, an aspartate pushes the leaving group out from an axial orientation. A water molecule then binds to the site vacated by the GlcNAc leaving group, where the basic glutamate residue enhances the nucleophilicity of a water molecule through deprotonation. The glutamate might deprotonate the water as it attacks as part of a concerted mechanism; however, if we doubt that the breaking and forming bonds vibrate in perfect synchrony, then we would be better off depicting attack of water and deprotonation as individual mechanistic steps.

Inhibiting glycosylhydrolase enzymes fight influenza

Inhibition of pathogenic glycosidases represents a rich area for treatment of infectious diseases, including viral infections. For example, the newly packaged virions of human influenza virus display two unique proteins on the viral coat: a hemagglutinin that affects fusion with target cells, and a neuraminidase that affects release from the host cell. *N*-Acetylneuraminic acid (Neu5Ac) is the most important member of a broader class of nine-carbon monosaccharides called sialic acids but *N*-Acetylneuraminic acid is the only sialic acid we will discuss in this book. New influenza virions are bound to host cells through the interaction of *N*-acetylneuraminic acid-binding proteins on the virion coat with *N*-acetylneuraminic acid residues on mammalian cells (**Figure 7.31**). The viral enzyme neuraminidase cleaves these *N*-acetylneuraminic acid residues, releasing the infectious virion.

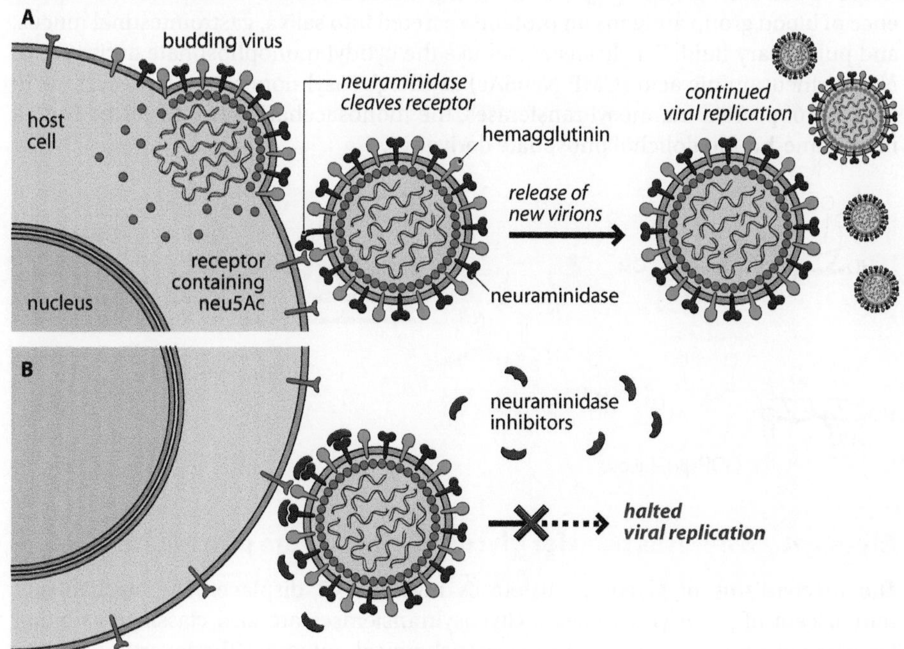

Figure 7.31 Interfering with release of virions. (A) The enzyme neuraminidase affects release of influenza virions by cleaving Neu5Ac linkages. (B) Inhibition of viral neuraminidase prevents release. (Adapted from A. Moscona, *N. Eng. J. Med.* 353:1363–1373, 2005.)

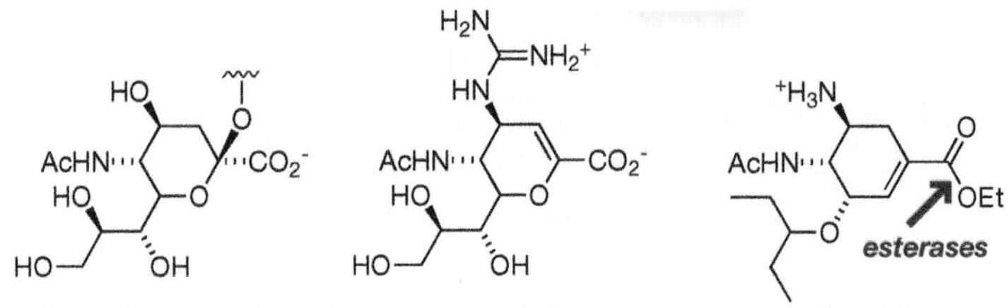

Figure 7.32 Flu fighters. Structures of substrates and inhibitors of influenza neuraminidase.

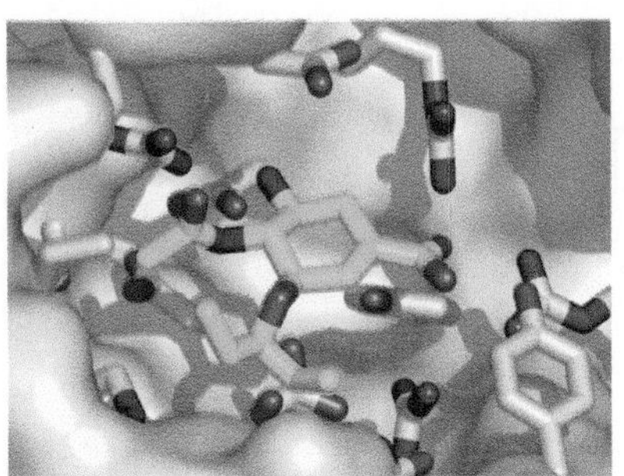

Figure 7.33 The active site of influenza neuraminidase N1. In this crystal structure oseltamivir carboxylic acid is bound to the same site as the neuraminic acid substrate (PDB 2HU0). See PDB 1NNC or 2HTQ for zanamivir.

Influenza α-neuraminidase is a retaining glycosidase in which a tyrosyl hydroxyl group is believed to form a β glycoside with Neu5Ac. The mechanism has been proposed to involve a carbocation, in contrast with other glycosidases. Because neuraminic acids are the only components of the human glycome that are connected through ketal rather than acetal linkages, the mechanism may indeed involve discrete carbocations rather than concerted S_N2 displacements (described in the previous section). Two neuraminidase inhibitors are available for the prevention of influenza and in early-stage treatment: zanamivir and oseltamivir. Both of these neuraminidase inhibitors, developed independently, include an sp^2-hybridized atom at the anomeric center (**Figure 7.32**). Zanamivir is a relatively close structural analog of *N*-acetylneuraminic acid, and inhibits the enzyme as a substrate mimic. However, this zwitterionic compound is not orally available and must be delivered by an inhaler. Oseltamivir is an orally available ester prodrug that is cleaved by human esterases to generate a carboxylic acid that ultimately binds to neuraminidase (**Figure 7.33**).

Glycosyltransferases transfer monosaccharides from glycosyl phosphate donors

In humans, glycosidic bonds are generated by glycosyltransferases that use glycosyl donors with phosphate leaving groups at the anomeric position. Most glycosyltransferases, whether transferring glycosyl groups to amino acid side chains, lipids, or other sugars, use glycosyl donors with nucleoside phosphate leaving groups. Uridyl diphosphate (UDP) is the most common leaving group and is present on the donors UDP-α-D-glucose, UDP-α-D-GlcNAc, UDP-α-D-GalNAc, UDP-α-D-xylose, and UDP-α-D-glucuronic acid. Mannosyltransferases and fucosyltransferases use GDP-α-D-mannose and GDP-β-L-fucose, respectively, as substrates. For example, fucosyltransferase 2 transfers α-fucosyl residues specifically to the 2-hydroxyl group of certain Galβ1−3GalNAcβ groups (**Figure 7.34**). This modification correlates with the presence of blood group antigens on proteins secreted into saliva, gastrointestinal mucus, and pulmonary fluid. Sialyltransferases use the cytidyl monophosphate derivative of *N*-acetylneuraminic acid (CMP-Neu5Ac) as the glycosyl donor. In a few cases, as in the case of protein mannosyltransferases, the monosaccharide is transferred from a membrane-bound dolichol phosphate derivative.

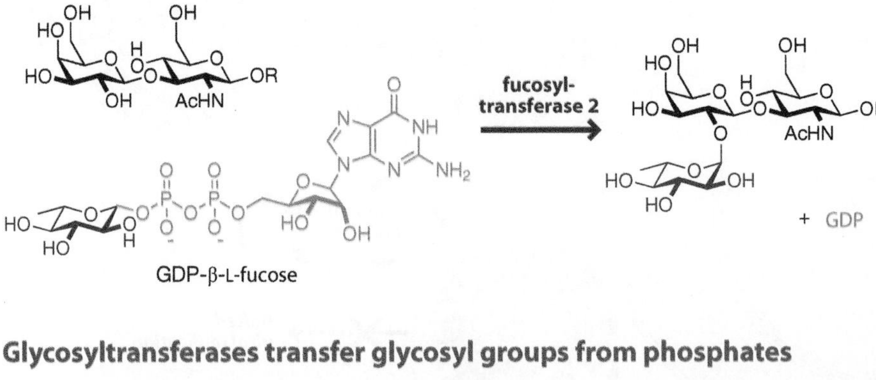

Figure 7.34 Enzymatic glycosylation. Fucosyltransferase 2 transfers an α-fucosyl group from GDP-β-L-fucose.

Glycosyltransferases transfer glycosyl groups from phosphates

The mechanisms of glycosyltransferases involve S_N2 displacement mechanisms, reminiscent of glycosylhydrolases. Glycosyltransferases are also classified as either inverting or retaining based on the stereochemical outcome. The inverting enzyme

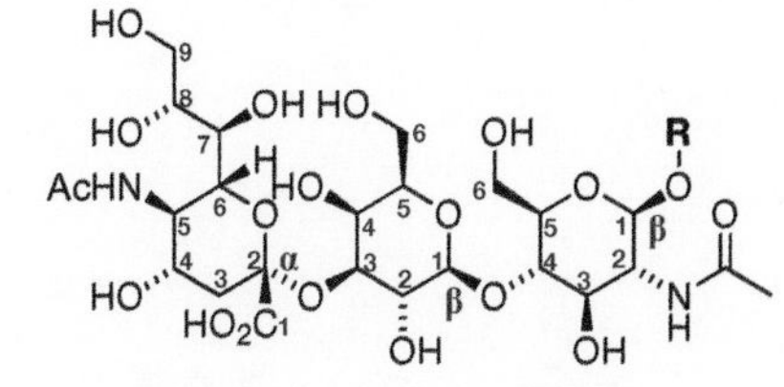

Figure 7.35 Shorthand notation for oligosaccharides. Unbranched oligosaccharides can be described with a compact nomenclature that uses the abbreviation for each monosaccharide unit, followed by the anomeric stereochemistry, followed by the connectivity (based on numbering of the monosaccharide backbone). The trisaccharide moiety shown would be Neu5Acα2–3Galβ1–4GlcNAcβ; alternatively, arrows or other symbols can be used to designate the connectivity Neu5Acα2→3Galβ1→4GlcNAcβ.

β1–4 galactosyltransferase adds β-galactosyl residues to the 4-hydroxyl group of GlcNAc to generate Galβ1–4GlcNAc. The retaining enzyme α1–3 galactosyltransferase then adds α-galactosyl groups to the 3-hydroxyl group of the galactose to generate Neu5Acα2–3Galβ1–4GlcNAc (**Figure 7.35**). Both the inverting and retaining galactosyltransferases use the same UDP-Gal substrate.

The structures of glycosylhydrolase enzymes are diverse, but the structures of glycosyltransferases that use nucleotidyl diphosphate donors are limited to two basic folds, GT-A and GT-B (**Figure 7.36**). Unfortunately there is no correlation between the protein fold and the stereochemical outcome of the reaction. For example, the retaining glycosyltransferase α1–3 *N*-acetylgalactosaminyltransferase and the inverting glycosyltransferase β1–4 galactosyltransferase both have GT-A folds.

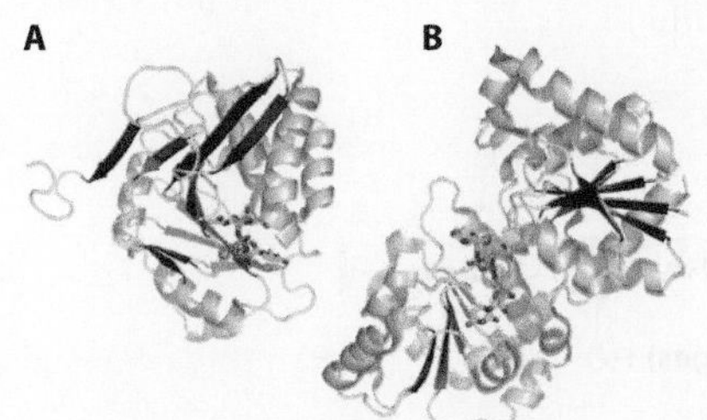

Figure 7.36 Two canonical structures for glycosyltransferases. (A) Mouse α1–4 *N*-acetylhexosaminyl-transferase complexed with UDP-GalNAc has a GT-A fold (PDB 1OMZ). (B) *E. coli* MurG complexed with UDP-GlcNAc has a GT-B fold (PDB 1NLM).

Glycosyltransferase enzymes have a catalytic base, usually a carboxylate side chain, that can remove the proton from the nucleophilic hydroxyl at some point during the reaction. In the active site, the phosphate leaving group is usually coordinated to divalent ions such as Mg^{2+} or Mn^{2+} (**Figure 7.37**). Divalent Mg^{2+} ions have no ligand field and can accept ligands with any coordination geometry; in contrast, a divalent Mn^{2+} ion prefers to adopt octahedral geometries. In a few cases, amino acid side chains act as Brønsted acids to facilitate loss of the anomeric phosphate.

Retaining glycosyltransferases, such as UDP-GalNAc:polypeptide transferase 1, catalyze **glycosylation** with retention of stereochemistry. Retaining enzymes are often presumed to proceed via a covalent glycosyl-enzyme intermediate, but even with several good crystal structures it is still unclear whether retaining glycosyltransferases proceed through a double inversion mechanism or through an attack on a discrete oxocarbenium ion with retention of anomeric configuration. In support of a glycosyl-enzyme intermediate, when the glutamate carboxylate of E137 of the retaining glycosyltransferase, α1–3 galactosyltransferase, was mutated to alanine, the E317A enzyme was essentially inactive; however, when azide anion was present at concentrations higher than 100 mM in the reaction medium, it was able to rescue the activity of the enzyme by forming a temporary β-azido intermediate (**Figure 7.38**).

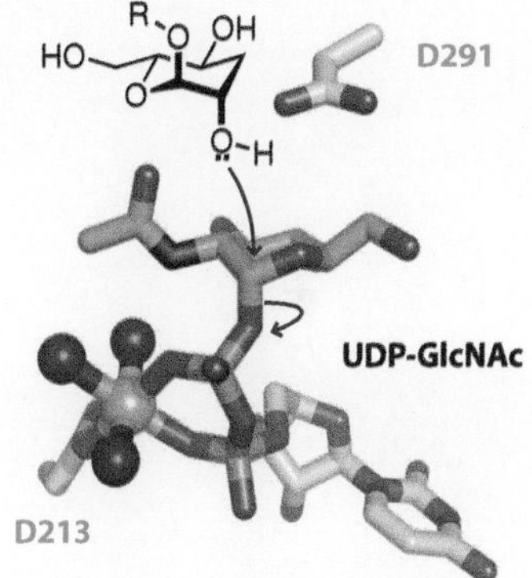

Figure 7.37 Functional groups in a glycosyltransferase active site. An aspartate side chain (D291) in the inverting glycosyltransferase *N*-acetylglucosaminyltransferase I is poised to deprotonate the 2-hydroxyl group of the mannose acceptor. A manganese ion (gray) coordinates to the diphosphate, making it a better leaving group. (PDB 1FOA)

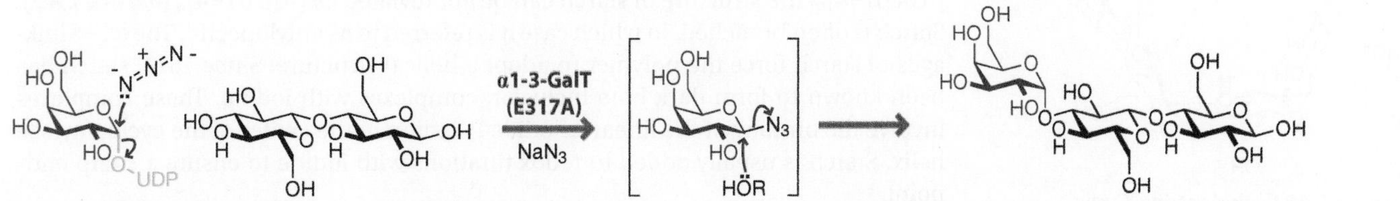

Figure 7.38 Chemical assistance. Azide anion rescues the catalytic activity of a retaining glycosyltransferase from which the nucleophilic glutamate has been removed.

Problem 7.8

Suggest a plausible S_N2 arrow-pushing mechanism for the transfer of α-UDP-GlcNAc to the 3 position of galactose. What is the preferred conformation of the GlcNAc pyranose ring in the transition state?

7.3 POLYSACCHARIDES

Diastereomers of glucose polymers have very different properties

From the standpoint of human glycobiology, oligomers of monosaccharides tend to be more interesting than repeating polymers, known as polysaccharides. However, polysaccharides such as starch, cellulose, and chitin are highly familiar carbohydrates, even to scientific lay persons (**Figure 7.39**). Starch is commonly found in potatoes, bread, and other foodstuffs that are nutritionally labeled as carbohydrates. The cell walls of plants are composed of cellulose, a fiber found in paper, wood, and cotton. Chitin forms the exoskeletons of insects and shellfish. In humans the important polysaccharides are glycogen, an oligomer of glucose, and polysialic acid [-Neu5Acα2–8]$_n$, which is found in developing embryonic tissues. Glycogen is analogous in structure to the starch found in foods.

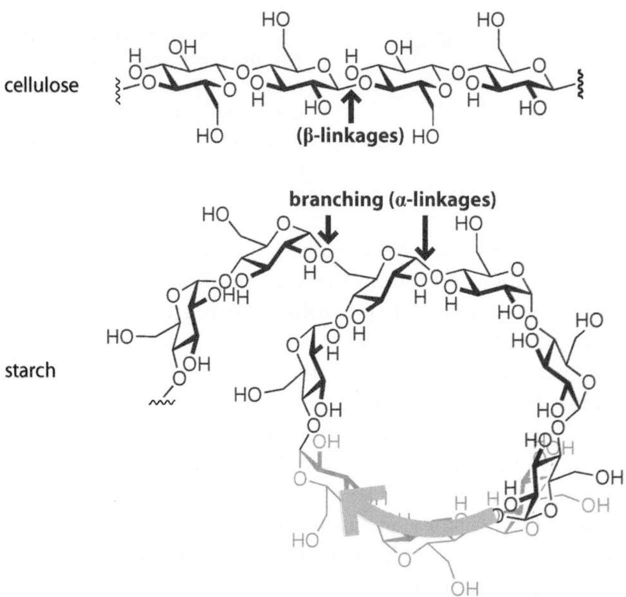

Figure 7.39 Polymers. Structures of the common polysaccharides cellulose and starch.

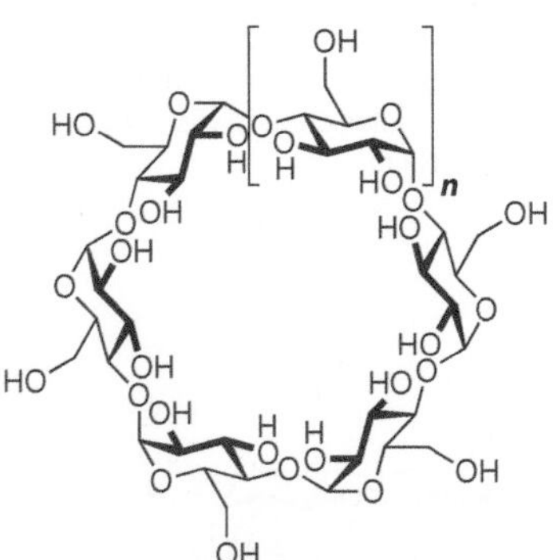

Figure 7.40 Cyclodextrins. Cyclic oligomers of glucose, [Glcα1–4]$_n$, are called cyclodextrins.

Cellulose and starch are polymers of D-glucose linked through the 4-hydroxyl group. The principal structural difference between starch and cellulose is that cellulose is linked through β-linkages, whereas starch, also known as α-amylose, is linked through α-linkages. The structure of cellulose can be succinctly abbreviated as [-Glcβ1–4]$_n$; the structure of starch can be abbreviated as [-Glcα1–4]$_n$ (**Figure 7.40**). Starch is often branched, in which case it is referred to as amylopectin. The α1–4 linkages of starch force the polymer to adopt a helical structure. Since 1814, starch has been known to form dark blue inclusion complexes with iodine. These complexes involve the binding of the linear triiodide I_3^- anion in the center of the cycloamylose helix. Starch is usually added to redox titrations with iodine to ensure a sharp endpoint.

If you wet a piece of bread with your saliva and then taste it later, it will taste sweet. Human saliva contains the enzyme **α-amylase**, which catalyzes the hydrolysis

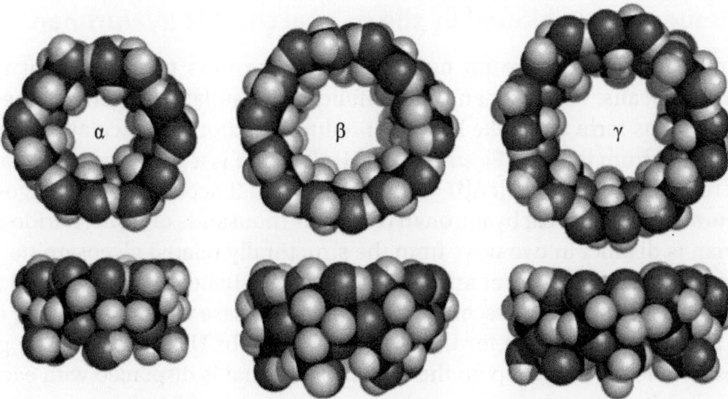

Figure 7.41 Crystal structures of α-, β-, and γ-cyclodextrins.

of amylose into sweet-tasting glucose. α-Amylose is also a substrate for the enzyme cyclodextrin glycosyltransferase. Cyclodextrin glycosyltransferase converts partly hydrolyzed starch helices into rings with six, seven, or eight glucose subunits. These cyclic molecules are referred to as α-, β-, and γ-**cyclodextrins**, respectively (**Figure 7.41**). The individual ringlets in medieval chain mail armor were made through an analogous process of cutting wire spirals into small wire rings. The hydrophobic cavity of cyclodextrins can bind to hydrophobic molecules. Cyclodextrins are adept at binding benzene rings, which are about 7 Å in diameter. For example, 4-bromopropiophenone binds to α-cyclodextrin with a K_d of 1.3 mM. β-Cyclodextrin binds aromatics with even higher affinity. The ability to encapsulate aromatic molecules has been commercialized by Procter & Gamble in two products. Bounce® dryer sheets have perfumes encapsulated that are slowly released in the heat of a dryer. Febreze® deodorizer traps foul-smelling molecules in cyclodextrins.

Problem 7.9

Draw each of the following structures as dash–wedge drawings. Do not worry about the conformations, but clearly indicate the configuration of each stereogenic center.

A Fucα1–2Galβ1–3GlcNAcβ1–3Galβ-OR

B Galβ1–4GlcNAcβ1–2Manα-Ser

Chitin is a resilient polymer in insect cuticles

Chitin is a β1–4 linked polymer of *N*-acetyl-D-glucosamine (**Figure 7.42**). It is the highly resilient material upon which arthropod (insects, crabs, shrimp) exoskeletons are based. When the acetyl groups of chitin are chemically hydrolyzed, the resulting polymer, known as chitosan, has a number of materials applications. For example, the ability of chitosan to strongly induce blood clotting has led to its application in battlefield dressings.

From the perspective of human nutrition, mammals seem to have missed out on a rich potential food source. All mammals have enzymes that can hydrolyze the α1–4 linkages of starch. However, no mammals have the enzymes necessary to hydrolyze the β1–4 linkages of cellulose or chitin. Fortunately, grazing animals have developed a symbiotic relationship with microorganisms that can hydrolyze cellulose into digestible fragments.

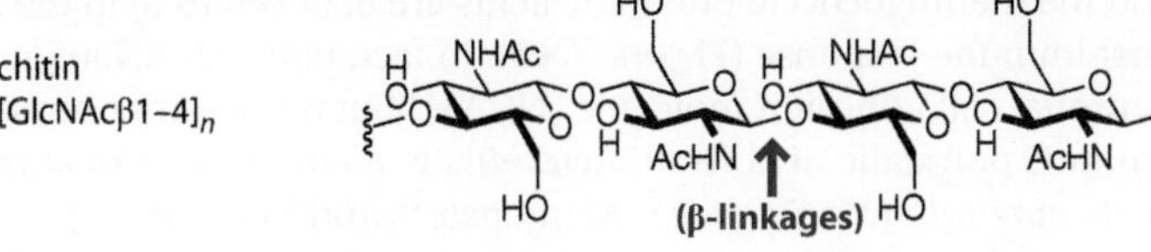

Figure 7.42 Chitin. A polymer of *N*-acetylglucosamine gives the shells of insects and crustaceans great resiliency.

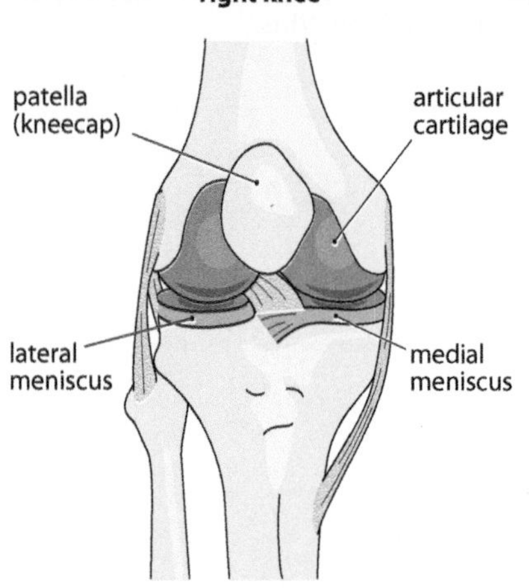

Figure 7.43 Glycosaminoglycans. The meniscal and articular cartilages of the knee joint are made up of glycosaminoglycans.

Some tissues are cushioned by the polysaccharide hyaluronan

The chondrocytes that make up human cartilage express two important types of glycosaminoglycans: hyaluronan and chondroitin sulfate. Together, these glycosaminoglycans form a gel-like layer that reduces friction in articulating joints and cushions joints from impact (**Figure 7.43**). Hyaluronan is a glycosaminoglycan composed of repeating units of [GlcAβ1–3GlcNAcβ1–4] that occurs as a free oligosaccharide (**Figure 7.44**). A typical hyaluronan chain has thousands of disaccharide repeats. Hyaluronan is distinct in two ways from the structurally related glycosaminoglycans. First, the oligosaccharide is not attached to a protein. Instead, it is extruded directly into the extracellular milieu. Second, hyaluronan synthase is a polymerizing enzyme that has two active sites: one for UDP-GlcA and one for UDP-GlcNAc. The growing chain has a UDP leaving group on the reducing end that is displaced with each addition of a subunit. As the chain grows it bounces from one active site to the other.

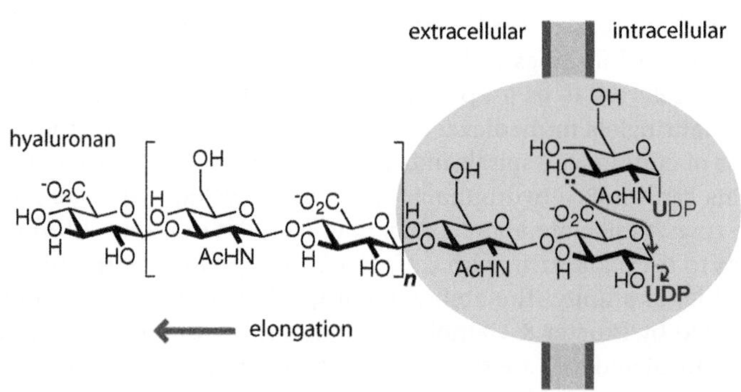

Figure 7.44 Extrusion polymerization. Hyaluronan synthase is a polymerase that extrudes the alternating copolymer hyaluronan into the extracellular medium.

Osteoarthritis involves a loss of cartilage and commonly affects the knee joint. Even though cartilage is composed of chondroitin sulfate and hyaluronan (both of which contain glucosamine) it would be naive to think that the disease is caused by a simple dietary deficiency. However, by 2007, sales of chondroitin and glucosamine in the United States exceeded half a billion dollars. Unfortunately, recent clinical trials have shown that oral chondroitin sulfate and glucosamine supplements are no better than placebos at reducing pain or damage to osteoarthritic knees.

Meningococci are coated with polysialic acids like those found on neurons

Meningococci are one of many types of infectious agents that can lead to meningitis, an inflammation of the membranes that surround the central nervous system, called the meninges. Young toddlers are at greatest risk for the disease, which can present in various ways, often as quickly spreading necrotic lesions. The cellular capsules of *Neisseria meningitidis* are coated with polymers that contain *N*-acetylneuraminic acids either as repeating polymers or as repeating copolymers; the type of polymer varies with the strain of *N. meningitidis*. For example, serogroup B is coated with $[\text{Neu5Ac}\alpha2\text{–}8]_n$; serogroup C is coated with $[\text{Neu5Ac}\alpha2\text{–}9]_n$; serogroup Y is coated with $[\text{Glc}\alpha1\text{–}4\text{Neu5Ac}\alpha2\text{–}6]_n$; serogroup W135 is coated with $[\text{Gal}\alpha1\text{–}4\text{Neu5Ac}\alpha2\text{–}6]_n$ (**Figure 7.45**). The branches of many human glycans terminate in sialic acid residues, and the meningococcal polysialic acids are believed to help the bacterium evade the host immune response (**Figure 7.46**). In fact, poly-α2–8Neu5Ac, is found on human neuronal cell adhesion molecule (NCAM) but on no other human protein. The serogroup C polysialic acid $[\text{Neu5Ac}\alpha2\text{–}9]_n$ is a component of several vaccines against *N. meningitidis* (Meningitec®, Menjugate®, and NeisVac-C®).

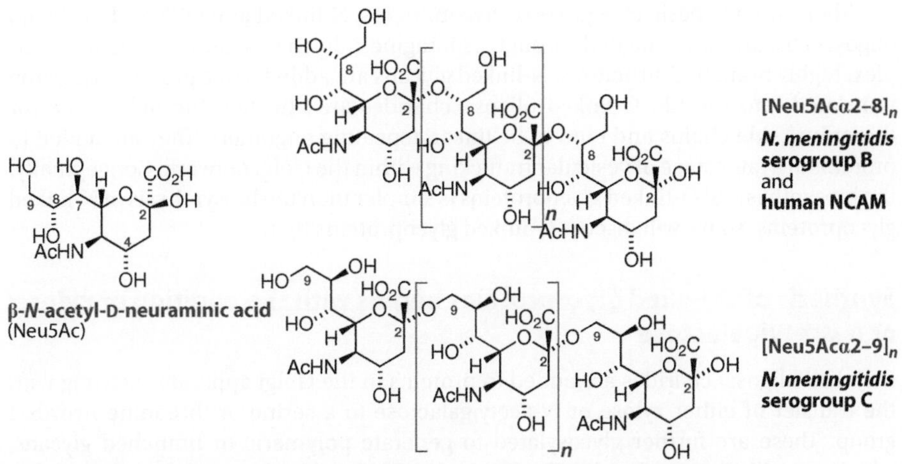

β-*N*-acetyl-D-neuraminic acid (Neu5Ac)

[Neu5Acα2-8]$_n$
***N. meningitidis* serogroup B and human NCAM**

[Neu5Acα2-9]$_n$
***N. meningitidis* serogroup C**

Figure 7.45 Polysialic acids. Two different types of Neu5Ac polymers are found in the capsular coats of meningococci.

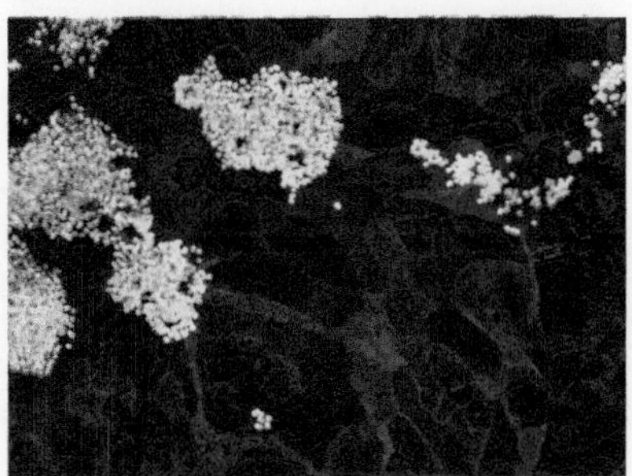

Figure 7.46 Sleeper cells. *Neisseria meningitidis* live in the nasopharynx of 5–10% of the population. Infections are problematic when *N. meningitidis* crosses the blood-brain barrier or blood-cerebrospinal fluid barrier. The fluorescence image above shows adherence of *N. meningitidis* (stained yellow) to human choroid plexus papilloma cells which serve as a model of the blood-cerebrospinal fluid barrier. (From C. Schwerk et al., *PLoS One* 7(1): e30069, 2012.)

7.4 GLYCOPROTEINS

Glycosylation of human proteins occurs in the vesicles of the secretory pathway

Glycoproteins are rare in prokaryotes but common in eukaryotic organisms. Virtually all glycosylation of human proteins occurs in the vesicles of the secretory pathway, the endoplasmic reticulum (ER), and the Golgi apparatus, which are not found in prokaryotic cells (**Figure 7.47**). The ER is a dynamic network of organelles. The contents of the ER are isolated from the cytosol by a membrane, and the surfaces of the rough ER are distinguished by the presence of membrane-bound ribosomes that "squirt" proteins directly into the ER during translation. The lamellae of the ER are dynamic structures that bud off as ribosome-free organelles, referred to as the Golgi apparatus. Interestingly, this process can be inhibited by the natural product brefeldin A. The Golgi vesicles migrate toward the plasma membrane and ultimately fuse, releasing soluble proteins from the cell. This process has two additional topological consequences. First, soluble proteins in the secretory vesicles are exported from the cell. Second, the cytosolic domain of proteins bound to the ER membrane remains on the cytosolic side of the plasma membrane; conversely, the domain of membrane-bound proteins that is directed into the ER ends up on the extracellular side of the plasma membrane. Because glycosylation of proteins occurs inside the ER and Golgi apparatus, the attached oligosaccharides are destined to be presented on the extracellular surface and never within the cytosol. Proteins and lipids decorated with complex glycans are meant for communication between cells, not for communication within cells.

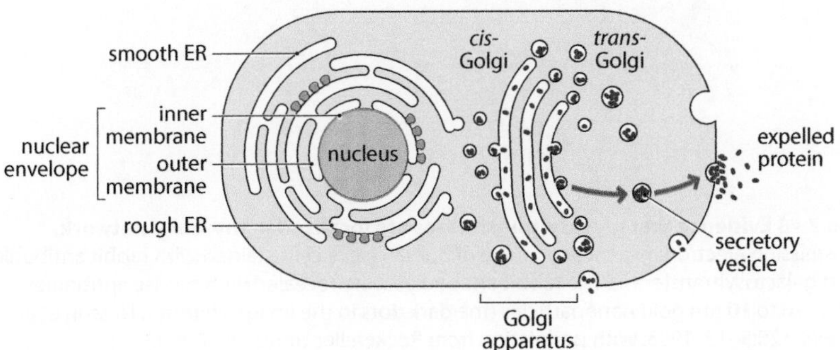

Figure 7.47 N-glycosylation of proteins occurs in the endoplasmic reticulum (ER). The glycans are further modified by enzymes in the Golgi vesicles. (Adapted from B. Alberts et al., Molecular Biology of the Cell, 5th ed. New York: Garland Science, 2008.)

There are two basic categories of glycoproteins: N-linked and O-linked. N-linked oligosaccharides are attached through asparagine side chains, and are usually complex, highly branched structures. N-linked glycans are added while proteins are being translated into the ER. O-linked oligosaccharides are attached through serine (or threonine) side chains and tend to be linear, repeating oligomers. They are added to proteins at a later stage of vesicular trafficking within the Golgi network. Conceptually, the biosynthesis of O-linked glycoproteins is simpler than the biosynthesis of N-linked glycoproteins, so we will discuss O-linked glycoproteins first.

Synthesis of O-linked glycoproteins begins with the addition of xylose or *N*-acetylgalactose

O-linked oligosaccharides are added to proteins in the Golgi apparatus starting with the transfer of either xylose or *N*-acetylgalactose to a serine or threonine hydroxyl group. These are further glycosylated to generate polymeric or branched glycans, which we will discuss shortly. The biosynthesis of proteoglycans is initiated by the enzyme UDP-D-xylose:protein β-D-xylosyltransferase. In this common form of enzyme nomenclature the name of the enzyme is preceded by the two substrates, which are separated by a colon. Humans have genes for two different isoforms of UDP-D-xylose:protein β-D-xylosyltransferase; the expression levels depend on the cell type. The biosynthesis of mucins, and that of many other O-linked glycoproteins, is initiated by UDP-*N*-acetyl-D-galactosamine:polypeptide *N*-acetylgalactosaminyltransferase. Fourteen of these polypeptide GalNAc transferases have been enumerated in humans (ppGalNAc T1–T14) and there are a handful of other genes that could code for isoforms. The expression of these various polypeptide GalNAc transferases is dependent on cell type and other factors. For example, the mucin protein MUC1 is initially glycosylated by the ubiquitous polypeptide ppGalNAc T1 and T2. In fibroblasts, the extracellular matrix protein fibronectin is glycosylated by ppGalNAc T3. In neurons, the post-translational modification of syndecan is initiated by the glycosyltransferase ppGalNAc T13. The GalNAcβ1–Ser group is further modified in healthy cells. Unmodified GalNAcα1–Ser groups are often present on the surface of cancer cells and leukemias and are referred to as the Tn antigen.

There are no simple peptide consensus sequences that allow one to identify O-glycosylation sites in proteins, but such sites can be predicted with complex computer algorithms. Localization of the substrate and the glycosyltransferase can be as important as any overt structural feature on the substrate (**Figure 7.48**). For example, human β1–4 galactosyltransferase 3 attaches galactosyl residues to any terminal GlcNAc residue in a membrane-bound acceptor, regardless of whether it is an O-linked glycoprotein, an N-linked glycoprotein, or a glycolipid (**Figure 7.49**).

Less commonly, extracellular human proteins are O-glycosylated by other monosaccharides. The α-dystroglycan component of the dystrophin glycoprotein complex is O-mannosylated by *O*-mannosyltransferase, which transfers an α-mannosyl residue from dolichol phosphate β-D-mannose to serine and threonine residues on the protein. In all mammals, this is further elaborated to oligosaccharides with a common trisaccharide core: Galβ1–4GlcNAcβ1–2Manα–Ser/Thr. The enzyme GDP-L-fucose:polypeptide fucosyltransferase transfers α-fucosyl residues to serine and threonine residues in the EGF domains of extracellular proteins like Notch.

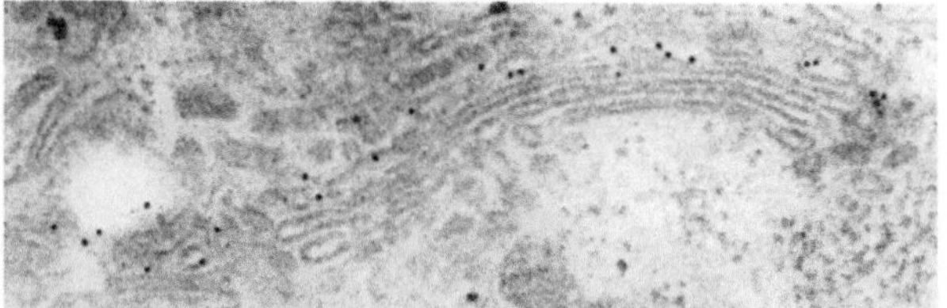

Figure 7.48 Evidence that glycosyltransferases are localized in the Golgi network. A transmission electron microscopy image of human HeLa cells stained with rabbit antibodies against galactosyltransferase. The rabbit antibodies were revealed with mouse antibodies conjugated to 10 nm gold nanoparticles (the dark dots in the image). (From T. Nilsson et al., *J. Cell Biol.* 120:5–13, 1993. With permission from Rockefeller University Press.)

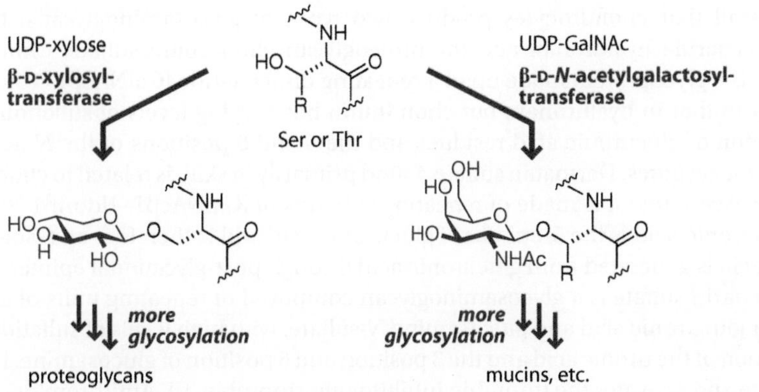

UDP-xylose
β-D-xylosyl-transferase
Ser or Thr
UDP-GalNAc
β-D-*N*-acetylgalactosyl-transferase
more glycosylation
proteoglycans
more glycosylation
mucins, etc.

Figure 7.49 Creating the O-linkage. The biosynthesis of O-linked glycoproteins with long oligosaccharides begins with the addition of β-D-xylosyl or β-D-*N*-acetylgalactosyl residues.

O-linked proteoglycans are polyanions

Proteoglycans are proteins with long O-linked glycosaminoglycan chains. Some examples include chondroitin sulfate, heparin/heparan sulfate, and keratan sulfate. Proteoglycans are generated through the initial addition of β-xylose to serine, followed by elaboration to a trisaccharide core: Galβ1–3Galβ1–4Xylβ-OSer (**Figure 7.50**). This trisaccharide core is then elaborated to various longer oligosaccharides.

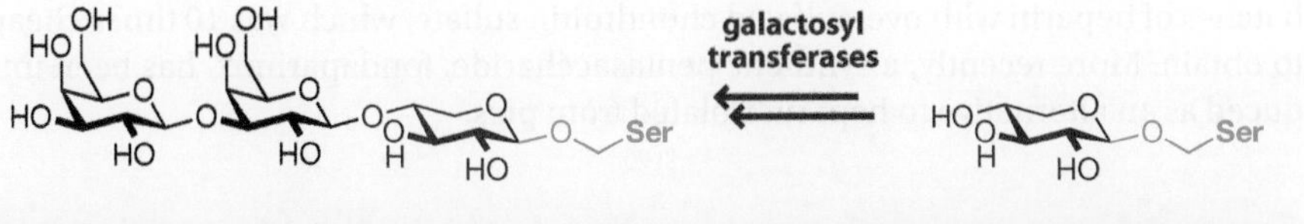

galactosyl transferases

Figure 7.50 Making the trisaccharide core. The trisaccharide core of O-linked proteoglycans is assembled one monosaccharide at a time, prior to an uncontrolled polymerization.

Problem 7.10

Using chair conformations, draw the structure of the following pentasaccharide segment of a proteoglycan: GlcNAcβ1–4GlcAβ1–3Galβ1–3Galβ1–4Xylβ-OSer.

Glycosaminoglycans are hygroscopic proteoglycans that are formed from repeating disaccharide units: a uronic acid such as D-glucuronic or L-iduronic acid, and an acetamido sugar such as *N*-acetylglucosamine or *N*-acetylgalactosamine. Found on the surface of nearly all cell types, glycosaminoglycans determine the physicochemical and mechanical properties of the tissue (**Figure 7.51**). Glycosaminoglycans always occur as heterogeneous mixtures as a result of varying levels of sulfation. Heterogeneity is a general property of many resilient materials, because it avoids the brittleness associated with crystallinity.

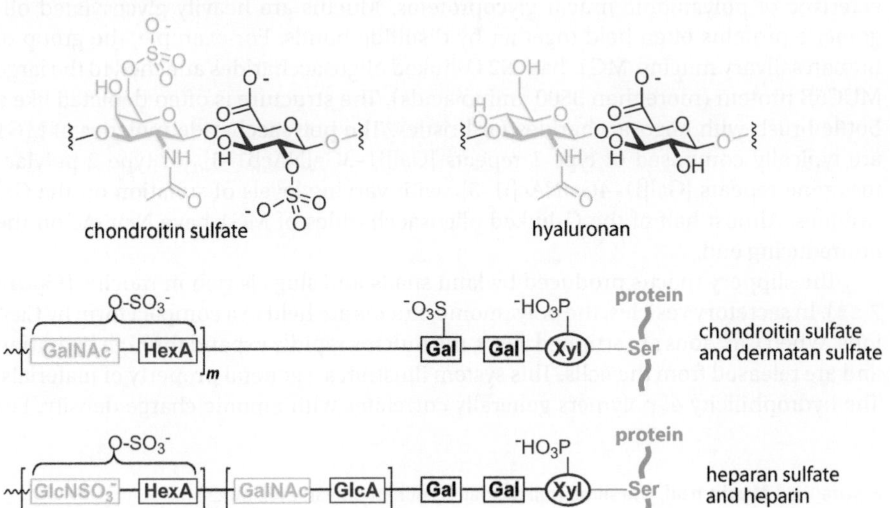

chondroitin sulfate
hyaluronan

GalNAc — HexA ... Gal — Gal — Xyl — Ser (protein): chondroitin sulfate and dermatan sulfate

GlcNSO$_3^-$ — HexA ... GalNAc — GlcA ... Gal — Gal — Xyl — Ser (protein): heparan sulfate and heparin

Figure 7.51 Glycosaminoglycans. The glycosaminoglycans in human cartilage (HexA = GlcA or IdoA) are made up of repeating disaccharide subunits.

Recall that chondrocytes produce two types of glycosaminoglycans: the free polysaccharide hyaluronan and the proteoglycan chondroitin sulfate. Chondroitin sulfate is a glycoprotein made up of a repeating disaccharide [GalNAcβ1–4GlcAβ1–3] similar to that in hyaluronan, but chondroitin has varying levels of sulfation on the 2 position of glucuronic acid residues and the 4 and 6 positions of the *N*-acetylglucosamine residues. Dermatan sulfate, found primarily in skin, is related to chondroitin sulfate except that it is made of repeating subunits of [GalNAcβ1–4IdoAβ1–3]. Recall that iduronic acid is the 5-epimer of glucuronic acid and is therefore an L-sugar. Iduronic acid is generated from glucuronic acid through post-glycational epimerization.

Heparin sulfate is a glycosaminoglycan composed of repeating units of iduronic acid or glucuronic acid and glucosamine *N*-sulfate, with high levels of sulfation at the 2 position of the uronic acid and the 3 position and 6 position of glucosamine. Heparin binds to and activates the thrombin inhibitor antithrombin-III. Antithrombin-III then forms a covalent complex with various serine proteases in the clotting cascade (Section 6.3). Antithrombin-III is best described as a very poor substrate that is turned over slowly by the protease. Antithrombin-III is most strongly activated by a specific pentasaccharide sequence of heparin (**Figure 7.52**). Low-molecular-weight heparin, isolated from pig intestinal mucosa, is used clinically as an injectable anticoagulant during surgery when there is a high risk of unwanted blood clotting. In 2008, a major pharmaceutical company recalled batches of heparin that were contaminated with a related glycosaminoglycan. Allegedly, the original supplier had augmented the batches of heparin with oversulfated chondroitin sulfate, which was 10 times cheaper to obtain. More recently, a synthetic pentasaccharide, fondaparinux, has been introduced as an alternative to heparin isolated from pigs.

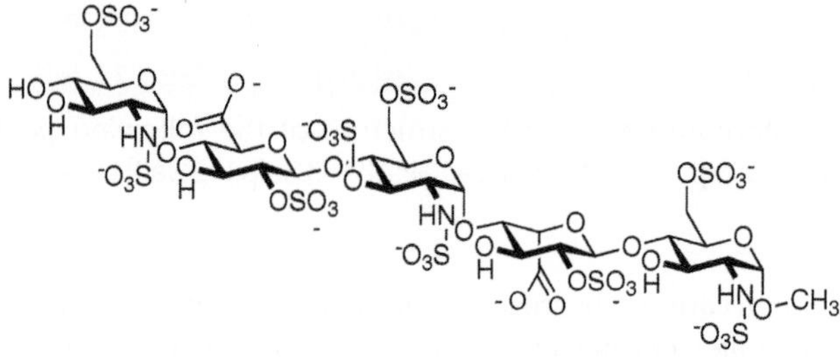

Figure 7.52 A heparin-like drug. Fondaparinux is based on the antithrombin-III binding sequence of heparin.

The slippery hydrophilic properties of mucous secretions such as snot are characteristic of polyanionic mucin glycoproteins. Mucins are heavily glycosylated oligomeric proteins often held together by disulfide bonds. For example, the group of human salivary mucins, MG1, has 292 O-linked oligosaccharides attached to the large MUC5B protein (more than 3500 amino acids). The structure is often depicted like a bottle brush with oligosaccharides for bristles. The polysaccharide moieties of MG1 are typically composed of type 1 repeats $[\text{Gal}\beta 1\text{–}3\text{GalNAc}\beta 1\text{–}3]_n$ or type 2 polylactosamine repeats $[\text{Gal}\beta 1\text{–}4\text{GalNAc}\beta 1\text{–}3]_n$ with varying levels of sulfation on the Gal residues. Almost half of the O-linked oligosaccharides of MG1 have Neu5Ac on the nonreducing end.

The slippery mucus produced by land snails and slugs is rich in mucins (**Figure 7.53**). In secretory vesicles, the polyanionic mucins are held in a compact form by Ca^{2+} ions. When Ca^{2+} ions are stripped away, the mucins rapidly expand through hydration and are released from the cells. This system illustrates a general property of materials. The hydrophilicity of polymers generally correlates with anionic charge density. For

Figure 7.53 Mucin trail. The slime trail of snails is rich in polyanionic mucin glycoproteins. (Courtesy of Photo Researchers.)

example, the superabsorbant polymers found in disposable diapers are simply polyacrylates, $[CH_2CH(CO_2Na)]_n$. They absorb up to 500 times their weight in water.

Hagfish are one of the most notorious sources of mucin-based slime (**Figure 7.54**). When stressed, hagfish produce voluminous quantities of a thick slime composed of approximately 99.996% seawater, 0.0015% mucin, and 0.002% protein threads. The protein imparts elasticity, and the mucin imparts viscosity. The slime has been hypothesized to clog the gills of potential predators.

Figure 7.54 Instant slime. Dried hagfish slime, composed of polyanionic mucins, can be reconstituted from seawater. (Courtesy of Jamie Miller, University of Guelph.)

The carbohydrate moiety of N-linked glycoproteins is initially added as an oligosaccharide

The N-glycosylation of proteins takes place on the asparagine residue of Asn-Xxx-Ser and Asn-Xxx-Thr sequences as the proteins are translated into the endoplasmic reticulum (**Figure 7.55**). Initially, the branched tetradecasaccharide $Glc_3Man_9GlcNAc_2$ is transferred from a membrane-bound precursor by the enzyme oligosaccharyltransferase (**Figure 7.56**). The structure of the initial mannose-rich oligosaccharide varies among eukaryotes, but the $Man_3GlcNAc_2$ core is common to all of them.

As the N-linked glycoprotein is trafficked through the vesicular network of the endoplasmic reticulum and Golgi apparatus, various glycosylhydrolases trim the

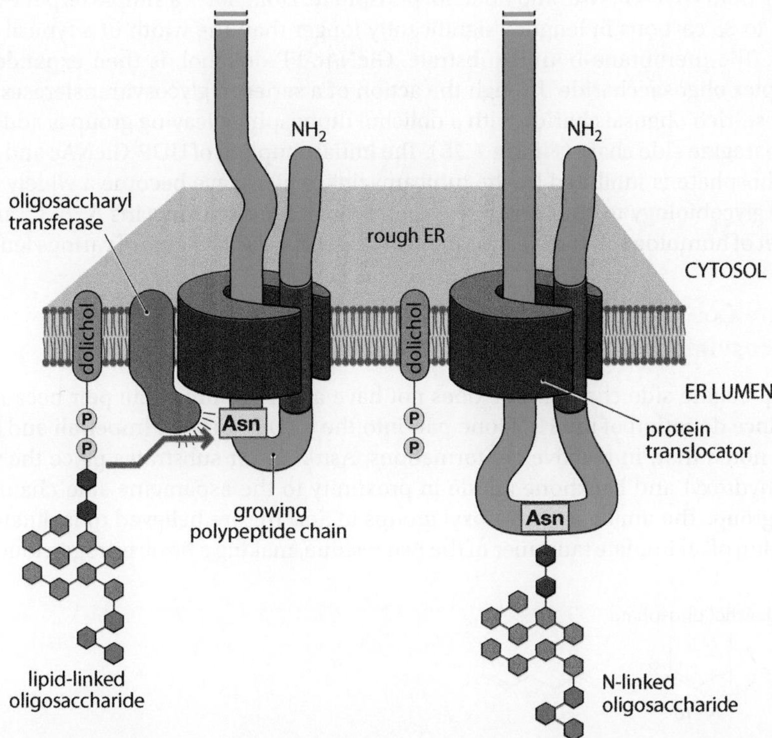

Figure 7.55 Grafting the scion onto the stock. Transfer of a complex oligosaccharide to the tripeptide sequence AsnXxxSer occurs as the ribosome extrudes the protein chain into the endoplasmic reticulum. (Adapted from B. Alberts et al., Molecular Biology of the Cell, 5th ed. New York: Garland Science, 2008.)

common to all eukaryotes

Manα1→2Manα1→6
Manα1→2Manα1→3 ⟩ Manα1→6
Glcα1→2Glcα1→3Glcα1→3Manα1→2Manα1→2Manα1→3 ⟩ Manβ1→4GlcNAcβ1→4GlcNAcβ1→Asn

Figure 7.56 The initial N-glycan core. The core oligosaccharide that is added to asparagine side chains of human proteins is composed of 14 monosaccharide subunits.

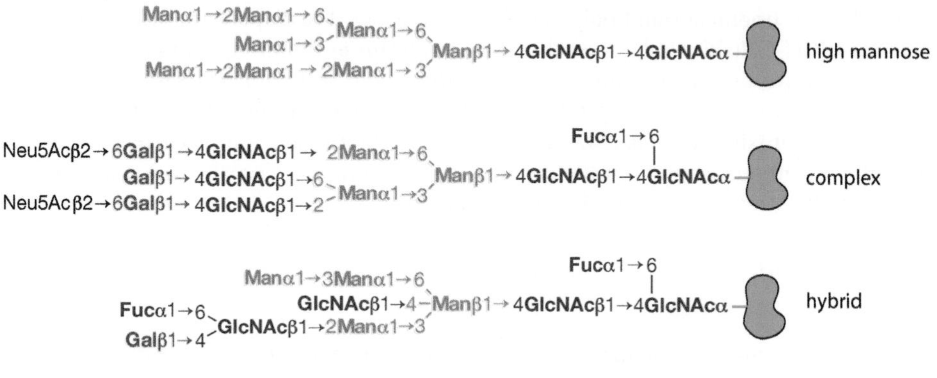

Figure 7.57 Man kind. The three types of N-glycans that ultimately emerge after vesicular trafficking are classified on the basis of the mannose core: high mannose (Man_8), complex (Man_3), and hybrid (Man_{3-8}).

structure, and various glycosyltransferases graft on additional monosaccharide units. The resulting N-linked glycoproteins can be subdivided into three types: oligomannose (high mannose), hybrid, and complex (**Figure 7.57**). Oligomannose structures are found in all eukaryotes, including yeast, whereas hybrid and complex N-linked glycoproteins are characteristic of multicellular organisms.

The oligosaccharide substrate for the oligosaccharyltransferase is synthesized starting from UDP-GlcNAc and dolichol phosphate. Dolichol is a simple terpene alcohol up to 85 carbons in length—significantly longer than the width of a typical lipid bilayer. The membrane-bound substrate, GlcNAc-PP-dolichol, is then expanded to a complex oligosaccharide through the action of a series of glycosyltransferases. The mannose-rich oligosaccharide with a dolichol diphosphate leaving group is added to the asparagine side chain (**Figure 7.58**). The initial coupling of UDP-GlcNAc and dolichol phosphate is inhibited by the tunicamycins, which have become a widely used tool in glycobiology to shut down N-glycosylation. The tunicamycins A, B, C, and D are a set of homologous natural products with lipophilic acyl chains of various lengths.

An Asn-Xxx-Ser motif adopts a reactive conformation in the N-glycosylation of proteins

The asparagine side chain amide does not have a nucleophilic lone pair because of resonance donation of the H_2N lone pair into the C=O π^* orbital. Imperiali and coworkers noted that, in reactive conformations, Asn-Xxx-Ser substrates place the side-chain hydroxyl and backbone amide in proximity to the asparagine side-chain carbonyl group. The amide and hydroxyl groups of Ser/Thr are believed to facilitate the formation of an imidate tautomer of the Asn residue, making a basic nitrogen lone pair

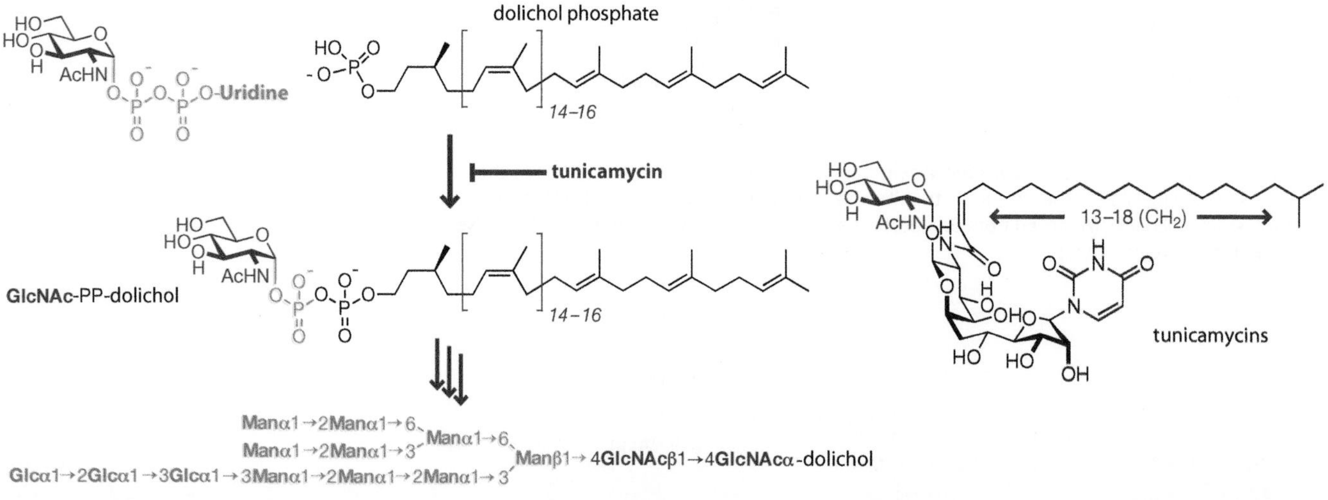

Figure 7.58 Blocking N-glycan synthesis. Tunicamycin blocks the synthesis of N-linked glycoproteins by inhibiting the initial attachment of a lipophilic dolichol chain to the first GlcNAc residue in the glycan core.

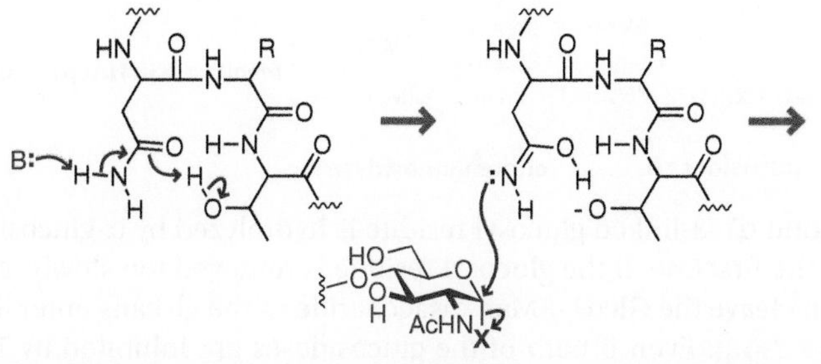

Figure 7.59 The mechanism for N-glycan attachment. The mechanism of asparagine glycosylation by oligosaccharyl transferase is believed to proceed through an imidate tautomer.

available for glycosylation (**Figure 7.59**). This example of **substrate-assisted catalysis** illustrates how sequence specificity for post-translational processes results from both the preferences of modifying enzymes and the functional groups of the enzyme's substrate placed in close proximity to the targeted side chain.

The processing of glycans occurs during vesicular trafficking

The processing of N-linked oligosaccharides begins with trimming of the branches within the endoplasmic reticulum. Further removal and addition of individual monosaccharides occurs as the glycans are trafficked outward toward the cell surface (see Figure 7.47), first through the endoplasmic reticulum and then through the Golgi network toward the cell surface. At no time are the glycans exposed to the cytoplasm, and most of the trimming reactions are performed by membrane-bound glycosylhydrolases with their active sites oriented toward the interior of the vesicular compartments. To appreciate why N-linked glycans are so heterogeneous and complex, it is helpful to imagine an automobile assembly line staffed by workers who perform their operation only when certain parts are missing. Now imagine speeding up the assembly line so that workers do not have sufficient time to work on every vehicle.

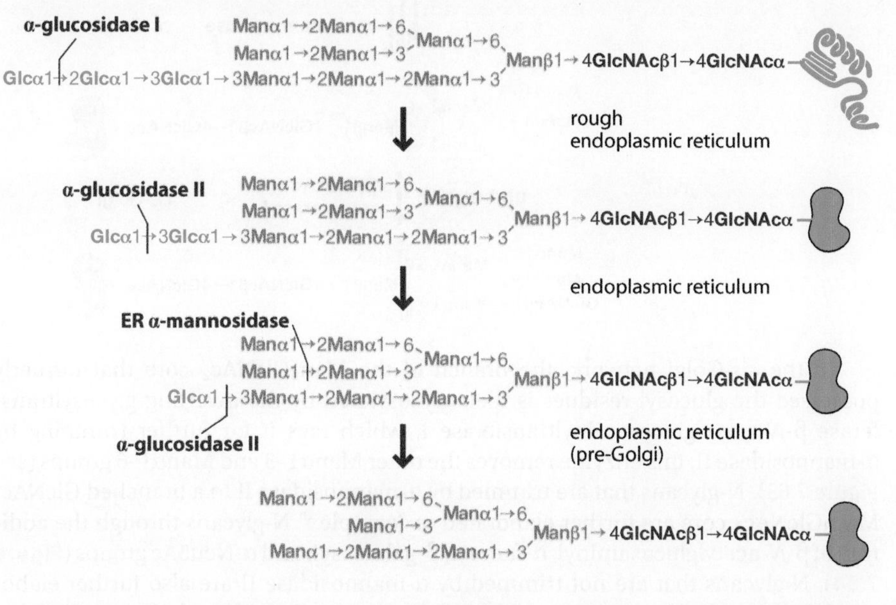

Figure 7.60 Early stages of trimming. Trimming reactions of N-linked oligosaccharides that take place within the endoplasmic reticulum.

Trimming starts within the rough endoplasmic reticulum (covered with ribosomes), as proteins are being translated by ribosomes, where α-glucosidase I removes the outermost α1-2-linked glucosyl residue (**Figure 7.60**). The next two α-1-3-linked glucosyl residues are removed by α-glucosidase II, which is weakly bound to the membranes of the smooth endoplasmic reticulum and the pre-Golgi vesicles. α-Glucosidases I and II are inhibited by the natural products 1-deoxynojirimycin and castanospermine (**Figure 7.61**).

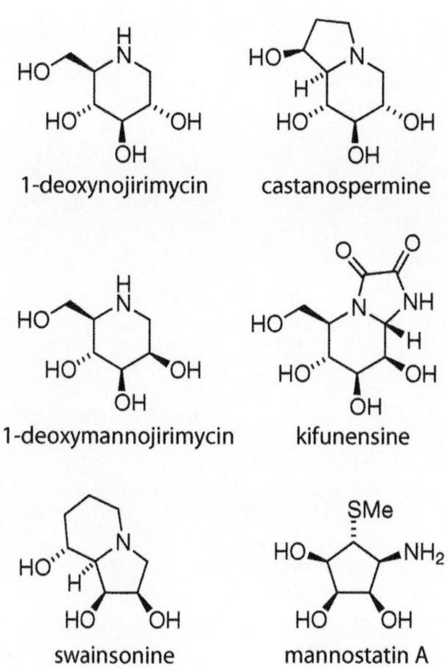

1-deoxynojirimycin

castanospermine

1-deoxymannojirimycin

kifunensine

swainsonine

mannostatin A

Figure 7.61 Tools from nature. A number of natural product inhibitors of glycosidases exhibit sufficient selectivity against various glycosidases to be useful as tools for studying N-glycan processing in human cells.

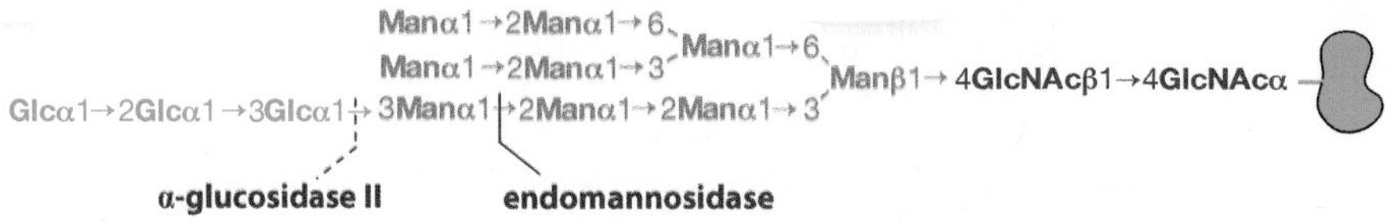

Manα1→2Manα1→6
Manα1→6
Manα1→2Manα1→3
Manβ1→4GlcNAcβ1→4GlcNAcα
Glcα1→2Glcα1→3Glcα1→3Manα1→2Manα1→2Manα1→3
α-glucosidase II
endomannosidase

Figure 7.62 Alternative early stage trimming. If any of the outer glucosidase residues remain then endomannosidase takes care of any unfinished business by cleaving an internal glycosidic bond.

The second α1-3-linked glucosyl residue is hydrolyzed by α-glucosidase II more slowly than the first one. If the glucosyl residue is removed too slowly, an endomannosidase can cleave the Glcα1-3Man disaccharide as the glycans enter the Golgi network (**Figure 7.62**). Even if both of the glucosidases are inhibited by 1-deoxynojirimycin or castanospermine, endomannosidase can still cleave off the entire Glc_3Man tetrasaccharide. Endomannosidase is the only mammalian trimming enzyme that cleaves an internal glycosidic bond.

As the pre-Golgi vesicles fuse with the Golgi vesicular network, they are set upon by α1-2 mannosidases IA, IB, or IC (**Figure 7.63**). The face of the Golgi network that is closest to the nucleus is referred to as *cis*; the face of the Golgi network that is closest to the cell surface is referred to as *trans*. Since Golgi vesicles slowly bubble outward toward the cell surface, glycosidases in the *cis*-Golgi act before glycosidases in the *trans*-Golgi. The three α1-2 mannosidases IA, IB, and IC seem to have the same function, cleaving the three α1-2-linked mannosyl residues, but their expression is tissue-specific. α-Mannosidases are inhibited by the natural products 1-deoxymannojirimycin and kifunensine. Glycans that are trimmed back to the $Man_5GlcNAc_2$ core are elaborated by glycosyltransferases. Those "high-mannose" N-glycans that are not completely trimmed by α1-2 mannosidases within the *cis*-Golgi may eventually make their way to the cell surface without further modification.

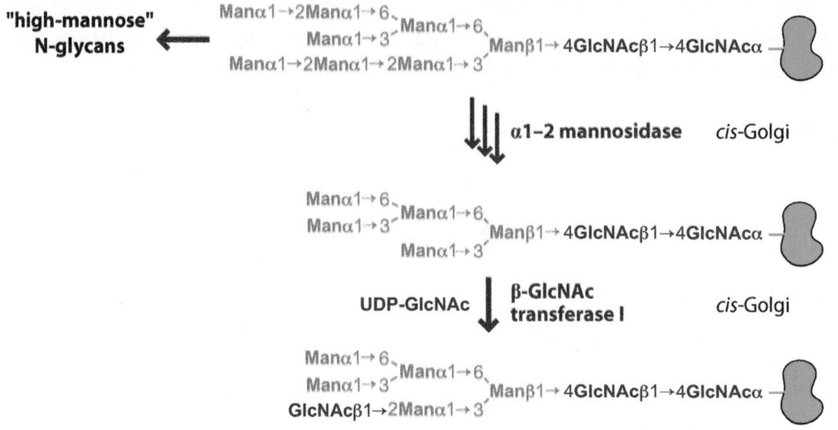

"high-mannose" N-glycans ←
Manα1→2Manα1→6
Manα1→6
Manα1→3
Manβ1→4GlcNAcβ1→4GlcNAcα
Manα1→2Manα1→2Manα1→3
α1-2 mannosidase *cis*-Golgi
Manα1→6
Manα1→6
Manα1→3
Manβ1→4GlcNAcβ1→4GlcNAcα
Manα1→3
UDP-GlcNAc β-GlcNAc transferase I *cis*-Golgi
Manα1→6
Manα1→6
Manα1→3
Manβ1→4GlcNAcβ1→4GlcNAcα
GlcNAcβ1→2Manα1→3

Figure 7.63 Addition or subtraction. As they enter the *cis*-Golgi N-glycans are either elaborated to high-mannose glycans or trimmed further. Small molecules (shown in Figure 7.61) can inhibit N-glycan modifying enzymes.

In the *cis*-Golgi network, the branch of the $Man_5GlcNAc_2$ core that formerly possessed the glucosyl residues is then glycosylated by the inverting glycosyltransferase β-*N*-acetylglucosaminyltransferase I, which tags it for further trimming by α-mannosidase II; this enzyme removes the outer Manα1-3 and Manα1-6 groups (see Figure 7.63). N-glycans that are trimmed by α-mannosidase II to a branched GlcNAc-$Man_3GlcNAc_2$ core are further elaborated to "complex" N-glycans through the addition of β-*N*-acetylglucosaminyl, α-fucosyl, β-galactosyl, and α-Neu5Ac groups (**Figure 7.64**). N-glycans that are not trimmed by α-mannosidase II are also further elaborated, but because they have more than three mannosyl residues they are referred to as "hybrid" N-glycans because they have the extra mannose residues of a high-mannose N-glycan and the glycosylated branches of a complex N-glycan.

Golgi α-mannosidase II can be selectively inhibited by the natural products swainsonine and mannostatin A, diverting N-glycans toward hybrid cores rather than complex cores. The Man_3 and Man_5 cores are glycosylated by a number of β-*N*-acetylglucosaminyltransferases, which add up to six GlcNAc branches. For example, β-GlcNAc transferase II adds GlcNAc residues to the 2-hydroxyl groups of mannosyl residues (those that were not glycosylated by β-*N*-acetylglucosaminyltransferase II); β-GlcNAc transferases IV and V generate branches by adding a second GlcNAc residue to the 6-hydroxyl groups of mannosyl residues; β-GlcNAc transferase I adds a

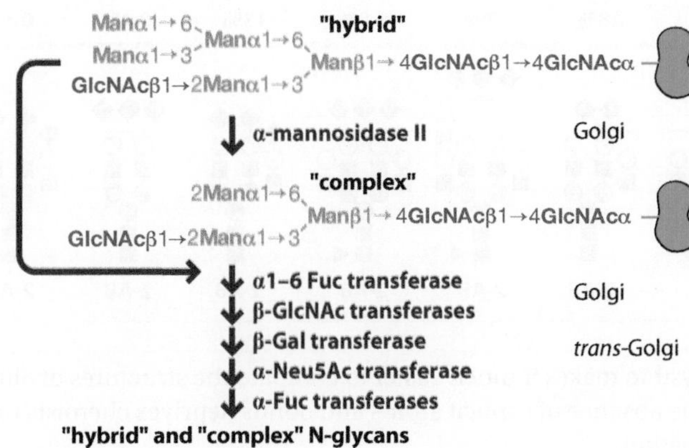

Figure 7.64 Late-stage glycosylation in the Golgi network. Further glycosylation reactions add additional sugars to the mannosyl branches and to the outer GlcNAc residue.

third GlcNAc residue to the 4-hydroxyl groups of mannosyl residues. As the N-glycans reach the *trans* face of the Golgi network, α1–6 fucosyltransferase VIII adds a fucosyl residue to the *N*-acetylglucosamine residue that is directly connected to the protein.

The outer GlcNAc branches of the N-glycans are then further adorned with β-galactosyl residues by two of the seven human β1–4 galactosyltransferases, II or V (**Figure 7.65**). The 3-hydroxyl or 6-hydroxyl group of the galactosyl residues can be modified by any of 20 different human α-sialyltransferases. There are 11 different fucosyltransferases in the human genome; they selectively add α-fucosyl groups to the 2-hydroxyl group of β-galactosyl residues, whereas other fucosyltransferases selectively add α-fucosyl groups to the 3-hydroxyl or 4-hydroxyl group of β-GlcNAc residues. More than 100 different glycosyltransferases are known to be expressed in Golgi compartments of different human cells. Most of the late-stage glycosylation reactions are performed by glycosyltransferases that recognize just a few monosaccharide subunits in the substrate rather than the full-length oligosaccharide. Thus a glycosyltransferase might recognize a GlcNAc residue in the acceptor, but the enzyme is unable to tell whether the acceptor residue is attached to a glycolipid or a glycoprotein. For example, β1–4 galactosyltransferase V targets GlcNAc residues in both glycolipids and proteins. Similarly, α1–2 fucosyltransferase II adds fucosyl groups to galactosyl residues on both glycolipids and glycoproteins.

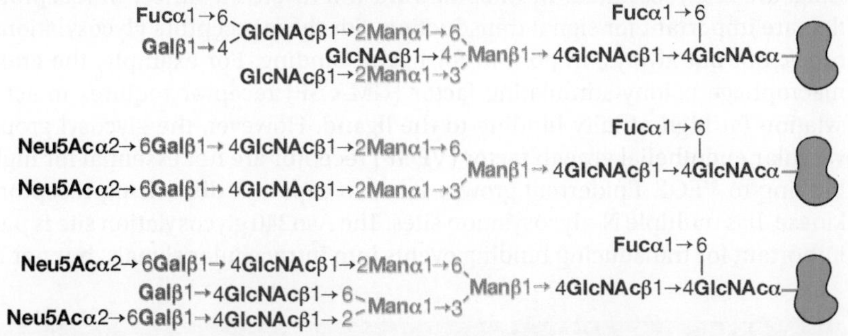

Fucα1→6
Galβ1→4
GlcNAcβ1→2Manα1→6
GlcNAcβ1→4–Manβ1→ 4GlcNAcβ1→4GlcNAcα
GlcNAcβ1→2Manα1→3
Fucα1→6

Neu5Acα2→ 6Galβ1→ 4GlcNAcβ1→2Manα1→6
Manβ1→ 4GlcNAcβ1→4GlcNAcα
Neu5Acα2→ 6Galβ1→ 4GlcNAcβ1→2Manα1→3
Fucα1→6

Neu5Acα2→ 6Galβ1→ 4GlcNAcβ1→2Manα1→6
Galβ1→4GlcNAcβ1→ 6
Manβ1→ 4GlcNAcβ1→4GlcNAcα
Manα1→3
Neu5Acα2→6Galβ1→4GlcNAcβ1→ 2
Fucα1→6

Figure 7.65 Examples of complex N-glycans. Protein glycans are further elaborated one monosaccharide at a time.

The time spent in transit through the Golgi network is limited, and there is not sufficient time for all of the potential glycosylation reactions to reach completion. As a result, N-linked glycans are always produced as heterogeneous combinatorial mixtures, making it difficult to purify them, characterize them, or even accurately represent the structures on paper. To understand the degree of heterogeneity in glycosylated proteins it is useful to look at an example of the glycans that are present on just one type of protein. When the soluble form of intracellular adhesion molecule-1 (sICAM-1) is expressed in Chinese hamster ovary (CHO) cells it is produced as a mixture of N-glycans, most with a complex $Man_3GlcNAc_2$ core (**Figure 7.66**). Increasingly, glycobiologists are turning to a symbolic representation system, using colored shapes to represent the individual monosaccharide subunits. The symbolic representation

relative amount	N/A	15%	6.5%	48%	2%	6.5%	13%	0.4%	0.5%	3%	5%
main N-glycan	2-AB	2-AB	2-AB	2-AB	2-AB	2-AB	2-AB	2-AB	2-AB	2-AB	2-AB

Figure 7.66 Heterogeneity. In order to reveal the wide range of glycan structures on a single type of protein, glycans were enzymatically cleaved from the expressed glycoprotein sICAM-1 and then fluorescently labeled with 2-aminobenzamide (2-AB) to facilitate chromatographic analysis. The connectivity of the monosaccharide subunits in each glycan was determined by mass spectrometry (blue square, GlcNAc; green circle, mannose; red triangle, fucose; yellow circle, galactose; purple diamond, *N*-acetylneuraminic acid).

system makes it much easier to compare the structures of oligosaccharides. However, the absence of explicit atoms and bonds deprives chemists of potentially useful information.

A few human proteins are C-mannosylated on tryptophan residues

Surprisingly, a few secreted human proteins have been found to be post-translationally glycosylated on tryptophan residues within the endoplasmic reticulum. The modification is an α-mannosyl group attached to the 2-position of the indole ring. Ribonuclease 2 and interleukin-12 are C-mannosylated on the first tryptophan in a WXXW tetrapeptide motif. Several proteins in the complement system have been found to be C-mannosylated enzymatically. The generic name of the complement system belies its ruthless lethality toward foreign cells that have been targeted for destruction. The complement system is a set of proteolytically activated proteins that self-assemble into gaping channels in the membranes of pathogenic organisms. These channels allow the contents of the cell to leak out. The complement proteins C6, C7, C8α, C8β, and C9 are multiply C-mannosylated: six in C6, four in C7, C8α, and C8β, and two in C9. All of these proteins are glycosylated on WXXWXXW sequences in a common thrombospondin type I repeat domain; however, other tryptophan residues that are not part of this domain are also C-mannosylated.

Glycosylation of proteins sometimes, but not always, affects the intrinsic function of the protein

Most membrane-bound proteins that traffic through the endoplasmic reticulum and Golgi are N-glycosylated. In Chapter 9 we will discuss a variety of receptor proteins that are important for signal transduction. For these receptors glycosylation is sometimes, but not always, important for ligand binding. For example, the granulocyte/macrophage colony-stimulating factor (GM-CSF) receptor requires intact N-glycosylation for high-affinity binding to the ligand. However, the glycosyl groups of the vascular endothelial growth factor (VEGF) receptor are not essential for high-affinity binding to VEGF. Epidermal growth factor receptor, a dimerizing receptor tyrosine kinase, has multiple N-glycosylation sites. The Asn240 glycosylation site is particularly important for transducing binding events into intracellular signals, but not for ligand

Problem 7.11

Suggest a plausible arrow-pushing mechanism for the enzymatic transfer of a mannosyl residue from β-mannosyl dolichol phosphate to tryptophan.

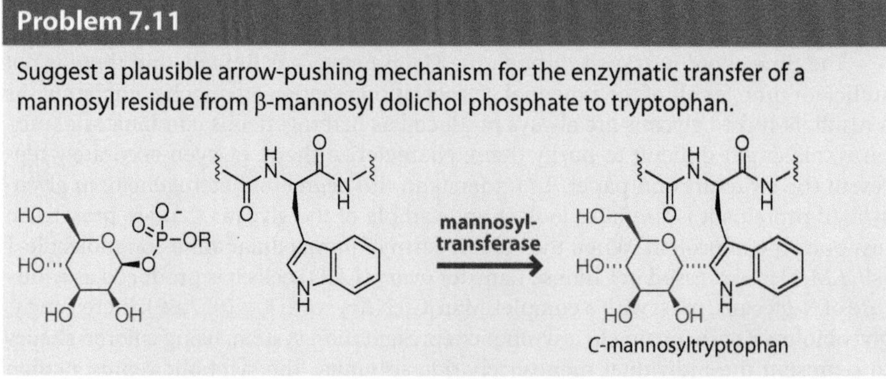

binding. The N240Q mutant of the EGF receptor, which lacks glycosylation at position 240, undergoes spontaneous autophosphorylation even in the absence of the ligand; a mutation with this phenotype would be termed constitutively active.

More than 95% of 7TM receptors are glycosylated. About two-thirds of the 7TM receptors that have been analyzed have glycosylation sites in the second extracellular loop; the remaining one-third have glycosylation sites in the first or third extracellular loop. The role of these oligosaccharides is still a subject of investigation. In many cases, the oligosaccharides on cell-surface receptors appear to contribute to membrane localization and ligand binding. For example, when two of the three glycosylation sites (Asn4 and Asn188) of the angiotensin II receptor were mutated to non-glycosylatable threonine residues, most of the membrane expression and ligand affinity of the receptors were retained. However, mutation of the third glycosylation site Asn176 led to a significant reduction in both membrane expression of the receptor and affinity of the receptor for the peptide ligand.

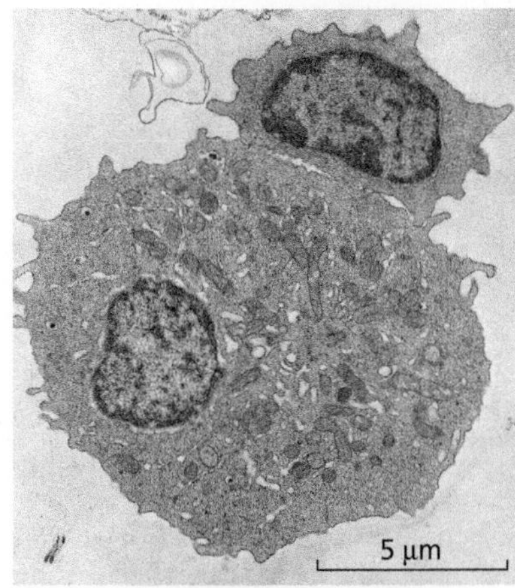

Figure 7.67 An intimate chemical conversation. Cytotoxic T lymphocyte (top) interacting with an antigen-presenting cell (bottom). (From B. Alberts et al., Molecular Biology of the Cell, 5th ed. New York: Garland Science, 2008.)

During a viral infection, human cells fragment viral proteins and present the peptide fragments bound to major histocompatibility complex (MHC) receptors on the cell surface for recognition by T cells. A Braille-like communication takes place between the two cells, as proteins touch each other. Productive binding results from protein–protein molecular recognition at the interface between the two cells, termed the immunological synapse. Through this mechanism, killer T cells can distinguish between friendly cells and enemy cells, and between healthy cells and infected cells. A key interaction between the antigen-presenting cell and the cytotoxic T lymphocyte is mediated by the B7-1 protein of the antigen-presenting cell and the cytotoxic T-lymphocyte-associated antigen 4 (CTLA-4) (**Figure 7.67** and **Figure 7.68**). Both proteins are heavily glycosylated (CTLA-4 has two glycosylation sites, and B7-1 has ten). However, the carbohydrates do not make key interactions responsible for specificity. Instead, protein–protein recognition triggers a T-cell response. Although the glycan moieties on CTLA-4 do not make contacts with B7-1, they are still important. The drug abatacept, which is a dimeric form of an antibody Fc domain fused to CTLA-4, has a longer half-life when produced in CHO cells than in other cell lines, primarily because of the presence of terminal Neu5Ac residues.

Most extracellular signaling proteins are glycosylated with oligosaccharides

The majority of soluble signaling proteins are glycosylated. Most of the cytokine signaling proteins that we discuss in Chapter 9 (such as interleukins, interferons,

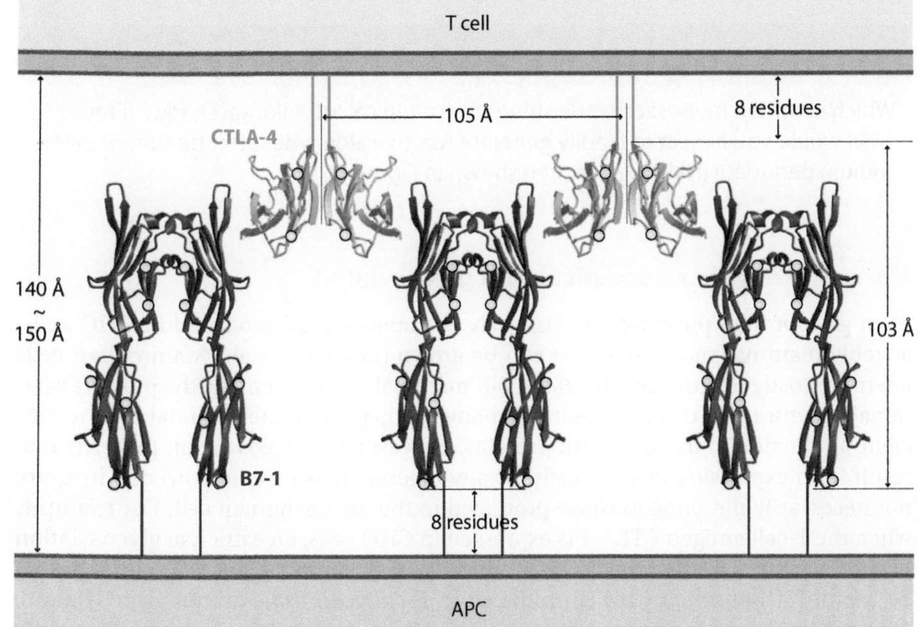

Figure 7.68 Sweet greeting. The recognition interface between a T cell and an antigen-presenting cell (APC) has many glycosylation sites (yellow), but none are important for binding. (Courtesy of S. Ikemizu and S.J. Davis.)

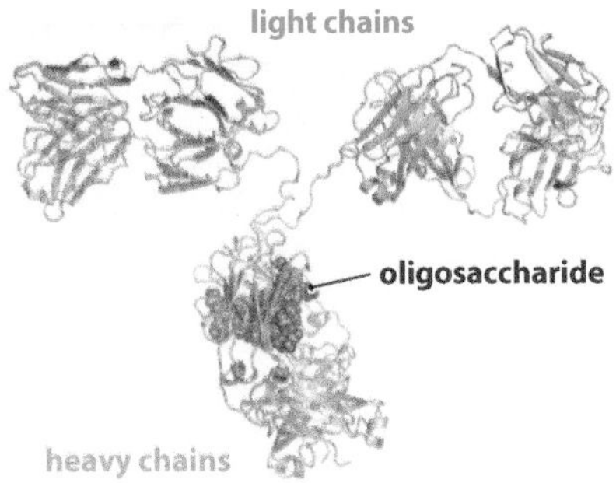

Figure 7.69 An inside out glycoprotein. The oligosaccharides of most glycoproteins are attached to the outside, but the opposite is true for antibodies. In this crystal structure of a full-length type G immunoglobulin the oligosaccharide is clearly embedded deeply within the constant region of the heavy chains: two heavy chains (green), two light chains (blue), and two N-linked oligosaccharides (purple). The sialic acid residues are missing from the oligosaccharide. (PDB 1IGT)

and colony-stimulating factors) are glycosylated, which extends the half-life of the cytokine in the blood. Such glycosylation usually has no effect on protein binding. With respect to serum half-life, the structure of the oligosaccharide fusion matters. For example, glycoproteins stripped of Neu5Ac residues to reveal terminal galactosyl residues (asialoglycoproteins) are taken up by the liver within minutes.

In some cases the oligosaccharide is also important for function. For example, β-human chorionic gonadotropin (Pregnyl®) induces ovulation. When deglycosylated, the protein binds to its cognate receptor with similar affinity, but the binding fails to stimulate adenylate cyclase activity inside the target cell. The most common antibodies, gamma immunoglobulins (IgG), are glycosylated on Asn297 of the heavy chain fragment (**Figure 7.69**). The oligosaccharides are present in various related forms (**Figure 7.70**). Surprisingly, although the added oligosaccharides are heterogeneous, the addition of the carbohydrates is still important for the function of the protein. Glycosylation is essential for binding to immunoglobulin Fc receptors and to the C1q complement protein that leads to lysis of foreign cells.

Neu5Acα2→6Galβ1→4GlcNAcβ1→2Manα1→6 Fucα1→6
GlcNAcβ1→4–Manβ1→4GlcNAcβ1→4GlcNAcα→Asn
Neu5Acα2→6Galβ1→4GlcNAcβ1→2Manα1→3

Figure 7.70 Sequence of an IgG oligosaccharide. Truncated forms of this oligosaccharide are also observed.

Aldehydes condense with acyl hydrazides to form stable hydrazone linkages, even in water. A common method for creating aldehyde functional groups is to cleave oligosaccharides oxidatively with sodium periodate (pronounced per-iodate) (**Figure 7.71**). Because the cleavage reaction involves a cyclic five-membered ring periodate ester, the vicinal hydroxyl groups must be able to adopt a *syn* orientation. Acyclic 1,2-diols can adopt an appropriate conformation through rotation about a carbon-carbon bond. However, cyclic 1,2-diols will only react if the hydroxyl groups are on the same face of the ring (*syn*); cyclic *anti* 1,2-diols are much less susceptible to cleavage by periodate.

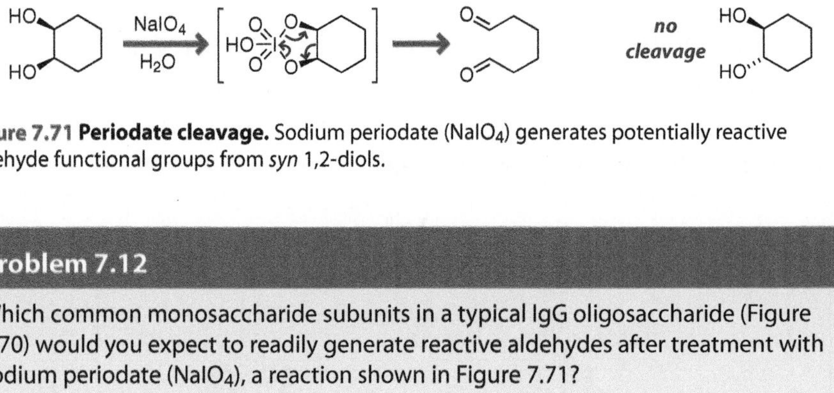

Figure 7.71 Periodate cleavage. Sodium periodate ($NaIO_4$) generates potentially reactive aldehyde functional groups from *syn* 1,2-diols.

Problem 7.12

Which common monosaccharide subunits in a typical IgG oligosaccharide (Figure 7.70) would you expect to readily generate reactive aldehydes after treatment with sodium periodate ($NaIO_4$), a reaction shown in Figure 7.71?

Many protein pharmaceuticals are glycosylated

Most glycoprotein pharmaceuticals, such as Aranesp®, are produced in CHO cells, a stable mammalian cell line that can be grown on a large scale. Mammalian cells are more costly to grow than bacteria, but mammalian cells provide the post-translational enzymes and compartmentalized environments that are essential for the efficient production of many human proteins. However, the glycosylation patterns that result from expression in CHO cells, or any heterologous (nonhuman) cell line, are not necessarily the same as those produced in the native, human cell. For example, when the T-cell antigen CTLA-4 is expressed in CHO cells, an extra Asn glycosylation occurs. Avonex® (β-interferon 1a) is produced as a glycoprotein in CHO cells, whereas Betaseron® (β-interferon 1b) is produced in *Escherichia coli* without glycosylation. Avonex® has been reported to be more than 10 times as potent as Betaseron®, mainly as a result of the presence of an N-linked oligosaccharide rather than the small differences in sequence between the two pharmaceuticals. The carbohydrate group of Avonex® confers greater resistance to protein aggregation. The glycosylation patterns of proteins expressed in cultured cells such as CHO cells are not necessarily the same as those found in proteins isolated from humans. For example, the interferon β1, isolated from humans, is a mixture of four sialylated triantennary and tetraantennary glycoproteins (**Figure 7.72**). CHO cells produce the same mixture of glycans, but in slightly different ratios. Fortunately, these differences do not seem to have adverse effects on the therapeutic benefit. Significant advances have been made in the engineering of yeast (*Saccharomyces cerevisiae* and *Pichia pastoris*) cell lines that can produce glycoproteins. Through screening it is possible to identify cell lines that produce essentially homogeneous glycoforms of biologically active erythropoietin, although not the major glycoform produced in humans. There is every reason to expect that yeast will become a major source of glycoprotein pharmaceuticals in the future.

	human	CHO
(Neu5Acα2→3) Galβ1→4GlcNAcβ1→2Manα1→6 (Neu5Acα2→3) Galβ1→4GlcNAcβ1→2Manα1→3 } Manβ1→	74%	68%
(Neu5Acα2→3) Galβ1→4GlcNAcβ1→2Manα1→6 (Neu5Acα2→3) Galβ1→4GlcNAcβ1→4 (Neu5Acα2→3) Galβ1→4GlcNAcβ1→2 } Manβ1→3 } Manβ1→	10%	0%
(Neu5Acα2→3) Galβ1→4GlcNAcβ1→4 (Neu5Acα2→3) Galβ1→4GlcNAcβ1→2 } Manβ1→6 (Neu5Acα2→3) Galβ1→4GlcNAcβ1→2Manα1→3 } Manβ1→	8%	27%
(Neu5Acα2→3) Galβ1→4GlcNAcβ1→2Manα1→6 (Neu5Acα2→3) Galβ1→4GlcNAcβ1→3 Galβ1→4GlcNAcβ1→2Manα1→3 } Manβ1→	8%	0%
(Neu5Acα2→3Galβ1→4GlcNAcβ1→3)Galβ1→4GlcNAcβ1→4 (Neu5Acα2→3Galβ1→4GlcNAcβ1→3)Galβ1→4GlcNAcβ1→2 } Manβ1→6 (Neu5Acα2→3Galβ1→4GlcNAcβ1→3)Galβ1→4GlcNAcβ1→2Manα1→3 } Manβ1→	0%	4%

Figure 7.72 Different glycoforms from different cells. The glycoforms of human interferon β1 expressed in CHO cells are slightly different from the glycoforms isolated from humans. Each of the N-glycans shown is attached to interferon β1 through 4GlcNAcβ1–4GlcNAcα-Asn (not shown) with varying amounts of α-fucosylation on the first GlcNAc. The outer branches exhibit varying degrees of sialylation with Neu5Ac on the 3- and 6-positions of the terminal Gal residues.

Cell–cell recognition is often mediated by glycoproteins

The spikes that project from an infectious HIV virion bind selectively to helper T cells (**Figure 7.73**). These spikes are composed of trimers of a heavily glycosylated protein gp120, which undergoes a drastic rearrangement when the spikes bind to the CD4 receptor of T cells (**Figure 7.74**). HIV has evolved several different tricks to disguise this unique spike glycoprotein from the human immune system. Rapid mutation is the chief mechanism. After infection with HIV, the viral population rapidly diversifies, and

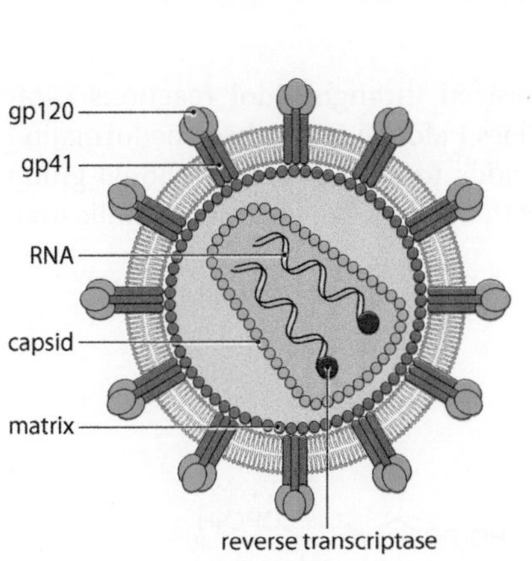

Figure 7.73 Structure of the HIV virion.

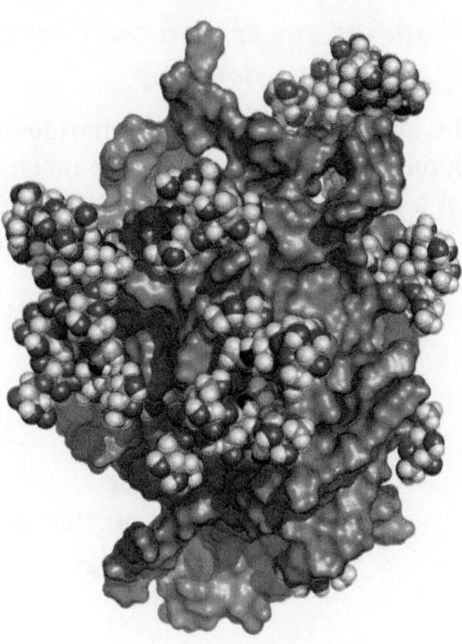

Figure 7.74 Many glycosylation sites. A crystal structure of gp120 from simian immunodeficiency virus, a monkey homolog of HIV, shows the extent of glycosylation. (PDB 2BF1)

viruses that are no longer recognized by neutralizing antibodies emerge. When these mutations were analyzed, the sequences often corresponded to changes in N-glycosylation patterns.

Introduction of N-glycosylation sites can improve protein pharmaceuticals

The protein pharmaceutical erythropoietin is a cytokine that induces the proliferation of red blood cells. More red blood cells translate into more oxygen intake, a key parameter for wound healing and aerobic sports. In the late 1990s, the pharmaceutical erythropoietin was used by athletes to gain a competitive advantage over rivals. Many organizations now test athletes for recombinant erythropoietin. Eythropoietin is a simple four-helix bundle protein with three N-glycosylation sites and one O-glycosylation site; 40% of the total mass is carbohydrate. Ironically, glycosylation of erythropoietin decreases the affinity of the protein for its receptor by a factor of 10. However, the carbohydrate increases serum half-life. The mixture of N-linked oligosaccharides on recombinant human erythropoietin expressed in CHO cells is similar to the pattern found on human interferon β1 (**Figure 7.75**); note the presence of Gal-GlcNAc disaccharide repeats, sometimes abbreviated as LacNAc because the disaccharide is Galβ1-4Glc.

Neu5Acα2→3Galβ1→4GlcNAcβ1→3Galβ1→4GlcNAcβ1→4
Neu5Acα2→3Galβ1→4GlcNAcβ1→3Galβ1→4GlcNAcβ1→2 Manβ1→6
(Fucα1→)6
Manβ1→4GlcNAcβ1→4GlcNAcα→Asn
Neu5Acα2→3Galβ1→4GlcNAcβ1→3Galβ1→4GlcNAcβ1→4
Neu5Acα2→3Galβ1→4GlcNAcβ1→3Galβ1→4GlcNAcβ1→2 Manβ1→3

Figure 7.75 Glycosylation of Epogen™. The most abundant oligosaccharide on erythropoietin expressed in CHO cells.

The O-linked oligosaccharide on Ser126 of erythropoietin is a mixture of O-glycosides: Neu5Acα2-3Galβ1-4GalNAc and Neu5Acα2-6[Neu5Acα2-3]Galβ1-4GalNAc. When the terminal Neu5Ac residues (up to 14 of them) of erythropoietin are removed to reveal terminal galactosyl residues, the modified erythropoietin is rapidly cleared. An improved version of erythropoietin, sold as Aranesp®, has been engineered to include two additional AsnXxxSer N-glycosylation sites. The additional glycosylation results in slightly weaker binding to the erythropoietin receptor, but the weaker binding is more than made up for by a threefold increase in serum half-life.

N-linked oligosaccharides can be selectively released from proteins by peptide: N-glycosidase F (PNGase F) as long as the asparagine-linked GlcNAc lacks a Fucα1-3 group. The enzyme cleaves the side-chain amide of asparagine and does not affect O-linked oligosaccharides. O-linked oligosaccharides can be selectively cleaved with an alkaline solution of sodium borohydride.

Modified sugars can carry reactive groups through the glycoprotein biosynthesis pathway

In human cells, monosaccharides are synthesized through aldol reactions catalyzed by class I aldolases. The mechanism of class I aldolases involves the formation of a nucleophilic enamine intermediate that adds to an aldehyde carbonyl group (**Figure 7.76**). Enamines are more nucleophilic than enols, but less nucleophilic than

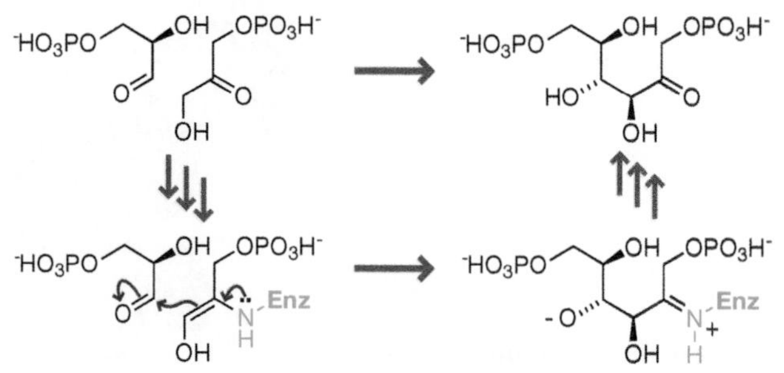

Figure 7.76 Aldolase mechanism. The biosynthesis of monosaccharides by class I aldolases involves enamines.

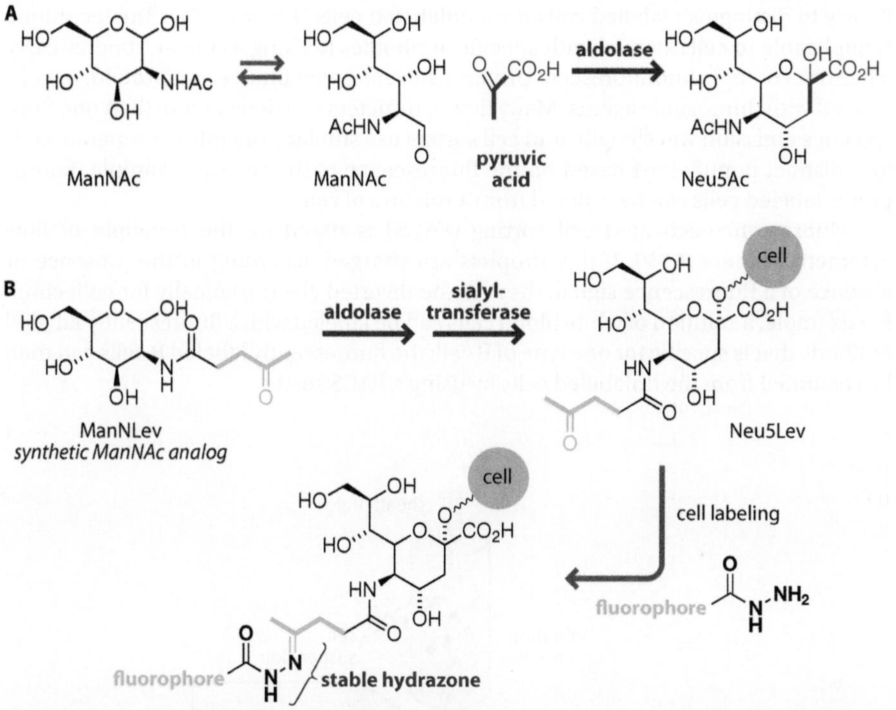

Figure 7.77 Metabolic incorporation of reactive ketone groups. (A) Biosynthesis of *N*-acetylneuraminic acid (Neu5Ac) from ManNAc. (B) Incorporation of levulinamide groups into cell-surface glycoproteins.

enolates. Fungi and bacteria also have class II aldolases that catalyze the addition of a zinc enolate to an aldehyde carbonyl group.

In human cells, *N*-acetylneuraminic acid is biosynthesized through an aldol reaction of a two-carbon fragment to *N*-acetylmannosamine (**Figure 7.77**). The enzymes that catalyze *N*-acetylneuraminic acid biosynthesis and glycosyl transfer are tolerant enough to allow additional substituents on the *N*-acetyl group. For example, when a synthetic ManNAc analog, ManNLev, is taken up by cells, it is converted first into a neuraminic acid derivative Neu5Lev by an aldolase and then into an activated CMP derivative. The CMP-Neu5Lev serves as a substrate for sialyltransferases, which incorporate the Neu5Ac analog into N-glycans, including N-glycans on membrane-bound cell-surface proteins. Free ketones are more reactive as electrophiles than esters or amides are. Like aldehydes generated through periodate cleavage of glycans, ketones can be selectively condensed with hydrazine derivatives in water to generate stable hydrazone linkages. Thus, after cells are allowed to display the levulinic derivatives of Neu5Ac on their surface, the free ketone groups can be selectively conjugated to biotin, fluorophores, or other synthetic molecules with acylhydrazide functional groups. This powerful approach developed by Bertozzi and coworkers allows the specific chemical decoration of cell-surface glycans.

Problem 7.13

Lys165 has a key role in the mechanism of sialic acid aldolase. Suggest a plausible arrow-pushing mechanism for the transformation of ManNAc to Neu5Ac by sialic acid aldolase.

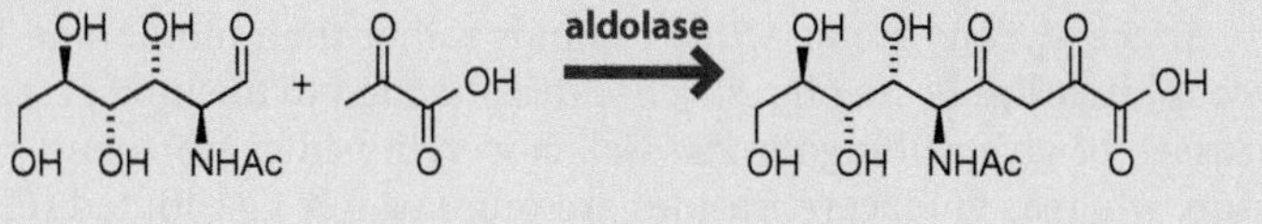

Cells displaying fluorescently labeled oligosaccharides can be identified by using **flow cytometry**. Flow cytometers use ultra-fast electronics to measure the fluorescence of cells in a hydrodynamically focused stream of medium containing cells. A typical flow cytometer can easily measure the fluorescence of tens of thousands of cells per second. When fluorescence intensity is plotted against the number of cells, it

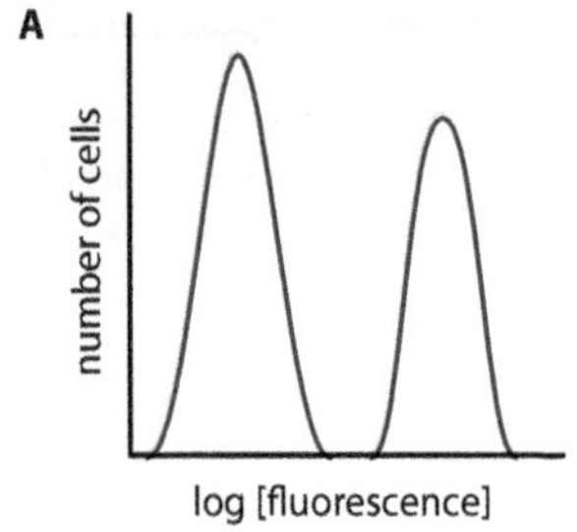

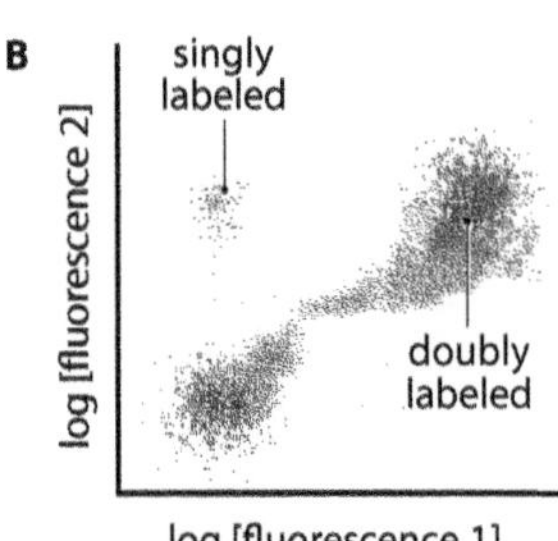

Figure 7.78 Flow cytometry. (A) Flow cytometry can distinguish between populations of cells with low fluorescence and high fluorescence. (B) Two-dimensional flow cytometry allows more than one type of fluorescent label to be read on each cell.

is easy to distinguish labeled cells from unlabeled cells (**Figure 7.78**). The technique is applicable to cells stained with specific antibodies (as long as the antibodies carry a fluorescent tag), autofluorescent proteins such as green fluorescent protein, or cells labeled with fluorogenic agents. Many flow cytometers can detect more than one fluorescence emission wavelength, and cell sorters use similar principles to separate cells into distinct populations based on the fluorescence of the cell (for example, fluorophore-labeled cells can be isolated from a mixture of cells).

Fluorescence-activated cell sorting (FACS) is based on the principle of flow cytometry (**Figure 7.79**). If tiny droplets are charged according to the presence or absence of a fluorescence signal, they can be diverted electrostatically for collection. For example, a solution of white blood cells can be labeled with a fluorescently labeled antibody that is specific for one type of B cell; the fluorescently labeled B cells can then be separated from the unlabeled cells by using a FACS sorter.

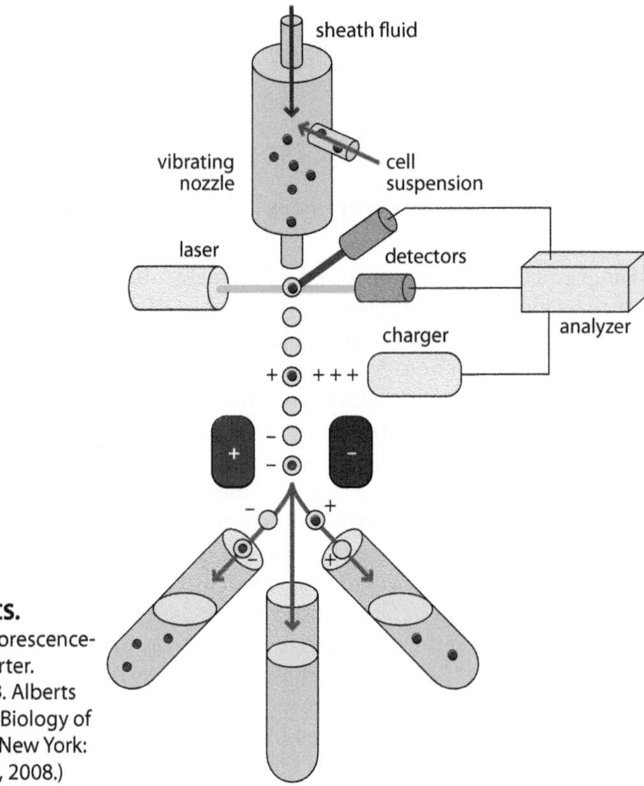

Figure 7.79 FACS. Diagram of a fluorescence-activated cell sorter. (Adapted from B. Alberts et al., Molecular Biology of the Cell, 5th ed. New York: Garland Science, 2008.)

7.5 GLYCOLIPIDS

Glycosphingolipids are lipid-like glycoconjugates

Some cell-surface glycans are attached to lipids rather than membrane-bound proteins. **Glycosphingolipids** are oligosaccharides attached to the lipid ceramide. There are two classes of glycosphingolipids. Galactocerebrosides are found primarily in neural tissue, whereas glucocerebrosides are more widely distributed (**Figure 7.80**). A deficiency in a specific glycosidase, glucocerebrosidase, leads to the accumulation of fatty glucocerebrosides in various tissues, a condition known as Gaucher's disease.

The hemagglutinin protein of the influenza virus recognizes a particular glucocerebroside called a ganglioside. Gangliosides are glucocerebrosides with the tetrasaccharide Galβ1-3GalNAcβ1-4Galβ1-4Glcβ1 bound to ceramide. Gangliosides usually have one or more Neu5Ac residues on the oligosaccharide.

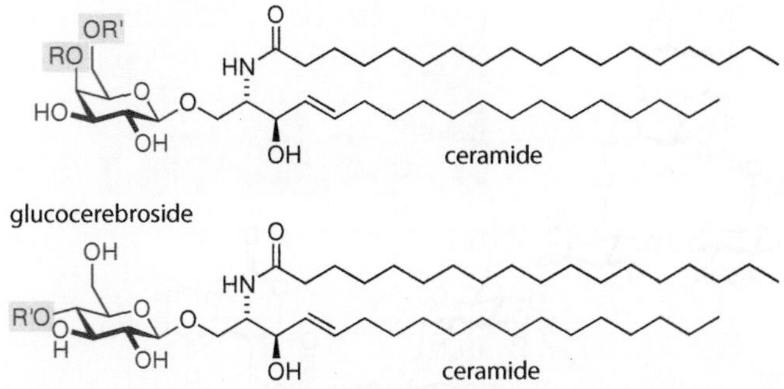

galactocerebroside

glucocerebroside

ceramide

ceramide

Figure 7.80 Cerebrosides. Cerebrosides are an important class of glycosphingolipids in brain tissue. The ceramide moieties anchor the glycans to cell membranes. In galactocerebrosides, glycan chains are attached to either the 4- or 6-hydroxyl of the galactosyl residue. In glucocerebrosides, glycan chains are attached to the 4-hydroxyl of the glucosyl residue.

Globosides are glucocerebrosides with a GalNAcβ1–3Galα1–4Galβ1–4Glcβ1 tetrasaccharide bound to ceramide. The globo H antigen is a hexasaccharide that is heavily expressed on globosides and glycoproteins on the surface of prostate cancer cells. When covalently linked to nonhuman proteins (such as keyhole limpet hemocyanin), the carbohydrate can be used to raise an immune response in humans; there is significant interest in using globo H in cancer vaccines (**Figure 7.81**).

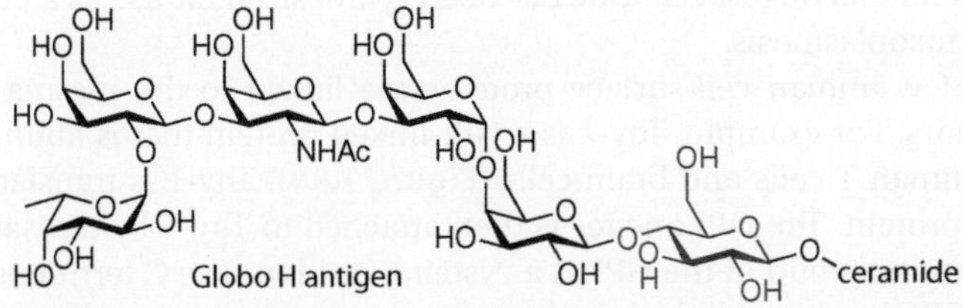

Globo H antigen

Figure 7.81 Anticancer carbohydrate vaccines. A hexasaccharide known as the globo H antigen is characteristic of human cancer cells. Vaccines based on globo H may help the human body fight cancers.

Glycosylphosphatidylinositols from pathogens are potential vaccines

The human immune system is optimized to respond to pathogens through two unique molecular signatures: their oligosaccharide coats and their polypeptide debris. Sometimes the immune system needs help in learning to recognize the molecular signatures of pathogens. Glycosylphosphatidylinositols (GPIs) serve as anchors that link proteins and carbohydrates to the cell membrane of eukaryotes, but the glycosylated lipids are more common in single-celled eukaryotes than in humans. The core of GPI is the cyclohexane derivative inositol. The inositol of GPIs is linked to a lipid moiety. The GPI core resembles the lipids that make up the cell membrane (**Figure 7.82**).

Less common on human cells, GPI anchors are often distinctive features of infectious microorganisms such as *Toxoplasma gondii, Leishmania, Trypanosoma brucei,* and *Plasmodium falciparum* (the causative agent of malaria). The pathogen *T. gondii* displays a set of oligosaccharides with a unique conserved GPI core. *T. gondii* is

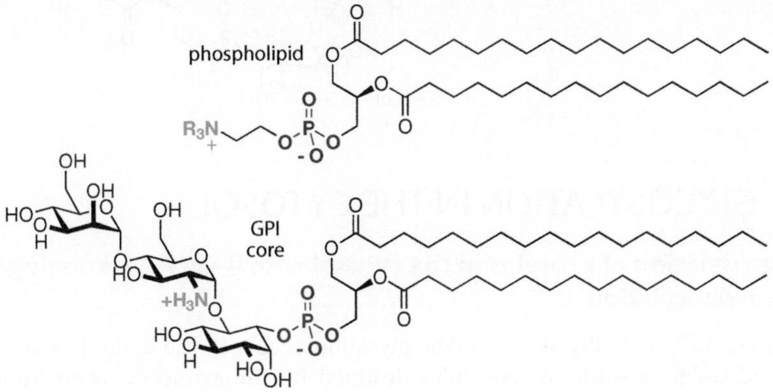

phospholipid

GPI core

Figure 7.82 Glycosylphosphatidylinositols. The zwitterionic glycan moiety of a glycosylphosphatidylinositol (GPI) resembles the polar headgroup of a phospholipid.

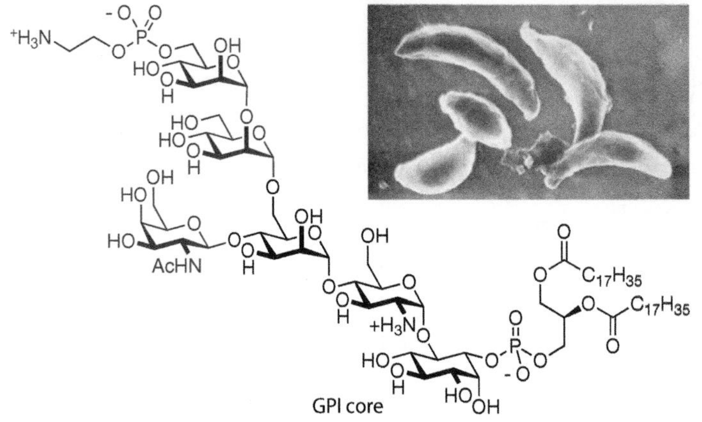

Figure 7.83 Targeting the glycans in infectious diseases. Some of the glycosylphosphatidylinositols on the surface of *Toxoplasma gondii* (inset), which causes toxoplasmosis, are unique to the pathogen. These nonhuman glycans can be used as vaccines against the disease. (Inset, courtesy of Michael J. Cuomo. With permission from Public Health Source.)

carried by cats; pregnant women and individuals with a compromised or suppressed immune system are at risk of infection (toxoplasmosis) from cat feces (**Figure 7.83**). In advanced stages of infection, *T. gondii* forms cysts inside human skeletal muscle and brain cells, which are not easily accessible to the immune system. However, the human immune system can mount an initial response toward a GPI oligosaccharide of *T. gondii*, and that oligosaccharide has been synthesized for use as a potential vaccine against toxoplasmosis.

Quite a few human cell-surface proteins are linked to the plasma membrane by GPI anchors. For example, Thy-1 is a GPI-linked protein that is abundant on the surface of human T cells and brain cells (**Figure 7.84**). Thy-1 is translated as a 161 amino acid protein. The GPI anchor is then attached to Thy-1 by transamidation of the ethanolamine group of the GPI at a cysteine residue. The C terminus is cleaved, leaving the GPI covalently linked to Thy-1. The bacterial toxin aerolysin binds to the carbohydrate portion of human GPI anchors, such as those found on Thy-1, with high affinity. As the concentration of aerolysin increases on the surface of the human cell, the aerolysin molecules form a heptameric pore in the cell membrane, killing the cell.

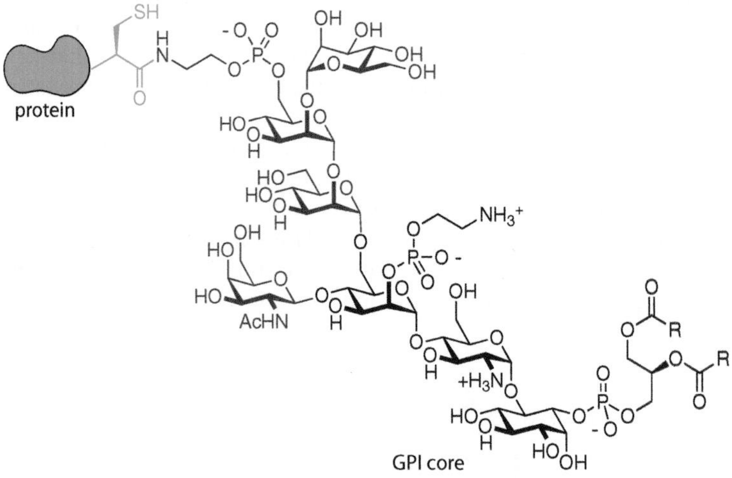

Figure 7.84 A GPI anchor. The glycoprotein Thy-1 is anchored to the membrane of T cells by a glycosylphosphatidylinositol.

7.6 GLYCOSYLATION IN THE CYTOSOL

O-glycosylation of proteins in the cytosol with β-GlcNAc is analogous to phosphorylation

In human cells, virtually all enzymatic glycation occurs in the endoplasmic reticulum and Golgi network on molecules destined for the extracellular environment.

One type of protein glycosylation occurs in the cytosol of eukaryotic cells: the addition of GlcNAc to specific serine and threonine residues. The glycosylation of serine and threonine residues by GlcNAc bears a strong functional analogy to protein phosphorylation. For example, the modification is highly specific. Indeed, many of the glycosylated residues are also sites of protein phosphorylation. Phosphorylation and glycosylation are mutually exclusive modification reactions. Thus, the actual structure of the *O*-GlcNAc residue may be less important for protein function than the fact that it prevents the addition of a phosphate group. Phosphorylation of proteins is controlled by many enzymes, but glycosylation of intracellular proteins is mediated by just one enzyme, UDP-*O*-GlcNAc transferase, using the substrate UDP-GlcNAc (**Figure 7.85**). Dephosphorylation of proteins is controlled by several phosphatases, but the hydrolytic removal of *O*-GlcNAc groups from proteins is mediated by just one enzyme: *O*-GlcNAcase.

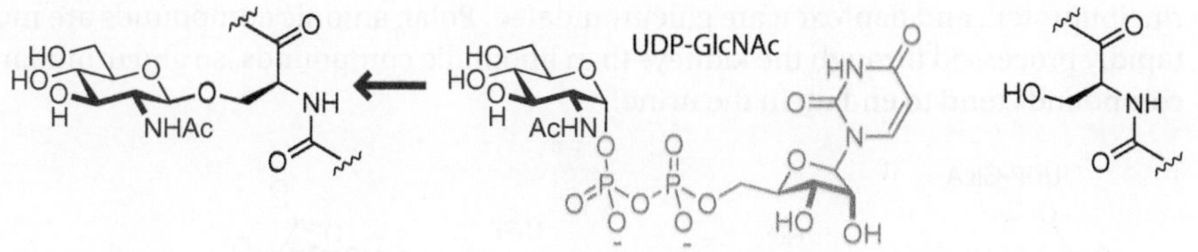

Figure 7.85 Glycation within the human cytosol. *O*-GlcNAc groups are transferred from UDP-GlcNAc to serine sidechains.

Nuclear pore proteins are extensively modified with *O*-GlcNAc residues. Other proteins known to be modified include transcription factors, polymerases, RNA-binding proteins, kinases, and cytoskeletal proteins. Glycosylation with *O*-GlcNAc tends to correlate with extracellular glucose concentrations. Thus, *O*-GlcNAc modification may be a cellwide signal to ramp up the metabolism and storage of glucose.

A fluorescence polarization assay has been developed to identify chemical inhibitors of *O*-GlcNAc transferase (OGT) that could compete with UDP-GlcNAc (**Figure 7.86**). When fluorophores are excited with light, it takes a while (10^{-7}–10^{-9} second) for them to relax to the ground state and emit a photon. When fluorescent molecules are irradiated with polarized light, they emit light with the same polarization, as long as they do not change orientation. Small molecules rapidly change orientation in solution by tumbling, but small molecules bound to macromolecules such as DNA or proteins tumble slowly. With the use of fluorescence polarization it is easy to tell when fluorophores are bound to macromolecules or free in solution. In the absence of a suitable glycosyl acceptor, UDP-GlcNAc forms a stable complex with OGT. A fluorescein derivative was synthesized and shown to bind to the transferase enzyme in the same way as unmodified UDP-GlcNAc. A fluorescence polarization assay was used to identify inhibitors that could displace the protein-bound fluoresceinylated UDP-GlcNAc derivative.

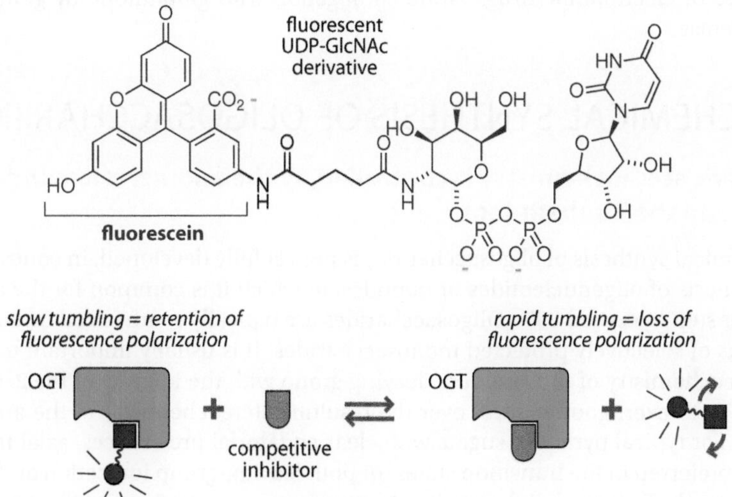

Figure 7.86 Fluorescent O-GlcNAc probe. Emission of polarized fluorescent light is greater when molecules tumble slowly. A fluorescence polarization assay was developed to screen for compounds that could prevent the binding of UDP-GlcNAc.

Drugs are targeted for export by glucuronidation

The human liver is a major processing center for the elimination of foreign molecules. Recall that the liver, for example, is rich in cytochrome P450 (CYP) enzymes that catalyze the oxidation of benzo[*a*]pyrene, aflatoxin, cyclophosphamide, and other xenobiotics (that is, foreign molecules). The liver is also rich in UDP-glucuronosyltransferases (UGTs) that transfer β-D-glucuronic acid groups to various compounds, a process that targets them for excretion. These enzymes transfer glucuronic acid to various nucleophilic functional groups, but hydroxyl groups and carboxylic acids are the most common substrates. Some endogenous alcohols, such as estradiol, testosterone, and bile acids, are good substrates. Lipophilic carboxylic acids such as retinoic acid are also glucuronidated. Glucuronidation is a major clearance mechanism for many drugs. For example, the phenolic hydroxyls of morphine and acetominophen (paracetamol) are glucuronidated (**Figure 7.87**). Similarly, the carboxylates of aspirin, ibuprofen, and naproxen are glucuronidated. Polar, anionic compounds are more rapidly processed through the kidneys than lipophilic compounds, so glucuronidated compounds tend to end up in the urine.

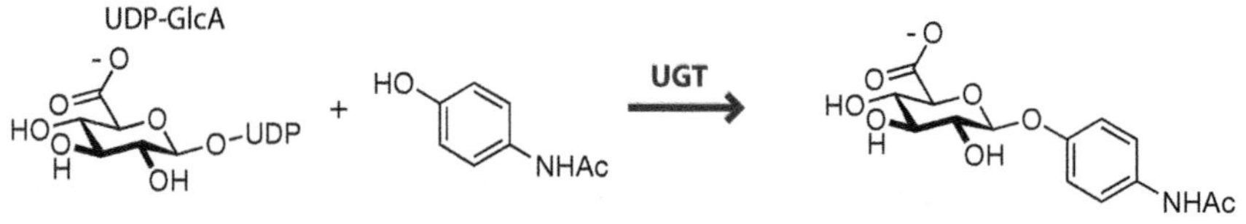

UDP-GlcA + HO–C₆H₄–NHAc —UGT→ glucuronide

Figure 7.87 Glucuronidation. Drugs with phenolic groups, like acetominophen (also known as paracetamol) are targeted for cellular export by glucuronidation.

Problem 7.14

Find the structures of ibuprofen, naproxen, and morphine and draw the structure of the glucuronide derivatives that would be generated by UDP-glucuronosyltransferase.

There are at least 13 UDP-glucuronosyltransferases in the human genome, and the selectivity of each isoform is different. For example, morphine is rapidly glucuronidated by UGT 2B7, slowly glucuronidated by UGTs 1A3 and 1A8, and not glucuronidated at all by other UGTs. Some UGTs are expressed in other tissues instead of the liver. For example, UGT 1A7 is expressed in the stomach, UGT 1A8 is expressed in the small intestine and in the colon, but neither of these UGTs are expressed in the liver.

There are related mechanisms for the clearance of unwanted compounds. For example, *O*-sulfation of steroids facilitates excretion. A related mechanism for the clearance of electrophilic drugs is the conjugation with glutathione by glutathione S-transferase.

7.7 CHEMICAL SYNTHESIS OF OLIGOSACCHARIDES

Anomeric stereochemistry is controlled by the anomeric leaving group and the 2 substituent

The chemical synthesis of oligosaccharides is not yet fully developed, in contrast with the synthesis of oligonucleotides or peptides, in which it is common for the average yield per step to exceed 98%. Oligosaccharides are typically synthesized through S_N1 reactions of selectively protected monosaccharides. It is usually important to match the stereochemistry of the anomeric leaving group with the identity of the 2′ substituent so as to exert good control over the resulting stereochemistry of the anomeric carbon. For typical pyranose sugars with clear equatorial preferences, axial trajectories are preferred in the transition states for both leaving-group ionization and nucleophilic attack. To favor α-linkages, it is best to use a glycosyl donor with a protected

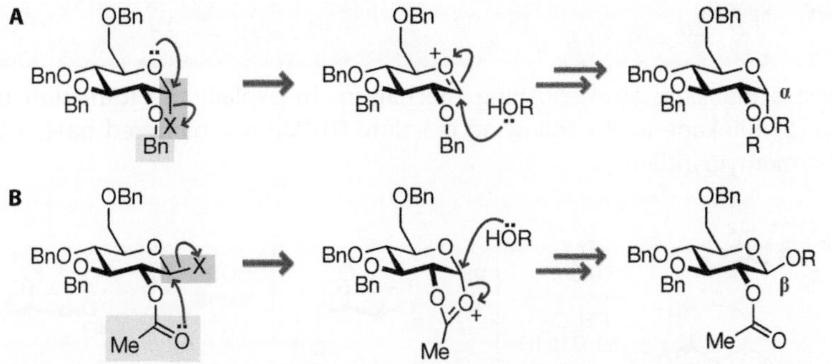

Figure 7.88 Careful choice of glycosyl donor can lead to good stereocontrol. (A) A non-participatory protecting group at the 2′ position directs the displacement to proceed through an S_N1 reaction, and favors α-linkages. (B) Neighboring-group displacement at the 2′ position and β leaving groups favors β-linkages.

2′ substituent such as benzyloxy that cannot displace the anomeric leaving group (**Figure 7.88**A). Under polar conditions that favor S_N1 over S_N2 with inversion of configuration, good selectivity of 10–20:1 α/β can be expected. To favor the β anomer, it is best to use a glycosyl donor with a 2′ substituent such as an ester or an amide that can generate a five-membered ring oxonium intermediate (see Figure 7.88B). The nucleophile then attacks, resulting in equatorial selectivity. Such glycosylations usually proceed with high selectivity (more than 20:1 β/α). Interestingly, mechanistic studies of some enzymes that hydrolyze *N*-acetylglucosamine residues have revealed the same type of neighboring-group participation.

Neighboring-group participation can also be used to synthesize α-mannosyl linkages because an axial 2′ nucleophile can push out the axial anomeric leaving group in an analogous manner to the mechanism in **Figure 7.89**. Efficient incorporation of β-mannosyl sugars into complex oligosaccharides on a solid phase represents an unmet synthetic challenge. Because the axial 2′ substituent of a mannose sugar is on top, it cannot push out an equatorial leaving group from the bottom. If the nucleophile pushes out the leaving group directly, it generates the α anomer. If the leaving group ionizes without assistance to generate an oxacarbenium ion, the nucleophile will still prefer an axial trajectory. The β-mannosyl problem seems close to resolution. Mannosyl triflates have been shown to favor β-mannosyl adducts in solid-phase glycosylations as long as the glycosyl donor is protected by a 4,6-*O*-benzylidene protecting group.

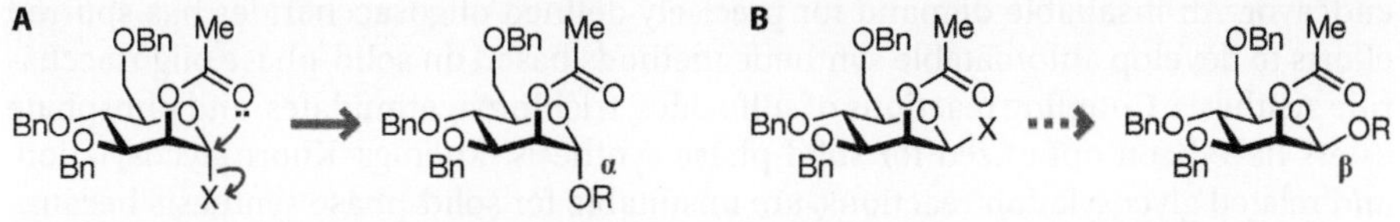

Figure 7.89 The β-mannosyl problem. (A) α-Mannosyl linkages are not a problem because neighboring-group participation leads to substitution with retention of anomeric configuration. (B) β-Mannosyl linkages, however, present a big synthetic problem.

Problem 7.15

Suggest a plausible arrow-pushing mechanism to explain the formation of the β-mannosyl linkage in the following reaction.

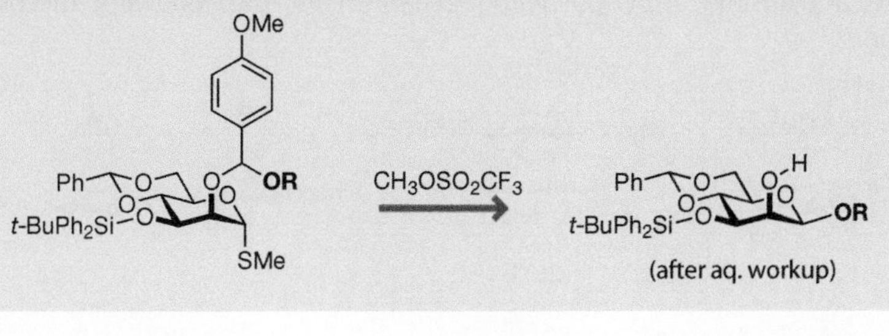

$CH_3OSO_2CF_3$

(after aq. workup)

Problem 7.16

Suggest a plausible arrow-pushing mechanism to explain the formation of the β-mannosyl linkage in the following reaction. DTBMP is a hindered base, 2,6-di-*t*-butyl-4-methylpyridine.

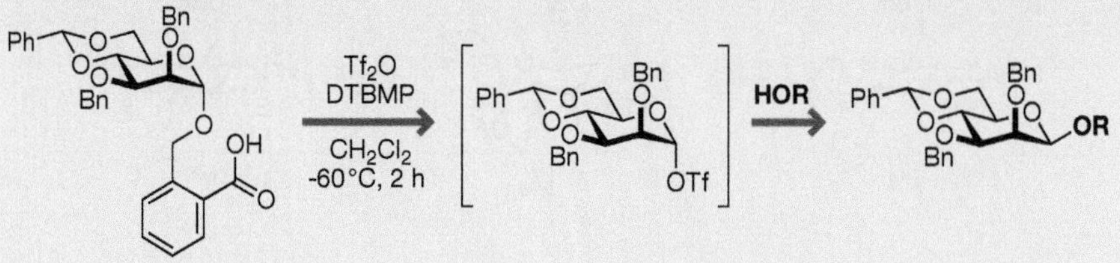

Tf$_2$O, DTBMP, CH$_2$Cl$_2$, -60 °C, 2 h; HOR

Modern oligosaccharide synthesis takes advantage of activatable leaving groups

The earliest chemical glycosylation method was developed, shortly on the heels of Emil Fischer's landmark studies of monosaccharides, by his contemporary Wilhelm Koenigs. The Koenigs–Knorr glycosylation method is based on a combination of halide leaving groups and halophilic silver ions. Pyranosyl halides are extremely reactive, even in the absence of silver ions, and are prone to degradation. A wide range of stable glycosyl donors have been developed that are stable until they are activated *in situ* (**Figure 7.90**). The most common examples of activatable anomeric functional groups include fluorides, thioethers, sulfoxides, imidates, phosphates, and *n*-pentenyloxy groups.

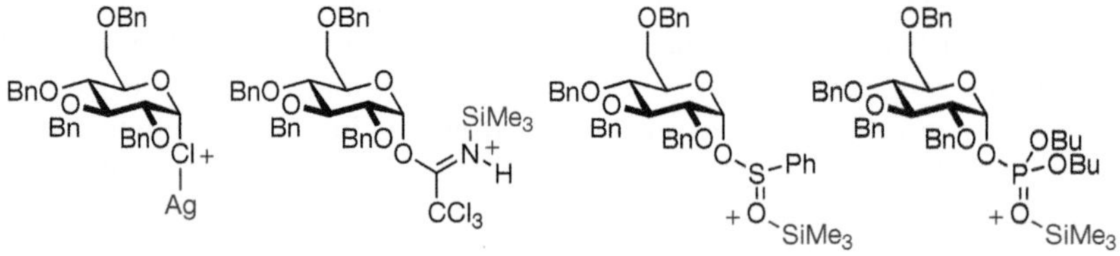

Figure 7.90 Chemical activation of glycosyl donors. In chemical glycosylation reactions, leaving groups are activated by reagents that confer a positive charge.

Until recently, glycosyl donor building blocks were only useful to skilled artisans of organic synthesis, and the synthesis of any pure oligosaccharide was a harrowing endeavor. An insatiable demand for precisely defined oligosaccharides has spurred efforts to develop automatable synthetic methods based on solid-phase oligosaccharide synthesis. Coupling reactions of sulfoxides, trichloroacetimidates, and phosphate esters have been optimized for solid-phase synthesis. Koenigs–Knorr glycosylation, and related glycosylation reactions, are unsuitable for solid-phase synthesis because they involve insoluble precipitates. Trichloroacetimidates, sulfoxides, and phosphates are usually activated by silyl trifluoromethanesulfonates (triflates), which ultimately form very stable silicon–oxygen bonds in the by-products. Protic substrates or adventitious water react aggressively with trimethylsilyl triflate to generate triflic acid. In the absence of proton-specific bases (such as 2,6-di-*t*-butylpyridine), the silyl group of silyl triflates drives the reaction thermodynamically, while the proton of triflic acid drives the reaction kinetically.

Problem 7.17

Suggest a plausible arrow-pushing mechanism for the following glycosylation reaction.

Me$_3$SiOTf, CH$_2$Cl$_2$, 23 °C

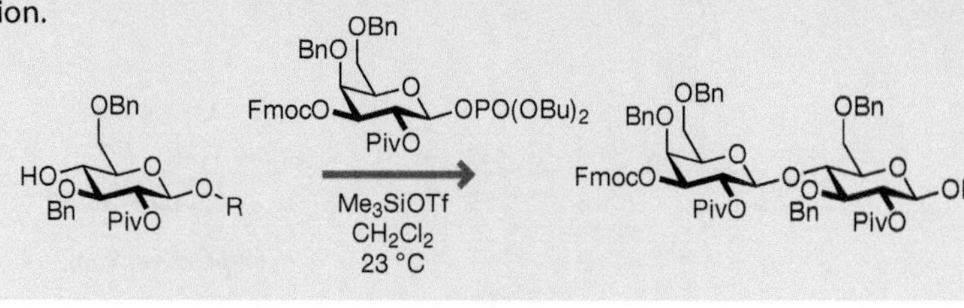

Synthesis of oligosaccharides still requires a skilled synthetic organic chemist

Monosaccharide building blocks for oligosaccharide synthesis are highly specific. For example, two different classes of galactosyl donors are needed for the addition of either Galα or Galβ residues. For each of these classes of galactosyl donors, four different protecting-group schemes are required to allow the selective deprotection of one of the four hydroxyl groups at the 2, 3, 4, or 6 position. Even more protecting-group combinations are needed for the addition of a branching galactosyl residue. The synthesis of a single selectively protected glycosyl donor requires about 10 steps from the unprotected monosaccharide. Compare this with the synthesis of a typical amino acid building block for peptide synthesis, which takes on average two steps and usually does not require chromatography. As of 2009, few selectively protected monosaccharide building blocks for oligosaccharide synthesis were commercially available. However, as the demand for custom-designed oligosaccharides increases, commercial starting materials may find their way to the market.

Benzyl protecting groups are usually removed at the end of the synthesis by hydrogenolysis with hydrogen and catalytic Pd-C. The deprotected oligosaccharide is easily removed from the catalyst by filtration, and the hydrogenolysis by-product, toluene, by rotary evaporation. Fmoc groups can be selectively removed from alkoxy substituents in the presence of hindered esters such as pivalates (*t*-BuCOO) by using piperidine.

Problem 7.18

Levulinoyl groups can be selectively removed by using hydrazine buffered with acetic acid. Suggest a plausible arrow-pushing mechanism for the deprotection, and show the structure of the by-product.

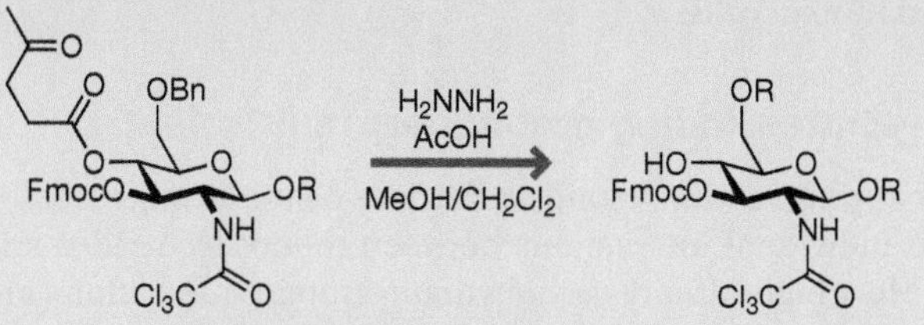

The first automated oligosaccharide synthesizer was designed to use glycosyl phosphate donors. Starting from five monosaccharide building blocks, the Lewis^y–Lewis^x nonasaccharide tumor antigen (**Figure 7.91**) was prepared in 23 hours. Each

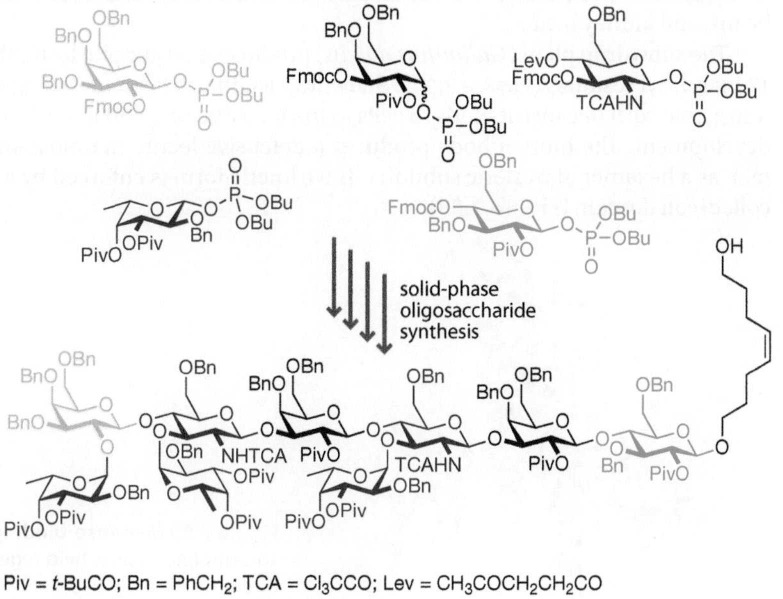

Figure 7.91 Solid-phase oligosaccharide synthesis. Automated synthesis of the branched Lewis^y–Lewis^x tumor antigen involves carefully orchestrated reactions with protected glycosyl donors. (Adapted from P.H. Seeberger and D.B. Werz, *Nature Rev. Drug Disc.* 4:751–763, 2005. With permission from Macmillan Publishers Ltd.)

coupling was performed with five equivalents of glycosyl phosphate donor and five equivalents of trimethylsilyl triflate. After washing, the coupling was repeated to effect the glycosylation of any unreacted hydroxyl acceptors. After purification by high-performance liquid chromatography, the protected oligosaccharide was isolated in 6.5% overall yield. The hydroxyolefin tail is usually not removed from the oligosaccharide product. It is a useful linker for conjugation to biomolecules, and it protects the terminal anomeric hydroxyl from side reactions of the open-chain form under basic deprotection conditions. The use of excess reagents and double couplings requires a large amount of the glycosyl donor. However, it is easier to synthesize glycosyl donors than to purify a full-length oligosaccharide from a mixture of incomplete products. For typical applications such as vaccine testing, only a small amount of the final product is necessary.

7.8 PROTEINS THAT BIND TO CARBOHYDRATE LIGANDS

Glycans differentiate the surfaces of human cells

At the molecular level, the language of glycobiology is spoken through carbohydrate–protein interactions. A wide variety of oligosaccharide structures are now known to differentiate different human cells not only within an individual but also at different stages of development of an individual and between different individuals. In some cases, the determinants are distinguished by multivalent lectin proteins and in other cases the determinants are recognized by antibodies, which are bivalent. Richard D. Cummings recently tabulated some of the prominent determinants of identity and compatibility in the human glycome. This impressive list is reproduced in **Table 7.1**, not for memorization but to impress on the reader the great diversity of oligosaccharides that adorn human cells.

Most carbohydrate-binding proteins are multivalent

To a first-order approximation, monosaccharides bear a strong resemblance to aqueous solvent, so individual interactions between monosaccharides and proteins are usually weak. Most high-affinity carbohydrate–protein interactions are mediated by multivalent interactions. Protein receptors for carbohydrates are referred to as lectins; all known lectins bind carbohydrates through multivalent interactions (**Table 7.2**). Plants are a rich source of lectins. Lectins are often toxic because they crosslink cells, leading to aggregation. Many uncooked beans contain lectins that can be used to hemagglutinate red blood cells; examples are fava beans, lima beans, lentils, string beans, and kidney beans.

The snowdrop plant (*Galanthus nivalis*) produces a tetrameric lectin that binds to 12 mannosyl residues (**Figure 7.92**). Snowdrop lectin has been investigated for use in transgenic corn because it binds to cells in the intestine of aphid nymphs and inhibits development. The human body produces a defensive lectin, mannose-binding protein, as a hexamer of trimeric subunits. The trimeric form is enforced by a three-helix coiled-coil domain (**Figure 7.93**).

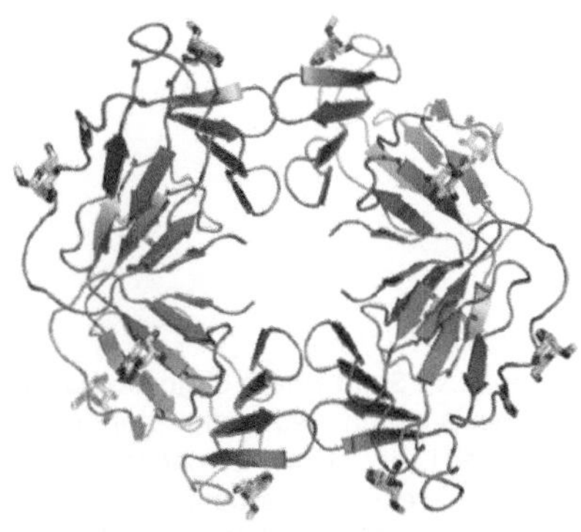

Figure 7.92 Lectins bind glycans. All lectins are polyvalent. This lectin from the snowdrop plant is bound to 12 molecules of methyl α-D-mannose. (PDB 1MSA)

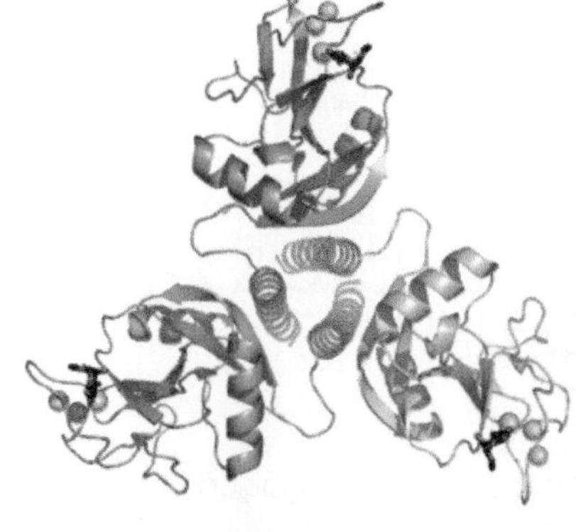

Figure 7.93 Mannose-binding protein. This trimeric lectin is held together by a three-helix bundle.

Table 7.1 Human glycan determinants in N-linked glycoproteins, O-linked glycoproteins and glycolipids.

Glycan structure	Trivial name*
Galβ1–4GlcNAcβ1–R	Type 2 LN (*N*-acetyllactosamine)
Galβ1–4GlcNAcβ1–3Galβ1–4GlcNAcβ1–R	Type 2 LN_2 ("I" antigen)
Galβ1–4GlcNAcβ1–3(Galβ1–4GlcNAcβ1–6)Galβ1–4GlcNAcβ1–R	"I" antigen
Galβ1–4GlcNAcβ1–3(Galβ1–4GlcNAcβ1–3)Galβ1–4GlcNAcβ1–R	"I" antigen
Galβ1–3GlcNAcβ1–R	Type 1 LN
Galβ1–3GlcNAcβ1–3Galβ1–3GlcNAcβ1–R	Type 1 LN_2
Galβ1–3GlcNAcβ1–3Galβ1–3GlcNAcβ1–3Galβ1–3GlcNAcβ1–R	Type 1 LN_3
GalNAcα1–Ser/Thr	Tn antigen
Galβ1–3GalNAcα1–Ser/Thr	T (also TF) antigen
Fucα1–2Galβ1–3GalNAcα1–Ser/Thr	H antigen (type 3)
GalNAcα1–3(Fucα1–2)Galβ1–3GalNAcα1–Ser/Thr	A antigen (type 3)
Neu5Acα2–6GalNAcα1–R	Sialyl Tn antigen
Neu5Acα2–3Galβ1–3GalNAcα1–Ser/Thr	Sialyl T antigen
Neu5Acα2–3Galβ1–3(Neu5Acα2–6)GalNAcα1–Ser/Thr	Di-sialyl T antigen
Galα1–3Galβ1–R	α-Gal antigen
GalNAcβ1–4GlcNAc–R	LacdiNAc (LDN)
GalNAcβ1–4(Fucα1–3)GlcNAc–R	Fucosylated LDN (LDNF)
Galβ1–4(Fucα1–3)GlcNAc–R	Lewis x (Le^x) (SSEA-1)
Fucα1–2Galβ1–4(Fucα1–3)GlcNAc–R	Lewis y (Le^y)
Galβ1–3(Fucα1–4)GlcNAc–R	Lewis a (Le^a)
Fucα1–2Galβ1–3(Fucα1–4)GlcNAc–R	Lewis b (Le^b)
Neu5Acα2–3Galβ1–4(Fucα1–3)GlcNAcβ1–R	Sialyl Lewis x (SLe^x)
Cyclic Neu5Acα2–3Galβ1–4(Fucα1–3)(Su–6)GlcNAcβ1–R†	Cyclic sialyl 6-sulfo Lewis X
Neu5Acα2–3Galβ1–3(Fucα1–4)GlcNAcβ1–R	Sialyl Lewis a (SLe^a) (CA19-9 antigen)
Neu5Acα2–3(Su–6)Galβ1–4(Fucα1–3)GlcNAcβ1–R	6′-Sulfo-sialyl Lewis x (6′-sulfo SLe^x)
Neu5Acα2–6Galβ1–4(Su–6)GlcNAcβ1–R	2,6-sialyl-Sulfo-LN (6-sialyl-6-sulfo LN)
Neu5Acα2–3Galβ1–4(Fucα1–3)(Su–6)GlcNAcβ1–R	6-Sulfo-sialyl Lewis x (6-sulfo SLe^x)
Neu5Acα2–3(Su–6)Galβ1–4(Fucα1–3)(Su–6)GlcNAcβ1–R	6,6′-bisSulfo-Lewis x (6,6′-bissulfo Le^x)
Su–3Galβ1–4(Fucα1–3)GlcNAcβ1–R	3′-Sulfo-Lewis x (3′-sulfo Le^x)
Su–3Galβ1–3(Fucα1–4)GlcNAcβ1–R	3′-Sulfo-Lewis a (3′-Sulfo Le^a)
Galβ1–4(Fucα1–3)GlcNAcβ1–3Galβ1–4(Fucα1–3)GlcNAcβ1–R	Le^x–Le^x
Neu5Acα2–3Galβ1–4(Fucα1–3)GlcNAcβ1–3Galβ1–4(Fucα1–3)GlcNAcβ1–R	$SDLe^x$
Neu5Acα2–3Galβ1–4GlcNAcβ1–3Galβ1–4(Fucα1–3)GlcNAcβ1–R	VIM-2
Neu5Acα2–3Galβ1–4GlcNAc–R	3-Sialyl-LN (type 2)
Neu5Acα2–6Gal(NAc)β1–R	6-Sialyl-GalNAc or 6-Sialyl LN (type 1 or 2)
Fucα1–2Galβ1–4GlcNAcβ1–3Galβ1–R	Blood group H (type 2)
Galα1–3(Fucα1–2)Galβ1–4GlcNAcβ1–3Galβ1–R	Blood group B (type 2)
GalNAcα1–3(Fucα1–2)Galβ1–4GlcNAcβ1–R	Blood group A (type 2)
GalNAcα1–3(Fucα1–2)Galβ1–4(Fucα1–3)GlcNAcβ1–R	Blood group A (type 2) (A-Le^y)
GalNAcα1–3(Fucα1–2)Galβ1–3GalNAcα1–3(Fucα1–2)Galβ1–4GlcNAcβ1–R	Blood group A1 (type 3)
Fucα1–2Galβ1–3GalNAcα1–3(Fucα1–2)Galβ1–4GlcNAcβ1–R	Blood group A2 (A-associated H type 3)
Fucα1–2Galβ1–3GlcNAcβ1–3Galβ1–R	Blood group H (type 1)
Galα1–3(Fucα1–2)Galβ1–3GlcNAcβ1–R	Blood group B (type 1)
GalNAcα1–3(Fucα1–2)Galβ1–3GlcNAcβ1–R	Blood group A (type 1)

Glycan structure	Trivial name*
Fucα1–2Galβ1–4(Fucα1–3)GlcNAcβ1–R	Blood group H (type 2) (Le^y)
Galα1–3(Fucα1–2)Galβ1–3(Fucα1–4)GlcNAcβ1–R	Blood group B (type 1) (B-Le^y)
GalNAcα1–3(Fucα1–2)Galβ1–3(Fucα1–4)GlcNAcβ1–R	Blood group A (type 1) (A-Le^b)
Fucα1–2Galβ1–4(Fucα1–3)GlcNAcβ1–R	Lewis b (Le^b)
Su–4GalNAcβ1–4GlcNAc–R	4′-sulfated LDN
Neu5Acα2–6GalNAcβ1–4GlcNAc–R	Sialylated LDN
Neu5Acα2–3(GalNAcβ1–4)Galβ1–4GlcNAcβ1–R	Sd^a/CT antigen
Neu5Acα2–8(Neu5Acα2–8)*n*Neu5Acα2–3Galβ1–4GlcNAcβ1–R	Polysialic acid
(Neu5Acα2–3Galβ1–4(Fucα1–3)GlcNAcβ1–6)(Neu5Acα2–3Galβ1–3)GalNAcα1–Ser/Thr	SLe^x Core 2 *O*-glycan glycan
P–6–Manα1–2Manα1–3(Manα1–3)(P–6–Manα1–6)Manα1–6Manβ1–4GlcNAcβ–R	Diphosphorylated Man_6
Neu5Acα2–3Galβ1–4GlcNAcβ1–2Manα1–Ser/Thr	O-linked mannose
Neu5Acα2–3Galβ1–4GlcNAcβ1–2(Neu5Acα2–3Galβ1–4GlcNAcβ1–6)Manα1–Ser/Thr	2,6-Branched *O*–mannose
Galβ1–4(Fucα1–3)GlcNAcβ1–2Manα1–Ser/Thr	*O*-Mannose Le^x
Su–3GlcAβ1–3Galβ1–4GlcNAcβ1–R	HNK-1 antigen
Su–3GlcAβ1–3Galβ1–4GlcNAcβ1–2Manα1–Ser/Thr	HNK-1 on *O*-mannose
Glcβ1–Cer	Glucosylceramide
Galβ1–Cer	Galactosylceramide
Su–3Galβ1–alkyl–2–acyl–s–glycerol	Seminolipid
Su–3Galβ1–Cer	Sulfatide
Galβ1–4Glcβ1–Cer	Lactosylceramide
Su–3Galβ1–4Galβ1–Cer	Ceramide dihexosyl sulfate
Su–3Galβ1–3GalNAcβ1–4Galβ1–4Glcβ1–Cer	Monosulfated gangliotetraosylceramide
GlcNAcβ1–3Galβ1–4Glcβ1–Cer	Lactotriaosylceramide (Lc_3)
Galα1–4Galβ1–4Glcβ1–Cer	P^k antigen (Gb_3, globotriaosylceramide)
Galα1–4Galβ1–4GlcNAcβ1–3Galβ1–4Glcβ1–Cer	P1 antigen
GalNAcβ1–3Galα1–4Galβ1–4Glcβ1–Cer	Globoside (P antigen) (Gb_4)
GalNAcβ1–3Galα1–3Galβ1–4Glcβ–Cer	Isoglobotetraosylceramide
Su–GalNAcβ1–3Galα1–3Galβ1–4Glcβ1–Cer	Monosulfated globopentaosylceramide
Su–3GalNAcβ1–3Galα1–4Galβ1–4Glcβ–Cer	Monosulfated globotetraosylceramide
Su–3GalNAcβ1–3Galα1–3Galβ1–4Glcβ–Cer	Sulfo-isogloboside
GalNAcβ1–4(GlcNAcβ1–3)Galβ1–4Glcβ1–Cer	LcGg4
Galβ1–3GalNAcβ1–3Galα1–4Galβ1–4Glcβ1–Cer	Gb_5
Galβ1–4GlcNAcβ1–3Galβ1–4Glcβ1–Cer	Paragloboside
Fucα1–2Galβ1–3GalNAcβ1–3Galα1–4Galβ1–4Glcβ1–Cer	GL-6 fucosylated (globoH)
Galβ1–3GalNAcβ1–3Galα1–4Galβ1–4Glcβ1–Cer	SSEA-3
Neu5Acα2–3Galβ1–3GalNAcβ1–3Galα1–4Galβ1–4Glcβ1–Cer	GL-7 globoseries ganglioside (SSEA-4)
Neu5Acα2–3GalNAcβ1–3Galβ1–4GlcNAcβ1–3Galβ1–4Glcβ1–Cer	Sialosyl paragloboside
Neu5Acα2–3Galβ1–3(Neu5Acα2–6)GalNAcβ1–3Galα1–4Galβ1–4Glcβ1–Cer	Disialosyl globopentaosylceramide
GalNAcβ1–3Galα1–3Galβ1–4Glcβ1–Cer	Cytolipin R
GalNAcα1–3GalNAcβ1–3Galα1–4Galβ1–4Glcβ1–Cer	Forssman glycolipid
GalNAcβ1–3GalNAcβ1–3Galα1–4Galβ1–4Glcβ1–Cer	*para*-Forssman glycolipid
Fucα1–2Galβ1–3GalNAcβ1–3Galα1–4Galβ1–4Glcβ1–Cer	Blood group H (type 4)
Galα1–3(Fucα1–2)Galβ1–3GalNAcβ1–3Galα1–4Galβ1–4Glcβ1–Cer	Blood group B (type 4)
GalNAcα1–3(Fucα1–2)Galβ1–3GalNAcβ1–3Galα1–4Galβ1–4Glcβ1–Cer	Blood group A (type 4)
Neu5Acα2–3Galβ1–Cer	GM4

Glycan structure	Trivial name*
Neu5Acα2–3Galβ1–4Glcβ1–Cer	GM3
Neu5Acα2–8Neu5Acα2–3Galβ1–4Glcβ1–Cer	GD3
Neu5Acα2–8Neu5Acα2–8Neu5Acα2–3Galβ1–4Glcβ1–Cer	GT3
9-O–Neu5Acα2–8Neu5Acα2–3Galβ1–4Glcβ1–Cer	9-*O*-Acetyl GD3
GalNAcβ1–4(Neu5Acα2–8Neu5Acα2–3)Galβ1–4Glcβ1–Cer	GD2
GalNAcβ1–4(Neu5Acα2–3)Galβ1–4Glcβ1–Cer	GM2
GalNAcβ1–4(Neu5Acα2–3)Galβ1–4Glcβ1–Cer	*N*-Glycolyl-GM2
GalNAcβ1–4(Neu5Acα2–3)Galβ1–4GlcNAcβ1–3Galβ1–4Glcβ1–Cer	Sialopentaosylceramide
Galβ1–3GalNAcβ1–4(Neu5Acα2–3)Galβ1–4Glcβ1–Cer	GM1
Galβ1–3GalNAcβ1–4Galβ1–4Glcβ1–Cer	Asialo-GM1
Neu5Acα2–3Galβ1–3GalNAcβ1–4Galβ1–4Glcβ1–Cer	*cis*GM1 (GM1b)
Fucα1–2Galβ1–3GalNAcβ1–4(Neu5Acα2–3)Galβ1–4Glcβ1–Cer	2-Fucosyl-GM1
Galα1–3(Fucα1–2)Galβ1–3GalNAcβ1–4(Neu5Acα2–3)Galβ1–4Glcβ1–Cer	B-GM1
Fucα1–2Galβ1–3GalNAcβ1–4(Neu5Acα2–8Neu5Acα2–3)Galβ1–4Glcβ1–Cer	2-Fucosyl-GD1b
Galα1–3(Fucα1–2)Galβ1–3GalNAcβ1–4(Neu5Acα2–8Neu5Acα2–3)Galβ1–4Glcβ1–Cer	B-GD1b
Neu5Acα2–8Neu5Acα2–3Galβ1–3GalNAcβ1–4Galβ1–4Glcβ1–Cer	GD1 (GD1c)
Neu5Acα2–3Galβ1–3(Neu5Acα2–6)GalNAcβ1–4Galβ1–4Glcβ1–Cer	GD1α
Neu5Acα2–3Galβ1–3GalNAcβ1–4(Neu5Acα2–3)Galβ1–4Glcβ1–Cer	GD1a
Galβ1–3GalNAcβ1–4(Neu5Acα2–8Neu5Acα2–3)Galβ1–4Glcβ1–Cer	GD1b
GalNAcβ1–4(Neu5Acα2–8Neu5Acα2–8Neu5Acα2–3)Galβ1–4Glcβ1–Cer	GT2
Neu5Acα2–8Neu5Acα2–3Galβ1–3GalNAcβ1–4(Neu5Acα2–3)Galβ1–4Glcβ1–Cer	GT1a
Neu5Acα2–3Galβ1–3(Neu5Acα2–6)GalNAcβ1–4(Neu5Acα2–3)Galβ1–4Glcβ1–Cer	GT1aα
Neu5Acα2–3Galβ1–3GalNAcβ1–4(Neu5Acα2–8Neu5Acα2–3)Galβ1–4Glcβ1–Cer	GT1b
Galβ1–3GalNAcβ1–4(Neu5Acα2–8Neu5Acα2–8Neu5Acα2–3)Galβ1–4Glcβ1–Cer	GT1c
Neu5Acα2–8Neu5Acα2–3Galβ1–3(Neu5Acα2–6)GalNAcβ1–4(Neu5Acα2–3)Galβ1–4Glcβ1–Cer	GQ1aα
Neu5Acα2–8Neu5Acα2–3Galβ1–3GalNAcβ1–4(Neu5Acα2–8Neu5Acα2–3)Galβ1–4Glcβ1–Cer	GQ1b
Neu5Acα2–3Galβ1–3(Neu5Acα2–6)GalNAcβ1–4(Neu5Acα2–8Neu5Acα2–3)Galβ1–4Glcβ1–Cer	GQ1bα
Neu5Acα2–3Galβ1–3GalNAcβ1–4(Neu5Acα2–8Neu5Acα2–8Neu5Acα2–3)Galβ1–4Glcβ1–Cer	GQ1c
Neu5Acα2–8Neu5Acα2–3Galβ1–3GalNAcβ1–4(Neu5Acα2–8Neu5Acα2–8Neu5Acα2–3) Galβ1–4Glcβ1–Cer	GP1c
Neu5Acα2–3Galβ1–3(Neu5Acα2–6)GalNAcβ1–4(Neu5Acα2–8Neu5Acα2–8Neu5Acα2–3) Galβ1–4Glcβ1–Cer	GP1cα
Neu5Acα2–8Neu5Acα2–3Galβ1–3(Neu5Acα2–6)GalNAcβ1–4(Neu5Acα2–8Neu5Acα2–8Neu5Acα2–3)Galβ1–4Glcβ1–Cer	GH1cα
Galβ1–4GlcNAcβ1–3Galβ1–4(Fucα1–3)GlcNAcβ1–3Galβ1–4GlcNAcβ1–3Galβ1–4GlcNAcβ1–3Galβ1–4Glcβ1–Cer	Mono-fucosyl LN_5
Galβ1–4GlcNAcβ1–3Galβ1–4(Fucα1–3)GlcNAcβ1–3Galβ1–4Glcβ1–Cer	Mono-fucosyl LN_3
(Neu5Acα2–3Galβ1–4GlcNAcβ1–6)(Neu5Acα2–3Galβ1–4GlcNAcβ1–3)Galβ1–4GlcNAcβ1–R	Disialyl-branched Type 2
Fucα1–2Galβ1–3(Fucα1–4)GlcNAcβ1–3(Fucα1–2)Galβ1–3(Fucα1–4)GlcNAcβ1–3Galβ1–4Glcβ1–Cer	Extended tetrafucosyl-Le^b
Fucα1–2Galβ1–3(Fucα1–4)GlcNAcβ1–3Galβ1–4(Fucα1–4)GlcNAcβ1–3Galβ1–4Glcβ1–Cer	Extended trifucosyl-Le^b
Galβ1–3(Fucα1–4)GlcNAcβ1–3Galβ1–3(Fucα1–4)GlcNAcβ1–3Galβ1–4Glcβ1–Cer	Dimeric Le^a
Neu5Acα2–3Galβ1–3(Fucα1–4)GlcNAcβ1–3Galβ1–4Glcβ1–Cer	Sialyl Le^a glycolipid
Neu5Acα2–6Galβ1–4GlcNAcβ1–3Galβ1–4Glcβ1–Cer	Monosialylganglioside LSTb
Su–3GlcAβ1–3Galβ1–4GlcNAcβ1–3Galβ1–4GlcNAcβ1–3Galβ1–4Glcβ1–Cer	HNK-expressing SGLPG

*Common names and symbols for different glycans are used.
†Su = sulfate.
(Data from R. Cummings, *Molecular BioSystems* 5:1087–1104, 2009.)

Table 7.2 Lectin examples.

Lectin group	Typical ligands	Examples of functions
Calnexin	Glc_1Man_9	Protein sorting in the endoplasmic reticulum
M-type lectins	Man_8	Endoplasmic reticulum-associated degradation of glycoproteins
L-type lectins	Various	Protein sorting in the endoplasmic reticulum
P-type lectins	Man-6-phosphate	Protein sorting post Golgi
C-type lectins	Various	Cell adhesion (selectins); glycoprotein clearance; innate immunity (collectins)
Galectins	β-Galactosyl	Glycan crosslinking in the extracellular matrix
I-type lectins	Neu5Ac	Cell adhesion (siglecs)
R-type lectins	Various	Enzyme targeting; glycoprotein hormone turnover

(Data from Kurt Drickamer, Imperial College London, 2012. A Genomics Resource for Animal Lectins. http://www.imperial.ac.uk/research/animallectins)

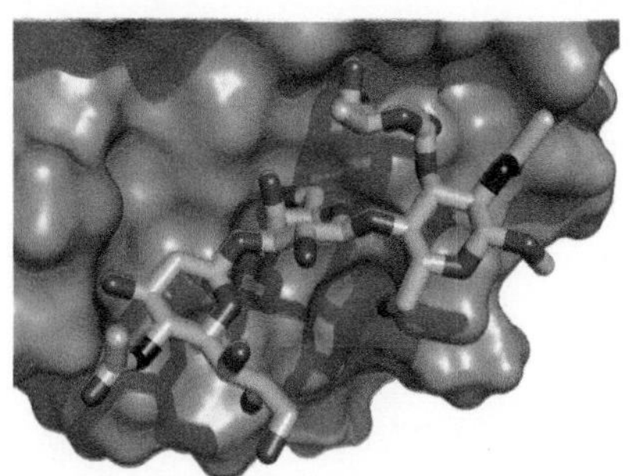

Figure 7.94 Protein–glycan recognition. Sialyl-Lewis^X binds selectively to P-selectin without being fully enveloped by the protein. (PDB 1G1R)

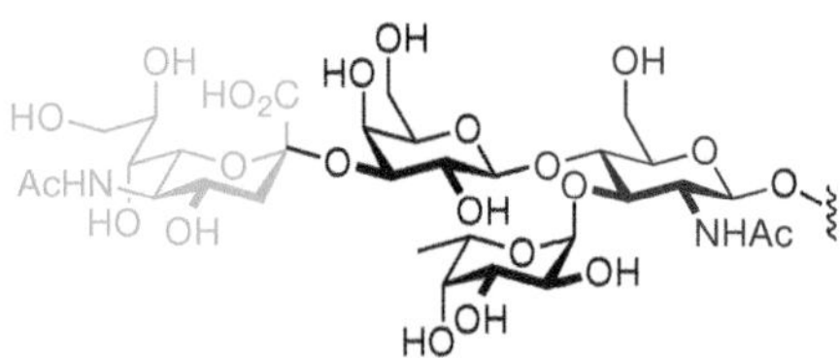

Figure 7.95 The sialyl-Lewis^X antigen. This tetrasaccharide mediates the adhesion of leukocytes to the P-selectins on vascular endothelium.

Human lectins mediate selective adhesion of leukocytes

One well-known human glycan determinant is important for the spread of cancer cells. On a cellular scale, cells in the bloodstream move at dizzying speeds, so the ability of blood cells to engage in meaningful molecular recognition while in transit is truly remarkable. The interaction of blood cells with the walls of blood vessels is mediated by a group of lectins referred to as selectins: L-selectins are found on leukocytes (white blood cells), P-selectins are found on platelets and endothelial cells, and E-selectins are found on endothelial cells. As part of the inflammatory response, cells in the vascular endothelium display P-selectin and E-selectin on their surface. Leukocytes display P-selectin glycoprotein ligand, which has a tetrasaccharide carbohydrate antigen, sialyl-Lewis^X, on its O-linked glycans (**Figure 7.94** and **Figure 7.95**). The interaction between P-selectin glycoprotein ligand and P-selectin allows leukocytes to slow down and roll along the vessel wall before finally squeezing between endothelial cells to invade the surrounding tissue, where they can provide assistance. Unfortunately, many metastatic tumors have been found to express glycoproteins with the sialyl-Lewis^X antigen. The sialyl-Lewis^X antigen allows them to invade tissues other than those in which they are normally localized.

The lymph nodes express the proteoglycan GlyCAM-1, which is a ligand for L-selectin on the surface of leukocytes. The interaction between GlyCAM-1 and L-selectin allows leukocytes to localize to the lymph nodes. The key epitope of GlyCAM-1 is believed to be a sulfated form of a sialyl-Lewis^X with a sulfate group on the 6′ position of galactose.

Human blood group antigens are found on glycolipids and glycoproteins

The most familiar antigens are those that distinguish the ABO blood groups. When persons with type A blood receive a transfusion from persons with type B blood, antibodies of the person with type A blood attack the type B blood cells. The converse is also true. When persons with type B blood receive a transfusion from persons with type A blood, antibodies of the person with type B blood attack the type A blood cells. Neither type A nor type B blood is compatible with persons with type O blood. However, persons with type A or B blood can receive transfusions from persons with type O blood. For this reason, persons with type O blood are referred to as universal donors.

The basis for these compatibilities rests in a particular class of oligosaccharide structures, known as the ABH antigens, present on the surface of erythrocytes. The ABH antigens are found on the ends of polylactosamines (e.g., $[Gal\beta1\text{-}3GlcNAc\beta1\text{-}3]_n$ or $[Gal\beta1\text{-}4GlcNAc\beta1\text{-}3]_n$) present on three types of membrane-bound molecules: N-linked glycoproteins, O-linked glycoproteins, and GPIs. The two most abundant carriers of the ABH antigens on erythrocytes are the oligosaccharides attached to

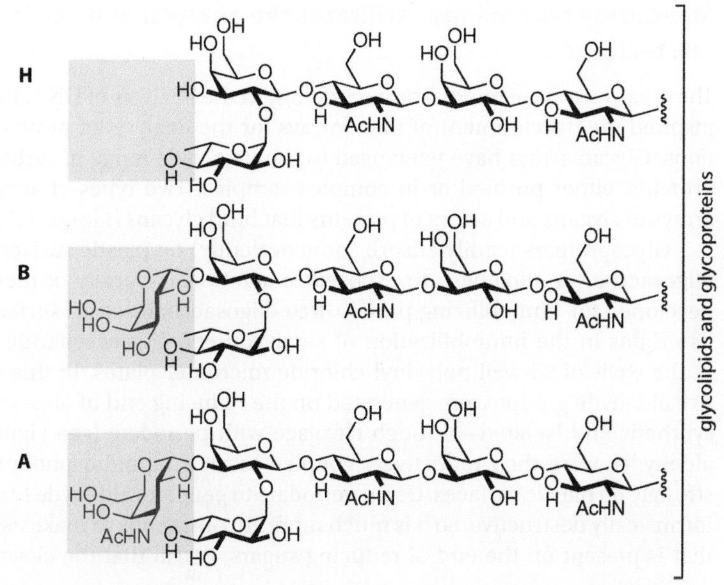

Figure 7.96 Glycans as determinants of ABO blood type. Subtle differences in the ABH antigens distinguish various blood types.

the anion transporter and the glucose transporter. The ABH antigenic determinants are only subtly different (**Figure 7.96**). The immunogenic response to these antigens rests on the presence or absence of a single sugar or acetamido group. All of the ABH antigens share a common pentasaccharide core. Persons with type O blood have the type H pentasaccharide antigen. Persons with type B blood have the type B antigen characterized by one additional Galα1–3 residue. Persons with type A blood have the type A antigen characterized by one additional GalNAcα1–3 residue.

Problem 7.19

Draw a diagram of each of the ABH antigens, depicted in Figure 7.96, using abbreviated notation.

Some toxins enter cells through multivalent carbohydrate recognition

Cholera is an intestinal disease mediated by the microorganism *Vibrio cholerae*. This organism produces a potent toxin composed of a catalytic A subunit and a pentameric B subunit that acts as a delivery vehicle (**Figure 7.97**). The pentameric B subunit acts as a lectin that binds selectively to ganglioside GM1 on the surface of intestinal cells (**Figure 7.98**). The catalytic subunit enters the intestinal cell, where the protein then begins to conjugate the ADP moiety of NAD^+ molecules to the α subunit of heterotrimeric $G\alpha_s$ proteins, which will be discussed in Chapter 9. The persistently activated $G\alpha_s$ proteins then generate cAMP without control. The massive diarrhea that results has two consequences. First, the affected individual suffers from lethal dehydration. Second, the cholera-rich effluent is highly infectious and can easily contaminate water supplies. Several other catalytic toxins attach to cellular targets through a carbohydrate-binding domain: pertussis toxin, Shiga toxin, ricin, and *E. coli* heat-labile enterotoxin A.

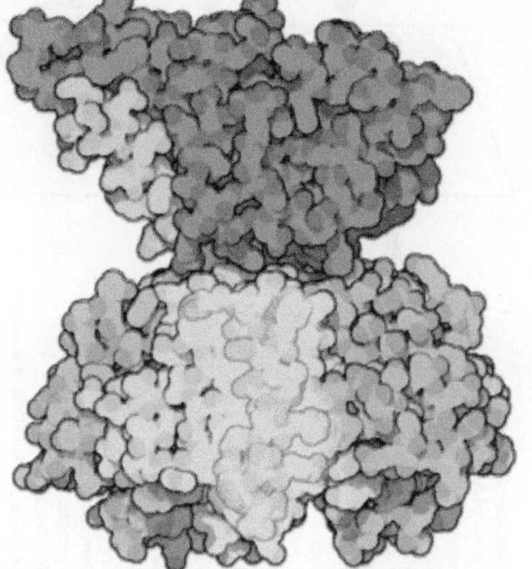

Figure 7.97 Cholera toxin. The catalytic A subunit is rendered in red and the pentameric B, which binds to the ganglioside GM1, is rendered in blue. (Courtesy of David Goodsell and RCSB Protein Data Bank.)

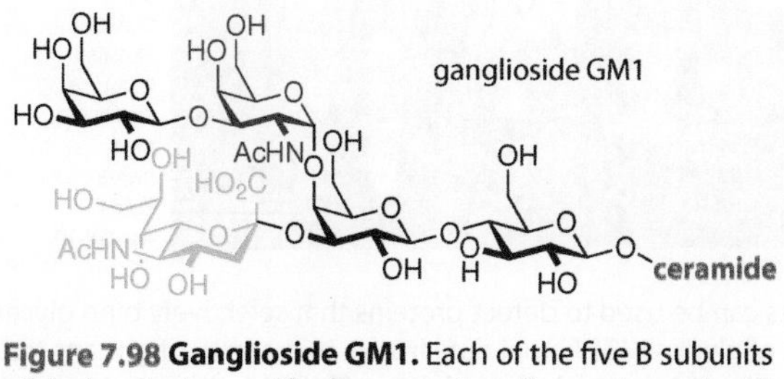

Figure 7.98 Ganglioside GM1. Each of the five B subunits of cholera toxin specifically recognizes the pentasaccharide moiety of GM1.

Microarray technology facilitates the analysis of protein–glycan interactions

The great success of microarray technology in the analysis of DNA, RNA, and proteins inspired the development of microarrays for the analysis of protein–glycan interactions. Glycan arrays have been used to profile a wide range of carbohydrate-binding proteins, either purified or in complex samples. Two types of arrays are common: arrays of glycans and arrays of proteins that bind glycans (**Figure 7.99**).

Glycoproteins readily adsorb (noncovalently) on plastic surfaces such that their oligosaccharide moieties are exposed to solution. A variety of methods have been developed for immobilizing protein-free oligosaccharides on surfaces. The idea has its origins in the immobilization of semisynthetic oligosaccharide lipid conjugates in the wells of 96-well polyvinyl chloride microtiter plates. In this early work, reactive aldehyde groups were generated on the reducing end of oligosaccharides—both synthetic and isolated—through cleavage with periodate (see Figure 7.71), and the aldehydes were then reductively coupled to an aminophospholipid that adsorbed strongly on plastic surfaces. Using periodate to generate aldehyde functional groups is intrinsically destructive, so it is much more advantageous to make use of the aldehyde that is present on the end of reducing sugars. Recall that the closed form of sugars

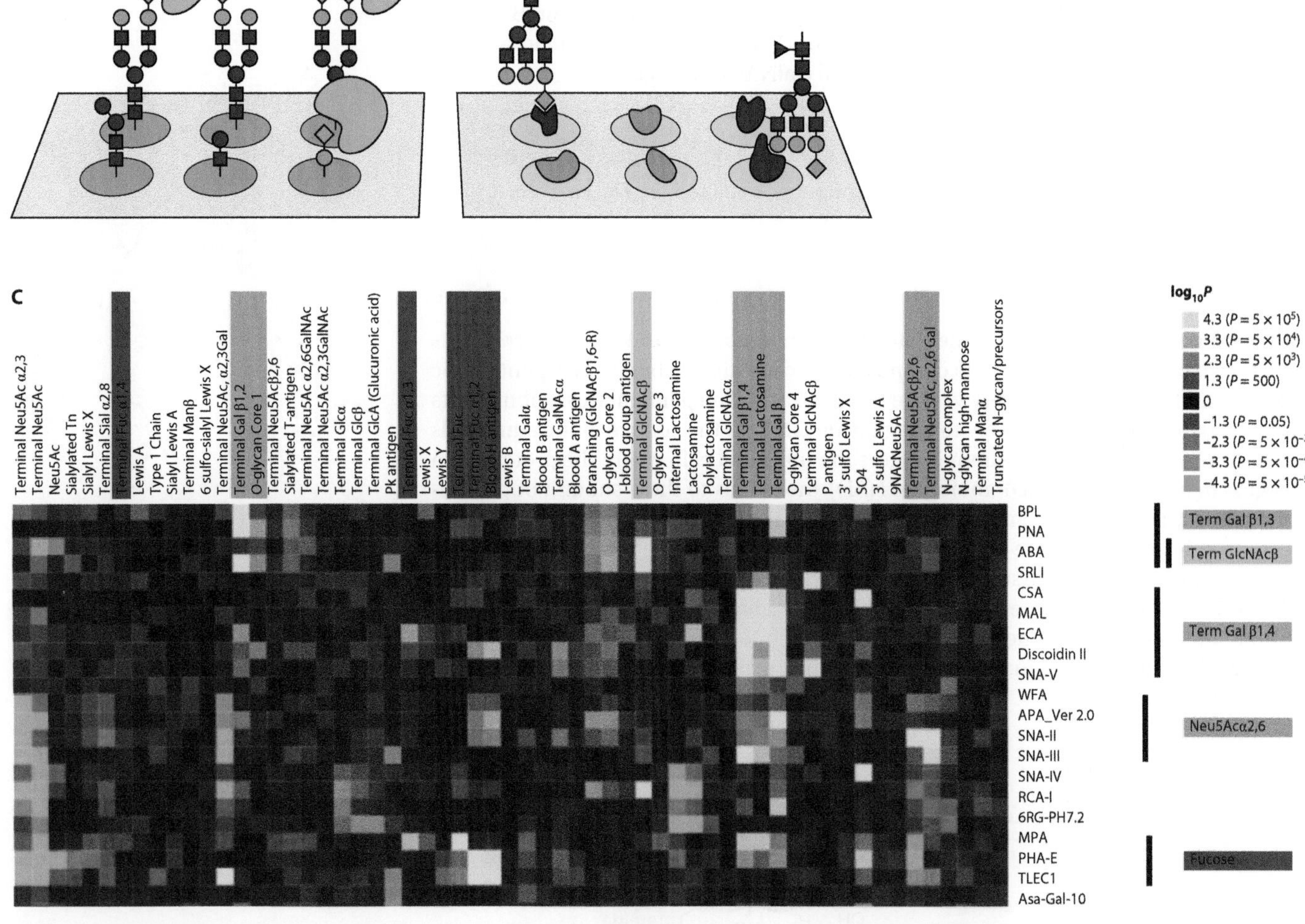

Figure 7.99 Glycan microarray. (A) Microarrays of oligosaccharides can be used to detect proteins that selectively bind glycans. (B) Microarrays of lectins can be used to identify structural features of glycans. (C) Actual data from a microarray of glycans treated with various fluorescently labeled lectins. Yellow correlates with high-affinity binding; blue correlates with low-affinity binding. (C, from A. Porter et al., *Glycobiology* 20:369–380, 2010. With permission from Oxford University Press.)

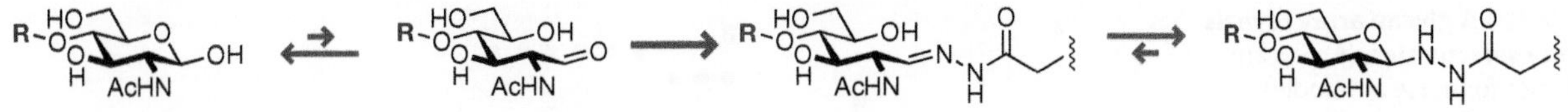

predominates over the open-chain aldehyde form; nevertheless, alkoxyamines and hydrazides react so aggressively with aldehydes that they can be used to form N-glycosides of reducing sugars, even in water. Such reactions have been used to attach complex oligosaccharides covalently to spot surfaces on microarrays (**Figure 7.100**). Glycans can also be immobilized by using surfaces derivatized with reactive epoxides or *N*-hydroxysuccinimido esters, but the linking reaction is not selective among the many hydroxyl groups that exist on typical glycans. The ability to synthesize complex oligosaccharides, with or without linkers for immobilization, has been a major advance.

Figure 7.100 Glycan capture. The reducing end of sugars reacts readily with acylhydrazides.

The attachment of polyfluorinated alkyl chains to glycans facilitates interactions that are neither hydrophilic nor hydrophobic but involve interactions between fluorocarbon functional groups. In fact, when fluorocarbons are mixed with oil and water, the fluorocarbons form a third phase that is distinct from the aqueous and organic layers (**Figure 7.101**). The nonstick polymer Teflon® is the most common example of a perfluoroalkane that repels aqueous and organic compounds. Compounds with polyfluorinated alkyl chains (for example -$(CF_2)_7CF_3$), often referred to as fluorous tags, can be extracted from organic solvents into fluorocarbon solvents or adsorbed on fluorous surfaces when making spot arrays. Various types of fluorinated media are commercially available for fluorous phase applications: solvents, supports for chromatography, and fluorocarbon-coated glass slides for microarray applications. Intermediates in the synthesis of complex oligosaccharides can be separated from reaction by-products through solid-phase fluorous extraction when one of the monosaccharide units (usually the first) bears a perfluoroalkane tag. Fluorous-tagged synthesis of biooligomers offers the advantages of solution phase and solid phase. Reactions can be conducted in solution phase, where yields are high, and separated from by-products on a solid phase, facilitating rinse-based purification.

Figure 7.101 Separation of liquid phases. Each phase contains a different dye that is soluble only in that phase. (Courtesy of Fluorous Technologies.)

To gain a better appreciation for the power of microarrays it is useful to consider an example in which a glycan microarray was used to study how mammalian immune systems target the glycans of the SARS (severe acute respiratory syndrome) coronavirus. A glycan array was made by spotting, in triplicate, 51 structurally characterized glycoproteins and glycolipids onto nitrocellulose-coated glass slides; proteins and lipids adhere strongly to nitrocellulose. Horse antibodies against SARS coronavirus were covalently modified with fluorescein isothiocyanate (FITC), which reacts randomly with lysine amino groups (**Figure 7.102**); only a small number of the lysine side chains are modified on the antibodies. As a control experiment, antibodies were also isolated from a horse immunized with a bacterial glycan and fluorescently labeled with FITC. When the array was treated with the anti-SARS antibodies, the antibodies bound to an asialo-orosomucoid glycan with a weaker binding to lacto-*N*-tetraose, Galβ(1,3) GlcNAcβ1-3Galβ1-4Glc. The antibodies against the bacterial glycan did not bind to

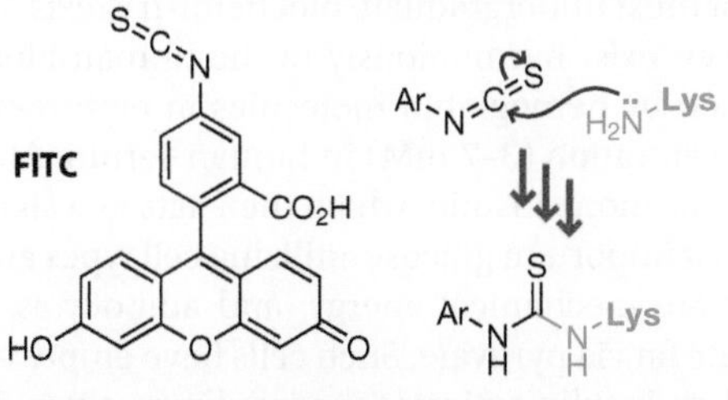

Figure 7.102 Lysine side chains react with the isothiocyanate of FITC. Isocyanates, R–N=C=O, hydrolyze too quickly for use in protein labeling.

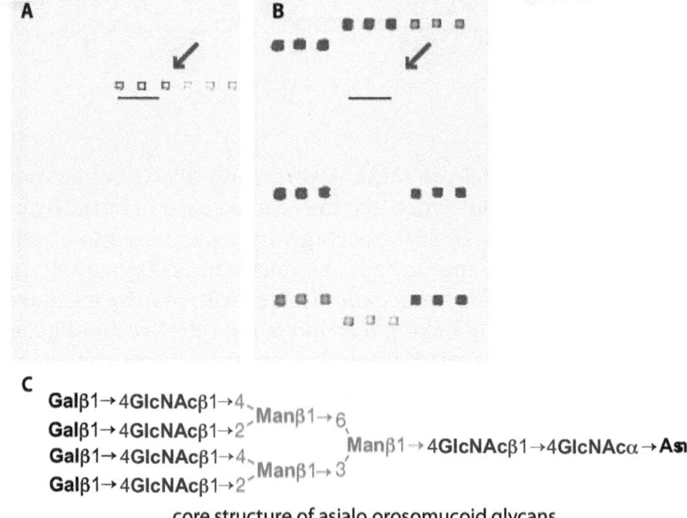

Figure 7.103 A glycan array reveals antibody selectivities. (A) Glycan selectivities for horse antibodies against the SARS coronavirus. (B) Glycan selectivities for horse antibodies raised against a bacterial glycan from *Streptococcus pneumoniae* serotype 18. (C) The anti-SARS antibodies seem to target the Galβ1–4GlcNAc branches of the orosomucoid glycans. (A, from D. Wang and J. Lu, *Physiol. Genomics* 18:245–248, 2004. With permission from the American Physiological Society.)

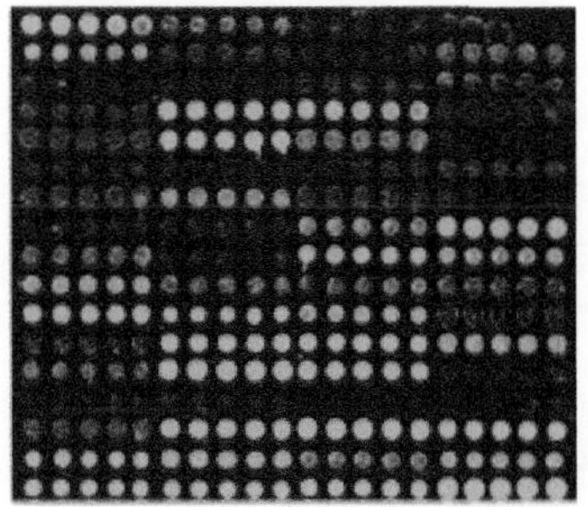

Figure 7.104 A lectin microarray reveals glycan epitopes. This array contains about 70 discrete carbohydrate-binding proteins with known selectivity. After treatment with fluorescently labeled HIV, one can determine which glycan epitopes are present on the virus. (From L. Krishnamoorthy et al., *Nat. Chem. Biol.* 5:244–250, 2009. With permission from Macmillan Publishers Ltd.)

the orosomucoid glycan (**Figure 7.103**). Orosomucoid is a serum glycoprotein, about 1-3% of total plasma protein, that is produced mainly by hepatocytes; it binds nonselectively to cationic and/or lipophilic drugs. Further characterization of the anti-SARS antibodies with glycan arrays (not shown) suggested that the antibodies recognize the Galβ1-4GlcNAc branches of the asialo-orosomucoid glycans, implying that related glycans are present on the coat of the SARS virus.

Lectin microarrays are used to reveal the carbohydrate composition of a biological sample—either pure oligosaccharides or complex mixtures of oligosaccharides isolated from, or displayed on, a cell membrane (**Figure 7.104**). A wide range of lectins (and antibodies) with known carbohydrate binding selectivity are commercially available, readily expressed, or readily isolated from plant sources. These lectins can be readily spotted onto glass slides coated with reactive aldehyde or epoxide groups. Lectins usually recognize only a small part, often the outer branches, of large oligosaccharides; however, these are the same parts that any other biological entity, like a cell or virus, would see.

7.9 GLUCOSE HOMEOSTASIS AND DIABETES

Human metabolism and paper-burning are related transformations

Aerobic organisms combine oxygen with glucose to produce carbon dioxide and water. The transformation seems nothing short of miraculous until we realize that a burning piece of paper achieves the same effect with even greater brilliance. The human biochemical process of glucose-burning, referred to as respiration, is miraculous because it harnesses enthalpy and subdues entropy. One result of this process, the production of ATP, is the subject of most undergraduate biochemistry texts.

Glucose and oxygen exist harmoniously in the human bloodstream. Oxygen is sequestered and carried by hemoglobin molecules in erythrocytes. Glucose is kept at a relatively low concentration (3–7 mM) in human serum. After meals, pancreatic beta cells produce the hormone insulin, which then acts as a signal for cells to take in glucose. Two of the most important glucose-utilizing cell types are muscle cells, which burn glucose to generate mechanical energy, and adipocytes, in which glucose is metabolized to generate fat via pyruvate. Such cells have empty vesicles that fuse with the cell membrane when insulin activates the insulin receptor. These empty vesicles have GLUT4 glucose transporters directed into the cytosol so that, when fused with the cell membrane, the transporters are directed outside where glucose is plentiful. In this way, insulin signaling can rapidly notify cells to take in glucose (**Figure 7.105**).

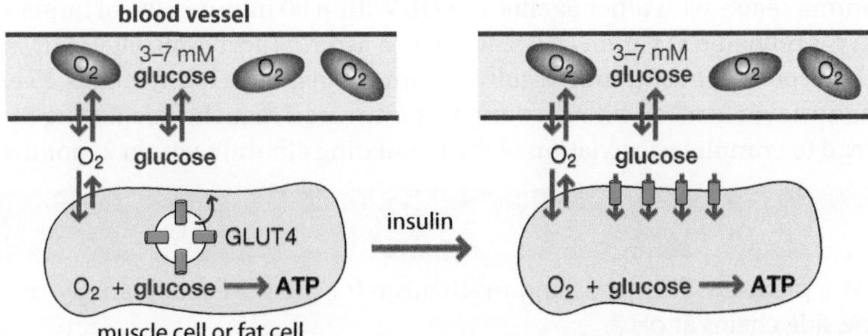

Figure 7.105 **Glucose transporters.** An important aspect of glucose homeostasis involves the import of glucose into cells. Glucose import is increased when vesicles lined with glucose transporters fuse with the cell membrane.

Patients with type I diabetes do not produce functional insulin, and must take insulin supplements. Patients with type II diabetes produce functional insulin, but have other problems in the glucose homeostasis pathway. The protein hormone incretin induces the glucose-dependent release of insulin from the pancreatic cells. Interestingly, exenatide, a 39 amino acid peptide toxin from the venom of the Gila monster, was shown to mimic the effect of incretin (**Figure 7.106**). It has been approved in the United States for the treatment of type II diabetes.

Any mutation that interferes with glucose uptake is likely to lead to diabetes. For example, insufficient insulin, insulin mutations, and insulin receptor mutations can all cause diabetes with dangerous consequences. Under hypoglycemic conditions (not enough glucose) cells cannot get the glucose required to produce energy. Glucose concentrations rise drastically after meals, and high levels of glucose are passed out in the urine. The most prevalent condition is referred to as diabetes mellitus, a Greek term referring to the flow of honey.

Figure 7.106 Venom of the Gila monster is a source for the drug exenatide. (Courtesy of Digital West Media, Inc., 2012. Desert USA, http://www.desertusa.com)

Glucose reacts with proteins over time

The chronic effects of diabetes are attributable to hyperglycemia (too much glucose in the bloodstream). Glucose is dangerously reactive. The open-chain form of glucose is an aldehyde, which can condense with amines to form highly reactive iminium ions. This nonenzymatic reaction of proteins with glucose is referred to as glycation and should be distinguished from the enzymatic process of glycosylation. Every serum protein that has been studied has been shown to be glycated by glucose, usually on lysine side chains. For example, more than 11% of human serum albumin, which makes up 67% of the serum protein, is glycated (**Figure 7.107**). About half of the glycation of human serum albumin occurs on one particular lysine side chain, Lys525 (**Figure 7.108**). The high concentration of serum albumin in the blood (about 40 g l^{-1}) makes it important in the pharmacokinetics of many drugs. The protein readily binds neutral or anionic compounds, and most drugs that remain in human plasma for extended periods do so by binding to albumin. The corresponding albumin from cow's blood, bovine serum albumin (BSA), is one of the least expensive soluble proteins available to chemical biologists. It is often used as a generic protein platform for the conjugation of biomolecules and for blocking nonspecific hydrophobic interactions through the addition of high concentrations of BSA to experimental conditions.

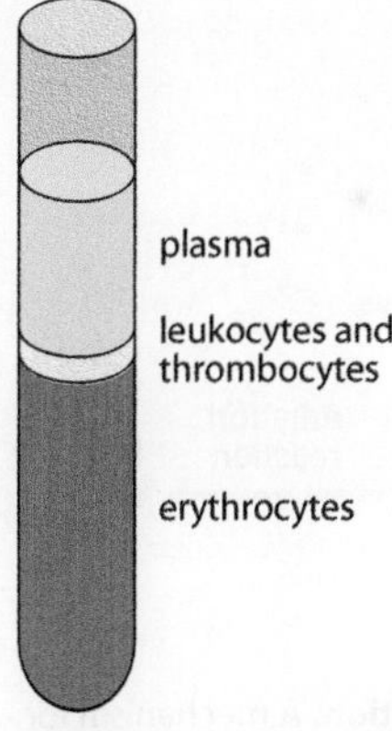

Figure 7.107 Blood components. When whole blood is spun down in a centrifuge the plasma (yellow) is separated from the white and red blood cells. Serum is plasma that has been depleted of fibrinogen and clotting factors.

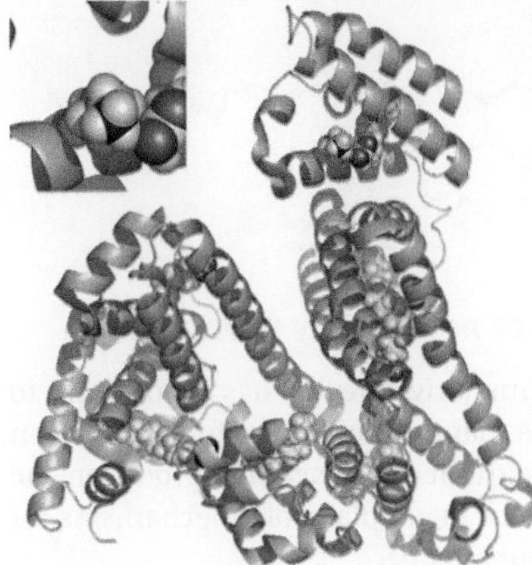

Figure 7.108 A protein sponge. This crystal structure of human serum albumin is bound to five molecules of the fatty acid myristate (yellow). Lys525 of human serum albumin (inset) is readily modified by glucose.

Albumin reacts with other agents as well. Within 90 minutes, 2% of human serum albumin is acetylated at various sites by 100 μM aspirin (acetylsalicylic acid), which is less than the peak concentration resulting from two regular aspirin tablets. Even when albumin turnover is taken into account, it is estimated that chronic dosing of aspirin would lead to complete acetylation of the circulating albumin within 2 months.

Problem 7.20

Suggest a plausible arrow-pushing mechanism for the acid-catalyzed glycosylation of lysine side chains at pH 7.

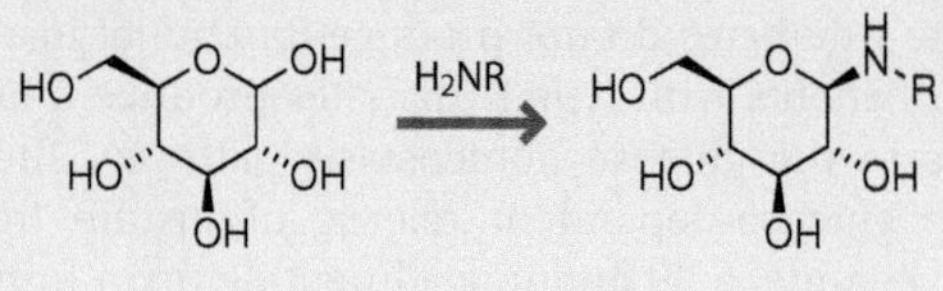

H_2NR

Nonenzymatic glycation has two detrimental effects: first, the modification creates immunogenic structures on proteins capable of raising a deleterious immune response; and second, glycation can generate undesirable protein crosslinks. Covalent crosslinking reduces the rate of protein degradation, which is a necessary part of the ongoing process of renewal that prevents the buildup of damaged proteins. The most common protein–protein crosslinks involve lysine (**Figure 7.109**). Retro-aldol reactions lead to smaller fragments of glucose (glyoxal and methylglyoxal) that can also crosslink proteins. Nonenzymatic glycation ultimately produces many chemical structures, referred to as advanced glycation end-products. A variety of cells in the human body express a receptor (RAGE) that binds to advanced glycation end-products. A variety of ligands have been identified for RAGE. Carboxymethyllysine ($LysNHCH_2CO_2$-) is a prominent end-product of glycation that activates RAGE. RAGE activation initiates an inflammatory response and has been implicated in the pathogenesis of diabetes, Alzheimer's disease, and aging.

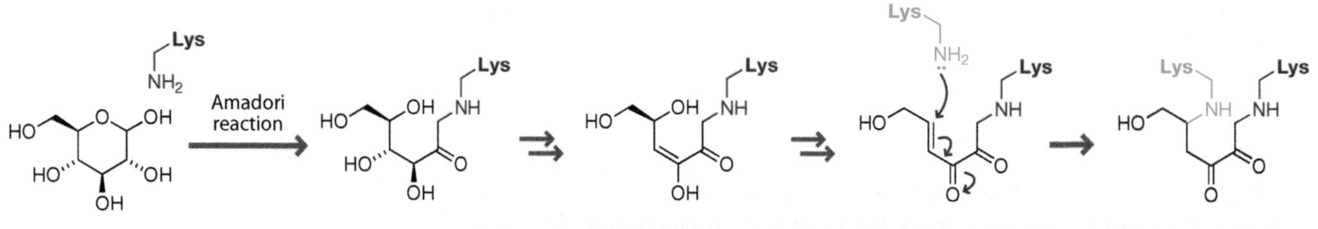

Amadori reaction

Figure 7.109 Glycation. A mechanism for protein crosslinking by glucose.

Problem 7.21

Glucose can form a fructosyl adduct with lysine side chains through the Amadori reaction. Suggest a plausible arrow-pushing mechanism for the reaction at pH 7.

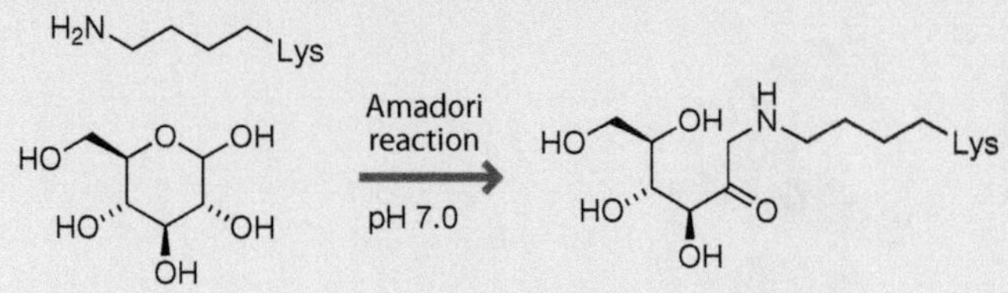

Amadori reaction

pH 7.0

Glucose-derived protein crosslinks are not necessarily permanent

The investigational drug alagebrium is a thiazolium derivative that has been shown to catalytically cleave 1,2-dicarbonyls such as those that are sometimes found between proteins. Unlike thiamine, alagebrium has an extra acidic functional group on the nitrogen that can form a second C–C bond. One of many plausible mechanisms for the cleavage of protein crosslinks is shown in **Figure 7.110**.

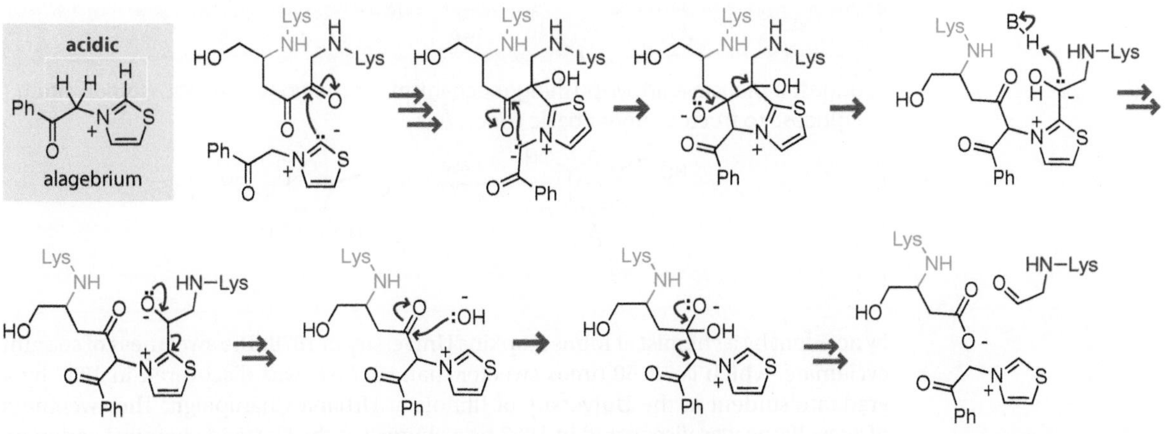

Figure 7.110 One of many possible mechanisms for 1,2-dicarbonyl cleavage by alagebrium.

Mature erythrocytes have no nucleus and do not actively produce proteins, but they last a long time in circulation, with a lifespan of between 3 and 5 months. Because the proteins inside erythrocytes are not turned over, those proteins are subject to glycation, just as proteins in the serum are. One of the most common tests for diabetes analyzes the fraction of a glycated form of hemoglobin, referred to as hemoglobin A1c (HbA1c).

There is a big market for artificial ligands for human taste receptors

In part as a result of the reactivity of glucose described here, the demand for sweet foods in the American diet is a contributory factor in obesity and diabetes. Much of the dietary sugar is consumed as the sweetener sucrose, or table sugar. Most sucrose comes from sugar cane, but sugarbeet now accounts for almost one-third of the world's table sugar production. Sucrose is a 1,1′-disaccharide derived from glucose and the isomeric ketosugar fructose (**Figure 7.111**). Mixtures of glucose and fructose are available in grocery stores under the name corn syrup (**Figure 7.112**). The 1,1′-α,α′-disaccharide of glucose called trehalose is the blood sugar found in insects, much like glucose in humans.

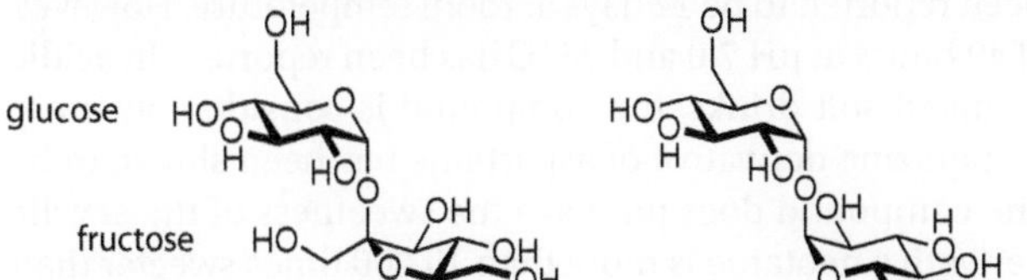

Figure 7.111 Two common 1,1′-disaccharides.

Fructose is slightly sweeter than sucrose. Thanks to farm subsidies and the efficiency of the US agricultural–industrial complex, high-fructose corn syrup has become a much cheaper alternative to sucrose. High-fructose corn syrup has become ubiquitous in packaged foods, particularly soda. The human GLUT5 transporter is selective for fructose, which is also taken in by the nonselective GLUT2 transporter. However, the effects of fructose and glucose on human metabolism are essentially the same.

In humans, sweet-tasting molecules are sensed by three different 7TM G protein-coupled receptors, T1R1, T1R2, and T1R3, which act in pairs. Most of the sweet-tasting compounds are sensed by cells that co-express T1R2 and T1R3 receptors. These G protein-coupled receptors are believed to act as a heterodimeric complex; its structure has not yet been fully defined.

The earliest artificial sweeteners to be discovered were the sulfamates, the oldest of which is saccharin. The anionic salts have good solubility and dissolve readily in cold beverages. The sweetness of saccharin, 300 times that of sucrose, was discovered

Figure 7.112 Cooking with glucose. This 1909 advertisement for Karo corn syrup, a glucose-rich sweetener made from corn starch, describes some of the culinary applications.

Problem 7.22

Suggest a plausible arrow-pushing mechanism for the nonenzymatic isomerization of glucose to fructose under basic conditions.

glucose ⇌ (base) fructose

by accident by a chemist at Johns Hopkins University in 1878. The sweetness of sodium cyclamate, which is 30–50 times sweeter than sucrose, was discovered in 1937 by a graduate student at the University of Illinois at Urbana-Champaign. The sweetness of acesulfame was discovered in 1967 by a chemist at the German chemical company Hoechst AG. Acesulfame potassium is up to 200 times sweeter than sugar. There is widespread public concern over the safety of these sweeteners as a result of some early animal studies. There is no worldwide consensus on the safety of these sulfamates in humans. For example, saccharin is banned in Canada, whereas cyclamate is banned in the United States. The pharmaceutical company Abbott Laboratories has petitioned the US Food and Drug Administration to approve cyclamate (**Figure 7.113**).

calcium saccharin (Ca^{2+}) | sodium cyclamate (Na^+) | acesulfame potassium (K^+)

Figure 7.113 Sulfamate sweeteners.

The dipeptide sweetener aspartame was discovered at G.D. Searle & Company in 1965. With a sweetness 180 times that of sucrose, aspartame looks almost like the natural fragment of a protein except that it has a C-terminal methyl ester (**Figure 7.114**). Under basic conditions, the N terminus rapidly cyclizes onto the ester to form a diketopiperazine. Diketopiperazine formation can complicate the solid-phase synthesis of peptides linked through C-terminal ester linkages. At pH 7, the half-life for this cyclization has been reported to be 12 days at room temperature. However, in other studies a half-life of 49 hours at pH 7.0 and 25 °C has been reported. In acidic medium, such as cold carbonated soft drinks, the compound is considerably more stable. Although the diketopiperazine derivative of aspartame has been shown to be safe in numerous studies, the compound does not have the sweetness of the acyclic dipeptide. The aspartame derivative neotame is more than 10,000 times sweeter than sucrose, and only reluctantly forms a diketopiperazine. Several naturally occurring proteins are known to have an intensely sweet taste: thaumatin, monellin, mabinlin, pentadin, brazzein, and curculin (**Figure 7.115**). Monellin was discovered in 1972 in

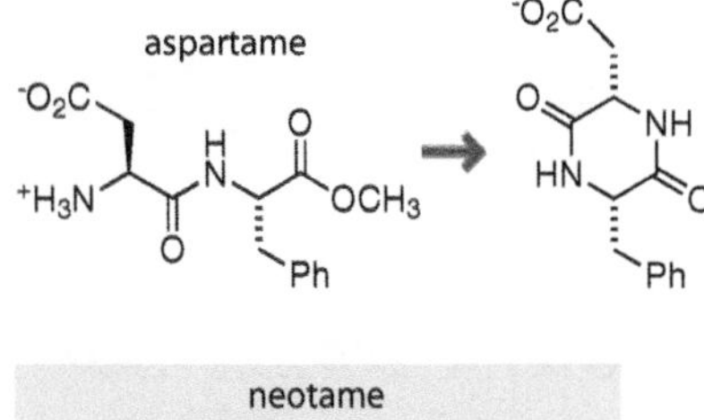

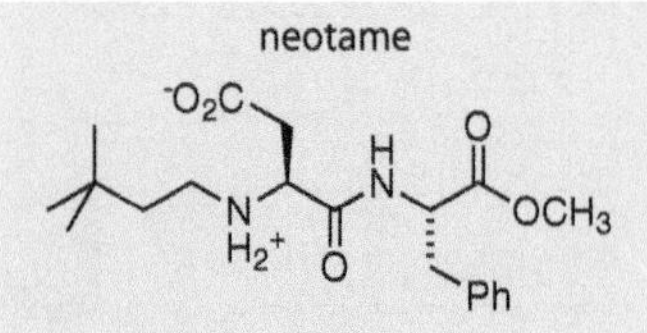

Figure 7.114 Aspartame cyclization. The dipeptide aspartame spontaneously cyclizes to form a diketopiperazine. The more recently developed analog neotame resists cyclization.

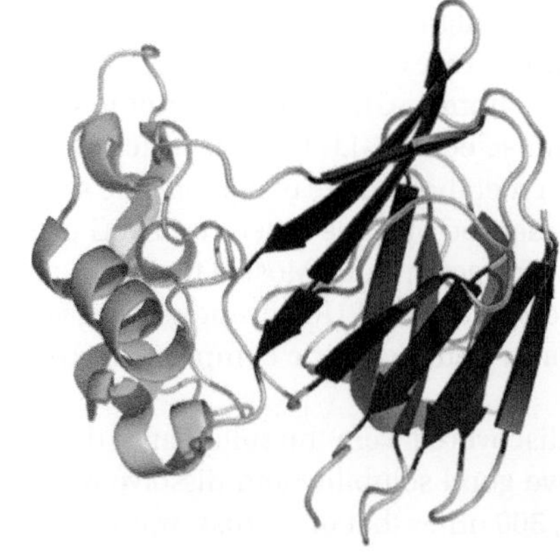

Figure 7.115 Thaumatin. This protein from the fruit of the African katemfe plant is sold as the sweetener Talin®. (PDB 1RQW)

the fruit of the west African shrub called the serendipity berry, also known as miracle fruit (**Figure 7.116**). The glycoprotein miraculin from the miracle fruit does not itself taste sweet. Instead, the name reflects the compound's remarkable ability to cause sour foods to taste sweet long after exposure to miraculin, a result of strong binding to taste receptors. Thaumatin is marketed under the trade name Talin®, but proteins have limited application as low-calorie sweeteners because of their tendency to denature at elevated temperatures.

Figure 7.116 Miracle fruit. A glycoprotein in the miracle fruit changes taste perception for hours after consumption. (Courtesy of Hamale Lyman.)

Sucralose, a chlorodeoxysucrose derivative, is 600 times sweeter than sucrose and is more stable than aspartame. Many chlorinated compounds are toxic to the liver because the compounds are metabolized into DNA-alkylating agents. However, extensive studies of sucralose have revealed no long-term exposure risks. The compound is poorly absorbed in the gastrointestinal tract, and because of its hydrophilicity it is rapidly cleared.

Sucralose is more economical to synthesize than aspartame because the compound can be made directly from sucrose (**Figure 7.117**). An unusual chlorodehydration of sucrose is performed with sulfuryl chloride (SO_2Cl_2), which is not to be confused with thionyl chloride ($SOCl_2$). Sulfuryl chloride resembles Cl_2 in its reactivity and is rarely used for substitution of hydroxyl groups. Sulfonate esters of sucrose have differential reactivity: 6-$OSO_2R \approx$ 6′-OSO_2R >> 4-OSO_2R > 1′-OSO_2R. Thus, practical routes for the synthesis of sucralose always involve a selective protection of the 6-hydroxyl group of sucrose so that the 6′-OH is not sulfonated and converted to a chloride.

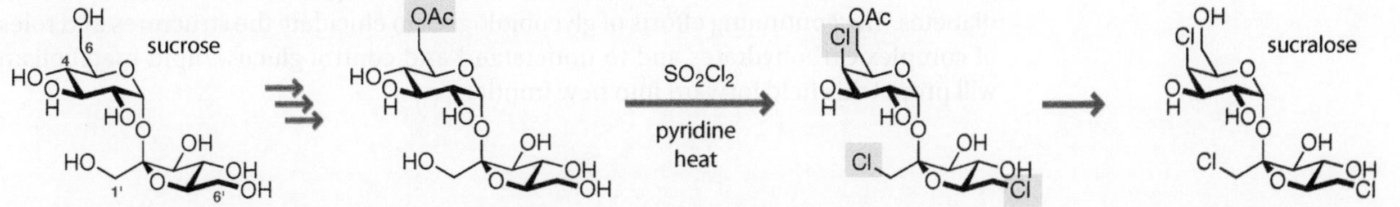

sucrose
SO_2Cl_2
pyridine
heat
sucralose

Figure 7.117 Sucralose. Sucralose is a chlorinated compound that is chemically synthesized from sucrose.

The South American plant known as sweetleaf (*Stevia rebaudiana* Bertoni) produces a group of terpene glycosides with an intensely sweet taste. Extracts of this plant are sold as a low-calorie sweetener. The most abundant of these sweet-tasting terpenes, stevioside, makes up about 5–10% of the steviol glycosides; the sweetest derivative, rebaudioside A, is up to 300 times sweeter than the equivalent weight of sucrose (**Figure 7.118**).

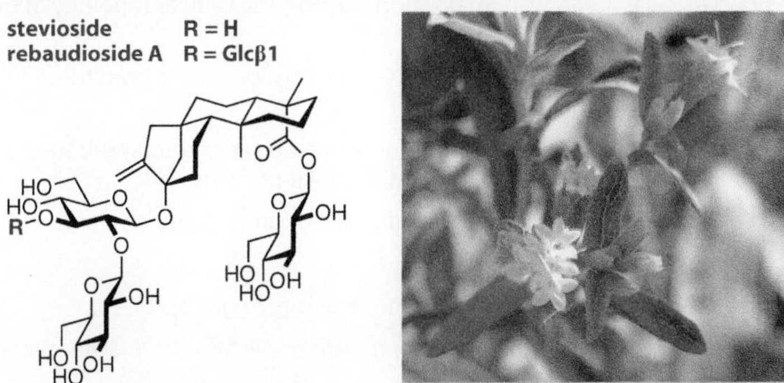

Figure 7.118 Sweet terpenes. Steviol glycosides (left) account for the intensely sweet properties of the sweetleaf plant (right). (Right, courtesy of Ethel Aardvark.)

7.10 SUMMARY

Since the 1930s, most students of biochemistry have been indoctrinated with a myopic view of glycobiology, focusing on the role of glucose as an energy source and as a molecular building block. However, by the end of the past millennium, the confluence of biological problems and emerging techniques has galvanized the field of glycobiology toward large, complex oligosaccharides.

Human oligosaccharides are constructed from about half as many building blocks as proteins, yet oligosaccharides are structurally more complex than proteins. The structural complexity of each oligosaccharide is easily traced to the potential for branching, and is further complicated by the fact that the formation of glycosidic bonds creates a new stereogenic center. These challenges are further exacerbated by the fact that mammalian cells are designed to produce mixtures of oligosaccharides. The oligosaccharides are assembled by enzymes with varying levels of selectivity, and there is no proofreading system to ensure the fidelity of glycosylation reactions.

Human cells have evolved to use glucose as a carbon source and as the molecular currency of energy in the bloodstream. They wear a coat of glycoproteins and glycolipids that mediate cell–cell interactions. Cells speak with glycans and listen with lectins. Many pathogenic organisms recognize their human cellular targets through protein–oligosaccharide interactions. Chemists can exploit the differences between human oligosaccharides and those of pathogenic organisms to design vaccines and therapeutic drugs. Many of the proteins that circulate in the bloodstream are glycosylated, and the oligosaccharide structure is important; for example, the half-life of proteins in the serum can be altered by glycosylation.

The confectionery of evolution has produced some sweet treats indeed, and our view of carbohydrate chemistry and biology has also evolved. In a little more than 100 years, the glycobiologist's perspective has progressed from a malt-ball view of human cells, with sugar on the inside and fat on the outside, to a gumdrop model, with protein on the inside and sugar on the outside. Unfortunately, the importance of glucose metabolism is on the rise, driven by an alarming increase in the incidence of diabetes. The continuing efforts of glycobiologists to elucidate the structures and roles of complex carbohydrates and to understand and control glucose/lipid metabolism will propel this field forward into new frontiers.

LEARNING OUTCOMES

- Draw the structures of the 10 monosaccharide subunits found in human oligosaccharides, including anomeric stereochemistry and conformation.
- Draw mechanisms for inverting and retaining glycosylhydrolases.
- Distinguish polysaccharides from oligosaccharides.
- Explain how vesicular compartmentalization controls the cellular topology of protein glycosylation processes.
- Distinguish the synthesis and structure of N-linked glycoproteins from those of O-linked glycoproteins.
- Contrast the presence of oligosaccharides on extracellular/membrane-bound proteins with the absence of oligosaccharides on intracellular proteins.
- Describe the functional/physiological significance of protein glycosylation.
- Draw the lipid anchors of glycolipids.
- Contrast the structure of glycolipids with the structure of glycoproteins.
- Explain why vaccines are an important therapeutic application of oligosaccharides.
- Draw the mechanisms for the chemical formation of glycosidic bonds.
- Understand the importance of multivalency in lectin–carbohydrate interactions.
- Describe some of the physiological functions of human lectins.
- Explain how insulin affects glucose homeostasis.
- Draw mechanisms for nonenzymatic protein glycation.
- Draw the structures of common artificial sweeteners.

PROBLEMS

***7.23** Suggest a plausible base-catalyzed arrow-pushing mechanism for the formation of 3-deoxyglucosone from glucose.

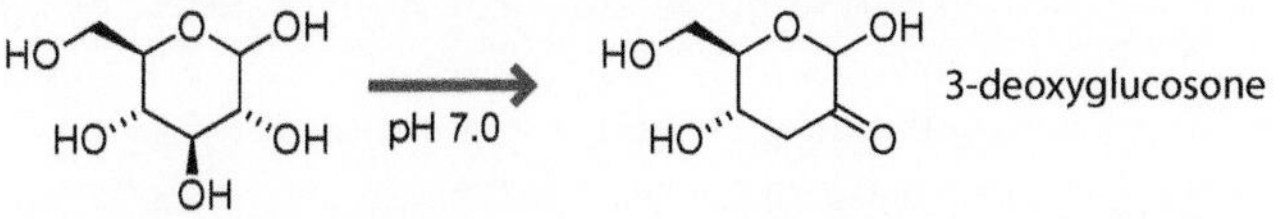

pH 7.0 → 3-deoxyglucosone

***7.24** Methyl α-L-fucoside selectively forms a single bicyclic bisketal because it minimizes 1,3-diaxial interactions and maximizes anomeric effects. Draw the structure of the preferred ketal, using chair conformations.

3-MP; esterases

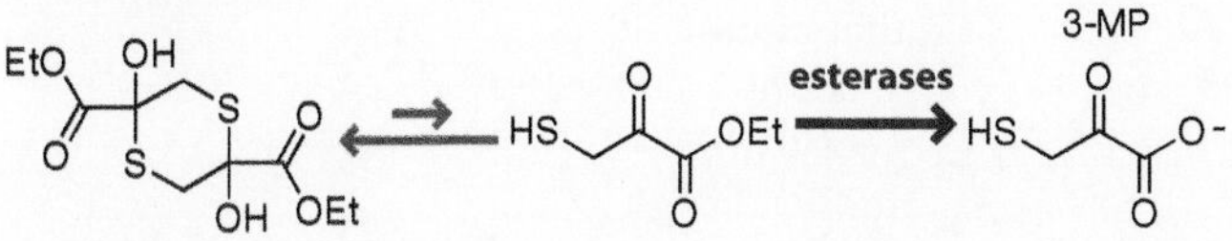

7.25 3-Deoxyglucosones are highly reactive and can form crosslinks between lysine side chains. A developmental anti-diabetic drug called pimagedine (aminoguanidine) reacts with 3-deoxyglucosones to form two different stable aromatic products. Suggest a plausible structure for either aromatic product and a mechanism for its formation.

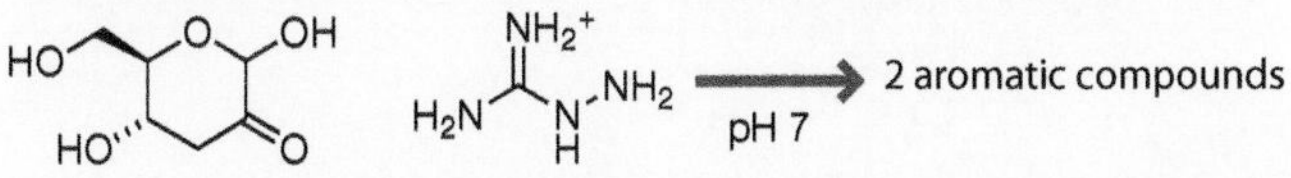

pH 7 → 2 aromatic compounds

***7.26** Draw the structure of the following glucosphingolipid with the Lewis[a] antigen, using chair conformations.

Fucα1→4\
Galβ1→3 **GlcNAcβ1→3 Galβ1→4 Glcβ1→Cer**

7.27 One strategy for the synthesis of asparagine-linked oligosaccharides involves the use of the Ritter reaction. Suggest a plausible arrow-pushing mechanism for the reaction. Do not worry about the identity of the leaving group.

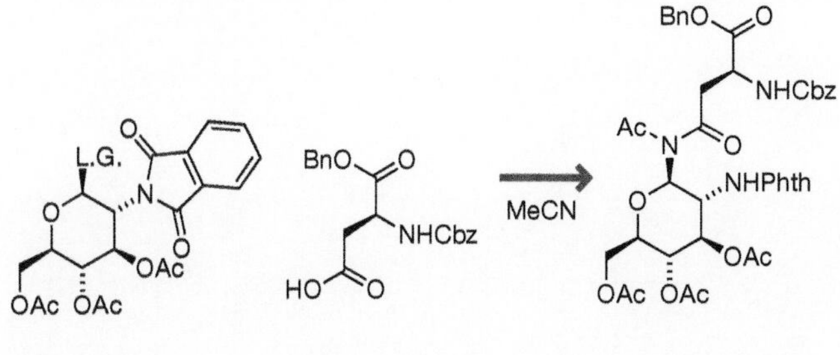

MeCN

7.28 The human pancreas produces insulin—a hormone that controls serum glucose levels. Streptozotocin is a glucose derivative isolated from the soil microorganism *Streptomyces achromogenes* that exhibits selective cytotoxicity toward pancreatic beta cells. It has been approved for the treatment of pancreatic cancer (Zanosar®) and is even used to induce diabetes in rats. Streptozotocin exerts selective cytotoxicity against pancreatic beta cells through a mechanism that includes DNA alkylation. Suggest a plausible arrow-pushing mechanism (base-catalyzed) for the methylation of guanine bases in DNA by streptozotocin.

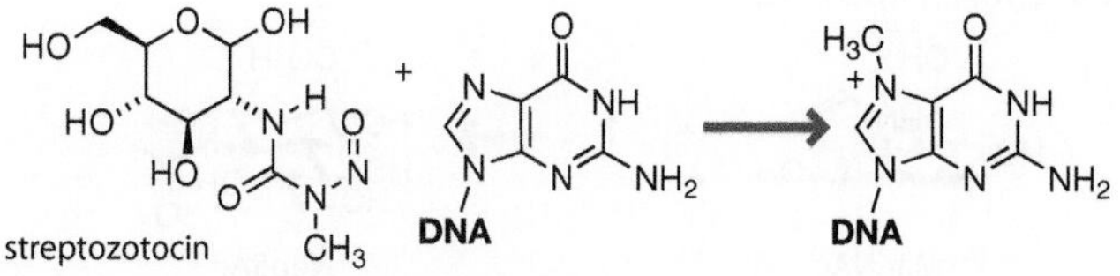

streptozotocin + DNA → DNA

***7.29** O-linked oligosaccharides can be selectively released from glycopeptides (derived from digestion with the peptidase pronase) under mild basic conditions in the presence of excess sodium borohydride without affecting N-linked oligosaccharides. Suggest a plausible arrow-pushing mechanism.

0.05 M NaOH, 1 M $NaBH_4$, H_2O, 45 °C, 10 h

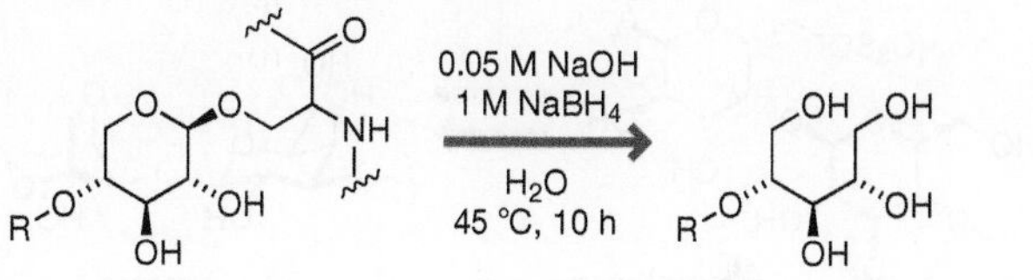

7.30 The enzyme chitinase is selectively inhibited by the natural product inhibitor allosamidin. On the basis of the analogy between allosamidin and a key mechanistic intermediate, suggest a plausible arrow-pushing mechanism for chitinase.

allosamidin

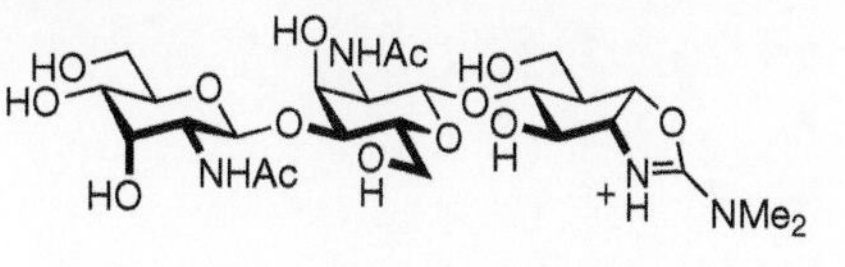

7.31 The natural product acarbose is an inhibitor of human pancreatic α-amylase, which is a retaining glucosidase. The mechanism of inhibition involves the hydrolysis and transglycosylation of acarbose to produce a pseudopentasaccharide that binds tightly in the active site. Suggest a plausible arrow-pushing mechanism for the formation of the tightly bound inhibitor.

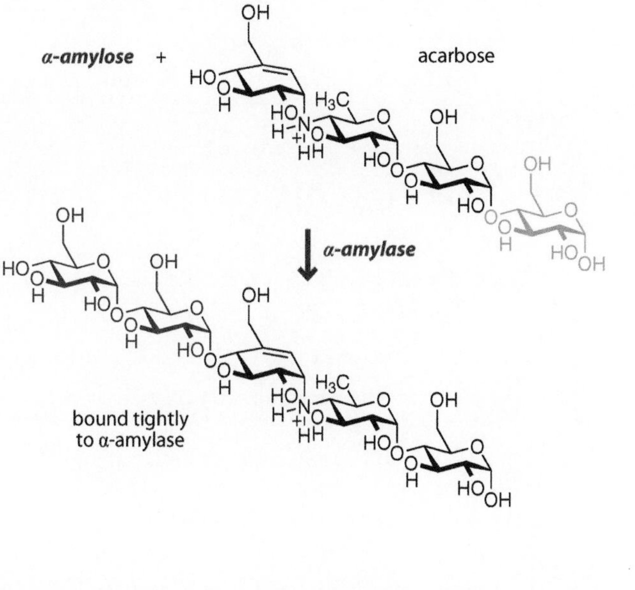

α-amylose + acarbose

↓ *α-amylase*

bound tightly to α-amylase

***7.32** Predict the structure of the missing starting material for the aldolase-catalyzed biosynthesis of *N*-acetylneuraminic acid from *N*-acetylmannosamine.

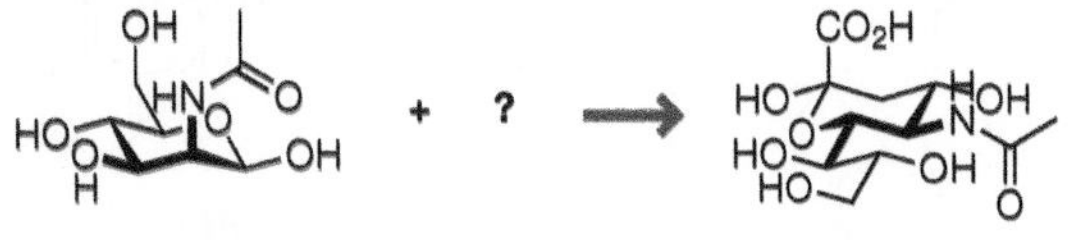

ManNAc + ? → Neu5Ac

7.33 The agarose gel used for the electrophoresis of DNA is seaweed polysaccharide containing an unusual bicyclic subunit. An enzyme isolated from the red seaweed that produces the related polysaccharide porphyran has been shown to convert galactose 6-sulfate residues into the anhydrogalactosyl residues of agarose. Suggest a plausible arrow-pushing mechanism for the enzymatic reaction.

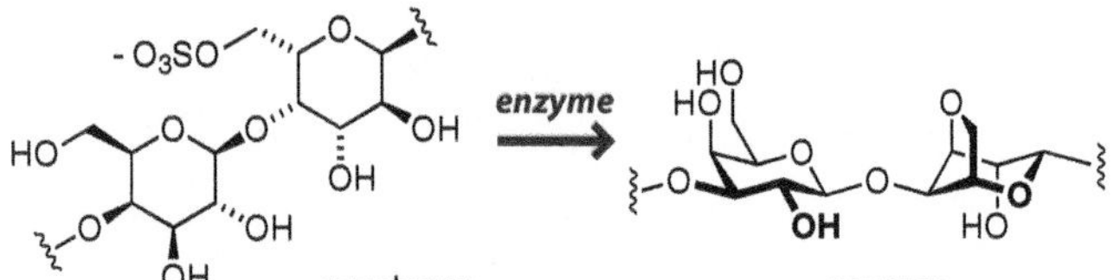

porphyran —enzyme→ agarose

7.34 Draw the following xylose derivative in a chair conformation and then consider the conformation of the ester substituents. On the basis of $A_{1,3}$ strain, which carbonyl groups do you expect to point above the plane of the pyran ring, and which do you expect to point below the plane of the pyran ring?

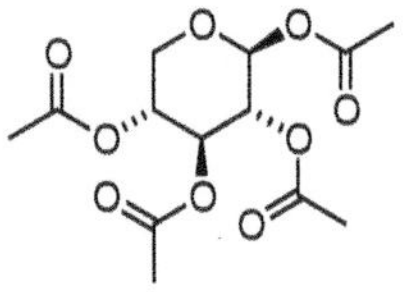

8 Polyketides and Terpenes

LEARNING OBJECTIVES

- Distinguish between the chemical mechanisms for iterative carbon–carbon bond formation in polyketide biosynthesis and terpene biosynthesis.
- Recognize the limited structural range of human polyketides and human terpenes.
- Appreciate the unlimited structural range of polyketides and terpenes found in lower organisms and plants.
- Describe the origin and effects of eicosanoid and sphingoid signaling molecules.
- Understand the origin of diversity in polyketide metabolites from lower organisms.
- Explain the structure and reactivity of nonclassical carbocations.
- Describe how cationic cyclizations and rearrangements amplify the diversity of terpene natural products.

Carbon–carbon bonds are as ubiquitous in organic chemistry as the notes in music, the words in English, and the mathematics in physics. Yet, despite their foundational centrality, carbon–carbon bonds seem slighted by the central dogma, which emphasizes the roles of bonds to and between heteroatoms, such as the phosphorus–oxygen bonds in DNA and RNA, the carbon–oxygen bonds in glycans, and the carbon–nitrogen bonds in proteins. Why is carbon–carbon bond formation relegated to a relatively minor role in the central dogma?

Naive answers to this question, right or wrong, might point to an incompatibility between carbon nucleophiles or carbon electrophiles, and the aqueous environment, but that kind of thinking loses traction when the effortless nature of enzymatic carbon–carbon bond formation is revealed. There is also no reason to impugn the stability of carbon–carbon bonds. Aliphatic carbon–carbon bonds are exceptionally robust—perhaps too robust. They are not readily cleaved by any single reaction, either chemical or enzymatic, whereas DNA, RNA, proteins, and glycans are easily hydrolyzed by enzymes into their molecular components for eventual reuse by the cell. The durability of aliphatic biooligomers makes them ideal for the lipid chainmail armor that protects all living cells, the long-distance signals that coordinate the actions of human cells, and the advanced weaponry used by the fittest microorganisms.

Polyketides and **terpenes** range in complexity and form, from simple linear hydrocarbons (**Figure 8.1**) to ornate polycyclic architectures; however, the iterative reactions used to construct these two classes of natural products are quite different. Polyketides are built up from two-carbon and three-carbon subunits through the reactions of enolates with carboxylic esters and thioesters. Terpenes, in contrast, are built up from five-carbon subunits through the addition of carbocations to olefins. We have

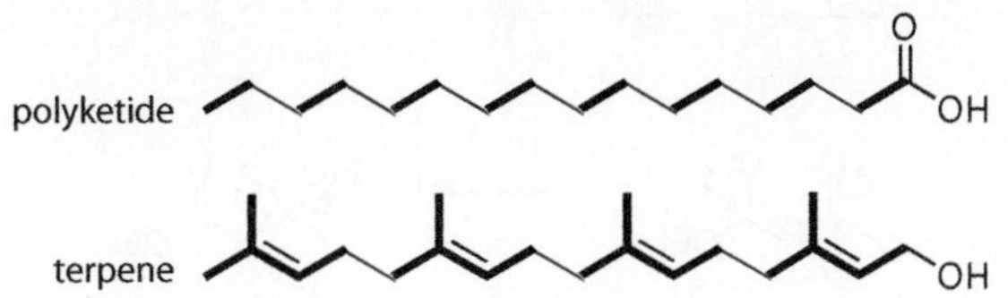

polyketide O OH

terpene OH

Figure 8.1 Rudimentary examples of a polyketide and a terpene. The repeating subunits are shown in bold, and the bonds formed through oligomerization are highlighted in red.

already discussed a wide range of polyketide and terpene natural products, such as daunomycin, aflatoxin, ptaquiloside, erythromycin, rapamycin, parthenolide, and steviol. These natural products are so complex that it can sometimes be difficult to discern the monomeric units from which they are assembled.

Cellular biosynthesis produces both primary and secondary metabolites. **Primary metabolites** are the molecules considered essential for normal cellular function, such as ATP, glyceraldehyde 3-phosphate, or cholesterol. Signaling molecules such as dopamine and estrogen are also considered primary metabolites. In contrast, **secondary metabolites** are nonessential for the immediate survival of a cell; instead, cells produce secondary metabolites for their chemical defense. Bacteria favor the polyketide pathway for the production of bioactive secondary metabolites. Plants, in contrast, tend to favor the terpene biosynthesis pathway for the production of secondary metabolites.

The evolutionary arms race has led to some truly ingenious polyketide and terpene weaponry. All organisms use polyketide and terpene biosynthesis to make primary metabolites, but humans seem to have missed out on the defensive potential of polyketide and terpene secondary metabolites. Instead, we developed agile physiques to fight with or flee from predators, and sophisticated protein and cellular defenses to fend off pathogens. This powerful combination of defenses has ensured the survival of our species, but not the survival of individuals who must battle infectious diseases, metabolic diseases, or cancer. After millions of years of evolutionary neglect, humans have finally started to harness the power of polyketide and terpene biosynthesis pathways, borrowing natural products from lower organisms and weaponizing them against human afflictions through a combination of molecular biology and chemical synthesis. Thus, over the past hundred years, we have moved beyond the molecular hunter–gatherer phase to the domestication phase. Perhaps over the next hundred years, through harnessing such remarkably complex biosynthetic technology, we might reach a level of molecular sophistication on a par with bacteria, fungi, and plants.

8.1 THE CLAISEN REACTION IN POLYKETIDE BIOSYNTHESIS

The diverse structures of polyketide natural products belie their iterative construction

Polyketides were first defined by the mountaineer–chemist J. Norman Collie in 1907 (**Figure 8.2**) as formal polymers of ketene ($H_2C{=}C{=}O$), which he described as a ketide group. Collie did not investigate any specific biochemical pathways, but he recognized that a wide range of chemical structures were accessible through further functionalization of the basic polyketide backbone $(CH_2CO)_n$.

Figure 8.2 Mountaineer and chemist. J. Norman Collie coined the term "ketide." (Courtesy of the Whyte Museum of the Canadian Rockies.)

Polyketides are derived from two-carbon and three-carbon building blocks

Polyketides are built up through repeated Claisen reactions. Each enzyme-catalyzed Claisen reaction of either two-carbon acetate esters or three-carbon propionate esters generates a β-ketoester (**Figure 8.3**). Such an intermolecular dimerization is usually referred to as a Claisen condensation, even though the reaction generates an alcohol, rather than water, as its by-product. In a base-catalyzed Claisen condensation, the

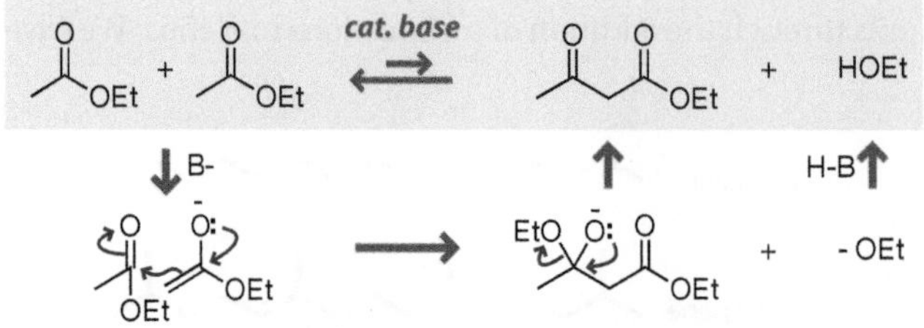

Figure 8.3 Claisen condensation mechanism. The base-catalyzed Claisen condensation involves the addition of an enolate to a carboxylic acid ester.

base—usually an alkoxide—deprotonates the ester, generating low concentrations of a highly reactive ester enolate. The enolate then attacks one of the unreacted esters, and the alkoxide anion that is ejected from the tetrahedral intermediate recaptures a proton from the protonated base.

Unfortunately, the Claisen condensation of ethyl acetate is energetically uphill by over 20 kcal mol^{-1}. In this reaction, one of the ester functional groups in the starting material, stabilized by resonance, is converted to a ketone, which does not enjoy such resonance stabilization. The retro-Claisen condensation is facile in the presence of alkoxides, so the yield of β-ketoester at equilibrium is low in an alkoxide-catalyzed Claisen condensation. The successful laboratory version of this reaction, developed by Rainer Ludwig Claisen in 1887, involves the use of *stoichiometric* ethoxide anion ($pK_a \approx 15.9$) to help drive the reaction toward a highly stabilized enolate of ethyl acetoacetate ($pK_a \approx 10.7$). Even when a stoichiometric base is employed, the reaction is not high-yielding with respect to the ester, but the yield is decent with respect to the sodium alkoxide base. Fortunately, ethyl acetate is an inexpensive solvent. This and/or other special tricks are needed to make the Claisen condensation efficient in the laboratory. One notable trick introduced by Walter Dieckmann was to apply the reaction to tethered esters that can cyclize to form highly favorable five- and six-membered rings.

Problem 8.1

Draw all of the stereoisomeric products that are expected from the following synthetic sequence.

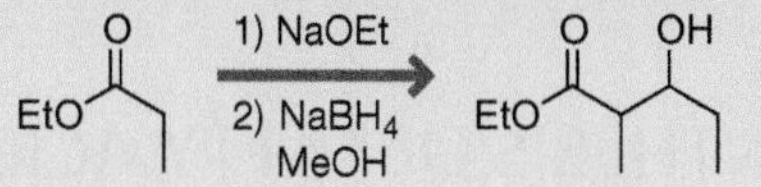

Nature favors the Claisen condensation for the formation of carbon–carbon bonds, but uses thioesters instead of carboxylic acid esters. The thioesters used by Nature are derivatives of **coenzyme A** (**Figure 8.4**). Coenzyme A is composed of two modular subunits: a phosphopantetheinyl unit and adenosine 3,5-diphosphate. To emphasize the sulfur atom, coenzyme A is often abbreviated as HS-CoA or CoA-SH. CoA is readily recognized by enzymatic machinery, and it may eventually turn out to be a ligand for microRNA regulation like many of the other vitamin cofactors. As far as we know, none of the rococo elements of the phosphopantetheinyl linker are important in the Claisen condensations of acyl-CoA esters. Nevertheless, enzymes that assemble polyketides use phosphopantetheinyl linkers just like the one in coenzyme A.

Thioester-based Claisen condensations are different in a couple of ways from Claisen condensations of carboxylic acid esters. First, the condensation of thioesters is energetically unfavorable, but less unfavorable than the corresponding condensation of carboxylic acid esters, by a very large amount, 11.7 kcal mol^{-1}. Second, the α-protons of thioesters are more acidic than the α-protons of traditional esters. For example, ethyl thioacetate has a pK_a of 21, which is closer to the pK_a of a ketone

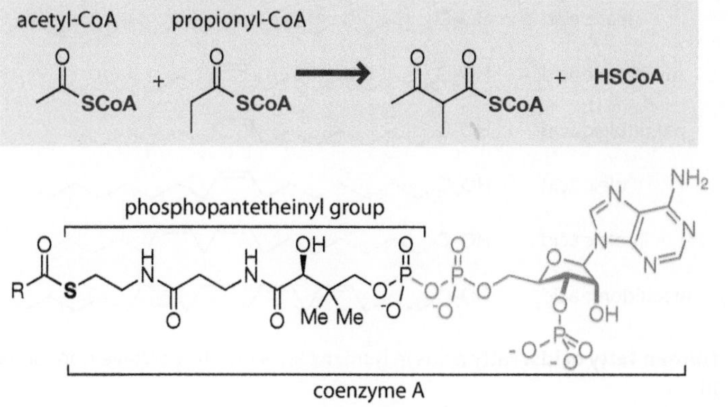

Figure 8.4 Carbon-carbon bond formation with thioesters. Nature performs Claisen condensations with thioesters derived from the thiol coenzyme A.

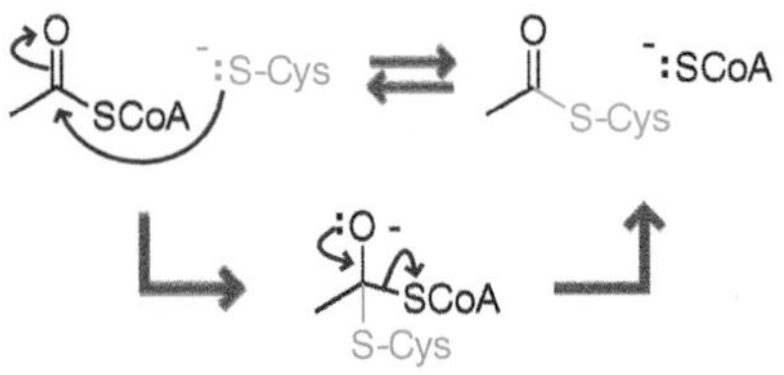

(for example, pK_a = 20 for acetone) than a carboxylic ester (for example, pK_a = 24 for ethyl acetate). Enolates of thioesters are commonly used in organic synthesis, but the Claisen condensation of thioesters, usually requiring Grignard reagents as bases, has no synthetic advantages over the Claisen condensation of the much more readily available carboxylic acid esters. As we will see, the method of enolate formation in biology is much less direct than deprotonation.

Most enzymatic Claisen condensations involving acyl-CoA esters involve the formation of an acyl-enzyme intermediate through acyl transfer to a cysteine thiolate (**Figure 8.5**). The transesterification of thioesters is more facile than the transesterification of carboxylic esters. Thiol ester exchange between acyl-CoA and glutathione (GSH) occurs with a second-order rate constant of 0.02 M^{-1} s^{-1} (**Figure 8.6**), which is about 100 times faster than exchange between regular carboxylic acid esters and GSH. This rate of transesterification is significantly faster than thioester hydrolysis, which is slow at physiological pH. The latter attribute also explains why the native chemical ligation of peptide thioesters works. Thiols are more reactive than water in large part because a significant fraction (about 1%) of thiols (pK_a ≈ 9) are deprotonated at physiological pH.

Figure 8.5 Thioester exchange mechanism. Unlike carboxylic esters derived from alcohols, acyl group exchange rapidly between thiols under physiological conditions.

Figure 8.6 Water resistant. Chemical studies reveal that thioester exchange (right) is much faster than hydrolysis (left).

8.2 THE BIOSYNTHESIS OF FATTY ACIDS IS A PARADIGM FOR POLYKETIDE BIOSYNTHESIS

Fatty acids have varying levels of unsaturation

Fatty acids are the prototypical polyketides. They have the simplest chemical structures, yet their biosyntheses involve most of the steps common to the synthesis of more complex polyketide natural products, both human and nonhuman. Human fatty acids have even numbers of carbons. The saturated fatty acids myristic acid, palmitic acid, and stearic acid, for example, have 14, 16, and 18 carbons, respectively (**Figure 8.7**). Fatty acids are important in chemical biology because they are essential components of the **lipid** molecules that make up cellular membranes and are precursors to signaling molecules. Human fatty acids have *cis*, but never *trans*, double bonds. Sometimes fatty acids have many *cis* double bonds, but they are never in conjugation with each other. Notice in Figure 8.7, for example, that there is a methylene group between the double bonds of both linoleic and arachidonic acids.

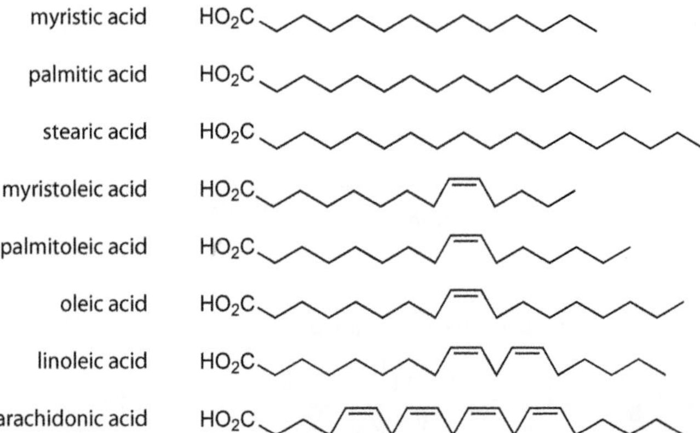

Figure 8.7 Human fatty acids. Fatty acids in humans are saturated or have nonconjugated *cis* double bonds.

Figure 8.8 ω-n nomenclature. There are a number of alternatives to IUPAC chemical nomenclature. Nutritional labels tend to favor the ω-n system in which positions of unsaturation are numbered from the last (ω) carbon. α-linolenic acid above is an ω-3 fatty acid, specifically, 3ω-3 octadecatrienoic acid.

Traditional rules of IUPAC nomenclature enumerate the carbons starting from the carboxylic acid end. Lipid chemists, in contrast, have devised a specialized naming system, which numbers the carbon atoms of lipids starting from the end farthest from the carboxylic acid group (**Figure 8.8**). The last position is named ω-1 (pronounced "omega one"). This naming system is used to categorize ω-3 and ω-6 **fatty acids**, the two types of fatty acids that cannot be biosynthesized by humans. Because these fatty acids must be obtained from the diet, they are called essential fatty acids. In ω-3 fatty acids, there is a *cis* double bond three carbons away from the ω-terminus, regardless of the overall length of the fatty acid. Fish oil is a common source of ω-3 fatty acids. The ratio of ω-6 to ω-3 fatty acids in the human diet is considered critical, because the ω-6 fatty acids can displace ω-3 fatty acids in cell membranes. An ideal human diet would have a 4:1 ratio, but very high levels of ω-6 fatty acids in safflower (50:1 ratio), corn (250:1), and other common plant seed oils make the average Western diet closer to an overall ratio of 10:1 to 20:1. Cooking with canola oil (2:1) can help maintain the balance of ω-6 to ω-3 fatty acids.

Fatty acid/polyketide synthases are categorized on the basis of their supramolecular structure

The enzymatic machinery that constructs polyketides through iterative carbon–carbon bond-forming reactions works in a highly cooperative fashion, passing synthetic intermediates from one enzymatic subunit to another. Most commonly, the growing polyketide chain is shuttled between clustered enzymatic subunits by a carrier protein. In some cases, all of the enzymatic subunits are ribosomally synthesized as a single polypeptide chain, and these assemblies are referred to as **type I polyketide synthases.** In other cases, the enzymatic subunits are ribosomally synthesized as individual peptide chains that fold independently and then associate through noncovalent interactions; such complexes are referred to as **type II polyketide synthases. Type III polyketide synthases** are smaller, more compact enzymes, which lack the domain organization of type I and type II polyketide synthases.

Human **fatty acid synthase** is a type I polyketide synthase, so the catalytic domains are tethered together, as shown schematically in **Figure 8.9**. These multidomain catalytic factories pair up as symmetrical dimers. Mammalian fatty acid synthase is optimized to produce palmitic acid, a saturated aliphatic acid with 16 carbon atoms. This synthesis requires repetitive cycles of aliphatic chain extension by a series of catalytic domains surveyed briefly here and in more detail below.

The **transacylase domain** (MAT) catalyzes the transfer of malonyl (HO2CCH2C(=O)) and acetyl (CH3C(=O)) groups from coenzyme A to the **acyl carrier protein** (ACP), which carries the synthetic intermediates between the enzymatic subunits. The **ketosynthase domain** (KS) catalyzes the Claisen condensation of malonyl and acetyl thioesters, which results in the formation of a β-ketoester (RC(=O)CH2C(=O)SR′) (hence the name of the enzyme). The **ketoreductase domain** (KR) catalyzes the reduction of the β-keto group in RC(=O)CH2C(=O)SR′, following the Claisen condensation, to a β-hydroxyl group. The **dehydratase domain** (DH) catalyzes a dehydration to yield an enone (RCH=CHC(=O)SR′). The **enoyl reductase** (ER) domain catalyzes conjugate reduction of the enone to give a saturated alkanoyl group (RCH2CH2C(=O)SR′. Finally, when the fatty acid reaches an appropriate length after iterative reactions catalyzed by the enzymatic domains, the **thioesterase** (TE) **domain** catalyzes the hydrolysis of the enzyme-bound acyl chain to generate a free fatty acid. In the X-ray crystal structure of the porcine fatty acid synthase, the ACP and TE domains were not well resolved, suggesting a high degree of mobility, which would be expected for the ACP in its role as a transporter. Most bacteria and plants synthesize fatty acids through type II synthases. The domains of fungal fatty acid synthases are less conveniently categorized because the various catalytic domains are present on two polypeptide chains.

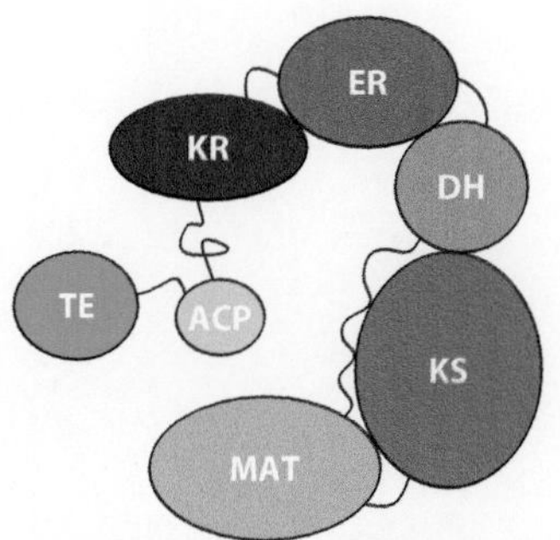

Figure 8.9 Type I domain organization of mammalian fatty acid synthase. For clarity, only half of the homodimeric complex is shown. The fully assembled dimeric complex resembles a large "X." TE, thioesterase; ACP, acyl carrier protein; KR, ketoreductase; ER, enoyl reductase; DH, dehydratase; KS, ketosynthase; MAT, malonyl-CoA/acetyl-CoA transacylase domain.

The acyl carrier protein shuttles the growing polyketide chain from one catalytic domain to another

The chemical transformations catalyzed by polyketide synthases take place exclusively while acyl chains are covalently bound as thioesters to protein subunits. Type I and type II polyketide synthases use an acyl carrier protein (ACP) to shuttle the thioester intermediate between various catalytic subunits. The acyl group of the nascent polyketide is not attached directly to a cysteine thiol of the ACP; instead, the acyl intermediate forms a thioester with a phosphopantetheinyl group that is added post-translationally to a serine residue of the ACP by the enzyme phosphopantetheinyl transferase. In type I polyketide synthases, the acyl carrier protein is part of the same long polypeptide chain composing all of the catalytic domains. It is a wonder that such large multidomain proteins can fold correctly! In such polyketide synthases, the ACP domain is attached by a long flexible peptide linker, which allows the domain to carry the growing polyketide chain to each of the catalytic domains (**Figure 8.10**).

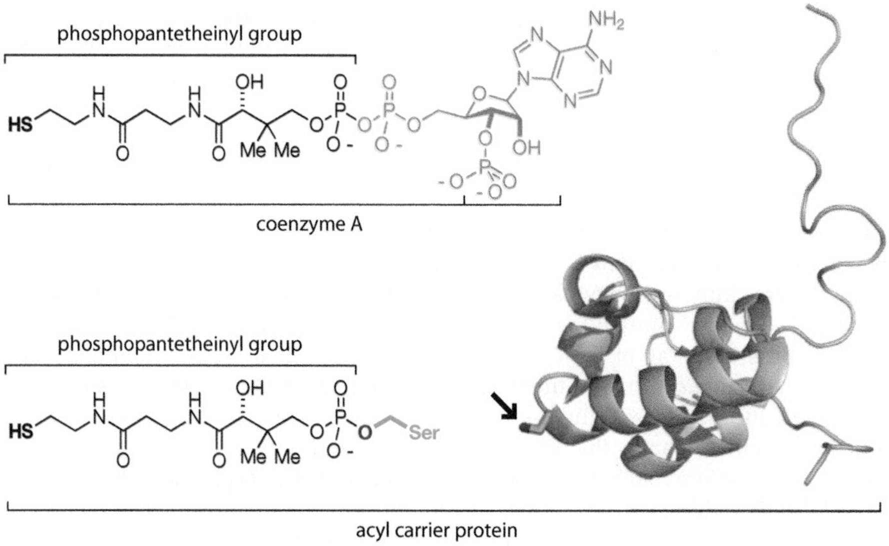

phosphopantetheinyl group

coenzyme A

phosphopantetheinyl group

acyl carrier protein

Figure 8.10 The thiol leash. The polyketide chain is bound to the acyl carrier protein via a phospho-pantetheinyl linker attached to a serine hydroxyl group (red). The free thiol at the end of the phosphopantetheinyl group provides a reactive site for conjugation to the acyl group of the growing polyketide. (PDB 2PNG)

A transacylase loads monomeric subunits onto the carrier protein

The biosynthesis of polyketides begins with the loading of the first thioester subunit onto the acyl carrier protein by a transacylase domain. For fatty acids, the first subunit is always acetyl-CoA (**Figure 8.11**). Chemically, this is an unremarkable reaction because thioester exchange is usually facile. However, during each subsequent round of elongation, malonyl-CoA/acetyl-CoA transacylase (MAT) then loads a malonyl subunit, not an acetyl subunit, onto the carrier protein.

transacylase

transacylase

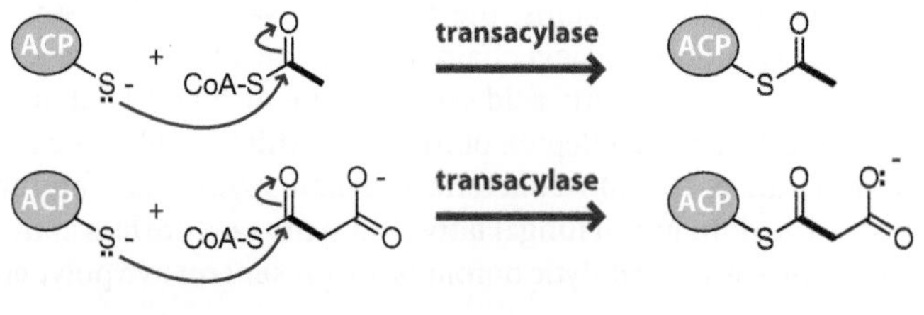

Figure 8.11 Batter up. The transacylase loads one of two possible acyl groups onto the carrier protein ACP. Acetyl-CoA (top) is used for the first subunit in fatty acid synthases, and malonyl-CoA (bottom) is used for subsequent rounds of elongation.

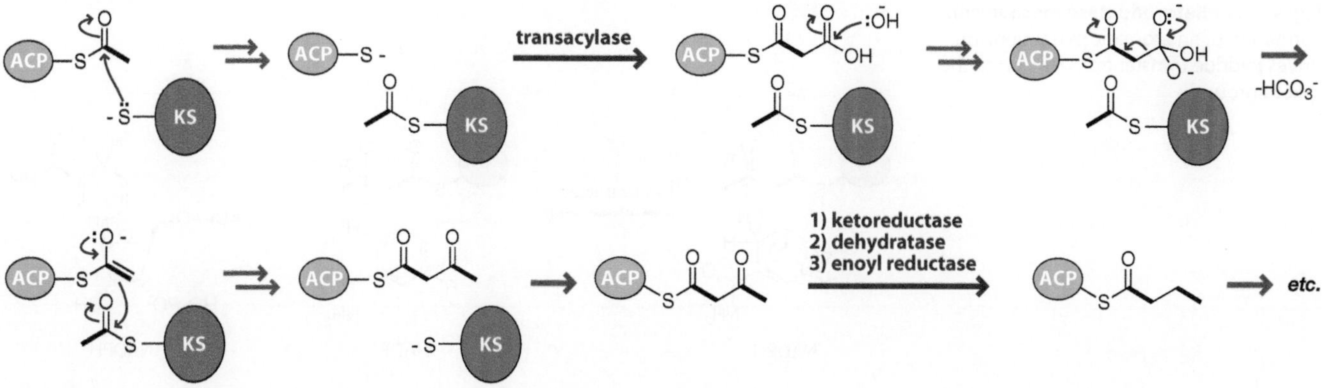

Figure 8.12 Forming a carbon-carbon bond. The ketosynthase domain (KS) catalyzes the Claisen condensation during each cycle of polyketide biosynthesis.

Ketosynthases catalyze a decarboxylative Claisen condensation

Our understanding of the key mechanisms for the Claisen condensation in polyketide biosynthesis comes primarily from studies of FabH, a bacterial fatty acid ketosynthase enzyme. Both the electrophilic thiol ester and the nucleophilic thiol ester enolate are bound to proteins via cysteine side chains (**Figure 8.12**). The electrophilic thiol ester is transferred to the ketosynthase enzyme through thiol ester exchange. The thiol enolate precursor is transferred to an acyl carrier protein as a malonate derivative. Decarboxylation of the malonate in the active site of the ketosynthase enzyme leaves a nucleophilic enolate. Studies of a bacterial ketosynthase strongly support the idea that decarboxylation is initiated by the addition of hydroxide to the carboxylic acid moiety of the malonate, followed by a loss of bicarbonate, in contrast to the more conventional decarboxylation resulting in a loss of carbon dioxide. The resulting enolate is stabilized by two imidazolium ions from histidine residues that can form hydrogen bonds with the enolate oxygen. This stabilization strongly suggests that the ketosynthase mechanism can best be drawn as the reaction of an enol with a thioester. The tetrahedral intermediate generated in the key carbon-carbon bond-forming event is stabilized by two amide N-H bonds—one from the backbone of the ketosynthase and the other from the nearby phosphopantetheine.

Cerulenin, platensimycin, and thiolactomycin (**Figure 8.13**) are natural-product inhibitors of ketosynthases. Cerulenin is a general inhibitor of type I and type II ketosynthases. It covalently reacts with one of the thiolates in the active site. Cerulenin leads to profound weight loss and feeding inhibition in mice, and it garnered a lot of interest as a potential weight-control medication. Although the mechanisms of ketosynthase domains in type I and type II fatty acid synthases are believed to operate through the same mechanisms, the type II ketosynthase used to synthesize fatty acids in bacteria can be selectively inhibited by platensimycin and thiolactomycin. Both platensimycin and thiolactomycin possess chemically reactive functional groups, but they have been shown to be reversible inhibitors.

Problem 8.2

Cerulenin has four potentially electrophilic sites for reaction with a thiol in the ketosynthase active site (numbered in the structure shown in Figure 8.13). Assuming that the thiol of the ketosynthase active site reacts irreversibly, provide an arrow-pushing mechanism for the reaction that indicates the correct stereochemistry of the product. How did you choose the more chemically reactive carbon of the epoxide?

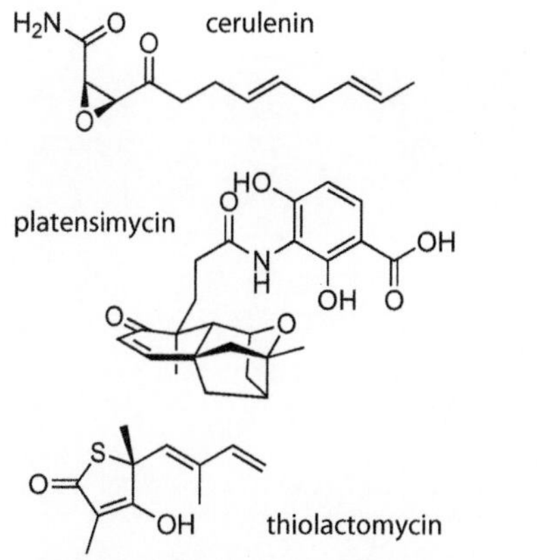

Figure 8.13 Fat fighters. A variety of natural products inhibit the ketosynthase activity of fatty acid synthases.

Ketoreductases catalyze hydride transfer from NADPH

After the Claisen condensation, the second step in fatty acid synthesis is the reduction of the β-ketone to an alcohol. This reaction is catalyzed by the ketoreductase (KR) domain, using NADPH as a hydride donor. The hydride transfer from the

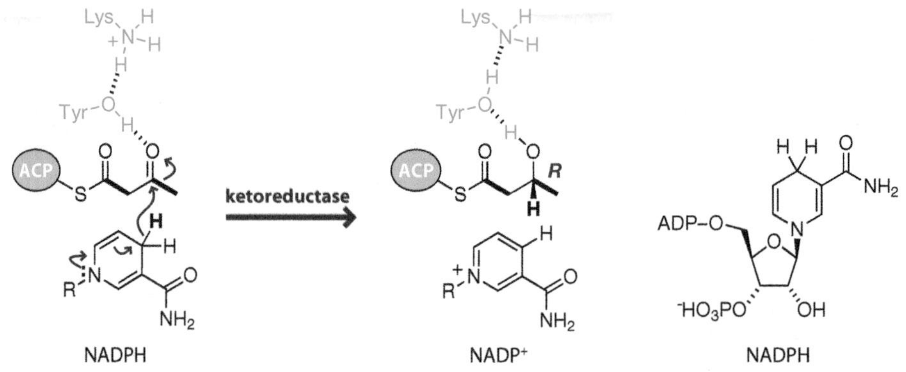

Figure 8.14 Ketoreductase mechanism. Formation of an aromatic pyridinium ion drives hydride transfer from NADPH to the β-keto group.

dihydropyridine to the carbonyl group is essentially the reverse mechanism for alcohol dehydrogenase discussed in Chapter 6, except that the carbonyl group undergoing hydride addition is activated not by a Zn^{2+} ion but by a tyrosyl group that is acidified by hydrogen bonding to a lysine ammonium group (**Figure 8.14**). As expected for an enzymatically catalyzed reaction, the hydride addition reaction stereoselectively adds to only one side of the carbonyl carbon; the two faces of planar, sp^2-hybridized carbons are designated either *re* or *si* for a clockwise or a counter-clockwise arrangement, respectively, of functional groups, which are prioritized by the Cahn–Ingold–Prelog rules. In mammalian fatty acid synthase, the hydride adds from the *si* face of the carbonyl group to generate an *R* stereocenter.

NADPH differs from NADH by the presence of a single phosphate on the 3′-hydroxyl group. It is important for cells to have two independent pools of reversible hydride donor/acceptors. In humans, NADPH is generally used as a hydride donor/acceptor in biosynthetic transformations, whereas NADH is generally involved in redox equilibria. In a healthy mammalian cell, NADPH is usually used as a hydride donor because the ratio of [NADPH] to [$NADP^+$] is about 200:1. When a hydride acceptor is needed, as it is for alcohol dehydrogenase, it makes more sense for an enzyme to use NAD^+ because the ratio of [NADH] to [NAD^+] is closer to 1.

Dehydratases catalyze β-elimination

The third step in fatty acid synthesis involves elimination of the β-hydroxyl group. Two residues have a key role in this reaction, an aspartic acid and a histidine residue. The aspartic acid protonates the hydroxyl group. Then the histidine catalyzes an elimination reaction. In a solution-phase reaction, such an elimination would certainly proceed via a two-step E1cB mechanism involving an enolate as the conjugate base. However, there are no trivial experiments that allow one to rule out a fully concerted E2 mechanism. A useful pattern is revealed by highlighting the individual two-carbon subunits—mainly that the double bonds generated by dehydratases are generally found between the acetyl and propionyl subunits rather than within them (**Figure 8.15**).

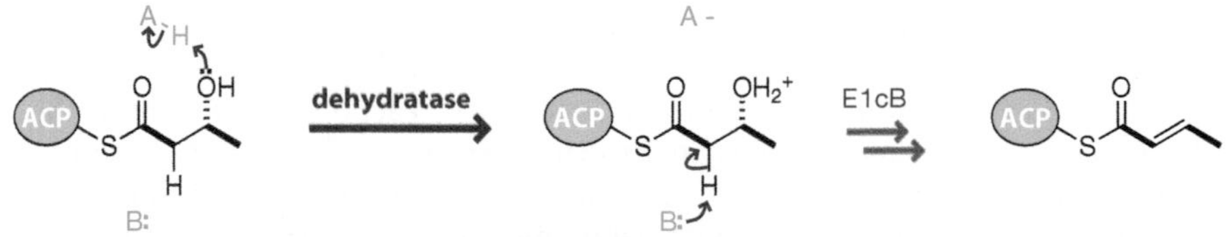

Figure 8.15 Going and gone. The β-elimination involves acid and base catalysis. The darkened bonds emphasize the two-carbon units derived from acetyl groups.

Enoyl reductases catalyze a conjugate reduction

The final step in a single elongation cycle by the fatty acid synthase involves the conjugate reduction of the enone by the enoyl reductase (ER) domain. NADPH again serves as the hydride donor, and the resulting enolate is protonated by a neighboring residue

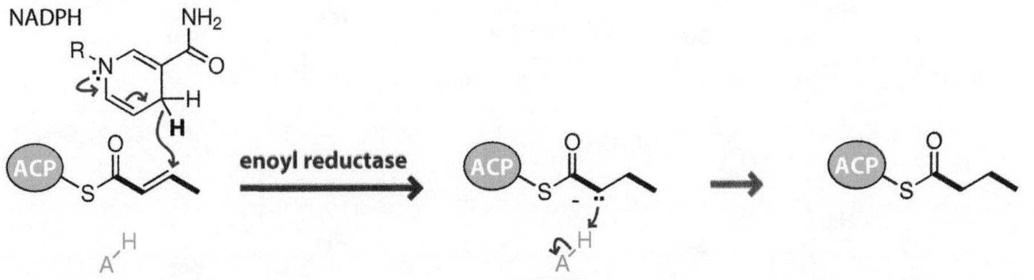

Figure 8.16 **Hydrogenation by enoyl reductase.** In the final step of the elongation of fatty acids, the enoyl reductase domain uses a hydride from NADPH to reduce the conjugated double bond.

(**Figure 8.16**). Both an aspartic acid and a lysine ammonium group are in close proximity in the active site, and it is difficult to know which of these residues actually protonates the enolate intermediate.

Several different enzymes help to maintain the equilibrium between NADPH and $NADP^+$. In *Mycobacterium tuberculosis*, the enzyme KatG oxidizes NADH and NADPH to their oxidized forms. The reaction is slow with oxygen and much faster with peroxide or superoxide. The powerful antibiotic isoniazid, used against tuberculosis, is oxidatively activated by KatG (**Figure 8.17**). The resulting reactive intermediate is believed to be an acyl radical, which then undergoes a Minisci radical addition reaction with the pyridinium ion of $NADP^+$. The resulting adduct is a potent inhibitor of the enoyl reductase in *M. tuberculosis*. Thus, patients taking isoniazid are tricking the bacterium into synthesizing the instruments of its own destruction.

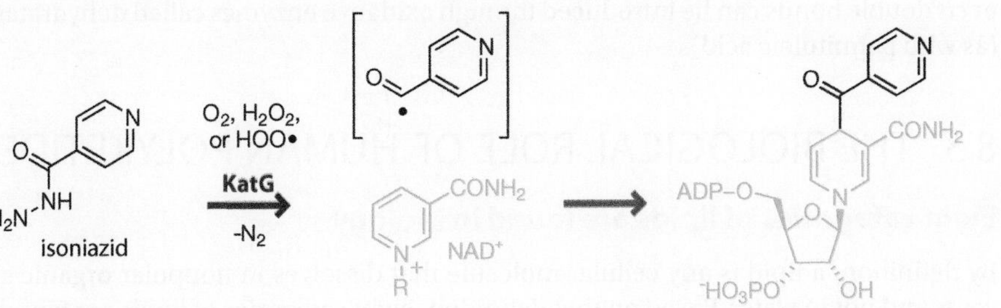

Figure 8.17 **Booby trap.** The bacterial enzyme KatG unwittingly synthesizes a deadly inhibitor from isoniazid.

Problem 8.3

Suggest a plausible arrow-pushing mechanism (using fishhook half-arrows) for the Minisci reaction of acyl radicals with pyridinium ions.

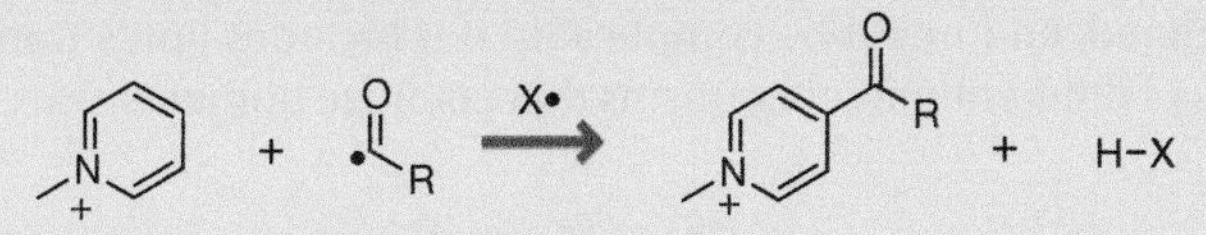

A thioesterase uses a catalytic triad to cleave the acyl group from the acyl carrier protein

Synthesis of the C_{16} palmitic acid requires eight cycles through the fatty acid synthase machinery described above. Once the chain has reached the intended length (as determined by the size of the active site), the acyl chain is directed to the active site of the thioesterase domain. The thioesterase uses a catalytic triad—composed of aspartate, histidine, and serine—like those found in serine proteases (**Figure 8.18**). The nucleophilic serine residue attacks the thioester linkage that attaches the acyl chain to the carrier protein. Water then replaces the thiol in the active site and the mechanistic sequence is reversed.

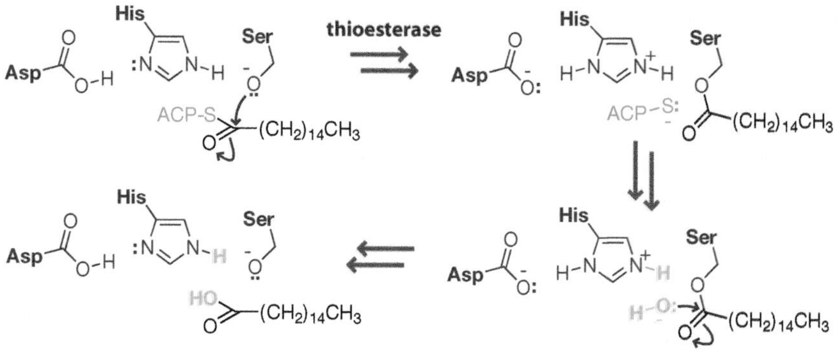

Figure 8.18 **Release.** Thioesterase cleaves the finished fatty acid from acyl carrier protein by using a catalytic triad analogous to that found in a serine protease.

Enzymes associated with the endoplasmic reticulum put the finishing touches on fatty acids

Molecules of palmitic acid that are released from the acyl carrier protein of human fatty acid synthase can be incorporated directly into lipids through an enzymatic esterification reaction. However, such intermediates are also subject to further transformations by enzymes associated with the endoplasmic reticulum. They can be extended by two carbons to stearic acid through a series of reactions analogous to those just described, or *cis* double bonds can be introduced through oxidative enzymes called dehydratases (as with palmitoleic acid).

8.3 THE BIOLOGICAL ROLE OF HUMAN POLYKETIDES

Eight categories of lipids are found in biology

By definition, a lipid is any cellular molecule that dissolves in nonpolar organic solvents and not in water. Based on that definition, eight categories of lipids are found in living organisms—namely, fatty acids, glycerolipids, glycerophospholipids, sphingolipids, saccharolipids, sterols, prenols, and polyketides. Most molecules that are present in the human **lipidome** (the totality of lipids in cells) derive their hydrophobic character from polyketide chains, particularly fatty acids, and we will focus on such lipids in this section. Saccharolipids are comparatively rare, with only about a dozen bacterial examples known and characterized (**Figure 8.19**). The broadest class of lipids, polyketides, includes around 7000 examples. The field of lipidomics relies largely on mass spectrometry and other tools to characterize the complete set of lipids from an organism, tissue, or cell type. As yet, the complete lipidome for humans has not been completely defined, and might be complicated if it includes lipids from the tremendous numbers of different microorganisms that colonize humans.

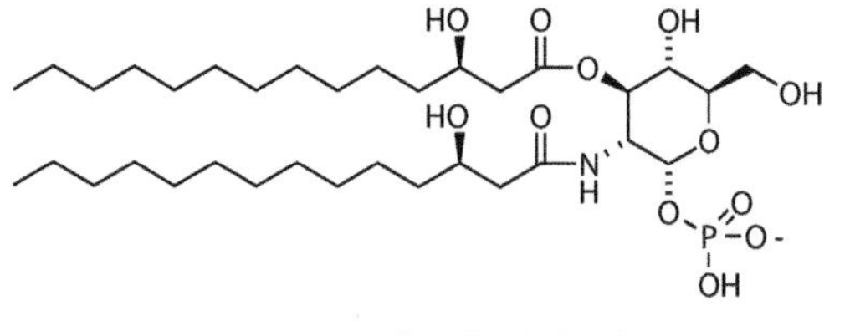

saccharolipids: lipid X
(2,3-bis-(3*R*-hydroxy-tetradecanoyl)-α-D-glucosamine 1-phosphate)

Figure 8.19 **Molecular anchor bolt.** Lipid X is one of the basic building blocks of larger bacterial saccharolipids, which anchor long polysaccharide chains to the outer bacterial membrane.

Lipid membranes are composed of lipids with a polar head group and a nonpolar tail

From the simplest category of lipids, fatty acids are present in two of the three general classes of molecules that make up mammalian lipid bilayers (**Figure 8.20** and **Figure 8.21**): phospholipids and glycosphingolipids, but not cholesterol. Lipids are present

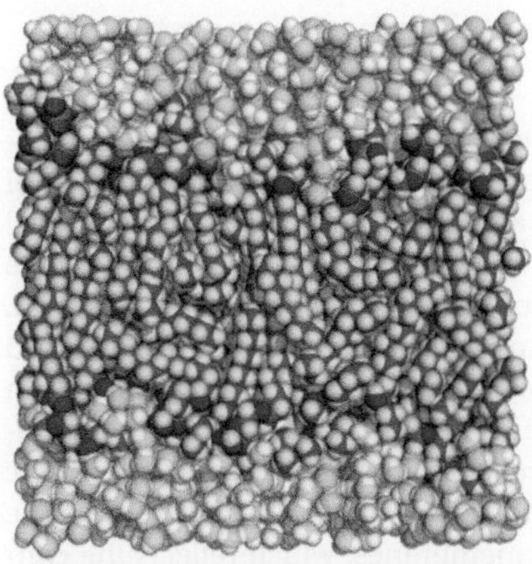

Figure 8.20 Lipid barrier. A model of a lipid bilayer generated through molecular dynamics reveals substantial disorder within the lipid chains, consistent with the fluid-like nature of the lipid bilayer. Water molecules are shown in green.

Table 8.1 Lipid composition of an average mammalian cell membrane.

Lipid	Percentage
Phosphatidylcholine	45–55
Phosphatidylethanol-amine	15–25
Phosphatidylinositol	10–15
Cholesterol	10–20
Sphingomyelin	5–10
Phosphatidylserine	5–10
Cardiolipin	2–5
Phosphatidic acid	1–2

in membranes as combinatorial mixtures, with different combinations of head groups and fatty acids. Phospholipids have a phosphatidic acid tail with two fatty acid ester groups. Sphingolipids have a ceramide tail, with one fatty acid attached to the amino group of the aminoalcohol sphingosine. There is significant variation in the polar head groups of phospholipids. Choline is the most common head group of mammalian phospholipids, but ethanolamine, inositol, and serine are also common (**Table 8.1**).

The two leaflets of the plasma membrane have distinctly different phospholipid compositions. The leaflet facing the cytoplasm, for example, has an excess of anionic phospholipid head groups. The appearance of such anionic head groups on the outside of the cell is one indicator of apoptosis. Maintaining this asymmetry requires energy, because the hydrophilic and charged head group cannot readily diffuse through the aliphatic-packed center of the membrane. Enzymes called flippases catalyze the flip-flop of phospholipids from one side of the membrane to the other. These P-type ATP-binding transporters both maintain the membrane and have a role in the synthesis of the glycoconjugated phospholipids.

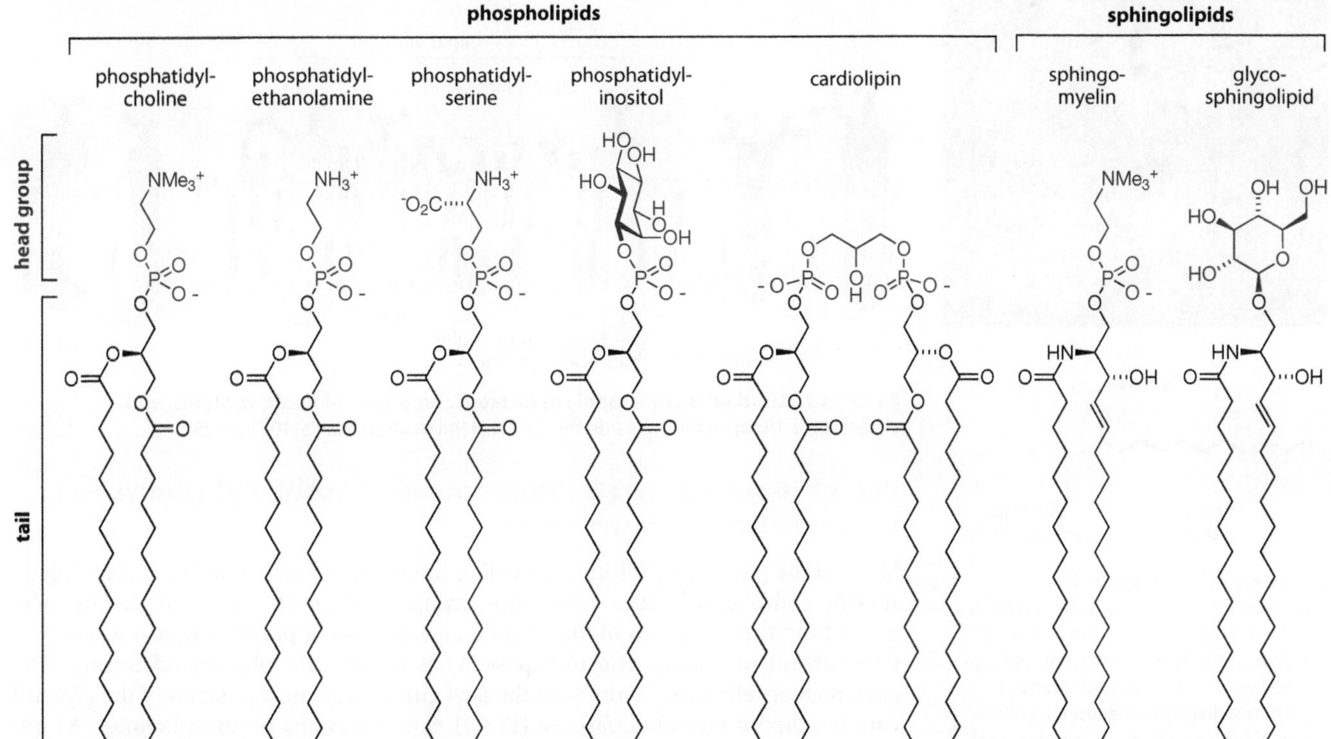

Figure 8.21 Common lipid components of human cell membranes. These lipids exist as mixtures with acyl groups of various lengths and various degrees of unsaturation. Most of the head groups are anions or zwitterions, with the exception of the glycosyl groups of glycosphingolipids.

The lipid bilayer entropically favors interactions between embedded molecules

Proteins embedded within lipid bilayers have fewer degrees of freedom than molecules in solution. Molecules bound to lipid bilayers have essentially two degrees of translational freedom (side-to-side movement) and at most one degree of rotational freedom (spinning perpendicular to the membrane). Membrane-bound proteins are translated directly into the membrane with a precise orientation, which remains fixed. Glycans always point outward from the cell, SH2 protein domains always point inward into the cytoplasm, and so on. With fewer degrees of freedom, interactions between membrane-bound molecules should have a natural entropic advantage over solution-phase reactions. However, the complexity of cell membranes makes this entropic advantage difficult to quantify. Lipids diffuse up to two orders of magnitude more slowly in real cell membranes than in artificial membranes, which reflects the considerable heterogeneity of the plasma membrane and contacts to the actin skeleton.

A wide range of signaling proteins, such as Hedgehog, are anchored to the plasma membrane through ester, thioester, and amide linkages to long-chain fatty acids such as C_{14} myristic acid and C_{16} palmitic acid. Other signaling proteins, such as Ras and Raf, are anchored to membranes by terpenes. These signaling proteins usually lose their function when the lipid anchor is removed and the protein is allowed to float aimlessly within the cytoplasm.

Until the late 1980s, the fluid mosaic model of membranes held that phospholipids, glycolipids, cholesterol, and proteins were randomly and evenly distributed throughout the lipid bilayer. However, most membrane-bound proteins associate with proteins or microfilaments on the cytosolic side of the membrane. Those structures impose order on the components of the membrane. Furthermore, it is now known that the outer membranes of human cells possess microdomains called **lipid rafts** that are particularly rich in glycosphingolipids and cholesterol (**Figure 8.22** and **Figure 8.23**). Many transmembrane receptors, such as immunoglobulin E receptors, T-cell antigen receptors, and B-cell receptors, are clustered in lipid rafts. Thus, lipid rafts are often the focal points for cell signaling.

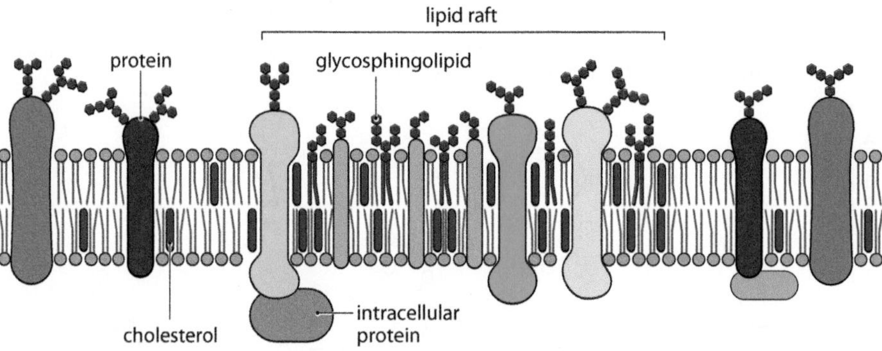

Figure 8.22 Lipid rafts are strongly associated complexes of lipids, proteins, and cholesterol. (Adapted from Inside the Cell, NIH Publication No. 05-1051, 2005.)

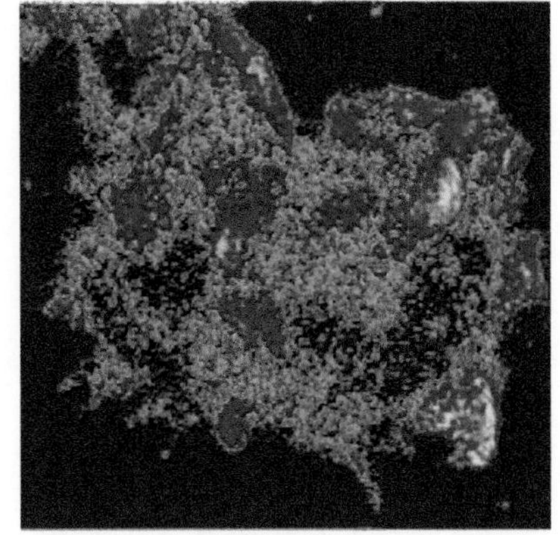

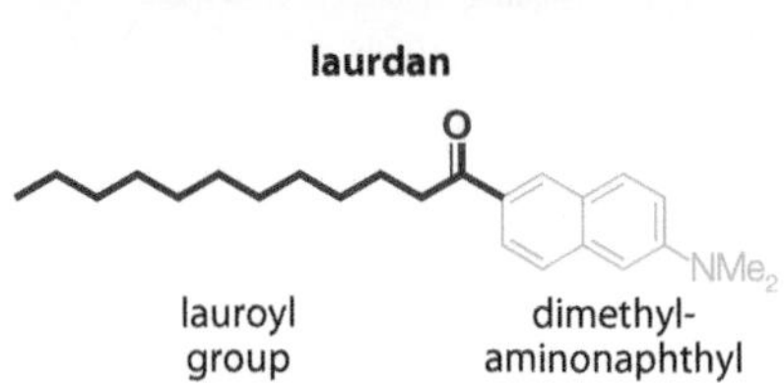

Figure 8.23 Paint the cell. The fluorescence of the lauric acid derivative laurdan is dependent on the polarity of the environment. A human macrophage stained with laurdan reveals areas of high generalized polarization in yellow. Areas of high polarization correlate with lipid rafts. (Top, from K. Gaus et al., *Proc. Natl. Acad. Sci. USA* 100:15554–15559, 2003. With permission from the National Academy of Sciences.)

Phospholipases generate distinct chemical signals by hydrolyzing various bonds of phospholipids

Many of the phospholipids in mammalian membranes serve as substrates for highly specific hydrolases (**Figure 8.24**). Any enzyme that selectively hydrolyzes the acyl group from the 1 position of the glycerol core is called a **phospholipase** A1 (PLA1). One version of human phospholipase A1 is specific for phosphatidylserine. Any enzyme that selectively hydrolyzes the acyl group from the 2 position of the glycerol core is called a phospholipase A2 (PLA2). Some versions of phospholipase A2 are secreted from human cells, whereas others are localized in the cytosol. Some types of phospholipase A2 cleave arachidonoyl esters from the C2 position of phospholipids. The arachidonic acids are converted to various signaling molecules that are important in inflammatory immune responses. Wasp, hornet, and snake venoms contain various

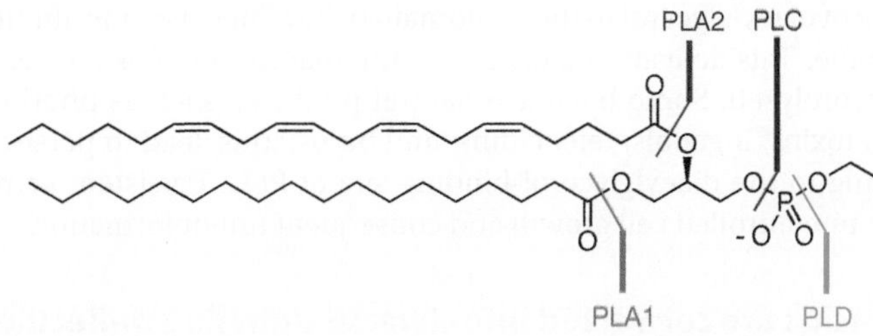

Figure 8.24 Regioselectivity of various classes of phospholipases acting on a generic lipid substrate. PLA1 is phospholipase A1, PLA2 is phospholipase A2, PLC is phospholipase C, and PLD is phospholipase D.

phospholipases (A1 and A2) that act on all kinds of phospholipids. Phospholipases C and D (PLC and PLD, respectively) target phosphodiester bonds. Phospholipase C cleaves the polar lipid head group from the hydrophobic lipid tail, whereas phospholipase D cleaves the other phosphodiester bond from phosphatidylcholine to generate phosphatidic acid and choline. The science of lipid signaling molecules has been slow to evolve. Simple ideas like concentration, measured in mol L^{-1}, aren't readily applied to molecules that reside in two-dimensional membranes.

Phospholipase Cβ generates two signaling molecules

Recall from our discussion of protein kinase regulation in Chapter 6 that the activity of protein **kinase** C (PKC) is controlled by two regulatory domains. The C2 domain acts as a calcium sensor, whereas the C1 domain is activated by 1,2-diacyl-*sn*-glycerol (often referred to as diacylglycerol). Don't worry about the *sn* (*stereospecific numbering*) nomenclature, which is specific for lipids based on the prochiral molecule glycerol (**Figure 8.25**). Although most lipids are made through the attachment of phosphoryl head groups to diacylglycerols, the concentration of diacylglycerols is low in a resting cell. Activation of some 7TM G protein-coupled receptors coupled to the heterotrimeric G protein G_q leads to the activation of phospholipase Cβ. Phospholipase Cβ cleaves phosphatidylinositol 4,5-bisphosphate (PIP_2) to give 1,2-diacyl-*sn*-glycerol and inositol 1,4,5-triphosphate (**Figure 8.26**). Inositol triphosphate leads to the opening of calcium storage vesicles, whereas diacylglycerol directly binds to PKC. PKC is generally active in cells that are actively performing their physiological function, such as the contraction of smooth muscle cells; secretion by stomach, salivary, and lacrimal cells; and the generation of glucose by fat and liver cells.

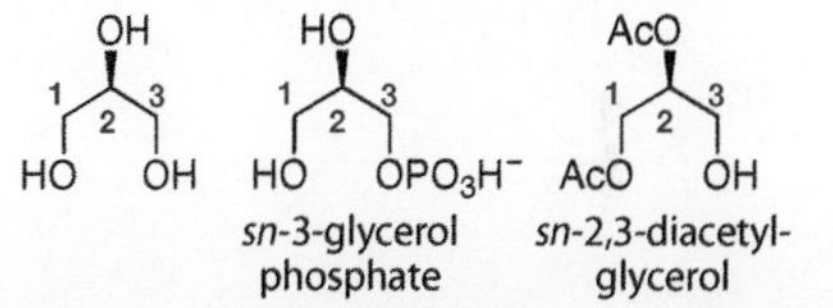

Figure 8.25 Numbering glycerol. With stereospecific numbering (*sn*), the positions of glycerol are assigned fixed numbers, regardless of the Cahn–Ingold–Prelog priorities of the substituents.

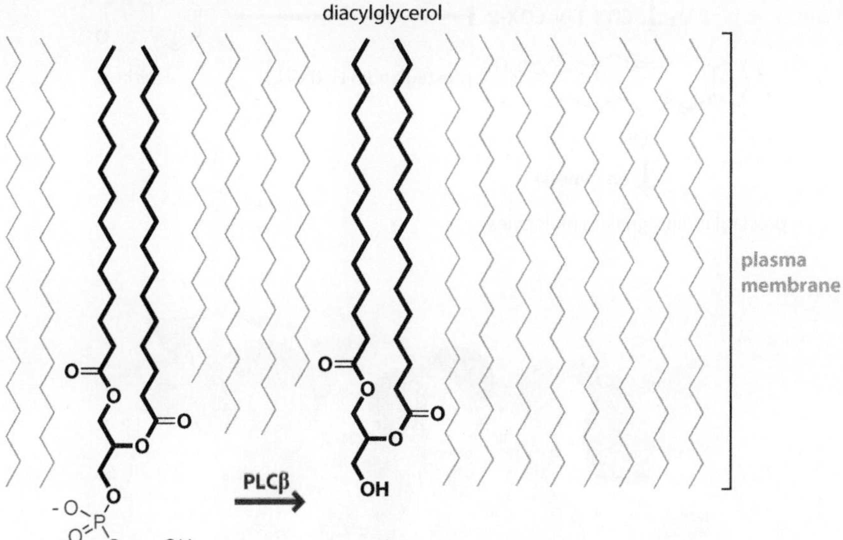

Figure 8.26 PLCβ initiates a cytosolic calcium burst. The activation of some seven-transmembrane receptors leads to the dissociation of the heterotrimeric G protein subunits. Depending on the cell, either the Gα or Gβγ subunit activates phospholipase Cβ (PLCβ), which generates diacylglycerol (DAG), an activator of protein kinase C.

The binding of diacylglycerol to the C1 domain of PKC increases the affinity for the plasma membrane. This activation subsides as the diacylglycerol is converted back into lipids or hydrolyzed. Some bioactive natural products, such as phorbol esters, debromoaplysiatoxins, ingenols, teleocidins, and bryostatins, lead to persistent activation by binding to the diacylglycerol-binding site of PKC. Persistent activation is associated with uncontrolled cell growth and consequent tumor formation.

Arachidonic acids are converted into diverse signaling molecules during inflammation

A wide range of human cells convert the fatty acid arachidonic acid into eicosanoid signaling molecules, usually in response to inflammatory signals (**Figure 8.27**). There are four general classes of eicosanoid signaling molecules: thromboxanes, prostacyclins, prostaglandins, and leukotrienes. These signaling molecules usually act locally, on nearby tissues. Most eicosanoids affect two types of cells—namely, smooth muscle cells and cells derived from hematopoiesis, such as platelets, white blood cells, and bone-forming cells. The first step in the biosynthesis of thromboxanes, prostacyclins, and prostaglandins is the conversion of arachidonic acid (released by phospholipase A2) into prostaglandin H_2 (PGH_2) by the enzyme cyclooxygenase. Humans have two variants of cyclooxygenase, COX-1 and COX-2, both of which are the target of aspirin and other nonsteroidal anti-inflammatory drugs such as ibuprofen and naproxen. COX-1 is expressed constitutively in all cells, whereas the expression of COX-2 is upregulated in response to proinflammatory signals. Aspirin is an irreversible inhibitor of both cyclooxygenases, because aspirin acylates a serine hydroxyl group in the active site (**Figure 8.28**). In some patients, the high doses of aspirin needed for chronic pain relief lead to bleeding in the stomach as a result of COX-1 inhibition, so new selective inhibitors of COX-2 were developed (for example rofecoxib, valdecoxib, and celecoxib). Unfortunately, these selective COX-2 inhibitors were unexpectedly found to cause cardiovascular injuries in some patients.

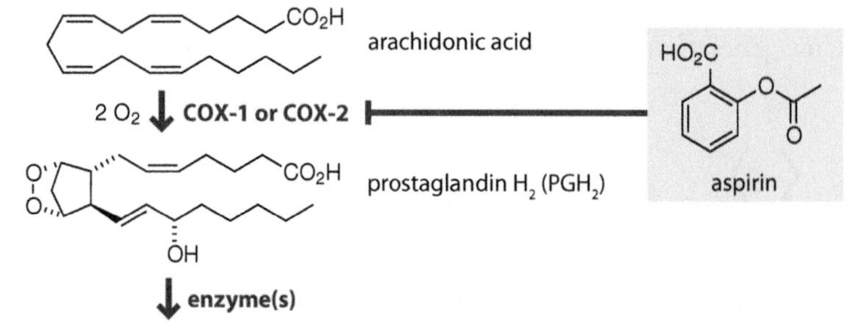

CO_2H arachidonic acid

$2\ O_2$ ↓ **COX-1 or COX-2**

HO_2C aspirin

CO_2H prostaglandin H_2 (PGH_2)

OH

↓ **enzyme(s)**

prostaglandin signaling molecules

Figure 8.27 The start of prostaglandin signaling. The first step in prostaglandin biosynthesis is the conversion of arachidonic acid to prostaglandin H_2 (PGH_2). The red line with a bar at the end implies inhibition.

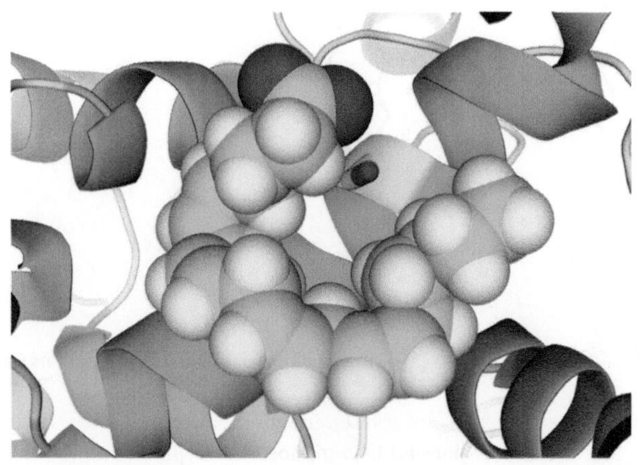

Figure 8.28 Active site for inflammation. Arachidonic acid (space-filling) nestles within the active site of cyclooxygenase-2 (COX-2). Ser530 (sticks) in the active site of COX-2 is susceptible to acylation by aspirin. (PDB 1CVU)

arachidonic acid → PGH_2 → TXA_2 → smooth muscle, platelets

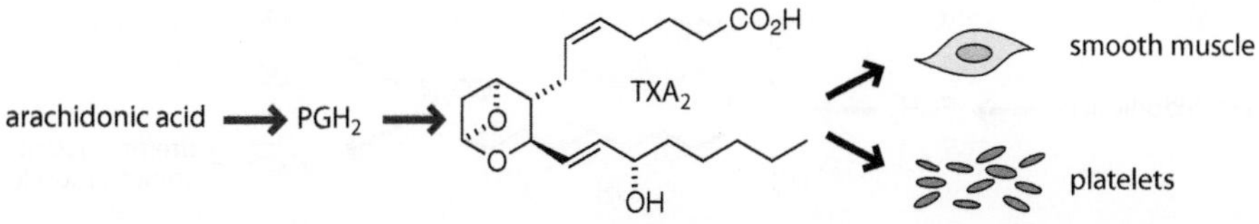

Figure 8.29 Signaling by a strained bicyclic acetal. Thromboxane A_2 (TXA_2) is produced by activated platelets.

The PGH_2 is further converted into other prostaglandins, dependent on cell type, that can signal neighboring cells. Most of these prostaglandin signaling molecules, or chemical analogs, are sold as prescription pharmaceuticals. When platelets are activated as part of a blood clotting response, they convert prostaglandin H_2 into thromboxane A_2 (TXA_2; **Figure 8.29**), which has a half-life of 32 seconds, as a result of the hydrolysis of the strained bicyclic acetal. The hydrolysis product, thromboxane B_2, has no known biological activity. Because of the brief half-life, TXA_2 acts only on nearby smooth muscle, causing vasoconstriction, and on other nearby platelets, resulting in a localized autocatalytic effect that amplifies clot formation.

Problem 8.4

Thromboxane A_2 is unstable at physiological pH, with a half-life of less than 1 minute. Predict the product of thromboxane A_2 hydrolysis.

Cells of the vascular endothelium that line blood vessels convert prostaglandin H_2 into prostaglandin I_2, more commonly called prostacyclin (**Figure 8.30**), which counters the effects of TXA_2. Like TXA_2, prostacyclin is unstable at physiological pH because of the hydrolyzable enol ether, and has a half-life of under 1 minute at physiological pH. It acts on platelets to inhibit aggregation and also acts on nearby smooth muscle cells, causing vasodilation.

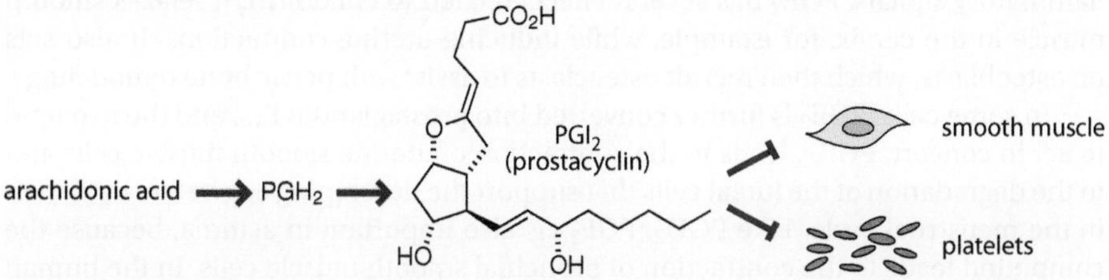

arachidonic acid → PGH_2 → PGI_2 (prostacyclin) → smooth muscle, platelets

Figure 8.30 Signaling by an enol ether. Prostacyclin (PGI_2) is produced by vascular endothelial cells.

Problem 8.5

Prostacyclin is unstable at physiological pH, with a half-life of less than 1 minute. Predict the product of prostacyclin hydrolysis.

Mast cells are a type of white blood cell that have an important role in allergic responses. In response to proinflammatory signals, the cells release the allergic signaling molecule histamine and convert prostaglandin H_2 into prostaglandin D_2 (PGD_2; **Figure 8.31**). To mimic this effect in the laboratory, lipopolysaccharides derived from bacterial cell walls are commonly used as a proinflammatory signal, which can

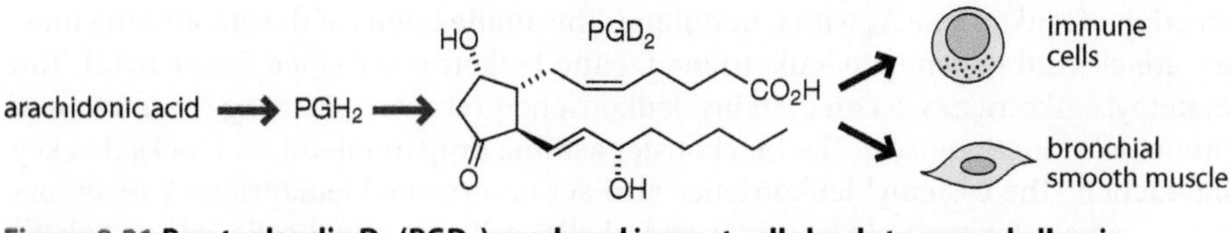

arachidonic acid → PGH_2 → PGD_2 → immune cells, bronchial smooth muscle

Figure 8.31 Prostaglandin D_2 (PGD_2) produced in mast cells leads to several allergic responses.

Figure 8.32 Prostaglandin signaling. Prostaglandin E_2 (PGE_2) induces rapid inflammatory responses and slower responses affecting smooth muscle.

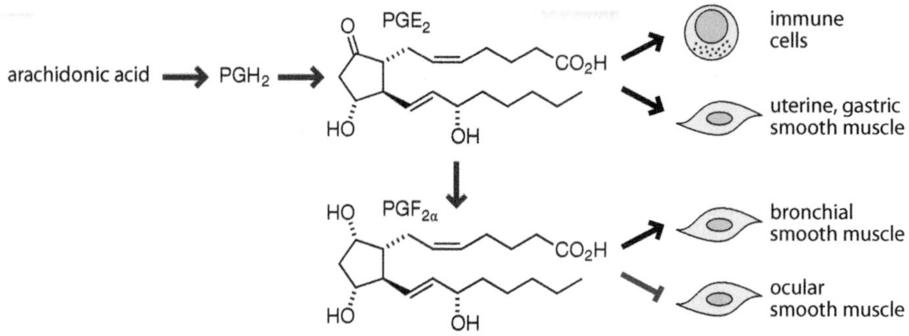

be added to the cell culture. PGD_2 activates several types of immune cells, such as T helper cells, eosinophils, and basophils. It also induces the contraction of bronchial smooth muscle, which causes the shortness of breath associated with asthma. PGD_2 acts on cells within the nervous system and on mucus-releasing goblet cells.

Prostaglandin E_2 (PGE_2) is the most abundant prostanoid in the human body and is generated from prostaglandin H_2 by prostaglandin E_2 synthase (**Figure 8.32**). Depending on the cell type and location, PGE_2 can affect inflammation, the integrity of the gastric mucosa, fertility, and parturition (labor and delivery). A cytosolic version of prostaglandin E_2 synthase is primarily associated with rapid inflammatory responses. In stomach, uterine, and other cells, a microsomal (that is, derived from vesicles of the endoplasmic reticulum) form of prostaglandin E_2 synthase, usually associated with inducible COX-2 activity, is associated with slower smooth muscle responses such as stomach distress, the menstrual cycle, and childbirth. However, the expression of microsomal PGE_2 synthase is also upregulated in response to proinflammatory signals. PGE_2 has several effects related to childbirth. It relaxes smooth muscle in the cervix, for example, while inducing uterine contractions. It also acts on osteoblasts, which then recruit osteoclasts to assist with pelvic bone remodeling.

In some cells, PGE_2 is further converted into prostaglandin $F_{2\alpha}$, and the two tend to act in concert. $PGF_{2\alpha}$ leads to the contraction of uterine smooth muscle cells and to the degradation of the luteal cells that support the developing egg before ovulation in the menstrual cycle. Like PGD_2, $PGF_{2\alpha}$ is also important in asthma, because the compound leads to the contraction of bronchial smooth muscle cells. In the human eye, PGE_2 and $PGF_{2\alpha}$ inhibit the contraction of smooth muscle cells, lowering the intraocular pressure.

Leukotrienes represent the fourth class of eicosanoid signaling molecules. They generally act to sustain the rapid initial inflammatory response and are important in diseases such as asthma. Leukotrienes are produced by inflammatory immune cells—namely mast cells, macrophages, eosinophils, and neutrophils. Leukotrienes do not arise through the action of cyclooxygenase or the intermediation of prostaglandin H_2. Instead, the leukotrienes are generated by the action of 5-lipoxygenase on arachidonic acid to generate 5-hydroperoxyeicosatetraenoic acid (5-HPETE) (**Figure 8.33**). 5-HPETE is converted into leukotriene A_4, which has two potential fates. In some mast cells and monocytes, leukotriene A_4 is hydrolyzed to leukotriene B_4; in other mast cells, eosinophils, and basophils, leukotriene A_4 is converted into cysteinyl leukotrienes. Leukotriene B_4 exerts a powerful chemotactic response, attracting leukocytes and some cytotoxic T lymphocytes to sites of inflammation.

The synthesis of the cysteinyl leukotrienes begins with an enzyme-catalyzed S_N2 reaction of leukotriene A_4 with glutathione. The amide bonds of the glutathione moiety are cleaved to generate leukotriene D_4 and leukotriene E_4 (see Figure 8.33). The cysteinyl leukotrienes act on cysteinyl leukotriene 1 receptors, causing contraction of bronchial smooth muscle. The blockbuster asthma drug montelukast blocks this key interaction. The cysteinyl leukotrienes also act on cysteinyl leukotriene 2 receptors expressed on, for example, vascular endothelial cells, neuronal cells, adrenal gland cells, and cardiac Purkinje cells.

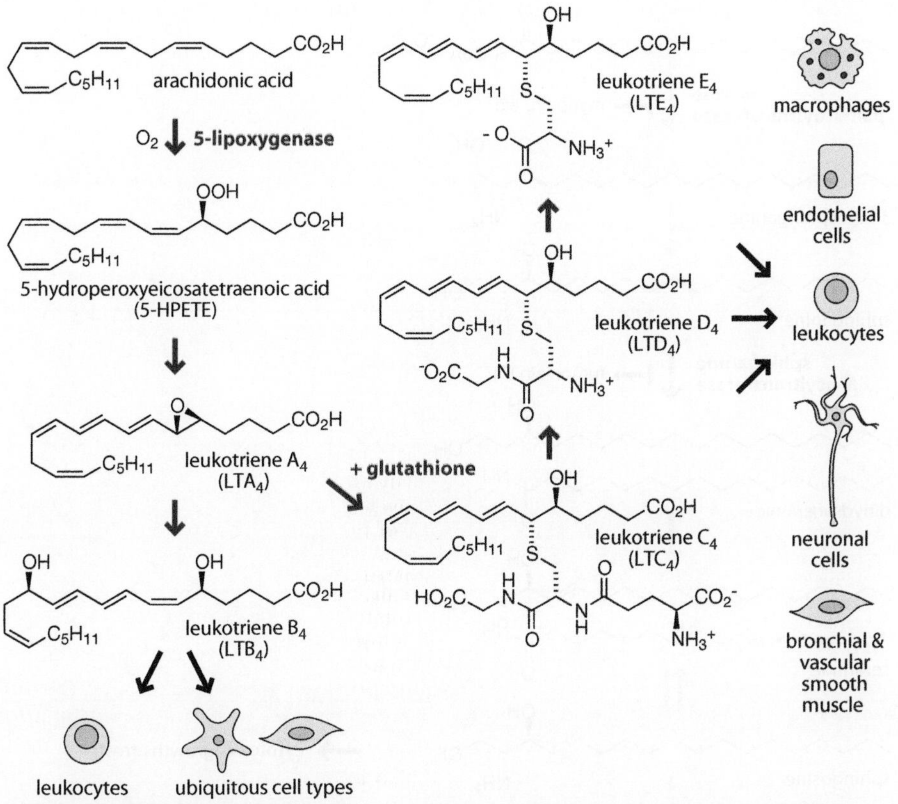

arachidonic acid

O_2 5-lipoxygenase

5-hydroperoxyeicosatetraenoic acid (5-HPETE)

leukotriene A_4 (LTA_4)

+ glutathione

leukotriene B_4 (LTB_4)

leukocytes ubiquitous cell types

leukotriene C_4 (LTC_4)

leukotriene D_4 (LTD_4)

leukotriene E_4 (LTE_4)

macrophages

endothelial cells

leukocytes

neuronal cells

bronchial & vascular smooth muscle

Figure 8.33 Signaling by leukotrienes. Leukotrienes derived from arachidonic acid mediate sustained inflammatory responses.

Problem 8.6

Suggest a plausible arrow-pushing mechanism for the enzymatic conversion of 5-HPETE into leukotriene A_4, assuming that the enzyme 5-lipoxygenase has acidic and basic residues in its active site. The following is an abbreviated version of the reaction shown in Figure 8.33.

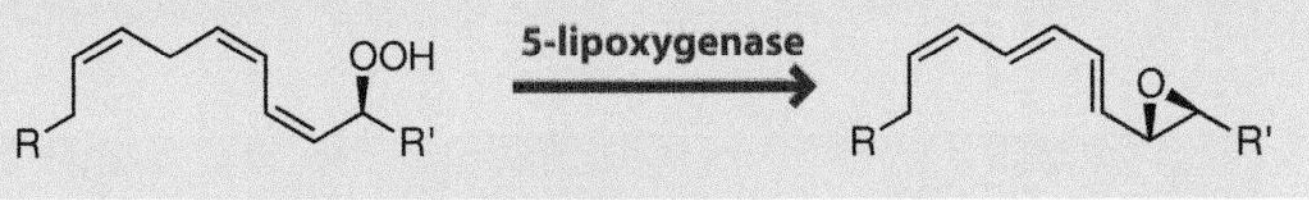

5-lipoxygenase

Sphingosine derivatives are important in intracellular signaling

Sphingolipids are essential components of human cell membranes, yet the intermediates in sphingolipid biosynthesis also have important effects on cell function (**Figure 8.34**). Many of these sphingoid molecules exhibit profound biological activities when added to cells. Ceramide halts cell growth, for example, and is cytotoxic as a result of interference with cell cycle regulation via binding to protein phosphatases and the protease cathepsin D. Sphingosine has similar effects on cells. Interestingly, sphingosine 1-phosphate has the opposite effect, inducing the inhibition of apoptosis and causing cell growth. The enzyme sphingosine kinase 1 required to phosphorylate sphingosine is recruited to the plasma membrane by phosphatidic acid, a product of phospholipase D.

A variety of natural mycotoxins synthesized by microfungi target early steps in sphingoid biosynthesis (**Figure 8.35**). Myriocin, sphingofungins, and lipoxamycin inhibit the first step in sphingoid biosynthesis, which is a decarboxylative C-acylation reaction of serine with palmitoyl-CoA. These compounds exhibit immunosuppressant activity and have been shown to inhibit the proliferation of mouse T cells.

Figure 8.34 Stop and go. The biosynthesis of sphingolipid signaling molecules involves several enzyme-catalyzed steps. Mycotoxin inhibitors of this pathway (red) are immunosuppressants.

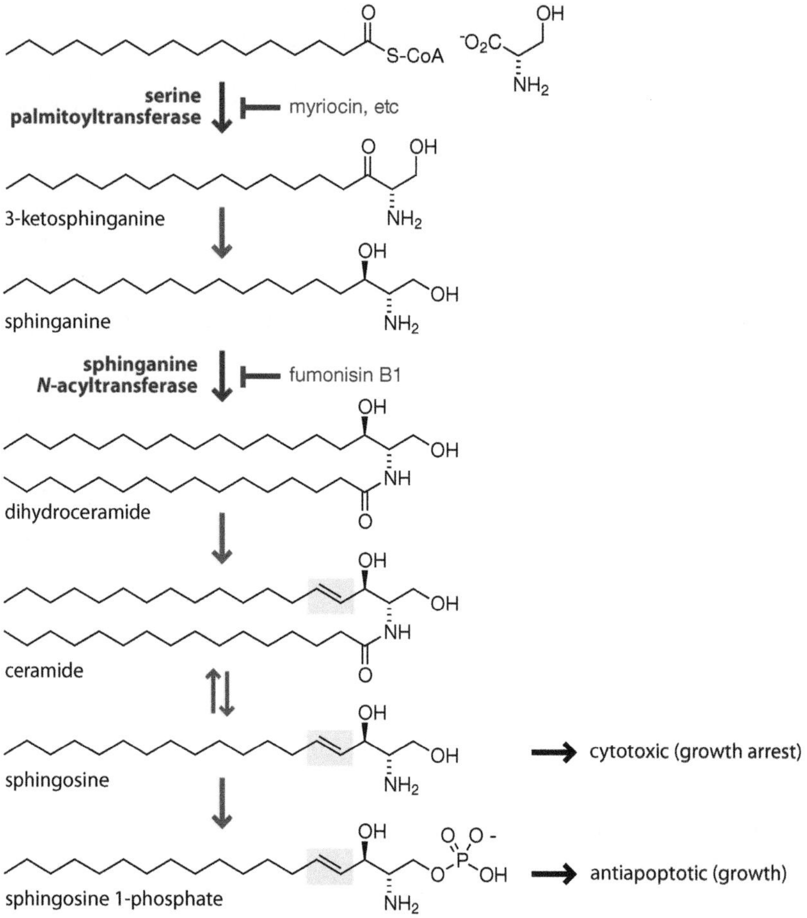

Fumonisin B1 inhibits the further acylation of sphinganine to produce dihydroceramide. The toxin is produced by a fungus (*Fusarium moniliforme*) that infects corn. The fungus has been shown to cause pulmonary edema in pigs, leukoencephalomalacia in horses, and tumors in rats.

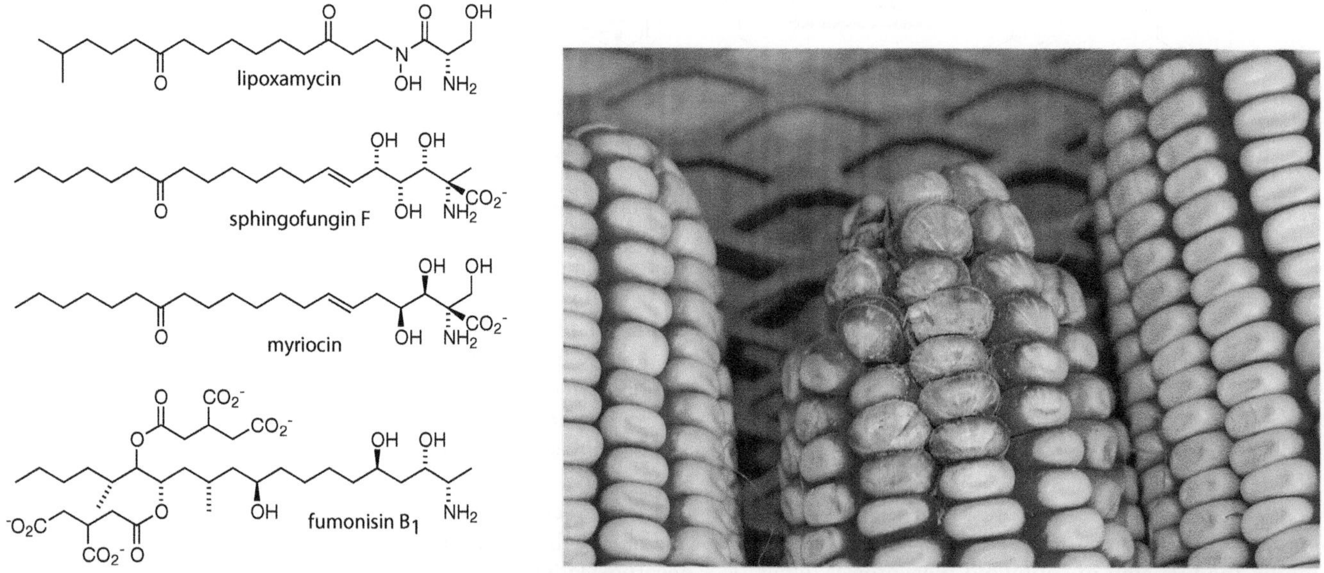

Figure 8.35 Mycotoxins produced by the mold *Fusarium moniliforme*. This mold infects corn, leading to the ear rot shown in the photograph. (Photograph courtesy of Alison Robertson, Iowa State University.)

Metal-catalyzed hydrogenation of unsaturated fats changed the human diet

Obesity and diabetes are two of the most challenging health problems facing industrialized nations. Whereas most diseases of developing countries are diseases of deficiency (that is, too little food, water, and vitamins), most diseases of developed countries are diseases of excess (that is, too much food, cholesterol, sugar, and fat). The major problem with the modern human diet is not merely the quantity of fat, but the type of fat. Polyunsaturated fatty acids and *cis* fatty acids are an essential part of the human diet, and they are healthier than *trans* fatty acids and saturated fatty acids, which contribute to blockage of the coronary arteries (**Figure 8.36**).

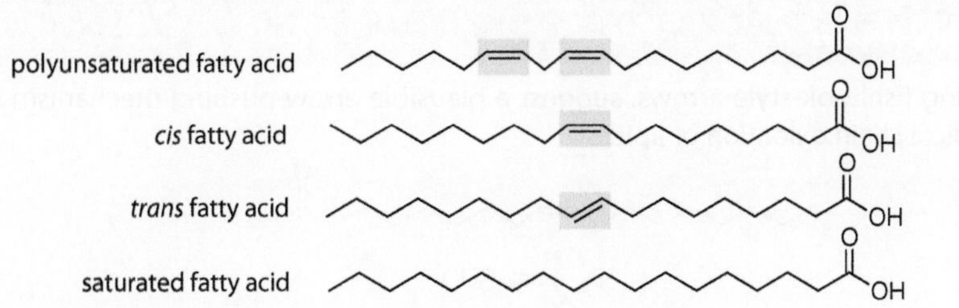

polyunsaturated fatty acid

cis fatty acid

trans fatty acid

saturated fatty acid

Figure 8.36 Classifying fatty acids. The number of double bonds and stereochemistry (*cis* or *trans*) distinguish the types of fatty acids used in nutritional labeling.

Saturated fatty acids and *trans* fatty acids can pack tightly together and tend to be solids at room temperature. In contrast, *cis* fatty acids cannot readily pack together and tend to be liquids. Unfortunately, highly liquid fats based on polyunsaturated fatty acids and *cis* fatty acids readily undergo autoxidation. In 1820, the Swiss chemist Nicolas-Théodore de Saussure noted that a thin layer of walnut oil was able to absorb 150 times its volume of oxygen over a one-year period. Back in the days of horse-drawn wagons, oils and fats often went rancid owing to a combination of short shelf life and slow transport. Solid fats derived from saturated fats tend to have longer shelf lives, however, and they give foods a smooth velvety texture that coats the tongue.

In 1909 Procter & Gamble began to market Crisco®, a form of partly hydrogenated cottonseed oil with a long shelf life (**Figure 8.37**). Partial hydrogenation converts

Figure 8.37 Cooking cottonseed oil. Partly hydrogenating vegetable oil turns *cis* fats into *trans* fats and saturated fats.

good, healthy, *cis* fats to unhealthy *trans* fats and unhealthy saturated fats. Before the development of partly hydrogenated vegetable oils, the only sources of *trans* fats in a natural diet were meat and dairy products from cows and sheep (less than 5%). Partly hydrogenated vegetable oils have become a cheap staple of western diets. Before 2008, a large order of french fries from the world's largest hamburger chain contained 8 g of *trans* fat (out of 30 g), but since then the company has started making its french fries with oil that is free of *trans* fats. In modern times, it is unnecessary to rely on partly hydrogenated oils, because transportation is fast and shelf life is not an issue. Healthy diets tend to be made up of oils such as olive oil, in which 55–85% of the fatty acids are oleic acid, a *cis* fatty acid.

Problem 8.7

Using fishhook-style arrows, suggest a plausible arrow-pushing mechanism for the radical chain oxidation of lipids.

O, OR

O_2 ↓ cat. ROO•

HOO, O, OR

Some lipids from lower organisms contain cyclopropane rings

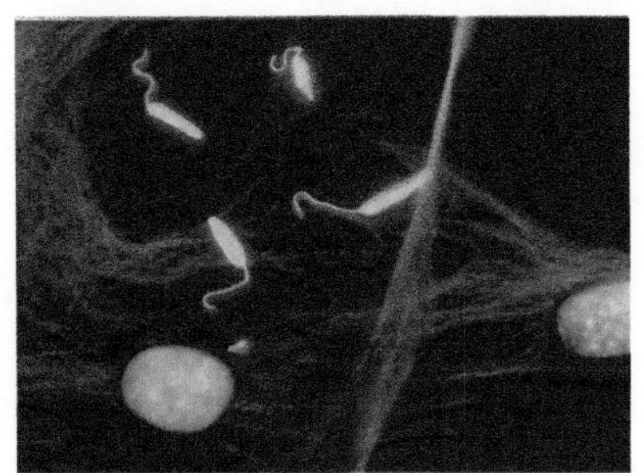

Figure 8.38 The free-swimming form of *Leishmania* (green), called a promastigote, infecting a macrophage. (Courtesy of the Beverley Laboratory, Washington University School of Medicine, St. Louis.)

In the 1950s, the bacteria in the genus *Lactobacillus* were found to contain lipids that incorporated cyclopropane rings. More than 16% of the fatty acid content of lipids from the common strain *Lactobacillus casei* is composed of the cyclopropane-containing fatty acid lactobacillic acid. *Leishmania* cells produce another cyclopropane fatty acid, dihydrosterculic acid. Sandflies transmit the protozoal parasite *Leishmania*, which causes leishmaniasis, a disease characterized by disfiguring open sores. The *Leishmania* cells enter and proliferate inside macrophages, which normally kill microorganisms (**Figure 8.38**).

Cyclopropane fatty acids are biosynthesized in two steps from saturated fatty acids (**Figure 8.39**). The first step involves the oxidative introduction of a *cis* double bond by a desaturase enzyme, and the second step involves cyclopropanation by the sulfonium ion *S*-adenosylmethionine. Interestingly, the biosynthesis of dihydrosterculic acid was shown to be most strongly inhibited by 9-thiastearic acid, although actual growth inhibition of *Leishmania* was strongest for the 10-isomer. In similar studies against the parasite *Trypanosoma cruzi*, it was speculated that the thia fatty acids were inhibiting the desaturase enzyme.

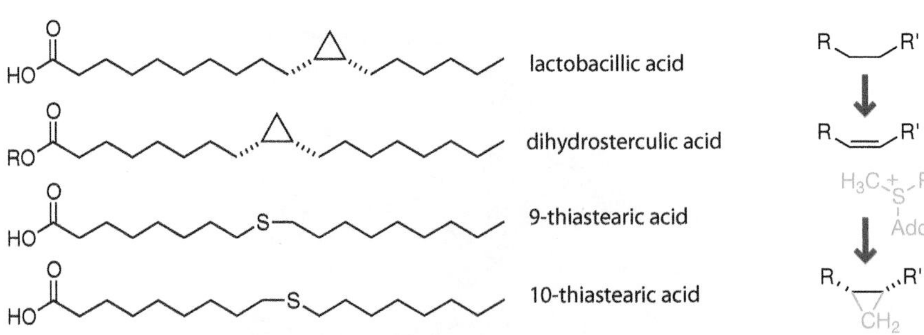

Figure 8.39 Cyclopropane- and thia-based lipids. Some lipids from microorganisms, such as lactobacillic acid and dihydrosterculic acid, contain cyclopropane rings generated through the two-step process shown on the right. 9-Thiastearic acid and 10-thiastearic acid, in contrast, inhibit the biosynthesis of dihydrosterculic acid.

Problem 8.8

Suggest a plausible arrow-pushing mechanism for the following base-promoted reaction.

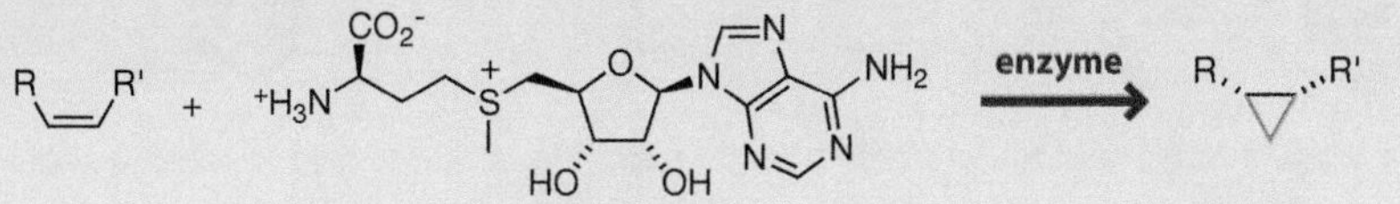

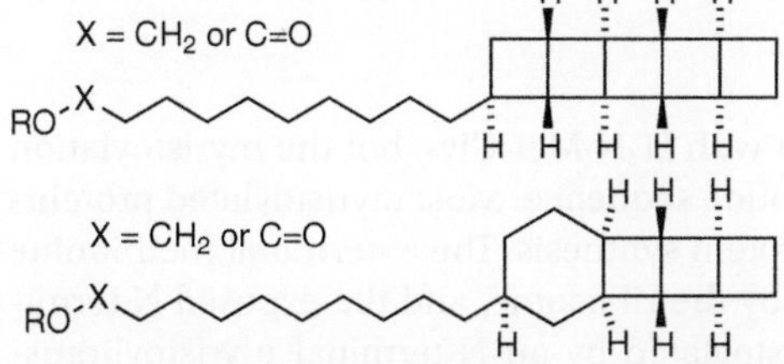

Figure 8.40 Lipid ladders. Ladderane fatty acids from *Candidatus Brocadia anammoxidans* contain cyclobutane rings.

An amazing set of lipid structures comes from the ammonotrophic bacterium *Candidatus Brocadia anammoxidans*. About half of the total lipid content in these bacteria is composed of ethers and fatty acids containing concatenated four-membered rings (**Figure 8.40**). Such compounds are named ladderanes for their resemblance to ladders in a two-dimensional depiction.

Problem 8.9

Cyclopropane fatty acids can be distinguished from saturated fatty acids by hydrogenating mixtures of fatty acid esters and then subjecting them to gas chromatography (GC) before and after treatment with bromine; after the treatment, the peaks for the cyclopropane-containing molecules (green arrows) disappear from the chromatogram. Suggest a plausible arrow-pushing mechanism for the reaction of lactobacillic acid with bromine to yield two diastereomeric dibromides.

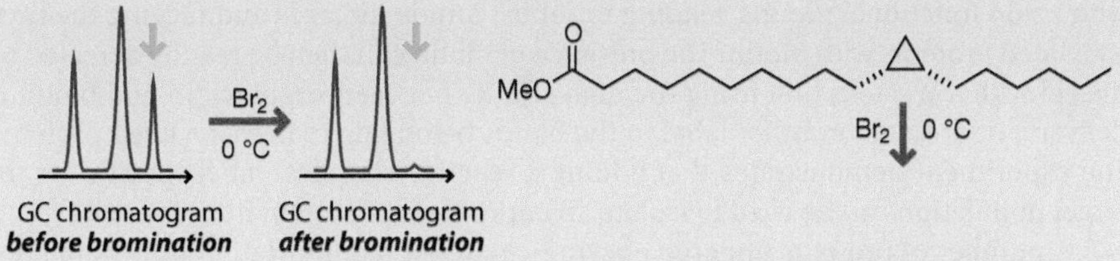

Acylation of human proteins induces membrane localization

Proteins participating in signal transduction pathways interact with each other through specific high-affinity interactions. Association with membranes is often a key requirement for these kinds of interactions; such binding confines the protein to a particular subcellular location and can cause a change in conformation. For many cell-surface receptors, hydrophobic residues confer a permanent affinity for the plasma membrane. For other proteins, membrane affinity is a conditional property that initiates, propagates, or terminates a signaling pathway. Reversible association with lipophilic molecules can control membrane affinity. Recall that the binding of diacylglycerol to the C1 domain of PKC confers membrane affinity. Later we shall see how vitamin K brings about the binding of phosphatidylserine to Gla protein domains, thus conferring membrane affinity.

Other proteins are covalently modified with fatty acids, which provides a greasy tail for binding to the membrane. For example, some human proteins possess an N-terminal signal that causes the N terminus to be acylated with the 14-carbon fatty

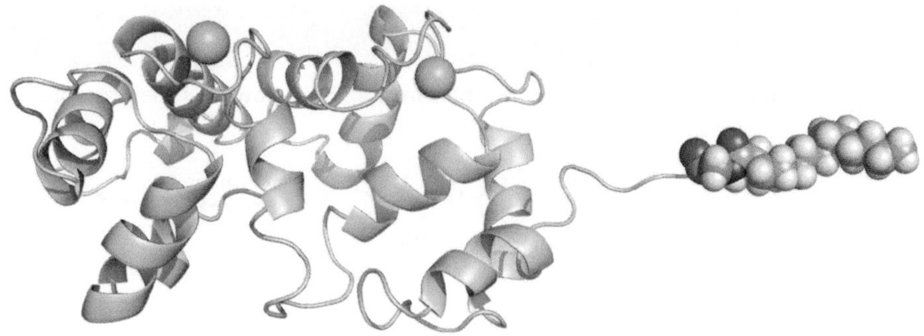

Figure 8.41 Unmasked with calcium. When recoverin binds Ca^{2+} ions (green spheres), the N-terminal myristoyl group (van der Waals spheres) is available to interact with membranes.

acid myristic acid. All such proteins start with H_2N-Met-Gly-, but the myristoylation signal is not simply a particular short peptide sequence. Most myristoylated proteins are modified cotranslationally, during protein synthesis. The N-terminal methionine is cleaved as the proteins are translated by the ribosome, and the exposed N terminus of the glycine residue is then myristoylated by an N-terminal myristoyltransferase enzyme. The protein recoverin, which is important for human vision, normally sequesters its N-terminal myristoyl group; however, when it binds calcium, the myristoyl group is extruded, thus allowing it to interact with membranes (**Figure 8.41**).

Less commonly, myristoylation is a post-translational event. For example, recall that the protein Bid associates with mitochondria as part of caspase-mediated apoptosis (see Figure 6.37). Myristoylation is induced when the cleavage of Bid by caspase-8 reveals an N-terminal glycine residue. The acylation of serine and threonine residues is rare in human proteins. One of the few examples is the 28-amino-acid gastric peptide hormone ghrelin. It is enzymatically acylated on Ser3 with dietary octanoyl groups and then released into the circulation, where it stimulates a hunger response.

The bio-orthogonal Staudinger ligation has been used to identify myristoylated proteins in cells (**Figure 8.42**). The azido derivative, 12-azidododecanoic acid, can replace myristic acid during protein myristoylation. After treatment with this compound, the cells are lysed, and their proteins are run on a polyacrylamide gel. The myristoylated proteins can be selectively detected by treating the gel with a 2-phosphinobenzoate ester carrying a biotin handle. 2-Phosphinobenzoates react selectively with azido functional groups, leading to robust amide linkages and tagging the myristoylated proteins with biotin. The presence of biotin can then be readily detected on the gel with a Western blot using the high affinity between streptavidin and biotin to deliver a fluorescent or other label to the bands belonging to myristoylated proteins. The experiment demonstrates that adding a reactive chemical label specific for the target population can be used to isolate an entire class of protein from the cell.

A number of proteins undergo enzyme-catalyzed palmitoylation of cysteine residues. The cysteines are often located at an N terminus or C terminus that already has a lipophilic modification. For example, the protein tyrosine kinase $p56^{lck}$ is

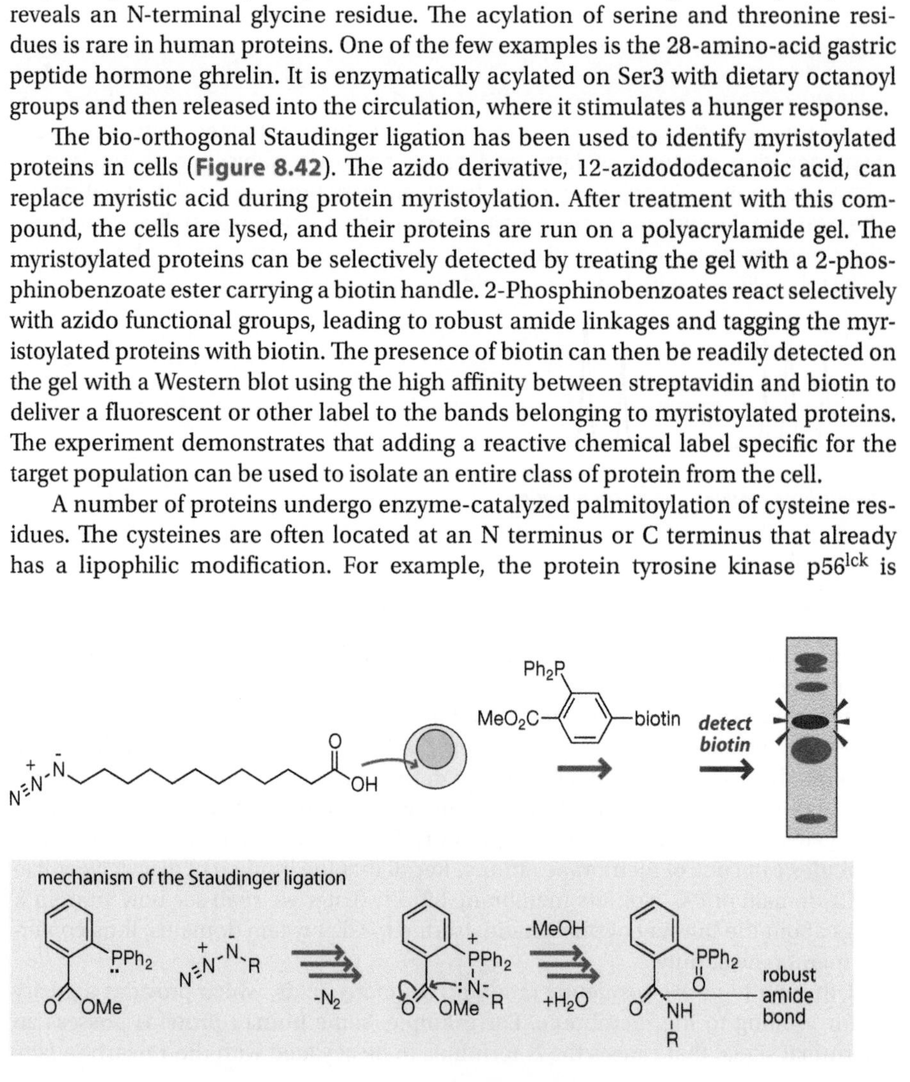

Figure 8.42 Pinned by a ligation reaction. The Staudinger ligation can be used to detect proteins that are acylated with an azido analog of myristate.

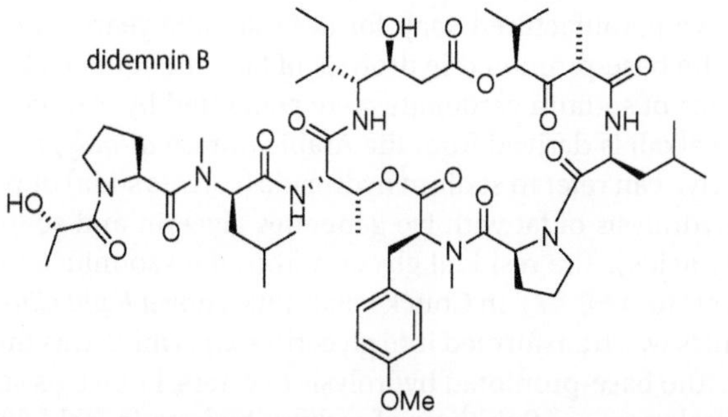

Figure 8.43 Didemnin B, an inhibitor of palmitoyl protein thioesterase.

palmitoylated at two of the cysteines near the N-terminal myristoyl group. In a protein with the sequence myristoyl-GCGCSS, for example, the underlined cysteine residues are modified. The *S*-palmitoylation reaction is catalyzed by the enzyme palmitoyl protein transferase, which catalyzes a transacylation with palmitoyl-CoA. In some instances (for example mannose-6-phosphate receptor) depalmitoylation occurs on a reasonably fast timescale ($t_{1/2}$ = 2 hours). Depalmitoylation can be catalyzed by the enzyme palmitoyl protein thioesterase. The cyclic peptide natural product didemnin B (**Figure 8.43**), produced by a primitive ocean organism called a tunicate, is a potent inhibitor (K_i = 92 nM) of palmitoyl protein thioesterase. Inhibition of this enzyme confers strong cytotoxic activity against cells infected with RNA viruses and against tumor cells. Unfortunately, the compound failed to progress beyond phase II clinical trials because of anaphylactic toxicity.

Problem 8.10

Most *S*-palmitoyl groups are readily cleaved by thiolates, but the palmitoyl group at the N-terminal cysteine of Sonic hedgehog protein is not readily cleaved. Suggest a plausible reason for this stability of the Sonic hedgehog palmitoyl group.

Chemical transformation of fats generates useful compounds

Humans store fatty acids as triesters of glycerol, more commonly referred to as triacylglycerols, or triglycerides. Such compounds are stored in adipocytes for subsequent retrieval as energy sources. Triacylglycerols are the component of cooked foods that we typically associate with the word fat. Soybean oil (marketed as vegetable oil) and other cooking oils, chicken fat, and bacon grease are mainly composed of triacylglycerols. Because of their high boiling points, such fats are convenient solvents for frying foods.

In the search for a more economical use of natural resources, there has been a growing trend toward the replacement of fossil fuels with fuels based directly on renewable resources. Early in this millennium, in response to rapidly escalating fuel prices, there was a push toward the reclamation of used cooking oils for use as automotive fuels. A century earlier, Rudolf Diesel demonstrated a diesel engine that operated on peanut oil, but high-viscosity triacylglycerols are not well suited for modern diesel engines; however, it is simple to convert triacylglycerols to esters of simple alcohols through base-catalyzed transesterification (**Figure 8.44** and **Figure 8.45**). Methyl and ethyl esters of fatty acids have lower viscosities and lower flash points and are conveniently burned in diesel automobiles with little or no modification to the engine.

Figure 8.44 Biodiesel in various stages of preparation. (Courtesy of Luc S. Blais.)

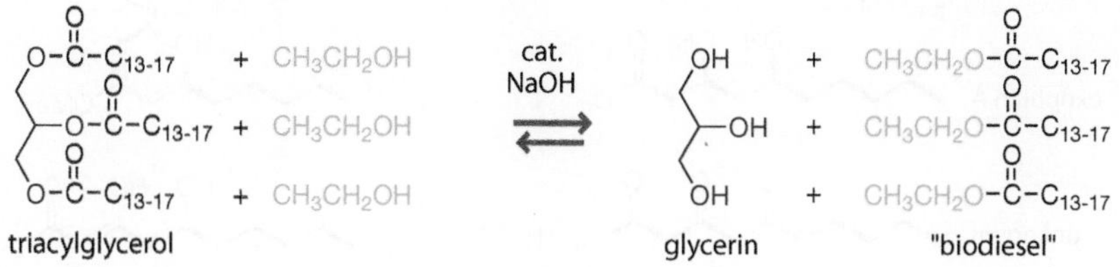

Figure 8.45 The synthesis of biodiesel through transesterification.

Figure 8.46 A woman from the Mississippi valley, ca. 1911, making lye soap. (From C. Johnson, Highways and Byways of the Mississippi Valley. New York: Macmillan, 1913.)

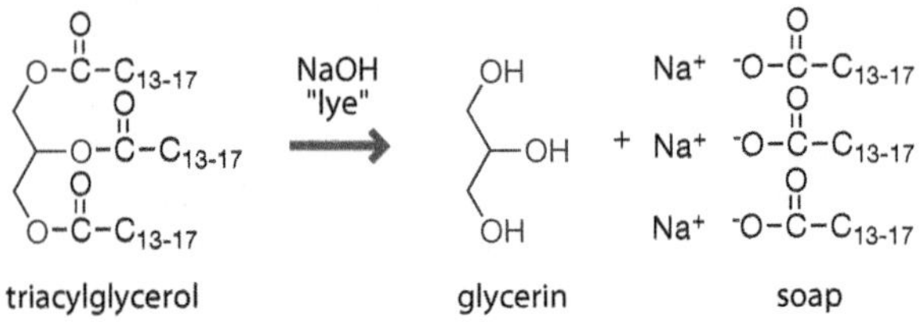

Humans have manufactured soap for at least 2000 years. Traditionally, soap is made through the base-promoted hydrolysis of fat with aqueous base (**Figure 8.46**). Alkaline solutions of sodium carbonate were generated by extracting wood ash with water. The term alkali is derived from the Arabic phrase *al-qaly*, referring to ash. The shortened term lye can refer to sodium hydroxide (caustic soda) or potassium hydroxide (potash). Hydrolysis of fat with lye generates glycerin and soap (sodium salts of long-chain fatty acids). The residual glycerin from the saponification of fat is valued as a moisturizer (**Figure 8.47**). In Chuck Palahniuk's novel *Fight Club*, human fat from liposuction clinics was transformed into glycerin soap, which was then resold in high-end boutiques. The base-promoted hydrolysis of esters, known as saponification, is a rapid and general reaction. Solutions of sodium hydroxide feel "slippery" when you rub them between your fingers because the hydroxide is saponifying the lipids in your outer layer of dead skin cells.

Figure 8.47 The synthesis of glycerin soap by the hydrolysis of fat.

8.4 NONHUMAN POLYKETIDE NATURAL PRODUCTS

Several tricks amplify the potential diversity of polyketide natural products

Unlike simpler organisms that produce a structurally diverse range of polyketides, humans never really tapped the full biosynthetic potential of polyketide biosynthesis in their metabolism. A variety of tricks are used by some organisms to expand the potential diversity of polyketides. One trick that expands the diversity of polyketides is to skip some of the steps that lead to an aliphatic backbone. Recall that fatty acid synthase follows each Claisen condensation with a carbonyl reduction, a β-elimination, and a conjugate enone reduction. Omitting any of these steps generates an alternative functional group. For example, leaving out the conjugate reduction leaves a double bond between the ketide subunits (**Figure 8.48**). Many mammalian fatty acids have double bonds at specific positions, but such double bonds are introduced by a dehydratase domain, long after the fatty acid synthase has completed its work. Recall that the ketoreductase in human fatty acid synthase creates a stereogenic carbinol center during each cycle, and then destroys it through a β-elimination. What a waste! Leaving out both the conjugate reduction and the β-elimination leaves a hydroxyl group with a defined configuration. Repeatedly skipping these steps generates 1,3-diols, which are a distinguishing feature of many polyketide natural products. Leaving out all of the steps after the Claisen condensations would generate a polymer of ketide subunits. Such a polymer, containing skipped carbonyl groups, would be unstable because of

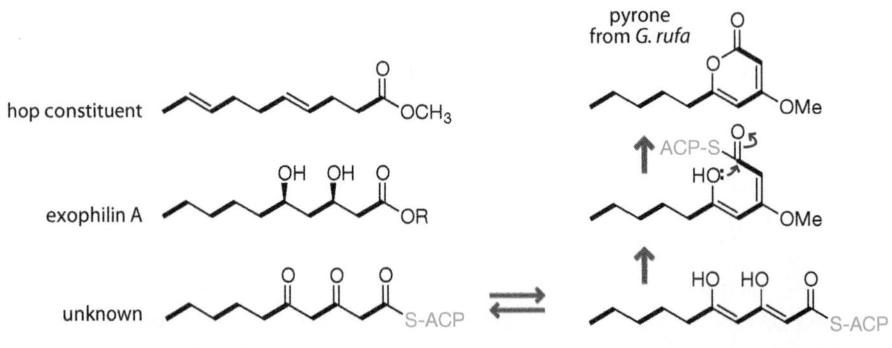

Figure 8.48 Diversifying polyketides. Leaving out conjugate reductions (hop constituent), β-eliminations (exophilin A), and carbonyl reductions (unknown) increases the potential diversity of polyketides. Reactive compounds resulting from such omissions can undergo rearrangements (for example, to give the pyrone from *Galiella rufa*).

the potential for cyclization reactions. The enolic forms of 3,5-diketoesters are susceptible to cyclization to generate pyrones, and some polyketide synthases harness these cyclization reactions. For example, the fungus *Galiella rufa* produces a 3-methoxypyrone through a route involving a 3,5-diketoester intermediate.

Polymers containing skipped carbonyl groups are also highly susceptible to intramolecular aldol condensations. For example, the naturally occurring benzoic acid derivative orsellinic acid is a tetraketide derived from an aldol condensation followed by double tautomerization to an aromatic compound (**Figure 8.49**). Similarly, curvulinic acid is derived from a pentaketide through an aldol condensation and aromatization.

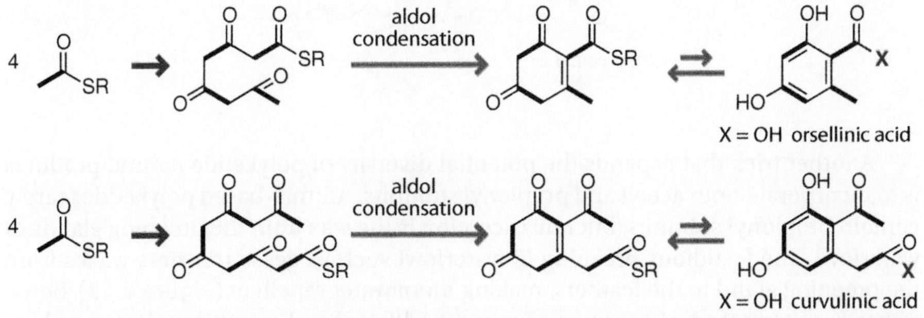

aldol condensation

X = OH orsellinic acid

aldol condensation

X = OH curvulinic acid

Figure 8.49 Synthesis of aromatic polyketides. Polyketide synthases use intramolecular aldol condensations to generate aromatic compounds.

Problem 8.11

Suggest a plausible arrow-pushing mechanism for the following transformation using a catalytic base.

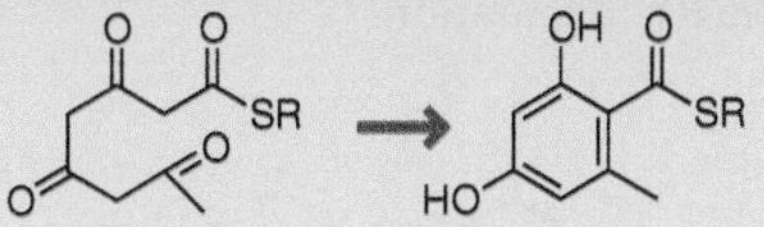

A wide range of aromatic natural products can arise through this type of aldol condensation/aromatization cascade. Usually, the synthases that catalyze the formation of aromatic polyketides perform additional steps both before and after the aromatization steps. The reduction of ketone groups before aromatization leads to the β-elimination of water during the aromatization step, leaving the aromatic carbon without a hydroxyl substituent (**Figure 8.50**). Many of the anthraquinone antibiotics that bind to DNA are generated by the oxidation of electron-rich anthracyclines (**Figure 8.51**). During the biosynthesis of nogalamycin, for example, two of the aromatic positions are further oxidized by enzymes coded for by the nogalamycin biosynthesis gene cluster.

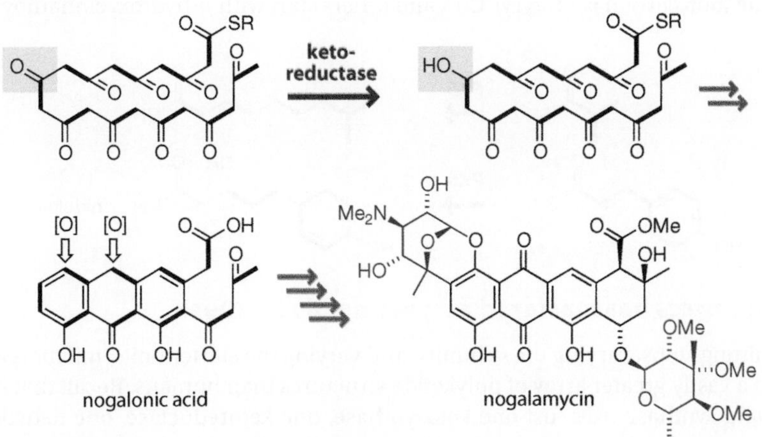

Figure 8.50 Making nogalamycin. Nogalamycin biosynthesis involves additional steps before and after aromatization of the decaketide.

Figure 8.51 Nogalamycin in action. Two molecules of nogalamycin (depicted as stick models) are intercalated between the base pairs of double-stranded DNA with the sequence CGTACG (rendered as spheres). (PDB 182D)

Figure 8.52 Water proofing. Waterfowl such as greylag geese wax their feathers by transferring water-repellent waxes from their preening gland. (Courtesy of Alan Weaver.)

Problem 8.12

For the following two aromatic polyketides, identify the backbone resulting from aldol condensation reactions, highlighting the two-carbon building blocks.

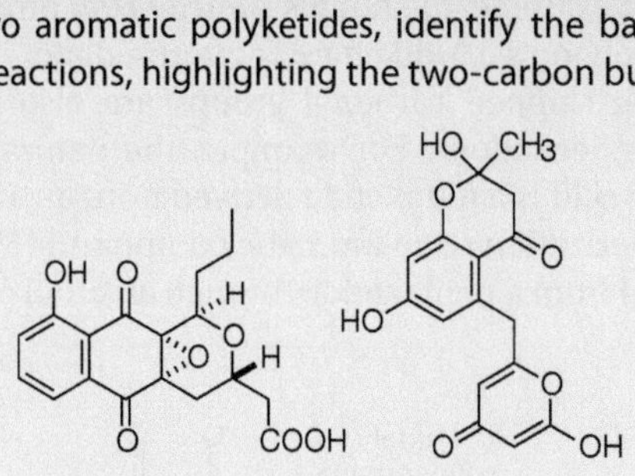

frenolicin B SEK4

Another trick that expands the potential diversity of polyketide natural products is to incorporate both acetyl and propionyl subunits. Animal-based polyketides rarely contain propionyl subunits, but one exception is the wax from the preening glands of waterfowl. The fastidious preening of waterfowl such as geese transfers waxes from the preening gland to the feathers, making them water repellent (**Figure 8.52**). Some waxes are composed of mixtures of acetyl and propionyl subunits, whereas others contain fatty acids composed solely of propionyl subunits (**Figure 8.53**). The incorporation of propionyl subunits can lead to stereocenters with *R* or *S* configurations. Each of the methyl groups in goose preening wax is attached to a stereogenic center with the *R* configuration.

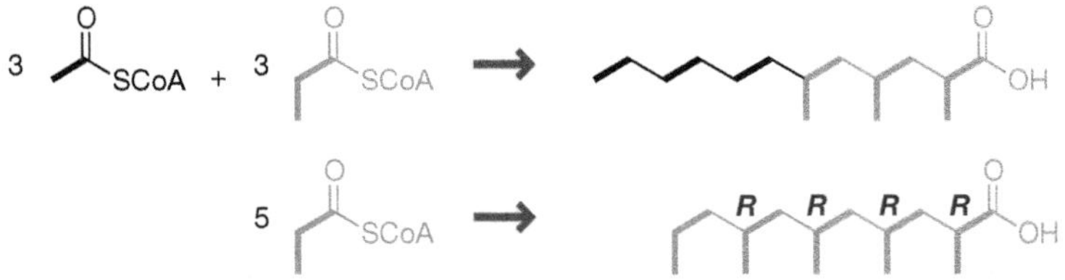

Figure 8.53 Waxy fatty acids from geese are made up from both acetyl and propionyl subunits.

Polyketide synthases rarely construct polyketides from subunits with more than three carbon atoms. However, there is considerable variability in the starter unit that is loaded by the loading module (**Figure 8.54**). For example, some polyketides start with the four-carbon isobutyryl-CoA and others start with *p*-hydroxycinnamoyl-CoA.

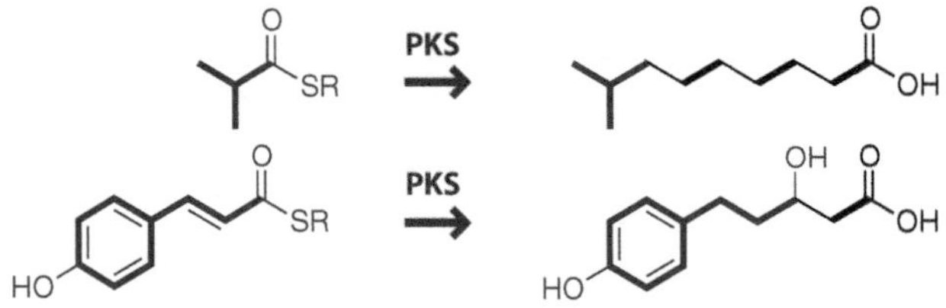

niduloic acid

Figure 8.54 Variation in the starter units of polyketides: isobutyryl-CoA (top) and *p*-hydroxycinnamoyl-CoA (bottom). PKS, polyketide synthase.

Streptomyces has mastered polyketide biosynthesis

By omitting steps, varying the subunits, and varying the starter units, microorganisms access a vastly greater array of polyketide structures than humans. Recall that human fatty acid synthase uses just one ketosynthase, one ketoreductase, one dehydratase, and one enoyl reductase for each of the two-carbon ketide subunits in the resulting oligomer. The resulting short polymers are monotonous. The most significant trick exploited by microorganisms to produce diverse polyketides is to devote a unique set

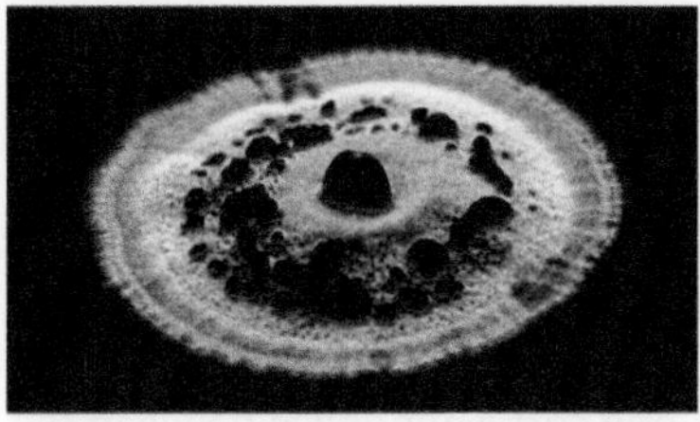

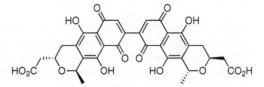

Figure 8.55 Colonies of *Streptomyces coelicolor* A3 excreting droplets containing the blue antibiotic actinorhodin, depicted on the right. (Photograph courtesy of Sir David Hopwood, John Innes Centre.)

of enzymatic subunits to the synthesis of each and every ketide unit. This makes it possible to expand polyketide biosynthesis beyond repeating polymers to oligomers with a defined sequence of functionalities and structures.

Soil bacteria in the phylum Actinobacteria are some of the premier architects of polyketide natural products. Microbiologists have had considerable success in culturing Actinobacteria in the laboratory and isolating bioactive natural products, often polyketides. Among the various species and strains of Actinobacteria that have been collected around the world, those of the genus *Streptomyces* are particularly common sources of bioactive polyketide natural products (**Figure 8.55**). Using the same basic reactions as those employed to make fatty acids, *Streptomyces* produces a dazzling array of bioactive polyketide natural products, such as amphotericin B, tetracycline, tylosin, and avermectin B_{1a} (**Figure 8.56**). Because the backbone is generated through iterative Claisen condensations, vestigial repeating patterns of 1,3-oxygenation are apparent, particularly in natural products such as amphotericin B.

amphotericin B
Streptomyces nodosus

tetracycline
Streptomyces aureofaciens

tylosin
Streptomyces fradiae

avermectin B_{1a}
Streptomyces avermitilis

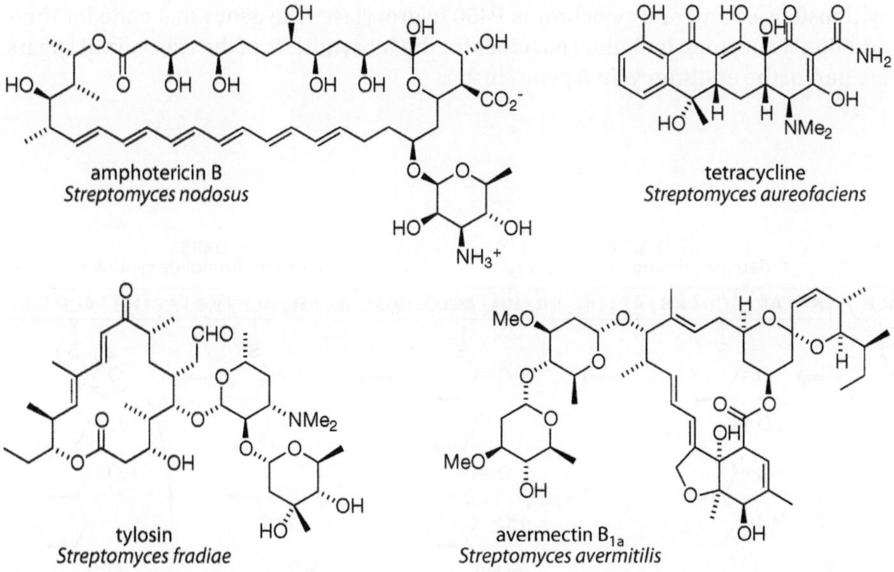

Figure 8.56 *Streptomyces* and their natural products. Clinically useful natural products from *Streptomyces* illustrate the power of polyketide biosynthesis to access a wide range of structures.

Problem 8.13

Highlight each of the individual two-carbon acetate and three-carbon propionate subunits in linearmycin A.

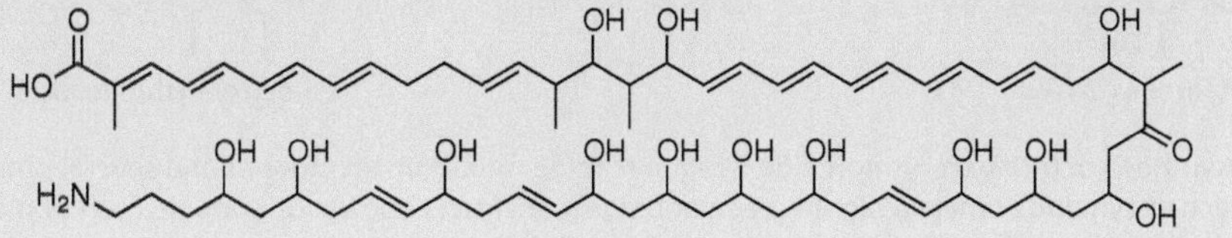

The modular genetic organization of type I polyketide synthases facilitates genetic reprogramming

In RNA transcription there is largely a one-to-one sequence correspondence between the DNA template and the RNA transcript (not counting splice variants or other alterations to the RNA sequence). In protein translation there is also a one-to-one correspondence between the nascent protein and the mRNA sequence. Similarly, in the biosynthesis of many polyketides there is a one-to-one correspondence between the polyketide structure and the sequence of enzymatic domains involved in biosynthesis. In theory, the modular genetic organization of many polyketide synthases offers the potential to (re)program the biosynthesis of any polyketide by deleting or replacing genetic elements. It is easier and more efficient to cut and splice DNA than to re-engineer the total synthesis of complex polyketide natural products.

The best-studied example of a modular type I polyketide synthase is that involved in the synthesis of the macrolide antibiotic erythromycin. Several thousand tons of erythromycin are produced annually through microbial fermentation. The key macrocycle, 6-deoxyerythronolide B, is generated by three very large, multisubunit proteins—namely, 6-deoxyerythronolide B synthases 1, 2, and 3 (DEBS1, DEBS2, and DEBS3). Each of the DEBS proteins is composed of enzymatic domains that perform the same reactions used to produce fatty acids in the order required to program the erythronolide structure. The single exception is the thiolesterase domain, which catalyzes a macrolactonization rather than a hydrolysis of the acyl-enzyme linkage that terminates fatty acid synthesis (**Figure 8.57**). The 6-hydroxyl group is added by a cytochrome P450 hydroxylase, eryF, to produce the aglycone of erythromycin B. Three more enzymes are required to convert erythronolide B to erythromycin A: two glycosyltransferases and one cytochrome P450 hydroxylase. The genes that code for these additional enzymes, including enzymes for the biosynthesis of the specialized sugars, are part of the erythromycin A gene cluster.

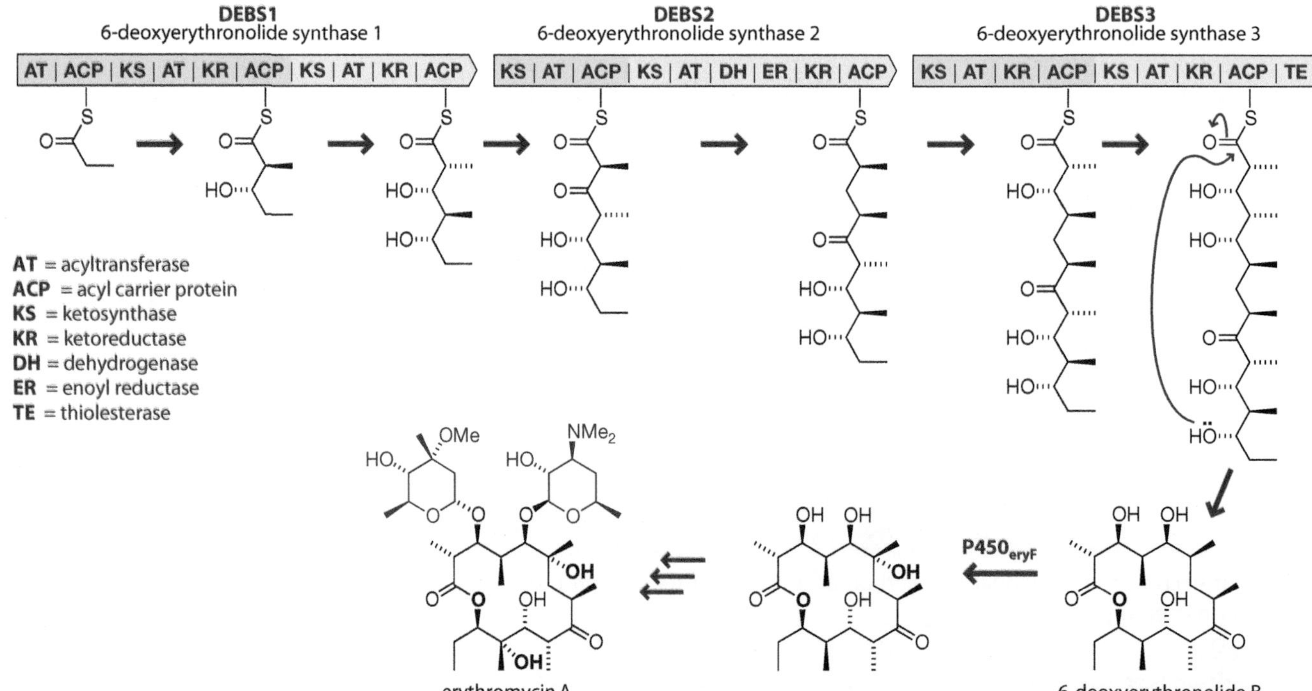

Figure 8.57 A polyketide assembly line. The biosynthesis of 6-deoxyerythronolide B requires three multidomain proteins and several other enzymes such as hydroxylases and glycosylases. Each enzymatic domain (colored green or beige) constructs one ketide subunit. Each domain is composed of multiple enzymatic subunits, each with an independent catalytic function (see the key).

Problem 8.14

Highlight each of the acetate and propionate subunits in halstoctacosanolide A. Which of the carbon atoms in halstoctacosanolide A requires additional oxidation reactions by an oxidase?

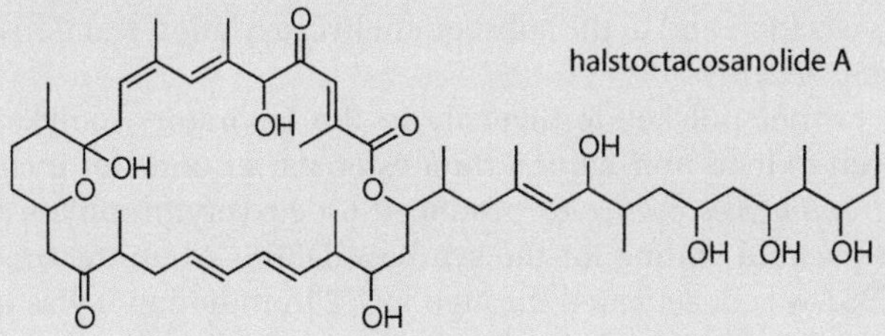

halstoctacosanolide A

When the modules are deleted or exchanged through genetic engineering, they can often still produce functional multienzyme complexes. Thus, the size, functionality, and stereochemistry of polyketides can be controlled through reliable genetic methods. For example, the thiolesterase subunit that cyclizes the final 6-deoxyerythronolide chain can be moved to an alternative location, resulting in a smaller ring (**Figure 8.58**).

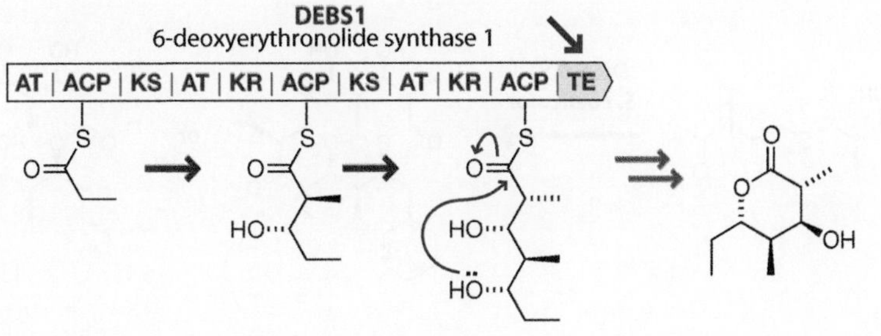

DEBS1
6-deoxyerythronolide synthase 1
AT | ACP | KS | AT | KR | ACP | KS | AT | KR | ACP | TE

Figure 8.58 Changed outcome. Moving the thiolesterase domain (TE) from DEBS3 to DEBS1 leads to a new lactone natural product through premature cyclization.

When additional modules are genetically introduced without optimization, the biosynthetic machinery happily introduces the new ketide units but the results are unpredictable. For example, when an acetate genetic module was transferred from the operon for rapamycin synthase to the operon for erythromycin synthase, new macrolide products were produced with an additional ketide unit; surprisingly, the original heptaketide **1** was still the major product of the reaction (**Figure 8.59**). Two

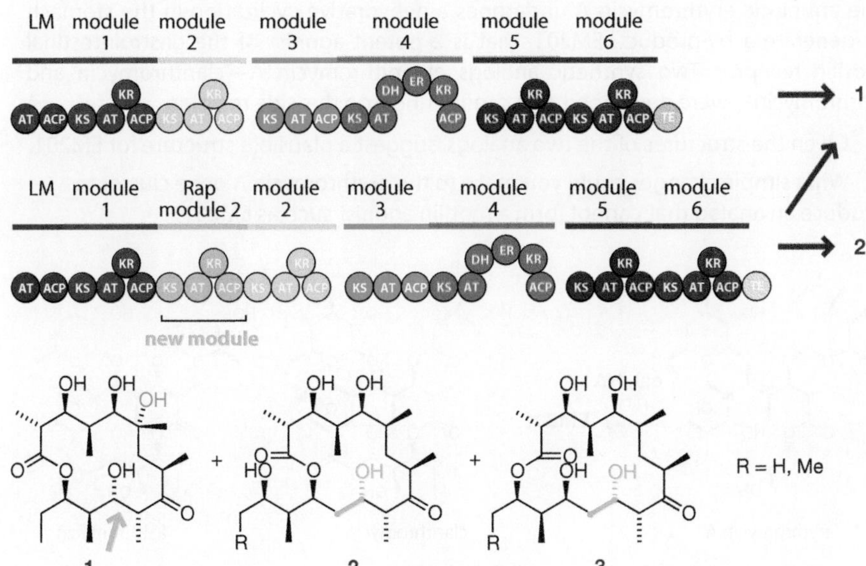

Figure 8.59 New natural products through genetic engineering. A new module was introduced into the operon for erythromycin synthase through genetic engineering, resulting in the biosynthesis of new "unnatural products." (Adapted from C.J. Rowe et al., *Chem. Biol.* 8:475–485, 2001. With permission from Elsevier.)

new types of octaketides (structures **2** and **3** in Figure 8.59) were detected, differing in the site of macrolactonization. It seems as though the synthase struggled to adapt to the newly introduced module. On the basis of the structures of the products, the loading module was frequently misloading acetyl-CoA instead of propionyl-CoA. Furthermore, the two new cyclic products were not being hydroxylated at the C6 position. Thus, the size and shape of the intermediates are probably important, both in the thiolesterase cyclization and in the subsequent hydroxylation reactions conducted by cytochrome P450 enzymes.

To access further polyketide diversity in the laboratory, polyketide synthases have been driven to load non-natural thiol ester starter units for incorporation into polyketides. *Streptomyces coelicolor* produced 6-deoxyerythronolide B when transformed with a plasmid coding for the synthase DEBS. When the first ketosynthase domain of DEBS1 was deactivated through a C729A mutation it was unable to initiate the assembly of 6-deoxyerythronolide B. However, when the growth medium was supplemented with chemically synthesized *N*-acetylcysteamine thiol esters (which mimic CoA thiol esters), those thiol esters were picked up by the first acyl carrier protein domain and transformed into macrolactone products. Through this combination of selective genetic deactivation and chemical rescue it was possible to create semisynthetic analogs of erythronolide B that would have been challenging to create with biological or chemical techniques alone (**Figure 8.60**).

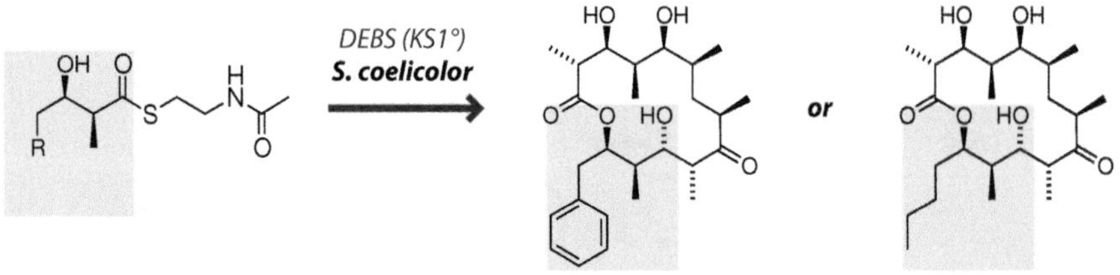

DEBS (KS1°)
S. coelicolor
or

Figure 8.60 Nonnatural starter units. When *S. coelicolor* is transformed with a plasmid encoding the C279A variant of the DEBS ketosynthase domain [*DEBS (KS1°)*], the ketosynthase will accept non-natural thioesters, including synthetic analogs.

Problem 8.15

The antibiotic erythromycin A undergoes a dehydrative cyclization in the stomach to generate a by-product, EM201, that is a potent agonist of the gastrointestinal motilin receptor. Two synthetic analogs of erythromycin A—clarithromycin and azithromycin—were developed that cannot undergo this side reaction.

A Given the structures of the two analogs, suggest a plausible structure for EM201.

B What simple changes could you make to the erythromycin A gene cluster to produce an analog that cannot form a motilin agonist such as EM201?

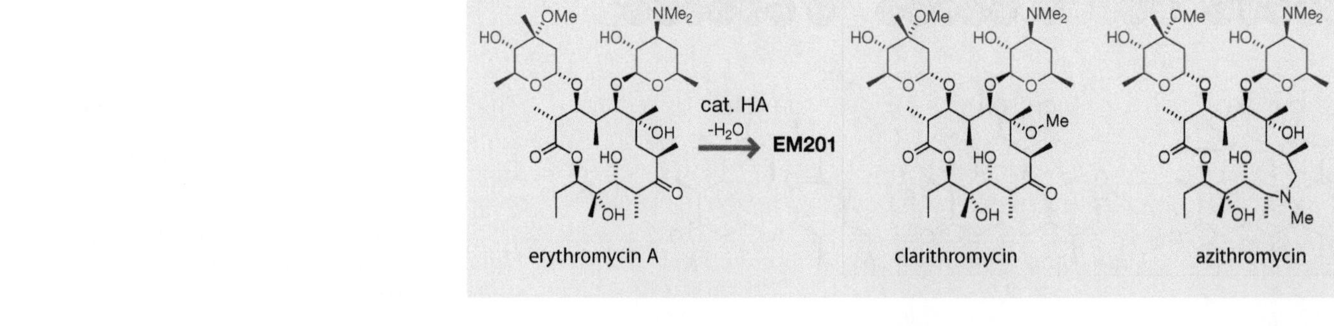

cat. HA
$-H_2O$
EM201

erythromycin A clarithromycin azithromycin

Microbial expression has obvious advantages over chemical synthesis, particularly in the ease of scale-up and the costs of starting materials. However, mixing and matching polyketide synthase modules in novel combinations often results in low yields of the polyketide natural products. It is expected that the expression of worthwhile polyketides can be optimized through combinations of random mutation and selection, much like the expression of erythromycin A. In reality, most newer versions of erythromycin (such as clarithromycin and azithromycin) are produced through a combination of optimized microbial expression and chemical modification.

Sometimes additional methyl groups are added to the polyketide backbone

Sometimes polyketides possess additional methyl groups whose presence cannot be rationalized on the basis of acetate or propionate subunits. Often, these methyl groups are added through alkylation of the enzyme-bound polyketide with the sulfonium ion *S*-adenosylmethionine. For example, in the biosynthesis of the microtubule-stabilizing agent epothilone C, alkylation of a β-ketothioester intermediate by a methyltransferase domain generates the quaternary center that appears in the natural product (**Figure 8.61**). *S*-Adenosylmethionine is not used merely to create quaternary centers; sometimes methyl groups that seem to derive from propionate subunits are actually derived from alkylation with *S*-adenosylmethionine. For example, all three carbon atoms that dangle from the chain of alternaric acid are derived from *S*-adenosylmethionine (**Figure 8.62**). Although it is true that methyl substituents on polyketide backbones sometimes arise from *S*-adenosylmethionine, this is not always so. In the absence of specific data, such as feeding organisms isotope-labeled propionate and observing the incorporation of the isotope, you should assume that methyl groups are introduced as a part of a propionate subunit.

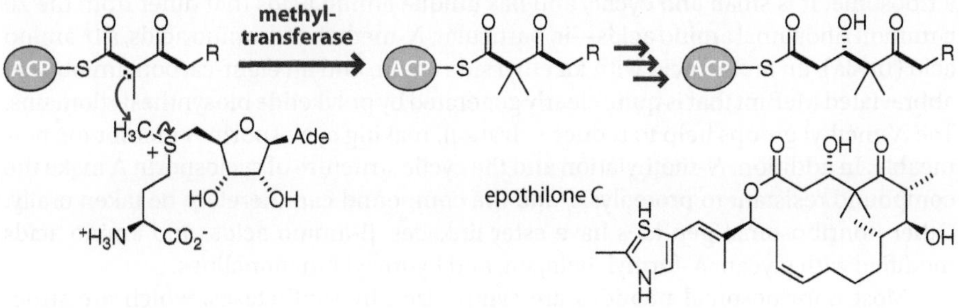

Figure 8.61 Give me a CH_3. The quaternary center in epothilone C arises through alkylation of a β-ketothioester with S-adenosylmethionine.

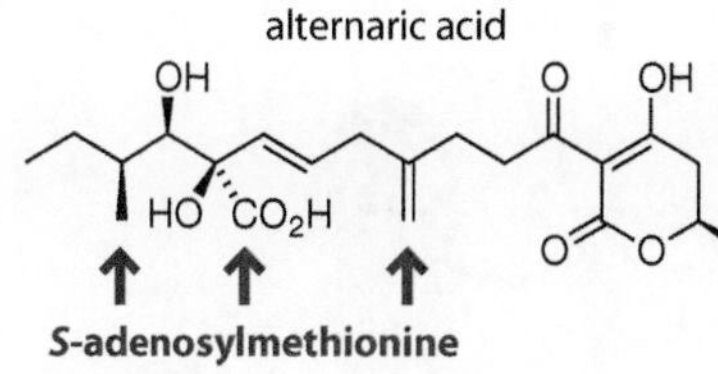

Figure 8.62 One methyl at a time. Methyl groups derived from *S*-adenosylmethionine masquerade as parts of propionate subunits.

8.5 NONRIBOSOMAL PEPTIDE SYNTHASES

Ribosomal translation is suited to the production of large proteins, not short peptides

The human body contains a wide variety of bioactive peptides that act as signaling molecules. All of these short peptides are produced by ribosomal translation of a full-length protein followed by some type of proteolytic cleavage reaction. For example, the hormone ghrelin (28 amino acids) is derived from a much larger protein called proghrelin, which has 99 amino acids. The short opioid signaling peptides enkephalin, melanocortin, and dynorphin are generated from the differential cleavage of the large protein precursors proenkephalin, proopiomelanocortin, and prodynorphin, respectively. The protein precursors to the enkephalins, preproenkephalins A and B, each have an N-terminal signal peptide sequence that directs localization to places such as the Golgi vesicles. Depending on the tissue of origin, the full-length protein (211 amino acids) can be cleaved to yield Met-enkephalin (YGGFM), Met-enkephalin variants, and/or Leu-enkephalin (YGGFL) (**Figure 8.63**). Enkephalins bind to δ-opioid receptors in the pons, the amygdala, the olfactory bulb, and the cortex areas of the brain.

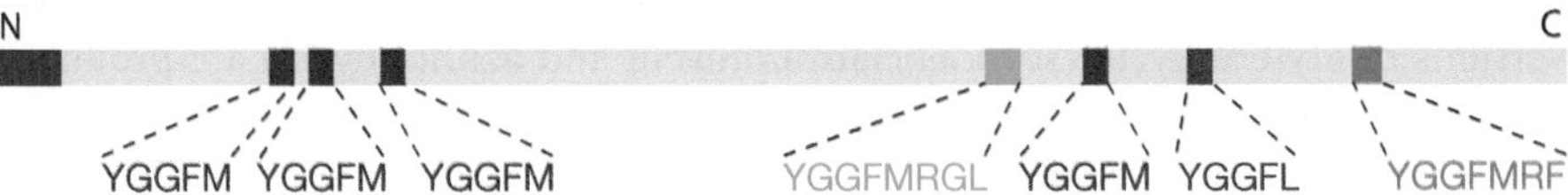

Figure 8.63 Short human peptide hormones, snipped from long peptide chains. Proteolytic processing of preproenkephalin A involves proteolytic cleavage of an N-terminal signal peptide followed by tissue-specific cleavage to generate various signaling peptides, such as Leu-enkephalin, Met-enkephalin, and variants of Met-enkephalin. The preproenkephalin A, depicted schematically here as the blue sequence from N to C terminus, is translated and proteolyzed into the indicated short peptide sequences.

Most bioactive peptide secondary metabolites are generated by peptide synthases, not by ribosomes

Bioactive peptides in humans are always derived from ribosomal translation. In contrast, many of the bioactive peptide metabolites found in fungi and microorganisms arise from nonribosomal pathways. The immunosuppressant peptide drug cyclosporin A was isolated from the soil fungus *Tolypocladium inflatum*. More so than any surgical technique, cyclosporin A made it possible to successfully transplant human organs between nonidentical patients by inhibiting the rejection of the foreign tissue mediated by human T cells (**Figure 8.64**). The 1:1 complex of cyclosporin A and cyclophilin (**Figure 8.65**) binds to, and inhibits, the phosphatase calcineurin in T cells. Various features of cyclosporin A suggest that the compound was not synthesized by a ribosome. It is small and cyclic, and has unique amino acids that differ from the 20 common ribosomal amino acids—in particular, *N*-methylated amino acids, a D-amino acid (D-Ala), an amino acid with an ethyl side chain, and an eight-carbon amino acid abbreviated MeBmt that is quite clearly generated by polyketide biosynthesis domains. The *N*-methyl groups help to reduce solvation, making cyclosporin A membrane permeable. In addition, *N*-methylation and the cyclic structure of cyclosporin A make the compound resistant to proteolysis, and the compound can therefore be taken orally. Other nonribosomal peptides have ester linkages, β-amino acids, and amino acids modified with glycan, *N*-formyl, halogen, and hydroxyl functionalities.

Most nonribosomal peptides are synthesized by synthetases, which are structurally and functionally akin to modular polyketide synthases. As in the case of cyclosporin A, nonribosomal peptide synthetases sometimes include modules for polyketide and peptide assembly. In the biosynthesis of cyclosporin A, the first amino acid that is loaded onto the peptidyl carrier protein (PCP) domain of the synthetase is the D-Ala residue. The peptidyl carrier protein serves the same role as the acyl carrier protein in polyketide synthases. Both the amino acid subunits and the growing peptide chain are bound to the PCP through a phosphopantetheinyl linker. The amino

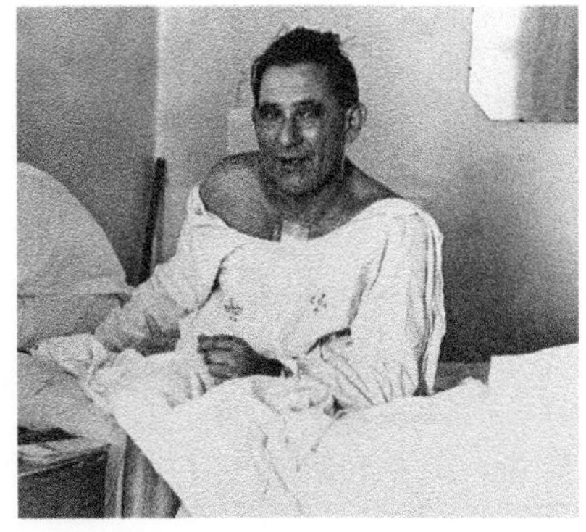

Figure 8.64 Immunosuppression for organ transplantation. The first successful heart transplant recipient survived for 18 days before succumbing to pneumonia. The state-of-the-art immunosuppressant azathioprine was administered to prevent rejection. Azathioprine is a DNA synthesis inhibitor that kills all rapidly dividing cells, which includes the entire immune system. Cyclosporin A is a much more effective immunosuppressant than azathioprine for heart transplant patients. (Courtesy of the Christiaan Barnard Division of Cardiothoracic Surgery at the University of Cape Town.)

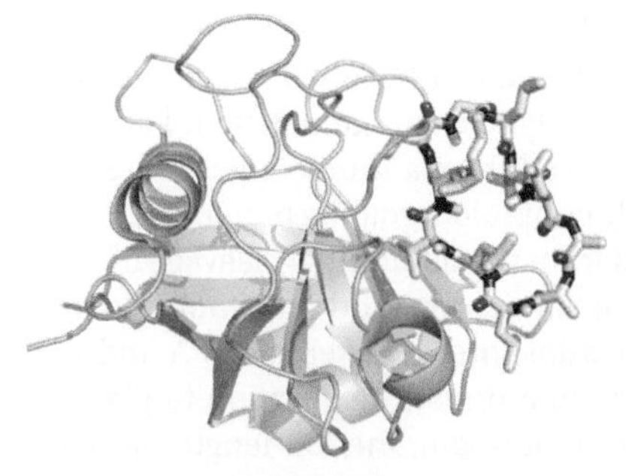

Figure 8.65 Cyclophilin D (stick model) bound to cyclosporin A (ribbon model). (PDB 2Z6W)

acids of cyclosporin are added one at a time through aminolysis reactions, including the unusual polyketide amino acid MeBmt. The final cyclic structure is generated by a thiolesterase domain, which catalyzes a macrolactamization (**Figure 8.66**).

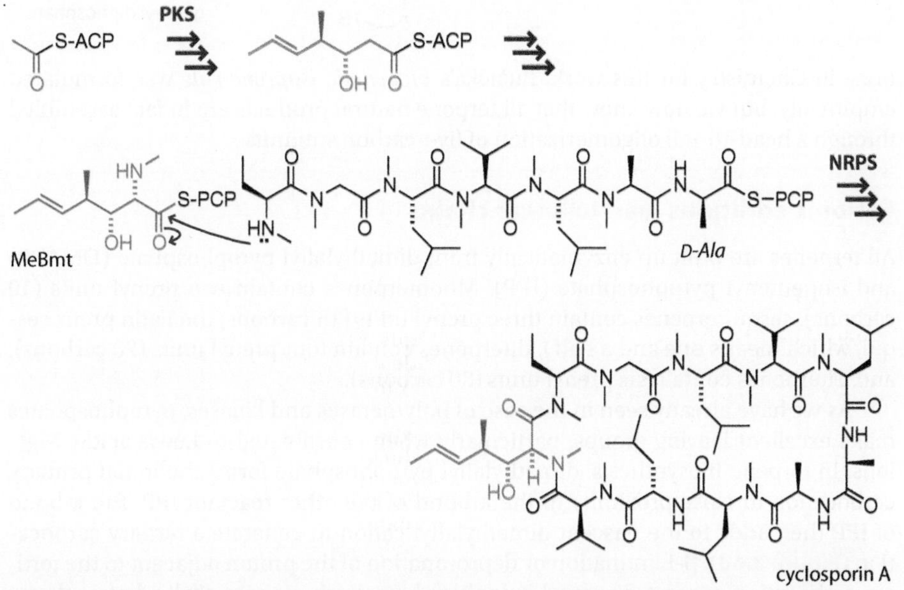

Figure 8.66 Forging the ring. The biosynthesis of cyclosporin A involves the aminolysis of thioesters attached to carrier proteins. PKS, polyketide synthase; NRPS, nonribosomal peptide synthetase.

Problem 8.16

Highlight each of the acetate, propionate, or amino acid subunits in carmabin A.

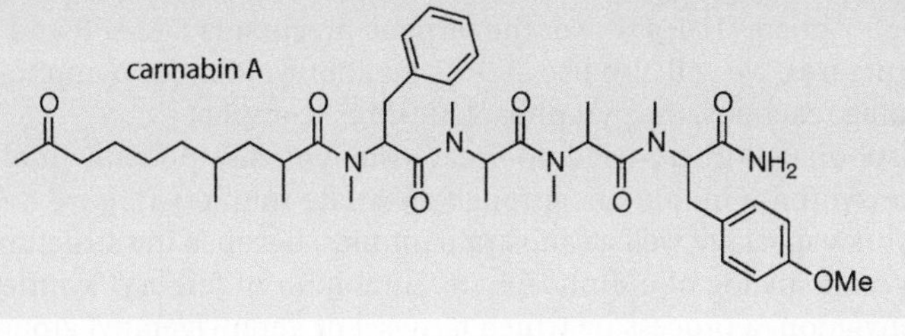

8.6 HUMAN TERPENES

Early chemists recognized terpenes as oligomers of isoprene

In the 1880s, the chemist Otto Wallach recognized that a wide range of natural products seemed, conceptually, to be oligomers of the five-carbon building block isoprene, $H_2C{=}C(CH_3)CH{=}CH_2$. Leopold Ružička expanded on this hypothesis, pointing out that the arrangement of isoprene units was not random; those natural products were best formulated as arising through the cyclization of linear oligomers of isoprene connected in a head-to-tail fashion (**Figure 8.67**). Ružička was awarded the 1939 Nobel

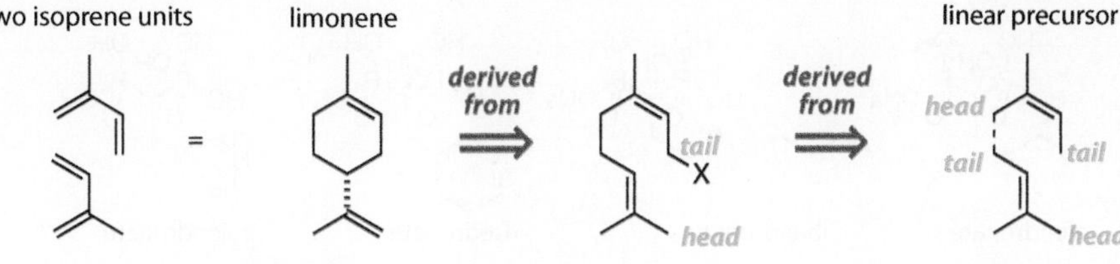

Figure 8.67 Head-to-tail. According to the biogenetic isoprene rule, terpene natural products are most likely to derive from a *linear* precursor composed of prenyl units joined in a head-to-tail fashion.

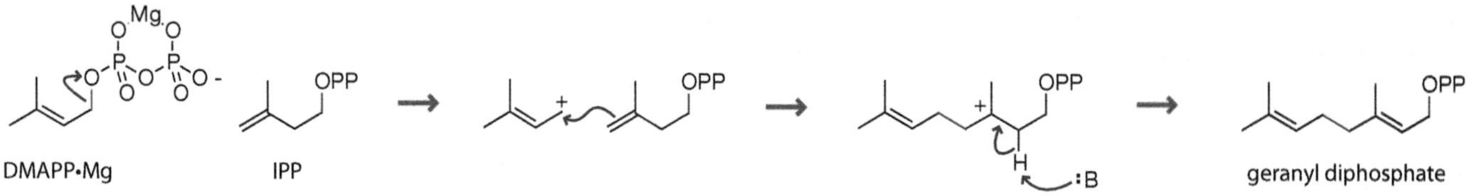

DMAPP•Mg IPP geranyl diphosphate

Figure 8.68 Connecting the prenyl units. The mechanism for enzyme-catalyzed carbon–carbon bond formation in terpene biosynthesis resembles the Friedel-Crafts alkylation reaction. DMAPP, dimethylallyl pyrophosphate; IPP, isopentenyl pyrophosphate.

Prize in Chemistry for this work. Ružička's *biogenetic isoprene rule* was formulated empirically, but we now know that all terpene natural products are in fact assembled through a head-to-tail oligomerization of five-carbon subunits.

Cationic additions lead to linear chains

All terpenes are built up enzymatically from dimethylallyl pyrophosphate (DMAPP) and isopentenyl pyrophosphate (IPP). Monoterpenes contain two prenyl units (10 carbons), sesquiterpenes contain three prenyl units (15 carbons; the Latin prefix sesqui, which means one and a half), diterpenes contain four prenyl units (20 carbons), and triterpenes contain six prenyl units (30 carbons).

As we have already seen in the case of polymerases and kinases, pyrophosphates make excellent leaving groups, particularly when coordinated to Lewis acidic Mg^{2+} ions. In terpene biosynthesis, dimethylallyl pyrophosphate forms the initial primary carbocation in close proximity to the π bond of the other reactant IPP. The π bond of IPP then adds to the nascent dimethylallyl cation to generate a tertiary carbocation (**Figure 8.68**). β-Elimination or deprotonation of the proton adjacent to the tertiary carbocation generates geranyl pyrophosphate. Each of these allylic diphosphates can serve as a substrate in further, iterative enzyme-catalyzed reactions with IPP. In all of these reactions, the growing oligomer generates a cation, and IPP serves as a cation trap. The formation of geranyl diphosphate illustrates the key steps in terpene assembly—namely, Lewis-acid-catalyzed loss of the diphosphate, formation of a carbocation, a series of carbocation-based reactions or rearrangements, and then a β-elimination step to quench the carbocation.

The suffix pyrophosphate is slowly giving way to the simpler suffix diphosphate. In this book, the term "diphosphate" will be favored, except in the case of the inorganic dimer pyrophosphate ($H_2P_2O_7^{2-}$) or the terpene precursors DMAPP and IPP. To simplify our structures, we will also use OPP as an abbreviation for pyrophosphate or, in cation-forming reactions, the pyrophosphate•Mg^{2+} complex.

Bisphosphonate drugs, used to treat osteoporosis, potently inhibit farnesyl diphosphate synthase in humans through substrate mimicry (**Figure 8.69**). Biphosphonates work especially well as substrate mimics, because the structure provides a nonhydrolyzable analog of a diphosphate. Inhibition of farnesyl synthesis prevents protein prenylation, a process by which farnesyl or geranylgeranyl groups are transferred to key proteins, such as small GTPase signaling proteins (described further below). The loss of protein prenylation interferes with the normal cell signaling processes in osteoclasts, eventually causing cell death. Bisphosphonates bind strongly to hydroxyapatite, the calcium phosphate mineral that makes up bones. Thus, the compounds are selectively taken up by osteoclasts, which are the only cells actively involved in bone resorption.

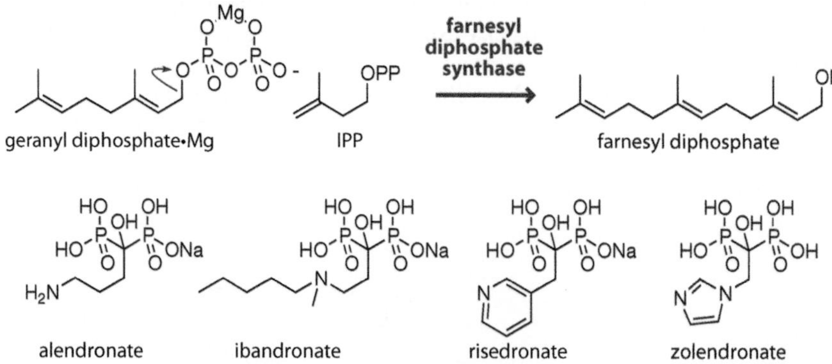

farnesyl diphosphate synthase

geranyl diphosphate•Mg IPP farnesyl diphosphate

alendronate ibandronate risedronate zolendronate

Figure 8.69 Bisphosphonates. The enzyme farnesyl diphosphate synthase (top) promotes ionization of the magnesium pyrophosphate leaving group to generate an allylic carbocation. Bisphosphonates (bottom) used to treat osteoporosis mimic essential features of the reactive intermediates in the enzymatic reaction.

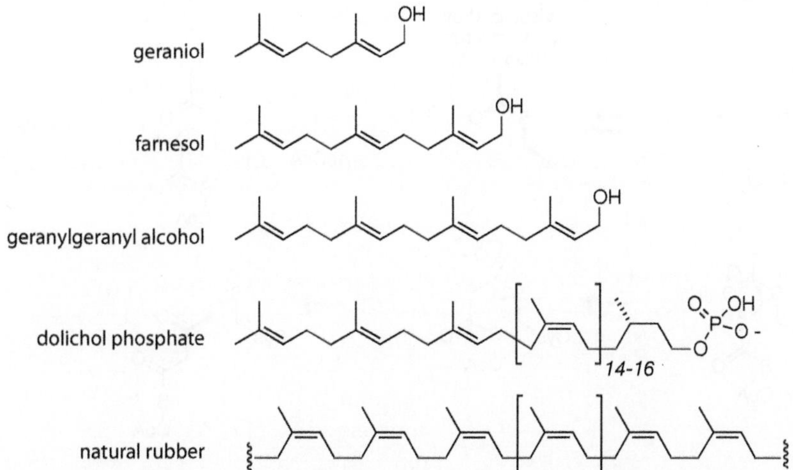

Figure 8.70 Structures and names of common linear terpenes. Note the saturation of the terminal isoprene group in dolichol. The five-carbon prenyl subunits shown in black and the bonds linking the subunits are shown in red.

The alcohols derived from the linear terpene diphosphates are found in a variety of fragrant essential plant oils (**Figure 8.70**). Geraniol is found in rose oil. Farnesol is present in citronella, lemon grass, rose, and neroli oil. Farnesol is active against some pests, and has been shown to inhibit transcription in the bacterial pathogen *Pseudomonas aeruginosa*. In humans, the most important terpenes are the prenyl chains dolichol, prenylated quinones, retinoids, and steroids. Farnesyl and geranylgeranyl groups are added post-translationally to protein cysteinyl groups. Recall that dolichol phosphate is the leaving group in the initial *N*-glycosylation of proteins. The dolichol keeps the glycosyl donor firmly anchored in the membrane of the endoplasmic reticulum, where it can be best utilized by the membrane-bound enzyme oligosaccharyltransferase. Dolichol and natural rubber are much longer oligomers of prenyl subunits. The first three prenyl units of dolichol have an *E* configuration, but the remaining prenyl units have the *Z* configuration like rubber. This suggests that dolichol is biosynthesized from an initial farnesyl diphosphate precursor. The terminal prenyl unit in dolichol is saturated, making it resistant to the formation of carbocations and the spontaneous loss of the phosphate or other leaving group.

Natural latex rubber is a polymer of prenyl units with the *Z* configuration. Synthetic rubber can be made by the polymerization of the diene isoprene, with butyllithium as an initiator. When anionic polymerization is performed in aliphatic solvents, 97% of the double bonds have the *Z* configuration, because the *Z*-allyllithium intermediate reacts faster than the *E*-allyllithium intermediate (**Figure 8.71**).

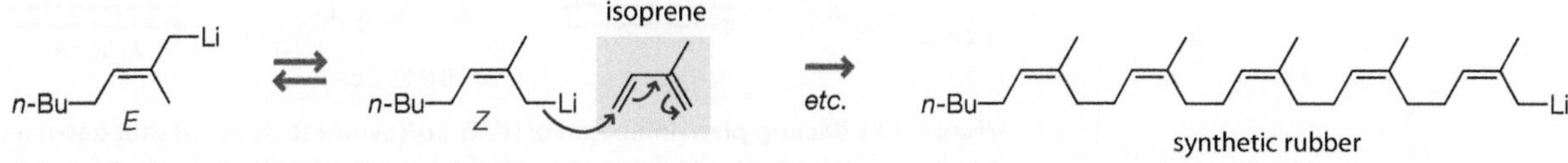

Figure 8.71 Synthesis of rubber. The anionic polymerization of isoprene is stereoselective because the *Z* isomer of the allyllithium intermediate reacts faster than the *E* isomer.

Prenyl subunits arise through enolate chemistry

In humans, the biosynthesis of terpenes initially begins with enolate chemistry, which looks a lot like polyketide biosynthesis. The biosynthesis of isopentenyl pyrophosphate from three molecules of acetyl-CoA (**Figure 8.72**) involves a six-step pathway. Each step is catalyzed by an enzyme.

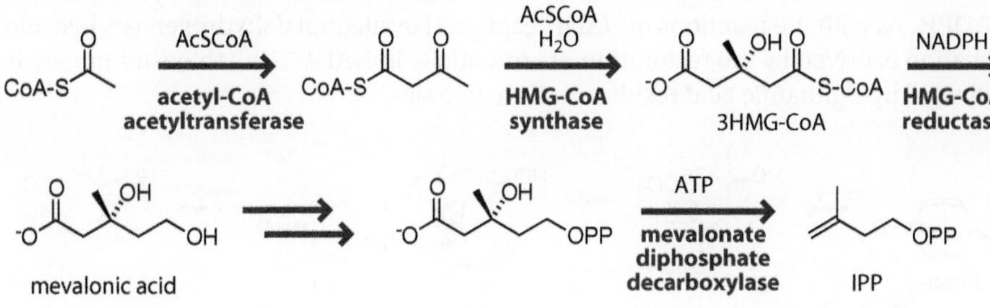

Figure 8.72 Building the building blocks. The biosynthesis of prenyl monomers involves an aldol addition reaction. HMG, hydroxymethylglutaryl.

Figure 8.73 The mechanism of the Claisen condensation catalyzed by acetyl-CoA acetyltransferase. The initial acetyl thioester intermediate was captured in a crystal structure as shown in Figure 8.74.

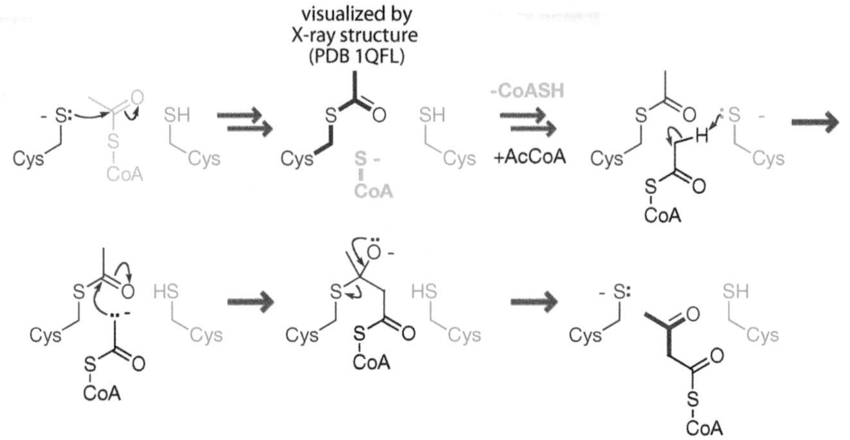

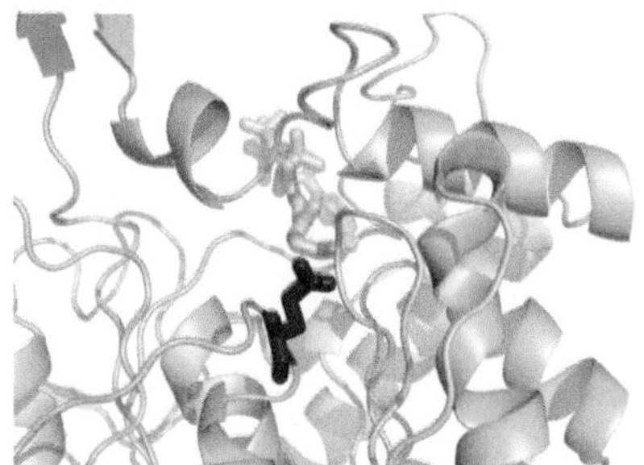

Figure 8.74 A bacterial thiolase that catalyzes the Claisen condensation of acetyl-CoA. To solve this crystal structure of a tetrameric thiolase from *Zoogloea ramigera*, the crystals were soaked briefly in a solution of acetyl-CoA before being flash-frozen. The acetylated residue Cys89 is rendered in red and the CoA leaving group is rendered in green. (PDB 1QFL)

The first two reactions in the synthesis of IPP involve enzymes called thiolases. The first thiolase, acetyl-CoA acetyltransferase, catalyzes the Claisen condensation of two acetyl-CoA esters via a covalently bound thiol ester intermediate (**Figure 8.73**). Initially, the acetyl group of acetyl-CoA is transferred to Cys89 in the active site, followed by Claisen condensation with an enolate derived from acetyl-CoA. When crystals of the thiolase from the bacterium *Zoogloea ramigera* (**Figure 8.74**) were soaked with substrate, X-ray diffraction revealed the presence of the acetylated cysteine residue. Unlike the ketosynthases involved in polyketide biosynthesis, thiolases generate enolates directly by deprotonation of the thiol ester. A second key difference between thiolases and ketosynthases is that the substrates for the thiolases involved in terpene biosynthesis (namely acetyl-CoA and acetoacetyl-CoA) are free in solution, whereas the substrates for modular polyketide synthases remain covalently attached to the ACP domain of the multienzyme complex at all times.

Acetyl-CoA, and hence acetoacetyl-CoA, arises from the well-known breakdown of glucose to produce energetic molecules such as ATP and NADH. In a key step, the enzyme 3-hydroxy-3-methylglutaryl-CoA synthase catalyzes an aldol reaction and hydrolysis to generate 3-hydroxymethylglutaryl-CoA (3HMG-CoA; **Figure 8.75**). However, 3HMG-CoA can also be produced by HMG-CoA lyase, which catalyzes the aldol reaction of acetyl-CoA to acetoacetate, a product of fat metabolism. Thus, the fundamental building block 3HMG-CoA can be generated from two different molecular feedstocks.

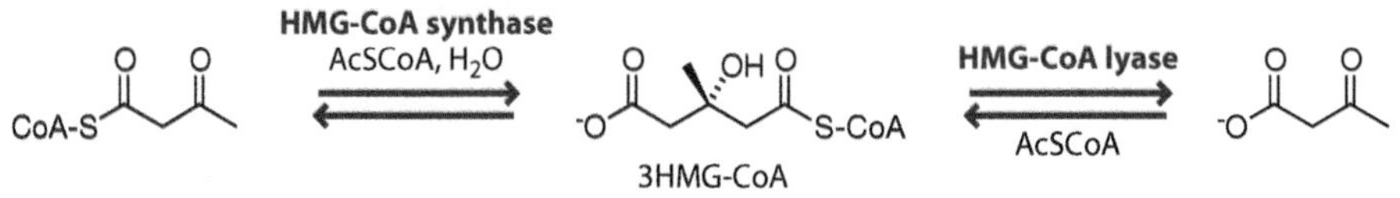

Figure 8.75 Backup plan. Inhibition of HMG-CoA synthase does not shut down terpene production in humans. The liver can generate 3-hydroxymethylglutaryl-CoA using an alternative enzyme.

The thioester group of 3HMG-CoA is reduced to a primary alcohol by the enzyme HMG-CoA reductase (**Figure 8.76**). As described in the next section, this humble transformation was enormously important in the ascendancy of the pharmaceutical industry at the end of the previous millennium and well into the current one. The mechanism of HMG-CoA reductase involves two reductions by the hydride donor NADPH. As with the reactions of NADH catalyzed by alcohol dehydrogenase, hydride donation is driven by the restoration of aromaticity in NADP+. The thioester moiety is activated by a glutamic acid residue in the active site.

Figure 8.76 The reduction of a thioester to a primary alcohol by HMG-CoA reductase.

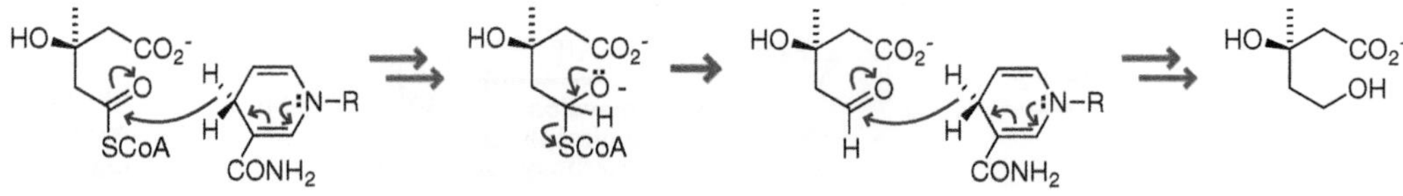

Inhibition of terpene biosynthesis is the number one treatment for heart disease

HMG-CoA reductase catalyzes the committed, rate-determining step in terpene biosynthesis and ultimately in the biosynthesis of cholesterol. High levels of cholesterol and lipids in the serum are associated with the occlusion of arteries, referred to as atherosclerosis, which eventually leads to heart attacks and/or strokes (**Figure 8.77**). Unhealthy diets contribute significantly to the prevalence of high cholesterol levels in western populations. However, a normal diet (300 mg cholesterol per day) contributes far less cholesterol than biosynthesis by the liver (800 mg per day). Thus, the inhibition of cholesterol biosynthesis is as essential as a decrease in dietary cholesterol in the treatment of hypercholesterolemia.

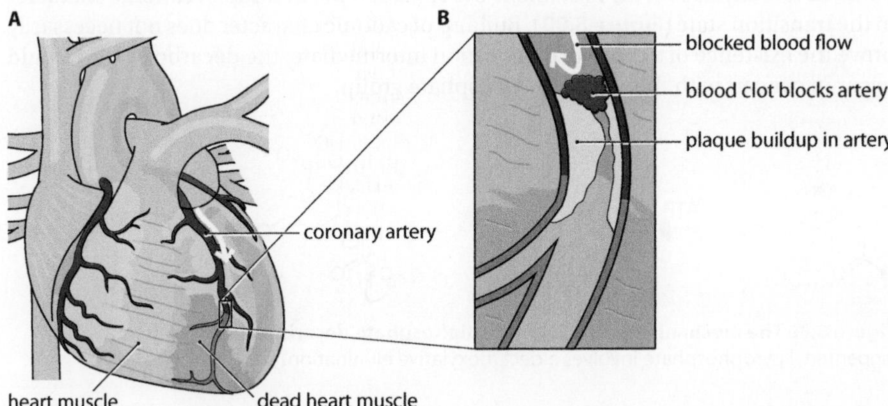

Figure 8.77 Heart attack! Hypercholesterolemia can result in blockage of coronary arteries and ultimately to heart attacks. (Adapted with permission from the National Heart Lung and Blood Institute.)

In humans, both cholesterol and fats are transported through the bloodstream as lipoprotein complexes. The key lipoproteins in cardiovascular disease are formed through the association of one molecule of apoprotein B-100 with many triacylglycerols and many cholesteryl esters. Lipoproteins can have varying ratios of protein to fats and cholesteryl esters, and such ratios determine their overall density. Low-density lipoproteins (LDLs), with low ratios of protein to cholesterol and fat, are particularly prone to forming plaques in arteries. Those plaques can grow until they occlude blood flow through the arteries. Both cholesterol biosynthesis and lipoprotein processing take place in the liver. When cholesterol biosynthesis is inhibited, the low levels of cholesterol lead to proteolytic release of the transcription factor sterol regulatory element binding protein from the endoplasmic reticulum membrane, allowing it to translocate to the nucleus, where it initiates the production of LDL receptors. The LDL receptors cause liver cells to take in low-density lipoproteins. Thus, the inhibition of cholesterol biosynthesis directly decreases serum levels of cholesterol and, more importantly, low-density lipoproteins.

In the 1970s, a screen for natural products that inhibit HMG-CoA reductase led to the discovery of the polyketide natural product mevastatin (**Figure 8.78**). The lactone

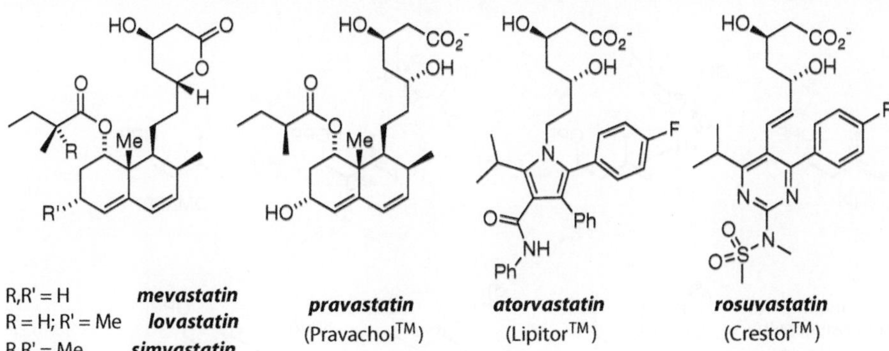

Figure 8.78 The evolution of statin drugs. Over several decades, the statin natural products mevastatin and lovastatin were carefully crafted into potent medicines that inhibit the biosynthesis of terpenes, and ultimately cholesterol.

Figure 8.79 Food as a pharmaceutical. Red yeast rice contains lovastatin. It is prepared by fermenting rice with the fungus *Monascus purpureus* and has been used in Chinese traditional medicine for more than 1000 years. In 2007 the US Food and Drug Administration declared that red yeast rice preparations containing significant quantities of lovastatin must be regulated as drugs. (Courtesy of Photo Researchers.)

moiety in mevastatin and similar structures found in all other members of the "statin" class of drugs resemble the cyclic form of mevalonic acid. Following on from this work, two pharmaceutical companies isolated related natural products for the treatment of cardiovascular disease—namely lovastatin (Mevacor™) at Merck and pravastatin (Pravachol™) at Sankyo (**Figure 8.79**). Merck improved upon lovastatin by replacing the ester with one possessing an additional methyl group, resulting in simvastatin (Zocor™). Chemists at Warner-Lambert showed that the decalin ring system could be replaced with a substituted heterocycle, resulting in the wildly successful pharmaceutical Lipitor™. These efforts culminated in the development of rosuvastatin (Crestor™) by AstraZeneca (see Figure 8.78). The introduction of cheaper generic versions of these effective pharmaceuticals may dissuade further incremental advances in statin development.

Catalyzing the last step in the synthesis of IPP, the enzyme mevalonate diphosphate decarboxylase seems to stabilize the considerable buildup of cationic character in the transition state (**Figure 8.80**). Buildup of cationic character does not necessarily prove the existence of a discrete carbocation intermediate; the decarboxylation could occur in concert with the loss of the phosphate group.

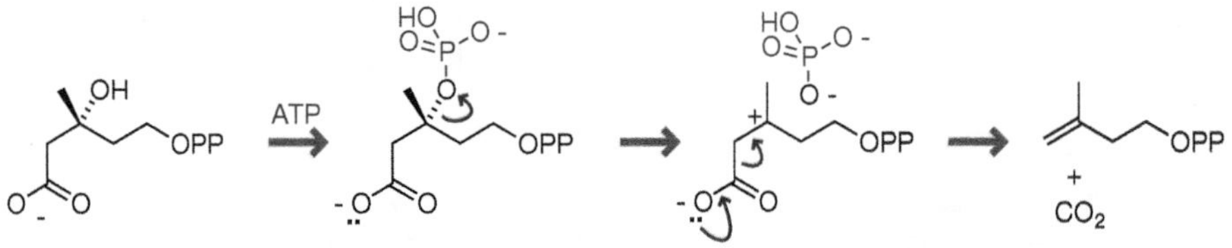

OH, OPP, ATP, HO, O=P-O-, O, OPP, O-, OPP, OPP, + CO_2

Figure 8.80 The mechanism of mevalonate diphosphate decarboxylase. The formation of isopentenyl pyrophosphate involves a decarboxylative elimination.

Problem 8.17

Suggest a plausible structure for the intermediate in the following two-step decarboxylation of β-hydroxy acids.

OH, BnO, HO, O → ($PhSO_2Cl$, pyridine, 0 °C, 1.5 h) → $C_{13}H_{14}O_3$ (ν 1840 cm^{-1}) → (100 °C, 2 h) → BnO

Isopentenyl pyrophosphate isomerase isomerizes the double bond from a less stable terminal position to a more stable internal position (**Figure 8.81**). A Mg^{2+} ion is coordinated to two glutamate residues in the active site. One of these glutamic acids protonates the double bond of IPP. On the opposite face, a cysteine thiolate abstracts a proton from the resulting tertiary carbocation. When the methylene carbon of IPP is labeled with ^{13}C, most of the isotopic label ends up in the *trans* methyl group of DMAPP; however, a small amount ends up in the *cis* methyl group of DMAPP. The origin of this stereochemical infidelity is unclear. Either the substrate binds, and is protonated, in a minor conformation, or else the carbocation undergoes sigma bond rotation on the timescale of enzymatic turnover.

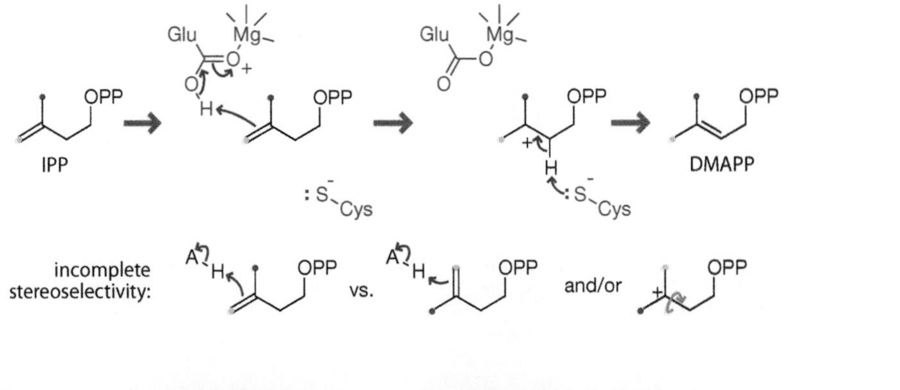

OPP, IPP, Glu, Mg, O, H, OPP, :S⁻ Cys, Glu, Mg, O, OPP, H, :S⁻ Cys, OPP, DMAPP

incomplete stereoselectivity: A, H, OPP vs. A, H, OPP and/or OPP

Figure 8.81 A critical double bond isomerization. Mammalian isopentenyl pyrophosphate isomerase is mostly stereoselective. The carbon atom labeled with a green dot is ^{13}C; the carbon atom labeled with a red dot is the normal isotope ^{12}C.

Problem 8.18

If you assume that isopentenyl pyrophosphate isomerase is completely stereo-selective, which of the carbon–carbon bonds in farnesyl diphosphate arises through cationic additions, and which arises through enolate reactions of acetyl thioesters?

ubiquinone (coenzyme Q10)

menaquinone-4 (vitamin K_2)

phylloquinone (vitamin K_1)

vitamin E

	R_1	R_2	R_3
α-tocopherol	Me	Me	Me
β-tocopherol	Me	H	Me
γ-tocopherol	Me	Me	H
δ-tocopherol	H	H	H

Figure 8.82 Prenylated quinones in humans.

Prenylated quinones serve important roles in redox chemistry

The human body uses two types of prenylated quinones (**Figure 8.82**): ubiquinones and vitamin K. The ubiquinones are a group of prenylated benzoquinones, which are important as redox carriers in mitochondrial membranes. The length of the prenyl tail varies from six to ten prenyl units. The variant with ten prenyl units, called coenzyme Q10 or ubiquinone, is most important in humans. The redox equilibrium between quinones and dihydroquinones is balanced and facile (**Figure 8.83**). The quinone form is favored by the exceptional stability of C=O π bonds relative to C=C π bonds. The benzoquinone form of coenzyme Q10 is further stabilized by resonance donation from the methoxy groups. The dihydroquinone form is favored by aromaticity.

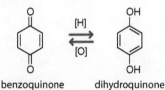

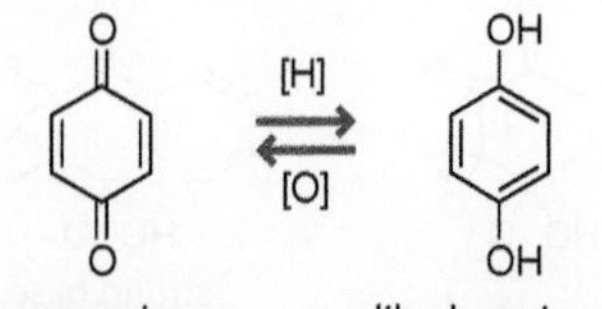

Figure 8.83 Redox messenger. Benzoquinone is easily reduced to dihydroquinone and reoxidized back to benzoquinone.

The term "vitamin K" describes a group of naphthoquinones with terpenoid tails that includes both menaquinones and phylloquinones. Humans do not possess genes for the biosynthesis of these quinones. Menaquinones are produced by obligate gastrointestinal bacteria. Phylloquinones are produced by plants. Thus, humans have two possible sources of vitamin K, which is essential for coagulation. The vitamin K group of naphthoquinones serve as cofactors in the enzyme-catalyzed carboxylation of glutamate residues in the Gla domains of some proteins. Once the glutamate residues of Gla domains are converted to γ-carboxyglutamate residues, the side chains can bind Ca^{2+}, leading to a change in protein conformation. For example, the carboxylated Gla domain of prothrombin binds to Ca^{2+} and creates a binding pocket for

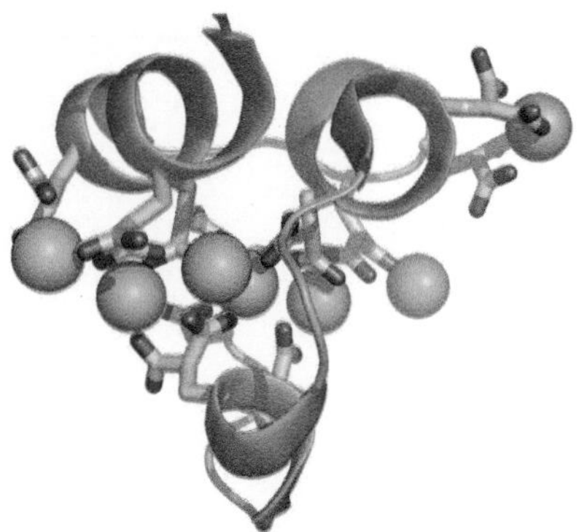

Figure 8.84 Ionic glue. The Gla domain of prothrombin binds to Ca^{2+} ions (green spheres). The Ca^{2+} ions serve as the glue makes prothrombin stick to cell membranes where it can be proteolytically activated by factor Xa. (PDB 1NL1)

phosphatidylserine, which confers membrane affinity (**Figure 8.84**). Thus, the post-translational generation of γ-carboxyglutamate residues is essential for normal blood clotting.

A series of reactions transfer carbon dioxide to the glutamic acid side chain and then recycle the then-modified vitamin K (**Figure 8.85**). The carboxylation of glutamate residues by γ-glutamyl carboxylase requires the initial reduction of the cofactor to the dihydroquinone form by a reductase enzyme. Each enzyme-catalyzed carboxylation event consumes a molecule of oxygen and converts the dihydroquinone to an epoxyquinone called vitamin K oxide. A consensus motif (EXXXEXC) seems to be essential for the initial recognition of the substrate protein by the carboxylase enzyme. The vitamin K cofactor is regenerated from vitamin K oxide by another reductase. Vitamin K oxide reductase is essential for the regeneration of vitamin K. The rodenticide warfarin (sold as the anticoagulant Coumadin®) acts by inhibiting vitamin K oxide reductase. Given the potential for uncontrolled bleeding due to the inhibition of blood clotting, the therapeutic index for warfarin is frighteningly narrow.

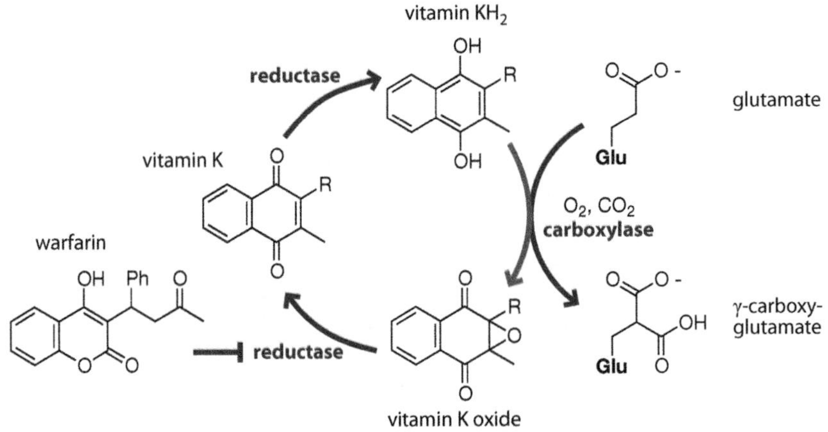

Figure 8.85 Cycling of vitamin K. The vitamin K-catalyzed carboxylation of glutamate residues in Gla domains leads to the binding of calcium (not shown).

The mechanism of action for γ-glutamyl carboxylase has been proposed to involve intramolecular epoxidation of the quinone by a hydroperoxy hemiketal (**Figure 8.86**). This transformation strongly resembles the base-catalyzed epoxidation of enones by hydrogen peroxide. The hemiketal alkoxide is proposed to deprotonate the glutamate side chain before it can eject hydroxide. The highly reactive enolate of the glutamate side chain then reacts with carbon dioxide to generate the γ-carboxyglutamate.

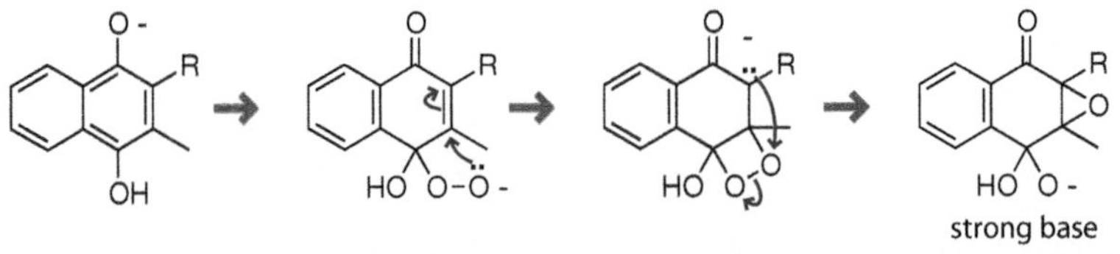

Figure 8.86 The mechanism for the formation of a powerful alkoxide base by γ-glutamyl carboxylase. The resultant strong base can then deprotonate α to the carboxylic group on the side chain of glutamic acid to generate a highly reactive enolate for reaction with carbon dioxide.

Vitamin E is not a quinone, but it is structurally related to ubiquinone and vitamin K (see Figure 8.82). Vitamin E includes any or all of the tocopherols, which are prenylated benzopyrans with various methyl substituents. Vitamin E is generally regarded as an antioxidant. Like all phenol derivatives, the aromatic hydroxyl group of the tocopherols serves as a hydrogen atom donor that can quench peroxy radical intermediates in lipid peroxidation. Among the tocopherols, only the trimethyl isomer, α-tocopherol, is nutritionally important. Not surprisingly, only α-tocopherol has substituents at both *ortho* positions, preventing radical dimerization (see Problem 5.41).

Prenylation of proteins confers membrane affinity

Recall that the acylation of proteins on their N termini or cysteine residues confers membrane affinity, which can be essential for function. An alternative, related post-translational modification involves the enzymatic prenylation of cysteine residues with 15-carbon farnesyl groups by farnesyltransferase or with 20-carbon geranylgeranyl groups by one of two geranylgeranyltransferases. The most prominent class of proteins that are prenylated are the small GTPases such as Ras and Rho, which control the duration of individual biochemical signaling events. These proteins are discussed more fully in Chapter 9.

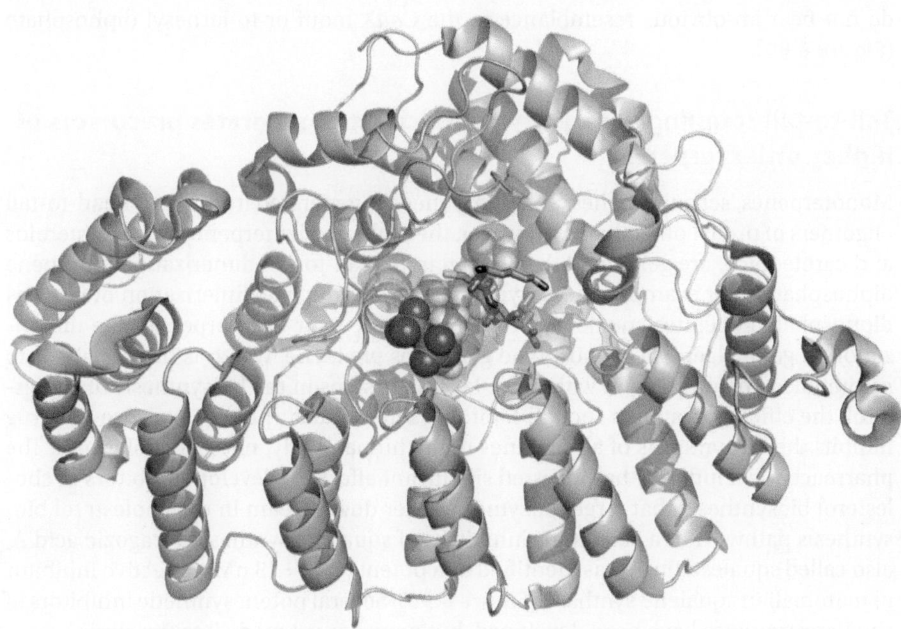

Figure 8.87 Bioconjugal visit. Two substrates bind in the active site of the enzyme geranylgeranyltransferase I (cyan): the electrophile geranylgeranyl diphosphate (spheres) and the nucleophilic tail of a protein substrate. In this crystal structure, the protein fragment KCVIL (sticks) is bound in the active site. The cysteine thiolate (yellow) is poised to displace the diphosphate leaving group in an SN2 reaction. (PDB 1N4Q)

Farnesyltransferase and geranylgeranyltransferase I (**Figure 8.87**) catalyze the prenylation of proteins with a C*aa*X tetrapeptide motif at the C terminus, where *a* tends to be a hydrophobic amino acid such as valine, leucine, isoleucine, or alanine, and where X is any amino acid except proline. After formation of the robust thioether linkage, several enzymatic modification reactions take place to rearrange the linkage to the lipid (**Figure 8.88**). First, an endopeptidase removes the three terminal amino acids. Then, the C terminus of the prenylated cysteine is enzymatically alkylated with

C*aa*X recognition site
a = Leu, Val, Ile, Ala
X ≠ Pro

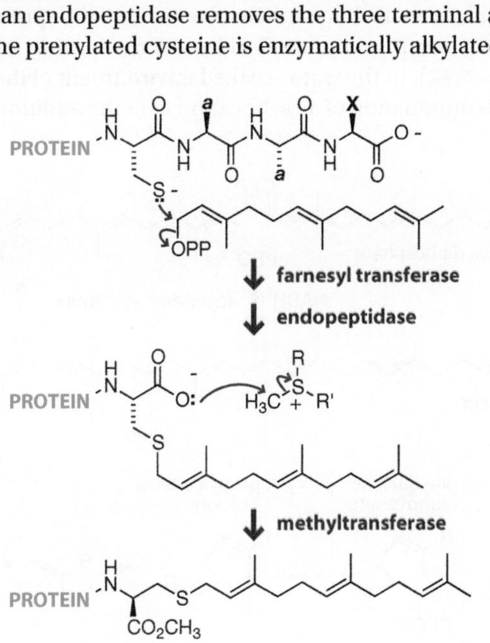

PROTEIN
OPP
farnesyl transferase
endopeptidase
PROTEIN
H_3C + R'
methyltransferase
PROTEIN
CO_2CH_3

Figure 8.88 Adding a membrane anchor. Two modifications increase the lipophilicity of the protein C terminus. The cysteine sidechain is prenylated and the tripeptide-carboxylate terminus is replaced with a carboxymethyl ester.

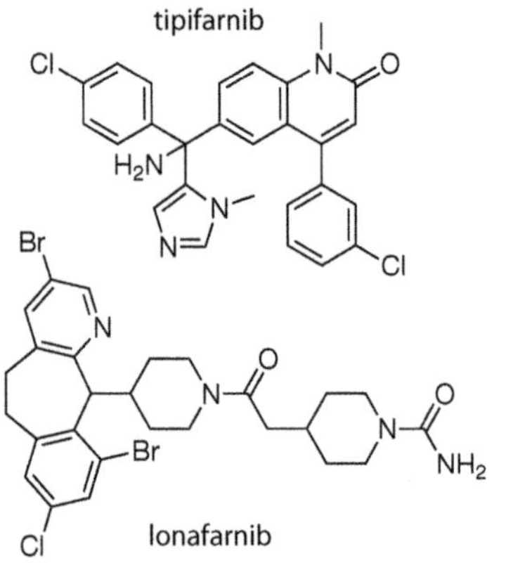

Figure 8.89 Farnesyltransferase inhibitors. Synthetic inhibitors of protein farnesyltransferases prevent prenylation of signal transduction proteins like ras. Without a membrane anchor, Ras is unable to associate with the cell membrane where it can transduce extracellular signals into intracellular instructions for cell proliferation.

S-adenosylmethionine. The resultant highly modified C-terminal cysteine is quite hydrophobic. It is common to find that S-palmitoylated cysteines are close to the farnesylated cysteine at the C terminus. Geranylgeranyltransferase II is highly specific and is only known to target the small GTPases in the Rab family, which have CC or CXC at the C terminus.

The prenylation of Ras is particularly important, because the protein amplifies signals for growth and differentiation. About one-quarter of all human tumors include mutations to the Ras protein, which leave the protein permanently stuck in an "on" position. Prenylation is essential for the function of Ras, which must interact with proteins in the plasma membrane. Inhibitors of farnesyltransferase have shown promise as anticancer drugs. Two of the leading clinical candidates, lonafarnib and tipifarnib, do not bear an obvious resemblance to the C*aa*X motif or to farnesyl diphosphate (**Figure 8.89**).

Tail-to-tail coupling of terpenyl diphosphates generates precursors of higher-order terpenes

Monoterpenes, sesquiterpenes, and diterpenes are generated from linear head-to-tail oligomers of prenyl phosphates. However, the higher-order terpenes, such as steroids and carotenoids, are generated through an initial tail-to-tail dimerization of terpene diphosphates. For example, the enzyme-catalyzed reductive dimerization of farnesyl diphosphate generates squalene, which is the precursor to triterpenes. The dimerization of geranylgeranyl diphosphate generates phytoene, which is the precursor to carotenoids. In comparison with these late-stage steps in the biosynthesis of cholesterol, the efficacy of statins such as Lipitor is somewhat surprising, because the drug inhibits the biosynthesis of all terpenes in the human body, not just cholesterol. The pharmaceutical industry has invested significant efforts to develop inhibitors of cholesterol biosynthesis that target enzymes further downstream in the cholesterol biosynthesis pathway. In a screen for inhibitors of squalene synthase, zaragozic acid A, also called squalestatin 1, was identified as a potent (IC_{50} = 13 nM), selective inhibitor of mammalian squalene synthase (**Figure 8.90**). Several potent synthetic inhibitors of squalene synthase have been developed, but none has yet made it to the clinic.

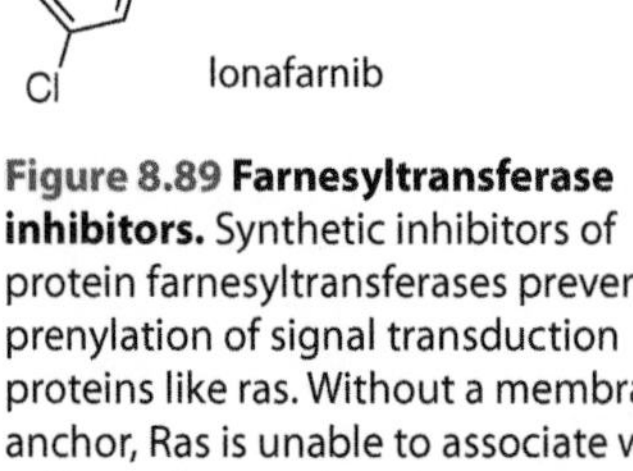

The dimerization of farnesyl diphosphate to generate squalene is deceptively simple, but the mechanism for this reductive coupling is complex. The first stage of the reaction involves a cyclopropanation reaction to generate presqualene diphosphate (**Figure 8.91**). In the synthetic chemistry laboratory, cyclopropanes can be formed by carbene attack on olefins. Such carbenes are generated through an α-elimination reaction, such as base-promoted cyclopropanations with chloroform. However, α-elimination mechanisms involve an initial deprotonation followed by the loss of a leaving group, as opposed to loss of the leaving group followed by deprotonation (**Figure 8.92**). In the water-soaked environment of the cell, imagining conditions that favor deprotonation of a carbocation over the addition of water or other quencher to a

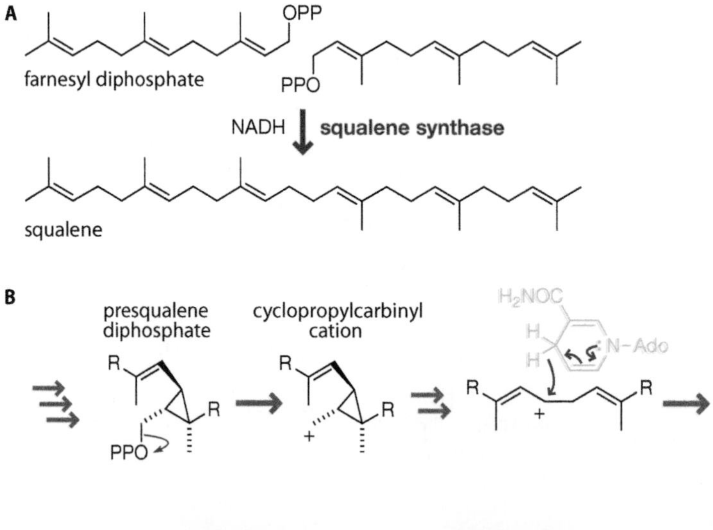

Figure 8.90 Shutting down cholesterol synthesis. Zaragozic acid A, a polyketide natural product, is a potent inhibitor of squalene synthase.

Figure 8.91 Mechanism of squalene formation. (A) Squalene synthase catalyzes the tail-to-tail reductive dimerization of farnesyl diphosphate. (B) The mechanism involves a cyclopropane intermediate.

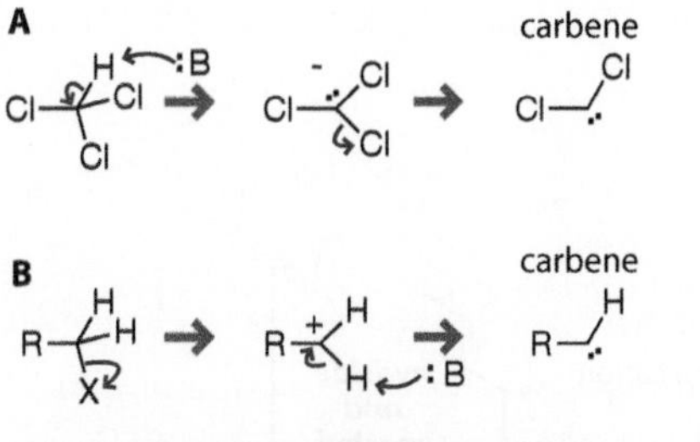

Figure 8.92 Generating free carbenes. α-Elimination mechanisms generally involve a carbanion (A), not a carbocation (B).

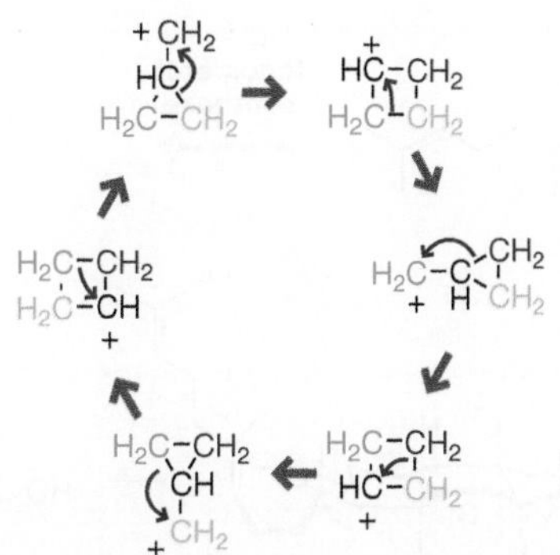

Figure 8.93 Rapid scrambling. Facile rearrangements between cyclopropylcarbinyl and cyclobutyl cations.

carbocation requires some creativity. However, the active site of an enzyme provides exactly the necessary environment where such processes can be controlled. Thus, the formation of presqualene diphosphate involves the loss of the diphosphate group, followed by deprotonation to generate a carbene. This carbene can then add to the olefin in the nearby farnesyl diphosphate to form the cyclopropane of presqualene diphosphate.

Presqualene diphosphate is poised to ionize, and in so doing to generate a cyclopropylcarbinyl cation. Recall that cyclopropylcarbinyl cations are exquisitely stable, as a result of the nucleophilicity of the strained carbon–carbon bonds in the three-membered ring. An unsubstituted cyclopropylcarbinyl cation rearranges effortlessly to a cyclobutyl cation, which is only slightly higher in energy (**Figure 8.93**). The energy barrier for these interconversions has been calculated to be less than 2 kcal mol^{-1}, and at equilibrium the ratio of cyclopropylcarbinyl cation to cyclobutyl cation should be 1.6:1.0. These facile rearrangements lead to immediate scrambling of the CH_2 positions. Undoubtedly, such rearrangements are occurring in the active site of squalene synthase. The formation of acyclic products is understood by looking at chlorination reactions of cyclopropylcarbinyl alcohol or the reaction of cyclobutanol with $SOCl_2$. Both reactions give very similar mixtures of products—namely cyclopropylcarbinyl chloride and cyclobutyl chloride in about a 2:1 ratio, with a much smaller amount of the acyclic homoallyl chloride. The acyclic homoallyl chloride arises not from an acyclic primary carbocation but from attack on one of the two cyclic cations (**Figure 8.94**). The situation is different when the cyclopropylcarbinyl cation can open to generate a stable carbocation, however, as occurs with presqualene, which can open to generate a stable, secondary allylic carbocation.

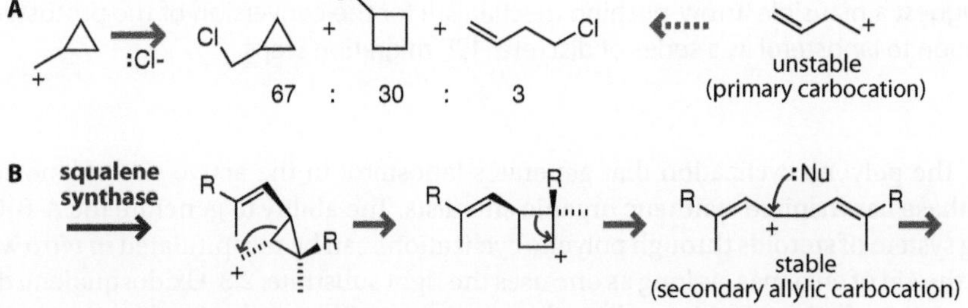

Figure 8.94 The different fates of cyclopropylcarbinyl cations. (A) Unsubstituted cyclopropylcarbinyl cations primarily give strained ring products. (B) The presqualene cation generates a ring-opened product.

Polyene cyclizations generate many rings in a single reaction

All steroids in humans are generated from squalene. The enzyme squalene monooxygenase catalyzes the epoxidation of squalene to generate 2,3-oxidosqualene, using oxygen and NADPH. Then, lanosterol synthase catalyzes a dramatic cascade cyclization of 2,3-oxidosqualene to produce four rings and six stereogenic centers

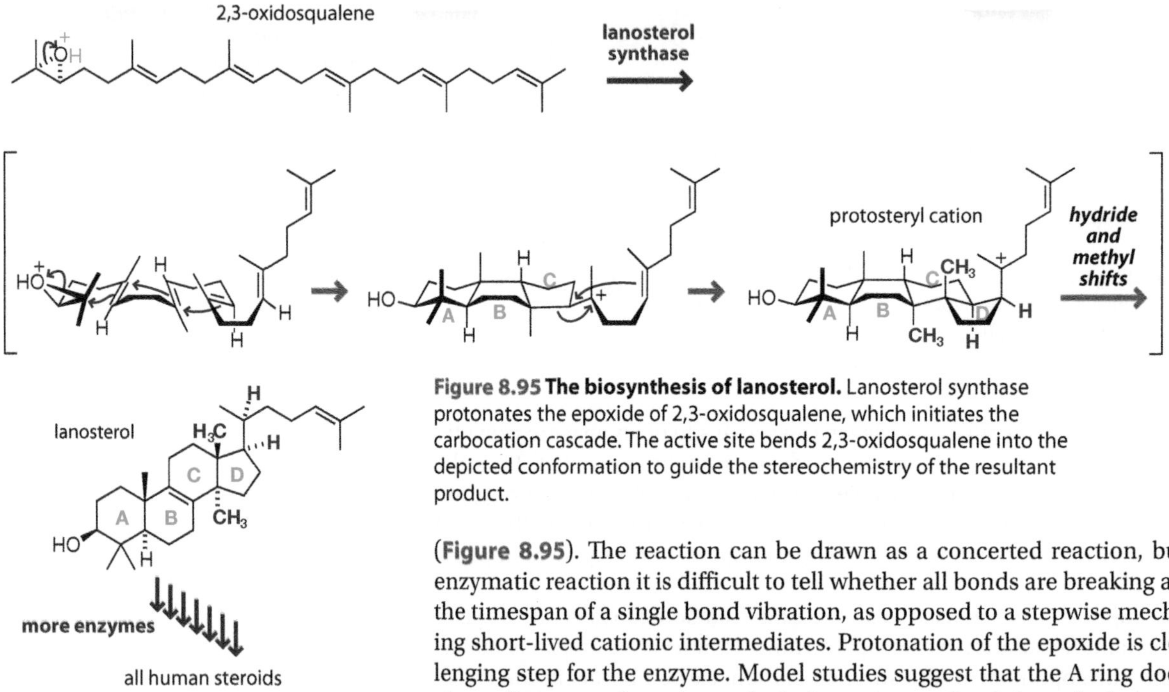

Figure 8.95 The biosynthesis of lanosterol. Lanosterol synthase protonates the epoxide of 2,3-oxidosqualene, which initiates the carbocation cascade. The active site bends 2,3-oxidosqualene into the depicted conformation to guide the stereochemistry of the resultant product.

(**Figure 8.95**). The reaction can be drawn as a concerted reaction, but as with any enzymatic reaction it is difficult to tell whether all bonds are breaking and cleaving in the timespan of a single bond vibration, as opposed to a stepwise mechanism involving short-lived cationic intermediates. Protonation of the epoxide is clearly the challenging step for the enzyme. Model studies suggest that the A ring does not open to give a discrete carbocation; calculations suggest that it is *pushed* open by the C6–C7 π bond, generating ring A, followed effortlessly by the B ring and a five-membered C ring. Formation of the D ring is accompanied by a migratory ring expansion of the C ring to generate the full A-B-C-D ring system of the protolanosteryl cation. In these types of ultra-fast reactions, one cannot predict whether the individual steps are concerted or stepwise, and for practitioners of arrow-pushing the distinction is not very useful. However, experiments conducted on oxidosqualene analogs, with oxygen in place of methylene groups, support the idea of discrete carbocations (see Problem 8.34). The most fascinating aspect of the enzyme mechanism is that the protosteryl cation is generated such that ring B favors a boat conformation, and in this conformation the bonds to the bridgehead substituents are aligned to facilitate a precise series of hydride and methyl shifts that culminate in the formation of lanosterol with the correct stereochemistry.

Problem 8.19

Suggest a plausible arrow-pushing mechanism for the conversion of the protosteryl cation to lanosterol as a series of discrete [1,2] migration steps.

The polyene cyclization that generates lanosterol in the active site of lanosterol synthase has inspired synthetic organic chemists. The ability to generate the A-B-C-D ring system of steroids through polyene cyclizations can be recapitulated *in vitro* without the aid of enzymes, as long as one uses the right substrate. 2,3-Oxidosqualene does not cyclize efficiently without the aid of an enzyme. The synthesis of progesterone is highly efficient when a five-membered ring is used to initiate the polyene cyclization and an alkyne is used as the nucleophile (**Figure 8.96**). The product in this reaction is of course racemic.

Figure 8.96 Borrowing from nature. Many of the most impressive synthetic achievements have been inspired by reactions in nature. A highly efficient chemical synthesis of progesterone takes advantage of a biomimetic polyene cyclization.

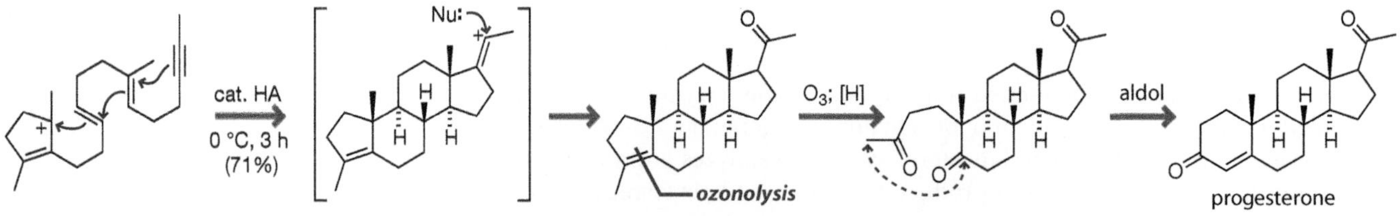

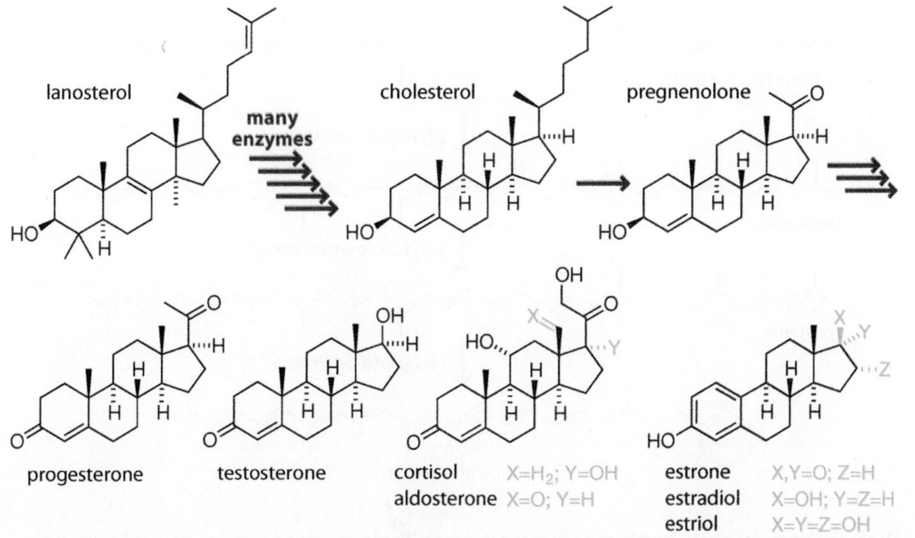

Figure 8.97 The wrong steroid. Humans don't need lanosterol; they need cholesterol. A host of enzymes is required to convert lanosterol into cholesterol and pregnenolone. All human steroid hormones are produced biosynthetically from pregnenolone.

The human conversion of oxidosqualene to lanosterol seems a wondrous trick—until one realizes that it is the wrong steroid. Nineteen more enzymes are required to convert lanosterol into the right steroid—cholesterol—as it is trafficked within the endoplasmic reticulum (**Figure 8.97**). Those enzymes perform hydroxylations, redox hydride transfers catalyzed by nicotinamide-dependent enzymes, decarboxylations, and hydrogenations. These same kinds of transformations are used by microorganisms and plants in the biosynthesis of terpene natural products. The removal of three methyl groups from lanosterol makes it difficult to trace the squalene backbone within the structure of cholesterol. Similarly, enzymatic modifications to other terpene natural products can also mask the location of the prenyl units within the structure. Most cholesterol winds up in cell membranes, but a small amount is converted into signaling molecules. The side chain of cholesterol is oxidatively cleaved to generate pregnenolone, the common precursor to all of the steroids that target human nuclear receptors (see Figure 8.97).

Humans lack genes for retinoid biosynthesis

Retinoids are an important class of terpenes used by the human body (**Figure 8.98**). *Cis* and *trans* retinoic acids serve as signaling molecules for nuclear receptors. Retinal serves as the covalently bound ligand for the vision receptor rhodopsin. These receptors are discussed in Chapter 9. Humans lack enzymes for the *de novo* synthesis of retinoids. Instead, human retinoids are derived from two dietary sources—β-carotene and retinyl esters, which are hydrolyzed to generate retinol, also known as vitamin A. Plants produce β-carotene from the tail-to-tail dimer phytoene. Phytoene synthase catalyzes the generation of phytoene from two molecules of geranylgeranyl diphosphate. The mechanism of phytoene synthase is believed to involve a cyclopropane intermediate, like the mechanism of squalene synthase. The penultimate carbocation is deprotonated to generate an alkene linkage, however, and does not involve hydride addition from NADH (**Figure 8.99**). To solve nutritional deficiency caused by diets low in β-carotene, scientists have genetically engineered a strain of rice, called Golden

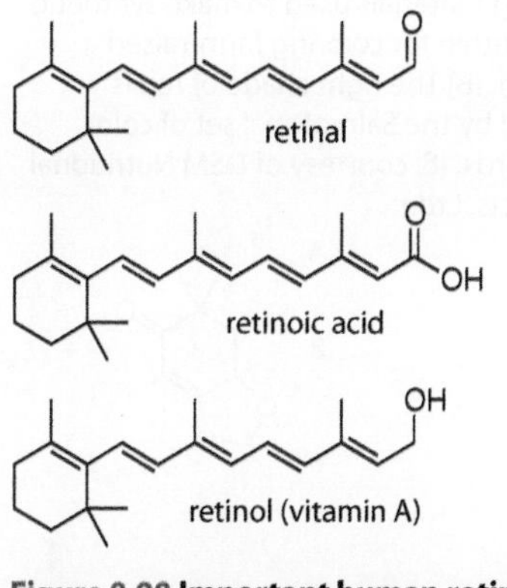

Figure 8.98 Important human retinoids. Retinal, retinoic acid, and retinol (vitamin A) are essential retinoids in humans, but we don't have enzymes that can construct the six-membered ring.

Problem 8.20

Suggest a plausible arrow-pushing mechanism for the enzyme-catalyzed cyclization of lycopene to β-carotene (for the necessary structures, see Figure 9.99). Assume that the active site of the enzyme, lycopene β-cyclase, has a catalytic acid residue.

Figure 8.99 Key compounds from vegetables. Carotenoids are derived from the tail-to-tail dimer phytoene.

geranylgeranyl diphosphate

OPP

PPO

phytoene synthase

phytoene

phytoene desaturase

lycopene

lycopene β-cyclase

β-carotene

Figure 8.100 Golden rice. Golden rice (right) has been genetically modified to include the genes required for the biosynthesis and accumulation of β-carotene in the edible grains. (Courtesy of Golden Rice Humanitarian Board.)

Rice, that produces β-carotene (**Figure 8.100**). To do so, the relevant genes from maize and a bacterium found in soil were transfected into rice, which was then bred into different varieties of rice important to vitamin A-deficient communities. In humans, β-carotene is oxidatively cleaved to retinal with O_2 by a monooxygenase enzyme. Retinoic acids are generated through the oxidation of retinal.

The red hues of many animals arise from dietary carotenoids. Wild salmon has a deep red hue due to the presence of the carotenoid pigment astaxanthin, which is accumulated from the oceanic diet of krill and shrimp (**Figure 8.101**).

Figure 8.101 Red shrimp. The red color of krill results from the accumulation of the dye astaxanthin. (Courtesy of Joseph Warren, Stony Brook University.)

Similarly, the brilliant pink hue of flamingos is attributable to their natural diet of brine shrimp. Farm-raised salmon do not naturally acquire the deep red color of wild oceanic salmon, because they are raised on a diet of other fish. To increase the value of farm-raised salmon, synthetic astaxanthin (DSM's Carophyll® Pink 10% CWS or BASF's Lucantin® Pink) is added to the fish food shortly before harvest; a color gauge called the SalmoFan^{TM} is used to judge the coloration of the fish (**Figure 8.102**).

Figure 8.102 Painting salmon with synthetic astaxanthin. (A) A set of starting materials used to make synthetic astaxanthin for coloring farm-raised salmon. (B) The right shade of red is judged by the SalmoFan™ set of color standards. (B, courtesy of DSM Nutritional Products, Ltd.)

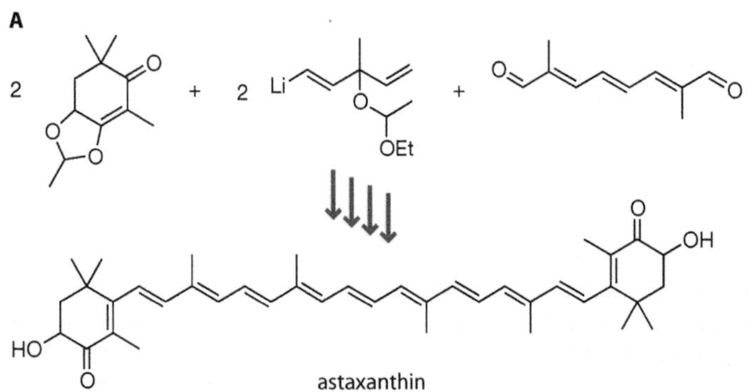

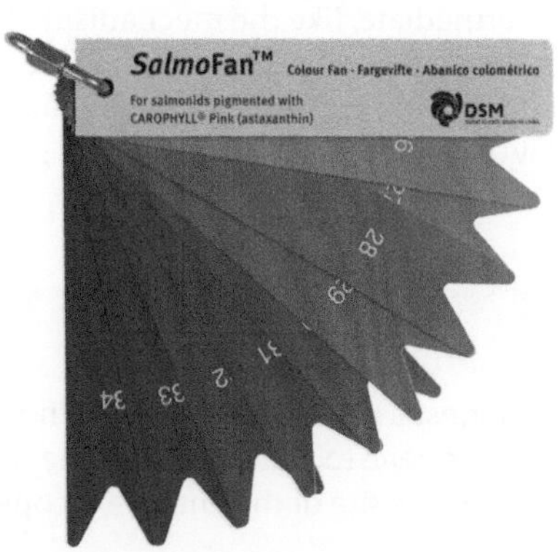

8.7 NONHUMAN TERPENE NATURAL PRODUCTS

Plants and microorganisms produce a much wider range of terpene natural products than humans

Terpene natural products have a much wider range of polycyclic structures than polyketides. Most terpene natural products are produced by, and isolated from, plants. In contrast, polyketide natural products are produced by a variety of organisms, but are most commonly isolated from readily cultured microorganisms. Terpenes were among the first natural products to be structurally studied. The name is associated with turpentine, which is the distillate from the resin of certain trees. Turpentine was originally distilled from the resin of the terebinth tree (*Pistacia terebinthus*) but is now more commonly obtained from pine trees. Of the 101 components in turpentine from *P. terebinthus*, the major components are α-pinene (39.6%), β-pinene (19.5%), sabinene (6.5%), terpinen-4-ol (3.8%), and δ-3-carene (3.3%). Modern turpentine produced from pine trees is more than 90% α- and β-pinenes (**Figure 8.103**).

Figure 8.103 Collecting terpenes. Turpentine is distilled from tree sap. (Left, courtesy of Wikimedia Commons; right, courtesy of Speedball Art Products.)

Terpene natural products inspire an irresistible allure in chemists, because a limited range of linear precursors generates an astounding number of complex cyclic structures. Higher diterpenes, such as triterpenes and tetraterpenes, tend to arise from tail-to-tail dimers of squalene or phytoene. Recall that the formation of lanosterol began with the opening of an epoxide to give a tertiary carbocation. Monoterpenes, sesquiterpenes, and diterpenes arise from the ionization of linear diphosphates to yield allyl cations (**Figure 8.104**). Polycyclic terpene natural products challenge the ingenuity of organic chemists who seek to demonstrate the efficiency of chemical synthesis.

The biosynthesis of terpenes by terpene synthase enzymes fascinates organic chemists because the synthetic steps involve the rich chemistry of carbocations, ionization reactions, S_N1 substitutions, rearrangements, and migrations. All known terpene synthases share a common structural fold (**Figure 8.105**) and must perform

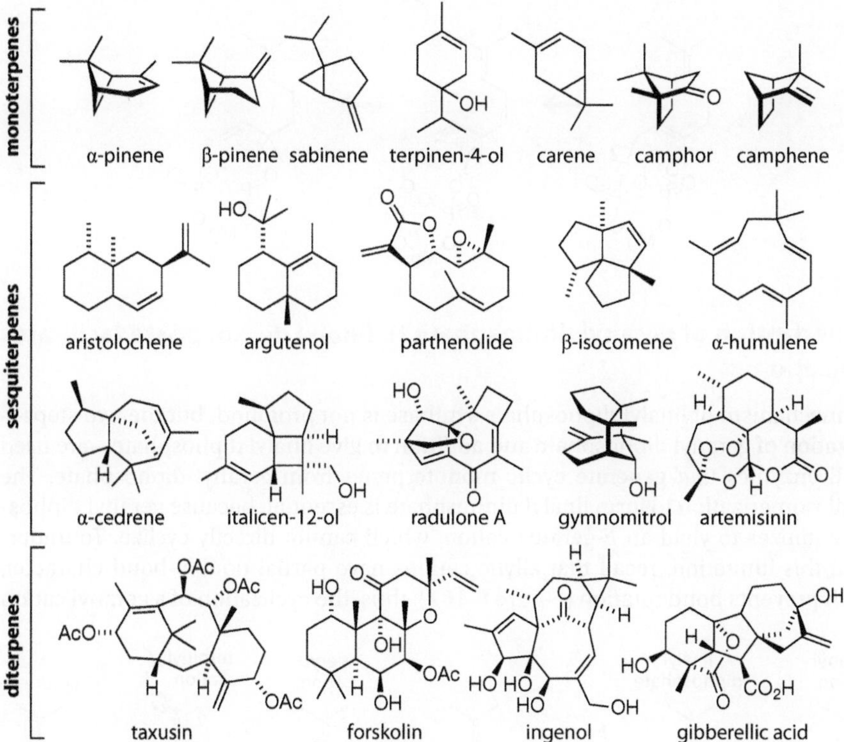

Figure 8.104 Examples of the many cyclic terpene natural products synthesized by plants and microorganisms.

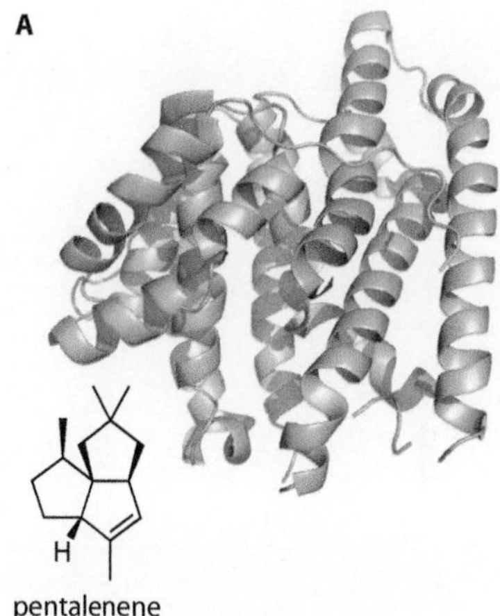

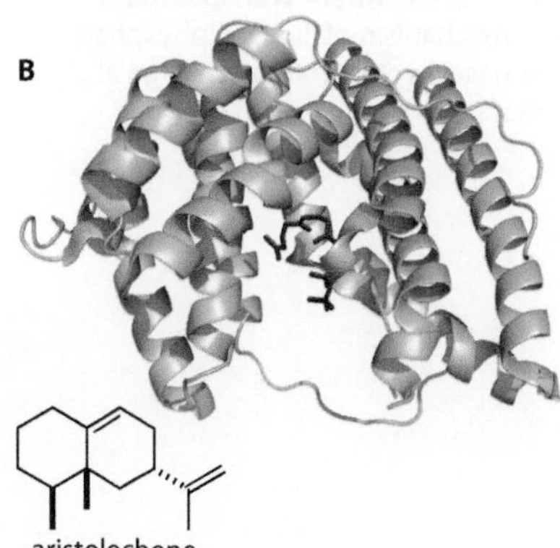

Figure 8.105 The structure of terpene synthases. Pentalenene synthase (A) (PDB 1HM7) and aristolochene synthase (B) (PDB 2BNY) have a common structural fold, despite their very low sequence homology and different products. The substrate mimic 2-fluorofarnesyl diphosphate (red) highlights the location of the enzyme's active site.

the following steps. First, the enzymes bind to the diphosphate substrate. During this binding event, water must be squeezed out of the binding site and the active site must be sealed shut; any water present during the cyclization would otherwise react quickly with any carbocation intermediates, shutting down the reaction. Thus, the active sites of terpene synthases resemble a hydrophobic deep cave. Binding to the back of this cave, the isoprenoid tail of the diphosphate allows the walls of the cave to collapse around it. The active site pushes the substrate into an approximate shape of the product, which helps to guide the product specificity. The mouth of the cave includes Mg^{2+} ions chelated by the carboxylate side chains of aspartic and glutamic acid residues. As Lewis acids, the Mg^{2+} ions bind to the diphosphate of the substrate, which seals the mouth of the cave shut.

Next, the Lewis acidic Mg^{2+} ions promote loss of diphosphate, which results in the generation of the carbocation intermediate. This intermediate requires careful shepherding to avoid premature quenching and the enzyme guides the formation of the correct product. Through cation–π and other interactions, the terpene synthase active site can stabilize specific intermediates, favoring particular reaction pathways and thus products. Illustrative of the power of this class of enzymes, examples of terpene synthases favoring a secondary over a tertiary carbocation are known. In the penultimate step, the reactive carbocation is quenched, either by enzyme-directed reaction with a nucleophile such as water, or through β-elimination to yield an olefin. Finally, the hydrophobic active site must unseal, allow the entry of water, and resolvate the hydrophobic product. Given the resistance of such hydrophobic compounds to dissolve readily in water, this last step, although perhaps the least complex from the view of chemical reactivity, can be the rate-determining step of the reaction.

The simplest terpene natural products are the linear monoterpenes. Geranyl diphosphate is the universal substrate among plant terpene synthases. Linalyl diphosphate synthase is among the simplest terpene synthase. The enzyme catalyzes the S_N1 isomerization of geranyl diphosphate to linalyl diphosphate (**Figure 8.106**). Although the mechanistic details are not yet available, it is tempting to suggest that the oxygen that attacks the allyl cation is different from the one that was initially bound, making it possible for the reaction to proceed through a highly facile transposition.

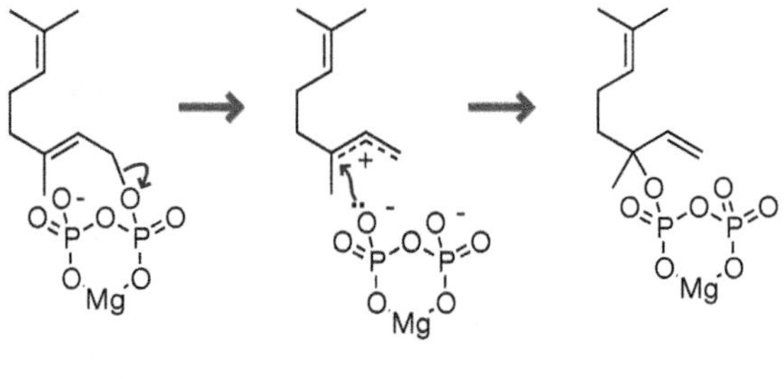

Figure 8.106 Allylic transposition. The mechanism of linalyl diphosphate synthase involves formation of an allyl cation.

Isomerization of geranyl diphosphate to linalyl diphosphate facilitates cyclization

The mechanism of linalyl diphosphate synthase is not profound, but the two steps—ionization of geranyl diphosphate and addition to give linalyl diphosphate—are used by all enzymes that generate cyclic monoterpenes from geranyl diphosphate. The initial isomerization to form linalyl diphosphate is essential, because geranyl diphosphate ionizes to yield an *E*-geranyl cation, which cannot directly cyclize. To understand this limitation, recall that allylic cations have partial double-bond character, which prevents bond rotation (**Figure 8.107**); thus, the cyclization of a geranyl cation

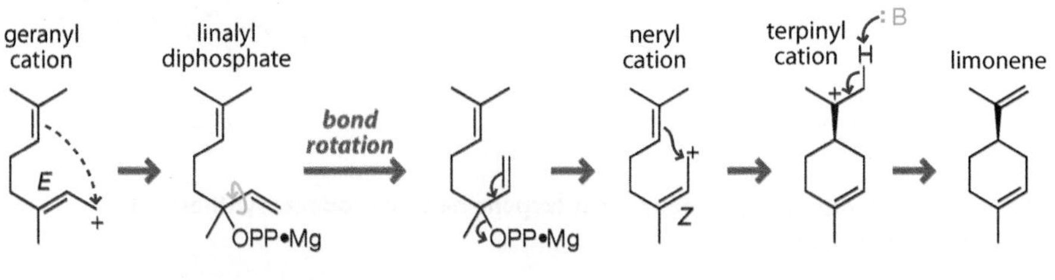

Figure 8.107 The mechanism of limonene synthase. The first step shown on the left is disallowed, because it would result in the formation of a *trans* (*E*) olefin in a six-membered ring.

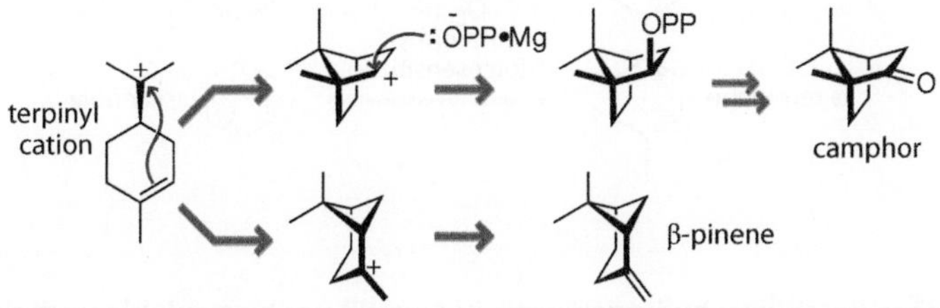

Figure 8.108 Two ways to cyclize. The transannular cyclization of a terpinyl cation can give two different ring systems.

would generate an impossibly strained six-membered ring containing a *trans* double bond. Trapping the geranyl cation to generate linalyl diphosphate provides a freely rotatable vinyl group. Linalyl diphosphate can ionize to form a neryl cation with a *Z* configuration. The allylic terminus of the neryl cation is poised to react with the trisubstituted olefin. After the cyclization of the neryl cation, the resulting terpinyl cation, common to many monoterpene biosynthetic pathways, can undergo many possible fates. In the enzyme limonene synthase, a base deprotonates one of the methyl groups to generate limonene, a common component in the oil of citrus fruits.

All terpenes are generated through enzymatic reactions, but the mechanistic pathways that lead to the diverse terpene skeletons can generally be explained by simple arrow-pushing mechanisms composed of cationic additions and [1,2] migrations. A large number of different terpene natural products can arise from a single precursor. For example, the terpinyl cation that generates limonene in limonene synthase can further cyclize to bicyclic structures in other enzymes (**Figure 8.108**), thus generating a [2.2.1] bornyl cation that is trapped by pyrophosphate or the [3.3.1] ring system of α-pinene and β-pinene.

The mechanism for the formation of (+)-sabinene hydrate by the cyclase enzyme from sweet marjoram involves an initial isomerization to (*R*)-linalyl diphosphate, followed by ionization to generate a terpinyl cation in a boat conformation (**Figure 8.109**). In theory this cation could undergo a **transannular** cyclization to generate a bornyl or pinanyl ring system. Instead, the cation undergoes a hydride migration to generate a homoallyl cation that then cyclizes to generate a cyclopropylcarbinyl cation. Ultimately, the cyclopropylcarbinyl cation is trapped by water to generate sabinene hydrate as a mixture of two epimers.

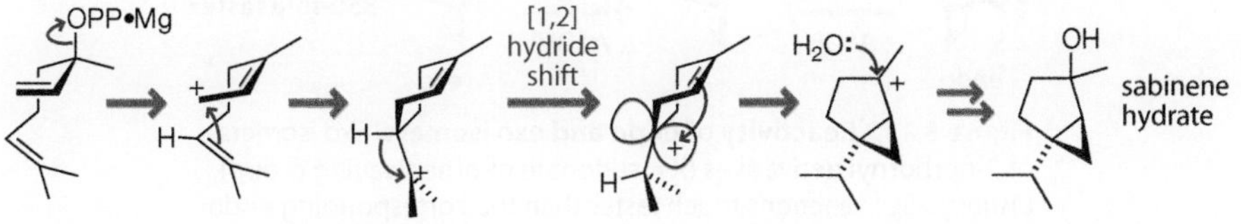

Figure 8.109 Cyclopropane formation. The formation of sabinene hydrate involves a hydride shift, facilitating formation of a highly stable tertiary cyclopropylcarbinyl cation.

While on vacation in Florence, Italy, the biochemist Konrad Bloch noticed the honey blonde hair of women depicted in the paintings of fifteenth-century Europe (**Figure 8.110**). Bloch found their hair color puzzling, especially since the women lacked the blue eyes expected for Nordic blondes. Italian women today, as then, typically have brunette hair, and the most common method for artificially lightening hair

Figure 8.110 Blondes in Venetian paintings. These selections from *Primavera* (1482, left) and *The Birth of Venus* (1486, right), both by Sandro Botticelli, illustrate that Italian women had access to blonde hair colors well before the invention of hydrogen peroxide.

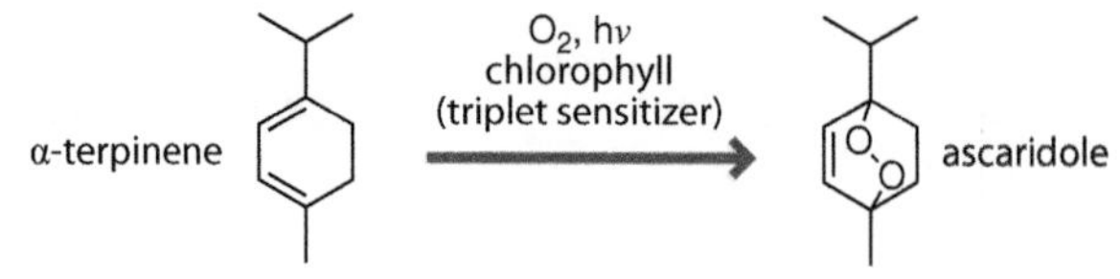

Figure 8.111 Diels-Alder reactions with air. Sensitized photooxidation has been proposed as a source of alkyl peroxides used in early hair bleaches. For example, the natural product ascaridole can be formulated as the product of a [4+2] cycloaddition with singlet oxygen.

α-terpinene O_2, hν chlorophyll (triplet sensitizer) ascaridole

color requires peroxides; hydrogen peroxide would not be available until three centuries later. Awarded the 1964 Nobel Prize in Physiology or Medicine for his investigations of cholesterol biosynthesis, Bloch was well aware of the chemical properties of terpenes. On further investigation, he learned that the women depicted in Botticelli's paintings would comb plant extracts containing terpenes into their hair and sit in sunlight. Through sensitized photooxygenation reactions, alkyl peroxides could be generated from the terpenes in pine oil (**Figure 8.111**). However, the inefficient process probably required multiple treatments and hours spent in sunlight.

The 2-norbornyl cation exhibits exceptional behavior

If the methyl groups are removed from the [2.2.1] ring system of borneol, the compound is said to possess a norbornyl ring system. 2-Norbornyl cations derived from terpenes and related synthetic substrates have exceptional properties that helped chemists understand the properties of carbocations and what makes them stable (**Figure 8.112**). There are two possible diastereomers of 2-norbornyl derivatives. When the substituent is on the same side as the *methylene* bridge, the compound is called the exo isomer. Conversely, when the substituent is on the same side as the *ethylene* bridge, it is called the endo isomer. Surprisingly, when an optically pure endo isomer of a 2-norbornyl sulfonate was subjected to a substitution reaction with the acetate ion, the exo isomer was isolated *in racemic form*. The formation of a racemic mixture could not be explained by a simple S_N2 reaction with inversion of configuration. When the optically pure exo isomer was subjected to the same conditions, the reaction was 350 times faster, and the exo acetate was formed with retention of configuration.

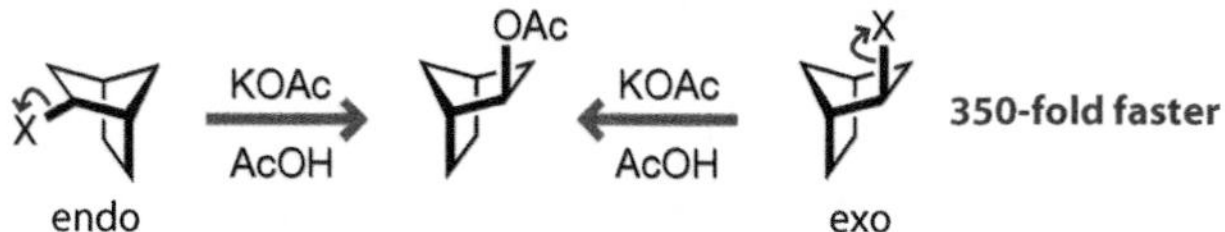

Figure 8.112 Reactivity of endo and exo isomers. Exo isomers of 2-norbornyl derivatives (X = sulfonate or other leaving group) undergo S_N1 reactions much faster than the corresponding endo isomers.

We now know that the exo isomer ionizes faster as a result of hyperconjugative donation by a perfectly aligned axial C–C bond, which is antiperiplanar to the leaving group. In fact, this alignment is so perfect that the 2-norbornyl cation actually exists as a symmetrical system with a three-center two-electron bond, as opposed to two carbocations that equilibrate through a higher-energy transition state (**Figure 8.113**). The exo isomer facilitates the formation of this three-center two-electron bond owing to the alignment of the axial C–C bond, which is not aligned correctly in the endo isomer. Bond alignment is also a critical factor in determining the ease by which [1,2] migration reactions can take place. If we recall the mechanism of lanosterol synthase, the impressive aspect of this reaction is not just the four rings and many stereocenters generated: an equally impressive achievement by lanosterol synthase is the grip that the enzyme exerts on the protosteryl cation to hold intermediates in a conformation that facilitates four precisely controlled [1,2] shifts.

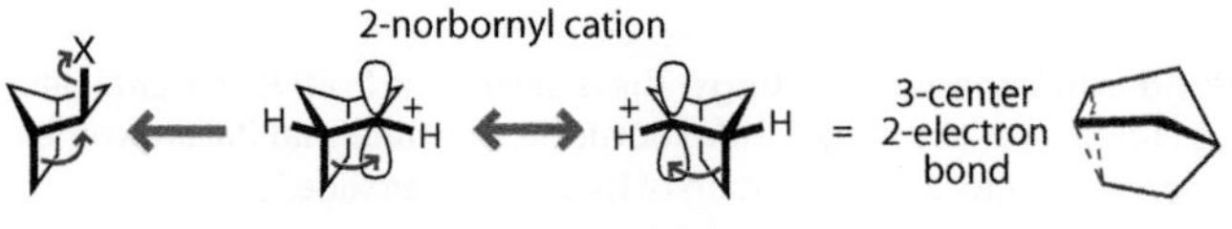

Figure 8.113 The importance of bond alignment. Precise alignment of a C–C bond facilitates ionization of the leaving group in exo-2-norbornyl derivatives and in the two resonance depictions of the 2-norbornyl cation.

2-norbornyl cation X H H H H = 3-center 2-electron bond

Problem 8.21

When α-pinene is treated with anhydrous HCl, it generates an adduct called pinene hydrochloride that is structurally analogous to camphor. When pinene hydrochloride is treated with sodium acetate at high temperature, camphene is produced. Suggest a plausible arrow-pushing mechanism for the formation of pinene hydrochloride from α-pinene and the subsequent conversion to camphene.

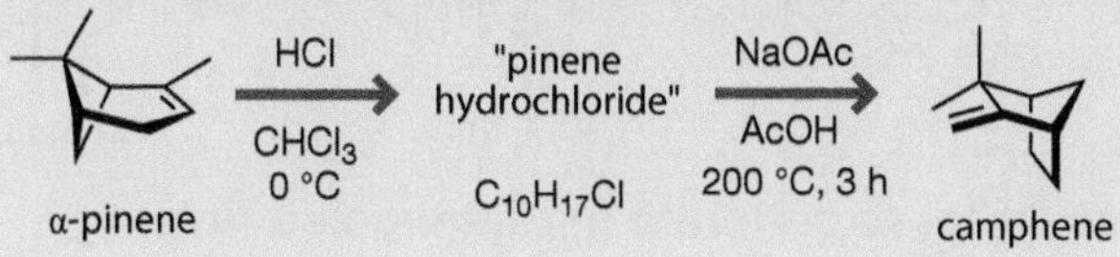

α-pinene → (HCl, $CHCl_3$, 0 °C) → "pinene hydrochloride" $C_{10}H_{17}Cl$ → (NaOAc, AcOH, 200 °C, 3 h) → camphene

Minor products offer clues to the enzymatic mechanisms of terpene cyclases

Terpene cyclases are not always perfect. Recombinant enzymes are often shown to generate mixtures of terpenes when supplied with an appropriate substrate. For example, *epi*-cedrol synthase was shown to convert farnesyl diphosphate into *epi*-cedrol along with cedrol in a 96:4 ratio. About 3% of the product mixture consists of other olefinic products, such as α-cedrene, β-cedrene, acoradiene, (*E*)-α-bisabolene, and (*E*)-β-farnesene (**Figure 8.114**). This molecular debris trail offers clues to the enzyme's mechanism.

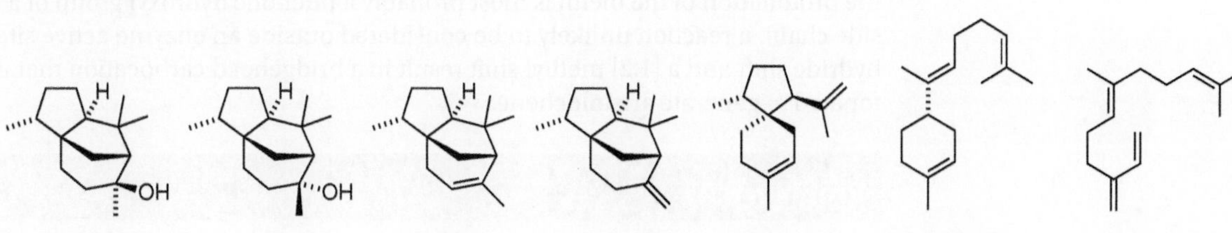

epi-cedrol | cedrol | α-cedrene | β-cedrene | α-acoradiene | (E)-α-bisabolene | (E)-β-farnesene

Figure 8.114 Debris trail. Major and minor products generated by *epi*-cedrol synthase.

For example, the mechanism of *epi*-cedrol synthase is believed to involve an initial isomerization to nerolidyl diphosphate that ionizes to yield a *Z*-allylic cation (**Figure 8.115**). Deprotonation of nerolidyl cation gives (*E*)-β-farnesene. Cyclization of the nerolidyl cation generates a six-membered ring and deprotonation of this cation generates (*E*)-α-bisabolene, but the preferred enzymatic pathway involves a [1,2] hydride shift to generate a cyclohexyl cation. Cyclization of the pendant olefin generates a spirocyclic [5.4.0] acorane ring system. The isopropyl cation is poised to cyclize with the cyclohene olefin, but a small amount of the isopropyl cation is deprotonated to generate acoradiene. The appropriate cyclization generates the cedrane ring system. Stereoselective attack by water favors the formation of *epi*-cedrol, but a small amount undergoes elimination to give α- and β-cedrene.

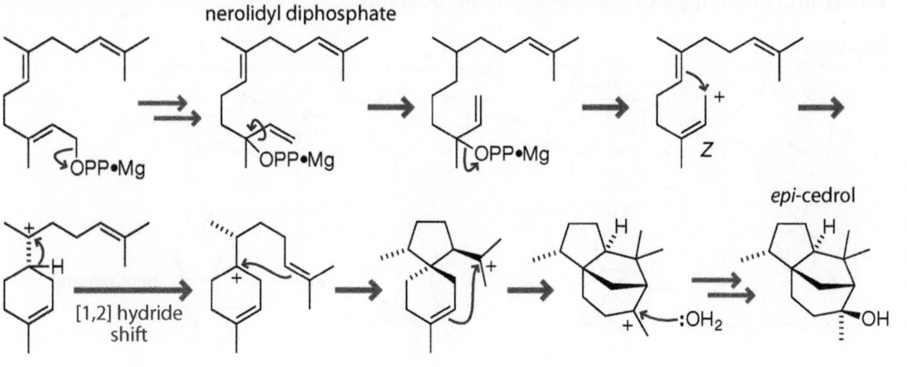

nerolidyl diphosphate

OPP•Mg | OPP•Mg | OPP•Mg | Z

[1,2] hydride shift | :OH_2 | epi-cedrol | OH

Figure 8.115 *epi*-Cedrol synthase. The mechanism of *epi*-cedrol synthase accounts for the other terpene natural products produced by the enzyme from nerolidyl diphosphate.

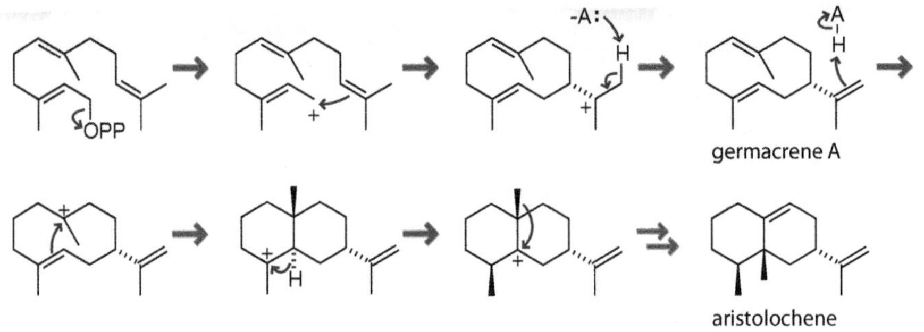

Figure 8.116 Aristolochene synthase. Medium-sized rings with 8–11 carbon atoms are among the most difficult to synthesize due to transannular steric interactions. Aristolochene synthase initially catalyzes the cyclization of farnesyl diphosphate to produce the 10-membered ring intermediate garmacrene A.

Some terpene cyclases generate medium-sized rings

Eight-, nine-, and ten-membered rings are among the most difficult to generate through chemical cyclization, because of the strain produced by transannular steric interactions. Recall that the direct cyclization of geranyl diphosphate, without isomerization to linalyl diphosphate, is virtually impossible because it would place a *trans* double bond in a small ring. Some enzymes generate cyclic sesquiterpenes without first isomerizing the *E*-allylic diphosphate to a tertiary allylic diphosphate. For example, germacrene A synthase catalyzes the cyclization of farnesyl diphosphate leading to a strained 10-membered ring (**Figure 8.116**). Aristolochene synthase performs this same cyclization to yield germacrene, but takes the transformation even further. An acidic functional group protonates one of the trisubstituted double bonds in the ring, leading to a tertiary carbocation that undergoes a transannular cyclization to generate a decalin ring system. The acidic functionality in the active site responsible for the protonation of the olefin is most probably a phenolic hydroxyl group of a tyrosine side chain, a reaction unlikely to be considered outside an enzyme active site. A [1,2] hydride shift and a [1,2] methyl shift result in a bridgehead carbocation that is deprotonated to generate aristolochene.

Problem 8.22

Why is 2-fluorofarnesyl diphosphate an inhibitor of aristolochene synthase?

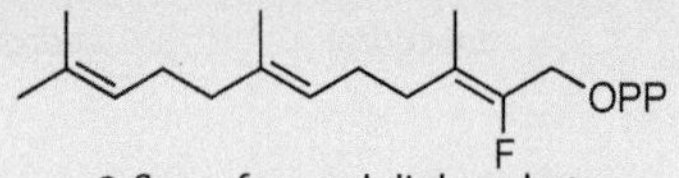

2-fluorofarnesyl diphosphate

The enzyme pentalenene synthase generates pentalenene from farnesyl diphosphate. The reaction proceeds via initial cyclization to an 11-membered ring humulyl cation (**Figure 8.117**). A [1,2] hydride shift would generate a secondary carbocation poised for transannular cyclization. For many years it was speculated that the humulyl cation was deprotonated to generate the well-known natural product α-humulene, and then reprotonated to generate an isomeric cation. However, there is no evidence for a catalytic base in the enzyme. Transannular cyclization generates a fused 5–8 ring system. A [1,2] hydride shift would generate the bridgehead carbocation needed for the final cyclization to generate the tricyclic skeleton of pentalenene. When isotopically labeled [8-^{3}H]-farnesyl diphosphate was transformed by pentalenene synthase, the tritium label appeared at the expected position.

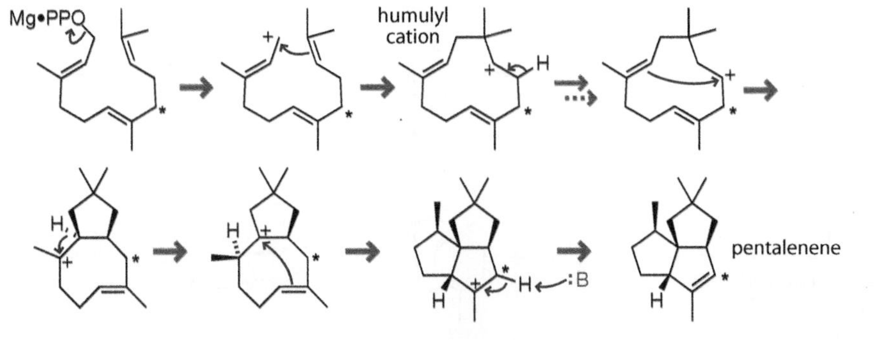

Figure 8.117 Isotoping labeling. The mechanism for the formation of pentalenene from farnesyl pyrophosphate is supported by tritium labeling. The tritium-substituted carbon is indicated with an asterisk.

The biosynthesis of some terpenes involves nontraditional [1,3] hydride shifts

Hydride and alkyl shifts are highly responsive to bond alignment. Thus, terpene synthases can facilitate certain [1,2] migrations by holding cationic intermediates in specific conformations. Similarly, terpene synthases can also disfavor [1,2] shifts by holding carbocations in specific conformations. In such cases, less obvious cation reactions can occur. [1,2] Migration reactions dominate in solution-phase carbocation chemistry, but in a few systems there is clear evidence for hydride transfer over longer distances. In solvolytic reactions of strained cycloalkyl tosylates, for example, the initial secondary carbocation undergoes a transannular [1,5] hydride transfer to generate a more stable tertiary carbocation (**Figure 8.118**).

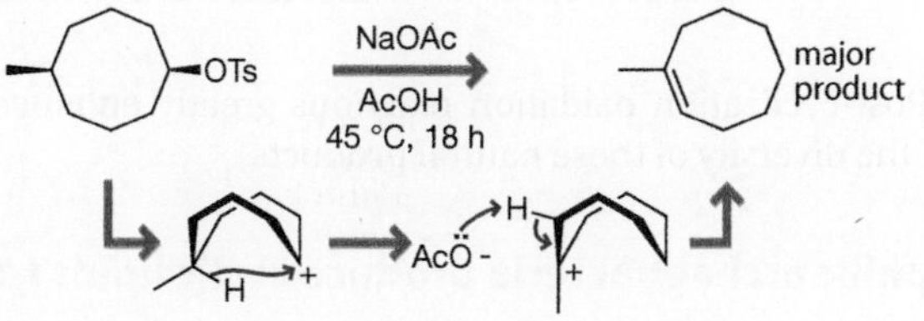

NaOAc
AcOH
45 °C, 18 h
major product

Figure 8.118 Proximity effect. Transannular [1,5] hydride migrations are observed in eight-membered rings.

The mechanisms of some terpene cyclases are believed to involve [1,3] hydride migration. In the case of longifolene synthase, for example, the mechanism is believed to involve an initial ionization to generate the tertiary allylic diphosphate nerolidyl diphosphate, which can ionize to form a *Z*-allylic cation (**Figure 8.119**). Contrary to chemical intuition, the cation cyclizes to generate a 13-membered ring with a secondary carbocation. A [1,3] hydride shift next generates a more favorable allylic cation. The allylic cation could undergo transannular cyclization to generate the [5.4.0] ring system. A second transannular cyclization would generate a norbornyl cation, which could be deprotonated to generate longifolene. Thus, terpene synthases have numerous powerful points of control over the reaction pathway: such enzymes can control the stereochemistry of the initial allylic cation, which atoms form the initial ring, which atoms are aligned to migrate, and which protons are set to eliminate.

nerolidyl diphosphate
[1,3] hydride shift
longifolene

Figure 8.119 Mechanism of longifolene biosynthesis. Longifolene biosynthesis is believed to involve a [1,3] hydride shift. To follow the final cyclization to form the three-ring system, note the formation of a new bond (shown in green) between two carbons attached to methyl groups.

Plants can also make complex triterpenes from squalene

The cyclization of oxidosqualene to lanosterol is a truly amazing reaction. However, the range of possible structures is far greater than the repertoire of human steroids would suggest. Plants also make terpenes from tail-to-tail dimers such as oxidosqualene, and triterpenes have been isolated with more than 90 different skeletons

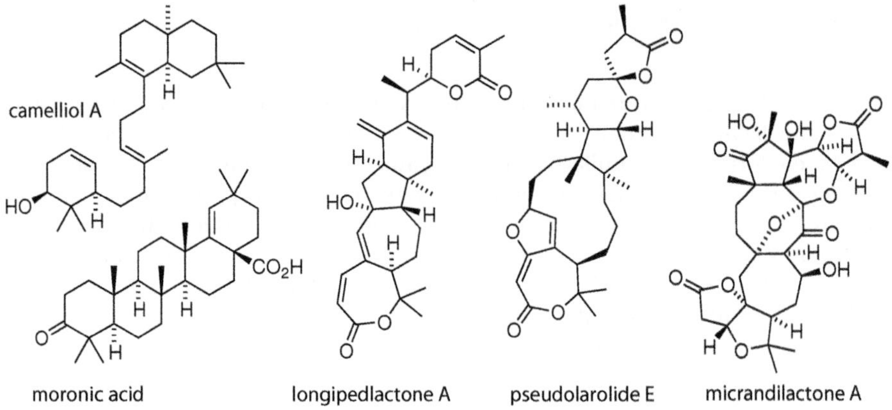

camelliol A

moronic acid

longipedlactone A

pseudolarolide E

micrandilactone A

Figure 8.120 Dazzling complexity. Plant triterpenes, such as the five shown here, have a much wider range of structures than human steroids.

(**Figure 8.120**). Post-cyclization oxidation reactions greatly enhance the degree of functionality and the diversity of these natural products.

Hyperthermophilic archaebacteria produce cyclic lipids from terpenes

Figure 8.121 The heat is on. This boiling volcanic spring in Campi Flegrei, Italy, is home to the hyperthermophile *Sulfolobus solfataricus.* (Courtesy of Science@NASA.)

Heat disrupts lipid bilayers, yet many microorganisms in the primitive domain Archaea—which is distinct from Eukaryotes and prokaryotic Eubacteria—thrive at high temperatures (**Figure 8.121**). How do they maintain the integrity of their membranes? Hyperthermophilic bacteria that thrive at temperatures above 60 °C have been found to produce a structurally complex set of lipids, relative to the diacylglycerol motif of human lipids. Some hyperthermophiles prosper within the withering heat of an autoclave. The membranes of hyperthermophiles are predominantly cyclic glycerol dialkyl glycerol tetraether lipids, some containing additional five-membered rings (**Figure 8.122**). These lipids arise from terpene biosynthetic pathways rather than fatty acid synthases. Interestingly, the tetraterpene chains are linked head-to-head, whereas squalene and phytoene are linked tail-to-tail. Only a few strains of traditional Eubacteria have been found to contain glycerol dialkyl glycerol tetraether lipids, which are abundant in hyperthermophilic archaebacteria.

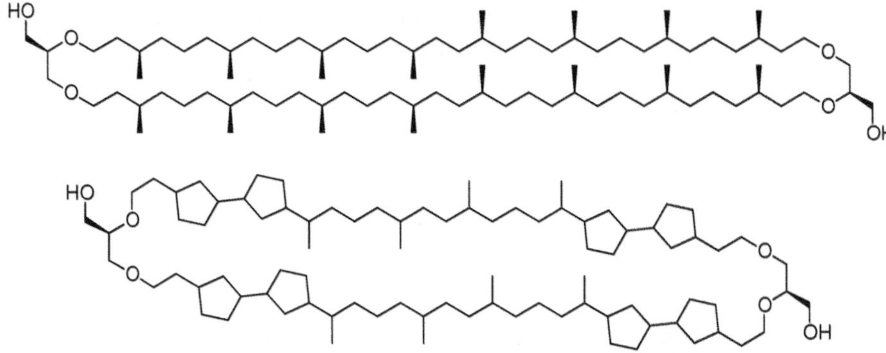

Figure 8.122 Super-sized rings. Some hyperthermophilic archaebacteria contain macrocyclic tetraether lipids that are structurally and biosynthetically distinct from traditional lipids based on fatty acids.

Problem 8.23

Identify each of the terpene subunits in this diol portion of an archaeal cyclic ether lipid.

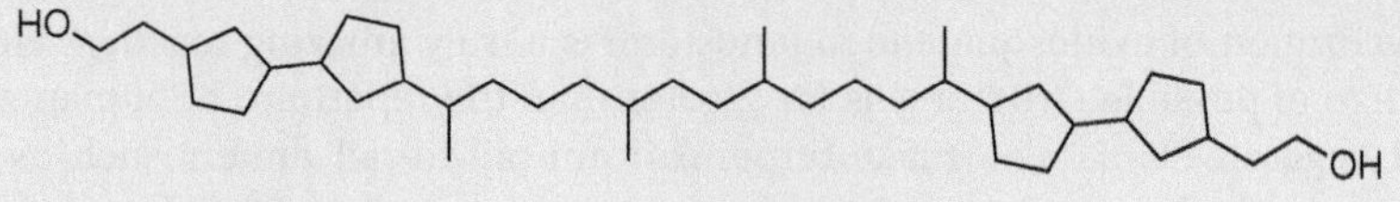

8.8 SUMMARY

We humans pride ourselves on the ingenuity of our species. We take simple resources and craft them into ingenious devices such as axles, slingshots, bicycles, telescopes, guns, engines, generators, pharmaceuticals, airplanes, computers, and spacecraft. Given our ability to invent so prolifically, it is disappointing that the catalog of human polyketides should be such a meager affair—namely, straight-chain fatty acids, linear sphingoid molecules, and some mildly interesting eicosanoids. The best that might be said of human polyketides is that they fulfill their intended function as primary metabolites. How can one not be awestruck by the spectacular range of polyketides produced by "primitive" organisms—namely, polycyclic aromatics, polyenes, polyynes, enediynes, heterocycles, spirocycles, strained rings, and macrocycles, which bristle with stereogenic centers and glisten with novel sugars? Our artless implementation of polyketide biosynthesis is even more embarrassing when we realize that microorganisms generate their resplendent polyketides by using the same reactions that we use to generate boring saturated fatty acids.

Humans have done an equally minimal job when it comes to the design and production of terpenes. Our cells generally tinker with simple prenyl chains, using them to alkylate proteins and quinones. We seem to have chanced upon one exciting cyclization—the conversion of squalene epoxide into lanosterol—but rather than develop new cyclization reactions our bodies simply modified lanosterol, investing 19 laborious steps to convert it to cholesterol, and even more effort to make the various steroid hormones with a monotonous A-B-C-D ring system. Vitamin D is a notable departure from this theme because it lacks a B ring. A quick glance at the range of monoterpene secondary metabolites found in plants inspires envy. Plants have managed to make every possible ring system that can be accessed from the 10 carbons in geranyl diphosphate. Starting with longer precursors—that is, with 15, 20, and 30 carbons—Nature has accessed a boundless array of structures.

Evolution has clearly pushed otherwise defenseless organisms to maximize the potential of polyketide and terpene biosynthesis. Somehow, the human combination of physical agility, a complex immune system, and mental computing power has obviated the need for chemical weaponry to ensure the survival of our species. However, we need these kinds of sleek, membrane-permeable chemical weapons to wage a war of survival against infectious diseases, metabolic diseases, and even cancer. Humans aspire to reach a level of synthetic sophistication on a par with plants and microorganisms. We are slowly catching up, not through human genetic evolution but through human intellect, by harnessing the combined power of organic synthesis, biochemistry, microbiology, and molecular biology.

LEARNING OUTCOMES

- Understand the chemical mechanisms of each step in fatty acid biosynthesis—namely, the Claisen condensation, carbonyl reduction, β-elimination, and conjugate reduction.
- Appreciate the limited range of human fatty acids.
- Recognize the presence of polyketides in human phospholipids and sphingolipids.
- Understand how lipases can generate signaling molecules from lipids.
- Recognize the four classes of eicosanoid signaling molecules derived from arachidonic acid—namely, thromboxanes, prostacyclin, prostaglandins, and leukotrienes.
- Draw the structure of ceramide and understand how it is formed from palmitoyl-CoA.
- Understand the effects of N-myristoylation and S-palmitoylation on protein localization.
- Understand the chemical transformation of fats into partly hydrogenated vegetable oil, soap, and biofuels.
- Understand the origin of 1,3-diol motifs in polyketide natural products.
- Draw the mechanism for aldol condensations that lead to aromatic polyketides.
- Understand how the incorporation of propionate subunits generates methyl substituents on polyketide chains.
- Identify the acetate and propionate subunits in polyketide natural products.
- Understand how the modular genetic organization of type I polyketide synthases facilitates genetic reprogramming.
- Recognize the similarities between polyketide biosynthesis and nonribosomal peptide synthesis.

- Draw the structures of the linear terpenes geraniol, farnesol, geranylgeranyl alcohol, and squalene.
- Identify which step of cholesterol biosynthesis is inhibited by statins.
- Draw the sequence of post-translational modifications that accompany prenylation.
- Recognize the prenylated quinones used by humans.
- Understand the structure and reactivity of cyclopropylcarbinyl cations.
- Draw the arrow-pushing mechanism of the transformation catalyzed by lanosterol synthase.
- Recognize human steroid signaling molecules.
- Appreciate the dietary origin of human retinoids.
- Understand how cationic cyclizations and rearrangements amplify the diversity of terpene natural products.
- Describe the key steps catalyzed by a terpene synthase in the biosynthesis of terpenes.
- Understand the structure and reactivity of 2-norbornyl cations.
- Recognize the complexity of cyclic ether lipids from hyperthermophilic archaebacteria.

PROBLEMS

8.24 Suggest a plausible arrow-pushing mechanism for the retro-Claisen reaction of 6-oxo-camphor to (2*R*,4*S*)-α-campholinic acid by the enzyme β-diketone hydrolase. Assume that the active site contains basic and acidic residues.

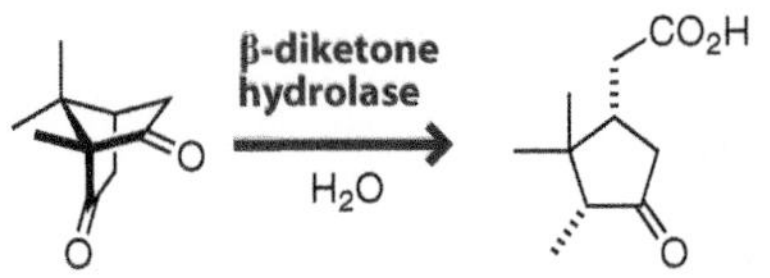

β-diketone hydrolase, H_2O; CO_2H

***8.25** The autoxidation of lipids leads to unstable allylic peroxides that can oxidatively cleave via the Criegee mechanism. Suggest a plausible arrow-pushing mechanism for this reaction.

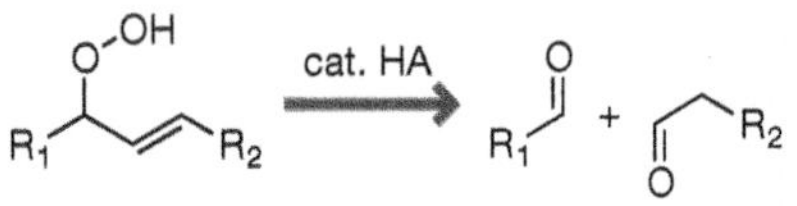

O–OH; R_1; R_2; cat. HA; O; R_1; +; R_2; O

8.26 Humans and other organisms metabolize fatty acids back into acetyl-CoA through a process called β-oxidation, which is the reverse of the steps used to make fatty acids.

A Draw the three intermediates in the first cycle of β-oxidation of stearic acid.

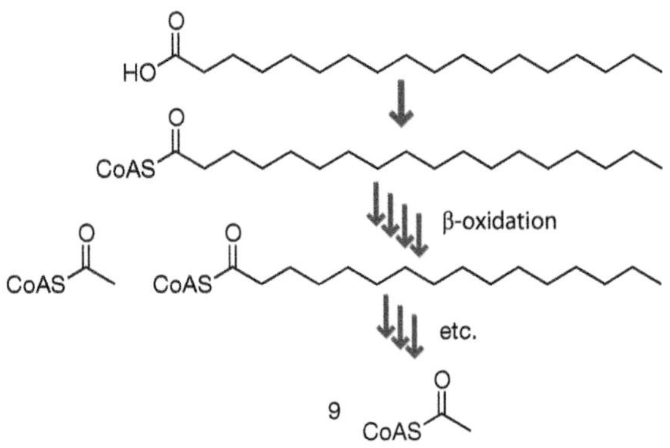

HO; O; CoAS; O; β-oxidation; CoAS; O; CoAS; O; etc.; 9 CoAS; O

B Thia-fatty acids, which have sulfur atoms at odd-numbered positions, cannot be broken down all the way by β-oxidation. Draw the intermediate derived from 9-thiastearic acid that is resistant to further reaction.

9-thiastearic acid; HO; O; S

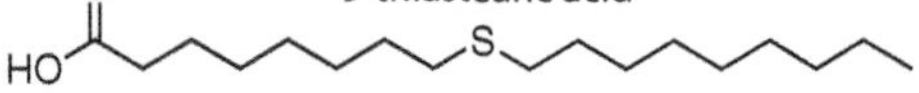

***8.27** Suggest a plausible acyclic polyketide precursor to SEK34, assuming that it arises through aldol reactions, dehydrations, and enolizations.

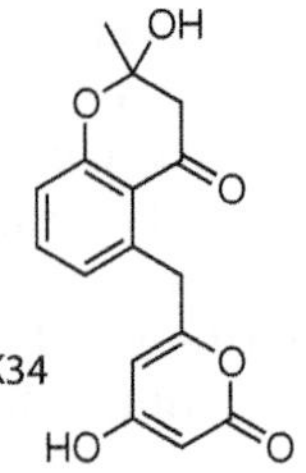

SEK34; OH; O; O; O; HO; O

example:

OH O; HO; O; O; O; SR; O; O; =; O O O O O; SR

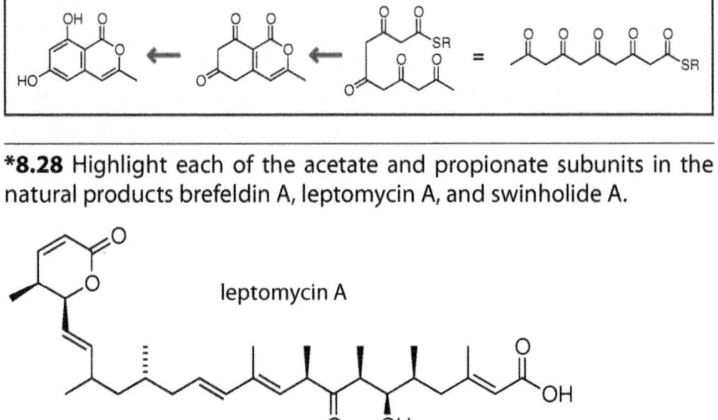

***8.28** Highlight each of the acetate and propionate subunits in the natural products brefeldin A, leptomycin A, and swinholide A.

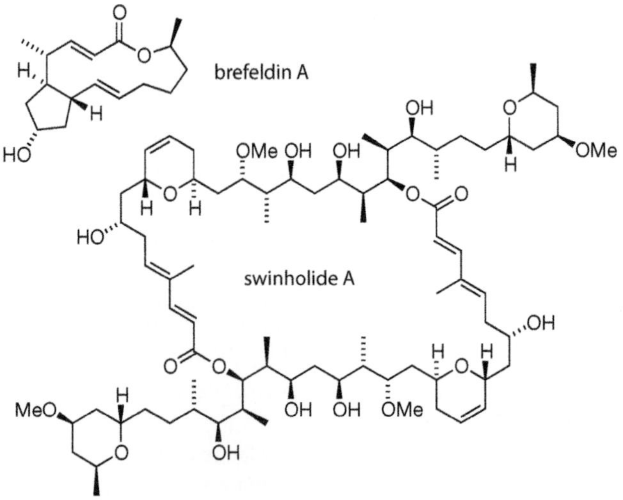

leptomycin A; O; O; O; OH; O; OH

brefeldin A; O; O; H; H; HO

swinholide A; OH; O; OMe; OMe OH OH; H; O; O; H; H; HO; OH; O; O; H; H; O; MeO; H; OH OH OMe; O; OH

8.29 Which part of tulearin A is the most difficult to rationalize on the basis of an iterative Claisen condensation of acetyl and propionyl subunits?

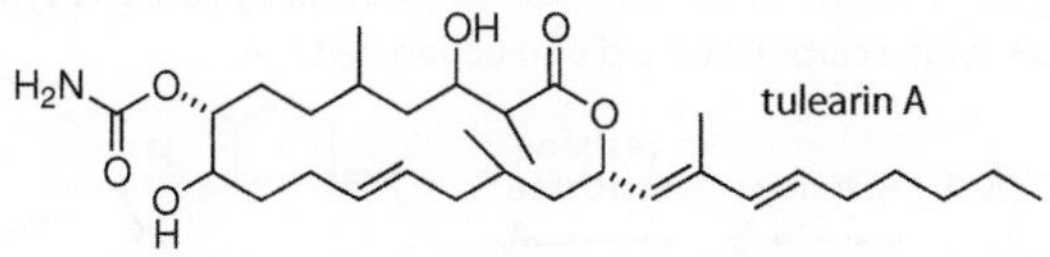

tulearin A

8.30 The polyketide spongistatin 1 is active at femtomolar concentrations. Highlight each of the acetate and propionate subunits in spongistatin 1 and circle the carbon atoms derived from the carbonyl carbon of each subunit. Note which methyl groups do not seem to arise from a propionate subunit.

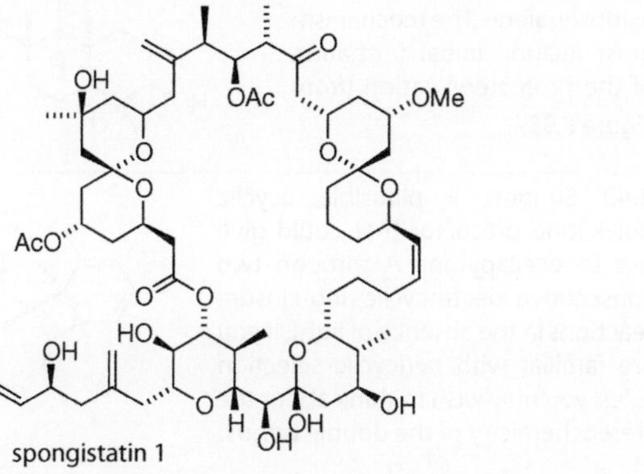

spongistatin 1

8.31 Which of the amino acids in the antibiotic gramicidin S is not normally incorporated into proteins by a ribosome?

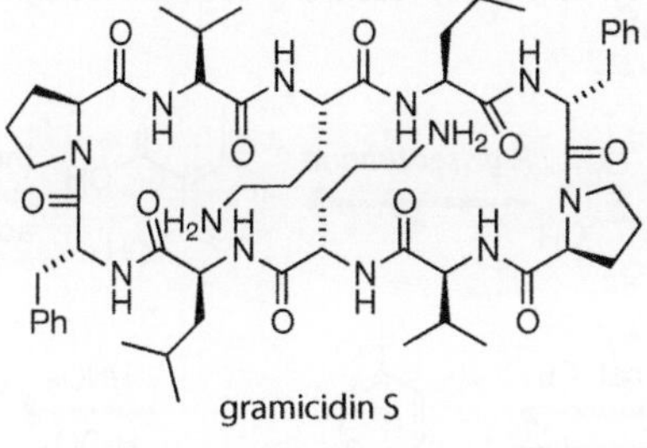

gramicidin S

***8.32** For each of the following compounds, predict whether it most plausibly arises through a polyketide pathway or a terpene pathway.

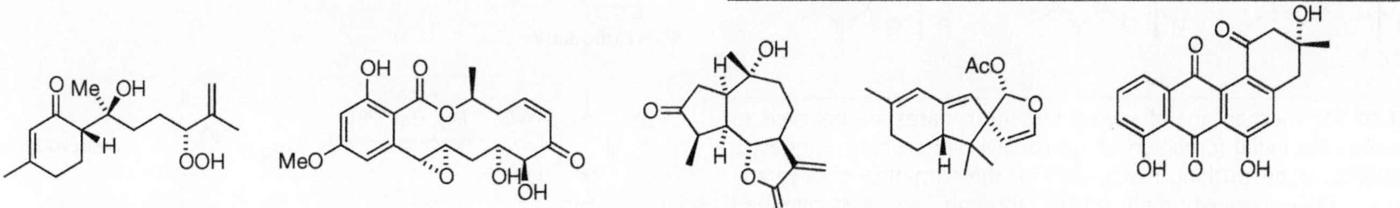

peroxylippidulcine B | hypothemycin | chinensiolide D | haterumadysin A | rabelomycin

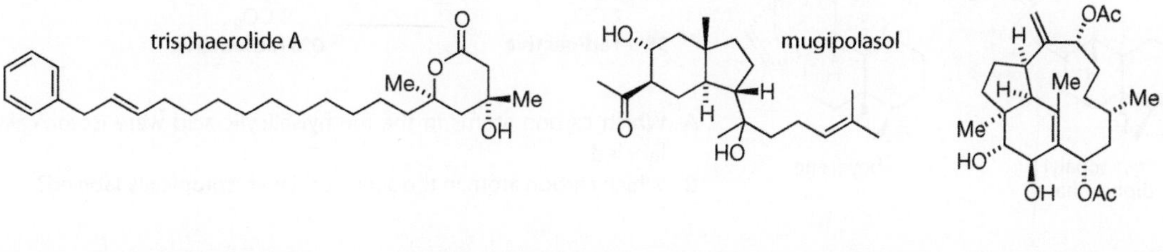

trisphaerolide A | mugipolasol | 9α,14α-diacetoxy-2β,3α-dihydroxy-1(15),8(19)-trinervitadiene

***8.33** Identify the five-carbon terpene subunits in the natural products (–)-limonene, α-cedrene, illudin M, and camphor. Illudin M is extra difficult; for experts only.

example:

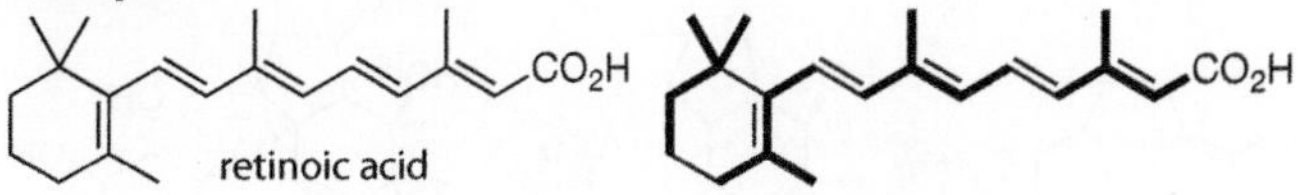

retinoic acid

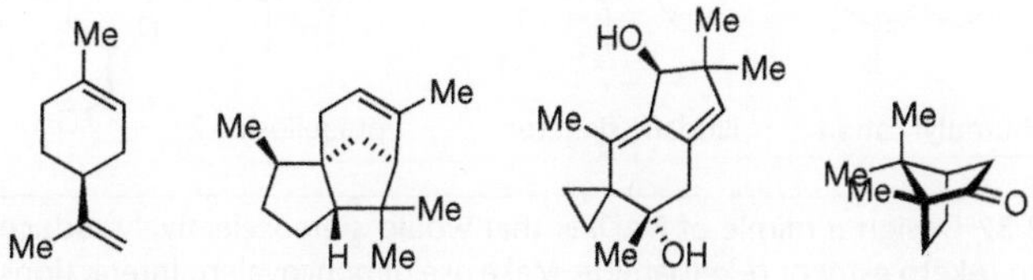

(–)-limonene | α-cedrene | illudin M | camphor

***8.34** Propose an arrow-pushing mechanism for the following transformation of the 20-oxa analog of 2,3-oxidosqualene.

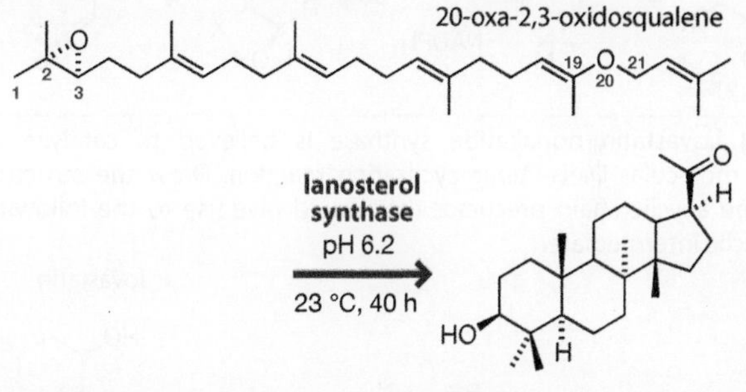

20-oxa-2,3-oxidosqualene

lanosterol synthase, pH 6.2, 23 °C, 40 h

8.35 The natural products dammarenediol-I and lupeol are both generated from oxidosqualene via the same cationic intermediates.

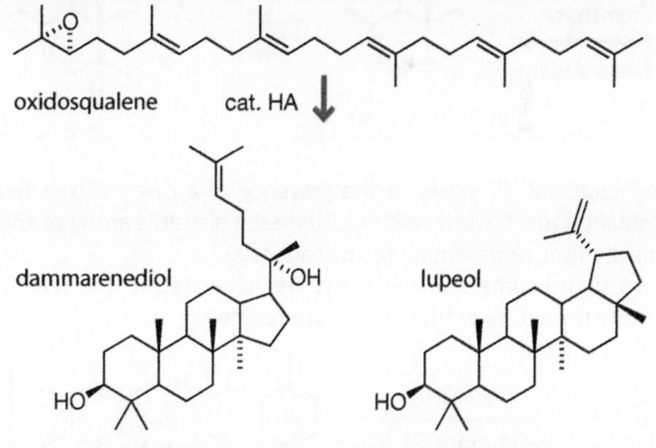

oxidosqualene → cat. HA → dammarenediol, lupeol

Suggest a plausible arrow-pushing mechanism for the biosynthesis of each triterpene.

***8.36** Suggest a plausible arrow-pushing mechanism for the enzymatic synthesis of illudane terpenes via the humulyl cation. Assume that either catalytic acids or bases are present.

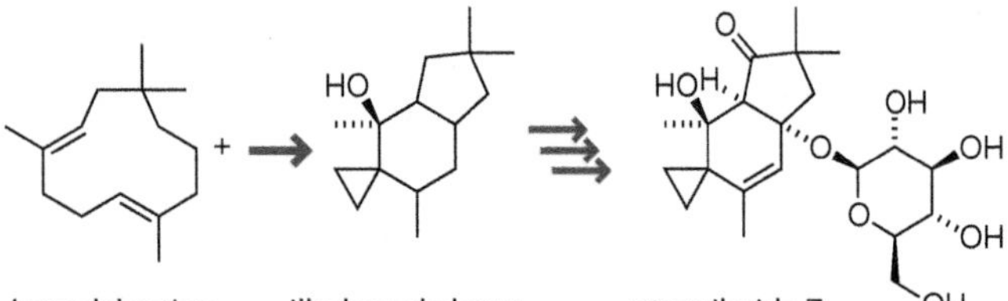

humulyl cation | illudane skeleton | ptaquiloside Z

8.37 Design a mimic of NADPH that would stereoselectively reduce an α-keto ester or α-ketoamide. Make use of noncovalent interactions such as π stacking, hydrogen bonding, and base pairing.

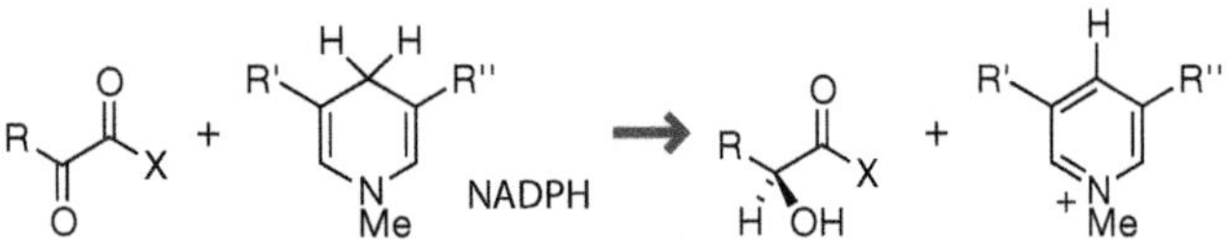

NADPH

8.38 Lovastatin nonaketide synthase is believed to catalyze an intramolecular Diels–Alder cyclization reaction. Draw the structure of the acyclic chain precursor that would give rise to the following bicyclic intermediate.

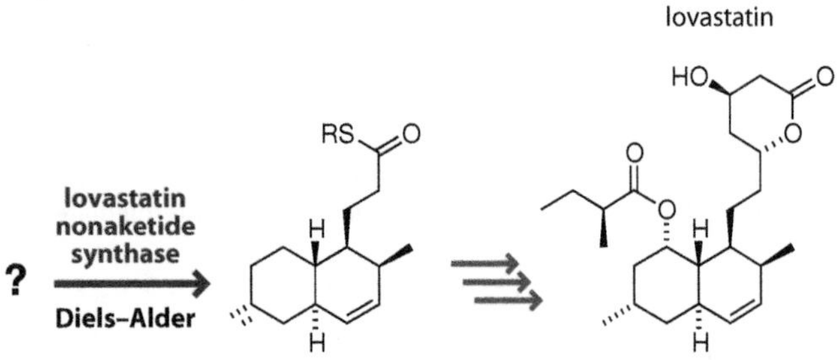

? → lovastatin nonaketide synthase, Diels–Alder → ⇒ lovastatin

***8.39**

A γ-Tocopherol dimerizes in the presence of a one-electron ferric oxidant under basic conditions. Suggest a plausible arrow-pushing mechanism, representing the oxidant as X•.

B Two diastereomers of the biaryl are generated in the reaction. Clarify the nature of these two diastereomers.

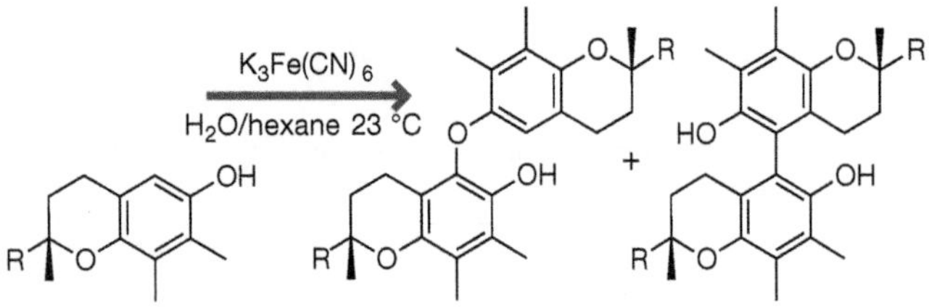

$K_3Fe(CN)_6$, H_2O/hexane 23 °C

***8.40** The mechanisms of several terpene cyclases are believed to involve the initial formation of *syn*-copalyl diphosphate. Suggest a plausible arrow-pushing mechanism for the formation of beyerene from geranylgeranyl diphosphate through an acid-catalyzed mechanism that involves *syn*-copalyl diphosphate.

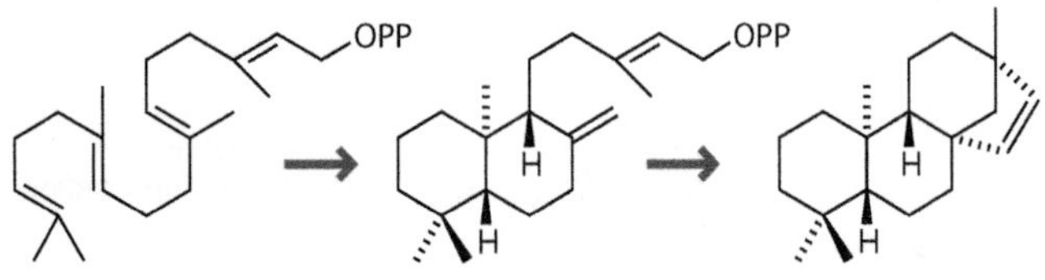

geranylgeranyl diphosphate → *syn*-copalyl diphosphate → beyerene

8.41 When ^{13}C-labeled isopentenyl pyrophosphate was used in a cell-free extract containing casbene synthase and other enzymes, eight carbons in casbene synthase were labeled. Suggest a plausible arrow-pushing mechanism for the synthesis of casbene by casbene synthase and show which carbons would end up labeled.

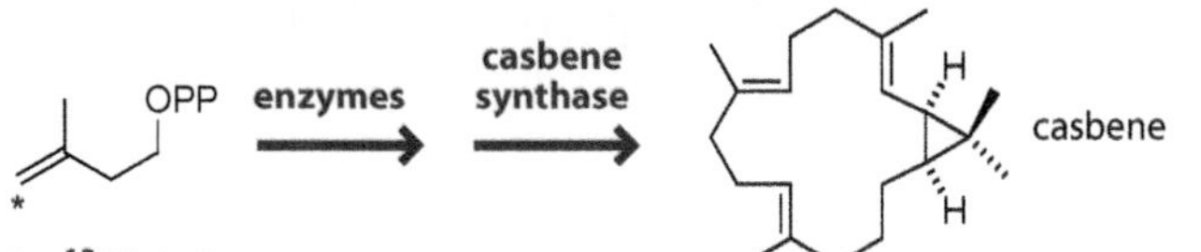

OPP → enzymes → casbene synthase → casbene

* = ^{13}C label

8.42 Suggest a plausible arrow-pushing mechanism for the formation of the triterpene betulinic acid skeleton from oxidosqualene. The mechanism must include initial formation of the protosteryl cation (from Figure 8.95).

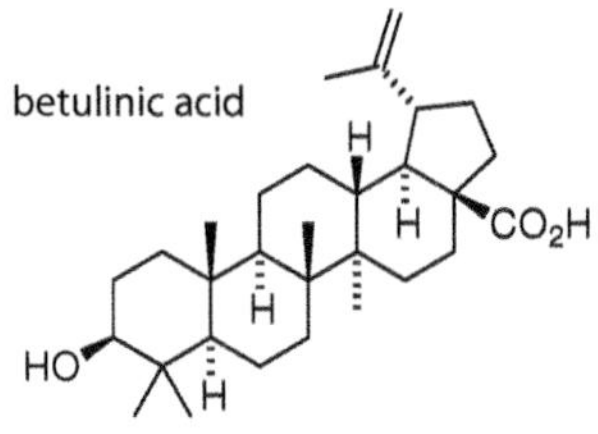

betulinic acid

8.43 Suggest a plausible acyclic polyketide precursor that could give rise to ocellapyrone A through two consecutive electrocyclic ring-closure reactions in the absence of light. If you are familiar with pericyclic selection rules, you may wish to think about the stereochemistry of the double bonds.

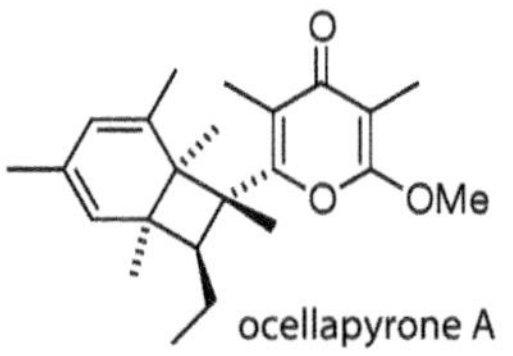

ocellapyrone A

***8.44** In the earliest isotopic labeling study of polyketide biosynthesis, ^{14}C-labeled acetic acid was fed to *Penicillium griseofulvum*. The resulting methylsalicylic acid was isolated and subjected to various chemical degradation and combustion reactions. On the basis of the radioactivity in the liberated CO_2, one can deduce the distribution of radioactive label in the methylsalicylic acid and each of the degradation products.

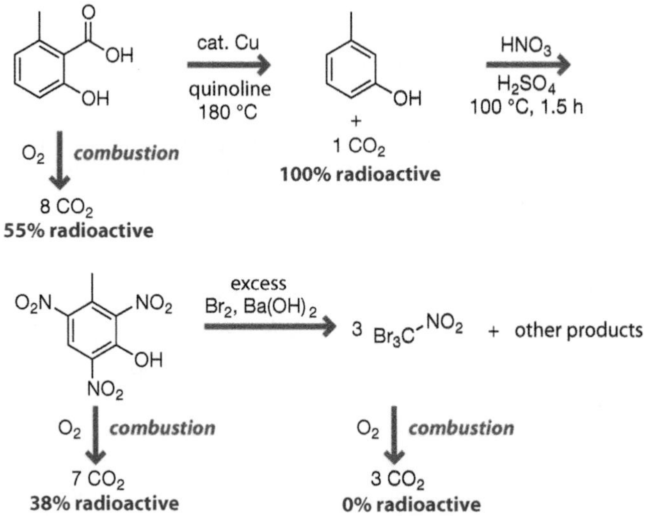

$H_3\overset{*}{C}-C(O)OH$ or $H_3C-\overset{*}{C}(O)OH$ → *P. griseofulvum* → methylsalicylic acid

cat. Cu, quinoline 180 °C → + 1 CO_2 **100% radioactive**; HNO_3, H_2SO_4 100 °C, 1.5 h

O_2 *combustion* → 8 CO_2 **55% radioactive**

excess Br_2, $Ba(OH)_2$ → 3 Br_3C-NO_2 + other products

O_2 *combustion* → 7 CO_2 **38% radioactive**

O_2 *combustion* → 3 CO_2 **0% radioactive**

A Which carbon atoms in the methylsalicylic acid were isotopically labeled?

B Which carbon atom in the acetic acid was isotopically labeled?

Chemical Control of Signal Transduction

9

LEARNING OBJECTIVES

- Recognize primary signal transduction pathways in the scientific literature.
- Predict the effects of activators and inhibitors on signaling outcomes.
- Describe two rapid nontranscriptional processes that involve Ca^{2+} ions.
- Identify and describe the seven major signal transduction pathways that control transcription in human cells.
- Distinguish the three major subpathways affected by receptor tyrosine kinases.
- Outline the two types of pathways affected by G protein-coupled receptors.
- Recognize the ionic imbalance maintained by human cells.

At the outset of this book, we proposed to use the central dogma of molecular biology as a framework for understanding how human cells work. Genes are transcribed into RNA; RNA is translated into enzymes; enzymes catalyze the formation of glycans, lipids, and terpenes. Our relentless focus on the central dogma led us to skirt *the* fundamental question: *Which genes*? Gene expression determines how cells differentiate, allowing each cell to have a precise role in the human organism. Gene expression also determines the way in which fully differentiated cells respond to their environment for the maximum benefit of the organism. Ultimately, extracellular cues determine which genes are expressed both during and after differentiation. In this concluding chapter, we focus on the biochemical pathways that transduce signals from outside the cell to changes in gene expression (**Figure 9.1**). Chemists who understand and control these pathways are empowered to remediate errant cells that cause disease and to harness cells for new applications.

The human quest to control cellular signaling with small molecules is an ancient one. The earliest molecular tools came from natural sources such as plants, and their use pre-dates written records. Archaeological evidence suggests that residents of the Nanchoc valley in Peru were orally freebasing cocaine in the eighth millennium BC by chewing on coca leaves along with lime (calcium carbonate). Furthermore, poppies, a source of bioactive compounds, were cultivated on the Eurasian continent as early as the fourth millennium BC, and were prominent in Egyptian art by the Eighteenth Dynasty (1549–1298 BC).

The earliest comprehensive written pharmacopeia is contained within the Ebers papyrus, an ancient Egyptian papyrus written around 1550 BC. Many formulations in the Ebers papyrus involve plants with plausible medicinal value—such as pomegranate root bark, poppy plants, and juniper berries; and many more of the formulations

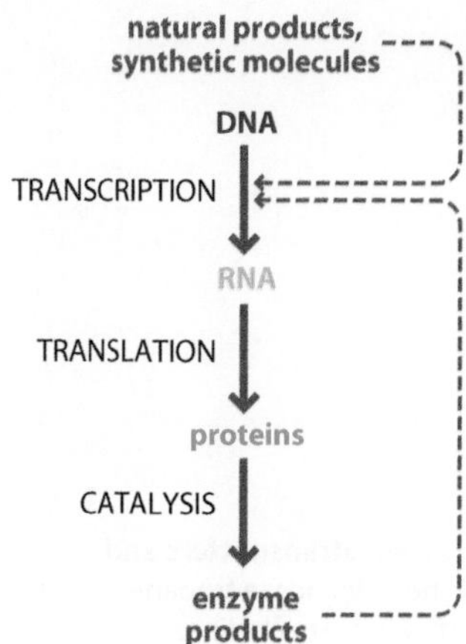

Figure 9.1 Small molecules at work. RNA, proteins, and enzyme products exert control over transcription from within the cell. Natural products and small synthetic molecules that originate outside the cell can also affect transcription.

Figure 9.2 The ancient Egyptian pharmacopeia. The Ebers medical papyrus is written in hieratic script. Page 33, column 89, line 13 describes a treatment for tooth pain involving the netherworld plant.

Figure 9.3 A Babylonian painkiller. A cuneiform tablet from the library of King Ashurbanipal describes an ancient treatment for toothache involving a plant believed to be henbane (right). The exact date of origin of the treatment cannot be assigned because the scribes (*ca* 650 BC) were copying and recording much earlier writings dating back many thousands of years. (Left, courtesy of Wikimedia; right, courtesy of Martin Vidner.)

involve plants that cannot be identified. For example, the Ebers papyrus describes a recipe for dental maladies involving celery, beer, and the netherworld plant (**Figure 9.2**). A similar prescription was found on a cuneiform tablet, penned a thousand years later, from the library of Ashurbanipal in Nineveh: a mixture of oil, beer, and the *sa-kil-bir* plant (**Figure 9.3**). The Babylonian *sa-kil-bir* plant has been interpreted as henbane, which contains large amounts of the tropane alkaloid hyoscyamine (L-atropine). Wisely, these treatments were not intended for ingestion, because hyoscyamine is toxic at high concentrations. Unfortunately, plausible formulations in the Ebers papyrus are listed side by side with questionable treatments involving, for example, incantations and animal excrement. Even today, desperate patients are faced with an equally wide variation in choices, ranging from clinically proven drugs to fake medicines with deceptive labels such as "natural" and "herbal."

The ancient formulations of Chinese traditional medicine are many thousands of years old and, unlike ancient Egyptian medicines, have endured to the present day. The precise age of the formulations is difficult to ascertain. The original copies of *Shénnóng Běn Cǎo Jīng*—the ancient materia medica of Chinese traditional medicine—have never been found, and the oldest copies are only about 2000 years old.

The earliest examples of natural medicines and poisons involve compounds that interfere with neuronal signaling, either between pairs of neurons or between neurons and muscle cells. Neuronal signaling is rapid, making it easy for primitive cultures to establish cause-and-effect relationships. For example, plants in the family Solanaceae (henbane, mandrake, belladonna, and nightshade, among others) produce tropane alkaloids such as hyoscyamine and scopolamine (**Figure 9.4**). The racemic form of (-)-hyoscyamine, produced by the plant belladonna, is referred to as (±)-atropine, but only the (-) enantiomer is active. Hyoscyamine binds to a specific class of receptors in parasympathetic neurons by mimicking the structural features of the neurotransmitter acetylcholine. Parasympathetic neuronal pathways cause the body to "rest and

acetylcholine

(-)-hyoscyamine
(±)-atropine

(-)-scopolamine

Figure 9.4 Neurotransmitters and poisons. The Solonaceae tropane alkaloids mimic acetylcholine.

digest." Hyoscyamine has many systemic effects. It is a local anesthetic, and causes dilation of the pupils, increased heart rate, decreased perspiration, decreased salivation, and cessation of bowel movement.

The most ancient of medicines—for example cocaine, morphine, atropine, and scopolamine—clearly have profound effects on cell function, but the immediate effects of these drugs do not involve the transcription of DNA into RNA. Medicines that exert their primary effects on transcription are much more recent. For example, *Croton tiglium*, mentioned in the *Shénnóng Běn Cǎo Jīng*, is a source of phorbol esters that activate protein kinase C, an enzyme that controls many transcriptional processes.

Our ability to "see" the inner workings of human cells is only a few decades old and is still evolving. The most common methods in use today still involve either tools that report on only one gene at a time or on post-mortem analysis of cellular debris such as DNA, RNA, and proteins. Chemical biologists do not yet have the ability to design potent selective inhibitors by simply looking at the structure of a protein target; however, by screening diverse collections of natural and synthetic compounds, we are discovering molecular tools that control the pathways that direct the transcription of human genes. Most of those tools that come from screening campaigns are rather blunt, but some can be sharpened for pinpoint accuracy. As we progress in the new millennium, one can expect to see new tools for inhibiting disease pathways, enhancing beneficial pathways, and directing the differentiation of stem cells to generate a tissue of choice.

9.1 SIGNAL TRANSDUCTION

Chemical signaling is universal

Communication is fundamental to all animals on the planet Earth and involves many different types of signals: visual, auditory, chemical, and tactile. Male Indian peafowl boast with colorful displays. Siberian huskies growl out warnings. Avocado seed moths beckon with alluring pheromones (**Figure 9.5**). African elephant mothers reassure their calves with gentle caresses from their massive trunks.

All examples of signaling between organisms involve the relay of information from the outside of a cell to the inside. Although the types of signals used by various organisms may differ, all cells in all organisms—even single-celled organisms such as bacteria—communicate using chemical signals. The ability to receive signals is essential for cells because they must be able to respond to changes in their environmental conditions: conditions such as crowding, famine, and safety. The ability to send signals is just as important as receiving signals because even the simplest unicellular organism needs a way to allow its clonal brethren to distinguish it from predators, competitors, and prey. Intercellular communication is particularly important between human cells, because each human cell has a distinct function. Muscle cells contract. Neurons forge connections. Goblet cells spew mucus. Epithelial cells proliferate. B cells differentiate. Such cellular actions are conditional and occur only when the cells receive the correct chemical signals. Understanding how molecular signals control cell behavior, or misbehavior, is a central goal of molecular biology. Using small molecules as precise tools to control these processes is a major goal of chemical biology. Honing these small molecules into safe, effective medicines is the goal of the pharmaceutical industry.

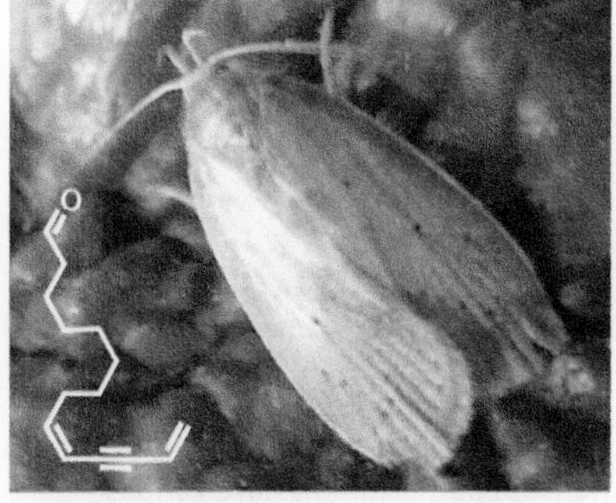

Figure 9.5 Chemical sex appeal. The avocado seed moth *Stenoma catenifer* responds to the pheromone 9(*Z*),13-tetradecadien-11-ynal. The larvae devastate crops by feeding on avocado pits and pulp. Adult moths can be lured into traps laced with the pheromone. (Courtesy of Mark S. Hoddle, University of California, Riverside.)

The field of biology is full of cryptic acronyms and ambiguous symbols

The conversion of one type of chemical signal into another type of chemical signal is called **signal transduction**. The student interested in understanding the signal transduction networks that control cells faces a seemingly impenetrable morass of cryptic acronyms. For example, SOS1 stands for "son of sevenless homolog 1" and PAK1 is an abbreviation for "p21/Cdc42/Rac1-activated kinase 1." Often, the names have nothing to do with our current understanding of the role of the protein, like the ryanodine receptor—named after a molecule not found in the human body. Human signal transduction is further complicated by the redundancy in naming that proliferated in the late 1990s. Researchers often christened their newly identified proteins with exotic new names, even when closely related proteins were already known in the literature. To understand how multiple names arise, consider the enzyme Akt. In 1991 the protein kinase Akt was identified in a virus (designated Akt-8) that causes cancer in the Ak strain of mouse. Around the same time, a nearly identical kinase was discovered independently in two different human cell lines and given two different names: protein kinase B and RAC protein kinase. Scientists published papers for many years on this kinase without agreeing on which name to use! Fortunately, one can now search vast online databases for any protein sequence to prevent redundant naming. However, the tendency to infer the function of a protein from a name is perilous. If you find yourself becoming disheartened by the large number of new names, then you might take comfort from the fact that you have learned the names of your friends and family members and developed a predictive understanding of their relationships. The names of human beings are just as arbitrary as the names of human proteins. Clearly time and patience will breed an easy familiarity with complex names and complex relationships.

Problem 9.1

Look at the most recent table of contents for any of the top scientific journals or any specialized journal of biological chemistry, chemical biology, or biochemistry. List the biological molecules that are found in the titles. When molecules are designated by acronyms, use the Internet to find the full name on which the acronym is based.

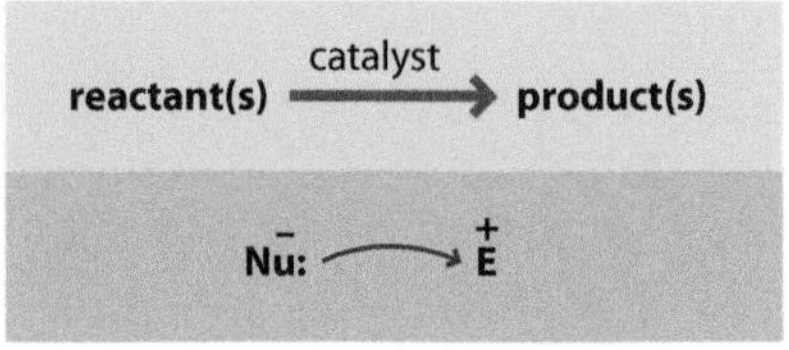

reactant(s) $\xrightarrow{\text{catalyst}}$ **product(s)**

Nu:$^{-}$ → **E**$^{+}$

Figure 9.6 Semantics of arrows. The two types of arrows used by chemists have precise meanings. Straight, horizontal arrows indicate a chemical reaction. Curved arrows indicate the interaction of filled orbitals with unfilled orbitals.

New students in chemical biology tend to confuse the arrows in a signal transduction diagram with the arrows used in reaction diagrams. Organic chemists have agreed on a precise system of representation to depict chemical reactions. They use straight, horizontal arrows to show the transformation of reactants to products, and curved arrows depict the interaction of filled orbitals with unfilled orbitals (**Figure 9.6**). A signal transduction diagram makes use of arrows to depict the known and hypothesized molecular interactions that are responsible for translating extracellular signals into cellular responses. Arrows with a single arrowhead start from some molecule and end on either a function that is facilitated or a second molecule with a function resulting from the first molecule. Arrows that end with a small perpendicular bar generally indicate that the function of the second molecule is inhibited by the presence of the first. These relationships can be established through genetic deletion or chemical inhibition studies.

In signal transduction diagrams, an arrow can have many chemical meanings (**Figure 9.7**). An arrow from Abc to Xyz could mean that proenzyme Abc is converted

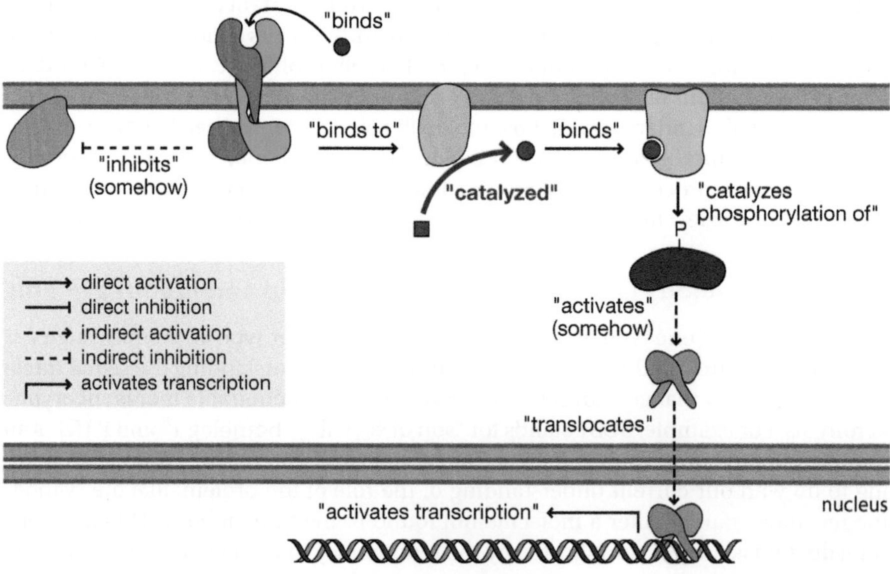

Figure 9.7 Deciphering signal transduction diagrams. Arrows in signal transduction diagrams have various chemical meanings, rarely corresponding to the transformation of starting materials to products. It is up to the reader to intuit the type of interaction that is being depicted. In this book we will use solid arrows to indicate processes that involve direct interactions between molecules (for example binding, chemical reactions, catalytic transformation, and covalent modification).

to the active form Xyz, or it could mean that enzyme Abc binds to a co-catalyst Xyz; alternatively, it might mean that enzyme Abc acts on substrate Xyz; of course, one cannot exclude the possibility that it means that enzyme Abc generates an active product Xyz. Arrows in signal transduction diagrams do not always imply a direct interaction; they are sometimes used to indicate spatial translocation, for example across a membrane. In other cases signal transduction arrows represent a series of complex biochemical phenomena in which the key components have been omitted. For example, an arrow from *Abc* to Xyz could mean that gene *Abc* is expressed as protein Xyz and necessarily involves a large number of intermediate steps. Note that the names of genes are usually italicized. Recently, there have been attempts to develop unambiguous notation for signal transduction diagrams, but these notations have not yet been widely adopted.

Fast cellular responses do not involve the production of proteins

Before we learn about human signaling pathways that affect the transcription of genes (and ultimately the translation of proteins), we need to distinguish those pathways that affect transcription from those that do not. Some signals lead to responses by the biochemical machinery already in place in the cell; such signaling does not involve transcriptional regulation (**Figure 9.8**A). For example, after acetylcholine receptors in intestinal goblet cells bind to acetylcholine, the cells rapidly release the oligosaccharide mucin from storage vesicles (**Figure 9.9**). The mucin helps to ensure a slippery pathway for debris traveling down the gastrointestinal tract. Like it or not, you resemble a slug turned inside out.

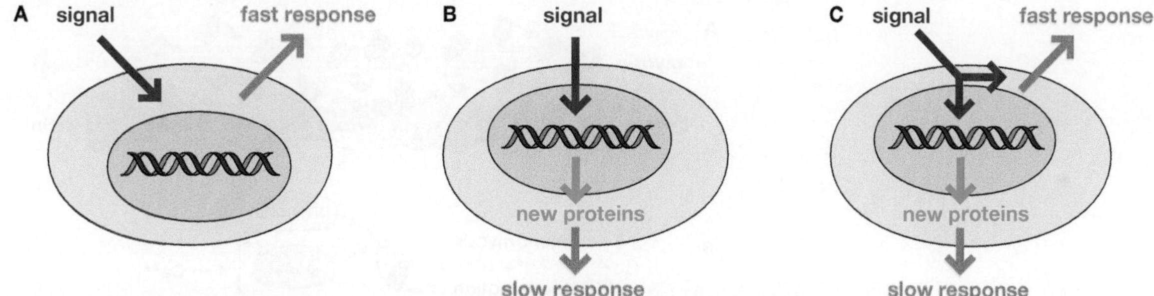

Figure 9.8 Scenarios for signaling. (A) Some signaling pathways lead to a response without affecting transcription. (B) Other signaling pathways involve the transcription of new proteins, and are necessarily slow on a cellular timescale. (C) Many signaling pathways affect both immediate nontranscriptional responses and slower responses that involve the transcription of new proteins.

In other cases, extracellular signals turn on signaling networks that penetrate the nucleus, and either activate or repress transcription of specific genes (Figure 9.8B). Processes that require the production of new proteins are generally slower than processes that do not. For example, when you cut your skin the binding of fibroblast growth factor to the fibroblast growth factor receptor leads to mitotic proliferation of skin cells; cell division is heavily reliant on the translation of new proteins. The transcriptional processes leading to mitosis take many hours.

In other cases, signaling pathways can bifurcate, leading to both a fast response and a slow response involving transcription (Figure 9.8C). For example, insulin exerts both fast and slow responses in liver cells. When insulin binds to the insulin receptor, it leads to the fusion of vesicles adorned with glucose transporters, allowing liver cells to rapidly vacuum up glucose from the medium. In addition, insulin induces slower responses through the regulation of genes. These slower effects include the upregulation of genes responsible for lipid biosynthesis: genes encoding fatty acid synthase, acetyl-CoA carboxylase, and others.

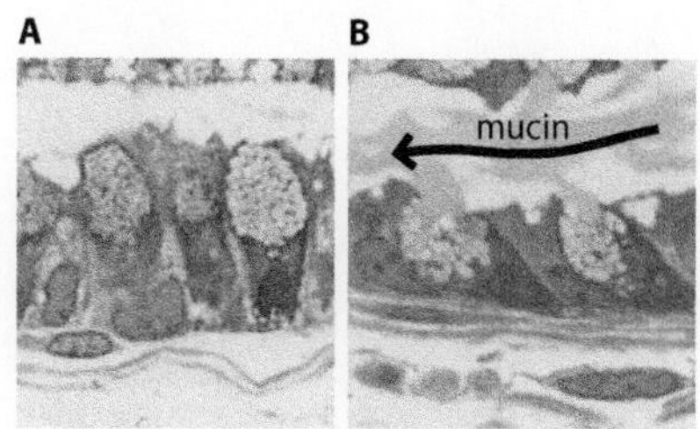

Figure 9.9 Spilling their guts. (A) Normally, mucin is released from goblet cells at a very slow rate. (B) After treatment with acetylcholine and physostigmine (an acetylcholinesterase inhibitor), the goblet cells rapidly release their mucus, which streams into the colon. (Adapted from R.D. Specian and M.R. Neutra, *J. Cell Biol.* 85:626–640, 1980. With permission from Rockefeller University Press.)

Problem 9.2

Which of the following compounds would greatly reduce the acetylcholine-induced release of mucin by intestinal goblet cells? Cycloheximide, a ribosome inhibitor; physostigmine, an inhibitor of acetylcholinesterases; atropine, an acetylcholine receptor antagonist.

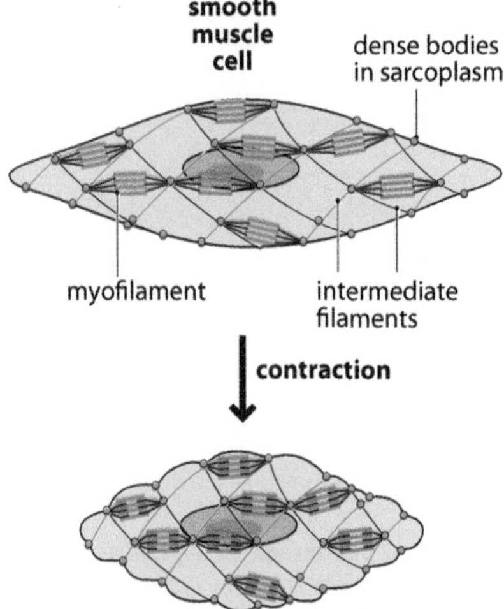

Figure 9.10 Smooth operator. The contractile state of smooth muscle cells arises from the contraction of myofilaments containing myosin, actin, and other proteins. (Adapted from E. Marieb and K. Hoehn, Human Anatomy and Physiology, 7th ed. Benjamin Cummings, 2007.)

Cell contraction and vesicle fusion: fast calcium-dependent responses that do not involve changes in transcription

Two of the most common examples of fast cellular responses (that do not involve transcription) are muscle cell contraction and vesicular export. Both responses are triggered by Ca^{2+} ions. These calcium-triggered cellular responses account for most of the physiological effects that can be monitored by a physician, including heart rate, blood pressure, breathing, pupil dilation, knee reflexes, lifting your arm, seeing, and speaking. Most fast-acting natural toxins act on nontranscriptional pathways. If you want to kill a potential predator before it eats you, or eat prey before it escapes, you would not want to target transcription. It is too slow. Both muscle contraction and the fusion of secretory vesicles are fast and involve Ca^{2+} ions; we will focus on these mechanisms now, and the origins of those Ca^{2+} ions later in this chapter.

The contraction of all muscle cells is due to the crawling of myosin filaments along actin filaments (**Figure 9.10**)—which is mediated by an increase in cellular Ca^{2+} ion concentration. The human body has two classes of muscle cells. Smooth muscle cells, such as those present in bronchial airways and vascular endothelia, are important in asthma and hypertension, respectively. The other class of muscle cells includes cardiac and skeletal muscle cells, which are important in heart disease and muscular dystrophy, respectively. Myosin proteins resemble dimeric bean sprouts; myosin dimers exist as filaments that are grouped into much larger bundles. The "heads" of the myosin proteins protrude from the myosin bundles. The myosin head is an ATP-driven motor that crawls along actin filaments within the cell through a rapid, hinge-like motion. Each stroke is powered by the hydrolysis of one ATP molecule (**Figure 9.11**).

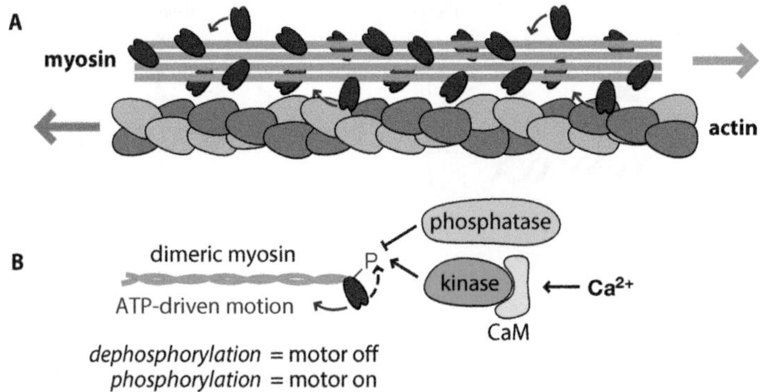

Figure 9.11 Powering muscles. (A) The head domains that protrude from the myosin bundle serve as ATP-driven motors that drag the myosin filament along an actin fiber. (B) Each myosin motor is controlled through the competing actions of two species: a phosphatase and a Ca^{2+}-dependent kinase complex. CaM, calmodulin. The magenta letter P represents a phosphate group.

Each individual myosin motor is controlled through phosphorylation; the motor is "on" when it is phosphorylated and "off" when it is dephosphorylated. The ratio between the phosphorylated and dephosphorylated states is determined by a kinase and a phosphatase, respectively. The phosphatase is always active. The activity of the kinase depends indirectly on the concentration of Ca^{2+} ions. When cytosolic Ca^{2+} levels increase, the Ca^{2+} binds to the calcium sensor protein calmodulin (CaM), leading to a marked change in conformation. The distorted Ca^{2+}•CaM complex binds to and activates the kinase, forming an active complex that phosphorylates myosin proteins, turning on the motors. The ratio of on-motors to off-motors is controlled with high gain, because each kinase and phosphatase acts on many myosin substrates. Controlling states through competing enzyme levels is common among all organisms from bacteria to humans.

The molecular mechanisms that control vesicle fusion are not as well understood as the mechanisms that lead to contraction. Vesicle fusion is most closely associated with the release of neurotransmitters such as acetylcholine, dopamine, and

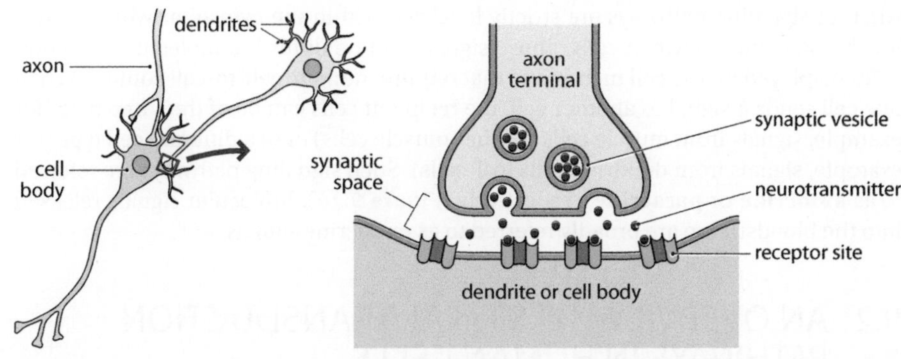

Figure 9.12 Signaling across the synapse. Neurotransmitters are released on one side of the synaptic gap and detected by 7TM GPCRs on the other side. (Adapted from US National Institutes of Health, National Institute on Aging.)

5-hydroxytryptamine from neurons (**Figure 9.12**), but it is also important for the secretion of signaling molecules from non-neuronal cells such as the release of histamine from mast cells. In addition, recall from Chapter 7 that vesicle fusion leads to rapid insertion of fully assembled glucose transporters (GLUT4) into the plasma membrane after stimulation with insulin. In neurons, neurotransmitters are prepackaged within vesicles that dock with the synaptic membrane (**Figure 9.13**). A complex of several proteins holds the vesicle close to the cell membrane, but not close enough to fuse accidentally. When the neuronal signal reaches the synaptic cleft, a flood of Ca^{2+} ions binds to synaptotagmin. Ca^{2+} ions have a high affinity for phosphate groups, such as those found on lipids; up to 50% of bone is a mineralized form of calcium phosphate. Once the vesicle has been drawn close to the plasma membrane, fusion occurs, releasing neurotransmitters into the synaptic space.

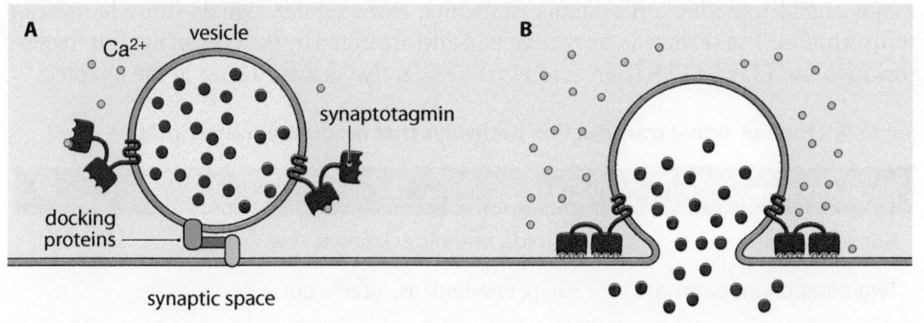

Figure 9.13 Release the hounds. (A) Neurons release neurotransmitters from vesicles that are docked with synaptic membranes. (B) When Ca^{2+} binds to synaptotagmin, the Ca^{2+} helps pull the vesicle into the cell membrane.

Cell signaling can involve pathways within cells and/or between cells

When you add a selective enzyme inhibitor to a cell or delete the gene for the enzyme, it will usually affect many other proteins within the cell. The great challenge is to distinguish *direct* interactions—mediated by binding, catalysis, or other chemical phenomena—from *indirect* interactions. A typical cell contains thousands of different proteins; how can one tell which proteins are responsible for the effects of an inhibitor on a cell? *In vitro* assays allow one to assess the effects of one protein on another directly, without the complications of complex cellular mixtures. However, there are many challenges to setting up unambiguous *in vitro* assays that can truly mimic the situation inside a cell. First, it is difficult—and sometimes impossible—to express and purify stably folded proteins for *in vitro* studies, particularly in the case of membrane-bound proteins. Second, highly sensitive functional assays are required to detect changes at physiological concentrations, usually in the low micromolar or nanomolar range. Third, many proteins exist as subunits in multiprotein complexes; once removed from the complex, the actions of an isolated protein might be physiologically irrelevant. Given the challenges inherent in assays *in vitro*, it is not surprising that so many studies are conducted in living cells.

Cell biologists are mainly interested in understanding how cells work and how they manage to work together. In many cases, the distinction between these two goals is blurry. Most studies of signal transduction are performed on thousands of cells in culture, often in plastic multiwell plates. In such circumstances, it is difficult to know

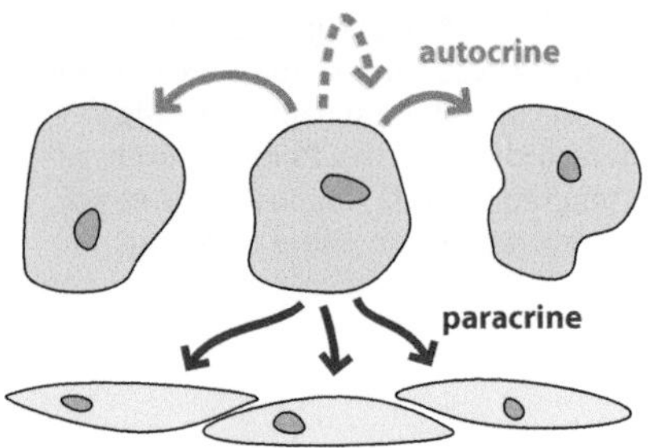

Figure 9.14 ***Entre nous.*** Autocrine signaling occurs between cells of the same type. Paracrine signaling occurs between cells of different types.

whether signaling pathways are strictly localized within the cytosol or whether they involve signaling between cells. Those signals can be diffusible molecules or molecules displayed on the cell membrane that require intimate cell-to-cell contact. When one cell sends a signal to another cell, the recipient cells can be of the same type (for example, signals from muscle cells to other muscle cells) or of a different cell type (for example, signals from dendritic cells to T cells). Such signaling pathways are referred to as **autocrine** or **paracrine**, respectively (**Figure 9.14**). Molecular signals released into the bloodstream are broadly referred to as endocrine signals.

9.2 AN OVERVIEW OF SIGNAL TRANSDUCTION PATHWAYS IN HUMAN CELLS

There are seven major signal transduction pathways in humans

If you have taken a traditional course in biochemistry, you probably experienced the harrowing process of memorizing biosynthetic pathways and metabolic pathways such as the citric acid cycle, amino acid biosynthesis, gluconeogenesis, and glycolysis. These were cutting-edge pathways 50–100 years ago—but no longer. The chemical mechanisms involved in metabolism are fascinating, but they are not the focus of this book. Instead, we choose to focus on how cells receive signals, particularly those signals that affect transcription—not just any cells, but *human* cells. Transcription is the start of the central dogma: DNA is transcribed to RNA; RNA is translated into enzymes; enzymes catalyze reactions of small (and big) molecules.

A quick glance at the literature would suggest that there are thousands of unique signaling pathways in human cells; the complexity is misleading. In humans, seven major signal transduction systems transduce extracellular signals into changes in transcription. The seven can be recognized and arranged by the type of ligands/receptors involved (**Table 9.1**). Each acronym will be fully discussed later in the chapter.

Table 9.1 Human signal transduction pathways that modulate transcription.

Receptor type	Ligands
Nuclear receptors	Steroids, retinoic acid, thyroxine
Two-component pathways	TGF-β, interleukins, interferons
Receptor tyrosine kinases	Growth factors (EGF, FGF, VEGF, NGF)
Trimeric death receptors	TNFα, FasL
G-protein-coupled receptors	Neurotransmitters, hormones, odorants, tastes, photons, enzymes
Ion channel receptors	Glutamate, Na^+ ions
Diffusible gas receptors	O_2, NO

In **Figure 9.15** we summarize the seven principal signal transduction pathways that transduce signals from outside human cells into changes in transcription inside human cells. The reader should commit these to memory. The dashed lines imply the involvement of multiple steps, and we will explain the details in the following sections. As you read those sections, return to this figure so that you are able to keep the big picture in perspective. Figure 9.15 is the forest; keep it in sight as we introduce you to the major trees.

Our understanding of human signal transduction is historically biased toward pathways associated with cancer as opposed to, for example, metabolic diseases. Cancer-derived cell lines are easily grown in the laboratory, whereas primary or normal human tissue is difficult to obtain and culture. Furthermore, the lethality of compounds toward cells is easily assayed, whereas more subtle effects on cell function and human physiology can be challenging to assess. Empowered by new tools from

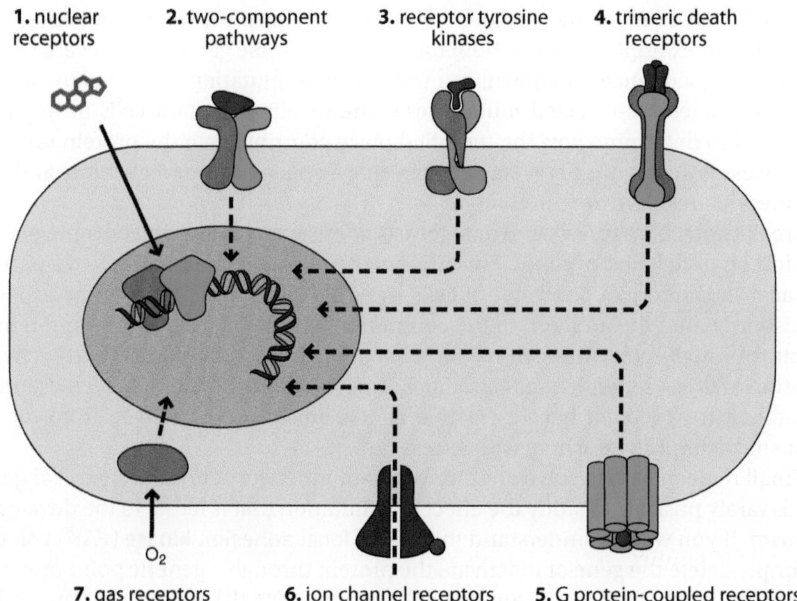

Figure 9.15 Seven pathways. Seven major signal transduction pathways control transcription of human genes. The dashed arrows imply multiple chemical steps.

Figure 9.16 The color of their eyes. The *white* (eye) gene of *Drosophila* codes for a protein that transports molecules such as guanine and tryptophan—precursors of red eye pigments called drosopterins—across membranes. (Courtesy of Wikimedia.)

genomics, cell biology, and chemical biology, the field of cell biology is slowly overcoming the bias toward cell growth pathways. Thus, the view of which pathways are important continues to evolve.

Problem 9.3

Which of the seven major signal transduction pathways involve ligands that pass through the cell membrane and enter the cell?

Chemical genetics involves the use of small molecules to understand gene function

Drosophila has been a fruitful system (pun intended) for the field of genetics, which is primarily concerned with the hunt for genes linked to phenotypic mutations. Phenotypic mutations can enhance the function of a protein, but in most cases the phenotype results from a debilitating effect of the mutation. For example, the *Drosophila white* mutation impairs the transporter that brings pigment precursors into cells of the eye (**Figure 9.16**). It is easy to generate mutations in *Drosophila* through irradiation, and then breed the mutant flies to sort out the effects of diploidy (two copies of each gene per fly). In **forward genetics**, such stochastic mutations are introduced into the organism or cells, and then DNA sequencing of the mutant phenotype identifies the genotype. The corollary, **reverse genetics**, involves examining a specific protein by making mutations to its encoding gene, and then examining the resultant function.

In **chemical genetics**, one seeks to trace the phenotypic effects of small molecules back to a protein target and, ultimately, the gene that controls expression of the protein target. A well-known example is the antifungal natural product rapamycin, discovered in a soil microorganism on Rapa Nui, also known as Easter Island. Interest in rapamycin was raised when it was shown to inhibit the activation of T cells. Rapamycin binds FKBP (FK506-binding protein), and the resultant composite surface then binds the prosaically named protein mTOR (mammalian target of rapamycin) (**Figure 9.17**). Interfering with mTOR suppresses the immune system, and rapamycin has been used both as a pharmaceutical immunosuppressant (for preventing the rejection of transplanted organs) and as an important tool in the laboratory for taking apart the related immune system pathways, a technique called reverse chemical genetics..

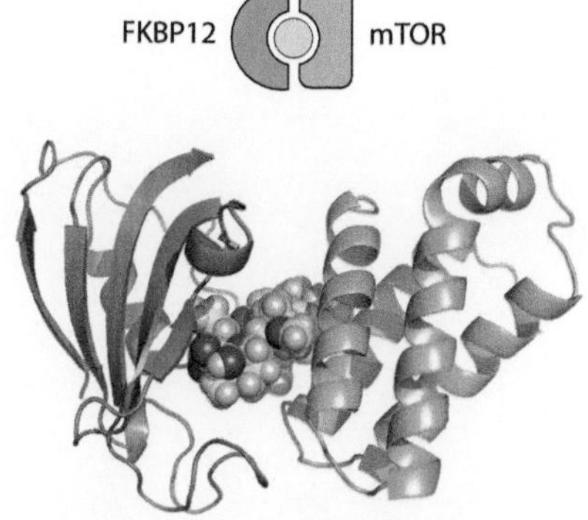

Figure 9.17 A small molecule dimerizer. The polyketide natural product rapamycin (yellow) induces the dimerization of FKBP12 (blue) and mTOR (green). Rapamycin is used clinically as an immunosuppressant. (PDB 1FAP)

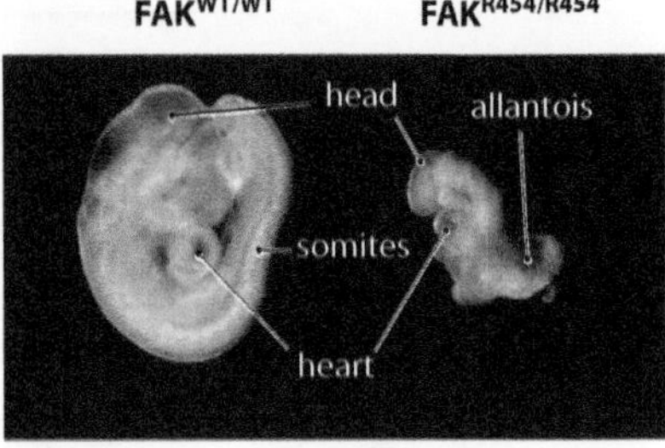

selective FAK inhibitor

Figure 9.18 Small-molecule control over embryo development. You cannot study mice with a K454R mutation in focal adhesion kinase (FAK) because they are never born. On the left is a wild-type embryo (WT/WT); on the right is a nonviable deformed embryo resulting from the mutation of K454 to R454. Bottom: a selective FAK inhibitor could be applied *after* the wild-type mouse has developed into a healthy adult. (Adapted from S.T. Lim et al., *J. Biol. Chem.* 285:21526–21536, 2010. With permission from The American Society for Biochemistry and Molecular Biology.)

Rapid DNA-sequencing methods have revolutionized the field of genetics. Such sequences, for example, facilitate both forward and reverse genetics. In reverse genetics, the gene (sequence) is known, but the effects of mutations to the gene are not. After oligonucleotide-directed mutagenesis, the resultant mutant cells or organism are studied to determine how the modified piece contributes to the protein function. This process is analogous to reverse engineering—change one part at a time and then examine what results from the change.

Small-molecule agonists and antagonists offer several important advantages over the addition or deletion of genes. First, the small molecules can be added to a particular place—say an organ or cluster of cells—at a set time point, and can be expected to act shortly after the addition of the compounds. In addition to spatial and temporal control, small molecules can be added at specific concentrations for the characterization of dose-dependent effects. Furthermore, small diffusible molecules, unlike genetic constructs, often have reversible effects and can be removed from the cell under study simply by washing with excess buffer.

Small molecules also solve another problem inherent in reverse chemical genetics. It is rarely possible to study the effect of a mutation that is lethal to the developing organism. If you wish to understand the role of focal adhesion kinase (FAK) you cannot simply delete the gene or inactivate the protein through a genetic point mutation. Genetic mutations that deactivate FAK (for example K454R) are lethal at the embryonic stage (**Figure 9.18**). However, one can assess the importance of FAK in a mouse at any stage of development by treating it with a selective small-molecule inhibitor of FAK such as the synthetic small molecule PF-562,271 (see Figure 9.18). Thus, chemical genetics provides a powerful toolkit of reagents for dissecting the tremendous complexity inherent in the overlapping, cross-communicating, and feedback-regulated cellular networks.

Screening identifies small molecules for use in chemical genetics

A major challenge in chemical biology and pharmaceutical development is finding small molecules that are selective for a particular protein target. We have seen throughout this book that Nature generates an abundance of small molecules that target cellular machinery, but natural inhibitors are rarely potent and selective against the full array of proteins in a human. Finding selective small-molecule modulators usually involves setting up a high-throughput assay and testing many thousands of compounds in a single screening campaign.

To illustrate a typical workflow, let us take an example of a screen to identify compounds that inhibit mitosis without affecting microtubules. With regard to the latter criterion, we now have many potent anticancer drugs that affect both microtubule assembly and disassembly; we do not need more examples in this class of compounds. The antimitotic assay was based on the use of an enzyme-linked immunosorbent assay (ELISA) to detect mitosis in whole cells, as opposed to solutions of proteins. Epithelial cells growing in 384-well plates were screened with a diverse library of 16,320 chemical compounds. Cells were fixed with 70% ethanol to wash away lipids and displace water, leaving most proteins insoluble; the fixed cells were then subjected to an ELISA with an antibody selective for phosphonucleolin, which is present only in mitotic cells. Fewer than 1% of the compounds were active (**Figure 9.19**A). When these 139 hits were individually assayed for the ability to disrupt microtubule formation, 86 were found to have no effect on microtubules (Figure 9.19B). The effects of each of these 86 compounds on various phases of the cell cycle were then assessed individually by fluorescence microscopy. Kidney cells were treated with each of the compounds, and then the cells were fixed and stained with fluorescent antibodies against microtubules (stained green) and chromatin (stained blue). Various effects were observed by microscopy. Among the five hits that affected only mitosis, monastrol led to drastic aberrations in the mitotic spindle (Figure 9.19C). Whereas normal cells align chromosomes along a scrimmage line during mitosis, in cells treated with monastrol chromosomes emanate from a central point, like an exploding star. Further studies revealed the true target of monastrol to be kinesin, a motor protein that drags DNA along microtubules (**Figure 9.20**). It is not uncommon for pharmaceutical companies or large academic

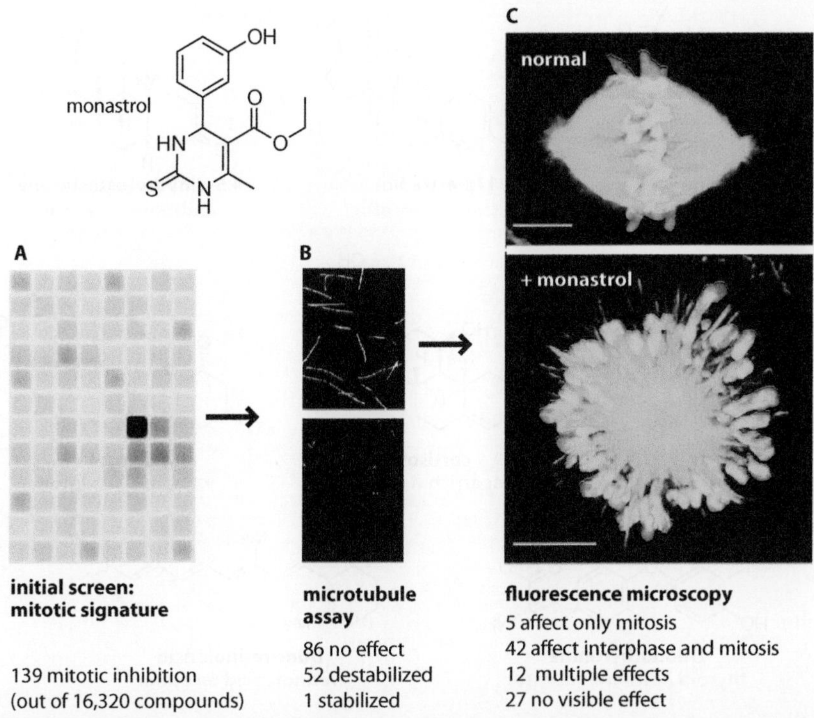

Figure 9.19 High-throughput screening for a cellular phenotype. (A) Screening of 16,320 chemical compounds for antimitotic effects led to 139 hits. (B) The antimitotic hits were further assayed for activity against microtubules. (C) Of the 86 compounds that had no effect on microtubules, five were selective for mitosis. Fluorescent staining of microtubules (green) and chromatin (blue) revealed an aberrant mitotic spindle induced by one of the compounds, which was dubbed monastrol. (From T.U. Mayer et al., *Science* 286:971–974, 1999. With permission from AAAS.)

centers to screen hundreds of thousands of compounds in a single screening campaign. These screens are a rich source of leads for drug development and for new tools for biologists.

How important are small-molecule screens? Recall from Chapter 4 that RNA interference makes it possible to stop the expression of proteins solely on the basis of mRNA sequence without the need to carry out high-throughput screens. Small interfering RNAs against any protein can be easily designed and synthesized, but RNA interference has two drawbacks. First, RNA is not membrane permeable. At present, RNA molecules lack the oral availability and pharmacokinetics of typical pharmaceuticals. Second, stopping the expression of a protein is not the same as inhibiting the function of protein that has already been expressed. Small molecules can solve these problems, and thus provide an invaluable tool in chemical biology studies.

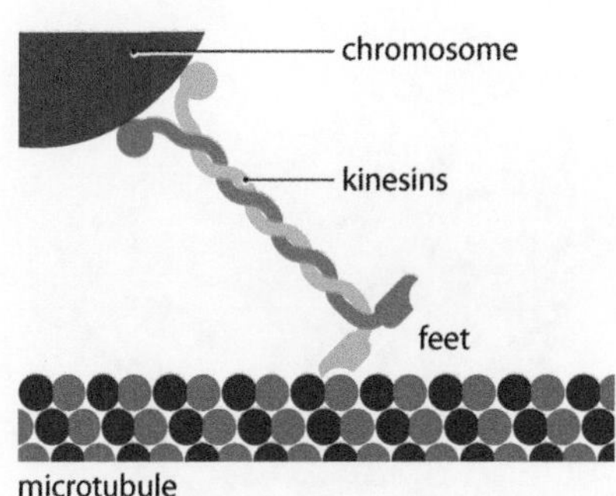

Figure 9.20 The longest journey begins with a single step. ATP hydrolysis drives a conformational change that prevents the kinesin from stepping backward. As the kinesin dimers waddle along the microtubule, they drag the massive chromosomes along with them. (Courtesy of C. Asbury and Steven M. Block.)

9.3 NUCLEAR RECEPTORS

Binding of small-molecule ligands activates nuclear receptor transcription factors

We will begin our discussion of signal transduction pathways by starting with the simplest: transcription factors that bind directly to signaling molecules. Bacteria use transcription factor receptors for quorum sensing, engaging in activities such as luminescence when they reach a high population density. In humans, transcription factor receptors are commonly referred to as **nuclear receptors**. The nuclear receptor pathways are simple but the ligands are structurally complex and interesting, so we will spend more time talking about the ligands than the pathway itself.

Analysis of the human genome has revealed 28 types of nuclear receptors (48 if you include subtypes such as α, β, γ, that bind to the same ligand). Naturally

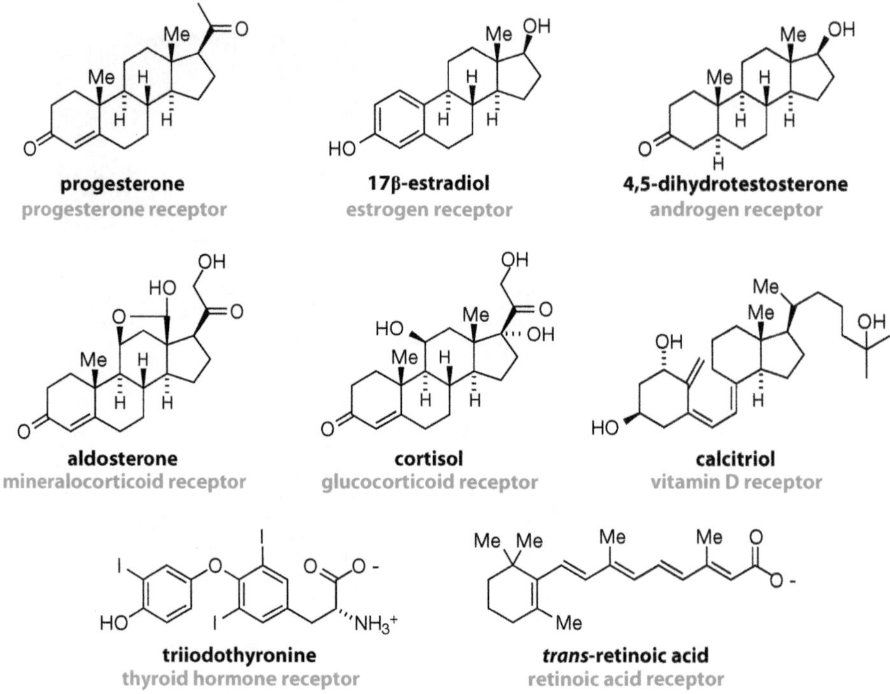

Figure 9.21 Ligands and their receptors. These examples of nuclear receptor ligands (black labels) bind to human nuclear receptors (blue labels) with high affinity and high specificity.

occurring high-affinity ligands have been identified for only 8 of these 28 types (**Figure 9.21**). Steroid hormones are the best-known ligands for nuclear receptors: progesterone, estrogen, testosterone, aldosterone, and cortisol. In addition, vitamin D, triiodothyronine, and *trans*-retinoic acid are also high-affinity ligands. Nuclear receptors bind hormones with high affinity (nanomolar K_d values) and high specificity. Crystal structures of nuclear receptor–hormone complexes show that the ligand is completely sequestered within the ligand-binding domain (**Figure 9.22**).

Receptors for which the endogenous ligand has not yet been identified are referred to as orphan nuclear receptors. Extensive efforts have not revealed high-affinity, high-specificity ligands for 20 of the 28 human nuclear receptors. It has been speculated that the orphan nuclear receptors bind with relatively low affinity to abundant intracellular metabolites. Examples of nuclear receptors that are believed to bind to abundant intracellular metabolites include the liver X receptor (LXR), the farnesoid X receptor (FXR), the retinoid X receptor (RXR), the peroxisome proliferator activator receptor (PPAR), the aromatic hydrocarbon receptor, the prostanoid X receptor, and the constitutive androstane receptor (CAR). The liver X receptor can be activated by cholesterol metabolites, referred to as *oxysterols*, at concentrations below their physiological levels (**Figure 9.23**). There is no nuclear receptor for cholesterol, which is a highly abundant component of lipid bilayers, and the liver X receptor is believed to

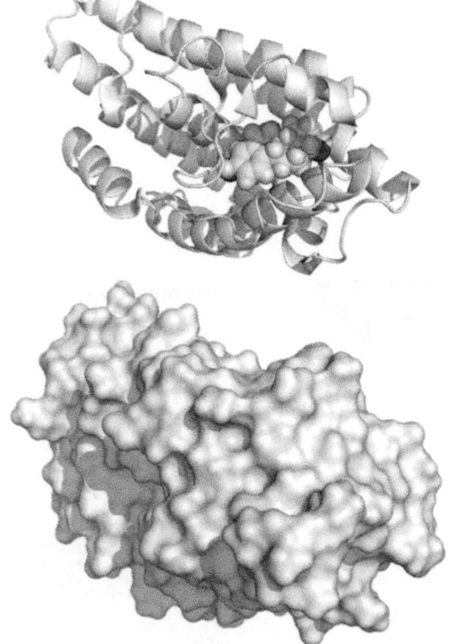

Figure 9.22 Trapped! The ligand-binding domain of human estrogen receptor-α (PDB 1QKU) rendered in ribbon mode (top) and as a solvent-accessible surface (bottom) reveals that the ligand, 17β-estradiol, is completely encapsulated by the receptor.

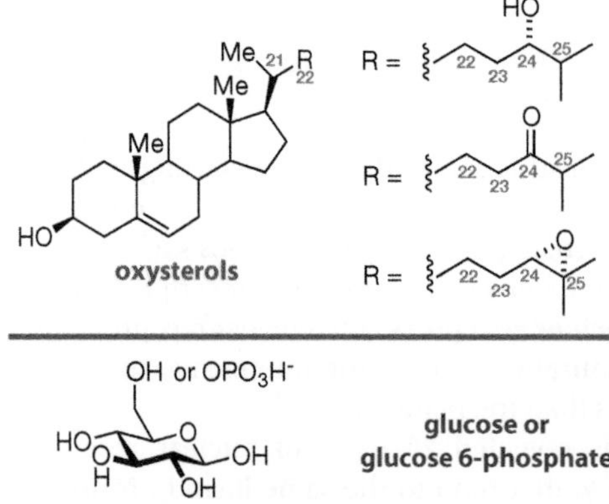

Figure 9.23 Ligands for the liver X receptor. Two very different types of ligands, sterols and monosaccharides, activate the liver X receptor at physiological concentrations.

serve as an indirect sensor for cholesterol levels. Surprisingly, the liver X receptor has also been shown to be activated by glucose and glucose 6-phosphate at physiological levels. Because the liver X receptor activates genes for lipid metabolism and binds to oxysterols, the receptor seems to connect lipid metabolism with glucose levels which could have relevance to diabetes.

Nuclear receptors are intracellular, and the ligands for nuclear receptors must be either membrane permeable or intracellular. The known high-affinity ligands for nuclear receptors are lipophilic, and therefore have low solubility in water. These ligands are shuttled through the bloodstream by proteins with lipophilic pockets, such as retinol-binding protein and thyroxine-binding globulin. Some cells contain retinoic-acid-binding proteins that assist in transport to the nucleus.

Some nuclear receptors translocate from cytoplasm to the nucleus, and bind DNA as homodimers

The signal transduction pathway for 17β-estradiol illustrates how nuclear receptors work. The ligand for the estrogen receptor, estradiol, is most abundant in women. During reproductive years it is produced primarily within the ovaries, but smaller amounts are produced by the brain in both sexes and in the testes in men. Estradiol has a key role in the menstrual cycle, targeting cells in both the uterus and the brain. Estradiol diffuses across the cell membrane of hypothalamic cells, binding tightly to the estrogen receptor-α, which is sequestered within the cytoplasm as a dimeric complex with heat shock protein 90 (Hsp90). Subsequent dissociation of Hsp90 allows the estrogen receptor dimer to be actively transported into the nucleus. The estrogen receptor possesses a short peptide sequence, called a **nuclear localization signal** (NLS), that can bind to karyopherin β. This binding event allows the protein to be ferried through a nuclear pore complex. Once in the nucleus, the ligand-bound form of the estrogen receptor-α binds tightly to a specific DNA sequence, referred to as the estrogen response element, and enhances transcription of the gene for the protein hormone gonadotropin-releasing hormone (**Figure 9.24**). The signal transduction pathways for other ligands to nuclear receptors, including progesterone, 4,5-dihydrotestosterone, cortisol, and aldosterone, follow similar pathways, but the tissue distribution and target genes are different.

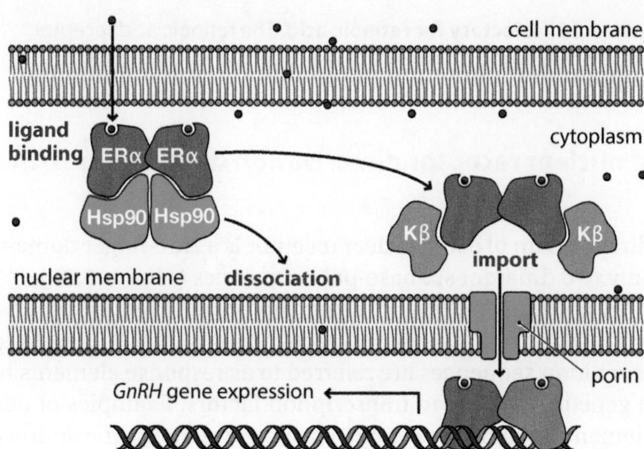

Figure 9.24 Road map for estrogen. The estrogen receptor pathway in hypothalamic neurons involves dissociation from Hsp90, association with karyopherin β (Kβ), and import through a nuclear porin. The estrogen receptor dimer (ERα) binds to DNA and promotes transcription of the *GnRH* gene.

Some nuclear receptors are localized in the nucleus and bind to DNA as heterodimers

The second major variant of the nuclear receptor signal transduction pathway involves the heterodimerization of nuclear receptors that remain within the nucleus. This pathway is best appreciated by considering the signal transduction pathway for the endogenous ligand retinoic acid. *Trans*-retinoic acid (isotretinoin) is prescribed for the treatment of severe cystic acne, but like other nuclear receptor ligands it has profound

Figure 9.25 Profound birth defects by retinoids. Treatment of pregnant mice with massive doses of retinol, a precursor to retinoic acid, causes embryonic malformation. (Courtesy of the Nobel Foundation.)

effects on cell growth, particularly in the developing fetus (**Figure 9.25**). Retinoic acid is a potent teratogen, known to cause human birth defects including hydrocephalus (an accumulation of cerebrospinal fluid in the brain) and malformed or missing ears, so prescriptions of retinoic acid are carefully controlled. In the USA, women must submit two negative pregnancy tests before the initial prescription, and all patients are advised not to donate blood while taking retinoic acid.

Retinoic acid diffuses across both the cell membrane and nuclear membrane. In the nucleus of developing embryonic neuronal cells, retinoic acid binds to the retinoic acid receptor α (RARα), which binds as a heterodimer with the retinoid X receptor (RXR). Together, the RARα–RXR heterodimer binds to specific DNA sequences enhancing the transcription of some genes and repressing the transcription of others. For example, in developing neuronal cells, the RARα–RXR heterodimer enhances the expression of the *homeobox b-1* gene (*Hoxb-1*). Hox genes code for transcription factors that further control cell growth and differentiation (**Figure 9.26**). This type of heterodimerization pathway is also found with the thyroid receptor, vitamin D receptor, the peroxisome proliferator activator receptor, liver X receptor, and the constitutive androstane receptor.

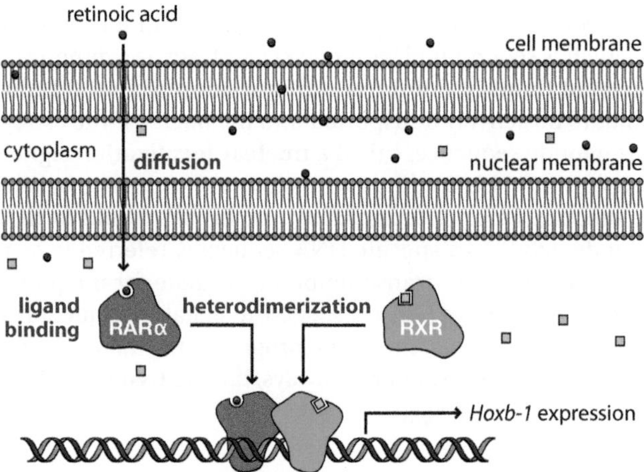

Figure 9.26 Trajectory for retinoic acid. The retinoic acid receptor pathway in hypothalamic neurons is a canonical nuclear receptor pathway.

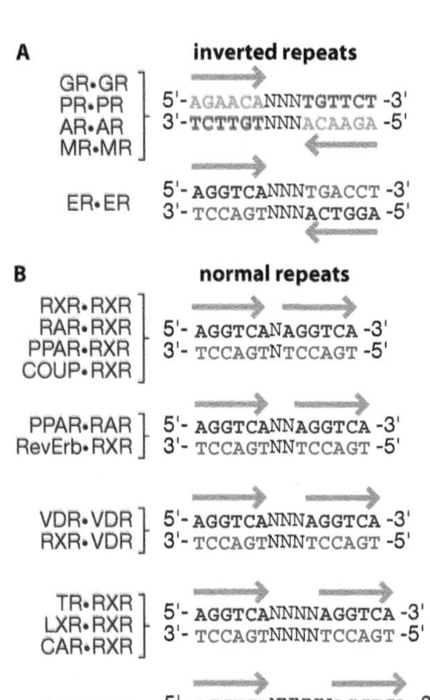

Figure 9.27 DNA binding by nuclear receptors. Unique DNA sequences are recognized by homodimers or heterodimers of nuclear receptors. Some DNA response elements are inverted repeats (A); others are simply repeats (B) separated by one to five base pairs. See Figure 9.21 for the full names of the high affinity nuclear receptors. Some orphan nuclear receptors have cryptic acronyms: COUP, chicken ovalbumin upstream promoter; RevErb, reverse orientation of the c-erbA gene.

The mode of nuclear receptor dimerization determines DNA sequence selectivity

The DNA-binding domain of each nuclear receptor is a zinc-finger domain that recognizes one of only two different six-base-pair sequences (**Figure 9.27**): 5′-AGAACA-3′ or 5′-AGGTCA-3′. A significant level of selectivity in gene regulation arises from the fact that nuclear receptors bind as dimers recognizing a total of 12 unique base pairs. These DNA recognition sequences are referred to as response elements because they determine the genetic response to transcription factors. Examples of nuclear receptor response elements include the estrogen response element, the androgen response element, and the vitamin D response element.

Some nuclear receptors form head-to-head homodimers that bind to inverted repeats of the six-base-pair recognition sequences. For example, the glucocorticoid receptor, the progesterone receptor, the androgen receptor, and the mineralocorticoid receptor form homodimers that bind to the inverted repeat 5′-AGAACA*NNN*TGTTCT-3′. Note that the complementary strand of this sequence is the same when read in the 5′ to 3′ direction. Estrogen receptors bind as homodimers to the inverted repeat 5′-AGGTCA*NNN*TGACCT-3′. Other nuclear receptors bind six-base-pair repeats such as 5′-AGGTCA$(N)_{1-5}$AGGTCA-3′, for example retinoid X receptor, retinoic acid receptor, peroxisome proliferator activator receptor (PPAR), vitamin D receptor, thyroxine

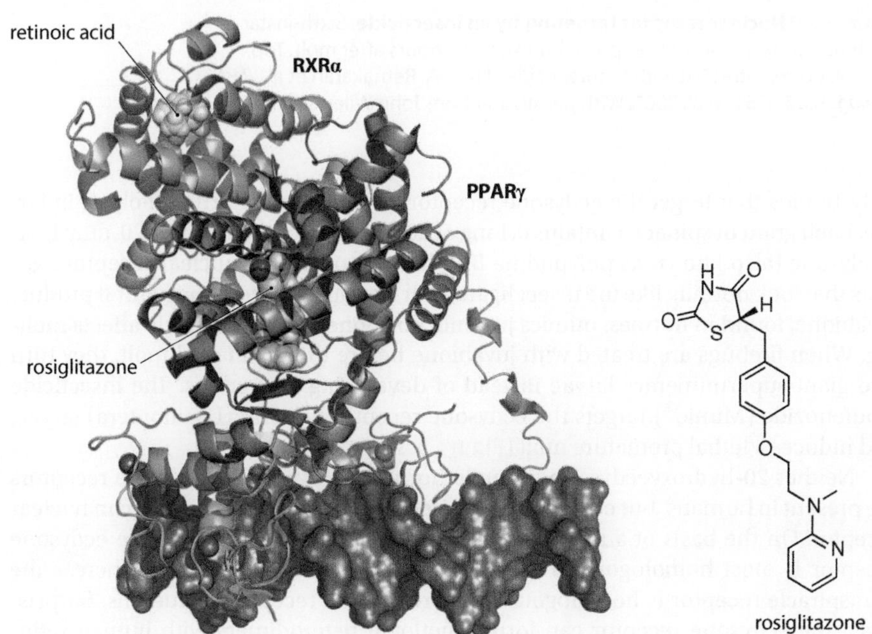

Figure 9.28 Structure of a full-length nuclear receptor bound to DNA. The structure of the PPARγ–RXRα complex bound to the ligand rosiglitazone and retinoic acid and DNA. The DNA includes the following consensus binding sequence (underlined) AGGTCAAAGGTCA. (PDB 3DZY)

receptor, and liver X receptor. The number of intervening base pairs matters; a span of five base pairs represents a 180° twist of the helical axis of B-form DNA.

A wide range of partial structures of nuclear receptors have been solved through X-ray crystallography of either the ligand-binding domain or the DNA-binding domain, but not both. So far, there is only one crystal structure of full-length nuclear receptors bound to DNA—the RXRα-PPARγ heterodimer. In the crystal structure, retinoic acid is bound to the RXRα and the pharmaceutical (*S*)-rosiglitazone is bound to PPARγ. As expected, the zinc fingers of each nuclear receptor bind to the sequence 5′-AGGTCA-3′ (**Figure 9.28**).

Human cells can be rewired for control by *Drosophila* nuclear receptors

Nuclear receptors have a common structure and mechanism across a range of different organisms. However, they do not bind to precisely the same ligands in different organisms. For example, in humans, steroid hormones such as testosterone and estrogen have a key role in development during puberty. However, in *Drosophila*, progression from nonreproductive juveniles to reproductive adults is controlled by the binding of the steroid 20-hydroxyecdysone to the ecdysone receptor and the binding of juvenile hormone III to the ultraspiracle receptor. These insect nuclear receptors form a heterodimer that controls differentiation (**Figure 9.29**). Nearly all plants produce

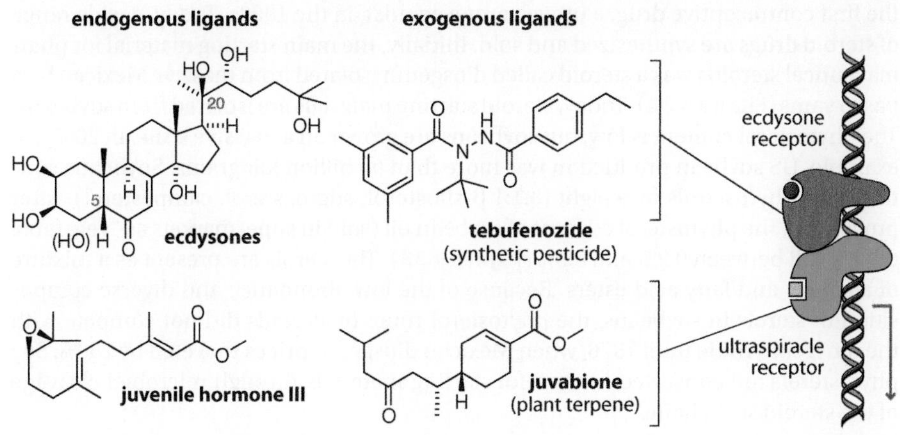

Figure 9.29 Controlling insects with ligands to nuclear receptors. Endogenous ligands like ecdysones and juvenile hormones activate nuclear receptors in insects. Exogenous ligands like synthetic and natural pesticides also target these insect nuclear receptors, inducing a premature molt.

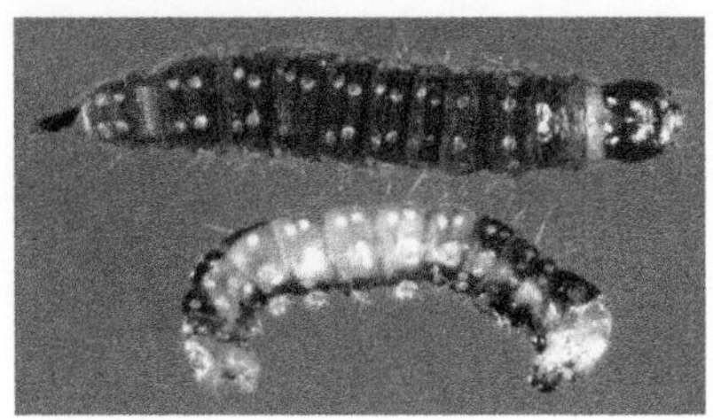

Figure 9.30 Nuclear receptor targeting by an insecticide. Sixth-instar (molting phase) larvae of the spruce budworm 48 hours after molt. Top, control; bottom, treated with tebufenozide. (From A. Retnakaran et al., *Pest Manag. Sci.* 57:951–957, 2001. With permission from John Wiley & Sons.)

ecdysteroids that target the ecdysone receptor and induce abnormal molting in larvae. Each gram of spinach contains 0.1 mg of 20-hydroxyecdysone and 5,20-dihydroxyecdysone (also known as polypodine B). Plants also contain nuclear receptor agonists that look nothing like the insect ligand. For example, the terpene natural product juvabione, found in fir trees, mimics juvenile hormone III and adversely affects molting. When firebugs are treated with juvabione before the final larval molt, they turn into giant supernumerary larvae instead of developing into adults. The insecticide tebufenozide (Mimic®) targets the ecdysone receptor of moth (Lepidoptera) larvae, and induces a lethal premature molt (**Figure 9.30**).

Neither 20-hydroxyecdysone, juvenile hormone III, nor their cognate receptors are present in humans, but each *Drosophila* receptor is a homolog of a human nuclear receptor. On the basis of amino acid sequence and protein structure, the ecdysone receptor is most homologous to the farnesoid X receptor in humans, whereas the ultraspiracle receptor is homologous to the retinoid X receptor in humans. Surprisingly, the ecdysone receptor can form functional heterodimers with human retinoid X receptor. When human cells are transfected with plasmids expressing ecdysone receptor and then treated with ecdysone, the ecdysone receptors associate with human retinoid X receptors, and bind to DNA (**Figure 9.31**).

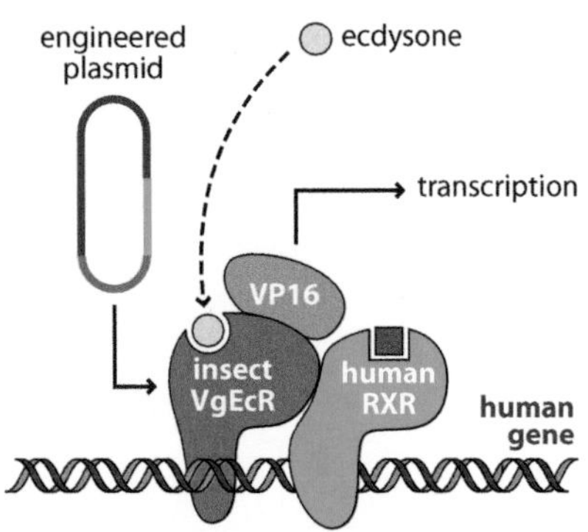

Figure 9.31 Re-engineering human cells to respond to an insect hormone. Human cell lines can be transformed with a plasmid encoding a special form of the ecdysone receptor (bearing the human transcription flag VP16).

The ecdysone receptor has been re-engineered to include a mammalian VP16 domain that recruits human transcriptional machinery (as opposed to *Drosophila* transcriptional machinery), and three point mutations that confer the same DNA binding selectivity as the glucocorticoid receptor: 5′-AGAACA-3′. When expressed in human cells, this re-engineered nuclear receptor VgEcR forms a heterodimer with the retinoid X receptor and can be used to turn on the expression of genes driven by a novel response element, 5′-AGGTCANAGAACA-3′. This combination of *Drosophila* nuclear receptor and novel response element can be used to turn on transcription in human cells by using an insect hormone, without affecting other human genes.

Problem 9.4

Draw a cartoon depiction showing how you could use the DNA-binding domains of both the androgen receptor and the estrogen receptor to engineer a new pair of proteins that would bind to the DNA sequence 5′-AGGTCATAGTGTTCT-3′ in the presence of rapamycin.

Steroids make highly potent pharmaceuticals

Chemical interest in steroid hormones drove the field of nuclear receptors, leading to the first contraceptive drug, a progesterone agonist, in the 1960s. Today, a wide range of steroid drugs are synthesized and sold. Initially, the main starting material for pharmaceutical steroids was a steroid called diosgenin isolated from massive Mexican barbasco yams (**Figure 9.32**). Today, steroid starting materials are isolated from soybeans. The phytosterol content is low, but soybeans are grown on a massive scale. In 2009, for example, US soybean production was more than 80 billion kilograms. Soybeans contain 0.3% phytosterols by weight (3:1:1 β-sitosterol, stigmasterol, campesterol). After processing, the phytosterol content of soybean oil (sold in supermarkets as "vegetable oil") is still between 0.2% and 0.3% (**Figure 9.33**). The sterols are present as a mixture of alcohols and fatty acid esters. Because of the low abundance and diverse composition of sterols in soybeans, the phytosterol route to steroids did not compete with the diosgenin route until 1976, when Mexican diosgenin prices rocketed by 250%. Soy phytosterols are converted into useful starting materials through microbial cleavage of the steroid side chain.

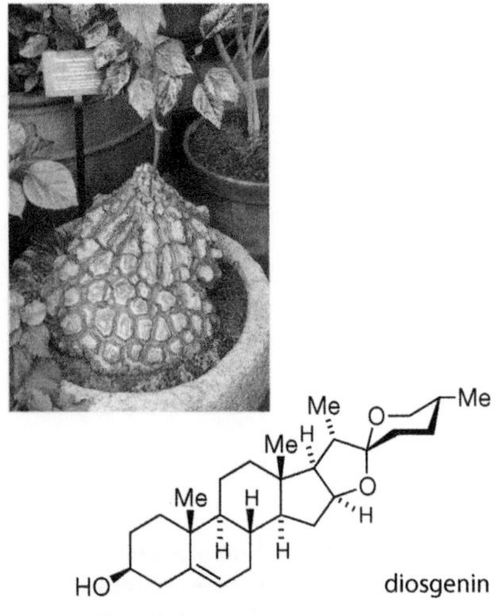

Figure 9.32 The massive tubers of the barbasco yam. Two percent of the tuber is a single steroid, diosgenin. (Courtesy of Nhu Nguyen, University of California, Berkeley.)

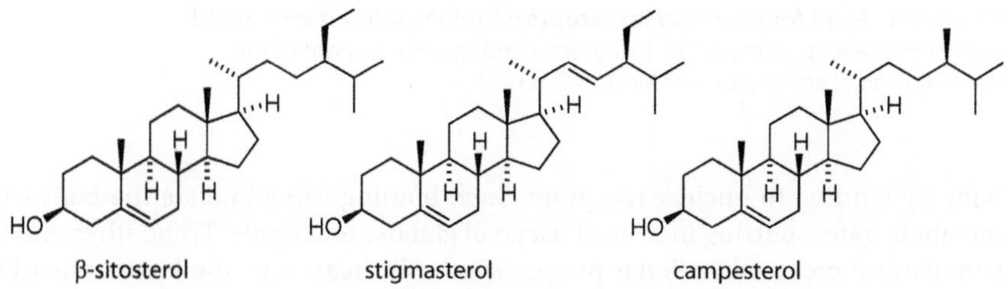

Figure 9.33 Steroids from soybeans. A mixture of three phytosterols derived from soybean oil is now the most common starting material for pharmaceutical steroid synthesis.

Steroids are still widely prescribed for a range of indications beyond birth control (**Figure 9.34**). For many years, a mixture of estrone conjugates, mainly *O*-sulfate esters, isolated from the urine of pregnant mares (sold as Premarin™) was widely used for hormone replacement therapy to ameliorate post-menopausal symptoms such as hot flashes. At its peak in 2002, US sales of Premarin were more than $2 billion a year. After 2002, sales declined due to mounting evidence that long-term use of estrogen and progestin increases the risk of cardiovascular disease and cancer.

Glucocorticoids are widely used as anti-inflammatory drugs. Hydrocortisone creams (1% hydrocortisone) are widely available over the counter as anti-itch medications. The powerful synthetic glucocorticoid fluticasone is a component of Flonase™ nasal spray and the two-component asthma medication Advair™. US sales of Advair™ were almost $4 billion in 2008.

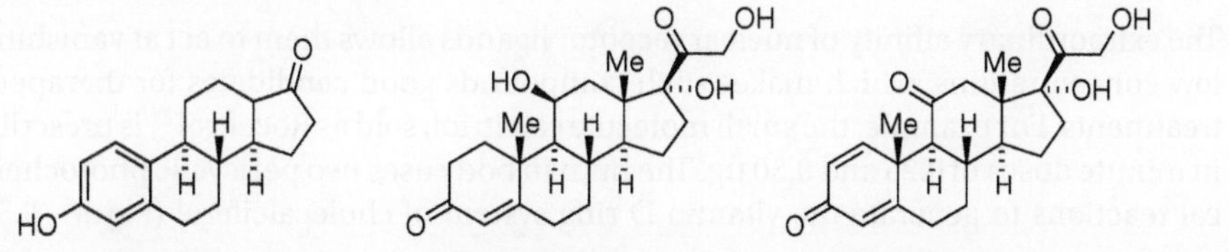

Figure 9.34 Steroids as therapeutics. Estrone derivatives and glucocorticoids (such as hydrocortisone) have been widely prescribed as medicines.

Problem 9.5

Many steroid hormones and steroid pharmaceuticals possess enone functional groups that can react with thiols such as glutathione or cysteine side chains. The equilibrium is exquisitely sensitive to sterics. Given the equilibrium constants below and a thiol concentration of 5 mM, what percentage of progesterone would be present as the thiol adduct at equilibrium? What percentage of prednisone would be present as the thiol adduct at equilibrium?

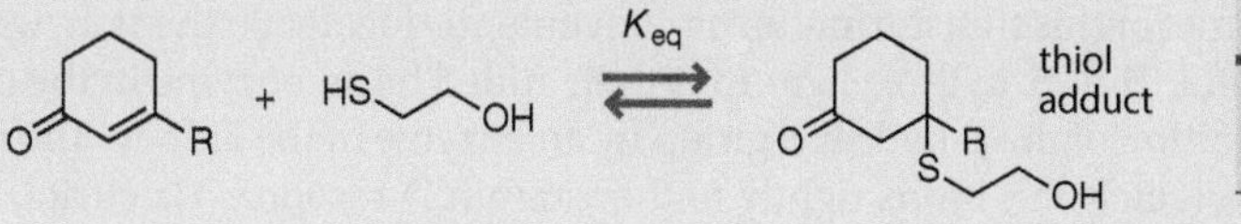

R	K_{eq}
H	7000
CH_3	2

Nonsteroidal ligands for nuclear receptors are also widely used as drugs

Nonsteroidal ligands for nuclear receptors are both biosynthesized and prescribed widely as drugs. The thyroid gland in the neck produces triiodothyronine and tetraiodothyronine, which is converted into the more potent triiodothyronine by enzymes in the target cells (**Figure 9.35**). Triiodothyronine controls metabolism throughout the

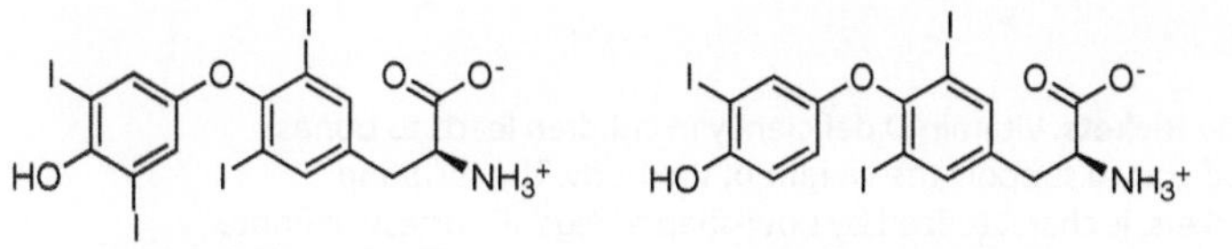

Figure 9.35 Regulators of basal metabolic rate. Synthesized by the thyroid, the iodo-substituted, nonsteroidal thyronine derivatives are derivatives of tyrosine, and exert control over diverse physiological responses.

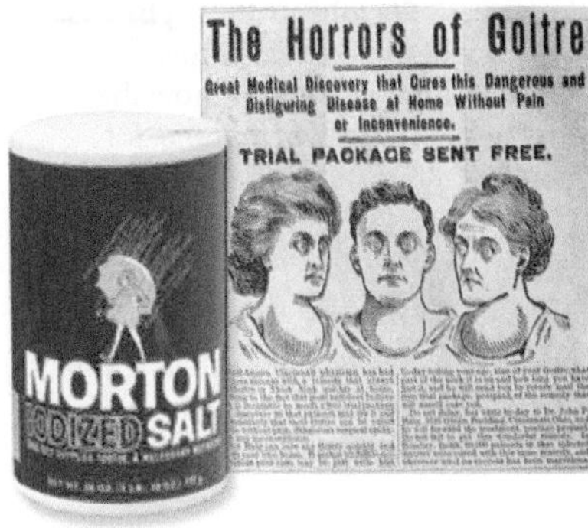

Figure 9.36 Food for your nuclear receptors. Iodized salt was introduced in 1924 to prevent goiter, which is characterized by enlargement of the thyroid gland. (Left, courtesy of Morton Salt.)

body by binding to nuclear receptors. Such binding can stimulate the body's basal metabolic rate, spurring increased usage of glucose and lipids. Triiodothyronine also stimulates increases in cellular protein synthesis, heart rate, the force of heart contractions, the systolic component of blood pressure, and body temperature. Taken together, the two compounds mediate a broad range of physiological responses through direct binding to nuclear receptors.

When the human diet is deficient in iodine, the cells in the thyroid gland proliferate, leading to a condition known as goiter (**Figure 9.36**). Since 1924, grocery salt has been "iodized" with the addition of 0.006–0.010% KI. Tetraiodothyronine is an important prescription pharmaceutical, with almost $1 billion a year in US sales.

Nuclear accidents generate radioactive ^{131}I, which ends up being concentrated in the thyroid gland, leading to thyroid cancer. Taking potassium iodide pills overwhelms the body with the nonradioactive natural isotope ^{127}I, preventing the accumulation of ^{131}I in the thyroid gland.

Drugs can be designed to target specific mutations of nuclear receptors

The extraordinary affinity of nuclear receptor ligands allows them to act at vanishingly low concentrations, which makes such compounds good candidates for therapeutic treatments. For example, the small molecule calcitriol, sold as Rocaltrol®, is prescribed in minute doses of 0.25 and 0.50 μg. The human body uses two pericyclic photochemical reactions to generate the vitamin D ring system of cholecalciferol (**Figure 9.37**):

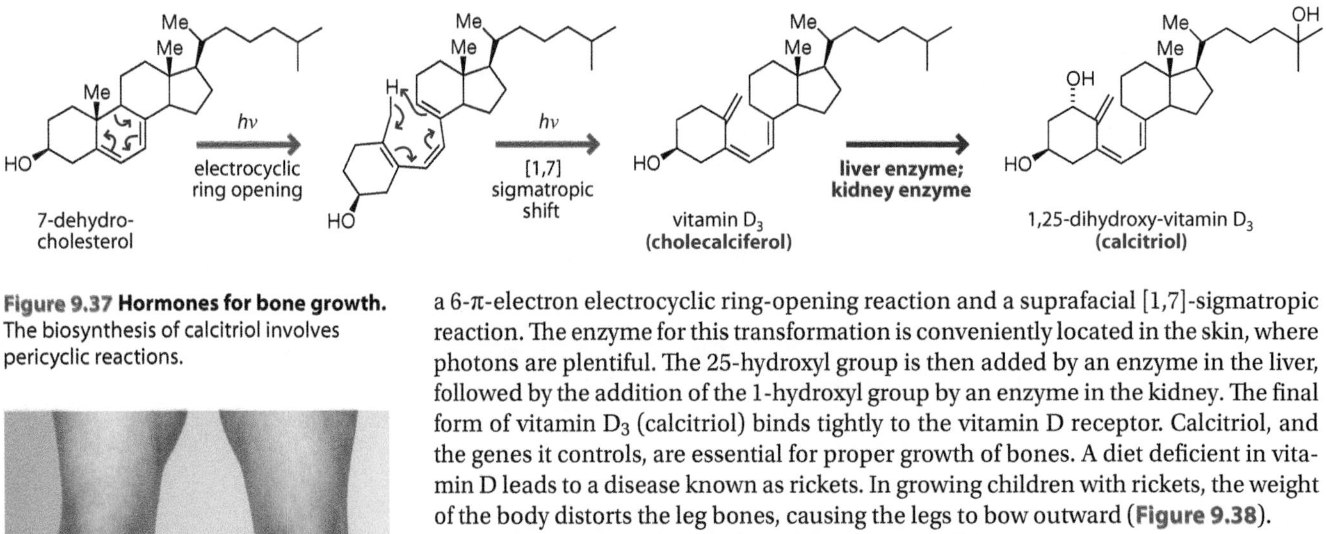

7-dehydro-cholesterol
$h\nu$ electrocyclic ring opening
$h\nu$ [1,7] sigmatropic shift
vitamin D_3 **(cholecalciferol)**
liver enzyme; kidney enzyme
1,25-dihydroxy-vitamin D_3 **(calcitriol)**

Figure 9.37 Hormones for bone growth. The biosynthesis of calcitriol involves pericyclic reactions.

a 6-π-electron electrocyclic ring-opening reaction and a suprafacial [1,7]-sigmatropic reaction. The enzyme for this transformation is conveniently located in the skin, where photons are plentiful. The 25-hydroxyl group is then added by an enzyme in the liver, followed by the addition of the 1-hydroxyl group by an enzyme in the kidney. The final form of vitamin D_3 (calcitriol) binds tightly to the vitamin D receptor. Calcitriol, and the genes it controls, are essential for proper growth of bones. A diet deficient in vitamin D leads to a disease known as rickets. In growing children with rickets, the weight of the body distorts the leg bones, causing the legs to bow outward (**Figure 9.38**).

Mutations that affect the ligand-binding site of the vitamin D receptor are associated with vitamin D-resistant rickets. For example, the R274L mutation makes the vitamin D receptor more than 1000 times less responsive to vitamin D because it removes a hydrogen bond and creates a void that cannot be filled by the natural ligand (**Figure 9.39**). However, the addition of *O*-benzyl groups to the 1-hydroxyl group of vitamin D generates derivatives that are capable of restoring up to 80% of the inducible activity.

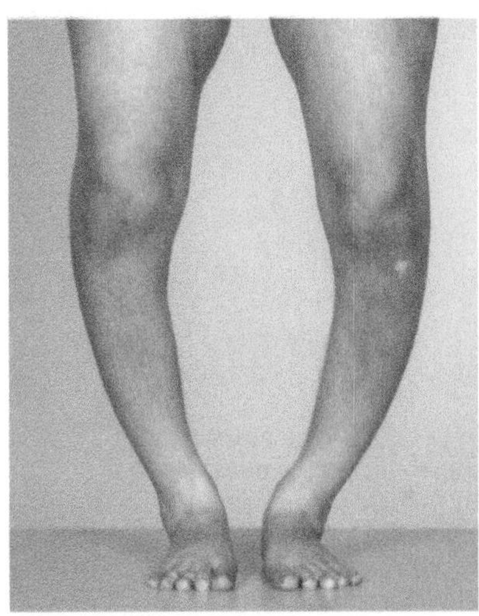

Figure 9.38 Rickets. Vitamin D deficiency in children leads to bones that are too soft to support the weight of the body. The resulting disease, rickets, is characterized by bow-shaped legs. (Courtesy of Photo Researchers.)

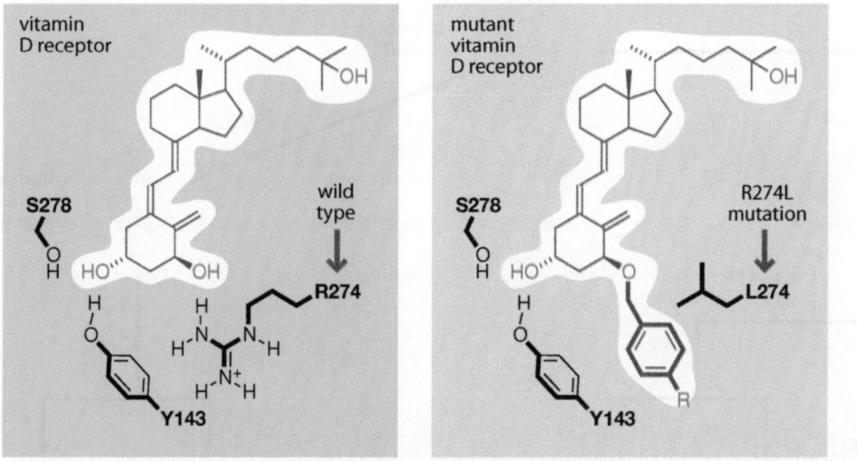

Figure 9.39 A synthetic ligand to restore biological function. Calcitriol binds tightly to the wild-type vitamin D receptor (left) by making numerous contacts, including a key hydrogen bond with Arg274; however, calcitriol cannot fill the void left by an R274L mutation. A synthetic *O*-benzyl derivative of calcitriol fills the void in the R274L variant (right), rescuing patients from vitamin D-resistant rickets.

Problem 9.6

Triiodothyronine (T_3) binds tightly to the wild-type thyroid hormone receptor, forming a stabilizing hydrogen bond with His435 (H435) in the ligand-binding site. An orthogonal, non-metabolizable analog (QH2) was designed to be inactive against the natural receptor, but highly active against a compensatory mutant. Which thyroid receptor mutant would be most likely to respond to QH2: H435A, H435F, H435G, H435K, H435W, or H435R?

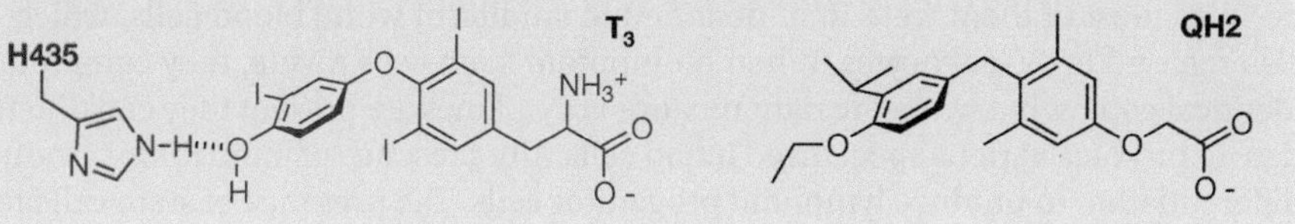

9.4 CELL-SURFACE RECEPTORS THAT INTERACT DIRECTLY WITH TRANSCRIPTION FACTORS

Hematopoietic proliferation and differentiation are controlled by molecular signals

The ability to respond to membrane-impermeable ligands requires cell-surface receptors. Bacteria such as *Staphylococcus aureus* use peptides, called auto-inducing peptides, for quorum sensing. These impermeable peptides activate a two-component signal transduction system that involves a membrane-bound receptor and a phosphorylating protein in addition to the transcription factor. The bacterial phosphorylating protein acts on histidine residues as opposed to serine, threonine, or tyrosine. Humans use a two-component signal transduction system to control the proliferation and differentiation of blood cells—a process referred to as hematopoiesis (**Figure 9.40**). The bone marrow is the central factory for hematopoiesis. It contains a family of cells that originate from a self-reproducing cell line known as multipotent hematopoietic stem cells. Multipotent stem cells can proliferate to produce more cells of the same type or, as they divide, differentiate to become a variety of essential cell types: red blood cells, platelets, antibody-producing B lymphocytes, killer T lymphocytes, bacteria-killing macrophages, bone-forming osteoclasts, and bone-remodeling osteoblasts. If you suffer a traumatic injury you probably need to clot the wound with platelets, fight off infection with white blood cells, and replace the lost erythrocytes. However, you do not always need equal amounts of the various blood types. For example, in response to a viral infection you need B cells, not platelets; to assist with clotting you need platelets, not T cells. Pluripotent stem cells can differentiate to produce any type of cell in the organism: for example, blood, brain, muscle.

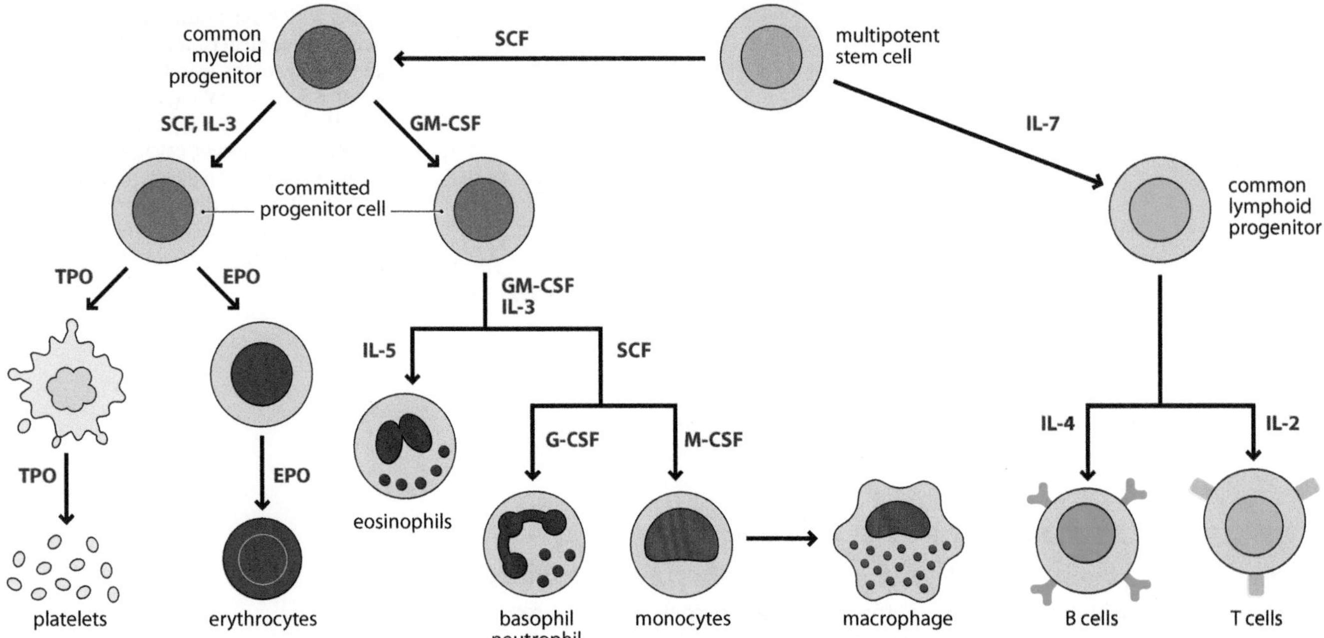

Figure 9.40 Cytokines that control the differentiation of blood cells. Proteins called cytokines (for example erythropoietin (EPO), interleukin (IL)-3, and IL-6) control which types of blood cells are being made. (SCF, stem cell factor; TPO, thrombopoietin; GM-CSF, granulocyte/macrophage colony-stimulating factor; G-CSF, granulocyte colony-stimulating factor; M-CSF, macrophage colony-stimulating factor.)

The proliferation and differentiation of hematopoietic cell types are induced by endogenous protein signals called **cytokines** or by extracellular cues from pathogenic organisms. Many of the cytokines that control hematopoiesis are named **interleukins** because most of them were first identified in studies of white blood cells, which are also referred to as leukocytes. When multipotent stem cells divide, they can produce identical copies, but when the right mixture of cytokines are present they can differentiate to produce various specialized blood cells. The presence of interleukin-7 induces differentiation to produce lymphoid progenitor cells. The presence of stem cell factor induces differentiation to produce myeloid progenitor cells. Depending on the mixture of cytokines in the environment, myeloid progenitors can further differentiate as they divide. For example, the presence of interleukin-5 induces differentiation to produce eosinophils, which help to fight off infections that include multicellular parasites.

Most cytokines mediate their intracellular effects through a closely related set of pathways that involve the activation of a kinase and the phosphorylation of a transcription factor. As we will see, the organization of the receptor assemblies varies widely from cytokine to cytokine, and this variation can permit activation of and cross-talk with more than one signaling pathway.

Problem 9.7

Which cytokines would be necessary to ensure the production of (a) eosinophils, (b) B cells?

Human cytokines can be used as pharmaceuticals

Naturally occurring ligands for cytokine receptors have proven to be a rich source of protein pharmaceuticals. Whereas most pharmaceuticals inhibit biochemical or cellular processes, cytokines can be used to induce cellular processes, such as differentiation. Erythropoietin, sold as Epogen® and Aranesp®, induces growth of red blood cells, and is used to treat anemia. Athletes have taken injections of erythropoietin to increase their supply of red blood cells, and hence the oxygen-carrying capacity of their blood. Thus, the use of erythropoietin is banned by most international sporting organizations. Colony-stimulating factor induces the proliferation of a wide range of white blood cells. Granulocyte/macrophage colony-stimulating factor, sold as

Neupogen®, is particularly important for patients undergoing chemotherapy, whose immune systems have been decimated. Interferon-α (IFN-α), sold as Roferon® and Intron A®, is used to boost the immune system of patients with various forms of hepatitis and various forms of cancer. Interferon-β, sold as Avonex® and Betaseron®, is used to modulate the overactive immune system in multiple sclerosis. Interferon-γ, sold as Actimmune®, is used to treat chronic granulomatous disease, in which neutrophils take pathogens into granulocytic vesicles but are unable to destroy them—rather like a garbage disposal unit without a blade.

Other hormones, not related to hematopoietic cell lines, also act through receptor pathways that resemble those of cytokines. Recombinant human growth hormone 1, sold as Nutropin®, is used to promote growth in children who grow slowly as the result of a deficiency of the natural human growth hormone. Excess growth hormone, taken inappropriately by athletes seeking faster recovery from injury, can cause acromegaly (**Figure 9.41**).

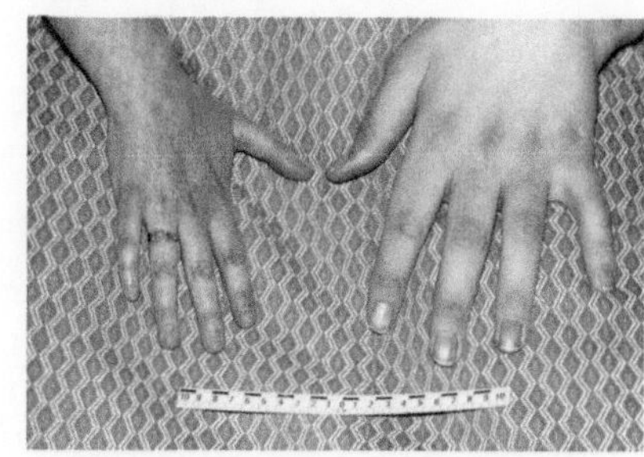

Figure 9.41 Excess growth hormone. This photo depicts acromegaly caused by genetic overproduction of human growth hormone. (Courtesy of Biophoto Associates. With permission from Photo Researchers.)

The JAK–STAT pathway involves a receptor, a kinase, and a transcription factor

The signal transduction pathway involved in cytokine signaling is the closest human analogy to the primitive bacterial signal transduction pathway that mediates quorum proliferation. The human kidneys monitor oxygen and extracellular volume; when more erythrocytes are needed, they release the cytokine erythropoietin. The signal transduction pathway by which erythropoietin induces the differentiation and proliferation of erythrocytes is representative of all the other intracellular cytokine signal transduction pathways. Erythropoietin binds to a pair of erythropoietin receptors in the membrane of erythrocyte progenitor cells found in the bone marrow. Cytokines such as erythropoietin bind to two receptor molecules in a 1:2 complex. Other cytokines, such as interleukin-6, bind to their receptors in a 2:2 stoichiometry. Because erythropoietin lacks two-fold symmetry, each half of the dimeric erythropoietin receptor is binding to a different part of the erythropoietin molecule (**Figure 9.42**). Each receptor protein in the dimer is constitutively complexed to a Janus kinase, abbreviated JAK (Figure 9.42 and **Figure 9.43**). There are four different Janus kinases—JAK1, JAK2, JAK3, and TYK2—in the human genome. A key conformational change occurs upon binding of the cytokine ligand that causes each Janus kinase to catalyze the transfer

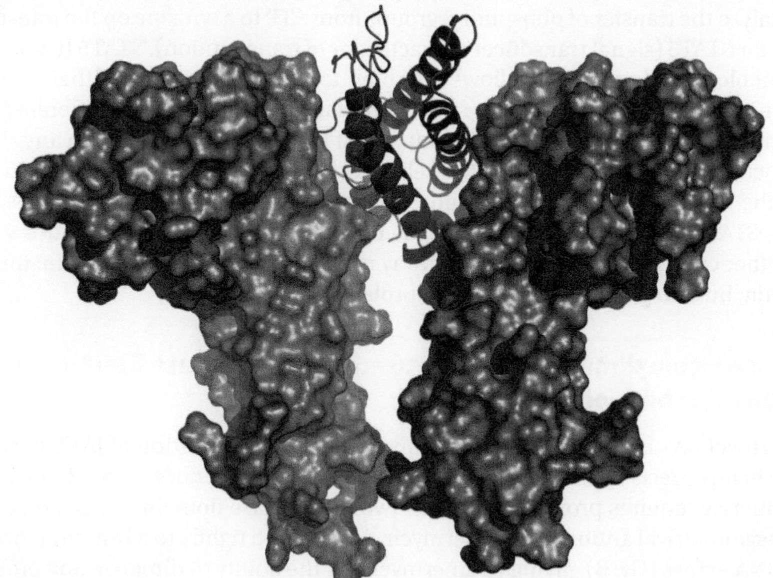

Figure 9.42 A receptor–ligand sandwich. A 2:1 complex forms between the extracellular domain of the erythropoietin receptor (blue) and the aglycone of erythropoietin (violet); the extracellular domain represents only about 40% of the full-length receptor. (PDB 1CN4)

Figure 9.43 Two-faced. Janus, the god of doorways, has two faces. Janus kinases are part of the doorway for cytokine signaling.

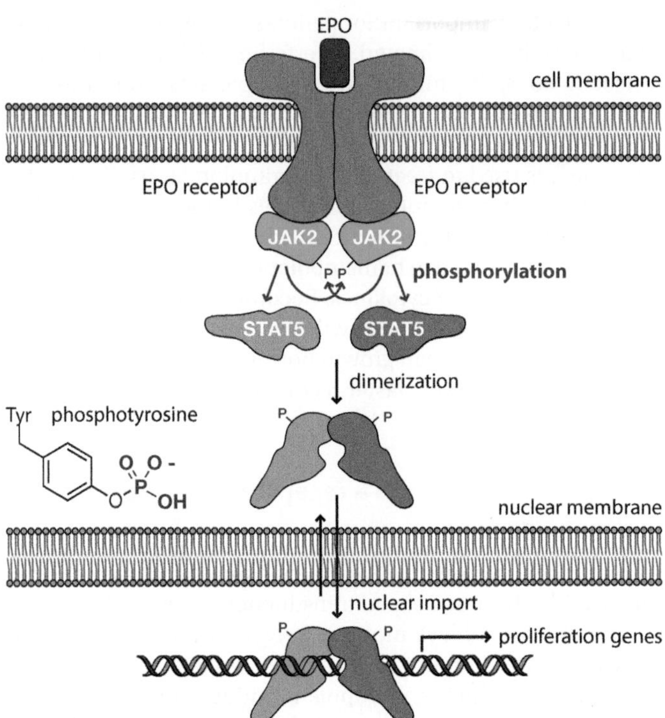

Figure 9.44 JAK–STAT signaling. The erythropoietin receptor–Janus kinase complex and the STAT transcription factor constitute a two-component signal transduction system. The red symbol P represents a phosphorylated tyrosine, serine, or threonine side chain. The red arrows indicate phosphorylation reactions.

of phosphate groups from ATP to a specific tyrosine residue of the opposing Janus kinase. This type of reciprocal phosphorylation is referred to as **autophosphorylation** because the Janus kinases seem to have phosphorylated themselves when they are analyzed by electrophoresis; however, the phosphorylation is actually catalyzed by the other half of the receptor pair.

Once the Janus kinases have become phosphorylated, the erythropoietin receptors catalyze the transfer of phosphoryl groups from ATP to a tyrosine on the transcription factor STAT5 (signal transducer and activator of transcription). STAT5 has an SH2 (Src homology 2) domain that allows it to bind to phosphotyrosines, either on JAK2 or on another STAT5 protein (**Figure 9.44**). Humans have genes for six different STAT transcription factors. The precise combination of JAK and STAT proteins in any signal transduction pathway depends on the cell type. STAT5 is actively transported in and out of the nucleus both before and after phosphorylation; however, once phosphorylated, STAT5 accumulates in the nucleus by binding tightly to DNA (**Figure 9.45**). Three other cytokines act through a pathway much like that of erythropoietin: thrombopoietin, human growth hormone, and prolactin.

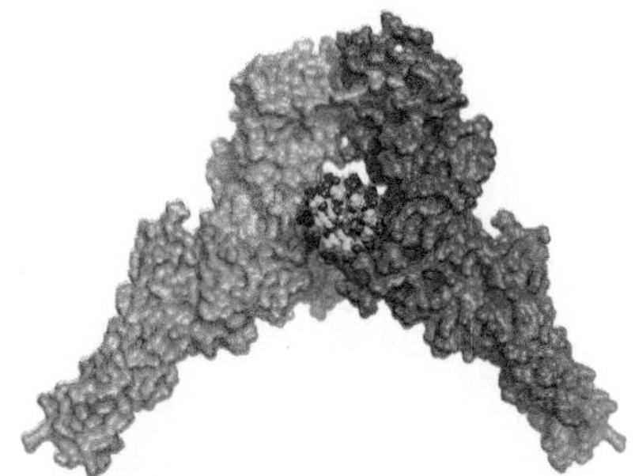

Figure 9.45 STAT-DNA binding. The STAT3B dimer straddles double-stranded DNA. (PDB 1BG1)

Small-molecule dimerizers can be used to demonstrate functional relationships between proteins

Coumermycin A1 has been used to demonstrate that the activation of JAK2 does not require highly precise positioning by the cytokine receptor (**Figure 9.46**). Rather, activation merely requires proximity between two JAK2 kinase domains. Each end of the pseudosymmetrical antibiotic coumermycin A1 can bind tightly to a B subunit of bacterial DNA gyrase (GyrB), giving coumermycin A1 the ability to dimerize any proteins that are attached to GyrB. A non-natural conjugate between JAK2 and GyrB did not exhibit autophosphorylation. However, in the presence of coumermycin A1, autophosphorylation of the GyrB–JAK2 conjugate was observed, as was phosphorylation of STAT5.

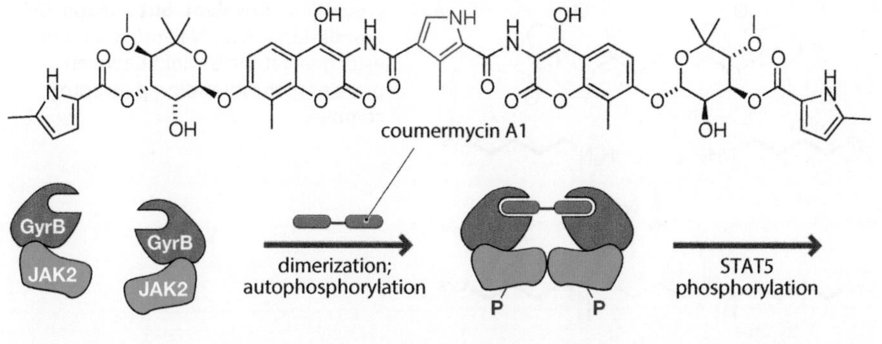

Figure 9.46 Dimerization by a pseudosymmetrical antibiotic. Chemically induced dimerization of GyrB–Jak2 fusion proteins with coumermycin A1 leads to autophosphorylation and downstream signaling.

Other interferons bind to heterodimeric and higher-order receptor assemblies

Interferons are cytokines that were first noted for their ability to interfere with viral and bacterial infections by activating cells of the immune system. However, these cytokines are now known to have more diverse functions. Interferons like interferon-α and inteferon-β are five-helix bundles that bind to heterodimeric receptors on human cells. Interferon-α binds tightly to the interferon-α receptor 2 subunit (IFN-αR2) and ultimately as a ternary complex including the interferon-α receptor 1 subunit (**Figure 9.47**). Each of these two receptor subunits binds to a different Janus kinase: IFN-αR2 binds to the Janus kinase JAK1, whereas IFN-αR1 binds to the Janus kinase TYK2. The two Janus kinases activate each other through transphosphorylation. TYK2 then phosphorylates Tyr466 on the IFN-αR1 and those phosphotyrosines serve as a recognition site for the SH2 domain of STAT2 (see Figure 9.47). Once STAT2 has bound to the receptor, it is phosphorylated by TYK2 and recruits STAT1 to the receptor complex. JAK1 then phosphorylates STAT1, leading to a heterodimeric STAT1–STAT2 complex. Other cytokines act through higher-order ligand–receptor assemblies. The interleukin-6 ligand–receptor complex is rather like a dimer of the interferon-α receptor complex, composed of interleukin-6, the receptor gp130, and the interleukin-6 co-receptor in a 2:2:2 ratio. Other interferons, such as interferon-γ, bind to heterotrimeric receptor assemblies involving one cytokine and three different receptor proteins.

Synthetic *N*-hydroxysuccinimidyl esters can acylate proteins in aqueous solution

N-Hydroxysuccinimidyl esters (NHS esters) are activated esters that react rapidly with amines and much more slowly with hydroxyl groups. The difference in reactivity is sufficient that the functional group can be used to acylate protein amines even in buffered aqueous solution. With two reactive esters connected by a short linker, bis-NHS esters can provide a tool for spot-welding together complexes found in cells; addition of the bis-NHS linker can form covalent crosslinks between closely associated proteins. The approach, however, can also turn up nonspecific interactions (for example proteins at high concentration that bind only weakly to each other), and the results need to be validated by additional techniques.

A bis-NHS ester was used to demonstrate the formation of a ternary complex between IFN-β, IFN-β receptor 2, and IFN-β receptor 1. When treated with a large excess of the crosslinking agent, random crosslinks were formed within and between proteins in the ternary receptor complex even when the proteins were present at relatively low concentrations (**Figure 9.48**). The covalently crosslinked complex between IFN-β, IFN-β receptor 2, and IFN-β receptor 1 was stable enough to be observed by denaturing SDS-polyacrylamide gel electrophoresis.

The chemical yield of such protein acylation reactions is usually low with respect to the NHS ester, as a result of hydroxide-dependent hydrolysis by water and other OH groups. The half-life for hydrolysis of an NHS ester is about 1 hour at pH 8.0 and 25 °C. pH is a critical variable in the acylation of lysine side chains. If the pH is too low, lysine side chains are protonated and unreactive. If the pH is too high, the concentration of hydroxide anions is so high that the NHS esters will be rapidly hydrolyzed.

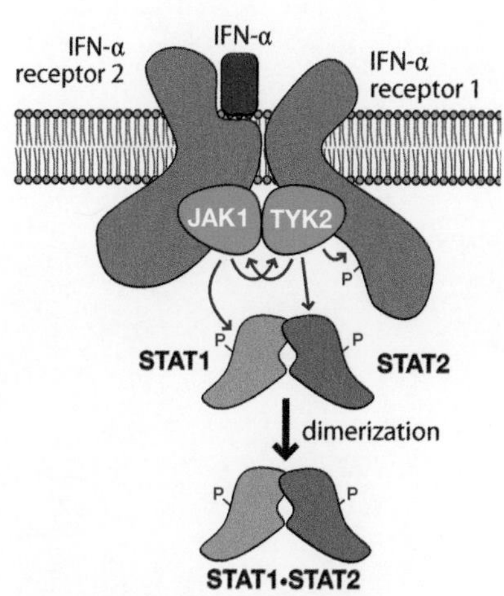

Figure 9.47 Heterodimerization during signal transduction. Type I interferon (IFN) pathways involve two different receptor subunits, two different Janus kinases, and two different STAT transcription factors.

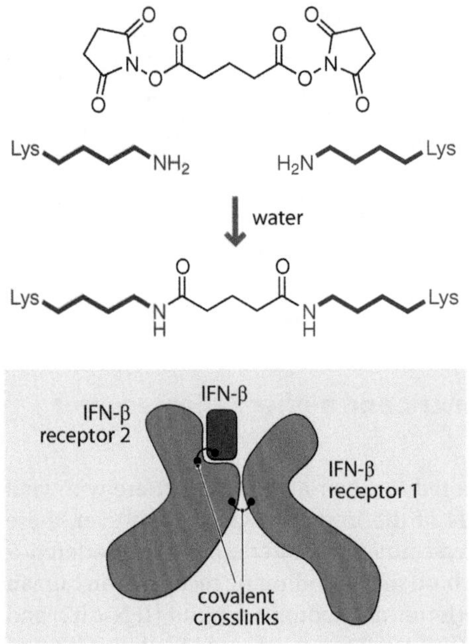

Lys NH_2 H_2N Lys

water

Lys N H N H Lys

IFN-β
IFN-β receptor 2
IFN-β receptor 1
covalent crosslinks

Figure 9.48 Covalent, but nonspecific, crosslinking. A bis-*N*-hydroxysuccinimidyl ester generates crosslinks between proteins in the interferon-β receptor complex.

NHS esters are used extensively in chemical biology to form covalent bonds to proteins in aqueous solutions. NHS esters can provide a straightforward route to modify protein surfaces through covalent bond formation. For example, recombinant human interferon-α_{2b} boosts the activity of the immune system and is prescribed to decrease the possibility of relapse from malignant melanoma. The half-life in serum is significantly increased by acylating interferon-α_{2b} with polyethylene glycol (PEG) chains. In general, covalent modification with PEG increases the molecular mass of the protein, thus slowing excretion from the body; in addition, modification with PEG often increases the solubility of the modified protein.

Polyethylene glycols are not readily available in monodisperse form, because the polymers are synthesized through the polymerization of oxirane. Instead, polyethylene glycols are used as mixtures with a known average molecular mass. When human interferon-α_{2b} is acylated in pH 6.5 phosphate buffer with a mixture of polyethylene glycols (average molecular mass 12 kDa; 273 oxyethylene units) bearing an NHS carbonate ester, the reaction generates a mixture of mono-pegylated proteins, of which 95% possess a single polyethylene glycol chain. This "pegylated," long-half-life form of interferon-α_{2b} is sold commercially as PegIntron® (**Figure 9.49**). The acylation reaction targets lysine, histidine, serine, threonine, and tyrosine side chains, but more than 45% of the pegylated species are acylated on a single residue, His34.

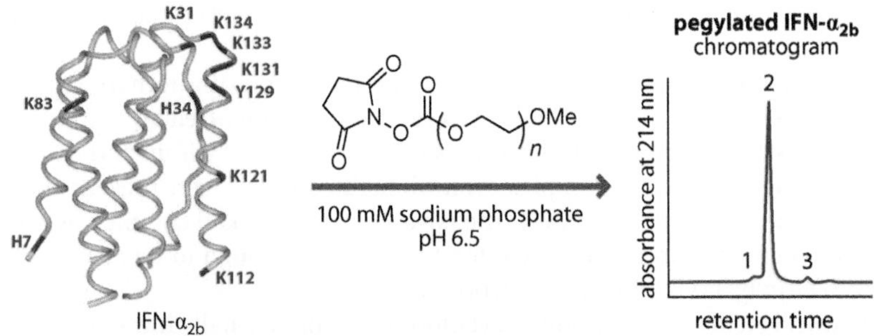

Figure 9.49 A mixture of different products from a nonspecific PEGylation reaction. Acylation of interferon-α_{2b} with PEG–*N*-hydroxysuccinimidyl carbonate esters generates a mixture of many different species, acylated on different nucleophilic side chains. Following chromatography, the main product consists of a mixture of many different monoacylated species (peak 2), along with a small amount of diacylated species (peak 1) and unacylated IFN-α_{2b}.

Problem 9.8

The small molecule "stattic" selectively blocks the binding of the 5-carboxyfluorescein-G(pY)LPQTV-NH_2 to the SH2 domain of STAT3. Surprisingly, the saturated analog had no activity and the degree of inhibition was found to increase at longer incubation times of STAT3 with stattic. What do the data suggest about the interaction between stattic and STAT3?

"stattic" analog

Transforming growth factor-β receptors possess built-in serine/threonine kinase domains

Transforming growth factor-β (TGF-β) is a key signaling protein that exerts opposing effects on the cell's balance between proliferation and apoptosis, depending on the cell type. Mechanistically, the pathway resembles that of interleukin-6 (with two ligands, two receptors, and two co-receptors) except that the kinase domain is built into the receptor protein (**Figure 9.50**). The transcription factor in the TGF-β pathway is named SMAD (or RSMAD) instead of STAT. Whereas phosphotyrosines mediate dimerization of STAT transcription factors, it is a phosphoserine that mediates interactions between RSMAD and its partner co-SMAD. As we will see shortly, built-in kinase domains are usually associated with **receptor tyrosine kinases** (discussed in the next section) and not cytokine receptors.

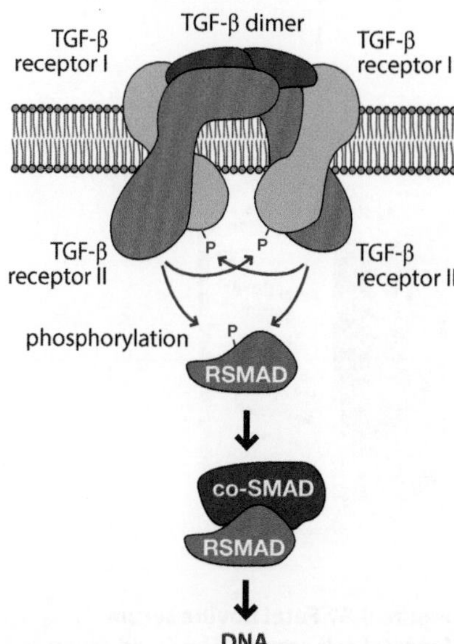

Figure 9.50 Ser/Thr kinase. The transforming growth factor-β receptor has a serine/threonine kinase built into the receptor. The red arrows in this pathway indicate phosphorylation reactions.

The DNA consensus sequence for the binding of SMAD was determined through affinity selection. A library of oligonucleotides, with 20 random nucleotides, was prepared using a DNA synthesizer. The random region in each nucleotide was flanked by constant regions with a known sequence, permitting amplification by PCR. The library of oligonucleotides was mixed with the DNA-binding domain of SMAD4 and the mixture was subjected to polyacrylamide gel electrophoresis. The larger protein-oligonucleotide complexes moved more slowly on the gel than uncomplexed oligonucleotides. The area of the gel containing DNA complexes was then removed, and the oligonucleotides were extracted and amplified by PCR (**Figure 9.51**). The sequences were determined and analyzed, revealing an idealized consensus sequence for SMAD binding: 5′-GTCTAGAC-3′.

5'-TAGTAAACACTCTATCAATTGGNNNNNNNNNNNNNNNNNNNNGGCTGTAAACGATACTGGAC-3'
3'-ATCATTTGTGAGATAGTTAACCNNNNNNNNNNNNNNNNNNNNCCGACATTTGCTATGACCTG-5'

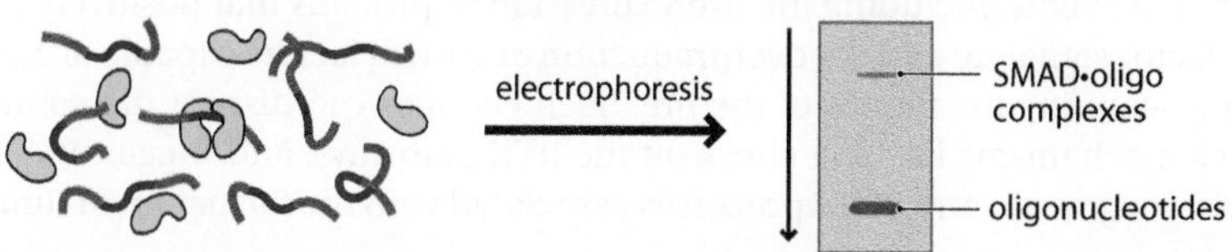

Figure 9.51 Selecting DNA sequences that bind to a protein. When a library of oligonucleotides is mixed with a tightly binding transcription factor, the DNA•protein complexes can be selected by electrophoresis, amplified by PCR, and sequenced to reveal the DNA target sequence. This selection method was used to identify the DNA sequence of genetic SMAD response elements.

9.5 RECEPTOR TYROSINE KINASES

Receptor tyrosine kinases control tissue growth

Growth of human tissues is carefully controlled. During fetal development, tissue growth is exquisitely organized to allow development of the human body plan. As we grow from infancy to adulthood, tissue growth is directed (mostly) upward and outward. In adulthood, tissue growth is mostly restricted to and triggered by response to injury. If you cut your finger, you need immune cells to fight off bacterial infection. You need new skin, made up of epidermal cells and fibroblasts, to protect the body. You need new blood vessels to supply the growing tissue and new nerves to innervate the tissue. To support rapid proliferation, these cells must also metabolize nutrients at enhanced rates. These needs are generally communicated by growth factor proteins

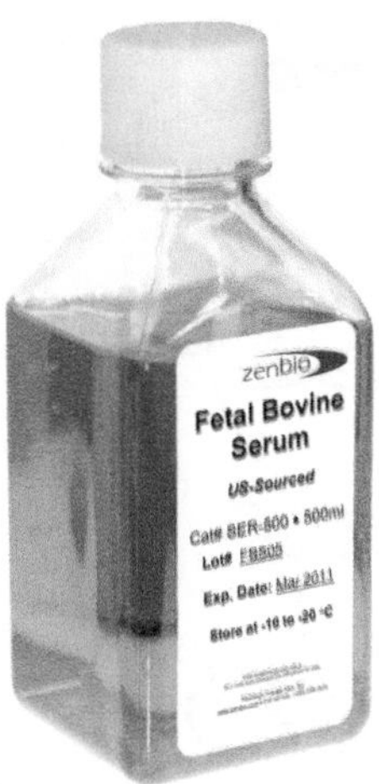

Figure 9.52 Fetal bovine serum (or fetal calf serum). The liquid serum from the blood of cow fetuses is a rich source of protein growth factors and cytokines. Once purified, this relatively inexpensive mixture of protein hormones is usually used in culturing mammalian cells. (Courtesy of ZenBio, Inc.)

Table 9.2 Subclasses of receptor tyrosine kinases with known ligands.

RTK class name	Ligand(s)
EGF receptor	Epidermal growth factor
Insulin receptor	Insulin
PDGF receptor	Platelet-derived growth factor, stem cell factor
FGF receptor	Fibroblast growth factor
VEGF receptor	Vascular endothelial growth factor
HGF receptor	Hepatocyte growth factor
Trk receptor	Nerve growth factor
Ephrin receptor	Ephrin
AXL receptor	Vitamin K-dependent growth-arrest specific gene 6 product
TIE receptor	Angiopoietin
DDR receptor	Collagen?
RET receptor	Glial cell line-derived neurotrophic factor
MuSK receptor	Agrin

(**Figure 9.52**). Growth factors act on a class of receptors that function in ways similar to JAK-STAT receptors, but activate very different downstream pathways. Analysis of the human genome has revealed 20 subclasses of these growth factor receptors, referred to as receptor tyrosine kinases (RTKs), but ligands have not yet been identified for all of these receptors (**Table 9.2**). The names can be misleading or uninformative. For example, tropomyosin receptor kinase (Trk) receptor is activated by nerve growth factor, and the muscle-specific kinase (MuSK) receptor is activated by the proteoglycan agrin.

Given the central role of RTKs in tissue growth, it should not be surprising to find that the receptors and their associated signaling pathways contribute critically to all steps in the progression of cancer. Many different cancer types are associated with mutations in the RTK pathway, including breast, lung, and skin cancers. RTK pathways can go awry as a result of any mutations that decouple growth factor signals from downstream events, including the overexpression of proteins that positively regulate growth factor signaling and the overproduction of growth factors capable of autocrine signaling—fanning the flames of the fire. Such changes can disrupt the control and feedback mechanisms keeping check on the RTK pathways, and trigger undesirable cell division, growth, and other processes associated with the formation of tumors.

Figure 9.53 Growing a bladder. Top: transitional epithelial cells of the bladder. Bottom: forming an artificial bladder through the application of growth factors to bladder cells growing on a bladder-shaped form. (Top, courtesy of Wikimedia; bottom, courtesy of Photo Researchers.)

Growth factors have a role in proliferation of urothelial cells

The human bladder has an amazing ability to expand and contract as urine accumulates and is eliminated. A specialized type of endothelial cell, termed the transitional epithelium (also known as the urothelium), is specific to the bladder, and can change from flat for an empty bladder to cubical when the bladder fills. Biomedical engineers have recently succeeded in growing artificial bladders in the laboratory by culturing a mixture of urothelial cells and smooth muscle cells around a bladder-shaped mold. Growth factors, which stimulate growth through RTKs, are an essential component of the growth media used to culture these artificial bladders (**Figure 9.53**).

The propensity for the urothelium to grow has a more sinister side—90% of bladder cancers are carcinomas of the urothelial cells, not fibroblasts or other cells found in bladder tissue. Many of the mutagens that enter our body ultimately end up in the bladder, where they can contribute to the formation of bladder cancers. Smoking is one of the biggest risk factors for bladder cancer. Several easily cultured urothelial carcinoma cell lines have been isolated from humans and each of those cell lines has a unique set of mutations that permit uncontrolled growth. Most urothelial cell lines exhibit mutations in the fibroblast growth factor receptor (FGFR). There are four genes for FGFR in the human genome, namely *FGFR1–4*; urothelial cells express *FGFR3*. Postnatal mutations in FGFR3 contribute to cancer. Three FGFR3 mutations appear with relatively high frequency in urothelial carcinomas: S249C (67%), Y375C (15%), and R248C (10%). Congenital mutations of FGFR3 lead to problems in embryological development. Mutations to FGFR3 that contribute to the cancer phenotype usually arise spontaneously after birth, usually through lifelong exposure to carcinogens that collect in the urine.

Problem 9.9

Suggest a plausible rationale for why an S249C mutation in the extracellular domain might lead to constitutive activation of the FGFR3 receptor.

Comparing receptor tyrosine kinases and cytokine receptors reveals useful commonalities

Receptor tyrosine kinases share many features with the cytokine receptors discussed in the previous section. First, all receptor tyrosine kinases form dimers in response to ligand binding. Second, the cytoplasmic domain of each receptor molecule is coupled to a kinase that transphosphorylates tyrosine residues on the partner. The phosphotyrosines then serve as beacons for the binding of proteins with SH2 domains, alerting the cell to gear up for big changes in protein expression. Third, some receptor tyrosine kinases act as homodimers, whereas others act as heterodimers. All growth factor ligands bind with RTKs in a 2:2 ratio.

There are also significant differences between RTKs and cytokine receptors. The first difference is that the tyrosine kinase is a *domain* of the receptor as opposed to an independent protein; we have already seen an example of a kinase domain embedded in the TGF-β receptor, but it was a serine/threonine kinase. A second difference between cytokine receptors and receptor tyrosine kinases is that the phosphotyrosine residues are coupled to four subpathways: mitogen-activated protein (MAP) kinase, phospholipase Cγ (PLCγ), phosphoinositide 3-kinase (PI3K), and STAT (**Figure 9.54**). From our discussion of TGF-β, you are already familiar with the recruitment of STAT transcription factors by phosphotyrosines on membrane-bound receptors. So we will focus on the other three subpathways, all of which involve interactions with the cell membrane.

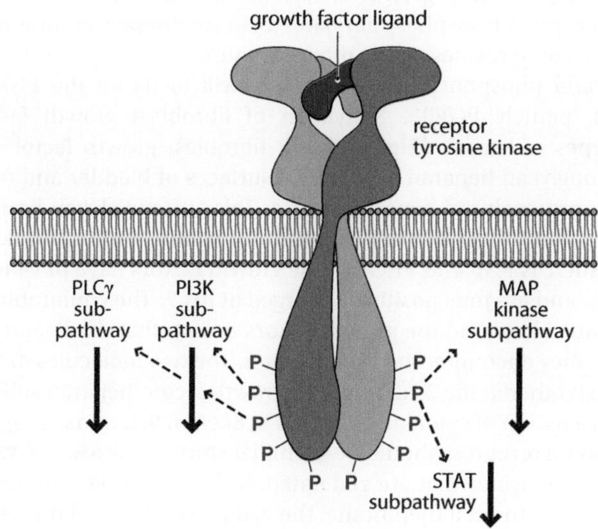

Figure 9.54 Four subpathways. Receptor tyrosine kinases activate four different subpathways. Dimerization of receptor tyrosine kinases leads to the sequential autophosphorylation of tyrosine residues, each marked with a P. Some of these phosphotyrosines serve as beacons that turn on subpathways.

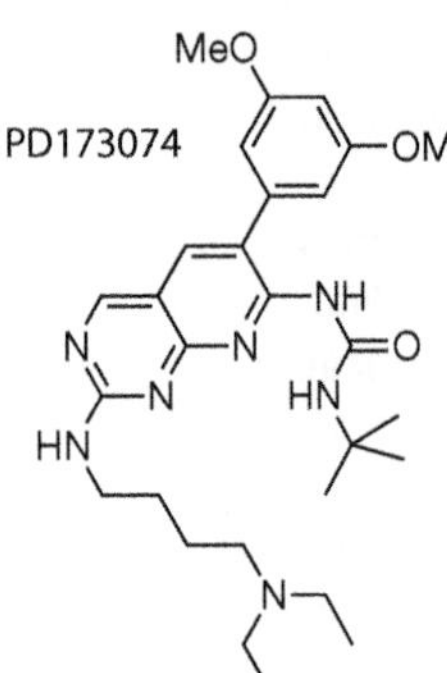

A

RTK	IC_{50} (nM)
FGFR1	4
FGFR2	3
FGFR3	5
VEGFR2	100
PDGFR	17,600
c-SRC	19,800
EGFR	>50,000
InsR	>50,000

B

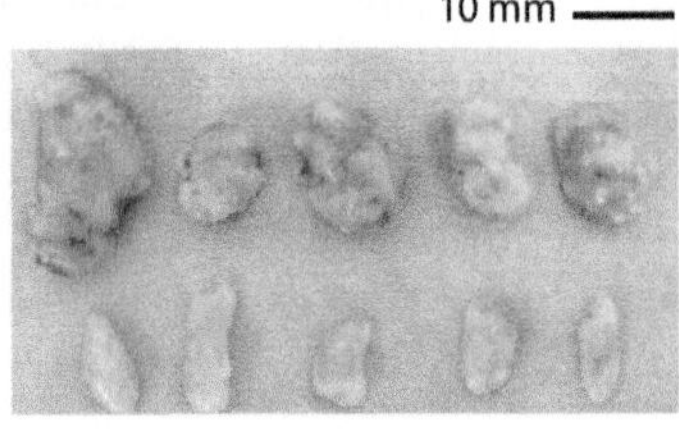

Figure 9.55 Selective inhibition of RTKs. (A) The synthetic inhibitor PD173074 (left) shows significant selectivity among receptor tyrosine kinases. (B) The inhibitor slows the growth of urothelial carcinomas. (B, from M. Miyake et al., *J. Pharmacol. Exp. Ther.* 332:795, 2010. With permission from American Society for Pharmacology and Experimental Therapeutics.)

The ATP-binding sites of receptor tyrosine kinases are sufficiently different that they can be selectively inhibited by small molecules

All protein kinase catalytic domains use Mg•ATP as a phosphoryl donor and have similar folded structures. Yet the ATP-binding pockets are not identical. The subtle differences between the various tyrosine kinases are sufficient to allow the development of selective kinase inhibitors through a combination of empirical screening and targeted synthesis. Chemists at Pfizer developed a selective inhibitor of the tyrosine kinase domain of the fibroblast growth factor receptor (**Figure 9.55**). When screened against 244 different protein kinases, PD173074 was shown to be active against only FGFR1, FGFR2, FGFR3 (but not FGFR4), and the γ isoform of CaM kinase II (see Figure 9.55). When administered orally, PD173074 significantly decreases the growth of the human transitional cell carcinoma cell line MGHU3 transplanted into nude mice, which lack an immune system to reject human cells. Other cell lines (T24, KU7, UM-UC-2, UM-UC-3, UM-UC-6, and J82) were completely unaffected by PD173074. In these cell lines, the aberrant signaling is not due to mutations in an FGF receptor. For example, a GGC (Gly) to GTC (Val) mutation in the Ras protein downstream of Ras in T24 cells causes persistent activation of the FGFR3 pathway.

Problem 9.10

Which tyrosine kinases in Figure 9.55 would be inhibited by more than 90% at 1 μM PD173074?

Transphosphorylation of tyrosine residues is sequential

The various growth factor receptors show strong mechanistic similarities and subtle structural differences in the way in which they bind the growth factor ligands. All growth factor receptors bind growth factors as 2:2 ligand–receptor complexes. All growth factor receptors transphosphorylate: a kinase domain on one receptor molecule phosphorylates tyrosines on the other receptor.

To understand phosphorylation better, we will focus on the FGF receptor in urothelial cells, namely FGFR3. Activation of fibroblast growth factor receptor requires two types of ligands (**Figure 9.56**): fibroblast growth factor and the anionic glycosaminoglycan heparan sulfate. The surfaces of bladder and other cells are coated with the proteoglycan heparan sulfate. This anionic glycan limits adherence by bacteria. Heparan sulfate sifts the extracellular milieu, binding weakly to cationic growth factors such as FGF and VEGF; these growth factors have p*I* values of 8.2 and 7.7, respectively, and are thus positively charged at pH 7. The coulombic interactions between heparan sulfate and the growth factors ensure that the ligands are in close proximity when they encounter the FGF receptor. The two molecules of FGF make no contact with each other in the 2:2:2 receptor–growth factor–heparan sulfate complex.

After our discussion of cytokine receptors in Section 9.4, transphosphorylation of tyrosine residues on receptors should be a familiar story (**Figure 9.57**). Within urothelial cells, phosphotyrosines that are generated on FGFR3 beckon proteins equipped with SH2 domains to the cell membrane; the cytoplasmic side of the cell membrane

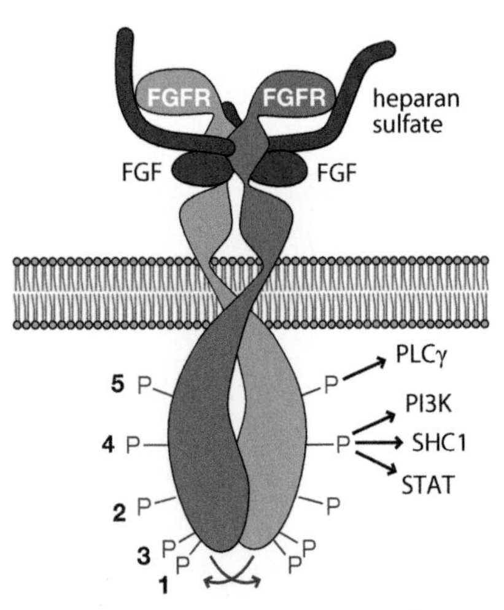

Figure 9.56 Sequential phosphorylation. Tyrosine residues on FGFR3 are phosphorylated in a precise order (1–5). The fourth and fifth phosphotyrosines activate specific subpathways. The actual locations of the phosphotyrosines are different from that shown in this schematic diagram.

is where the signal transduction pathway begins. Much of our understanding of FGFR3 comes from studies of the closely related receptor FGFR1. The kinase domain of FGFR3 is kept at bay by an inhibitory loop with the sequence $Y_{647}Y_{648}KK$. The net positive charge confers a strong affinity for the active site, but this effect is abrogated by phosphorylation of Y647. Phosphorylation of Y647 allows the kinase to phosphorylate Y577 on the opposing receptor, followed by the second tyrosine (Y648) in the inhibitory loop. The effect of phosphorylating tyrosines in the inhibitory loop has been quantified for FGFR1: phosphorylation of the first tyrosine increases activity 50-fold, and phosphorylation of the second tyrosine further increases activity 500-fold. Once both tyrosines in the inhibitory loop are phosphorylated, the affinity of the loop for the active site is fully reversed and two additional tyrosines (Y724 and Y760) are then phosphorylated. These two residues have a critical role in recruiting signaling proteins with SH2 domains. Tyr724 recruits STAT3, PI3K, and SH2-containing transforming protein 1 (SHC1). Tyr760 recruits the enzyme phospholipase Cγ.

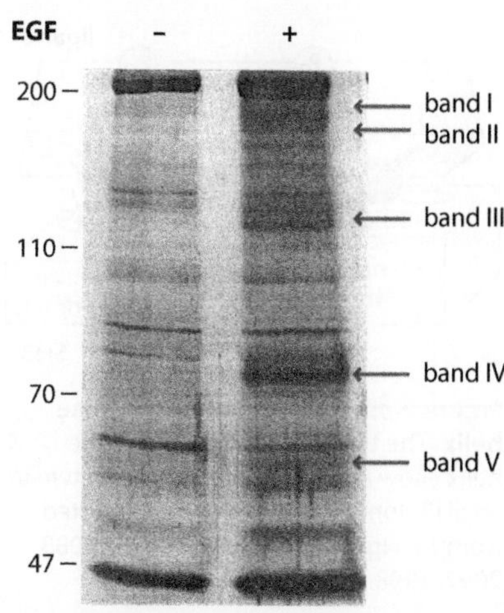

Figure 9.57 Immunoblot. Antibodies that specifically bind to phosphotyrosine residues can be used to label proteins in an SDS–PAGE gel. Cells treated (+) with epidermal growth factor show new patterns of tyrosine phosphorylation compared with untreated cells (–), as indicated by the presence of new bands in the (+) lane. (From H. Steen et al., *J. Biol. Chem.* 277:1031–1039, 2002. With permission from The American Society for Biochemistry and Molecular Biology.)

Receptor tyrosine kinases signal growth via a MAP kinase cascade

Early studies of mitogens pointed to a connection between protein growth factors (such as FGF) and protein kinases. Whereas some protein kinases seemed to phosphorylate many proteins (protein kinase C, protein kinase A), the protein kinases that were activated by mitogenic growth factors seemed to be highly specific. Studies of mitogen-activated protein kinases (MAP kinases) gave birth to the idea that mitogenic signals are transduced through a cascade of successive protein phosphorylations. In these MAP kinase cascades, which we discussed in Chapter 6, one protein kinase activates, through phosphorylation, a second protein kinase, and the second protein kinase activates a third kinase, and so on.

In urothelial cells, FGFR3 transduces growth signals to a MAP kinase cascade through a complex of three adaptor proteins: SHC1, Grb2, and SOS. The ubiquitous adaptor protein SHC1 (SH2 domain-containing protein 1) binds the phosphorylated form of Tyr724 on the FGFR3 receptor, and undergoes tyrosine phosphorylation. The adaptor protein Grb2 has an SH2 domain that binds to the phosphotyrosine on SHC1 (**Figure 9.58**). Grb2 also has two SH3 domains (Chapter 5) that bind specifically to

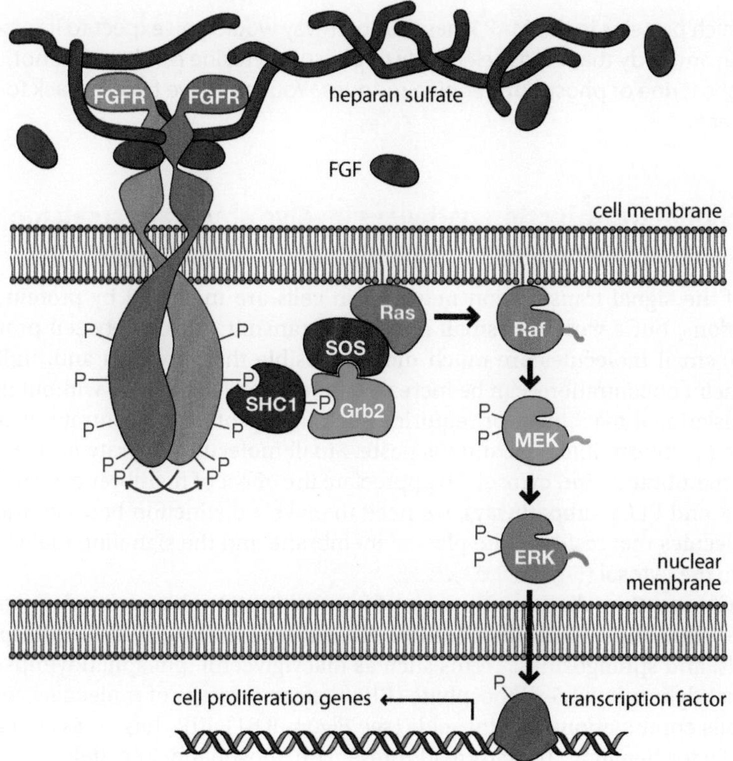

Figure 9.58 The MAP kinase subpathway. Receptor tyrosine kinases such as FGFR promote genes for proliferation via a cascade of protein phosphorylations by the kinases Raf, MEK, and ERK.

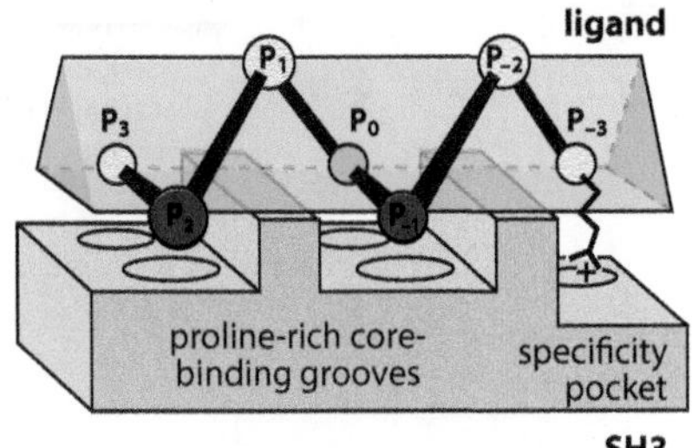

Figure 9.59 Binding to a polyproline helix. The tight turns of a polyproline helix allow extensive interactions between an SH3 domain and its ligand. (Adapted from J.T. Nguyen et al., *Science* 282:2088–2092, 1998.)

helical polyproline segments (PPXPPR) on SOS (**Figure 9.59**). SOS catalytically activates a G protein called Ras, which is held in the plasma membrane by a farnesyl group.

The kinases in the MAP kinase pathways phosphorylate serine and threonine residues. Ras activates the serine/threonine protein kinase called Raf, which specifically phosphorylates the kinase MEK. MEK phosphorylates the protein ERK (which stands for extracellular signal regulated kinase). ERK translocates to the nucleus, where it phosphorylates transcription factors, which turns on genes for cell proliferation and for inhibition of the FGFR signaling pathway (*Sprouty*, *Sef*, and *MKP3* phosphatase). In this pathway, ERK is considered the mitogen-activated protein kinase. The kinase that phosphorylates a MAP kinase (such as ERK) is called a MAP kinase kinase. The kinase that phosphorylates a MAP kinase kinase (such as MEK) is called a MAP kinase kinase kinase. Thus, Raf is a MAP kinase kinase kinase. There are two other homologous MAP kinases, JNK and p38 MAP kinase, and each of these kinases is activated by a specific series of MAP kinase kinases and MAP kinase kinase kinases.

The exquisite specificity of kinases in MAP kinase pathways is due to the presence of docking grooves in the kinase, and corresponding docking peptide segments in either the upstream or downstream kinase. The interaction of the docking peptide on the MAP kinase kinase MEK with the docking groove on the MAP kinase ERK is essential for specificity and binding. The devious enzyme anthrax lethal factor, produced by *Bacillus anthracis*, specifically castrates the docking peptide from MEK and other MAP kinase kinases, leaving them impotent (**Figure 9.60**).

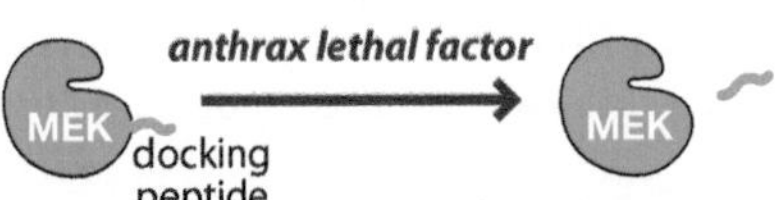

Figure 9.60 Chemical castration. Anthrax lethal factor is a protease that cleaves the docking peptide from MAP kinase kinases such as MEK.

Problem 9.11

A Which proteins in the MAP kinase pathway (Figure 9.58) are phosphorylated when the FGF is added to urothelial cells?

B Which proteins in the MAP kinase pathway would be phosphorylated if both FGF and vemurafenib, a selective inhibitor of Raf, were added to urothelial cells?

C Which proteins in the MAP kinase subpathway would you expect to interact with an antibody that binds selectively to phosphotyrosine residues, but not phosphoserine or phosphothreonine residues? You may have to refer back to Chapter 6.

Many signal transduction pathways involve abundant small molecules and scarce proteins

Most of the signal transduction networks in cells are mediated by protein–protein interactions, but a variety of small molecules transmit signals between proteins. In general, small molecules are much more diffusible than proteins and, unlike proteins, their concentrations can be increased to relatively high levels without involving the translational machinery or requiring such high protein concentrations that they turn the cytoplasm into a gelatinous paste. Small molecules operate in two environments: membranes and cytosol. To appreciate the effect of membrane localization in the PI3K and PLCγ subpathways, we need to make a distinction between the signaling molecules that reside in the plasma membrane and the signaling molecules that reside in the cytosol (**Figure 9.61**).

In Chapter 8 we discussed a range of fatty acids and sphingosine derivatives that mediate signaling within membranes: for example diacylglycerols, phosphatidylinositols, and sphingosines. Terms such as diacylglycerol, phosphatidylinositol, and phosphatidylinositol 4,5-diphosphate (PIP_2) refer to classes of molecules composed of various combinations of fatty acids (see Figure 9.61). PIP_2 has an essential role in growth factor signaling. The fatty acids present on phosphatidylinositols vary in length

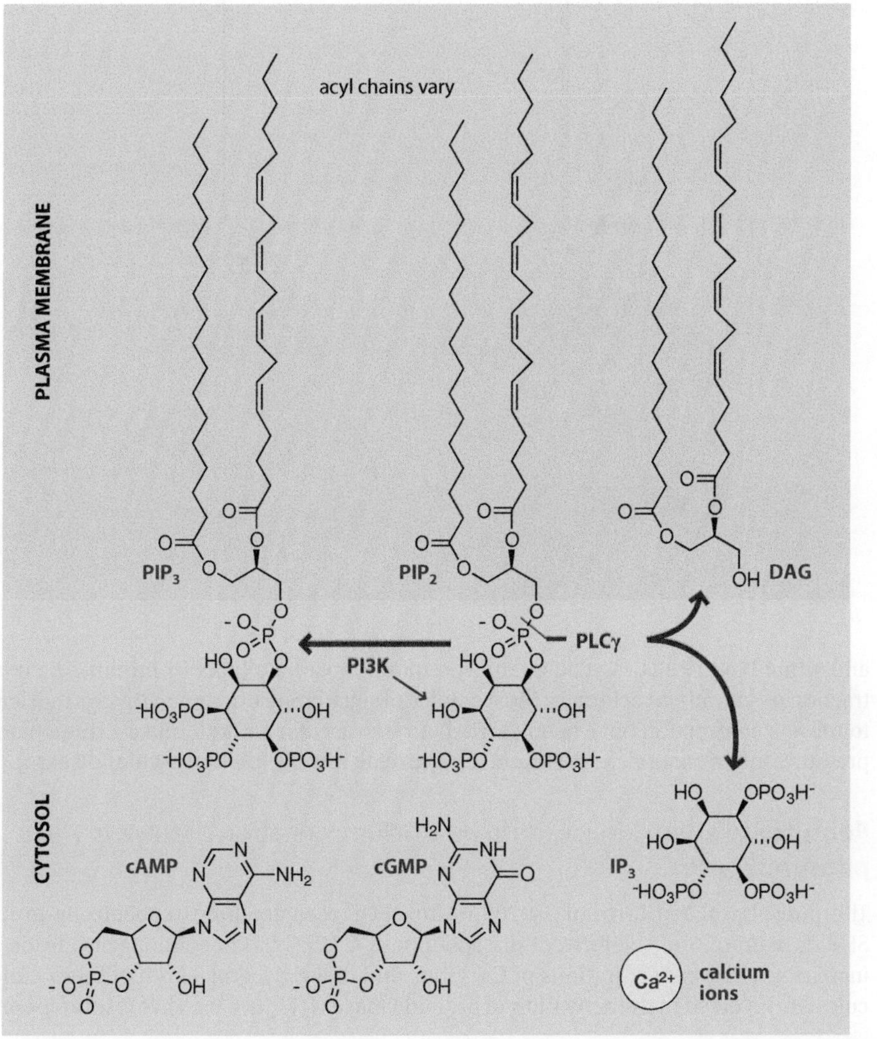

Figure 9.61 Small molecules, big effects. Abundant small, nonprotein molecules are important in signal transduction. Fatty acid derivatives such as PIP_2, PIP_3, and DAG are localized in the plasma membrane, whereas polar ionic compounds are localized in the cytosol.

(16–20 carbons) and unsaturation (0–4 double bonds), but an arachidonyl group is most common at the 2 position of glycerol, and a stearoyl group is most common at the 1 position. There is some selectivity among the various enzymes that process phosphatidylinositols; for example, some prefer arachidonyl groups at the S-2 position of glycerol; however, there seems to be a wide tolerance for different fatty acyl chains. Given the heterogeneity of lipids, we cannot adequately represent all phosphatidylinositols with a single structure, any more than we can represent a voter with a single human face. Each time a phosphatidylinositol is drawn it is up to the reader to envision a collection of molecules with variations in the fatty acid chains.

Diacylglycerol remains localized within the plasma membrane, where it interacts with membrane-bound proteins such as protein kinase C. Similarly, phosphatidylinositol 1,4,5-triphosphate (PIP_3) also remains localized within the membrane, but it dangles the inositol subunit—like bait on a hook. The phosphoinositol head group of PIP_3 is irresistible to proteins with pleckstrin homology domains, such as the kinase PDK1 (**Figure 9.62**).

Small molecules that diffuse through the cytosol include cyclic adenosine monophosphate (cAMP), inositol 1,4,5-triphosphate (IP_3), and metal cations, particularly Ca^{2+}. Several of these small molecules are associated with common effects among the various types of human cells. cAMP usually promotes cell growth via the relatively promiscuous kinase cAMP-dependent protein kinase (PKA) and a transcription factor called the cAMP response element binding protein. IP_3 opens calcium storage vessels, resulting in an immediate burst in Ca^{2+} ion concentration. An increase in intracellular Ca^{2+} levels induces the contraction of muscle cells (such as skeletal, cardiac, vascular, bronchial, and ciliary) and the release of messengers from vesicles (such as neurons

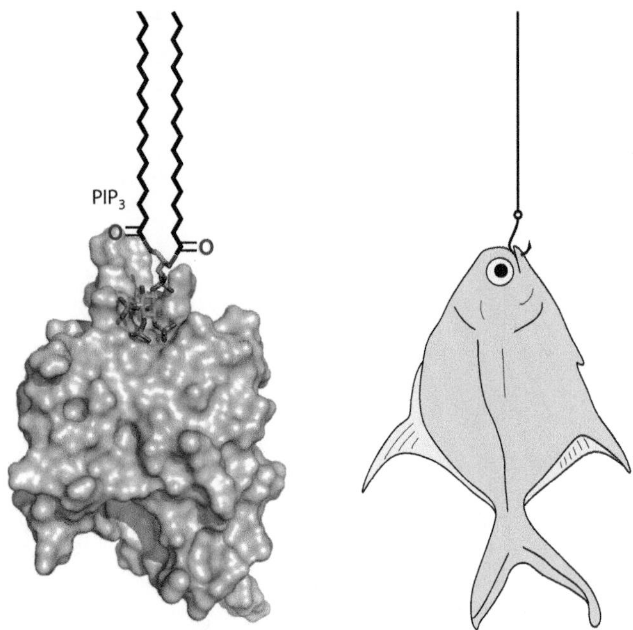

Figure 9.62 Fishing for proteins. The pleckstrin homology domain of PDK1 binds to the phosphoinositide head group of PIP_3 (PDB 1W1D). As a result, the entire kinase is localized at the cell membrane.

and white blood cells). Cyclic guanosine monophosphate (cGMP) inhibits the contraction of smooth muscle cells. Most studies of signal transduction pathways that lead to muscle contraction have been carried out on vascular smooth muscle (high blood pressure, inflammation, and asthma) and cardiac muscle (cardiovascular disease).

Receptor tyrosine kinases turn on calcium signaling pathways via phospholipase C

The phosphorylated form of Tyr760 on the FGF receptor binds specifically to the SH2 domain of the γ isoform of phospholipase C (PLCγ). Phospholipase C leads to increases in the concentrations of Ca^{2+} ions and diacylglycerol—both of these molecules are involved in the activation of protein kinase C (**Figure 9.63**). Phospholipase C

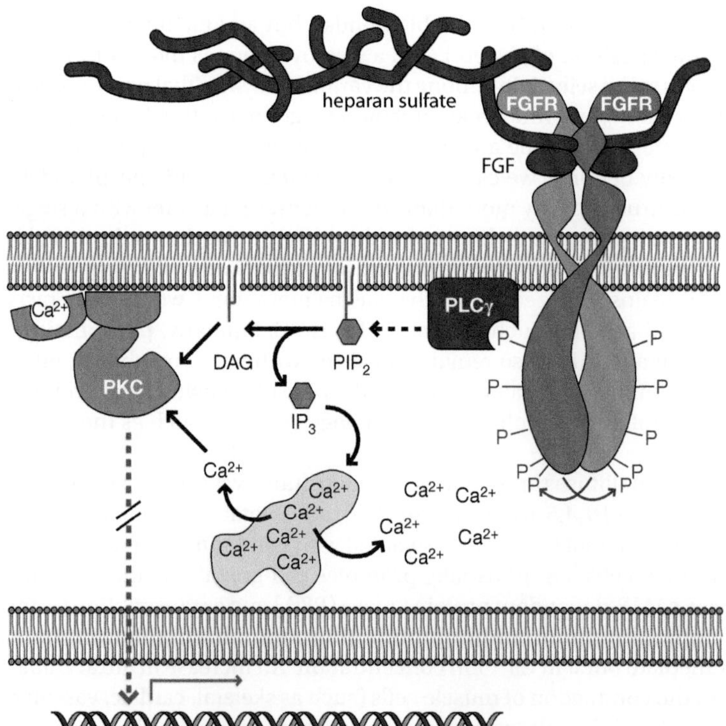

Figure 9.63 The PLCγ subpathway. PLCγ catalyzes the hydrolysis of PIP_2 to give DAG and IP_3. DAG directly and selectively activates protein kinase C, whereas IP_3 indirectly activates PKC by increasing cellular Ca^{2+} levels.

cleaves the diacylglycerol moiety of PIP_2, generating the membrane-bound diacylglycerol and cytosolic IP_3. IP_3 opens Ca^{2+} channels in calcium storage vesicles, resulting in a burst of cytosolic Ca^{2+} ions. Recall from Chapter 6 that Ca^{2+} ions bind to a domain of protein kinase C, which otherwise blocks a membrane-localization domain. Once Ca^{2+} unblocks the membrane localization domain, PKC binds to the plasma membrane, where the concentration of diacylglycerol is high. The binding of diacylglycerol to PKC leads to a fully activated form of the kinase localized at the cell membrane.

Problem 9.12

The active site of phospholipase Cγ contains three key catalytic residues: two histidines and one arginine. Suggest a plausible arrow-pushing mechanism for the enzyme-catalyzed hydrolysis of PIP_2 to IP_3 and DAG that includes a role for these three key residues and that involves neighboring-group participation by analogy with ribonuclease A.

Receptor tyrosine kinases broadcast both proliferative and anti-apoptotic signals via Akt

The phosphotyrosines that result from activation of the growth factor receptors are recognized by proteins with SH2 domains. Grb2 (pronounced "grab two") is a key mediator of growth factor receptor signals. In urothelial cells, the SH2 domain of Grb2 binds to phosphotyrosine 724 of the FGF receptor, further recruiting the proteins Gab1 and PI3K to the cell membrane (**Figure 9.64**). Once PI3K is localized at the cell membrane it can catalyze the transfer of a phosphate group from ATP to PIP_2 to generate PIP_3. As the concentrations of PIP_3 increase, the PIP_3 molecules recruit two protein kinases to the membrane, PDK1 and Akt, each of which possesses a pleckstrin homology domain (**Figure 9.65**). Binding to PIP_3 in the plasma membrane fixes both the location and orientation of proteins, allowing PDK1 to phosphorylate Thr308 and Ser473 in Akt, leading to a fully activated form of Akt that targets many proteins.

Most of the proteins that are phosphorylated by Akt are inhibited, but because those proteins inhibit mitosis and differentiation, Akt ultimately *activates* pathways that permit cell proliferation. The result of this boolean logic is reminiscent of the

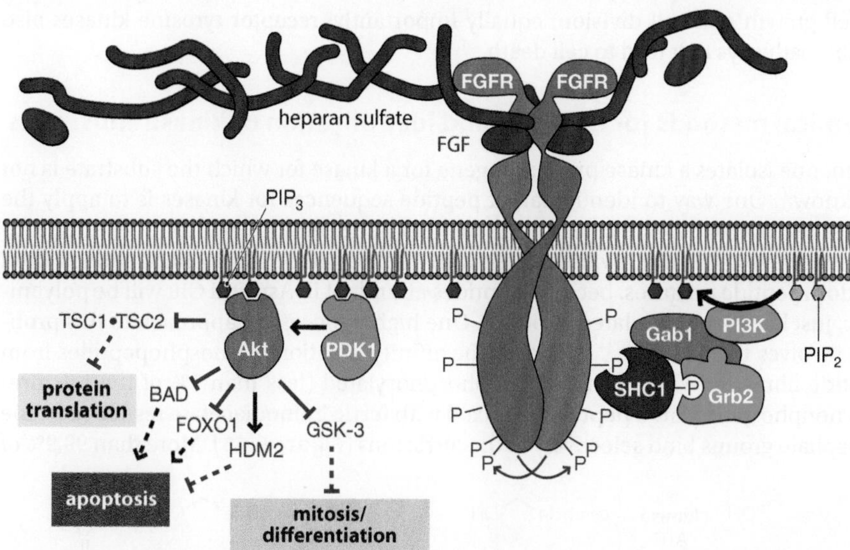

Figure 9.64 The PI3K/Akt subpathway. The kinase Akt binds to a lipid derivative PI3K. Once localized to the plasma membrane Akt exerts three effects: (1) enhancement of protein translation, (2) inhibition of apoptosis, and (3) promotion of proliferation and differentiation. The dashed arrows suggest that we have omitted many of the intervening steps.

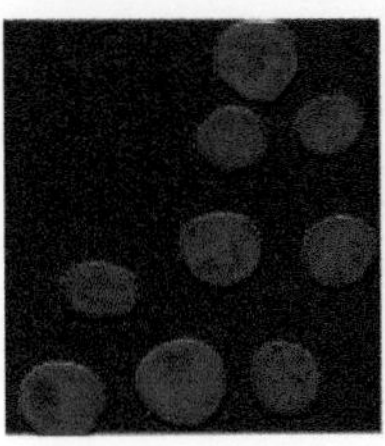
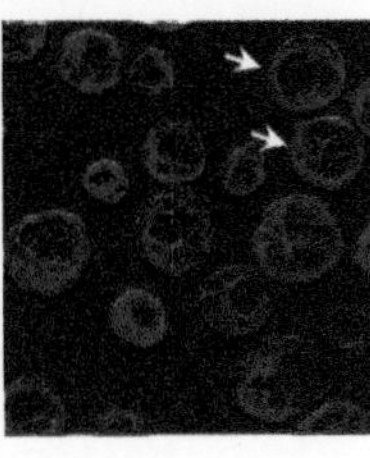

Figure 9.65 Akt at the membrane. The localization of Akt in a fibroblast cell line can be determined by staining fixed cells with a red fluorescent antibody. In the absence of mitogenic signals (top) Akt is distributed throughout the cytoplasm. In the presence of mitogenic signaling (bottom), Akt is localized mainly at the plasma membrane. (From G. Guo et al., *Nature* 29:3845–3853, 2010. With permission from Macmillan Publishers Ltd.)

ancient proverb "the enemy of my enemy is my friend." Akt phosphorylates the proteins TSC1 and TSC2, preventing them from inhibiting protein translation. By releasing the brakes on protein translation, the cell is ready for rapid growth. Akt suppresses the suicidal tendencies of the cell by deactivating BAD and FOXO1 and by activating HDM2. Left unchecked, these three proteins would otherwise couple any attempts at cell division with apoptosis. Finally, Akt deactivates glycogen synthase kinase 3 (GSK-3) through phosphorylation. When GSK-3 is active, it suppresses cell division. By releasing the brakes, Akt allows cell division and differentiation. In the next section, we discuss how GSK-3 leads to the degradation of β-catenin, ultimately affecting transcription.

Problem 9.13

Wortmannin is a highly specific covalent inhibitor of PI3K, forming a covalent bond with Lys802 in the ATP-binding pocket. Suggest a plausible arrow-pushing mechanism for an acid-catalyzed reaction of wortmannin with PI3K.

wortmannin

cat. HA

PI3K

The differences between various receptor tyrosine kinase pathways are less important than the similarities

We began our discussion of receptor tyrosine kinases by listing the many types of growth factor signaling pathways (EGF, VEGF, PDGF, NGF, and so on) and contrasting the different structures of the receptor–ligand complexes. Our decision to focus on the FGF receptor was inconsequential (for readers of this book). We would have discussed the same common downstream pathways—STAT, MAP kinase, and PLCγ and PI3K—regardless of which receptor tyrosine kinase was chosen. Given the complexities of those divergent pathways, it is more profitable to focus on the similarities rather than the differences between the various receptor tyrosine kinase pathways. The overall message is that receptor tyrosine kinases activate the transcription of genes leading to cell growth and cell division; equally importantly, receptor tyrosine kinases also inhibit pathways that lead to cell death.

Chemical methods for isolation and identification of kinase substrates

Often, one isolates a kinase protein or gene for a kinase for which the substrate is not yet known. One way to identify target peptide sequences for kinases is to apply the kinase reaction to peptide libraries and then sequence only the phosphopeptide products. However, it is nontrivial to selectively remove anionic phosphopeptides from random peptide mixtures, because peptides abundant in Asp and Glu will be polyanionic, just like phosphorylated peptides. One highly successful approach to this problem involves the use of Fe^{3+} resins for the affinity isolation of phosphopeptides from peptide libraries. When a mixture of phosphorylated (less than 1% of the mixture) and nonphosphorylated peptides is mixed with ferric iminodiacetate resin beads, the phosphate groups bind selectively to the ferric ions (**Figure 9.66**). More than 99.9% of

peptide library

kinase ATP

Fe^{3+}-IDA resin pH ≤ 6

Figure 9.66 Fishing for phosphoproteins, part 1. The Cantley soluble oriented peptide library method allows one to isolate phosphorylated peptides from large peptide libraries on the basis of the affinity of phosphate anions for Fe^{3+}. IDA, imino diacetic acid.

Figure 9.67 Fishing for phosphoproteins, part 2. The Shokat strategy for the isolation of kinase products from complex mixtures of proteins involves an ATP analog, ATP-γ-S.

the nonphosphorylated peptides can be washed off by elution with buffers at or below pH 6.0. With a concentrated high-pH buffer (500 mM NH_4HCO_3, pH 8.0), more than 90% of the phosphopeptides can be eluted. The entire phosphopeptide mixture is then sequenced to determine which amino acids are preferred at each of the positions of the preferred substrates.

An alternative approach for identifying substrates for kinases is to phosphorylate protein mixtures by using a mixture of kinase and ATP-γ-S. The sulfur atom in a phosphorothioate prefers the anionic form, even at neutral pH, and like the cysteine thiol it is highly nucleophilic. Both cysteine and phosphorothioserines are alkylated, but antibodies against the anionic nitrobenzylphosphorothio group readily distinguish between an alkylated cysteine and an alkylated phosphorothioserine. This method can even be applied to mixtures of full-length proteins from cell lysates. The protein can be digested and the sequences of the peptides determined by mass spectrometry (**Figure 9.67**). Once you know the sequence of the peptide, you can search through databases to identify the full-length protein.

9.6 G PROTEIN-COUPLED RECEPTORS

Seven-transmembrane domain G protein-coupled receptors can respond to a wide range of ligands with high dynamic range

Seven-transmembrane domain G protein-coupled receptors (7TM GPCRs, or GPCRs for short) are prosaically named. Recall from Chapter 5 that they consist of a membrane-bound receptor, with seven membrane-spanning α helices, coupled to a GTP hydrolase (called a G protein) that serves as a timing device. The more apt term "heptahelical" is slowly entering the vernacular as a replacement for the clumsy term "seven-transmembrane domain," but we will continue to use the 7TM description to reinforce a connection to the large body of GPCR literature (**Figure 9.68**). There are more than 900 different genes for GPCRs in the human genome, about 5% of the total number of genes. Almost half of the genes for 7TM GPCRs code for olfactory receptors that recognize diverse odorants. It is common to group G proteins into superfamilies based on genetic sequence. In order of population, the human GPCR families are: Rhodopsin (R), Adhesion (A), Frizzled/Taste2 (F), Glutamate (G), and Secretin (S) (**Table 9.3**).

Table 9.3 Families of human 7TM GPCRs.

Family	Number of receptors
Rhodopsin	284 non-olfactory
Rhodopsin	388 olfactory
Adhesion	33
Taste2	25
Glutamate	22 (includes Taste1)
Secretin	15
Frizzled	11

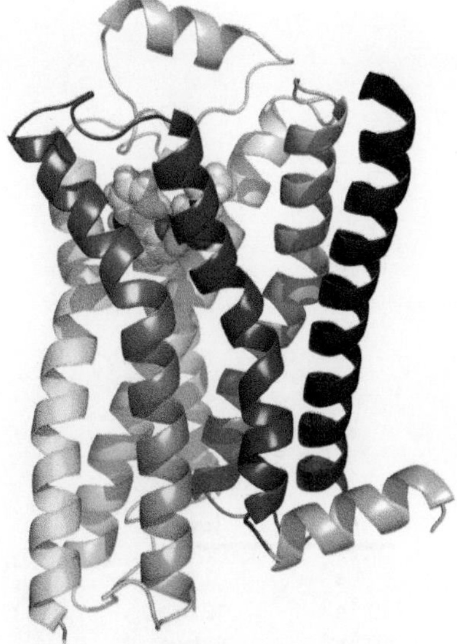

Figure 9.68 Structure of a GPCR. The β_2-adrenergic receptor with the synthetic antagonist carazolol (van der Waals spheres) bound in place of epinephrine. The transmembrane helices are rendered in color. The structure is lacking the lower cytoplasmic loop between the yellow and orange helices. (PDB 2RH1)

It is not surprising that the majority of pharmaceutical drugs that were developed in the previous millennium were ultimately found to target GPCRs. Small molecules have the best oral availability, and 7TM receptors are the most common receptor for small-molecule signals in humans. Furthermore it is relatively easy to identify small-molecule ligands (for example histamine, adrenaline, and thyrotropin releasing hormone) and to design medicinally useful analogs, even if the structure of the protein target is unknown. Major advances were made in genomics and structural biology in the 1990s that emboldened pharmaceutical chemists to target signal transduction pathways inside cells. Thus, the focus of pharmaceutical chemists has broadened from membrane-bound receptors such as GPCRs to include many intracellular targets as well.

High-affinity ligand–receptor interactions lead to slow response times and low dynamic range

An ideal sensor system should have high sensitivity, fast response time, and wide dynamic range. Designing highly sensitive receptors using reversible ligand–receptor interactions leads to slow response times and low dynamic range. To understand the origin of these long response times, let us return to the high-affinity streptavidin-biotin interaction that we discussed in Chapter 6. Recall that the half-life for the avidin-biotin complex is over half a year long. The half-life for the closely related streptavidin-biotin is several days. Thus, streptavidin could be a highly sensitive receptor for biotin and bind to a minuscule amount of biotin. However, the streptavidin receptor cannot respond to *changes* in concentration on a timescale that is useful for a cell.

To understand the origin of the slow off-rates of high-affinity receptors, we need to return to a kinetic picture of equilibrium binding that was introduced in Chapter 6. Recall that smaller dissociation constants correlate with tighter binding. The dissociation constant, like any equilibrium constant, can be expressed as the ratio of the rate constant for the off-rate and the rate constant for the on-rate (**Figure 9.69**). There is usually little variation in the on-rate constants for small-molecule ligands and receptors (this simplifying assumption only works well for small-molecule ligands). However, off-rate constants vary considerably. The half-life for dissociation is determined by the equation $t_{1/2} = (\ln 2)/k_{off}$ (**Figure 9.70**). At pH 7.4, a streptavidin–biotin interaction with an off-rate of $3 \times 10^{-5}\ s^{-1}$ has a half-life of more than 6 hours.

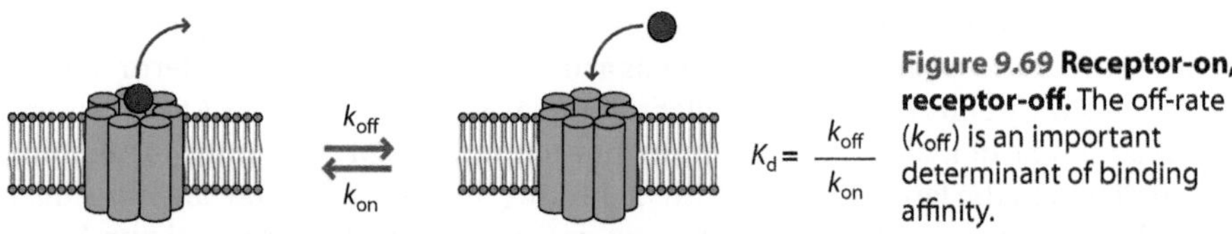

Figure 9.69 Receptor-on, receptor-off. The off-rate (k_{off}) is an important determinant of binding affinity.

The limited dynamic range or lack of responsiveness to different ligand conditions by high-affinity receptors such as streptavidin is more difficult to appreciate. Recall that biological responses correlate with the percentage of bound receptor. A larger response results from a higher percentage of receptors bound to the ligand. Thus, reversible ligand–receptor interactions are highly sensitive to changes in ligand concentration only when the ligand concentration is close to the K_d for the receptor–ligand interaction. At this concentration, 50% of the receptors are bound to ligand, and 50% of the receptors remain empty; this situation provides empty receptors to react to an influx of new ligand, and also bound receptors ready to dissociate in response to a decrease in the concentration of the ligand. For the streptavidin–biotin interaction, only concentrations of biotin in the subpicomolar range lead to differences in the percentage of biotin bound to streptavidin.

When the ligand concentration is close to the K_d for the receptor-ligand interaction, 10-fold changes in ligand concentration lead to marked changes in the percentage of bound receptor; however, outside that narrow window, 10-fold changes in ligand concentrations lead to very small changes in the percentage of bound receptors (**Figure 9.71**). For example, if the K_d for a receptor–ligand interaction is 1 μM, the

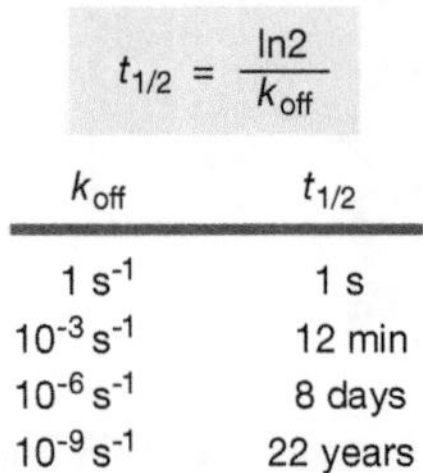

$$t_{1/2} = \frac{\ln 2}{k_{off}}$$

k_{off}	$t_{1/2}$
$1\ s^{-1}$	1 s
$10^{-3}\ s^{-1}$	12 min
$10^{-6}\ s^{-1}$	8 days
$10^{-9}\ s^{-1}$	22 years

Figure 9.70 Binding and half-lives. Dissociation of a ligand from the receptor–ligand complex is a first-order or unimolecular reaction. The top equation relates the rate constant to the half-life for a reaction (lower table).

percentage of bound receptor is close to 100% regardless of whether the concentration of ligand is 100 μM, 1000 μM, or 10,000 μM; similarly, the percentage of bound receptor is close to 0% regardless of whether the concentration of ligand is 0.01 μM, 0.001 μM, or 0.0001 μM.

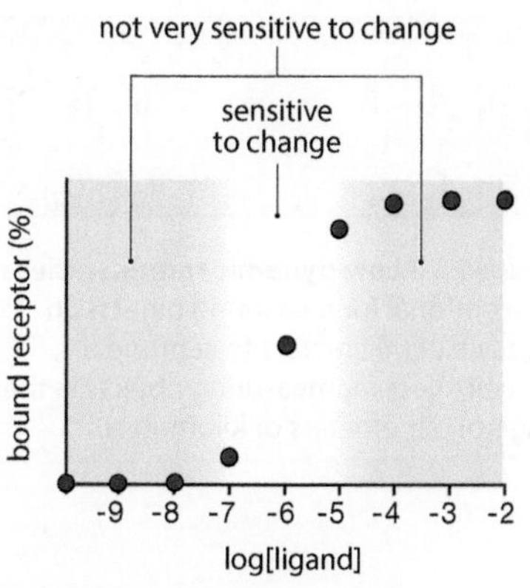

Figure 9.71 Limited window of sensitivity. As seen in this dose–response curve, reversible receptor–ligand interactions provide high dynamic sensitivity to changes in ligand concentration only when the concentration of the ligand is close to the K_d for the receptor–ligand interaction (the unshaded portion of the graph).

G proteins allow low-affinity receptors to have high sensitivity

G proteins are common elements of signal transduction pathways that must modulate the strength of the incoming signal. G proteins slowly catalyze the hydrolysis of GTP to GDP (**Figure 9.72**). The GTP-bound state turns on downstream signaling; in the GDP-bound state, signaling is shut off. G proteins are not enzymes in a traditional sense, because the catalytic rate is low and product dissociation is slow. However, each of these phases of the catalytic cycle is accelerated by a specific class of GTP-binding-protein regulators. **Guanine nucleotide exchange factors** (GEFs) catalyze the exchange of GDP for GTP; GEFs therefore increase signal output. **GTPase-activating proteins** (GAPs) accelerate hydrolysis of the bound GTP; GAPs therefore decrease signal output. Thus, the ratio of GEF to GAP determines the strength of the signal (**Figure 9.73**). In addition, regulation through binding to other proteins and lipids, along with phosphorylation and subcellular localization, controls the activity of GEFs and GAPs. Such regulation can effectively tie the G protein signaling to other cell signaling pathways.

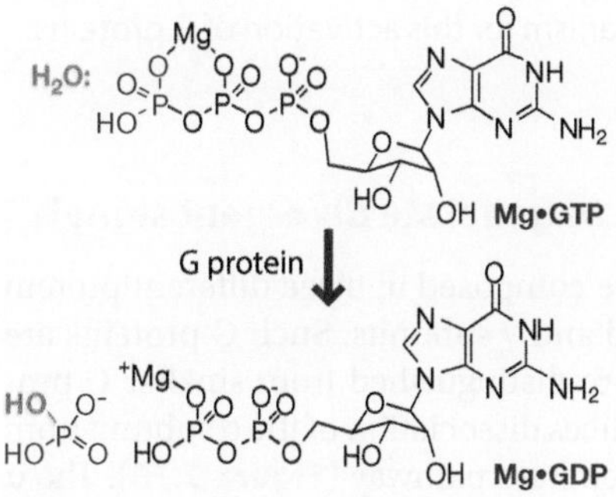

Figure 9.72 Shutting off a G protein. G proteins are deactivated when they hydrolyze Mg•GTP to GDP.

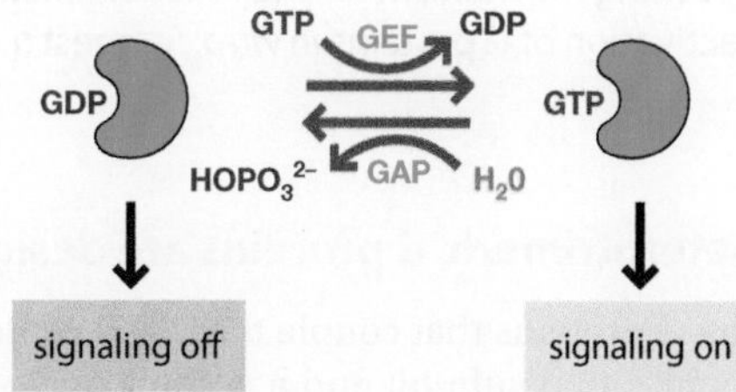

Figure 9.73 The volume knob. The ratio of GEF to GAP, both of which are catalysts, determines the strength of G protein signaling. (Adapted from D.P. Siderovski and F.S. Willard, *Int. J. Biol. Sci.* 1:51–66, 2005.)

Given their control over cell growth pathways, G proteins are a common site of oncogenic mutations. For example, mutations to the G protein Ras (Section 9.5) can prevent GTP hydrolysis. The resultant cells have growth signals permanently turned on. A third of all human cancers have been found to have such mutations in Ras. Furthermore, the small-molecule analogs of genetic mutations can alter the course of signal transduction pathways. For example, the small molecule brefeldin A binds to the G protein Arf1 and glues the protein to a GEF, effectively preventing the dissociation of GDP from Arf1. The activity prevents the binding of Arf1 to coat protein found in the membrane of the Golgi complex (**Figure 9.74**). Thus, brefeldin A blocks the transport of vesicles between the endoplasmic reticulum and the Golgi complex; this ability makes brefeldin A a widely used tool for the study of vesicle trafficking in the cell.

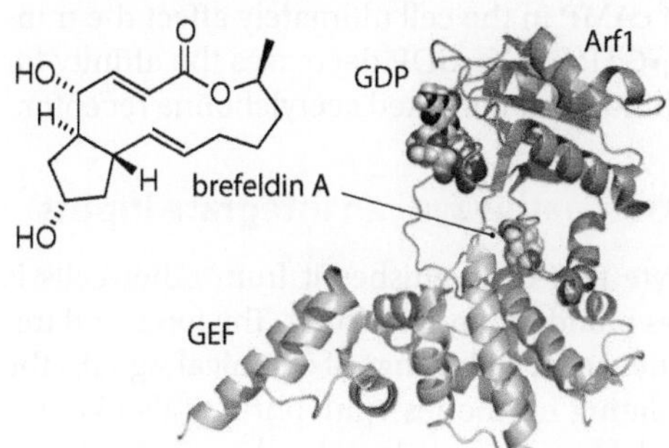

Figure 9.74 Blocking anterograde transport of vesicles between the ER and the Golgi. The G protein Arf1 has an essential role in Golgi vesicle trafficking. The polyketide natural product brefeldin A binds to and stabilizes the complex between a GEF and Arf1, inhibiting the release of GDP. (PDB 1RE0)

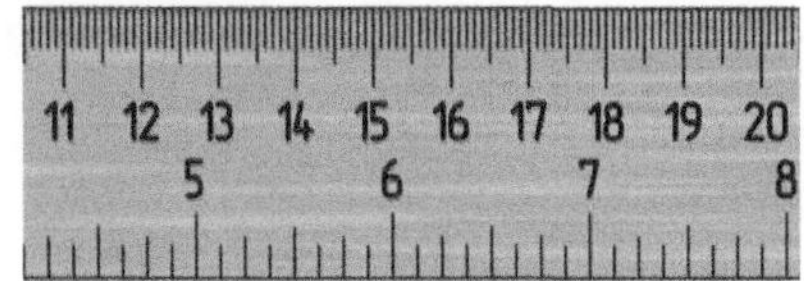

Figure 9.75 Low dynamic range. A ruler is exceptional for measuring objects on the scale of millimeters to centimeters, but worthless for measuring objects in the range of nanometers or kilometers.

Seven-transmembrane domain G protein-coupled receptors can respond to a wide range of ligands with high dynamic range

G protein-coupled receptors respond with high sensitivity, fast response times, and high dynamic range. They can bind ligands with modest affinity—briefly—and then release them. During the short time that 7TM receptors bind to their cognate ligands, they act as guanine nucleotide exchange factors (GEFs), and release active G proteins. The power of a G protein-coupled receptor is that the lifetime of the signal is not limited by the amount of time that the receptor is bound! Instead, the lifetime of the signal is determined by the ratio of GEFs to GAPs, which the cell can control. This powerful concept bears repeating: cells can modulate the sensitivity of G protein signaling pathways. A detector that can accurately distinguish responses with both high and low sensitivity is said to have high (or wide) dynamic range (**Figure 9.75**). For example, your eyes rely on 7TM G protein-coupled receptors to detect light. Having photon receptors with a wide dynamic range allows you to distinguish objects in a dimly lit room as easily as you distinguish them under sunlight that is a billion times brighter.

Problem 9.14

G proteins hydrolyze a bound Mg•GTP, using water as a nucleophile. Draw an arrow-pushing mechanism for the reaction. Mixtures of AlF_3 and GDP can lead to persistent activation of G proteins *in vitro*. Suggest a mechanism for this activation of G proteins.

Heterotrimeric G proteins are designed to generate divergent signals

The G proteins that couple to 7TM receptors are composed of three different protein chains: an α subunit and a complex between β and γ subunits. Such G proteins are referred to as heterotrimeric G proteins, and are distinguished from smaller G proteins such as Ras. Activation of a 7TM GPCR induces dissociation of the α subunit from the βγ subunit, and each of these turns on a separate pathway (**Figure 9.76**). The α subunit often modulates pathways that generate small molecules (cAMP or DAG and IP_3), which ultimately affect transcription. The βγ subunit modulates rapid cellular responses (such as the contraction of muscle cells or the firing of neurons), usually based on **ion channels**: for example GIRK1 potassium channels (cardiac myocytes) and CaCn calcium channels (Purkinje neurons).

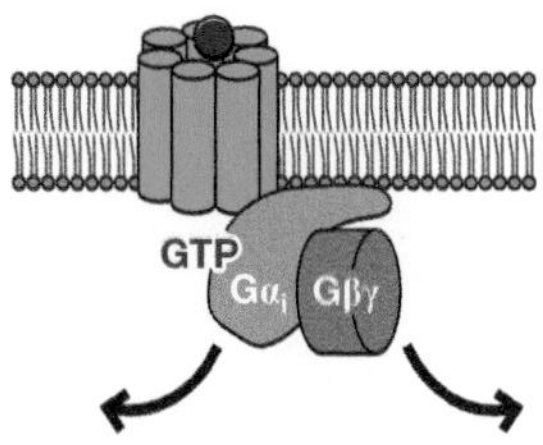

Figure 9.76 Fork in the road. Dissociation of Gα and Gβγ subunits allows 7TM GPCRs to activate two different pathways.

To understand the divergence of signals from heterotrimeric G proteins, it is instructive to examine how the vagus nerve decreases the contractile force of heart muscle cells. Vagal neurons release the neurotransmitter acetylcholine, which binds to a 7TM GPCR, the M2 acetylcholine receptor, on cardiac myocytes (**Figure 9.77**). Those acetylcholine receptors are coupled to a heterotrimeric G protein, leading to an exchange of GDP for GTP within the α subunit. The α subunit, $G\alpha_i$•GTP, dissociates from the βγ subunit, Gβγ. Gβγ activates GIRK1 potassium channels in the cell membrane, and the leakage of K^+ ions from the cell decreases the force of contractions. The K^+ gradient is essential for myocyte function; in the United States, death by lethal injection is performed by infusion of potassium chloride, which stops the heart. The $G\alpha_i$•GTP subunit inhibits adenylyl cyclase, slowing the rate at which it cyclizes ATP into cAMP (see Figure 9.77). The levels of cAMP in the cell ultimately affect the transcription of genes. The conversion of $G\alpha_i$•GTP to $G\alpha_i$•GDP decreases the affinity for adenylyl cyclase, freeing the $G\alpha_i$•GDP to bind to an unligated acetylcholine receptor.

Some elements of signal transduction pathways can integrate inputs

The principal feature of a cardiac myocyte that distinguishes it from other cells is ceaseless rhythmic contraction that moves blood around the body. The force and frequency of myocyte contraction are dependent on a wide range of chemical signals: the neurotransmitter acetylcholine, the circulating hormones epinephrine (also known as adrenaline) and norepinephrine (also known as noradrenaline), and circulating

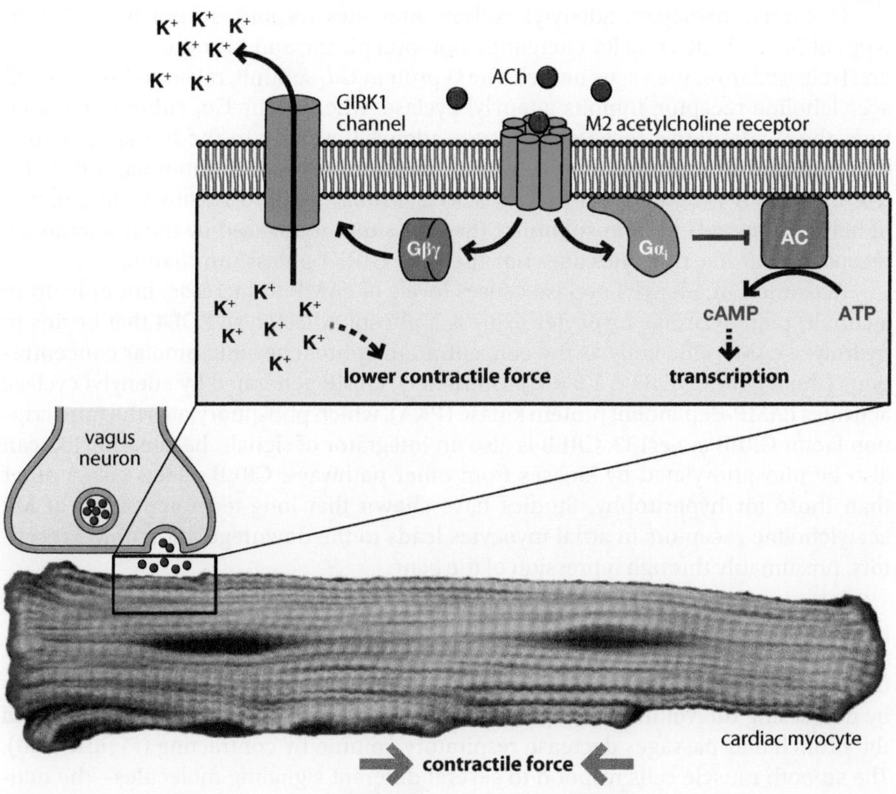

Figure 9.77 **Dual effects.** The M2 acetylcholine receptor exerts two effects on the cell via dissociation of the heterotrimeric G protein. It decreases the contractile force and affects transcription. AC, adenylyl cyclase. (Cardiac myocyte, courtesy of Nicholas Geisse, Harvard University. With permission from Kevin Kit Parker, Harvard University.)

peptide hormones such as endothelin I and angiotensin II. An overworked heart cell has relatively few ways in which to ease the burden. Cell proliferation is not an option because cardiac myocytes are terminally differentiated, without the capacity for proliferation. Instead of proliferating to share the workload, individual cardiac myocytes grow larger, leading to a visibly enlarged heart (**Figure 9.78**). The genes for myocyte hypertrophy are controlled by a transcription factor called the cyclic AMP response element binding protein (CREB). Thus, any signaling pathways that affect levels of cyclic AMP can affect transcription by CREB. The diterpene natural product forskolin activates adenylyl cyclase, mimicking agonists of 7TM GPCRs coupled to $G\alpha_s$ (**Figure 9.79**).

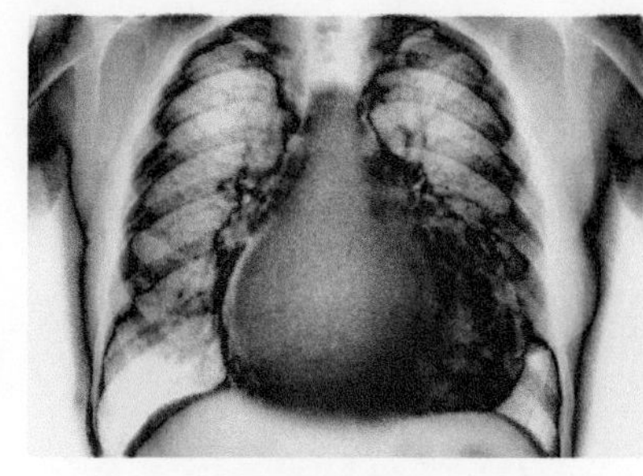

Figure 9.78 **Enlarged heart.** Alcoholism and high blood pressure overwork the heart, causing it to enlarge. (Courtesy of Photo Researchers.)

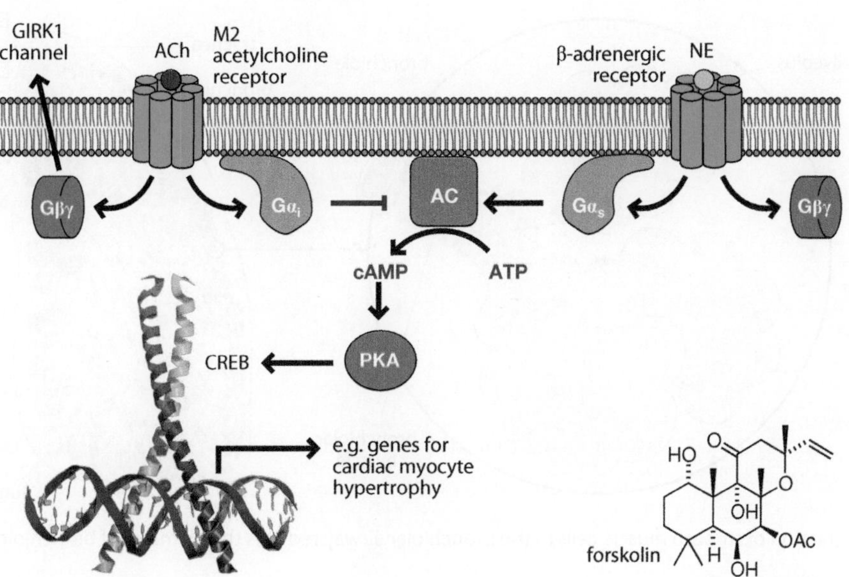

Figure 9.79 **Two paths converge.** Adenylyl cyclase (AC) integrates input from two different 7TM GPCRs. Inset: the small molecule forskolin activates adenylyl cyclase. NE, norepinephrine; AC, adenyl cyclase.

In cardiac myocytes, adenylyl cyclase integrates responses from two different types of 7TM GPCRs: one for circulating norepinephrine and the other for acetylcholine (released from the vagus nerve). The G protein $G\alpha_i$ subunit, released from the M2 acetylcholine receptor, inhibits adenylyl cyclase, whereas the $G\alpha_s$ subunit, released from the β-adrenergic receptor, activates adenylyl cyclase. In the human genome, there are 23 different genes for α subunits, 7 different genes for β subunits, and 12 different genes for γ subunits; each GPCR in a cell usually binds to a distinct combination of heterotrimeric G protein subunits. The Gβγ subunit released by the β-adrenergic receptor in cardiac myocytes does not affect the GIRK1 potassium channel.

Activation of adenylyl cyclase causes levels of cAMP to increase, but only up to a certain point. Cardiac myocytes express a phosphodiesterase PDE4 that begins to hydrolyze cAMP efficiently as the concentration approaches micromolar concentrations (the K_m for PDE4D is 1.5 μM). Ultimately, cAMP generated by adenylyl cyclase activates cAMP-dependent protein kinase (PKA), which phosphorylates the transcription factor CREB at Ser133. CREB is also an integrator of signals, because Ser133 can also be phosphorylated by kinases from other pathways. CREB affects genes other than those for hypertrophy. Studies have shown that long-term activation of M2 acetylcholine receptors in atrial myocytes leads to the downregulation of M2 receptors, presumably through repression of the gene.

Contraction of endothelial smooth muscle is controlled by $G\alpha_q$

On inhalation of noxious compounds or dangerous pathogens, the body responds by decreasing the volume of inspired air. Airway smooth muscle cells that surround the bronchiolar passages decrease respiratory volume by contracting (**Figure 9.80**). The smooth muscle cells respond to several different signaling molecules—the neurotransmitter acetylcholine, the cysteinyl leukotriene LTD4, the peptide endothelin, and the inflammatory signal histamine—each through a specific 7TM GPCR. These GPCRs are coupled to heterotrimeric G proteins containing $G\alpha_q$ subunits. Whereas $G\alpha_i$ and $G\alpha_s$ subunits affect adenylyl cyclase, $G\alpha_q$ activates the membrane-bound enzyme phospholipase Cβ, and no Gα subunits inhibit PLCβ. Recall from Chapter 8 that PLCβ cleaves the lipid phosphatidylinositol 4,5-bisphosphate to generate inositol triphosphate, which opens ion channels in calcium vesicles, and diacylglycerol, which remains associated with the membrane, where it activates protein kinase C.

$G\alpha_q$ affects the transcription of genes in several ways. Protein kinase C phosphorylates a wide range of signaling proteins that ultimately affect transcription, often associated with cell proliferation. Ca^{2+} ions exert both rapid and slow effects on muscle cells. The Ca^{2+}-induced contraction of muscle cells is a rapid effect that does not

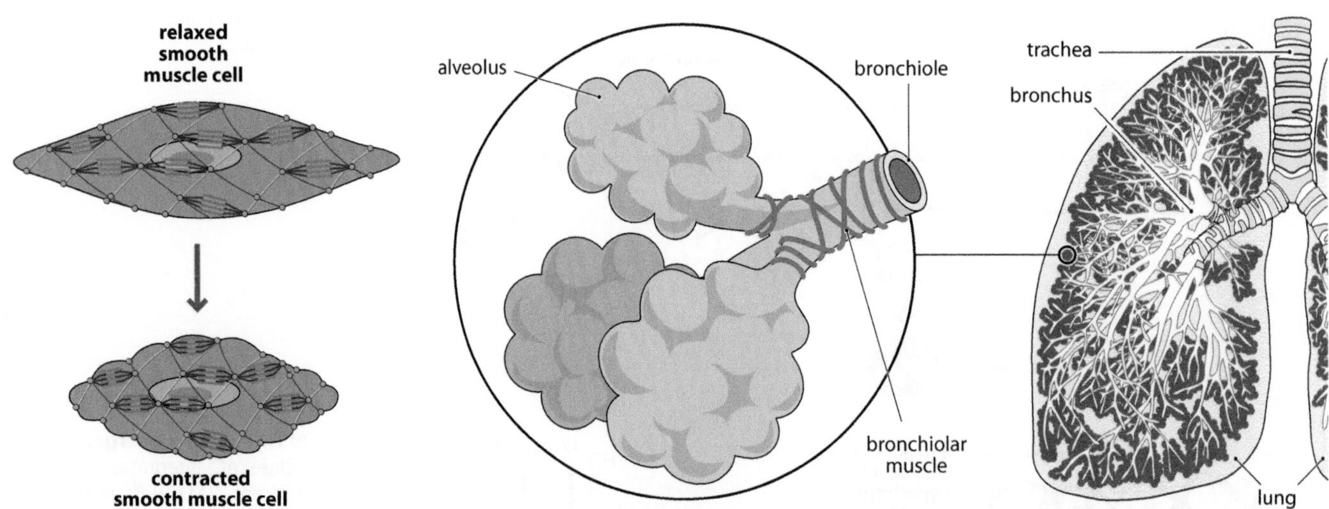

Figure 9.80 Response to noxious stimuli. Contraction of smooth muscle cells in the bronchiolar airway reduces the diameter of bronchioles and restricts airflow.

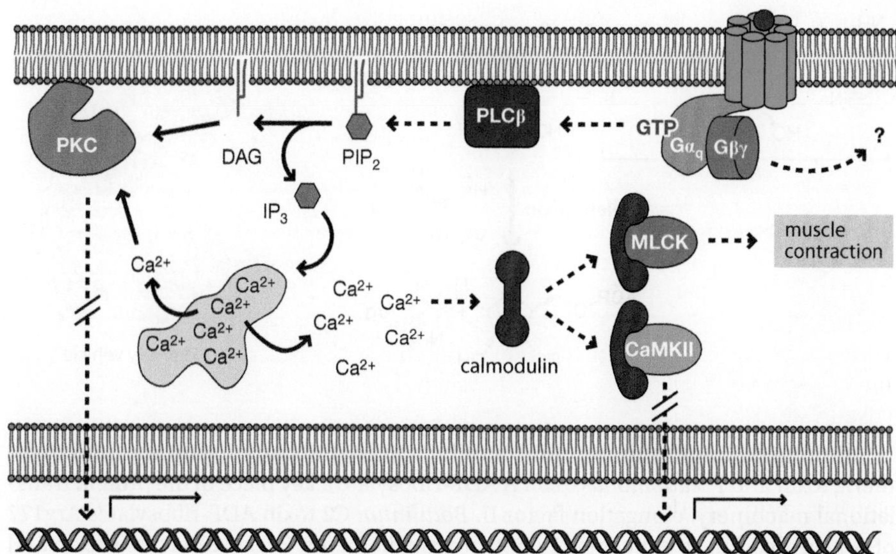

Figure 9.81 Signaling contraction. Histamine induces the contraction of smooth muscle cells in the airway by increasing the levels of Ca^{2+} ions in the cell. MLCK, myosin light chain kinase.

involve transcription. Ca^{2+} ions also affect transcription via protein kinases. Recall from our discussion of protein kinase C in Chapter 6 that Ca^{2+} ions promote the association of protein kinase C with diacylglycerol groups in the plasma membrane. Once localized at the plasma membrane, protein kinase C phosphorylates proteins that ultimately lead to cell growth, including various subdomains of G proteins, and other regulatory proteins and receptors. Furthermore, Ca^{2+} levels are monitored within the cell by a protein called calmodulin, which has binding sites for four Ca^{2+} ions. In the presence of Ca^{2+}, calmodulin undergoes a marked conformational change that allows the protein to activate Ca^{2+}/calmodulin-dependent protein kinase II (CaMKII). CaMKII phosphorylates cytoskeletal proteins and other proteins that ultimately affect the transcription of genes such as endothelial nitric oxide synthase (**Figure 9.81**).

Some bacterial toxins reprogram Gα subunits, with deadly results

Although humans live in harmony with many strains of bacteria, some bacteria are highly pathogenic and produce sophisticated protein toxins that are responsible for the most overt disease symptoms. Often these bacterial toxins are composed of two components: an A subunit that catalyzes a modification of an intracellular protein, and a B subunit that acts as a delivery vehicle for getting across the host membrane. The bacterium *Vibrio cholerae* and some strains of *Escherichia coli* produce a similar protein toxin (**Figure 9.82**). The B subunit of cholera toxin is a hexameric complex that binds multivalently to GM1 gangliosides on the surface of human cells. The A subunit of the cholera toxin is a glycosyltransferase that uses the high concentrations of NAD^+ in the cell as a substrate for modifying the G protein $G\alpha_s$ at Arg201; the by-product is the pyridine derivative nicotinamide. The modified Gα subunit cannot hydrolyze GTP; this leads to constitutive activation of adenylyl cyclase (**Figure 9.83**). In the epithelial cells of the gastrointestinal tract, constitutive $G\alpha_s$ activation leads to the opening of chloride channels. The massive export of Cl^- ions is accompanied by a loss of water through uncontrollable diarrhea. The latter incident ensures that *V. cholerae* re-enters the environment and often the drinking water, where it can find more victims to infect.

Pertussis toxin, produced by the bacterium responsible for whooping cough, ADP-ribosylates a specific cysteine residue, four residues from the C terminus in $G\alpha_i$, $G\alpha_o$, and transducin, the G protein coupled to rhodopsin. In the ciliary cells of the respiratory tract, ADP-ribosylation of this cysteine residue of G proteins prevents the exchange of GDP, leaving the G protein in an inactive form. Without the ability to inhibit **adenylate cyclase**, the levels of cAMP and the activity of cAMP-dependent protein kinase are maintained at uncontrollably high levels. Ultimately, the effect on airway epithelial cells leads to a characteristic coughing pattern and ultimately to a whooping sound when the victim finally inhales.

Figure 9.82 Death's dispensary. In 1854, Dr. John Snow founded the field of epidemiology by tracking incidents during a cholera outbreak to a single pump providing water in the Soho neighborhood of London. Removing the handle from the pump ended the epidemic. (George J. Pinwell, Death's Dispensary, 1866.)

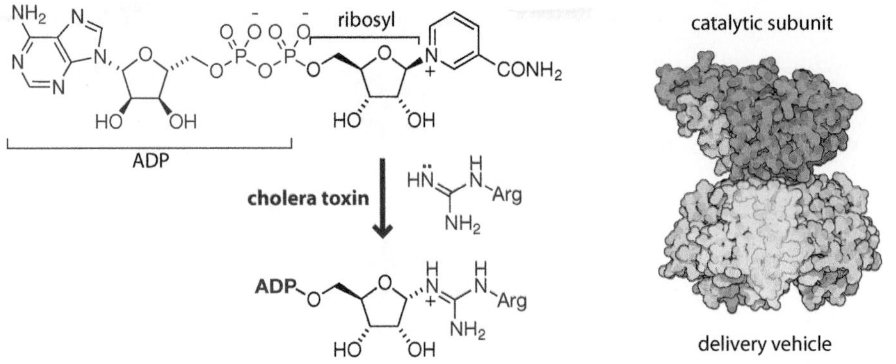

Figure 9.83 G protein assassin. Left: cholera toxin catalyzes the transfer of ADP-ribose from NAD^+ to a specific arginine side chain of G proteins. Right: the structure of cholera toxin with the B subunit rendered in blue and the catalytic A subunit rendered in red. (Right, courtesy of David S. Goodsell and RCSB Protein Data Bank.)

G proteins are not the sole substrates for bacterial ribosylating toxins. Both diphtheria toxin and *Pseudomonas* toxin A ADP-ribosylate a key piece of the protein translational machinery, elongation factor II. *Botulinum* C2 toxin ADP-ribosylates Arg177 of G-actin, leading to depolymerization of actin filaments. Because pertussis toxin and cholera toxin have orthogonal substrate selectivity, the toxins served for many years as diagnostic reagents for characterizing various G protein signaling pathways.

Problem 9.15

Forskolin induces the formation of new synapses between rat hippocampal neurons growing in cell culture. The effect is inhibited by the active ingredient in marijuana (Δ^9-tetrahydrocannabinol) which acts on another type of GPCR called the CB1 receptor. Which type of Gα subunit is coupled to the CB1 receptor: $G\alpha_i$, $G\alpha_q$, or $G\alpha_s$? How could you use an ADP-ribosylation toxin to confirm your answer?

Adenylyl cyclase and phospholipase Cβ are the most common mediators of 7TM GPCRs

7TM GPCRs have a significant role in the function of all human cells, regardless of whether the cell is an vascular smooth muscle cell, a cardiac muscle cell, a neuron, or a white blood cell. If you understand how 7TM GPCRs modulate the activity of adenylyl cyclase (via $G\alpha_s$ or $G\alpha_i$) or phospholipase Cβ (via $G\alpha_q$), you understand how most cells in the human body respond to small-molecule ligands. In a few cell types, Gα subunits modulate the phosphodiesterase PDE6 or a small GTPase called Rho, but we will not allow ourselves to be distracted by these fascinating exceptions. For many 7TM GPCRs, no one has yet determined what kind of Gα subtype transduces the receptor signal. Much less is known about the Gβγ subunits. It is tempting to assume that Gβγ subunits serve as buffers that sequester Gα subunits, but the ability of Gβγ to activate GIRK1 potassium ion channels suggests that they have much more interesting roles.

The remainder of our discussion of 7TM-GPCR signaling will focus on the 7TM receptors and their ligands. Because the ligands are often small molecules, they offer the best opportunity to exploit our knowledge of chemistry at the level of atoms and bonds. Before proceeding, consider that all 7TM GPCRs have the capacity for fast response times, wide dynamic range, and high sensitivity. In all cases, signals from GPCRs are usually being integrated with signals from other GPCRs and even other types of receptors.

Many pharmaceuticals act on 7TM GPCRs that respond to ligands derived from amino acids

Historically, 7TM GPCRs have been the most popular target for the development of pharmaceuticals. Even if nothing is known about the structure of the receptor, the endogenous ligand is usually a sensible lead for analog design. Drugs that target GPCRs do not have to enter cells; however, the pill-a-day paradigm for the development of

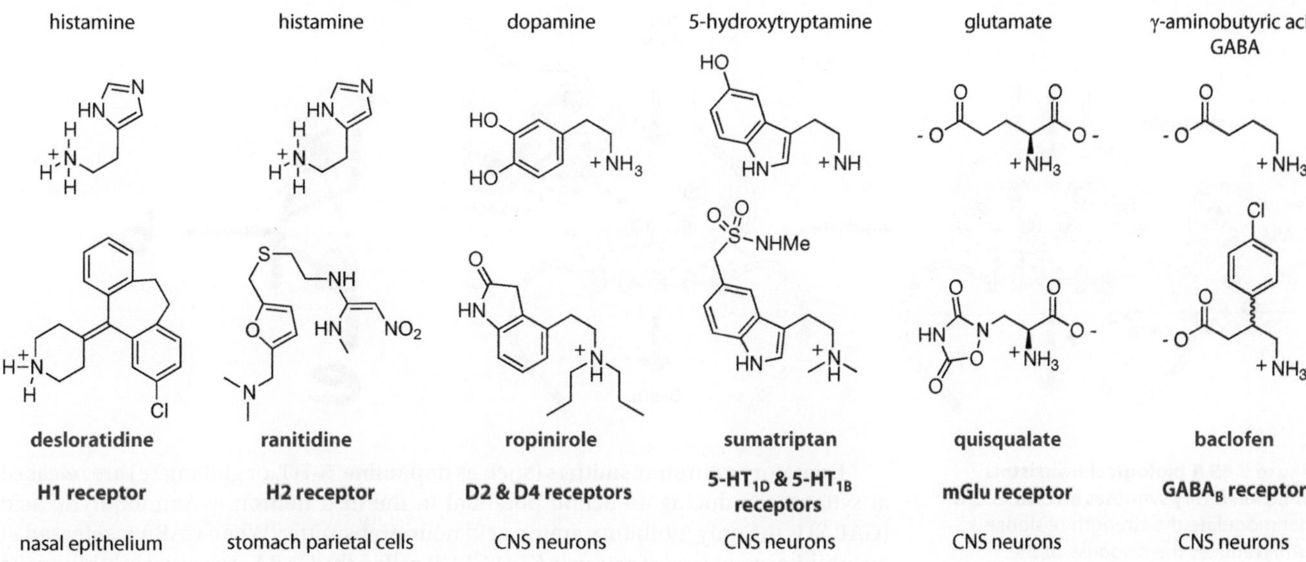

Figure 9.84 7TM GPCR ligands and their drug analogs. A variety of signaling molecules that act on 7TM GPCRs are derived from amino acids (upper). From the structures you cannot tell which drugs activate the GPCRs and which drugs inhibit them (lower).

blockbuster drugs requires one to develop a pharmaceutical that can enter the bloodstream by first crossing the gastrointestinal lumen. Such drugs require a low molecular mass and a balance between hydrophobic character and polarity. Neurotransmitters derived from amino acids are particularly amenable to analog synthesis. A large number of pharmaceuticals target neurotransmitter GPCRs.

Another significant challenge to targeting GPCRs is that one usually wants to target just a single subtype. For example, both stomach cells and cells in the nasal epithelium respond to histamine via GPCRs; however, the stomach cells express the H2 histamine receptor subtype, whereas the cells in the nasal epithelium express the H1 histamine receptor subtype. These two different GPCRs have different protein sequences and different ligand-binding sites. Thus, the H1 receptor antagonist desloratidine decreases the allergic response in the nasal epithelium (and other tissues) but not in the stomach. In contrast, the anti-ulcer drug ranitidine acts on H2 receptors in parietal cells in the stomach, preventing the release of acid. GPCRs also have an important role in neuronal signaling by neurotransmitters, including dopamine, 5-hydroxytryptamine, glutamate, and γ-aminobutyric acid (**Figure 9.84**).

The neurotransmitters dopamine, 5-hydroxytryptamine (5-HT; also known as serotonin), glutamate, and glycine also act on 7TM GPCRs within the central nervous system. Most of these receptors are located within synapses. Synapses usually transmit neuronal signals in one direction. One neuron releases neurotransmitters into the synaptic cleft and the other neuron has 7TM GPCRs for that neurotransmitter. For synapses in the brain, the signal is usually terminated by specialized proteins that pump the neurotransmitter back into the neuron that released it. Neuroscientists have developed some drugs that act on 7TM GPCRs, but the blockbuster drugs for depression and schizophrenia inhibit reuptake pumps rather than activating the receptors. If the reuptake pump is blocked, when a neuron releases a neurotransmitter the concentration of neurotransmitter within the synaptic cleft stays high for a longer period of time.

Structurally, dopamine is a phenethylamine, and many phenethylamine derivatives interfere with signaling at dopaminergic synapses (see Figure 9.84). Methamphetamine, sold legally as Adderall® but commonly known as "speed," is the archetype of these drugs, which include legal drugs such as Ritalin™ and drugs of abuse such as ecstasy. LSD (lysergic acid diethylamide) is the most notorious drug that interferes with 5-HT signaling, but many drugs for schizophrenia and migraines also target 5-HT receptors. A food additive, the amino acid glutamate (monosodium glutamate, or MSG) tastes like meat, and imparts a pleasant savory flavor, which coats the tongue with umami taste. Some individuals get headaches after ingesting large amounts of MSG, but because the compound is a common flavoring agent in foods, particularly Asian cuisine, restaurants sometimes advertise with signs that say "no MSG."

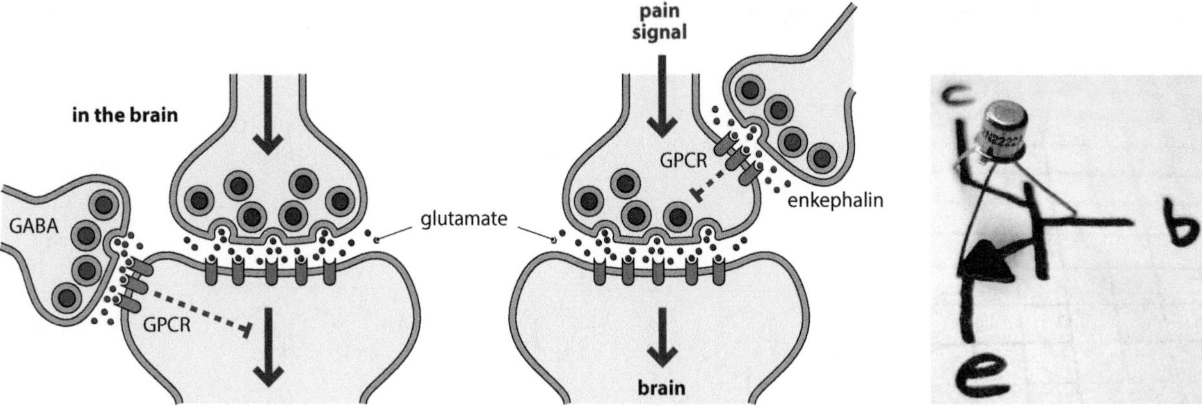

Figure 9.85 A biological transistor. Some inhibitory synapses involve GPCRs that modulate the strength of signals. GABA reduces the response of the postsynaptic neuron that receives the signal. The neuropeptide enkephalin reduces the response of the presynaptic neuron that sends the signal. These types of modulated neuronal connections resemble a transistor. (Right image, from Coltecnica, Ltda, 2012. http://www.coltecnica.com.)

Excitatory neurotransmitters (such as dopamine, 5-HT, or glutamate) are released at synapses, inducing an action potential in the next neuron. γ-Aminobutyric acid (GABA) is the only inhibitory amino acid neurotransmitter. When GABA is released at an inhibitory synapse, it acts on a 7TM GPCR called the $GABA_B$ receptor, inhibiting the effect of excitatory amino acids on that same neuron. By modulating neuronal signaling, GABA controls the flow of signals much like a transistor does (**Figure 9.85**). The GABA derivative baclofen also inhibits synaptic transmission and is used to diminish uncontrolled jerky movements (spasticity) (see Figure 9.84).

Opioids act on 7TM GPCRs that bind to neuropeptides

Morphine is the most enduring example of a small molecule that is used to activate signal transduction pathways. The Sumerians described the collection of opium, which contains morphine, in cuneiform script on tablets dating back to 5000 BC. The immediacy and potency of the effect made it easy to assign a direct cause-and-effect relationship between the latex of the poppy and analgesia (**Figure 9.86**). Morphine acts on μ-opioid receptors, GPCRs that normally respond to neuropeptides called enkephalins. The analgesic properties of morphine are due to μ-opioid receptors found in sensory neurons in the spinal cord that lead to pain centers in the brain. Like GABA, opioids inhibit synaptic transmission by other neurons, but whereas GABA acts on the neuron that receives the signal, opioids act on neurons that send the signal. Opioids do not act through pathways involving transcriptional control. Enkephalins

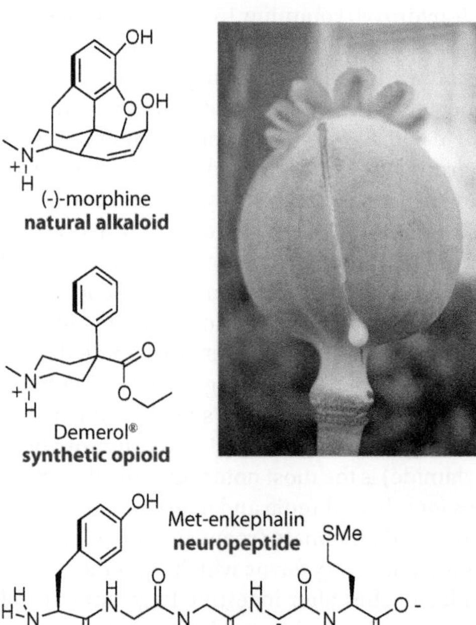

Figure 9.86 Ligands for the μ-opioid receptor. Opioids act upon the μ-opioid receptor. Morphine is present in the latex of the opium poppy, *Papaver somniferum*. Morphine and synthetic opioids such as Demerol mimic the effects of enkephalin. Met-enkephalin has a methionine at the C terminus; Leu-enkephalin has a leucine at the C terminus. (Image courtesy of Wikimedia.)

have a tyrosine residue at the N terminus of the peptide, and the aryl ring of nonpeptidic opioids mimics this tyrosine residue. A wide range of other peptides also act via 7TM GPCRs, for example secretin, vasoactive intestinal peptide, neuropeptide Y, and substance P.

Smell and taste involve 7TM GPCRs

All of the human senses involve signal transduction through 7TM GPCRs. The aroma of cooked foods is usually associated with a large number of compounds. For example, the enticing aroma of freshly roasted peanuts can be attributed to at least 27 volatile compounds (**Figure 9.87**). The most significant contributor is methanethiol, not because it is most abundant but because we have odorant receptors that are extremely sensitive to thiols. Each of the 5 million olfactory neurons in the nasal epithelium encodes just one of the 900 different odorant receptors, all of which are 7TM GPCRs. If you inhale a pure compound, such as camphor, it binds tightly to some receptors and weakly to others. The olfactory bulb processes this pattern of signals and your brain interprets it as a single odor (**Figure 9.88**). This kind of multiplexed receptor mapping allows you to distinguish more than 10,000 different odors with only 900 odorant receptors.

	conc. (μg kg^{-1})	odor threshold (μg kg^{-1})	odor activity value
—SH	113	0.06	1889
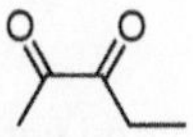	83	0.3	286
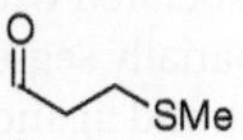 SMe	40	0.2	200
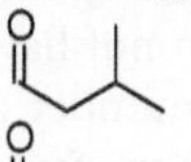	637	5.4	118
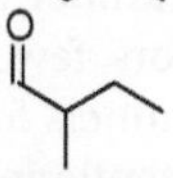	971	10.0	97
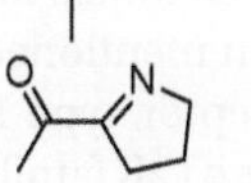 N	8.9	0.1	89
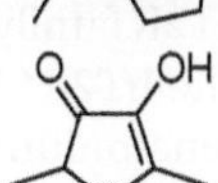 OH	1953	25	78

Figure 9.87 Peanuts! The relative intensities of odors in fresh roasted peanuts are due to both the concentration of the odorants and our sensitivity to those odorants.

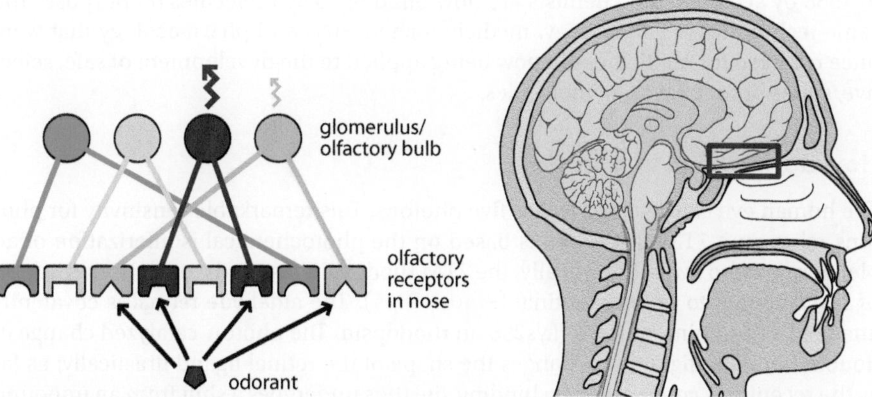

Figure 9.88 Odorant receptor mapping. You distinguish individual odors through spatial mapping of olfactory neurons in the nasal epithelium. Your brain decodes the pattern of neuronal signals as an odor. The olfactory region of the brain is highlighted. (Adapted from K. Touhara, *Microsc. Res. Tech.* 58:135–141, 2002. With permission from John Wiley & Sons.)

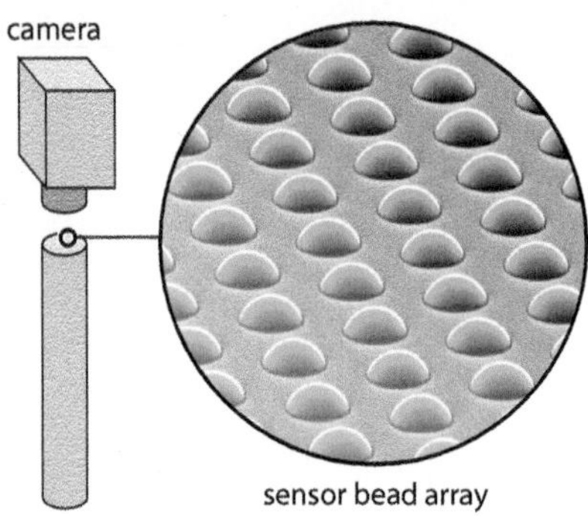

Figure 9.89 Robo-smell. An artificial nose can be made by spatially mapping the response of random sensor bead arrays on the ends of optical fiber bundles to various analytes, followed by computer decoding. (From M. Wadman, *Nature* 444:256, 2006. With permission from Macmillan Publishers Ltd.)

Multiplexed spatial mapping has been used to create an artificial nose capable of detecting a wide range of analytes. By randomly adsorbing a library of different beads, each coated with a different fluorescence-based sensor, on the ends of an optical fiber bundle, it is possible to individually monitor each of the beads in response to a pure analyte or a complex mixture of analytes. The beads can be derivatized with small molecules, peptides, oligonucleotides—anything. A computer can "memorize" the image of the pure analyte and then, using pattern recognition algorithms, look for the pattern produced by the analyte even when the array is exposed to a complex mixture. This technique works with proteins, DNA, and odor molecules (**Figure 9.89**).

Problem 9.16

There is no crystal structure for an odorant receptor, but modeling studies suggest that the I7 odorant receptor tightly binds aliphatic aldehydes through hydrogen bonding with the ammonium ion on K164. Suggest an alternative mode of binding that would explain the high affinity of the ligand.

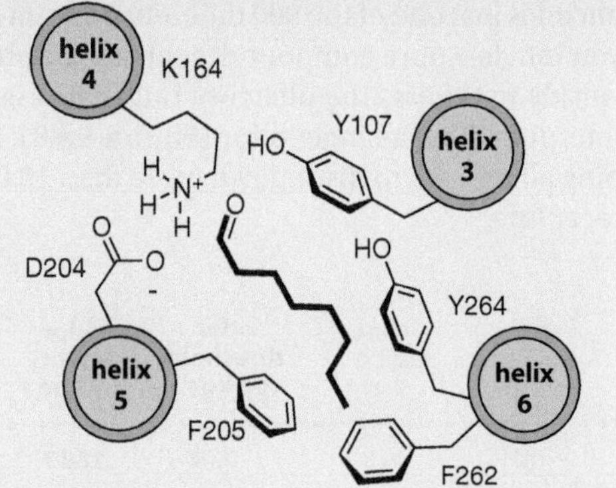

At one time, the ability to taste different flavors was thought to be associated with different regions of the tongue. However, the receptors for taste are not spatially segregated. The ability to distinguish tastes is based on a mixture of 7TM GPCRs and ligand-gated ion channels; however, the taste receptors in the human tongue do not have the level of diversity of the odorant receptors. In contrast with the large number of odorant receptors, humans have a relatively small repertoire of taste receptors, fewer than 50 7TM GPCRs for savory, sweet, and bitter tastes, and a few ion channels for salty and sour tastes. The sweet-tasting proteins thaumatin and monellin mentioned in Chapter 7 bind to the human taste receptor hT1R2 (human taste receptor, type 1, member 2). There are more than 25 members of the human taste receptor hT2R family that respond to bitter flavors. The ammonium ion denatonium binds to the hT2R44 receptor and is the bitterest compound known to man (**Figure 9.90**). Denatonium benzoate is added to alcohol to denature it, and is used to make household chemicals unpalatable to young children. The bitter properties of denatonium were discovered in 1958 by accident, but chemists are now building tasty molecules on purpose. The same techniques of cell biology, medicinal chemistry, and pharmacology that were once reserved for medicines are now being applied to the development of safe, selective, super-potent flavoring molecules.

How do you bind to a photon?

The human eye can sense as few as five photons. This remarkable sensitivity for photons relies on a 7TM GPCR and is based on the photochemical isomerization of an olefin from *cis* to *trans*. Specifically, the 7TM rhodopsin responds to the isomerization of 11-*cis*-retinal to 11-*trans*-retinal (**Figure 9.91**). The aldehyde retinal is covalently attached as an iminium ion to Lys296 on rhodopsin. The photon-catalyzed change in double-bond configuration changes the shape of the retinal ligand drastically; as far as the receptor is concerned, the binding site thus undergoes a shift from an unbound state to a bound state.

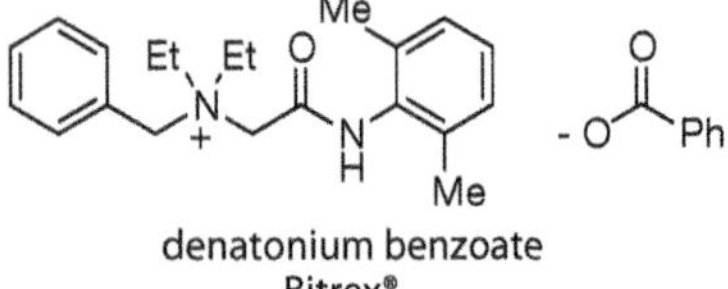

Figure 9.90 Super bitter. Denatonium benzoate is the most bitter known compound, binding potently to the hT2R44 receptor. When added to household chemicals, it deters curious toddlers from ingesting them.

Most studies on the molecular basis of vision have been conducted with bacteriorhodopsin from the bacterium *Halobacterium halobium*. Bacteriorhodopsin differs from mammalian rhodopsin in two important ways. In bacteriorhodopsin, retinal is

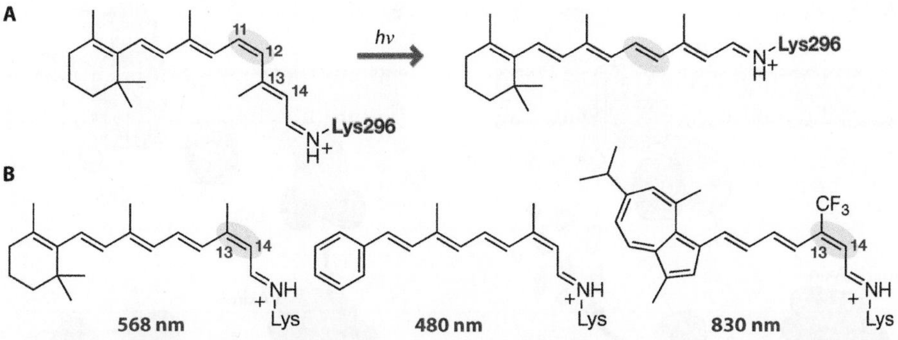

Figure 9.91 Retinal and its analogs. (A) In mammalian rhodopsin, the photoisomerization of 11-*cis*-retinal iminium ion into the 11-*trans* isomer drives human vision and leads to the same effect as binding a new ligand. (B) Chemical analogs of retinal in bacteriorhodopsin allow bacteria to "see" at different wavelengths.

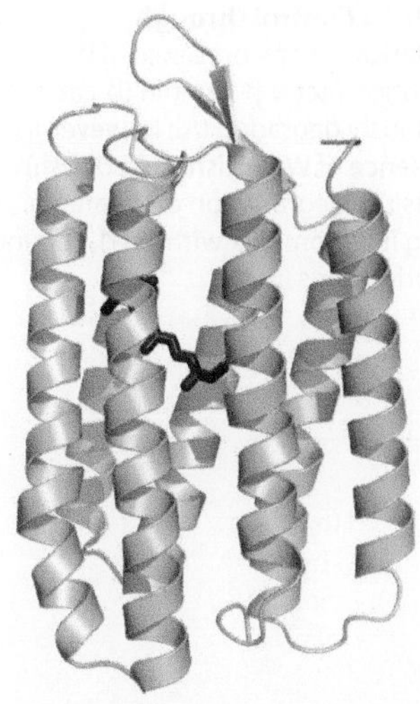

Figure 9.92 Bacteriorhodopsin. The 7TM protein bacteriorhodopsin (blue) opens a proton channel in response to light, causing the *cis–trans* isomerization of retinal (red). (PDB 1C3W)

bound as the 13-*cis* isomer instead of the 11-*cis* isomer of mammalian rhodopsin. Furthermore, bacteriorhodopsin is a light-driven proton pump, not a G protein-coupled receptor. For many years, the crystal structure of bacteriorhodopsin was the only precise model for a 7TM receptor that was available to chemical biologists (**Figure 9.92**). Chemical analogs of retinal have been synthesized and incorporated into bacteriorhodopsin, leading to receptors with altered sensitivity to light of different wavelengths (see Figure 9.91).

Problem 9.17

The azulene ring system of the retinoid analog below interacts favorably with cation-stabilizing residues in bacteriorhodopsin. The neutral resonance structure for azulene is nonaromatic as a result of cross-conjugation. Draw six charge-separated aromatic resonance structures for azulene, and determine which ring bears the positive charge.

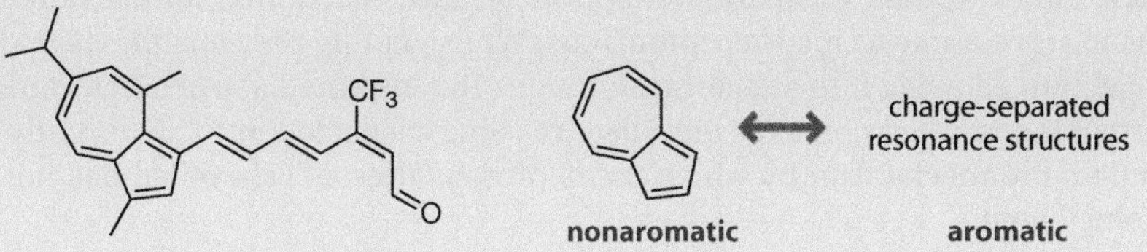

nonaromatic ⟷ charge-separated resonance structures aromatic

The decision between immortality and destiny involves the protein Wnt and the β-catenin pathway

In the classic children's story *The 500 Hats of Bartholomew Cubbins*, a young boy finds that each time he removes his feathered hat a new one appears in its place. After the 450th hat, each of the following hats becomes more ornate than the previous one. The process is exhausted with the 500th and most splendorous hat, leaving him bareheaded (**Figure 9.93**). Human stem cells follow a similar trend. They have unlimited capacity for replication, but once they have received the right set of signals they begin to differentiate. Differentiation pushes the cell toward a useful function at a significant cost—loss of immortality. The inevitable fate of all fully differentiated cells (for example cells found in the leg muscle, heart, kidney, and brain) is death, because such cells have a limited capacity for mitotic cell division. Thus, each mitotic event involves a weighty decision to produce identical daughter cells or differentiated daughter cells. The Wnt pathway has a key role in that choice. The choice of immortality is made by cancer cells; the choice to fulfill a mortal destiny is made by cells in developing embryos.

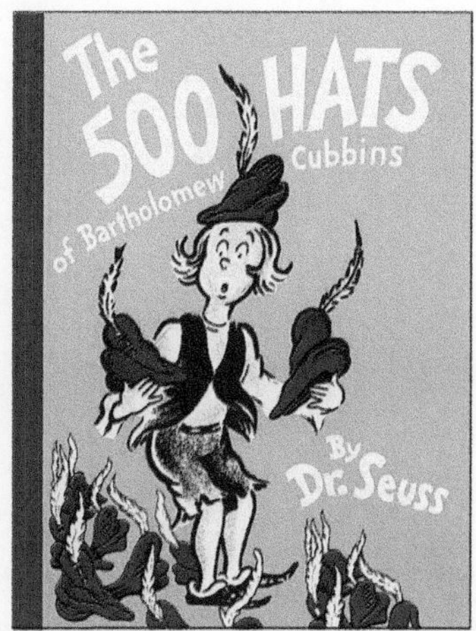

Figure 9.93 Regenerative hats. Developmental biologists hope to control proliferation and differentiation, much like Bartholomew Cubbins in the classic work by Dr. Seuss. (Courtesy of Dr. Seuss Enterprises, L.P. With permission from Random House, Inc.)

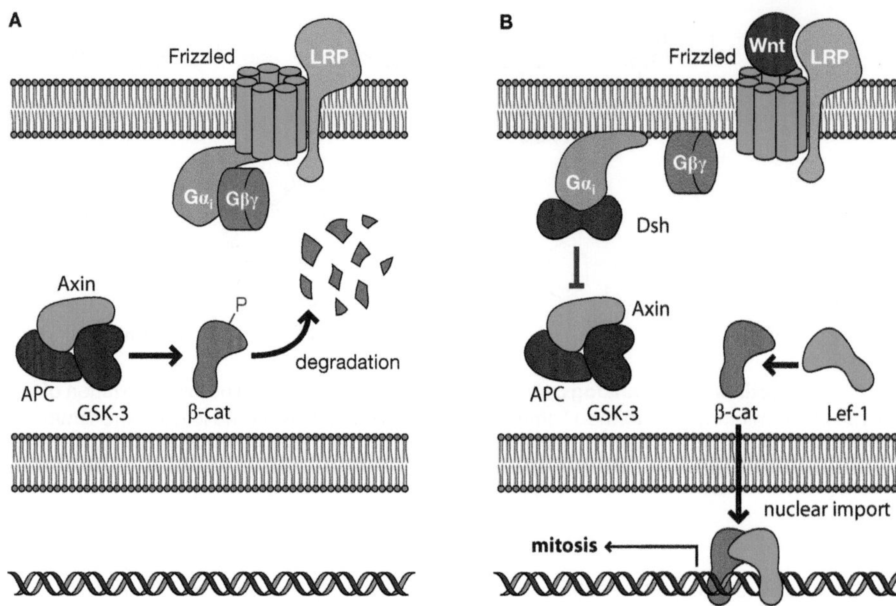

Figure 9.94 Control through degradation. In the presence of Wnt, the transcription factor β-catenin (β-cat) is continuously degraded (A); however, in the presence of Wnt, Dishevelled (Dsh) prevents the degradation of β-catenin, allowing it to complex with Lef-1, turning on mitotic genes (B).

To understand the effect of Wnt signaling on cells, it is essential to understand the basal metabolism of the transcription factor β-catenin. In the absence of Wnt, β-catenin is continuously marked for proteolytic degradation by phosphorylation. The phosphorylation of β-catenin is performed by a trimeric complex composed of axin, APC, and glycogen synthase kinase 3 (**Figure 9.94A**). In the presence of Wnt, the trimeric kinase complex is inhibited, and concentrations of β-catenin increase until it affects transcription by forming a complex with Lef-1 (Figure 9.94B). In colon cancers, the β-catenin–Lef-1 complex turns on the transcription of genes such as those encoding c-Myc and cyclin D1, both related to mitosis. How does Wnt inhibit the trimeric kinase complex? The signaling protein Wnt binds to a 7TM GPCR called Frizzled (instead of Wnt receptor) which causes the protein Dishevelled to inhibit the trimeric kinase complex. An additional protein, LRP, is required for activation, and seems to serve a role as a co-receptor. Thus, Wnt signaling prevents the degradation of β-catenin, allowing it to affect transcription. The intermediacy of a heterotrimeric G protein in the activation of Dishevelled was long suspected but only recently demonstrated; the mechanism by which the G protein affects Dishevelled has not been fully elucidated.

A seven-transmembrane receptor that controls development does not bind to an extracellular ligand

The cyclops was a frightening figure in Greek and Roman mythology, with just one eye positioned in the middle of its head. The inspiration for the cyclops may come from cyclopia, which is a specific form of a rare birth defect called holoprosencephaly. The first clues to the molecular origin of holoprosencephaly came from Idaho sheep ranchers who noted sheep born with cyclopia (**Figure 9.95**). The congenital deformities were later linked to the consumption of *Veratrum californicum*, also known as the California corn lily or California false hellebore. The magical properties of medicinal hellebores, easily confused with false hellebores, date back to ancient times. *Veratrum californicum* produces a group of steroidal alkaloids such as cyclopamine, with potent teratogenic (causing birth defects) activity. Cyclopamine does not kill cells indiscriminately; it interferes with a specific GPCR that controls cell differentiation during fetal development. Steroidal ligands have an important role in this cell differentiation pathway.

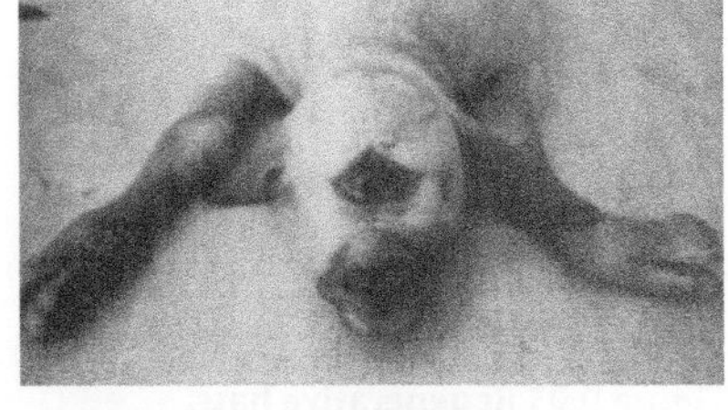

Figure 9.95 Cyclops. This lamb was born to a ewe that had fed on California false hellebore, which contains the alkaloid cyclopamine. (From R.F. Keeler, *Lipids* 13:708–715, 1978. With permission from Springer.)

In *Drosophila*, mutations to a gene named *Hedgehog* result in larvae with a profusion of spiky tentacles. At a molecular level, *Veratrum* alkaloids such as cyclopamine exert teratogenic effects similar to the *Hedgehog* mutations in fruit flies. The Hedgehog

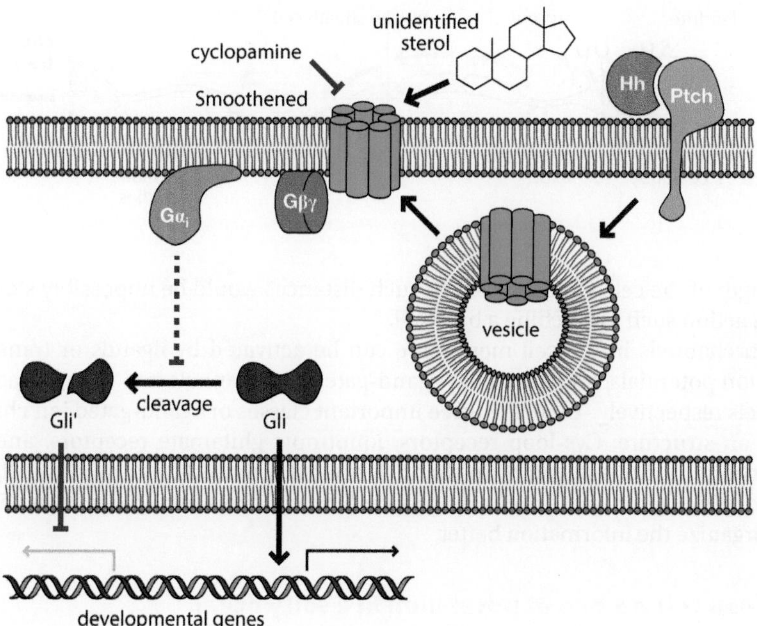

Figure 9.96 Cyclopamine repression of genes required for normal development. Gli can affect two different sets of genes required for normal development. Full-length Gli activates one set of genes. Truncated Gli' represses another set of genes. When Smoothened is activated, one set of developmental genes is activated and another is not repressed. Hh, Hedgehog; Ptch, Patched.

protein is modified post-translationally with a cholesterol molecule, and then packaged into lipoprotein complexes for transport around the developing organism. When Hedgehog binds to the receptor Patched, it allows the GPCR Smoothened to move from endocytic vesicles to the cell membrane (**Figure 9.96**). Once in the cell membrane, Smoothened acts as a receptor for some kind of steroidal alcohol (sterol) that has not yet been identified; cyclopamine binds to Smoothened, which ultimately prevents the cleavage of a transcription factor called Gli. At the time of writing of this book, many details of this pathway are incomplete. Full-length Gli activates the transcription of genes for differentiation. In the absence of Hedgehog, Gli is cleaved and the truncated form of Gli acts as a repressor of gene transcription.

There are vague similarities between the β-catenin pathway and the Hedgehog signaling pathway. Activation of the 7TM receptors Frizzled and Smoothened requires proteins other than heterotrimeric G proteins. Both pathways involve the constitutive proteolysis of a transcription factor in the absence of ligand. Both pathways affect the differentiation of cells.

9.7 ION CHANNEL RECEPTORS

Ion channel receptors provide an ultra-fast response to stimuli

Human cells maintain an imbalance of ion concentrations between the inside and the outside of the cell. At a moment's notice, ion channels are poised to open in rapid succession, allowing a flood of ions to permeate the cytoplasm. This type of ionic flooding is the fastest mechanism for communicating an event to all parts of the cell. Neuronal signaling is the most significant physiological event that involves ion channel receptors. As you read the words on this page, hundreds of millions of ion channels are opening and closing. If you choose to memorize the ion concentrations in **Table 9.4** to the right, you will probably have to look at the data several times ... ruminate on it and savor it. Repetitive neuronal signaling will eventually change the neurons in those pathways at a transcriptional level.

Neurons come in many shapes and sizes, but a common morphological theme is that they have a cell body that receives chemical signals, and long extensions called axons that transmit the cellular signal toward a synaptic junction (**Figure 9.97**). The axons in the nerves in your arms and legs can be up to 1 meter in length. However, in communicating the signal from one end of the cell to the other, no molecules travel

Table 9.4 Approximate ion concentrations inside and outside human cells.

Ion	Concentration (mM)	
	Inside	Outside
Na^+	10	145
K^+	140	5
Cl^-	10	110
HCO_3^-	10	25
$H_2PO_4^-$	–	1
Mg^{2+}	0.5	2
Ca^{2+}	0	2

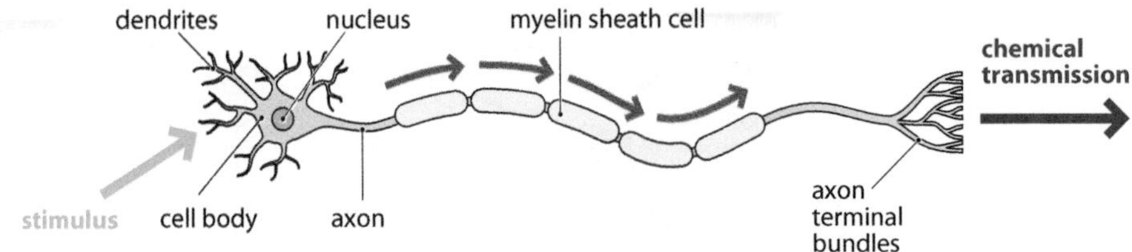

Figure 9.97 Anatomy of a neuron. Dendrites at the cell body receive signals that are propagated along the axons.

the length of the cell. Diffusion across such distances would be impossibly slow for a simple action such as catching a baseball.

Ion channels in the cell membrane can be activated by ligands or transmembrane ion potentials, and are called ligand-gated ion channels and voltage-gated ion channels respectively. There are three important classes of ligand-gated ion channels based on structure: Cys-loop receptors, ionotropic glutamate receptors, and ATP-gated channels. Voltage-gated ion channels are not activated by extracellular ligands, but grouping ligand-gated and voltage-gated ion channels into one section will help us to organize the information better.

A human cell is a bag of potassium in a salty ocean

Human serum, the liquid that surrounds the cells in whole blood, is a reasonable approximation to liquid that surrounds the other cells in the human body (see Table 9.4). Serum has many of the same ions as seawater, with about one-third of the salinity (sodium chloride concentration). In contrast, the cytoplasm of human cells has very little sodium, very little chloride, and a relatively high concentration of potassium. Experiments meant to mimic the extracellular environment are performed in sodium chloride buffers such as phosphate-buffered saline (PBS). Experiments meant to mimic intracellular conditions should be performed in low-saline, potassium-rich buffers.

To maintain this ionic imbalance between the inside of the cell and the outside of the cell, energy-driven pumps pull K^+ ions in, and drive Na^+ and Cl^- ions out (**Figure 9.98**). Hydrolysis of ATP to ADP provides the energy for Na^+/K^+ pumps. Recall that contraction of muscle cells involves a burst of cytosolic Ca^{2+} ions. Pumps also actively pump Ca^{2+} ions out of the cell and/or into cellular compartments such as the endoplasmic reticulum (or the sarcoplasmic reticulum of muscle cells). In these compartments, the concentration of Ca^{2+} ions (about 0.1 mM) is 1000 times the concentration in the cytosol. Various stimuli can open calcium channels in calcium vesicles and in the plasma membrane, leading to a rapid influx of Ca^{2+} that triggers the release of neurotransmitters, for example.

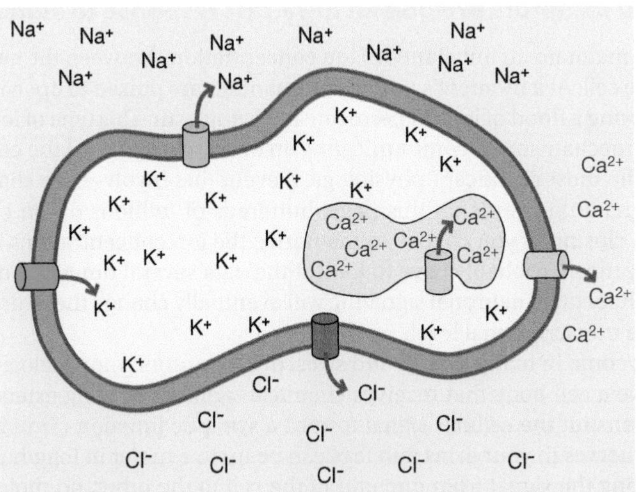

Figure 9.98 Ions in, ions out. Pumps maintain the concentration of anions and cations inside the cell. The imbalance of ions creates a tension that can be rapidly relieved through the opening of membrane ion channels (not shown).

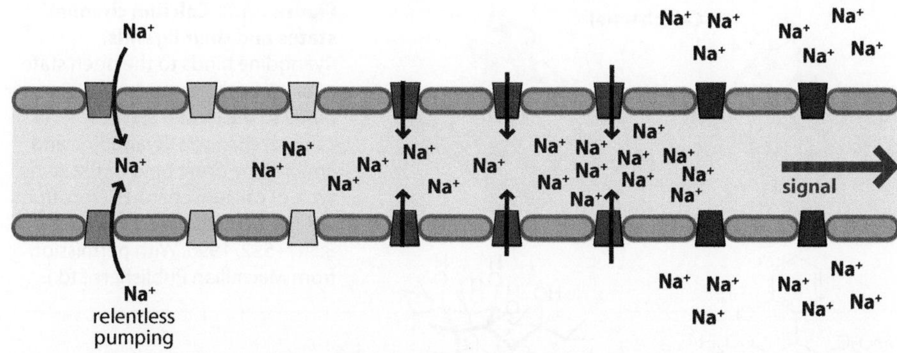

Figure 9.99 **Firing the neuron.** Sodium ion channels open in response to intracellular Na^+ ions, but only for a brief period. Energy-driven pumps (left) along the length of the axon then expel the Na^+ ions from the axon.

We measure the differences in ion concentration using the same units used by electricians to measure an imbalance in electrons—volts—but try not to confuse the chemical potential created by an imbalance of cations with the electrician's concept of an electromagnetic potential.

Voltage-gated ion channels are activated by transmembrane differences in ion concentrations

How does a signal travel along an axon? Energy-driven export pumps relentlessly pump Na^+ ions out of the axon. As a result of this concentration difference, Na^+ ions yearn to enter the axon from outside the cell. At intervals along the axon, voltage-gated sodium channels are poised to open and allow a rush of Na^+ ions into the axon. Ironically, the opening of these sodium channels is initiated by a high concentration of intracellular Na^+. Thus, after one channel allows Na^+ ions to enter, neighboring sodium channels follow suit. The opening of these voltage-gated sodium channels is temporary, and they rapidly close. Once the voltage-gated sodium channels close, the intracellular concentration of Na^+ begins to decrease, as the energy-driven Na^+ pumps continue their relentless expulsion of Na^+ ions out of the cell. The region of high intracellular Na^+ concentration travels along the axon at impressive rates, up to 1 m s^{-1}, even though each individual Na^+ ion travels only a microscopic distance into the cell and then back out again (**Figure 9.99**).

At the end of the axon, voltage-gated calcium channels allow a wave of Ca^{2+} ions to enter, causing the release of neurotransmitters into a slender gap between neurons called the synapse. Grayanotoxin, found in rhododendrons, is an inhibitor of voltage-gated sodium channels, keeping them in an open state. Significant amounts of grayanotoxin can accumulate in the honey of bees that feed on rhododendron nectar (**Figure 9.100**). Eating honey contaminated with grayanotoxin leads to a variety of symptoms, including altered mental states. Xenophon, an Athenian author and military commander, first described the phenomenon of "mad honey poisoning" in 401 BC.

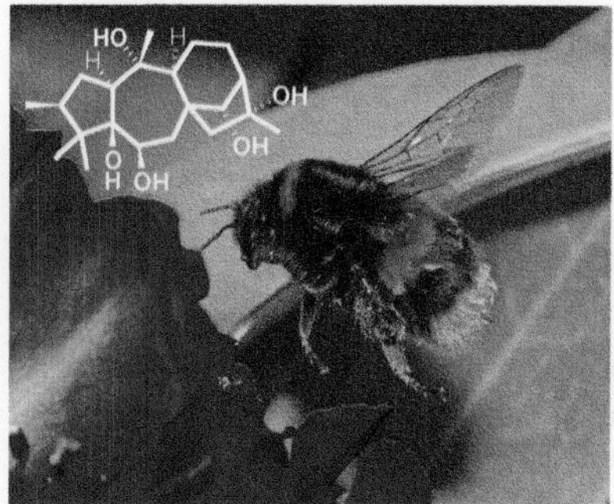

Figure 9.100 **Dangerous honey.** Honey produced by bees feeding on rhododendrons can become contaminated with grayanotoxin 1, an inhibitor of voltage-gated sodium channels in humans. (Courtesy of W.P. Coombs, Jr., Western New England University.)

Problem 9.18

Fura-2 is non-fluorescent in the absence of Ca^{2+} because a key nitrogen lone pair quenches the fluorescence through resonance. However, fura-2 tightly grips Ca^{2+} ions as an octadentate ligand, and in so doing becomes highly fluorescent. Suggest a plausible structure for the fura-2•Ca^{2+} complex.

resonance quenches fluorescence

fura-2

Voltage-gated ion channels control processes other than neuronal firing. In muscle cells, voltage-gated calcium channels on the membranes of storage vesicles contribute to the burst of Ca^{2+} initiated by inositol triphosphate. These voltage-gated ion channels are usually classified on the basis of the types of non-native small molecules

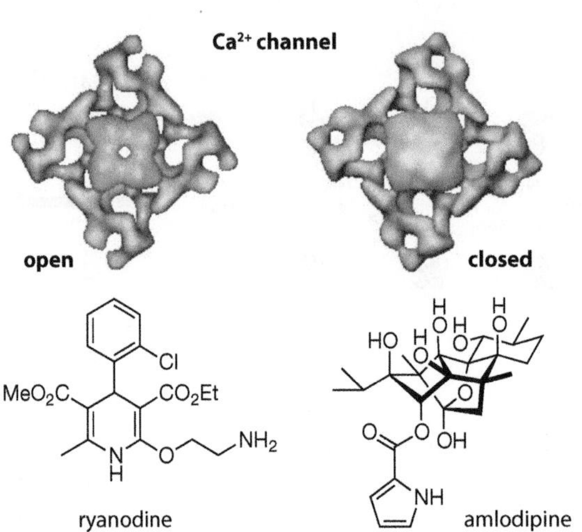

Figure 9.101 Calcium channel states and their ligands. Ryanodine binds to the open state of calcium channels. Amlodipine binds to the closed state of calcium channels. Ryanodine and amlodipine don't bind to the same type of calcium channel. (Top, from E.V. Orlova et al., *Nat. Struct. Biol.* 3:547–552, 1996. With permission from Macmillan Publishers Ltd.)

that can be used by pharmacologists to open them. Ryanodine receptors, are ion channels which allow Ca^{2+} to escape from intracellular storage vesicles, and are named for their response to the plant alkaloid ryanodine (**Figure 9.101**). Under normal circumstances, the ryanodine receptors open in response to Ca^{2+}. Dihydropyridine receptors are so named because they respond to synthetic dihydropyridine drugs like the blockbuster amlodipine. Channels with similar functions account for the monogamous relationship between egg and sperm. Once a spermatozoon penetrates the exterior of the ovum, IP_3 initiates a wave of Ca^{2+} ions that, with blinding speed, makes the egg impenetrable to all other suitors.

Voltage-gated potassium channels also have a role in cell signaling. For example, immunological T cells have voltage-gated potassium channels. Forty out of the 78 human genes for potassium channels are voltage-gated. A variety of protein toxins, from scorpions, sea anemones, snakes, and cone snails, block voltage-gated potassium channels, often by plugging the K^+ exit site by inserting a cationic lysine side chain into the narrow orifice of the channel (**Figure 9.102**). One such toxin, charybdotoxin from the death stalker scorpion, blocks Ca^{2+}-activated potassium channels in aortic smooth muscle. Ca^{2+}-activated potassium channels are, in a functional sense, a hybrid between voltage-gated ion channels and ligand-gated ion channels because the interactions with Ca^{2+} are highly specific (see Figure 9.102). Another toxin that targets the Ca^{2+}-activated potassium channel is lolitrem B, a small molecule from the fungus *Epichloë*. Lolitrem B is a potent inhibitor (IC_{50} = 4 nM) of the Ca^{2+}-activated potassium channel that is present on the presynaptic membrane of Purkinje neurons.

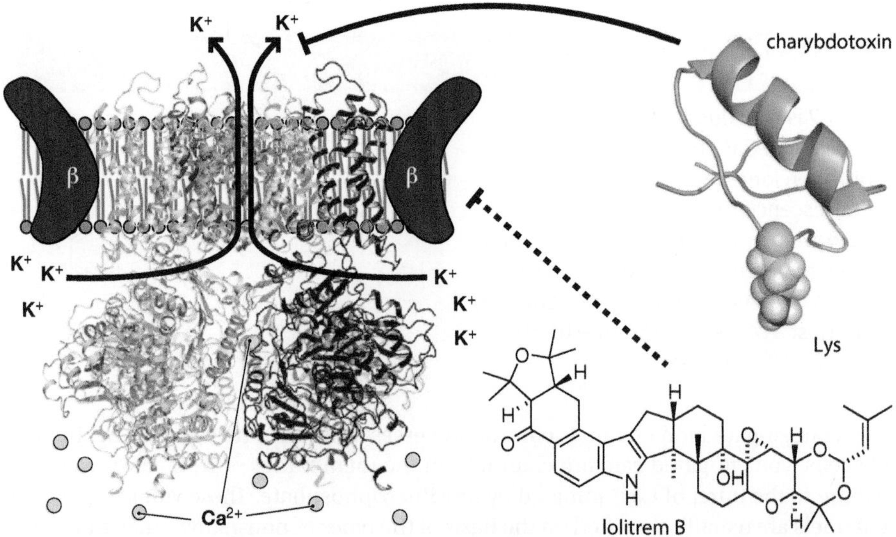

Figure 9.102 Control over the calcium-activated potassium channel. The binding of Ca^{2+} ions to the calcium-activated potassium channel allows K^+ ions to leak out of cells. The effect of Ca^{2+} is modulated by β subunits. The channels are inhibited by both protein toxins, such as charybdotoxin (PDB 2A9H), and small molecules, such as lolitrem B. Lolitrem B acts on the modulatory β subunits. (Adapted from P. Yuan et al., *Science* 329:182–186, 2010.)

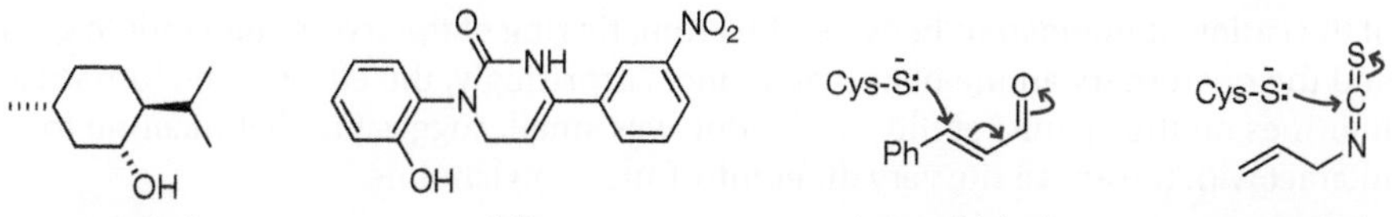

menthol icilin cinnamaldehyde mustard oil

Figure 9.103 Compounds that activate the TRPA1 cation channel. Cinnamaldehyde and mustard oil are electrophiles, which covalently modify TRP channels for long-lasting effects.

Hapless sheep that feed on ryegrass infected with *Epichloë* develop a neurological condition known as ryegrass staggers, characterized by tremors and a lack of coordination. The action of lolitrem B is dependent on a regulatory protein that modulates the activity of Ca^{2+}-activated potassium channels.

Transient receptor potential (TRP) cation channels are stimulated by cold, and directly activate neurons in pain pathways. Menthol, for example, creates a cooling sensation by activating TRP channels. The synthetic compound icilin is 200-fold more potent in producing these effects. Surprisingly, electrophiles such as enones, isothiocyanates, α-iodoacetamide, and thiosulfonates act at the TRP channel TRPA1. These compounds covalently modify cysteines on the channel, leading to persistent activation (**Figure 9.103**).

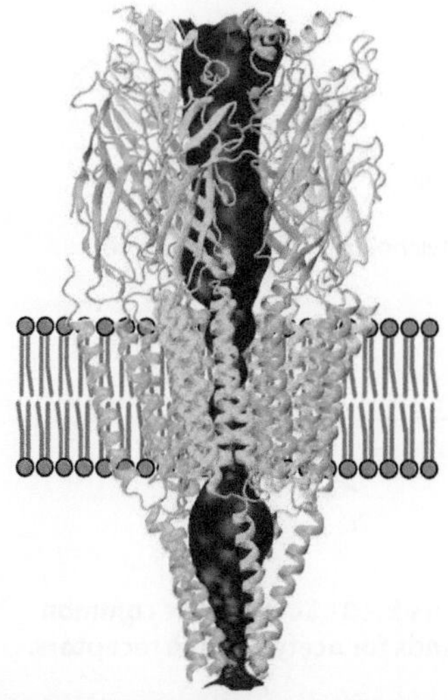

Figure 9.104 Pentameric nicotinic acetylcholine receptor. The five subunits of the nicotinic acetylcholine receptor are rendered in cyan and green. The cation channel is rendered in blue. (PDB 2BG9) (Adapted from S.M. Sine and A.G. Engel, *Nature* 440:448–455, 2006. With permission from Macmillan Publishers Ltd.)

Pentameric Cys-loop receptors are gated by neurotransmitters

Cys-loop receptors are ligand-gated ion channels that have a common structure based on five protein subunits. These veritable skyscrapers extend far above and below the cell membrane (**Figure 9.104**). The excitatory neurotransmitter 5-HT acts on both 7TM GPCRs and on a ligand-gated cation channel called the 5-HT3 receptor (**Figure 9.105**). GABA modulates synaptic transmission by acting on the 7TM GPCR called the $GABA_B$ receptor. However, in other neurons it acts on a ligand-gated chloride channel called the $GABA_A$ receptor.

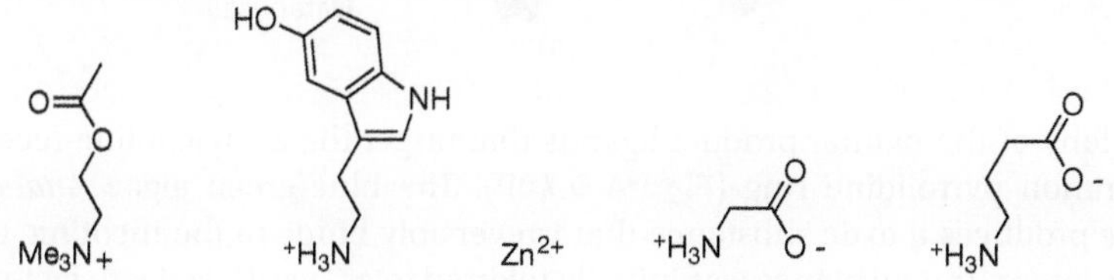

acetylcholine 5-HT zinc glycine GABA

Figure 9.105 Examples of ligands that act at Cys-loop receptor ion channels.

The binding site for the nicotinic acetylcholine receptor has no anionic side chains. Instead, acetylcholine binds to an active site composed of three tyrosine side chains and two tryptophan side chains. One, and only one, of these aromatic groups is important for the ability of acetylcholine to open the channel. The importance of a cation–π interaction between the quaternary ammonium ion and tryptophan was demonstrated through the creation and assay of a series of fluorotryptophan mutants by loading tryptophan analogs onto suppressor tRNAs (Chapter 4). By this method a series of fluorinated tryptophans were synthesized and attached to mimics of tRNA that recognize the rare amber stop codon, UAG. The amber stop codon TAG was easily introduced into the muscle-type nicotinic acetylcholine receptor gene through oligonucleotide-directed mutagenesis. Fluorine substituents on the key tryptophan residue were found to strongly decrease the ability of acetylcholine to open the channel (**Figure 9.106**). Such electron-withdrawing functionalities can decrease the strength

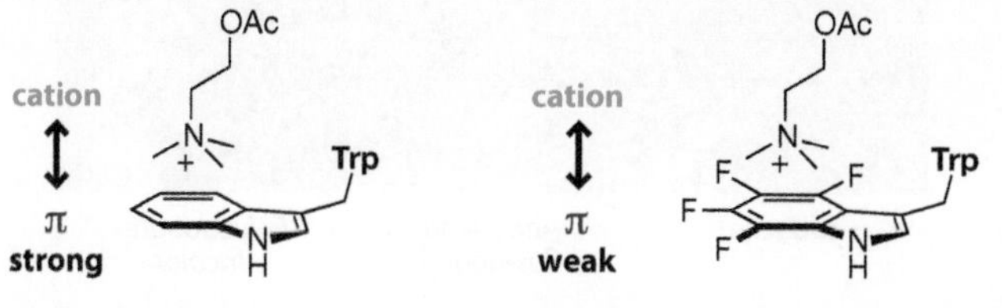

cation ↕ π strong | cation ↕ π weak

Figure 9.106 Exploring the biological implications of a cation–π interaction with synthetic chemistry and mutagenesis. A cation–π interaction is important in the binding of acetylcholine, as shown by much weaker binding after replacement in the nicotinic acetylcholine receptor with a fluorinated tryptophan. The fluorinated tryptophan substitution, however, has only a weak effect on nicotine binding.

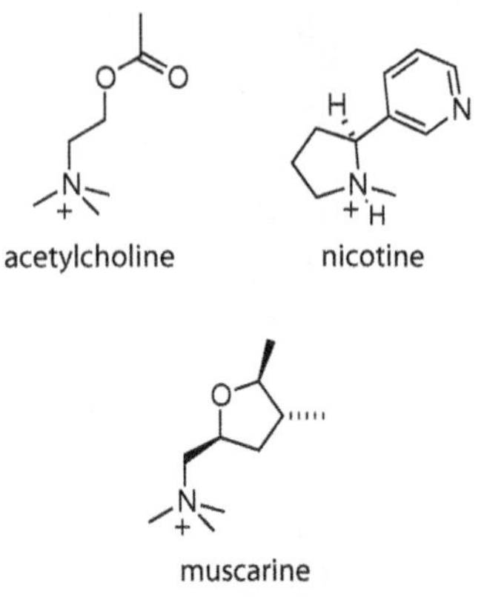

acetylcholine nicotine

muscarine

Figure 9.107 Some of the common ligands for acetylcholine receptors.

of the cation–π interaction between the aromatic ring of the tryptophan indole group and the quaternary amine of acetylcholine. Surprisingly, the effect of the introduced fluorines on the agonist ability of nicotine was small, suggesting that nicotine might interact with the active site very differently from acetylcholine.

The nicotinic acetylcholine receptor is a popular target for toxins

Before the ready availability of sequencing methods, pharmacologists used non-natural ligand susceptibility to distinguish between various subtypes of receptors for neurotransmitters (analogous to the naming of the ryanodine-sensitive voltage-gated ion channels described above). Acetylcholine receptors can be distinguished by their sensitivity to nicotine or muscarine (**Figure 9.107**). Nicotinic acetylcholine receptors are found between neurons in the autonomic ganglia and at the neuromuscular junction; such receptors control primitive responses such as "fight or flight" and "rest and digest." The nicotinic acetylcholine receptor has two binding sites for the neurotransmitter acetylcholine and a wide range of natural toxins target these binding sites. For example, curare, the poison used by Amazonian hunters to coat the tips of hunting darts, targets the acetylcholine-binding site. A variety of animal venoms contain proteins that jam the nicotinic acetylcholine receptor into its open conformation (**Figure 9.108**): erabutoxin (sea snake), α-bungarotoxin (krait, *Bungarus caeruleus*), α-conotoxin (*Conus* snails), and α-cobratoxin (cobra, *Naja naja kaouthia).*

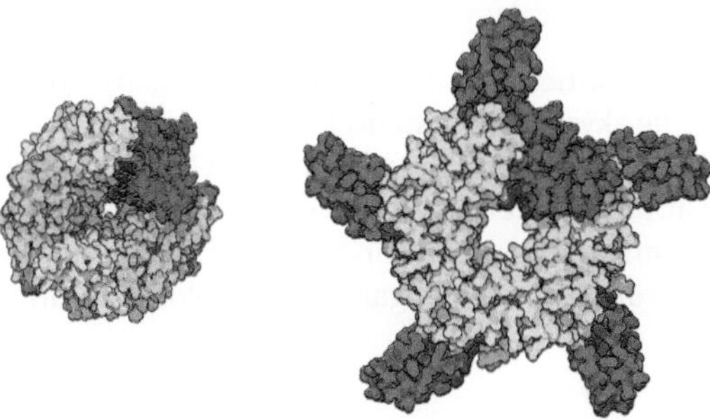

Figure 9.108 Open wide. These structures of the nicotinic acetylcholine receptor depict the closed form (left) with acetylcholine-binding sites in red, and the open form (right) bound to five molecules of α-cobratoxin. (Courtesy of David S. Goodsell and the RCSB Protein Database.)

Many of the natural product ligands that target the acetylcholine receptor have a common pyrrolidine ring (**Figure 9.109**). The blue-green algae *Anabaena flos-aquae* produces a toxic substance that irreversibly binds to the nicotinic acetylcholine receptor. The substance was initially referred to as "very fast death factor" but was later named anatoxin A. Surprisingly, other toxins that act at nicotinic receptors have potentially beneficial applications. For example, the compound epibatidine, isolated from the Ecuadoran frog *Epipedobates tricolor*, is a potent analgesic, 200 times more powerful than morphine in alleviating pain. Sadly, when *E. tricolor* is raised in the laboratory, the frogs no longer produce epibatidine. The production of epibatidine is a consequence of the frog's diet or a symbiotic relationship with a microorganism unique to its native habitat.

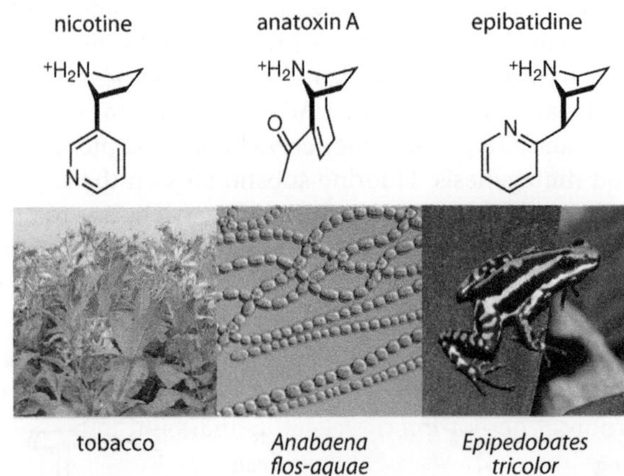

Figure 9.109 Nicotine and structural variants. Natural products that can target the nicotinic acetylcholine receptor share common structural features. The source of each compound is shown below its structure. (Bottom left, courtesy of Joachim Mullerchen, Wikimedia; middle, courtesy of the Cultural Collection of Autotrophic Organisms; right, courtesy of H. Krisp, Wikimedia.)

Tetrameric glutamate receptors are defined by their specificity for glutamate analogs

You may recall that glutamic acid is a ligand for some 7TM GPCRs, imparting umami flavor to food. More commonly, glutamic acid is a ligand for ion channel receptors. Twenty different subtypes of glutamate ion channel receptors are known, and each has four protein subunits. The major classes of glutamate receptors are categorized on the basis of their susceptibility to the glutamate analogs *N*-methyl-D-aspartate (NMDA), α-amino-3-hydroxy-5-methyl-4-isoxazole propionic acid (AMPA), and kainate (**Figure 9.110**). The glutamate receptors include AMPA, kainate, and NMDA receptors. Such receptors all respond to the excitatory amino acids glutamic acid and aspartic acid. The popular flavoring agent MSG (monosodium glutamate) is actually a pure, crystalline form of the neurotransmitter glutamate.

Most of the computing power of the brain is controlled by excitatory amino acid receptor pathways. Activation of ligand-gated ion channels does not always lead to changes in transcription. Often, such activation merely induces a fast response like neuronal depolarization. However, many processes that result in transcriptional activation, such as neuronal changes associated with memory, are initiated by ligand-gated ion channels. Repeated stimulation of NMDA receptors (by glutamate) in hippocampal neurons leads to Ca^{2+} influx. In these neurons, Ca^{2+} ions lead to the activation of protein kinase C (PKC) and CaMKII. These kinases affect neuronal function in two ways. First, PKC phosphorylates the AMPA receptors, making them more sensitive to glutamate. Second, the kinases PKC and CaMKII eventually activate the ERK pathway in these neurons, leading to changes in gene expression.

Kainic acid is named after the red algae *Digenea*, referred to as kaininso in Japan. *Digenea* has been used for centuries as an antihelminthic, or treatment for parasitic worms. Recently, a structurally related natural product, domoic acid, was discovered as the active agent in amnesic shellfish poisoning (ASP) (see Figure 9.110). Domoic acid is produced not by shellfish but by the diatom *Pseudonitzschia*, which can be accumulated by shellfish. Domoic acid activates AMPA and kainate receptors, leading to persistent Ca^{2+} influx. In serious cases, domoic acid poisoning results in the permanent loss of short-term memory.

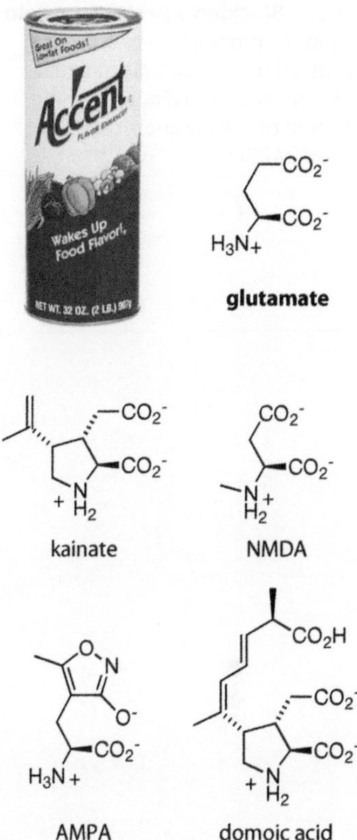

Figure 9.110 Glutamate and structural variants. Glutamate is a popular flavoring agent. A variety of nonendogenous ligands are selective for different subtypes of glutamate ion channel receptors. (Top, courtesy of B&G Foods, Inc.)

Problem 9.19

The following "caged" glutamate derivative can be induced to release glutamate on activation with light. Suggest a plausible arrow-pushing mechanism for the reaction.

9.8 TRIMERIC DEATH RECEPTORS

Tumor necrosis factor binding to TNF receptors triggers diverse, cell-dependent responses

Recall from Chapter 3 that rapidly dividing cells are poised to self-destruct if they fail to pass through various mitotic checkpoints. Such apoptotic pathways are mediated by the tetrameric transcription factor p53. However, even non-mitotic cells can be induced to self-destruct on receipt of the correct signals. Such signals are mainly mediated by a group of homotrimeric receptors that bind to homotrimeric ligands, either expressed on cell surfaces or in soluble forms. This three-fold symmetry of

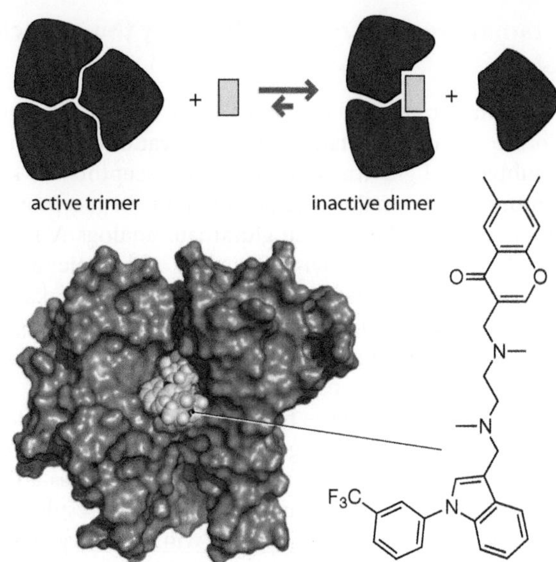

Figure 9.111 Blocking a protein–protein interaction. A combination of screening and design led to a small molecule that can bind to a dimer of TNFα, preventing the formation of the symmetrical TNFα trimer. (PDB 2AZ5)

trimeric death receptors is a distinctive feature of these ligands and receptors. The FasL-promoted protease cascade discussed in Chapter 6 is an example of such a pathway. A second important pathway leading to apoptosis is mediated by the ligand tumor necrosis factor (TNF). Of the two TNF isoforms, TNFα and TNFβ, TNFα has more potent effects.

Short bursts of TNFα induce acute-phase inflammatory responses in many cells, but chronic administration leads to nonspecific cytotoxicity. The **therapeutic index** (the window between beneficial and harmful doses) is so narrow that TNFα has not found use against human tumors. However, TNFα is highly cytotoxic toward many tumor cell lines grown in the laboratory; the cytokine does not operate via the mitotic checkpoints and p53-dependent pathways that we discussed in Chapter 3. From a clinical viewpoint, the most significant role of TNFα is the undesired proinflammatory effects seen in autoimmune diseases such as arthritis and Crohn's disease. Drugs that interfere with TNFα can prevent an overzealous immune system from attacking chondrocytes in the joints (causing arthritis) or tissue in the gastrointestinal tract (causing Crohn's disease). Small molecules have been identified that bind to and destabilize the TNFα trimer (**Figure 9.111**). However, the best drugs targeting TNFα on the market are proteins that sequester TNFα or block the receptor.

A wide range of cells produce TNFα, but the protein is mainly associated with cells of the immune system. For example, when macrophages recognize bacteria, the macrophages release large amounts of soluble TNFα. TNF is initially present as a homotrimer on the surface of cells. Then the protease TACE cleaves the membrane-bound form of TNFα to produce a soluble form. TNFα exists as the biologically active trimer only when its concentrations reach or exceed the nanomolar range.

There are two types of trimeric receptors for TNFα. Most cells express both TNFR-1 and TNFR-2 on their cell surfaces. TNFR-2 can only be activated by the membrane-bound form of TNF. TNFR-1 binds to FADD via an adaptor protein. Recall from the discussion of caspase cascades in Chapter 6 that FADD leads to DNA fragmentation via a mitochondrion-dependent caspase cascade.

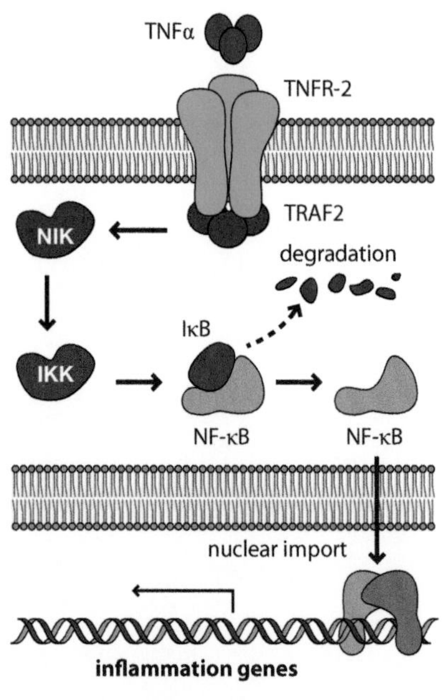

Figure 9.112 IκB sequestering the transcription factor NF-κB. When TNFα binds to TNFR-2 in T cells it leads to the degradation of IκB, releasing the transcription factor.

TNFα induces apoptosis via TNFR-1, but it has the opposite effect on immune cells via TNFR-2. TNFR-2 does not associate with proteins that display death domains. When activated by TNFα, TNFR-2 binds to the adaptor protein TRAF2, which activates the kinase NF-κB-inducing kinase (NIK). NIK phosphorylates another kinase, IκB kinase (IKK), which in turn phosphorylates IκB. IκB normally sequesters a transcription factor—from the NF-κB family—in the cytoplasm, but once phosphorylated, IκB is targeted for destruction by ubiquitination, releasing NF-κB. Free NF-κB is cleaved by a protease, allowing it to dimerize and translocate to the nucleus, upregulating proinflammatory pathways (**Figure 9.112** and **Figure 9.113**). In T cells, activation of NF-κB induces differentiation and proliferation—the exact opposite of the apoptotic effects mediated by the TNFR-1 receptor.

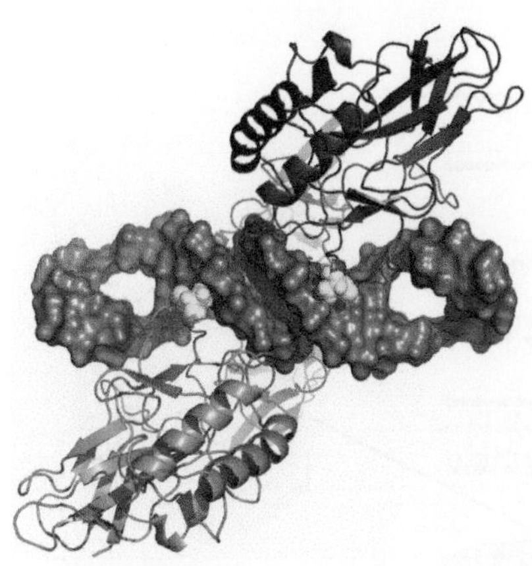

Figure 9.113 NF-κB (p50) binding to DNA as a dimer. In this structure of the C62A mutant (PDB 1SVC), the mutated residue is rendered as yellow spheres. The gaps in the DNA double helix arise from the fact that some of the DNA bases were not base-paired in the crystallized molecule.

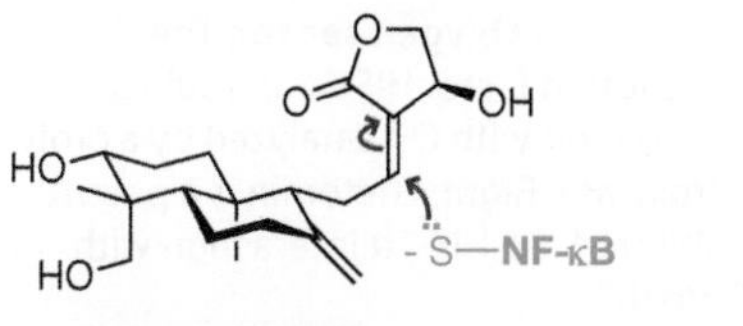

andrographolide

Figure 9.114 Covalent modification of NF-κB. The α,β-unsaturated lactone of andrographolide forms a covalent bond with NF-κB.

The natural product andrographolide was isolated from an *in vitro* screen of plant extracts for inhibitors of NF-κB transcriptional activity. Andrographolide was shown to form a covalent bond with Cys62 of the p50 variant of NF-κB (**Figure 9.114**). Electrophiles such as andrographolide can react with the many cysteines found inside the cell, but the reporter assay reveals only those that affect transcription of the gene under control of NF-κB (in this case, a luciferase reporter gene.)

Problem 9.20

The natural product kamebakaurin also targets the same reactive cysteine on NF-κB as andrographolide. Draw a structure of the cysteine adduct of kamebakaurin.

kamebakaurin

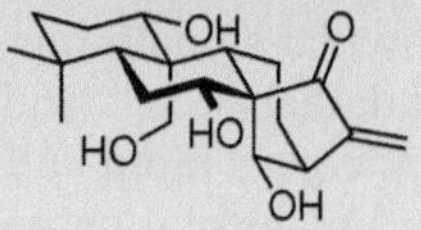

9.9 PATHWAYS CONTROLLED BY SMALL DIFFUSIBLE GAS MOLECULES

Oxygen levels are monitored through HIF-1α

Human cells need to monitor levels of oxygen. Under aerobic conditions, human cells degrade glucose oxidatively to generate carbon dioxide, like a slow flame except that the energy is harnessed through the production of ATP. Under anaerobic conditions, a different set of enzymes extract energy from glucose, ultimately producing lactic acid. The switch between genes for aerobic respiration and anaerobic respiration is controlled by a transcription factor called hypoxia inducible factor-1α (HIF-1α). Under aerobic conditions, HIF-1α has a half-life of about 5 minutes because Pro402 and Pro564 on HIF-1α are constitutively oxidized to hydroxyprolines by a prolyl hydroxylase enzyme. The hydroxylated prolines flag HIF-1α for enzymatic ubiquitination and proteolysis. Thus, the hydroxylase enzyme serves as a kind of oxygen receptor, because oxygen binds as a substrate. When oxygen levels are low, the oxidation of HIF-1α is prevented and concentrations of unmodified HIF-1α build up, eventually becoming sufficient to form a transcriptional complex involving many proteins, including a key protein CBP/p300.

The genes controlled by HIF-1α depend on the cell type. In muscle cells, HIF-1α controls genes for aerobic respiration. In kidney cells, HIF-1α promotes expression of the cytokine erythropoietin, which leads to the production of oxygen-carrying red blood cells. HIF-1α inhibits genes for cell proliferation; cell division does not make sense in the absence of nutrients. This effect is particularly important for solid tumors, which are chronically deficient in oxygen. A screen for small molecules that could inhibit the interaction of CBP/p300 with HIF-1α revealed that the natural product

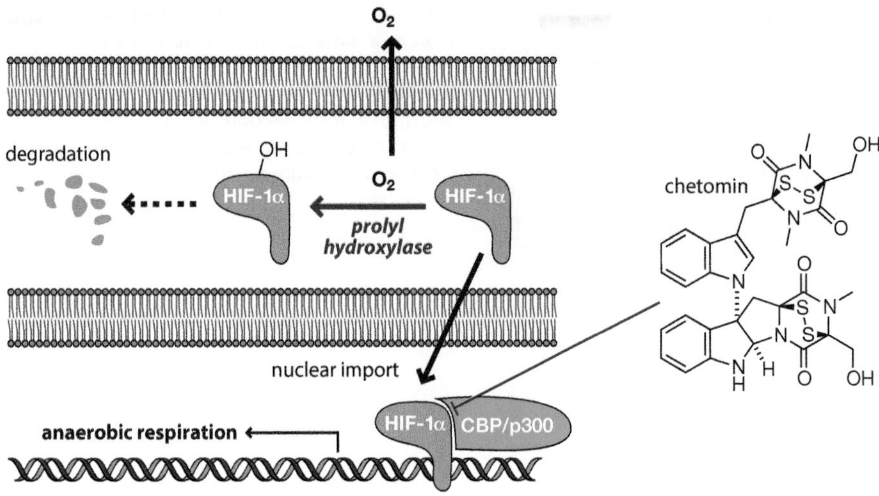

Figure 9.115 Oxygen sensor. The transcription factor HIF-1α is a substrate for oxidation with O_2, catalyzed by a prolyl hydroxylase. Right: chetomin is a potent inhibitor of the HIF-1α interaction with CBP/p300.

chetomin is a potent inhibitor of the interaction (**Figure 9.115**). Chetomin tricks cells into thinking that there is no oxygen, and causes the cells to stop proliferating.

A nitric oxide receptor induces the production of cGMP

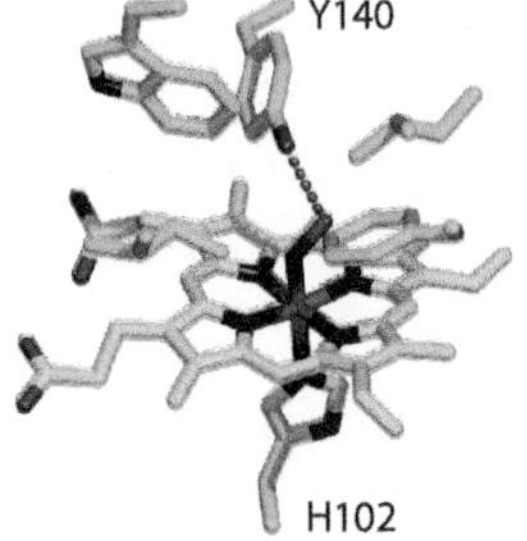

Figure 9.116 Dr. NO. NO binds to the iron porphyrin cofactor of guanylate cyclase. (Courtesy of C.S. Raman and Pierre Nioche, *Science* 306:1550–1553, 2004.)

Vascular smooth muscle cells control blood pressure by contracting or relaxing. Nonmuscle cells of the vascular endothelium can exert highly localized control over vascular smooth muscle by generating nitric oxide (NO), which diffuses freely out of endothelial cells and into nearby muscle cells; however, NO is highly reactive and undergoes rapid and promiscuous reactions with many molecules, including thiols and molecular oxygen. Thus, NO can only exert local effects, unlike hormones that circulate throughout the body. Many diatomic gas molecules (for example CO, O_2, and NO) form tight complexes with iron porphyrins. Vascular smooth muscle cells contain a form of **guanylate cyclase** that is regulated by an iron–porphyrin complex. When nitric oxide binds to the iron atom of the porphyrin (**Figure 9.116**), the enzyme undergoes an activating conformational change that allows it to convert GTP into cyclic guanosine monophosphate (cGMP), in direct analogy to the reaction catalyzed by adenylyl cyclase. cGMP binds to cGMP-dependent protein kinase I, which phosphorylates calcium channels in the cell membrane and in calcium storage vesicles. Phosphorylation inhibits these channels from opening even if other pathways are generating IP_3. The cGMP-dependent protein kinase also phosphorylates proteins that, in turn, affect phosphorylation of myosin, directly inhibiting muscle contraction. By inhibiting the contraction of vascular smooth muscles, NO leads to vasodilation. cGMP levels are diminished by the action of cGMP phosphodiesterase 5 (PDE5), which hydrolyzes cGMP to guanosine monophosphate (**Figure 9.117** and **Figure 9.118**).

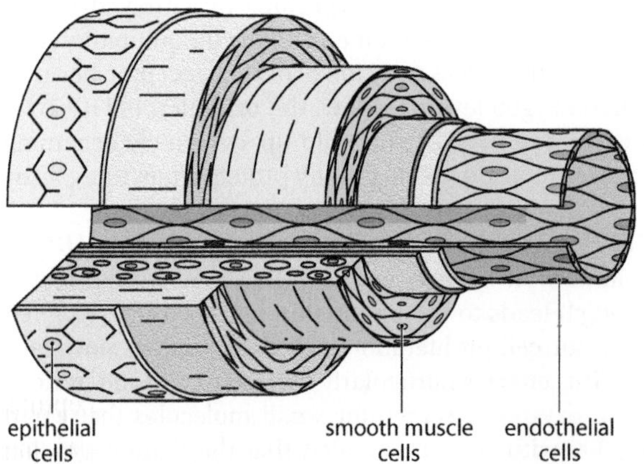

Figure 9.117 Cutaway view of an arterial blood vessel showing layers of cells. (Adapted from S. Fox, Human Physiology, 12th ed. McGraw-Hill, 2010.)

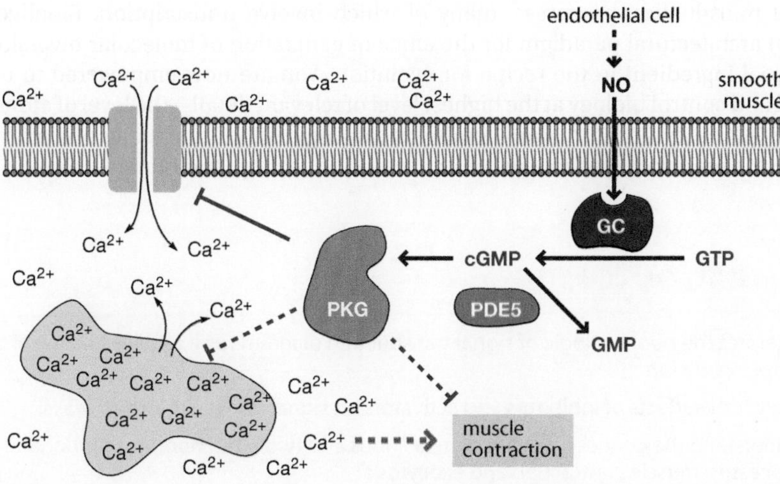

Figure 9.118 **Control by NO.** In blood vessels, endothelial cells inhibit the contraction of smooth muscle cells by releasing NO, which activates guanylate cyclase (GC) an enzyme that generates cGMP. cGMP-activated protein kinase (PKG) phosphorylates ion channels, preventing them from opening.

The vascular smooth muscle cells that supply blood to the penis are also controlled by cGMP levels. Medications for erectile dysfunction (such as Viagra®, Cialis®, and Levitra®) promote vasodilation by inhibiting PDE5; the resultant high levels of cGMP ultimately depress contraction of the blood vessels (**Figure 9.119**). Nitric oxide also acts as a diffusible signal for neurons and macrophages. Since 1879, long before its mechanism was understood, nitroglycerin has been used as a medication to increase blood flow to the heart. As shown more recently, aldehyde dehydrogenase generates nitric oxide from nitroglycerin, and the NO causes vasodilation of coronary arteries.

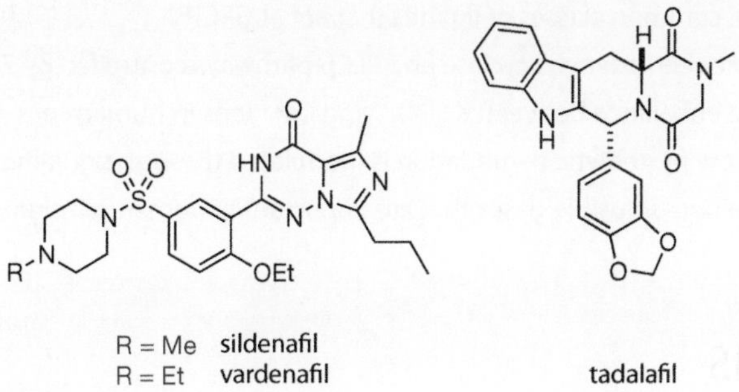

R = Me sildenafil
R = Et vardenafil

tadalafil

Figure 9.119 **The compound in the blue pill.** Inhibitors of PDE5 lead to relaxation of the vascular smooth muscle and are used to treat erectile dysfunction.

9.10 SUMMARY

The cells of the human body are engaged in an endless conversation mediated by secreted signals and displayed signals. Some signals initiate rapid responses, such as contraction or secretion; other signals initiate slower responses involving the synthesis of new proteins according to the central dogma: transcription of DNA into RNA, and translation of RNA into proteins. Those transcriptional processes can redefine the constitution and capabilities of a human cell. In this chapter we have taught you how extracellular signals are transduced into transcriptional changes, dropping you into a vast sea of acronyms. To help you chart your course we described signal transduction pathways mediated by seven types of receptors: nuclear receptors, cytokine receptors, receptor tyrosine kinases, trimeric death receptors, G protein-coupled receptors, ion channel receptors, and diffusible gas receptors. These seven flotillas of signaling proteins will help you to confidently navigate the primary research literature.

We have come full circle. We started this book by describing how the transcription of genes leads to all of the downstream biooligomers in the human cell. Filling in the details took up most of the book. With this chapter we have concluded our journey by describing how small molecules and macromolecules, ultimately produced by biooligomers, can control the transcription of other biooligomers in the cell. Moreover, we covered many examples where synthetic small molecules could be used to control signal transduction processes, many of which involve transcription. Biooligomers are an architectural paradigm for the efficient generation of molecular diversity—an essential ingredient in the recipe for evolution. You are now empowered to understand and control biology at the highest level of relevant detail—the level of atoms and bonds. At the level of atoms and bonds you can engineer the medicines of the future or probe the secrets of cells that make us healthy, unhealthy, and human.

LEARNING OUTCOMES

- Interpret the boolean logic of signal transduction diagrams as it applies to control of gene expression.
- Predict the effects of inhibitors and activators on signal transduction pathways.
- Understand the connection between Ca^{2+} ions and two rapid nontranscriptional processes: muscle contraction and exocytosis.
- Distinguish the seven major signal transduction pathways that control transcription in human cells.
- Recognize the high-affinity ligands for nuclear receptors.
- Know the three types of ligands for human two-component pathways: interleukins, interferons, and TGF-β.
- Recognize growth factors as the ligands for receptor tyrosine kinases.
- Distinguish between the four subpathways downstream of receptor tyrosine kinases: STAT, MAP kinases, PLCγ, and PI3K/Akt.
- Distinguish the death pathways (TNFα and FADD) from receptor tyrosine kinase pathways.
- Recognize the common classes of ligands that act at GPCRs.
- Distinguish between adenylyl cyclase and PLCβ pathways controlled by 7TM GPCRs.
- Recognize the imbalance between K^+, Na^+, and Ca^{2+} ions in human cells.
- Understand how proteolytic degradation is modulated through signaling.
- Recognize the two diffusible gases that are important in human cell signaling: NO and O_2.

PROBLEMS

9.21 Draw the Lewis dot structure for nitric oxide.

9.22 Using cell signaling diagrams throughout the chapter, rank each of the seven-signal transduction pathways according to the maximum number of steps required to affect transcription within the nucleus.

***9.23** Consider the action of the following small molecules on the following signal transduction pathway (see figure on the next page).
tyrphostin, a specific inhibitor of tyrosine kinases
AZD8330, a specific inhibitor of MEK1 and MEK2
wortmannin, a specific inhibitor of PI3K
rapamycin, a specific inhibitor of TOR
FR180204, a specific inhibitor of ERK
A-443654, a specific inhibitor of PKB
A Which compound would most effectively prevent uncontrolled proliferation?
B Which compound best inhibits apoptotic signals without inhibiting proliferative signals?
C Which compound would most selectively inhibit protein synthesis?

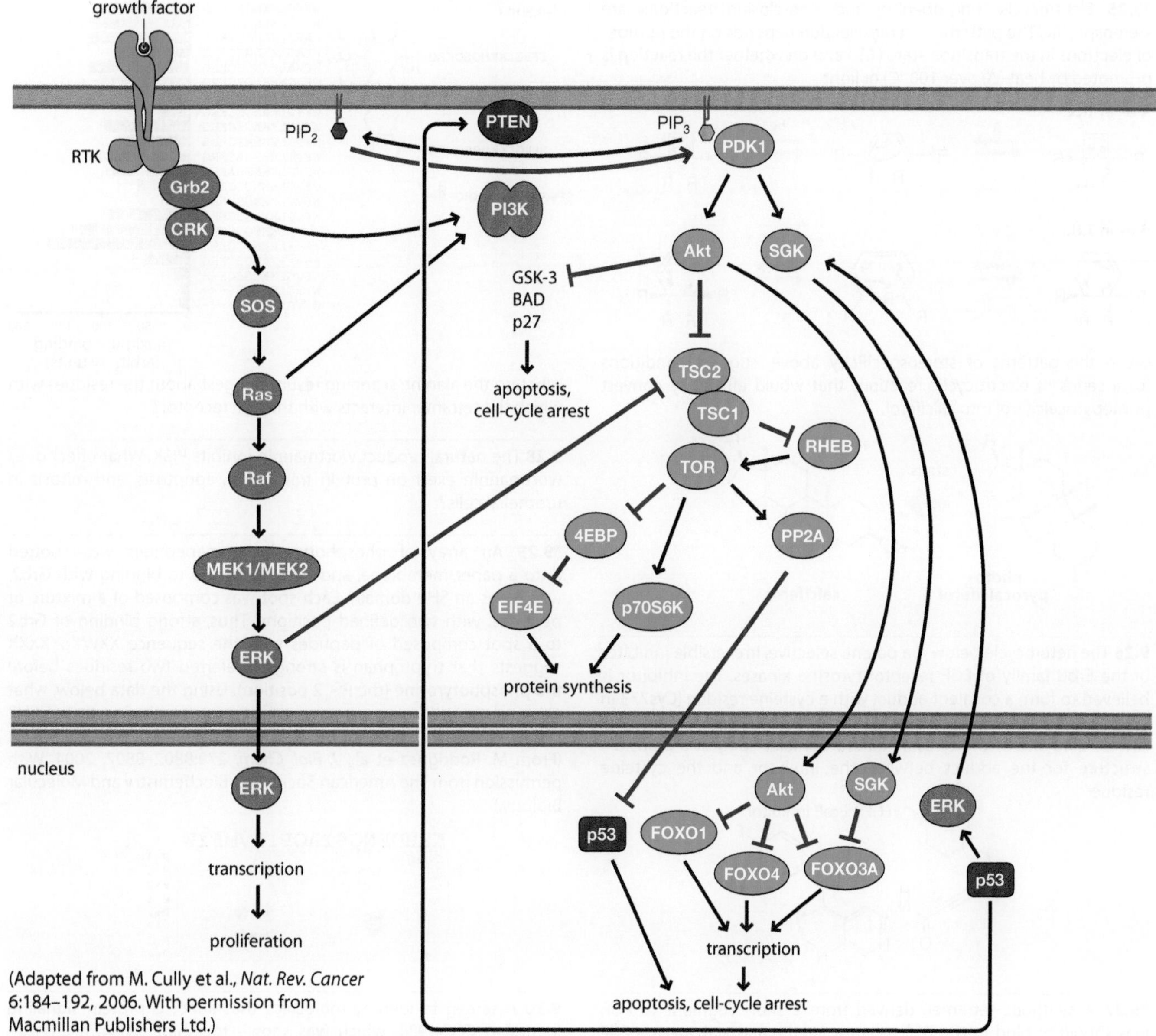

(Adapted from M. Cully et al., *Nat. Rev. Cancer* 6:184–192, 2006. With permission from Macmillan Publishers Ltd.)

9.24 The red natural product ammosamide B is produced by *Streptomyces* strain CNR-698, isolated from ocean sediments in the Bahamas. It is cytotoxic to various cultured cancer cells with an IC_{50} value as low as 20 nM. To identify a molecular target, an aminocoumarin fluorophore was added by using a carbodiimide coupling in low, but useful, yield. The carbodiimide EDC (1-ethyl-3-[dimethylaminopropyl] carbodiimide•HCl) is a water-soluble carbodiimide that works just like DCC and DIC. In cells treated with the probe reagent, fluorescence was shown to localize within lysosomal vesicles. The inability to wash away or compete away the fluorescence with free ammosamide B suggested a covalent or tight binding interaction between ammosamide and the molecular target. Suggest a plausible arrow-pushing mechanism for the carbodiimide coupling reaction.

aminocoumarin

OH, NMe$_2$, H$_2$N, Cl, N, CONH$_2$, NH$_2$ — **ammosamide B**

EDC
DMAP
60 °C

Me$_2$N, H, N, Cl, N, CONH$_2$, 10% NH$_2$ — **probe reagent**

***9.25** Electrocyclic ring-opening and ring-closing reactions are stereospecific. The pattern of stereoselection depends on the number of electrons in the transition state (T.S.) and on whether the reaction is promoted by heat (Δ; over 100 °C) or light.

4 e- in T.S.:

6 e- in T.S.:

Given the patterns of stereospecificity above, suggest conditions for a series of electrocyclic reactions that would efficiently convert photopyrocalciferol into calciferol.

photopyrocalciferol → steps → **calciferol**

9.26 The heterocycle below is a potent, selective, irreversible inhibitor of the ErbB family of EGF receptor tyrosine kinases. The inhibitor is believed to form a covalent adduct with a cysteine residue (Cys773 in ErbB1, Cys784 in ErbB2, and Cys778 in ErbB4) near the entrance to the ATP-binding site of the tyrosine kinase domain. Suggest a plausible structure for the adduct between the inhibitor and the cysteine residue.

EGFR/ErbB inhibitor

***9.27** A synthetic tetramer, derived from a short segment of FGF, was shown to bind to the FGF receptor *in vitro* and activate the FGF receptor pathway in neuronal cells. An alanine scan was used to assess the contribution of each of the residues to the interaction with the receptor.

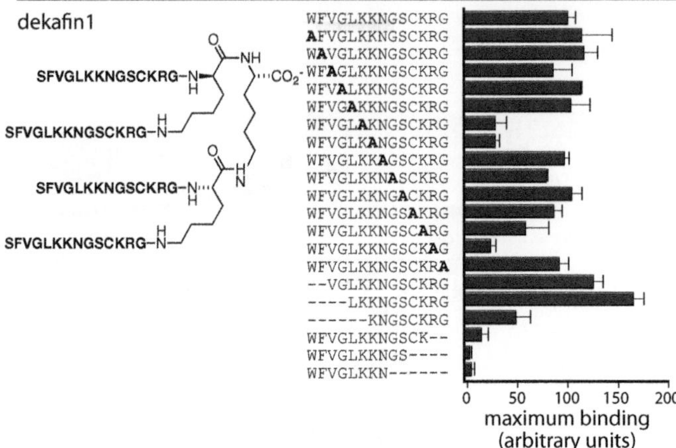

What do the alanine scanning results suggest about the residues with which the tetramer interacts with the FGF receptor?

9.28 The natural product wortmannin inhibits PI3K. What effect does wortmannin exert on protein translation, apoptosis, and mitosis in urothelial cells?

***9.29** An array of phosphotyrosine nonapeptides was spotted onto a paper membrane, and then subjected to binding with Grb2, which has an SH2 domain. Each spot was composed of a mixture of peptides, with two defined positions. Thus, strong binding of Grb2 to a spot composed of peptides with the sequence XXWXpYXXXX suggests that tryptophan is strongly preferred two residues before the phosphotyrosine (the P – 2 position). Using the data below, what consensus peptide sequence would be best for binding to the SH2 domain of Grb2? Use X to represent positions with no preference. (From M. Rodriguez et al., *J. Biol. Chem.* 279:8802–8807, 2004. With permission from The American Society for Biochemistry and Molecular Biology.)

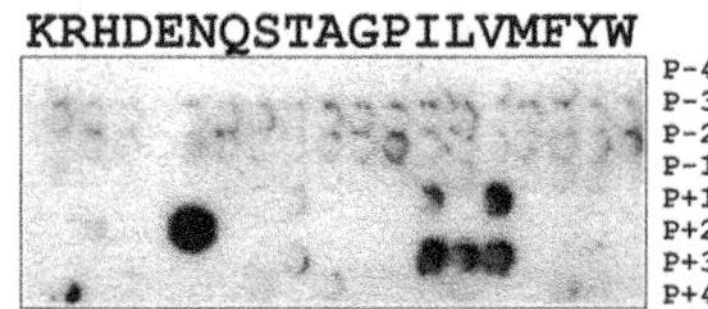

9.30 A screen to identify molecules that affect G protein signaling identified CCC-4986, which was shown to inhibit RGS4, a GTPase-activating protein. As summarized in the following signal transduction diagram, CCC-4986 was tested in a C6-glioma cell to determine the effect on cAMP production.

Figure for Problem 9.30

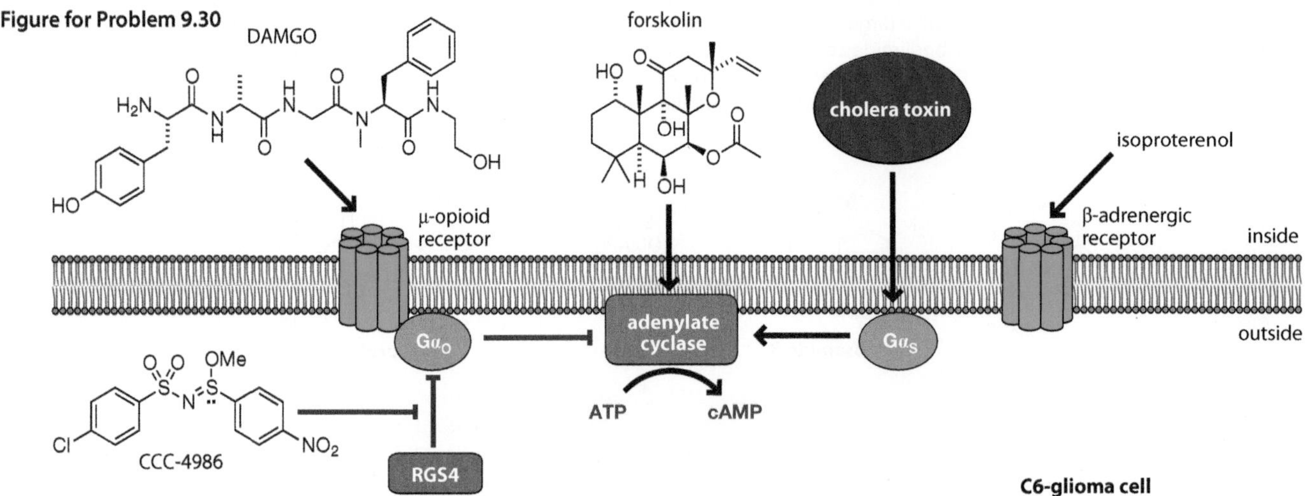

A For each of the following compounds, predict whether it will **increase** or **decrease** cAMP production:

i. isoproterenol (also known as isoprenaline)
ii. DAMGO
iii. forskolin
iv. cholera toxin
v. CCC-4986

B Subsequent studies showed that CCC-4986 covalently modifies RGS4 by reacting with the thiolate side chain of Cys132. On the basis of mass spectrometry, the adduct was proposed to be a disulfide. If you feel up to it, suggest a plausible arrow-pushing mechanism for the reaction.

RGS4 CCC-4986 50 mM Tris buffer pH 8.0 250 mM NaCl 5 mM DTT RGS4

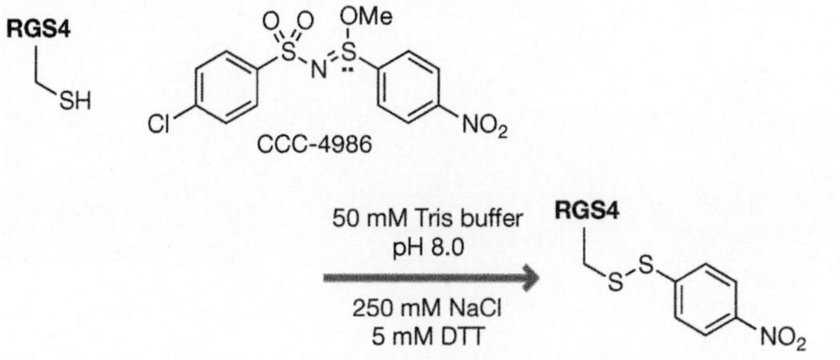

***9.31** Grb2 has two SH3 domains, one at the N terminus and one at the C terminus. Molecular modeling suggests that a simple dimer of a polyproline sequence VPPPVPPRRR would be able to bind simultaneously to both of the SH3 domains of Grb2. How would you synthesize the peptidimer ligand by solid-phase synthesis? Clearly indicate which protecting groups you would use in the synthesis.

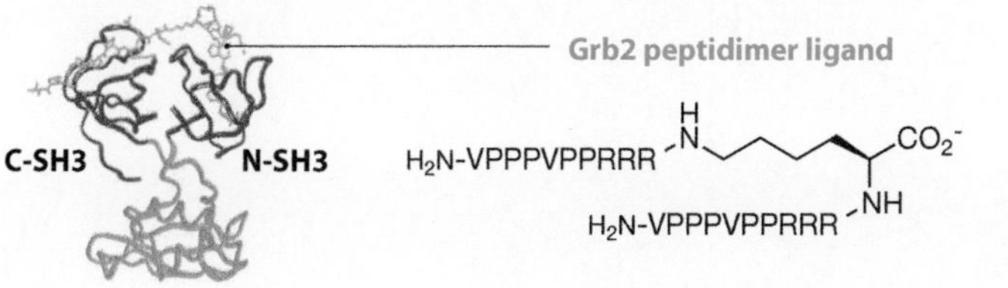

Grb2 peptidimer ligand

C-SH3 N-SH3 H_2N-VPPPVPPRRR H_2N-VPPPVPPRRR

9.32 The transcription factor Pap1 in fission yeast is normally exported from the nucleus because it has a nuclear export sequence—a 19 amino acid sequence near the N terminus (515–533). In response to hydrogen peroxide, export of Pap1 is inhibited, and it remains in the nucleus where it initiates transcription of a variety of genes. When two cysteine residues in the nuclear export sequence are mutated (C523A and C532D), nuclear export is abolished and the transcription factor accumulates in the nucleus, where it initiates transcription. Surprisingly, the non-oxidative reagent diethyl maleate exerts an effect similar to that of hydrogen peroxide. Suggest a plausible chemical mechanism accounting for the similar effect of hydrogen peroxide and diethyl maleate.

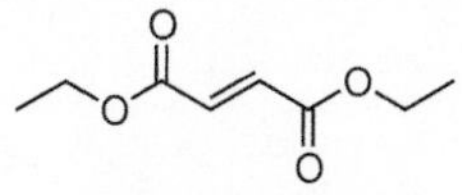

diethyl maleate

***9.33** The following "caged" glutamate derivative can be induced to release glutamate on activation by ultraviolet radiation. Suggest a plausible arrow-pushing mechanism for the reaction that goes through the intermediate shown.

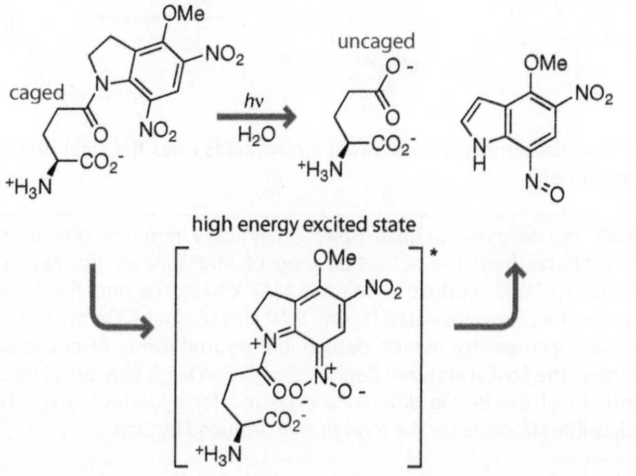

caged $h\nu$ H_2O uncaged

high energy excited state

9.34 Celastrol, isolated from the thunder god vine (*Tripterygium wilfordii*), forms an aromatic adduct by reacting with Cys179 in IKK. Suggest a plausible structure for the covalent adduct.

IKK

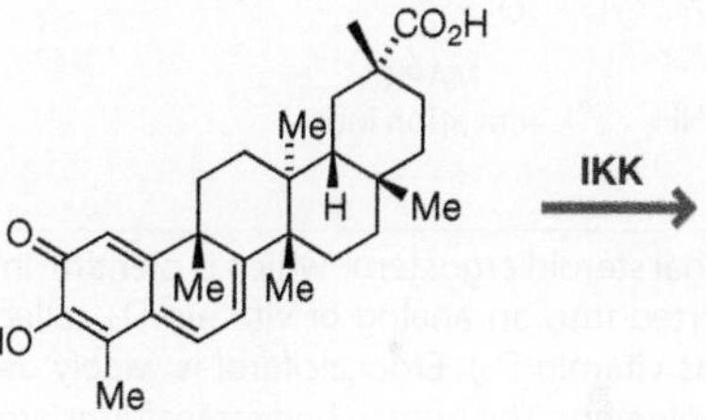

celastrol

***9.35** 17-Hydroxy-jolkinolide, isolated from *Euphorbia fischeriana* Steud, forms covalent crosslinks between cysteines in Janus kinases. Suggest a plausible structure for the covalent adduct between 17-hydroxy-jolkinolide and the cysteine residues of two Janus kinases.

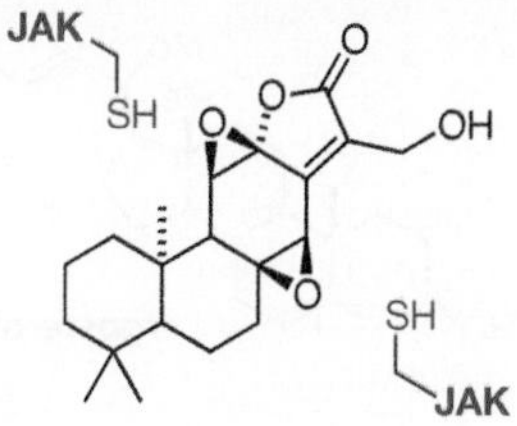

JAK JAK

9.36 The pyranonaphthoquinone antibiotic kalafungin inhibits the kinase Akt2 by reacting with a thiol in the active site. Which atom of kalafungin would you expect to be most reactive toward nucleophilic protein side chains (like cysteine thiolates) under bioreductive conditions?

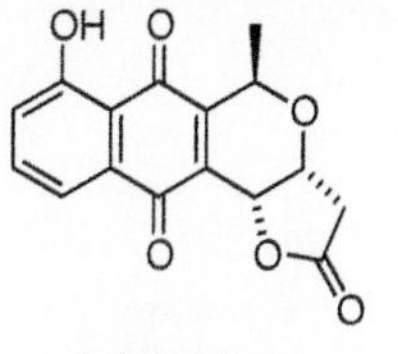

kalafungin
IC_{50} 70 nM in MDA468 cells

***9.37** A method for the detection of phosphoserine residues in proteins involves the two-step, one-pot reaction sequence below. Suggest a plausible arrow-pushing mechanism for each step. Which

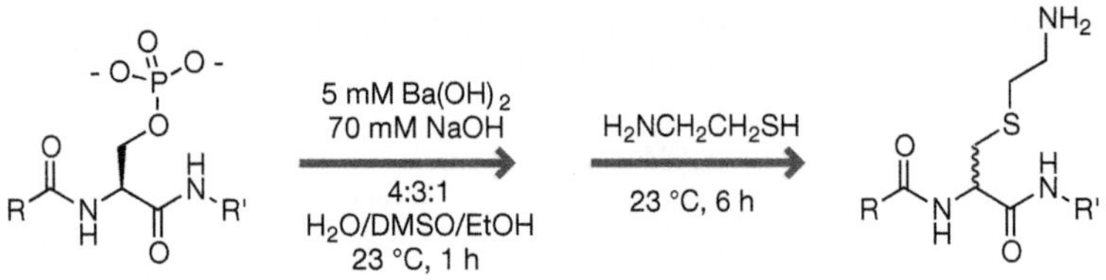

5 mM $Ba(OH)_2$
70 mM NaOH
4:3:1
H_2O/DMSO/EtOH
23 °C, 1 h

$H_2NCH_2CH_2SH$
23 °C, 6 h

of the ribosomally incorporated amino acids does the final product resemble?

9.38 The *Salmonella* toxin SpvC catalytically removes one of the phosphates from the activation loop of MAP kinases (for example FMTpEYpVA) to produce an inactive MAP kinase. The modified kinase cannot be phosphorylated by any MAP kinase kinase. On the basis of mass spectrometry (which detects the neutral forms of phosphate esters), the SpvC-catalyzed dephosphorylation leads to a loss of 98 Da, instead of the 80 Da difference expected for hydrolysis. Suggest a plausible structure for the modification induced by SpvC.

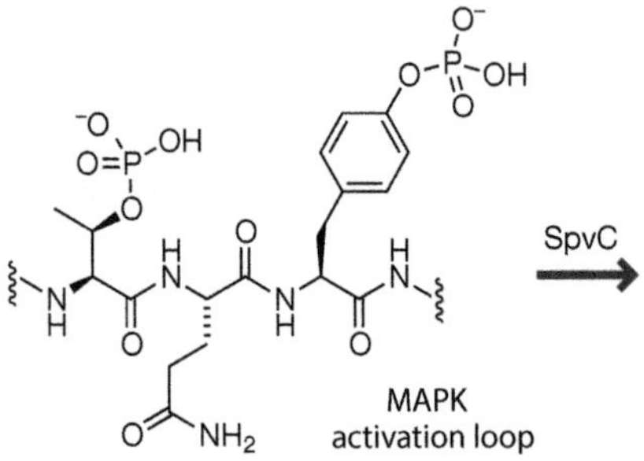

SpvC

MAPK activation loop

***9.39** The fungal steroid ergosterol, which is plentiful in brewer's yeast, can be converted into an analog of vitamin D_3 called ergocalciferol (also known as vitamin D_2). Ergocalciferol is widely used as a dietary vitamin D supplement. The human body transforms ergocalciferol into a ligand for the vitamin D receptor by using the same enzymes that metabolize vitamin D_3. Draw the structure of the tris-hydroxy form of vitamin D_2 that would bind to the vitamin D receptor.

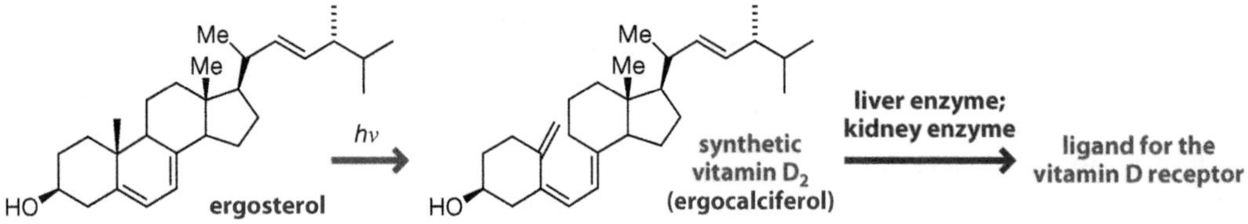

ergosterol

$h\nu$

synthetic vitamin D_2 (ergocalciferol)

liver enzyme; kidney enzyme

ligand for the vitamin D receptor

Glossary

α-amylase an enzyme that catalyzes the hydrolysis of amylose into glucose.

β barrel a cylindrical protein structural motif composed of antiparallel β strands.

β hairpin a protein structural motif composed of four amino acids and a hydrogen bond between the *i* and *i* + 3 residue. This type of turn structure leads to a 180° change in the direction of the peptide backbone.

activator a transcription factor responsible for initiating and promoting transcription of a gene or genes.

acyl carrier protein domain (ACP) the domain of a polyketide synthase that attaches to the nascent polyketide and carries the intermediate compound to different domains of the polyketide synthase.

adenylate cyclase an enzyme responsible for catalysis of the reaction converting ATP into cAMP.

alanine scanning systematic substitution of specific residues in a protein with the amino acid alanine. The approach is used to quantify contributions made by side chain atoms past the β-carbon.

annealing hybridization of strands of DNA by incubating the two at a high temperature (> 95 °C) for a short time and then slowly cooling the solution.

anomer a diastereomer of a monosaccharide with either an α or β configuration at the acetal or ketal carbon.

anomeric center (anomeric carbon) the acetal or ketal carbon in monosaccharides or glycoconjugates.

anomeric effect a stereoelectronic preference of anomeric alkoxy substituents in pyranose rings to adopt axial orientations over equatorial orientations.

antibody proteins of the immune system charged with binding to nonhuman molecules from pathogenic organisms.

antisense strand the DNA strand that provides the template for the sense strand and for RNA polymerase during transcription.

apoprotein proteins lacking a key cofactor.

apoptosis an orchestrated process leading to the death of the cell.

apoptosome a complex formed by cytochrome *c* with Apaf-1 and dATP, which activates caspase-9 by cleaving procaspase-9 during apoptosis.

aptamer a DNA or RNA sequence with affinity for specific target molecules.

Argonaute a ribonuclease that binds to siRNA and destroys any RNA transcripts complementary to the siRNA sequence.

association constant (K_a) the equilibrium constant for the association of two molecules, such as a ligand and receptor, to give a complex.

autocrine signaling hormonal communication between cells of the same type.

autophosphorylation in a dimer of protein kinases, reciprocal phosphorylation of side chains on the partner kinase.

bacteriophage viruses that infect only bacteria.

bait protein *see* yeast two-hybrid.

base the heterocyclic substituent that uniquely distinguishes each of the four DNA (or RNA) subunits. Each DNA or RNA molecule is composed of many nucleotide subunits and each subunit has a distinctive nitrogen heterocycle substituent. The term base is often used to refer to the entire nucleotide subunit. In the context of chemistry, the term base is used to mean a proton acceptor.

base pairs two DNA or RNA bases held together by hydrogen bonds, such as A•T or G•C.

biooligomers polymers produced by a cell that are constructed through the connection of subunits to form chain-like molecules, including proteins, DNA, RNA, glycans, lipids, and terpenes.

bioreductive activation a mechanism for converting an unreactive prodrug into an active compound through a reduction reaction that takes place inside the cell.

C terminus the end of the protein backbone with the carboxy functional group.

carbohydrates "hydrates of carbon" or compounds with the empirical formula $C_n(H_2O)_n$. This term is reserved for molecules composed exclusively of one or more monosaccharides, and excludes glycoproteins and other glycans.

central dogma of molecular biology an organizing principle to describe the encoding of sequence information in biooligomers. In the classical definition, it states that the DNA provides a blueprint for the molecules synthesized by the cell; the DNA is transcribed into RNA, which is then translated into proteins.

checkpoint during eukaryotic cell division, a conditional test that determines whether the cell is allowed to continue the process of cell division.

chemical biology the use of tools from chemistry to understand and manipulate biology at the level of atoms and bonds.

chemical genetics using small molecules as probes of protein function in the cell. For example, protein activity can be shut off through a specific inhibitor of the target protein.

chitin a β(1–4)-linked polymer of *N*-acetyl-D-glucosamine, which provides the exoskeletons for arthropods.

cloning making identical copies of a cell or gene. In the context of molecular biology, this process involves placing the gene into a plasmid for expression of the encoded protein or the production of multiple copies of the plasmid that contains the gene.

coenzyme A an enzyme co-factor attached through a thioester to provide a handle for enzyme binding to two- and three-carbon building blocks (acetyl and propionyl, respectively).

combinatorial chemistry a method for the synthesis of collections of molecules each containing the same number of subunits, typically accomplished using a modular approach to join subunits together.

complementation assay an approach for testing an interaction based on the restoration of a wild-type phenotype. For example, a split GFP assay will result in functional, fluorescent GFP if the two pieces of the protein can reassemble as a result of a mediating interaction.

covalent (bond) a type of bond in which the interaction energy is dominated by the interaction of filled orbitals with unfilled orbitals, the third term in Equation 2.2, for example $H_3C—H$ or $H_3C—CH_3$.

cross conjugated (cross conjugation) two or more π bonds that cannot participate in resonance due to an intervening C=C bond.

cyclodextrins cyclic oligomers of glucose, $[\text{-Glc}\alpha(1,4)\text{-}]_n$.

cystine a dimeric, oxidized form of cysteine linked by a sulfur–sulfur disulfide bond.

cytokines proteins used in extracellular cell–cell signaling. Inside the cell, their signals are transduced by Janus kinases and STAT transcription factors.

deacetylases enzymes responsible for removing an acetyl functionality from a protein or oligosaccharide.

dehydratase domain (DH) the domain of a polyketide synthase that catalyzes a dehydration reaction on the intermediate polyketide, which results in the formation of an enone intermediate polyketide.

developmental biology the study of how cells differentiate to grow from a fertilized egg to a complete organism.

dicer a nuclease charged with hydrolyzing pre-miRNAs to yield siRNAs.

dispersion force the attractive interactions between molecules or functional groups that is embodied by the third term of Equation 2.3.

dissociation constant (K_d) the equilibrium constant for the dissociation of a molecular complex to give the individual components, such as a ligand and receptor.

DNA ligase an enzyme that can repair nicked DNA by forming a phosphate ester bond from the 3′ hydroxyl of one strand and the 5′ phosphate of another.

DNA primase an enzyme charged with synthesizing short strands of RNA to provide a primer for DNA polymerization of the lagging strand during replication.

DNA shuffling a method for engineering new genes based on the hybridization of single-stranded fragments of homologous genes from different organisms or other sources during PCR-based amplification.

domains small regions of proteins (typically with fewer than 100 amino acid residues) that are capable of folding independently when removed from the larger protein structure.

dose-response curve a plot of drug concentration versus biological response, used to determine EC_{50} values.

DTT dithiothreitol. A synthetic reducing agent used to cleave disulfide bonds in peptides and proteins.

Edman degradation a chemical method for determining the amino acid sequence of a protein, based upon chemical cleavage and identification of the N-terminal amino acid.

efflux pump a membrane-bound protein that harnesses the hydrolysis of ATP to push small molecules out of the cell. This class of proteins provides a mechanism for antibacterial drug resistance.

enoyl reductase domain (ER) the domain of a polyketide synthase that catalyzes a conjugate reduction reaction on an enone polyketide intermediate.

enzyme a protein capable of catalyzing a chemical transformation.

epigenetics the study of heritable factors that are not encoded by the organism's DNA sequence. For example, DNA methylation or modification of the histones can influence otherwise identical sequences of DNA.

error-prone PCR a technique for generating a collection of DNA sequences based upon the addition of DMSO, Mn^{2+} or other factors, which can disrupt the fidelity of DNA polymerase during PCR.

exons the parts of a eukaryotic RNA transcript that remain after the removal of introns. The remaining part of the transcript is translated into a protein.

exonuclease an enzyme or subdomain of a larger enzyme responsible for the hydrolysis of DNA or RNA strands by cleaving one base at a time from the end of the strand.

fatty acid synthase the type I polyketide synthase protein complex responsible for the synthesis of fatty acids. It iteratively introduces each two-carbon subunit through the processive action of each catalytic domain.

fatty acid an unbranched alkanoic acid with an even number of carbons produced by a polyketide synthase. They vary in length and level of unsaturation.

flow cytometry a method for analyzing a population of cells. The method involves passing a hydrodynamically focused stream of the cells through a fluorimeter. The fluorescence of each cell, introduced by fluorescently tagged antibodies, fluorogenic agents, or fluorescent proteins, can be quantified. In FACS, the flow cytometry is used to separate out a population of cells using a fluorescence-based criterion.

fluorophore a fluorescent chromophore that absorbs photons and emits photons with lower energy and higher wavelength.

forward genetics the process of introducing random mutations into an organism to identify interesting phenotypes. DNA sequencing can then identify the associated genotype, thus implicating a specific gene required for biological activity. In forward chemical genetics, collections of small molecules are used in place of genetic mutations. Specific phenotypes from treatment with the small molecule can be identified, and the target of the small molecule can reveal a contribution to a cellular pathway.

frontier molecular orbitals the most reactive orbitals in a chemical species, usually corresponding to the highest occupied molecular orbital and the lowest occupied molecular orbital.

G proteins a class of proteins that bind and hydrolyze GTP. Many such proteins interact with 7-transmembrane domain receptors to transduce signaling information from the extracellular milieu to the cytoplasm.

G protein-coupled receptors (GPCRs) a common class of 7-transmembrane domain receptors that transduce extracellular hormone signals into intracellular actions like contraction, secretion, or mitotic proliferation.

gene a DNA sequence encoding a protein and a promoter.

genetic code the correlation between DNA or RNA triplets (codons) and the resultant amino acid in the protein.

genome the genetic material that is necessary and sufficient to encode an organism.

genotype a gene or combination of genes.

glycans molecules that contain carbohydrate functional groups, such as glycolipids, glycoproteins, and polysaccharides.

glycome the collection of glycans in a cell, tissue, or organism.

glycoproteins proteins post-translationally modified through the attachment of oligosaccharides.

glycosphingolipids oligosaccharides attached to the lipid ceramide.

glycosylation the enzyme-catalyzed transfer of carbohydrates to the surface of proteins, lipids and oligosaccharides.

glycosylhydrolases a class of enzymes that break glycosidic bonds through the addition of water.

glycosyltransferases the class of enzymes charged with forming glycosidic bonds during the synthesis of glycans and carbohydrates. The glycosyl donor substrates for this enzyme have nucleotide phosphates as the leaving group at the anomeric position.

GSH a symbolic abbreviation for the tripeptide glutathione that emphasizes the thiol functional group.

GSSG a symbolic abbreviation for the oxidized dimer of glutathione that emphasizes the disulfide functional group.

GTPase-activating proteins (GAPs) a regulatory protein that accelerates the hydrolysis of GTP to GDP by G proteins.

guanine nucleotide exchange factors (GEFs) a regulatory protein that catalyzes the ejection of the spent guanine nucleotide GDP from a G protein. The binding of a new GTP substrate regenerates an activated G protein–GTP complex.

guanylate cyclase an enzyme that converts GTP into cyclic guanosine monophosphate (cGMP).

heat shock raising the temperature of an organism to provoke transcriptional changes in response.

histone acetyltransferase (HAT) an enzyme responsible for transferring an acetyl group to specific lysine side chains of histones.

histone deacetylase (HDAC) an enzyme that hydrolyzes the acetyl amide functionalities from the lysine side chains of modified histones.

hotspot of binding energy found on the surface of a protein, a cluster of amino acid side chains that provide the bulk of binding energy to mediate the protein's noncovalent binding.

hybridization forming double-stranded DNA by incubating two complementary single-stranded DNA sequences.

hydrolase an enzyme that catalyzes the hydrolysis of bonds, such as amides and glycosides, through the addition of water.

hydrophobic effect the tendency of water molecules to maximize interactions between themselves while minimizing interactions with solutes.

inhibitory constant (K_i) the dissociation constant for an inhibitor-enzyme complex. For typical inhibitors that bind at the same site as the substrate, this equilibrium constant is independent of the inhibitor concentration. In contrast, IC_{50} is dependent on substrate concentration. When the substrate concentration is much lower than K_m, IC_{50} is equal to K_m.

inteins internal segments of proteins that are removed autocatalytically from a protein through a reaction involving an *N* to *S* acyl shift at a cysteine residue. The remaining proteins are referred to as exteins.

intercalation a sandwich-like interaction between DNA bases and small molecules involving an aromatic compound sliding between the π-stacked DNA bases.

interleukins a class of cytokines that are important in the differentiation and development of blood cells in the hematopoietic lineage.

introns the parts of a eukaryotic RNA transcript that are removed from the transcript and are not translated into proteins.

ion channels a class of proteins that form pores in cell membranes. After opening in response to a regulatory signal, these proteins allow the passage of ions through the pore. This class of proteins includes both ion-specific and nonspecific channels.

ionic (bond) a type of bond in which the interaction energy is dominated by the Coulombic terms of Equation 2.2 or 2.3. For example K–Br, Cs–OH.

isomerase an enzyme that catalyzes a reaction leading to the isomerization of a starting material without changing the molecular formula of the substrate.

ketoreductase domain (KR) the domain of a polyketide synthase that catalyzes the reduction of a β-ketone to an alcohol.

ketosynthase domain (KS) the domain of a polyketide synthase that catalyzes the Claisen condensation of malonyl and acetyl thioesters, which results in the formation of a new carbon-carbon bond.

kinase a class of transferase enzymes responsible for transferring a phosphoryl group to a hydroxyl group, using ATP as the phosphate donor.

lability refers to chemical reactivity, and is the opposite of chemical stability. In chemical biology, the term implies susceptibility to reactions that lead to loss of biological function.

ligase an enzyme that catalyzes the joining (ligation) of two independent molecules, typically in an energy-dependent process. For example, subtiligase is a variant of the enzyme subtilisin, and joins together two peptide fragments.

lipid a cellular molecule that can dissolve in nonpolar organic solvent and not in water.

lipid raft microdomains within the cell membrane that are particularly rich in glycosphingolipids and cholesterol. Diverse transmembrane receptors and membrane-associated proteins cluster with this microdomain, which thus provides a focal site for organizing cell signaling.

lipidome the collection of all lipids found in the cell.

lyase an enzyme that cleaves bonds through mechanisms other than hydrolysis or redox reactions.

monosaccharides monomeric sugars used as the fuel for metabolism and the building blocks for glycans and other structures in the cell (for example RNA and DNA).

mutation a change to a DNA sequence. Silent mutations result in no change to the protein sequence encoded by the DNA.

N terminus the end of the protein backbone with the amino functional group.

native chemical ligation a method for joining together unprotected peptide fragments.

neighboring group participation adjacent functionalities or atoms contributing to the reactivity of a functional group.

nick in double-stranded DNA, a site where adjacent nucleotides are not covalently bonded.

nonenzymic glycation post-translational reactions between carbohydrates and proteins. This reaction does not require catalysis by enzymes, and can lead to protein crosslinking and other deleterious products. For example, the open-chain (aldehyde) form of glucose condenses with the amino functionality of lysine side chains.

nuclear localization sequence a short peptide fused to a protein, which directs the protein into the nucleus, through binding to karyopherin β.

nuclear receptors the class of receptors that bind directly to hydrophobic small molecules, including steroids and thyroid hormones. The resultant receptor–ligand complex can enter the nucleus and act as a transcription factor through binding to DNA.

nucleoside a ribose or deoxyribose sugar covalently linked through a glycosidic bond to a purine or pyrimidine base.

nucleotide a nucleoside with one or more phosphate groups attached to a ribose or deoxyribose moiety through a phosphate ester bond.

oligosaccharides linear or branched chains of monosaccharides covalently linked through glycosidic bonds. These glycan chains are distinct from polysaccharides, which are composed of one or more monosaccharides connected as repeating units.

oncoprotein a protein associated with cancer as a result of mutations in its encoding gene.

operator a region of a DNA responsible for binding to transcription factors.

operon a cluster of genes meant to be transcribed together to provide a complete set of transcripts to encode proteins that then work together.

origin of replication (ORI or ori) a DNA sequence within a plasmid or chromosome instructing the cell to make copies of the plasmid or chromosome.

oxidoreductase an enzyme that catalyzes oxidation/reduction processes.

oxocarbenium ion a structure with a carbocation adjacent to an oxygen atom.

oxonium ion a species containing a positively charged oxygen atom with three bonds.

paracrine signaling hormonal communication between cells of a different type.

PEG polyethylene glycol, a polymer with the chemical structure $[CH_2CH_2O]_n$.

peptide an oligomer of α-amino acids, distinguished from proteins by their short length and lack of complex folded structure.

peptide nucleic acids (PNAs) synthetic analogs of DNA with the bases linked together by amide bonds in place of the phosphodiester and deoxyribose backbone of DNA.

phenotype the observable characteristics resulting from a genotype.

phospholipase an enzyme that catalyzes the hydrolysis of the acyl group from the glycerol core of triacylglycerols.

phosphoramidite a key reagent for oligonucleotide synthesis possessing a trivalent phosphorus atom with a highly reactive P–N bond.

plasma membrane the lipid barrier between the extracellular milieu and the cytoplasm of the cell.

plasmid a relatively short, circular piece of DNA with an origin of replication and one or more genes.

polar covalent (bond) a type of bond between atoms with substantially different electronegativities and in which the interaction energy is dependent on both the first and third terms of equation 2.2, for example HO–Na or H_3C–F.

polyketide a class of natural products biosynthesized from two- and three-carbon building blocks through a series of Claisen condensations.

polysaccharides polymers of monosaccharides covalently linked through glycosidic bonds (for example starch, cellulose, and amylose).

post-translational modifications covalent alterations to proteins that occur after the protein has been synthesized by the ribosome. These protein modifications include trimming, splicing, phosphorylation, glycosylation, oxidation, addition of membrane anchors, fusion with other proteins, alkylation, and acetylation.

prebiotic chemistry a subfield of chemistry that attempts to identify the most probable chemical reactions that led from inanimate molecules to living organisms.

prey protein *see* yeast two-hybrid.

primary metabolite molecules considered essential for normal cellular function, such as ATP, glyceraldehyde 3-phosphate, or cholesterol.

probe an oligonucleotide designed to hybridize to a specific sequence of DNA or RNA.

promoter a DNA sequence found near the beginning of an open reading frame that can recruit a transcription factor to initiate the transcription of a gene into RNA.

proteasome a large complex of proteins found in the cell that hydrolyzes proteins marked for destruction, usually by ubiquitinylation.

protein an oligomer of α-amino acids translated by the ribosome. Structurally, they are distinguished from peptides by the presence of polypeptides with at least two elements of interacting secondary structure.

proteolysis hydrolysis of amide bonds to degrade a protein to its constituent peptides and/or amino acids.

proteome the collection of all proteins found in a cell, tissue, organ, or organism.

purine the nitrogen-containing bicyclic aromatic ring system found in adenosine or guanosine.

pyrimidine the nitrogen containing aromatic ring found in cytidine, thymidine, or uridine.

pyrosequencing a technique for sequencing DNA based on the sensitive detection of pyrophosphate, which is produced by the successful incorporation of a dNTP into the nascent DNA strand during synthesis by DNA polymerase.

quaternary structure a level of protein structure defined by the three-dimensional arrangement of multiple folded protein chains.

Ramachandran plot a plot of the ϕ versus ψ angles for a polypeptide, protein, or group of proteins. Certain regions of the plot are associated with secondary structures such as α helices and β sheets.

receptor tyrosine kinases a class of cell-surface receptors that usually bind to growth factor proteins. These proteins undergo dimerization upon ligand binding, and then autophosphorylate the dimerizing partners.

replication the process by which a copy of a DNA molecule is made.

repressor a transcription factor responsible for shutting down RNA transcription.

retrovirus an RNA-based virus.

reverse genetics examining a specific protein by making mutations to its encoding gene, and then examining the resultant function. In reverse chemical genetics, a small molecule with a known cellular target is used as a probe to understand contributions to a cellular phenotype by the target.

reverse transcriptase an enzyme that converts RNA to DNA.

riboswitches structured regions of RNA that can exert control over transcription, splicing or translation.

ribozymes RNA molecules capable of catalyzing biological reactions.

RNA interference RNA molecules that elicit a destructive response toward homologous RNA transcripts.

secondary metabolite molecules considered nonessential for the immediate survival of a cell. Such compounds are often used for the chemical defense of the cell.

secondary structure elements of biooligomer structure with well-defined patterns of hydrogen bonds: in proteins, helices and sheets formed through hydrogen bonds with the peptide backbone; in DNA, stem loops and other structures formed through base pairing.

selectant a biooligomer of defined sequence (DNA, RNA, or protein) that is selected from a library of molecules on the basis of a unique chemical characteristic such as binding, reactivity, or catalysis.

sense strand the DNA strand that resembles the resultant RNA transcript. This strand of DNA hybridizes with the antisense strand.

sequencing determining the order of subunits in a biooligomer such as DNA or a protein.

seven-transmembrane domain receptors (7TM receptors) *see* G protein-coupled receptors (GPCRs).

signal transduction the conversion of one type of signal in a chemical or physical form into another type of chemical signal (for example a photon or small molecule converted into a protein conformational change or the conversion of extracellular information into the transcription of genes).

single nucleotide polymorphism (SNP) pronounced "snip." In humans, SNPs are the single nucleotide differences that exist between otherwise identical homologous genes, like pairs of homologous genes within a chromosome or homologous genes in genetically related individuals.

solid-phase peptide synthesis (SPPS) a method for the chemical synthesis of peptides, based upon the attachment of the C terminus of the peptide to an insoluble polymer followed by iterative amino acid coupling reactions.

spliceosome a multiprotein–RNA complex that is responsible for the removal of introns in RNA transcripts.

splicing removing an internal segment from a biopolymer.

stem cell a type of cell capable of differentiating into most or all different tissue types.

stereoelectronic effect a difference in conformation or reactivity that is attributable to the donation of filled orbitals into unfilled orbitals.

sticky end a segment of single-stranded DNA that extends from one of the strands of the duplex.

substrate the starting material for an enzyme-catalyzed transformation.

substrate-assisted catalysis catalysis by an enzyme that includes functional groups from the reaction starting material or substrate.

sulfotransferases a class of enzymes that adds sulfate (SO^{3-}) groups to hydroxyl groups or amines. Modifies glycans through the addition of a sulfate functionality.

supercoil (supercoiling) in DNA, the coils that form as the double helix of DNA wraps upon itself.

synthase an enzyme that catalyzes the synthesis of a biomolecule, such as terpenes by terpene synthase.

***t*-Boc protecting group** the tert-butoxycarbonyl protecting group used to mask amino functionalities.

TCEP tris-carboxyethylphosphine. A synthetic reducing agent used to cleave disulfide bonds in peptides and proteins. The reagent has three carboxylate groups that confer solubility in water.

terminator a DNA sequence that appears at the end of a gene to end transcription.

terpene a class of natural products biosynthesized from isoprene building blocks through the addition of carbocations to olefins.

tertiary structure elements of biooligomer structure resulting from the three-dimensional interactions of secondary structure with each other. In RNA, this level of structure results from the interactions of stems and loops with each other.

therapeutic index a comparison of the dosage of a drug that has a beneficial effect to the dosage that induces toxic effects.

thioesterase (TE) domain the domain of a polyketide synthase that catalyzes the hydrolysis of the thioester bond responsible for attaching the intermediate polyketide to the acyl carrier protein.

topoisomerase an enzyme that changes the topology of DNA but not the bond connectivity. Topology is a characteristic that distinguishes the vast number of unique knots that can be tied with a single piece of string.

transacylase domain (MAT) the domain of a polyketide synthase that catalyzes the transfer of malonyl and acetyl groups from coenzyme A to the acyl carrier protein.

transannular meaning "across the ring;" refers to reactions or interactions between atoms on opposite sides of a ring. Such interactions are particularly prominent in medium-sized rings composed of eight to eleven atoms.

transcribe to make a complementary strand of RNA based on a single-stranded DNA template by the enzyme RNA polymerase.

transcription factor a protein binding to a DNA sequence necessary to initiate or suppress RNA transcription by recruiting or blocking binding by RNA polymerase.

transcriptome the collection of all RNA transcripts from a cell, tissue, organ, or organism.

transferase an enzyme that catalyzes the transfer of a functional group from one compound to another.

transient receptor potential (TRP) cation channels channels for cationic ions that respond to heat, cold, or compounds that bind noncovalently or covalently.

transition-state analog a stable small molecule that mimics the transition state for an enzymatic reaction.

translate to synthesize a protein of defined sequence based on the sequence in a messenger RNA.

trimeric death receptors found at the cell surface, a class of homotrimeric receptors that transduce signals leading to apoptotic pathways.

type I polyketide synthase composed of a single, long polypeptide, a multidomain complex responsible for the synthesis of polyketides.

type II polyketide synthase composed of multiple, individual proteins, a multidomain complex responsible for the synthesis of polyketides.

type III polyketide synthase a protein responsible for the synthesis of polyketides, which lacks the multidomain organization of the type I and II polyketide synthases.

van der Waals interaction non-Coulombic interactions between nonbonded molecules.

yeast two-hybrid a method for investigating protein interactions inside yeast cells, based upon splitting the GAL4 transcription factor into DNA-binding and activation domains for fusion to the bait and prey proteins, respectively.

Index

For entries with no discussion in the main text, page numbers appear with an F to refer to discussion within a figure (for example, 123F) and with a T to refer to a table (for example 45T); *vs.* means compare/comparison. The numbered problems have been treated as normal text.

A

B

C

D

E

I

J

K

N

Q

R

S

For Product Safety Concerns and Information please contact our EU representative GPSR@taylorandfrancis.com Taylor & Francis Verlag GmbH, Kaufingerstraße 24, 80331 München, Germany

Printed and bound by CPI Group (UK) Ltd, Croydon, CR0 4YY

01/05/2025

01858603-0001